# Optimal Control

## An Introduction to the Theory and Its Applications


Michael Athans

*Professor of Electrical Engineering (emeritus)*
*Massachusetts Institute of Technology,*
*Cambridge, Massachusetts*
*and*
*Invited Research Professor, Instituto Superior Tecnico,*
*Lisbon, Portugal*

Peter L. Falb

*Professor of Applied Mathematics*
*Brown University,*
*Providence, Rhode Island*



Dover Publications, Inc.
Mineola, New York

*Bibliographical Note*

This Dover edition, first published in 2007, is an unabridged republication of the work published by McGraw-Hill, Inc., New York, in 1966. A new Preface to the Dover Edition has been provided for this edition.

*Library of Congress Cataloging-in-Publication Data*

Athans, Michael.
Optimal control : an introduction to the theory and its applications / Michael Athans and Peter L. Falb.
p. cm.
New York : Dover Publications, 2007. Originally published: New York, McGraw-Hill, 1966.
ISBN 0-486-45328-6 (pbk.)
1. Automatic control. I. Athans, Michael. II. Falb, Peter L.

TJ213.A8 2007
629.8 22 0701 2006050500

Manufactured in the United States of America
Dover Publications, Inc., 31 East 2nd Street, Mineola, N.Y. 11501

*To our children.*

# PREFACE TO THE DOVER EDITION

It has been forty years since the original version of *Optimal Control* was published. Actually, the book was ready in 1965, but McGraw-Hill delayed its publication to have a 1966 copyright. Its selling price was $19.95, a princely sum in those days. Subsequently, the price rose to over $80.

*Optimal Control* was written between 1962 and 1965 while we were staff members at the MIT Lincoln Laboratory. We thought we could write it in six months, but it took a bit longer. It was the first published textbook in the field, followed by the Bryson & Ho and Marcus & Lee texts a couple of years later.

Looking at the field forty years later, we marvel at the truly phenomenal growth of ideas, theories, methodologies, applications and design software. Optimal control, together with state-space representations and stochastic Kalman filtering concepts, are the foundations of what people still call "Modern Control and Systems Theory." It would be impossible to summarize the exponential growth of discoveries in this exciting field; all of them are based upon rigorous mathematical principles. Surely, nobody ever could have predicted in the early '60s the powerful multivariable feedback design methods we have today, such as *H2, H∞* and robust μ-synthesis accompanied by reliable MATLAB software.

Advances in microprocessors have helped the application of Modern Control and Systems Theory to a variety of disciplines: MPC in chemical process control; aerial, underwater and ground autonomous robotic vehicles; numerous automotive control systems; civilian and defense aerospace and marine vehicles; etc.

It is also very encouraging that, during the last decade, we see optimal control being used in novel multi-disciplinary applications such as immunology, biological systems, environmental systems, biochemical processes, communication networks, intelligent transportation systems and socio-economic systems. The Pontryagin Maximum Principle and the Hamilton-Jacobi-Bellman equation remain just as important today as they were forty years ago.

Our book *Optimal Control* has been described by some as a "classic." It is for this reason that we have not changed it over the years. Clearly, some of the chapters in the book are "old-fashioned" by today's standards. When we were writing the book, just after the translation of the Pontryagin *et al* research monograph, we believed that the future of optimal control would be primarily directed toward nonlinear systems (hence the emphasis on the time-optimal and fuel-optimal examples in Chapters 7, 8, and 10). The much briefer material in Chapter 9, dealing with the Linear-Quadratic problem, was added for the sake of completeness and at the urging of Prof. R. E. Kalman. In retrospect, it was the linear dynamic optimization methods that have blossomed. Solving nonlinear two-point boundary-value problems and/or computing approximate solutions to the Hamilton-Jacobi-Bellman partial differential equation still remain challenging issues; and let us not forget singular optimal control problems. We hope that, in the decades to come, the field will routinely deal with nonlinear optimal robust feedback control problems with similar unified methodologies and software as it does for their linear counterparts.

*Michael Athans*
Professor of Electrical Engineering (emeritus), Massachusetts Institute of Technology, Cambridge, Massachusetts, United States
and
Invited Research Professor, Instituto Superior Tecnico, Lisbon, Portugal

*Peter L. Falb*
Professor of Applied Mathematics, Brown University, Providence, Rhode Island, United States

# PREFACE

Recent advances in control science have created the need for an introductory account of the theory of optimal control and its applications which will provide both the student and the practicing engineer with the background necessary for a sound understanding of these advances and their implications. This book is directed toward the fulfillment of that need.

The book is primarily an introductory text for engineering students at approximately the first-year graduate level. We hope that the basic material provided in this text will enable the student, at an early stage in his career, to evaluate, on the one hand, the engineering implications of theoretical control work, and to judge, on the other hand, the merits and soundness of the numerous papers published in the engineering control literature. We have tried to make this book a bridge between the beginning student and the control literature, and to indicate the inestimable value to the student of careful thinking and disciplined intuition. In view of these hopes and aims, we have included a large number of examples and exercises.

Structurally, the book is divided into the following three major parts:

1. Mathematical preliminaries related to the description and analysis of dynamical systems (Chapters 2 to 4)
2. Aspects of the theory of optimal control, including the maximum principle of Pontryagin (Chapters 5 and 6)
3. Application of the optimal-control theory to the design of optimal feedback systems with respect to several performance criteria (Chapters 7 to 10)

More specifically, we review the basic concepts of linear algebra and develop the vector notation used throughout the book in Chapter 2. In Chapter 3, we discuss the elementary topological properties of $n$-dimensional Euclidean space, the basic theory of vector functions, and several aspects of vector differential equations. In Chapter 4, we provide an introduction to the state-space representation of dynamical systems and we define the control problem. We derive and study the basic conditions for optimality, including the maximum principle of Pontryagin, in Chapter 5. The structure and general properties of optimal systems with respect to several specific performance criteria are examined in Chapter 6, which provides a buffer between the theoretical material of previous chapters and the design problems considered in subsequent chapters. Chapter 7 is devoted to the solution of the time-optimal control problem for a number of specific systems. In Chapter 8, we apply the theory to the design of several minimum-fuel systems. We present the general results available for an important class of optimization problems, namely, the class of control problems involving linear plants and quadratic performance criteria, in Chapter 9. In the final chapter, Chapter 10, we study a class of control problems which are more easily solved by direct methods than by methods based on the maximum principle of Pontryagin.

We owe a significant debt of gratitude to many of our colleagues and students for their invaluable assistance in the preparation of this book. In particular, we wish to thank Prof. C. A. Desoer of the University of California at Berkeley, Dean J. G. Truxal of the Polytechnic Institute of Brooklyn, Prof. E. Polak of the University of California at Berkeley, Dr. H. K. Knudsen of the Lincoln Laboratory of the Massachusetts Institute of Technology, and Dr. S. J. Kahue of the Air Force Cambridge Research Laboratories for their careful reading of the manuscript and their numerous helpful suggestions and comments. We also wish to express our appreciation to Dr. H. Halkin of the Bell Telephone Laboratories at Whippany for his comments on Chapters 4 and 5, to Prof. R. W. Brockett of the Massachusetts Institute of Technology for his suggestions relating to Chapter 4, and to Prof. R. E. Kalman of Stanford University for his very helpful comments on Chapter 9. A number of students at the Massachusetts Institute of Technology have aided us with their suggestions and their careful proofreading, and we should like to especially thank D. Gray, D. Kleinman, W. Levine, J. Plant, and H. Witsenhausen for their exceptional assistance. R. A. Carroll of the Lincoln Laboratory provided us with the computer results in Chapter 9, for which we are very grateful. Finally, we should like to thank Miss J. M. Kelley, who did an outstanding job in typing the major portion of the manuscript, and Mrs. S. M. McKay, who also did an excellent job of typing for us.

*Michael Athans*
*Peter L. Falb*

# CONTENTS†



† See page 14 for an explanation of the starred sections.

# Chapter 1
# INTRODUCTION

## 1-1 Introduction

Considerable interest in optimal control has developed over the past decade, and a broad, general theory based on a combination of variational techniques, conventional servomechanism theory, and high-speed computation has been the result of this interest. We feel that, at this point, there is a need for an introductory account of the theory of optimal control and its applications which will provide both the student and the practicing engineer with the background and foundational material necessary for a sound understanding of recent advances in the theory and practice of control-system design. We attempt to fill this need in this book.

We briefly indicate some of our aims and philosophy and describe the contents of the book in this introductory chapter. In particular, we set the context by discussing the system design problem in Sec. 1-2 and by specifying the particular type of system design problem which generates the control problem in Sec. 1-3. We then discuss (Sec. 1-4) the historical development of control theory, thus putting a proper temporal perspective on the book. Next, we indicate (Sec. 1-5) the aims of the book. Following our statement of purposes, we make some general comments on the structure of the book in Sec. 1-6, and we present a chapter-by-chapter description of its contents in Sec. 1-7. We conclude this chapter in Sec. 1-8 with a statement of the necessary prerequisites and some study suggestions based on our teaching experience.

## 1-2 The System Design Problem

A system design problem begins with the statement (sometimes vague) of a task to be accomplished either by an existing physical process or by a physical process which is to be constructed. For example, the systems engineer may be asked to improve the yield of a chemical distillation column or to design a satellite communication system. As an integral part of this task statement, the engineer will usually be given:

1. A set of goals or objectives which broadly describe the desired performance of the physical process; for example, the engineer may be asked to

design a rocket which will be able to intercept a specified target in a reasonable length of time.

2. A set of constraints which represent limitations that either are inherent in the physics of the situation or are artificially imposed; for example, there are almost always requirements relating to cost, reliability, and size.

The development of a system which accomplishes the desired objectives and meets the imposed constraints is, in essence, the system design problem.

There are basically two ways in which the system design problem can be approached: the *direct* (or ad hoc) approach and the *usual* (or standard) approach.

Approaching the system design problem directly, the engineer combines experience, know-how, ingenuity, and the results of experimentation to produce a prototype of the required system. He deals with specific components and does not develop mathematical models or resort to simulation. In short, assuming the requisite hardware is available or can be constructed, the engineer simply builds a system which does the job. For example, if an engineer is given a specific turntable, audio power amplifier, and loudspeaker and is asked to design a phonograph system meeting certain fidelity specifications, he may, on the basis of direct experimentation and previous experience, conclude that the requirements can be met with a particular preamplifier, which he orders and subsequently incorporates ("hooks up") in the system. The direct approach is, indeed, often suitably referred to as the "art of engineering."

Unfortunately, for complicated systems and stringent requirements, the direct approach is frequently inadequate. Moreover, the risks and costs involved in extensive experimentation may be too great. For example, no one would attempt to control a nuclear reactor simply by experimenting with it. Finally, the direct approach, although it may lead to a sharpening of the engineer's intuition, rarely provides broad, general design principles which can be applied in a variety of problems. In view of these difficulties, the systems engineer usually proceeds in a rather different way.

The usual, or standard, approach to a system design problem begins with the replacement of the real-world problem by a problem involving mathematical relationships. In other words, the first step consists in formulating a suitable model of the physical process, the system objectives, and the imposed constraints. The adequate mathematical description and formulation of a system design problem is an extremely challenging and difficult task. Desirable engineering features such as reliability and simplicity are almost impossible to translate into mathematical language. Moreover, mathematical models, which are idealizations of and approximations to the real world, are not unique.

Having formulated the system design problem in terms of a mathematical model, the systems engineer then seeks a pencil-and-paper design which represents the solution to the mathematical version of his design problem.

Simulation of the mathematical relationships on a computer (digital, analog, or hybrid) often plays a vital role in this search for a solution. The design obtained will give the engineer an idea of the number of interconnections required, the type of computations that must be carried out, the mathematical description of subsystems needed, etc.

When the mathematical relations that specify the overall system have been derived, the engineer often simulates these relations to obtain valuable insight into the operation of the system and to test the behavior of the model under ideal conditions. Conclusions about whether or not the mathematics will lead to a reasonable physical system can be drawn, and the sensitivity of the model to parameter variations and unpredictable disturbances can be tested. Various alternative pencil-and-paper designs can be compared and evaluated.

After completing the mathematical design and evaluating it through simulation and experimentation, the systems engineer builds a prototype, or "breadboard." The process of constructing a prototype is, in a sense, the reverse of the process of modeling, since the prototype is a physical system which must adequately duplicate the derived mathematical relationships. The prototype is then tested to see whether or not the requirements are met and the constraints satisfied. If the prototype does the job, the work of the systems engineer is essentially complete.

Often, for economic and esthetic reasons, the engineer is not satisfied with a system which simply accomplishes the task, and he will seek to improve or optimize his design. The process of optimization in the pencil-and-paper stage is quite useful in providing insight and a basis for comparison, while the process of optimization in the prototype building stage is primarily concerned with the choice of best components. The role of optimization in the control-system design problem will be examined in the next section.

## 1-3 The Control Problem

A particular type of system design problem is the problem of "controlling" a system. For example, the engineer may be asked to design an autopilot with certain response characteristics or a "fast" tracking servo or a satellite attitude control system which does not consume too much fuel. The translation of control-system design objectives into the mathematical language of the pencil-and-paper design stage gives rise to what will be called the *control problem.*

The essential elements of the control problem are:

1. A mathematical model (system) to be "controlled"
2. A desired output of the system
3. A set of admissible inputs or "controls"
4. A performance or cost functional which measures the effectiveness of a given "control action"

How do these essential elements arise out of the physical control-system design problem?

The mathematical model, which represents the physical system, consists of a set of relations which describe the response or output of the system for various inputs. Constraints based upon the physical situation are incorporated in this set of relations.

In translating the design problem into a control problem, the engineer is faced with the task of describing desirable physical behavior in mathematical terms. The objective of the system is often translated into a requirement on the output. For example, if a tracking servo is being designed, the desired output is the signal being tracked (or something close to it).

Since "control" signals in physical systems are usually obtained from equipment which can provide only a limited amount of force or energy, constraints are imposed upon the inputs to the system. These constraints lead to a set of admissible inputs (or "control" signals).

Frequently, the desired objective can be attained by many admissible inputs, and so the engineer seeks a measure of performance or cost of control which will allow him to choose the "best" input. The choice of a mathematical performance functional is a highly subjective matter, as the choice of one design engineer need not be the choice of another. The experience and intuition of the engineer play an important role in his determination of a suitable cost functional for his problem. Moreover, the cost functional will depend upon the desired behavior of the system. For example, if the engineer wishes to limit the oscillation of a system variable, say $x(t)$, he may assign a cost of the form $[\dot{x}(t)]^6$ to this variable and try to make the integral of this cost over a period of time, say $t_1 \leq t \leq t_2$, small. Most of the time, the cost functional chosen will depend upon the input and the pertinent system variables.

When a cost functional has been decided upon, the engineer formulates his control problem as follows: Determine the (admissible) inputs which generate the desired output and which, in so doing, minimize (optimize) the chosen performance measure. At this point, optimal-control theory enters the picture to aid the engineer in finding a solution to his control problem. Such a solution (when it exists) is called an optimal control.

To recapitulate, a control problem is the translation of a control-system design problem into mathematical terms; the solution of a control problem is an (idealized) pencil-and-paper design which serves to guide the engineer in developing the actual working control system.

## 1-4 Historical Perspective

Before World War II, the design of control systems was primarily an art. During and after the war, considerable effort was expended on the design of closed-loop feedback control systems, and negative feedback was used to improve performance and accuracy. The first theoretical tools used were based upon the work of Bode and Nyquist. In particular, concepts such as frequency response, bandwidth, gain (in decibels), and phase margin were used to design servomechanisms in the frequency domain in a more or less trial-and-error fashion. This was, in a sense, the beginning of modern automatic-control engineering.

The theory of servomechanisms developed rapidly from the end of the war to the beginning of the fifties. Time-domain criteria, such as rise time, settling time, and peak overshoot ratio, were commonly used, and the introduction of the "root-locus" method by Evans in 1948 provided both a bridge between the time- and frequency-domain methods and a significant new design tool. During this period, the primary concern of the control engineer was the design of linear servomechanisms. Slight nonlinearities in the plant and in the power-amplifying elements could be tolerated since the use of negative feedback made the system response relatively insensitive to variations and disturbances.

The competitive era of rapid technological change and aerospace exploration which began around mid-century generated stringent accuracy and cost requirements as well as an interest in nonlinear control systems, particularly relay (bistable) control systems. This is not surprising, since the relay is an exceedingly simple and rugged power amplifier. Two approaches, namely, the *describing-function* and *phase-space* methods, were used to meet the new design challenge. The describing-function method enabled the engineer to examine the stability of a closed-loop nonlinear system from a frequency-domain point of view, while the phase-space method enabled the engineer to design nonlinear control systems in the time domain.

Minimum-time control laws (in terms of switch curves and surfaces) were obtained for a variety of second- and third-order systems in the early fifties. Proofs of optimality were more or less heuristic and geometric in nature. However, the idea of determining an optimum system with respect to a specific performance measure, the response time, was very appealing; in addition, the precise formulation of the problem attracted the interest of the mathematician.

The time-optimal control problem was extensively studied by mathematicians in the United States and the Soviet Union. In the period from 1953 to 1957, Bellman, Gamkrelidze, Krasovskii, and LaSalle developed the basic theory of minimum-time problems and presented results concerning the existence, uniqueness, and general properties of the time-optimal control. The recognition that control problems were essentially problems in the calculus of variations soon followed.

Classical variational theory could not readily handle the "hard" constraints usually imposed in a control problem. This difficulty led Pontryagin to first conjecture his celebrated *maximum principle* and then, together with Boltyanskii and Gamkrelidze, to provide a proof of it. The maximum principle was first announced at the International Congress of Mathematicians held at Edinburgh in 1958.

While the maximum principle may be viewed as an outgrowth of the Hamiltonian approach to variational problems, the method of dynamic programming, which was developed by Bellman around 1953–1957, may be viewed as an outgrowth of the Hamilton-Jacobi approach to variational problems. Considerable use has been made of dynamic-programming techniques in control problems.

Simultaneous with the rapid development of control theory was an almost continuous revolution in computer technology, which provided the engineer with vastly expanded computational facilities and simulation aids. The ready availability of special- and general-purpose computers greatly reduced the need for closed-form solutions and the demand that controllers amount to simple (network) compensation.

Modern control theory (and practice) can thus be viewed as the confluence of three diverse streams: the theory of servomechanisms, the calculus of variations, and the development of the computer.

At present, control theory is primarily a design aid which provides the engineer with insight into the structure and properties of solutions to the optimal-control problem. Specific design procedures and "rules of thumb" are rather few in number. Moreover, since optimal feedback systems are, in the main, complicated and nonlinear, it is difficult to analyze the effects of variations and disturbances. In addition, the need for accurate measurement of the relevant (state) variables and the computational difficulties associated with the determination of an optimal control often prevent the economical implementation of an optimal design.

We believe, at any rate, that the present theory will become increasingly useful to the engineer. There are several reasons for this belief. First of all, pencil-and-paper or computer designs of optimum systems can serve as comparison models in the evaluation of alternative designs. Secondly, knowledge of the optimal solution to a given problem provides the engineer with valuable clues to the choice of a suitable suboptimal design. Thirdly, improvements in computer technology will serve to mitigate some of the current on-line computational difficulties in applying the theory. Finally, although optimal designs may rarely be implemented, the theory has expanded the horizon of the engineer and has, thus, allowed the engineer to tackle complex and difficult problems which he would not have previously considered attacking.

## 1-5 Aims of This Book

We have two major aims in writing this book. First of all, we wish to provide an introductory text on the theory and applications of optimal control for engineering students at approximately the first-year graduate level. Secondly, we wish to provide the student and the practicing engineer with the background and foundational material necessary to make advances in the theory and practice of control-system design both accessible and meaningful.

We did not attempt or desire to write an exhaustive treatise, but rather we have concentrated on developing introductory material in a relatively careful and detailed way. In particular, we have tried to:

1. Develop the basic mathematical background needed for a sound understanding of the theory
2. Give a careful formulation of the control problem
3. Explicate the basic necessary conditions for optimality, paying particular attention to the maximum principle of Pontryagin and the basic sufficiency conditions in terms of the Hamilton-Jacobi equation
4. Illustrate the application of the theory to several relatively simple problems involving various "standard" performance criteria
5. Examine the structure, properties, and engineering realizations of several optimal feedback control systems

In keeping with the introductory nature of the book, we have not developed many of our topics "in depth," and we have omitted many important topics. We have frequently used heuristic arguments, referring the reader to the relevant literature for a rigorous treatment.

We hope that the basic material provided in this book will enable the student, at an early stage in his career, to evaluate, on the one hand, the engineering implications of theoretical control work and to judge, on the other hand, the merits and soundness of the numerous papers published in the engineering control literature (which unfortunately contains many unwarranted claims of generality). We hope also to indicate the inestimable value to the student of careful thinking and disciplined intuition, for these are essential to his development.

We have found that engineering students cannot really learn control theory without examining many examples and doing a number of exercises, for only thus can they appreciate the uses and shortcomings of the theory. Examples are also quite valuable to the applied mathematician who wishes to make contributions to the theory, since these examples often help him to distinguish between physically meaningful and purely mathematical problems. We have, therefore, included a great many examples and exercises which we hope will be useful to the student, the practicing engineer, and the applied mathematician.

## 1-6 General Comments on the Structure of This Book

The book can be divided into the following three major parts:

1. Mathematical preliminaries related to the description and analysis of dynamical systems (Chaps. 2 to 4)

2. Aspects of the theory of optimal control, including the maximum principle of Pontryagin, for continuous, finite-dimensional, deterministic systems (Chaps. 5 and 6)

3. Application of the theory to the design of optimal feedback systems with respect to several performance criteria (Chaps. 7 to 10)

We have attempted throughout the book to state definitions, theorems, and problems in a careful, precise manner; on the other hand, we have often only sketched proofs, and we have either omitted or left as exercises for the reader many proofs, giving suitable references to the literature. We have found that such a procedure is quite appealing to engineering students.

We use the compact notation of set theory and vector algebra to avoid obscuring general concepts in a cloud of complicated equations. We have found that students, after an initial transition period, adjust to this terminology and become accustomed to thinking "physically" in terms of sets, vectors, and matrices. We attempt to stress general ideas, and we try to illustrate these ideas with a number of examples. We do not, however, attempt to gloss over the difficulties, particularly those of a computational nature, associated with the analysis and design of systems involving many variables.

Many carefully worked examples are included in the book to provide the reader with concrete illustrations of the general theory. We believe that these examples serve to deepen the understanding and strengthen the intuition. In addition, we have included numerous exercises. The exercises fall into three categories:

1. Exercises which are routine.

2. Exercises which involve and illustrate "fine points" of the theory as well as some of the computational difficulties associated with the determination of an optimal control.†

3. Exercises which challenge the student to discover new aspects of the theory and its applications. Several of these are of the "almost impossible" variety.

We feel that the educational value of exercises is very great.

We cross-reference frequently, and in addition, we pinpoint the references which are relevant to a particular discussion so that the interested student can profitably consult the literature. We include an extensive bibliography with this in mind. However, we do not claim that our bibliography is complete, for we have simply included the papers, reports, and books with which we are familiar.

Since our purpose is to provide an introduction to the theory of optimal control and its applications, we have not discussed a number of important topics which are of an advanced nature or which require additional preparation. In particular, the following topics are not included:

† Several exercises in Chap. 9 require the use of a digital computer.

1. Computational algorithms for the determination of optimal controls for complex systems
2. Problems with state-space constraints
3. Optimal-control theory for discrete (or sampled-data) systems
4. Problems involving distributed-parameter systems
5. The optimal control of stochastic systems
6. The design of optimal filters, predictors, or smoothers

Briefly, some of our specific reasons for omitting these important topics are as follows:

1. Although considerable research effort has been (and is) devoted to the development of convergent computational algorithms, there have been (with some notable exceptions) few general results which guarantee convergence or contain information regarding the speed of convergence of an algorithm.

2. State-space constraints are more difficult to handle than control constraints. Necessary conditions for problems involving state-space constraints are available in the literature but do not, in our opinion, represent material of an introductory nature.

3. Currently, the theory for discrete, distributed-parameter, and stochastic systems is being developed. Since the study of distributed-parameter systems requires a knowledge of partial differential equations and since the study of stochastic systems requires a knowledge of advanced probability theory (including stochastic differential equations), these topics are clearly not suitable for an introductory text.

4. The Wiener-Kalman-Bucy theory can be used to design optimal filters for linear systems with Gaussian noise processes. However, general results pertaining to nonlinear systems and non-Gaussian noise are not currently available; moreover, the interested reader can readily follow the Wiener-Kalman-Bucy theory after digesting the deterministic material included in this book.

We shall present a chapter-by-chapter description of the contents of the book in the next section.

## 1-7 Chapter Description

We present a brief description of the contents of each chapter, taking care to point out the significant interrelationships between chapters.

### *Chapter 2 Mathematical Preliminaries: Algebra*

We review the basic concepts of linear algebra and develop the vector and matrix notation used throughout the book in this chapter. After a brief discussion of the familiar notions of set theory, we introduce the concept of a vector space and then consider linear transformations. We treat matrices from the point of view of linear transformations rather than with the more familiar array-of-numbers approach. We stress that a matrix is associated with a linear transformation and given coordinate systems, thus indicating that linear transformations are intrinsic whereas matrices are not. In the remainder of Chap. 2, we discuss eigenvalues and eigenvectors, similarity of matrices, inner and scalar products, Euclidean vector spaces, and some properties of symmetric matrices. The material in this chapter is used in every other chapter of the book.

*Chapter 3 Mathematical Preliminaries: Analysis*

We discuss the elementary topological (metric) properties of $n$-dimensional Euclidean space, the basic theory of vector functions, and several aspects of vector differential equations in this chapter. We begin by studying the notion of distance and the related concepts of convergence, open and closed sets, and compactness. We then define and examine the notion of convexity. Next, we consider functions of several variables, developing the definitions of continuity, piecewise continuity (or regularity), derivative, gradient, and integral for such functions. We introduce the important concepts of a function space and the distance between elements of a function space. In the remainder of Chap. 3, we are concerned with vector differential equations. We prove the basic existence-and-uniqueness theorem, and we develop the method of solution of linear differential equations by means of the fundamental matrix. We make frequent use of the material in this chapter in the remaining chapters of the book.

*Chapter 4 Basic Concepts*

We provide an introduction to the state-space representation of dynamical systems, and we define and discuss the control problem in this chapter. We start with an examination of the concept of state for some simple network systems. After an informal introduction to the mathematical description of physical systems, we present an axiomatic definition of a dynamical system. We then consider finite-dimensional continuous-time differential systems, and we use the material of Chap. 3 to establish the relationship between the state and the initial conditions of the differential equations representing the system. We describe the technique for finding a state-variable representation of single-input–single-output constant systems in Secs. 4-9 and 4-10. We indicate the physical significance of the state variables by means of an analog-computer simulation of the differential equations. We complete the chapter with the definition of the control problem and a discussion of some of the natural consequences of that definition. In particular, we consider the set of reachable states and the qualitative notions of controllability, observability, and normality. We also develop some of the important implications of these concepts. The basic material of this chapter is the cornerstone of the theory and examples which we develop in subsequent chapters.

*Chapter 5 Conditions for Optimality: The Minimum Principle*† *and the Hamilton-Jacobi Equation*

We derive and study the basic conditions for optimality in this chapter. We include a review of the theory of minima for functions defined on $n$-dimensional Euclidean space, an indication of the application of the techniques of the calculus of variations to control problems, a statement and heuristic proof of the minimum principle of Pontryagin, and a discussion of the Hamilton-Jacobi equation, based on a combination of Bellman's "principle of optimality" and a lemma of Caratheodory in a manner suggested by Kalman. To bolster the intuition of the student, we start with ordinary minima (Secs. 5-2 to 5-4), and we review the Lagrange multiplier technique for constrained minimization problems. We then introduce the variational approach to the control problem (Secs. 5-5 to 5-10), indicating the step-by-step procedure that can be used to derive some necessary and some sufficient conditions for optimality. We next develop the necessary conditions for optimality embodied in the minimum principle of Pontryagin (Secs. 5-11 to 5-17). We carefully state several control problems and the versions of the minimum principle corresponding to each of these problems. We give a heuristic and rather geometric proof of the minimum principle in Sec. 5-16, and we comment on some of the implications of the minimum principle in Sec. 5-17. We conclude the chapter with a discussion of the Hamilton-Jacobi equation and the sufficiency conditions associated with it. The minimum principle is the primary tool used to find optimal controls in Chaps. 7 to 10.

*Chapter 6 Structure and Properties of Optimal Systems*

We discuss the structure and general properties of optimal systems with respect to several specific performance criteria in this chapter. We indicate the methods which may be used to deduce whether or not optimal and extremal controls are unique, and we briefly consider the question of singular controls. In developing the chapter, we aim to provide a buffer between the theoretical material of the previous chapters and the more specific design problems considered in subsequent chapters. We begin with an examination of the time-optimal control problem for both linear and nonlinear systems (Secs. 6-2 to 6-10). After discussing the time-optimal problem geometrically, we use the minimum principle to derive suitable necessary conditions. We then define the notions of singularity and normality and state the "bang-bang" principle for normal time-optimal systems. We present a series of theorems dealing with existence, uniqueness, and number of switchings for linear time-invariant systems in Sec. 6-5. The material

† We speak of the *minimum principle* rather than the *maximum principle* in this book. The only difference in formulation is a change of sign which makes no essential difference in the results obtained.

in this portion of the chapter is used in Chap. 7. We continue with a discussion of the minimum-fuel control problem (Secs. 6-11 to 6-16). We again use the minimum principle to derive necessary conditions, and we again follow this with a definition of the singular and normal fuel-optimal problems. We establish the "on-off" principle for normal fuel-optimal problems, and we develop uniqueness theorems for linear time-invariant systems. We consider several additional problem formulations involving both time and fuel in Sec. 6-15. The material in this part of the chapter is used in Chap. 8. We next study minimum-energy problems (Secs. 6-17 to 6-20). We derive an analytic solution for the fixed-terminal-time–fixed-terminal-state problem for time-invariant linear systems, and we illustrate the general results with a simple example. The results of this segment of the chapter are applied and expanded upon in Chap. 9. We conclude the chapter with a brief introduction to singular problems (Secs. 6-21 and 6-22) and some comments on the existence and uniqueness of optimal and extremal controls.

*Chapter 7 The Design of Time-optimal Systems*

We apply the theory of Chaps. 5 and 6 to the time-optimal control problem for a number of specific systems in this chapter. We illustrate the construction of time-optimal controls as functions of the state of the system, and we discuss the structure of time-optimal feedback systems, indicating the types of nonlinearities required in the feedback loop. We also comment on the design of suboptimal systems. We begin, in Secs. 7-2 to 7-6, with a consideration of plants whose transfer functions have only real poles. The complexity of the time-optimal feedback system is illustrated by comparing the time-optimal systems for second-order plants (Secs. 7-2 and 7-3) with that of a third-order plant (Sec. 7-4). Next, we examine the time-optimal problem for harmonic-oscillator-type plants (Secs. 7-7 to 7-9). We show (Secs. 7-10 and 7-11) how the concepts developed for linear systems can be applied to the time-optimal control problem for a class of nonlinear systems, using experimental data and graphical constructions. We conclude the chapter with an introductory discussion of plants having both zeros and poles in their transfer function (Secs. 7-12 to 7-15). We indicate the effects of minimum and non-minimum-phase zeros on the character and structure of the time-optimal system.

*Chapter 8 The Design of Fuel-optimal Systems*

We apply the theory of Chaps. 5 and 6 (and some results of Chap. 7) to the design of a number of minimum-fuel systems. We show that conservative systems often have nonunique minimum-fuel controls. We illustrate ways in which the response time can be taken into account in minimum-fuel problems, indicating the engineering implications of the resultant optimal system. We also point out that nonlinear systems often admit singular fuel-optimal solutions (in marked contrast to the time-optimal case). We begin the chapter by illustrating the nonuniqueness of fuel-optimal controls (Secs. 8-2 to 8-5). Different methods for including the response time in the performance functional are examined in Secs. 8-6 to 8-10. The possibility of singular solutions is also discussed in Sec. 8-10, and we conclude with some comments on the graphical methods that are frequently used to determine the optimal control as a function of the state in such minimum-fuel problems.

*Chapter 9 The Design of Optimum Linear Systems with Quadratic Criteria*

In this chapter, we present the general results available for an important class of optimization problems, namely, the class of control problems involving linear time-varying plants and quadratic performance criteria. In view of some results of Kalman on the inverse problem, the material in this chapter may be construed as, in a sense, a generalization of conventional control theory. We begin with the state-regulator problem (Secs. 9-2 to 9-6); then we consider the output-regulator problem (Secs. 9-7 and 9-8); and finally, we conclude the chapter with an examination of the tracking problem (Secs. 9-9 to 9-13). Since linear time-invariant systems are studied in Secs. 9-5, 9-8, and 9-11, we obtain a degree of correlation between the properties of optimal systems and well-designed servo systems, as conventional servomechanism theory can be applied to the design of linear time-invariant feedback systems. However, the theoretical results developed in this chapter can easily be applied to the control of processes (e.g., multivariable or time-varying plants) for which the conventional servomechanism theory cannot be used directly.

*Chapter 10 Optimal-control Problems when the Control Is Constrained to a Hypersphere*

In this chapter, we study a certain broad class of control problems which can be solved more easily by direct methods than by methods based on the minimum principle. The basic ideas are discussed in Secs. 10-2 to 10-6. We also illustrate, in a specific example, the effect of constraints upon the design and properties of an optimal system with respect to several performance criteria.

## 1-8 Prerequisites and Study Suggestions

We suppose that the engineering student using this book as a text has, in general, the following background:

1. A knowledge of basic conventional servomechanism theory, including such ideas as the system response, the transfer function of a system, feedback, and linear compensation
2. A knowledge of the basics of ordinary differential equations and calculus, including the Laplace transform method for the solution of linear

differential equations with constant coefficients and the notion of a matrix, together with some skill in performing matrix manipulation.

We consider this background adequate for an understanding of the major portion of the book. We do not expect the student to be an expert in linear algebra or in vector differential equations; consequently, we incorporate in the text the specific mathematical material that we need.

The subject matter contained in the book has been covered in a one-semester course in the Department of Electrical Engineering at the Massachusetts Institute of Technology. Sections indicated with a star in the table of contents were treated in the classroom, and the remaining sections were assigned reading. No text should be followed verbatim, and different teachers will expand upon or delete various topics, depending on their own inclinations and the level of their students' preparation. In this regard, we note that Chaps. 8 to 10 are independent of each other and so can be treated in any order. Also, for students with a more advanced background (say, for example, a course based on the book "Linear System Theory," by Zadeh and Desoer), Chaps. 2 to 4 can be omitted from the classroom and simply assigned as review reading and the material in Chaps. 5 and 6 developed in greater detail.

## Chapter 2

# MATHEMATICAL PRELIMINARIES: ALGEBRA

### 2-1 Introduction

We shall use a number of mathematical results and techniques throughout this book. In an attempt to make the book reasonably self-contained, we collect, in this chapter and the next, the basic definitions and theorems required for an understanding of the main body of the text. Our treatment will be fairly rapid, as we shall assume that the reader has been exposed to most of the mathematical notions we need. Also, our treatment will be incomplete in the sense that we shall present only material which is used in the sequel.

We shall discuss sets, functions, vector spaces, linear algebra, and Euclidean spaces in this chapter. In essence, most of our discussion will be aimed at translating the familiar physical notion of linearity into mathematical terms. As a consequence of this aim, we shall view the concepts of a vector space and of a linear transformation rather than the concepts of an $n$-tuple of numbers and of a matrix as basic.

We shall suppose that the reader is familiar with the notion of determinant and its relation to the solution of linear equations. The particular information that is needed can be found in Ref. B-9,† chap. 10, or in Ref. B-6, app. A.

The material in this chapter, in whole or in part, together with various extensions and ramifications of it, can be found, for example, in Refs. B-6, B-9, D-5, H-4, K-20, R-10, S-4, and Z-1.

### 2-2 Sets

We consider objects which have various properties and which can be related to one another. A collection of objects which have common and distinguishing properties is called a *set*. We use $\in$ to denote membership in a set; in other words, if $A$ is a set, then $a \in A$ means that $a$ is a member

† Letter-and-number citations are keyed to the References at the back of the book.

(or element) of $A$. Two sets are the same if and only if they have the same elements.

If $A$ and $B$ are sets, then we shall say that $A$ is *contained in* $B$ or is a *subset of* $B$ if every element of $A$ is an element of $B$. We write this inclusion relationship in the form

$$A \subset B \tag{2-1}$$

If $A$ is a set and if $\mathcal{P}$ is a property which elements of $A$ may have, then we let $\{a: \mathcal{P}(a)\}$ stand for the set of elements of $A$ which actually have the property $\mathcal{P}$. For example, if $A$ is the set of all girls in New York and $\mathcal{P}$ is the property of being a blonde, then $\{a: \mathcal{P}(a)\}$ is the set of all blonde girls in New York.

**Example 2-1** If $R$ is the set of real numbers and if $\mathcal{P}$ is the property "$|\ \ | \leq 1$" (i.e., the absolute value is less than or equal to 1), then $\{r: |r| \leq 1\}$ is the set of real numbers whose absolute value is less than or equal to 1.

We let $\emptyset$ denote the set which has *no* elements. $\emptyset$ is called the *empty*, or *null*, set. We observe that $\emptyset$ may be defined by $\{a: \mathcal{P}(a)\}$, where $\mathcal{P}$ is any property which is never satisfied; for example, $\mathcal{P}$ is the property "$a \neq a$." We note also that $\emptyset \subset A$ for every set $A$.

A set $A$ whose elements may be enumerated in a sequence $a_1, a_2, \ldots$ which may or may not terminate is called a *countable set*. We note that if a set has only a finite number of elements, then it is a countable set, but that not every countable set has a finite number of elements. For example, the set $\{1, 2, 3\}$ is a finite and therefore countable set, while the set $\{1, \frac{1}{2}, \frac{1}{3}, \ldots, 1/n, \ldots\}$ is a countable set which is not finite. The set $\{r: |r| \leq 1\}$ of Example 2-1 is *not* countable.

## 2-3 Operations on Sets

We consider several operations on sets, namely, *union*, *intersection*, *complement*, and *product*. If $A$ and $B$ are sets, then the *union* of $A$ and $B$, denoted by $A \cup B$, is the set of all elements in either $A$ or $B$. We note that

$$A \cup B = \{x: x \in A \textit{ or } x \in B\} \tag{2-2}$$†

The *intersection* of two sets $A$ and $B$, denoted by $A \cap B$, is the set of all elements common to both $A$ and $B$. We note that

$$A \cap B = \{x: x \in A \textit{ and } x \in B\} \tag{2-3}$$

If $A$ and $B$ are sets and if $A \subset B$, then the set of elements of $B$ which are *not* in $A$ is called the *complement of* $A$ *in* $B$ and is denoted by $B - A$ (or, simply, $-A$ when $B$ is fixed by the context). If we write $x \notin A$ for "$x$ is *not* an

† The "or" here does not exclude the possibility that $x$ is in both $A$ and $B$.

element of $A$," then we note that

$$B - A = \{x: x \in B \textit{ and } x \notin A\} \tag{2-4}$$

Finally, the *product* of two sets $A$ and $B$, which we denote by $A \times B$, is the set of all *ordered* pairs $(a, b)$ where $a \in A$ and $b \in B$. We point out that $A \times B$ and $B \times A$ are *not* the same, and we note that

$$A \times B = \{(a, b): a \in A, b \in B\} \tag{2-5}$$

We shall sometimes write $\begin{bmatrix} a \\ b \end{bmatrix}$ instead of $(a, b)$. Let us examine some examples which illustrate these operations. *We shall always use $R$ to denote the set of real numbers in this book,* and we shall often refer to $R$ as the *real line* or, simply, the *reals.*

**Example 2-2** Let $A = \{r: 0 \le r \le 1\}$, where $r$ represents elements of $R$, and let $B = \{r: \frac{1}{2} < r \le 2\}$. Then

$$A \cup B = \{r: 0 \le r \le 2\}$$
$$A \cap B = \{r: \tfrac{1}{2} < r \le 1\}$$

**Example 2-3** Let $A = \{r: 0 \le r \le 1\}$, where $r$ represents elements of $R$, and let $B = \{r: 0 \le r \le 2\}$. Then $A \subset B$, and

$$B - A = \{r: 1 < r \le 2\}$$

**Example 2-4** Let $A = \{r: 0 \le r \le 1\}$ and let $B = \{s: 1 \le s \le 2\}$, where $r$, $s$ represent elements of $R$. Then

$$A \times B = \{(r, s): 0 \le r \le 1, 1 \le s \le 2\} \quad \text{Fig. 2-1}a$$
$$B \times A = \{(s, r): 1 \le s \le 2, 0 \le r \le 1\} \quad \text{Fig. 2-1}b$$

**Exercise 2-1** Let $R^2 = R \times R$, let $A$ be the subset of $R^2$ given by $A = \{(x, y): x^2 + y^2 \le 1\}$, and let $B$ be the subset of $R^2$ given by

$$B = \{(x, y): (x - 1)^2 + y^2 \le 1\}$$

What are $A \cap B$, $A \cup B$, and $R^2 - A$? Draw figures in the plane which represent these sets.

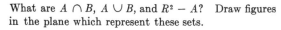

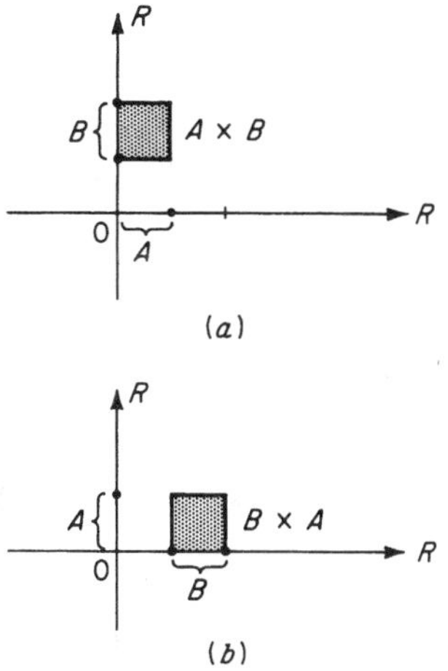

**Fig. 2-1** $A \times B$ is not the same as $B \times A$.

If $A_1, \ldots, A_n$ are sets, then we may define the union, intersection, and product of these sets. In particular, the *union* of $A_1, \ldots, A_n$, which we denote by $\bigcup_{i=1}^{n} A_i$, is defined by

$$\bigcup_{i=1}^{n} A_i = \left(\bigcup_{i=1}^{n-1} A_i\right) \cup A_n \tag{2-6}$$

the *intersection* of $A_1, \ldots, A_n$, which we denote by $\bigcap_{i=1}^{n} A_i$, is defined by

$$\bigcap_{i=1}^{n} A_i = \left(\bigcap_{i=1}^{n-1} A_i\right) \cap A_n \tag{2-7}$$

and, finally, the *product* of $A_1, \ldots, A_n$, which we denote by $\prod_{i=1}^{n} A_i$ or $A_1 \times A_2 \times \cdots \times A_n$, is defined by

$$\prod_{i=1}^{n} A_i = \{(a_1, a_2, \ldots, a_n): a_i \in A_i \text{ for } i = 1, \ldots, n\} \tag{2-8}$$

The elements of $\prod_{i=1}^{n} A_i$ are sometimes called *ordered n-tuples* and are sometimes written as columns,

$$\begin{bmatrix} a_1 \\ a_2 \\ \cdot \\ \cdot \\ \cdot \\ a_n \end{bmatrix} \tag{2-9}$$

We observe that

$$\begin{aligned} \bigcup_{i=1}^{n} A_i &= \{a: a \in A_1 \textit{ or } a \in A_2 \textit{ or } \cdots \textit{ or } a \in A_n\} \\ \bigcap_{i=1}^{n} A_i &= \{a: a \in A_1 \textit{ and } a \in A_2 \textit{ and } \cdots \textit{ and } a \in A_n\} \end{aligned} \tag{2-10}$$

**Exercise 2-2** Prove Eqs. (2-10).

We may also define these notions for infinite collections $A_i$, $i = 1, 2, \ldots$, of sets. For example, the *union* of the $A_i$, which we denote by $\bigcup_{i=1}^{\infty} A_i$, is given by

$$\bigcup_{i=1}^{\infty} A_i = \{a: \text{there is an } i_0 \text{ with } a \in A_{i_0}\} \tag{2-11}$$†

and the *intersection* of the $A_i$, which we denote by $\bigcap_{i=1}^{\infty} A_i$, is given by

$$\bigcap_{i=1}^{\infty} A_i = \{a: a \in A_i \text{ for every } i\} \tag{2-12}$$

We omit the definition of the infinite product, as we shall have no occasion to use it.

**Example 2-5** Let $A_i = \{r: -i \leq r \leq i\}$, for $i = 0, 1, 2, \ldots$, where $r$ represents elements of $R$. Then $\bigcup_{i=0}^{\infty} A_i = R$ and $\bigcap_{i=0}^{\infty} A_i = \{0\}$ (i.e., the subset of $R$ consisting of 0 alone).

† This means that there is at least one $i_0$ with $a$ in $A_{i_0}$; there may be more.

**Example 2-6** Let $A_i$, for $i = 1, 2, \ldots$, be the subset of $R^2$ ($= R \times R$) given by $A_i = \{(x, y): i \leq x < i + 1\}$. Then $A_i$ is the strip (Fig. 2-2) parallel to the $y$ axis between $x = i$ and $x = i + 1$, including the line $x = i$ but excluding the line $x = i + 1$, and

$$\bigcup_{i=1}^{\infty} A_i = \{(x, y): 1 \leq x\}$$

$$\bigcap_{i=1}^{\infty} A_i = \emptyset \text{ (the empty set)}$$

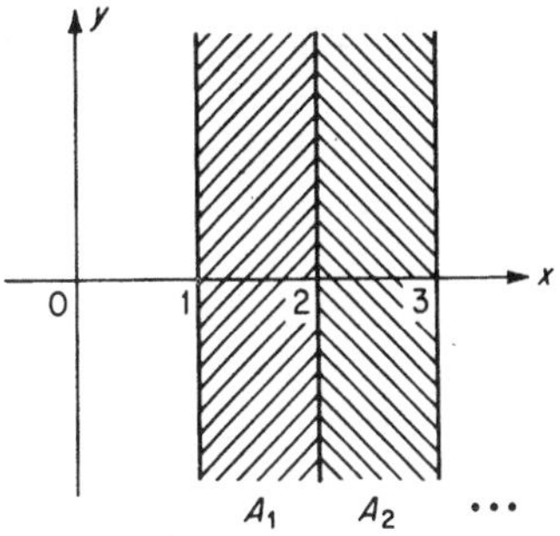


**Fig. 2-2** The "strips" $A_i = \{(x, y): i \leq x < i + 1\}$.

We relate these various operations on sets to one another in the following rather simple theorem.

***Theorem 2-1***

*a. If A, B, and C are sets, then*

1. $A - A = \emptyset$ *and* $A - \emptyset = A$
2. $A \cup A = A$ *and* $A \cap A = A$
3. $A \subset C$ *and* $B \subset C$ *if and only if* $A \cup B \subset C$
4. $C \subset A$ *and* $C \subset B$ *if and only if* $C \subset A \cap B$
5. $A \cup (B \cap C) = (A \cup B) \cap (A \cup C)$
6. $A \cap (B \cup C) = (A \cap B) \cup (A \cap C)$

*b. If* $A_i$, $i = 1, 2, \ldots$, *are subsets of a set A, then*

1. $A - (\bigcup_{i=1}^{\infty} A_i) = \bigcap_{i=1}^{\infty} (A - A_i)$
2. $A - (\bigcap_{i=1}^{\infty} A_i) = \bigcup_{i=1}^{\infty} (A - A_i)$

*c. If A, B, C, and D are sets, then*

1. $A \times B = \emptyset$ *if and only if* $A = \emptyset$ or $B = \emptyset$
2. $(A \times B) \cup (C \times B) = (A \cup C) \times B$
3. $(A \times B) \cap (C \times D) = (A \cap C) \times (B \cap D)$

PROOF Most of the assertions of the theorem are immediate consequences of the definitions. As an illustration of this, we shall prove part $b1$.

Suppose that $a \in A - (\bigcup_{i=1}^{\infty} A_i)$; then $a$ is not in $A_1$, for if $a$ were in $A_1$, then $a$ would be in $\bigcup_{i=1}^{\infty} A_i$. In other words, $a \in A - A_1$. In a similar way, we see that $a \in A - A_i$ for every $i$ and so $a \in \bigcap_{i=1}^{\infty} (A - A_i)$. On the other hand, if $a \in \bigcap_{i=1}^{\infty} (A - A_i)$, then $a \in A - A_i$ for every $i$, and so $a$ is *not* in any $A_i$. Therefore, $a \in A - (\bigcup_{i=1}^{\infty} A_i)$.

**Exercise 2-3** Prove part $b2$ of Theorem 2-1.

## 2-4 Functions

Heuristically speaking, a function is a rule for associating elements of one set with those of another. More precisely, let us suppose that $A$ and $B$ are sets and that $G$ is a subset of $A \times B$ such that there is *at most* one element $(a, b)$ of $G$ for each $a$ in $A$. [Note that for some elements $a$ of $A$, there may not be any pair $(a, b)$ in $G$.] We then call $G$ a *graph*. If $(a, b)$ is an element of the graph $G$, then we call $b$ the *value of $G$ at $a$*, and we write

$$b = G(a) \tag{2-13}$$

The relationship (2-13) between $A$ and $B$ is called a *function* (or *transformation* or *mapping*) from $A$ into $B$ and is often written simply as $G(a)$ or as $G$. The set of elements $a$ in $A$ for which there is a pair $(a, b)$ in $G$ is called the *domain of $G(a)$* (or of $G$), and the set of elements $b$ in $B$ for which there is a pair $(a, b)$ in $G$ is called *the range of $G(a)$* (or of $G$).

**Example 2-7** Suppose that $A = B = R$ and that $G$ is the subset of $R^2$ defined by $G = \{(a, b): b = a^2\}$ (see Fig. 2-3). Then $G$ is a graph, the relationship $b = G(a) = a^2$ is a function, the domain of $G(a)$ is all of $R$, and the range of $G(a)$ is the set of all nonnegative real numbers.

If $G$ is a graph and if $A_1 \subset A$, then the subset $G(A_1)$ of $B$ defined by

$$G(A_1) = \{b: (a_1, b) \text{ is in } G \text{ for some } a_1 \text{ in } A_1\} \tag{2-14}$$

is called the *image of $A_1$ under $G$*. If $B_1 \subset B$, then the subset $G^{-1}(B_1)$ of $A$ defined by

$$G^{-1}(B_1) = \{a: (a, b_1) \text{ is in } G \text{ for some } b_1 \text{ in } B_1\} \tag{2-15}$$

is called the *inverse image of $B_1$ under $G$*. We note that $G(A)$ is the range of $G$ and that $G^{-1}(B)$ is the domain of $G$.

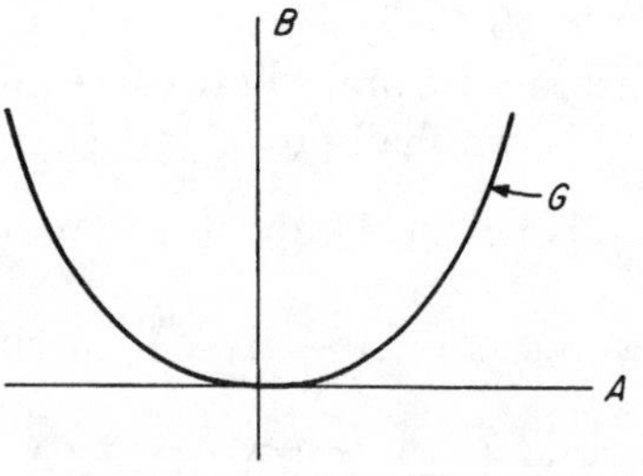


**Fig. 2-3** A typical graph $G$.

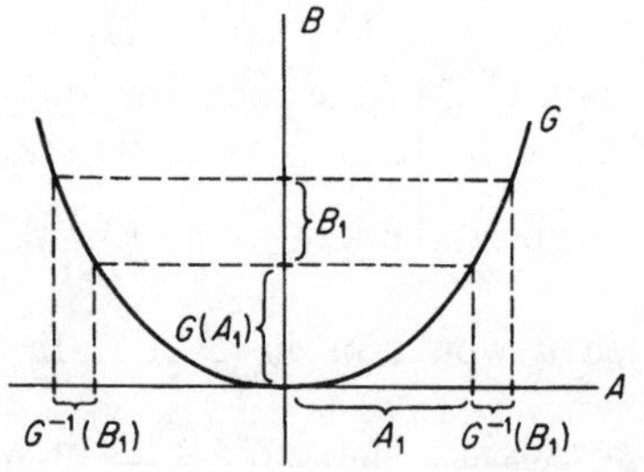


**Fig. 2-4** The graph $G$ and the sets $G(A_1)$ and $G^{-1}(B_1)$. Note that $G^{-1}(B_1)$ consists of two parts.

**Example 2-8** Suppose that $G$ is the graph of Example 2-7 and let $A_1 = \{a: 0 \le a \le 1\}$ and $B_1 = \{b: 1 \le b \le 4\}$. Then $G(A_1) = \{b: 0 \le b \le 1\}$ and $G^{-1}(B_1) = \{a: -2 \le a \le -1 \text{ or } 1 \le a \le 2\}$ (see Fig. 2-4).

**Exercise 2-4** Show that $G^{-1}(B_1 \cup B_2) = G^{-1}(B_1) \cup G^{-1}(B_2)$ and $G^{-1}(B_1 \cap B_2) = G^{-1}(B_1) \cap G^{-1}(B_2)$.

If, for every $b$ in the range of $G$, $b = G(a)$ and $b = G(a_1)$ imply that $a = a_1$, then we say that the function $G$ is *one-one*. In other words, if $b \in G(A)$ implies that there is a *unique* $a$ in $A$ such that $G(a) = b$, then $G(a)$ is one-one. On the other hand, if the range of $G$ is all of $B$, that is, if $G(A) = B$, then we say that the function $G(a)$ is *onto*. A function $G(a)$ which is both one-one and onto will sometimes be called a *correspondence* (or a one-one correspondence), and $b = G(a)$ and $a$ will be called *corresponding elements*.

**Example 2-9** Let $G$ be the subset of $R^2$ defined by $G = \{(a, b): b = a\}$. Then the function $G(a)$ is both one-one and onto. On the other hand, the function of Example 2-8 is neither one-one nor onto.

Suppose that $A$, $B$, and $C$ are sets, that $G$ is a graph contained in $A \times B$, and that $H$ is a graph contained in $B \times C$. We let $H \circ G$ be the subset of $A \times C$, defined by

$$H \circ G = \{(a, c)\text{: there is a } b \in B \text{ such that } (a, b) \in G \text{ and } (b, c) \in H\} \tag{2-16}$$

Then $H \circ G$ is a graph in $A \times C$ and is called the *composition of H with G*. The function $(H \circ G)(a)$, which we often write $H[G(a)]$, is called a *composite function* (and is sometimes referred to as a function of a function). We note that the domain of $H \circ G$ is $G^{-1}\{[H^{-1}(C)]\}$ and that the range of $H \circ G$ is $H\{[G(A)]\}$.

**Example 2-10** Let $G$ be the graph of Example 2-7 and let $H$ be the subset of $R^2$, defined by $H = \{(b, c): c = \sin b\}$. Then $H \circ G = \{(a, c): c = \sin a^2\}$ and $H[G(a)] = \sin a^2$.

Suppose that $G$ is a graph contained in $A \times B$ and that $A_1 \subset A$. Then the set $G_1 = \{(a_1, b): a_1 \in A_1 \text{ and } (a_1, b) \in G\}$ is also a graph. We call the function $G_1(a_1)$ the *restriction of G to $A_1$*, and we note that $G_1(a_1) = G(a_1)$ for $a_1 \in A_1$. We observe that $G_1$ maps the subset $A_1$ of $A$ into a subset of $B$. We often refer to $G_1$ as the *segment of G over $A_1$*, and we often write $G_{A_1}$ in place of $G_1$. We also note that $G$ is sometimes called an *extension of $G_1$*.

**Example 2-11** Let $G$ be the graph of Example 2-8 and let $A_1 = \{a: a \geq 0\}$. Then $G_1 = \{(a_1, b): a_1 \geq 0 \text{ and } b = a_1^2\}$ is a graph, and $G_1(a_1)$ is the restriction of $G$ to $A_1$. Is $G_1$ one-one?

## 2-5 Vector Spaces

We come now to a basic concept of this book, namely, the notion of a (real) vector space. *Loosely speaking, a vector space is a set whose elements can be added to one another and can be multiplied by (real) numbers.* Let us make this idea more precise. Suppose that $V$ is a set and that $G_+$ is a graph whose domain is *all* of $V \times V$ and whose range is contained in $V$. In other words, the function $G_+$ maps $V \times V$ into $V$. If $(v_1, v_2) \in V \times V$, then we write $v_1 + v_2$ for the elements $G_+((v_1, v_2))$ of $V$; that is,

$$v_1 + v_2 = G_+((v_1, v_2)) \tag{2-17}$$

We call $v_1 + v_2$ the *sum of* $v_1$ *and* $v_2$. Let us suppose also that $G.$ is a graph whose domain is *all* of $R \times V$ and whose range is contained in $V$. In other words, the function $G.$ maps $R \times V$ into $V$. If $(r, v) \in R \times V$, then we write $r \cdot v$ for the element $G.((r, v))$ of $V$; that is,

$$r \cdot v = G.((r, v)) \tag{2-18}$$

We call $r \cdot v$ the *product of* $r$ *and* $v$. We say that $V$ is a (real) *vector space* if the following conditions are satisfied by sums and products:

*Sums*

V1 $v_1 + v_2 = v_2 + v_1$ for all $v_1$ and $v_2$ in $V$

V2 $v_1 + (v_2 + v_3) = (v_1 + v_2) + v_3$ for all $v_1, v_2, v_3$ in $V$

V3 There is a unique element 0 of $V$ such that $v + 0 = 0 + v = v$ for all $v$ in $V$

V4 For every $v$ in $V$, there is a unique element $-v$ of $V$ such that $v + (-v) = (-v) + v = 0$

*Products*

V5 $r_1 \cdot (r_2 \cdot v) = (r_1 r_2) \cdot v$ for all $r_1, r_2$ in $R$ and for all $v$ in $V$

V6 $1 \cdot v = v$ for all $v$ in $V$

V7 $r \cdot (v_1 + v_2) = r \cdot v_1 + r \cdot v_2$ for all $r$ in $R$ and for all $v_1, v_2$ in $V$

V8 $(r_1 + r_2) \cdot v = r_1 \cdot v + r_2 \cdot v$ for all $r_1, r_2$ in $R$ and for all $v$ in $V$

If $V$ is a vector space, we shall call the elements of $V$ "vectors," and we shall *always use lowercase boldface type for vectors in the remainder of this book*. Because of this, we shall usually write $r\mathbf{v}$ in place of $r \cdot \mathbf{v}$.

We shall now give an important example which illustrates the concept of a vector space. This example will play a crucial role in the rest of the book and should be carefully scrutinized by the reader. Let $V$ be the set of all $n$-tuples of real numbers written as columns:

$$\begin{bmatrix} r_1 \\ r_2 \\ \cdot \\ \cdot \\ \cdot \\ r_n \end{bmatrix} \tag{2-19}$$

We sometimes refer to $r_i$ as the $i$th entry of column (2-19). If

$$\mathbf{v} = \begin{bmatrix} r_1 \\ r_2 \\ \cdot \\ \cdot \\ \cdot \\ r_n \end{bmatrix} \quad \text{and} \quad \mathbf{w} = \begin{bmatrix} s_1 \\ s_2 \\ \cdot \\ \cdot \\ \cdot \\ s_n \end{bmatrix}$$

are two elements of $V$, then we set

$$\mathbf{v} + \mathbf{w} = \begin{bmatrix} r_1 + s_1 \\ r_2 + s_2 \\ \cdot \\ \cdot \\ \cdot \\ r_n + s_n \end{bmatrix} \tag{2-20}$$

and if $r \in R$, then we set

$$r \cdot \mathbf{v} = \begin{bmatrix} rr_1 \\ rr_2 \\ \cdot \\ \cdot \\ \cdot \\ rr_n \end{bmatrix} = r\mathbf{v} \tag{2-21}$$

$V$ is then a vector space which we denote by $R_n$, and we call the elements of $V$ *n-column vectors.*

**Exercise 2-5** Verify conditions V1 to V8 for $R_n$. Show also that if $\mathbf{v} \in R_n$, then $-\mathbf{v} = (-1)\mathbf{v}$ and $0\mathbf{v} = \mathbf{0}$. Is this true for any vector space?

**Example 2-12** Let $V$ be the set of all $n$-tuples of real numbers written as rows $(r_1, r_2, \ldots, r_n)$. If $\mathbf{v} = (r_1, r_2, \ldots, r_n)$ and $\mathbf{w} = (s_1, s_2, \ldots, s_n)$ are elements of $V$, then set $\mathbf{v} + \mathbf{w} = (r_1 + s_1, r_2 + s_2, \ldots, r_n + s_n)$; and if $r \in R$, set $r \cdot \mathbf{v} = (rr_1, rr_2, \ldots, rr_n)$. $V$ is then a vector space which is denoted by $R^n$, and the elements of $V$ are called *n-row vectors.*

**Exercise 2-6** Show that $R^n$ is a vector space.

Suppose that $V$ is a vector space and that $W$ is a subset of $V$. We shall say that $W$ is a *subspace* of $V$ if:

1. $\mathbf{w}_1, \mathbf{w}_2 \in W$ implies that $\mathbf{w}_1 + \mathbf{w}_2 \in W$.
2. $r \in R$, $\mathbf{w} \in W$ implies that $r\mathbf{w} \in W$.

In other words, the subset $W$ is a subspace of $V$ if $W$ is itself a vector space with respect to the operations $+$ and $\cdot$. We observe that if $W_1$ and $W_2$ are subspaces of $V$, then $W_1 \cap W_2$ is a subspace of $V$.

If $W_1$ and $W_2$ are subsets of $V$, then we define a subset $W_1 + W_2$ of $V$ by setting

$$W_1 + W_2 = \{\mathbf{v}\text{: there is a } \mathbf{w}_1 \text{ in } W_1 \text{ and a } \mathbf{w}_2 \text{ in } W_2 \text{ such that } \mathbf{v} = \mathbf{w}_1 + \mathbf{w}_2\} \tag{2-22}$$

and we call $W_1 + W_2$ the *sum of* $W_1$ and $W_2$. In particular, if $W_1$ and $W_2$ are *subspaces* of $V$, then $W_1 + W_2$ is also a subspace of $V$. On the other hand, if $W$ is a subset of $V$ and if $r$ is a real number, then we may define a subset $r \cdot W$ of $V$ by setting

$$rW = \{\mathbf{v}\colon \text{there is a } \mathbf{w} \text{ in } W \text{ such that } \mathbf{v} = r\mathbf{w}\}$$

and we call $rW$ the *product* of $r$ and $W$. In particular, if $W$ is a *subspace* of $V$, then $rW$ is a subspace of $V$.

**Example 2-13** Let $W$ be the subset of $R_n$, defined by

$$W = \left\{ \mathbf{v}\colon \mathbf{v} = \begin{bmatrix} 0 \\ r_2 \\ \cdot \\ \cdot \\ \cdot \\ r_n \end{bmatrix} \right\}$$

Then $W$ is a subspace of $R_n$.

**Example 2-14** If $W$ is the subset of $V$ consisting of $\mathbf{0}$ alone, then $W$ is a subspace of $V$.

**Exercise 2-7** Show that if $W_1$ and $W_2$ are subspaces of $V$, then $W_1 \cap W_2$ and $W_1 + W_2$ are subspaces of $V$. Show that if $W$ is a subspace of $V$, then $rW$ is a subspace of $V$.

We shall deal with real vector spaces exclusively throughout this book. However, the notion of a vector space can be defined in a considerably more general way by replacing the set of real numbers $R$ with a set having essentially the same algebraic properties, such as, for example, the set $\mathcal{C}$ of complex numbers. In other words, a set $V$ for which the notions of sum and product with an element of the complex numbers $\mathcal{C}$ are defined and satisfy conditions analogous to VI to V8 is called a complex vector space. A discussion of the general notion of a vector space can be found in Ref. B-9.

We shall see many instances of the usefulness of the notion of a vector space in control problems later in the book. There are many other areas, such as circuit theory, communication theory, and electromagnetic theory, in which the concept of a vector space plays a crucial role and pervades much of the manipulation with which the reader may be familiar. In point of fact, the mathematization of the familiar physical idea of linearity is precisely the motivation for the definition of a vector space.

## 2-6 Linear Combinations and Bases

Suppose that $V$ is a vector space and that $\mathbf{v}_1, \mathbf{v}_2, \ldots, \mathbf{v}_n$ are elements of $V$. Then we say that a vector $\mathbf{v}$ in $V$ is a *linear combination* of the $\mathbf{v}_i$ (or *depends linearly* on the $\mathbf{v}_i$) if there are real numbers $r_1, r_2, \ldots, r_n$ such that

$$\mathbf{v} = r_1\mathbf{v}_1 + r_2\mathbf{v}_2 + \cdots + r_n\mathbf{v}_n\dagger \tag{2-23}$$

† The sum of $n$ elements $\mathbf{v}_1, \mathbf{v}_2, \ldots, \mathbf{v}_n$ in a vector space $V$ can be defined by induction [that is, $\mathbf{v}_1 + \cdots + \mathbf{v}_n = (\mathbf{v}_1 + \cdots + \mathbf{v}_{n-1}) + \mathbf{v}_n$] and is well defined in view of conditions V1 and V2.

We sometimes write $\sum_{i=1}^{n} r_i\mathbf{v}_i$ in place of $r_1\mathbf{v}_1 + r_2\mathbf{v}_2 + \cdots + r_n\mathbf{v}_n$; that is,

$$\sum_{i=1}^{n} r_i\mathbf{v}_i = r_1\mathbf{v}_1 + r_2\mathbf{v}_2 + \cdots + r_n\mathbf{v}_n\dagger \tag{2-24}$$

We say that the set $\{\mathbf{v}_1, \mathbf{v}_2, \ldots, \mathbf{v}_n\}$ of elements of $V$ is a *linearly dependent set* (or that the $\mathbf{v}_i$, $i = 1, 2, \ldots, n$, are *linearly dependent*) if $\mathbf{0}$ is a linear combination of the $\mathbf{v}_i$ in which not all of the $r_i$ are 0. In other words, the $\mathbf{v}_i$, $i = 1, 2, \ldots, n$, are *linearly dependent* if

$$\mathbf{0} = \sum_{i=1}^{n} r_i\mathbf{v}_i = r_1\mathbf{v}_1 + r_2\mathbf{v}_2 + \cdots + r_n\mathbf{v}_n \tag{2-25}$$

and there is some $r_i \neq 0$. If, on the other hand,

$$\mathbf{0} = \sum_{i=1}^{n} r_i\mathbf{v}_i \tag{2-26}$$

implies that $r_i = 0$ for all $i$, $i = 1, 2, \ldots, n$, then the set $\{\mathbf{v}_1, \mathbf{v}_2, \ldots, \mathbf{v}_n\}$ is called a *linearly independent set*, and the $\mathbf{v}_i$, $i = 1, 2, \ldots, n$, are said to be *linearly independent*.

We observe that the vectors $\mathbf{e}_1, \mathbf{e}_2, \ldots, \mathbf{e}_n$ of $R_n$, defined by

$$\mathbf{e}_1 = \begin{bmatrix} 1 \\ 0 \\ \cdot \\ \cdot \\ \cdot \\ 0 \\ 0 \end{bmatrix} \quad \mathbf{e}_2 = \begin{bmatrix} 0 \\ 1 \\ \cdot \\ \cdot \\ \cdot \\ 0 \\ 0 \end{bmatrix} \quad \cdots \quad \mathbf{e}_n = \begin{bmatrix} 0 \\ 0 \\ \cdot \\ \cdot \\ \cdot \\ 0 \\ 1 \end{bmatrix} \tag{2-27}$$

(that is, $\mathbf{e}_i$ has $i$th entry 1 and all other entries 0), are *linearly independent*.

**Exercise 2-8** Show that if $\mathbf{v} \neq \mathbf{0}$, then the set $\{\mathbf{v}\}$ is linearly independent. HINT: Use condition V6.

We note that the set of *all* linear combinations of the vectors $\mathbf{v}_1, \mathbf{v}_2, \ldots, \mathbf{v}_n$ of $V$ is a subspace of $V$, which we call the *span* of the set $\{\mathbf{v}_1, \mathbf{v}_2, \ldots, \mathbf{v}_n\}$. It is easy to show that the span of $\{\mathbf{e}_1, \mathbf{e}_2, \ldots, \mathbf{e}_n\}$ is all of $R_n$.

**Exercise 2-9** Show that the span of $\{\mathbf{v}_1, \mathbf{v}_2, \ldots, \mathbf{v}_n\}$ is a subspace of $V$.

*A set $\{\mathbf{v}_1, \mathbf{v}_2, \ldots, \mathbf{v}_n\}$ of vectors in $V$ is called a (finite) basis of $V$ if:*

1. *The $\mathbf{v}_i$, $i = 1, 2, \ldots, n$, are linearly independent.*
2. *Every element of $V$ is a linear combination of the $\mathbf{v}_i$ (that is, $V$ is the span of $\mathbf{v}_1, \ldots, \mathbf{v}_n$).*

† See previous footnote.

We note that $\mathbf{e}_1, \mathbf{e}_2, \ldots, \mathbf{e}_n$ [see (2-27)] is a basis of $R_n$, which we shall call the *natural* (or *canonical*) basis of $R_n$.

If $\{\mathbf{v}_1, \mathbf{v}_2, \ldots, \mathbf{v}_n\}$ is a basis of $V$, then $\mathbf{v} \in V$ implies that

$$\mathbf{v} = \sum_{i=1}^{n} \alpha_i \mathbf{v}_i \tag{2-28}$$

where the $\alpha_i$ are *unique* elements of $R$. The $\alpha_i$ are called the *coordinates of* $\mathbf{v}$ *with respect to the basis* $\{\mathbf{v}_1, \mathbf{v}_2, \ldots, \mathbf{v}_n\}$. It is easy to see that if

$$\mathbf{v} = \begin{bmatrix} r_1 \\ r_2 \\ \cdot \\ \cdot \\ \cdot \\ r_n \end{bmatrix} \in R_n$$

then $r_1, r_2, \ldots, r_n$ are the coordinates of $\mathbf{v}$ with respect to the basis $\{\mathbf{e}_1, \mathbf{e}_2, \ldots, \mathbf{e}_n\}$. We also observe that if $\{\mathbf{v}_1, \mathbf{v}_2, \ldots, \mathbf{v}_n\}$ is a (fixed) basis of $V$ and if

$$\mathbf{v} = \sum_{j=1}^{n} \alpha_j \mathbf{v}_j$$

is an element of $V$, then we may view the coordinates $\alpha_1, \alpha_2, \ldots, \alpha_n$ of $\mathbf{v}$ as defining a vector $\boldsymbol{\alpha}$ in $R_n$ whose components, with respect to the natural basis $\{\mathbf{e}_1, \mathbf{e}_2, \ldots, \mathbf{e}_n\}$ of $R_n$, are $\alpha_1, \alpha_2, \ldots, \alpha_n$; that is,

$$\boldsymbol{\alpha} = \begin{bmatrix} \alpha_1 \\ \alpha_2 \\ \cdot \\ \cdot \\ \cdot \\ \alpha_n \end{bmatrix}$$

If $V$ has a basis $\{\mathbf{v}_1, \mathbf{v}_2, \ldots, \mathbf{v}_n\}$, then $n$ is called the *dimension of* $V$. We note that it can be shown that if $V$ has a basis, then *any* two bases of $V$ have the same number of elements, so that the notion of dimension is well defined. Moreover, it can be shown that any linearly independent set $\{\mathbf{w}_1, \mathbf{w}_2, \ldots, \mathbf{w}_m\}$ of vectors in $V$ can be extended to a basis of $V$ (that is, there are elements $\mathbf{x}_1, \ldots, \mathbf{x}_{n-m}$ of $V$ such that the set $\{\mathbf{w}_1, \ldots, \mathbf{w}_m, \mathbf{x}_1, \ldots, \mathbf{x}_{n-m}\}$ is a basis of $V$). It follows that the dimension of any proper† subspace $W$ (that is, $W \neq V$) of an $n$-dimensional space $V$ is less than $n$.

**Example 2-15** If $W_1$ and $W_2$ are subspaces of the $n$-dimensional space $V$, then

$$\dim [W_1] + \dim [W_2] = \dim [W_1 + W_2] + \dim [W_1 \cap W_2]$$

† "Proper" here means that $W \subset V$ but $W \neq V$.

where dim [ ] denotes *the dimension of*. For example, in $R_n$, if $W_1$ is the span of $\{\mathbf{e}_1\}$ and $W_2$ is the span of $\{\mathbf{e}_1 + \mathbf{e}_2\}$, then dim $[W_1]$ + dim $[W_2] = 2 =$ dim $[W_1 + W_2]$ + dim $[W_1 \cap W_2]$ since $W_1 + W_2$ is the span of $\{\mathbf{e}_1, \mathbf{e}_2\}$ and $W_1 \cap W_2 = \{\mathbf{0}\}$.

## 2-7 Linear Algebra: Linear Transformations and Matrices

Once the mathematician has defined a notion (as we have done for the concept of a vector space), he then looks for a natural class of transformations which "preserve" the notion. For example, *if $V$ and $W$ are vector spaces and if $\mathcal{A}$ is a transformation (function) of $V$ into $W$, then we can ask the question:* What properties must $\mathcal{A}$ possess in order to preserve or be consistent with the vector-space structure of $V$ and $W$? First of all, $\mathcal{A}$ should preserve the notion of sum; in other words, for all elements $\mathbf{v}_1$ and $\mathbf{v}_2$ of $V$

LT1 $$\mathcal{A}(\mathbf{v}_1 + \mathbf{v}_2) = \mathcal{A}(\mathbf{v}_1) + \mathcal{A}(\mathbf{v}_2) \tag{2-29}$$

Secondly, $\mathcal{A}$ should preserve the notion of product; in other words, if $r$ is in $R$ and $\mathbf{v}$ is in $V$, then

LT2 $$\mathcal{A}(r\mathbf{v}) = r\mathcal{A}(\mathbf{v}) \tag{2-30}$$

A transformation $\mathcal{A}$ of $V$ into $W$ whose domain is all of $V$ and which satisfies conditions LT1 and LT2 is called a *linear transformation*. We shall *always use boldface script capital letters to denote linear transformations in the remainder of this book unless explicit mention to the contrary is made.*

Suppose that $V$ is $n$-dimensional with $\{\mathbf{v}_1, \mathbf{v}_2, \ldots, \mathbf{v}_n\}$ as basis, that $W$ is $m$-dimensional with $\{\mathbf{w}_1, \mathbf{w}_2, \ldots, \mathbf{w}_m\}$ as basis, *and that $\boldsymbol{\mathcal{A}}$ is a linear transformation of $V$ into $W$.* Since every element of $W$ is a linear combination of the $\mathbf{w}_i$ and since $\boldsymbol{\mathcal{A}}(\mathbf{v}_j)$ is an element of $W$, we may write

$$\boldsymbol{\mathcal{A}}(\mathbf{v}_j) = \sum_{i=1}^{m} a_{ij}\mathbf{w}_i \qquad \text{for } j = 1, 2, \ldots, n \tag{2-31}$$

Furthermore, if $\mathbf{v} = \sum_{j=1}^{n} \alpha_j \mathbf{v}_j$ is an element of $V$, then, in view of conditions LT1 and LT2, we see that

$$\boldsymbol{\mathcal{A}}(\mathbf{v}) = \sum_{j=1}^{n} \sum_{i=1}^{m} \alpha_j a_{ij} \mathbf{w}_i \tag{2-32}$$

The set of numbers $\{a_{ij}: i = 1, 2, \ldots, m; j = 1, 2, \ldots, n\}$ may be written in an array, denoted by $\mathbf{A}$, containing $m$ rows and $n$ columns, as

$$\mathbf{A} = \begin{bmatrix} a_{11} & a_{12} & \cdots & a_{1n} \\ a_{21} & a_{22} & \cdots & a_{2n} \\ \cdots & \cdots & \cdots & \cdots \\ a_{m1} & a_{m2} & \cdots & a_{mn} \end{bmatrix} \tag{2-33}$$

Such an array is called an $m \times n$ *matrix*. The numbers $a_{ij}$ are called the *entries* (or coefficients or elements) of the matrix $\mathbf{A}$. We sometimes write

$$\mathbf{A} = (a_{ij})$$

as a shorthand for (2-33). We note that $i$ indicates the row and $j$ the column of $\mathbf{A}$ in which $a_{ij}$ is. We shall *always use boldface capital letters for matrices in the remainder of this book.* We observe that the matrix $\mathbf{A}$ is "associated with" the linear transformation $\mathcal{A}$ *and* the bases $\{\mathbf{v}_1, \mathbf{v}_2, \ldots, \mathbf{v}_n\}$ and $\{\mathbf{w}_1, \mathbf{w}_2, \ldots, \mathbf{w}_m\}$. On the other hand, if

$$\mathbf{B} = \begin{bmatrix} b_{11} & b_{12} & \cdots & b_{1n} \\ b_{21} & b_{22} & \cdots & b_{2n} \\ \cdots & \cdots & \cdots & \cdots \\ b_{m1} & b_{m2} & \cdots & b_{mn} \end{bmatrix} \tag{2-34}$$

is an $m \times n$ matrix, then we can define a linear transformation $\mathcal{B}$ of $V$ into $W$ by setting

$$\mathcal{B}(\mathbf{v}) = \sum_{j=1}^{n} \sum_{i=1}^{m} \alpha_j b_{ij} \mathbf{w}_i \tag{2-35}$$

for $\mathbf{v} = \sum_{j=1}^{n} \alpha_j \mathbf{v}_j$ in $V$. $\mathbf{B}$ is then the matrix of $\mathcal{B}$ with respect to the bases $\{\mathbf{v}_1, \mathbf{v}_2, \ldots, \mathbf{v}_n\}$ and $\{\mathbf{w}_1, \mathbf{w}_2, \ldots, \mathbf{w}_m\}$ of $V$ and $W$, respectively.

**Exercise 2-10** Show that $\mathcal{B}$ as defined by (2-35) is indeed a linear transformation of $V$ into $W$.

**Example 2-16** Let $V = R_2$ and $W = R_3$ and let $\mathcal{A}$ be the linear transformation of $V$ into $W$ defined by $\mathcal{A}(\mathbf{e}_1) = \mathbf{e}_1 + 2\mathbf{e}_2 + 3\mathbf{e}_3$, $\mathcal{A}(\mathbf{e}_2) = \mathbf{e}_3$ [see (2-27)]. Then the matrix $\mathbf{A}$ of $\mathcal{A}$ is given by

$$\mathbf{A} = \begin{bmatrix} 1 & 0 \\ 2 & 0 \\ 3 & 1 \end{bmatrix}$$

If $\mathbf{B}$ is the matrix given by

$$\mathbf{B} = \begin{bmatrix} 1 & 2 \\ 0 & 2 \\ 1 & 0 \end{bmatrix}$$

then $\mathcal{B}(\mathbf{e}_1) = \mathbf{e}_1 + \mathbf{e}_3$ and $\mathcal{B}(\mathbf{e}_2) = 2\mathbf{e}_1 + 2\mathbf{e}_2$.

We note that if $\mathcal{A}$ is the linear transformation of $V$ into $W$, with $\mathcal{A}(\mathbf{v}) = \mathbf{0}$ for all $\mathbf{v}$ in $V$, then the matrix $\mathbf{A}$ of $\mathcal{A}$ has all 0 entries; and conversely, if the matrix $\mathbf{B}$ has all 0 entries, then the linear transformation $\mathcal{B}$ of Eq. (2-35) has the property that $\mathcal{B}(\mathbf{v}) = \mathbf{0}$ for all $\mathbf{v}$ in $V$.

## 2-8 Linear Algebra: Operations on Linear Transformations and Matrices

Let us suppose that $\mathcal{A}$ and $\mathcal{B}$ are linear transformations of $V$ into $W$ and that $\mathbf{A}$ and $\mathbf{B}$ are the matrices of $\mathcal{A}$ and $\mathcal{B}$, respectively, with respect to the bases $\{\mathbf{v}_1, \mathbf{v}_2, \ldots, \mathbf{v}_n\}$ and $\{\mathbf{w}_1, \mathbf{w}_2, \ldots, \mathbf{w}_m\}$ of $V$ and $W$, respectively. Let us find the matrices of the transformations $\mathcal{A} + \mathcal{B}$ and $r\mathcal{A}$, $r \in R$, which are defined by

$$(\mathcal{A} + \mathcal{B})(\mathbf{v}) = \mathcal{A}(\mathbf{v}) + \mathcal{B}(\mathbf{v}) \qquad \text{for all } \mathbf{v} \in V \tag{2-36}$$

and

$$(r\mathcal{A})(\mathbf{v}) = r[\mathcal{A}(\mathbf{v})] \qquad \text{for all } \mathbf{v} \in V \tag{2-37}$$

It is easy to see that $\mathcal{A} + \mathcal{B}$ and $r\mathcal{A}$ are again linear transformations of $V$ into $W$, and we call $\mathcal{A} + \mathcal{B}$ the *sum* of $\mathcal{A}$ and $\mathcal{B}$, and $r\mathcal{A}$ the *product* of $r$ and $\mathcal{A}$. If $\mathbf{A} = (a_{ij})$ and $\mathbf{B} = (b_{ij})$, then

$$(\mathcal{A} + \mathcal{B})(\mathbf{v}_j) = \mathcal{A}(\mathbf{v}_j) + \mathcal{B}(\mathbf{v}_j) = \sum_{i=1}^{m} a_{ij}\mathbf{w}_i + \sum_{i=1}^{m} b_{ij}\mathbf{w}_i = \sum_{i=1}^{m} (a_{ij} + b_{ij})\mathbf{w}_i \tag{2-38}$$

and

$$(r\mathcal{A})(\mathbf{v}_j) = r[\mathcal{A}(\mathbf{v}_j)] = r \sum_{i=1}^{m} a_{ij}\mathbf{w}_i = \sum_{i=1}^{m} (ra_{ij})\mathbf{w}_i \tag{2-39}$$

In other words, the matrix of $\mathcal{A} + \mathcal{B}$ is $(a_{ij} + b_{ij})$, and the matrix of $r\mathcal{A}$ is $(ra_{ij})$. We denote these matrices by $\mathbf{A} + \mathbf{B}$ and $r\mathbf{A}$, respectively. In explicit form,

$$\mathbf{A} + \mathbf{B} = \begin{bmatrix} a_{11} + b_{11} & a_{12} + b_{12} & \cdots & a_{1n} + b_{1n} \\ a_{21} + b_{21} & a_{22} + b_{22} & \cdots & a_{2n} + b_{2n} \\ \cdots & \cdots & \cdots & \cdots \\ a_{m1} + b_{m1} & a_{m2} + b_{m2} & \cdots & a_{mn} + b_{mn} \end{bmatrix} \tag{2-40}$$

and

$$r\mathbf{A} = \begin{bmatrix} ra_{11} & ra_{12} & \cdots & ra_{1n} \\ ra_{21} & ra_{22} & \cdots & ra_{2n} \\ \cdots & \cdots & \cdots & \cdots \\ ra_{m1} & ra_{m2} & \cdots & ra_{mn} \end{bmatrix} \tag{2-41}$$

$\mathbf{A} + \mathbf{B}$ is called the *sum* of $\mathbf{A}$ and $\mathbf{B}$, and $r\mathbf{A}$ is called the *product* of $r$ and $\mathbf{A}$. It can be shown that the set of *all* linear transformations of $V$ into $W$ and the set of *all* $m \times n$ matrices are vector spaces with respect to these notions of sum and product. We shall denote the set of all linear transformations of $V$ into $W$ by $\mathcal{L}(V, W)$, and we shall denote the set of all $m \times n$ matrices by $\mathfrak{M}(m, n)$.

**Exercise 2-11** Prove that $\mathcal{L}(V, R)$ is a vector space; i.e., verify axioms V1 to V8 for $\mathcal{L}(V, R)$. We denote $\mathcal{L}(V, R)$ by $\mathbf{V}^*$, and we call $V^*$ the *dual space* of $V$.

**Exercise 2-12** Let $V = R_n$ and let $\mathcal{A} \in R_n^*$ (see Exercise 2-11). Show that if $\mathbf{v} \in R_n$, then $\mathcal{A}(\mathbf{v}) = \sum_{j=1}^{n} \alpha_j \mathcal{A}(\mathbf{e}_j)$, where the $\alpha_j$ are the coordinates of $\mathbf{v}$ with respect to the $\mathbf{e}_j$ (see Sec. 2-6). Show that the transformations $\mathcal{A}_i$ of $R_n$ into $R$, defined by $\mathcal{A}_i(\mathbf{v}) = \alpha_i \left(\text{where } \mathbf{v} = \sum_{j=1}^{n} \alpha_j \mathbf{e}_j\right)$ for $i = 1, 2, \ldots, n$, are in $R_n^*$ and that $\mathcal{A}_i(\mathbf{e}_j) = \delta_{ij}$, where $\delta_{ij} = 0$ if $i \neq j$ and $\delta_{ii} = 1$. $\left(\text{HINT: } \mathbf{e}_i = \sum_{j=1}^{n} \delta_{ij}\mathbf{e}_j.\right)$ Show that $\{\mathcal{A}_1, \mathcal{A}_2, \ldots, \mathcal{A}_n\}$ is a basis of $R_n^*$. We call $\{\mathcal{A}_1, \mathcal{A}_2, \ldots, \mathcal{A}_n\}$ the *dual basis* to $\{\mathbf{e}_1, \mathbf{e}_2, \ldots, \mathbf{e}_n\}$. What is the matrix of $\mathcal{A}_i$ with respect to the bases $\{\mathbf{e}_1, \mathbf{e}_2, \ldots, \mathbf{e}_n\}$ of $R_n$ and 1 of $R$? Can you find a *one-one* linear transformation of $R_n^*$ *onto* $R^n$ (the space of $n$-row vectors)? (HINT: Let $\mathbf{f}_i$ be the $n$-row vector, with 1 in the $i$th place and 0s elsewhere.) If you can find such a linear transformation, what is the matrix "associated with" it and the bases $\{\mathcal{A}_1, \mathcal{A}_2, \ldots, \mathcal{A}_n\}$ and $\{\mathbf{f}_1, \mathbf{f}_2, \ldots, \mathbf{f}_n\}$?

Now let us suppose that $V$, $W$, and $X$ are vector spaces with bases $\{\mathbf{v}_1, \mathbf{v}_2, \ldots, \mathbf{v}_n\}$, $\{\mathbf{w}_1, \mathbf{w}_2, \ldots, \mathbf{w}_m\}$, and $\{\mathbf{x}_1, \mathbf{x}_2, \ldots, \mathbf{x}_p\}$, respectively. If $\mathcal{A}$ is a linear transformation of $W$ into $X$ and if $\mathcal{B}$ is a linear transformation of $V$ into $W$, then $\mathcal{A} \circ \mathcal{B}$ is a linear transformation of $V$ into $X$. For example, let us verify condition LT1. If $\mathbf{v}$ and $\mathbf{v}'$ are elements of $V$, then $(\mathcal{A} \circ \mathcal{B})(\mathbf{v} + \mathbf{v}') = \mathcal{A}[\mathcal{B}(\mathbf{v} + \mathbf{v}')] = \mathcal{A}[\mathcal{B}(\mathbf{v}) + \mathcal{B}(\mathbf{v}')] = \mathcal{A}[\mathcal{B}(\mathbf{v})] + \mathcal{A}[\mathcal{B}(\mathbf{v}')] = (\mathcal{A} \circ \mathcal{B})(\mathbf{v}) + (\mathcal{A} \circ \mathcal{B})(\mathbf{v}')$. Let us now try to find the matrix of $\mathcal{A} \circ \mathcal{B}$, where we suppose that $\mathbf{A} = (a_{ik})$ and $\mathbf{B} = (b_{kj})$ are the matrices of $\mathcal{A}$ and $\mathcal{B}$. We have

$$(\mathcal{A} \circ \mathcal{B})(\mathbf{v}_j) = \mathcal{A}[\mathcal{B}(\mathbf{v}_j)] = \mathcal{A}\Big[\sum_{k=1}^{m} b_{kj}\mathbf{w}_k\Big]$$

$$= \sum_{k=1}^{m} b_{kj}\mathcal{A}(\mathbf{w}_k) = \sum_{k=1}^{m} b_{kj} \sum_{i=1}^{p} a_{ik}\mathbf{x}_i = \sum_{i=1}^{p} \Big(\sum_{k=1}^{m} a_{ik}b_{kj}\Big)\mathbf{x}_i \tag{2-42}$$

In other words, the matrix $\mathbf{C} = (c_{ij})$ of $\mathcal{A} \circ \mathcal{B}$ has entries of the form

$$c_{ij} = \sum_{k=1}^{m} a_{ik}b_{kj} \tag{2-43}$$

for $i = 1, 2, \ldots, p$ and $j = 1, 2, \ldots, n$. We call $\mathbf{C}$ the *product* of $\mathbf{A}$ and $\mathbf{B}$, and we write

$$\mathbf{C} = \mathbf{AB} \tag{2-44}$$

We note that the number of columns of $\mathbf{A}$ must be the same as the number of rows of $\mathbf{B}$ in order that the product $\mathbf{AB}$ may be defined. In general, $\mathbf{AB} \neq \mathbf{BA}$.

**Example 2-17** Let $\mathbf{A} = \begin{bmatrix} 1 & 0 \\ -1 & 2 \end{bmatrix}$ and let $\mathbf{B} = \begin{bmatrix} 0 & 0 & 1 \\ 1 & 0 & 2 \end{bmatrix}$. Then $\mathbf{AB} = \begin{bmatrix} 0 & 0 & 1 \\ 2 & 0 & 3 \end{bmatrix}$.

**Exercise 2-13** Let $\mathbf{A} = \begin{bmatrix} 0 & -1 \\ 1 & 0 \end{bmatrix}$ and $\mathbf{B} = \begin{bmatrix} 1 & 2 \\ 0 & -1 \end{bmatrix}$. What is $\mathbf{AB}$? What is $\mathbf{BA}$?

If $\mathbf{A} = (a_{ij})$ is an $m \times n$ matrix, then the $n \times m$ matrix $\mathbf{A}'$, given by

$$\mathbf{A}' = \begin{bmatrix} a_{11} & a_{21} & \cdots & a_{m1} \\ a_{12} & a_{22} & \cdots & a_{m2} \\ \cdots & \cdots & \cdots & \cdots \\ a_{1n} & a_{2n} & \cdots & a_{mn} \end{bmatrix} \tag{2-45}$$

is called the *transpose of* $\mathbf{A}$. We note that the rows of $\mathbf{A}'$ are the columns of $\mathbf{A}$ and that the columns of $\mathbf{A}'$ are the rows of $\mathbf{A}$. If we represent the transpose of **A** by

$$\mathbf{A}' = (a'_{ji}) \qquad j = 1, 2, \ldots, n; i = 1, 2, \ldots, m \tag{2-46}$$

then $a'_{ji} = a_{ij}$.

**Example 2-18** If $\mathbf{A} = \begin{bmatrix} 0 & 1 & 0 \\ 2 & 3 & 4 \end{bmatrix}$, then

$$\mathbf{A}' = \begin{bmatrix} 0 & 2 \\ 1 & 3 \\ 0 & 4 \end{bmatrix}$$

If $V$ has $\{\mathbf{v}_1, \mathbf{v}_2, \ldots, \mathbf{v}_n\}$ as basis, if $W$ has $\{\mathbf{w}_1, \mathbf{w}_2, \ldots, \mathbf{w}_m\}$ as basis, and if $\mathcal{A}$ is a linear transformation of $V$ into $W$ with **A** as matrix with respect to these bases, then $\mathbf{v} = \sum_{i=1}^{n} \alpha_j \mathbf{v}_j$ implies that

$$\mathcal{A}(\mathbf{v}) = \sum_{j=1}^{n} \alpha_j \sum_{i=1}^{m} a_{ij}\mathbf{w}_i = \sum_{i=1}^{m} \Big( \sum_{j=1}^{n} a_{ij}\alpha_j \Big) \mathbf{w}_i = \sum_{i=1}^{m} \Big( \sum_{j=1}^{n} a'_{ji}\alpha_j \Big) \mathbf{w}_i \tag{2-47}$$

In other words, **A**′ may be viewed as defining a transformation from the coordinates $\alpha_j$ of **v** to the coordinates of $\mathcal{A}(\mathbf{v})$. We note also that

$$(\mathbf{A}')' = \mathbf{A} \qquad (\mathbf{A} + \mathbf{B})' = \mathbf{A}' + \mathbf{B}' \qquad (\mathbf{AB})' = \mathbf{B}'\mathbf{A}' \tag{2-48}$$

Let us look at the transpose of a matrix in another way. Suppose that $\mathbf{A} = (a_{ij})$ is an $m \times n$ matrix and that **w** is an element of $R_m$, that is, that

$$\mathbf{w} = \begin{bmatrix} s_1 \\ s_2 \\ \cdot \\ \cdot \\ \cdot \\ s_m \end{bmatrix} \tag{2-49}$$

Then we can view **w** as an $m \times 1$ matrix, and we can consider the product **A**′**w** with

$$\mathbf{A}'\mathbf{w} = \begin{bmatrix} \sum_{i=1}^{m} a'_{1i}s_i \\ \sum_{i=1}^{m} a'_{2i}s_i \\ \cdot \\ \cdot \\ \cdot \\ \sum_{i=1}^{m} a'_{ni}s_i \end{bmatrix} \tag{2-50}$$

$\mathbf{A}'\mathbf{w}$ is an element of $R_n$, and we can define a transformation $\mathcal{A}'$ of $R_m$ into $R_n$ by setting

$$\mathcal{A}'(\mathbf{w}) = \mathbf{A}'\mathbf{w} \tag{2-51}$$

for $\mathbf{w}$ in $R_m$. It is easy to see that $\mathcal{A}'$ is a linear transformation of $R_m$ into $R_n$. Since the basis elements $\mathbf{e}_1, \mathbf{e}_2, \ldots, \mathbf{e}_m$ of $R_m$ are given by

$$\mathbf{e}_i = \begin{bmatrix} 0 \\ 0 \\ \cdot \\ \cdot \\ \cdot \\ 1 \\ \cdot \\ \cdot \\ \cdot \\ 0 \end{bmatrix} \leftarrow i\text{th row} \tag{2-52}$$

we see that

$$\mathcal{A}'(\mathbf{e}_i) = \begin{bmatrix} a'_{1i} \\ a'_{2i} \\ \cdot \\ \cdot \\ \cdot \\ a'_{ni} \end{bmatrix} = \sum_{j=1}^{n} a'_{ji}\mathbf{e}_j \tag{2-53}$$

In other words, $\mathbf{A}'$ is the matrix of $\mathcal{A}'$ with respect to the natural bases† of $R_m$ and $R_n$.

## 2-9 Linear Algebra: Linear Transformations of $V$ into $V$

We shall now turn our attention to linear transformations of a vector space $V$ into itself and study them in detail. We suppose that $\{\mathbf{v}_1, \mathbf{v}_2, \ldots, \mathbf{v}_n\}$ is a given basis of $V$, and we observe that if $\mathcal{A}$ is a linear transformation of $V$ into $V$, then the matrix $\mathbf{A}$ of $\mathcal{A}$ with respect to this basis has $n$ rows and $n$ columns (that is, $\mathbf{A}$ is an $n \times n$ matrix). A particularly important linear transformation of $V$ into $V$ is the *identity* transformation $\mathcal{I}$, which is given by

$$\mathcal{I}(\mathbf{v}) = \mathbf{v} \tag{2-54}$$

The matrix $\mathbf{I}$ of $\mathcal{I}$ has the form

$$\mathbf{I} = \begin{bmatrix} 1 & 0 & \cdots & 0 \\ 0 & 1 & \cdots & 0 \\ \cdots & \cdots & \cdots & \cdots \\ 0 & 0 & \cdots & 1 \end{bmatrix} \tag{2-55}$$

that is,

$$\mathbf{I} = (\delta_{ij}) \tag{2-56}$$

† That is, the bases $\{\mathbf{e}_1, \mathbf{e}_2, \ldots, \mathbf{e}_m\}$ and $\{\mathbf{e}_1, \mathbf{e}_2, \ldots, \mathbf{e}_n\}$.

where $\delta_{ij} = 0$ if $i \neq j$ and $\delta_{ii} = 1$. $\delta_{ij}$ is called the *Kronecker delta*. **I** is called the *identity* matrix. We observe that $\mathcal{A} \circ \mathcal{I} = \mathcal{I} \circ \mathcal{A} = \mathcal{A}$ for any linear transformation $\mathcal{A}$ of $V$ into $V$ and hence that

$$\mathbf{IA} = \mathbf{AI} = \mathbf{A} \tag{2-57}$$

for any $n \times n$ matrix **A**.

**Exercise 2-14** Prove (2-57) by using formula (2-43).

**Exercise 2-15** Show in two ways that if

$$\mathbf{v} = \begin{bmatrix} r_1 \\ r_2 \\ \cdot \\ \cdot \\ \cdot \\ r_n \end{bmatrix}$$

is in $R_n$ and if **I** is the $n \times n$ identity matrix, then $\mathbf{Iv} = \mathbf{v}$ [see (2-50)]. HINT: Use formula (2-43) and definition (2-51).

Suppose that $\mathcal{A}$ is a linear transformation of $V$ into itself and that there is a linear transformation $\mathcal{B}$ of $V$ into itself such that

$$\mathcal{A} \circ \mathcal{B} = \mathcal{B} \circ \mathcal{A} = \mathcal{I} \tag{2-58}$$

Then $\mathcal{A}$ is said to be *nonsingular* and $\mathcal{B}$, which we denote by $\mathcal{A}^{-1}$, is called the *inverse of* $\mathcal{A}$. We observe that if **A** and **B** are the matrices of $\mathcal{A}$ and $\mathcal{B}$, respectively, then Eq. (2-58) implies that

$$\mathbf{BA} = \mathbf{AB} = \mathbf{I} \tag{2-59}$$

On the other hand, if **A** is an $n \times n$ matrix and if there is an $n \times n$ matrix **B** which satisfies (2-59), then **A** is said to be *nonsingular* and **B**, which we denote by $\mathbf{A}^{-1}$, is called the *inverse of* **A**. We note that

$$\mathbf{A}^{-1}\mathbf{A} = \mathbf{AA}^{-1} = \mathbf{I} \tag{2-60}$$

The following theorem (whose proof is outlined in Exercise 2-16) gives us more insight into the notion of a nonsingular linear transformation.

***Theorem 2-2*** *If $V$ is n-dimensional and if $\mathcal{A}$ is a linear transformation of $V$ into itself, then*

*a. $\mathcal{A}$ is nonsingular if and only if $\mathcal{A}(\mathbf{v}) = \mathbf{0}$ implies that $\mathbf{v} = \mathbf{0}$.*

*b. $\mathcal{A}$ is nonsingular if and only if the range of $\mathcal{A}$ is all of $V$.*

An important consequence of this theorem is: $\mathcal{A}$ is nonsingular if and only if "$\{\mathbf{v}_1, \mathbf{v}_2, \ldots, \mathbf{v}_n\}$ is a basis of $V$" implies that $\{\mathcal{A}(\mathbf{v}_1), \mathcal{A}(\mathbf{v}_2), \ldots, \mathcal{A}(\mathbf{v}_n)\}$ is a basis of $V$. To put it another way, *nonsingular transformations correspond to changes in the basis of $V$*. We note that the identity $\mathcal{I}$ is always nonsingular. It can also be shown that an $n \times n$ *matrix* **A** *is nonsingular if and only if the determinant of* **A**, *denoted by* det **A**, *is nonzero*, i.e., $\det \mathbf{A} \neq 0$.

**Example 2-19** Let $\mathcal{A}$ be the linear transformation of $R_2$ into itself defined by $\mathcal{A}(\mathbf{e}_1) = a\mathbf{e}_1 + b\mathbf{e}_2$, $\mathcal{A}(\mathbf{e}_2) = c\mathbf{e}_1 + d\mathbf{e}_2$. Then the matrix $\mathbf{A}$ of $\mathcal{A}$ with respect to $\{\mathbf{e}_1, \mathbf{e}_2\}$ is given by

$$\mathbf{A} = \begin{bmatrix} a & c \\ b & d \end{bmatrix}$$

and $\mathbf{A}$ is nonsingular if and only if $ad - bc \neq 0$. If $\mathcal{A}$ is nonsingular, then $\mathcal{A}^{-1}$ is defined by

$$\mathcal{A}^{-1}(\mathbf{e}_1) = \frac{d}{ad - bc}\,\mathbf{e}_1 + \frac{-b}{ad - bc}\,\mathbf{e}_2 \qquad \mathcal{A}^{-1}(\mathbf{e}_2) = \frac{-c}{ad - bc}\,\mathbf{e}_1 + \frac{a}{ad - bc}\,\mathbf{e}_2$$

and $\mathbf{A}^{-1}$ is given by

$$\mathbf{A}^{-1} = \begin{bmatrix} d/ad - bc & -c/ad - bc \\ -b/ad - bc & a/ad - bc \end{bmatrix}$$

For example, if $\mathbf{A} = \begin{bmatrix} 1 & 2 \\ -1 & 3 \end{bmatrix}$, then $\mathbf{A}^{-1} = \begin{bmatrix} \frac{3}{5} & -\frac{2}{5} \\ \frac{1}{5} & \frac{1}{5} \end{bmatrix}$.

**Exercise 2-16 Proof of Theorem 2-2** Let $\mathcal{A}$ be a linear transformation of $V$ into $V$ such that (1) if $\mathbf{v}_1 \neq \mathbf{v}_2$, then $\mathcal{A}(\mathbf{v}_1) \neq \mathcal{A}(\mathbf{v}_2)$ and (2) the range of $\mathcal{A}$ is *all* of $V$. Show that $\mathcal{A}$ is nonsingular. [HINT: If $\mathbf{v} \in V$, then $\mathbf{v} = \mathcal{A}(\mathbf{w})$ for some $\mathbf{w}$. Set $\mathcal{A}^{-1}(\mathbf{v}) = \mathbf{w}$. Is $\mathbf{w}$ unique?] Show that if $\mathcal{B}$ is a nonsingular linear transformation of $V$ into $V$, then $\mathcal{B}$ satisfies conditions 1 and 2. Now prove Theorem 2-2. [HINT: Use properties 1 and 2. For example, show that "$\mathcal{A}(\mathbf{v}) = \mathbf{0}$ implies that $\mathbf{v} = \mathbf{0}$" ensures that if $\{\mathbf{v}_1, \mathbf{v}_2, \ldots, \mathbf{v}_n\}$ is a basis of $V$, then $\mathcal{A}(\mathbf{v}_1), \mathcal{A}(\mathbf{v}_2), \ldots, \mathcal{A}(\mathbf{v}_n)$ are linearly independent.]

Let us suppose that $\mathcal{C}$ is a linear transformation of $V$ into itself and that $\{\mathbf{v}_1, \mathbf{v}_2, \ldots, \mathbf{v}_n\}$ and $\{\mathbf{w}_1, \mathbf{w}_2, \ldots, \mathbf{w}_n\}$ are bases of $V$. Then we may write

$$\mathcal{C}(\mathbf{v}_j) = \sum_{i=1}^{n} a_{ij}\mathbf{v}_i \qquad j = 1, 2, \ldots, n$$

and

$$\mathcal{C}(\mathbf{w}_j) = \sum_{i=1}^{n} b_{ij}\mathbf{w}_i \qquad j = 1, 2, \ldots, n \tag{2-61}$$

In other words, $\mathbf{A} = (a_{ij})$ is the matrix of $\mathcal{C}$ with respect to the basis $\{\mathbf{v}_1, \mathbf{v}_2, \ldots, \mathbf{v}_n\}$, and $\mathbf{B} = (b_{ij})$ is the matrix of $\mathcal{C}$ with respect to the basis $\{\mathbf{w}_1, \mathbf{w}_2, \ldots, \mathbf{w}_n\}$. Let us determine the relationship between $\mathbf{A}$ and $\mathbf{B}$. Since the $\mathbf{w}_i$ are in $V$ and $\{\mathbf{v}_1, \mathbf{v}_2, \ldots, \mathbf{v}_n\}$ is a basis,

$$\mathbf{w}_j = \sum_{i=1}^{n} p_{ij}\mathbf{v}_i \qquad j = 1, 2, \ldots, n \tag{2-62a}$$

and since the $\mathbf{v}_k$ are in $V$ and $\{\mathbf{w}_1, \mathbf{w}_2, \ldots, \mathbf{w}_n\}$ is a basis, we have

$$\mathbf{v}_k = \sum_{j=1}^{n} q_{jk}\mathbf{w}_j \qquad k = 1, 2, \ldots, n \tag{2-62b}$$

If we let $\mathbf{P}$ be the matrix $(p_{ij})$ and $\mathbf{Q}$ be the matrix $(q_{jk})$, then we can easily see that $\mathbf{PQ} = \mathbf{QP} = \mathbf{I}$ and hence that $\mathbf{Q} = \mathbf{P}^{-1}$. Now,

$$\mathcal{C}(\mathbf{w}_j) = \mathcal{C}\left(\sum_{i=1}^{n} p_{ij}\mathbf{v}_i\right) = \sum_{i=1}^{n} p_{ij}\mathcal{C}(\mathbf{v}_i) = \sum_{i=1}^{n}\sum_{k=1}^{n} (p_{ij}a_{ki})\mathbf{v}_k$$
$$= \sum_{i=1}^{n}\sum_{k=1}^{n}\sum_{r=1}^{n} (p_{ij}a_{ki}q_{rk})\mathbf{w}_r = \sum_{r=1}^{n}\left(\sum_{k=1}^{n}\sum_{i=1}^{n} q_{rk}a_{ki}p_{ij}\right)\mathbf{w}_r = \sum_{r=1}^{n} b_{rj}\mathbf{w}_r \tag{2-63}$$

It follows that

$$\mathbf{B} = \mathbf{P}^{-1}\mathbf{A}\mathbf{P} = \mathbf{Q}\mathbf{A}\mathbf{Q}^{-1} \tag{2-64}$$

The matrices $\mathbf{A}$ and $\mathbf{B}$ are then called *similar matrices*, and the linear transformations defined by (2-62*a*) and (2-62*b*) are called *similarity transformations*. In general, if $\mathbf{A}$ and $\mathbf{B}$ are any two $n \times n$ matrices for which there is a nonsingular matrix $\mathbf{P}$ such that Eq. (2-64) is satisfied, then $\mathbf{A}$ and $\mathbf{B}$ are called *similar matrices*. We note that similar matrices correspond to the same linear transformation but with a different choice of basis.

## 2-10 Linear Algebra: Eigenvectors and Eigenvalues

Suppose that $\mathcal{A}$ is a linear transformation of $V$ into itself and that $\mathbf{A}$ is the matrix of $\mathcal{A}$ with respect to the basis $\{\mathbf{v}_1, \mathbf{v}_2, \ldots, \mathbf{v}_n\}$ of $V$. A vector $\mathbf{v} \neq \mathbf{0}$ in $V$ is called an *eigenvector* of $\mathcal{A}$ if there is a $\lambda$ *in* $R$ such that

$$\mathcal{A}(\mathbf{v}) = \lambda\mathbf{v} \tag{2-65}$$

$\lambda$ is then called an *eigenvalue* of $\mathcal{A}$. Let us see what this means. To begin with, we observe that since $\mathbf{v} \neq \mathbf{0}$, the span of $\mathbf{v}$ (see Sec. 2-6) is a one-dimensional subspace $V_1$ of $V$. It is easy to see that $\mathcal{A}(V_1) \subset V_1$ (note that $\mathbf{v}$ is a basis of $V_1$). On the other hand, if $W_1$ is a one-dimensional subspace of $V$ with $\mathbf{w}_1$ as basis and if $\mathcal{A}(W_1) \subset W_1$, then $\mathcal{A}(\mathbf{w}_1) \in W_1$ implies that

$$\mathcal{A}(\mathbf{w}_1) = \lambda\mathbf{w}_1 \tag{2-66}$$

so that $\mathbf{w}_1$ is an eigenvector of $\mathcal{A}$. In different terms, eigenvalues of $\mathcal{A}$ correspond to one-dimensional subspaces of $V$ which $\mathcal{A}$ maps into themselves.

**Example 2-20** Suppose that $\mathcal{A}$ is the linear transformation of $R_2$ into itself for which $\mathcal{A}(\mathbf{e}_1) = \mathbf{e}_1 + \mathbf{e}_2$ and $\mathcal{A}(\mathbf{e}_2) = \mathbf{e}_1 + \mathbf{e}_2$; then $\mathcal{A}(\mathbf{e}_1 + \mathbf{e}_2) = 2(\mathbf{e}_1 + \mathbf{e}_2)$, so that $\mathbf{e}_1 + \mathbf{e}_2$ is an eigenvector of $\mathcal{A}$ and 2 is an eigenvalue of $\mathcal{A}$. Also, $\mathcal{A}(\mathbf{e}_1 - \mathbf{e}_2) = \mathbf{0} = 0(\mathbf{e}_1 - \mathbf{e}_2)$, so that $\mathbf{e}_1 - \mathbf{e}_2$ is an eigenvector of $\mathcal{A}$ and 0 is an eigenvalue of $\mathcal{A}$. The matrix $\mathbf{A}$ of $\mathcal{A}$ with respect to $\{\mathbf{e}_1, \mathbf{e}_2\}$ is given by

$$\mathbf{A} = \begin{bmatrix} 1 & 1 \\ 1 & 1 \end{bmatrix}$$

This matrix is similar to the matrix **B** given by

$$\mathbf{B} = \begin{bmatrix} 0 & 0 \\ 0 & 2 \end{bmatrix}$$

In fact,

$$\mathbf{B} = \begin{bmatrix} 1/\sqrt{2} & -1/\sqrt{2} \\ 1/\sqrt{2} & 1/\sqrt{2} \end{bmatrix} \begin{bmatrix} 1 & 1 \\ 1 & 1 \end{bmatrix} \begin{bmatrix} 1/\sqrt{2} & 1/\sqrt{2} \\ -1/\sqrt{2} & 1/\sqrt{2} \end{bmatrix}$$

What are the effects of the similarity transformation here?

We may look at eigenvalues in another way. If $\mathbf{v}$ is an eigenvector of $\mathfrak{A}$ with $\lambda$ as eigenvalue, then $\mathfrak{A}(\mathbf{v}) - \lambda\mathbf{v} = \mathbf{0}$, and noting that $\mathbf{v} = \sum_{j=1}^{n} \alpha_j \mathbf{v}_j$, since $\{\mathbf{v}_1, \mathbf{v}_2, \ldots, \mathbf{v}_n\}$ is a basis of $V$, we can see that

$$\begin{aligned} \mathfrak{A}(\mathbf{v}) - \lambda\mathbf{v} &= \sum_{j=1}^{n} \alpha_j \mathfrak{A}(\mathbf{v}_j) - \lambda \Big( \sum_{j=1}^{n} \alpha_j \mathbf{v}_j \Big) \\ &= \sum_{j=1}^{n} \sum_{i=1}^{n} \alpha_j a_{ij} \mathbf{v}_i - \sum_{j=1}^{n} (\lambda\alpha_j)\mathbf{v}_j = \mathbf{0} \end{aligned} \tag{2-67}$$

However, we know that $\mathbf{v}_j = \sum_{i=1}^{n} \delta_{ij}\mathbf{v}_i$, where $\delta_{ij}$ is the Kronecker delta, and so we may write Eq. (2-67) in the form

$$\begin{aligned} \mathfrak{A}(\mathbf{v}) - \lambda\mathbf{v} &= \sum_{j=1}^{n} \sum_{i=1}^{n} \alpha_j a_{ij} \mathbf{v}_i - \sum_{j=1}^{n} \sum_{i=1}^{n} \lambda\alpha_j \delta_{ij} \mathbf{v}_i \\ &= \sum_{i=1}^{n} \Big( \sum_{j=1}^{n} \alpha_j a_{ij} - \lambda\alpha_j\delta_{ij} \Big) \mathbf{v}_i = \mathbf{0} \end{aligned} \tag{2-67a}$$

Since the $\mathbf{v}_i$ are a basis of $V$, we may conclude that

$$\begin{aligned} (a_{11} - \lambda)\alpha_1 + a_{12}\alpha_2 + \cdots + a_{1n}\alpha_n &= 0 \\ a_{21}\alpha_1 + (a_{22} - \lambda)\alpha_2 + \cdots + a_{2n}\alpha_n &= 0 \\ \cdots\cdots\cdots\cdots\cdots\cdots\cdots\cdots\cdots & \\ a_{n1}\alpha_1 + a_{n2}\alpha_2 + \cdots + (a_{nn} - \lambda)\alpha_n &= 0 \end{aligned} \tag{2-68}$$

It follows that the system of $n$ linear equations in $n$ unknowns $x_1, x_2, \ldots, x_n$, given by

$$\begin{aligned} (a_{11} - \lambda)x_1 + a_{12}x_2 + \cdots + a_{1n}x_n &= 0 \\ a_{21}x_1 + (a_{22} - \lambda)x_2 + \cdots + a_{2n}x_n &= 0 \\ \cdots\cdots\cdots\cdots\cdots\cdots\cdots\cdots\cdots & \\ a_{n1}x_1 + a_{n2}x_2 + \cdots + (a_{nn} - \lambda)x_n &= 0 \end{aligned} \tag{2-69}$$

has the *nonzero* solution $x_1 = \alpha_1, x_2 = \alpha_2, \ldots, x_n = \alpha_n$ (inasmuch as $\mathbf{v} = \sum_{j=1}^{n} \alpha_j\mathbf{v}_j \neq \mathbf{0}$). We may then conclude that

$$\det(\mathbf{A} - \lambda\mathbf{I}) = 0 \tag{2-70}$$

If we expand det $(\mathbf{A} - \lambda\mathbf{I})$ as a function of $\lambda$, then we obtain a polynomial relation of the form

$$\det(\mathbf{A} - \lambda\mathbf{I}) = (-1)^n\lambda^n + c_1\lambda^{n-1} + \cdots + c_n = 0 \tag{2-71}$$

where $c_1, c_2, \ldots, c_n$ depend upon the $a_{ij}$. In other words, $\lambda$ is a *root* of the $n$th-degree polynomial $p(z)$ in the variable $z$, given by

$$p(z) = \det(\mathbf{A} - z\mathbf{I}) = (-1)^n z^n + c_1 z^{n-1} + \cdots + c_n \tag{2-72}$$

The polynomial $p(z)$ is sometimes called the *characteristic polynomial* of the matrix $\mathbf{A}$. If, on the other hand, $\mu \in R$ is a root of the polynomial $p(z)$, then the system of equations

$$\begin{aligned}(a_{11} - \mu)x_1 + a_{12}x_2 + \cdots + a_{1n}x_n &= 0\\ a_{21}x_1 + (a_{22} - \mu)x_2 + \cdots + a_{2n}x_n &= 0\\ \cdots\cdots\cdots\cdots\cdots\cdots\cdots\cdots\cdots\cdots&\\ a_{n1}x_1 + a_{n2}x_2 + \cdots + (a_{nn} - \mu)x_n &= 0\end{aligned} \tag{2-73}$$

has a *nonzero* solution $x_1 = \beta_1$, $x_2 = \beta_2, \ldots, x_n = \beta_n$, and the vector $\mathbf{w} = \sum_{j=1}^{n} \beta_j \mathbf{v}_j$ is an eigenvector of $\mathcal{A}$ with $\mu$ as eigenvalue. We observe that the characteristic polynomial $p(z)$ [Eq. (2-72)] of $\mathbf{A}$ has $n$ *complex* roots, and *we shall agree to call any root of* $p(z)$ *an eigenvalue of the matrix* $\mathbf{A}$. If the matrix $\mathbf{B}$ is similar to $\mathbf{A}$, with $\mathbf{B} = \mathbf{PAP}^{-1}$ ($\mathbf{P}$ being nonsingular), then

$$\begin{aligned}\det(\mathbf{B} - z\mathbf{I}) &= \det(\mathbf{PAP}^{-1} - \mathbf{P}z\mathbf{IP}^{-1})\\ &= \det(\mathbf{A} - z\mathbf{I})\end{aligned} \tag{2-74}$$

That is, the similar matrices $\mathbf{A}$ and $\mathbf{B}$ have the same characteristic polynomial and hence the same eigenvalues. We note that the eigenvalues of the linear transformation $\mathcal{A}$ are indeed eigenvalues of the matrix $\mathbf{A}$ and that every *real* eigenvalue of $\mathbf{A}$ is an eigenvalue of $\mathcal{A}$. However, the roots of the characteristic polynomial of $\mathbf{A}$ which have imaginary parts (and are sometimes called complex eigenvalues of $\mathbf{A}$) are *not* eigenvalues of the linear transformation $\mathcal{A}$.

Another interesting and useful property of the characteristic polynomial $p(z)$ of the matrix $\mathbf{A}$ is that $\mathbf{A}$ is a "root" of this polynomial. In other words, the following theorem is true.

***Theorem 2-3 The Cayley-Hamilton Theorem*** *If* $\mathbf{A}$ *is an* $n \times n$ *matrix with characteristic polynomial* $p(z)$ *given by Eq.* (2-72), *then* $\mathbf{A}$ *is a "root" of* $p(z)$ *in the sense that*

$$p(\mathbf{A}) = (-1)^n\mathbf{A}^n + c_1\mathbf{A}^{n-1} + \cdots + c_n\mathbf{I} = \mathbf{0}$$

A proof of this theorem may be found in Ref. B-9 or B-6.

**Example 2-21** Suppose that $\mathcal{A}$ is the linear transformation of $R_3$ into itself whose matrix $\mathbf{A}$ with respect to $\{\mathbf{e}_1, \mathbf{e}_2, \mathbf{e}_3\}$ is

$$\mathbf{A} = \begin{bmatrix} 1 & -1 & 0 \\ -1 & 1 & \sqrt{3} \\ 0 & \sqrt{3} & 1 \end{bmatrix}$$

Then the characteristic polynomial $p(z)$ of $\mathbf{A}$ is $-z^3 + 3z^2 + z - 3$ and the eigenvalues of $\mathbf{A}$ are the numbers 1, $-1$, and 3. Moreover, $\sqrt{3}\,\mathbf{e}_1 + \mathbf{e}_3$ is an eigenvector of $\mathcal{A}$ with 1 as eigenvalue, $\mathbf{e}_1 + 2\mathbf{e}_2 - \sqrt{3}\,\mathbf{e}_3$ is an eigenvector of $\mathcal{A}$ with $-1$ as eigenvalue, and $-\mathbf{e}_1 + 2\mathbf{e}_2 + \sqrt{3}\,\mathbf{e}_3$ is an eigenvector of $\mathcal{A}$ with 3 as eigenvalue. We note that $\mathbf{A}$ is similar to the matrix $\mathbf{B}$, given by

$$\mathbf{B} = \begin{bmatrix} 1 & 0 & 0 \\ 0 & -1 & 0 \\ 0 & 0 & 3 \end{bmatrix}$$

**Exercise 2-17** Verify the statements made in Example 2-21. Show that $\mathbf{A}$ satisfies the equation $-\mathbf{A}^3 + 3\mathbf{A}^2 + \mathbf{A} - 3\mathbf{I} = \mathbf{0}$.

Suppose that the transformation $\mathcal{A}$ has the property that

$$\mathcal{A}(\mathbf{v}_i) = \lambda_i \mathbf{v}_i \qquad i = 1, 2, \ldots, n \tag{2-75}$$

Then we say that $\mathcal{A}$ is a *diagonal* transformation (or is *diagonal*, for short) with respect to the basis $\{\mathbf{v}_1, \mathbf{v}_2, \ldots, \mathbf{v}_n\}$. In that case, the matrix $\mathbf{A}$ of $\mathcal{A}$ with respect to the $\mathbf{v}_i$ has the form

$$\mathbf{A} = \begin{bmatrix} \lambda_1 & 0 & \cdots & 0 \\ 0 & \lambda_2 & \cdots & 0 \\ \cdots & \cdots & \cdots & \cdots \\ 0 & 0 & \cdots & \lambda_n \end{bmatrix} \tag{2-76}$$

It is clear that $\lambda_1, \lambda_2, \ldots, \lambda_n$ are the eigenvalues of $\mathbf{A}$ (or of $\mathcal{A}$). If $\mathbf{B}$ is the matrix of $\mathcal{A}$ with respect to another basis $\{\mathbf{w}_1, \mathbf{w}_2, \ldots, \mathbf{w}_n\}$ of $V$, then we know [(2-64)] that $\mathbf{B}$ is similar to $\mathbf{A}$ (i.e., that there is a nonsingular matrix $\mathbf{P}$ with $\mathbf{B} = \mathbf{PAP}^{-1}$) and therefore that $\lambda_1, \lambda_2, \ldots, \lambda_n$ are also the eigenvalues of $\mathbf{B}$. (Take a look at Example 2-21 again, with this discussion in mind.)

Now let us consider a linear transformation $\mathcal{A}$ of $V$ into itself which has $n$ distinct eigenvalues $\lambda_1, \lambda_2, \ldots, \lambda_n$. Suppose that $\mathbf{w}_1, \mathbf{w}_2, \ldots, \mathbf{w}_n$ are the corresponding eigenvectors of $\mathcal{A}$ [that is, $\mathcal{A}(\mathbf{w}_i) = \lambda_i \mathbf{w}_i$]; then it can be shown that the set $\{\mathbf{w}_1, \mathbf{w}_2, \ldots, \mathbf{w}_n\}$ is actually a *basis* of $V$. It follows that the matrix $\mathbf{B}$ of $\mathcal{A}$ with respect to the $\mathbf{w}_i$ is diagonal, i.e., that

$$\mathbf{B} = \begin{bmatrix} \lambda_1 & 0 & \cdots & 0 \\ 0 & \lambda_2 & \cdots & 0 \\ \cdots & \cdots & \cdots & \cdots \\ 0 & 0 & \cdots & \lambda_n \end{bmatrix} \tag{2-77}$$

If $\mathfrak{A}$ happens to be a linear transformation of $R_n$ into itself with *n distinct* eigenvalues $\lambda_1, \lambda_2, \ldots, \lambda_n$, then we shall usually let $\mathbf{\Lambda}$ denote the $n \times n$ diagonal matrix of the form

$$\mathbf{\Lambda} = \begin{bmatrix} \lambda_1 & 0 & \cdots & 0 \\ 0 & \lambda_2 & \cdots & 0 \\ \cdots & \cdots & \cdots & \cdots \\ 0 & 0 & \cdots & \lambda_n \end{bmatrix} \tag{2-78}$$

and we shall call $\mathbf{\Lambda}$ the *matrix of the eigenvalues*.

Finally, if $\mathbf{A}$ is an $n \times n$ matrix and if we let $\mathbf{A}_k(\lambda)$ denote the $k \times k$ matrix given by

$$\mathbf{A}_k(\lambda) = \begin{bmatrix} \lambda & 1 & 0 & \cdots & 0 \\ 0 & \lambda & 1 & \cdots & 0 \\ \cdots & \cdots & \cdots & \cdots & \cdots \\ 0 & 0 & 0 & \cdots & 1 \\ 0 & 0 & 0 & \cdots & \lambda \end{bmatrix} \tag{2-79}$$

then it can be shown that if the eigenvalues of $\mathbf{A}$ are *all* real, then $\mathbf{A}$ is similar to a matrix of the form

$$\mathbf{J}(\mathbf{A}) = \left[\begin{array}{c|c|c|c} \mathbf{A}_{m_1}(\lambda_1) & \mathbf{0} & \cdots & \mathbf{0} \\ \hline \mathbf{0} & \mathbf{A}_{m_2}(\lambda_2) & \cdots & \mathbf{0} \\ \hline \cdots & \cdots & \cdots & \cdots \\ \hline \mathbf{0} & \mathbf{0} & \cdots & \mathbf{A}_{m_p}(\lambda_p) \end{array}\right] \tag{2-80}$$

where $m_1 + m_2 + \cdots + m_p = n$ and $\lambda_1, \lambda_2, \ldots, \lambda_p$ are the eigenvalues (not necessarily distinct) of $\mathbf{A}$. $\mathbf{J}(\mathbf{A})$ is called the *Jordan canonical form of* $\mathbf{A}$.

**Example 2-22** Consider the $3 \times 3$ matrix $\mathbf{A}$, where

$$\mathbf{A} = \begin{bmatrix} 1 & 1 & 0 \\ 1 & 1 & -2 \\ \sqrt{3}/9 & 0 & 1 \end{bmatrix}$$

Then it can be shown that either

$$\mathbf{J}(\mathbf{A}) = \begin{bmatrix} 1 + \sqrt{3}/3 & 1 & 0 \\ 0 & 1 + \sqrt{3}/3 & 0 \\ 0 & 0 & -\sqrt{3}/6 + 3\sqrt{3} \end{bmatrix}$$

or

$$\mathbf{J}(\mathbf{A}) = \begin{bmatrix} 1 + \sqrt{3}/3 & 0 & 0 \\ 0 & 1 + \sqrt{3}/3 & 0 \\ 0 & 0 & -\sqrt{3}/6 + 3\sqrt{3} \end{bmatrix}$$

**Example 2-23** If $\mathbf{A}$ is a $2 \times 2$ matrix and if $\lambda$ is a double root of $\det(\mathbf{A} - z\mathbf{I}) = 0$, then $\mathbf{A}$ is similar to a matrix either of the form $\begin{bmatrix} \lambda & 0 \\ 0 & \lambda \end{bmatrix}$ or of the form $\begin{bmatrix} \lambda & 1 \\ 0 & \lambda \end{bmatrix}$.

## 2-11 Euclidean Spaces: Inner Products

Suppose that $V$ is a vector space and suppose that $P$ is a function whose domain is *all* of $V \times V$ and whose range is contained in $R$. We say that $P$ is an *inner, or scalar, product on* $V$ if the following conditions are satisfied:

IP1

$$P(\mathbf{v}_1 + \mathbf{v}_2, \mathbf{v}_3) = P(\mathbf{v}_1, \mathbf{v}_3) + P(\mathbf{v}_2, \mathbf{v}_3) \qquad \text{for all } \mathbf{v}_1, \mathbf{v}_2, \mathbf{v}_3 \text{ in } V$$

IP2

$$P(r\mathbf{v}_1, \mathbf{v}_2) = rP(\mathbf{v}_1, \mathbf{v}_2) \qquad \text{for all } r \text{ in } R \text{ and all } \mathbf{v}_1, \mathbf{v}_2 \text{ in } V$$

IP3

$$P(\mathbf{v}_1, \mathbf{v}_2) = P(\mathbf{v}_2, \mathbf{v}_1) \qquad \text{for all } \mathbf{v}_1, \mathbf{v}_2 \text{ in } V$$

$P$ is sometimes referred to as a *symmetric bilinear form on* $V$.† The function $P(\mathbf{v}, \mathbf{v})$ is often referred to as the *quadratic form* on $V$ induced by $P$.

**Exercise 2-18** Show that if $P$ is an inner product on $V$, then (*a*) $P(\mathbf{v}_1, \mathbf{v}_2 + \mathbf{v}_3) = P(\mathbf{v}_1, \mathbf{v}_2) + P(\mathbf{v}_1, \mathbf{v}_3)$ for all $\mathbf{v}_1, \mathbf{v}_2, \mathbf{v}_3$ in $V$, and (*b*) $P(\mathbf{v}_1, r\mathbf{v}_2) = rP(\mathbf{v}_1, \mathbf{v}_2)$ for all $r$ in $R$ and all $\mathbf{v}_1, \mathbf{v}_2$ in $V$. HINT: Use condition IP3 again and again.

Suppose that $P$ is an inner product on $V$ and that $\{\mathbf{v}_1, \mathbf{v}_2, \ldots, \mathbf{v}_n\}$ is a basis of $V$. Let us set

$$P(\mathbf{v}_i, \mathbf{v}_j) = a_{ij} \qquad i = 1, 2, \ldots, n; j = 1, 2, \ldots, n \tag{2-81}$$

and let $\mathbf{A}$ be the $n \times n$ matrix $(a_{ij})$. $\mathbf{A}$ is called the *matrix of* $P$ with respect to the basis $\{\mathbf{v}_1, \mathbf{v}_2, \ldots, \mathbf{v}_n\}$. First of all, we note that condition IP3 implies that

$$a_{ij} = P(\mathbf{v}_i, \mathbf{v}_j) = P(\mathbf{v}_j, \mathbf{v}_i) = a_{ji} \tag{2-82}$$

so that the matrix $\mathbf{A}$ has the property that $a_{ij} = a_{ji}$ for $i = 1, 2, \ldots, n$ and $j = 1, 2, \ldots, n$. In other words, $\mathbf{A}$ and its transpose $\mathbf{A}'$ are the same; that is,

$$\mathbf{A} = \mathbf{A}' \tag{2-82a}$$

In general, if $\mathbf{B}$ is an $n \times n$ matrix which is the same as its transpose (i.e., if $\mathbf{B} = \mathbf{B}'$), then we call $\mathbf{B}$ a *symmetric matrix*. For example, the identity matrix $\mathbf{I}$ is a symmetric matrix. We have just shown that if $P$ is an inner product on $V$, then the matrix $\mathbf{A} = (a_{ij})$, where $a_{ij} = P(\mathbf{v}_i, \mathbf{v}_j)$, is symmetric. Now suppose that $\mathbf{B} = (b_{ij})$ is a symmetric $n \times n$ matrix. Can we define an inner product $Q$ on $V$ for which $Q(\mathbf{v}_i, \mathbf{v}_j) = b_{ij}$? We certainly can, and

† This is the more common terminology, as the term *inner product* is usually reserved for what we call a *positive definite inner product*. See Eq. (2-84).

in fact $Q$ is given by the formula

$$Q(\mathbf{v}, \mathbf{w}) = \sum_{i=1}^{n} \sum_{j=1}^{n} \alpha_i b_{ij} \beta_j \tag{2-83}$$

where $\mathbf{v} = \sum_{i=1}^{n} \alpha_i \mathbf{v}_i$ and $\mathbf{w} = \sum_{j=1}^{n} \beta_j \mathbf{v}_j$.

**Exercise 2-19** Show that Eq. (2-83) does indeed define an inner product $Q$ on $V$. If $\mathbf{B}$ is the identity matrix $\mathbf{I}$, what does the formula for $Q$ become?

**Example 2-24** The matrix

$$\mathbf{B} = \begin{bmatrix} 1 & 2 & -1 \\ 2 & 0 & 3 \\ -1 & 3 & 1 \end{bmatrix}$$

is symmetric and defines an inner product $Q$ on $R_3$ with

$$Q\left(\begin{bmatrix} \alpha_1 \\ \alpha_2 \\ \alpha_3 \end{bmatrix}, \begin{bmatrix} \beta_1 \\ \beta_2 \\ \beta_3 \end{bmatrix}\right) = \alpha_1\beta_1 + 2\alpha_1\beta_2 - \alpha_1\beta_3 + 2\alpha_2\beta_1 + 3\alpha_2\beta_3 - \alpha_3\beta_1 + 3\alpha_3\beta_2 + \alpha_3\beta_3$$

Note that

$$Q\left(\begin{bmatrix} 1 \\ 0 \\ 1 \end{bmatrix}, \begin{bmatrix} 1 \\ 0 \\ 1 \end{bmatrix}\right) = 0$$

and that

$$Q\left(\begin{bmatrix} 0 \\ 1 \\ -1 \end{bmatrix}, \begin{bmatrix} 0 \\ 1 \\ -1 \end{bmatrix}\right) = -5$$

In other words, there is a nonzero vector

$$\mathbf{v} = \begin{bmatrix} 1 \\ 0 \\ 1 \end{bmatrix}$$

such that $Q(\mathbf{v}, \mathbf{v}) = 0$, and there is a nonzero vector

$$\mathbf{w} = \begin{bmatrix} 0 \\ 1 \\ -1 \end{bmatrix}$$

such that $Q(\mathbf{w}, \mathbf{w}) < 0$.

We say that the inner product $P$ on $V$ is *positive* if $P$ has the property that

$$P(\mathbf{v}, \mathbf{v}) \geq 0 \qquad \text{for all } \mathbf{v} \text{ in } V \tag{2-84}$$

and we say that $P$ is *definite* (or nondegenerate) if "$P(\mathbf{v}, \mathbf{v}_0) = 0$ for all $\mathbf{v}$ in $V$" implies that $\mathbf{v}_0 = \mathbf{0}$. If $\mathbf{B}$ is a symmetric matrix for which the corresponding inner product on $V$ [as defined by Eq. (2-83)] is positive or definite, then we say that $\mathbf{B}$ is a *positive* or *definite* matrix. If $P$ (or $\mathbf{B}$) is both positive *and* definite, then we say that $P$ (or $\mathbf{B}$) is *positive definite.* The significance of being positive definite will be examined in Secs. 2-12,

2-13, and 2-15. A vector space $V$ together with a given *positive definite inner product* $P$ on $V$ will be called a *Euclidean space.*

For example, if we consider the inner product $P$ on $R_n$ whose matrix with respect to the natural basis $\{\mathbf{e}_1, \mathbf{e}_2, \ldots, \mathbf{e}_n\}$† is the identity $\mathbf{I}$, that is,

$$P\left(\begin{bmatrix}\alpha_1\\\alpha_2\\\cdot\\\cdot\\\cdot\\\alpha_n\end{bmatrix}, \begin{bmatrix}\beta_1\\\beta_2\\\cdot\\\cdot\\\cdot\\\beta_n\end{bmatrix}\right) = \sum_{i=1}^{n}\sum_{j=1}^{n} \alpha_i\delta_{ij}\beta_j = \sum_{i=1}^{n} \alpha_i\beta_i \tag{2-85}$$

then this inner product is positive definite and $R_n$ becomes a Euclidean space. We shall distinguish this inner product on $R_n$ by calling it the *scalar product* on $R_n$, and we shall use the notation $\langle\mathbf{v}, \mathbf{w}\rangle$ for $P(\mathbf{v}, \mathbf{w})$; in other words, if

$$\mathbf{v} = \begin{bmatrix}\alpha_1\\\alpha_2\\\cdot\\\cdot\\\cdot\\\alpha_n\end{bmatrix} \quad \text{and} \quad \mathbf{w} = \begin{bmatrix}\beta_1\\\beta_2\\\cdot\\\cdot\\\cdot\\\beta_n\end{bmatrix}$$

are elements of $R_n$, then the scalar product of $\mathbf{v}$ and $\mathbf{w}$, denoted by $\langle\mathbf{v}, \mathbf{w}\rangle$, is given by

$$\langle\mathbf{v}, \mathbf{w}\rangle = \sum_{i=1}^{n} \alpha_i\beta_i \tag{2-86}$$

This particular inner product will be used very frequently in the remainder of this book.

**Example 2-25** If $\mathbf{v} = \begin{bmatrix}1\\-1\end{bmatrix}$ and $\mathbf{w} = \begin{bmatrix}\sqrt{2}\\3\end{bmatrix}$ are elements of $R_2$, then $\langle\mathbf{v}, \mathbf{w}\rangle = \sqrt{2} - 3$, $\langle\mathbf{v}, \mathbf{v}\rangle = 2$, and $\langle\mathbf{w}, \mathbf{w}\rangle = 11$.

**Exercise 2-20** Show that the scalar product on $R_n$, $\langle\mathbf{v}, \mathbf{w}\rangle$, is positive definite.

## 2-12 Euclidean Spaces: The Schwarz Inequality

Suppose that $V$ is a Euclidean space with $P$ as inner product. Then we wish to prove the inequality, which is known as the *Schwarz inequality,*

$$|P(\mathbf{v}, \mathbf{w})|^2 \leq P(\mathbf{v}, \mathbf{v})P(\mathbf{w}, \mathbf{w}) \tag{2-87}$$

for all $\mathbf{v}$, $\mathbf{w}$ in $V$. We shall actually require only that $P$ be positive (not positive definite) for the proof of (2-87). Now, if $\lambda$ is *any* real number, then

$$P(\mathbf{v} + \lambda\mathbf{w}, \mathbf{v} + \lambda\mathbf{w}) = P(\mathbf{v}, \mathbf{v}) + 2\lambda P(\mathbf{v}, \mathbf{w}) + \lambda^2 P(\mathbf{w}, \mathbf{w}) \geq 0 \tag{2-88}$$

† See Eq. (2-27).

since $P$ is positive. If $P(\mathbf{w}, \mathbf{w}) \neq 0$, then substituting $\lambda = -P(\mathbf{v}, \mathbf{w})/P(\mathbf{w}, \mathbf{w})$ in Eq. (2-88) will produce the desired result (2-87). If $P(\mathbf{w}, \mathbf{w}) = 0$ but $P(\mathbf{v}, \mathbf{v}) \neq 0$, then a consideration of $P(\lambda\mathbf{v} + \mathbf{w}, \lambda\mathbf{v} + \mathbf{w})$ will again lead us to (2-87). Finally, if $P(\mathbf{w}, \mathbf{w}) = 0$ and $P(\mathbf{v}, \mathbf{v}) = 0$, then, setting $\lambda = -P(\mathbf{v}, \mathbf{w})$ in Eq. (2-88), we obtain the inequality

$$-2[P(\mathbf{v}, \mathbf{w})]^2 \geq 0 \tag{2-89}$$

which implies that $P(\mathbf{v}, \mathbf{w}) = 0$ and again establishes (2-87). We shall use the Schwarz inequality quite frequently in the sequel.

We observe that if $P$ is the scalar product on $R_n$ [see (2-86)], then the Schwarz inequality becomes

$$|\langle \mathbf{v}, \mathbf{w} \rangle|^2 \leq \langle \mathbf{v}, \mathbf{v} \rangle \langle \mathbf{w}, \mathbf{w} \rangle \tag{2-90}$$

or, letting

$$\mathbf{v} = \begin{bmatrix} \alpha_1 \\ \alpha_2 \\ \cdot \\ \cdot \\ \cdot \\ \alpha_n \end{bmatrix} \quad \text{and} \quad \mathbf{w} = \begin{bmatrix} \beta_1 \\ \beta_2 \\ \cdot \\ \cdot \\ \cdot \\ \beta_n \end{bmatrix}$$

we have

$$\left| \sum_{i=1}^{n} \alpha_i \beta_i \right|^2 \leq \left( \sum_{i=1}^{n} \alpha_i^2 \right) \left( \sum_{i=1}^{n} \beta_i^2 \right) \tag{2-91}$$

## 2-13 Euclidean Spaces: Orthogonality and Norm

Suppose that $V$ is a Euclidean space with $P$ as inner product; then we say that two vectors $\mathbf{v}$ and $\mathbf{w}$ in $V$ are *orthogonal* (or *perpendicular*) if

$$P(\mathbf{v}, \mathbf{w}) = 0 \tag{2-92}$$

For example, the vectors $\mathbf{e}_1$ and $\mathbf{e}_2$ in $R_2$ are orthogonal with respect to the scalar product on $R_2$; that is, $\langle \mathbf{e}_1, \mathbf{e}_2 \rangle = 0$. If $V$ is $n$-dimensional, then it can be shown that there is a basis $\{\mathbf{v}_1, \mathbf{v}_2, \ldots, \mathbf{v}_n\}$ of $V$ such that $\mathbf{v}_i$ is orthogonal to $\mathbf{v}_j$ if $i \neq j$ and such that $P(\mathbf{v}_i, \mathbf{v}_i) = 1$. In other words, the matrix of $P$ with respect to the basis $\{\mathbf{v}_1, \mathbf{v}_2, \ldots, \mathbf{v}_n\}$ is precisely the identity matrix $\mathbf{I}$. Such a basis is called an *orthonormal basis* of $V$. For example, the natural basis $\{\mathbf{e}_1, \mathbf{e}_2, \ldots, \mathbf{e}_n\}$ is an orthonormal basis of $R_n$ for the scalar product on $R_n$; that is,

$$\langle \mathbf{e}_i, \mathbf{e}_j \rangle = \delta_{ij} \qquad i = 1, 2, \ldots, n; j = 1, 2, \ldots, n \tag{2-93}$$

**Example 2-26** The matrix

$$\mathbf{A} = \begin{bmatrix} 1 & 1 & 0 \\ 1 & 3 & 2 \\ 0 & 2 & 4 \end{bmatrix}$$

is symmetric and defines an inner product $P$ on $R_3$ with

$$P\left(\begin{bmatrix}\alpha_1\\\alpha_2\\\alpha_3\end{bmatrix}, \begin{bmatrix}\beta_1\\\beta_2\\\beta_3\end{bmatrix}\right) = \alpha_1\beta_1 + \alpha_1\beta_2 + \alpha_2\beta_1 + 3\alpha_2\beta_2 + 2\alpha_2\beta_3 + 2\alpha_3\beta_2 + 4\alpha_3\beta_3$$

$$= (\alpha_1 + \alpha_2)(\beta_1 + \beta_2) + \alpha_2\beta_2 + (\alpha_2 + 2\alpha_3)(\beta_2 + 2\beta_3)$$

$P$ is positive definite, and if we set $\mathbf{w}_1 = \mathbf{e}_1$, $\mathbf{w}_2 = \frac{1}{2}\mathbf{e}_3$, and $\mathbf{w}_3 = -\mathbf{e}_1 + \mathbf{e}_2 - \frac{1}{2}\mathbf{e}_3$, then the vectors $\mathbf{w}_1$, $\mathbf{w}_2$, $\mathbf{w}_3$ form an orthonormal basis of $R_3$ with respect to the inner product $P$. We observe that $P(\boldsymbol{\alpha}, \boldsymbol{\beta}) = \langle\boldsymbol{\alpha}, \mathbf{A}\boldsymbol{\beta}\rangle$ and that $P(\mathbf{w}_i, \mathbf{w}_j) = \langle\mathbf{w}_i, \mathbf{A}\mathbf{w}_j\rangle = \delta_{ij}$ for $i, j = 1, 2, 3$.

**Exercise 2-21** Prove that the inner product $P$ in Example 2-26 is indeed positive definite. HINT: Use the orthonormal basis $\{\mathbf{w}_1, \mathbf{w}_2, \mathbf{w}_3\}$, but show that it is an orthonormal basis.

**Exercise 2-22** Show that $|(\alpha_1\beta_1 + 4\alpha_3\beta_3)|^2 \le (\alpha_1^2 + 4\alpha_3^2)(\beta_1^2 + 4\beta_3^2)$. HINT: Use (2-87) and the inner product $P$ of Example 2-26.

If $V$ is a Euclidean space with $P$ as inner product, then we call $\sqrt{P(\mathbf{v}, \mathbf{v})}$ the *norm* of the element $\mathbf{v}$ of $V$, and we write

$$\|\mathbf{v}\|_P = \sqrt{P(\mathbf{v}, \mathbf{v})} \tag{2-94}$$

In the case of the scalar product on $R_n$, we simply write $\|\mathbf{v}\|$ for the norm of $\mathbf{v}$, that is,

$$\|\mathbf{v}\| = \sqrt{\langle\mathbf{v}, \mathbf{v}\rangle} \qquad \text{for all } \mathbf{v} \text{ in } R_n \tag{2-95}$$

The norm has the following important properties:

N1

$$\|\mathbf{v}\|_P \ge 0 \qquad \text{for all } \mathbf{v} \text{ in } V \text{ with } \|\mathbf{v}\|_P = 0 \text{ if and only if } \mathbf{v} = \mathbf{0}$$

N2

$$\|\mathbf{v}_1 + \mathbf{v}_2\|_P \le \|\mathbf{v}_1\|_P + \|\mathbf{v}_2\|_P \qquad \text{for all } \mathbf{v}_1, \mathbf{v}_2 \text{ in } V$$

N3

$$\|r\mathbf{v}\|_P = |r|\,\|\mathbf{v}\|_P \qquad \text{for all } r \text{ in } R \text{ and for all } \mathbf{v} \text{ in } V$$

We shall have more to say about norms in the next chapter. We shall show in Sec. 3-2 that the norm allows us to define a notion of the distance between elements of $V$. Moreover, we can define the "angle" $\theta(\mathbf{v}_1, \mathbf{v}_2)$ between the vectors $\mathbf{v}_1$ and $\mathbf{v}_2$ by the formula

$$\cos\theta(\mathbf{v}_1, \mathbf{v}_2) = \frac{P(\mathbf{v}_1, \mathbf{v}_2)}{\|\mathbf{v}_1\|_P\,\|\mathbf{v}_2\|_P} \tag{2-96}$$

so that there are notions of length and angle in $V$. This is why $V$ is called a Euclidean space.

## 2-14 Euclidean Spaces: Some Properties of the Scalar Product on $R_n$

Let us suppose that $\mathbf{C} = (c_{ij})$ is an $n \times m$ matrix. Then, if $\mathbf{u}$ is an element of $R_m$, we may consider $\mathbf{Cu}$ (see Sec. 2-8) as an element of $R_n$. In other words, if

$$\mathbf{u} = \begin{bmatrix} u_1 \\ u_2 \\ \cdot \\ \cdot \\ \cdot \\ u_m \end{bmatrix} \qquad \text{then } \mathbf{Cu} = \begin{bmatrix} \sum_{j=1}^{m} c_{1j}u_j \\ \sum_{j=1}^{m} c_{2j}u_j \\ \cdot \\ \cdot \\ \cdot \\ \sum_{j=1}^{m} c_{nj}u_j \end{bmatrix}$$

Let us try to determine $\langle \mathbf{Cu}, \mathbf{x} \rangle$, where $\mathbf{x}$ is an element of $R_n$. We have

$$\begin{aligned} \langle \mathbf{Cu}, \mathbf{x} \rangle &= \left\langle \begin{bmatrix} \sum_{j=1}^{m} c_{1j}u_j \\ \sum_{j=1}^{m} c_{2j}u_j \\ \cdot \\ \cdot \\ \cdot \\ \sum_{j=1}^{m} c_{nj}u_j \end{bmatrix}, \begin{bmatrix} x_1 \\ x_2 \\ \cdot \\ \cdot \\ \cdot \\ x_n \end{bmatrix} \right\rangle = \sum_{i=1}^{n} \left( \sum_{j=1}^{m} c_{ij}u_j \right) x_i \\ &= \sum_{j=1}^{m} \left( \sum_{i=1}^{n} c_{ij}x_i \right) u_j = \sum_{j=1}^{m} \left( \sum_{i=1}^{n} c'_{ji}x_i \right) u_j \\ &= \left\langle \begin{bmatrix} u_1 \\ u_2 \\ \cdot \\ \cdot \\ \cdot \\ u_m \end{bmatrix}, \begin{bmatrix} \sum_{i=1}^{n} c'_{1i}x_i \\ \sum_{i=1}^{n} c'_{2i}x_i \\ \cdot \\ \cdot \\ \cdot \\ \sum_{i=1}^{n} c'_{mi}x_i \end{bmatrix} \right\rangle = \langle \mathbf{u}, \mathbf{C}'\mathbf{x} \rangle \end{aligned} \tag{2-97}$$

where $\mathbf{C}' = (c'_{ji})$ is the transpose of $\mathbf{C}$ and $c'_{ji} = c_{ij}$ for $i = 1, 2, \ldots, n$ and $j = 1, 2, \ldots, m$. To reiterate for emphasis,

$$\langle \mathbf{Cu}, \mathbf{x} \rangle = \langle \mathbf{u}, \mathbf{C}'\mathbf{x} \rangle \tag{2-98}$$

We shall use (2-98) repeatedly in the remainder of this book.

**Example 2-27** Let C be the 3 × 2 matrix

$$\begin{bmatrix} 1 & 0 \\ 0 & 1 \\ 1 & 2 \end{bmatrix}$$

If $\mathbf{u} = \begin{bmatrix} u_1 \\ u_2 \end{bmatrix}$ and if

$$\mathbf{x} = \begin{bmatrix} x_1 \\ x_2 \\ x_3 \end{bmatrix}$$

are elements of $R_2$ and $R_3$, respectively, then

$$\mathbf{Cu} = \begin{bmatrix} u_1 \\ u_2 \\ u_1 + 2u_2 \end{bmatrix} \qquad \mathbf{C'x} = \begin{bmatrix} x_1 + x_3 \\ x_2 + 2x_3 \end{bmatrix}$$

and $\langle \mathbf{Cu}, \mathbf{x} \rangle = u_1x_1 + u_2x_2 + u_1x_3 + 2u_2x_3 = \langle \mathbf{u}, \mathbf{C'x} \rangle = u_1(x_1 + x_3) + u_2(x_2 + 2x_3)$. (Try putting in some numbers for the $u_j$ and $x_i$ if you are still not convinced.)

From now on in this section, we shall consider only the space $R_n$. *All* vectors will be assumed to be elements of $R_n$, and *all* matrices will be assumed to have $n$ rows and $n$ columns. Suppose that $\mathbf{A} = (a_{ij})$ satisfies the relation

$$\mathbf{A} = -\mathbf{A}' \tag{2-99}$$

or, equivalently,

$$a_{ij} = -a_{ji} \qquad \text{for } i = 1, 2, \ldots, n; j = 1, 2, \ldots, n \tag{2-100}$$

Then we say that **A** is a *skew-symmetric matrix* (or is, simply, *skew-symmetric*).

What is the significance of this property with regard to the scalar product? According to (2-98), we have

$$\langle \mathbf{Ax}, \mathbf{x} \rangle = \langle \mathbf{x}, \mathbf{A'x} \rangle \tag{2-101}$$

for every **x**. This implies that

$$\langle \mathbf{Ax}, \mathbf{x} \rangle = \langle \mathbf{x}, -\mathbf{Ax} \rangle = -\langle \mathbf{Ax}, \mathbf{x} \rangle \tag{2-102}$$

for every **x** if **A** is skew-symmetric. In other words,

$$\langle \mathbf{Ax}, \mathbf{x} \rangle = 0 \tag{2-103}$$

for every **x** if **A** is skew-symmetric. The converse is also true; that is, if a matrix **A** satisfies Eq. (2-103) for every **x**, then **A** is skew-symmetric. Equation (2-103) implies that the vectors **Ax** and **x** are orthogonal; thus, a skew-symmetric matrix **A** operating on any vector **x** creates a vector **Ax** which is orthogonal to **x**.

**Example 2-28** Let

$$\mathbf{A} = \begin{bmatrix} 0 & a & b \\ -a & 0 & -c \\ -b & c & 0 \end{bmatrix}$$

Then $\mathbf{A}$ is skew-symmetric. In particular,

$$\begin{bmatrix} 0 & 1 & 1 \\ -1 & 0 & 0 \\ -1 & 0 & 0 \end{bmatrix}$$

is skew-symmetric.

**Example 2-29** If $\mathbf{B}$ is any $n \times n$ matrix, then the matrix $(\mathbf{B} - \mathbf{B}')/2$ is skew-symmetric (why?) and the matrix $(\mathbf{B} + \mathbf{B}')/2$ is symmetric. Note that $\mathbf{B} = (\mathbf{B} + \mathbf{B}')/2 + (\mathbf{B} - \mathbf{B}')/2$.

**Exercise 2-23** Show that if $\langle \mathbf{Ax}, \mathbf{x} \rangle = 0$ for every $\mathbf{x}$, then $\mathbf{A} = -\mathbf{A}'$. HINT: $\langle \mathbf{A}(\mathbf{e}_i - \mathbf{e}_j), \mathbf{e}_i - \mathbf{e}_j \rangle = 0$.

We know that $\{\mathbf{e}_1, \mathbf{e}_2, \ldots, \mathbf{e}_n\}$ is an orthonormal basis of $R_n$. Suppose that $\mathbf{\Phi}$ is a nonsingular matrix; then the set $\{\mathbf{\Phi e}_1, \mathbf{\Phi e}_2, \ldots, \mathbf{\Phi e}_n\}$ is a basis of $R_n$ (why?). Let us see what conditions $\mathbf{\Phi}$ must satisfy for this basis to be orthonormal. We have

$$\langle \mathbf{\Phi e}_i, \mathbf{\Phi e}_j \rangle = \langle \mathbf{e}_i, \mathbf{\Phi}'\mathbf{\Phi e}_j \rangle \tag{2-104}$$

If $\mathbf{\Phi} = (\varphi_{ij})$, then we may write

$$\langle \mathbf{e}_i, \mathbf{\Phi}'\mathbf{\Phi e}_j \rangle = \sum_{p=1}^{n} \varphi'_{ip}\varphi_{pj} \tag{2-105}$$

where $\mathbf{\Phi}' = (\varphi'_{ji})$, with $\varphi'_{ji} = \varphi_{ij}$. It follows that $\langle \mathbf{\Phi e}_i, \mathbf{\Phi e}_j \rangle = \delta_{ij}$ if and only if

$$\mathbf{\Phi}'\mathbf{\Phi} = \mathbf{I} \tag{2-106}$$

In a similar way, we can show that $\langle \mathbf{\Phi e}_i, \mathbf{\Phi e}_j \rangle = \delta_{ij}$ if and only if

$$\mathbf{\Phi\Phi}' = \mathbf{I} \tag{2-107}$$

The result of this is that the basis $\{\mathbf{\Phi e}_1, \mathbf{\Phi e}_2, \ldots, \mathbf{\Phi e}_n\}$ is again orthonormal if and only if the inverse of $\mathbf{\Phi}$ is the same as the transpose of $\mathbf{\Phi}$, that is, if and only if

$$\mathbf{\Phi}^{-1} = \mathbf{\Phi}' \tag{2-108}$$

or, equivalently,

$$\mathbf{\Phi}'\mathbf{\Phi} = \mathbf{\Phi\Phi}' = \mathbf{I} \tag{2-109}$$

Any matrix which satisfies (2-109) is called an *orthogonal matrix* (and the transformation defined by $\mathbf{\Phi}$, as in Sec. 2-7, is called an *orthogonal transformation*). We observe that if $\mathbf{\Phi}$ is an orthogonal matrix, then

$$\langle \mathbf{\Phi x}, \mathbf{\Phi x} \rangle = \langle \mathbf{x}, \mathbf{x} \rangle \tag{2-110}$$

for every vector $\mathbf{x}$, and so [see (2-95)]

$$\|\mathbf{\Phi x}\| = \|\mathbf{x}\| \tag{2-111}$$

for every $\mathbf{x}$ in $R_n$ whenever $\mathbf{\Phi}$ is an orthogonal matrix.

We can think of an orthogonal matrix (or transformation) $\mathbf{\Phi}$ as preserving the Euclidean length $\|\mathbf{x}\|$ of a vector $\mathbf{x}$ but changing its "direction."

**Example 2-30** Let $\mathbf{\Phi}$ be the matrix $\begin{bmatrix} \cos\theta & \sin\theta \\ -\sin\theta & \cos\theta \end{bmatrix}$ (for some angle $\theta$). Then

$$\mathbf{\Phi}' = \begin{bmatrix} \cos\theta & -\sin\theta \\ \sin\theta & \cos\theta \end{bmatrix} \qquad \mathbf{\Phi}'\mathbf{\Phi} = \begin{bmatrix} \cos^2\theta + \sin^2\theta & 0 \\ 0 & \sin^2\theta + \cos^2\theta \end{bmatrix} = \begin{bmatrix} 1 & 0 \\ 0 & 1 \end{bmatrix} = \mathbf{I}$$

so that $\mathbf{\Phi}$ is orthogonal.

**Exercise 2-24** Is the $3 \times 3$ matrix

$$\begin{bmatrix} \cos\theta & \sin\theta & 0 \\ -\sin\theta & \cos\theta & 0 \\ 0 & 0 & 1 \end{bmatrix}$$

orthogonal? What about

$$\begin{bmatrix} 1 & 0 & 0 \\ 0 & \cos\phi & \sin\phi \\ 0 & -\sin\phi & \cos\phi \end{bmatrix}$$

What about the product of these two matrices?

## 2-15 Euclidean Spaces: Some Properties of Symmetric Matrices

We have already noted in Sec. 2-11 that there is an intimate connection between inner products and symmetric matrices. We shall exploit this connection in this section to obtain some properties of symmetric matrices which will be used quite frequently in later chapters of the book.

We shall consider only the space $R_n$ in this section. Consequently, all vectors will be assumed to be elements of $R_n$, and all matrices will be assumed to have $n$ rows and $n$ columns. Moreover, we let $\mathbf{e}_1, \mathbf{e}_2, \ldots, \mathbf{e}_n$ denote the natural basis of $R_n$ [see Eqs. (2-27)], and we let $\mathbf{Q} = (q_{ij})$ be a symmetric $n \times n$ matrix so that

$$\mathbf{Q} = \mathbf{Q}' \tag{2-112}$$

where $\mathbf{Q}'$ is the transpose of $\mathbf{Q}$.

Now, we know that the function $Q(\mathbf{v}, \mathbf{w})$, defined by setting

$$Q(\mathbf{v}, \mathbf{w}) = \langle \mathbf{v}, \mathbf{Q}\mathbf{w} \rangle \tag{2-113}$$

for all $\mathbf{v}$ and $\mathbf{w}$ in $R_n$, is an inner product on $R_n$ since $\mathbf{Q}$ is symmetric. In point of fact, if $\mathbf{v} = \sum_{i=1}^{n} \alpha_i \mathbf{e}_i$ and $\mathbf{w} = \sum_{j=1}^{n} \beta_j \mathbf{e}_j$, then we have

$$\begin{aligned} Q(\mathbf{v}, \mathbf{w}) &= \left\langle \sum_{i=1}^{n} \alpha_i \mathbf{e}_i, \mathbf{Q} \sum_{j=1}^{n} \beta_j \mathbf{e}_j \right\rangle \\ &= \sum_{i=1}^{n} \sum_{j=1}^{n} \alpha_i \langle \mathbf{e}_i, \mathbf{Q}\mathbf{e}_j \rangle \beta_j \end{aligned} \tag{2-114}$$

However, we know that

$$\mathbf{Q}\mathbf{e}_j = \sum_{k=1}^{n} q_{kj} \mathbf{e}_k \tag{2-115}$$

and hence that

$$\langle \mathbf{e}_i, \mathbf{Q}\mathbf{e}_j \rangle = \sum_{k=1}^{n} q_{kj} \langle \mathbf{e}_i, \mathbf{e}_k \rangle = q_{ij} \tag{2-116}$$

since $\langle \mathbf{e}_i, \mathbf{e}_k \rangle = \delta_{ik}$ (where $\delta_{ik} = 0$ if $i \neq k$ and $\delta_{ii} = 1$). It follows that

$$Q(\mathbf{v}, \mathbf{w}) = \langle \mathbf{v}, \mathbf{Q}\mathbf{w} \rangle = \sum_{i=1}^{n} \sum_{j=1}^{n} \alpha_i q_{ij} \beta_j \tag{2-117}$$

and that $\mathbf{Q}$ is the matrix of the inner product $Q$ with respect to the natural basis $\{\mathbf{e}_1, \mathbf{e}_2, \ldots, \mathbf{e}_n\}$ of $R_n$.

We now state a very important theorem, which says that there is an orthonormal basis $\{\mathbf{f}_1, \mathbf{f}_2, \ldots, \mathbf{f}_n\}$ of $R_n$ (that is, $\langle \mathbf{f}_j, \mathbf{f}_j \rangle = \delta_{ij}$) with the property that

$$Q(\mathbf{f}_i, \mathbf{f}_j) = \lambda_i \delta_{ij} \qquad i, j = 1, 2, \ldots, n \tag{2-118}$$

where $\lambda_1, \lambda_2, \ldots, \lambda_n$ are $n$ real numbers. In other words, the matrix of the inner product $Q$ with respect to the basis $\{\mathbf{f}_1, \mathbf{f}_2, \ldots, \mathbf{f}_n\}$ is a *diagonal matrix* with entries $\lambda_1, \lambda_2, \ldots, \lambda_n$ on the diagonal. More precisely, we have:

***Theorem 2-4*** *If* $\mathbf{Q}$ *is a symmetric* $n \times n$ *matrix and if* $Q$ *is the inner product on* $R_n$ *defined by Eq.* (2-113), *then there is an orthonormal basis* $\{\mathbf{f}_1, \mathbf{f}_2, \ldots, \mathbf{f}_n\}$ *of* $R_n$, *that is,*

$$\langle \mathbf{f}_i, \mathbf{f}_j \rangle = \delta_{ij} \qquad i, j = 1, 2, \ldots, n \tag{2-119}$$

*such that the matrix of* $Q$ *with respect to the* $\mathbf{f}_i$ *is a diagonal matrix* $\mathbf{\Lambda}$ *with*

$$\mathbf{\Lambda} = \begin{bmatrix} \lambda_1 & 0 & \cdots & 0 \\ 0 & \lambda_2 & \cdots & 0 \\ \cdots & \cdots & \cdots & \cdots \\ 0 & 0 & \cdots & \lambda_n \end{bmatrix} \tag{2-120}$$

*In other words, there is an orthogonal matrix* $\mathbf{\Phi}$ [*that is,* $\mathbf{\Phi}'\mathbf{\Phi} = \mathbf{I}$; *see Eq.* (2-109)] *with the property that*

$$\langle \mathbf{\Phi}\mathbf{e}_i, \mathbf{Q}\mathbf{\Phi}\mathbf{e}_j \rangle = \lambda_i \delta_{ij} \qquad i, j = 1, 2, \ldots, n \tag{2-121}$$

*or, equivalently, with the property that*

$$\mathbf{\Phi}'\mathbf{Q}\mathbf{\Phi} = \mathbf{\Lambda} \tag{2-122}$$

*Since* $\mathbf{\Phi}$ *is orthogonal,* $\mathbf{\Phi}'$ *is the inverse of* $\mathbf{\Phi}$, *and so Eq.* (2-122) *implies that* $\mathbf{Q}$ *and* $\mathbf{\Lambda}$ *are similar matrices (see Sec.* 2-9). *It follows (see Sec.* 2-10) *that the real numbers* $\lambda_1, \lambda_2, \ldots, \lambda_n$ *are the eigenvalues of* $\mathbf{Q}$.†

† A proof of this theorem can be found in Ref. B-6 or B-9.

We can immediately deduce the following two corollaries of the theorem:

*Corollary 2-1* *If* $\mathbf{Q}$ *is a (real) symmetric* $n \times n$ *matrix, then the eigenvalues* $\lambda_1, \lambda_2, \ldots, \lambda_n$ *of* $\mathbf{Q}$ *are all real.*

*Corollary 2-2* *If* $Q(\mathbf{v}, \mathbf{v}) = \langle \mathbf{v}, \mathbf{Q}\mathbf{v} \rangle$ *is the quadratic form induced by the inner product* $Q$ *(see Sec. 2-11), then there is an orthonormal basis* $\{\mathbf{f}_1, \mathbf{f}_2, \ldots, \mathbf{f}_n\}$ *of* $R_n$ *such that*

$$Q(\mathbf{v}, \mathbf{v}) = \langle \mathbf{v}, \mathbf{Q}\mathbf{v} \rangle = \sum_{i=1}^{n} \lambda_i \beta_i^2 \tag{2-123}$$

*for all* $\mathbf{v}$ *in* $R_n$*, where the* $\lambda_i$ *are the eigenvalues of* $\mathbf{Q}$ *and the* $\beta_i$ *are the coordinates of* $\mathbf{v}$ *with respect to the* $\mathbf{f}_i$*, that is,* $\mathbf{v} = \sum_{i=1}^{n} \beta_i \mathbf{f}_i$.

**Example 2-31** Let $\mathbf{Q}$ be the $2 \times 2$ symmetric matrix given by

$$\mathbf{Q} = \begin{bmatrix} a & b \\ b & c \end{bmatrix}$$

so that $\langle \mathbf{e}_1, \mathbf{Q}\mathbf{e}_1 \rangle = a$, $\langle \mathbf{e}_1, \mathbf{Q}\mathbf{e}_2 \rangle = \langle \mathbf{e}_2, \mathbf{Q}\mathbf{e}_1 \rangle = b$, and $\langle \mathbf{e}_2, \mathbf{Q}\mathbf{e}_2 \rangle = c$. Then the theorem states that there is an orthonormal basis $\{\mathbf{f}_1, \mathbf{f}_2\}$ of $R_2$ such that $\langle \mathbf{f}_1, \mathbf{Q}\mathbf{f}_1 \rangle = \lambda_1$, $\langle \mathbf{f}_1, \mathbf{Q}\mathbf{f}_2 \rangle = \langle \mathbf{f}_2, \mathbf{Q}\mathbf{f}_1 \rangle = 0$, and $\langle \mathbf{f}_2, \mathbf{Q}\mathbf{f}_2 \rangle = \lambda_2$, where $\lambda_1$ and $\lambda_2$ are the eigenvalues of $\mathbf{Q}$. Since

$$\mathbf{Q} - \lambda\mathbf{I} = \begin{bmatrix} a - \lambda & b \\ b & c - \lambda \end{bmatrix}$$

we can see that $\det(\mathbf{Q} - \lambda\mathbf{I}) = \lambda^2 - (a + c)\lambda + (ac - b^2)$ and hence that the eigenvalues of $\mathbf{Q}$ are

$$\lambda_1 = \frac{(a + c) + \sqrt{(a - c)^2 + 4b^2}}{2} \qquad \lambda_2 = \frac{(a + c) - \sqrt{(a - c)^2 + 4b^2}}{2}$$

As $(a - c)^2 + 4b^2 \geq 0$, the eigenvalues $\lambda_1$ and $\lambda_2$ must be real. In particular, if $a = b = c = 1$, so that

$$\mathbf{Q} = \begin{bmatrix} 1 & 1 \\ 1 & 1 \end{bmatrix}$$

then (see Example 2-20) the eigenvalues of $\mathbf{Q}$ are 0 and 2 and there is an orthogonal matrix $\mathbf{\Phi}$, namely,

$$\mathbf{\Phi} = \begin{bmatrix} 1/\sqrt{2} & 1/\sqrt{2} \\ -1/\sqrt{2} & 1/\sqrt{2} \end{bmatrix}$$

such that

$$\mathbf{\Phi}'\mathbf{Q}\mathbf{\Phi} = \begin{bmatrix} 0 & 0 \\ 0 & 2 \end{bmatrix}$$

We can see that

$$\mathbf{f}_1 = \mathbf{\Phi}\mathbf{e}_1 = \begin{bmatrix} 1/\sqrt{2} \\ -1/\sqrt{2} \end{bmatrix}$$

and

$$\mathbf{f}_2 = \mathbf{\Phi}\mathbf{e}_2 = \begin{bmatrix} 1/\sqrt{2} \\ 1/\sqrt{2} \end{bmatrix}$$

are the new orthonormal basis vectors in $R_2$.

In view of Corollary 2-2, we shall introduce the following terminology for a symmetric $n \times n$ matrix $\mathbf{Q}$; we shall say that $\mathbf{Q}$ is:

1. *Positive definite* if and only if all the eigenvalues $\lambda_1, \lambda_2, \ldots, \lambda_n$ of $\mathbf{Q}$ are positive, that is,

$$\lambda_i > 0 \qquad \text{for all } i \tag{2-124}$$

or, equivalently, if and only if

$$Q(\mathbf{v}, \mathbf{v}) = \langle \mathbf{v}, \mathbf{Q}\mathbf{v} \rangle > 0 \tag{2-125}$$

for all $\mathbf{v}$ in $R_n$ with $\mathbf{v} \neq \mathbf{0}$

2. *Positive semidefinite* if and only if all the eigenvalues $\lambda_1, \lambda_2, \ldots, \lambda_n$ of $\mathbf{Q}$ are nonnegative and at least one eigenvalue of $\mathbf{Q}$ is zero, that is,

$$\lambda_i \geq 0 \qquad \text{for all } i \tag{2-126}$$

and

$$\lambda_{i_1} = 0 \qquad \text{for some } i_1 \in \{1, 2, \ldots, n\} \tag{2-127}$$

or, equivalently, if and only if

$$Q(\mathbf{v}, \mathbf{v}) = \langle \mathbf{v}, \mathbf{Q}\mathbf{v} \rangle \geq 0 \tag{2-128}$$

for all $\mathbf{v}$ in $R_n$ and there is a $\hat{\mathbf{v}} \neq \mathbf{0}$ in $R_n$ for which

$$Q(\hat{\mathbf{v}}, \hat{\mathbf{v}}) = \langle \hat{\mathbf{v}}, \mathbf{Q}\hat{\mathbf{v}} \rangle = 0 \tag{2-129}$$

3. *Negative definite* if and only if all the eigenvalues $\lambda_1, \lambda_2, \ldots, \lambda_n$ of $\mathbf{Q}$ are negative, that is,

$$\lambda_i < 0 \qquad \text{for all } i \tag{2-130}$$

or, equivalently, if and only if

$$Q(\mathbf{v}, \mathbf{v}) = \langle \mathbf{v}, \mathbf{Q}\mathbf{v} \rangle < 0 \tag{2-131}$$

for all $\mathbf{v}$ in $R_n$ with $\mathbf{v} \neq \mathbf{0}$

4. *Negative semidefinite* if and only if all the eigenvalues $\lambda_1, \lambda_2, \ldots, \lambda_n$ of $\mathbf{Q}$ are nonpositive and at least one eigenvalue of $\mathbf{Q}$ is zero, that is,

$$\lambda_i \leq 0 \qquad \text{for all } i \tag{2-132}$$

and

$$\lambda_{i_1} = 0 \qquad \text{for some } i_1 \in \{1, 2, \ldots, n\} \tag{2-133}$$

or, equivalently, if and only if

$$Q(\mathbf{v}, \mathbf{v}) = \langle \mathbf{v}, \mathbf{Q}\mathbf{v} \rangle \leq 0 \tag{2-134}$$

for all $\mathbf{v}$ in $R_n$ and

$$Q(\hat{\mathbf{v}}, \hat{\mathbf{v}}) = \langle \hat{\mathbf{v}}, \mathbf{Q}\hat{\mathbf{v}} \rangle = 0 \tag{2-135}$$

for some $\hat{\mathbf{v}}$ in $R_n$ with $\hat{\mathbf{v}} \neq \mathbf{0}$

We observe that if $\mathbf{Q}$ is either positive or negative definite, then $\mathbf{Q}$ must be *nonsingular*, since $\mathbf{Q}$ is similar to a diagonal matrix which has no zero entries on the main diagonal. On the other hand, if $\mathbf{Q}$ is either positive or negative *semi*definite, then $\mathbf{Q}$ must be a singular matrix, since $\mathbf{Q}$ has a zero eigenvalue. Particular use of these notions will be made in Chaps. 6 and 9.

**Example 2-32** The matrix

$$\mathbf{Q} = \begin{bmatrix} 1 & 1 \\ 1 & 1 \end{bmatrix}$$

is positive semidefinite since its eigenvalues are 0 and 2.

Since we shall frequently want to know whether or not a given symmetric $n \times n$ matrix $\mathbf{Q} = (q_{ij})$ is positive definite, we now exhibit two criteria for positive definiteness in the following theorem.

***Theorem 2-5*** *Let* $\mathbf{Q} = (q_{ij})$ *be a symmetric* $n \times n$ *matrix. Then*

*a.* $\mathbf{Q}$ *is positive definite if and only if there is a* $k > 0$ *such that*

$$\langle \mathbf{v}, \mathbf{Q}\mathbf{v} \rangle \geq k\|\mathbf{v}\|^2 \tag{2-136}$$

*for all* $\mathbf{v}$ *in* $R_n$, *where* $\|\mathbf{v}\| = \sqrt{\langle \mathbf{v}, \mathbf{v} \rangle}$ *is the Euclidean norm of* $\mathbf{v}$ *[see Eq. (2-95)].*

*b.* $\mathbf{Q}$ *is positive definite if and only if the relations*

$$\det(\mathbf{Q}_r) > 0 \qquad r = 1, 2, \ldots, n \tag{2-137}$$

*hold, where* $\det(\mathbf{Q}_r)$ *is the determinant of the* $r \times r$ *matrix* $\mathbf{Q}_r$, *given by*

$$\mathbf{Q}_r = (q_{\alpha\beta}) \qquad \alpha = 1, 2, \ldots, r;\ \beta = 1, 2, \ldots, r \tag{2-138}$$

*that is,*

$$\mathbf{Q}_r = \begin{bmatrix} q_{11} & q_{12} & \cdots & q_{1r} \\ q_{21} & q_{22} & \cdots & q_{2r} \\ \cdots & \cdots & \cdots & \cdots \\ q_{r1} & q_{r2} & \cdots & q_{rr} \end{bmatrix} \tag{2-139}$$

We shall prove part *a* of this theorem, and we note that a proof of part *b* of this theorem can be found in Ref. B-6, pp. 73–74.

Now let us prove part *a*. We first suppose that Eq. (2-136) is satisfied. Since $k > 0$ and since $\|\mathbf{v}\| \neq 0$ if $\mathbf{v} \neq \mathbf{0}$ (see Sec. 2-13), we can see that

$$\langle \mathbf{v}, \mathbf{Q}\mathbf{v} \rangle \geq k\|\mathbf{v}\|^2 > 0 \tag{2-140}$$

for all $\mathbf{v}$ in $R_n$ with $\mathbf{v} \neq \mathbf{0}$. In view of Eq. (2-125), we may then conclude that $\mathbf{Q}$ is positive definite. On the other hand, if we suppose that $\mathbf{Q}$ is positive definite, then we know from Corollary 2-2 that there is an *orthonormal* basis $\mathbf{f}_1, \mathbf{f}_2, \ldots, \mathbf{f}_n$ of $R_n$ such that

$$\langle \mathbf{v}, \mathbf{Q}\mathbf{v} \rangle = \sum_{i=1}^{n} \lambda_i \beta_i^2 \tag{2-141}$$

for all $\mathbf{v}$ in $R_n$, where the $\beta_i$ are the coordinates of $\mathbf{v}$ with respect to the $\mathbf{f}_i$ and where

$$\lambda_i > 0 \qquad \text{for } i = 1, 2, \ldots, n \tag{2-142}$$

(since $\mathbf{Q}$ is positive definite). However, since

$$\langle \mathbf{f}_i, \mathbf{f}_j \rangle = \delta_{ij} \qquad i, j = 1, 2, \ldots, n \tag{2-143}$$

we have

$$\|\mathbf{v}\|^2 = \langle \mathbf{v}, \mathbf{v} \rangle = \left\langle \sum_{i=1}^{n} \beta_i \mathbf{f}_i, \sum_{j=1}^{n} \beta_j \mathbf{f}_j \right\rangle = \sum_{i=1}^{n} \beta_i^2 \tag{2-144}$$

and so, if we let $k$ be a positive number (that is, $k > 0$) such that

$$k \leq \lambda_i \qquad \text{for } i = 1, 2, \ldots, n \tag{2-145}$$

then we can deduce from Eqs. (2-141) and (2-144) that

$$\langle \mathbf{v}, \mathbf{Q}\mathbf{v} \rangle = \sum_{i=1}^{n} \lambda_i \beta_i^2 \geq k \sum_{i=1}^{n} \beta_i^2 = k\|\mathbf{v}\|^2 \tag{2-146}$$

Thus, we have established part $a$ of Theorem 2-5.

**Example 2-33** Suppose that

$$\mathbf{Q} = \begin{bmatrix} q_{11} & q_{12} \\ q_{21} & q_{22} \end{bmatrix}$$

is a symmetric 2 × 2 matrix. Then **Q** is positive definite if and only if

$$q_{11} > 0 \qquad \text{and} \qquad q_{11}q_{22} - q_{12}^2 > 0$$

If

$$\mathbf{Q} = \begin{bmatrix} q_{11} & q_{12} & q_{13} \\ q_{21} & q_{22} & q_{23} \\ q_{31} & q_{32} & q_{33} \end{bmatrix}$$

is a symmetric 3 × 3 matrix, then **Q** is positive definite if and only if

$$q_{11} > 0 \qquad \text{and} \qquad q_{11}q_{22} - q_{12}^2 > 0 \qquad \text{and} \qquad \det \mathbf{Q} > 0$$

This example illustrates part $b$ of the theorem in the special cases where $n = 2$ and $n = 3$.

# Chapter 3

# MATHEMATICAL PRELIMINARIES: ANALYSIS

## 3-1 Introduction

We continue the development of the basic mathematical concepts necessary for an understanding of the main body of the book in this chapter. Previously, we examined ideas which depended upon such algebraic notions as addition and multiplication by a real number. We shall now turn our attention to ideas which depend upon the notions of distance and magnitude. In particular, we shall study distance, convexity, vector functions, differential equations, and linear systems in this chapter. Moreover, we shall, for the sake of completeness, develop some of the familiar results of what is generally known as advanced calculus.

We again note that our treatment will be fairly rapid, as we assume that the reader has met with most of the ideas we need. We also observe that our treatment will be incomplete in that we shall present only material used in the sequel. Moreover, we do not develop the concepts any further than is necessary for our purposes. The interested reader will find the material of this chapter, in whole or in part, together with various generalizations of it, in Refs. B-6, B-10, C-7, D-5, E-3, K-20, P-6, R-10, S-4, V-1, and Z-1 (and many other books).

## 3-2 Distance and Related Notions: Definition

We begin now to study a very basic concept, the notion of distance, and some of its consequences. Let us examine our intuitive idea of distance and try to abstract the essential elements from it. First of all, we observe that we usually speak of the distance between two points as in "the distance between New York and Boston is 208 miles." Secondly, we note that the distance between two points is zero if and only if the points are the same. For example, the distance between Boston and Boston is zero miles. Thirdly, we usually think of the distance between two points as being independent of the order in which the points are mentioned; e.g., the distance between New York and Boston is the same as the distance between Boston and New York. Finally, our intuitive idea of distance is triangular in the sense that if $A$, $B$, and $C$ are three points, the distance between $A$ and $C$ is no greater than the sum of the distance between $A$ and $B$ and the distance between $B$ and $C$. For example, the distance between New York and Boston is less than the sum of the distance between New York and Albany and the distance between Albany and Boston. These observations lead us to the following precise formulation of the notion of distance:

***Definition 3-1*** *If $X$ is a set, then a real-valued function $d$, which is defined on all of $X \times X$, is called a distance function (or distance or metric) on $X$ if the following conditions are satisfied:*

D1 $d(x, y) \geq 0$ *for all $x$, $y$ in $X$ and* $d(x, y) = 0$ *if and only if* $x = y$.
D2 $d(x, y) = d(y, x)$ *for all $x$, $y$ in $X$.*
D3 $d(x, z) \leq d(x, y) + d(y, z)$ *for all $x$, $y$, $z$ in $X$.*

*We often refer to $d(x, y)$ as "the distance between $x$ and $y$," and we call* D3 *the triangle inequality. A set $X$ together with a given distance function $d$ on $X$ is called a metric space.*

If $X$ is a Euclidean space with $P$ as inner product (see Secs. 2-11 and 2-13), then we may define a distance $d$ on $X$ by setting

$$d(\mathbf{x}, \mathbf{y}) = \|\mathbf{x} - \mathbf{y}\|_P = \sqrt{P(\mathbf{x} - \mathbf{y}, \mathbf{x} - \mathbf{y})} \tag{3-1}$$

where $\|\mathbf{x} - \mathbf{y}\|_P$ represents the norm of $\mathbf{x} - \mathbf{y}$ (see Sec. 2-13) and $\mathbf{x}$, $\mathbf{y}$ are in $X$. In view of properties N1 and N2 (Sec. 2-13) of the norm and property IP3 of $P$ (Sec. 2-11), we see that Eq. (3-1) does indeed define a distance on $X$. In the particular case where $X$ is the space $R_n$ and $P$ is the scalar product on $R_n$ [Eq. (2-86)], we have, for

$$\mathbf{x} = \begin{bmatrix} x_1 \\ x_2 \\ \cdot \\ \cdot \\ \cdot \\ x_n \end{bmatrix} \quad \text{and} \quad \mathbf{y} = \begin{bmatrix} y_1 \\ y_2 \\ \cdot \\ \cdot \\ \cdot \\ y_n \end{bmatrix}$$

$$d(\mathbf{x}, \mathbf{y}) = \|\mathbf{x} - \mathbf{y}\| = \sqrt{\langle \mathbf{x} - \mathbf{y}, \mathbf{x} - \mathbf{y} \rangle} \tag{3-2}$$

or, equivalently,

$$d\left(\begin{bmatrix} x_1 \\ x_2 \\ \cdot \\ \cdot \\ \cdot \\ x_n \end{bmatrix}, \begin{bmatrix} y_1 \\ y_2 \\ \cdot \\ \cdot \\ \cdot \\ y_n \end{bmatrix}\right) = \sqrt{\sum_{i=1}^{n} (x_i - y_i)^2} = \|\mathbf{x} - \mathbf{y}\| \tag{3-3}$$

We call this distance function the *natural, or Euclidean, distance* (or *metric*) on $R_n$.

**Example 3-1** If $X = R$ is the set of real numbers, then the natural distance between $x$ and $y$, $d(x, y)$, is simply the absolute value of $x - y$, that is, $d(x, y) = |x - y|$. For example, $d(5, 3) = |5 - 3| = 2$.

**Example 3-2** If

$$\mathbf{x} = \begin{bmatrix} 1 \\ 0 \\ 2 \end{bmatrix} \quad \text{and} \quad \mathbf{y} = \begin{bmatrix} 3 \\ 4 \\ 0 \end{bmatrix} \quad \text{and} \quad \mathbf{z} = \begin{bmatrix} 1 \\ 2 \\ 1 \end{bmatrix}$$

then $d(\mathbf{x}, \mathbf{z}) = \|\mathbf{x} - \mathbf{z}\| = \sqrt{5}$, $d(\mathbf{x}, \mathbf{y}) = \|\mathbf{x} - \mathbf{y}\| = 2\sqrt{6}$, and $d(\mathbf{y}, \mathbf{z}) = \|\mathbf{y} - \mathbf{z}\| = 3$. We note that $\sqrt{5} < 2\sqrt{6} + 3$, that $2\sqrt{6} < \sqrt{5} + 3$, and that $3 < \sqrt{5} + 2\sqrt{6}$.

**Example 3-3** If $X$ is any set and if we set $d(x, y) = 1$ if $x \neq y$ and $d(x, x) = 0$ for $x, y$ in $X$, then $d$ is a distance function on $X$.

**Example 3-4** Suppose that $X = R_2$ and that $\mathbf{x} = \begin{bmatrix} x_1 \\ x_2 \end{bmatrix}$ and $\mathbf{y} = \begin{bmatrix} y_1 \\ y_2 \end{bmatrix}$. Then the function $d(\mathbf{x}, \mathbf{y})$ defined by $d(\mathbf{x}, \mathbf{y}) = |x_1 - y_1| + |x_2 - y_2|$ is a distance function on $R_2$.

**Exercise 3-1** Show that the function $d$ of Example 3-4 is indeed a distance on $R_2$. Generalize this distance to $R_n$. Hint: In Example 3-4, $d(\mathbf{x}, \mathbf{y}) = \sum_{i=1}^{2} |x_i - y_i|$.

**Exercise 3-2** If we let $\max \{a, b\}$ denote the maximum of $a$ and $b$, then the function $d$ defined by $d\left(\begin{bmatrix} x_1 \\ x_2 \end{bmatrix}, \begin{bmatrix} y_1 \\ y_2 \end{bmatrix}\right) = \max \{|x_1 - y_1|, |x_2 - y_2|\}$ is a distance on $R_2$. Prove this and generalize this distance to $R_n$. Hint: $d\left(\begin{bmatrix} x_1 \\ x_2 \end{bmatrix}, \begin{bmatrix} y_1 \\ y_2 \end{bmatrix}\right) = \max_{i=1,2} \{|x_i - y_i|\}$.

## 3-3 Distance and Related Notions: Spheres and Limits

Let us suppose that $X$ is a set, that $d$ is a distance on $X$, and that $x_0$ is a fixed element of $X$. We are often interested in the points which are within a given distance of the fixed point $x_0$. If $\rho > 0$ is a real number, then the set $S(x_0, \rho)$ defined by

$$S(x_0, \rho) = \{x \in X : d(x_0, x) < \rho\} \tag{3-4}$$

is called the *open sphere of radius* $\rho$ *about* $x_0$. In other words, $S(x_0, \rho)$ is the set of elements of $X$ whose distance from $x_0$ is *less than* $\rho$. For example, if $X = R$ and if we are using the natural distance on $R$, then $S(0, 1) = \{r: |r| < 1\}$. We note that if $\rho < \sigma$, then $S(x_0, \rho) \subset S(x_0, \sigma)$.

**Example 3-5** Let $X$ be the space $R_2$ and let $\mathbf{x}_0 = \begin{bmatrix} 0 \\ 0 \end{bmatrix}$. Then, using the natural distance [Eq. (3-3)], we see that $S(\mathbf{x}_0, 1)$ is the set $\left\{\mathbf{x} = \begin{bmatrix} x_1 \\ x_2 \end{bmatrix} : x_1^2 + x_2^2 < 1\right\}$. (See Fig. 3-1*a*.) If we use the distance of Exercise 3-2, then $S(\mathbf{x}_0, 1)$ is the set $\left\{\mathbf{x} = \begin{bmatrix} x_1 \\ x_2 \end{bmatrix} : \max \{|x_1|, |x_2|\} < 1\right\}$. (See Fig. 3-1*b*.)

In a similar way, if $\rho > 0$ is a real number, then the set $\overline{S(x_0, \rho)}$ defined by

$$\overline{S(x_0, \rho)} = \{x \in X : d(x_0, x) \leq \rho\} \tag{3-5}$$

*is called the closed sphere of radius* $\rho$ *about* $x_0$. In other words, $\overline{S(x_0, \rho)}$ is the set of elements of $X$ whose distance from $x_0$ is *not greater than* $\rho$. For example, if $X = R$ and if we are using the natural distance on $R$, then $\overline{S(0, 1)} = \{r: |r| \leq 1\}$. The reason for using the terms *open* and *closed* will appear in the next section.

Finally, we observe that, in the particular case of $R_n$ with the distance $\|\mathbf{x} - \mathbf{y}\|$, we have

$$S(\mathbf{x}_0, \rho) = \{\mathbf{x} \in R_n: \|\mathbf{x} - \mathbf{x}_0\| < \rho\} \tag{3-6a}$$

and

$$\overline{S(\mathbf{x}_0, \rho)} = \{\mathbf{x} \in R_n: \|\mathbf{x} - \mathbf{x}_0\| \leq \rho\} \tag{3-6b}$$

Having defined the notions of open and closed sphere about a point, we are now ready to develop the idea of convergence. So we suppose that $x_n$, $n = 1, 2, \ldots$, is a sequence of elements of $X$. What does it mean to say that the sequence approaches $x_0$ or tends to $x_0$ or converges to $x_0$? Intuitively, we view the sequence as tending to $x_0$ if the elements $x_n$ get closer and closer to $x_0$ as $n$ increases. More precisely, we have:

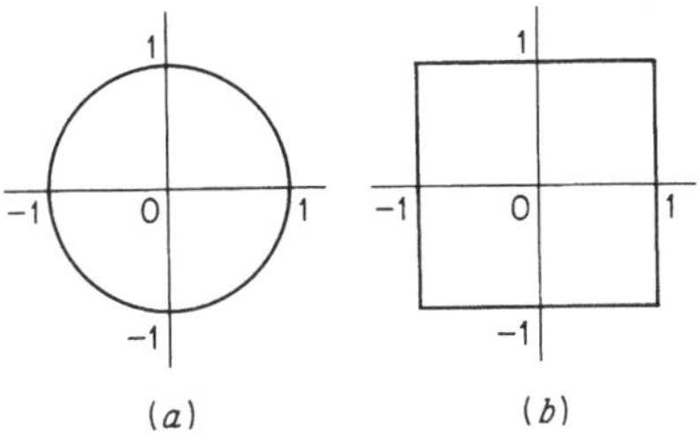


**Fig. 3-1** (*a*) $S(\mathbf{0}, 1)$ is the interior of the disk, and $\overline{S(\mathbf{0}, 1)}$ is the interior of the disk together with the circle which is the boundary of the disk. (*b*) $S(\mathbf{0}, 1)$ is the interior of the square, and $\overline{S(\mathbf{0}, 1)}$ is the entire square.

***Definition 3-2 Convergence*** *The sequence* $\{x_n: n = 1, 2, \ldots\}$ *is said to converge to* $x_0$ *if, given any real number* $\epsilon > 0$, *there is an* $N(\epsilon)$ *such that if* $n > N(\epsilon)$, *then* $x_n$ *is in* $S(x_0, \epsilon)$, *that is,* $d(x_0, x_n) < \epsilon$ *for all* $n$ *bigger than* $N(\epsilon)$. *In other words, if we take any open sphere about* $x_0$, *then, from some element on, all the members of the sequence are in that open sphere.*

We often write $x_n \to x_0$ to represent the statement "$x_n$ converges to $x_0$." For example, the sequence $x_n = 1/n$, $n = 1, 2, \ldots$, in $R$ converges to 0. It is easy to see that the sequence $x_n$ converges to $x_0$ if and only if the sequence of real numbers $d(x_0, x_n)$ converges to 0. In the particular case of $R_n$ with the distance $\|\mathbf{x} - \mathbf{y}\|$, we see that a sequence $\mathbf{x}_n$ converges to $\mathbf{x}_0$ if and only if $\epsilon > 0$ implies that there is an $N(\epsilon)$ such that $\|\mathbf{x}_n - \mathbf{x}_0\| < \epsilon$ for all $n > N(\epsilon)$.

**Example 3-6** The points $\mathbf{x}_n = \begin{bmatrix} 1/n \\ 1/n \end{bmatrix}$ in $R_2$ converge to the origin $\begin{bmatrix} 0 \\ 0 \end{bmatrix}$ in $R_2$ with respect to the natural distance in $R_2$ and also with respect to the distances of Exercises 3-1 and 3-2.

**Example 3-7** If $X$ is any set and if $d$ is the distance of Example 3-3 [that is, $d(x, y) = 1$ if $x \neq y$, and $d(x, x) = 0$ for $x$, $y$ in $X$], then the sequence $x_n = x_0$ for $n = 1, 2, \ldots$ converges to $x_0$.

***Definition 3-3 Limit of a Sequence*** *If the sequence $x_n$, $n = 1, 2, \ldots$, converges to $x_0$, then we say that $x_0$ is the limit of the sequence, and we often write $x_0 = \lim_{n \to \infty} x_n$.*

Can a sequence $x_n$, $n = 1, 2, \ldots$, have more than one limit? Well, let us first observe that if $x_0$ and $y_0$ are two *distinct* elements of $X$, then $d(x_0, y_0) = \rho > 0$, and consequently, $S(x_0, \rho/2)$ and $S(y_0, \rho/2)$ do not meet, that is, $S(x_0, \rho/2) \cap S(y_0, \rho/2) = \emptyset$ (Fig. 3-2). For if $x$ were in both $S(x_0, \rho/2)$ and $S(y_0, \rho/2)$, then $d(x_0, y_0) = \rho \leq d(x_0, x) + d(x, y_0)$ would give the contradiction $\rho < \rho$ (why?). If we take $\epsilon = \rho/2$, then we see that, for $n > N(\epsilon)$, $x_n$ is not in $S(y_0, \epsilon)$, so that $y_0$ is not a limit of the sequence.

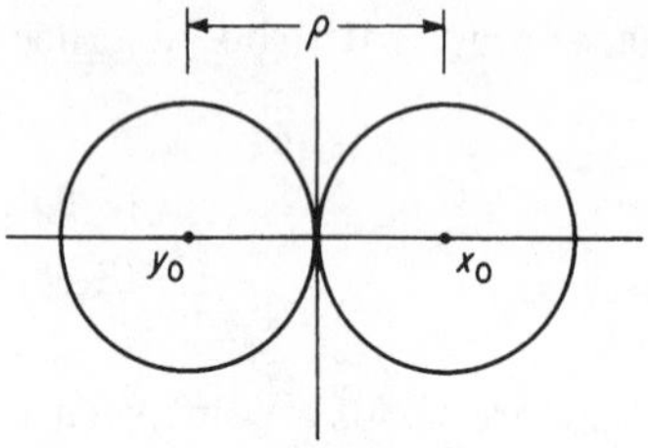

**Fig. 3-2** The two *open* [see Eq. (3-4)] spheres $S(x_0, \rho/2)$ and $S(y_0, \rho/2)$ do not meet.

Finally, let us suppose that $\mathbf{x}_n$, $n = 1, 2, \ldots$, is a sequence of elements of $R_N$, and let us denote the components of $\mathbf{x}_n$ by $x_{n,i}$, that is,

$$\mathbf{x}_n = \begin{bmatrix} x_{n,1} \\ x_{n,2} \\ \cdot \\ \cdot \\ \cdot \\ x_{n,N} \end{bmatrix}$$

Then an $N$ vector $\mathbf{x}$ with components $x_1, x_2, \ldots, x_N$ is the limit of the sequence $\{\mathbf{x}_n\}$ if and only if each $x_i$ is the limit of the sequence $\{x_{n,i}\}$. In other words,

$$\mathbf{x}_n \to \mathbf{x} \qquad \text{if and only if } x_{n,i} \to x_i \text{ for all } i \tag{3-6c}$$

To see this, we simply observe that

$$|x_{n,i} - x_i| \leq \|\mathbf{x}_n - \mathbf{x}\| \leq \sum_{i=1}^{N} |x_{n,i} - x_i| \tag{3-6d}$$

since

$$\begin{aligned} |x_{n,i} - x_i| &= \sqrt{|x_{n,i} - x_i|^2} = \sqrt{(x_{n,i} - x_i)^2} \\ &\leq \sqrt{\sum_{i=1}^{N} (x_{n,i} - x_i)^2} \leq \sqrt{\left(\sum_{i=1}^{N} |x_{n,i} - x_i|\right)^2} = \sum_{i=1}^{N} |x_{n,i} - x_i| \end{aligned}$$

## 3-4 Distance and Related Notions: Open and Closed Sets

Again let us suppose that $X$ is a set and that $d$ is a distance on $X$. In this section, we are going to investigate some important properties which subsets of $X$ may have.

***Definition 3-4 Interior Point*** *If $A$ is a subset of $X$ and if $x \in A$, then we say that $x$ is an interior point of $A$ if there is a $\rho > 0$ such that the open sphere $S(x, \rho)$ is contained in $A$, that is, $S(x, \rho) \subset A$.*

For example, if $A$ is itself an open sphere [say $A = S(x_0, \rho)$], then every element of $A$ is an interior point of $A$.

**Exercise 3-3** Show that if $A = S(x_0, \sigma)$ and $x \in A$, then $x$ is an interior point of $A$. HINT: Consider the open sphere $S(x, \rho)$ with $\rho = \sigma - d(x_0, x)$.

**Example 3-8** Let $X = R$, with the natural distance $d(r, s) = |r - s|$, and let $A$ be the subset of $X$ defined by $A = \{r: 0 < r \leq 1\}$. Then the point $r = \frac{1}{2}$ is an interior point of $A$, but the point $r = 1$ is not.

**Example 3-9** Let $X$ be any set and let $d$ be the distance function of Example 3-3 [that is, $d(x, y) = 1$ if $x \neq y$ and $d(x, x) = 0$]. If $A$ is *any* subset of $X$, then *every* element of $A$ is an interior point of $A$.

***Definition 3-5 Open Set*** *We say that $A$ is an open subset of $X$ (or, simply, is open) if every element of $A$ is an interior point of $A$.*

In other words, $A$ is open if, for each $x$ in $A$, there is a $\rho > 0$ (which may well depend upon $x$) such that $S(x, \rho)$ is contained in $A$. We observe that an open sphere is indeed an open subset of $X$ (Exercise 3-3). If $X = R$ and $a$ and $b$ are elements of $R$ with $a < b$, then the set $\{r: a < r < b\}$ is open and is called the *open interval with end points $a$ and $b$*. We sometimes denote this set by $(a, b)$,† that is, $(a, b) \triangleq \{r \in R: a < r < b\}$.

If $A_1$ and $A_2$ are open, then it can be shown that $A_1 \cap A_2$ is open. Moreover, if $A_i$, $i = 1, 2, \ldots$, are open, then it is easy to see that $\bigcup_{i=1}^{\infty} A_i$ is also open. Finally, we note that the whole space $X$ is open and that the null set $\emptyset$ is also open.

**Exercise 3-4** Show that if $A_1$ and $A_2$ are open, then $A_1 \cap A_2$ is open. [HINT: If $x \in A_1 \cap A_2$, then there is a $\rho_1 > 0$ and a $\rho_2 > 0$ such that $S(x, \rho_1) \subset A_1$ and $S(x, \rho_2) \subset A_2$. What is $S(x, \rho_1) \cap S(x, \rho_2)$?] Does it follow that if $A_1, A_2, \ldots, A_n$ are open, then $A_1 \cap A_2 \cap \cdots \cap A_n$ is open? Is the set $\left\{\begin{bmatrix} x_1 \\ x_2 \end{bmatrix}: x_1^2 + x_2^2 < 1 \text{ and } x_1 > 0\right\}$ open in $R_2$?

***Definition 3-6 Limit Point*** *Now suppose that $B$ is a subset of $X$ and that $x$ is an element of $X$. We shall say that $x$ is a limit point of $B$ if there is a sequence $x_n$, $n = 1, 2, \ldots$, of elements of $B$ (that is, $x_n \in B$ for $n = 1, 2, \ldots$) which converges to $x$.*

We note that $x$ need *not* be in $B$ in order to be a limit point of $B$. For example, if $B$ is the subset of $R$ given by $B = \{r: 0 < r \leq 1\}$, then 0 is a limit point of $B$ which is not in $B$, since the sequence $1/n$, $n = 1, 2, \ldots$, of elements of $B$ converges to 0, which is not in $B$. If $y$ is any point in $B$,

† This should not be confused with the pair $(a, b)$ in $R \times R$, and it will always be clear from the context which meaning is to be given to the symbol $(a, b)$.

then $y$ is a limit point of $B$, since the sequence $x_n = y$, $n = 1, 2, \ldots$, converges to $y$.

***Definition 3-7 Closed Set*** *We shall say that $B$ is a closed subset of $X$ (or, simply, is closed) if every limit point of $B$ is actually in $B$.*

Let us now prove the following:

***Theorem 3-1*** *If $A$ is open, then $X - A$ (the complement of $A$; see Sec. 2-3) is closed; and, conversely, if $B$ is closed, then $X - B$ is open.*

PROOF Suppose that $A$ is open and that $x \in A$. Then there is a $\rho > 0$ such that $S(x, \rho) \subset A$, and therefore $S(x, \rho)$ does *not* meet $X - A$, that is, $S(x, \rho) \cap (X - A) = \emptyset$. Clearly, $x$ cannot be a limit point of $X - A$. This shows that every limit point of $X - A$ is in $X - A$, and so $X - A$ is closed.

On the other hand, suppose that $B$ is closed and that $x \in X - B$. Consider the open spheres $S(x, 1/n)$ for $n = 1, 2, \ldots$. We claim that there is an $N$ for which $S(x, 1/N) \subset X - B$. If not, then each of the spheres $S(x, 1/n)$ meets $B$, and so there is an $x_n$ in $S(x, 1/n) \cap B$ for $n = 1, 2, \ldots$. The sequence $\{x_n\}$, $n = 1, 2, \ldots$, converges to $x$, since $m > n$ implies that $S(x, 1/m) \subset S(x, 1/n)$ and since $\rho > 0$ implies that there is an $m$ with $\rho > 1/m$. Since $B$ is closed and each $x_n$ is in $B$, we obtain the contradiction $x \in B$. So the open sphere $S(x, 1/N)$ is contained in $X - B$ for some $N$, and therefore $x$ is an interior point of $X - B$. It follows that $X - B$ is open.

**Exercise 3-5** Show that the closed sphere $\overline{S(x_0, \rho)}$ is a closed set. HINT: If $x_n$, $n = 1, 2, \ldots$, are in $\overline{S(x_0, \rho)}$ and $\{x_n\}$ converges to $x$, then $d(x_n, x)$ converges to 0, so that $d(x_0, x) \leq d(x_0, x_n) + d(x_n, x) \leq \rho + d(x_n, x)$.

If $B_1$ and $B_2$ are *closed*, then it can be shown that $B_1 \cup B_2$ is *closed*. Moreover, if $B_i$, $i = 1, 2, \ldots$, are closed, then it is easy to see that $\bigcap_{i=1}^{\infty} B_i$ is also closed. Finally, we note that the whole space $X$ is closed and that the null set $\emptyset$ is also closed.

If $X = R$ and $a$ and $b$ are elements of $R$, with $a < b$, then the set $\{r: a \leq r \leq b\}$ is closed and is called the *closed interval with end points a and b*. We sometimes denote this set by $[a, b]$, that is, $[a, b] \triangleq \{r \in R: a \leq r \leq b\}$. We shall also use the notations $[a, b)$ and $(a, b]$ for the sets $\{r: a \leq r < b\}$ and $\{r: a < r \leq b\}$, respectively. Summarizing, we have

$$\begin{aligned} (a, b) &= \{r: a < r < b\} \\ [a, b] &= \{r: a \leq r \leq b\} \\ [a, b) &= \{r: a \leq r < b\} \\ (a, b] &= \{r: a < r \leq b\} \end{aligned} \tag{3-7}$$

We note that the sets $[a, b)$ and $(a, b]$ are neither open nor closed and are often called *semiclosed* (or *half-closed*) or *semiopen* (or *half-open*) intervals.

Let us suppose now that $C$ is a subset of $X$. We are going to define three subsets of $X$ which are called the *interior*, the *closure*, and the *boundary* of $C$.

***Definition 3-8 Interior of a Set*** *Given a set $C$, then the interior of $C$, denoted by $i(C)$, is the set of all interior points of $C$; that is,*

$$i(C) = \{x\text{: there is a } \rho > 0 \text{ with } S(x, \rho) \subset C\} \tag{3-8}$$

We observe that a set $A$ is open if and only if it is the same as its interior [in other words, $i(A) = A$]. For example, the interior of the set $[a, b)$ [Eq. (3-7)] is the open interval $(a, b)$.

***Definition 3-9 Closure of a Set*** *Given a set $C$, the closure of $C$, denoted by $c(C)$ (or sometimes by $\bar{C}$), is the set of all limit points of $C$; that is,*

$$c(C) = \{x\text{: there is a sequence } x_n,\ n = 1, 2, \ldots, x_n \in C, \text{ such that } x_n \text{ converges to } x\} \tag{3-9}$$

It can be shown (Exercise 3-6) that $c(C)$ is a *closed* set. Moreover, *a set $B$ is closed if and only if it coincides with its closure, that is, $B = c(B)$.* For example, the closure of the set $[a, b)$ [Eq. (3-7)] is the closed interval $[a, b]$.

***Definition 3-10 Boundary of a Set*** *Given a set $C$, then the boundary of $C$, denoted by $b(C)$ (or sometimes by $\partial C$), is the set of limit points of $C$ which are not interior points of $C$; that is,*

$$b(C) = c(C) - i(C) \tag{3-10}$$

It can be shown that the boundary of $C$, $b(C)$, is a closed set. In fact,

$$b(C) = c(C) \cap c(X - C) \tag{3-11}$$

In other words, *the boundary of $C$ is the intersection of the closure of $C$ and the closure of $X - C$.* For example, the boundary of the set $[a, b)$ is the set consisting of the two points $a$ and $b$.

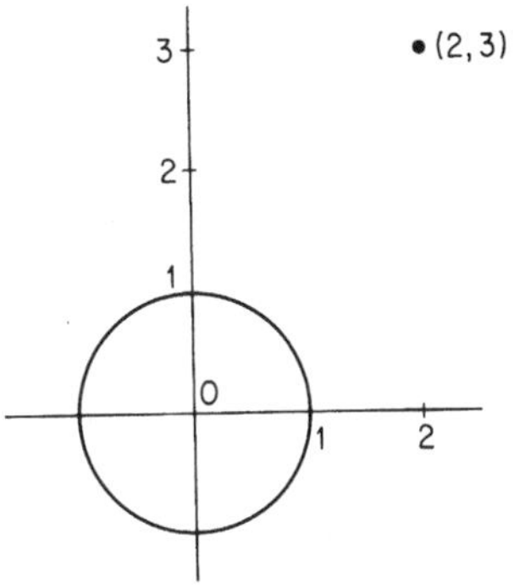


**Fig. 3-3** The set $C$ consists of the interior of the disk and the single point (2, 3). The interior $i(C)$ of $C$ is the interior of the disk; the boundary $b(C)$ of $C$ is the boundary of the disk *and* the point (2, 3); the closure $c(C)$ of $C$ is the entire disk and the point (2, 3).

**Example 3-10** Consider the set $C$ in $R_2$ given by $C = \left\{\begin{bmatrix} x_1 \\ x_2 \end{bmatrix}: x_1^2 + x_2^2 < 1 \text{ or } x_1 = 2, x_2 = 3\right\}$ (Fig. 3-3). In other words, $C = S(\mathbf{0}, 1) \cup \left\{\begin{bmatrix} 2 \\ 3 \end{bmatrix}\right\}$.† Then $i(C) = S(\mathbf{0}, 1)$, $c(C) = \overline{S(\mathbf{0}, 1)} \cup \left\{\begin{bmatrix} 2 \\ 3 \end{bmatrix}\right\}$, and $b(C) = \left\{\begin{bmatrix} x_1 \\ x_2 \end{bmatrix}: x_1^2 + x_2^2 = 1 \text{ or } x_1 = 2, x_2 = 3\right\}$. We note that no two of the four sets $C$, $i(C)$, $c(C)$, and $b(C)$ are the same.

† $\{(2, 3)\}$ is the set consisting of the single point $\begin{bmatrix} 2 \\ 3 \end{bmatrix}$.

**Exercise 3-6** Prove that $c(C)$ is closed. HINT: Suppose that $x \in c\{c(C)\}$; then there is a sequence $y_n$, $y_n \in c(C)$, which converges to $x$. But $y_n \in c(C)$ implies that there is an $x_n$ in $C$ with $d(x_n, y_n) < 1/n$. Show that the sequence $x_n$ converges to $x$.

**Example 3-11** Let $X = R$ be the real line and let $C$ be the set of rational numbers, that is, $C = \{r \in R\colon$ there are integers $p$ and $q$ with $r = p/q\}$. Then $i(C)$ is empty, $c(C)$ is all of $R$, and $b(C)$ is all of $R$.

The concepts we have been discussing can be generalized quite a bit. In fact, we have been talking about metric spaces and what is called the *topology* induced by the metric (or distance function). The general notion of a topology and of a topological space is based upon the concept of open sets rather than upon the notion of distance. These more general ideas can be found, for example, in Ref. S-4.

## 3-5 Distance and Related Notions: Completeness and Contractions

Let us suppose that $X$ is a set with $d$ as distance function and that $x_n$, $n = 1, 2, \ldots$, is a sequence which converges to $x_0$. If $\epsilon > 0$, then there is an $N(\epsilon/2)$ such that if $n > N(\epsilon/2)$, then $d(x_0, x_n) < \epsilon/2$. It follows from the triangle inequality (Definition 3-1) that if $n, m > N(\epsilon/2)$, then $d(x_n, x_m) < \epsilon$. This property of convergent sequences is quite important and leads us to the following notion:

***Definition 3-11 Cauchy Sequence*** *A sequence $y_n$, $n = 1, 2, \ldots$, of elements of $X$, whether or not it converges to an element of $X$, is called a Cauchy sequence if, for every $\epsilon > 0$, there is an $N(\epsilon)$ such that if $n$ and $m$ are bigger than $N(\epsilon)$ [that is, $n > N(\epsilon)$, $m > N(\epsilon)$], then $d(x_n, x_m) < \epsilon$. If every Cauchy sequence in $X$ converges to an element of $X$, we say that $X$ is complete. For example, the Euclidean space $R_n$ with the Euclidean distance function is complete.*†

**Example 3-12** Let $X = (0, 1]$ and let $d(x, y) = |x - y|$ for $x, y$ in $(0, 1]$. Then the sequence $x_n = 1/n$, $n = 1, 2, \ldots$, is a Cauchy sequence in $X$ but does not converge to an element of $X$. In other words, $X = (0, 1]$ is not complete.

**Exercise 3-7** Prove that if $X = [1, 2)$ and if $d(x, y) = |x - y|$ for $x, y$ in $[1, 2)$, then $X$ is not complete.

***Definition 3-12 Contraction*** *Assume that $X$ is complete and that $T$ is a transformation (function) of $X$ into itself; that is, $T$ is a function whose domain is all of $X$ and whose range is contained in $X$. If there is a real number $k$ with $0 \le k < 1$ such that, for all $x, y$ in $X$,*

$$d[T(x), T(y)] \le kd(x, y) \tag{3-12}$$

† In fact, $R_n$ is complete with respect to the distance functions $d(\mathbf{x}, \mathbf{y}) = \|\mathbf{x} - \mathbf{y}\| = \sqrt{\sum_{i=1}^{n} (x_i - y_i)^2}$ (the Euclidean distance), $d(\mathbf{x}, \mathbf{y}) = \sum_{i=1}^{n} |x_i - y_i|$, and $d(\mathbf{x}, \mathbf{y}) = \max \{|x_i - y_i|\}$.

*then the transformation T is called a contraction on X (or, simply, a contraction). In other words, T is a contraction if T decreases the distance between points. Observe that the number k in Eq.* (3-12) *does not depend on x and y.*

**Example 3-13** Let $X$ be the closed interval $[0, 1]$ [with $d(x, y) = |x - y|$ for $x, y$ in $[0, 1]$] and let $f$ be the transformation of $[0, 1]$ into itself defined by $f(x) = x^2/3$. Then $|f(x) - f(y)| = |x^2 - y^2|/3 \leq |x + y|\,|x - y|/3 \leq \frac{2}{3}|x - y|$, so that $f$ is a contraction on $[0, 1]$.

**Example 3-14** Suppose that $X = R_n$ and that $\mathcal{A}$ is the linear transformation of $R_n$ into itself given by $\mathcal{A}(\mathbf{v}) = \mathbf{A}\mathbf{v}$, where $\mathbf{A} = (a_{ij})$ is an $n \times n$ matrix and $\mathbf{v} \in R_n$ [see Eq. (2-51)]. Suppose also that $\sum_{i=1}^{n} \sum_{j=1}^{n} a_{ij}^2 = k < 1$. Then we claim that $\mathcal{A}$ is a contraction. We note that if $\mathbf{v}, \mathbf{w} \in R_n$, then

$$d(\mathbf{A}\mathbf{v}, \mathbf{A}\mathbf{w}) = \|\mathbf{A}\mathbf{v} - \mathbf{A}\mathbf{w}\| = \|\mathbf{A}(\mathbf{v} - \mathbf{w})\| = \sqrt{\sum_{i=1}^{n} \left\{\sum_{j=1}^{n} a_{ij}(v_j - w_j)\right\}^2}$$

and, in view of the Schwarz inequality [Eq. (2-87) or (2-91)],

$$d(\mathbf{A},\mathbf{v}\ \mathbf{A}\mathbf{w}) \leq \sqrt{\sum_{i=1}^{n} \sum_{j=1}^{n} a_{ij}^2 \|\mathbf{v} - \mathbf{w}\|^2}$$

It follows that $d(\mathbf{A}\mathbf{v}, \mathbf{A}\mathbf{w}) \leq kd(\mathbf{v}, \mathbf{w})$, which establishes our claim. For example, if $n = 2$ and if $\mathbf{A} = \begin{bmatrix} 1/\sqrt{3} & -\frac{1}{2}\sqrt{6} \\ \frac{1}{2}\sqrt{6} & -1/\sqrt{3} \end{bmatrix}$, then $a_{11}^2 = \frac{1}{3}$, $a_{12}^2 = \frac{1}{24}$, $a_{21}^2 = \frac{1}{24}$, $a_{22}^2 = \frac{1}{3}$, so that $a_{11}^2 + a_{12}^2 + a_{21}^2 + a_{22}^2 = \frac{3}{4} = k < 1$; and hence the transformation $\mathcal{A}$ given by $\mathcal{A}(\mathbf{e}_1) = \begin{bmatrix} 1/\sqrt{3} \\ \frac{1}{2}\sqrt{6} \end{bmatrix}$, $\mathcal{A}(\mathbf{e}_2) = \begin{bmatrix} -\frac{1}{2}\sqrt{6} \\ -1/\sqrt{3} \end{bmatrix}$ $\left(\text{where } \mathbf{e}_1 = \begin{bmatrix} 1 \\ 0 \end{bmatrix}, \mathbf{e}_2 = \begin{bmatrix} 0 \\ 1 \end{bmatrix}\right)$ is a contraction.

We now prove an important theorem which shows how contractions can be related to the solution of equations.

***Theorem 3-2*** *If T is a contraction on X, then there is a unique element $x^*$ of X such that*

$$T(x^*) = x^* \tag{3-13}$$

*Moreover, if $x_1$ is an arbitrary element of X and if $x_2 = T(x_1)$, $x_3 = T(x_2)$, . . . , $x_n = T(x_{n-1})$, . . . , then the sequence $\{x_n\}$ converges to $x^*$.*

PROOF First, let us show that such an $x^*$ must be unique. Suppose that $x^*$ and $y^*$ are elements of $X$ which satisfy Eq. (3-13), that is, $T(x^*) = x^*$ and $T(y^*) = y^*$. Then $d[T(x^*), T(y^*)] \leq kd(x^*, y^*)$ since $T$ is a contraction, and therefore $d(x^*, y^*) \leq kd(x^*, y^*)$, which implies that $d(x^*, y^*) = 0$, since $0 \leq k < 1$. Hence, $x^* = y^*$.

Now let us prove that there is an $x^*$ which satisfies Eq. (3-13). Let $x_1$ be any element of $X$ and let us set $x_2 = T(x_1)$, $x_3 = T(x_2)$, and, in general,

$$x_n = T(x_{n-1}) \qquad n = 1, 2, \ldots \tag{3-14}$$

We claim that the sequence $x_n$, $n = 1, 2, \ldots$, is a Cauchy sequence. If $\epsilon > 0$, then there is an $N(\epsilon)$ such that if $n > N(\epsilon)$, then

$$k^{n-1}\, d(x_1, x_2) \left\{\frac{1}{1-k}\right\} < \epsilon \tag{3-15}$$

Suppose that $n, m > N(\epsilon)$ (with, say, $m \geq n$); then

$$\begin{aligned} d(x_n, x_m) &= d[T(x_{n-1}), T(x_{m-1})] \leq k\, d[T(x_{n-2}), T(x_{m-2})] \\ &\leq k^{n-1}\, d[x_1, T(x_{m-n-1})] \\ &\leq k^{n-1}\{d(x_1, x_2) + d(x_2, x_3) + \cdots + d(x_1, x_{m-n})\} \\ &\leq k^{n-1}\, d(x_1, x_2)\{1 + k + k^2 + \cdots + k^{m-n-1}\} \\ &\leq k^{n-1}\, d(x_1, x_2) \left\{\frac{1}{1-k}\right\} < \epsilon \qquad \text{by Eq. (3-15)} \end{aligned} \tag{3-16}$$

In other words, the sequence $x_n$, $n = 1, 2, \ldots$, is a *Cauchy sequence* and therefore has a limit $x^*$ (since $X$ is complete).

We assert that $T(x^*) = x^*$. Consider the sequence $y_n = T(x_n) = x_{n+1}$ for $n = 1, 2, \ldots$. Clearly, $y_n$ converges to $x^*$, and we shall show that $y_n$ converges to $T(x^*)$. This will establish the theorem. Since $x^*$ is the limit of the sequence $\{x_n\}$, we know that if $\epsilon > 0$, then there is an $N(\epsilon)$ such that if $n > N(\epsilon)$, then $x_n$ is in $S(x^*, \epsilon)$, that is, $d(x^*, x_n) < \epsilon$. It follows that $d[T(x^*), T(x_n)] = d[T(x^*), y_n] \leq k\, d(x^*, x_n) < \epsilon$, and therefore we see that $y_n$ is in $S(T(x^*), \epsilon)$ for all $n > N(\epsilon)$. Thus we have shown that $y_n$ converges to $T(x^*)$.

We shall apply this theorem in later sections to prove an existence theorem for differential equations.

## 3-6 Properties of Sets in $R_n$: Compactness

In this section and the next, we shall deal exclusively with the Euclidean space $R_n$ and with the natural distance on $R_n$ [see Eq. (3-2)]. We shall be interested in studying properties of subsets of $R_n$ which will be crucial in our later work.

***Definition 3-13 Bounded Set*** *Suppose that $C$ is a subset of $R_n$. Then we shall say that $C$ is bounded if there is a finite $\rho > 0$ such that $C$ is contained in the sphere $S(\mathbf{0}, \rho)$ of radius $\rho$ about the origin.*

***Definition 3-14 Compactness*** *Let $C$ be a subset of $R_n$. If the set $C$ is both closed and bounded, then we say that $C$ is a compact subset of $R_n$ (or, simply, is compact).*

For example, the closed interval $[0, T]$, with $T > 0$, is a compact subset of $R = R_1$.

There are several properties which may serve as alternative definitions of compactness and which are often useful in proving things which involve compact sets. These properties are as follows:

C1 $C$ is compact if $C$ is closed and bounded.

C2 $C$ is compact if, given any collection $A_1, A_2, \ldots$ of *open* sets such that $C \subset \bigcup_{i=1}^{\infty} A_i$, there is a finite set of integers $n_1, n_2, \ldots, n_N$ such that $C \subset A_{n_1} \cup A_{n_2} \cup \cdots \cup A_{n_N}$.† (This is sometimes called the Heine-Borel property.)

C3 $C$ is compact if, given any collection $B_1, B_2, \ldots$ of *closed* sets such that $C \cap (\bigcap_{j=1}^{n} B_j) \neq \emptyset$ for every $n$, then $C \cap (\bigcap_{j=1}^{\infty} B_j) \neq \emptyset$. (This is sometimes called the finite-intersection property.)

C4 $C$ is compact if, given any sequence $x_n$, $n = 1, 2, \ldots$, in $C$, there is a subsequence $x_{n_1}, x_{n_2}, \ldots$ which converges to an element of $C$. (This is sometimes called the Bolzano-Weierstrass property.)

It can be shown that if a subset $C$ of $R_n$ has any one of properties C1 to C4, then $C$ has them all. (See Ref. D-5 or S-4.)

**Example 3-15** If $\mathbf{x}$ is an element of $R_n$ and if $\rho > 0$, then the closed sphere $\overline{S(\mathbf{x}, \rho)}$ is compact but the open sphere $S(\mathbf{x}, \rho)$ is not. In fact, if $A$ is a bounded subset of $R_n$, then

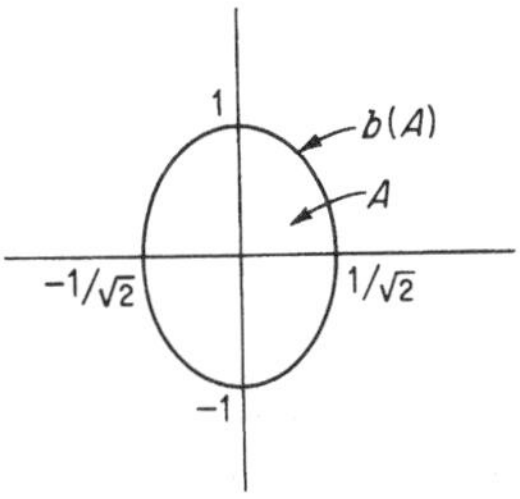


**Fig. 3-4** The closure $c(A)$ and the boundary $b(A)$ of the bounded open set $A$ are compact subsets of $R_2$.

the closure of $A$, $c(A)$ [Eq. (3-9)], is compact and the boundary $b(A)$ [Eq. (3-11)] is also compact. For example, if $A$ is the subset of $R_2$ defined by $A = \left\{\begin{bmatrix} x_1 \\ x_2 \end{bmatrix} : 2x_1^2 + x_2^2 < 1\right\}$, then $c(A) = \left\{\begin{bmatrix} x_1 \\ x_2 \end{bmatrix} : 2x_1^2 + x_2^2 \leq 1\right\}$ and $b(A) = \left\{\begin{bmatrix} x_1 \\ x_2 \end{bmatrix} : 2x_1^2 + x_2^2 = 1\right\}$ are compact subsets of $R_2$. (See Fig. 3-4.)

## 3-7 Properties of Sets in $R_n$: Hyperplanes and Cones

We shall examine two special classes of subsets in the Euclidean space $R_n$ in this section, preparatory to our study of the important notion of convexity in the next section.

† This is sometimes phrased as follows: "Every open covering of $C$ has a finite subcovering."

Loosely speaking, a hyperplane in $R_n$ is the analog of a line in the plane $R_2$ and may be viewed as the translate of an $(n-1)$-dimensional subspace of $R_n$. More precisely:

***Definition 3-15 Hyperplanes in $R_n$*** *Let $L(\mathbf{x})$ be the real-valued function on $R_n$, defined by*

$$L(\mathbf{x}) = \langle \mathbf{a}, \mathbf{x} \rangle - b = \left\langle \begin{bmatrix} a_1 \\ a_2 \\ \cdot \\ \cdot \\ \cdot \\ a_n \end{bmatrix}, \begin{bmatrix} x_1 \\ x_2 \\ \cdot \\ \cdot \\ \cdot \\ x_n \end{bmatrix} \right\rangle - b = \sum_{i=1}^{n} a_i x_i - b \tag{3-17}$$

*where $\mathbf{a}$ is a given nonzero element of $R_n$ and $b$ is a given real number; then we call the subset $L$ of $R_n$, given by*

$$L = \left\{ \mathbf{x} \colon L(\mathbf{x}) = \sum_{i=1}^{n} a_i x_i - b = 0 \right\} \tag{3-18}$$

*a hyperplane in $R_n$.*

For example, the set $L$ in $R_3$, given by

$$L = \left\{ \begin{bmatrix} x_1 \\ x_2 \\ x_3 \end{bmatrix} : 1 \cdot x_1 + 1 \cdot x_2 + 1 \cdot x_3 - \sqrt{2} = 0 \right\}$$

is a hyperplane in $R_3$.

We let

$$L^+ = \left\{ \mathbf{x} \colon L(\mathbf{x}) = \sum_{i=1}^{n} a_i x_i - b > 0 \right\} \tag{3-19}$$

and

$$L^- = \left\{ \mathbf{x} \colon L(\mathbf{x}) = \sum_{i=1}^{n} a_i x_i - b < 0 \right\} \tag{3-20}$$

$L^+$ and $L^-$ are called the *open half-spaces determined by $L$*. We observe that $L^+$ and $L^-$ are open sets and that $R_n = L^+ \cup L \cup L^-$. We shall sometimes refer to the sets $L^+ \cup L$ and $L^- \cup L$ as the *closed half-spaces determined by $L$*. It is easy to see that $L^+ \cup L$ is the closure of $L^+$, that $L^- \cup L$ is the closure of $L^-$, and that $L$ is the boundary of both $L^+$ and $L^-$. (See Fig. 3-5.)

If $A$ and $B$ are any two subsets of $R_n$, then we say that the hyperplane *$L$ separates $A$ and $B$* if $A$ is contained in one closed half-space determined by $L$ and $B$ is contained in the other closed half-space determined by $L$, that is, if

$$A \subset L^+ \cup L \quad \text{and} \quad B \subset L^- \cup L \tag{3-21a}$$

or

$$A \subset L^- \cup L \quad \text{and} \quad B \subset L^+ \cup L \tag{3-21b}$$

If it should happen that $A$ is contained in one of the open half-spaces determined by $L$ and that $B$ is contained in the other, then we shall say that *$L$ separates $A$ and $B$ strictly*. For example, in $R_3$, the hyperplane $x_1 - 1 = 0$

separates the sets $\overline{S(\mathbf{0}, 1)}$ and

$$\overline{S\left(\begin{bmatrix}2\\0\\0\end{bmatrix}, 1\right)}$$

and strictly separates the sets $S(\mathbf{0}, 1)$ and

$$S\left(\begin{bmatrix}2\\0\\0\end{bmatrix}, 1\right)$$

Now let $L$ be a hyperplane in $R_n$ and let $A$ be a subset of $R_n$. It is trivial to observe that either $L$ does not meet the closure of $A$, that is,

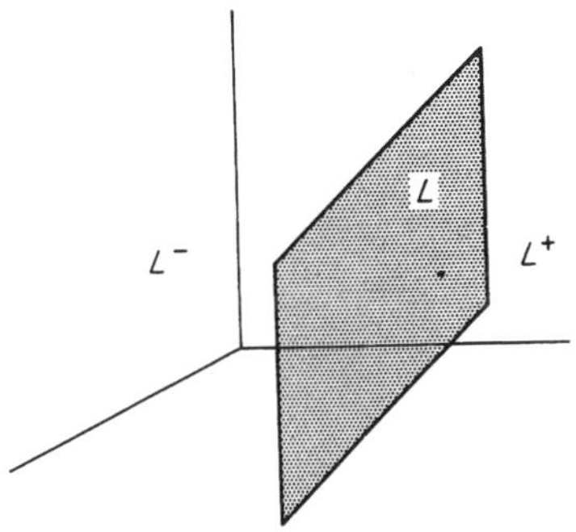


**Fig. 3-5** The hyperplane $L$ and the half-spaces $L^+$ and $L^-$.

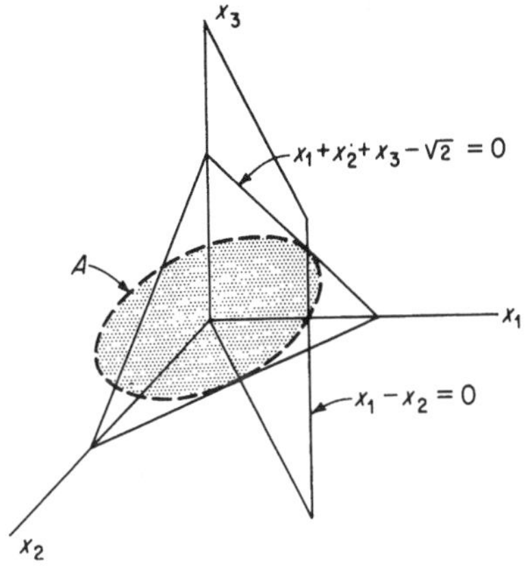


**Fig. 3-6** The sets of Example 3-16.

$L \cap c(A) = \emptyset$, or $L$ meets the closure of $A$, that is, $L \cap c(A) \neq \emptyset$. If $L$ meets the closure of $A$, one of the following three conditions must be satisfied:

1. $L$ contains $A$, that is, $A \subset L$.
2. Both $L^+ \cap A$ and $L^- \cap A$ are not empty.
3. $L$ does not contain $A$, and $A$ is contained in either $L^+ \cup L$ or $L^- \cup L$; that is, $L^+ \cap A$ or $L^- \cap A$ is empty.

If $L$ satisfies condition 2, then we say that $L$ *cuts* $A$; if $L$ satisfies condition 3, then we say that $L$ is a *support hyperplane of* $A$.

**Example 3-16** Let $A$ be the subset of $R_3$, defined by

$$A = \left\{\begin{bmatrix}x_1\\x_2\\x_3\end{bmatrix} : x_1^2 + x_2^2 \leq 1,\ x_3 = 0\right\}$$

Then the hyperplane $x_3 - 1 = 0$ does not meet the closure of $A$, the hyperplane $x_3 = 0$ contains $A$, the hyperplane $x_1 - x_2 = 0$ cuts $A$, and the hyperplane $x_1 + x_2 + x_3 - \sqrt{2} = 0$ is a support hyperplane of $A$ (meeting the closure of $A$ at the point $x_1 = \sqrt{2}/2$, $x_2 = \sqrt{2}/2$, $x_3 = 0$). (See Fig. 3-6.)

**Example 3-17** Let $A$ be the subset $S(\mathbf{0}, 1)$ of $R_2$; that is, $A = \left\{\begin{bmatrix} x_1 \\ x_2 \end{bmatrix}: x_1^2 + x_2^2 < 1\right\}$. If $L$ is a line in $R_2$ (in other words, $L$ is a hyperplane in $R_2$) with $L = \left\{\begin{bmatrix} x_1 \\ x_2 \end{bmatrix}: a_1x_1 + a_2x_2 - b = 0\right\}$, then $L$ satisfies one of three conditions: ($a$) $L \cap A = \emptyset$, ($b$) $L$ cuts $A$, ($c$) $L$ is "tangent" to $A$; that is, $L \cap c(A)$ consists of a single point on the boundary of $A$. (See Fig. 3-7.) The lines which satisfy condition $a$ or condition $c$ are the support lines of $A$.

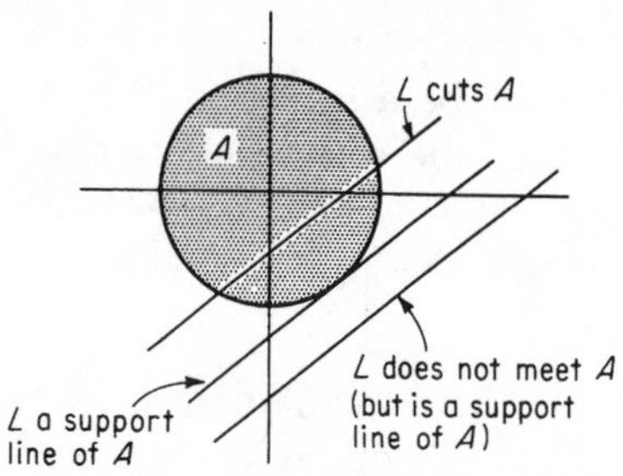


**Fig. 3-7** The three possible conditions for a line $L$ and a set $A$.

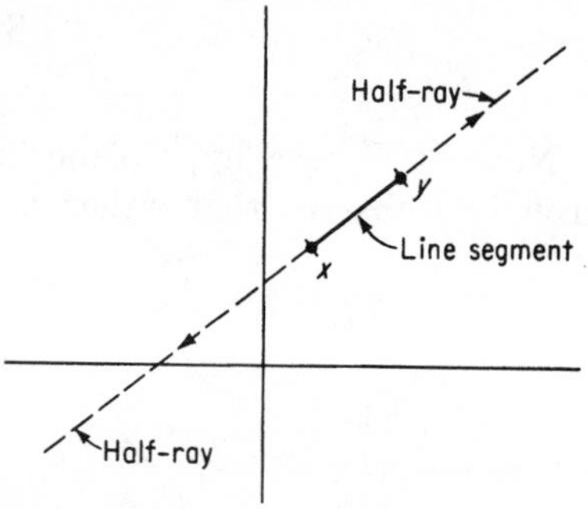


**Fig. 3-8** The line segment joining $\mathbf{x}$ and $\mathbf{y}$ and the two half-rays determined by $\mathbf{x}$ and $\mathbf{y}$.

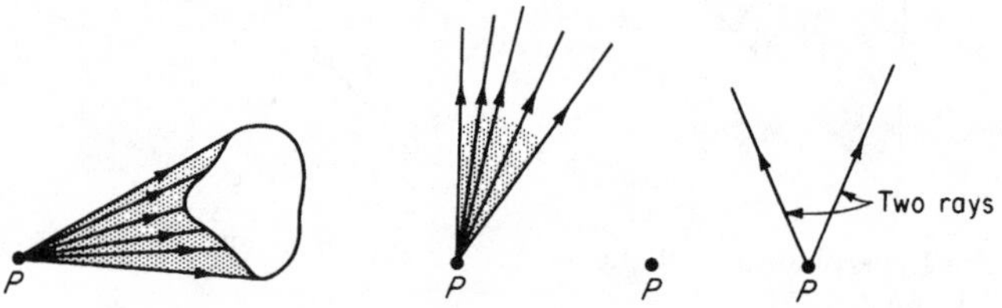


**Fig. 3-9** Cones with vertex $P$.

We shall conclude this section by defining the notions of a line segment and a half-ray and then the notion of a cone. We have:

***Definition 3-16*** *If* $\mathbf{x}$ *and* $\mathbf{y}$ *are two elements of* $R_n$, *then the line segment joining* (*or connecting*) $\mathbf{x}$ *and* $\mathbf{y}$ *is the subset of* $R_n$, *given by*

$$\{\mathbf{z} \in R_n: \mathbf{z} = r\mathbf{x} + s\mathbf{y},\ r \geq 0,\ s \geq 0,\ r + s = 1\} \tag{3-22a}$$

*and the half-ray joining* $\mathbf{x}$ *and* $\mathbf{y}$, *emanating from* $\mathbf{x}$, *is the subset of* $R_n$, *given by*

$$\{\mathbf{z} \in R_n: \mathbf{z} = \mathbf{x} + r(\mathbf{y} - \mathbf{x}),\ r \geq 0\} \tag{3-22b}$$

We observe that the line segment joining $\mathbf{x}$ and $\mathbf{y}$ is the *same* as the line segment joining $\mathbf{y}$ and $\mathbf{x}$ but that the half-ray joining $\mathbf{x}$ and $\mathbf{y}$, emanating from $\mathbf{x}$, is not the same as the half-ray joining $\mathbf{y}$ and $\mathbf{x}$, emanating from $\mathbf{y}$. (See Fig. 3-8.)

***Definition 3-17 Cones*** *A subset $K$ of $R_n$ is called a cone with vertex $\mathbf{x}_0$ if, for any given point $\mathbf{x}$ in $K$ with $\mathbf{x} \neq \mathbf{x}_0$, all points on the half-ray joining $\mathbf{x}_0$ and $\mathbf{x}$, emanating from $\mathbf{x}_0$, belong to $K$.*

Several cones are illustrated in Fig. 3-9.

## 3-8 Properties of Sets in $R_n$: Convexity

We now turn our attention to a very important concept, namely, the notion of convexity. Loosely speaking, a subset of $R_n$ is convex if, for *any* two points in the set, the line segment between those two points is contained within the set. More precisely:

***Definition 3-18 Convexity*** *A subset $A$ of $R_n$ is convex if, for any $\mathbf{x}$ and $\mathbf{y}$ in $A$ and $r$, $s$ in $R$ with $r \geq 0$, $s \geq 0$, $r + s = 1$, the point $r\mathbf{x} + s\mathbf{y}$ is in $A$.*

For example, the open sphere $S(\mathbf{0}, \rho)$ in $R_n$ is convex since we may deduce from $\|\mathbf{x}\| < \rho$ and $\|\mathbf{y}\| < \rho$ that $\|r\mathbf{x} + s\mathbf{y}\| \leq \|r\mathbf{x}\| + \|s\mathbf{y}\| = r\|\mathbf{x}\| + s\|\mathbf{y}\| < r\rho + s\rho = \rho$ if $r \geq 0$, $s \geq 0$, and $r + s = 1$.

**Example 3-18** In the space $R_3$, the set

$$A = \left\{ \begin{bmatrix} x_1 \\ x_2 \\ x_3 \end{bmatrix} : |x_1| \leq 1, |x_2| \leq 1, |x_3| \leq 1 \right\} \qquad \text{a cube}$$

is convex; the set

$$L = \left\{ \begin{bmatrix} x_1 \\ x_2 \\ x_3 \end{bmatrix} : x_1 + x_2 + x_3 - 1 = 0 \right\} \qquad \text{a plane}$$

is convex; and any line segment or ray is convex.

**Example 3-19** The set $A$ of points in $R_2$, defined by $A = \left\{ \begin{bmatrix} x_1 \\ x_2 \end{bmatrix} : x_1 \geq 0,\ x_2 \geq 0\ (x_1^{\frac{2}{3}} + x_2^{\frac{2}{3}})^{\frac{3}{2}} \leq 1 \right\}$, is *not convex* (see Fig. 3-10). The set $K$ of points in $R_2$, given by $K = \left\{ \begin{bmatrix} x_1 \\ x_2 \end{bmatrix} : \text{ either } x_1 = 0 \text{ or } x_2 = 0 \right\}$, is not convex although it is a cone with $\begin{bmatrix} 0 \\ 0 \end{bmatrix}$ as vertex.

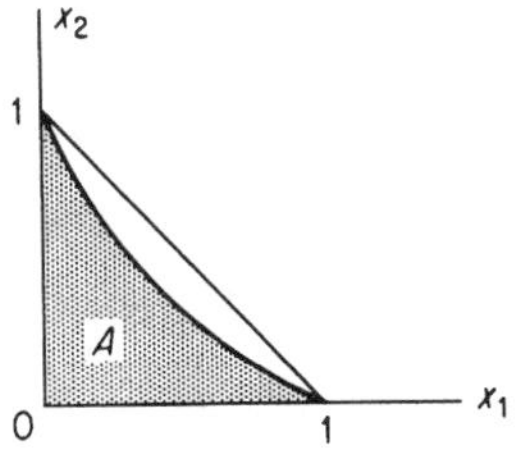


**Fig. 3-10** The set $A$ is not convex, and the convex cover of $A$ is the triangle.

**Example 3-20** If $a$, $b$ are elements of $R$, with $a < b$, then the intervals $(a, b]$, $(a, b)$, $[a, b)$, and $[a, b]$ are all convex.

**Exercise 3-8** Prove that the closed sphere $\overline{S(\mathbf{0}, \rho)}$ in $R_n$ is convex.

**Exercise 3-9** Show that the intervals $[0, 2]$ and $[4, 5]$ are convex but that $[0, 2] \cup [4, 5]$ is not convex.

***Definition 3-19 Convex Cone*** *A cone $K$ in $R_n$ which is also a convex set is often called a convex cone.*

We note that not every cone is convex (see Example 3-19). Convex cones will be discussed in Chaps. 5 and 6.

***Definition 3-20 Convex Combination*** *Suppose that $\mathbf{x}_1, \mathbf{x}_2, \ldots, \mathbf{x}_N$ are $N$ elements of the Euclidean space $R_n$; then we say that $\mathbf{x}$ is a convex com-*

*bination of* $\mathbf{x}_1, \mathbf{x}_2, \ldots, \mathbf{x}_N$ *if there exist real numbers* $r_1, r_2, \ldots, r_N$ *such that*

$$\mathbf{x} = \sum_{i=1}^{N} r_i \mathbf{x}_i \tag{3-23}$$

*and* $r_i \geq 0$ *for* $i = 1, 2, \ldots, N$ *and* $\sum_{i=1}^{N} r_i = 1$.

We claim that if $A$ is a convex subset of $R_n$ and $\mathbf{x}_1, \mathbf{x}_2, \ldots, \mathbf{x}_N$ are elements of $A$, then *every* convex combination of the $\mathbf{x}_i$ is in $A$. Since $A$ is convex, our claim is true for $N = 2$, and so we shall use induction on $N$. Assuming that our claim is true for $N - 1$, let us suppose that $\mathbf{x} = \sum_{i=1}^{N} r_i \mathbf{x}_i$ is a convex combination of the $\mathbf{x}_i$. We may suppose that $r_N \neq 1$ (why?), in which case

$$r_1 + r_2 + \cdots + r_{N-1} = 1 - r_N \neq 0 \tag{3-24}$$

It follows that

$$\mathbf{x} = (r_1 + r_2 + \cdots + r_{N-1}) \left\{ \sum_{i=1}^{N-1} \frac{r_i}{r_1 + r_2 + \cdots + r_{N-1}} \mathbf{x}_i \right\} + r_N \mathbf{x}_N \tag{3-25}$$

But the term in braces in Eq. (3-25) represents an element of $A$ by our induction assumption, and so we see that, in view of the convexity of $A$, $\mathbf{x}$ is in $A$.

***Definition 3-21 Convex Cover or Hull*** *Let* $B$ *be any subset of* $R_n$ *and let* co$(B)$ *denote the set of all convex combinations of elements of* $B$, *that is,*

co$(B)$ = {$\mathbf{x}$: *there are elements* $\mathbf{x}_1, \mathbf{x}_2, \ldots, \mathbf{x}_N$† *in* $B$ *such that* $\mathbf{x}$ *is a convex combination of the* $\mathbf{x}_i$} (3-26)

co$(B)$ *is called the convex cover (or hull) of* $B$. *We observe that the set* $B$ *is convex if and only if it coincides with its convex cover, that is,* $B =$ co$(B)$. *In fact, the convex cover (or hull) of* $B$ *is the smallest convex set containing* $B$.

**Exercise 3-10** Show that the set co$(B)$ is convex. HINT: Suppose, first, that $\mathbf{x} = \sum_{i=1}^{N} r_i \mathbf{x}_i$ and $\mathbf{y} = \sum_{i=1}^{N} s_i \mathbf{x}_i$ are in co$(B)$ and note that $r\mathbf{x} + s\mathbf{y} = \sum_{i=1}^{N-1} (rr_i + ss_i)\mathbf{x}_i + \left\{r\left(1 - \sum_{i=1}^{N-1} r_i\right) + s\left(1 - \sum_{i=1}^{N-1} s_i\right)\right\} \mathbf{x}_N$. Then show that if $\mathbf{x} = \sum_{i=1}^{N} r_i \mathbf{x}_i$ and $\mathbf{y} = \sum_{j=1}^{M} s_j \mathbf{y}_j$ and $\mathbf{z}_1 = \mathbf{x}_1, \ldots, \mathbf{z}_N = \mathbf{x}_N, \mathbf{z}_{N+1} = \mathbf{y}_1, \ldots, \mathbf{z}_{N+M} = \mathbf{y}_M$, then $\mathbf{x} = \sum_{i=1}^{N} r_i \mathbf{z}_i + \sum_{N+1}^{N+M} 0\mathbf{z}_j$ and $\mathbf{y} = \sum_{i=1}^{N} 0\mathbf{z}_i + \sum_{N+1}^{N+M} s_j \mathbf{z}_{N+j}$.

† $N$ may depend upon $\mathbf{x}$.

**Example 3-21** Suppose that $B$ is the set of Example 3-19, that is, $B = \{(x_1, x_2): x_1 \geq 0, x_2 \geq 0, (x_1^{2/3} + x_2^{2/3})^{3/2} \leq 1\}$. Then the convex cover of $B$ is given by $\text{co}(B) = \{(x_1, x_2): x_1 \geq 0, x_2 \geq 0, x_1 + x_2 \leq 1\}$. (See Fig. 3-10.)

**Example 3-22** Let $B = B_1 \cup B_2$, where $B_1 = \{(x_1, x_2): x_1^2 + x_2^2 \leq 1, x_1 \leq 0\}$ and $B_2 = \{(x_1, x_2): x_1^2 + x_2^2 \leq 1, x_2 \leq 0\}$. In other words, $B$ consists of the three quadrants of a unit circle, as shown in Fig. 3-11. Then the convex cover of $B$, $\text{co}(B)$, is

$$\text{co}(B) = B \cup \{(x_1, x_2): x_1 > 0, x_2 > 0, x_1 + x_2 \leq 1\}$$

We shall now state and prove several important theorems about convex sets in $R_n$.

***Theorem 3-3*** *If $A$ is a convex subset of $R_n$, then the closure of $A$, $c(A)$ [Eq. (3-9)], is also convex.*

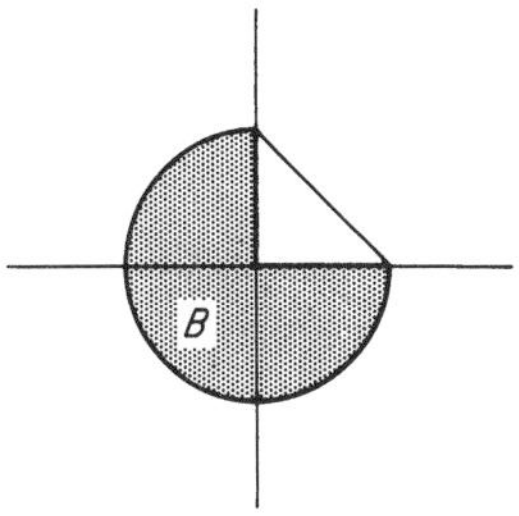


**Fig. 3-11** The convex cover of the set $B$ is $B$ together with the triangle.

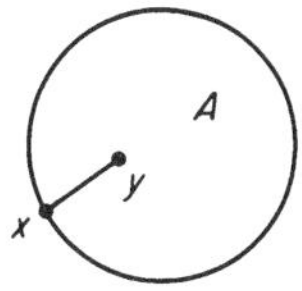


**Fig. 3-12** Every point on the line segment joining $\mathbf{x}$ and $\mathbf{y}$, with the possible exception of $\mathbf{x}$, is an interior point of the convex set $A$.

PROOF Suppose that $\mathbf{x}$ and $\mathbf{y}$ are elements of $c(A)$ and that $r$, $s$ are elements of $R$, with $r \geq 0$, $s \geq 0$, and $r + s = 1$. We want to show that $r\mathbf{x} + s\mathbf{y}$ is in $c(A)$; that is, we want to find a sequence $\mathbf{z}_n$ of elements of $A$ which converges to $r\mathbf{x} + s\mathbf{y}$. Well, since $\mathbf{x}$ and $\mathbf{y}$ are in $c(A)$, there are sequences $\mathbf{x}_n$ and $\mathbf{y}_n$ of elements of $A$ such that $\mathbf{x}_n$ converges to $\mathbf{x}$ and $\mathbf{y}_n$ converges to $\mathbf{y}$. We claim that we may take $\mathbf{z}_n = r\mathbf{x}_n + s\mathbf{y}_n$. First of all, $\mathbf{z}_n$ is in $A$ since $A$ is convex. Now let us show that the sequence $\mathbf{z}_n$ converges to $r\mathbf{x} + s\mathbf{y}$. If $\epsilon > 0$, then there is an $M(\epsilon)$ such that $n > M(\epsilon)$ implies that $\|\mathbf{x}_n - \mathbf{x}\| < \epsilon$ and $\|\mathbf{y}_n - \mathbf{y}\| < \epsilon$. It follows that, for all $n > M(\epsilon)$,

$$\begin{aligned} \|r\mathbf{x}_n + s\mathbf{y}_n - (r\mathbf{x} + s\mathbf{y})\| &= \|r(\mathbf{x}_n - \mathbf{x}) + s(\mathbf{y}_n - \mathbf{y}_n)\| \\ &\leq r\|\mathbf{x}_n - \mathbf{x}\| + s\|\mathbf{y}_n - \mathbf{y}\| < \epsilon \qquad (3\text{-}27) \end{aligned}$$

This shows that the sequence $\mathbf{z}_n$ converges to $r\mathbf{x} + s\mathbf{y}$ and completes the proof of the theorem.

***Theorem 3-4*** *If $A$ is a convex subset of $R_n$, then the interior of $A$, $i(A)$ [Eq. (3-8)], is either convex or empty. In fact, if $A$ is convex and if $i(A)$ is not empty, then the following is true: Given two points $\mathbf{x}$ and $\mathbf{y}$ in $A$, with $\mathbf{y}$ in the interior of $A$, then every point on the line segment between $\mathbf{x}$ and $\mathbf{y}$ (with the possible exception of $\mathbf{x}$ itself) is an interior point of $A$. (See Fig. 3-12.)*

PROOF Since $\mathbf{y}$ is in the interior of $A$, there is a $\rho > 0$ with $S(\mathbf{y}, \rho) \subset A$. Let $\mathbf{z}$ be a point on the line segment joining $\mathbf{x}$ and $\mathbf{y}$ which is distinct from $\mathbf{x}$. Then

$$\mathbf{z} = r\mathbf{x} + s\mathbf{y} \qquad r + s = 1 \qquad r \geq 0 \qquad s > 0 \tag{3-28}$$

We shall show that the open sphere $S(\mathbf{z}, s\rho)$ is contained in $A$ and, therefore, that $\mathbf{z}$ is an interior point of $A$. (See Fig. 3-13.) If $\mathbf{w} \in S(\mathbf{z}, s\rho)$, then

$$\|\mathbf{w} - \mathbf{z}\| = \|\mathbf{w} - (r\mathbf{x} + s\mathbf{y})\| < s\rho \tag{3-29}$$

which implies that

$$\left\| \frac{s}{s}(\mathbf{w} - r\mathbf{x}) - s\mathbf{y} \right\| = s \left\| \frac{1}{s}(\mathbf{w} - r\mathbf{x}) - \mathbf{y} \right\| < s\rho \tag{3-30}$$

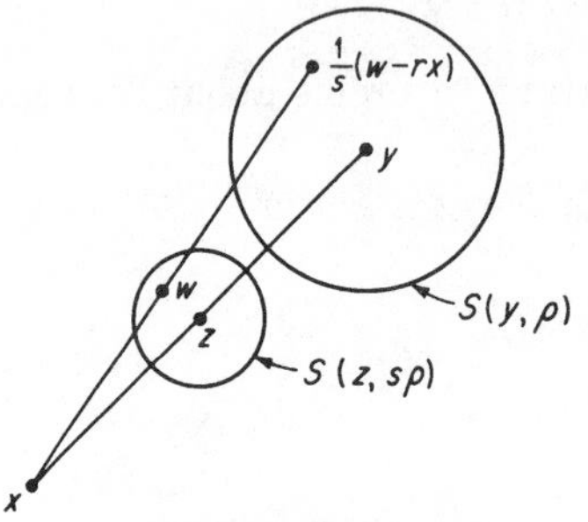


**Fig. 3-13** Illustration for the proof of Theorem 3-4.

In other words,

$$\left\| \frac{1}{s}(\mathbf{w} - r\mathbf{x}) - \mathbf{y} \right\| < \rho \tag{3-31}$$

which shows that $(1/s)(\mathbf{w} - r\mathbf{x})$ is in $S(\mathbf{y}, \rho)$. Since $S(\mathbf{y}, \rho)$ is contained in $A$ and $A$ is convex, we see that $\mathbf{w} = s(1/s)(\mathbf{w} - r\mathbf{x}) + r\mathbf{x}$ is an element of $A$. Thus $S(\mathbf{z}, s\rho) \subset A$, and the proof is complete.

The next theorem, whose proof we shall omit (see Ref. E-3), will be used in Chap. 5. We shall first state the theorem and then indicate its intuitive content.

***Theorem 3-5***

a. *If $A$ is an open convex subset of $R_n$ and $B$ is a convex subset of $R_n$ which does not meet $A$, then there is a hyperplane $L$ which separates $A$ and $B$.*
b. *If $A$ is a convex subset of $R_n$ and if $\mathbf{x}$ is an element of the boundary of $A$, then there is at least one support hyperplane of $A$ which contains $\mathbf{x}$.*
c. *If $A$ is a closed set in $R_n$ whose interior, $i(A)$, is not empty and if there is a support hyperplane of $A$ for each $\mathbf{x}$ in the boundary of $A$, then $A$ is convex.*

Part $a$ of the theorem is illustrated in Fig. 3-14$a$ and $b$. We note that in Fig. 3-14$a$ the hyperplane $L$ strictly separates $A$ and $B$ but that in Fig. 3-14$b$ there is *no* hyperplane which *strictly* separates $A$ and $B$. Part $b$ of the theorem, which is illustrated in Fig. 3-15, may be rephrased as follows: "If $\mathbf{x}$ is a boundary point of the convex set $A$, then there is a hyperplane $L$ passing through $\mathbf{x}$ such that $A$ is contained in either $L^+ \cup L$ or $L^- \cup L$." We shall use this form of part $b$ in Chap. 5. Part $c$ of the theorem provides an indirect means for determining whether or not a closed set is convex.

Finally, let us take a closer look at points on the boundary of a convex set and at support hyperplanes. We have:

1. *A point $\mathbf{x}$ on the boundary of the convex set $A$ in $R_n$ is a regular point if it is an element of only one support hyperplane of $A$.*

2. *A support hyperplane L of the convex set A is a regular support hyperplane of A if it meets A in a single point.*

3. *The convex set A is a regular convex set (or, simply, is regular) if every boundary point of A is a regular point and if every support hyperplane of A is a regular support hyperplane of A.*

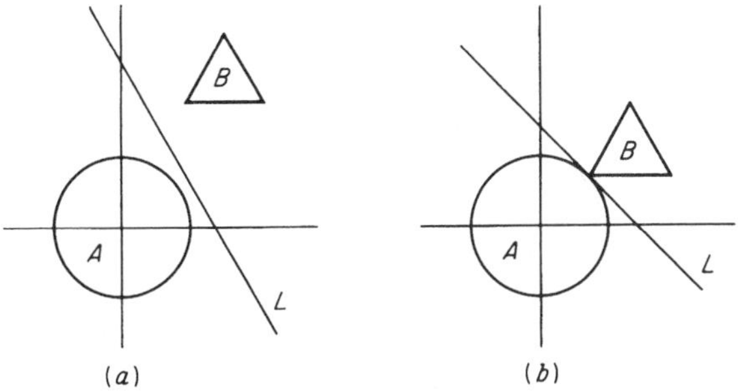


**Fig. 3-14** (*a*) $L$ strictly separates $A$ and $B$. (*b*) $L$ separates $A$ and $B$, and there is no line which strictly separates $A$ and $B$.

The sphere $S(\mathbf{0}, \rho)$ in $R_n$ is the archetype of a regular convex set. The term *regular* applied to a convex set means roughly that the set has *no* "corners." The set $A_1$ in Fig. 3-16 is regular; the set $A_2$ in Fig. 3-16 has two nonregular points (or "corners"), but every support hyperplane of $A_2$

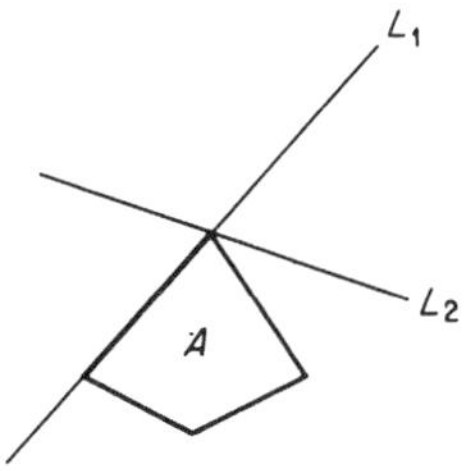


**Fig. 3-15** Support hyperplanes $L_1$ and $L_2$ through boundary points of the convex set $A$.

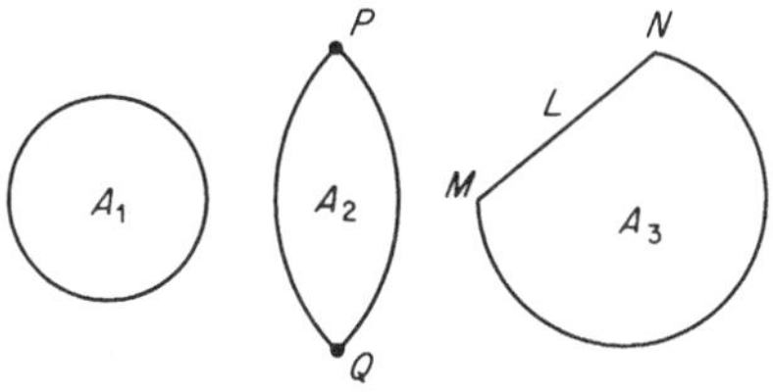


**Fig. 3-16** $A_1$ is a regular convex set. $P$ and $Q$ are nonregular points (or "corners") of the convex set $A_2$; $M$ and $N$ are nonregular points of the convex set $A_3$, and $L$ is a nonregular support hyperplane of $A_3$.

is regular; the set $A_3$ in Fig. 3-16 has both nonregular points and a nonregular support hyperplane. We note also that convex sets like $\{(x_1, x_2): x_2 = 0, -1 \le x_1 \le 1\}$ or the unit cube in $R_3$ are also nonregular convex sets. Finally, we note that it is not possible to define a unique normal to a convex set at a nonregular point (or "corner") of the set.

## 3-9 Vector Functions: Introductory Remarks

We are now going to study vector functions, i.e., functions taking on values in a vector space, and several important properties which such functions may have. We shall suppose that the reader has met in his calculus courses many of the notions that we shall discuss, and so we shall often proceed quite rapidly over familiar ground. However, we shall establish notations which will be used throughout the remainder of the book, and we shall try to be somewhat more precise than basic calculus courses are apt to be. Moreover, we shall pay particular attention to those concepts which are important to the main body of the book. We shall examine such ideas as continuity, the derivative and integral of a function, function spaces, and functionals.†

We shall often begin by talking about the familiar case of functions from $R$ into $R$, that is, real-valued functions of one variable; then we shall usually discuss functions from $R$ into $R_n$, that is, vector-valued functions of one variable; and finally, we shall consider functions from $R_m$ into $R_n$, that is, vector-valued functions of several (real) variables. Although we shall usually introduce concepts such as continuity for functions of one variable with reference only to $R$ and not to an interval [see Eqs. (3-7)], we note that if the reader substitutes an interval for $R$ as the domain (see Sec. 2-4) of the functions in the definitions, he will readily obtain the proper formulation of the concept in question and will be able to interpret the theorems for this situation.

Let us now recall that if $\mathbf{x}$ is an element of $R_n$, with

$$\mathbf{x} = \begin{bmatrix} x_1 \\ x_2 \\ \cdot \\ \cdot \\ \cdot \\ x_n \end{bmatrix} \tag{3-32}$$

then $x_1, x_2, \ldots, x_n$ are called the *components of* $\mathbf{x}$. If $\mathbf{f}$ is a function from $R$ into $R_n$, then we have, for $t$ in $R$,

$$\mathbf{f}(t) = \begin{bmatrix} f_1(t) \\ f_2(t) \\ \cdot \\ \cdot \\ \cdot \\ f_n(t) \end{bmatrix} \tag{3-33}$$

† References D-5, R-10, and S-4 are of particular interest with regard to these ideas.

where $f_1, f_2, \ldots, f_n$ are real-valued functions which are called the *components of* **f**. For example, the function $\mathbf{f}(t)$, from $R$ to $R_3$,

$$\mathbf{f}(t) = \begin{bmatrix} t \\ e^{-t} \\ \cos t \end{bmatrix}$$

is of the form (3-33), with $f_1(t) = t, f_2(t) = e^{-t}, f_3(t) = \cos t$. We shall use *boldface type for vector-valued functions.* In a similar way, we see that if **g** is a function from $R_m$ into $R_n$ and if **u** is an element of $R_m$, then

$$\mathbf{g}(\mathbf{u}) = \begin{bmatrix} g_1(\mathbf{u}) \\ g_2(\mathbf{u}) \\ \cdot \\ \cdot \\ \cdot \\ g_n(\mathbf{u}) \end{bmatrix} \tag{3-34}$$

where the $g_i$ are real-valued functions on $R_m$ which are called the *components of* **g**. For example, if **g** is a function from $R_2$ into $R_3$ and if $\mathbf{u} = \begin{bmatrix} u_1 \\ u_2 \end{bmatrix}$ is an element of $R_2$, then a typical function **g** from $R_2$ into $R_3$ might have the form

$$\mathbf{g}(\mathbf{u}) = \begin{bmatrix} u_1 + 3u_2^2 \\ \log u_1 + e^{-u_2} \\ \sin u_1 + u_2^3 \end{bmatrix}$$

where $g_1(\mathbf{u}) = u_1 + 3u_2^2$, $g_2(\mathbf{u}) = \log u_1 + e^{-u_2}$, and $g_3(\mathbf{u}) = \sin u_1 + u_2^3$. In view of Eqs. (3-33) and (3-34), we observe that there is a very close connection between vector-valued and real-valued functions. We shall exploit this connection later on.

If we let $\mathfrak{F}(R_m, R_n)$ denote the set of *all* functions from $R_m$ into $R_n$, that is,

$$\mathfrak{F}(R_m, R_n) = \{\mathbf{f}: \mathbf{f} \text{ is a function from } R_m \text{ into } R_n\} \tag{3-35}$$

and if we define the notions of *sum* and *product* (by an element of $R$) for elements of $\mathfrak{F}(R_m, R_n)$ by setting

$$[\mathbf{f} + \mathbf{g}](\mathbf{u}) = \mathbf{f}(\mathbf{u}) + \mathbf{g}(\mathbf{u}) \qquad \text{sum} \tag{3-36}$$

$$[r\mathbf{f}](\mathbf{u}) = r\mathbf{f}(\mathbf{u}) \qquad \text{product} \tag{3-37}$$

for **f**, **g** in $\mathfrak{F}(R_m, R_n)$, **u** in $R_m$, and $r$ in $R$, then $\mathfrak{F}(R_m, R_n)$ becomes a vector space with respect to these notions of sum and product (see Sec. 2-5). We shall make frequent use of these notions of sum and product in the sequel.

## 3-10 Vector Functions: Continuity

In Sec. 2-7, we observed that once the mathematician has defined a concept, he tries to find out what sort of transformations "preserve" that concept. We defined the notion of limit point in Sec. 3-4, and we shall now consider transformations which preserve that notion. To begin with, let us introduce the idea of a limit point of a transformation. We have:

***Definition 3-22*** *If $X_1$ and $X_2$ are sets with distance functions $d_1$ and $d_2$, respectively, if $f$ is a transformation from $X_1$ into $X_2$ whose domain is all of $X_1$, and if $x_0$ is an element of $X_1$, then we say that an element $y$ of $X_2$ is a limit point of $f$ at $x_0$, if, for every sequence $x_n$, $n = 1, 2, \ldots$, of elements of $X_1$ which converges to $x_0$, with $x_n \neq x_0$ for every $n$, the corresponding sequence $f(x_n)$, $n = 1, 2, \ldots$, converges to $y$. We write, in this case,*

$$y = \lim_{x \to x_0} f(x) \tag{3-38}$$

We note that a transformation may *not* have a limit point at a given $x_0$ but that if it does, then this limit point is unique. We also note that $y = \lim_{x \to x_0} f(x)$ does *not* mean that $y = f(x_0)$. In fact, we have:

***Definition 3-23 Continuity*** *If $X_1$ and $X_2$ are sets with distance functions $d_1$ and $d_2$, respectively, if $f$ is a transformation from $X_1$ into $X_2$ whose domain is all of $X_1$, and if $x_0$ is a point in $X_1$, then we say that $f$ is continuous at $x_0$ if $f(x_0)$ is a limit point of $f$ at $x_0$, that is, if*

$$f(x_0) = \lim_{x \to x_0} f(x) \tag{3-39}$$

*If $f$ is continuous at every point $x_0$, then we say that $f$ is a continuous transformation (or mapping or function) or, simply, that $f$ is continuous.*

We note that the continuity of $f$ means that if $x_0$ is the limit of the sequence $x_n$, $n = 1, 2, \ldots$, then $f(x_0)$ is the limit of the sequence $f(x_n)$, $n = 1, 2, \ldots$.

We are familiar with the so-called $\epsilon$, $\delta$ notion of continuity for real-valued functions of one variable. Let us see how this relates to our Definition 3-23. We shall, in fact, relate the idea of a limit point of a transformation to spheres in the sets $X_1$ and $X_2$. The following theorem provides us with the desired relation.

***Theorem 3-6*** *$y$ is a limit point of the transformation $f$ at $x_0$ if and only if, for each $\epsilon > 0$, there is a $\delta > 0$ such that*

$$f[S(x_0, \delta)] \subset S(y, \epsilon) \cup \{f(x_0)\} \tag{3-40}$$

*or, equivalently, if and only if*

$$0 < d_1(x, x_0) < \delta \textit{ implies that } d_2(f(x), y) < \epsilon \tag{3-41}$$

*In particular, if $f$ is a function from $R$ into itself, then $y$ is a limit point of $f$ at $x_0$ if and only if, for each $\epsilon > 0$, there is a $\delta > 0$ such that*

$$0 < |x - x_0| < \delta \textit{ implies that } |f(x) - y| < \epsilon \tag{3-42}$$

*More generally, if* $\mathbf{f}$ *maps* $R_m$ *into* $R_n$*, then* $\mathbf{y}$ *is a limit point of* $\mathbf{f}$ *at* $\mathbf{x}_0$ *if and only if, for each* $\epsilon > 0$*, there is a* $\delta > 0$*, which may depend on* $\epsilon$*, such that*

$$0 < \|\mathbf{x} - \mathbf{x}_0\| < \delta \textit{ implies that } \|\mathbf{f}(\mathbf{x}) - \mathbf{y}\| < \epsilon \tag{3-43}$$

*where* $\|\ \ \|$ *is the Euclidean norm in* $R_n$*. Finally, it follows from Eq.* (3-40) *that* $f$ *is continuous at* $x_0$ *if and only if, for each* $\epsilon > 0$*, there is a* $\delta > 0$ *such that*

$$f[S(x_0, \delta)] \subset S(f(x_0), \epsilon) \tag{3-44}$$

(*This final assertion is illustrated in Fig.* 3-17 *for a transformation of* $R_2$ *into itself.*)

PROOF It is clear from the definition of limit point that if $f$ satisfies (3-40), then $y$ is a limit point of $f$ at $x_0$. On the other hand, if $y$ is a limit point of $f$ at $x_0$ and if $\epsilon > 0$ is given, then we claim that there is a $\delta > 0$ with

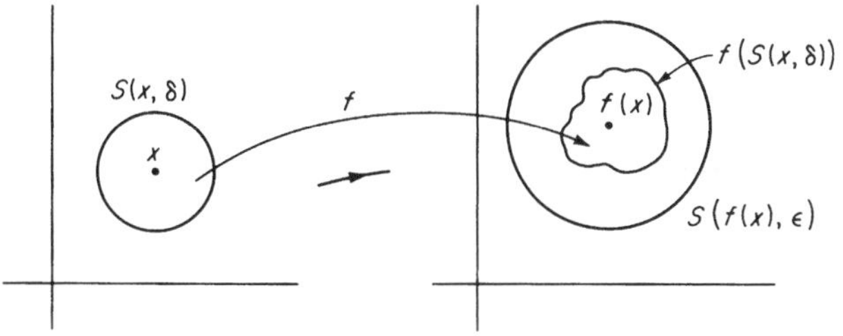


**Fig. 3-17** The $\epsilon$, $\delta$ property of continuity.

$f[S(x_0, \delta)] \subset S(y, \epsilon) \cup \{f(x_0)\}$. If this were not true, then, for $n = 1, 2, \ldots$, there would be an $x_n \neq x_0$ in $X_1$ with $d_1(x_n, x_0) < 1/n$ and $d_2(f(x_n), y) \geq \epsilon$ and $f(x_n) \neq f(x_0)$. But this would mean that $x_n$ converged to $x_0$ while $f(x_n)$ did *not* converge to $y$, which would contradict the fact that $y$ was a limit point of $f$ at $x_0$. Thus, our claim is established, and with it the theorem.

It is easy to see that this theorem has the immediate consequence that *$f$ is continuous if and only if the inverse image* [*see Eq.* (2-15)] *of every open subset of* $X_2$ *is an open subset of* $X_1$*; that is,* $f^{-1}(A)$ *is open in* $X_1$ *if* $A$ *is open in* $X_2$. In view of Theorem 3-1, we may also observe that *$f$ is continuous if and only if the inverse image of every closed subset of* $X_2$ *is a closed subset of* $X_1$*; that is,* $f^{-1}(B)$ *is closed in* $X_1$ *if* $B$ *is closed in* $X_2$.

If $\mathbf{g}$ is a function from $R_m$ into $R_n$, then we claim that it is easy to see from Eq. (3-44) that $\mathbf{g}$ is continuous if and only if each of the components $g_1, g_2, \ldots, g_n$ of $\mathbf{g}$ is a continuous function from $R_m$ into $R$ [see Eq. (3-34)]. In fact, we note that if $\mathbf{u}$ and $\mathbf{v}$ are elements of $R_m$, then, for $i = 1, 2, \ldots, n$,

$$|g_i(\mathbf{u}) - g_i(\mathbf{v})| \leq \|\mathbf{g}(\mathbf{u}) - \mathbf{g}(\mathbf{v})\| \leq \sum_{i=1}^{n} |g_i(\mathbf{u}) - g_i(\mathbf{v})| \tag{3-45}†$$

which implies the validity of our claim.

† See Eq. (3-6$d$).

Furthermore, we can readily see that if $\mathbf{g}$ and $\mathbf{h}$ are continuous functions from $R_m$ into $R_n$, then the sum of $\mathbf{g}$ and $\mathbf{h}$, $\mathbf{g} + \mathbf{h}$ [see Eq. (3-36)], and the product of $\mathbf{g}$ and an element $r$ of $R$, $r\mathbf{g}$ [see Eq. (3-37)], are again continuous functions from $R_m$ into $R_n$. In other words, the set

$$\{\mathbf{g}: \mathbf{g} \text{ is a continuous function from } R_m \text{ into } R_n\} \tag{3-46}$$

is a vector space which is a subspace of the set of all functions from $R_m$ into $R_n$, $\mathfrak{F}(R_m, R_n)$ [see Eq. (3-35)]. We shall have more to say about spaces of this type in Sec. 3-15. Finally, we note that if $\mathbf{g}$ is a continuous function from $R_m$ into $R_n$ and if $\mathbf{h}$ is a continuous function from $R_n$ into $R_p$, then the composite function $\mathbf{h} \circ \mathbf{g}$ [see Eq. (2-16)] is a continuous function from $R_m$ into $R_p$.

**Example 3-23** Let $f_n$ be the transformation of $R$ into itself, defined by $f_n(x) = x^n$ for $n = 0, 1, \ldots$. Then $f_n$ is continuous, and it follows that every polynomial function is continuous. On the other hand, if $U$ is the transformation of $R$ into itself, defined by $U(x) = 0$ if $x \leq 0$ and $U(x) = 1$ if $x > 0$, then $U$ is not continuous at $x = 0$ but is continuous at every point $x \neq 0$.

**Example 3-24** Let $\mathcal{A}$ be a linear transformation of $R_m$ into $R_n$, with $\mathbf{A} = (a_{ij})$ as matrix with respect to the natural bases in $R_m$ and $R_n$ (see Sec. 2-6). We claim that $\mathcal{A}$ is continuous. To see this, we observe that if

$$\mathbf{v} = \begin{bmatrix} r_1 \\ r_2 \\ \cdot \\ \cdot \\ \cdot \\ r_m \end{bmatrix} \quad \text{and} \quad \mathbf{w} = \begin{bmatrix} s_1 \\ s_2 \\ \cdot \\ \cdot \\ \cdot \\ s_m \end{bmatrix}$$

are elements of $R_m$, then $\mathcal{A}(\mathbf{v}) - \mathcal{A}(\mathbf{w}) = \mathcal{A}(\mathbf{v} - \mathbf{w})$ and

$$\langle \mathcal{A}(\mathbf{v}-\mathbf{w}), \mathcal{A}(\mathbf{v}-\mathbf{w})\rangle = \left\langle \begin{bmatrix} \sum_{j=1}^{m} a_{1j}(r_j - s_j) \\ \sum_{j=1}^{m} a_{2j}(r_j - s_j) \\ \cdot \\ \cdot \\ \cdot \\ \sum_{j=1}^{m} a_{nj}(r_j - s_j) \end{bmatrix}, \begin{bmatrix} \sum_{j=1}^{m} a_{1j}(r_j - s_j) \\ \sum_{j=1}^{m} a_{2j}(r_j - s_j) \\ \cdot \\ \cdot \\ \cdot \\ \sum_{j=1}^{m} a_{nj}(r_j - s_j) \end{bmatrix} \right\rangle$$

[using Eq. (2-32)], from which it follows that

$$\begin{aligned} \langle \mathcal{A}(\mathbf{v}-\mathbf{w}), \mathcal{A}(\mathbf{v}-\mathbf{w})\rangle &= \sum_{i=1}^{n} \left\{ \sum_{j=1}^{m} a_{ij}(r_j - s_j) \right\}^2 \\ &\leq \sum_{i=1}^{n} \left\{ \sum_{j=1}^{m} (r_j - s_j)^2 \right\} \left\{ \sum_{j=1}^{m} a_{ij}^2 \right\} \qquad \text{using Eq. (2-91)} \\ &\leq \|\mathbf{v} - \mathbf{w}\|^2 \left\{ \sum_{i=1}^{n} \sum_{j=1}^{m} a_{ij}^2 \right\} \end{aligned}$$

and hence that $\mathcal{A}$ is continuous.

**Example 3-25** Let $f$ be the function from $R$ into itself, defined by $f(x) = |x|$ (see Fig. 3-18). Then $f$ is continuous. In fact, the function $f$ from $R_m$ into $R$, defined by setting $f(\mathbf{x}) = \|\mathbf{x}\|$ for $\mathbf{x}$ in $R_m$, is a continuous function.

**Example 3-26** If

$$\mathbf{x} = \begin{bmatrix} x_1 \\ x_2 \\ \cdot \\ \cdot \\ \cdot \\ x_m \end{bmatrix}$$

is an element of $R_m$ and if we define a function $f$ from $R_m$ into $R$ by setting

$$f(\mathbf{x}) = \max_{i=1,2,\ldots,m} \{|x_i|\}$$

[that is, $f(\mathbf{x})$ is the maximum of the absolute values of the components of $\mathbf{x}$], then $f$ is continuous, since $|x_i| \le \left(\sum_{j=1}^{m} x_j^2\right)^{\frac{1}{2}}$ for $i = 1, 2, \ldots, m$.

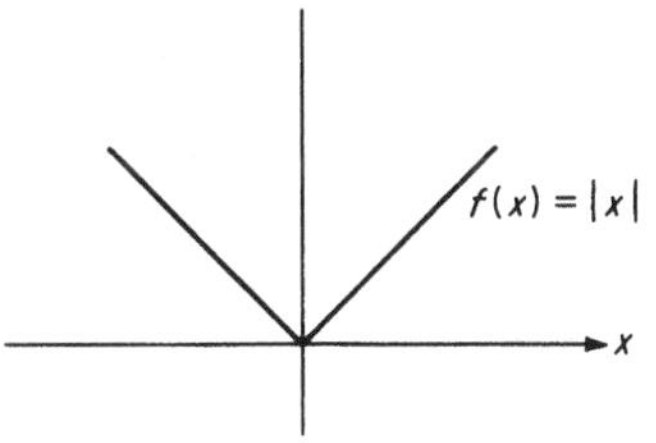


**Fig. 3-18** The continuous function $f(x) = |x|$.

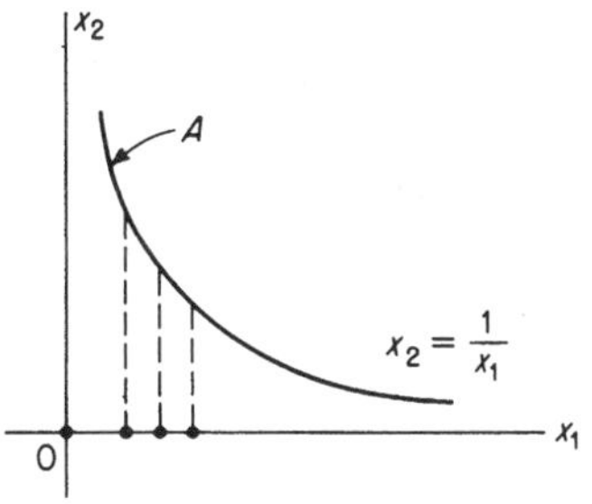


**Fig. 3-19** The set $A$ is closed, but the projection of $A$ on the $x_1$ axis is not closed.

**Example 3-27** If

$$\mathbf{x} = \begin{bmatrix} x_1 \\ x_2 \\ \cdot \\ \cdot \\ \cdot \\ x_m \end{bmatrix}$$

is an element of $R_m$ and if we define functions $\phi_i$, $i = 1, 2, \ldots, m$, from $R_m$ into $R$ by setting $\phi_i(\mathbf{x}) = x_i$, then the $\phi_i$ are continuous. In particular, if we are considering $R_2$ and if we set $\phi_1\left(\begin{bmatrix} x_1 \\ x_2 \end{bmatrix}\right) = x_1$, then $\phi_1$ is continuous. If $A$ is the subset of $R_2$, defined by $A = \left\{\begin{bmatrix} x_1 \\ x_2 \end{bmatrix} : x_2 = 1/x_1,\ x_1 > 0\right\}$ (see Fig. 3-19), then $A$ is *closed* in $R_2$ but $\phi_1(A)$ is *not closed* in $R$. Consider the remarks after Theorem 3-6 in the light of this example.

**Exercise 3-11** Show that if $\mathbf{g}$ and $\mathbf{h}$ are continuous functions from $R_m$ into $R_n$, then $\mathbf{g} + \mathbf{h}$ is a continuous function from $R_m$ into $R_n$. HINT: If $\mathbf{u}$ and $\mathbf{v}$ are in $R_m$, then $\|(\mathbf{g} + \mathbf{h})(\mathbf{u}) - (\mathbf{g} + \mathbf{h})(\mathbf{v})\| \le \|\mathbf{g}(\mathbf{u}) - \mathbf{g}(\mathbf{v})\| + \|\mathbf{h}(\mathbf{u}) - \mathbf{h}(\mathbf{v})\|$.

Now let us suppose that $\mathbf{f}$ is a continuous function from $R_m$ into $R_n$ and that $C$ is a compact subset of $R_m$ (see Definition 3-14). We shall soon show (Theorem 3-7) that $\mathbf{f}(C)$ [see Eq. (2-14)] is a compact subset of $R_n$. A particularly important consequence of this is the fact that *a real-valued continuous function actually attains both a maximum and a minimum value when restricted to a compact set.*

***Theorem 3-7*** *If* $\mathbf{f}$ *is a continuous transformation from* $R_m$ *into* $R_n$ *and if* $C$ *is a compact subset of* $R_m$, *then* $\mathbf{f}(C)$ *is a compact subset of* $R_n$.

PROOF We shall use property C4 (see Sec. 3-6) to prove that $\mathbf{f}(C)$ is compact.† Suppose that $\mathbf{y}_n$, $n = 1, 2, \ldots$, is a sequence of elements of $\mathbf{f}(C)$; then there are elements $\mathbf{x}_n$, $n = 1, 2, \ldots$, of $C$ such that $\mathbf{f}(\mathbf{x}_n) = \mathbf{y}_n$ for $n = 1, 2, \ldots$. Since $C$ is compact, there is a subsequence $\mathbf{x}_{n_1}, \mathbf{x}_{n_2}, \ldots$ of the sequence $\{\mathbf{x}_n\}$ which has a limit $\mathbf{x}$ *in* $C$. However, it follows from the continuity of $\mathbf{f}$ that $\mathbf{y}_{n_1} = \mathbf{f}(\mathbf{x}_{n_1})$, $\mathbf{y}_{n_2} = \mathbf{f}(\mathbf{x}_{n_2})$, . . . converges to $\mathbf{f}(\mathbf{x})$, which is an element of $\mathbf{f}(C)$. In other words, $\mathbf{y}_{n_1}, \mathbf{y}_{n_2}, \ldots$ is the desired convergent subsequence of the sequence $\{\mathbf{y}_n\}$.

## 3-11 Vector Functions: Piecewise Continuity

We examined the notion of continuity at a point in the previous section. Let us now see what can happen at a point where a transformation is not continuous. Such a point is called a *discontinuity* of the transformation. To begin with, let us consider two examples.

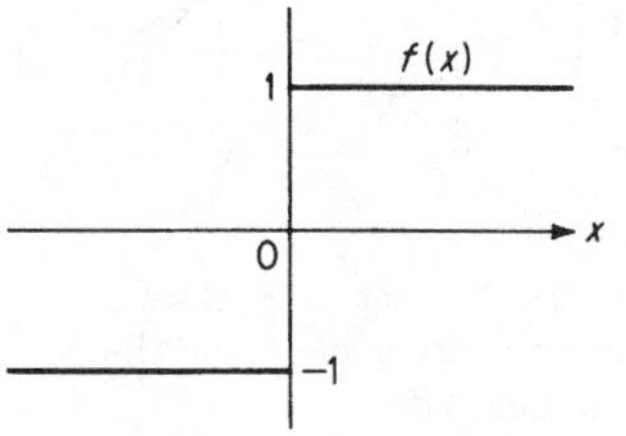


**Fig. 3-20** The function $f(x)$ is not continuous at $x = 0$.

**Example 3-28** Let $f$ be the function from $R$ into itself, defined by setting $f(x) = -1$ if $x \leq 0$ and $f(x) = 1$ if $x > 0$. (See Fig. 3-20.) Then $f$ is not continuous at 0; however, we note that if $x_n$, $n = 1, 2, \ldots$, is a sequence converging to 0, with either $x_n \leq 0$ for every $n$ or $x_n > 0$ for every $n$, then the limit of the sequence $f(x_n)$, $n = 1, 2, \ldots$, exists. In fact, if $x_n \leq 0$ for every $n$, then $\lim_{n\to\infty} f(x_n) = -1$, and if $x_n > 0$ for every $n$, then $\lim_{n\to\infty} f(x_n) = 1$.

**Example 3-29** Let $g$ be the function from $R$ into itself, defined by setting $g(x) = -1$ if $x$ is an irrational number and $g(x) = 1$ if $x$ is a rational number. Then $g$ is not continuous at any $x$ and, in particular, is not continuous at 0. Moreover, we note that there are sequences $x_n$, $n = 1, 2, \ldots$, converging to 0, with (say) $x_n > 0$ for every $n$ for which either $\lim_{n\to\infty} g(x_n) = -1$ or $\lim_{n\to\infty} g(x_n) = 1$ or for which the sequence $g(x_n)$, $n = 1, 2, \ldots$, has *no* limit.

† Recall that property C4 states that a set $K$ is compact if, given any sequence $x_n$, $n = 1, 2, \ldots$, of elements of $K$, there is a subsequence $x_{n_1}, x_{n_2}, \ldots$ of $\{x_n\}$ which converges to an element of $K$.

In a sense, the function $f$ of Example 3-28 is much "closer" to being continuous at 0 than the function $g$ of Example 3-29. We shall try to make this idea more precise in the remainder of this section.

Now let $f$ be a transformation from $R$ into a set $X$, with $d$ as distance on $X$.

***Definition 3-24 Limits on the Right and Left*** *If $t$ is an element of $R$, then we say that $f$ has a limit on the right at $t$ (or has a limit from above at $t$) if there is an element $x$ in $X$ such that, given any sequence $t_n$, $n = 1, 2, \ldots$, which converges to $t$ with $t_n > t$ for every $n$, $\lim_{n\to\infty} f(t_n) = x$; that is, $x$ is the limit of the sequence $\{f(t_n)\}$. We often write*

$$x = f(t+) \qquad or \qquad \lim_{t\to t+} f(t) = x \tag{3-47}$$

*In a similar way, we say that $f$ has a limit on the left at $t$ (or has a limit from below at $t$) if there is an element $y$ in $X$ such that, given any sequence $t_n$, $n = 1, 2, \ldots$, which converges to $t$ with $t_n < t$ for every $n$, $\lim_{n\to\infty} f(t_n) = y$; that is, $y$ is the limit of the sequence $\{f(t_n)\}$. We often write*

$$y = f(t-) \qquad or \qquad \lim_{t\to t-} f(t) = y \tag{3-48}$$

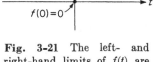

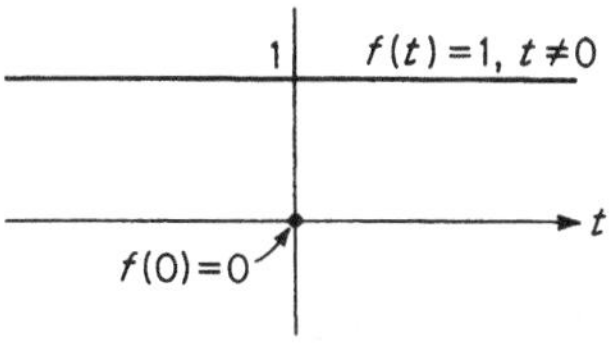


**Fig. 3-21** The left- and right-hand limits of $f(t)$ are the same at $t = 0$, but $f(t)$ is not continuous at $t = 0$.

For example, the function $f$ of Example 3-28 has the limit 1 from the right at 0 and the limit $-1$ from the left at 0. We note that even if $x = f(t+)$ and $y = f(t-)$ exist and are equal, $f$ may still *not* be continuous at $t$. This is illustrated in Example 3-30.

**Example 3-30** Let $f$ be the function from $R$ into itself, defined by $f(t) = 1$ if $t \neq 0$ and $f(0) = 0$. Then $\lim_{t\to 0+} f(t) = 1 = \lim_{t\to 0-} f(t)$, but $f$ is not continuous at 0. (See Fig. 3-21.)

***Definition 3-25 Piecewise Continuity*** *If $f$ is a transformation from $R$ into a set $X$, with $d$ as distance on $X$, then we say that $f$ has a simple (or jump) discontinuity at $t$ if (1) $f$ is not continuous at $t$ and (2) $f$ has both a limit from the right and a limit from the left at $t$, that is, both $f(t+)$ and $f(t-)$ exist. If $\mathfrak{D}(f) = \{t\colon t$ is a discontinuity of $f$; that is, $f$ is not continuous at $t\}$ is a countable set† [i.e., if the elements of $\mathfrak{D}(f)$ can be enumerated in a sequence $t_n$, $n = 1, 2, \ldots$ (which may terminate)] and consists of only simple discontinuities, then we shall say that $f$ is a piecewise continuous (or regulated)‡ function. In other words, $f$ is piecewise continuous if there are only a countable number of points at which $f$ is not continuous and if, at each of these points, $f$ has a limit both on the left and on the right.*

† See Sec. 2-2.
‡ See Ref. D-5.

**Example 3-31** Let $f$ be the function from $R$ into itself which is defined by setting $f(t) = 0$ if $t < 0$, $f(t) = 1$ if $0 \le t \le 1$, $f(t) = -1$ if $1 < t \le 2$, and $f(t) = 0$ if $t > 2$. Then $f$ is piecewise continuous and, in fact, is what is called a *step*, or *piecewise constant*, function. We note that there is a finite sequence $t_1, t_2, t_3$, with $t_1 = 0$, $t_2 = 1$, $t_3 = 2$, such that $f$ is constant on the intervals $(t_1, t_2)$, $(t_2, t_3)$ and on the sets $\{t: t < t_1\}$ and $\{t: t > t_3\}$. (See Fig. 3-22.) We shall have more to say about step functions in Sec. 3-15.

We observe that every continuous function is also piecewise continuous and that $\mathbf{f}$ is a piecewise continuous function from $R$ into $R_n$ if and only if the components $f_1, f_2, \ldots, f_n$ of $\mathbf{f}$ [see Eq. (3-33)] are piecewise continuous. Moreover, if we define the notions of sum and product (with an element of $R$) for piecewise continuous functions as in Eqs. (3-36) and (3-37), then we can see that the set of all piecewise continuous functions from $R$ into $R_n$ forms a vector space which is a subspace of $\mathfrak{F}(R, R_n)$. We shall have more to say about this space in Sec. 3-15. We note finally that it is *not* true that the composition of piecewise continuous functions is piecewise continuous; however, if $f$ is a *continuous* function from $R_n$ into $R$ and if $\mathbf{g}$ is a piecewise continuous function from $R$ into $R_n$, then $f \circ \mathbf{g}$ is a piecewise continuous function from $R$ into $R$.†

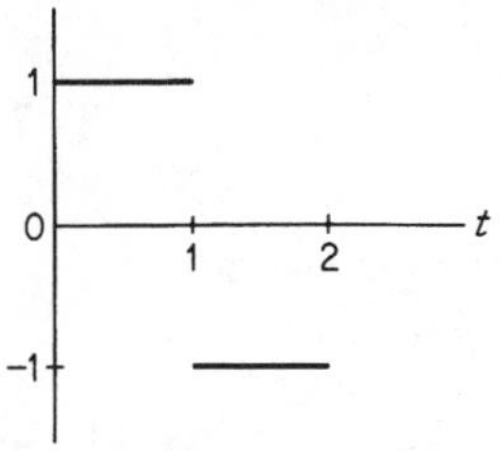

**Fig. 3-22** A piecewise constant function.

**Example 3-32** Let $I$ be the closed interval with end points 0 and 1, that is, $I = [0, 1]$ [see the second of Eqs. (3-7)], and let $\mathbf{f}$ be a function from $I$ into $R_n$ with components $f_1, f_2, \ldots, f_n$. Then we say that $\mathbf{f}$ is piecewise continuous on $I$ (compare with the remarks at the end of Sec. 3-9) if the only points of $I$ at which $\mathbf{f}$ is not continuous are simple discontinuities of $\mathbf{f}$ and if these are countable in number. For example, the function $\mathbf{f}$ from $I$ into $R_2$, defined by $\mathbf{f}(t) = \begin{bmatrix} 1 \\ 0 \end{bmatrix}$ for $t$ in $[0, \frac{1}{4})$, $\mathbf{f}(t) = \begin{bmatrix} 0 \\ -1 \end{bmatrix}$ for $t$ in $[\frac{1}{4}, \frac{1}{2})$, and $\mathbf{f}(t) = \begin{bmatrix} -1 \\ 0 \end{bmatrix}$ for $t$ in $[\frac{1}{2}, 1]$, is piecewise continuous. It is easy to see that the set of *all* piecewise continuous functions from $I$ into $R_n$ with sum and product defined as in (3-36) and (3-37) forms a vector space.

## 3-12 Vector Functions: Derivatives

We are going to study a very important notion in this section, namely, the notion of the derivative of a function. Loosely speaking, the derivative of a function represents a "local" approximation to the function by means of a linear function. As such, the derivative may be thought of as measuring the rate of change of the function. Let us attempt to make these ideas more precise.

First, let us suppose that $f$ is a function from $R$ into itself and that $t_0$ is an element of $R$. Now, if $t$ is "near" $t_0$, we may try to approximate $f(t)$ by

† This is so since $\lim_{t \to t+} \mathbf{g}(t) = \mathbf{v}$ (say) implies $\lim_{t \to t+} f[\mathbf{g}(t)] = f(\mathbf{v})$. We shall use this remark in Sec. 3-15.

a function $f^a(t)$ of the following type:

$$f(t_0) + \mathfrak{A}(t - t_0) = f^a(t) \tag{3-49}$$

where $\mathfrak{A}$ is a linear transformation of $R$ into itself. We may write (3-49) in the form

$$f^a(t) = f(t_0) + a(t - t_0) \tag{3-50}$$

where $a$ is an element of $R$.† It is easy to see that $f^a(t)$ represents a line which passes through $f(t_0)$. (See Fig. 3-23.) We should like to choose $a$ in such a way that $f^a(t)$ is tangent to $f(t)$ at $t_0$, that is, in such a way that the limit of $|f^a(t) - f(t)|/|t - t_0|$ as $t$ approaches $t_0$ is 0. In other words, we want to choose $a$ so that

$$\lim_{t \to t_0} \left| \frac{f(t) - f(t_0)}{t - t_0} - a \right| = 0 \tag{3-51}$$

If there is an $a$ in $R$ for which Eq. (3-51) is satisfied, then we say that $f$ is *differentiable at* $t_0$ and that $a$ is the *derivative of* $f$ *at* $t_0$. In this case, we often write

$$a = \left.\frac{df}{dt}\right|_{t_0} = Df(t_0) = \dot{f}(t_0) \tag{3-52}$$

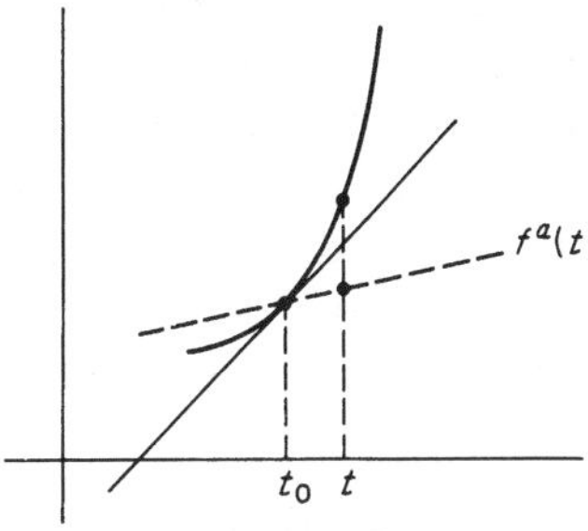


**Fig. 3-23** An approximating function $f^a(t)$.

We say simply that $f$ is *differentiable* if $f$ has a derivative at every $t_0$ in $R$, and we write

$$\frac{df}{dt} \qquad \text{or} \qquad Df(t) \qquad \text{or} \qquad \dot{f}(t) \tag{3-53}$$

for the function from $R$ into itself whose value at $t$ is the derivative of $f$ at $t$. We call this function the *derivative* of $f$. We note that it can be shown that if $f$ has a derivative, then $f$ is continuous. Furthermore, it can be shown that if $f$ is differentiable and if $a$, $b$ are in $R$ with $a < b$, then there is a $\theta$ in the interval $(a, b)$ with

$$f(b) - f(a) = \dot{f}(\theta)(b - a) = \left.\frac{df}{dt}\right|_{\theta} (b - a)$$

This is often called the *mean-value theorem.*

If $f$ is a differentiable function from $R$ into itself, then we may consider the function $\dot{f}(t) = df/dt$ and ask whether or not this function has a deriva-

† We note that $a$ is the matrix of $\mathfrak{A}$ (which has a $1 \times 1$ matrix). (See Sec. 2-8.) Furthermore, we may also write (3-50) in the form

$$f^a(t) = f(t_0) + \langle a, t - t_0 \rangle$$

where $\langle\ ,\ \rangle$ denotes the scalar product on $R$ [see Eq. (2-86)].

tive at a given $t_0$ in $R$. If $\dot{f}(t)$ has a derivative at $t_0$, then we say that $f(t)$ is *twice-differentiable at* $t_0$ or that $f(t)$ has a *second derivative at* $t_0$, and we write

$$\left.\frac{d[\dot{f}(t)]}{dt}\right|_{t_0} = \ddot{f}(t_0) = \left.\frac{d^2 f}{dt^2}\right|_{t_0}$$

We may repeat this procedure to obtain higher-order derivatives of $f(t)$, and we shall write

$$\left.\frac{d^n f(t)}{dt^n}\right|_{t_0} = f^{(n)}(t_0) = \left.\frac{d}{dt}\left[\frac{d^{n-1}f(t)}{dt^{n-1}}\right]\right|_{t_0}$$

for these higher-order derivatives. We shall not bother to define the notion of higher derivative, which corresponds to the concepts of derivative soon to be discussed, but rather shall leave this particular matter to the reader.

We may repeat this procedure to define the notion of derivative for a vector-valued function $\mathbf{f}$ from $R$ into $R_n$. In other words, if $t_0$ is an element of $R$ and if $\mathcal{A}$ is a linear transformation of $R$ into $R_n$, then we consider functions $\mathbf{f}^a(t)$ which are given by

$$\mathbf{f}^a(t) = \mathbf{f}(t_0) + \mathcal{A}(t - t_0) \tag{3-54}$$

or, equivalently, by

$$\mathbf{f}^a(t) = \mathbf{f}(t_0) + \mathbf{a}(t - t_0) \tag{3-55}$$

where $\mathbf{a}$ is a vector in $R_n$.† As in (3-51), we want to choose $\mathbf{a}$ in such a way that

$$\lim_{\substack{t \to t_0 \\ t \neq t_0}} \left\| \frac{\mathbf{f}(t) - \mathbf{f}(t_0)}{t - t_0} - \mathbf{a} \right\| = 0 \tag{3-56}$$

If there is an $\mathbf{a}$ in $R_n$ for which Eq. (3-56) is satisfied, then we say that $\mathbf{f}$ is *differentiable at* $t_0$ and that $\mathbf{a}$ is the *derivative of* $\mathbf{f}$ *at* $t_0$. In this case, we write

$$\mathbf{a} = D\mathbf{f}(t_0) = \dot{\mathbf{f}}(t_0) \tag{3-57}$$

We note that $\mathbf{f}$ is differentiable at $t_0$ if and only if the components $f_1, f_2, \ldots, f_n$ of $\mathbf{f}$ (see Sec. 3-9) are differentiable at $t_0$ and that

$$\dot{\mathbf{f}}(t_0) = \begin{bmatrix} \dot{f}_1(t_0) \\ \dot{f}_2(t_0) \\ \cdot \\ \cdot \\ \cdot \\ \dot{f}_n(t_0) \end{bmatrix} \tag{3-58}$$

† $\mathbf{a}$ is the $n$-column vector which is $\mathcal{A}(1)$ and is the matrix of the linear transformation $\mathcal{A}$ [see (2-31)].

We say simply that $\mathbf{f}$ is *differentiable* if $\mathbf{f}$ has a derivative at every $t_0$ in $R$, and we write

$$\frac{d\mathbf{f}}{dt} \qquad \text{or} \qquad D\mathbf{f}(t) \qquad \text{or} \qquad \dot{\mathbf{f}}(t) \tag{3-59}$$

for the function from $R$ into $R_n$ whose value at $t$ is the derivative of $\mathbf{f}$ at $t$. We call this function the *derivative of* $\mathbf{f}$. We note that

$$\dot{\mathbf{f}}(t) = \begin{bmatrix} \dot{f}_1(t) \\ \dot{f}_2(t) \\ \cdot \\ \cdot \\ \cdot \\ \dot{f}_n(t) \end{bmatrix} \tag{3-60}$$

Furthermore, we can see that if $\mathbf{f}$ and $\mathbf{g}$ are differentiable, then the sum $\mathbf{f} + \mathbf{g}$ is differentiable and the product $r\mathbf{f}$ is differentiable ($r$ being an element of $R$), with

$$\frac{d[\mathbf{f} + \mathbf{g}]}{dt} = \dot{\mathbf{f}}(t) + \dot{\mathbf{g}}(t) \tag{3-61}$$

and

$$\frac{d(r\mathbf{f})}{dt} = r\dot{\mathbf{f}}(t) \tag{3-62}$$

An immediate consequence of Eqs. (3-61) and (3-62) is the fact that *the set of all differentiable functions from $R$ into $R_n$ forms a vector space which is a subspace of the space of all functions from $R$ into $R_n$, $\mathfrak{F}(R, R_n)$.* Finally, we note that if $g$ is a differentiable function from $R$ into $R$ and if $\mathbf{f}$ is a differentiable function from $R$ into $R_n$, then the composite function $\mathbf{f} \circ g$ is differentiable and

$$\frac{d[\mathbf{f} \circ g]}{dt} = \dot{\mathbf{f}}[g(t)]\dot{g}(t) = \left.\frac{d\mathbf{f}}{dt}\right|_{g(t)} \frac{dg}{dt} \tag{3-63}$$

**Exercise 3-12** Show that a function $\mathbf{f}$ from $R$ into $R_n$ is differentiable at $t_0$ if and only if each component $f_i$ of $\mathbf{f}$ is differentiable at $t_0$ and that

$$\dot{\mathbf{f}}(t_0) = \begin{bmatrix} \dot{f}_1(t_0) \\ \cdot \\ \cdot \\ \cdot \\ \dot{f}_n(t_0) \end{bmatrix}$$

HINT: Make use of Eq. (3-6c).

**Example 3-33** Let $I$ be the closed interval with end points 0 and 1, that is, $I = [0, 1]$ [see (3-7)], and let $\mathbf{f}$ be a function from $I$ into $R_n$, with components $f_1, f_2, \ldots, f_n$. Then we say that $\mathbf{f}$ is *differentiable on I* if, for each $t_0$ in $I$, there is an $\mathbf{a}$ in $R_n$ (which may depend on $t_0$) such that

$$\lim_{\substack{t \to t_0 \\ t \in I}} \left\| \frac{\mathbf{f}(t) - \mathbf{f}(t_0)}{t - t_0} - \mathbf{a} \right\| = 0$$

where the notation

$$\lim_{\substack{t \to t_0 \\ t \in I}} \left\| \frac{\mathbf{f}(t) - \mathbf{f}(t_0)}{t - t_0} - \mathbf{a} \right\| = 0$$

means that, for *every* sequence $t_n$, $n = 1, 2, \ldots$, of elements of $I$, with $t_n \neq t_0$ for all $n$, which converges to $t_0$, the corresponding sequence

$$\left\| \frac{\mathbf{f}(t_n) - \mathbf{f}(t_0)}{t_n - t_0} - \mathbf{a} \right\|$$

converges to 0. It is easy to see that $\mathbf{f}$ is differentiable on $I$ if and only if each component $f_i$ of $\mathbf{f}$ is differentiable on $I$. For example, the function $\mathbf{f}$ from $I$ into $R_2$, defined by $\mathbf{f}(t) = \begin{bmatrix} \sin \pi t \\ \cos \pi t \end{bmatrix}$, is differentiable on $I$ and $\dot{\mathbf{f}}(t) = \begin{bmatrix} \pi \cos \pi t \\ -\pi \sin \pi t \end{bmatrix}$.

So far, we have considered functions of one (real) variable; now let us suppose that $f$ is a function from $R_m$ into $R$. There are several senses in which we may speak of the derivative of $f$: First of all, we may repeat the procedure used to obtain the derivative for functions of one variable; secondly, we may view $f$ as a function of the components of vectors in $R_m$ and obtain derivatives of $f$ with respect to those components (i.e., partial derivatives); and finally, we may consider the behavior of $f$ along a given "direction" in $R_m$ and obtain a derivative of $f$ with respect to this direction. We shall now make these different notions of derivative more precise.

If $\mathbf{u}_0$ is an element of $R_m$ and if $\mathbf{u}$ is "near" $\mathbf{u}_0$, we may try to approximate $f(\mathbf{u})$ by a function $f^a(\mathbf{u})$ of the type

$$f(\mathbf{u}_0) + \mathcal{A}(\mathbf{u} - \mathbf{u}_0) = f^a(\mathbf{u}) \tag{3-64}$$

where $\mathcal{A}$ is a linear transformation from $R_m$ into $R$ [that is, $\mathcal{A}$ is an element of the dual space $R_m^*$ of $R_m$ (see Exercises 2-11 and 2-12)]. We may write Eq. (3-64) in the form

$$f^a(\mathbf{u}) = f(\mathbf{u}_0) + \langle \mathbf{a}, \mathbf{u} - \mathbf{u}_0 \rangle \tag{3-65}$$ †

where $\mathbf{a}$ is an element of $R_m$. Again, we should like to choose the $m$ vector $\mathbf{a}$ so that $f^a(\mathbf{u})$ is tangent to $f(\mathbf{u})$ at $\mathbf{u}_0$, that is, in such a way that the limit of $|f^a(\mathbf{u}) - f(\mathbf{u})|/\|\mathbf{u} - \mathbf{u}_0\|$ as $\mathbf{u}$ approaches $\mathbf{u}_0$ is 0. In other words, we want

$$\lim_{\mathbf{u} \to \mathbf{u}_0} \left| \frac{f(\mathbf{u}) - f(\mathbf{u}_0)}{\|\mathbf{u} - \mathbf{u}_0\|} - \frac{\langle \mathbf{a}, \mathbf{u} - \mathbf{u}_0 \rangle}{\|\mathbf{u} - \mathbf{u}_0\|} \right| = 0 \tag{3-66}$$

If there is an $\mathbf{a}$ in $R_m$ for which Eq. (3-66) is satisfied, then we say that $f$ is *differentiable* at $\mathbf{u}_0$ and that $\mathbf{a}$ is the *gradient* (or *derivative*) *of $f$ at* $\mathbf{u}_0$. In this case, we often write

$$\mathbf{a} = \nabla f(\mathbf{u}_0) = \operatorname{grad} f(\mathbf{u}_0) = \left. \frac{\partial f}{\partial \mathbf{u}} \right|_{\mathbf{u}_0} \tag{3-67}$$

† Where $\langle \ , \ \rangle$ denotes the scalar product on $R_m$ [see Eq. (2-86)].

We say simply that $f$ is *differentiable* if $f$ has a gradient at every $\mathbf{u}_0$ in $R_m$, and we write

$$(\nabla f)(\mathbf{u}) \qquad \text{or} \qquad \text{grad}\, f(\mathbf{u}) \qquad \text{or} \qquad \frac{\partial f}{\partial \mathbf{u}} \tag{3-68}$$

for the function from $R_m$ into itself whose value at $\mathbf{u}$ is the gradient of $f$ at $\mathbf{u}$. We call this function the *gradient of $f$*. We observe that the gradient of $f$ is a vector-valued function, and we shall denote the components of it by

$$(D_i f)(\mathbf{u}) \qquad \text{or} \qquad \frac{\partial f}{\partial u_i} \tag{3-69}$$

that is,

$$\frac{\partial f}{\partial \mathbf{u}} = \begin{bmatrix} (D_1 f)(\mathbf{u}) \\ (D_2 f)(\mathbf{u}) \\ \cdot \\ \cdot \\ \cdot \\ (D_m f)(\mathbf{u}) \end{bmatrix} = \begin{bmatrix} \partial f/\partial u_1 \\ \partial f/\partial u_2 \\ \cdot \\ \cdot \\ \cdot \\ \partial f/\partial u_m \end{bmatrix} \tag{3-70}$$

For example, if $f$ is the function from $R_2$ into $R$, given by

$$f\left(\begin{bmatrix} u_1 \\ u_2 \end{bmatrix}\right) = u_1^2 + \sin u_2$$

then

$$\frac{\partial f}{\partial \mathbf{u}} = \begin{bmatrix} 2u_1 \\ \cos u_2 \end{bmatrix} \qquad \text{and} \qquad \frac{\partial f}{\partial u_1} = 2u_1 \qquad \frac{\partial f}{\partial u_2} = \cos u_2$$

We note, finally, that if $f$ is differentiable and if $\mathbf{u}_1$ and $\mathbf{u}_2$ are points of $R_m$, then it can be shown that there is a point $\mathbf{u}^*$ on the line segment joining $\mathbf{u}_1$ and $\mathbf{u}_2$ (see Definition 3-16) such that

$$f(\mathbf{u}_1) - f(\mathbf{u}_2) = \left\langle \frac{\partial f}{\partial \mathbf{u}}(\mathbf{u}^*), \mathbf{u}_1 - \mathbf{u}_2 \right\rangle \tag{3-70a}$$

We shall use this relation, which is often referred to as the *mean-value theorem*, in Sec. 3-18.

Now let us look at $f$ as a function of the components of vectors in $R_m$. In other words, if

$$\mathbf{u} = \begin{bmatrix} u_1 \\ u_2 \\ \cdot \\ \cdot \\ \cdot \\ u_m \end{bmatrix} = \sum_{i=1}^{m} u_i \mathbf{e}_i \dagger$$

then we may write

$$f(\mathbf{u}) = f\left(\sum_{i=1}^{m} u_i \mathbf{e}_i\right) = f(u_1, u_2, \ldots, u_m) \tag{3-71}$$

† Where the $\mathbf{e}_i$ are the elements of the natural basis in $R_m$ [see Eqs. (2-27)].

Now let us suppose that $\mathbf{u}_0$ is a given element of $R_m$, with components $u_{01}, u_{02}, \ldots, u_{0m}$, and let us examine the behavior of $f$ as we change only one of the components of $\mathbf{u}_0$—say the first. In other words, let us consider the function $g_1$ from $R$ into itself, which is given by

$$g_1(u) = f(u, u_{02}, \ldots, u_{0m}) \tag{3-72}$$

If $g_1$ has a derivative at $u_{01}$, then we say that $f$ has a *partial derivative with respect to $u_1$ at $\mathbf{u}_0$*. We denote this partial derivative by $(D_1f)(\mathbf{u}_0)$ or $(\partial f/\partial u_1)\,|_{\mathbf{u}_0}$, that is,

$$(Dg_1)(u_{01}) = (D_1f)(\mathbf{u}_0) = \left.\frac{\partial f}{\partial u_1}\right|_{\mathbf{u}_0} \tag{3-73}$$

and we note that

$$(D_1f)(\mathbf{u}_0) = \lim_{u \to u_{01}} \frac{f(u, u_{02}, \ldots, u_{0m}) - f(u_{01}, u_{02}, \ldots, u_{0m})}{u - u_{01}} \tag{3-74}$$

We may define the partial derivatives of $f$ with respect to the other components $u_2, u_3, \ldots, u_m$ of $\mathbf{u}$ in an entirely similar way; i.e., we say that $f$ has a *partial derivative with respect to $u_i$ at $\mathbf{u}_0$* if there is a number $a_i$ such that

$$\lim_{u \to u_{0i}} \left| \frac{f(u_{01}, \ldots, u_{0i-1}, u, u_{0i+1}, \ldots, u_{0m}) - f(u_{01}, u_{02}, \ldots, u_{0m})}{u - u_{0i}} - a_i \right| = 0 \tag{3-75}$$

and in that case, we write

$$a_i = (D_if)(\mathbf{u}_0) = \left.\frac{\partial f}{\partial u_i}\right|_{\mathbf{u}_0} \tag{3-76}$$

If $f$ has a partial derivative with respect to $u_i$ at every point $\mathbf{u}_0$, then we say that $f$ has a *partial derivative with respect to $u_i$* and we denote the function from $R_m$ into $R$ whose value at $\mathbf{u}_0$ is $(D_if)(\mathbf{u}_0)$ by

$$(D_if)(\mathbf{u}) \qquad \text{or} \qquad \frac{\partial f}{\partial u_i} \tag{3-77}$$

But we have used this notion before for the components of the gradient of $f$ [see Eq. (3-70)]. We shall leave it to the reader to verify that we are being consistent in our use of this notation.

We are now ready to consider the notion of a derivative of the function $f$ from $R_m$ into $R$ along a given "direction" in $R_m$. Let us suppose that $\mathbf{u}_0$ is an element of $R_m$ and that $\mathbf{e}$ is a given vector in $R_m$ of norm 1, that is,

$$\langle \mathbf{e}, \mathbf{e} \rangle = 1 \tag{3-78}$$

We shall call the unit vector $\mathbf{e}$ a *direction,* and we shall say that $f$ has a *directional derivative at* $\mathbf{u}_0$ *in the direction* $\mathbf{e}$ if the function $g_\mathbf{e}$ from $R$ into itself, defined by

$$g_\mathbf{e}(t) = f(\mathbf{u}_0 + t\mathbf{e}) \tag{3-79}$$

has a derivative at $t = 0$. We call the derivative of $g_\mathbf{e}$ at 0, $\dot{g}_\mathbf{e}(0)$, the *directional derivative of* $f$ *at* $\mathbf{u}_0$ *in the direction* $\mathbf{e}$. (See Fig. 3-24.) We remark that

$$\left.\frac{\partial f}{\partial u_i}\right|_{\mathbf{u}_0} = \dot{g}_{\mathbf{e}_i}(0) \tag{3-80}$$

where $\mathbf{e}_1, \mathbf{e}_2, \ldots, \mathbf{e}_m$ is the natural basis in $R_m$; that is, the partial derivatives of $f$ are the directional derivatives of $f$ in the directions $\mathbf{e}_i$ determined by

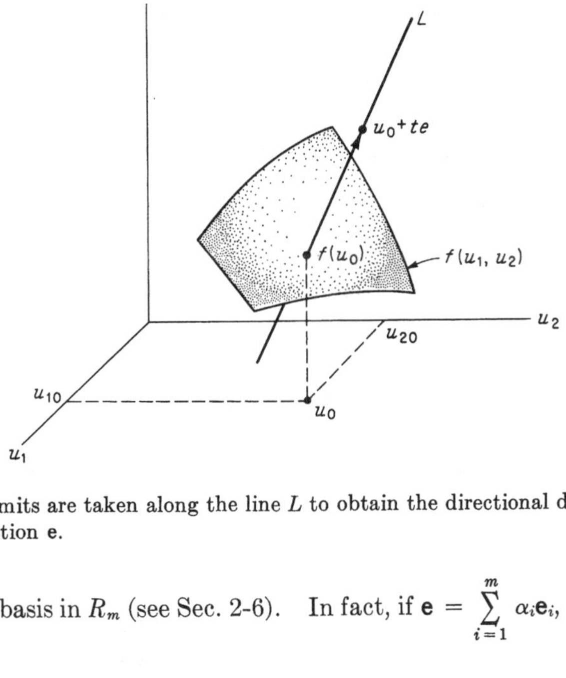


**Fig. 3-24** Limits are taken along the line $L$ to obtain the directional derivative of $f$ at $\mathbf{u}_0$ in the direction $\mathbf{e}$.

the natural basis in $R_m$ (see Sec. 2-6). In fact, if $\mathbf{e} = \sum_{i=1}^{m} \alpha_i \mathbf{e}_i$, then it can be shown that

$$\dot{g}_\mathbf{e}(0) = \sum_{i=1}^{m} \alpha_i \left.\frac{\partial f}{\partial u_i}\right|_{\mathbf{u}_0} \tag{3-81}$$

$$= \left\langle \begin{bmatrix} \alpha_1 \\ \alpha_2 \\ \cdot \\ \cdot \\ \cdot \\ \alpha_m \end{bmatrix}, \begin{bmatrix} (D_1 f)(\mathbf{u}_0) \\ (D_2 f)(\mathbf{u}_0) \\ \cdot \\ \cdot \\ \cdot \\ (D_m f)(\mathbf{u}_0) \end{bmatrix} \right\rangle \tag{3-82}$$

$$= \langle \mathbf{e}, \operatorname{grad} f(\mathbf{u}_0) \rangle \tag{3-83}$$

For example, if $f$ is the function from $R_2$ into $R$, given by

$$f\left(\begin{bmatrix} u_1 \\ u_2 \end{bmatrix}\right) = u_1^2 + u_2$$

and if

$$\mathbf{e} = \begin{bmatrix} 1/\sqrt{2} \\ 1/\sqrt{2} \end{bmatrix}$$

then the directional derivative of $f$ at $\mathbf{0}$ in the direction $\mathbf{e}$ is

$$\frac{1}{\sqrt{2}}\left(= \left\langle \begin{bmatrix} 1/\sqrt{2} \\ 1/\sqrt{2} \end{bmatrix}, \begin{bmatrix} 0 = (D_1f)(\mathbf{0}) \\ 1 = (D_2f)(\mathbf{0}) \end{bmatrix} \right\rangle\right)$$

Finally, let us consider a transformation $\mathbf{f}$ of $R_m$ into $R_n$ with components $f_1, f_2, \ldots, f_n$ and let us suppose that $\mathbf{u}_0$ is an element of $R_m$. Now, if $\mathbf{u}$ is "near" $\mathbf{u}_0$, we may try to approximate $\mathbf{f}(\mathbf{u})$ by a function $\mathbf{f}^a(\mathbf{u})$ of the type

$$\mathbf{f}^a(\mathbf{u}) = \mathbf{f}(\mathbf{u}_0) + \mathcal{A}(\mathbf{u} - \mathbf{u}_0) \tag{3-84}$$

where $\mathcal{A}$ is a linear transformation of $R_m$ into $R_n$. We should like to choose $\mathcal{A}$ in such a way that $\mathbf{f}^a(\mathbf{u})$ is tangent to $\mathbf{f}(\mathbf{u})$ at $\mathbf{u}_0$, that is, in such a way that

$$\lim_{\mathbf{u}\to\mathbf{u}_0} \frac{\|\mathbf{f}^a(\mathbf{u}) - \mathbf{f}(\mathbf{u})\|}{\|\mathbf{u} - \mathbf{u}_0\|} = 0 \tag{3-85}$$

In other words, if $\mathbf{A} = (a_{ij})$ is the matrix of $\mathcal{A}$ (see Sec. 2-7), then we want to choose the $a_{ij}$ so that

$$\lim_{\mathbf{u}\to\mathbf{u}_0} \left\| \frac{\mathbf{f}(\mathbf{u}) - \mathbf{f}(\mathbf{u}_0)}{\|\mathbf{u} - \mathbf{u}_0\|} - \frac{\mathbf{A}\cdot\mathbf{u} - \mathbf{u}_0}{\|\mathbf{u} - \mathbf{u}_0\|} \right\| = 0 \tag{3-86}$$

If there are $a_{ij}$ in $R$ for which Eq. (3-86) is satisfied, then we say that $\mathbf{f}$ is *differentiable at* $\mathbf{u}_0$ and that the *matrix* $\mathbf{A}$ is the *derivative of* $\mathbf{f}$ *at* $\mathbf{u}_0$. $\mathbf{A}$ is often called the *Jacobian matrix* of $\mathbf{f}$ at $\mathbf{u}_0$, and it can be shown that the $n \times m$ matrix $\mathbf{A}$ has the form

$$\mathbf{A} = \begin{bmatrix} (D_1f_1)(\mathbf{u}_0) & (D_2f_1)(\mathbf{u}_0) & \cdots & (D_mf_1)(\mathbf{u}_0) \\ (D_1f_2)(\mathbf{u}_0) & (D_2f_2)(\mathbf{u}_0) & \cdots & (D_mf_2)(\mathbf{u}_0) \\ \cdots & \cdots & \cdots & \cdots \\ (D_1f_n)(\mathbf{u}_0) & (D_2f_n)(\mathbf{u}_0) & \cdots & (D_mf_n)(\mathbf{u}_0) \end{bmatrix} \tag{3-87}$$

where $f_1, f_2, \ldots, f_n$ are the components of $\mathbf{f}$ and the $D_i$ represent partial derivatives. In other words, the entry in the $i$th row and $j$th column of $\mathbf{A}$ is precisely

$$D_jf_i(\mathbf{u}_0) \tag{3-88}$$

We often write $(\partial\mathbf{f}/\partial\mathbf{u})|\mathbf{u}_0$ for the matrix $\mathbf{A}$. In the case where $\mathbf{f}$ is a transformation of $R_n$ into itself, we call the determinant of the matrix $\mathbf{A}$ the

*Jacobian of* **f** *at* $\mathbf{u}_0$, and we write

$$\det \mathbf{A} = \det [D_j f_i(\mathbf{u}_0)] = \left.\frac{\partial(f_1, f_2, \ldots, f_n)}{\partial(u_1, u_2, \ldots, u_n)}\right|_{\mathbf{u}_0} = \det \left.\frac{\partial \mathbf{f}}{\partial \mathbf{u}}\right|_{\mathbf{u}_0} \tag{3-89}$$

We say simply that **f** is *differentiable* if **f** has a derivative at every $\mathbf{u}_0$ in $R_m$, and we write

$$\frac{\partial \mathbf{f}}{\partial \mathbf{u}} \quad \text{or} \quad [D_j f_i(\mathbf{u})] \tag{3-90}$$

for the function from $R_m$ into the set of $n \times m$ matrices $\mathfrak{M}(n, m)$ (see Sec. 2-8) whose value at **u** is the Jacobian matrix of **f** at **u**. We call this function the *derivative of* **f**. We observe that if

$$\mathbf{f}(\mathbf{u}) = \begin{bmatrix} f_1(u_1, u_2, \ldots, u_m) \\ f_2(u_1, u_2, \ldots, u_m) \\ \cdots\cdots\cdots\cdots \\ f_n(u_1, u_2, \ldots, u_m) \end{bmatrix} \tag{3-91}$$

then the derivative of **f**, $\partial \mathbf{f}/\partial \mathbf{u}$, is given by

$$\frac{\partial \mathbf{f}}{\partial \mathbf{u}} = \begin{bmatrix} \dfrac{\partial f_1}{\partial u_1} & \dfrac{\partial f_1}{\partial u_2} & \cdots & \dfrac{\partial f_1}{\partial u_m} \\ \dfrac{\partial f_2}{\partial u_1} & \dfrac{\partial f_2}{\partial u_2} & \cdots & \dfrac{\partial f_2}{\partial u_m} \\ \cdots & \cdots & \cdots & \cdots \\ \dfrac{\partial f_n}{\partial u_1} & \dfrac{\partial f_n}{\partial u_2} & \cdots & \dfrac{\partial f_n}{\partial u_m} \end{bmatrix} \tag{3-92}$$

For example, if **f** is the transformation of $R_3$ into $R_2$, given by

$$\mathbf{f}\left(\begin{bmatrix} u_1 \\ u_2 \\ u_3 \end{bmatrix}\right) = \begin{bmatrix} u_1^2 + u_2 \\ \\ u_3^2 - u_2 \end{bmatrix}$$

then

$$\frac{\partial \mathbf{f}}{\partial \mathbf{u}} = \begin{bmatrix} 2u_1 & 1 & 0 \\ 0 & -1 & 2u_3 \end{bmatrix}$$

**Example 3-34** Suppose that $\mathfrak{A}$ is a linear transformation from $R_m$ into $R_n$ with $\mathbf{A} = (a_{ij})$ as matrix. Then we know that $\mathfrak{A}(\mathbf{u}) = \mathbf{A}\mathbf{u}$ for **u** in $R_m$, and hence we can assert that $\partial\mathfrak{A}/\partial\mathbf{u} = \mathbf{A}$. In particular, if $\mathfrak{A}$ is the linear transformation from $R_3$ into $R_2$, with matrix **A** given by

$$\mathbf{A} = \begin{bmatrix} 1 & -1 & 0 \\ 0 & 2 & 3 \end{bmatrix}$$

then

$$\frac{\partial \mathfrak{A}}{\partial \mathbf{u}} = \mathbf{A} = \begin{bmatrix} 1 & -1 & 0 \\ 0 & 2 & 3 \end{bmatrix}$$

**Example 3-35** Suppose that $\mathbf{f}$ is a function from $R_2$ into $R_3$ which is given by

$$\mathbf{f}(\mathbf{u}) = \begin{bmatrix} u_1^2 + u_2^2 \\ -u_1u_2 \\ u_1 \sin u_2 \end{bmatrix}$$

Then

$$\frac{\partial \mathbf{f}}{\partial \mathbf{u}} = \begin{bmatrix} 2u_1 & 2u_2 \\ -u_2 & -u_1 \\ \sin u_2 & u_1 \cos u_2 \end{bmatrix}$$

and

$$\left.\frac{\partial \mathbf{f}}{\partial \mathbf{u}}\right|_{\mathbf{0}} = \begin{bmatrix} 0 & 0 \\ 0 & 0 \\ 0 & 0 \end{bmatrix}$$

We note that

$$\mathbf{f}\left(\begin{bmatrix} 0 \\ \pi \end{bmatrix}\right) = \mathbf{f}\left(\begin{bmatrix} \pi \\ 0 \end{bmatrix}\right) = \begin{bmatrix} \pi^2 \\ 0 \\ 0 \end{bmatrix}$$

so that $\mathbf{f}$ is not one-one.

## 3-13 Vector Functions: Smooth Sets in $R_n$

We are now going to use the notions of derivative and gradient introduced in the previous section to study the concept of "smoothness" for subsets of $R_n$. Roughly speaking, a subset of $R_n$ is "smooth" if it has a tangent plane at every point. Let us now make this idea more precise.

We begin with a series of definitions.

***Definition 3-26*** *Suppose that $f(\mathbf{x})$ is a continuous real-valued function on $R_n$. Then we let $S(f)$ denote the subset of $R_n$ given by*

$$S(f) = \{\mathbf{x}: f(\mathbf{x}) = 0\} \tag{3-93}$$

*In other words, $S(f)$ is the set of points $\mathbf{x}$ in $R_n$ at which $f$ is zero. We call $S(f)$ the (continuous) hypersurface in $R_n$ determined by $f$.*

***Definition 3-27*** *Suppose that $S(f)$ is a hypersurface in $R_n$ and that $\mathbf{x}_0$ is an element of $S(f)$ [that is, that $f(\mathbf{x}_0) = 0$]. Then we shall say that $\mathbf{x}_0$ is a regular point of $S(f)$ if the gradient of $f$ with respect to $\mathbf{x}$ at $\mathbf{x}_0$ exists and is not the zero vector. In other words, $\mathbf{x}_0$ is a regular point of $S(f)$ if*

$$\frac{\partial f}{\partial \mathbf{x}}(\mathbf{x}_0) \neq \mathbf{0} \tag{3-94}$$

We observe that if $\mathbf{x}_0$ is a regular point of $S(f)$, then the set $L(\mathbf{x}_0)$ given by

$$L(\mathbf{x}_0) = \left\{\mathbf{x}: \left\langle \frac{\partial f}{\partial \mathbf{x}}(\mathbf{x}_0), \mathbf{x} - \mathbf{x}_0 \right\rangle = 0\right\} \tag{3-95}$$

is a well-defined hyperplane (see Definition 3-15) which passes through the point $\mathbf{x}_0$. This observation leads us to the following definition:

***Definition 3-28*** *If $\mathbf{x}_0$ is a regular point of the hypersurface $S(f)$, then we call $L(\mathbf{x}_0)$ the tangent hyperplane of $S(f)$ at $\mathbf{x}_0$. We also say that a vector*

*$\mathbf{n}(\mathbf{x}_0)$ is normal to $S(f)$ at $\mathbf{x}_0$ if*

$$\mathbf{n}(\mathbf{x}_0) = c\frac{\partial f}{\partial \mathbf{x}}(\mathbf{x}_0) \qquad c \neq 0 \tag{3-96}$$

*In other words, $\mathbf{n}(\mathbf{x}_0)$ is a nonzero vector which is collinear with the gradient vector of $f$ at $\mathbf{x}_0$.*

We note that if $\mathbf{n}(\mathbf{x}_0)$ is normal to $S(f)$ at the (regular) point $\mathbf{x}_0$, then the tangent hyperplane $L(\mathbf{x}_0)$ is also given by

$$L(\mathbf{x}_0) = \{\mathbf{x}: \langle \mathbf{n}(\mathbf{x}_0), \mathbf{x} - \mathbf{x}_0 \rangle = 0\} \tag{3-97}$$

Now let us suppose that $f$ is a real-valued function defined on $R_n$. Then we shall say that $f$ is *smooth* if:

1. $f$ is differentiable [i.e., if $\partial f/\partial \mathbf{x}$ exists; see (3-68)].
2. The gradient $\partial f/\partial \mathbf{x}$ is a *continuous* function from $R_n$ into itself. We have:

***Definition 3-29*** *If $f$ is a smooth function and if every point $\mathbf{x}_0$ of $S(f)$ is regular, then we say that $S(f)$ is a smooth hypersurface.*

We observe that a smooth hypersurface has a well-defined tangent hyperplane at every point.

**Example 3-36** Let $f(x) = \langle \mathbf{a}, \mathbf{x} \rangle$, where $\mathbf{a}$ is some nonzero vector in $R_n$. Then $f$ is a smooth function, the set $S(f)$ is the hyperplane $L$ through the origin given by $\langle \mathbf{a}, \mathbf{x} \rangle = 0$, and the tangent hyperplane $L(\mathbf{x}_0)$ at any point $\mathbf{x}_0$ of $S(f)$ is precisely the set $S(f)$ ($= L$) itself.

**Example 3-37** Let $f$ be the real-valued function on $R_2$ given by $f\left(\begin{bmatrix} x_1 \\ x_2 \end{bmatrix}\right) = x_1^2 + x_2^2 - 1$. Then $f$ is smooth, the set $S(f)$ is the unit circle (see Fig. 3-25), and, for example, the tangent hyperplane $L\left(\begin{bmatrix} 1/\sqrt{2} \\ 1/\sqrt{2} \end{bmatrix}\right)$ of $S(f)$ at the point $\begin{bmatrix} 1/\sqrt{2} \\ 1/\sqrt{2} \end{bmatrix}$ is the line $x_1 + x_2 - \sqrt{2} = 0$ (as illustrated in Fig. 3-25). The reader should consider this example in the light of Definition 3-22.

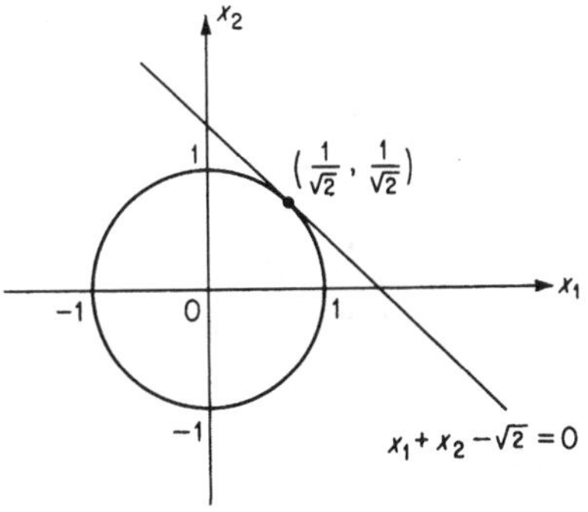


**Fig. 3-25** The set $S(f)$ of Example 3-37.

We are now ready to define the notion of a "smooth" set in $R_n$. We let $f_1, f_2, \ldots, f_{n-k}$ be $n - k$ distinct smooth functions on $R_n$, where $1 \leq k \leq n - 1$. We let $S(f_1, f_2, \ldots, f_{n-k})$ denote the intersection [see Eqs. (2-10)] of the hypersurfaces $S(f_1), S(f_2), \ldots, S(f_{n-k})$, that is,

$$S(f_1, f_2, \ldots, f_{n-k}) = S(f_1) \cap S(f_2) \cap \cdots \cap S(f_{n-k}) \tag{3-98}$$

In other words, $S(f_1, f_2, \ldots, f_{n-k})$ is the set of points $\mathbf{x}$ in $R_n$ at which *all* the $f_i$, $i = 1, 2, \ldots, n-k$, are zero; thus,

$$S(f_1, f_2, \ldots, f_{n-k}) = \{\mathbf{x}: f_1(\mathbf{x}) = 0 \text{ and } f_2(\mathbf{x}) = 0 \text{ and } \cdots \text{ and } f_{n-k}(\mathbf{x}) = 0\} \tag{3-99}$$

We now have:

***Definition 3-30*** *We shall say that* $S(f_1, f_2, \ldots, f_{n-k})$ *is a smooth k-fold in* $R_n$ *if, for every point* $\mathbf{x}_0$ *in* $S(f_1, f_2, \ldots, f_{n-k})$, *the* $n-k$ *vectors* $(\partial f_1/\partial \mathbf{x})(\mathbf{x}_0)$, $(\partial f_2/\partial \mathbf{x})(\mathbf{x}_0), \ldots, (\partial f_{n-k}/\partial \mathbf{x})(\mathbf{x}_0)$ *are linearly independent* [*see Eq.* (2-26)].

We observe that if $S(f_1, f_2, \ldots, f_{n-k})$ is a smooth $k$-fold, then each of the hypersurfaces $S(f_i)$ is smooth. It follows that if $\mathbf{x}_0$ is an element of $S(f_1, f_2, \ldots, f_{n-k})$, then the tangent hyperplane $L_i(\mathbf{x}_0)$ of $S(f_i)$ at $\mathbf{x}_0$ is defined for $i = 1, 2, \ldots, n-k$. We let $M(\mathbf{x}_0)$ denote the intersection of the hyperplanes $L_i(\mathbf{x}_0)$, that is,

$$M(\mathbf{x}_0) = L_1(\mathbf{x}_0) \cap L_2(\mathbf{x}_0) \cap \cdots \cap L_{n-k}(\mathbf{x}_0) \tag{3-100}$$

$$= \left\{\mathbf{x}: \left\langle \frac{\partial f_1}{\partial \mathbf{x}}(\mathbf{x}_0), \mathbf{x} - \mathbf{x}_0 \right\rangle = 0 \text{ and } \left\langle \frac{\partial f_2}{\partial \mathbf{x}}(\mathbf{x}_0), \mathbf{x} - \mathbf{x}_0 \right\rangle = 0 \text{ and } \cdots \text{ and } \left\langle \frac{\partial f_{n-k}}{\partial \mathbf{x}}(\mathbf{x}_0), \mathbf{x} - \mathbf{x}_0 \right\rangle = 0 \right\} \tag{3-101}$$

and we call $M(\mathbf{x}_0)$ the *tangent plane of* $S(f_1, f_2, \ldots, f_{n-k})$ *at* $\mathbf{x}_0$. Thus, a smooth $k$-fold has a tangent plane at every point.

**Example 3-38** Let $f_1(\mathbf{x}) = \langle \mathbf{a}^1, \mathbf{x} \rangle$ and $f_2(\mathbf{x}) = \langle \mathbf{a}^2, \mathbf{x} \rangle$, where $\mathbf{a}^1$ and $\mathbf{a}^2$ are linearly independent vectors in $R_n$. Then $f_1$ and $f_2$ are smooth, the set $S(f_1, f_2)$ is an $(n-2)$-fold, and the tangent plane $M(\mathbf{x}_0)$ of $S(f_1, f_2)$ at any point $\mathbf{x}_0$ of $S(f_1, f_2)$ is precisely the set $S(f_1, f_2)$ itself.

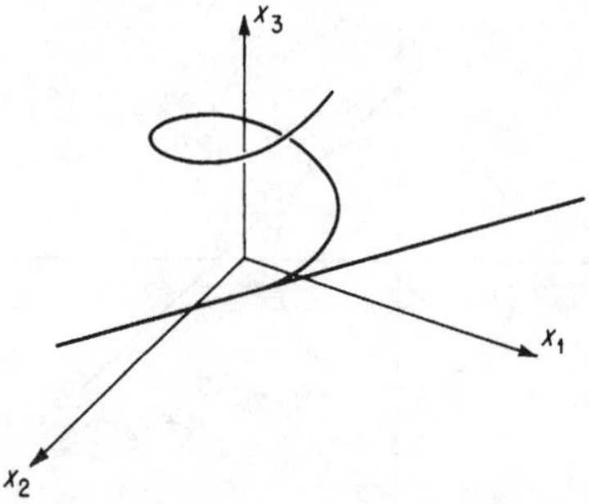


**Fig. 3-26** The set $S(f_1, f_2)$ of Example 3-39.

**Example 3-39** Let $f_1$ be the real-valued function on $R_3$ given by

$$f_1\left(\begin{bmatrix} x_1 \\ x_2 \\ x_3 \end{bmatrix}\right) = x_1 - \sin x_3$$

and let $f_2$ be the real-valued function on $R_3$ given by

$$f_2\left(\begin{bmatrix} x_1 \\ x_2 \\ x_3 \end{bmatrix}\right) = x_2 - \cos x_3$$

Then $f_1$ and $f_2$ are smooth, the 1-fold $S(f_1, f_2)$ in $R_3$ is the helix illustrated in Fig. 3-26, and, for example, the tangent line

$$M\left(\begin{bmatrix} 0 \\ 1 \\ 0 \end{bmatrix}\right)$$

of $S(f_1, f_2)$ at the point

$$\begin{bmatrix} 0 \\ 1 \\ 0 \end{bmatrix}$$

is the intersection of the planes $x_1 - x_3 = 0$ and $x_2 - 1 = 0$.

Finally, we have the following definition:

***Definition 3-31*** *Let $S(f_1, f_2, \ldots, f_{n-k})$ be a smooth k-fold in $R_n$ and let $\mathbf{x}_0$ be an element of $S(f_1, f_2, \ldots, f_{n-k})$. Then we shall say that a vector $\mathbf{p}$ is transversal to $S(f_1, f_2, \ldots, f_{n-k})$ at $\mathbf{x}_0$ if*

$$\langle \mathbf{p}, \mathbf{x} - \mathbf{x}_0 \rangle = 0 \qquad \textit{for all } \mathbf{x} \textit{ in } M(\mathbf{x}_0) \tag{3-102}$$

We observe that $\mathbf{p}$ is transversal to $S(f_1, f_2, \ldots, f_{n-k})$ at $\mathbf{x}_0$ if and only if $\mathbf{p}$ is a linear combination (see Sec. 2-6) of the $n - k$ vectors $(\partial f_1/\partial \mathbf{x})(\mathbf{x}_0)$, $(\partial f_2/\partial \mathbf{x})(\mathbf{x}_0), \ldots, (\partial f_{n-k}/\partial \mathbf{x})(\mathbf{x}_0)$. This notion of transversality will be quite important in Chap. 5. We also note that a vector $\mathbf{p}$ will be transversal to $S(f_1, f_2, \ldots, f_{n-k})$ at $\mathbf{x}_0$ if and only if $\mathbf{p}$ satisfies *the k relations*

$$\langle \mathbf{p}, \mathbf{x}^j - \mathbf{x}_0 \rangle = 0 \qquad j = 1, 2, \ldots, k \tag{3-103}$$

where the $\mathbf{x}^j$ are elements of $M(\mathbf{x}_0)$ such that the vectors $\mathbf{x}^j - \mathbf{x}_0$ are linearly independent. This remark will take on greater significance in Chap. 5.

## 3-14 Vector Functions: Integrals

We are now going to discuss a very important concept: the notion of the integral of a function. Loosely speaking, the integral of a function $f$ is a function $g$ whose derivative is the function $f$. We shall shortly make this idea more precise.

Let us suppose that $f$ is a real-valued function which is defined on the closed interval $[a, b]$ of $R$, and let us suppose that $g$ is a *continuous* function from $[a, b]$ into $R$. Then we shall say that $g$ is an *antiderivative of f* if there is a countable subset (see Sec. 2-2) $A$ of $[a, b]$ such that if $t$ is in $[a, b] - A$ [see Eq. (2-4)], then $g$ has a derivative at $t$ and $\dot{g}(t) = f(t)$. We remark that if $g$ and $h$ are two antiderivatives of $f$, then the difference $g - h$ is a constant on $[a, b]$.† If $f$ is a piecewise continuous function on $[a, b]$ (see Sec. 3-11), then it can be shown that there is an antiderivative $g$ of $f$. For example, if $f$ is a *step function* (see Example 3-31 and Sec. 3-15), that is, if there are a finite number of elements $t_0, t_1, \ldots, t_n$ of $[a, b]$ with $a = t_0 < t_1 < \cdots < t_n = b$ such that $f$ is a constant $c_i$ on each of the (open) intervals $(t_i, t_{i+1})$, then the function $g$, defined by setting

$$g(t) = c_i(t - t_i) + \sum_{j=0}^{i-1} c_j(t_{j+1} - t_j) \tag{3-104}$$

for $t$ in $[t_i, t_{i+1}]$, is an antiderivative of $f$. A function like $g(t)$ is often said to be *piecewise linear*. This leads us to the following definition:

***Definition 3-32 Integrals*** *If $f$ is a real-valued piecewise continuous (i.e., regulated) function defined on the (closed) interval $[a, b]$ and if $g$ is an antiderivative of $f$, then the difference $g(t_2) - g(t_1)$ for $t_2$, $t_1$ in $[a, b]$ is called*

† This follows from the mean-value theorem. (See Ref. D-5.)

*the (definite) integral of f from $t_1$ to $t_2$ (or, simply, the integral of f), and we write*

$$g(t_2) - g(t_1) = \int_{t_1}^{t_2} f(\tau)\, d\tau \tag{3-105}$$

We shall use the notation $\sup_{\tau \in I} |f(\tau)|$, where $I$ is an interval, to denote the maximum value of $|f(\tau)|$ on the interval $I$ if this maximum exists and to denote the greatest lower bound of the set of numbers $M$ such that $|f(\tau)| \le M$ for *all* $\tau$ in $I$ if this maximum does not exist. We call the number $\sup_{\tau \in I} |f(\tau)|$ the *supremum of* $|f(\tau)|$ *on* $I$. We shall show in Sec. 3-15 that if $f$ is piecewise continuous, then there are numbers $M$ such that $|f(\tau)| \le M$ for all $\tau$ in $I$.

We collect several important properties of the integral of a function in the next theorem, whose proof we shall omit (see Ref. D-5).

***Theorem 3-8*** *If $f$ and $h$ are piecewise continuous functions defined on $[a, b]$ and if $c$ is an element of $R$, then for $t_1$, $t_2$, $t$ in $[a, b]$,*

*a.*
$$\int_{t_1}^{t_2} [f + h](\tau)\, d\tau = \int_{t_1}^{t_2} f(\tau)\, d\tau + \int_{t_1}^{t_2} h(\tau)\, d\tau \tag{3-106}$$

*b.*
$$\int_{t_1}^{t_2} [cf](\tau)\, d\tau = c \int_{t_1}^{t_2} f(\tau)\, d\tau \tag{3-107}$$

*c.*
$$\left| \int_{t_1}^{t_2} f(\tau)\, d\tau \right| \le \int_{t_1}^{t_2} |f(\tau)|\, d\tau \le (t_2 - t_1) \sup_{\tau \in [a,b]} |f(\tau)| \tag{3-108}$$

*d. If $f$ is actually continuous, then the continuous function*

$$\int_{t_1}^{t} f(\tau)\, d\tau$$

*is differentiable for every $t$ in $[a, b]$ and*

$$\frac{d}{dt}\left[\int_{t_1}^{t} f(\tau)\, d\tau\right] = f(t)$$

*e.*
$$\int_{t_1}^{t} f(\tau)\, d\tau + \int_{t}^{t_2} f(\tau)\, d\tau = \int_{t_1}^{t_2} f(\tau)\, d\tau \tag{3-109}$$

*f. If*

$$\int_a^b |f(\tau)|\, d\tau = 0$$

*then there is a countable subset $A$, which may be empty, of $[a, b]$ such that $f(\tau) = 0$ if $\tau \in [a, b] - A$. In particular,*

$$\int_a^b |f(\tau)|\, d\tau = 0$$

*implies that $f(\tau) = 0$ for all $\tau$ in $[a, b]$ provided that $f$ is continuous and that $f(\tau) = 0$ for all $\tau$ in $[a, b)$ if $f$ is continuous on the right, that is, if $f(\tau+) = f(\tau)$ for all $\tau$ in $[a, b)$.*

*g. If there is a countable subset A of* $[a, b]$ *such that* $f(\tau) = h(\tau)$ *for all* $\tau$ *in* $[a, b] - A$, *then*

$$\int_a^b f(\tau)\, d\tau = \int_a^b h(\tau)\, d\tau$$

*In other words, two piecewise continuous functions f and h can differ at a countable number of points and still have the same integral. In particular, if* $f(\tau)$ *is a continuous function for all* $\tau \in [a, b]$ *and if* $h(\tau) = f(\tau)$ *for all* $\tau \in (a, b)$, *that is,* $f(a) \neq h(a)$ *and* $f(b) \neq h(b)$ *only, then*

$$\int_a^b f(\tau)\, d\tau = \int_a^b h(\tau)\, d\tau$$

The properties of the integral enumerated in this theorem will be used again and again in the remainder of the book.

Let us now suppose that $\mathbf{f}$ is a function from $[a, b]$ into $R_n$ with components $f_1, f_2, \ldots, f_n$. Then we may define the integral of $\mathbf{f}$ componentwise. In other words, we have:

***Definition 3-33*** *If* $\mathbf{f}$ *is a piecewise continuous function, then the integral of* $\mathbf{f}$ *from* $t_1$ *to* $t_2$ *(or, simply, the integral of* $\mathbf{f}$*) is the element of* $R_n$ *whose components are the integrals of the* $f_i$ *from* $t_1$ *to* $t_2$, *that is,*

$$\int_{t_1}^{t_2} \mathbf{f}(\tau)\, d\tau = \begin{bmatrix} \int_{t_1}^{t_2} f_1(\tau)\, d\tau \\ \int_{t_1}^{t_2} f_2(\tau)\, d\tau \\ \cdot \\ \cdot \\ \cdot \\ \int_{t_1}^{t_2} f_n(\tau)\, d\tau \end{bmatrix} \tag{3-110}$$

We note that the properties expressed in Theorem 3-8 have natural analogs for the integral of a vector-valued function. For example, property *c* [Eq. (3-108)] becomes

$$\left\| \int_{t_1}^{t_2} \mathbf{f}(\tau)\, d\tau \right\| \leq \int_{t_1}^{t_2} \|\mathbf{f}(\tau)\|\, d\tau \leq (t_2 - t_1) \sup_{\tau \in [a,b]} \|\mathbf{f}(\tau)\| \tag{3-111}$$†

and property *f* becomes: If

$$\int_a^b \|\mathbf{f}(\tau)\|\, d\tau = 0$$

then there is a countable subset $A$ of $[a, b]$ such that $\mathbf{f}(\tau) = \mathbf{0}$ if $\tau \in [a, b] - A$; in particular,

$$\int_a^b \|\mathbf{f}(\tau)\|\, d\tau = 0$$

† Where $\sup_{\tau \in [a,b]} \|\mathbf{f}(\tau)\|$ is the greatest lower bound of the set of numbers $M$ such that $\|\mathbf{f}(\tau)\| \leq M$ for all $\tau$ in $[a, b]$.

implies that $\mathbf{f}(\tau) = \mathbf{0}$ for all $\tau$ in $[a, b]$ if $\mathbf{f}$ is continuous and for all $\tau$ in $[a, b)$ if $\mathbf{f}$ is continuous on the right. Another important property of the integral of a vector-valued function relates to linear transformations. If $\mathcal{A}$ is a linear transformation of $R_n$ into $R_m$, then we have

$$\mathcal{A}\left[\int_{t_1}^{t_2} \mathbf{f}(\tau)\, d\tau\right] = \int_{t_1}^{t_2} \mathcal{A}[\mathbf{f}(\tau)]\, d\tau \tag{3-112}$$

which can also be written as

$$\mathbf{A}\begin{bmatrix} \int_{t_1}^{t_2} f_1(\tau)\, d\tau \\ \int_{t_1}^{t_2} f_2(\tau)\, d\tau \\ \cdot \\ \cdot \\ \cdot \\ \int_{t_1}^{t_2} f_n(\tau)\, d\tau \end{bmatrix} = \begin{bmatrix} \int_{t_1}^{t_2} \left\{\sum_{j=1}^{n} a_{1j} f_j(\tau)\right\} d\tau \\ \int_{t_1}^{t_2} \left\{\sum_{j=1}^{n} a_{2j} f_j(\tau)\right\} d\tau \\ \cdot \\ \cdot \\ \cdot \\ \int_{t_1}^{t_2} \left\{\sum_{j=1}^{n} a_{mj} f_j(\tau)\right\} d\tau \end{bmatrix} \tag{3-113}$$

where $\mathbf{A} = (a_{ij})$ is the $m \times n$ matrix which is the matrix of $\mathcal{A}$.

Finally, let us suppose that $\mathbf{F}$ is a function from the interval $[a, b]$ into the set $\mathfrak{M}(n, m)$ (see Sec. 2-8) of $n \times m$ matrices. In other words, $\mathbf{F}(t)$ for $t$ in $[a, b]$ is an $n \times m$ matrix whose entries we shall denote by $f_{ij}(t)$ for $i = 1, 2, \ldots, n$ and $j = 1, 2, \ldots, m$. Thus, we have

$$\mathbf{F}(t) = \begin{bmatrix} f_{11}(t) & f_{12}(t) & \cdots & f_{1m}(t) \\ f_{21}(t) & f_{22}(t) & \cdots & f_{2m}(t) \\ \cdots & \cdots & \cdots & \cdots \\ f_{n1}(t) & f_{n2}(t) & \cdots & f_{nm}(t) \end{bmatrix} \tag{3-114}$$

We observe that the $f_{ij}$ are functions from $[a, b]$ into $R$, and we shall say that the matrix-valued function $\mathbf{F}(t)$ is *piecewise continuous* (or *continuous*) if each of the functions $f_{ij}(t)$ is *piecewise continuous* (or *continuous*). This leads us to the following definition:

***Definition 3-34*** *If* $\mathbf{F}$ *is a piecewise continuous matrix-valued function, then the integral of* $\mathbf{F}$ *from* $t_1$ *to* $t_2$ *(or, simply, the integral of* $\mathbf{F}$*) is the element of* $\mathfrak{M}(n, m)$ *whose entries are the integrals of the* $f_{ij}$ *from* $t_1$ *to* $t_2$*, that is,*

$$\left(\int_{t_1}^{t_2} \mathbf{F}(\tau)\, d\tau\right)_{ij} = \int_{t_1}^{t_2} f_{ij}(\tau)\, d\tau \tag{3-115}$$

*or*

$$\int_{t_1}^{t_2} \mathbf{F}(\tau)\, d\tau = \begin{bmatrix} \int_{t_1}^{t_2} f_{11}(\tau)\, d\tau & \int_{t_1}^{t_2} f_{12}(\tau)\, d\tau & \cdots & \int_{t_1}^{t_2} f_{1m}(\tau)\, d\tau \\ \int_{t_1}^{t_2} f_{21}(\tau)\, d\tau & \int_{t_1}^{t_2} f_{22}(\tau)\, d\tau & \cdots & \int_{t_1}^{t_2} f_{2m}(\tau)\, d\tau \\ \cdots & \cdots & \cdots & \cdots \\ \int_{t_1}^{t_2} f_{n1}(\tau)\, d\tau & \int_{t_1}^{t_2} f_{n2}(\tau)\, d\tau & \cdots & \int_{t_1}^{t_2} f_{nm}(\tau)\, d\tau \end{bmatrix} \tag{3-116}$$

We note again that the properties of the integral expressed in Theorem 3-8 have natural analogs for the integral of a matrix-valued function. There are several other properties of this integral which we shall require in the main body of the book. These are given in the next theorem.

***Theorem 3-9*** *Let* $\mathbf{F}(t)$ *be a piecewise continuous function from* $[a, b]$ *into the set of* $n \times m$ *matrices, let* $\mathbf{P}$ *be a given (constant)* $p \times n$ *matrix and* $\mathbf{Q}$ *be a given (constant)* $m \times q$ *matrix, and let* $\mathbf{v}$ *be a given (constant) vector in* $R_m$. *Then*

*a.* $\mathbf{PF}(t)\mathbf{Q}$ *is a piecewise continuous function from* $[a, b]$ *into the set of* $p \times q$ *matrices, and*

$$\int_{t_1}^{t_2} [\mathbf{PF}(\tau)\mathbf{Q}]\, d\tau = \mathbf{P}\left\{\int_{t_1}^{t_2} \mathbf{F}(\tau)\, d\tau\right\}\mathbf{Q} \tag{3-117}$$

*b.* $\mathbf{F}(\tau)\mathbf{v}$ *is a piecewise continuous function from* $[a, b]$ *into* $R_n$, *and*

$$\int_{t_1}^{t_2} \{\mathbf{F}(\tau)\mathbf{v}\}\, d\tau = \left\{\int_{t_1}^{t_2} \mathbf{F}(\tau)\, d\tau\right\}\mathbf{v} \tag{3-118}$$

Let us now turn our attention to some examples which illustrate the notion of integral and some of its consequences.

**Example 3-40** Let $f$ be the function from $[0, T]$ into $R$ defined by $f(t) = e^t$. Then

$$\int_{t_1}^{t_2} e^\tau\, d\tau = e^{t_2} - e^{t_1}$$

and

$$\int_0^T e^\tau\, d\tau = e^T - 1$$

If $g$ is the function from $[0, T]$ into $R$ defined by $g(t) = t + 1$, then

$$\int_{t_1}^{t_2} g(\tau)\, d\tau = \frac{t_2^2 - t_1^2}{2} + t_2 - t_1$$

and

$$\int_0^T g(\tau)\, d\tau = \frac{T^2}{2} + T$$

It is clear that

$$\int_0^T \{e^\tau + (\tau + 1)\}\, d\tau = e^T - 1 + \frac{T^2}{2} + T = \int_0^T e^\tau\, d\tau + \int_0^T (\tau + 1)\, d\tau$$

**Example 3-41** Let $\mathbf{f}$ be the function from $[0, T]$ into $R_3$ defined by

$$\mathbf{f}(t) = \begin{bmatrix} e^t \\ \sin t \\ 1 \end{bmatrix}$$

Then

$$\int_{t_1}^{t_2} \mathbf{f}(\tau)\, d\tau = \begin{bmatrix} \int_{t_1}^{t_2} e^\tau\, d\tau \\ \int_{t_1}^{t_2} \sin \tau\, d\tau \\ \int_{t_1}^{t_2} 1\, d\tau \end{bmatrix} = \begin{bmatrix} e^{t_2} - e^{t_1} \\ -\cos t_2 + \cos t_1 \\ t_2 - t_1 \end{bmatrix}$$

**Example 3-42** Let $\mathbf{F}$ be the function from $[0, T]$ into the set of $2 \times 2$ matrices given by $\mathbf{F}(t) = \begin{bmatrix} e^t & 0 \\ -1 & t \end{bmatrix}$. Then

$$\int_0^T \mathbf{F}(\tau)\, d\tau = \begin{bmatrix} e^T - 1 & 0 \\ -T & T^2/2 \end{bmatrix}$$

If $\mathbf{P} = \begin{bmatrix} 0 & 1 \\ -1 & 0 \end{bmatrix}$ and $\mathbf{Q} = \begin{bmatrix} 1 & 0 \\ 1 & 0 \end{bmatrix}$, then $\mathbf{PF}(t)\mathbf{Q} = \begin{bmatrix} t-1 & 0 \\ -e^t & 0 \end{bmatrix}$ and

$$\int_0^T \mathbf{PF}(\tau)\mathbf{Q}\, d\tau = \begin{bmatrix} T^2/2 - T & 0 \\ 1 - e^T & 0 \end{bmatrix} = \mathbf{P}\left\{\int_0^T \mathbf{F}(\tau)\, d\tau\right\}\mathbf{Q}$$

If $\mathbf{v} = \begin{bmatrix} 1 \\ 1 \end{bmatrix}$, then $\mathbf{F}(t)\mathbf{v} = \begin{bmatrix} e^t \\ t-1 \end{bmatrix}$ and

$$\int_0^T \mathbf{F}(\tau)\mathbf{v}\, d\tau = \begin{bmatrix} e^T - 1 \\ T^2/2 - T \end{bmatrix} = \begin{bmatrix} e^T - 1 & 0 \\ -T & T^2/2 \end{bmatrix}\begin{bmatrix} 1 \\ 1 \end{bmatrix} = \left\{\int_0^T \mathbf{F}(\tau)\, d\tau\right\}\mathbf{v}$$

## 3-15 Vector Functions: Function Spaces

We have already come across the idea of a set of functions which have properties in common, as in the set of all piecewise continuous functions. Usually the sets of functions which we considered were vector spaces; now we should like to introduce into these sets of functions notions of distance compatible with the notions of sum and product. Loosely speaking, then, *a vector space of functions with a compatible distance function will be called a function space.* We shall make this concept more precise, and we shall examine several very important examples in the remainder of this section.

***Definition 3-35 Function Space*†** *Let $\mathfrak{F}$ be either a subspace of the vector space $\mathfrak{F}(R_m, R_n)$ of all functions from $R_m$ into $R_n$ or a subspace of the vector space $\mathfrak{F}([a, b], R_n)$ of all functions from the interval $[a, b]$ into $R_n$, and let $d$ be a distance on $\mathfrak{F}$ (see Definition 3-1). Then we shall say that $\mathfrak{F}$ is a function space if:*

FS1 *Given elements $\mathbf{f}$ and $\mathbf{g}$ of $\mathfrak{F}$ and sequences $\{\mathbf{f}_k\}$ and $\{\mathbf{g}_k\}$ in $\mathfrak{F}$ which converge to $\mathbf{f}$ and $\mathbf{g}$, respectively [that is, $d(\mathbf{f}_k, \mathbf{f}) \to 0$ and $d(\mathbf{g}_k, \mathbf{g}) \to 0$]; then $\mathbf{f}_k + \mathbf{g}_k$ converges to $\mathbf{f} + \mathbf{g}$, that is,*

$$d(\mathbf{f}_k + \mathbf{g}_k, \mathbf{f} + \mathbf{g}) \to 0 \tag{3-119}$$

FS2 *Given an element $\mathbf{f}$ of $\mathfrak{F}$ and an element $r$ of $R$ and sequences $\{\mathbf{f}_k\}$ in $\mathfrak{F}$ and $\{r_k\}$ in $R$ which converge to $\mathbf{f}$ and $r$, respectively; then the sequence $r_k\mathbf{f}_k$ converges to $r\mathbf{f}$, that is,*

$$d(r_k\mathbf{f}_k, r\mathbf{f}) \to 0 \tag{3-120}$$

† See Refs. S-4 and V-1. Actually, this definition is a special case of the definition of what is known as a topological vector space. The concept of a topological vector space is quite general.

Now let us turn our attention to some important examples of function spaces. We shall begin by considering the set of all bounded functions from the interval $[a, b]$ into $R_n$, where we have:

***Definition 3-36 Bounded Function*** *A function* $\mathbf{f}$ *from* $[a, b]$ *into* $R_n$ *is said to be bounded if there is an* $M$ *in* $R$, $M > 0$, *such that*

$$\|\mathbf{f}(t)\| \leq M \qquad \text{for all } t \text{ in } [a, b] \tag{3-121}$$

We let $\mathcal{B}([a, b], R_n)$ denote this set; that is,

$$\mathcal{B}([a, b], R_n) = \{\mathbf{f}: \mathbf{f} \text{ is a bounded function from } [a, b] \text{ into } R_n\} \tag{3-122}$$

It is easy to see that $\mathcal{B}([a, b], R_n)$ is a vector space which is a subspace of the set $\mathcal{F}([a, b], R_n)$ of all functions from $[a, b]$ into $R_n$. We should now like to introduce a distance into $\mathcal{B}([a, b], R_n)$ which makes it into a function space.

***Definition 3-37 Norm of a Function*** *Let* $\mathbf{f}$ *be an element of* $\mathcal{B}([a, b], R_n)$; *that is,* $\mathbf{f}$ *is a bounded function from* $[a, b]$ *into* $R_n$. *Then the (uniform) norm of* $\mathbf{f}$, *which we write* $\|\mathbf{f}\|$, *is, by definition, the greatest lower bound of the set of numbers* $M$ *such that* $\|\mathbf{f}(t)\| \leq M$ *for all* $t$ *in* $[a, b]$.† *In particular, if* $\mathbf{f}$ *is continuous, then*

$$\|\mathbf{f}\| = \sup_{t \in [a,b]} \|\mathbf{f}(t)\| = \max_{t \in [a,b]} \|\mathbf{f}(t)\| \tag{3-123}$$

*since* $\|\mathbf{f}(t)\|$ *actually attains its maximum value on* $[a, b]$ *(see Theorem 3-7). If* $\mathbf{f}$ *and* $\mathbf{g}$ *are elements of* $\mathcal{B}([a, b], R_n)$, *that is, if both* $\mathbf{f}$ *and* $\mathbf{g}$ *are bounded, then the distance between* $\mathbf{f}$ *and* $\mathbf{g}$, $d(\mathbf{f}, \mathbf{g})$, *is simply the norm of* $\mathbf{f} - \mathbf{g}$; *that is,*

$$d(\mathbf{f}, \mathbf{g}) = \|\mathbf{f} - \mathbf{g}\| = \sup_{t \in [a,b]} \|\mathbf{f}(t) - \mathbf{g}(t)\| \tag{3-124}‡$$

Let us now show that $\mathcal{B}([a, b], R_n)$ with this notion of distance is a function space. We have:

***Theorem 3-10*** *The function* $d$ *given by Eq. (3-124) is indeed a distance on the set* $\mathcal{B}([a, b], R_n)$. *The set* $\mathcal{B}([a, b], R_n)$ *is a function space with respect to this distance.*

PROOF We shall first verify properties D1 to D3 of Definition 3-1 for the function $d$ given by Eq. (3-124). Property D2 is clear, since $\|\mathbf{f}(t) - \mathbf{g}(t)\| = \|\mathbf{g}(t) - \mathbf{f}(t)\|$ for all $t$ in $[a, b]$. As for the triangle inequality D3, we note that if $\mathbf{f}$, $\mathbf{g}$, and $\mathbf{h}$ are in $\mathcal{B}([a, b], R_n)$, then

$$\|\mathbf{f}(t) - \mathbf{h}(t)\| \leq \|\mathbf{f}(t) - \mathbf{g}(t)\| + \|\mathbf{g}(t) - \mathbf{h}(t)\| \tag{3-125}$$

† $\|\mathbf{f}\|$ is often called the supremum of $\|\mathbf{f}(t)\|$ for $t$ in $[a, b]$.

‡ Where $\sup_{t \in [a,b]} \|\mathbf{f}(t) - \mathbf{g}(t)\|$ is to be construed as meaning the greatest lower bound of the set of numbers $M$ such that $\|\mathbf{f}(t) - \mathbf{g}(t)\| \leq M$ for all $t$ in $[a, b]$; that is, $\sup_{t \in [a,b]} \|\mathbf{f}(t) - \mathbf{g}(t)\|$ is the supremum of $\|\mathbf{f}(t) - \mathbf{g}(t)\|$.

for *all* $t$ in $[a, b]$, which implies that

$$\sup_{t\in[a,b]} \|\mathbf{f}(t) - \mathbf{h}(t)\| \le \sup_{t\in[a,b]} \{\|\mathbf{f}(t) - \mathbf{g}(t)\| + \|\mathbf{g}(t) - \mathbf{h}(t)\|\}$$
$$\le \sup_{t\in[a,b]} \|\mathbf{f}(t) - \mathbf{g}(t)\| + \sup_{t\in[a,b]} \|\mathbf{g}(t) - \mathbf{h}(t)\| \quad (3\text{-}126)\dagger$$

It is clear that $d(\mathbf{f}, \mathbf{g}) \ge 0$ and that $d(\mathbf{f}, \mathbf{f}) = 0$. Now, if $d(\mathbf{f}, \mathbf{g}) = 0$, then, for any given $t_0$ in $[a, b]$, we must have $\|\mathbf{f}(t_0) - \mathbf{g}(t_0)\| = 0$, so that $\mathbf{f}(t_0) = \mathbf{g}(t_0)$. Thus, we have shown that $d$ is a distance on $\mathcal{B}([a, b], R_n)$.

Now let us verify property FS1 of Definition 3-35. Suppose that we show that if $\mathbf{f}$ and $\mathbf{g}$ are elements of $\mathcal{B}([a, b], R_n)$, then

$$\|\mathbf{f} + \mathbf{g}\| \le \|\mathbf{f}\| + \|\mathbf{g}\| \quad (3\text{-}127)$$

It will then follow that

$$\|\mathbf{f}_k + \mathbf{g}_k - (\mathbf{f} + \mathbf{g})\| \le \|\mathbf{f}_k - \mathbf{f}\| + \|\mathbf{g}_k - \mathbf{g}\| \quad (3\text{-}128)$$

that is, that

$$d(\mathbf{f}_k + \mathbf{g}_k, \mathbf{f} + \mathbf{g}) \le d(\mathbf{f}_k, \mathbf{f}) + d(\mathbf{g}_k, \mathbf{g}) \quad (3\text{-}129)$$

Clearly, Eq. (3-129) implies that property FS1 is satisfied. Now, as for Eq. (3-127), we have

$$\|\mathbf{f} + \mathbf{g}\| = \sup_{t\in[a,b]} \|(\mathbf{f} + \mathbf{g})(t)\| = \sup_{t\in[a,b]} \|\mathbf{f}(t) + \mathbf{g}(t)\|$$
$$\le \sup_{t\in[a,b]} \|\mathbf{f}(t)\| + \sup_{t\in[a,b]} \|\mathbf{g}(t)\| = \|\mathbf{f}\| + \|\mathbf{g}\| \quad (3\text{-}130)\ddagger$$

We leave it to the reader to verify that property FS2 is satisfied.

We observe that $\mathcal{B}([a, b], R_n)$ is *complete* (see Sec. 3-5) with respect to the distance $d$ of Eq. (3-124). To see this, we note that if $\mathbf{f}_j$ is a Cauchy sequence (see Definition 3-11), then each $\mathbf{f}_j(t)$ is a Cauchy sequence in $R_n$ and so must converge to an element of $R_n$, which we denote by $\mathbf{f}(t)$. If $\epsilon > 0$ is given, then there is an $N(\epsilon/2)$ such that $\|\mathbf{f}_j - \mathbf{f}_k\| < \epsilon/2$ if $j, k > N(\epsilon/2)$. We claim that if $j > N(\epsilon/2)$, then $\|\mathbf{f}_j - \mathbf{f}\| < \epsilon$. But this follows from the relation $\|\mathbf{f}_j(t) - \mathbf{f}(t)\| \le \|\mathbf{f}_j(t) - \mathbf{f}_k(t)\| + \|\mathbf{f}_k(t) - \mathbf{f}(t)\|$, which holds for $t$ in $[a, b]$ [that is, if $t \in [a, b]$, then pick $k(t) > N(\epsilon/2)$ so that $\|\mathbf{f}_{k(t)}(t) - \mathbf{f}(t)\| < \epsilon/2$].

Now let us denote by $\mathcal{C}([a, b], R_n)$ the set of all *continuous* functions from $[a, b]$ into $R_n$; that is,

$$\mathcal{C}([a, b], R_n) = \{\mathbf{f}: \mathbf{f} \text{ is continuous on } [a, b] \text{ into } R_n\} \quad (3\text{-}131)$$

In view of Theorem 3-7, we observe that the set $\mathcal{C}([a, b], R_n)$ is contained in the set $\mathcal{B}([a, b], R_n)$. Moreover, it is actually a subspace. (See Sec. 3-10.)

† The basis for Eq. (3-126) is the following type of argument: If $M$ and $N$ are numbers such that $\|\mathbf{f}(t) - \mathbf{g}(t)\| \le M$ and $\|\mathbf{g}(t) - \mathbf{h}(t)\| \le N$ for all $t$ in $[a, b]$, then $\|\mathbf{f}(t) - \mathbf{g}(t)\| + \|\mathbf{g}(t) - \mathbf{h}(t)\| \le M + N$ for all $t$ in $[a, b]$ and, consequently, $\|\mathbf{f}(t) - \mathbf{h}(t)\| \le M + N$ for all $t$ in $[a, b]$. The reader should bear this in mind whenever we make statements similar to Eq. (3-126).

‡ Compare this with Eq. (3-126).

In other words, *continuous functions defined on a compact interval are bounded.* The distance $d$ given by Eq. (3-124) is, when restricted to $\mathcal{C}([a, b], R_n)$, a distance between continuous functions. Is $\mathcal{C}([a, b], R_n)$ a function space with respect to this distance? Well, we shall show in Theorem 3-11 that if $\mathbf{f}_k$ is a sequence of continuous functions which converges to $\mathbf{f}$, then $\mathbf{f}$ must be continuous. This will show that $\mathcal{C}([a, b], R_n)$ is a function space. In fact, $\mathcal{C}([a, b], R_n)$ is a *complete* function space with respect to this distance.

***Theorem 3-11*** *If $\mathbf{f}_k$ is a sequence of continuous functions which converges to $\mathbf{f}$, then $\mathbf{f}$ is continuous.*

PROOF Suppose that $t_0$ is in $[a, b]$ and that $\{t_m\}$ is a sequence in $[a, b]$ which converges to $t_0$. We shall show that $\mathbf{f}(t_m)$ converges to $\mathbf{f}(t_0)$. Now let $\epsilon > 0$ be given; then there is a $K$ such that $k > K$ implies that

$$\sup_{t \in [a,b]} \|\mathbf{f}_k(t) - \mathbf{f}(t)\| < \frac{\epsilon}{3}$$

Now we note that

$$\|\mathbf{f}(t_m) - \mathbf{f}(t_0)\| \leq \|\mathbf{f}(t_m) - \mathbf{f}_k(t_m)\| + \|\mathbf{f}_k(t_m) - \mathbf{f}_k(t)\| + \|\mathbf{f}_k(t) - \mathbf{f}(t_0)\| \quad (3\text{-}132)$$

Since $\mathbf{f}_k$ is continuous, there is an $M$ such that $m > M$ implies that

$$\|\mathbf{f}_k(t_m) - \mathbf{f}_k(t)\| < \frac{\epsilon}{3} \quad (3\text{-}133)$$

It follows from Eq. (3-121) that, for $m > M$,

$$\|\mathbf{f}(t_m) - \mathbf{f}(t_0)\| < \epsilon \quad (3\text{-}134)$$

and hence that $\mathbf{f}(t_m)$ converges to $\mathbf{f}(t_0)$. This shows that $\mathbf{f}$ is continuous.

**Exercise 3-13** Show that Theorem 3-11 implies that the set of continuous functions $\mathcal{C}([a, b], R_n)$ is a function space. HINT: Make use of Theorems 3-7, 3-11, and 3-10.

Theorem 3-11 is often stated as: *The limit of a uniformly convergent sequence of continuous functions is again continuous.* We observe also that the theorem remains true if "*continuous*" is replaced by "*piecewise continuous*" throughout; that is, we have:

***Theorem 3-11a*** *If $\mathbf{f}_k$ is a sequence of piecewise continuous functions which converges to $\mathbf{f}$ [with respect to the distance of Eq.* (3-124)], *then $\mathbf{f}$ is piecewise continuous.*

This will again imply that the set of *all* piecewise continuous functions from $[a, b]$ into $R_n$ will be a function space, provided that we can show that *every* piecewise continuous function from $[a, b]$ into $R_n$ is bounded. We shall do this by showing that if $\mathbf{f}$ is piecewise continuous, then there is a sequence $\mathbf{s}_m$ of *step functions* (see Sec. 3-14)† for which $\|\mathbf{s}_m - \mathbf{f}\|$ converges to 0. In other words, we have:

***Theorem 3-12*** *If* $\mathbf{f}$ *is a piecewise continuous function from* $[a, b]$ *into* $R_n$, *then there is a sequence* $\mathbf{s}_m$ *of step functions such that, given any* $\epsilon > 0$, *there is an* $M(\epsilon)$ *such that* $m > M(\epsilon)$ *implies that* $\|\mathbf{s}_m(t) - \mathbf{f}(t)\| < \epsilon$ *for all* $t$ *in* $[a, b]$.

PROOF (The idea of the proof is illustrated in Fig. 3-27.) Let $N$ be an integer. Suppose that $t$ is in $[a, b]$ and $t \neq a$ or $b$; then there are elements $\iota(t)$ and $r(t)$ in $[a, b]$ such that if $\sigma, \tau$ are in $(\iota(t), t)$ or in $(t, r(t))$, then $\|\mathbf{f}(\sigma) - \mathbf{f}(\tau)\| \leq 1/N$. This is so because $\mathbf{f}(t+)$ and $\mathbf{f}(t-)$ exist.‡ Similarly, there

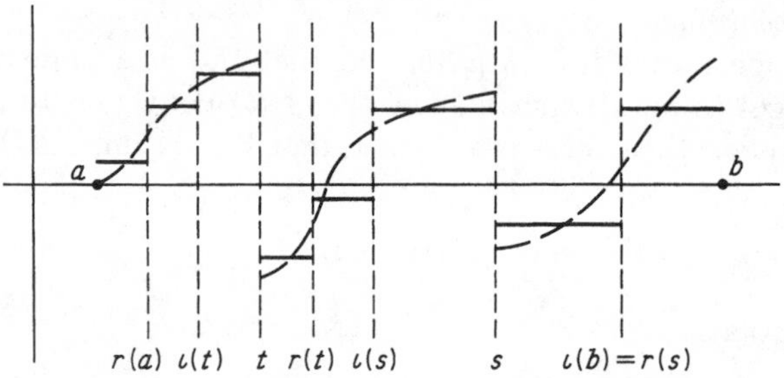


**Fig. 3-27** Illustration for the proof of Theorem 3-12. The dashed curve is the piecewise continuous function $f(t)$, and the solid curve is the approximating step function $s_N(t)$.

are elements $r(a)$ and $\iota(b)$ such that if, for example, $\sigma, \tau$ are in $(a, r(a))$, then $\|\mathbf{f}(\sigma) - \mathbf{f}(\tau)\| \leq 1/N$. Now pick elements $s_1, s_2, \ldots, s_p$ of $[a, b]$ such that every $t$ in $[a, b]$ is in at least one of the intervals $[a, r(a))$, $(\iota(s_1), r(s_1))$, $\ldots$, $(\iota(s_p), r(s_p))$, $(\iota(b), b]$. We may do this in view of the Heine-Borel property of compact sets (see Sec. 3-6, property C2). We now consider the set $\{a, b, s_1, s_2, \ldots, s_p, r(a), \iota(b), \iota(s_1), r(s_1), \ldots, \iota(s_p), r(s_p)\}$, and we define $a = t_0 < t_1 < \cdots < t_n = b$ to be the points of this set, arranged according to increasing size. As $t_i$ is in one of the intervals $[a, r(a))$, $\ldots$, etc., the point $t_{i+1}$ either is in the same interval as $t_i$ or is an $r(s_q)$ for some $q$. It follows that if $\sigma, \tau$ are in $(t_i, t_{i+1})$, then $\|\mathbf{f}(\sigma) - \mathbf{f}(\tau)\| \leq 1/N$. If we define $\mathbf{s}_N$ by

$$
\begin{aligned}
\mathbf{s}_N(t_i) &= \mathbf{f}(t_i) \\
\mathbf{s}_N(t) &= \mathbf{f}\left(\frac{t_i + t_{i+1}}{2}\right) \qquad \text{for } t \text{ in } (t_i, t_{i+1})
\end{aligned}
\tag{3-135}
$$

then the sequence of step functions $\mathbf{s}_N$ for $N = 1, 2, \ldots$ is our desired sequence.

† Recall that $\mathbf{s}$ is a step function if there are elements $t_0, t_1, \ldots, t_n$ of $[a, b]$ with $a = t_0 < t_1 < \cdots < t_n = b$ such that $\mathbf{s}$ is a constant $\mathbf{c}_i$ on each of the open intervals $(t_i, t_{i+1})$.

‡ In other words, $1/N > 0$ implies that there is an $\iota(t)$ (say) such that $\|\mathbf{f}(\tau) - \mathbf{f}(t-)\| \leq 1/2N$ for $\tau$ in $(\iota(t), t)$. It follows from the triangle inequality that $\|\mathbf{f}(\sigma) - \mathbf{f}(\tau)\| \leq \|\mathbf{f}(\sigma) - \mathbf{f}(t-)\| + \|\mathbf{f}(t-) - \mathbf{f}(\tau)\| \leq 1/2N + 1/2N = 1/N$.

We shall denote the function space of all piecewise continuous functions from $[a, b]$ into $R_n$ by $\mathcal{P}([a, b], R_n)$, that is,

$$\mathcal{P}([a, b], R_n) = \{\mathbf{f}: \mathbf{f} \text{ is a piecewise continuous function from } [a, b] \text{ into } R_n\} \tag{3-136}$$

We observe that one important consequence of Theorems 3-11, 3-11*a*, and 3-12 is the fact that the set of continuous functions $\mathcal{C}([a, b], R_n)$ and the set of piecewise continuous functions $\mathcal{P}([a, b], R_n)$ are *closed* subsets of the set of bounded functions $\mathcal{B}([a, b], R_n)$ (see Definition 3-7) and are, therefore, *complete* (see Sec. 3-5), since $\mathcal{B}([a, b], R_n)$ is. Thus, every continuous or piecewise continuous function from $[a, b]$ to $R_n$ is a bounded function. This observation will be of interest to us in Secs. 3-18 and 3-19.

Another important consequence of these theorems and the property of integrals given in Eq. (3-111) is the fact that if $\mathbf{f}_k$ is a sequence of elements of $\mathcal{P}([a, b], R_n)$ which converges to $\mathbf{f}$ [with regard to the distance given by Eq. (3-124)], then the sequence

$$\int_a^b \mathbf{f}_k(\tau)\, d\tau$$

converges to

$$\int_a^b \mathbf{f}(\tau)\, d\tau$$

This, of course, leads to the familiar result about integrating series term by term. In other words, the integral is a continuous linear function on $\mathcal{P}([a, b], R_n)$.

Now we shall consider the *set* of continuous functions $\mathcal{C}([a, b], R_n)$, and we shall introduce a different type of distance function into this set. We have:

***Definition 3-38*** *Let* $\mathbf{f}$ *be an element of* $\mathcal{C}([a, b], R_n)$. *Then the (one) norm of* $\mathbf{f}$, *which we write* $\|\mathbf{f}\|_1$,† *is simply the integral of the Euclidean norm of* $\mathbf{f}(t)$ *over the interval* $[a, b]$, *that is,*

$$\|\mathbf{f}\|_1 = \int_a^b \|\mathbf{f}(\tau)\|\, d\tau \tag{3-137}$$

*If* $\mathbf{f}$ *and* $\mathbf{g}$ *are elements of the set* $\mathcal{C}([a, b], R_n)$, *then the distance (one) between* $\mathbf{f}$ *and* $\mathbf{g}$, $d_1(\mathbf{f}, \mathbf{g})$, *is simply the (one) norm of* $\mathbf{f} - \mathbf{g}$, *that is,*

$$d_1(\mathbf{f}, \mathbf{g}) = \|\mathbf{f} - \mathbf{g}\|_1 = \int_a^b \|\mathbf{f}(\tau) - \mathbf{g}(\tau)\|\, d\tau \tag{3-138}$$

It is easy to show that Eq. (3-138) does indeed define a distance on $\mathcal{C}([a, b], R_n)$ and that $\mathcal{C}([a, b], R_n)$ becomes a function space with respect to this notion of distance.

**Exercise 3-14** Prove that (Eq. 3-138) defines a distance on $\mathcal{C}([0, 1], R)$ and that $\mathcal{C}([0, 1], R)$ is a function space with respect to this distance. HINT: Make use of Theorem 3-8; in particular, properties *c* [Eq. (3-108)] and *f* should be useful.

† The subscript 1 distinguishes this notion of norm from that of Definition 3-37.

Now Eq. (3-137) also has meaning if **f** is only piecewise continuous on $[a, b]$; that is,

$$\int_a^b \|\mathbf{f}(\tau)\| \, d\tau$$

exists. However, there are piecewise continuous functions **g**, which are not zero, such that

$$\int_a^b \|\mathbf{g}(\tau)\| \, d\tau = 0$$

This would prevent us from defining a distance on $\mathcal{P}([a, b], R_n)$ by means of Eq. (3-138), as property D1 of Definition 3-1 would be violated. In order to get around this difficulty, we shall say two piecewise continuous functions **f** and **g** are "similar," or *equivalent* (or agree almost everywhere), if the set $A$ of points $t$ in $[a, b]$ such that $\mathbf{f}(t) \neq \mathbf{g}(t)$ is countable.† It is easy to see that **f** and **g** are equivalent if and only if

$$\int_a^b \|\mathbf{f}(\tau) - \mathbf{g}(\tau)\| \, d\tau = 0$$

Let us denote for the moment by $[\mathbf{f}]$ the set of *all* **h** in $\mathcal{P}([a, b], R_n)$ which are "similar" to **f**, that is,

$$[\mathbf{f}] = \{\mathbf{h} \colon \mathbf{f} \text{ is similar to } \mathbf{h}\} \tag{3-139}$$

If we now define a function $d_1$ by setting

$$d_1([\mathbf{f}], [\mathbf{g}]) = \int_a^b \|\mathbf{h}(\tau) - \mathbf{k}(\tau)\| \, d\tau \tag{3-140}$$

where **h** is any element of $[\mathbf{f}]$ and **k** is any element of $[\mathbf{g}]$, then it is easy to see that $d_1$ is a distance on the set of *all* $[\mathbf{f}]$. Although we should, strictly speaking, use Eq. (3-140) to define $d_1$ and should consider the set of $[\mathbf{f}]$, we shall in the remainder of the book gloss over this point and shall consider Eq. (3-138) as defining the distance $d_1$ on $\mathcal{P}([a, b], R_n)$. Bearing this in mind, we shall view $\mathcal{P}([a, b], R_n)$ as a function space with respect to $d_1$.

We note that $\mathcal{C}([a, b], R_n)$ and $\mathcal{P}([a, b], R_n)$ are *not* complete (see Sec. 3-5) with respect to the distance $d_1$.

Once again, we turn our attention to the set of continuous functions $\mathcal{C}([a, b], R_n)$, and we shall introduce still another distance on this set. We have:

***Definition 3-39*** *Let* **f** *and* **g** *be elements of* $\mathcal{C}([a, b], R_n)$. *Then the scalar (or inner) product of* **f** *and* **g**, *which is denoted by* $[\mathbf{f}, \mathbf{g}]$, *is given by*

$$[\mathbf{f}, \mathbf{g}] = \int_a^b \langle \mathbf{f}(\tau), \mathbf{g}(\tau) \rangle \, d\tau \tag{3-141}$$

*where* $\langle \mathbf{f}(\tau), \mathbf{g}(\tau) \rangle$ *is the usual scalar product on* $R_n$ *[see Eq. (2-86)].*

We shall now show that this is indeed a meaningful definition and that the inner product $[\mathbf{f}, \mathbf{g}]$ is actually a positive definite inner product on the vector space $\mathcal{C}([a, b], R_n)$ in the sense of Sec. 2-11.

† See Sec. 2-2.

***Theorem 3-13*** $[\mathbf{f}, \mathbf{g}]$ *is a positive definite inner product on* $\mathcal{C}([a, b], R_n)$.

PROOF (See Sec. 2-11.) Properties IP1 to IP3 for $[\mathbf{f}, \mathbf{g}]$ follow readily from the corresponding properties of the scalar product $\langle\ ,\ \rangle$. For example, let us verify property IP2. If $r \in R$ and $\mathbf{f}$, $\mathbf{g}$ are in $\mathcal{C}([a, b], R_n)$, then

$$[r\mathbf{f}, \mathbf{g}] = \int_a^b \langle r\mathbf{f}(\tau), \mathbf{g}(\tau)\rangle\, d\tau = \int_a^b r\langle \mathbf{f}(\tau), \mathbf{g}(\tau)\rangle\, d\tau$$
$$= r\int_a^b \langle \mathbf{f}(\tau), \mathbf{g}(\tau)\rangle\, d\tau = r[\mathbf{f}, \mathbf{g}] \tag{3-142}$$

Now let us consider $[\mathbf{f}, \mathbf{f}]$ [see Eq. (2-84)] and show that it is $\geq 0$. Well,

$$[\mathbf{f}, \mathbf{f}] = \int_a^b \langle \mathbf{f}(\tau), \mathbf{f}(\tau)\rangle\, d\tau \tag{3-143}$$

But $\langle \mathbf{f}(\tau), \mathbf{f}(\tau)\rangle \geq 0$ for every $\tau$ in $[a, b]$, and hence, using Theorem 3-8*c*,

$$[\mathbf{f}, \mathbf{f}] = \int_a^b \langle \mathbf{f}(\tau), \mathbf{f}(\tau)\rangle\, d\tau = \int_a^b |\langle \mathbf{f}(\tau), \mathbf{f}(\tau)\rangle|\, d\tau$$
$$\geq \left| \int_a^b \langle \mathbf{f}(\tau), \mathbf{f}(\tau)\rangle\, d\tau \right| \geq 0 \tag{3-144}$$

This shows that $[\mathbf{f}, \mathbf{f}]$ is nonnegative.

Finally, suppose that $\mathbf{f}_0$ is an element of $\mathcal{C}([a, b], R_n)$ such that $[\mathbf{f}, \mathbf{f}_0] = 0$ for every $\mathbf{f}$ in $\mathcal{C}([a, b], R_n)$. In particular, $[\mathbf{f}_0, \mathbf{f}_0] = 0$. But

$$[\mathbf{f}_0, \mathbf{f}_0] = \int_a^b \langle \mathbf{f}_0(\tau), \mathbf{f}_0(\tau)\rangle\, d\tau = \int_a^b \|\mathbf{f}_0(\tau)\|^2\, d\tau = 0 \tag{3-145}$$

and it follows from Theorem 3-8*f* that $\|\mathbf{f}_0(\tau)\|^2 = 0$ for *all* $\tau$ in $[a, b]$. Clearly, then, $\mathbf{f}_0 = \mathbf{0}$ and the theorem is established.

In view of this theorem, we may define a notion of norm as in Sec. 2-13 [Eq. (2-94)] for elements of $\mathcal{C}([a, b], R_n)$. We shall call this the (two) *norm*, and we shall write $\|\mathbf{f}\|_2$ for this norm. In other words,

$$\|\mathbf{f}\|_2 = \sqrt{[\mathbf{f}, \mathbf{f}]} = \left[\int_a^b \langle \mathbf{f}(\tau), \mathbf{f}(\tau)\rangle\, d\tau\right]^{\frac{1}{2}} = \left[\int_a^b \|\mathbf{f}(\tau)\|^2\, d\tau\right]^{\frac{1}{2}} \tag{3-146}$$

We observe also that the Schwarz inequality [Eq. (2-87)] is valid for the inner product $[\mathbf{f}, \mathbf{g}]$, and so we have

$$|[\mathbf{f}, \mathbf{g}]| \leq \|\mathbf{f}\|_2 \|\mathbf{g}\|_2 \tag{3-147}$$

or, equivalently,

$$\left| \int_a^b \langle \mathbf{f}(\tau), \mathbf{g}(\tau)\rangle\, d\tau \right| \leq \left[\int_a^b \langle \mathbf{f}(\tau), \mathbf{f}(\tau)\rangle\, d\tau\right]^{\frac{1}{2}} \left[\int_a^b \langle \mathbf{g}(\tau), \mathbf{g}(\tau)\rangle\, d\tau\right]^{\frac{1}{2}} \tag{3-148}$$

We shall use Eq. (3-148) in Chap. 10.

We note that this notion of norm allows us to define a distance $d_2$ in $\mathcal{C}([a, b], R_n)$ [compare with Eq. (3-1)] with

$$d_2(\mathbf{f}, \mathbf{g}) = \|\mathbf{f} - \mathbf{g}\|_2 = \sqrt{[\mathbf{f} - \mathbf{g}, \mathbf{f} - \mathbf{g}]} = \left[\int_a^b \langle \mathbf{f}(\tau) - \mathbf{g}(\tau), \mathbf{f}(\tau) - \mathbf{g}(\tau)\rangle\, d\tau\right]^{\frac{1}{2}}$$

We leave it to the reader to verify that $\mathcal{C}([a, b], R_n)$ is a function space with respect to this notion of distance. We observe that $\mathcal{C}([a, b], R_n)$ is *not* complete with respect to the distance $d_2$. We also note that this notion of distance can be applied to the set of piecewise continuous functions where we agree to view "similar" functions as being the same [compare with Eqs. (3-139) and (3-140)].

**Example 3-43** Let $f(t) = \sin t$ for $t$ in $[0, \pi]$. Then $f(t)$ is an element of $\mathcal{C}([0, \pi], R)$, and we have

$$\|f\| = \sup_{t \in [0,\pi]} |\sin t| = 1$$

$$\|f\|_1 = \int_0^\pi |\sin t| \, dt = 2$$

$$\|f\|_2 = \sqrt{\int_0^\pi \sin^2 t \, dt} = \sqrt{\frac{\pi}{2}}$$

We note that $\|f\| < \|f\|_2 < \|f\|_1$.

**Example 3-44** Let $f(t) = 1$ for $t$ in $[0, 1]$ and let $g(t)$ be the "spike" (see Fig. 3-28) given by

$$g(t) = \begin{cases} 8t & 0 \le t \le \frac{1}{4} \\ 4 - 8t & \frac{1}{4} \le t \le \frac{1}{2} \\ 0 & \frac{1}{2} \le t \le 1 \end{cases}$$

Then $f(t)$ and $g(t)$ are elements of $\mathcal{C}([0, 1], R)$, and we have

$$\begin{array}{lll} \|f\| = 1 & \|g\| = 2 & \|f - g\| = 1 \\ \|f\|_1 = 1 & \|g\|_1 = \frac{1}{2} & \|f - g\|_1 = \frac{3}{4} \\ \|f\|_2 = 1 & \|g\|_2 = \sqrt{\frac{2}{3}} & \|f - g\|_2 = \sqrt{\frac{2}{3}} \end{array}$$

We observe that $\|f - g\|_2 < \|f - g\|_1 < \|f - g\|$ and that $\|f\| < \|g\|$ but $\|g\|_1 < \|f\|_1$.

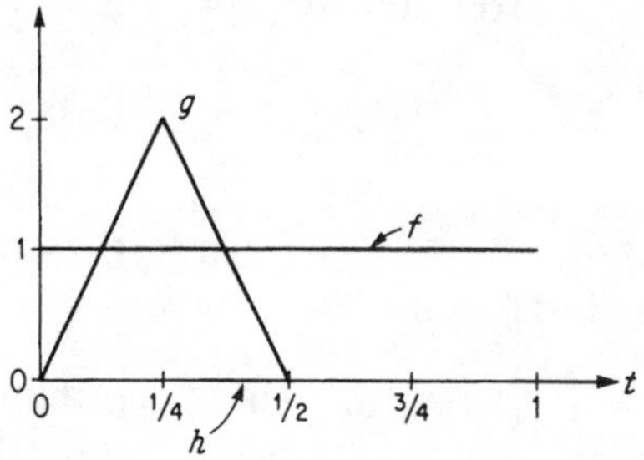


**Fig. 3-28** The functions $f$, $g$, and $h$ of Example 3-44. Note that $h(t) = 0$ for all $t$ in $[0, 1]$.

We let $h(t) = 0$ for all $t$ in $[0, 1]$. Now let us use the notion of the distance between elements of a function space to determine which of the functions $f$ and $g$ is closer to $h$. If we use the distance $d$, then

$$d(f, h) = \|f\| = 1 \qquad \text{and} \qquad d(g, h) = \|g\| = 2$$

so that $f$ is closer to $h$ than $g$ is. On the other hand, if we use the distance $d_1$, then

$$d_1(f, h) = \|f\|_1 = 1 \qquad \text{and} \qquad d_1(g, h) = \|g\|_1 = \tfrac{1}{2}$$

so that, with respect to this distance, $g$ is closer to $h$ than $f$ is. Similarly, $g$ is closer to $h$ than $f$ is with respect to the distance $d_2$. Thus, we can see that closeness of functions depends very much upon the distance used.

## 3-16 Vector Functions: Functionals

Loosely speaking, a functional is a real-valued function on a vector space (usually a function space). We shall, after a brief digression, give a somewhat more restrictive and precise definition of this notion.

Suppose that $L$ is a real-valued function on $R_N$. If $\mathbf{y}$ is an element of $R_N$ with components $y_1, y_2, \ldots, y_N$, then

$$L(\mathbf{y}) = L(y_1, y_2, \ldots, y_N) \tag{3-149}$$

Now, if $\mathbf{f}(t)$ is a function from $R$ into $R_N$ with components $f_1(t), f_2(t), \ldots, f_N(t)$, then $L[\mathbf{f}(t)]$ is a function from $R$ into $R$ (that is, is a function of $t$) and

$$L[\mathbf{f}(t)] = L[f_1(t), f_2(t), \ldots, f_N(t)] \tag{3-150}$$

If we suppose, for example, that $L$ is differentiable with respect to $y_i$ [see (3-77)], then $\partial L/\partial y_i$ is a function from $R_N$ into $R$, and so $(\partial L/\partial y_i)[\mathbf{f}(t)]$ is a function from $R$ into $R$, which we shall denote by $\partial L/\partial f_i(t)$. In other words,

$$\frac{\partial L}{\partial f_i(t)}[\mathbf{f}(t)] = \frac{\partial L}{\partial y_i}[\mathbf{f}(t)] = \frac{\partial L}{\partial y_i}[f_1(t), f_2(t), \ldots, f_N(t)] = (D_iL)[\mathbf{f}(t)] \tag{3-151}$$

We shall often be interested in determining the minimum (or maximum) of a real-valued function of the type

$$J(\mathbf{f}) = \int_a^b L[\mathbf{f}(t)]\, dt = \int_a^b L[f_1(t), f_2(t), \ldots, f_N(t)]\, dt \tag{3-152}$$

where $L$ is a *continuous* function from $R_N$ into $R$, $\mathbf{f}$ is a continuous or piecewise continuous function from $[a, b]$ into $R_N$, and $f_1, f_2, \ldots, f_N$ are the components of $\mathbf{f}$. We note that $J$ is a function from $\mathcal{P}([a, b], R_N)$ [see Eq. (3-136)] into $R$. We also observe that if $T \in (a, b]$, then we may define a real-valued function $J(T, \mathbf{f})$ by setting

$$J(T, \mathbf{f}) = \int_a^T L[\mathbf{f}(t)]\, dt = \int_a^T L[f_1(t), f_2(t), \ldots, f_N(t)]\, dt \tag{3-153}$$

We shall call $J(T, \mathbf{f})$ a *functional.*

Now, if we suppose that $L$ is *differentiable with respect to* $y_i$ and that the *derivative* of $L$ with respect to $y_i$, $\partial L/\partial y_i$ [see (3-77)], is *a continuous function* from $R_N$ into $R$, then we may define a new functional $\partial J(T, \mathbf{f})/\partial f_i(t)$ by setting [see Eq. (3-151)]

$$\frac{\partial J(T, \mathbf{f})}{\partial f_i(t)} = \int_a^T \frac{\partial L}{\partial f_i(t)}[\mathbf{f}(t)]\, dt = \int_a^T \frac{\partial L}{\partial f_i(t)}[f_1(t), f_2(t), \ldots, f_N(t)]\, dt \tag{3-154}$$

We call $\partial J(T, \mathbf{f})/\partial f_i(t)$ the *partial derivative of* $J(T, \mathbf{f})$ *with respect to* $f_i(t)$. We note that Eq. (3-154) will be used quite frequently in the sequel.

Finally, we observe that, for each fixed $\mathbf{f}$, the function $J(T, \mathbf{f})$ of $T$ has a derivative at all points except those of a countable subset $A$ of $(a, b]$. If $T$ is not in $A$, then the *derivative of* $J(T, \mathbf{f})$ *with respect to* $T$, $\partial J(T, \mathbf{f})/\partial T$, is given by

$$\frac{\partial J(T, \mathbf{f})}{\partial T} = L[\mathbf{f}(T)] = L[f_1(T), f_2(T), \ldots, f_N(T)] \tag{3-155}$$

In particular, if $\mathbf{f}$ is actually continuous, then

$$\frac{\partial J(T, \mathbf{f})}{\partial T} = L[\mathbf{f}(T)] \qquad \text{for all } T \text{ in } [a, b] \tag{3-156}$$

We shall often use Eq. (3-155) in the rest of the book.

**Example 3-45** Suppose that $L(\mathbf{y}) = 1$ for all $\mathbf{y}$ in $R_N$; then $L$ is continuous and $(\partial L/\partial y_i)(\mathbf{y}) = 0$ for $i = 1, 2, \ldots, N$ and $\mathbf{y}$ in $R_N$. If $\mathbf{f}$ is in $\mathcal{P}([a, b], R_N)$, then

$$J(T, \mathbf{f}) = \int_a^T 1\, dt = T - a$$

and $\partial J(T, \mathbf{f})/\partial T = 1$.

**Example 3-46** Suppose that

$$L(\mathbf{y}) = \sum_{i=1}^{N} c_i |y_i|$$

where $c_1, c_2, \ldots, c_N$ are positive constants; then $L$ is continuous and

$$L[\mathbf{f}(t)] = \sum_{i=1}^{N} c_i |f_i(t)|$$

for $\mathbf{f}$ in $\mathcal{P}([a, b], R_N)$. It follows that

$$J(T, \mathbf{f}) = \int_a^T \sum_{i=1}^{N} c_i |f_i(t)|\, dt = \sum_{i=1}^{N} c_i \int_a^T |f_i(t)|\, dt$$

and that
$$\frac{\partial J(T, \mathbf{f})}{\partial T} = \sum_{i=1}^{N} c_i |f_i(T)|$$

We note that $\partial J(T, \mathbf{f})/\partial f_i(t)$ is not, in general, defined.

**Example 3-47** Suppose that

$$L(\mathbf{y}) = \tfrac{1}{2} \langle \mathbf{y}, \mathbf{y} \rangle = \tfrac{1}{2} \sum_{i=1}^{N} y_i^2$$

Then $L$ is continuous and

$$L[\mathbf{f}(t)] = \tfrac{1}{2} \langle \mathbf{f}(t), \mathbf{f}(t) \rangle = \tfrac{1}{2} \sum_{i=1}^{N} f_i(t)^2$$

for $\mathbf{f}$ in $\mathcal{P}([a, b], R_N)$. It follows that

$$J(T, \mathbf{f}) = \tfrac{1}{2} \int_a^T \sum_{i=1}^{N} f_i(t)^2\, dt$$

$$\frac{\partial J(T, \mathbf{f})}{\partial T} = \tfrac{1}{2} \sum_{i=1}^{N} f_i(T)^2$$

and $\partial J(T, \mathbf{f})/\partial f_i(T) = f_i(T)$.

## 3-17 Differential Equations: Introductory Remarks

We are going to study ordinary differential equations in the remaining sections of this chapter. Since the behavior of all the physical systems which we shall later examine is described by systems of differential equations, the topics treated in these sections are of extreme importance and should be carefully scrutinized by the reader. In this introductory section, we shall define the concept of a system of ordinary differential equations, we shall say what we mean by a solution of the system, we shall discuss initial and boundary conditions, and we shall examine the role of parameters in the system. In subsequent sections, we shall prove a basic existence theorem and shall study linear systems in great detail.

Loosely speaking, a differential equation is an equation involving derivatives of the unknown functions. For example, the equations

$$\dot{x}(t) = -x(t) \tag{3-157}$$

and

$$\frac{\partial F(x, t)}{\partial t} = 1 + \frac{\partial F(x, t)}{\partial x} \tag{3-158}$$

are differential equations. In Eq. (3-157), the unknown function $x$ is a function of *one* real variable, and this equation is called an *ordinary differential equation.* In Eq. (3-158), the unknown function $F$ is a function of several variables, and this equation is called a *partial differential equation. We shall use the expression* "differential equation" *in place of the expression* "ordinary differential equation" *throughout the remainder of this book.* Now let us be more precise about the notion of a differential equation which we shall use in this book. We have:

***Definition 3-40*** *Let $F$ be a continuous† function from an open subset $A$ of $R_{n+2}$ (see Definition 3-5) into $R$. Then the equation*

$$F[x(t), \dot{x}(t), \ldots, x^{(n)}(t), t] = 0 \tag{3-159}$$

*is called an nth-order differential equation.‡ A real-valued function $\psi(t)$ whose domain is an open interval $(t_1, t_2)$ is called a solution of the differential equation (3-159) if:*

*a. $\psi(t)$ is continuous on the interval $(t_1, t_2)$.*

*b. The point*

$$\begin{bmatrix} \psi(t) \\ \dot{\psi}(t) \\ \cdot \\ \cdot \\ \cdot \\ \psi^{(n)}(t) \\ t \end{bmatrix}$$

*is in $A$ for $t$ in $(t_1, t_2)$.*

*c.* $$F[\psi(t), \dot{\psi}(t), \ldots, \psi^{(n)}(t), t] \equiv 0 \text{ for } t \text{ in } (t_1, t_2) \tag{3-160}$$

† Although weaker assumptions than this are usually made, continuity will suffice for our purposes.

‡ $x^{(n)}(t)$ is the $n$th derivative $dx^n/dt^n$ of $x(t)$.

If Eq. (3-159) may be written in the form

$$x^{(n)}(t) - G[x(t), \dot{x}(t), \ldots, x^{(n-1)}(t), t] = 0 \tag{3-161a}$$

or, equivalently,

$$x^{(n)}(t) = G[x(t), \dot{x}(t), \ldots, x^{(n-1)}(t), t] \tag{3-161b}$$

then we shall say that the equation is in *explicit form.* We shall always consider equations in explicit form in the sequel.

We observe that the order of the differential equation corresponds to the order of the highest derivative which occurs in the equation. We also note that solutions of the equation are associated with an interval of definition. For example, the function $ce^{-t}$ is a solution of Eq. (3-157) on *any* interval $(t_1, t_2)$.

**Example 3-48** Let $F$ be the function from $R_3$ into $R$, given by $F(y_1, y_2, y_3) = y_2 - 1/y_3$. Then $F$ is continuous on the open set

$$A^+ = \left\{ \begin{bmatrix} y_1 \\ y_2 \\ y_3 \end{bmatrix} : y_3 > 0 \right\}$$

and on the open set

$$A^- = \left\{ \begin{bmatrix} y_1 \\ y_2 \\ y_3 \end{bmatrix} : y_3 < 0 \right\}$$

We let $A = A^+ \cup A^-$; then $F$ is continuous on $A$ and we may consider the differential equation $\dot{x}(t) - 1/t = 0$. We note that if $t > 0$, then $x(t) = \log t + c$ ($c$ a constant) is a solution of the differential equation, and that if $t < 0$, then $x(t) = \log(-t) + c$ ($c$ a constant) is a solution of the differential equation.

Now let us define the concept of a system of first-order differential equations that will be used throughout this book.†

***Definition 3-41*** *Let $f_1, f_2, \ldots, f_n$ be continuous functions from the product of an open set $A_n$ of $R_n$ with an open interval $(T_1, T_2)$ (which may be all of $R$ or a set $\{t: t < T_2\}$, etc.), that is, from $A_n \times (T_1, T_2)$ [see Eq. (2-5)]‡ into $R$, and suppose that the partial derivatives*

$$\frac{\partial f_i(x_1, x_2, \ldots, x_n, t)}{\partial x_j} \qquad i, j = 1, 2, \ldots, n \tag{3-162}$$

*where $x_1, x_2, \ldots, x_n$ are the coordinates in $R_n$, are continuous functions from $A_n \times (T_1, T_2)$ into $R$. The system of equations*

$$\begin{aligned} \dot{x}_1(t) &= f_1[x_1(t), x_2(t), \ldots, x_n(t), t] \\ \dot{x}_2(t) &= f_2[x_1(t), x_2(t), \ldots, x_n(t), t] \\ &\cdots\cdots\cdots\cdots\cdots\cdots \\ \dot{x}_n(t) &= f_n[x_1(t), x_2(t), \ldots, x_n(t), t] \end{aligned} \tag{3-163}$$

† Again, greater generality is possible but will not be needed for our purposes.
‡ $A_n \times (T_1, T_2)$ may be viewed as a subset (open) of $R_{n+1}$.

*will be called a system of n first-order differential equations. The $x_i$ are often called dependent variables, and t is often called the independent variable. We often write Eqs.* (3-163) *in vector form as*

$$\dot{\mathbf{x}}(t) = \mathbf{f}[\mathbf{x}(t), t] \tag{3-164}$$

*where*

$$\mathbf{x}(t) = \begin{bmatrix} x_1(t) \\ x_2(t) \\ \cdot \\ \cdot \\ \cdot \\ x_n(t) \end{bmatrix} \quad \text{and} \quad \mathbf{f}[\mathbf{x}(t), t] = \begin{bmatrix} f_1[x_1(t), x_2(t), \ldots, x_n(t), t] \\ f_2[x_1(t), x_2(t), \ldots, x_n(t), t] \\ \cdots\cdots\cdots\cdots \\ f_n[x_1(t), x_2(t), \ldots, x_n(t), t] \end{bmatrix} \tag{3-165}$$

*An $R_n$-valued function $\boldsymbol{\psi}(t)$, defined on a subinterval $(t_1, t_2)$ of $(T_1, T_2)$, is called a solution of the system if:*

*a. $\boldsymbol{\psi}(t)$ is continuous on $(t_1, t_2)$.*

*b.*

$$\boldsymbol{\psi}(t) = \begin{bmatrix} \psi_1(t) \\ \psi_2(t) \\ \cdot \\ \cdot \\ \cdot \\ \psi_n(t) \end{bmatrix}$$

*is in $A_n$ for all t in $(t_1, t_2)$.*

*c. $\dot{\boldsymbol{\psi}}(t) = \mathbf{f}[\boldsymbol{\psi}(t), t]$ for t in $(t_1, t_2)$, except possibly for the elements of some countable subset A of $(t_1, t_2)$, that is,*

$$\begin{aligned} \dot{\psi}_1(t) &= f_1[\psi_1(t), \psi_2(t), \ldots, \psi_n(t), t] \\ \dot{\psi}_2(t) &= f_2[\psi_1(t), \psi_2(t), \ldots, \psi_n(t), t] \\ &\cdots\cdots\cdots\cdots \\ \dot{\psi}_n(t) &= f_n[\psi_1(t), \psi_2(t), \ldots, \psi_n(t), t] \end{aligned} \tag{3-166}$$

Systems of $n$ first-order differential equations will occur again and again in the main body of the book and represent a wide class of physical systems. We note that solutions of such systems correspond to functions from $R$ (or an interval) into $R_n$ and, as such, may be thought of as being represented by *trajectories* in the space $R_n$. We shall often "eliminate" $t$ in describing these trajectories; i.e., we shall write the equations of the trajectories in terms of the variables $x_i$ alone.

**Example 3-49** Let $f_1(x_1, x_2, t) = x_2$ and $f_2(x_1, x_2, t) = -x_1$. Then $f_1$ and $f_2$ are continuous on $R_n \times R$ $(= R_{n+1})$, as are $\partial f_1/\partial x_1 = 0$, $\partial f_1/\partial x_2 = 1$, $\partial f_2/\partial x_1 = -1$, and $\partial f_2/\partial x_2 = 0$. Thus, we have a system of two first-order differential equations

$$\dot{x}_1 = x_2 \qquad \dot{x}_2 = -x_1\dagger$$

† We shall often drop the "$t$" when writing the system equations.

The function $\boldsymbol{\psi}(t)$ with components $\psi_1(t) = c_1 \cos t + c_2 \sin t$, $\psi_2(t) = -c_1 \sin t + c_2 \cos t$, where $c_1$, $c_2$ are constants, is easily seen to be a solution of the system. We note that if $\boldsymbol{\Psi}(t)$ is the $2 \times 2$ matrix given by

$$\boldsymbol{\Psi}(t) = \begin{bmatrix} \cos t & \sin t \\ -\sin t & \cos t \end{bmatrix}$$

then $\mathbf{x}(t) = \boldsymbol{\Psi}(t) \begin{bmatrix} c_1 \\ c_2 \end{bmatrix}$ is the solution of the system. We also note that $\boldsymbol{\Psi}(t)$ is an orthogonal matrix (see Example 2-30) for every $t$. The trajectories of the system are circles in the $x_1x_2$ plane, whose equations may be written as $x_1^2 + x_2^2 = c_1^2 + c_2^2$.

In studying differential equations which describe the behavior of physical systems, we usually seek solutions which satisfy additional "boundary conditions." For example, Eq. (3-157) may be viewed as representing the radioactive decay of a substance, with $x(t)$ being the amount of material left after $t$ seconds. If $x(0) = C$ is the amount of material at the start, then we seek a solution $\psi(t)$ of Eq. (3-157) for which $\psi(0) = C$. In other words, the solution must satisfy an *initial condition* [$\psi(0) = C$]. More generally, we have:

***Definition 3-42*** *Suppose that* $\dot{\mathbf{x}}(t) = \mathbf{f}[\mathbf{x}(t), t]$ *is a system of n first-order differential equations (see Definition* 3-41), *with* $\mathbf{f}$ *defined on the set* $A_n \times (T_1, T_2)$. *Let* $(\mathbf{x}_0, t_0)$ *be a point in* $A_n \times (T_1, T_2)$. *We call* $(\mathbf{x}_0, t_0)$ *an initial point, and we call the numbers* $x_{10}, x_{20}, \ldots, x_{n0}$, *which are the components of* $\mathbf{x}_0$, *initial values. We say that the relations*

$$x_i(t_0) = x_{i0} \qquad i = 1, 2, \ldots, n \tag{3-167}$$

*are initial conditions. A solution* $\boldsymbol{\psi}(t)$ *of the system is said to satisfy the initial conditions or be a solution of the initial-value problem if*

$$\boldsymbol{\psi}(t_0) = \mathbf{x}_0 \tag{3-168}$$

*that is, if*

$$\psi_i(t_0) = x_{i0} \qquad i = 1, 2, \ldots, n \tag{3-169}$$

We shall prove in the next section that solutions to the initial-value problem exist under the assumptions we have made.

**Example 3-50** Consider the system of Example 3-49: $\dot{x}_1 = x_2$, $\dot{x}_2 = -x_1$. Let $\left(\begin{bmatrix} 1 \\ 0 \end{bmatrix}, 0\right)$ be an initial point, so that 1, 0 are the initial values. Then

$$\boldsymbol{\Psi}(t) = \begin{bmatrix} \cos t & \sin t \\ -\sin t & \cos t \end{bmatrix} \begin{bmatrix} 1 \\ 0 \end{bmatrix} = \begin{bmatrix} \cos t \\ -\sin t \end{bmatrix}$$

solves the initial-value problem.

We observe that there are $n$ initial conditions and that all these conditions must be satisfied at the single "time" $t_0$. In general, $n$ conditions are required to specify the solution of a system of $n$ first-order differential equations; we shall call these $n$ conditions *boundary conditions*. For example, suppose we have a system $\dot{\mathbf{x}}(t) = \mathbf{f}[\mathbf{x}(t), t]$ with $\mathbf{f}$ defined on $R_{n+1}$ and we require a solution $\boldsymbol{\psi}(t)$ such that $\psi_1(t_1) = x_{10}$, $\psi_2(t_2) = x_{20}$, $\ldots$, $\psi_n(t_n) =$

$x_{n0}$, where $t_1, t_2, \ldots, t_n$ are in $R$ and the $x_i$ are given; then the relations $x_i(t_i) = x_{i0}$ form the set of boundary conditions.

**Example 3-51** Again consider the system $\dot{x}_1 = x_2$, $\dot{x}_2 = -x_1$ of Examples 3-49 and 3-50. Then $x_1(0) = 1$, $x_2(\pi) = 0$ is a set of boundary conditions and $\psi_1(t) = \cos t$, $\psi_2(t) = -\sin t$ is a solution of the system satisfying these boundary conditions.

Now, in our later work we shall be concerned with systems which depend upon parameters; in other words, we consider systems of the form

$$\dot{\mathbf{x}}(t) = \mathbf{f}[\mathbf{x}(t), \mathbf{u}, t] \tag{3-170}$$

or, in components,

$$\dot{x}_i(t) = f_i[x_1(t), x_2(t), \ldots, x_n(t), u_1, u_2, \ldots, u_m, t] \tag{3-171}$$

where $\mathbf{f}$ is a continuous function on a set $A_n \times B_m \times (T_1, T_2)$, where $A_n$ and $B_m$ are open sets in $R_n$ and $R_m$, respectively, and where all the partial derivatives

$$\frac{\partial f_i}{\partial x_j} \quad \text{and} \quad \frac{\partial f_i}{\partial u_k} \tag{3-172}$$

are also continuous on $A_n \times B_m \times (T_1, T_2)$ (viewed as a subset of $R_{n+1+m}$). We call the system of Eq. (3-170) a *system of n first-order differential equations depending on the parameters* $u_1, u_2, \ldots, u_m$. Solutions of such a system, $\boldsymbol{\psi}(t, \mathbf{u})$, depend upon the parameters $\mathbf{u}$ and satisfy the equation

$$\dot{\boldsymbol{\psi}}(t, \mathbf{u}) = \mathbf{f}[\boldsymbol{\psi}(t, \mathbf{u}), \mathbf{u}, t] \tag{3-173}$$

We shall have more to say about such systems in the next section, and we shall indicate their importance in control in Chap. 4.

**Example 3-52** Let us consider the system $\dot{x}_1(t) = x_2(t)$, $\dot{x}_2(t) = -x_1(t) + u$, where $u$ is a parameter. In other words, $f_1(x_1, x_2, t, u) = x_2$ and $f_2(x_1, x_2, t, u) = -x_1 + u$; clearly, $f_1, f_2$ are continuous and have the required continuous partial derivatives. The vector-valued function $\boldsymbol{\psi}(t, u)$, given by

$$\boldsymbol{\psi}(t, u) = \begin{bmatrix} \cos t & \sin t \\ -\sin t & \cos t \end{bmatrix} \begin{bmatrix} c_1 + \int_0^t -(\sin \tau)u \, d\tau \\ c_2 + \int_0^t (\cos \tau)u \, d\tau \end{bmatrix}$$

is readily seen to be a solution of the system.

## 3-18 Differential Equations: The Basic Existence Theorem

We are going to prove in this section that solutions to the initial-value problem for systems of $n$ first-order equations exist and that they are unique. We shall first show that the system may be replaced by a system of integral equations; then we shall show how to apply Theorem 3-2 to obtain the desired result. The proof of the theorem is fairly difficult, and we wish to stress that, for our purposes, *it is very important that the reader understand the statement of the theorem but not essential that he follow the proof.* We shall begin by carefully stating the theorem; then we shall prove it. We shall then present an example illustrating the proof, and finally we shall comment on the theorem.

***Theorem 3-14*†** *Let $A_n$ be an open set in $R_n$ and let $(T_1, T_2)$ be an open interval (which may be all of $R$ or a set $\{t: t < T_2\}$, etc.). Suppose that*

H1 *$f_1(\mathbf{x}, \mathbf{u}, t), f_2(\mathbf{x}, \mathbf{u}, t), \ldots, f_n(\mathbf{x}, \mathbf{u}, t)$ are continuous functions from $A_n \times R_m \times (T_1, T_2)$ [see Eq. (2-5)] into $R$.*

H2 *The partial derivatives $\partial f_i(\mathbf{x}, \mathbf{u}, t)/\partial x_j$ are continuous functions from $A_n \times R_m \times (T_1, T_2)$ into $R$.*

H3 *$\mathbf{x}_0$ is an element of $A_n$, and $t_0$ is an element of $(T_1, T_2)$.*

H4 *$\mathbf{u}(\tau)$ is a piecewise continuous function from $(T_1, T_2)$ into $R_m$.*

*Then there is a function $\boldsymbol{\psi}(t)$ from an interval $(t_1, t_2)$ containing $t_0$ into $R_n$ with components $\psi_1(t), \psi_2(t), \ldots, \psi_n(t)$ such that*

C1 *$\boldsymbol{\psi}(t)$ is continuous on $(t_1, t_2)$, and $\boldsymbol{\psi}(t) \in A_n$ for $t$ in $(t_1, t_2)$.*

C2
$$\boldsymbol{\psi}(t_0) = \mathbf{x}_0 \tag{3-174}$$

C3 *$\boldsymbol{\psi}(t)$ is a solution of the system $\dot{\mathbf{x}}(t) = \mathbf{f}[\mathbf{x}(t), \mathbf{u}(t), t]$, that is,*

$$\dot{\psi}_i(t) = f_i[\boldsymbol{\psi}(t), \mathbf{u}(t), t] \tag{3-175}$$

*for $i = 1, 2, \ldots, n$ and for all but a countable set $A$ of points $t$ in $(t_1, t_2)$.*

*Moreover, if $\boldsymbol{\phi}(t)$ is another function which satisfies conditions* C1 *to* C3 *on an interval $(s_1, s_2)$ (which contains $t_0$ by condition* C3*), then*

$$\boldsymbol{\phi}(t) = \boldsymbol{\psi}(t) \tag{3-176}$$

*for all $t$ in $(s_1, s_2) \cap (t_1, t_2)$ [see Eq. (2-3)]; in other words, the solutions are unique.*

PROOF Let us suppose for the moment that we have found a solution $\boldsymbol{\psi}(t)$ with $\boldsymbol{\psi}(t_0) = \mathbf{x}_0$. Then we claim that

$$\boldsymbol{\psi}(t) = \mathbf{x}_0 + \int_{t_0}^{t} \mathbf{f}[\boldsymbol{\psi}(\tau), \mathbf{u}(\tau), \tau]\, d\tau \tag{3-177}$$

Since $\dot{\boldsymbol{\psi}}(\tau) = \mathbf{f}[\boldsymbol{\psi}(\tau), \mathbf{u}(\tau), \tau]$ except on a countable set and $\mathbf{f}$ is continuous and $\boldsymbol{\psi}$ and $\mathbf{u}$ are piecewise continuous,‡ we may integrate to obtain the relation of Eq. (3-177). On the other hand, suppose that $\boldsymbol{\psi}(t)$ is a function which satisfies the equation

$$\mathbf{x}(t) = \mathbf{x}_0 + \int_{t_0}^{t} \mathbf{f}[\mathbf{x}(\tau), \mathbf{u}(\tau), \tau]\, d\tau \tag{3-178}$$

† See Refs. C-7, D-5, P-6, and S-4.
‡ See the remarks at the end of Sec. 3-11.

that is, $\psi(t)$ satisfies Eq. (3-177). Then we see that $\psi(t_0) = \mathbf{x}_0$ and that $\dot{\psi}(t) = \mathbf{f}[\psi(t), \mathbf{u}(t), t]$ except for a countable set of $t$'s. Thus, we shall try to find a solution to the *integral equation* (3-178).

Now $\mathbf{x}_0$ is in $A_n$ and $t_0$ is in $(T_1, T_2)$. Since $A_n$ is open, there is a sphere $S(\mathbf{x}_0, \hat{\lambda})$ about $\mathbf{x}_0$ contained in $A_n$ (see Definition 3-4), and consequently, if $\lambda < \hat{\lambda}$, the *closed* sphere $\overline{S(\mathbf{x}_0, \lambda)}$ is contained in $A_n$. In a similar way, we can find a $\mu > 0$ such that the *closed* interval $[t_0 - \mu, t_0 + \mu]$ is contained in $(T_1, T_2)$. In other words, if

$$\|\mathbf{x} - \mathbf{x}_0\| \leq \lambda \qquad \text{and} \qquad |t - t_0| \leq \mu \tag{3-179}$$

then $\mathbf{x} \in A_n$ and $t \in (T_1, T_2)$. Now, $\mathbf{u}(\tau)$ is piecewise continuous on $[t_0 - \mu, t_0 + \mu]$, and hence there is a *closed* sphere $\overline{S(\mathbf{0}, \nu)}$ about the origin in $R_m$ such that $\mathbf{u}(\tau)$ is in $\overline{S(\mathbf{0}, \nu)}$ for *all* $\tau$ in $[t_0 - \mu, t_0 + \mu]$. Now we note that the set $\overline{S(\mathbf{x}_0, \lambda)} \times \overline{S(\mathbf{0}, \nu)} \times [t_0 - \mu, t_0 + \mu]$ is closed and bounded when viewed as a subset of $R_{n+m+1}$ and hence is *compact* (see Sec. 3-6). It follows that there are numbers $M$ and $N$ such that if $\mathbf{x}$ and $t$ satisfy Eq. (3-179), then

$$\|\mathbf{f}[\mathbf{x}, \mathbf{u}(t), t]\| \leq M \tag{3-180a}$$

and

$$\left| \frac{\partial f_i}{\partial x_j} [\mathbf{x}, \mathbf{u}(t), t] \right| \leq N \tag{3-180b}$$

for $i, j = 1, 2, \ldots, n$.

Now let us pick a $\rho \leq \mu$ such that for some given $k$, $0 < k < 1$, we have

$$\rho \leq \min \left\{ \frac{\lambda}{M}, \frac{k}{n^2 N} \right\} \tag{3-181}$$

We shall consider the interval $[t_0 - \rho, t_0 + \rho]$ and the function space $\mathcal{C}([t_0 - \rho, t_0 + \rho], R_n)$ of all *continuous* functions from this interval into $R_n$, with the uniform norm [see Eqs. (3-124) and (3-131)]. As we have observed in Sec. 3-15, $\mathcal{C}([t_0 - \rho, t_0 + \rho], R_n)$ is *complete*. If we let $X$ be the subset of $\mathcal{C}([t_0 - \rho, t_0 + \rho], R_n)$, defined by

$$X = \{\boldsymbol{\phi}: \|\boldsymbol{\phi}(\tau) - \mathbf{x}_0\| \leq \lambda \text{ for all } \tau \text{ in } [t_0 - \rho, t_0 + \rho]\} \tag{3-182}$$

then we can easily verify that $X$ is a *closed*† subset of $\mathcal{C}([t_0 - \rho, t_0 + \rho], R_n)$ and hence is *complete*.

Now *let $T$ be the transformation of $X$ into itself, which is given by*

$$T(\boldsymbol{\phi})(t) = \mathbf{x}_0 + \int_{t_0}^{t} \mathbf{f}[\boldsymbol{\phi}(\tau), \mathbf{u}(\tau), \tau]\, d\tau. \tag{3-183}$$

In other words, if $\boldsymbol{\phi}(t)$ is a continuous function on $[t_0 - \rho, t_0 + \rho]$ whose range is near $\mathbf{x}_0$, then $T(\boldsymbol{\phi})$ is the continuous function on $[t_0 - \rho, t_0 + \rho]$ whose value at $t$ is given by Eq. (3-183). To see that $T$ actually maps $X$

† If $\mathbf{f}_m \to \mathbf{f}$ and $\|\mathbf{f}_m - \mathbf{x}_0\| \leq \lambda$, then $\|\mathbf{f} - \mathbf{x}_0\| \leq \|\mathbf{f} - \mathbf{f}_m + \mathbf{f}_m - \mathbf{x}_0\| \leq \|\mathbf{f}_m - \mathbf{x}_0\| + \|\mathbf{f} - \mathbf{f}_m\| \leq \lambda + \|\mathbf{f} - \mathbf{f}_m\|$ and the assertion follows.

into itself, we note that

$$\|T(\boldsymbol{\phi})(t) - \mathbf{x}_0\| = \left\| \int_{t_0}^{t} \mathbf{f}[\boldsymbol{\phi}(\tau), \mathbf{u}(\tau), \tau]\, d\tau \right\| \tag{3-184}$$

$$\leq |t - t_0| \max \|\mathbf{f}[\boldsymbol{\phi}(\tau), \mathbf{u}(\tau), \tau]\| \tag{3-185†}$$

$$\leq \rho M \qquad \text{by Eq. (3-180}a\text{)} \tag{3-186}$$

$$\leq M \frac{\lambda}{M} = \lambda \qquad \text{by Eq. (3-181)} \tag{3-187}$$

Now let us prove that $T$ is a *contraction on X* [see Sec. 3-5 and Eq. (3-12)]. In other words, we shall show that if $\boldsymbol{\phi}_1$ and $\boldsymbol{\phi}_2$ are elements of $X$, then

$$\|T(\boldsymbol{\phi}_1) - T(\boldsymbol{\phi}_2)\| \leq k\|\boldsymbol{\phi}_1 - \boldsymbol{\phi}_2\| \tag{3-188}$$

where $k$ is the given number between 0 and 1 [see Eq. (3-181)]. Suppose that $t$ is in $[t_0 - \rho, t_0 + \rho]$; then

$$\|T(\boldsymbol{\phi}_1)(t) - T(\boldsymbol{\phi}_2)(t)\| = \left\| \int_{t_0}^{t} \{\mathbf{f}[\boldsymbol{\phi}_1(\tau), \mathbf{u}(\tau), \tau] - \mathbf{f}[\boldsymbol{\phi}_2(\tau), \mathbf{u}(\tau), \tau]\}\, d\tau \right\| \tag{3-189}$$

$$\leq \int_{t_0}^{t} \|\mathbf{f}[\boldsymbol{\phi}_1(\tau), \mathbf{u}(\tau), \tau] - \mathbf{f}[\boldsymbol{\phi}_2(\tau), \mathbf{u}(\tau), \tau]\|\, d\tau \tag{3-190}$$

$$\leq |t - t_0| \max \|\mathbf{f}[\boldsymbol{\phi}_1(\tau), \mathbf{u}(\tau), \tau] - \mathbf{f}[\boldsymbol{\phi}_2(\tau), \mathbf{u}(\tau), \tau]\| \tag{3-191}$$

$$\leq \rho n^2 N \|\boldsymbol{\phi}_1 - \boldsymbol{\phi}_2\|$$

$$\leq k\|\boldsymbol{\phi}_1 - \boldsymbol{\phi}_2\| \qquad \text{by Eq. (3-181)} \tag{3-192}$$

The step involved in going from Eq. (3-191) to Eq. (3-192) requires some justification. It can be justified by making use of Eq. (3-70$a$), the Schwarz inequality [Eq. (2-87)], and Eq. (3-180$b$). (See Exercise 3-15.) Since $T$ is a contraction, there is, by Theorem 3-2, a *unique* $\boldsymbol{\psi}(t)$ in $X$ such that

$$T(\boldsymbol{\psi})(t) = \boldsymbol{\psi}(t) \tag{3-193}$$

that is,

$$\boldsymbol{\psi}(t) = \mathbf{x}_0 + \int_{t_0}^{t} \mathbf{f}[\boldsymbol{\psi}(\tau), \mathbf{u}(\tau), \tau]\, d\tau \tag{3-194}$$

In view of Eq. (3-178) and the discussion which led to it, we see that $\boldsymbol{\psi}(t)$ is the desired solution on the interval $(t_0 - \rho, t_0 + \rho)$.

Now suppose that $\hat{\boldsymbol{\psi}}$ and $\hat{\boldsymbol{\phi}}$ are two solutions to the problem defined on intervals $(t_1, t_2)$ and $(s_1, s_2)$, respectively. Consider the set $(t_1, t_2) \cap (s_1, s_2)$, which is not empty since $t_0$ is in it. We note that

$$(t_1, t_2) \cap (s_1, s_2) = (\max\{t_1, s_1\}, \min\{t_2, s_2\}) \tag{3-195}$$

that is, $(t_1, t_2) \cap (s_1, s_2)$ is an interval $(m_1, m_2)$, with $m_1 = \max\{t_1, s_1\}$, $m_2 = \min\{t_2, s_2\}$. Suppose that $\tau$ were a point of $(m_1, m_2)$ such that

$$\hat{\boldsymbol{\psi}}(\tau) \neq \hat{\boldsymbol{\phi}}(\tau) \tag{3-196}$$

† Using Eq. (3-111).

and, say, $t_0 < \tau < m_2$. If we let $E$ be the set of points in $[t_0, \tau)$ at which $\hat{\boldsymbol{\psi}} = \hat{\boldsymbol{\phi}}$, then $E$ has an upper bound and therefore also a least upper bound, which we shall denote by $\sigma$.† Is $\sigma$ in $E$? Well, if $\sigma$ were in $E$, then there would be an open interval about $\sigma$, $(\sigma - \epsilon, \sigma + \epsilon)$, *contained in* $E$, since we could apply our proof of conditions C1 to C3 to the problem with $\hat{t}_0 = \sigma$ and $\hat{\mathbf{x}}_0 = \hat{\boldsymbol{\psi}}(\sigma) = \hat{\boldsymbol{\phi}}(\sigma)$. But this would give us elements of $E$ *bigger than* $\sigma$, which is impossible. Hence $\sigma$ is *not in* $E$. But there is a sequence of elements $\hat{t}_m$ of $E$ which converges to $\sigma$, since $\sigma$ is the least upper bound of $E$. Since $\hat{\boldsymbol{\phi}}$ is continuous, we have $\lim \hat{\boldsymbol{\phi}}(\hat{t}_m) = \hat{\boldsymbol{\phi}}(\sigma)$, and since $\hat{\boldsymbol{\psi}}$ is continuous, $\lim \hat{\boldsymbol{\psi}}(\hat{t}_m) = \hat{\boldsymbol{\psi}}(\sigma)$. But $\hat{\boldsymbol{\phi}}(\hat{t}_m) = \hat{\boldsymbol{\psi}}(\hat{t}_m)$ for all $m$, which implies that $\hat{\boldsymbol{\phi}}(\sigma) = \hat{\boldsymbol{\psi}}(\sigma)$, that is, that $\sigma$ is in $E$. This contradiction tells us that Eq. (3-196) cannot hold, and so the proof of the theorem is complete.

**Exercise 3-15** Verify the step involved in going from Eq. (3-191) to Eq. (3-192). HINT: Consider (say) $f_1[\hat{\boldsymbol{\phi}}_1(\tau), \mathbf{u}(\tau), \tau] - f_1[\hat{\boldsymbol{\phi}}_2(\tau), \mathbf{u}(\tau), \tau]$ and note that there is, by Eq. (3-70*a*), a point $r\hat{\boldsymbol{\phi}}_1(\tau) + s\hat{\boldsymbol{\phi}}_2(\tau)$, $r + s = 1$, $r, s \geq 0$ in $A_n$ such that $|f_1[\hat{\boldsymbol{\phi}}_1(\tau), \mathbf{u}(\tau), \tau] - f_1[\hat{\boldsymbol{\phi}}_2(\tau), \mathbf{u}(\tau), \tau]| = |\langle(\partial f_1/\partial \mathbf{x})[r\hat{\boldsymbol{\phi}}_1(\tau) + s\hat{\boldsymbol{\phi}}_2(\tau), \mathbf{u}(\tau), \tau], \hat{\boldsymbol{\phi}}_1(\tau) - \hat{\boldsymbol{\phi}}_2(\tau)\rangle|$. Now use the Schwarz inequality [Eq. (2-87)] and then use Eqs. (3-180*b*) and (3-45).

We shall now illustrate the proof of the theorem by carrying out the steps in a simple example.

**Example 3-53** Let $A_2$ be the entire space $R_2$ and let $(T_1, T_2)$ be the open interval $(0, 1)$. Take $f_1(x_1, x_2, u, t) = x_2 + u$ and $f_2(x_1, x_2, u, t) = -x_1$. Let $u(t)$ be the function $1/t(t-1)$. It is clear that $f_1$ and $f_2$ satisfy conditions H1 and H2 of Theorem 3-14. Let $\mathbf{x}_0 = 0$ and $t_0 = \frac{1}{2}$ be our initial conditions. Now we want to solve the system

$$\begin{aligned} \dot{x}_1(t) &= x_2(t) + \frac{1}{t(t-1)} \\ \dot{x}_2(t) &= -x_1(t) \end{aligned} \tag{3-197}$$

If $\boldsymbol{\psi}(t)$ is a solution of this system, with $\boldsymbol{\psi}(\frac{1}{2}) = \mathbf{0}$, then we have

$$\begin{aligned} \dot{\psi}_1(t) &= \psi_2(t) + \frac{1}{t(t-1)} \\ \dot{\psi}_2(t) &= -\psi_1(t) \end{aligned} \tag{3-198}$$

and, integrating, we obtain

$$\begin{aligned} \psi_1(t) &= 0 + \int_{t_0}^{t} \left\{\psi_2(\tau) + \frac{1}{\tau(\tau-1)}\right\} d\tau \\ \psi_2(t) &= 0 - \int_{t_0}^{t} \psi_1(\tau)\, d\tau \end{aligned} \tag{3-199}$$

In other words, we want to solve the system of integral equations

$$\begin{aligned} x_1(t) &= \int_{t_0}^{t} \left\{x_2(\tau) + \frac{1}{\tau(\tau-1)}\right\} d\tau \\ x_2(t) &= -\int_{t_0}^{t} x_1(\tau)\, d\tau \end{aligned} \tag{3-200}$$

† $\sigma$ satisfies the conditions (1) $\sigma \geq t$ for all $t$ in $E$ and (2) if $\sigma' \geq t$ for all $t$ in $E$, then $\sigma' \geq \sigma$.

Since $R_2$ is open, the sphere $S(\mathbf{0}, 2)$ (say) is contained in $R_2$, and so we may take $\lambda = 1$ for convenience. We shall take $\mu = \frac{1}{4}$, and we note that if $\tau \in [\frac{1}{4}, \frac{3}{4}]$, then $u(\tau) = 1/\tau(\tau - 1)$ is in $\overline{S(0, \frac{16}{3})} = [-\frac{16}{3}, \frac{16}{3}]$. So we consider the compact set $\overline{S(\mathbf{0}, 1)} \times [-\frac{16}{3}, \frac{16}{3}] \times [\frac{1}{4}, \frac{3}{4}] = \tilde{Y}$; on this set $\tilde{Y}$, we see that

$$\|\mathbf{f}(\mathbf{x}, u(t), t)\| \leq \sqrt{1 + (\tfrac{16}{3})^2 + 2(\tfrac{16}{3}) + 1} < 7 \qquad \text{and} \qquad \left|\frac{\partial f_i}{\partial x_j}[\mathbf{x}, u(t), t]\right| \leq 1$$

If we take $k = \frac{1}{2}$, then we want a $\rho \leq \frac{1}{4}$ such that

$$\rho \leq \min\{\tfrac{1}{7}, \tfrac{1}{8}\} = \tfrac{1}{8}$$

Well, we might as well take $\rho = \frac{1}{8}$, and so we consider the space $\mathcal{C}([\frac{3}{8}, \frac{5}{8}], R_2)$ and the closed subset $X$ of this space, defined by

$$X = \boldsymbol{\varphi}\{: \max_{t \in [\frac{3}{8}, \frac{5}{8}]} \sqrt{\varphi_1(t)^2 + \varphi_2(t)^2} \leq 1\} \tag{3-201}$$

Our transformation $T$ of $X$ into itself is given by

$$T(\boldsymbol{\varphi})(t) = \begin{bmatrix} \displaystyle\int_{\frac{1}{2}}^{t} \left\{\varphi_2(\tau) + \frac{1}{\tau(\tau - 1)}\right\} d\tau \\ \displaystyle -\int_{\frac{1}{2}}^{t} \varphi_1(\tau)\, d\tau \end{bmatrix} \tag{3-202}$$

We note that

$$\begin{aligned} \|T(\boldsymbol{\varphi})(t) - \mathbf{0}\| &\leq |t - \tfrac{1}{2}| \max_{t \in [\frac{3}{8}, \frac{5}{8}]} \|\mathbf{f}[\boldsymbol{\varphi}(\tau), u(\tau), \tau]\| \\ &\leq \tfrac{1}{8} \times 7 \leq 1 \end{aligned} \tag{3-203}$$

(Here $\rho = \frac{1}{8}$, $M = 7$, and $\lambda = 1$.) Now let us verify that $T$ is a *contraction*, i.e., that

$$\|T(\boldsymbol{\varphi}_1)(t) - T(\boldsymbol{\varphi}_2)(t)\| \leq \tfrac{1}{2}\|\boldsymbol{\varphi}_1 - \boldsymbol{\varphi}_2\|$$

for all $t$ in $[\frac{3}{8}, \frac{5}{8}]$. Well, we have

$$\begin{aligned} \|T(\boldsymbol{\varphi}_1)(t) - T(\boldsymbol{\varphi}_2)(t)\| &\leq \int_{\frac{1}{2}}^{t} \|\mathbf{f}[\boldsymbol{\varphi}_1(\tau), u(\tau), \tau] - \mathbf{f}[\boldsymbol{\varphi}_2(\tau), u(\tau), \tau]\|\, d\tau \\ &\leq |t - \tfrac{1}{2}| \max_{t \in [\frac{3}{8}, \frac{5}{8}]} \|\mathbf{f}[\boldsymbol{\varphi}_1(\tau), u(\tau), \tau] - \mathbf{f}[\boldsymbol{\varphi}_2(\tau), u(\tau), \tau]\| \\ &\leq \tfrac{1}{8} \times 4 \times 1 \times \|\boldsymbol{\varphi}_1 - \boldsymbol{\varphi}_2\| = \tfrac{1}{2}\|\boldsymbol{\varphi}_1 - \boldsymbol{\varphi}_2\| \end{aligned} \tag{3-204}$$

(Here $\rho = \frac{1}{8}$, $n^2 = 4$, and $N = 1$.) The reader should check the numerical computations for himself. The upshot of our argument is that there is a unique solution $\boldsymbol{\psi}(t)$ of the system (3-200), defined on the interval $(\frac{3}{8}, \frac{5}{8})$ with $\boldsymbol{\psi}(\frac{1}{2}) = \mathbf{0}$. It can be shown that

$$\boldsymbol{\psi}(t) = \begin{bmatrix} \cos(t - \frac{1}{2}) & \sin(t - \frac{1}{2}) \\ -\sin(t - \frac{1}{2}) & \cos(t - \frac{1}{2}) \end{bmatrix} \begin{bmatrix} \displaystyle\int_{\frac{1}{2}}^{t} \frac{\cos(\tau - \frac{1}{2})}{\tau(\tau - 1)}\, d\tau \\ \displaystyle\int_{\frac{1}{2}}^{t} \frac{\sin(\tau - \frac{1}{2})}{\tau(\tau - 1)}\, d\tau \end{bmatrix} \tag{3-205}$$

Now let us make some comments about Theorem 3-14 and its proof. We shall number our comments to facilitate future reference to them.

*Comment 1*

We observe that we make no assumptions about the existence or continuity of the partial derivatives of the $f_i$ with respect to the components $u_j$ of $\mathbf{u}$. This remark will be of interest to us in Chap. 5.

*Comment 2*

We note that $R_m$ may be replaced by any subset of $R_m$ which contains the closure (see Definition 3-9) of the range of $\mathbf{u}(\tau)$. In other words, the theorem will still be true if we weaken the hypotheses slightly by changing $R_m$ to the closure of the range of $\mathbf{u}(\tau)$ throughout. This comment will take on greater significance when we consider the control problem and constraints in Chap. 4.

*Comment 3*

We can see that the theorem is a *local* result, in that the existence and uniqueness of a solution are established only for an *interval* [namely, $(t_1, t_2)$] about the initial time $t_0$. If the reader reexamines Example 3-48, he will appreciate the significance of this comment. We shall see in the next section that a global result can be obtained for linear systems (see Theorem 3-15).

*Comment 4*

We have stated and proved the theorem in the form needed in the sequel. However, hypothesis H2 is somewhat stronger than is necessary. In fact, the theorem is usually stated with hypothesis H2 replaced by the so-called Lipschitz condition, which is as follows:

**Lipschitz Condition** There is a constant $K > 0$ with the property that

$$\|\mathbf{f}(\mathbf{x}_1, \mathbf{u}, t) - \mathbf{f}(\mathbf{x}_2, \mathbf{u}, t)\| \leq K\|\mathbf{x}_1 - \mathbf{x}_2\|$$

for all $\mathbf{x}_1$, $\mathbf{x}_2$ in $A_n$, $\mathbf{u}$ in $R_m$, and $t$ in $(T_1, T_2)$. A slight modification of the proof gives us the same result with this condition replacing hypothesis H2. (See for example, Ref. B-10 or C-7.)

*Comment 5*

We note that in the statement of the theorem we assume that $A_n$ is an *open* set and $(T_1, T_2)$ an *open* interval. We make these assumptions in order to avoid points both in space and in time at which the solution cannot be defined, i.e., to avoid so-called "singular" points. For example, if we consider the system

$$\frac{dx}{dt} = \frac{1}{x - 2}$$

then there is no solution which passes through the point $x = 2$, although there are unique solutions passing through points arbitrarily near $x = 2$. Similarly, there is no solution of the equation of Example 3-48 at $t = 0$, that is, of the equation

$$\frac{dx}{dt} = \frac{1}{t}$$

We shall use Theorem 3-14 repeatedly in the remainder of the book.

## 3-19 Linear Differential Equations: Basic Concepts

In the remaining sections of this chapter, we are going to examine in great detail a special class of systems of differential equations, called linear systems. These systems are of great importance because they are very frequently used to represent the dynamical behavior of the physical systems encountered in engineering practice. Moreover, it is possible to give explicit solutions to systems of this type, which greatly facilitates the analytical treatment of practical problems. Well, let us see what we mean by a linear system. We have:

***Definition 3-43*** *Let $A_n$ be an open set in $R_n$ and let $(T_1, T_2)$ be an open interval (which may be all of $R$ or a set $\{t: t > T_1\}$, etc.). Let $\mathbf{v}(t)$ be a piecewise continuous function on $(T_1, T_2)$ to $R_n$ and let $\mathbf{A}(t)$ be a continuous function from $(T_1, T_2)$ into the set $\mathfrak{M}(n, n)$ of all $n \times n$ matrices (compare with the end of Sec.* 3-14). *Then the system of equations*

$$\dot{\mathbf{x}}(t) = \mathbf{A}(t)\mathbf{x}(t) + \mathbf{v}(t) \tag{3-206}$$

*or, equivalently, the system of equations*

$$\dot{x}_i(t) = \sum_{j=1}^{n} a_{ij}(t)x_j(t) + v_i(t) \tag{3-207}$$

*for $i = 1, 2, \ldots, n$, where $\mathbf{A}(t) = [a_{ij}(t)]$ [that is, the $a_{ij}(t)$ are the entries of the matrix $\mathbf{A}(t)$] and $v_1(t), v_2(t), \ldots, v_n(t)$ are the components of $\mathbf{v}(t)$, is called a linear system with forcing function $\mathbf{v}(t)$. The system of equations*

$$\dot{\mathbf{x}}(t) = \mathbf{A}(t)\mathbf{x}(t) \tag{3-208}$$

*or, equivalently, the system*

$$\dot{x}_i(t) = \sum_{j=1}^{n} a_{ij}(t)x_j(t) \tag{3-209}$$

*is often called the homogeneous (or unforced or free) part of the linear system. If $\mathbf{A}(t)$ is a constant matrix, i.e., if*

$$\mathbf{A}(t) = \mathbf{A} = (a_{ij}) \qquad \text{for all } t \text{ in } (T_1, T_2) \tag{3-210}$$

*then we say that the linear system is time-invariant [or is a constant (coefficient) system]. Otherwise, we say that we have a time-varying linear system.*

**Example 3-54** Let $\mathbf{A}(t) = \mathbf{\Lambda}$ be the $n \times n$ diagonal matrix with distinct entries $\lambda_1, \lambda_2, \ldots, \lambda_n$ along the diagonal. Then the system

$$\begin{bmatrix} \dot{x}_1(t) \\ \dot{x}_2(t) \\ \cdots \\ \dot{x}_n(t) \end{bmatrix} = \begin{bmatrix} \lambda_1 & 0 & \cdots & 0 \\ 0 & \lambda_2 & \cdots & 0 \\ \cdots & \cdots & \cdots & \cdots \\ 0 & 0 & \cdots & \lambda_n \end{bmatrix} \begin{bmatrix} x_1(t) \\ x_2(t) \\ \cdots \\ x_n(t) \end{bmatrix} \tag{3-211}$$

is a homogeneous, time-invariant linear system. We observe that this system may also be written in the vector form

$$\dot{\mathbf{x}}(t) = \mathbf{A}\mathbf{x}(t) \tag{3-212}$$

or in the form

$$\dot{x}_i(t) = \lambda_i x_i(t) \qquad i = 1, 2, \ldots, n \tag{3-213}$$

From this last equation, we can readily see that the vector-valued function $\boldsymbol{\psi}(t)$, with components $\psi_i(t) = c_i e^{\lambda_i t}$, is a solution of this equation. We note that

$$\boldsymbol{\psi}(t) = \begin{bmatrix} e^{\lambda_1 t} & 0 & \cdots & 0 \\ 0 & e^{\lambda_2 t} & \cdots & 0 \\ \cdot & \cdot & \cdots & \cdot \\ 0 & 0 & \cdots & e^{\lambda_n t} \end{bmatrix} \begin{bmatrix} c_1 \\ c_2 \\ \cdots \\ c_n \end{bmatrix} \tag{3-214}$$

and that $\boldsymbol{\psi}(0) = \mathbf{c}$, where $\mathbf{c}$ has the components $c_1, c_2, \ldots, c_n$.

Now we shall turn our attention to recasting the basic existence theorem (Theorem 3-14) for linear systems. The crucial point we shall demonstrate is that the solution of a linear system is defined wherever the system is defined [i.e., the solution $\boldsymbol{\psi}(t)$ of Theorem 3-14 is defined on the entire interval $(T_1, T_2)$]. We have:

***Theorem 3-15*** *Let $(T_1, T_2)$ be an open interval and let $\mathbf{A}(t)$ be a continuous function from the interval $(T_1, T_2)$ into the set of $n \times n$ matrices $\mathfrak{M}(n, n)$. Suppose that $\mathbf{v}(t)$ is a piecewise continuous function from $(T_1, T_2)$ into $R_n$ and that $\mathbf{x}_0 \in R_n$, $t_0 \in (T_1, T_2)$. Then there is a function $\boldsymbol{\psi}(t)$ from all of $(T_1, T_2)$*† *into $R_n$ with components $\psi_1(t), \psi_2(t), \ldots, \psi_n(t)$ such that*

C1 *$\boldsymbol{\psi}(t)$ is continuous.*

C2 $\boldsymbol{\psi}(t_0) = \mathbf{x}_0$.

C3 *$\boldsymbol{\psi}(t)$ is a solution of the linear system $\dot{\mathbf{x}}(t) = \mathbf{A}(t)\mathbf{x}(t) + \mathbf{v}(t)$, that is,*

$$\dot{\boldsymbol{\psi}}(t) = \mathbf{A}(t)\boldsymbol{\psi}(t) + \mathbf{v}(t) \tag{3-215}$$

C4 *$\boldsymbol{\psi}(t)$ is unique; i.e., any function satisfying conditions* C1 *to* C3 *must coincide with $\boldsymbol{\psi}(t)$.*

PROOF If we set $\mathbf{f}(\mathbf{x}, \mathbf{v}, t) = \mathbf{A}(t)\mathbf{x} + \mathbf{v}$, then it is clear that $\mathbf{f}$ satisfies the hypotheses of Theorem 3-14. Thus, we can find a *unique* solution to Eq. (3-215) on some subinterval $(t_1, t_2)$, containing $t_0$, of $(T_1, T_2)$. *We want to show that $(t_1, t_2)$ can be taken to be all of $(T_1, T_2)$.*

So, let $[s_1, s_2]$ be any closed subinterval of $(T_1, T_2)$, with $t_0 \in [s_1, s_2]$. We shall show that Eq. (3-215) has a solution on $[s_1, s_2]$, and it will then follow that Eq. (3-215) has a solution defined at any point $t$ of $(T_1, T_2)$. Condition C4 will then be a consequence of Theorem 3-14. Well, set

$$\boldsymbol{\phi}^0(t) = \mathbf{x}_0 \qquad \text{for } t \text{ in } [s_1, s_2] \tag{3-216}$$

† This is the key point, because the general result, Theorem 3-14, was purely local.

Then $\boldsymbol{\phi}^0(t) \in \mathcal{C}([s_1, s_2], R_n)$; that is, $\boldsymbol{\phi}^0(t)$ is a continuous function. We define a sequence of elements $\boldsymbol{\phi}^j(t)$ of $\mathcal{C}([s_1, s_2], R_n)$ by setting

$$\boldsymbol{\phi}^j(t) = \mathbf{x}_0 + \int_{t_0}^{t} \{\mathbf{A}(\tau)\boldsymbol{\phi}^{j-1}(\tau) + \mathbf{v}(\tau)\} \, d\tau \tag{3-217}$$

for $j = 1, 2, \ldots$. We claim that $\boldsymbol{\phi}^j$ is a Cauchy sequence (see Definition 3-11) in this function space with respect to the uniform norm [see Eq. (3-124)]. Assuming for the moment that this claim is valid, we see that there is a $\boldsymbol{\phi}$ in $\mathcal{C}([s_1, s_2], R_n)$ such that $\boldsymbol{\phi}^j$ converges to $\boldsymbol{\phi}$ [since $\mathcal{C}([s_1, s_2], R_n)$ is complete (see Sec. 3-15)]. But then $\boldsymbol{\phi}(t_0) = \mathbf{x}_0$, and it is easy to see that the sequence

$$\mathbf{x}_0 + \int_{t_0}^{t} \{\mathbf{A}(\tau)\boldsymbol{\phi}^j(\tau) + \mathbf{v}(\tau)\} \, d\tau = \boldsymbol{\phi}^{j+1}(t)$$

converges to

$$\mathbf{x}_0 + \int_{t_0}^{t} \{\mathbf{A}(\tau)\boldsymbol{\phi}(\tau) + \mathbf{v}(\tau)\} \, d\tau$$

(See Exercise 3-16.) It follows that, for $t$ in $[s_1, s_2]$,

$$\boldsymbol{\phi}(t) = \mathbf{x}_0 + \int_{t_0}^{t} \{\mathbf{A}(\tau)\boldsymbol{\phi}(\tau) + \mathbf{v}(\tau)\} \, d\tau \tag{3-218}$$

and hence that $\boldsymbol{\phi}$ is a solution of Eq. (3-215) on this interval. Thus, all that remains to complete the proof is the demonstration that the $\boldsymbol{\phi}^j$ are a Cauchy sequence.

Now we have, for $t$ in $[s_1, s_2]$,

$$\|\boldsymbol{\phi}^1(t) - \boldsymbol{\phi}^0(t)\| = \left\| \int_{t_0}^{t} \{\mathbf{A}(\tau)\mathbf{x}_0 + \mathbf{v}(\tau)\} \, d\tau \right\| \le |t - t_0| M \tag{3-219}$$

where $M$ is some constant. Then

$$\begin{aligned} \|\boldsymbol{\phi}^2(t) - \boldsymbol{\phi}^1(t)\| &= \left\| \int_{t_0}^{t} \{\mathbf{A}(\tau)\}[\boldsymbol{\phi}^1(\tau) - \boldsymbol{\phi}^0(\tau)] \, d\tau \right\| \\ &\le Nn^2 \int_{t_0}^{t} M|\tau - t_0| \, d\tau & \text{(3-220)} \\ &\le NMn^2 \frac{|t - t_0|^2}{2!} & \text{(3-221)} \end{aligned}$$

where $N$ is a constant such that $|a_{ij}(t)| \le N$ for $t$ in $[s_1, s_2]$. Inductively, we can show that

$$\|\boldsymbol{\phi}^{j+1} - \boldsymbol{\phi}^j\| \le M \frac{[n^2N(s_2 - s_1)]^j}{j!} \tag{3-222}$$

But we note that

$$\sum_{j=0}^{\infty} \frac{M[n^2N(s_2 - s_1)]^j}{j!} = Me^{n^2N(s_2 - s_1)} \tag{3-223}$$

and hence that the sequence $\boldsymbol{\phi}^j$ is a Cauchy sequence, as

$$\|\boldsymbol{\phi}^{k+\nu} - \boldsymbol{\phi}^k\| \le \sum_{j=1}^{\nu-1} \frac{M[n^2N(s_2 - s_1)]^{k+j}}{k} + j! \tag{3-224}$$

This completes the proof of the theorem.

**Exercise 3-16** Show that the sequence $\boldsymbol{\phi}^{j+1}(t)$ converges to

$$\mathbf{x}_0 + \int_{t_0}^{t} \{\mathbf{A}(\tau)\boldsymbol{\phi}(\tau) + \mathbf{v}(\tau)\}\, d\tau$$

HINT:

$$\left\| \boldsymbol{\phi}^{j+1}(t) - \left[\mathbf{x}_0 + \int_{t_0}^{t} \{\mathbf{A}(\tau)\boldsymbol{\phi}(\tau) + \mathbf{v}(\tau)\}\, d\tau\right] \right\| = \left\| \int_{t_0}^{t} [\mathbf{A}(\tau)]\{\boldsymbol{\phi}^{j}(\tau) - \boldsymbol{\phi}(\tau)\}\, d\tau \right\|$$
$$\leq |t - t_0|\, Nn^2\, \|\boldsymbol{\phi}^j - \boldsymbol{\phi}\|$$

where $N$ is as in Eq. (3-221).

Although we have used $\boldsymbol{\psi}(t)$ to denote the solution of a system of differential equations in Secs. 3-17 to 3-19, *we shall in the remainder of this book often notationally slur over the distinction between the solution of a system and the variables of the system. In other words, we shall often write* $\mathbf{x}(t)$ *in place of* $\boldsymbol{\psi}(t)$ *for the solution of the system*

$$\dot{\mathbf{x}} = \mathbf{f}(\mathbf{x}, \mathbf{u}, t) \tag{3-225}$$

## 3-20 Linear Differential Equations: The Fundamental Matrix

We are now going to make use of Theorem 3-15 to give an "explicit" solution to the initial-value problem for linear systems. Let us begin by reexamining Example 3-54. We have the system

$$\begin{bmatrix} \dot{x}_1(t) \\ \dot{x}_2(t) \\ \cdots \\ \dot{x}_n(t) \end{bmatrix} = \begin{bmatrix} \lambda_1 & 0 & \cdots & 0 \\ 0 & \lambda_2 & \cdots & 0 \\ \cdots & \cdots & \cdots & \cdots \\ 0 & 0 & \cdots & \lambda_n \end{bmatrix} \begin{bmatrix} x_1(t) \\ x_2(t) \\ \cdots \\ x_n(t) \end{bmatrix} \tag{3-226}$$

where the $\lambda_i$ are distinct, and we have observed that the solution of this system $\mathbf{x}(t)$, with $\mathbf{x}(0) = \mathbf{c}$, is given by

$$\mathbf{x}(t) = e^{\mathbf{A}t}\mathbf{c} \tag{3-227}$$

where $e^{\mathbf{A}t}$ denotes the diagonal matrix with entries $e^{\lambda_i t}$ along the diagonal, that is,

$$e^{\mathbf{A}t} = \begin{bmatrix} e^{\lambda_1 t} & 0 & \cdots & 0 \\ 0 & e^{\lambda_2 t} & \cdots & 0 \\ \cdots & \cdots & \cdots & \cdots \\ 0 & 0 & \cdots & e^{\lambda_n t} \end{bmatrix} \tag{3-228}$$

Now suppose that we wanted to find the solution of Eq. (3-226) which satisfied the initial condition $\mathbf{x}(t_0) = \mathbf{x}_0$. Well, we can see immediately that it must be given by

$$\mathbf{x}(t) = e^{\mathbf{A}(t-t_0)}\mathbf{x}_0 \tag{3-229}$$
$$= e^{\mathbf{A}t}e^{-\mathbf{A}t_0}\mathbf{x}_0 \tag{3-230}$$

In other words, once we know the matrix $e^{\mathbf{A}t}$, we can find an explicit solution to the initial-value problem with ease. Thus, we see that the matrix $e^{\mathbf{A}t}$

is really fundamental to the solution of the initial-value problem. Our goals in this section will be:

1. To show that every linear system has associated with it a matrix of this type
2. To study the properties of these matrices

Now suppose that $(T_1, T_2)$ is an open interval, that $\mathbf{A}(t)$ is a continuous function from $(T_1, T_2)$ into the set $\mathfrak{M}(n, n)$ of all $n \times n$ matrices, and that $\mathbf{v}(t)$ is a piecewise continuous function from $(T_1, T_2)$ into $R_n$. We consider the linear system

$$\dot{x}_i(t) = \sum_{j=1}^{n} a_{ij}(t)x_j(t) + v_i(t) \qquad i = 1, 2, \ldots, n \tag{3-231}$$

where $\mathbf{A}(t) = (a_{ij}(t))$, which we also write in vector form as

$$\dot{\mathbf{x}}(t) = \mathbf{A}(t)\mathbf{x}(t) + \mathbf{v}(t) \tag{3-232}$$

Let $t_0$ be an element of $(T_1, T_2)$. Then we have:

***Theorem 3-16*** *Let $\mathcal{S}$ denote the set of all solutions of the homogeneous part of Eq.* (3-232) [*see Definition* 3-43, *Eq.* (3-209)], *i.e., of the equation*

$$\dot{\mathbf{x}}(t) = \mathbf{A}(t)\mathbf{x}(t)$$

*In other words,*

$$\mathcal{S} = \{\mathbf{x}(t) \colon \dot{\mathbf{x}}(t) = \mathbf{A}(t)\mathbf{x}(t) \text{ for } t \text{ in } (T_1, T_2)\} \tag{3-233}$$

*Then $\mathcal{S}$ is an n-dimensional vector space, and a basis $\{\mathbf{x}^1(t), \mathbf{x}^2(t), \ldots, \mathbf{x}^n(t)\}$ of $\mathcal{S}$ may be obtained by letting $\mathbf{x}^i(t)$ be the (unique) element of $\mathcal{S}$ which satisfies the condition*

$$\mathbf{x}^i(t_0) = \mathbf{e}_i \tag{3-234}$$

*where $\{\mathbf{e}_1, \mathbf{e}_2, \ldots, \mathbf{e}_n\}$ is the natural basis of $R_n$.*†

PROOF Clearly, the function $\mathbf{x}(t) = \mathbf{0}$ for all $t$ is in $\mathcal{S}$. This function is called the *trivial solution.* Moreover, if $r$ and $s$ are in $R$ and $\mathbf{x}(t)$ and $\mathbf{y}(t)$ are in $\mathcal{S}$, then

$$\begin{aligned} \frac{d}{dt}[r\mathbf{x}(t) + s\mathbf{y}(t)] &= r\dot{\mathbf{x}}(t) + s\dot{\mathbf{y}}(t) \\ &= r\mathbf{A}(t)\mathbf{x}(t) + s\mathbf{A}(t)\mathbf{y}(t) \\ &= \mathbf{A}(t)[r\mathbf{x}(t) + s\mathbf{y}(t)] \end{aligned} \tag{3-235}$$

so that $r\mathbf{x}(t) + s\mathbf{y}(t)$ is in $\mathcal{S}$. We leave it to the reader to verify axioms V1 to V8 of Sec. 2-5 for $\mathcal{S}$.

† Recall that

$$\mathbf{e}_1 = \begin{bmatrix} 1 \\ 0 \\ \cdot \\ \cdot \\ \cdot \\ 0 \end{bmatrix} \qquad \mathbf{e}_2 = \begin{bmatrix} 0 \\ 1 \\ \cdot \\ \cdot \\ \cdot \\ 0 \end{bmatrix} \qquad \cdots \qquad \mathbf{e}_n = \begin{bmatrix} 0 \\ 0 \\ \cdot \\ \cdot \\ \cdot \\ 1 \end{bmatrix}$$

Now let us show that the $\mathbf{x}^i(t)$ are a basis of $\mathcal{S}$ (see Sec. 2-6). Suppose that

$$\sum_{i=1}^{n} c_i \mathbf{x}^i(t) = \mathbf{0} \qquad \text{for all } t \text{ in } (T_1, T_2) \tag{3-236}$$

Then

$$\sum_{i=1}^{n} c_i \mathbf{x}^i(t_0) = \sum_{i=1}^{n} c_i \mathbf{e}_i = \mathbf{0} \tag{3-237}$$

and so the $c_i$ are all 0. In other words, the $\mathbf{x}^i(t)$ are linearly independent functions. Now, if $\mathbf{x}(t)$ is any element of $\mathcal{S}$, then $\mathbf{x}(t_0)$ may be written as a linear combination of the $\mathbf{e}_i$, that is,

$$\mathbf{x}(t_0) = \sum_{i=1}^{n} \beta_i \mathbf{e}_i \tag{3-238}$$

We claim that

$$\mathbf{x}(t) = \sum_{i=1}^{n} \beta_i \mathbf{x}^i(t) \qquad \text{for all } t \text{ in } (T_1, T_2) \tag{3-239}$$

But the function

$$\sum_{i=1}^{n} \beta_i \mathbf{x}^i(t)$$

is in $\mathcal{S}$ and agrees with $\mathbf{x}(t)$ at $t_0$, that is, $\Sigma \beta_i \mathbf{x}^i(t_0) = \mathbf{x}(t_0)$. It follows from the uniqueness part of Theorem 3-15 that

$$\mathbf{x}(t) = \sum_{i=1}^{n} \beta_i \mathbf{x}^i(t)$$

which establishes Eq. (3-239).

This theorem leads us to the following definition:

***Definition 3-44 Fundamental Matrix*** *Let $\mathbf{\Phi}(t, t_0)$ be the $n \times n$ matrix whose jth column is the vector $\mathbf{x}^j(t)$ in $\mathcal{S}$, with $\mathbf{x}^j(t_0) = \mathbf{e}_j$. In other words, the columns of $\mathbf{\Phi}(t, t_0)$ are the solutions of the unforced part of the system* (3-232) *satisfying the initial condition $\mathbf{x}^j(t_0) = \mathbf{e}_j$. We say that $\mathbf{\Phi}(t, t_0)$ is the fundamental, or transition, matrix of the system* (3-232). *We note that*

$$\mathbf{\Phi}(t, t_0) = \begin{bmatrix} x_1{}^1(t) & x_1{}^2(t) & \cdots & x_1{}^n(t) \\ x_2{}^1(t) & x_2{}^2(t) & \cdots & x_2{}^n(t) \\ \cdots & \cdots & \cdots & \cdots \\ x_n{}^1(t) & x_n{}^2(t) & \cdots & x_n{}^n(t) \end{bmatrix} \tag{3-240}$$

*where $x_i{}^j(t)$ is the ith component of $\mathbf{x}^j(t)$, and we note that*

$$\mathbf{\Phi}(t_0, t_0) = \begin{bmatrix} 1 & 0 & \cdots & 0 \\ 0 & 1 & \cdots & 0 \\ \cdots & \cdots & \cdots & \cdots \\ 0 & 0 & \cdots & 1 \end{bmatrix} = \mathbf{I} \tag{3-241}$$

*where* **I** *is the identity matrix* [*see Eq.* (2-55)].

We observe that $\mathbf{\Phi}(t, t_0)$ is, in reality, a function from $(T_1, T_2)$ into the set of all $n \times n$ matrices. We claim that this function $\mathbf{\Phi}(t, t_0)$ is differentiable [i.e., each coefficient of $\mathbf{\Phi}(t, t_0)$ is a differentiable function of $t$] and, in fact, that

$$\frac{d}{dt}\mathbf{\Phi}(t, t_0) = \dot{\mathbf{\Phi}}(t, t_0) = \mathbf{A}(t)\mathbf{\Phi}(t, t_0) \tag{3-242}$$

To see this, we observe that the entry in the $i$th row and $j$th column of $\mathbf{\Phi}(t, t_0)$ is $x_i{}^j(t)$. Since $\mathbf{x}^j(t)$ is a solution of the homogeneous part of our system, we have

$$\dot{x}_i{}^j(t) = \sum_{k=1}^{n} a_{ik}(t)x_k{}^j(t) \tag{3-243}$$

where $\mathbf{A}(t) = (a_{ik}(t))$. But the right-hand side of Eq. (3-243) is precisely the entry in the $i$th row and $j$th column of the product $\mathbf{A}(t)\mathbf{\Phi}(t, t_0)$ [see Eq. (2-33)]; consequently, Eq. (3-242) has been established. We note that this implies that the fundamental matrix $\mathbf{\Phi}(t, t_0)$ may be viewed as the (unique) solution of the matrix differential equation [i.e., the variable $\mathbf{X}(t)$ is an $n \times n$ matrix]

$$\dot{\mathbf{X}}(t) = \mathbf{A}(t)\mathbf{X}(t) \tag{3-244}$$

which satisfies the initial condition

$$\mathbf{X}(t_0) = \mathbf{I} \tag{3-245}$$

Furthermore, if $\mathbf{x}_0$ is an element of $R_n$, then we can see that the solution $\mathbf{x}(t)$ of the homogeneous part of Eq. (3-232), which satisfies the initial condition $\mathbf{x}(t_0) = \mathbf{x}_0$, is given by

$$\mathbf{x}(t) = \mathbf{\Phi}(t, t_0)\mathbf{x}_0 \tag{3-246}$$

Let us now show that the matrix $\mathbf{\Phi}(t, t_0)$ is *nonsingular* for every $t$ (see Sec. 2-9). There are several ways to do this. For example, suppose that $\mathbf{\Phi}(\hat{t}, t_0)$ were not nonsingular for some $\hat{t}$ in $(T_1, T_2)$; then the determinant of $\mathbf{\Phi}(\hat{t}, t_0)$, det $\mathbf{\Phi}(\hat{t}, t_0)$, would be 0 and the columns $\mathbf{x}^j(\hat{t})$ of $\mathbf{\Phi}(\hat{t}, t_0)$ would be linearly dependent with

$$\sum_{j=1}^{n} \beta_j\mathbf{x}^j(\hat{t}) = \mathbf{0} \tag{3-247}$$

and not all the $\beta_j = 0$. But the function

$$\sum_{j=1}^{n} \beta_j\mathbf{x}^j(t)$$

is an element of the set of Eq. (3-233) which satisfies the condition of being zero at $\hat{t}$. It follows from the uniqueness of solutions that

$$\sum_{j=1}^{n} \beta_j \mathbf{x}^j(t)$$

must be the trivial solution; that is,

$$\sum_{j=1}^{n} \beta_j \mathbf{x}^j(t) = \mathbf{0} \qquad \text{for all } t \text{ in } (T_1, T_2) \tag{3-248}$$

This contradicts the fact that the $\mathbf{x}^j(t)$ are linearly independent functions. Hence,

$$\mathbf{\Phi}(t, t_0) \text{ is nonsingular} \tag{3-249}$$

for every $t$ in $(T_1, T_2)$.

Well, now that we know that $\mathbf{\Phi}(t, t_0)$ is nonsingular, let us try to find its inverse. We shall first show that

$$\mathbf{\Phi}(\hat{t}, t_1)\mathbf{\Phi}(t_1, t_0) = \mathbf{\Phi}(\hat{t}, t_0) \tag{3-250}$$

for any $\hat{t}$, $t$, $t_0$ in $(T_1, T_2)$. This is called the *transition property* of $\mathbf{\Phi}(t, t_0)$. Let $\mathbf{x}_0$ be an element of $R_n$. Then we know that

$$\mathbf{x}(t) = \mathbf{\Phi}(t, t_0)\mathbf{x}_0 \tag{3-251}$$

is the *unique* solution to the homogeneous part of Eq. (3-232), with $\mathbf{x}(t_0) = \mathbf{x}_0$. It follows that

$$\mathbf{x}(t_1) = \mathbf{\Phi}(t_1, t_0)\mathbf{x}_0 \tag{3-252}$$

and

$$\mathbf{x}(\hat{t}) = \mathbf{\Phi}(\hat{t}, t_0)\mathbf{x}_0 \tag{3-253}$$

If we set $\mathbf{x}_1 = \mathbf{x}(t_1)$, then we note that the *unique* solution $\mathbf{x}_1(t)$ of the homogeneous part of Eq. (3-232), with $\mathbf{x}_1(t_1) = \mathbf{x}_1$, is given by

$$\mathbf{x}_1(t) = \mathbf{\Phi}(t, t_1)\mathbf{x}_1 \tag{3-254}$$

It follows from the uniqueness that

$$\mathbf{x}(\hat{t}) = \mathbf{x}_1(\hat{t}) \tag{3-255}$$

and hence that

$$\mathbf{\Phi}(\hat{t}, t_0)\mathbf{x}_0 = \mathbf{\Phi}(\hat{t}, t_1)\mathbf{\Phi}(t_1, t_0)\mathbf{x}_0 \tag{3-256}$$

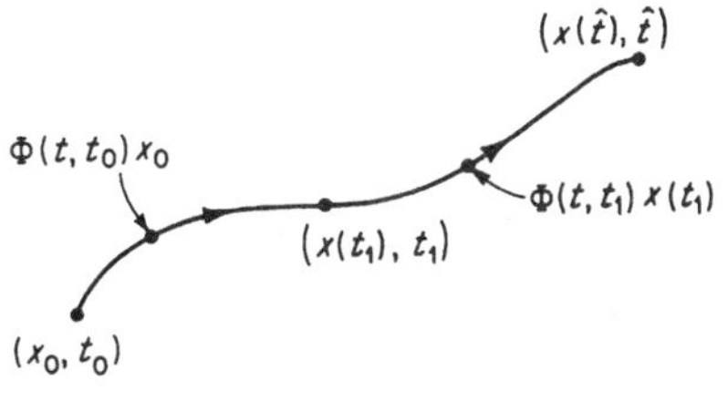


**Fig. 3-29** The transition property of the fundamental matrix.

(See Fig. 3-29.) Since Eq. (3-256) holds for every $\mathbf{x}_0$ in $R_n$, we see that Eq. (3-250) must be satisfied. Since $\mathbf{\Phi}(t_0, t_0) = \mathbf{I}$, we see that Eq. (3-250) implies that the inverse of $\mathbf{\Phi}(t, t_0)$ is $\mathbf{\Phi}(t_0, t)$, that is,

$$\mathbf{\Phi}^{-1}(t, t_0) = \mathbf{\Phi}(t_0, t) \tag{3-257}$$

With these properties of the fundamental matrix in mind, we are now ready to exhibit an "explicit" solution to the initial-value problem for the system of Eq. (3-232). We claim that

$$\mathbf{x}(t) = \mathbf{\Phi}(t, t_0)\left\{\mathbf{x}_0 + \int_{t_0}^{t} \mathbf{\Phi}^{-1}(\tau, t_0)\mathbf{v}(\tau)\, d\tau\right\} \tag{3-258}$$

$$= \mathbf{\Phi}(t, t_0)\left\{\mathbf{x}_0 + \int_{t_0}^{t} \mathbf{\Phi}(t_0, \tau)\mathbf{v}(\tau)\, d\tau\right\} \tag{3-259}$$

is the desired solution. First of all,

$$\mathbf{x}(t_0) = \mathbf{\Phi}(t_0, t_0)\left\{\mathbf{x}_0 + \int_{t_0}^{t_0} \mathbf{\Phi}^{-1}(\tau, t_0)\mathbf{v}(\tau)\, d\tau\right\} \tag{3-260}$$

$$= \mathbf{I}\mathbf{x}_0 = \mathbf{x}_0 \tag{3-261}$$

so that the initial condition is satisfied. Finally,

$$\begin{aligned}\dot{\mathbf{x}}(t) &= \dot{\mathbf{\Phi}}(t, t_0)\left\{\mathbf{x}_0 + \int_{t_0}^{t} \mathbf{\Phi}^{-1}(\tau, t_0)\mathbf{v}(\tau)\, d\tau\right\} \\ &\qquad + \mathbf{\Phi}(t, t_0)\frac{d}{dt}\left\{\mathbf{x}_0 + \int_{t_0}^{t} \mathbf{\Phi}^{-1}(\tau, t_0)\mathbf{v}(\tau)\, d\tau\right\} \\ &= \mathbf{A}(t)\mathbf{\Phi}(t, t_0)\left\{\mathbf{x}_0 + \int_{t_0}^{t} \mathbf{\Phi}^{-1}(\tau, t_0)\mathbf{v}(\tau)\, d\tau\right\} + \mathbf{\Phi}(t, t_0)\mathbf{\Phi}^{-1}(t, t_0)\mathbf{v}(t)\end{aligned} \tag{3-262}$$

$$= \mathbf{A}(t)\mathbf{x}(t) + \mathbf{v}(t)\dagger \qquad \text{using Eq. (3-242)} \tag{3-263}$$

which implies that $\mathbf{x}(t)$ is a solution of Eq. (3-232). We observe that $\mathbf{x}(t)$ is the sum of the two terms

$$\mathbf{\Phi}(t, t_0)\,\mathbf{x}_0 \qquad \text{and} \qquad \mathbf{\Phi}(t, t_0)\int_{t_0}^{t} \mathbf{\Phi}^{-1}(\tau, t_0)\mathbf{v}(\tau)\, d\tau \tag{3-264}$$

the first of which is a solution of the homogeneous part of the system and the second of which depends upon the forcing function. Now let us look at some examples.

**Example 3-55** Let us consider a system of the form

$$\begin{bmatrix}\dot{x}_1(t)\\ \dot{x}_2(t)\end{bmatrix} = \begin{bmatrix}0 & g(t)\\ -g(t) & 0\end{bmatrix}\begin{bmatrix}x_1(t)\\ x_2(t)\end{bmatrix} + \begin{bmatrix}v_1(t)\\ v_2(t)\end{bmatrix} \tag{3-265}$$

where $g(t)$ and the $v_i(t)$ are piecewise continuous. Let us try to find the solution of this system which satisfies the initial condition

$$\mathbf{x}(0) = \boldsymbol{\pi} \tag{3-266}$$

where $\boldsymbol{\pi}$ is a given element of $R_2$. The homogeneous part of the system has the form

$$\begin{aligned}\dot{x}_1(t) &= g(t)x_2(t)\\ \dot{x}_2(t) &= -g(t)x_1(t)\end{aligned} \tag{3-267}$$

† Note that this holds for $t$ in $(T_1, T_2) - A$, where $A$ is a countable set.

If we let $\mathbf{x}^1(t)$ and $\mathbf{x}^2(t)$ be given by

$$\mathbf{x}^1(t) = \begin{bmatrix} \cos \int_0^t g(\tau)\, d\tau \\ -\sin \int_0^t g(\tau)\, d\tau \end{bmatrix} \qquad \mathbf{x}^2(t) = \begin{bmatrix} \sin \int_0^t g(\tau)\, d\tau \\ \cos \int_0^t g(\tau)\, d\tau \end{bmatrix} \tag{3-268}$$

then we can see that $\mathbf{x}^1(t)$ and $\mathbf{x}^2(t)$ are solutions of Eqs. (3-267), with $\mathbf{x}^1(0) = \mathbf{e}_1$ and $\mathbf{x}^2(0) = \mathbf{e}_2$. Consequently, we take

$$\boldsymbol{\Phi}(t, 0) = \begin{bmatrix} \cos \int_0^t g(\tau)\, d\tau & \sin \int_0^t g(\tau)\, d\tau \\ -\sin \int_0^t g(\tau)\, d\tau & \cos \int_0^t g(\tau)\, d\tau \end{bmatrix} \tag{3-269}$$

and we see that the desired solution of Eq. (3-265) is given by

$$\mathbf{x}(t) = \boldsymbol{\Phi}(t, 0) \left\{ \boldsymbol{\pi} + \int_0^t \boldsymbol{\Phi}^{-1}(\tau, 0)\mathbf{v}(\tau)\, d\tau \right\} \tag{3-270}$$

**Exercise 3-17** Consider the system of Example 3-55. Show that
(*a*) $\boldsymbol{\Phi}(0, 0) = \mathbf{I}$.
(*b*) $\dot{\boldsymbol{\Phi}}(t, 0) = \begin{bmatrix} 0 & g(t) \\ -g(t) & 0 \end{bmatrix} \boldsymbol{\Phi}(t, 0)$.
(*c*) $\boldsymbol{\Phi}^{-1}(t, 0) = \boldsymbol{\Phi}(0, t)$.
(*d*) The Euclidean norm of $\boldsymbol{\Phi}(t, 0)\boldsymbol{\pi}$ is the same as the Euclidean norm of $\boldsymbol{\pi}$ (see Sec. 2-13).

Property *d* states that $\boldsymbol{\Phi}(t, 0)$ is an *orthogonal matrix* (see Sec. 2-14).

**Example 3-56** Let $\boldsymbol{\Psi}(t)$ be the matrix given by

$$\boldsymbol{\Psi}(t) = \begin{bmatrix} e^t & e^{2t} \\ 0 & 0 \end{bmatrix}$$

Then we note that the vector functions

$$\mathbf{x}^1(t) = \begin{bmatrix} e^t \\ 0 \end{bmatrix} \qquad \mathbf{x}^2(t) = \begin{bmatrix} e^{2t} \\ 0 \end{bmatrix}$$

are linearly independent functions but that $\mathbf{x}^1(0)$ and $\mathbf{x}^2(0)$ are linearly dependent elements of $R_2$. It follows that $\boldsymbol{\Psi}(t)$ cannot be the fundamental matrix of a linear system.

**Example 3-57** Let us consider the homogeneous system

$$\begin{bmatrix} \dot{x}_1(t) \\ \dot{x}_2(t) \end{bmatrix} = \begin{bmatrix} 0 & 1 \\ 0 & t \end{bmatrix} \begin{bmatrix} x_1(t) \\ x_2(t) \end{bmatrix} \tag{3-271}$$

It is easy to see that the fundamental matrix of this system has the form

$$\boldsymbol{\Phi}(t, t_0) = \begin{bmatrix} 1 & \int_{t_0}^t e^{-t_0^2/2} e^{\tau^2/2}\, d\tau \\ 0 & e^{-t_0^2/2} e^{t^2/2} \end{bmatrix} \tag{3-272}$$

and hence that the solution $\mathbf{x}(t)$ of the system with $\mathbf{x}(t_0) = \boldsymbol{\pi} = \begin{bmatrix} \pi_1 \\ \pi_2 \end{bmatrix}$ is given by

$$\mathbf{x}(t) = \boldsymbol{\Phi}(t, t_0)\boldsymbol{\pi} = \begin{bmatrix} \pi_1 + \pi_2 \int_{t_0}^t e^{-t_0^2/2} e^{\tau^2/2}\, d\tau \\ \pi_2 e^{-t_0^2/2} e^{t^2/2} \end{bmatrix} \tag{3-273}$$

**Exercise 3-18** Verify Eq. (3-250) for the matrix $\boldsymbol{\Phi}(t, t_0)$ given by Eq. (3-272).

Often we shall suppose that 0 is our initial time, i.e., that $t_0 = 0$. In this case, we shall simply denote the fundamental matrix of our system by $\mathbf{\Phi}(t)$ rather than by $\mathbf{\Phi}(t, 0)$. In other words, *from now on we shall write* $\mathbf{\Phi}(t)$ in place of $\mathbf{\Phi}(t, 0)$. This remark will be of particular importance in the next section.

## 3-21 Time-invariant Systems: The Exponential of A$t$

We shall now turn our attention to the special, but very important, case of time-invariant systems. In other words, we shall examine systems of the form

$$\dot{\mathbf{x}}(t) = \mathbf{A}\mathbf{x}(t) + \mathbf{v}(t) \tag{3-274}$$

where $\mathbf{A} = (a_{ij})$ is a (constant) $n \times n$ matrix, called the *system matrix* (see Definition 3-43). The system of Example 3-54 is such a time-invariant system, with $\mathbf{\Lambda}$ as system matrix. To begin with, we shall introduce some terminology which is useful in dealing with these systems. We have:

***Definition 3-44a*** *Given the linear system of Eq.* (3-274), *we shall say that the eigenvalues of the system matrix* $\mathbf{A}$ (*see Sec.* 2-10) *are the system eigenvalues* (*or the eigenvalues of the system*), *and we shall say that the fundamental matrix* $\mathbf{\Phi}(t)$ $[= \mathbf{\Phi}(t, 0)$ *in accordance with the concluding paragraph of Sec.* 3-20] *is the exponential of* $\mathbf{A}t$. *We then write* $e^{\mathbf{A}t}$ *in place of* $\mathbf{\Phi}(t)$, *that is,*

$$\mathbf{\Phi}(t) = e^{\mathbf{A}t} \tag{3-275}$$

(*compare with Example* 3-54).

We observe that if $\mathbf{B}$ is *any* $n \times n$ matrix, then the exponential of $\mathbf{B}t$, $e^{\mathbf{B}t}$, is defined, since it is the fundamental matrix of the homogeneous system

$$\dot{\mathbf{x}}(t) = \mathbf{B}\mathbf{x}(t) \tag{3-276}$$

Well, now let us see why we used the term "exponential of $\mathbf{A}t$" and, in so doing, determine the properties of $e^{\mathbf{A}t}$. We note, first of all, that, by virtue of Eq. (3-241),

$$\mathbf{\Phi}(0) = e^{\mathbf{A}0} = \mathbf{I} \tag{3-277}$$

Secondly, we observe that, in view of Eq. (3-242),

$$\frac{d}{dt} e^{\mathbf{A}t} = \mathbf{A}e^{\mathbf{A}t} \tag{3-278}$$

Thirdly, Eq. (3-250) tells us that

$$e^{\mathbf{A}(\hat{t}-t_1)}e^{\mathbf{A}t_1} = e^{\mathbf{A}\hat{t}} \tag{3-279}$$

so that

$$e^{\mathbf{A}(r+s)} = e^{\mathbf{A}r}e^{\mathbf{A}s} \tag{3-280}$$

Finally, we can see that the solution $\mathbf{x}(t)$ of Eq. (3-274) which satisfies the initial condition

$$\mathbf{x}(0) = \boldsymbol{\xi} \tag{3-281}$$

is given by

$$\mathbf{x}(t) = e^{\mathbf{A}t}\boldsymbol{\xi} + e^{\mathbf{A}t}\int_0^t e^{-\mathbf{A}\tau}\mathbf{v}(\tau)\,d\tau \tag{3-282a}$$

$$= e^{\mathbf{A}t}\boldsymbol{\xi} + \int_0^t e^{\mathbf{A}(t-\tau)}\mathbf{v}(\tau)\,d\tau \tag{3-282b}$$

Well, so far we have not really said anything new; now, however, let us show that $\mathbf{A}$ and $e^{\mathbf{A}t}$ commute and then that we may view $e^{\mathbf{A}t}$ as the sum of an infinite series of matrices. Suppose that $\mathbf{v}$ is *any* element of $R_n$; then we claim that both $\mathbf{A}e^{\mathbf{A}t}\mathbf{v}$ and $e^{\mathbf{A}t}\mathbf{A}\mathbf{v}$ are solutions of the system

$$\dot{\mathbf{x}}(t) = \mathbf{A}\mathbf{x}(t) \tag{3-283}$$

satisfying the initial condition

$$\mathbf{x}(0) = \mathbf{A}\mathbf{v} \tag{3-284}$$

Clearly,

$$\mathbf{A}e^{\mathbf{A}0}\mathbf{v} = \mathbf{A}\mathbf{I}\mathbf{v} = \mathbf{A}\mathbf{v} \tag{3-285}$$

and

$$e^{\mathbf{A}0}\mathbf{A}\mathbf{v} = \mathbf{I}\mathbf{A}\mathbf{v} = \mathbf{A}\mathbf{v} \tag{3-286}$$

so that both $\mathbf{A}e^{\mathbf{A}t}\mathbf{v}$ and $e^{\mathbf{A}t}\mathbf{A}\mathbf{v}$ satisfy Eq. (3-284). Moreover, by virtue of Eq. (3-278), we see that

$$\frac{d}{dt}(\mathbf{A}e^{\mathbf{A}t}\mathbf{v}) = \mathbf{A}\frac{d}{dt}(e^{\mathbf{A}t})\mathbf{v} = \mathbf{A}(\mathbf{A}e^{\mathbf{A}t}\mathbf{v}) \tag{3-287}$$

and that

$$\frac{d}{dt}(e^{\mathbf{A}t}\mathbf{A}\mathbf{v}) = \frac{d}{dt}(e^{\mathbf{A}t})\mathbf{A}\mathbf{v} = \mathbf{A}(e^{\mathbf{A}t}\mathbf{A}\mathbf{v}) \tag{3-288}$$

It follows from the uniqueness of solutions of Eq. (3-283) that, for all $t$,

$$\mathbf{A}e^{\mathbf{A}t}\mathbf{v} = e^{\mathbf{A}t}\mathbf{A}\mathbf{v} \tag{3-289}$$

and hence, since $\mathbf{v}$ was any element of $R_n$,† that $\mathbf{A}$ and $e^{\mathbf{A}t}$ commute, i.e., that, for all $t$,

$$\mathbf{A}e^{\mathbf{A}t} = e^{\mathbf{A}t}\mathbf{A} \tag{3-290}$$

Now let us show heuristically that $e^{\mathbf{A}t}$ may be viewed as the infinite series‡

$$\mathbf{I} + \mathbf{A}t + \mathbf{A}^2\frac{t^2}{2!} + \cdots + \mathbf{A}^n\frac{t^n}{n!} + \cdots \tag{3-291}$$

or, more compactly, that

$$e^{\mathbf{A}t} = \sum_{k=0}^{\infty} \mathbf{A}^k \frac{t^k}{k!} \tag{3-292}$$

† Note that $(\mathbf{A}e^{\mathbf{A}t} - e^{\mathbf{A}t}\mathbf{A})\mathbf{v} = \mathbf{0}$ for all $\mathbf{v}$ in $R_n$ and use the remark at the end of Sec. 2-7.

‡ If we define the norm of a matrix $\mathbf{A}$, $\|\mathbf{A}\|$, by setting $\|\mathbf{A}\| = \max\{|a_{ij}|: i, j = 1, 2, \ldots, n\}$, then we can show that the series of Eq. (3-291) converges in this norm for all $t$ and can be differentiated term by term. Bearing this in mind, our heuristic justification of Eq. (3-292) can be made rigorous. We warn the reader that the entries in $e^{\mathbf{A}t}$ are *not*, in general, the functions $e^{a_{ij}t}$.

We observe that at $t = 0$ the series becomes

$$\mathbf{I} + \mathbf{A}0 + \mathbf{A}^2 \frac{0}{2!} + \cdots = \mathbf{I} \tag{3-293}$$

Moreover, if we differentiate the series term by term, then we obtain the relation

$$\frac{d}{dt}\left(\sum_{k=0}^{\infty} \mathbf{A}^k \frac{t^k}{k!}\right) = \sum_{k=0}^{\infty} \frac{\mathbf{A}^k}{k!}\frac{dt^k}{dt} = \sum_{k=0}^{\infty} \mathbf{A}^k \frac{kt^{k-1}}{k!} \tag{3-294}$$

$$= \mathbf{A} \sum_{p=0}^{\infty} \mathbf{A}^p \frac{t^p}{p!} \tag{3-295}$$

which implies that the series satisfies the matrix differential equation

$$\dot{\mathbf{X}}(t) = \mathbf{A}\mathbf{X}(t) \tag{3-296}$$

[$\mathbf{X}(t)$ being a variable $n \times n$ matrix]. It follows that if $\mathbf{v}$ is any element of $R_n$, then $e^{\mathbf{A}t}\mathbf{v}$ and

$$\sum_{k=0}^{\infty} \mathbf{A}^k \frac{t^k}{k!} \mathbf{v}$$

are both solutions of Eq. (3-283) which satisfy the initial condition $\mathbf{x}(0) = \mathbf{v}$ and hence that Eq. (3-292) is satisfied.

Finally, let us suppose that $\mathbf{B}$ is an $n \times n$ matrix which is similar to $\mathbf{A}$ [see Eq. (2-64)], that is, that there is a nonsingular matrix $\mathbf{P}$ with

$$\mathbf{B} = \mathbf{P}^{-1}\mathbf{A}\mathbf{P} \tag{3-297}$$

and let us show that

$$e^{\mathbf{B}t} = \mathbf{P}^{-1}e^{\mathbf{A}t}\mathbf{P} \tag{3-298}$$

Again, let $\mathbf{v}$ be *any* element of $R_n$ and consider the functions $e^{\mathbf{B}t}\mathbf{v}$ and $\mathbf{P}^{-1}e^{\mathbf{A}t}\mathbf{P}\mathbf{v}$. We shall show that both of these functions are solutions of the system

$$\dot{\mathbf{x}}(t) = \mathbf{B}\mathbf{x}(t) \tag{3-299}$$

satisfying the initial condition $\mathbf{x}(0) = \mathbf{v}$. It will then follow that Eq. (3-298) is satisfied. Clearly, $e^{\mathbf{B}t}\mathbf{v}$ satisfies Eq. (3-299) and the condition $\mathbf{x}(0) = \mathbf{v}$. Now,

$$\mathbf{P}^{-1}e^{\mathbf{A}0}\mathbf{P}\mathbf{v} = \mathbf{P}^{-1}\mathbf{I}\mathbf{P}\mathbf{v} = \mathbf{P}^{-1}\mathbf{P}\mathbf{v} = \mathbf{I}\mathbf{v} = \mathbf{v} \tag{3-300}$$

and

$$\frac{d}{dt}(\mathbf{P}^{-1}e^{\mathbf{A}t}\mathbf{P})\mathbf{v} = \mathbf{P}^{-1}\frac{d}{dt}(e^{\mathbf{A}t})\mathbf{P}\mathbf{v} \tag{3-301}$$

$$= \mathbf{P}^{-1}\mathbf{A}e^{\mathbf{A}t}\mathbf{P}\mathbf{v} \tag{3-302}$$

$$= \mathbf{P}^{-1}\mathbf{A}(\mathbf{P}\mathbf{P}^{-1})e^{\mathbf{A}t}\mathbf{P}\mathbf{v} \tag{3-303}$$

$$= \mathbf{P}^{-1}\mathbf{A}\mathbf{P}\mathbf{P}^{-1}e^{\mathbf{A}t}\mathbf{P}\mathbf{v} \tag{3-304}$$

$$= \mathbf{B}(\mathbf{P}^{-1}e^{\mathbf{A}t}\mathbf{P}\mathbf{v}) \tag{3-305}$$

so that $\mathbf{P}^{-1}e^{\mathbf{A}t}\mathbf{P}\mathbf{v}$ also satisfies Eq. (3-299) and the condition $\mathbf{x}(0) = \mathbf{v}$. In other words, we have shown that if $\mathbf{A}$ and $\mathbf{B}$ are similar matrices, then $e^{\mathbf{A}t}$ and $e^{\mathbf{B}t}$ are similar matrices.

**Example 3-58** Let $\mathbf{A}$ be the $2 \times 2$ matrix $\begin{bmatrix} 1 & 1 \\ 0 & 0 \end{bmatrix}$. Then it is easy to see that $\mathbf{A}^2 = \mathbf{A}$ and hence that

$$e^{\mathbf{A}t} = \mathbf{I} + \mathbf{A}\left(t + \frac{t^2}{2!} + \cdots\right) = \mathbf{I} + (e^t - 1)\mathbf{A} = \begin{bmatrix} e^t & e^t - 1 \\ 0 & 1 \end{bmatrix}$$

We note that $e^{\mathbf{A}t}$ is very different from the matrix

$$\begin{bmatrix} e^t & e^t \\ 1 & 1 \end{bmatrix}$$

which is obtained by taking the exponentials of the entries in $\mathbf{A}$.

## 3-22 Time-invariant Systems: Reduction to Canonical Form

Consider the initial-value problem

$$\dot{\mathbf{x}}(t) = \mathbf{A}\mathbf{x}(t) \qquad \mathbf{x}(0) = \boldsymbol{\xi} \tag{3-306}$$

We know that the solution of this problem is given by the function

$$\mathbf{x}(t) = e^{\mathbf{A}t}\boldsymbol{\xi} \tag{3-307}$$

What we propose to do now is to relate this solution of our problem to the eigenvalues of the system (i.e., to the eigenvalues of $\mathbf{A}$).† We shall first examine the effects of a similarity transformation [see Eqs. (2-62*a*) and (2-62*b*)] and then consider the case where the eigenvalues of $\mathbf{A}$ are real and distinct.

Suppose that we consider the linear transformation $\mathcal{P}^{-1}$ of $R_n$ into itself, which is given by

$$\mathcal{P}^{-1}(\mathbf{v}) = \mathbf{P}^{-1}\mathbf{v} \qquad \mathbf{v} \in R_n \tag{3-308}$$

where $\mathbf{P}$ is a nonsingular $n \times n$ matrix [see Eq. (2-50)]. Then we can define a linear transformation $\mathcal{P}^{-1}$ on the vector space of all $R_n$-valued functions on $R$ by setting

$$\mathcal{P}^{-1}[\mathbf{x}(t)] = \mathbf{P}^{-1}\mathbf{x}(t) \tag{3-309}$$

We shall usually write

$$\mathbf{y}(t) = \mathcal{P}^{-1}[\mathbf{x}(t)] = \mathbf{P}^{-1}\mathbf{x}(t) \tag{3-310}$$

It is easy to see that

$$\dot{\mathbf{y}}(t) = \mathbf{P}^{-1}\dot{\mathbf{x}}(t) \tag{3-311}$$

† See Definition 3-44 and Sec. 2-10.

and hence that Eq. (3-306) is transformed into the equation

$$\dot{\mathbf{y}}(t) = \mathbf{P}^{-1}\mathbf{A}\mathbf{x}(t) \tag{3-312}$$
$$= \mathbf{P}^{-1}\mathbf{A}\mathbf{P}\mathbf{P}^{-1}\mathbf{x}(t) = \mathbf{P}^{-1}\mathbf{A}\mathbf{P}\mathbf{y}(t) \tag{3-313}$$

with the initial condition

$$\mathbf{y}(0) = \mathbf{P}^{-1}\mathbf{x}(0) = \mathbf{P}^{-1}\boldsymbol{\xi} \tag{3-314}$$

If we let **B** be the matrix, given by

$$\mathbf{B} = \mathbf{P}^{-1}\mathbf{A}\mathbf{P} \tag{3-315}$$

then **B** is similar to **A**, and we have

$$\dot{\mathbf{y}}(t) = \mathbf{B}\mathbf{y}(t) \qquad \mathbf{y}(0) = \mathbf{P}^{-1}\boldsymbol{\xi} \tag{3-316}$$

The solution to the initial-value problem of Eq. (3-316) is, by virtue of Eq. (3-282*a*),

$$\mathbf{y}(t) = e^{\mathbf{B}t}\mathbf{P}^{-1}\boldsymbol{\xi} \tag{3-317}$$

and it follows that

$$\mathbf{x}(t) = \mathbf{P}\mathbf{y}(t) = \mathbf{P}e^{\mathbf{B}t}\mathbf{P}^{-1}\boldsymbol{\xi} \tag{3-318}$$

We also note that, in view of Eq. (2-74), *the eigenvalues of the systems* (3-306) *and* (3-316) *are the same.*

Suppose now that the eigenvalues of **A** are *real* and *distinct.* We let $\lambda_1, \lambda_2, \ldots, \lambda_n$ denote these eigenvalues, and we let $\boldsymbol{\Lambda}$ be the matrix of the eigenvalues [see Eq. (2-78)]; that is,

$$\boldsymbol{\Lambda} = \begin{bmatrix} \lambda_1 & 0 & \cdots & 0 \\ 0 & \lambda_2 & \cdots & 0 \\ \cdots & \cdots & \cdots & \cdots \\ 0 & 0 & \cdots & \lambda_n \end{bmatrix} \tag{3-319}$$

Then we know (Sec. 2-10) that there is a nonsingular matrix **P** with

$$\boldsymbol{\Lambda} = \mathbf{P}^{-1}\mathbf{A}\mathbf{P} \tag{3-320}$$

It follows from Eq. (3-318) that the solution $\mathbf{x}(t)$ of Eqs. (3-306) is given by

$$\mathbf{x}(t) = \mathbf{P}e^{\boldsymbol{\Lambda}t}\mathbf{P}^{-1}\boldsymbol{\xi} \tag{3-321}$$

$$= \mathbf{P}\begin{bmatrix} e^{\lambda_1 t} & 0 & \cdots & 0 \\ 0 & e^{\lambda_2 t} & \cdots & 0 \\ \cdots & \cdots & \cdots & \cdots \\ 0 & 0 & \cdots & e^{\lambda_n t} \end{bmatrix}\mathbf{P}^{-1}\boldsymbol{\xi} \tag{3-322}$$

and it follows from Eq. (3-317) that the solution $\mathbf{y}(t)$ of Eqs. (3-316) is given by

$$\mathbf{y}(t) = e^{\mathbf{A}t}\boldsymbol{\eta} \qquad \text{where } \boldsymbol{\eta} = \mathbf{P}^{-1}\boldsymbol{\xi} \tag{3-323}$$

$$= \begin{bmatrix} e^{\lambda_1 t}\eta_1 \\ e^{\lambda_2 t}\eta_2 \\ \cdot \\ \cdot \\ \cdot \\ e^{\lambda_n t}\eta_n \end{bmatrix} \qquad \boldsymbol{\eta} = \begin{bmatrix} \eta_1 \\ \eta_2 \\ \cdot \\ \cdot \\ \cdot \\ \eta_n \end{bmatrix} \tag{3-324}$$

Thus, we have related the solution of our problem to the eigenvalues of **A**.

More generally, if we suppose only that the eigenvalues of **A** are real, then, letting

$$\mathbf{J} = \mathbf{J}(\mathbf{A}) \tag{3-325}$$

denote the Jordan canonical form of **A** [see Eq. (2-80)], we note that there is a nonsingular matrix **P** with

$$\mathbf{J} = \mathbf{P}^{-1}\mathbf{A}\mathbf{P} \tag{3-326}$$

and that we can again obtain $\mathbf{x}(t)$ by using the similarity transformation of Eq. (3-309), in the form

$$\mathbf{x}(t) = \mathbf{P}e^{\mathbf{J}t}\mathbf{P}^{-1}\boldsymbol{\xi} \tag{3-327}$$

It can be shown that if

$$\mathbf{J} = \begin{bmatrix} \mathbf{J}_0 & \mathbf{0} & \cdots & \mathbf{0} \\ \mathbf{0} & \mathbf{J}_1 & \cdots & \mathbf{0} \\ \cdots & \cdots & \cdots & \cdots \\ \mathbf{0} & \mathbf{0} & \cdots & \mathbf{J}_p \end{bmatrix} \tag{3-328}$$

with

$$\mathbf{J}_0 = \begin{bmatrix} \lambda_1 & 0 & \cdots & 0 \\ 0 & \lambda_2 & \cdots & 0 \\ \cdots & \cdots & \cdots & \cdots \\ 0 & 0 & \cdots & \lambda_k \end{bmatrix} \tag{3-329}$$

and

$$\mathbf{J}_m = \begin{bmatrix} \lambda_{k+m} & 1 & 0 & \cdots & 0 \\ 0 & \lambda_{k+m} & 1 & \cdots & 0 \\ \cdots & \cdots & \cdots & \cdots & \cdots \\ 0 & 0 & 0 & \cdots & 1 \\ 0 & 0 & 0 & \cdots & \lambda_{k+m} \end{bmatrix} \tag{3-330}$$

where $\lambda_1, \lambda_2, \ldots, \lambda_k, \lambda_{k+1}, \ldots, \lambda_{k+p}$ are the eigenvalues of **A**, then

$$e^{t\mathbf{J}} = \begin{bmatrix} e^{t\mathbf{J}_0} & \mathbf{0} & \cdots & \mathbf{0} \\ \mathbf{0} & e^{t\mathbf{J}_1} & \cdots & \mathbf{0} \\ \cdots & \cdots & \cdots & \cdots \\ \mathbf{0} & \mathbf{0} & \cdots & e^{t\mathbf{J}_p} \end{bmatrix} \tag{3-331}$$

with
$$e^{t\mathbf{J}_0} = \begin{bmatrix} e^{t\lambda_1} & 0 & \cdots & 0 \\ 0 & e^{t\lambda_2} & \cdots & 0 \\ \cdots & \cdots & \cdots & \cdots \\ 0 & 0 & \cdots & e^{t\lambda_k} \end{bmatrix} \tag{3-332}$$

and
$$e^{t\mathbf{J}_m} = e^{t\lambda_{k+m}} \begin{bmatrix} 1 & t & t^2/2! & \cdots & t^{\nu_m-1}/(\nu_m - 1)! \\ 0 & 1 & t & \cdots & t^{\nu_m-2}/(\nu_m - 2)! \\ \cdots & \cdots & \cdots & \cdots & \cdots \\ 0 & 0 & 0 & \cdots & 1 \end{bmatrix} \tag{3-333}$$

where $\mathbf{J}_m$ is a $\nu_m \times \nu_m$ matrix. (See Ref. C-7.) Thus, we have again related the solution of our problem to the eigenvalues of $\mathbf{A}$.

If we introduce a forcing function $\mathbf{v}(t)$ into our considerations—in other words, if we consider the problem

$$\dot{\mathbf{x}}(t) = \mathbf{A}\mathbf{x}(t) + \mathbf{v}(t) \qquad \mathbf{x}(0) = \boldsymbol{\xi} \tag{3-334}$$

then we can readily see that application of the similarity transformation

$$\mathbf{y}(t) = \mathbf{P}^{-1}\mathbf{x}(t) \tag{3-335}$$

will lead us to the problem

$$\dot{\mathbf{y}}(t) = \mathbf{P}^{-1}\mathbf{A}\mathbf{P}\mathbf{y}(t) + \mathbf{P}^{-1}\mathbf{v}(t) \qquad \mathbf{y}(0) = \mathbf{P}^{-1}\boldsymbol{\xi} \tag{3-336}$$

If we let $\mathbf{B} = \mathbf{P}^{-1}\mathbf{A}\mathbf{P}$, then we note that the solution of Eq. (3-336) is given by

$$\mathbf{y}(t) = e^{\mathbf{B}t}\left\{\mathbf{P}^{-1}\boldsymbol{\xi} + \int_0^t e^{-\mathbf{B}\tau}\mathbf{P}^{-1}\mathbf{v}(\tau)\, d\tau\right\} \tag{3-337}$$

and hence that the solution $\mathbf{x}(t)$ of Eqs. (3-334) is given by

$$\mathbf{x}(t) = \mathbf{P}e^{\mathbf{B}t}\left\{\mathbf{P}^{-1}\boldsymbol{\xi} + \int_0^t e^{-\mathbf{B}\tau}\mathbf{P}^{-1}\mathbf{v}(\tau)\, d\tau\right\} \tag{3-338}$$

In particular, if the eigenvalues $\lambda_1, \lambda_2, \ldots, \lambda_n$ of $\mathbf{A}$ are real and distinct and if $\boldsymbol{\Lambda}$ is the matrix of the eigenvalues [see Eqs. (2-78) and (3-319)], then

$$\mathbf{x}(t) = \mathbf{P}e^{\boldsymbol{\Lambda}t}\mathbf{P}^{-1}\boldsymbol{\xi} + \mathbf{P}\int_0^t e^{\boldsymbol{\Lambda}(t-\tau)}\,\mathbf{P}^{-1}\mathbf{v}(\tau)\, d\tau \tag{3-339}$$

We shall make frequent use of this formula and the similarity transformation of Eq. (3-335) in the remainder of this book.

## 3-23 Time-invariant Systems: Evaluation of the Fundamental Matrix by Means of the Laplace Transform

In this section, we shall show how the fundamental matrix of a time-invariant linear system can be evaluated by means of the Laplace transform.

We assume that the reader is familiar with the notion of the Laplace transform and its basic properties.

If $x(t)$ is a real-valued function of "time," then we let $x(s)$ denote the Laplace transform of $x(t)$ and we write

$$x(s) = \mathcal{L}[x(t)] \tag{3-340}$$

If we are given $x(s)$, then we may take the inverse Laplace transform to evaluate $x(t)$, and we write

$$x(t) = \mathcal{L}^{-1}[x(s)] \tag{3-341}$$

We note that there are extensive tables available for determining inverse Laplace transforms. The crucial property of the Laplace transform for our purposes can be stated as follows: Given the differentiable function $x(t)$ with $x(0) = \xi$, then the Laplace transform of the derivative $\dot{x}(t)$ of $x(t)$ is given by

$$\mathcal{L}[\dot{x}(t)] = s\mathcal{L}[x(t)] - \xi \tag{3-342}$$
$$= sx(s) - \xi \tag{3-343}$$

**Example 3-59** Let $x(t) = e^{at}$. Then $x(s) = \mathcal{L}[e^{at}] = 1/(s - a)$ and $\mathcal{L}[\dot{x}(t)] = sx(s) - 1 = s/(s - a) - 1 = a/s - a$.

Suppose now that $\mathbf{x}(t)$ is an $R_n$-valued function of "time" (that is, on $R$) with components $x_1(t), x_2(t), \ldots, x_n(t)$; then we may define the Laplace transform of $\mathbf{x}(t)$ as the $n$ vector whose components are the Laplace transforms of the components of $\mathbf{x}(t)$. We write

$$\mathbf{x}(s) = \mathcal{L}[\mathbf{x}(t)] \tag{3-344}$$

and we note that the components $x_i(s)$ of $\mathbf{x}(s)$ are given by

$$x_i(s) = \mathcal{L}[x_i(t)] \qquad i = 1, 2, \ldots, n \tag{3-345}$$

Furthermore, if $\mathbf{x}(t)$ is differentiable and if $\mathbf{x}(0) = \boldsymbol{\xi}$, then we have

$$\mathcal{L}[\dot{\mathbf{x}}(t)] = s\mathcal{L}[\mathbf{x}(t)] - \boldsymbol{\xi} \tag{3-346}$$
$$= s\mathbf{x}(s) - \boldsymbol{\xi} \tag{3-347}$$

Finally, if $\mathbf{C}(t)$ is an $n \times m$ matrix with entries $c_{ij}(t)$, then we may define the Laplace transform $\mathbf{C}(s)$ of $\mathbf{C}(t)$ to be the $n \times m$ matrix whose entries $c_{ij}(s)$ are the Laplace transforms of the $c_{ij}(t)$. We write

$$\mathbf{C}(s) = \mathcal{L}[\mathbf{C}(t)] \tag{3-348}$$

and we note that

$$c_{ij}(s) = \mathcal{L}[c_{ij}(t)] \tag{3-349}$$

**Example 3-60** Suppose that

$$\mathbf{x}(t) = \begin{bmatrix} e^{-t} \\ \cos \omega t \end{bmatrix} \quad \text{and} \quad \mathbf{C}(t) = \begin{bmatrix} k & \sin t \\ t & e^{-2t} \end{bmatrix} \tag{3-350}$$

Then

$$\mathbf{x}(s) = \mathcal{L}[\mathbf{x}(t)] = \begin{bmatrix} 1/s + 1 \\ s/s^2 + \omega^2 \end{bmatrix} \tag{3-351}$$

$$\mathcal{L}[\dot{\mathbf{x}}(t)] = s\mathbf{x}(s) - \boldsymbol{\xi} = \begin{bmatrix} s/s + 1 \\ s^2/s^2 + \omega^2 \end{bmatrix} - \begin{bmatrix} 1 \\ 1 \end{bmatrix} = \begin{bmatrix} -1/s + 1 \\ -\omega^2/s^2 + \omega^2 \end{bmatrix} \tag{3-352}$$

$$\mathbf{C}(s) = \mathcal{L}[\mathbf{C}(t)] = \begin{bmatrix} k/s & 1/s^2 + 1 \\ 1/s^2 & 1/s + 2 \end{bmatrix} \tag{3-353}$$

Now let us use these concepts to calculate the fundamental matrix for time-invariant systems. Suppose, first of all, that we consider the initial-value problem for the homogeneous equation

$$\dot{\mathbf{x}}(t) = \mathbf{A}\mathbf{x}(t) \qquad \mathbf{x}(0) = \boldsymbol{\xi} \tag{3-354}$$

Then we know that the solution is given by

$$\mathbf{x}(t) = e^{\mathbf{A}t}\boldsymbol{\xi} \tag{3-355}$$

and we want to evaluate the fundamental matrix $e^{\mathbf{A}t}$. Taking the Laplace transform of Eq. (3-354), we find that

$$s\mathbf{x}(s) - \boldsymbol{\xi} = \mathbf{A}\mathbf{x}(s) \tag{3-356}$$

and hence that

$$s\mathbf{x}(s) - \mathbf{A}\mathbf{x}(s) = \boldsymbol{\xi} \tag{3-357}$$

or, equivalently,

$$(s\mathbf{I} - \mathbf{A})\mathbf{x}(s) = \boldsymbol{\xi} \tag{3-358}$$

If we let

$$\mathbf{Q}(s) = s\mathbf{I} - \mathbf{A} \tag{3-359}$$

then we can assert that

$$\det \mathbf{Q}(s) = 0 \qquad \text{if and only if } s = \text{some } \lambda_i \tag{3-360}$$

where $\lambda_1, \lambda_2, \ldots, \lambda_n$ are the eigenvalues of $\mathbf{A}$, since

$$\det \mathbf{Q}(s) = \det (s\mathbf{I} - \mathbf{A}) = 0 \tag{3-361}$$

implies, and is implied by, the relation

$$\det (\mathbf{A} - s\mathbf{I}) = 0 \tag{3-362}$$

[see Eq. (2-70)]. It follows that $\mathbf{Q}(s)$ is nonsingular for all $s \neq \lambda_i$, $i = 1, 2, \ldots, n$. So we have

$$\mathbf{x}(s) = \mathbf{Q}^{-1}(s)\boldsymbol{\xi} = (s\mathbf{I} - \mathbf{A})^{-1}\boldsymbol{\xi} \tag{3-363}$$

and, taking the inverse Laplace transform,

$$\mathbf{x}(t) = \mathfrak{L}^{-1}[\mathbf{x}(s)] = \mathfrak{L}^{-1}[(s\mathbf{I} - \mathbf{A})^{-1}]\boldsymbol{\xi} \tag{3-364}$$

Since the solution of our system is unique, we may conclude that

$$e^{\mathbf{A}t} = \mathfrak{L}^{-1}[(s\mathbf{I} - \mathbf{A})^{-1}] \tag{3-365}$$

In other words, we must form the matrix $s\mathbf{I} - \mathbf{A}$, invert it, and then find the inverse Laplace transform of $(s\mathbf{I} - \mathbf{A})^{-1}$ in order to evaluate the fundamental matrix $e^{\mathbf{A}t}$.

**Example 3-61** Given the system

$$\begin{bmatrix} \dot{x}_1(t) \\ \dot{x}_2(t) \end{bmatrix} = \begin{bmatrix} 0 & 1 \\ 6 & 1 \end{bmatrix} \begin{bmatrix} x_1(t) \\ x_2(t) \end{bmatrix} \tag{3-366}$$

with system matrix

$$\mathbf{A} = \begin{bmatrix} 0 & 1 \\ 6 & 1 \end{bmatrix} \tag{3-367}$$

Then
$$\mathbf{Q}(s) = s\mathbf{I} - \mathbf{A} = \begin{bmatrix} s & -1 \\ -6 & s-1 \end{bmatrix} \tag{3-368}$$

and
$$\det \mathbf{Q}(s) = s(s-1) - 6 = (s-3)(s+2) \tag{3-369}$$

and
$$\mathbf{Q}^{-1}(s) = (s\mathbf{I} - \mathbf{A})^{-1} = \frac{1}{(s-3)(s+2)} \begin{bmatrix} s-1 & 1 \\ 6 & s \end{bmatrix} \tag{3-370}$$

$$= \begin{bmatrix} \dfrac{s-1}{(s-3)(s+2)} & \dfrac{1}{(s-3)(s+2)} \\ \dfrac{6}{(s-3)(s+2)} & \dfrac{s}{(s-3)(s+2)} \end{bmatrix} \tag{3-371}$$

It follows that

$$e^{\mathbf{A}t} = \mathfrak{L}^{-1}[(s\mathbf{I} - \mathbf{A})^{-1}] = \begin{bmatrix} \dfrac{2e^{3t} + 3e^{-2t}}{5} & \dfrac{e^{3t} - e^{-2t}}{5} \\ \dfrac{6e^{3t} - 6e^{-2t}}{5} & \dfrac{3e^{3t} + 2e^{-2t}}{5} \end{bmatrix} \tag{3-372}$$

is the fundamental matrix of the system (3-366).

**Example 3-62** Given the system

$$\begin{bmatrix} \dot{x}_1(t) \\ \dot{x}_2(t) \end{bmatrix} = \begin{bmatrix} -2 & 3 \\ -3 & -2 \end{bmatrix} \begin{bmatrix} x_1(t) \\ x_2(t) \end{bmatrix} \tag{3-373}$$

with system matrix

$$\mathbf{A} = \begin{bmatrix} -2 & 3 \\ -3 & -2 \end{bmatrix} \tag{3-374}$$

Then
$$\mathbf{Q}(s) = s\mathbf{I} - \mathbf{A} = \begin{bmatrix} s+2 & -3 \\ 3 & s+2 \end{bmatrix} \tag{3-375}$$

and
$$\det \mathbf{Q}(s) = (s+2)^2 + 9 \tag{3-376}$$

and
$$\mathbf{Q}^{-1}(s) = (s\mathbf{I} - \mathbf{A})^{-1} = \begin{bmatrix} \dfrac{s+2}{(s+2)^2+9} & \dfrac{3}{(s+2)^2+9} \\ \dfrac{-3}{(s+2)^2+9} & \dfrac{s+2}{(s+2)^2+9} \end{bmatrix} \tag{3-377}$$

It follows that

$$e^{\mathbf{A}t} = \mathfrak{L}^{-1}[(s\mathbf{I} - \mathbf{A})^{-1}] = \begin{bmatrix} e^{-2t}\cos 3t & e^{-2t}\sin 3t \\ -e^{-2t}\sin 3t & e^{-2t}\cos 3t \end{bmatrix} \tag{3-378}$$

Let us turn our attention to the problem

$$\dot{\mathbf{x}}(t) = \mathbf{A}\mathbf{x}(t) + \mathbf{v}(t) \qquad \mathbf{x}(0) = \boldsymbol{\xi} \tag{3-379}$$

with $\mathbf{v}(t)$ as our forcing function. We have shown that the solution $\mathbf{x}(t)$ of this problem is given by

$$\mathbf{x}(t) = e^{\mathbf{A}t}\boldsymbol{\xi} + e^{\mathbf{A}t}\int_0^t e^{-\mathbf{A}\tau}\mathbf{v}(\tau)\,d\tau \tag{3-380}$$

Now let us give an interpretation of this formula based upon the Laplace transform. We can compute $\mathbf{x}(t)$ by taking the Laplace transform of both sides of Eq. (3-379). Doing so, we obtain the relation

$$s\mathbf{x}(s) - \boldsymbol{\xi} = \mathbf{A}\mathbf{x}(s) + \mathbf{v}(s) \tag{3-381}$$

It follows that

$$(s\mathbf{I} - \mathbf{A})\mathbf{x}(s) = \boldsymbol{\xi} + \mathbf{v}(s) \tag{3-382}$$

Letting $\mathbf{Q}(s) = s\mathbf{I} - \mathbf{A}$ as before [Eq. (3-359)], we see that

$$\mathbf{x}(s) = \mathbf{Q}^{-1}(s)\boldsymbol{\xi} + \mathbf{Q}^{-1}(s)\mathbf{v}(s) \tag{3-383}$$

and hence, taking inverse Laplace transforms, that

$$\mathbf{x}(t) = \mathcal{L}^{-1}[\mathbf{Q}^{-1}(s)]\boldsymbol{\xi} + \mathcal{L}^{-1}[\mathbf{Q}^{-1}(s)\mathbf{v}(s)] \tag{3-384}$$

Now we know that

$$e^{\mathbf{A}t} = \mathcal{L}^{-1}[\mathbf{Q}^{-1}(s)] \tag{3-385}$$

but what is $\mathcal{L}^{-1}[\mathbf{Q}^{-1}(s)\mathbf{v}(s)]$? Well, let us recall that if $f_1(t)$ and $f_2(t)$ are real-valued functions of $t$, with

$$\mathcal{L}[f_1(t)] = f_1(s) \qquad \mathcal{L}[f_2(t)] = f_2(s) \tag{3-386}$$

then

$$f_1(s)f_2(s) = \mathcal{L}[(f_1 * f_2)(t)] \tag{3-387}$$

where $(f_1 * f_2)(t)$ is the *convolution of $f_1$ and $f_2$* and is given by

$$(f_1 * f_2)(t) = \int_0^t f_1(t - \tau)f_2(\tau)\,d\tau \tag{3-388}$$

Clearly, then,

$$\mathcal{L}^{-1}[f_1(s)f_2(s)] = (f_1 * f_2)(t) = \int_0^t f_1(t - \tau)f_2(\tau)\,d\tau \tag{3-389}$$

We leave it to the reader to verify that

$$\mathcal{L}^{-1}[\mathbf{Q}^{-1}(s)\mathbf{v}(s)] = \int_0^t e^{\mathbf{A}(t-\tau)}\mathbf{v}(\tau)\,d\tau \tag{3-390}$$

Thus, we may interpret the somewhat formidable expression

$$e^{\mathbf{A}t}\int_0^t e^{-\mathbf{A}\tau}\mathbf{v}(\tau)\,d\tau$$

in Eq. (3-380) as nothing more than the convolution of the fundamental matrix $e^{\mathbf{A}t}$ with the forcing function $\mathbf{v}(t)$.

**Exercise 3-19** Bearing in mind the results of this section, comment on the problem of evaluating the fundamental matrix $\Phi(t, t_0)$ of the time-varying homogeneous system $\dot{\mathbf{x}}(t) = \mathbf{A}(t)\mathbf{x}(t)$ by means of the Laplace transform.

**Exercise 3-20** Determine the matrix $e^{\mathbf{A}t}$, where $\mathbf{A}$ has the following forms:

$$(a)\ \mathbf{A} = \begin{bmatrix} 1 & 1 \\ 1 & 1 \end{bmatrix} \qquad (b)\ \mathbf{A} = \begin{bmatrix} 3 & 2 \\ -1 & 5 \end{bmatrix} \qquad (c)\ \mathbf{A} = \begin{bmatrix} -2 & 4 \\ -4 & -3 \end{bmatrix}$$

$$(d)\ \mathbf{A} = \begin{bmatrix} 0 & 1 & 0 \\ 0 & 0 & 1 \\ 0 & 0 & 0 \end{bmatrix} \qquad (e)\ \mathbf{A} = \begin{bmatrix} 0 & 1 & 0 \\ 0 & 0 & 1 \\ 0 & 0 & -1 \end{bmatrix}$$

## 3-24 Time-invariant Systems: The $n$th-order System

We are now going to consider the $n$th-order linear differential equation with constant coefficients, which we write in the form

$$\frac{d^n y(t)}{dt^n} + a_{n-1}\frac{d^{n-1}y(t)}{dt^{n-1}} + \cdots + a_1\frac{dy(t)}{dt} + a_0 y(t) = f(t) \tag{3-391}$$

We often use the notation $D = d/dt$ [see (3-53)], and we then represent Eq. (3-391) by

$$\{D^n + a_{n-1}D^{n-1} + \cdots + a_1 D + a_0\}y(t) = f(t) \tag{3-392}$$

We shall show that this system is equivalent to a linear system of $n$ first-order equations whose system eigenvalues are precisely the roots of the polynomial

$$x^n + a_{n-1}x^{n-1} + \cdots + a_1 x + a_0 \tag{3-393}$$

Moreover, we shall show how to find the matrix of the similarity transformation in the case where the system eigenvalues are real and distinct. Well, on to the task at hand.

Suppose that $y(t)$ is a solution of Eq. (3-391), and let us define an $R_n$-valued function $\mathbf{x}(t)$ by setting

$$\begin{aligned} x_1(t) &= y(t) \\ x_2(t) &= \dot{x}_1(t) = \dot{y}(t) \\ x_3(t) &= \dot{x}_2(t) = \frac{d^2y(t)}{dt^2} \\ &\cdots\cdots\cdots\cdots \\ x_n(t) &= \dot{x}_{n-1}(t) = \frac{d^{n-1}y(t)}{dt^n} \end{aligned} \tag{3-394}$$

where the $x_i(t)$ are to be the components of $\mathbf{x}(t)$. Then we can immediately deduce that

$$\begin{aligned} \dot{x}_1(t) &= x_2(t) \\ \dot{x}_2(t) &= x_3(t) \\ &\cdots\cdots \\ \dot{x}_{n-1}(t) &= x_n(t) \\ \dot{x}_n(t) &= -a_0x_1(t) - a_1x_2(t) - \cdots - a_{n-1}x_n(t) + f(t) \end{aligned} \tag{3-395}$$

which may be written in vector form as

$$\begin{bmatrix} \dot{x}_1(t) \\ \dot{x}_2(t) \\ \cdots \\ \dot{x}_{n-1}(t) \\ \dot{x}_n(t) \end{bmatrix} = \begin{bmatrix} 0 & 1 & 0 & \cdots & 0 & 0 \\ 0 & 0 & 1 & \cdots & 0 & 0 \\ \cdots & \cdots & \cdots & \cdots & \cdots & \cdots \\ 0 & 0 & 0 & \cdots & 0 & 1 \\ -a_0 & -a_1 & -a_2 & \cdots & -a_{n-2} & -a_{n-1} \end{bmatrix} \begin{bmatrix} x_1(t) \\ x_2(t) \\ \cdots \\ x_{n-1}(t) \\ x_n(t) \end{bmatrix} + \begin{bmatrix} 0 \\ 0 \\ \cdots \\ 0 \\ f(t) \end{bmatrix} \tag{3-396}$$

or, more succinctly, as

$$\dot{\mathbf{x}}(t) = \mathbf{A}\mathbf{x}(t) + \mathbf{v}(t) \tag{3-397}$$

where

$$\mathbf{A} = \begin{bmatrix} 0 & 1 & 0 & \cdots & 0 & 0 \\ 0 & 0 & 1 & \cdots & 0 & 0 \\ \cdots & \cdots & \cdots & \cdots & \cdots & \cdots \\ 0 & 0 & 0 & \cdots & 0 & 1 \\ -a_0 & -a_1 & -a_2 & \cdots & -a_{n-2} & -a_{n-1} \end{bmatrix} \tag{3-398}$$

and

$$\mathbf{v}(t) = \begin{bmatrix} 0 \\ 0 \\ \cdot \\ \cdot \\ \cdot \\ 0 \\ f(t) \end{bmatrix} \tag{3-399}$$

On the other hand, if $\mathbf{x}(t)$ is a solution of Eq. (3-396), then we can immediately conclude that the function $y(t) = x_1(t)$ [the first component of $\mathbf{x}(t)$] is a solution of Eq. (3-391). So we have exhibited a linear system of $n$ first-order equations which is equivalent to the $n$th-order equation.

Now let us show that the eigenvalues of the system matrix **A** given by Eq. (3-398) are precisely the roots of the polynomial

$$x^n + a_{n-1}x^n + \cdots + a_1 x + a_0 \tag{3-400}$$

In order to determine the eigenvalues of **A**, we must evaluate the determinant of $\mathbf{A} - \lambda\mathbf{I}$ and find the $\lambda$'s for which it is zero [see Eq. (2-71)]. Now

$$\mathbf{A} - \lambda\mathbf{I} = \begin{bmatrix} -\lambda & 1 & 0 & \cdots & 0 & 0 \\ 0 & -\lambda & 0 & \cdots & 0 & 0 \\ \cdots & \cdots & \cdots & \cdots & \cdots & \cdots \\ 0 & 0 & 0 & \cdots & -\lambda & 1 \\ -a_0 & -a_1 & -a_2 & \cdots & -a_{n-2} & -\lambda - a_{n-1} \end{bmatrix} \tag{3-401}$$

and so, expanding in terms of the minors of the first column, we have

$$\det(\mathbf{A}-\lambda\mathbf{I}) = -\lambda \det \begin{bmatrix} -\lambda & 1 & 0 & \cdots & 0 & 0 \\ 0 & -\lambda & 1 & \cdots & 0 & 0 \\ \cdots & \cdots & \cdots & \cdots & \cdots & \cdots \\ 0 & 0 & 0 & \cdots & -\lambda & 1 \\ -a_1 & -a_2 & -a_3 & \cdots & -a_{n-2} & -\lambda - a_{n-1} \end{bmatrix}$$

$$+ (-1)^{n+1}(-a_0) \det \begin{bmatrix} 1 & 0 & \cdots & 0 \\ -\lambda & 1 & \cdots & 0 \\ & -\lambda & & \\ \cdots & \cdots & \cdots & \cdots \\ 0 & 0 & \cdots & 1 \end{bmatrix} \tag{3-402}$$

Using an inductive argument, we may conclude from Eq. (3-402) that

$$\det(\mathbf{A}-\lambda\mathbf{I}) = (-1)^n\{\lambda^n + a_{n-1}\lambda^{n-1} + \cdots + a_1\lambda + a_0\} \tag{3-403}$$

and hence that $\lambda$ is an eigenvalue of **A** if and only if $\lambda$ is a root of the polynomial of Eq. (3-400), that is, if and only if

$$\lambda^n + a_{n-1}\lambda^{n-1} + \cdots + a_1\lambda + a_0 = 0 \tag{3-404}$$

**Example 3-63** Consider the system

$$\frac{d^3y(t)}{dt^3} - 3\frac{d^2y(t)}{dt^2} + 4y(t) = 1 \tag{3-405}$$

Then we set $x_1(t) = y(t)$, $x_2(t) = \dot{y}(t)$, and $x_3(t) = \ddot{y}(t)$, and we obtain the system

$$\begin{bmatrix} \dot{x}_1(t) \\ \dot{x}_2(t) \\ \dot{x}_3(t) \end{bmatrix} = \begin{bmatrix} 0 & 1 & 0 \\ 0 & 0 & 1 \\ -4 & 0 & 3 \end{bmatrix} \begin{bmatrix} x_1(t) \\ x_2(t) \\ x_3(t) \end{bmatrix} + \begin{bmatrix} 0 \\ 0 \\ 1 \end{bmatrix} \tag{3-406}$$

The system matrix **A** is given by

$$\mathbf{A} = \begin{bmatrix} 0 & 1 & 0 \\ 0 & 0 & 1 \\ -4 & 0 & 3 \end{bmatrix} \tag{3-407}$$

and the eigenvalues of **A** are the roots of the equation

$$\lambda^3 - 3\lambda^2 + 4 = 0 \tag{3-408}$$

Thus, the system eigenvalues are $\lambda = 2$, $\lambda = 2$, and $\lambda = -1$.

**Exercise 3-21** Consider the system

$$\frac{d^3y(t)}{dt^3} + \frac{d^2y(t)}{dt^2} = 1$$

Determine the equivalent system. What is its system matrix and what are the system eigenvalues? Find the solution $y(t)$. (See Exercise 3-20.)

Now let us suppose that the eigenvalues of the matrix $\mathbf{A}$ of Eq. (3-398) are *real* and *distinct*. We let $\lambda_1, \lambda_2, \ldots, \lambda_n$ denote these eigenvalues, and we let $\mathbf{\Lambda}$ be the matrix of the eigenvalues [see Eq. (2-78)], that is,

$$\mathbf{\Lambda} = \begin{bmatrix} \lambda_1 & 0 & \cdots & 0 \\ 0 & \lambda_2 & \cdots & 0 \\ \cdots & \cdots & \cdots & \cdots \\ 0 & 0 & \cdots & \lambda_n \end{bmatrix} \tag{3-409}$$

We know that there is a nonsingular matrix $\mathbf{P}$ such that

$$\mathbf{\Lambda} = \mathbf{P}^{-1}\mathbf{A}\mathbf{P} \tag{3-410}$$

We claim that $\mathbf{P}$ is the Vandermonde matrix of the $\lambda_i$, that is, that

$$\mathbf{P} = \begin{bmatrix} 1 & 1 & \cdots & 1 \\ \lambda_1 & \lambda_2 & \cdots & \lambda_n \\ \lambda_1^2 & \lambda_2^2 & \cdots & \lambda_n^2 \\ \cdots & \cdots & \cdots & \cdots \\ \lambda_1^{n-1} & \lambda_2^{n-1} & \cdots & \lambda_n^{n-1} \end{bmatrix} \tag{3-411}$$ †

In order to establish this claim, we shall show that

$$\mathbf{P}^{-1}\mathbf{A}\mathbf{P}\mathbf{e}_j = \lambda_j \mathbf{e}_j = \mathbf{\Lambda}\mathbf{e}_j \tag{3-412}$$

for $j = 1, 2, \ldots, n$, where the $\mathbf{e}_j$ are the elements of the natural basis of $R_n$ [see Eqs. (2-27)]. For example,

$$\mathbf{P}\mathbf{e}_1 = \begin{bmatrix} 1 & 1 & \cdots & 1 \\ \lambda_1 & \lambda_2 & \cdots & \lambda_n \\ \lambda_1^2 & \lambda_2^2 & \cdots & \lambda_n^2 \\ \cdots & \cdots & \cdots & \cdots \\ \lambda_1^{n-1} & \lambda_2^{n-1} & \cdots & \lambda_n^{n-1} \end{bmatrix} \begin{bmatrix} 1 \\ 0 \\ 0 \\ \cdots \\ 0 \end{bmatrix} = \begin{bmatrix} 1 \\ \lambda_1 \\ \lambda_1^2 \\ \cdots \\ \lambda_1^{n-1} \end{bmatrix} \tag{3-413}$$

and

$$\mathbf{A}\mathbf{P}\mathbf{e}_1 = \begin{bmatrix} 0 & 1 & 0 & \cdots & 0 & 0 \\ 0 & 0 & 1 & \cdots & 0 & 0 \\ \cdots & \cdots & \cdots & \cdots & \cdots & \cdots \\ 0 & 0 & 0 & \cdots & 0 & 1 \\ -a_0 & -a_1 & -a_2 & \cdots & -a_{n-2} & -a_{n-1} \end{bmatrix} \begin{bmatrix} 1 \\ \lambda_1 \\ \lambda_1^2 \\ \cdot \\ \cdot \\ \cdot \\ \lambda_1^{n-1} \end{bmatrix} \tag{3-414}$$

† We note that the determinant of $\mathbf{P}$ is given by

$$\det \mathbf{P} = \prod_{i>j} (\lambda_i - \lambda_j) = (\lambda_2 - \lambda_1) \cdots (\lambda_n - \lambda_1)(\lambda_3 - \lambda_2) \cdots (\lambda_n - \lambda_{n-1})$$

so that $\mathbf{P}$ is nonsingular if the $\lambda_i$ are all distinct.

$$= \begin{bmatrix} \lambda_1 \\ \lambda_1^2 \\ \cdot \\ \cdot \\ \cdot \\ \lambda_1^{n-1} \\ -a_0 - a_1\lambda_1 - \cdots - a_{n-1}\lambda_1^{n-1} \end{bmatrix} = \begin{bmatrix} \lambda_1 \\ \lambda_1^2 \\ \cdot \\ \cdot \\ \cdot \\ \lambda_1^{n-1} \\ \lambda_1^n \end{bmatrix} \tag{3-415}$$

since $\lambda_1$, being an eigenvalue of $\mathbf{A}$, is a root of Eq. (3-404). It follows that

$$\mathbf{APe}_1 = \lambda_1\mathbf{Pe}_1 \tag{3-416}$$

and, therefore, that

$$\mathbf{P}^{-1}\mathbf{APe}_1 = \mathbf{P}^{-1}\lambda_1\mathbf{Pe}_1 = \lambda_1\mathbf{P}^{-1}\mathbf{Pe}_1 = \lambda_1\mathbf{e}_1 \tag{3-417}$$

We may establish Eq. (3-402) for $j = 2, \ldots, n$ in an entirely similar way and, in so doing, establish Eq. (3-410).

## 3-25 The Adjoint System

Let us consider the homogeneous time-varying system

$$\dot{\mathbf{x}}(t) = \mathbf{A}(t)\mathbf{x}(t) \tag{3-418}$$

We shall often examine this system and the closely related system

$$\dot{\mathbf{z}}(t) = -\mathbf{A}'(t)\mathbf{z}(t) \tag{3-419}$$

[where $\mathbf{A}'(t)$ is the transpose of the matrix $\mathbf{A}(t)$; see Eq. (2-45)] together. The system of Eq. (3-419) is called the *adjoint system* to (3-418).

Now, we know that the solution of Eq. (3-418) which satisfies the initial condition $\mathbf{x}(0) = \boldsymbol{\xi}$ is given by

$$\mathbf{x}(t) = \boldsymbol{\Phi}(t)\boldsymbol{\xi} \tag{3-420}$$

where $\boldsymbol{\Phi}(t)$ is the fundamental matrix of Eq. (3-418), and we know that the solution of Eq. (3-419) which satisfies the initial condition $\mathbf{z}(0) = \boldsymbol{\pi}$ is given by

$$\mathbf{z}(t) = \boldsymbol{\Psi}(t)\boldsymbol{\pi} \tag{3-421}$$

where $\boldsymbol{\Psi}(t)$ is the fundamental matrix of Eq. (3-419). What is the relationship between $\boldsymbol{\Phi}(t)$ and $\boldsymbol{\Psi}(t)$? Well, we claim that

$$\boldsymbol{\Psi}'(t)\boldsymbol{\Phi}(t) = \mathbf{I} \qquad \text{for all } t \tag{3-422}$$

First of all, we note that

$$\boldsymbol{\Psi}'(0)\boldsymbol{\Phi}(0) = \mathbf{II} = \mathbf{I} \tag{3-423}$$

Secondly, if we let $\mathbf{v}$ be *any* element of $R_n$ and if we set

$$\mathbf{h}(t) = \boldsymbol{\Psi}'(t)\boldsymbol{\Phi}(t)\mathbf{v} \tag{3-424}$$

then

$$\dot{\mathbf{h}}(t) = \{\dot{\boldsymbol{\Psi}}'(t)\boldsymbol{\Phi}(t) + \boldsymbol{\Psi}'(t)\dot{\boldsymbol{\Phi}}(t)\}\mathbf{v} \tag{3-425}$$

But it follows from the definition of a fundamental matrix that

$$\dot{\mathbf{\Psi}}(t) = -\mathbf{A}'(t)\mathbf{\Psi}(t) \qquad \dot{\mathbf{\Phi}}(t) = \mathbf{A}(t)\mathbf{\Phi}(t) \tag{3-426}$$

Therefore, by virtue of Eqs. (2-48), we may conclude that

$$\dot{\mathbf{\Psi}}'(t) = -\mathbf{\Psi}'(t)\mathbf{A}(t) \tag{3-427}$$

and that

$$\dot{\mathbf{h}}(t) = \{-\mathbf{\Psi}'(t)\mathbf{A}(t)\mathbf{\Phi}(t) + \mathbf{\Psi}'(t)\mathbf{A}(t)\mathbf{\Phi}(t)\}\mathbf{v} \tag{3-428}$$

$$= \mathbf{0} \tag{3-429}$$

Since the function $\mathbf{h}(t)$ is a solution of the initial-value problem

$$\dot{\mathbf{h}}(t) = \mathbf{0} \qquad \mathbf{h}(0) = \mathbf{v} \tag{3-430}$$

which also has the solution $\mathbf{g}(t) = \mathbf{I}\mathbf{v}$, we see that

$$\mathbf{\Psi}'(t)\mathbf{\Phi}(t)\mathbf{v} = \mathbf{I}\mathbf{v} \tag{3-431}$$

and hence that Eq. (3-422) holds, as $\mathbf{v}$ was an arbitrary element of $R_n$.

An immediate and important consequence of Eq. (3-422) is the fact that

$$\langle \mathbf{z}(t), \mathbf{x}(t)\rangle = \langle \mathbf{\Psi}(t)\boldsymbol{\pi}, \mathbf{\Phi}(t)\boldsymbol{\xi}\rangle \tag{3-432}$$

$$= \langle \boldsymbol{\pi}, \mathbf{\Psi}'(t)\mathbf{\Phi}(t)\boldsymbol{\xi}\rangle \qquad \text{by Eq. (2-98)} \tag{3-433}$$

$$= \langle \boldsymbol{\pi}, \boldsymbol{\xi}\rangle \tag{3-434}$$

where $\langle\ ,\ \rangle$ is the scalar product on $R_n$ [see Eq. (2-86)].

Suppose that the system we are considering has the property that its system matrix is skew-symmetric [see Eq. (2-99)]; then we shall say that the system is *self-adjoint*. In other words, the system

$$\dot{\mathbf{x}}(t) = \mathbf{A}(t)\mathbf{x}(t) \tag{3-435}$$

is *self-adjoint* if

$$\mathbf{A}(t) + \mathbf{A}'(t) = \mathbf{0} \tag{3-436}$$

We observe that the adjoint system to a self-adjoint system is the *same* as the system itself. It follows that if $\mathbf{\Phi}(t)$ is the fundamental matrix of a self-adjoint system, then

$$\mathbf{\Phi}'(t)\mathbf{\Phi}(t) = \mathbf{I} \tag{3-437}$$

[since $\mathbf{\Phi}(t)$ is also the fundamental matrix of the adjoint system]. In other words, $\mathbf{\Phi}(t)$ is an orthogonal matrix [see Eq. (2-109)]. Therefore, if we let

$$\mathbf{x}(t) = \mathbf{\Phi}(t)\boldsymbol{\xi}$$

be the solution of a self-adjoint system satisfying the initial condition $\mathbf{x}(0) = \boldsymbol{\xi}$, we find that

$$\langle \mathbf{x}(t), \mathbf{x}(t)\rangle = \langle \mathbf{\Phi}(t)\boldsymbol{\xi}, \mathbf{\Phi}(t)\boldsymbol{\xi}\rangle = \langle \boldsymbol{\xi}, \boldsymbol{\xi}\rangle \tag{3-438}$$

and that

$$\|\mathbf{x}(t)\| = \|\boldsymbol{\xi}\| \tag{3-439}$$

Thus, if a homogeneous system is self-adjoint, the (Euclidean) length of the solution $\mathbf{x}(t)$ remains constant for all "time."

**Example 3-64** Consider the time-invariant system $\dot{\mathbf{x}}(t) = \mathbf{A}\mathbf{x}(t)$. Its adjoint system is $\dot{\mathbf{z}}(t) = -\mathbf{A}'\mathbf{z}(t)$. The fundamental matrix of the system is $e^{\mathbf{A}t}$, and the fundamental matrix of the adjoint system is $e^{-\mathbf{A}'t}$. Equation (3-422) becomes $(e^{-\mathbf{A}'t})'e^{\mathbf{A}t} = \mathbf{I}$, and it follows that $e^{-\mathbf{A}'t} = (e^{-\mathbf{A}t})'$. If the system is self-adjoint, that is, if $\mathbf{A} + \mathbf{A}' = \mathbf{0}$, then $e^{\mathbf{A}t}$ is an orthogonal matrix and $\|e^{\mathbf{A}t}\xi\| = \|\xi\|$ for all $\xi$ in $R_n$.

**Exercise 3-22** Let $\mathbf{A} = \begin{bmatrix} 0 & 1 \\ -1 & 0 \end{bmatrix}$. Show that the system $\dot{\mathbf{x}}(t) = \mathbf{A}\mathbf{x}(t)$ is self-adjoint, calculate its fundamental matrix $e^{\mathbf{A}t}$, and verify by direct computation that $e^{\mathbf{A}t}$ is an orthogonal matrix.

**Exercise 3-23** Show, by differentiating $\langle \mathbf{x}(t), \mathbf{x}(t)\rangle$ and applying Eq. (2-103), that if the system $\dot{\mathbf{x}}(t) = \mathbf{A}(t)\mathbf{x}(t)$ is self-adjoint, then $\|\mathbf{x}(t)\|$ is constant. Carry out the steps of your demonstration for the particular case $\mathbf{A}(t) = \begin{bmatrix} 0 & t \\ -t & 0 \end{bmatrix}$.

## 3-26 Stability of Linear Time-invariant Systems

Let us consider the homogeneous linear time-invariant system

$$\dot{\mathbf{x}}(t) = \mathbf{A}\mathbf{x}(t) \tag{3-440}$$

where $\mathbf{A} = (a_{ij})$ is a (constant) $n \times n$ matrix. Loosely speaking, the system (3-440) is *stable* if small deviations from equilibrium [i.e., the solution $\mathbf{x}(t) = \mathbf{0}$] remain small as time lapses and is *unstable* if arbitrarily small deviations from equilibrium can grow very large as time lapses. More precisely, we have:

***Definition 3-45*** *We say that the system* (3-440) *is stable if the* (*Euclidean*) *norm* $\|\mathbf{x}(t)\|$ *remains bounded as* $t \to \infty$ *for every solution* $\mathbf{x}(t)$ *of the system. We call the system strictly stable if it is stable and if*

$$\lim_{t\to\infty} \|\mathbf{x}(t)\| = 0 \tag{3-441}$$

*for every solution* $\mathbf{x}(t)$ *of the system. If the system is not stable or, in other words, if there is a solution* $\hat{\mathbf{x}}(t)$ *such that*

$$\lim_{t\to\infty} \|\hat{\mathbf{x}}(t)\| = \infty \tag{3-442}$$

*then we say that the system is unstable.*

Let us denote the eigenvalues of the matrix $\mathbf{A}$ by

$$\lambda_i + j\mu_i \qquad i = 1, 2, \ldots, n \tag{3-443}$$

where the $\lambda$'s and the $\mu$'s are *real* numbers and $j = \sqrt{-1}$. The following criterion relates the stability of the system (3-440) to the eigenvalues (3-443) of the system matrix $\mathbf{A}$.

**Stability Criterion**† *The system* (3-440) *is stable if and only if*

1. $$\lambda_i \le 0 \qquad \textit{for all } i \tag{3-444}$$

† See Refs. B-6, B-10, and C-7.

2. *If* $\lambda_k + j\mu_k$ *is a multiple root of the characteristic polynomial of* **A**, *then*

$$\lambda_k < 0 \tag{3-445}$$

*The system* (3-440) *is strictly stable if and only if*

$$\lambda_i < 0 \qquad \textit{for all } i \tag{3-446}$$

In other words, the system is strictly stable whenever *all* of its eigenvalues have negative real parts and is stable whenever *none* of its eigenvalues has a positive real part and *all* of its *multiple* eigenvalues have negative real parts.

**Example 3-65** Consider the system

$$\begin{bmatrix} \dot{x}_1(t) \\ \dot{x}_2(t) \end{bmatrix} = \begin{bmatrix} 0 & 1 \\ -a_0 & -a_1 \end{bmatrix} \begin{bmatrix} x_1(t) \\ x_2(t) \end{bmatrix}$$

Then the eigenvalues of this system are the roots of the equation

$$\lambda^2 + a_1\lambda + a_0 = 0$$

and the system will be strictly stable if and only if both $a_0$ and $a_1$ are positive, that is, $a_0 > 0$, $a_1 > 0$. If the system has a multiple eigenvalue, that is, if $a_1{}^2 = 4a_0$, then the system will be stable if and only if it is strictly stable. We observe that if $a_0 = a_1 = 0$, then the system is unstable.

We observe that the adjoint system of (3-440) is of the form

$$\dot{\mathbf{z}}(t) = -\mathbf{A}'\mathbf{z}(t) \tag{3-447}$$

where **A**′ is the transpose of **A**. If $\lambda$ is an eigenvalue of **A**, then

$$\det(\mathbf{A} - \lambda\mathbf{I}) = 0 \tag{3-448}$$

But $$\det(\mathbf{A} - \lambda\mathbf{I}) = -\det(-\mathbf{A} + \lambda\mathbf{I}) = -\det[(-\mathbf{A}') + \lambda\mathbf{I}'] \tag{3-449}$$

since the determinant of a matrix is the same as the determinant of its transpose. It follows that $-\lambda$ is an eigenvalue of $-\mathbf{A}'$ whenever $\lambda$ is an eigenvalue of **A**. Thus, in view of the stability criterion, we can see that if a system is strictly stable, then its adjoint is unstable.

We have only touched upon the notion of stability here, as this is all that we shall need in the sequel.

# Chapter 4

# BASIC CONCEPTS

## 4-1 Introduction

In Chaps. 2 and 3, we reviewed the basic mathematical notions that we need. In this chapter, we shall use these notions to define and develop the conceptual framework within which our subsequent studies will take place. In particular, we shall define the notion of a dynamical system, and we shall define the *control problem* which is the keystone of our later work.

We shall begin, in the next few sections, by examining some simple examples which will serve to motivate the formal definition of a dynamical system. We shall see that the essential elements in this definition are the notions of input, output, and state.

After examining the basic ideas associated with dynamical systems and after relating these ideas to the more familiar concept of the transfer function, we shall define the control problem and we shall consider some important special cases of it.

In the final portion of the chapter, we shall define and examine certain qualitative properties of control systems, such as accessibility, controllability, and observability. These qualitative properties play a very important role in control theory.

Most of the definitions and most of the development in this chapter are keyed to the requirements of the sequel. The basic concepts of input, output, and state are discussed in much greater detail and generality in the book of Zadeh and Desoer [Z-1]. The notion of dynamical system is examined in Refs. Z-1, K-7, K-9, and W-5, and the general control problem is discussed, for example, in Refs. B-8, H-3, K-2, and D-5.

## 4-2 An *RL* Network

We shall examine the very simple *RL* network of Fig. 4-1 in order to motivate several of the basic concepts which we shall precisely define somewhat later on.

Let us agree to call the voltage $u(t)$ the *input*, and let us denote the current by $i(t)$. Suppose that we can observe and measure the voltage $y(t)$ across the resistor $R$, and let us agree to call this voltage the *output*. We know that the relation between the current $i(t)$ and the input voltage $u(t)$ is the differential equation

$$L\frac{di(t)}{dt} + Ri(t) = u(t) \tag{4-1}$$

and, since $y(t) = Ri(t)$, that the relation between $y(t)$ and $u(t)$ can be written as

$$\frac{dy(t)}{dt} + \frac{R}{L}y(t) = \frac{R}{L}u(t) \tag{4-2}$$

Every self-respecting engineer knows that, *as long as the applied voltage is finite, the current through the inductor L cannot change instantaneously.*† Bearing this fact in mind, let us perform the following experiment: During the time interval $(0, T]$, we apply a known continuous voltage input $u(t)$, and we suppose that we know $y(0)$ and $u(0)$. To emphasize that the input is known over the entire semiclosed interval $(0, T]$, we shall denote it by

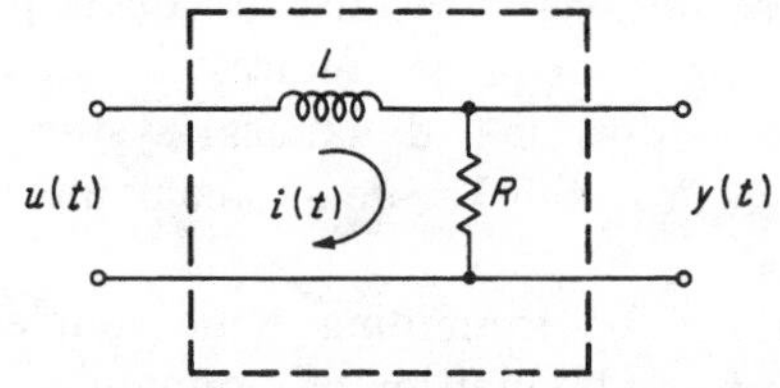

**Fig. 4-1** In this network, the output voltage $y(t)$ is measured across $R$.

$$u_{(0,T]} \tag{4-3}$$

In other words, $u_{(0,T]}$ is the function defined on $(0, T]$ by the relation

$$u_{(0,T]}(t) = u(t) \qquad \text{for all } t \text{ in } (0, T] \tag{4-4}$$

Now we measure the output $y(t)$ during this time interval, and let us denote by $y_{(0,T]}$ the function defined on $(0, T]$ by the relation

$$y_{(0,T]}(t) = y(t) \qquad \text{for all } t \text{ in } (0, T] \tag{4-5}$$

We shall call the pair

$$(u_{(0,T]}, y_{(0,T]}) \tag{4-6}$$

an *input-output pair* on the interval $(0, T]$. What we would like to do is *predict* the output voltage $y_{(0,T]}$. Thus, we are faced with the question:

*Given* $u_{(0,T]}$ *and the output differential equation* (4-2), *what additional information is needed to completely specify* $y_{(0,T]}$*?*

If we set $a = R/L$, then Eq. (4-2) takes the form

$$\dot{y}(t) + ay(t) = au(t) \tag{4-7}$$

and its solution may be written as

$$y(t) = y(0)e^{-at} + e^{-at}\int_0^t e^{a\tau}au(\tau)\,d\tau \tag{4-8}$$

† We assume that impulses do *not* occur in the physical world, so that $u(t)$ cannot be an impulse.

Since $u_{(0,T]}$ is known, the term

$$e^{-at}\int_0^t e^{a\tau}au(\tau)\,d\tau$$

is known, and so *the only additional quantity that we must know to predict* $y_{(0,T]}$ *is* $y(0)$, the value of the output voltage at $t = 0$. We note that we *do not need to know the value of the input* $u(0)$ *at* $t = 0$. We could have deduced this by the following reasoning: Suppose that $u(0+) \neq u(0)$, that is, that a jump in the input voltage occurred at $t = 0+$; then the current in $L$ would not change instantaneously, so that

$$i(0+) = i(0) \tag{4-9}$$

and hence we would find that

$$y(0+) = y(0) \tag{4-10}$$

independent of $u(0)$. *To recapitulate:* For the network of Fig. 4-1, we have found that knowledge of $u_{(0,T]}$ and $y(0)$ was sufficient to completely specify $y_{(0,T]}$.

If we make the trivial definition

$$x(t) = y(t) \qquad t \in [0, T] \tag{4-11}$$

we can see that $x(0)$ is the minimum amount of information required to determine $y_{(0,T]}$, given $u_{(0,T]}$. We shall call $x(0)$ the *state* of the system at $t = 0$, and we note that $x(t)$ satisfies the differential equation

$$\dot{x}(t) + ax(t) = au(t) \tag{4-12}$$

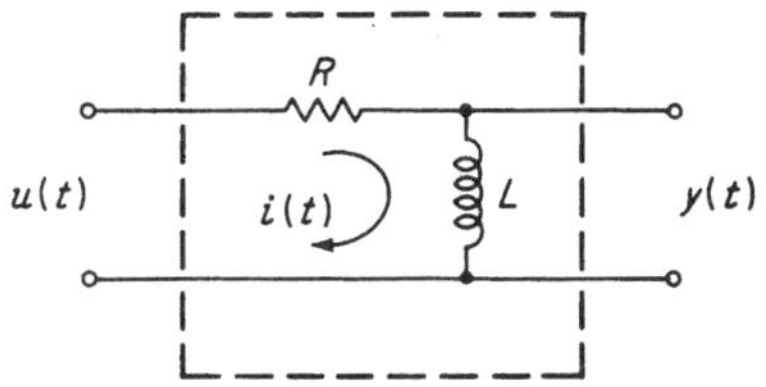


**Fig. 4-2** In this network, the output voltage $y(t)$ is measured across $L$.

Furthermore, we observe that if $\hat{t}$ is an element of $(0, T)$, that is, if $0 < \hat{t} < T$, then knowledge of $x(\hat{t})$ $[= y(\hat{t})]$ is sufficient to determine the output $y(t)$ on the interval $(\hat{t}, T]$, given the input $u(t)$ on that interval [cf. Eq. (3-250)]. Thus, we call $x(t)$ the *state* (variable) of the network.

Now let us examine the same network in a somewhat different way (see Fig. 4-2). Let us suppose that we now observe and measure the voltage across the inductor $L$ rather than across the resistor $R$. We shall now call this voltage *across* $L$ the output $y(t)$. The current $i(t)$ still satisfies Eq. (4-1); however, the output $y(t)$ is now described by the relation

$$y(t) + Ri(0) + \frac{R}{L}\int_0^t y(\tau)\,d\tau = u(t) \tag{4-13}$$

since

$$y(t) = L\frac{di(t)}{dt} \tag{4-14}$$

Now let us suppose that we apply a continuous input $u(t)$ which is *differentiable* on the time interval $(0, T]$. Then Eq. (4-13) may be differentiated to obtain

$$\frac{dy(t)}{dt} + \frac{R}{L}y(t) = \frac{du(t)}{dt} \qquad t \in (0, T] \tag{4-15}$$

which, letting $a = R/L$, becomes

$$\dot{y}(t) + ay(t) = \dot{u}(t) \qquad t \in (0, T] \tag{4-16}$$

If we know $y(0)$, $u(0)$, and $u_{(0,T]}$, then we would like to be able to predict the output $y_{(0,T]}$. Thus, we are faced with the question:

*Given* $u_{(0,T]}$ *and the output differential equation* (4-16), *what additional information is needed to completely specify* $y_{(0,T]}$*?*

Examining the network of Fig. 4-2 from a physical point of view, we observe that if the input jumps at 0, that is, if

$$u(0+) \neq u(0) \tag{4-17}$$

then the voltage $y(t)$ across the inductor $L$ must also jump at $t = 0+$ from $y(0)$ to $y(0+)$ since the current through $L$, and hence the voltage across the resistor $R$, does not change. In other words, the fact that the current through $L$ cannot change instantaneously implies that

$$Ri(0) = Ri(0+) \tag{4-18}$$

and hence that

$$y(0) - u(0) = y(0+) - u(0+) \tag{4-19}$$

or, equivalently,

$$y(0+) = y(0) + u(0+) - u(0) \tag{4-20}$$

However, the solution of Eq. (4-16) on the interval $(0, T]$ is given by

$$y(t) = e^{-at}y(0+) + e^{-at}\int_{0+}^{t} e^{a\tau}\dot{u}(\tau)\,d\tau \tag{4-21}$$

and so, in order to determine $y_{(0,T]}$, we must know $y(0+)$ or, equivalently, *both* $y(0)$ *and* $u(0)$.

If we define a new variable $x(t)$ by setting

$$x(t) = y(t) - u(t) \tag{4-22}$$

for $t$ in $[0, T]$, that is, $x(t)$ is the difference between the output $y(t)$ and the input $u(t)$, then

$$\dot{x}(t) = \dot{y}(t) - \dot{u}(t) \qquad t \in (0, T] \tag{4-23}$$

and so, by virtue of Eqs. (4-16) and (4-22),

$$\dot{x}(t) + ax(t) = -au(t) \tag{4-24}$$

for $t$ in $(0, T]$. Since Eq. (4-22) holds for all $t$ in $[0, T]$, we have

$$x(0) = y(0) - u(0) \tag{4-25}$$
$$x(0+) = y(0+) - u(0+) \tag{4-26}$$

and

$$x(0+) - x(0) = y(0+) - y(0) - u(0+) + u(0) \tag{4-27}$$
$$= 0 \qquad \text{by Eq. (4-20)} \tag{4-28}$$

Thus, the solution of Eq. (4-24) may be written as

$$x(t) = e^{-at}x(0) - ae^{-at}\int_0^t e^{a\tau}u(\tau)\,d\tau \tag{4-29}$$

It follows that knowledge of $x(0) = y(0) - u(0)$ and of $u_{(0,T]}$ is sufficient to completely specify $x_{(0,T]}$ and therefore, in view of Eq. (4-22), to completely specify $y_{(0,T]}$. We call $x(0)$ the *state at* $t = 0$, and we observe that if $\hat{t}$ is an element of $(0, T)$, that is, $0 < \hat{t} < T$, then knowledge of $x(\hat{t}) = y(\hat{t}) - u(\hat{t})$ is sufficient to determine the output $y(t)$ on the interval $(\hat{t}, T]$, given the input $u(t)$ on that interval. Thus, we call $x(t)$ the *state* (variable) of the system. We also note that Eqs. (4-22) and (4-29) provide us with a description of the behavior of the network of Fig. 4-2.

Let us summarize what we have done. For the system of Fig. 4-1, in which the output $y(t)$ was the voltage across the resistor $R$, we found that the equations

$$\dot{x}(t) + ax(t) = au(t) \qquad \text{state equation} \tag{4-30}$$
$$y(t) = x(t) \qquad \text{output equation} \tag{4-31}$$

where $a = R/L$, described the behavior of the network, in that knowledge of $x(0)$ and $u_{(0,T]}$ served to determine $y_{(0,T]}$. On the other hand, for the system of Fig. 4-2, in which the output $y(t)$ was the voltage across the inductor $L$, we found that the equations

$$\dot{x}(t) + ax(t) = -au(t) \qquad \text{state equation} \tag{4-32}$$
$$y(t) = x(t) + u(t) \qquad \text{output equation} \tag{4-33}$$

where $a = R/L$, described the behavior of the network.

**Example 4-1** Suppose that

$$L = R = 1 \quad a = 1 \quad T = 2\text{ sec} \quad y(0) = -0.5$$
$$u(0) = -1 \quad u(t) = e^{-t} \quad 0 < t \le 2 \tag{4-34}$$

For the network of Fig. 4-1, Eq. (4-30) becomes

$$\dot{x}(t) + x(t) = e^{-t} \tag{4-35}$$

and, since [Eq. (4-31)] $x(0) = y(0)$,

$$x(0) = -0.5 \tag{4-36}$$

so that

$$x(t) = -0.5e^{-t} + te^{-t} \tag{4-37}$$

and [Eq. (4-31)]

$$y(t) = -0.5e^{-t} + te^{-t} \tag{4-38}$$

The state $x(t)$, output $y(t)$, and input $u(t)$ are illustrated in Fig. 4-3.

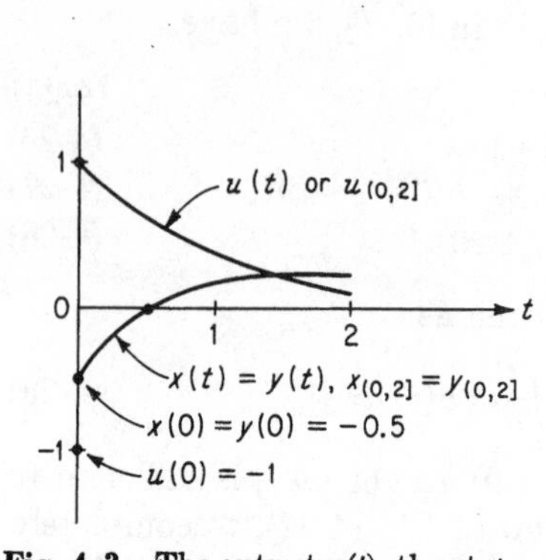


**Fig. 4-3** The output $y(t)$, the state $x(t)$, and the input $u(t)$ for the network of Fig. 4-1.

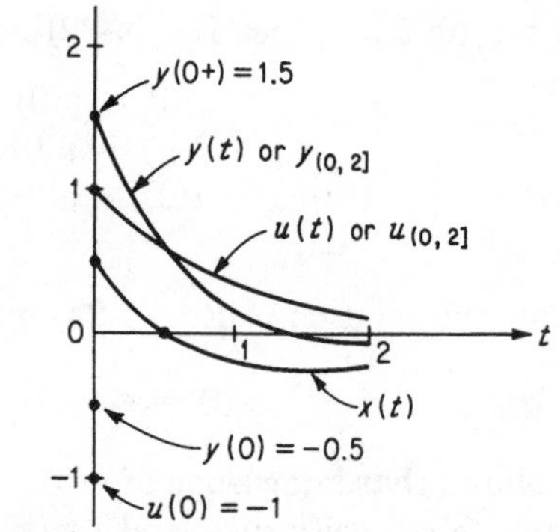


**Fig. 4-4** The output $y(t)$, the state $x(t)$, and the input $u(t)$ for the network of Fig. 4-2.

For the network of Fig. 4-2, Eq. (4-32) reduces to

$$\dot{x}(t) + x(t) = -e^{-t} \tag{4-39}$$

Since [Eq. (4-33)] $x(0) = y(0) - u(0)$, we have

$$x(0) = -0.5 + 1 = +0.5 \tag{4-40}$$

so that the solution of Eq. (4-39) is

$$x(t) = 0.5e^{-t} - te^{-t} \tag{4-41}$$

The output is, from Eq. (4-33), given by $y(t) = x(t) + u(t)$, so that

$$y(t) = 0.5e^{-t} - t + e^{-t} = 1.5e^{-t} - te^{-t} \tag{4-42}$$

The state $x(t)$, output $y(t)$, and input $u(t)$ are shown in Fig. 4-4.

**Exercise 4-1** Consider the *RLC* network shown in Fig. 4-5. Let $u(t)$ denote the input voltage and let $y_R(t)$, $y_L(t)$, and $y_C(t)$ denote the voltages across the resistor $R$, the inductor $L$, and the capacitor $C$, respectively. Suppose that $u_{(0,T]}$ is a differentiable voltage. Derive the input-output relationships between $u(t)$ and $y_R(t)$, $y_L(t)$ and $y_C(t)$. In each case, determine from physical considerations the signals that must be known at $t = 0$ so that the outputs $y_{R(0,T]}$, $y_{L(0,T]}$, and $y_{C(0,T]}$ are completely specified, given $u_{(0,T]}$.

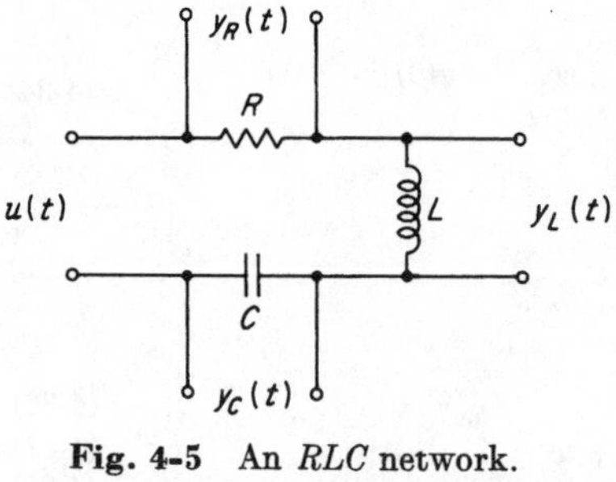


**Fig. 4-5** An *RLC* network.

**Exercise 4-2** Given a system (a differentiator) with input $u(t)$ and output $y(t)$ such that

$$y(t) = \dot{u}(t)$$

Show that if $u_{(0,T]}$ is differentiable, then $y_{[0,T]}$ is completely specified if $u(0)$ is also known. HINT: You may think of the system as a unit inductor, the input $u(t)$ being the current and the output $y(t)$ the voltage across the inductor.

## 4-3 A Multivariable System

We shall now turn our attention to a system, illustrated in Fig. 4-6, which is obtained by the interconnection of several subsystems (denoted by $S_1$, $S_2$, $S_3$, and $S_4$ in Fig. 4-6). The system has two inputs, $u_1(t)$ and $u_2(t)$, and has two *observable* outputs, $y_1(t)$ and $y_2(t)$. The input $u_1(t)$ is applied to the subsystems $S_1$ and $S_2$, generating outputs $x_1(t)$ and $x_2(t)$, respectively; the input $u_2(t)$ is applied to the subsystems $S_3$ and $S_4$, generating outputs $x_3(t)$ and

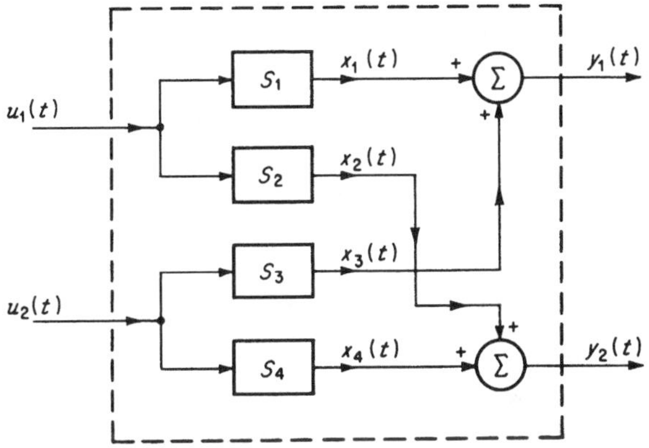


**Fig. 4-6** A multivariable system.

$x_4(t)$, respectively. We suppose that the output signals $y_1(t)$ and $y_2(t)$ *which we measure* are given by

$$\begin{aligned} y_1(t) &= x_1(t) + x_3(t) \\ y_2(t) &= x_2(t) + x_4(t) \end{aligned} \tag{4-43}$$

and that we *cannot* observe the $x_i(t)$.

We shall assume that each of the subsystems $S_i$, $i = 1, 2, 3, 4$, can be described by a simple first-order linear differential equation. In other words, we have

$$S_1\colon \dot{x}_1(t) = a_1x_1(t) + u_1(t) \tag{4-44}$$

$$S_2\colon \dot{x}_2(t) = a_2x_2(t) + u_1(t) \tag{4-45}$$

$$S_3\colon \dot{x}_3(t) = a_3x_3(t) + u_2(t) \tag{4-46}$$

$$S_4\colon \dot{x}_4(t) = a_4x_4(t) + u_2(t) \tag{4-47}$$

From this assumption and from Eqs. (4-43), we may deduce the following pair of differential equations relating the output signals $y_1(t)$ and $y_2(t)$ to the input signals $u_1(t)$ and $u_2(t)$:

$$\ddot{y}_1(t) - (a_1 + a_3)\dot{y}_1(t) + (a_1a_3)y_1(t) = \dot{u}_1(t) + \dot{u}_2(t) - a_3u_1(t) - a_1u_2(t) \tag{4-48}$$

$$\ddot{y}_2(t) - (a_2 + a_4)\dot{y}_2(t) + (a_2a_4)y_2(t) = \dot{u}_1(t) + \dot{u}_2(t) - a_4u_1(t) - a_2u_2(t) \tag{4-49}$$

We define, for convenience, the *input vector* $\mathbf{u}(t)$ and the *output vector* $\mathbf{y}(t)$ by setting

$$\mathbf{u}(t) = \begin{bmatrix} u_1(t) \\ u_2(t) \end{bmatrix} \qquad \mathbf{y}(t) = \begin{bmatrix} y_1(t) \\ y_2(t) \end{bmatrix} \tag{4-50}$$

Again, what we should like to do is *predict* the output (vector) $\mathbf{y}_{(0,T]}$, and thus we are faced with the following question:

*Given* $\mathbf{u}_{(0,T]}$ *and the output differential equations* (4-48) *and* (4-49), *what additional information is needed to completely specify* $\mathbf{y}_{(0,T]}$*?*

If we examine the differential equations (4-48) and (4-49), then we shall note that knowledge of the following initial conditions is sufficient to completely determine $\mathbf{y}_{(0,T]}$:

$$y_1(0),\ \dot{y}_1(0),\ y_2(0),\ \dot{y}_2(0),\ u_1(0),\ u_2(0)$$

or, equivalently,

$$\mathbf{y}(0),\ \dot{\mathbf{y}}(0),\ \mathbf{u}(0) \tag{4-51}$$

On the other hand, if we examine the block diagram of Fig. 4-6, then we may conclude that knowledge of

$$x_1(0),\ x_2(0),\ x_3(0),\ x_4(0) \tag{4-52}$$

is sufficient to completely specify $x_{1(0,T]}$, $x_{2(0,T]}$, $x_{3(0,T]}$, and $x_{4(0,T]}$, given $\mathbf{u}_{(0,T]}$ (see Sec. 4-2). In view of Eqs. (4-46) and (4-47) (i.e., in view of the fact that our system is the interconnection of the subsystems $S_1$, $S_2$, $S_3$, and $S_4$), we deduce that knowledge of the vector $\mathbf{x}(t)$ at $t = 0$, where

$$\mathbf{x}(t) = \begin{bmatrix} x_1(t) \\ x_2(t) \\ x_3(t) \\ x_4(t) \end{bmatrix} \tag{4-53}$$

is required to determine the output $\mathbf{y}_{(0,T]}$, given $\mathbf{u}_{(0,T]}$.

We shall call the vector $\mathbf{x}(0)$ the *state* of the system at $t = 0$ and the vector $\mathbf{x}(t)$ the *state* of the system at $t$. If $\mathbf{x}(t)$ could be measured, then the output could be predicted; however, we have assumed that $\mathbf{x}(t)$ cannot be measured, and so let us see if $\mathbf{x}(t)$ can be computed from the signals that we can observe, namely, $\mathbf{u}(t)$ and $\mathbf{y}(t)$. Now, it follows from Eqs. (4-43) that

$$\begin{aligned} y_1(0) &= x_1(0) + x_3(0) \\ y_2(0) &= x_2(0) + x_4(0) \end{aligned} \tag{4-54}$$

and it follows from differentiation of Eqs. (4-43) that

$$\begin{aligned} \dot{y}_1(0) &= \dot{x}_1(0) + \dot{x}_3(0) = a_1x_1(0) + a_3x_3(0) + u_1(0) + u_2(0) \\ \dot{y}_2(0) &= \dot{x}_2(0) + \dot{x}_4(0) = a_2x_2(0) + a_4x_4(0) + u_1(0) + u_2(0) \end{aligned} \tag{4-55}$$

Thus, if we know $\mathbf{y}(0)$, $\dot{\mathbf{y}}(0)$, and $\mathbf{u}(0)$ [that is, the same quantities as appear in (4-51)], then we can solve Eqs. (4-54) and (4-55) for $\mathbf{x}(0)$. If the $a_i$ are all distinct, then $\mathbf{x}(0)$ is *uniquely* determined; on the other hand, if (say)

$a_1 = a_3$, then $\mathbf{x}(0)$ is not unique (see Exercise 4-3). In any case, although we may not be able to measure the state $\mathbf{x}(t)$, we can compute it from the input and output signals.

**Exercise 4-3** Suppose that the subsystems $S_1$, $S_2$, $S_3$, and $S_4$ are identical, that is, $a_1 = a_2 = a_3 = a_4$ in Eqs. (4-44) to (4-47). Show that knowledge of $y_1(0)$ and $y_2(0)$ is sufficient to determine $y_1(t)$ and $y_2(t)$, given the inputs.

## 4-4 Dynamical Systems: Introductory Remarks

Knowledge of the physical world is based upon experiment and abstraction. The engineer examines specific physical systems with definite objectives in mind, while the theoretician attempts to discover the basic laws which govern and describe the behavior of physical systems in general.

Let us begin by playing the role of the engineer. We are given a physical system $\mathcal{P}$, the proverbial "black box," and we can apply certain input "signals" to $\mathcal{P}$ in order to observe and measure the resultant output "signals." Our ultimate objective is the determination of an input which will produce an output with certain desired characteristics and which will, in so doing, minimize our "cost" of operation. We might try to achieve our objective by a trial-and-error procedure, by trying first one input and then another. If we were (very) lucky, such a procedure might produce an answer; in general, it would not. However, we might look upon our efforts as experiments which could lead us to a description of the behavior of $\mathcal{P}$ and could indicate to us what output would result from the application of any input. Thus, we become faced with the problems of representing $\mathcal{P}$ and of finding a suitable model for the behavior of $\mathcal{P}$.

We generally attempt to solve these problems by developing mathematical models in which equations relate the output to the input. The types of models we choose are based upon the results of our experiments and upon assumptions about the fundamental laws which govern the behavior of all physical systems. These assumptions are usually taken as axioms which provide a precise definition for the theoretical concepts involved in constructing a mathematical model for the behavior of the physical system $\mathcal{P}$. Once the model of $\mathcal{P}$ has been developed, certain conclusions are then drawn about it and are tested by further experimentation on $\mathcal{P}$.

In this book, we shall be concerned with physical systems $\mathcal{P}$ whose behavior may be adequately described by mathematical models which are called *dynamical systems*. We shall, in Sec. 4-5, give an axiomatic definition of a dynamical system. Here, bearing the examples of Secs. 4-2 and 4-3 in mind, let us try to examine in a heuristic fashion some of the ideas which motivate our more formal approach.

To begin with, let us consider the following "gedanken" (i.e., thought) experiment. We start at some *initial time*, and we apply an input to our physical system $\mathcal{P}$ until some future time. We observe the output of $\mathcal{P}$

over this time interval, which we shall call the *observation interval.* Now, we know that our observed output will not only depend upon the applied input but will also depend upon the conditions in $\mathcal{P}$ at our initial time. In other words, if we imagine that $\mathcal{P}_1$ is an exact replica of $\mathcal{P}$, that is, if the two physical systems $\mathcal{P}$ and $\mathcal{P}_1$ are identical, then we may ask the following question:

*If the same input is applied to $\mathcal{P}$ and $\mathcal{P}_1$, starting at the same initial time, shall we observe the same output from $\mathcal{P}$ and $\mathcal{P}_1$?*

For example, if $\mathcal{P}$ and $\mathcal{P}_1$ are identical *RLC* networks and if we apply the *same* voltage signal, starting from a time $t_0$ (say) until some future time $t_1$, to both $\mathcal{P}$ and $\mathcal{P}_1$, then we are asking whether or not the observed output-voltage signals of $\mathcal{P}$ and $\mathcal{P}_1$ are the *same* over the entire observation interval $(t_0, t_1]$ [see Eqs. (3-7)]. However, we know that the output voltage will depend upon the charge on the capacitors at $t_0$ and the current through the inductors at $t_0$ as well as upon the input voltage. Thus, if the charges on the capacitors in $\mathcal{P}$ at $t_0$ are different from the charges on the capacitors in $\mathcal{P}_1$ at $t_0$ or if the currents through the inductors in $\mathcal{P}$ at $t_0$ are different from the currents through the inductors in $\mathcal{P}_1$ at $t_0$, then, applying the *same* input-voltage signal to $\mathcal{P}$ and $\mathcal{P}_1$ over the time interval $(t_0, t_1]$, we should expect to observe *different* output-voltage signals from $\mathcal{P}$ and $\mathcal{P}_1$. We can see then that, in order to predict the output, we must know both the input *and* the conditions in $\mathcal{P}$ at the starting time. These conditions, which represent the additional information we must know at the initial time $t_0$, are called the (initial) *state* of the system $\mathcal{P}$ at $t_0$. We thus observe that the essential elements in a description of $\mathcal{P}$ are the notions of *input, output,* and *state.*

Now let us play the role of the theoretician and attempt to develop a formal way of describing $\mathcal{P}$. To begin with, we establish some notation. We agree to let the symbol $\mathbf{u}$ denote a typical input time function and to let the symbol $\mathbf{y}$ denote the corresponding output time function. If $t$ is some time, then

$$\mathbf{u}(t) \text{ is the value of the input } \mathbf{u} \text{ at the time } t \tag{4-56}$$

and

$$\mathbf{y}(t) \text{ is the value of the output } \mathbf{y} \text{ at the time } t \tag{4-57}$$

The values $\mathbf{u}(t)$ and $\mathbf{y}(t)$ may be scalars or elements of $R_m$ and $R_p$, respectively [that is, $\mathbf{u}(t)$ may be an $m$-tuple of real numbers, and $\mathbf{y}(t)$ may be a $p$-tuple of real numbers], or considerably more general objects. For example, if $\mathcal{P}$ were a pattern-recognition device, then the input $\mathbf{u}(t)$ at time $t$ might be a two-dimensional picture and the output $\mathbf{y}(t)$ at time $t$ might be a decision on whether $\mathbf{u}(t)$ was a picture of a bird, of a plane, or of superman. Although the concepts we consider and the definitions we give in Sec. 4-5 are quite general, we shall, in the remainder of the book, consider only the situation in which the values $\mathbf{u}(t)$ of the input are elements of $R_m$ and the values $\mathbf{y}(t)$ of the output are elements of $R_p$.

Now we start our experiment at some *initial time* $t_0$, and we apply the input $\mathbf{u}$ until some *future time* $t_1$. Since we observe and measure the output $\mathbf{y}$ during the time interval $(t_0, t_1]$, we shall call the semiclosed interval [see Eqs. (3-7)] $(t_0, t_1]$ the *observation interval.* For convenience, we shall denote the segments (restrictions) of $\mathbf{u}$ and $\mathbf{y}$ over this interval by $\mathbf{u}_{(t_0,t_1]}$ and $\mathbf{y}_{(t_0,t_1]}$, respectively; that is, $\mathbf{u}_{(t_0,t_1]}$ is the function defined on $(t_0, t_1]$ by

$$\mathbf{u}_{(t_0,t_1]}(t) = \mathbf{u}(t) \qquad \text{for all } t \text{ in } (t_0, t_1] \tag{4-58}$$

and, similarly, $\mathbf{y}_{(t_0,t_1]}$ is the function defined on $(t_0, t_1]$ by

$$\mathbf{y}_{(t_0,t_1]}(t) = \mathbf{y}(t) \qquad \text{for all } t \text{ in } (t_0, t_1] \tag{4-59}$$

We shall call the pair of functions $(\mathbf{u}_{(t_0,t_1]}, \mathbf{y}_{(t_0,t_1]})$ an *input-output pair* on the observation interval $(t_0, t_1]$. Loosely speaking, our system $\mathcal{P}$ may be described as the set of *all* these input-output pairs taken over *all* observation intervals and over *all* conditions or states of $\mathcal{P}$. This view of $\mathcal{P}$ corresponds to the carrying out of all possible experiments on $\mathcal{P}$ and, as such, has little value. We may also view $\mathcal{P}$ as generating a relation between input, output, and state. This is the point of view we shall adopt.

Since we shall be concerned with "macroscopic" phenomena in this book, we shall suppose that our systems are completely deterministic and that our systems obey the classical cause-and-effect laws of physics. In other words, we shall suppose that:

1. All the quantities and functions with which we deal are exactly determined, so that *no random elements occur.*
2. Our systems are *nonanticipatory,* in that present values of the input, output, and state do not depend upon the future values of the input, output, and state.

Now let us suppose that $\mathcal{S}$ is to be an abstract mathematical model of our physical system $\mathcal{P}$. Then we must have, associated with $\mathcal{S}$, a certain set of *input* time functions $\mathbf{u}$, a corresponding set of *output* time functions $\mathbf{y}$, and a set of *state* time functions $\mathbf{x}$. Moreover, we must also have, associated with $\mathcal{S}$, a pair of equations, the output and state equations, which relate the input, output, and state of $\mathcal{S}$. We write these equations in the form

$$\mathbf{y}(t) = \mathbf{g}[\mathbf{x}(t_0); \mathbf{u}_{(t_0,t]}; (t_0, t]] \qquad \text{output equation} \tag{4-60}$$

and

$$\mathbf{x}(t) = \mathbf{f}[\mathbf{x}(t_0); \mathbf{u}_{(t_0,t]}; (t_0, t]] \qquad \text{state equation} \tag{4-61}$$

since we know that the output $\mathbf{y}(t)$ at time $t$ depends upon the state $\mathbf{x}(t_0)$ at the initial time $t_0$, the input applied over the observation interval $(t_0, t]$, and the observation interval $(t_0, t]$, and since we also know that application of the input $\mathbf{u}_{(t_0,t]}$ not only will generate the output $\mathbf{y}_{(t_0,t]}$ but also will produce a change in the conditions, or state, of our system and that the result of this change, i.e., the state $\mathbf{x}(t)$ at time $t$, will depend upon the state $\mathbf{x}(t_0)$ at the initial time $t_0$, the input applied over the observation interval $(t_0, t]$, and the observation interval itself.

Now, if our model $\mathcal{S}$ is to be consistent with our assumptions 1 and 2 and if $\mathcal{S}$ is to be an adequate representation of our experience with the physical system $\mathcal{P}$, then the state and output equations of $\mathcal{S}$ should have certain natural and desirable properties. For example, the functions **f** and **g** should be both *deterministic* (in that none of the quantities appearing in them is random or contains random elements) and *nonanticipatory* (in that future values of the input, output, or state do not influence the present values of these quantities). Moreover, the functions **f** and **g** should be chosen in such a way that the model $\mathcal{S}$ approximates the results of our experiments on the physical system $\mathcal{P}$. In other words, the functions **f** and **g** should be chosen to "fit" our experimental data.

In the next section, we shall give an axiomatic definition of a rather general class of system models, namely, the class of *dynamical systems*. The axioms will involve certain requirements on the set of inputs, the set of outputs, the set of states, and the functions **f** and **g**. In Sec. 4-6, we shall impose still more requirements in order to carefully delineate the class of systems with which the main portion of the book will be concerned. However, before beginning our formal development, we shall, in the remainder of this section, indicate some of the intuitive aspects of the general axioms and of the particular class of systems in which we shall be most interested.

Our first axiom will essentially assert that if we know the initial state at time $t_0$ and if we apply a known input over an interval $(t_0, t]$ with $t > t_0$, then we obtain a uniquely defined output. Our second axiom will ensure that there are "enough" states of the system so that every input-output pair can be accounted for. In fact, our second axiom will also assert that knowledge of the initial state and of the input applied over the observation interval will be sufficient to determine not only the output over the observation interval but also the *state* over the observation interval. This is crucial because it means that the *state at any time summarizes, in a sense, all the past information required to predict the future output signals and future states.* Our third axiom asserts a smoothness condition which guarantees that small changes in our input or in our initial state will cause correspondingly small changes in our output and in our future states. Our fourth (and final) axiom contains the conditions which the function representing the change of state must satisfy.

Now, the systems we shall study in the main portion of this book not only will satisfy our four axioms but also will satisfy certain additional conditions. In particular, we shall suppose that the states of our system are elements of some Euclidean space $R_n$, that is, are $n$ vectors or, equivalently, $n$-tuples of real numbers. Moreover, we shall demand that the values of our inputs be elements of a Euclidean space $R_m$ with $m \leq n$ and that the values of our outputs be elements of a Euclidean space $R_p$. We shall also assume that all the time-dependent quantities which we consider, such as inputs or outputs, are defined over some open interval $(T_1, T_2)$ [see Eqs. (3-7)] rather than at discrete times (i.e., we consider continuous-time, rather than discrete-time, systems). Finally, we shall suppose the state equation represents the solution of a differential equation which satisfies the conditions of the basic existence theorem (Theorem 3-14). Most of these remarks will become clearer as we progress.

## 4-5 Dynamical Systems: Formal Definition†

We shall define the notion of a dynamical system in a formal and rather general way in this section. Although we shall adopt an axiomatic approach, we shall not be quite as general as possible and we shall, in the next section, restrict the definition still further so as to carefully delineate the class of systems which we shall examine in the remainder of the book. We note again that our concept of a dynamical system will be based upon the notions of input, output, and state. We now proceed to the task at hand.

Let $T$ be a subset of the real numbers; let $\Sigma$ be a set, with $d$ as distance on $\Sigma$ (see Sec. 3-1); let $\Omega$ be a set, with $\hat{d}$ as distance on $\Omega$; and let $U$ be a set of piecewise continuous functions on $T$ with values in $\Omega$. Let $\mathbf{x}(t)$ be a variable which is defined on $T$ and which takes values in $\Sigma$, and let $\mathbf{g}$ be a function from $\Sigma \times \Omega \times T$ (see Sec. 2-3) into the Euclidean space $R_p$. If $t_0$ and $t$ are elements of $T$ with $t_0 \leq t$, then we shall, *in this section*, write $(t_0, t]$ in place of the set of elements of $T$ between $t_0$ and $t$, that is,

$$(t_0, t] = \{\tau : \tau \in T \text{ and } t_0 < \tau \leq t\} \tag{4-62}‡$$

We shall use the notation $\mathbf{u}_{(t_0,t]}$ to denote the segment (i.e., restriction) of the function $\mathbf{u}$ in $U$ over the set $(t_0, t]$. If $t$ is an element of $T$, if $\mathbf{u}$ is an element of $U$, and if $\mathbf{x}(t)$ is an element of $\Sigma$, then $\mathbf{g}[\mathbf{x}(t), \mathbf{u}(t), t]$ is a well-defined element of $R_p$, which is denoted by $\mathbf{y}(t)$; that is,

$$\mathbf{y}(t) = \mathbf{g}[\mathbf{x}(t), \mathbf{u}(t), t] \tag{4-63}$$

We shall use the notation $\mathbf{y}_{(t_0,t]}$ to denote the segment of the function $\mathbf{y}(t)$ of Eq. (4-63) over $(t_0, t]$, and we shall write

$$\mathbf{y}_{(t_0,t]} = \hat{\mathbf{g}}[\mathbf{x}(t_0); \mathbf{u}_{(t_0,t]}] \tag{4-64}$$

We are now ready to present the axioms.

***Axiom 4-1*** *For all* $\mathbf{x}(t_0)$ *in* $\Sigma$, *all* $t_0$ *in* $T$, *all* $t \geq t_0$ *with* $t \in T$, *and all* $\mathbf{u}_{(t_0,t]}$ *with* $\mathbf{u} \in U$, *knowledge of* $\mathbf{x}(t_0)$ *and* $\mathbf{u}_{(t_0,t]}$ *uniquely determines* $\mathbf{y}_{(t_0,t]}$. *In particular, if* $\mathbf{u}$ *and* $\mathbf{v}$ *are elements of* $U$ *such that*

$$\mathbf{u}_{(t_0,t]} = \mathbf{v}_{(t_0,t]} \tag{4-65}$$

*then*

$$\hat{\mathbf{g}}[\mathbf{x}(t_0); \mathbf{u}_{(t_0,t]}] = \hat{\mathbf{g}}[\mathbf{x}(t_0); \mathbf{v}_{(t_0,t]}] \tag{4-66}$$

† See Refs. Z-1, K-7, K-9, and W-5. Chapters 1 to 4 of Ref. Z-1 are of particular interest.

‡ Observe that $(t_0, t]$ in this sense is merely the intersection of the semiclosed interval $(t_0, t]$ with the set $T$ [see Eqs. (2-3) and (3-7)].

***Axiom 4-2*** *If $t_0 < \hat{t} < t$ are elements of $T$, if $\mathbf{x}(t_0)$ is an element of $\Sigma$, if $\Sigma[\mathbf{x}(t_0), \mathbf{u}, \hat{t}]$ denotes the set of all the elements $\mathbf{x}(\hat{t})$ of $\Sigma$ which satisfy the equation*

$$\mathbf{y}_{(\hat{t},t]} = \hat{\mathbf{g}}[\mathbf{x}(t_0); \mathbf{u}_{(t_0,t]}]_{(\hat{t},t]} = \hat{\mathbf{g}}[\mathbf{x}(\hat{t}); \mathbf{u}_{(\hat{t},t]}] \tag{4-67}$$

*and if $\mathbf{u}^*$ is a given element of $U$, then the intersection of the sets $\Sigma[\mathbf{x}(t_0), \mathbf{u}, \hat{t}]$, where $\mathbf{u}$ is an element of $U$ with $\mathbf{u}_{(t_0,\hat{t}]} = \mathbf{u}^*_{(t_0,\hat{t}]}$, is not empty; that is,*

$$\bigcap_{\substack{\mathbf{u} \in U \\ \mathbf{u}_{(t_0,\hat{t}]} = \mathbf{u}^*_{(t_0,\hat{t}]}}} \Sigma[\mathbf{x}(t_0), \mathbf{u}, \hat{t}] \neq \emptyset \tag{4-68}$$

*In particular, no set $\Sigma[\mathbf{x}(t_0), \mathbf{u}, \hat{t}]$ is empty. This axiom ensures that there is at least one element of $\Sigma$ which accounts for any pair $(\mathbf{u}_{(\hat{t},t]}, \mathbf{y}_{(\hat{t},t]})$.*

Now it can be shown† that Axioms 4-1 and 4-2 imply that there is a function $\boldsymbol{\phi}[t; \mathbf{u}_{(t_0,t]}, \mathbf{x}(t_0)]$ such that

$$\mathbf{x}(t) = \boldsymbol{\phi}[t; \mathbf{u}_{(t_0,t]}, \mathbf{x}(t_0)] \tag{4-69}$$

***Axiom 4-3*** *The functions $\mathbf{g}$, $\hat{\mathbf{g}}$, and $\boldsymbol{\phi}$ are continuous in all their arguments. As regards the dependence of $\mathbf{g}$ and $\boldsymbol{\phi}$ upon $\mathbf{u}_{(t_0,t]}$ this means that if $\mathbf{u}$ and $\mathbf{v}$ are elements of $U$ such that*

$$d_u(\mathbf{u}_{[t_0,t]}, \mathbf{v}_{[t_0,t]}) = \sup_{\tau \in [t_0,t] \cap T} \{\hat{d}(\mathbf{u}(\tau), \mathbf{v}(\tau))\} \tag{4-70}‡$$

*is small, then*

$$d\{\boldsymbol{\phi}[t; \mathbf{u}_{(t_0,t]}, \mathbf{x}(t_0)], \boldsymbol{\phi}[t; \mathbf{v}_{(t_0,t]}, \mathbf{x}(t_0)]\} \tag{4-71}$$

*and*

$$\sup_{\tau \in [t_0,t] \cap T} \{\|\hat{\mathbf{g}}[\mathbf{x}(t_0); \mathbf{u}_{(t_0,\tau]}] - \hat{\mathbf{g}}[\mathbf{x}(t_0); \mathbf{v}_{(t_0,\tau]}]\|\} \tag{4-72}$$

*are also small where $\hat{\mathbf{g}}[\mathbf{x}(t_0); \mathbf{u}_{(t_0,t_0]}] = \mathbf{g}[\mathbf{x}(t_0), \mathbf{u}(t_0+), t_0]$ and $\hat{\mathbf{g}}[\mathbf{x}(t_0); \mathbf{v}_{(t_0,t_0]}] = \mathbf{g}[\mathbf{x}(t_0), \mathbf{v}(t_0+), t_0]$ by definition.§*

***Axiom 4-4*** *The function $\boldsymbol{\phi}$ satifies the following conditions:*

*a. For all $t$, $t_0 \in T$, $\mathbf{u} \in U$, and $\mathbf{x}(t_0) \in \Sigma$, $\boldsymbol{\phi}[t_0; \mathbf{u}_{(t_0,t]}, \mathbf{x}(t_0)] = \mathbf{x}(t_0)$ in the sense that the limit of $\boldsymbol{\phi}[t; \mathbf{u}_{(t_0,t]}, \mathbf{x}(t_0)]$ as $t$ approaches $t_0$ from the right is $\mathbf{x}(t_0)$. This may also be written by abuse of notation as $\boldsymbol{\phi}[t_0; \mathbf{u}_{(t_0,t_0]}, \mathbf{x}(t_0)]$.*

*b. For all $t_0 < \hat{t} \leq t$ in $T$, $\mathbf{u} \in U$, and $\mathbf{x}(t_0) \in \Sigma$,*

$$\boldsymbol{\phi}[t; \mathbf{u}_{(t_0,t]}, \mathbf{x}(t_0)] = \boldsymbol{\phi}[t; \mathbf{u}_{(\hat{t},t]}, \boldsymbol{\phi}[\hat{t}; \mathbf{u}_{(t_0,\hat{t}]}, \mathbf{x}(t_0)]] \tag{4-73}$$

*where $t_0 < \hat{t} \leq t$. This is called the transition, or semigroup, condition.*

† See Ref. Z-1.

‡ Here "sup" means the greatest lower bound of the set of numbers $M$ such that $\hat{d}(\mathbf{u}(\tau), \mathbf{v}(\tau)) \leq M$ for all $\tau$ in $[t_0, t] \cap T$. We observe that $d_u$ is a distance on the set of functions $\mathbf{u}_{[t_0,t]}$ with $\mathbf{u}$ in $U$ (cf. Sec. 3-15).

§ Recall that $\hat{d}$ is the distance on $\Omega$, $d$ is the distance on $\Sigma$, and $\| \quad \|$ is the Euclidean norm on $R_p$.

c. *For all* $\tau$, $t_0$, $t$ *with* $\tau$ *in* $[t_0, t] \cap T$, *and for all* $\mathbf{x}(t_0)$ *in* $\Sigma$, *if* $\mathbf{u}$ *and* $\mathbf{v}$ *are elements of* $U$ *with* $\mathbf{u}_{[t_0,t]} = \mathbf{v}_{[t_0,t]}$, *then*

$$\boldsymbol{\phi}[\tau;\, \mathbf{u}_{(t_0,t]},\, \mathbf{x}(t_0)] = \boldsymbol{\phi}[\tau;\, \mathbf{v}_{(t_0,\tau]},\, \mathbf{x}(t_0)] \tag{4-74}$$

*for all* $\tau$ *in* $[t_0, t] \cap T$.

We can now make the following definition:

***Definition 4-1*** *A dynamical system* $\mathcal{S}$ *is the composite concept consisting of sets* $T$, $\Sigma$, $\Omega$, *and* $U$, *a variable* $\mathbf{x}(t)$, *and a function* $\mathbf{g}$ *such that Axioms* 4-1 *to* 4-4 *are satisfied. In that case,* $T$ *is called the domain of the system;* $\Sigma$ *is called the state space of the system;* $U$ *is called the input space of the system;* $\mathbf{x}(t)$ *is called the state (or state variable) of the system;* $\mathbf{u}_{(t_0,t]}$ *is called an input over the observation interval* $(t_0, t]$ *to the system, and* $\mathbf{y}_{(t_0,t]} = \mathbf{g}[\mathbf{x}(t_0);\, \mathbf{u}_{(t_0,t]}]$ *is called a (corresponding) output of the system, with* $(\mathbf{u}_{(t_0,t]}, \mathbf{y}_{(t_0,t]})$ *forming an input-output pair of the system; the function* $\boldsymbol{\phi}[t;\, \mathbf{u}_{(t_0,t]},\, \mathbf{x}(t_0)]$ *is called the transition function of the system; the subset* $\{\mathbf{x}(\tau): \mathbf{x}(\tau) = \boldsymbol{\phi}[\tau;\, \mathbf{u}_{(t_0,\tau]},\, \mathbf{x}(t_0)]$ *for* $\tau$ *in* $[t_0, t] \cap T\}$ *of the state space* $\Sigma$ *is called the trajectory, or motion, of the system during the time interval* $[t_0, t] \cap T$, *starting from the initial state* $\mathbf{x}(t_0)$ *and generated by the input* $\mathbf{u}_{(t_0,t]}$; *and, finally, Eqs.* (4-64) *and* (4-69) *are called the output equation and the state equation, respectively, of the system.*

Let us now briefly interpret the axioms.

Axiom 4-1 essentially states that if we know the initial state at time $t_0$ and if we apply a known input over $(t_0, t]$ with $t > t_0$, then we obtain a uniquely defined output. The essence of this axiom is that, in order to predict the output over $(t_0, t]$ when the state at $t_0$ is known, it is not necessary to know the inputs prior to $t_0$. The state at $t_0$ and $\mathbf{u}_{(t_0,t]}$ suffice. We note also that future values of the input do not affect $\mathbf{y}_{(t_0,t]}$, so that the axiom implies the nonanticipative character of the system.

Axiom 4-2 ensures that there are "enough" states of the system so that every input-output pair can be accounted for. Axiom 4-2 also implies that knowledge of the initial state $\mathbf{x}(t_0)$ and of the applied input $\mathbf{u}_{(t_0,t]}$ is sufficient to determine not only the output $\mathbf{y}(\hat{t})$, $t_0 < \hat{t} \le t$, but also the state at the time $\hat{t}$, that is, $\mathbf{x}(\hat{t})$, $t_0 < \hat{t} \le t$. This is a most important property because it means that the state at any time summarizes, in a sense, all the past information required to predict the future output signals and future states.

Axiom 4-3 is a smoothness condition which guarantees that small changes in the input or state of the system cause correspondingly small changes in the output and motion of the system.

Axiom 4-4 enumerates the conditions that changes in the state of the system must satisfy. In particular, we require that:

1. Our initial condition must correspond to the starting point of the motion (where we recall that "motion" means the trajectory in the state space).

2. If our input takes the system from a state $\mathbf{x}_0$ to a state $\mathbf{x}$ along a certain path and if $\hat{\mathbf{x}}$ is a state on that path, then application of our input must take us from $\hat{\mathbf{x}}$ to $\mathbf{x}$.

3. Our system must be nonanticipatory, in that future values of the input will have no effect on the present state.

We can begin to see that our axioms are indeed a reasonable abstraction of what we observe in the physical world.

The consequences of the axioms are illustrated in Fig. 4-7. Figure 4-7*a* illustrates two inputs, $\mathbf{u}^1_{(t_0,t_1]}$ and $\mathbf{u}^2_{(t_0,t_1]}$, with the property

$$\left.\begin{aligned} \mathbf{u}^1_{(t_0,\hat{t}]} &= \mathbf{u}^2_{(t_0,\hat{t}]} \\ \mathbf{u}^1_{(\hat{t},t_1]} &\neq \mathbf{u}^2_{(\hat{t},t_1]} \end{aligned}\right\} \qquad \hat{t} \in [t_0, t_1]$$

Let $\mathbf{x}(t_0)$ be the state at $t_0$, let $\mathbf{y}^1_{(t_0,t_1]}$ denote the unique output resulting from $\mathbf{u}^1_{(t_0,t_1]}$, and let $\mathbf{y}^2_{(t_0,t_1]}$ denote the output resulting from $\mathbf{u}^2_{(t_0,t_1]}$. The outputs are shown in Fig. 4-7*b*. According to our terminology

$$\begin{aligned} \mathbf{y}^1_{(t_0,t_1]} &= \hat{\mathbf{g}}[\mathbf{x}(t_0), \mathbf{u}^1_{(t_0,t_1]}] \\ \mathbf{y}^2_{(t_0,t_1]} &= \hat{\mathbf{g}}[\mathbf{x}(t_0), \mathbf{u}^2_{(t_0,t_1]}] \end{aligned}$$

Axiom 4-1 and, in particular, Eqs. (4-65) and (4-66) guarantee that

$$\mathbf{y}^1_{(t_0,\hat{t}]} = \mathbf{y}^2_{(t_0,\hat{t}]}$$

Let $\bar{t}$ be any time such that $\bar{t} \in [t_0, t_1]$. Then Axiom 4-2 and, in particular, Eq. (4-67) guarantee the existence of states $\mathbf{x}^1(\bar{t})$ and $\mathbf{x}^2(\bar{t})$ such that

$$\begin{aligned} \mathbf{y}^1_{(\bar{t},t_1]} &= \hat{\mathbf{g}}[\mathbf{x}^1(\bar{t}), \mathbf{u}^1_{(\bar{t},t_1]}] \\ \mathbf{y}^2_{(\bar{t},t_1]} &= \hat{\mathbf{g}}[\mathbf{x}^2(\bar{t}), \mathbf{u}^2_{(\bar{t},t_1]}] \end{aligned}$$

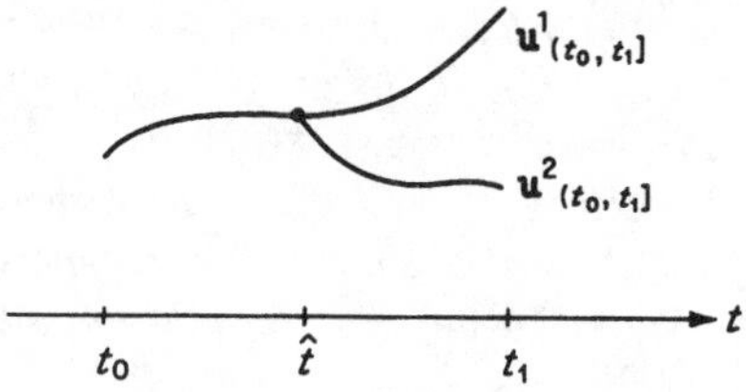

(*a*)

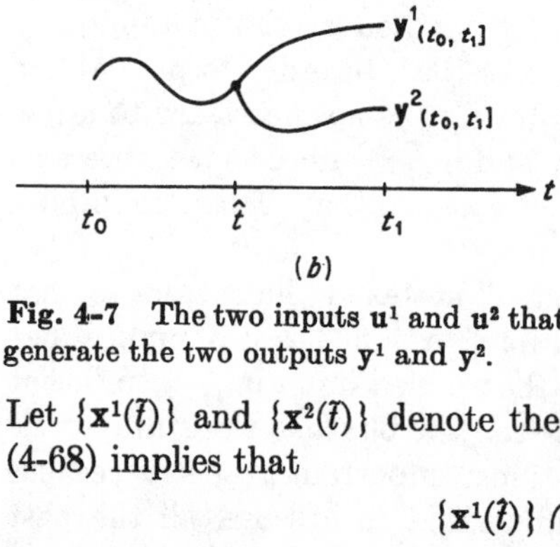

(*b*)

**Fig. 4-7** The two inputs $\mathbf{u}^1$ and $\mathbf{u}^2$ that generate the two outputs $\mathbf{y}^1$ and $\mathbf{y}^2$.

Let $\{\mathbf{x}^1(\bar{t})\}$ and $\{\mathbf{x}^2(\bar{t})\}$ denote the set of states $\mathbf{x}^1(\bar{t})$ and $\mathbf{x}^2(\bar{t})$. Then Eq. (4-68) implies that

$$\{\mathbf{x}^1(\hat{t})\} \cap \{\mathbf{x}^2(\hat{t})\} \neq \emptyset$$

This implies that there exists a state

$$\mathbf{x}(\hat{t}) \in \{\mathbf{x}^1(\hat{t})\} \cap \{\mathbf{x}^2(\hat{t})\}$$

such that

$$\begin{aligned} \mathbf{y}^1_{(\hat{t},t_1]} &= \hat{\mathbf{g}}[\mathbf{x}(\hat{t}), \mathbf{u}^1_{(\hat{t},t_1]}] \\ \mathbf{y}^2_{(\hat{t},t_1]} &= \hat{\mathbf{g}}[\mathbf{x}(\hat{t}), \mathbf{u}^2_{(\hat{t},t_1]}] \end{aligned}$$

Now let us consider the simple *RL* network of Fig. 4-1. We take $T$ to be all the real numbers, we let $\Sigma$ be the real numbers with the usual distance, and we let $\Omega = \Sigma$ and $U$ be the set of piecewise continuous functions from $\Omega$ (the real numbers) into $\Sigma$ (the real numbers). Finally, we let $\mathbf{g}$ be the function given by

$$\mathbf{g}(r, s, t) = r \tag{4-75}$$

where $r$, $s$, and $t$ are real numbers. Since $\mathbf{g}$ does not depend on $s$ and $t$, we shall write, for convenience of exposition,

$$\mathbf{g}(r) = r \tag{4-76}$$

in place of Eq. (4-75).

If $u$ is an element of $U$ and $x_0$ is an element of $\Sigma$, then we set

$$x(t) = e^{-at}x_0 + e^{-at}\int_0^t e^{a\tau}u(\tau)\,d\tau \tag{4-77}$$

where $a = R/L$. In other words,

$$\varphi(t;\, u_{(0,t]},\, x_0) = e^{-at}x_0 + e^{-at}\int_0^t e^{a\tau}u(\tau)\,d\tau \tag{4-78}$$

We also let

$$y(t) = \mathbf{g}[x(t)] = x(t) \tag{4-79}$$

Then we assert that we have defined a dynamical system with $\Sigma$ (= real numbers) as state space, with $y(t)$ as output, and with the state and output equations

$$\begin{aligned} x(t) &= \varphi(t;\, u_{(0,t]},\, x(0)) = e^{-at}x(0) + e^{-at}\int_0^t e^{a\tau}u(\tau)\,d\tau && \text{state equation} \\ y(t) &= x(t) && \text{output equation} \end{aligned} \tag{4-80}$$

It is clear that if we know $x(0)$ and $u_{(0,t]}$, then we know $x(t)$ and hence, in view of Eqs. (4-80), $y(t)$. Moreover, in view of Theorem 3-14, $x(t)$ and $y(t)$ are completely determined. As for Axiom 4-2, the element

$$x(\hat{t}) = e^{-a\hat{t}}x(0) + e^{-a\hat{t}}\int_0^{\hat{t}} e^{a\tau}u^*(\tau)\,d\tau$$

is in the intersection of all the sets $\Sigma[x(0), u, t]$ [see Eq. (3-259)]. We leave it to the reader to verify Axioms 4-3 and 4-4.

**Exercise 4-4** Verify Axiom 4-4 for the *RL* network.

**Exercise 4-5** Consider the *RL* network of Fig. 4-2 and determine a dynamical system which represents the network. Verify that the axioms are satisfied. HINT: The function $\mathbf{g}$ is given by $\mathbf{g}(r, s, t) = r + s$; $\varphi$ is given by

$$\varphi(t;\, u_{(0,t]},\, x_0) = e^{-at}x_0 - ae^{-at}\int_0^t e^{a\tau}u(\tau)\,d\tau$$

and the output and state equations are

$$\begin{aligned} y(t) &= x(t) + u(t) && \text{output equation} \\ x(t) &= e^{-at}x_0 - ae^{-at}\int_0^t e^{a\tau}u(\tau)\,d\tau && \text{state equation} \end{aligned}$$

Also, Theorem 3-8 is used in verifying Axiom 4-3.

## 4-6 Dynamical Systems: The Systems Considered in This Book

The systems we shall study in this book not only will be dynamical systems in the sense of Definition 4-1 but also will satisfy certain additional conditions. In point of fact, we shall only consider *finite-dimensional continuous-time differential* systems in this book. We shall define this class of systems and discuss some of its properties in this section.

We begin with some definitions.

***Definition 4-2*** *A dynamical system* $\mathcal{S}$ *is said to be finite-dimensional if* (1) *the state space* $\Sigma$ *of* $\mathcal{S}$ *is a Euclidean space* $R_n$, *that is,* $\Sigma = R_n$, *and* (2) *the set* $\Omega$ *in which the inputs take their values is a Euclidean space* $R_m$ *with* $m \leq n$, *that is,* $\Omega = R_m$. *In that case, we say that* $n$, *which is the dimension of* $\Sigma$, *is the order (or dimension) of the dynamical system.*

A vibrating string and a cavity resonator (either acoustic or electromagnetic) are examples of systems which are *not* finite-dimensional.

*We shall assume from now on that the term* "dynamical system" *means* "finite-dimensional dynamical system." We also note that the states of the system are $n$ vectors $\mathbf{x}$ and the inputs to the system are $m$ vectors $\mathbf{u}(t)$, with

$$m \leq n \tag{4-81}$$

***Definition 4-3*** *A dynamical system* $\mathcal{S}$ *is said to be a continuous-time (or, simply, a continuous) system if the set* $T$ *is an open interval* $(T_1, T_2)$ *(which may be all of* $R$*).* $(T_1, T_2)$ *is often called the domain, or interval of definition, of the system.*

Sampled-data systems provide us with a large class of systems which are not continuous.

*Again, we shall suppose from now on that the term* "dynamical system" *means* "continuous-time dynamical system."

***Definition 4-4*** *A dynamical system* $\mathcal{S}$ *is said to be a differential system if the state and output equations of the system are of the form*

$$\mathbf{x}(t) = \boldsymbol{\phi}[t; \mathbf{u}_{(t_0,t]}, \mathbf{x}(t_0)] \tag{4-82}$$

*where* $\boldsymbol{\phi}$ *is the solution of the system of differential equations*

$$\dot{\mathbf{x}}(t) = \mathbf{f}[\mathbf{x}(t), \mathbf{u}(t), t] \tag{4-83}$$

*with* $\mathbf{x}(t_0)$ *as initial point and with* $\mathbf{f}$ *satisfying the conditions of Theorem* 3-14, *and*

$$\mathbf{y}(t) = \mathbf{g}[\mathbf{x}(t), \mathbf{u}(t), t] \tag{4-84}$$

*where* $\mathbf{g}$ *is a continuous function of all its arguments. We shall often, in this case, refer to Eqs.* (4-83) *and* (4-84) *as the equations of the system, and we shall often call Eq.* (4-83) *the state equation.*

*So, from now on, we shall use the term* "dynamical system" *or* "system" *to mean a* "finite-dimensional continuous-time differential system."

*To recapitulate:* In this book, a dynamical system will be the composite concept consisting of an open interval $(T_1, T_2)$ which may be all of $R$, a set $U$ of piecewise continuous functions from this interval into $R_m$, a continuous function $\mathbf{f}$ from $R_n \times R_m \times (T_1, T_2)$ into $R_n$, and a continuous function $\mathbf{g}$ from $R_n \times R_m \times (T_1, T_2)$ into $R_p$ such that the following conditions are satisfied:

DS1 For all $\mathbf{x}_0$ in $R_n$, all $t_0$ in $(T_1, T_2)$, all $t$ in $(T_1, T_2)$ with $t \geq t_0$, and all $\mathbf{u}_{(t_0,t]}$ with $\mathbf{u}$ in $U$, there is a *unique* solution $\mathbf{x}(\tau)$ of the vector differential equation

$$\dot{\mathbf{x}}(\tau) = \mathbf{f}[\mathbf{x}(\tau), \mathbf{u}(\tau), \tau] \tag{4-85}$$

defined on the interval $[t_0, t]$† with

$$\mathbf{x}(t_0) = \mathbf{x}_0 \tag{4-86}$$

We write

$$\mathbf{x}(t) = \boldsymbol{\phi}(t; \mathbf{u}_{(t_0,t]}, \mathbf{x}_0) \qquad \text{state equation} \tag{4-87}$$

to represent this solution.

DS2 $\boldsymbol{\phi}$ is a continuous function of all its arguments which satisfies the conditions

*a.*
$$\mathbf{x}_0 = \boldsymbol{\phi}(t_0; \mathbf{u}_{(t_0,t]}, \mathbf{x}_0) \tag{4-88}$$

*b.*
$$\boldsymbol{\phi}(t; \mathbf{u}_{(t_0,t]}, \mathbf{x}_0) = \boldsymbol{\phi}(t; \mathbf{u}_{(\hat{t},t]}, \boldsymbol{\phi}(\hat{t}; \mathbf{u}_{(t_0,\hat{t}]}, \mathbf{x}_0)) \tag{4-89}$$

for all $\hat{t}$ in $[t_0, t]$.

*c.* If $\mathbf{u} = \mathbf{v}$ on $(t_0, t]$, then, for all $\tau$ in $[t_0, t]$,

$$\boldsymbol{\phi}(\tau; \mathbf{u}_{(t_0,\tau]}, \mathbf{x}_0) = \boldsymbol{\phi}(\tau; \mathbf{v}_{(t_0,\tau]}, \mathbf{x}_0) \tag{4-90}$$

DS3 If the output $\mathbf{y}(t)$ is defined by setting

$$\mathbf{y}(\tau) = \mathbf{g}[\boldsymbol{\phi}(\tau; \mathbf{u}_{(t_0,\tau]}, \mathbf{x}_0), \mathbf{u}(\tau), \tau] \qquad \text{output equation} \tag{4-91}$$

then $\mathbf{y}_{(t_0,t]}$ is *uniquely* determined by $\mathbf{x}_0 = \mathbf{x}(t_0)$ and $\mathbf{u}_{(t_0,t]}$. Moreover, if $t_0 \leq \hat{t} \leq t$ and if $\mathbf{x}(\hat{t}) = \boldsymbol{\phi}(\hat{t}; \mathbf{u}_{(t_0,\hat{t}]}, \mathbf{x}_0)$, then $\mathbf{y}_{(\hat{t},t]}$ is uniquely determined by $\mathbf{x}(\hat{t})$ and $\mathbf{u}_{(\hat{t},t]}$.

We shall be especially concerned with a particular class of dynamical systems, namely, the class of *linear systems*. We shall define and investigate this type of dynamical system in the next few sections. However, before concluding this section, we should like to briefly examine a notion of *equivalence* which will be useful to us later on, and we should like to examine several examples of dynamical systems.

Let us recall that our dynamical system $\mathcal{S}$ was to be a "model" of some physical system $\mathcal{P}$. The "motion" of $\mathcal{P}$ is intrinsic and geometric in the sense that it is independent of the way in which we represent it. Now, the trajectories of our dynamical system $\mathcal{S}$ may be viewed as curves in $R_n$, and the output $\mathbf{y}$ may be viewed as a function of the points on these curves.

† This actually means that there is a solution over some *open* interval containing $[t_0, t]$.

In other words, the trajectories and output of $\mathcal{S}$ are geometric objects *whose coordinates satisfy the equations of the system.* Thus, $\mathbf{x}(t)$ represents a vector $\mathbf{v}$ in $R_n$ with coordinates $x_1(t), x_2(t), \ldots, x_n(t)$ with respect to the natural basis $\mathbf{e}_1, \mathbf{e}_2, \ldots, \mathbf{e}_n$ of $R_n$ (see Sec. 2-6).† If $\mathbf{P}$ is a nonsingular $n \times n$ matrix [see Eq. (2-59)], then we may view $\mathbf{z}(t) = \mathbf{P}^{-1}\mathbf{x}(t)$ as representing the vector $\mathbf{v}$ with respect to the basis $\mathbf{Pe}_1, \mathbf{Pe}_2, \ldots, \mathbf{Pe}_n$ of $R_n$. It follows that if we consider the dynamical system $\mathcal{S}'$, with system equations

$$\dot{\mathbf{z}}(t) = \mathbf{P}^{-1}\mathbf{f}[\mathbf{Pz}(t), \mathbf{u}(t), t] \tag{4-92}$$
$$\mathbf{y}(t) = \mathbf{g}[\mathbf{Pz}(t), \mathbf{u}(t), t] \tag{4-93}$$

then the trajectories and output of this system are the *same* as those of $\mathcal{S}$. This leads us to the following definition:

***Definition 4-5*** *Two dynamical systems $\mathcal{S}$ and $\mathcal{S}'$ are said to be equivalent if there is a nonsingular $n \times n$ matrix $\mathbf{P}$ with constant entries such that*

$$\mathbf{P}^{-1}\mathbf{x}(t) = \mathbf{z}(t) \tag{4-94}$$

*where $\mathbf{x}(t)$ is the state variable of $\mathcal{S}$ and $\mathbf{z}(t)$ is the state variable of $\mathcal{S}'$.*‡

This notion of equivalence will be particularly useful to us in the study of linear systems (see Sec. 4-7).

**Example 4-2** Let $(T_1, T_2)$ be all of $R$ and let $U$ be the set of all piecewise continuous functions from $R$ into $R$. Let $\mathbf{f}$ be the continuous function from $R_2 \times R \times R$ into $R_2$, given by

$$\mathbf{f}\left(\begin{bmatrix} x_1 \\ x_2 \end{bmatrix}, u, t\right) = \begin{bmatrix} x_2 + u \\ tx_2 \end{bmatrix} \tag{4-95}$$
$$= \begin{bmatrix} 0 & 1 \\ 0 & t \end{bmatrix} \begin{bmatrix} x_1 \\ x_2 \end{bmatrix} + \begin{bmatrix} 1 \\ 0 \end{bmatrix} u \tag{4-96}$$

and let $\mathbf{g}$ be the continuous function from $R_2 \times R \times R$ into $R$, given by

$$\mathbf{g}\left(\begin{bmatrix} x_1 \\ x_2 \end{bmatrix}, u, t\right) = x_1 \tag{4-97}$$
$$= [1 \quad 0] \begin{bmatrix} x_1 \\ x_2 \end{bmatrix} + 0u \tag{4-98}$$

If we consider the vector differential equation

$$\dot{\mathbf{x}}(\tau) = \mathbf{f}[\mathbf{x}(\tau), u(\tau), \tau] \tag{4-99}$$
$$= \begin{bmatrix} 0 & 1 \\ 0 & \tau \end{bmatrix} \begin{bmatrix} x_1(\tau) \\ x_2(\tau) \end{bmatrix} + \begin{bmatrix} 1 \\ 0 \end{bmatrix} u(\tau) \tag{4-100}$$

† Recall that

$$\mathbf{e}_1 = \begin{bmatrix} 1 \\ 0 \\ \cdot \\ \cdot \\ \cdot \\ 0 \\ 0 \end{bmatrix} \quad \mathbf{e}_2 = \begin{bmatrix} 0 \\ 1 \\ \cdot \\ \cdot \\ \cdot \\ 0 \\ 0 \end{bmatrix} \quad \cdots \quad \mathbf{e}_n = \begin{bmatrix} 0 \\ 0 \\ \cdot \\ \cdot \\ \cdot \\ 0 \\ 1 \end{bmatrix}$$

‡ This is not the most general notion of equivalence, but it will be adequate for our purposes.

then we know that (see Example 3-51) this equation has the unique solution

$$\mathbf{x}(t) = \mathbf{\Phi}(t, t_0)\mathbf{x}_0 + \mathbf{\Phi}(t, t_0) \int_{t_0}^{t} \mathbf{\Phi}(t_0, \tau) \begin{bmatrix} 1 \\ 0 \end{bmatrix} u(\tau)\, d\tau \tag{4-101}$$

where $\mathbf{\Phi}(t, t_0)$ is given by

$$\mathbf{\Phi}(t, t_0) = \begin{bmatrix} 1 & \int_{t_0}^{t} e^{-t_0^2/2} e^{\tau^2/2}\, d\tau \\ 0 & e^{-t_0^2/2} e^{t^2/2} \end{bmatrix} \tag{4-102}$$

and where we require that

$$\mathbf{x}(t_0) = \mathbf{x}_0 \tag{4-103}$$

Thus, we shall write Eq. (4-101) in the form

$$\mathbf{x}(t) = \boldsymbol{\phi}(t;\, u_{(t_0,t]},\, \mathbf{x}_0) = \mathbf{\Phi}(t, t_0) \left\{ \mathbf{x}_0 + \int_{t_0}^{t} \mathbf{\Phi}(t_0, \tau) \begin{bmatrix} 1 \\ 0 \end{bmatrix} u(\tau)\, d\tau \right\} \tag{4-104}$$

It is clear from the properties of the fundamental matrix $\mathbf{\Phi}(t, t_0)$ (see Sec. 3-19) that the function $\boldsymbol{\phi}(t;\, u_{(t_0,t]},\, \mathbf{x}_0)$ satisfies condition DS2. Moreover, since $x_{1(t_0,t]}$ is *uniquely* determined by $\mathbf{x}_0$ and $u_{(t_0,t]}$, it follows that condition DS3 is also satisfied. As for condition DS1, that is a consequence of Theorem 3-14. Thus we have defined a second-order dynamical system with the equations

$$\dot{\mathbf{x}}(t) = \begin{bmatrix} 0 & 1 \\ 0 & t \end{bmatrix} \mathbf{x}(t) + \begin{bmatrix} 1 \\ 0 \end{bmatrix} u(t) \qquad \text{state} \tag{4-105}$$

$$y(t) = [1 \quad 0]\mathbf{x}(t) + 0u(t) \qquad \text{output} \tag{4-106}$$

## 4-7 Linear Dynamical Systems

Loosely speaking, a dynamical system is linear if the equations of the system are linear. More precisely, we have:

***Definition 4-6*** *A dynamical system is said to be a linear dynamical system or, simply, a linear system if the vector differential equation for the state* $\mathbf{x}(t)$ *is a linear differential equation (see Sec.* 3-19) *and if the output* $\mathbf{y}(t)$ *is a linear function of* $\mathbf{x}(t)$ *and* $\mathbf{u}(t)$. *In that case, the equations of the system are of the form*

$$\dot{\mathbf{x}}(t) = \mathbf{A}(t)\mathbf{x}(t) + \mathbf{B}(t)\mathbf{u}(t) \tag{4-107a}$$

$$\mathbf{y}(t) = \mathbf{C}(t)\mathbf{x}(t) + \mathbf{D}(t)\mathbf{u}(t) \tag{4-107b}$$

*where* $\mathbf{A}(t)$ *is an* $n \times n$ *matrix-valued function,* $\mathbf{B}(t)$ *an* $n \times m$ *matrix-valued function,* $\mathbf{C}(t)$ *a* $p \times n$ *matrix-valued function, and* $\mathbf{D}(t)$ *a* $p \times m$ *matrix-valued function. If all of these matrix-valued functions are constant, then we say that the system is time-invariant; otherwise, we say that the system is time-varying (cf. Definition* 3-43).

The *RL* networks of Figs. 4-1 and 4-2 and Example 4-2 provide us with some examples of linear systems.

We shall be particularly interested in linear systems whose equations are of the form

$$\begin{aligned} \dot{\mathbf{x}}(t) &= \mathbf{A}(t)\mathbf{x}(t) + \mathbf{B}(t)\mathbf{u}(t) \\ \mathbf{y}(t) &= \mathbf{C}\mathbf{x}(t) \end{aligned} \tag{4-108}$$

and we shall, in this case, often speak of the equation

$$\dot{\mathbf{x}}(t) = \mathbf{A}(t)\mathbf{x}(t) + \mathbf{B}(t)\mathbf{u}(t) \tag{4-109}$$

as either the *equation of the system* or, more simply, the *system*. This is indeed an abuse of language but should cause the reader no difficulty and will, in fact, foster economy of exposition in the sequel.

We observe that if we are considering a linear system with Eqs. (4-107*a*) and (4-107*b*), then the state and output equations of this system may be written in terms of the fundamental matrix $\mathbf{\Phi}(t, t_0)$ of Eq. (4-107*a*) as (see Sec. 3-20)

$$\mathbf{x}(t) = \boldsymbol{\phi}(t; \mathbf{u}_{(t_0,t]}, \mathbf{x}_0) = \mathbf{\Phi}(t, t_0)\left\{\mathbf{x}_0 + \int_{t_0}^{t} \mathbf{\Phi}^{-1}(\tau, t_0)\mathbf{B}(\tau)\mathbf{u}(\tau)\, d\tau\right\} \tag{4-110}$$

$$\begin{aligned}\mathbf{y}(t) &= \mathbf{C}(t)\boldsymbol{\phi}(t; \mathbf{u}_{(t_0,t]}, \mathbf{x}_0) + \mathbf{D}(t)\mathbf{u}(t)\\ &= \mathbf{C}(t)\mathbf{\Phi}(t, t_0)\left\{\mathbf{x}_0 + \int_{t_0}^{t} \mathbf{\Phi}^{-1}(\tau, t_0)\mathbf{B}(\tau)\mathbf{u}(\tau)\, d\tau\right\} + \mathbf{D}(t)\mathbf{u}(t)\end{aligned} \tag{4-111}$$

In particular, for a time-invariant linear system with $\mathbf{A}(t) = \mathbf{A}$, $\mathbf{B}(t) = \mathbf{B}$, $\mathbf{C}(t) = \mathbf{C}$, and $\mathbf{D}(t) = \mathbf{D}$, we have

$$\mathbf{x}(t) = e^{\mathbf{A}(t-t_0)}\left\{\mathbf{x}_0 + \int_{t_0}^{t} e^{-\mathbf{A}(\tau-t_0)}\mathbf{B}\mathbf{u}(\tau)\, d\tau\right\} \tag{4-112}$$

$$\mathbf{y}(t) = \mathbf{C}e^{\mathbf{A}(t-t_0)}\left\{\mathbf{x}_0 + \int_{t_0}^{t} e^{-\mathbf{A}(\tau-t_0)}\mathbf{B}\mathbf{u}(\tau)\, d\tau\right\} + \mathbf{D}\mathbf{u}(t) \tag{4-113}$$

We note that the properties of the transition function $\boldsymbol{\phi}(t; \mathbf{u}_{(t_0,t]}, \mathbf{x}_0)$ [see Eqs. (4-88) to (4-90)] are immediate consequences of the properties of the fundamental matrix (see Sec. 3-20), and conversely.

We also note that the notion of equivalence (see Definition 4-5) for linear systems corresponds to the use of the similarity transformation in Sec. 3-22. To put it another way, if we apply a similarity transformation to a linear system $\mathcal{S}$, we obtain an *equivalent linear* system $\mathcal{S}'$; and conversely, if the system $\mathcal{S}'$ is equivalent to the linear system $\mathcal{S}$ in the sense of Definition 4-5, then the equations of $\mathcal{S}'$ are obtained from the equations of $\mathcal{S}$ by a similarity transformation. This means, in effect, that the similarity transformation alters the equations of the system but does not change the geometric properties of the motion of the system. In particular, if $\mathbf{P}$ is a nonsingular $n \times n$ matrix [see Eq. (2-59)] and if we let

$$\mathbf{z}(t) = \mathbf{P}^{-1}\mathbf{x}(t) \tag{4-114}$$

then the system with state variable $\mathbf{z}(t)$, which is equivalent to the system with Eqs. (4-107*a*) and (4-107*b*), has the equations

$$\dot{\mathbf{z}}(t) = \mathbf{P}^{-1}\mathbf{A}(t)\mathbf{P}\mathbf{z}(t) + \mathbf{P}^{-1}\mathbf{B}(t)\mathbf{u}(t) \tag{4-115}$$

$$\mathbf{y}(t) = \mathbf{C}(t)\mathbf{P}\mathbf{z}(t) + \mathbf{D}(t)\mathbf{u}(t) \tag{4-116}$$

We shall make frequent use of these ideas in the sequel.

## 4-8 Input-Output Relationships and the Transfer Function†

So far, we have been concerned primarily with the basic ideas associated with the abstract notion of a dynamical system; now let us turn our attention to the relationship of these ideas to the transfer function, which is, perhaps, the most familiar concept to the control engineer and which represents the input-output relationship for a linear system. We shall review input-output relationships and the transfer function in this section, and we shall show, in the next two sections, how to determine the dynamical-system representation of a process from the transfer function.

One of the most difficult problems which confronts the engineer is the problem of obtaining an adequate mathematical description of a given physical process or system. If the process is as simple as the *RL* network of Sec. 4-2, equations which describe the process can be developed by inspection. However, in general, this is not possible, and a good mathematical description of the process can be obtained only after extensive experimentation. Usually, the experimenter applies known input signals and observes the resulting output signals. As a result of these experiments and from a priori theoretical knowledge, an input-output relationship tying the observable output signals to the available input signals is developed for the system. For example, the frequency-response method and the step-response method are used to obtain the transfer function for a linear time-invariant system.

Quite often, the result of the experiments on a single-input–single-output process is a linear time-invariant differential equation relating the output $y(t)$ to the input $u(t)$ as

$$\{D^n + a_{n-1}D^{n-1} + \cdots + a_1 D + a_0\}y(t) = b_0 u(t) \tag{4-117}$$

where $D = d/dt$ indicates differentiation with respect to $t$. If a system is adequately described by the differential equation (4-117), then it is common practice in servomechanism theory to say that the system has the *transfer function*

$$G(s) = \frac{y(s)}{u(s)} = \frac{b_0}{s^n + a_{n-1}s^{n-1} + \cdots + a_1 s + a_0} \tag{4-118}$$

where $y(s)$ is the Laplace transform of the output $y(t)$ and $u(s)$ is the Laplace transform of the input $u(t)$. Since the transfer function $G(s)$ of Eq. (4-118) has no finite zeros, it is said to contain only *poles*. The *poles* of $G(s)$ are the roots of the *denominator polynomial*

$$s^n + a_{n-1}s^{n-1} + \cdots + a_1 s + a_0 \tag{4-119}$$

of $G(s)$.

† See Refs. B-19, B-21, and Z-1 (chap. 9).

Frequently, the equation relating the output $y(t)$ to the input $u(t)$ for a single-input–single-output process is of the form

$$\{D^n + a_{n-1}D^{n-1} + \cdots + a_1D + a_0\}y(t) = \{b_mD^m + b_{m-1}D^{m-1} + \cdots + b_1D + b_0\}u(t) \quad (4\text{-}120)$$

where $D = d/dt$. Such a system is said to have the *transfer function*

$$H(s) = \frac{y(s)}{u(s)} = \frac{b_ms^m + b_{m-1}s^{m-1} + \cdots + b_1s + b_0}{s^n + a_{n-1}s^{n-1} + \cdots + a_1s + a_0} \quad (4\text{-}121)$$

where $y(s)$ is the Laplace transform of $y(t)$ and $u(s)$ is the Laplace transform of $u(t)$. The transfer function $H(s)$ of Eq. (4-121) contains both *zeros* and *poles*. The *zeros* of $H(s)$ are the roots of the *numerator polynomial*

$$b_ms^m + b_{m-1}s^{m-1} + \cdots + b_1s + b_0 \quad (4\text{-}122)$$

of $H(s)$, and the *poles* of $H(s)$, as before, are the roots of the *denominator polynomial* of $H(s)$.

Suppose now that the system we are considering has $n$ output signals $y_1(t), y_2(t), \ldots, y_n(t)$ and $m$ input signals $u_1(t), u_2(t), \ldots, u_m(t)$. Such systems are often called *multivariable systems.* If we experiment with a multivariable system which is linear and time-invariant, then we shall obtain a set of input-output relationships of the form

$$\begin{aligned} \mathcal{Q}_1(D)y_1(t) &= \mathcal{P}_{11}(D)u_1(t) + \mathcal{P}_{12}(D)u_2(t) + \cdots + \mathcal{P}_{1m}(D)u_m(t) \\ \mathcal{Q}_2(D)y_2(t) &= \mathcal{P}_{21}(D)u_1(t) + \mathcal{P}_{22}(D)u_2(t) + \cdots + \mathcal{P}_{2m}(D)u_m(t) \\ &\cdots\cdots\cdots\cdots\cdots\cdots \\ \mathcal{Q}_n(D)y_n(t) &= \mathcal{P}_{n1}(D)u_1(t) + \mathcal{P}_{n2}(D)u_2(t) + \cdots + \mathcal{P}_{nm}(D)u_m(t) \end{aligned} \quad (4\text{-}123)$$

where the $\mathcal{Q}_i(D)$ and the $\mathcal{P}_{ij}(D)$, $i = 1, 2, \ldots, n$ and $j = 1, 2, \ldots, m$, are linear differential operators with constant coefficients. For a multivariable system described by the set of differential equations (4-123), we can define the *transfer matrix* of the system as follows:

Let $\mathbf{y}(t)$ be the output vector whose components are $y_1(t), y_2(t), \ldots, y_n(t)$, and let $\mathbf{u}(t)$ be the input vector whose components are $u_1(t), u_2(t), \ldots, u_m(t)$. If the Laplace transforms of $\mathbf{y}(t)$ and $\mathbf{u}(t)$ are denoted by $\mathbf{y}(s)$ and $\mathbf{u}(s)$, respectively, then the $n \times m$ matrix $\mathbf{G}(s)$ is called the *transfer matrix* of the system if

$$\mathbf{y}(s) = \mathbf{G}(s)\mathbf{u}(s) \quad (4\text{-}124)$$

**Example 4-3** Consider the multivariable system of Sec. 4-3. The transfer matrix $\mathbf{G}(s)$ of this system is given by

$$\mathbf{G}(s) = \begin{bmatrix} 1/(s - a_1) & 1/(s - a_3) \\ 1/(s - a_2) & 1/(s - a_4) \end{bmatrix} \quad (4\text{-}125)$$

Finally, in concluding this section, we should like to point out that the problem of obtaining an adequate input-output relationship from experiment is particularly difficult for systems which are either linear but time-varying or nonlinear, and we should also like to stress the point that the process of experimentation yields a relationship (usually a set of differential equations) between the applied inputs and the observed outputs.

## 4-9 Finding the State (or Dynamical-system) Representation of a Plant Whose Transfer Function Contains Only Poles

The control engineer is very familiar with single-input–single-output systems which are described by $n$th-order linear differential equations with constant coefficients of the form

$$\{D^n + a_{n-1}D^{n-1} + \cdots + a_1D + a_0\}y(t) = b_0u(t) \tag{4-126}$$

where $y(t)$ is the (scalar) output and $u(t)$ is the (scalar) input. We have observed in the previous section [see Eq. (4-118)] that such a system has the transfer function

$$G(s) = \frac{y(s)}{u(s)} = \frac{b_0}{s^n + a_{n-1}s^{n-1} + \cdots + a_1s + a_0} \tag{4-127}$$

If we let $s_1, s_2, \ldots, s_n$ denote the (complex) roots of the denominator polynomial

$$s^n + a_{n-1}s^{n-1} + \cdots + a_1s + a_0 \tag{4-128}$$

of $G(s)$, then the $s_i$ are the poles of $G(s)$ and we can write

$$G(s) = \frac{b_0}{(s - s_1)(s - s_2) \cdots (s - s_n)} \tag{4-129}$$

In this section, we shall determine a state (or dynamical-system) representation of the system described by Eq. (4-126), and we shall attempt to attach some physical significance to the state.

Since we must know $n$ initial conditions, namely, $y(0), \dot{y}(0), \ldots, y^{(n-1)}(0)$, to obtain the solution of the differential equation (4-126), we can almost guess what the state will be. Bearing this in mind, let us mimic the development of Sec. 3-24 and define $n$ variables $z_1(t), z_2(t), \ldots, z_n(t)$ by setting

$$\begin{aligned} z_1(t) &= y(t) \\ z_2(t) &= \dot{y}(t) \\ &\cdots\cdots\cdots \\ z_k(t) &= y^{(k-1)}(t) \\ &\cdots\cdots\cdots \\ z_n(t) &= y^{(n-1)}(t) \end{aligned} \tag{4-130}$$

From Eqs. (4-130), we immediately deduce that

$$\begin{aligned} \dot{z}_1(t) &= z_2(t) \\ \dot{z}_2(t) &= z_3(t) \\ \cdots&\cdots \\ \dot{z}_{n-1}(t) &= z_n(t) \end{aligned} \tag{4-131}$$

and from Eq. (4-126), we deduce that

$$\dot{z}_n(t) = D^n y(t) = -a_0 y(t) - a_1 \dot{y}(t) - \cdots - a_{n-1} y^{(n-1)}(t) + b_0 u(t) \tag{4-132}$$

It follows that

$$\dot{z}_n(t) = -a_0 z_1(t) - a_1 z_2(t) - \cdots - a_{n-1} z_n(t) + b_0 u(t) \tag{4-133}$$

and hence that, if we define the $z_i$ by Eqs. (4-130), the $z_i$ satisfy a system of $n$ first-order linear differential equations which may be written in matrix form as

$$\begin{bmatrix} \dot{z}_1(t) \\ \dot{z}_2(t) \\ \cdots \\ \dot{z}_{n-1}(t) \\ \dot{z}_n(t) \end{bmatrix} = \begin{bmatrix} 0 & 1 & 0 & \cdots & 0 \\ 0 & 0 & 1 & \cdots & 0 \\ \cdots & \cdots & \cdots & \cdots & \cdots \\ 0 & 0 & 0 & \cdots & 1 \\ -a_0 & -a_1 & -a_2 & \cdots & -a_{n-1} \end{bmatrix} \begin{bmatrix} z_1(t) \\ z_2(t) \\ \cdots \\ z_{n-1}(t) \\ z_n(t) \end{bmatrix} + \begin{bmatrix} 0 \\ 0 \\ \cdot\cdot \\ 0 \\ b_0 \end{bmatrix} u(t) \tag{4-134}$$

The vector differential equation (4-134) is of the form

$$\dot{\mathbf{z}}(t) = \mathbf{A}\mathbf{z}(t) + \mathbf{b}u(t) \tag{4-135}$$

We note that Eq. (4-135) is (except for different symbols) identical to Eq. (3-396) and that the system matrix $\mathbf{A}$ of Eq. (4-135) is given by Eq. (3-398). Furthermore, the $n$ vector $\mathbf{b}$ in Eq. (4-135) is given by

$$\mathbf{b} = \begin{bmatrix} 0 \\ 0 \\ \cdot \\ \cdot \\ \cdot \\ 0 \\ b_0 \end{bmatrix} \tag{4-136}$$

Now let us verify that the vector $\mathbf{z}(t)$ qualifies as a state variable for the system of Eq. (4-126). In order to do this, we must show that knowledge of $\mathbf{z}(t_0)$ and $u_{(t_0,t]}$ completely specifies both $\mathbf{z}_{(t_0,t]}$ and the output $y_{(t_0,t]}$. Now, if $\mathbf{z}(t_0)$ and $u_{(t_0,t]}$ are known, then the *unique* solution of Eq. (4-135), with $\mathbf{z}(t_0)$ as initial point and $u_{(t_0,t]}$ as driving function, is given by

$$\mathbf{z}(t) = e^{\mathbf{A}(t-t_0)}\mathbf{z}(t_0) + e^{\mathbf{A}(t-t_0)} \int_{t_0}^{t} e^{-\mathbf{A}(\tau-t_0)}\mathbf{b}u(\tau)\, d\tau \tag{4-137}$$

where $e^{\mathbf{A}(t-t_0)}$ is the fundamental matrix of Eq. (4-135). It follows that $\mathbf{z}_{(t_0,t]}$ is completely determined by $\mathbf{z}(t_0)$ and $u_{(t_0,t]}$. Moreover, since $y(t) = z_1(t)$ [in view of Eqs. (4-130)], it is clear that $y_{(t_0,t]} = z_{1(t_0,t]}$ is completely determined by $\mathbf{z}(t_0)$ and $u_{(t_0,t]}$. The state equations of the system in terms of $\mathbf{z}(t)$ are

$$\begin{aligned} \dot{\mathbf{z}}(t) &= \mathbf{A}\mathbf{z}(t) + \mathbf{b}u(t) \\ y(t) &= z_1(t) \end{aligned} \tag{4-138}$$

We leave it to the reader to verify that we have indeed found a dynamical system representing Eq. (4-126).

Figure 4-8 illustrates the simulation of the vector differential equation (4-134) on an analog computer. We observe that $n$ integrators are required for the simulation and that the state variables $z_1(t)$, $z_2(t)$, . . . , $z_n(t)$

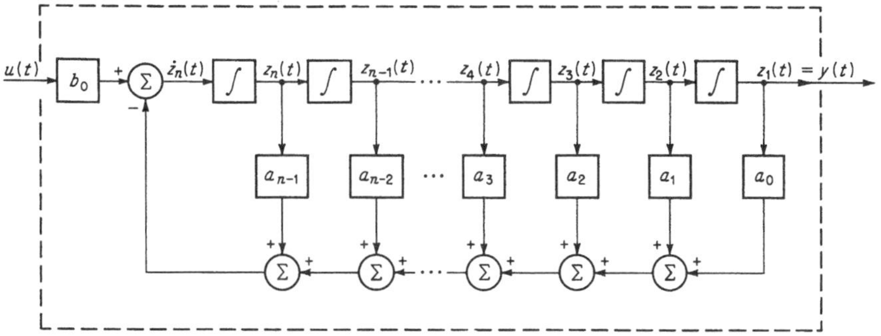


**Fig. 4-8** The analog-computer representation of the system (4-134).

correspond to the output signals of the integrators. The constant coefficients $a_0$, $a_1$, . . . , $a_{n-1}$ appear as gains in the feedback channels.

We have shown in Sec. 3-24 that the eigenvalues $\lambda_1$, $\lambda_2$, . . . , $\lambda_n$ of the system matrix given by Eq. (3-396) are the roots of the polynomial

$$\lambda^n + a_{n-1}\lambda^{n-1} + \cdots + a_1\lambda + a_0 \tag{4-139}$$

But the two polynomials given by Eqs. (4-128) and (4-139) are identical and, hence, must have the same roots. Therefore, *the eigenvalues of the system matrix* **A** *given by Eq.* (3-396) *are the same as the poles of the transfer function* $G(s)$ *given by Eq.* (4-127). This was to be expected, since the poles of the transfer function determine the system response and since Eqs. (4-138) represent another way of looking at the system. In other words, the eigenvalues of the system matrix **A** determine the response of the system given by Eqs. (4-138), and the poles of $G(s)$ determine the response of the system of Eq. (4-126); since these are both representations of the same physical process, the poles and eigenvalues must be related, and, in fact, they are the same, as we have just demonstrated.

We have shown in Sec. 2-10 that the eigenvalues of a linear transformation are an invariant property of the linear transformation and that the matrix associated with the linear transformation depends upon the particular bases of the vector spaces involved. In a similar way, we may view the system as a linear transformation and, consequently, see that the poles (or eigenvalues) represent invariant quantities while the system matrix and the state vector depend upon the basis chosen. This is just another way of saying that the eigenvalues of equivalent systems (see Definition 4-5) are the same, while the state variables and system matrices may be different.

In order to illustrate the nonuniqueness of the state vector, we shall perform the following change of variable [which corresponds to the replacing of the system of Eqs. (4-138) by a system equivalent to it, in the sense of Definition 4-5]: Let $\mathbf{Q}$ be any nonsingular $n \times n$ matrix, and let us define a vector $\mathbf{x}(t)$ by setting

$$\mathbf{x}(t) = \mathbf{Q}^{-1}\mathbf{z}(t) \tag{4-140}$$

where $\mathbf{z}(t)$ is the vector defined by Eqs. (4-130). Differentiating both sides of Eq. (4-140), we have

$$\dot{\mathbf{x}}(t) = \mathbf{Q}^{-1}\dot{\mathbf{z}}(t) \tag{4-141}$$

Now we premultiply both sides of Eq. (4-135) by the matrix $\mathbf{Q}^{-1}$ to obtain the equation

$$\mathbf{Q}^{-1}\dot{\mathbf{z}}(t) = \mathbf{Q}^{-1}\mathbf{A}\mathbf{z}(t) + \mathbf{Q}^{-1}\mathbf{b}u(t) \tag{4-142}$$

Substituting Eqs. (4-140) and (4-141) into Eq. (4-142), we find that $\mathbf{x}(t)$ satisfies the differential equation

$$\dot{\mathbf{x}}(t) = \mathbf{Q}^{-1}\mathbf{A}\mathbf{Q}\mathbf{x}(t) + \mathbf{Q}^{-1}\mathbf{b}u(t) \tag{4-143}$$

Let

$$\mathbf{S} = \mathbf{Q}^{-1}\mathbf{A}\mathbf{Q} \qquad \mathbf{c} = \mathbf{Q}^{-1}\mathbf{b} \tag{4-144}$$

so that Eq. (4-143) reduces to

$$\dot{\mathbf{x}}(t) = \mathbf{S}\mathbf{x}(t) + \mathbf{c}u(t) \tag{4-145}$$

We note that the matrices $\mathbf{S}$ and $\mathbf{A}$ are similar matrices† and, therefore, have the same eigenvalues (see Sec. 2-10). Since the eigenvalues of $\mathbf{A}$ are the same as the poles of $G(s)$, it follows that the eigenvalues of $\mathbf{S}$ are also the same as the poles of $G(s)$. Furthermore, we claim that the vector $\mathbf{x}(t)$ also qualifies as a state vector for the system of Eq. (4-126). To see this, we observe that, for any $t_0$, knowledge of $\mathbf{x}(t_0)$ and $u_{(t_0,t]}$ completely specifies $\mathbf{x}_{(t_0,t]}$, since

$$\mathbf{x}(t) = e^{\mathbf{S}(t-t_0)}\mathbf{x}(t_0) + e^{\mathbf{S}(t-t_0)} \int_{t_0}^{t} e^{-\mathbf{S}(\tau-t_0)}\mathbf{c}u(\tau)\, d\tau \tag{4-146}$$

However, if $\mathbf{x}(\tau)$ is known for all $\tau$ in $(t_0, t]$, then so is $\mathbf{z}(\tau)$, since $\mathbf{z}(\tau) = \mathbf{Q}\mathbf{x}(\tau)$, where $\mathbf{Q}$ is nonsingular, and consequently, so is $z_1(\tau) = y(\tau)$ for all $\tau$ in

† See Sec. 2-9.

$(t_0, t]$. We may thus conclude that $\mathbf{x}(t)$ is an appropriate state vector for the system of Eq. (4-126) and that Eqs. (4-135) and (4-145) define equivalent state representations of that system.

**Exercise 4-6** Construct an analog-computer simulation of the system (4-126) using the representation suggested by Eq. (4-145). Use $s_{ij}$ for the elements of $\mathbf{S}$, $q_{ij}$ for the elements of $\mathbf{Q}$, and $c_i$ for the components of $\mathbf{c}$. Show that the state variables $x_i(t)$ are again the outputs of $n$ integrators.

A particularly useful change of variables (or equivalence) can be found when the poles of the transfer function $G(s)$ (equivalently, the eigenvalues of the system matrix) are *real* and *distinct.* As we have seen in Sec. 3-24, we can find in this case a nonsingular matrix $\mathbf{P}$, which is the Vandermonde matrix given by Eq. (3-411), such that

$$\Lambda = \mathbf{P}^{-1}\mathbf{A}\mathbf{P} \tag{4-147}$$

where $\Lambda$ is the diagonal matrix of the eigenvalues given by Eq. (3-409). Thus, if we define the state vector $\mathbf{x}(t)$ by setting

$$\mathbf{x}(t) = \mathbf{P}^{-1}\mathbf{z}(t) \tag{4-148}$$

then $\mathbf{x}(t)$ satisfies the differential equation

$$\dot{\mathbf{x}}(t) = \Lambda\mathbf{x}(t) + \mathbf{P}^{-1}\mathbf{b}u(t) \tag{4-149}$$

We let the elements of $\mathbf{P}^{-1}$ be denoted by $r_{ij}$, $i = 1, 2, \ldots, n$ and $j = 1, 2, \ldots, n$. Since the vector $\mathbf{b}$ is defined by Eq. (4-136), it follows that Eq. (4-149) can be written in the form

$$\dot{x}_i(t) = \lambda_i x_i(t) + r_{in} b_0 u(t) \qquad i = 1, 2, \ldots, n \tag{4-150}$$

where the $\lambda_i$ are the eigenvalues of the system matrix and the $r_{in}$ are the elements of the last ($n$th) column of the matrix $\mathbf{P}^{-1}$.

Figure 4-9 illustrates the analog-computer simulation of the system represented by Eq. (4-149) or (4-150). Again, the state variables $x_i(t)$ are the output signals of the $n$ integrators. The output $y(t) = z_1(t)$ is the *sum* of the $x_i(t)$, that is,

$$y(t) = z_1(t) = \sum_{i=1}^{n} x_i(t) \tag{4-151}$$

since the elements of the first row of the Vandermonde matrix $\mathbf{P}$ are all 1 [see Eq. (3-411)].

We can directly evaluate the numbers $r_{in}$, $i = 1, 2, \ldots, n$, appearing in Eq. (4-150) by inverting the matrix $\mathbf{P}$; however, it is easier to compute them in a different manner, and in so doing, we shall obtain some additional physical insight. Since the poles of $G(s)$ are at $s = \lambda_i$, we can write

$$G(s) = b_0 \sum_{i=1}^{n} \frac{r_i}{s - \lambda_i} \tag{4-152}$$

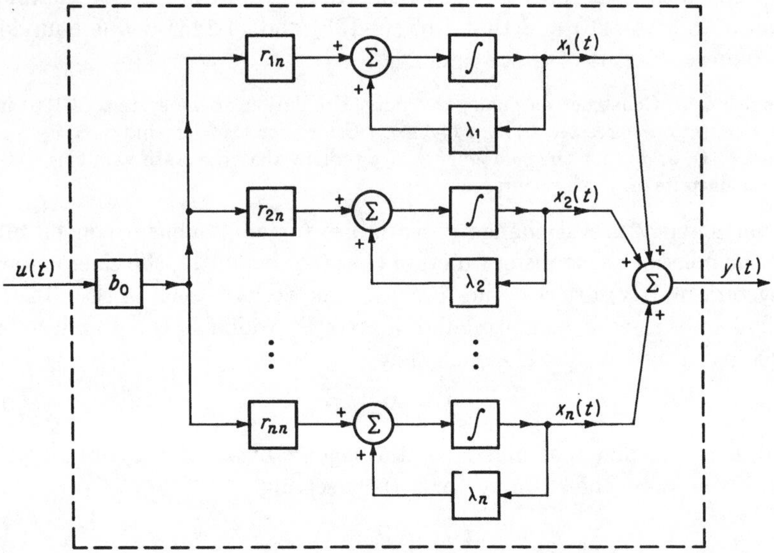


**Fig. 4-9** The analog-computer representation of the system (4-150).

where $r_i$ is the residue of $G(s)$ at the $i$th pole, $\lambda_i$. Since the $\lambda_i$ are distinct, we have

$$r_i = \frac{1}{(\lambda_i - \lambda_1)(\lambda_i - \lambda_2) \cdots (\lambda_i - \lambda_{i-1})(\lambda_i - \lambda_{i+1}) \cdots (\lambda_i - \lambda_n)} \tag{4-153}$$

Suppose that the system is initially at rest (i.e., the initial conditions are all zero) and that the input $u(t)$ is applied; then

$$y(s) = b_0 \sum_{i=1}^{n} \frac{r_i}{s - \lambda_i} u(s) \tag{4-154}$$†

But from Eq. (4-150), again assuming that the system is initially at rest, we find that

$$x_i(s) = \frac{b_0 r_{in} u(s)}{s - \lambda_i} \tag{4-155}$$

Since

$$y(t) = \sum_{i=1}^{n} x_i(t)$$

it follows that

$$y(s) = \sum_{i=1}^{n} x_i(s) = b_0 \sum_{i=1}^{n} \frac{r_{in}}{s - \lambda_i} u(s) \tag{4-156}$$

Comparison of Eqs. (4-154) and (4-156) leads us to the conclusion that

$$r_{in} = r_i \tag{4-157}$$

† Recall that $y(s)$ and $u(s)$ are the Laplace transforms of $y(t)$ and $u(t)$, respectively.

which means that the elements of the last column of $\mathbf{P}^{-1}$ are the *residues* of $G(s)$.

We observe that, in effect, the similarity transformation has changed the analog-computer representation of Fig. 4-8 to that of Fig. 4-9. The gains $r_{1i}, r_{2n}, \ldots, r_{nn}$ in Fig. 4-9 are precisely the residues of the transfer function $G(s)$ at its poles $\lambda_1, \lambda_2, \ldots, \lambda_n$ respectively.

**Example 4-4** Consider the single-input–single-output system described by the third-order differential equation

$$\{D^3 + 3D^2 + 2D\}y(t) = u(t) \tag{4-158}$$

This system has the transfer function

$$\frac{y(s)}{u(s)} = G(s) = \frac{1}{s^3 + 3s^2 + 2s} = \frac{1}{s(s+1)(s+2)} \tag{4-159}$$

with poles at $s = 0$, $s = -1$, $s = -2$. If we set

$$\begin{aligned} z_1(t) &= y(t) \\ z_2(t) &= \dot{y}(t) \\ z_3(t) &= \ddot{y}(t) \end{aligned} \tag{4-160}$$

then we may deduce that the vector $\mathbf{z}(t)$ with components $z_1(t)$, $z_2(t)$, $z_3(t)$ satisfies the vector differential equation

$$\begin{bmatrix} \dot{z}_1(t) \\ \dot{z}_2(t) \\ \dot{z}_3(t) \end{bmatrix} = \begin{bmatrix} 0 & 1 & 0 \\ 0 & 0 & 1 \\ 0 & -2 & -3 \end{bmatrix} \begin{bmatrix} z_1(t) \\ z_2(t) \\ z_3(t) \end{bmatrix} + \begin{bmatrix} 0 \\ 0 \\ 1 \end{bmatrix} u(t) \tag{4-161}$$

or, more succinctly,

$$\dot{\mathbf{z}}(t) = \mathbf{A}\mathbf{z}(t) + \mathbf{b}u(t) \tag{4-162}$$

Let us find the eigenvalues of the matrix $\mathbf{A}$. In order to do this, we must find the zeros of $\det(\lambda\mathbf{I} - \mathbf{A})$. Now, we have

$$\det(\lambda\mathbf{I} - \mathbf{A}) = \det \begin{bmatrix} \lambda & -1 & 0 \\ 0 & \lambda & -1 \\ 0 & 2 & \lambda + 3 \end{bmatrix} = \lambda^3 + 3\lambda^2 + 2\lambda \tag{4-163}$$

so that the eigenvalues of $\mathbf{A}$ are at $\lambda = 0$, $\lambda = -1$, and $\lambda = -2$. In other words, the poles of $G(s)$ and the eigenvalues of $\mathbf{A}$ are indeed the same. We note that these eigenvalues are real and distinct. If we let $\mathbf{P}$ be the Vandermonde matrix, given by

$$\mathbf{P} = \begin{bmatrix} 1 & 1 & 1 \\ 0 & -1 & -2 \\ 0 & 1 & 4 \end{bmatrix} \tag{4-164}$$

and if we apply the similarity transformation to Eq. (4-161), then we obtain the following system equivalent to our system:

$$\begin{bmatrix} \dot{x}_1(t) \\ \dot{x}_2(t) \\ \dot{x}_3(t) \end{bmatrix} = \begin{bmatrix} 0 & 0 & 0 \\ 0 & -1 & 0 \\ 0 & 0 & -2 \end{bmatrix} \begin{bmatrix} x_1(t) \\ x_2(t) \\ x_3(t) \end{bmatrix} + \begin{bmatrix} 1 & \frac{3}{2} & \frac{1}{2} \\ 0 & -2 & -1 \\ 0 & \frac{1}{2} & \frac{1}{2} \end{bmatrix} \begin{bmatrix} 0 \\ 0 \\ 1 \end{bmatrix} u(t) \tag{4-165}$$

$$= \begin{bmatrix} 0 & 0 & 0 \\ 0 & -1 & 0 \\ 0 & 0 & -2 \end{bmatrix} \begin{bmatrix} x_1(t) \\ x_2(t) \\ x_3(t) \end{bmatrix} + \begin{bmatrix} \frac{1}{2} \\ -1 \\ \frac{1}{2} \end{bmatrix} u(t) \tag{4-166}$$

where $\mathbf{x}(t) = \mathbf{P}^{-1}\mathbf{z}(t)$. We also note that

$$G(s) = \frac{1}{s(s+1)(s+2)} = \frac{1}{2s} - \frac{1}{s+1} + \frac{1}{2(s+2)} \tag{4-167}$$

which implies that the residues of $G(s)$ are precisely the elements of the last column of $\mathbf{P}^{-1}$ [that is, the residues are the components of the vector multiplying $u(t)$ in Eq. (4-166)].

## 4-10 Finding the State (or Dynamical-system) Representation of a Plant Whose Transfer Function Contains Poles and Zeros

Let us now consider a single-input–single-output plant which is described by the linear differential equation with constant coefficients

$$\{D^n + a_{n-1}D^{n-1} + \cdots + a_1D + a_0\}y(t) = \{b_nD^n + b_{n-1}D^{n-1} + \cdots + b_1D + b_0\}u(t) \tag{4-168}$$

where $y(t)$ is the (scalar) output and $u(t)$ is the (scalar) input. We have observed in Sec. 4-8 [see Eq. (4-121)] that such a system has the transfer function

$$H(s) = \frac{y(s)}{u(s)} = \frac{b_ns^n + b_{n-1}s^{n-1} + \cdots + b_1s + b_0}{s^n + a_{n-1}s^{n-1} + \cdots + a_1s + a_0} \tag{4-169}$$

If we let $s_1, s_2, \ldots, s_n$ denote the (complex) roots of the denominator polynomial of $H(s)$ and if we let $\sigma_1, \sigma_2, \ldots, \sigma_n$† denote the (complex) roots of the numerator polynomial of $H(s)$, then the $s_i$ are the poles of $H(s)$ and the $\sigma_j$ are the zeros of $H(s)$. We may thus write

$$H(s) = b_n\frac{(s-\sigma_1)(s-\sigma_2)\cdots(s-\sigma_n)}{(s-s_1)(s-s_2)\cdots(s-s_n)} \tag{4-170}$$

In this section, we shall determine a state (or dynamical-system) representation of the plant described by Eq. (4-168).

Since we must know $2n$ initial conditions, namely, $y(0), \dot{y}(0), \ldots, y^{(n-1)}(0)$ and $u(0), \dot{u}(0), \ldots, u^{(n-1)}(0)$, in order to obtain the solution of the differential equation (4-168), we shall attempt to determine the state vector as follows: Let $\mathbf{z}(t)$ be an $n$ vector with components $z_1(t), z_2(t), \ldots, z_n(t)$ which are given by

$$\begin{aligned}
z_1(t) &= y(t) - h_0u(t)\\
z_2(t) &= \dot{y}(t) - h_0\dot{u}(t) - h_1u(t)\\
z_3(t) &= \ddot{y}(t) - h_0\ddot{u}(t) - h_1\dot{u}(t) - h_2u(t)\\
&\cdots\cdots\cdots\cdots\cdots\cdots\\
z_n(t) &= y^{(n-1)}(t) - h_0u^{(n-1)}(t) - h_1u^{(n-2)}(t) - \cdots - h_{n-1}u(t)
\end{aligned} \tag{4-171}$$

† We shall suppose, for the moment, that the number of zeros is the same as the number of poles; this is not essential, since we may choose

$$b_n = b_{n-1} = \cdots = b_{m+1} = 0 \qquad b_m \neq 0$$

In that case, there are $m$ zeros and $n$ poles.

where $h_0, h_1, \ldots, h_{n-1}$ are $n$ constants whose values are still to be determined. We may write Eqs. (4-171) in the form

$$z_i(t) = y^{(i-1)}(t) - \sum_{k=0}^{i-1} u^{(k)}(t)h_{i-k-1} \qquad i = 1, 2, \ldots, n \qquad (4\text{-}172)\dagger$$

We should now like to determine the differential equation which the $z_i(t)$ satisfy. Now, we deduce from Eqs. (4-171) that

$$\dot{z}_1(t) = \dot{y}(t) - h_0\dot{u}(t) \qquad (4\text{-}173)$$
$$= z_2(t) + h_1u(t) \qquad (4\text{-}174)$$

Our deduction is based upon differentiation of the equation defining $z_1(t)$ and upon substitution of the result in the equation defining $z_2(t)$. We may, in a completely analogous way, deduce that, for $i = 1, 2, \ldots, n-1$,

$$\dot{z}_i(t) = y^{(i)}(t) - \sum_{k=0}^{i-1} u^{(k+1)}(t)h_{i-k-1} \qquad (4\text{-}175)$$
$$= z_{i+1}(t) + h_iu(t) \qquad (4\text{-}176)$$

since

$$z_{i+1}(t) = y^{(i)}(t) - \sum_{k=0}^{i} u^{(k)}(t)h_{i-k}$$
$$= y^{(i)}(t) - h_iu(t) - \sum_{k=1}^{i} u^{(k)}(t)h_{i-k} \qquad (4\text{-}177)$$

As for $\dot{z}_n(t)$, we note that differentiation of Eq. (4-172) for $i = n$ yields

$$\dot{z}_n(t) = y^{(n)}(t) - \sum_{k=1}^{n} u^{(k)}(t)h_{n-k} \qquad (4\text{-}178)$$

However, the system differential equation (4-168) gives us the relation

$$y^{(n)}(t) = -\sum_{i=0}^{n-1} a_iy^{(i)}(t) + b_nu^{(n)}(t) + \sum_{k=1}^{n-1} b_ku^{(k)}(t) + b_0u(t) \qquad (4\text{-}179)$$

It follows from Eq. (4-177) that

$$\sum_{i=0}^{n-1} a_iy^{(i)}(t) = \sum_{i=0}^{n-1} a_iz_{i+1}(t) + \sum_{i=0}^{n-1} a_i\left[\sum_{k=1}^{i} u^{(k)}(t)h_{i-k}\right] + u(t)\sum_{i=0}^{n-1} a_ih_i \qquad (4\text{-}180)$$

and hence that

$$\dot{z}_n(t) = -\sum_{i=0}^{n-1} a_iz_{i+1}(t) + b_nu^{(n)}(t) + \sum_{k=1}^{n-1} b_ku^{(k)}(t) + b_0u(t)$$
$$- \sum_{k=1}^{n} u^{(k)}(t)h_{n-k} - \sum_{i=0}^{n-1} a_i\left[\sum_{k=1}^{i} u^{(k)}(t)h_{i-k}\right] - u(t)\sum_{i=0}^{n-1} a_ih_i \qquad (4\text{-}181)$$

† Recall that $u^{(0)}(t) = u(t)$.

However, we can see by direct computation that

$$-\sum_{i=0}^{n-1} a_i \sum_{k=1}^{i} u^{(k)}(t)h_{i-k} = -\sum_{k=1}^{n-1} u^{(k)}(t) \sum_{i=0}^{n-k-1} h_i a_{i+k} \tag{4-182}$$

If we substitute in Eq. (4-181) and collect terms, we obtain the relation

$$\begin{aligned} \dot{z}_n(t) &= -\sum_{i=0}^{n-1} a_i z_{i+1}(t) + (b_n - h_0)u^{(n)}(t) \\ &+ \sum_{k=1}^{n-1} u^{(k)}(t) \left\{b_k - h_{n-k} - \sum_{i=0}^{n-k-1} h_i a_{i+k}\right\} + u(t) \left\{b_0 - \sum_{i=0}^{n-1} a_i h_i\right\} \end{aligned} \tag{4-183}$$

We are now ready to specify the values of the constants $h_0, h_1, \ldots, h_{n-1}$. In fact, we shall choose the $h$'s in such a fashion as to make $\dot{z}_n(t)$ independent of all the derivatives of $u(t)$. This can be done by taking

$$\begin{aligned} h_0 &= b_n \\ h_{n-k} &= b_k - \sum_{i=0}^{n-k-1} h_i a_{i+k} \qquad \text{for } k = 1, 2, \ldots, n-1 \end{aligned} \tag{4-184}$$

By substitution of Eqs. (4-184) into Eq. (4-183), we obtain the differential equation

$$\dot{z}_n(t) = -\sum_{i=0}^{n-1} a_i z_{i+1}(t) + h_n u(t) \tag{4-185}$$

where

$$h_n = b_0 - \sum_{i=0}^{n-1} a_i h_i \tag{4-186}$$

Thus, we have determined the differential equations satisfied by the $z_i(t)$, which may be written in vector form as

$$\begin{bmatrix} \dot{z}_1(t) \\ \dot{z}_2(t) \\ \cdots \\ \dot{z}_{n-1}(t) \\ \dot{z}_n(t) \end{bmatrix} = \begin{bmatrix} 0 & 1 & 0 & \cdots & 0 \\ 0 & 0 & 1 & \cdots & 0 \\ \cdots & \cdots & \cdots & \cdots & \cdots \\ 0 & 0 & 0 & \cdots & 1 \\ -a_0 & -a_1 & -a_2 & \cdots & -a_{n-1} \end{bmatrix} \begin{bmatrix} z_1(t) \\ z_2(t) \\ \cdots \\ z_{n-1}(t) \\ z_n(t) \end{bmatrix} + \begin{bmatrix} h_1 \\ h_2 \\ \cdots \\ h_{n-1} \\ h_n \end{bmatrix} u(t) \tag{4-187}$$

or, more succinctly, as

$$\dot{\mathbf{z}}(t) = \mathbf{A}\mathbf{z}(t) + \mathbf{h}u(t) \tag{4-188}$$

where $\mathbf{A}$ is the matrix given by Eq. (3-398) and the vector $\mathbf{h}$ has components $h_1, h_2, \ldots, h_n$ defined by Eqs. (4-184) and (4-186). We note also that computation of the constants $h_i$, $i = 0, 1, \ldots, n$, is not particularly difficult since Eqs. (4-184) are of the form

$$
\begin{aligned}
h_0 &= b_n \\
h_1 &= b_{n-1} - h_0 a_{n-1} \\
h_2 &= b_{n-2} - h_0 a_{n-2} - h_1 a_{n-1} \\
&\cdots\cdots\cdots\cdots\cdots\cdots
\end{aligned}
\tag{4-189}
$$

and the $h_i$ can be found by successive substitution.

We may readily verify that the vector $\mathbf{z}(t)$ qualifies as a state variable for the system of Eq. (4-168), since knowledge of $\mathbf{z}(t_0)$ and $u_{(t_0,t]}$ completely determines the solution of the differential equation (4-188) on the interval $(t_0, t]$ (that is, $\mathbf{z}_{(t_0,t]}$). Moreover, since the output $y(t)$ is given by

$$y(t) = z_1(t) + h_0 u(t) \tag{4-190}$$

it is clear that $y_{(t_0,t]}$ is indeed determined by $\mathbf{z}(t_0)$ and $u_{(t_0,t]}$. We leave it to the reader to determine the state equations and to verify that we have actually found a dynamical system which represents Eq. (4-168).

It is instructive to compare the state differential equation (4-187) for the system (4-168) with the state differential equation (4-135) for the system (4-126). We note that the system matrix **A** is the same in Eqs. (4-187) and (4-135); moreover, the only difference between these equations is the difference between the vectors **b** and **h** which multiply the control $u(t)$. It should also be clear that the eigenvalues of the system matrix **A** are the poles of the transfer function $H(s)$ of Eq. (4-169) as well as of the transfer function $G(s)$ of Eq. (4-127).

Figure 4-10 illustrates the simulation of the vector differential equation (4-187) on an analog computer. We observe that the state variables are the outputs of $n$ integrators. We can also compare Figs. 4-8 and 4-10 and observe that the only difference between them is that there are feedforward channels in Fig. 4-10. The feedback channels in Figs. 4-8 and 4-10 are identical.

So far, we have assumed that the number of poles and zeros of $H(s)$ is the same; now let us consider the situation in which the transfer function $H(s)$ has $m$ zeros and $n$ poles, with

$$m < n$$

In other words, we now suppose that

$$b_n = b_{n-1} = \cdots = b_{m+1} = 0 \qquad b_m \neq 0 \tag{4-191}$$

in Eq. (4-168) and that $H(s)$ is of the form

$$H(s) = \frac{b_m(s - \sigma_1)(s - \sigma_2) \cdots (s - \sigma_m)}{(s - s_1)(s - s_2) \cdots (s - s_n)} \tag{4-192}$$

We can, in a manner entirely analogous to that used for the $n$-zero $n$-pole system, show that the vector $\mathbf{z}(t)$ with components $z_1(t), z_2(t), \ldots, z_n(t)$,

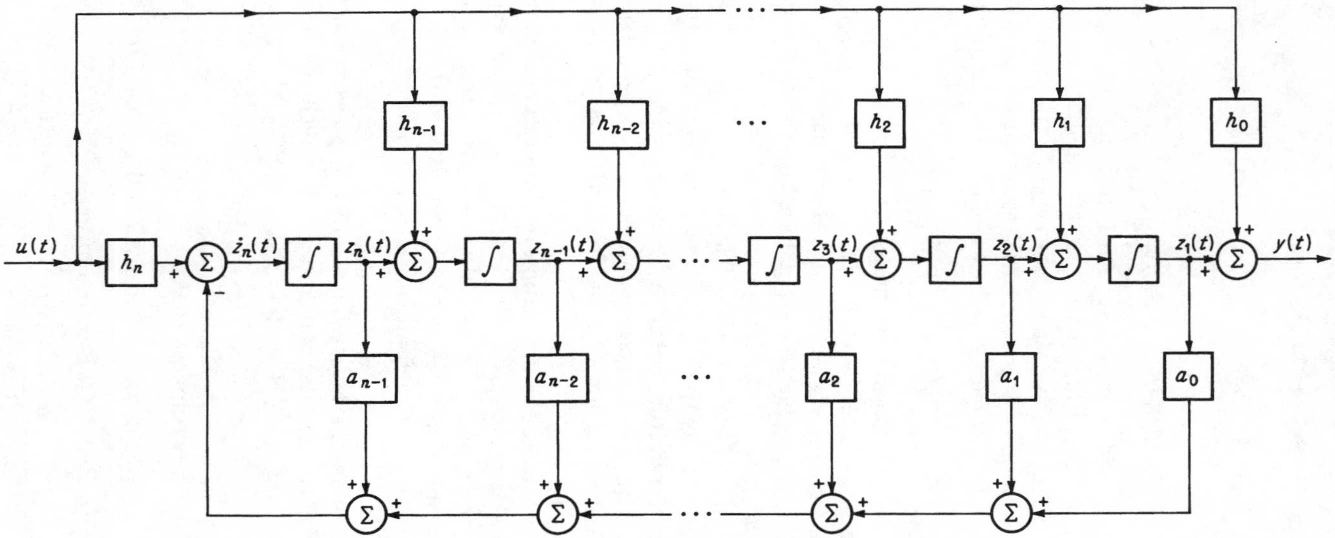


**Fig. 4-10** The analog-computer representation of the system (4-187).

given by

$$\begin{aligned} z_1(t) &= y(t) \\ z_2(t) &= \dot{y}(t) \\ &\cdots\cdots\cdots \\ z_{n-m}(t) &= y^{(n-m-1)}(t) \\ z_{n-m+1}(t) &= y^{(n-m)}(t) - h_{n-m}u(t) \\ &\cdots\cdots\cdots\cdots\cdots\cdots \\ z_n(t) &= y^{(n-1)}(t) - h_{n-m}u^{(m-1)}(t) - \cdots - h_{n-1}u(t) \end{aligned} \tag{4-193}$$

where
$$h_{n-k} = b_k - \sum_{j=n-m}^{n-k-1} a_{j+k}h_j \qquad k = 0, 1, \ldots, m \tag{4-194}$$

qualifies as a state variable for the system whose transfer function $H(s)$ is given by Eq. (4-192). We leave the verification of this to the reader, and we note that $\mathbf{z}(t)$ satisfies the vector differential equation

$$\begin{bmatrix} \dot{z}_1(t) \\ \dot{z}_2(t) \\ \cdots \\ \dot{z}_{n-1}(t) \\ \dot{z}_n(t) \end{bmatrix} = \begin{bmatrix} 0 & 1 & 0 & \cdots & 0 \\ 0 & 0 & 1 & \cdots & 0 \\ \cdots & \cdots & \cdots & \cdots & \cdots \\ 0 & 0 & 0 & \cdots & 1 \\ -a_0 & -a_1 & -a_2 & \cdots & -a_{n-1} \end{bmatrix} \begin{bmatrix} z_1(t) \\ z_2(t) \\ \cdots \\ z_{n-1}(t) \\ z_n(t) \end{bmatrix} + \begin{bmatrix} 0 \\ 0 \\ \cdot \\ \cdot \\ \cdot \\ 0 \\ h_{n-m} \\ \cdot \\ \cdot \\ \cdot \\ h_{n-1} \\ h_n \end{bmatrix} u(t) \tag{4-195}$$

We have previously discussed the effects of similarity transformations and their particular value when the system matrix has real and distinct eigenvalues (see Secs. 4-7 and 4-9). Let us see what interesting insights the similarity transformation leads us to now.

Supposing that the eigenvalues $\lambda_1, \lambda_2, \ldots, \lambda_n$ of the system matrix $\mathbf{A}$ of Eq. (4-188) are *real* and *distinct*, we define a new state vector $\mathbf{x}(t)$ by setting

$$\mathbf{x}(t) = \mathbf{P}^{-1}\mathbf{z}(t) \tag{4-196}$$

where $\mathbf{P}$ is the Vandermonde matrix given by Eq. (3-411). This new state vector $\mathbf{x}(t)$ satisfies the differential equation

$$\dot{\mathbf{x}}(t) = \mathbf{\Lambda}\mathbf{x}(t) + \mathbf{P}^{-1}\mathbf{h}u(t) \tag{4-197}$$

where $\mathbf{\Lambda}$ is the diagonal matrix of the eigenvalues [given by Eq. (3-409)]. If we let $x_1(t), x_2(t), \ldots, x_n(t)$ be the components of $\mathbf{x}(t)$ and if we let

$$\mathbf{v} = \mathbf{P}^{-1}\mathbf{h} = \begin{bmatrix} v_1 \\ v_2 \\ \cdot \\ \cdot \\ \cdot \\ v_n \end{bmatrix} \tag{4-198}$$

then we may write Eq. (4-197) in the form

$$\dot{x}_i(t) = \lambda_i x_i(t) + v_i u(t) \qquad i = 1, 2, \ldots, n \tag{4-199}$$

Figure 4-11 shows a simulation of the system of Eq. (4-199) under the assumption that $b_n \neq 0$ [that is, $H(s)$ has $n$ zeros and $n$ poles]. If we compare Fig. 4-11 with Fig. 4-9, we can see that the only difference in the structure of the two systems is the feedforward channel from the input $u(t)$. On

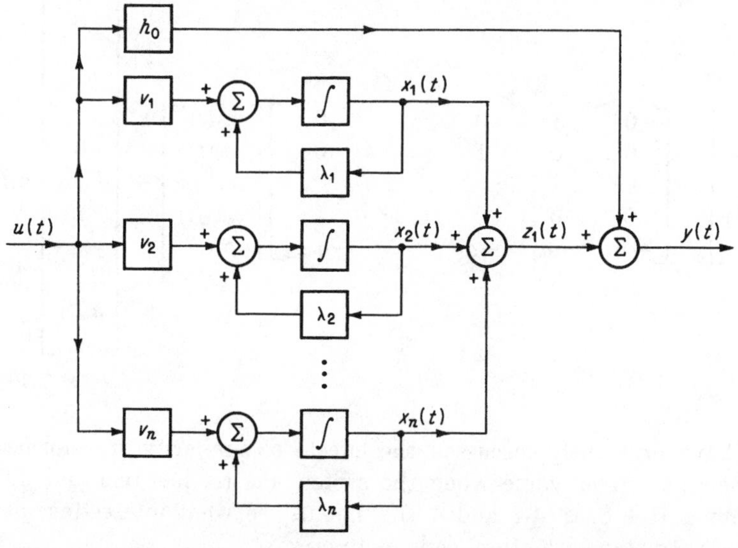


**Fig. 4-11** The analog-computer representation of the system (4-199).

the other hand, if $b_n = 0$ [that is, if the number of zeros of $H(s)$ is at least 1 less than the number of poles of $H(s)$], then $h_0 = 0$ and so Figs. 4-11 and 4-9 have the same structure.

If $b_n = 0$, then it is possible to calculate the gains $v_1, v_2, \ldots, v_n$ directly; however, it is also possible to identify the $v_i$ in terms of the residues of $H(s)$ at its poles $\lambda_i$. To accomplish this identification, we suppose that the system is initially at rest, i.e., that

$$y(0) = \dot{y}(0) = \cdots = y^{(n-1)}(0) = u(0) = \cdots = u^{(m-1)}(0) = 0 \tag{4-200}$$

and that the input $u(t)$ is applied; then we note that the Laplace transform $y(s)$ of the output $y(t)$ is given by

$$y(s) = H(s)u(s) = \sum_{i=1}^{n} \frac{\rho_i}{s - \lambda_i} u(s) \tag{4-201}$$

where $\rho_i$ is the residue of $H(s)$ at $\lambda_i$. Since the $\lambda_i$ are distinct, we have

$$\rho_i = \frac{b_m(\lambda_i - \sigma_1)(\lambda_i - \sigma_2) \cdots (\lambda_i - \sigma_m)}{(\lambda_i - \lambda_1)(\lambda_i - \lambda_2) \cdots (\lambda_i - \lambda_{i-1})(\lambda_i - \lambda_{i+1}) \cdots (\lambda_i - \lambda_n)} \tag{4-202}$$

where $H(s)$ has $m$ zeros $\sigma_1, \sigma_2, \ldots, \sigma_m$ [that is, $H(s)$ is given by Eq. (4-192)]. However, Eq. (4-200) implies [in view of Eqs. (4-193), which defines $\mathbf{z}(t)$] that $\mathbf{z}(0) = \mathbf{0}$ and hence, since $\mathbf{P}^{-1}$ is nonsingular, that

$$x_1(0) = x_2(0) = \cdots = x_n(0) = 0 \tag{4-203}$$

In view of this and Eq. (4-199), we observe that

$$x_i(s) = \frac{v_i}{s - \lambda_i} u(s) \tag{4-204}$$

Since the elements of the first row of the Vandermonde matrix $\mathbf{P}$ are all 1 [see Eq. (3-411)] and since $b_n = 0$ implies that $h_0 = 0$, we have

$$y(s) = z_1(s) = \sum_{i=1}^{n} x_i(s) = \sum_{i=1}^{n} \frac{v_i}{s - \lambda_i} u(s) \tag{4-205}$$

Comparison of Eqs. (4-205) and (4-201) leads us to the conclusion that

$$v_i = \rho_i \qquad i = 1, 2, \ldots, n \tag{4-206}$$

We have seen, in this section and in the previous section, how to determine suitable state-variable representations from the transfer function of a system. We have shown that the eigenvalues of the system matrix are the same as the poles of the transfer function and that the zeros of the transfer function affect only the control (in the sense that the zeros determine the "gain" vector $\mathbf{b}$).

We also noted that the state-variable representation corresponds to an analog-computer representation of the system in which the state variables are the outputs of integrators. We observed that the similarity transformation generated an alternate analog-computer representation of the system.

Since we have shown that the poles of the transfer function and the eigenvalues of the system matrix are the same, we can assert that the stability criterion of Sec. 3-26 can be translated into the familiar criteria for stability given in terms of the position of the poles of the transfer function in the complex plane.

**Example 4-5** Consider the system

$$\dddot{y}(t) + 3\ddot{y}(t) + 2\dot{y}(t) = \ddot{u}(t) + 7\dot{u}(t) + 12u(t)$$

For this system, $n = 3$, $a_0 = 0$, $a_1 = 2$, $a_2 = 3$, $b_3 = 0$, $b_2 = 1$, $b_1 = 7$, and $b_0 = 12$. Using Eq. (4-194), we find that

$$h_0 = 0 \qquad h_1 = 1 \qquad h_2 = 4 \qquad h_3 = -2$$

Thus, the state variables $z_1(t)$, $z_2(t)$, and $z_3(t)$ are given by [see Eqs. (4-193)]

$$\begin{aligned} z_1(t) &= y(t) \\ z_2(t) &= \dot{y}(t) - u(t) \\ z_3(t) &= \ddot{y}(t) - \dot{u}(t) - 4u(t) \end{aligned}$$

and they satisfy the vector differential equation [see Eq. (4-195)]

$$\begin{bmatrix} \dot{z}_1(t) \\ \dot{z}_2(t) \\ \dot{z}_3(t) \end{bmatrix} = \begin{bmatrix} 0 & 1 & 0 \\ 0 & 0 & 1 \\ 0 & -2 & -3 \end{bmatrix} \begin{bmatrix} z_1(t) \\ z_2(t) \\ z_3(t) \end{bmatrix} + \begin{bmatrix} 1 \\ 4 \\ -2 \end{bmatrix} u(t)$$

**Exercise 4-7** Determine the differential equations of the $z_i(t)$ state variables for the following systems:

(a) $\dddot{y}(t) = \dddot{u}(t)$ (b) $\dddot{y}(t) = \ddot{u}(t) + 8\dot{u}(t) + 15u(t)$

(c) $\dddot{y}(t) - \dot{y}(t) = \dot{u}(t)$ (d) $y^{(4)}(t) + 2y^{(3)}(t) + y^{(2)}(t) = u^{(2)}(t) + u^{(1)}(t)$

**Exercise 4-8** Consider the system

$$\ddot{y}(t) + a_1(t)\dot{y}(t) + a_0(t)y(t) = b_2(t)\ddot{u}(t) + b_1(t)\dot{u}(t) + b_0(t)u(t)$$

where the coefficients are continuous time functions. Derive the state-variable representation. HINT: Use $z_1(t) = y(t) - h_0(t)u(t)$, etc.

**Exercise 4-9** Generalize the result of Exercise 4-8 to the case of an $n$th-order linear time-varying system.

## 4-11 The Control Problem: Introductory Remarks

We have carefully defined the basic notion of a dynamical system and have examined the related concepts of input, output, and state. Thus, we are now prepared to discuss, in a rather general way, the problem of "controlling" a dynamical system. Before attempting a precise formulation, let us try to see, in an intuitive way, what this means.

When we consider physical systems, we usually have some definite objectives in mind. We want our physical systems to accomplish some given task as "cheaply" as possible. For example, we might want to design an autopilot which had certain response characteristics or a satellite attitude control system which responded rapidly but did not consume too much fuel or a "fast" tracking servo. In other words, we want to obtain a desired output from our system with an input which makes some cost (or performance) measure small. The translation of these engineering design objectives into the abstract language of dynamical systems gives rise to what we shall call the *control problem*. We can see that the essential elements of the control problem are:

1. A dynamical system which is to be "controlled"
2. A desired output or objective of the system
3. A set of allowable (or admissible) "controls" (i.e., inputs)

4. A performance functional which measures the effectiveness of a given "control action"

Since we have previously examined the concept of a dynamical system (1), let us briefly turn our attention to the other basic elements of the control problem: the desired objective (2), the set of admissible controls (3), and the performance functional (4).

We shall usually consider the desired objective of our system to be a given state or set of states which may vary with time. In other words, we shall replace the question of obtaining a certain output with the problem of "hitting" a certain target set in the state space of the system. For example, if we wished to reduce the output $\mathbf{y}(t)$ of a linear system such as

$$\begin{aligned} \dot{\mathbf{x}}(t) &= \mathbf{A}(t)\mathbf{x}(t) + \mathbf{B}(t)\mathbf{u}(t) \\ \mathbf{y}(t) &= \mathbf{C}\mathbf{x}(t) \end{aligned} \tag{4-207}$$

to zero, then we would attempt to reach the set of states $\hat{\mathbf{x}}$ for which $\mathbf{C}\hat{\mathbf{x}} = \mathbf{0}$.†

"Control" signals in physical systems are usually obtained from equipment which can provide only a limited amount of force or energy. This necessitates placing restrictions or constraints upon the set of inputs which may be used to "control" the system. We shall call the set $U$ of controls which satisfy the constraints, the *set of admissible controls.*

For most physical systems, the desired objective can be attained by many admissible inputs, each of which results in a different response. Thus, we would like to evaluate each response and, if possible, pick the "best" one. This requires the use of a performance criterion (or functional). The performance functionals we shall use to measure the "cost" of controlling our system will be real-valued and will depend upon the states at which we initiate and terminate our "control action," the path which the motion of the system takes in going from the initial to the terminal state, the time interval during which control is applied, and the particular control used.

We shall formally define the control problem in the next section, and then we shall examine some important special cases of it.

## 4-12 The Control Problem: Formal Definition

We shall define the control problem in a rather general way in this section and shall discuss several important special cases of it in the next section.

Let us suppose that we are given an $n$th-order dynamical system with $(T_1, T_2)$ as interval of definition (see Sec. 4-6) and with system equations

$$\begin{aligned} \dot{\mathbf{x}}(t) &= \mathbf{f}[\mathbf{x}(t), \mathbf{u}(t), t] \\ \mathbf{y}(t) &= \mathbf{g}[\mathbf{x}(t), \mathbf{u}(t), t] \end{aligned} \tag{4-208}$$

† If we let $\mathfrak{C}$ be the mapping of $R_n$ into $R_p$, given by $\mathfrak{C}\mathbf{x} = \mathbf{C}\mathbf{x}$, then this target set is the kernel $\mathfrak{C}^{-1}(\mathbf{0})$ (see Sec. 2-4) of $\mathfrak{C}$.

We shall denote the transition function of the system by

$$\boldsymbol{\phi}(t;\, \mathbf{u}_{(t_0,t]},\, \mathbf{x}_0) \tag{4-209}$$

We let $L$ be a continuous real-valued function on $R_n \times R_m \times (T_1, T_2)$,† that is, a continuous function of the form

$$L(x_1, x_2, \ldots, x_n, u_1, u_2, \ldots, u_m, t) \tag{4-210}$$

and we let $K$ be a real-valued function on $R_n \times (T_1, T_2)$, that is, a function of the form

$$K(x_1, x_2, \ldots, x_n, t) \tag{4-211}$$

We suppose that a subset $S$ of $R_n \times (T_1, T_2)$ is given, and we shall call $S$ the *target set.* We observe that the elements of the target set are pairs $(\mathbf{x}, t)$, consisting of a state $\mathbf{x}$ and a point $t$ in the interval of definition of the system.

Now, an important element in the definition of a dynamical system is the set $U$ of piecewise continuous functions (see Definition 4-4); we shall call the elements $\mathbf{u}$ of $U$ *controls*, and we shall now show how to relate constraints to the definition of $U$.

Let us assume that, for each $t$ in the interval of definition $(T_1, T_2)$ of our system, there is given a subset $U_t$ of $R_m$ (usually taken to be closed, bounded, and convex or all of $R_m$). Let us denote the collection of the sets $U_t$ by $\Omega$; that is,

$$\Omega = \{U_t\colon t \in (T_1, T_2)\} \tag{4-212}$$‡

We then have:

***Definition 4-7*** *$U_t$ is called the constraint set at time $t$, and $\Omega$ is called the constraint set or, simply, the constraint. If $U$ is the set of all bounded piecewise continuous functions $\mathbf{u}(t)$ on $(T_1, T_2)$ such that*

$$\mathbf{u}(t) \in U_t \qquad \textit{for all } t \textit{ in } (T_1, T_2) \tag{4-213}$$

*then we say that $U$ is the set of controls satisfying the constraint $\Omega$ or that $U$ is the set of admissible controls. Consequently, any element $\mathbf{u}$ of $U$ is called an admissible control.*

**Example 4-6** Let us denote by $\hat{U}$ the subset of $R_m$ given by

$$\hat{U} = \{u_1\mathbf{e}_1 + u_2\mathbf{e}_2 + \cdots + u_m\mathbf{e}_m\colon |u_i| \le M_i\} \tag{4-214}$$

where the $M_i$ are given constants and $\mathbf{e}_1, \mathbf{e}_2, \ldots, \mathbf{e}_m$ is the natural basis of $R_m$ [see Eqs. (2-27)]. Then a very common constraint set $\Omega$ is given by

$$\Omega = \{U_t = \hat{U}\colon t \text{ in } (T_1, T_2)\} \tag{4-215}$$

In other words, $\mathbf{u}(t)$ satisfies the constraint $\Omega$ if and only if, for all $t$ in $(T_1, T_2)$,

$$|u_i(t)| \le M_i \qquad i = 1, 2, \ldots, m \tag{4-216}$$

† See Sec. 2-3 and, in particular, Eq. (2-5).
‡ $\Omega$ is *not* the same as the set of Sec. 4-5. Here, the set of Sec. 4-5 is $R_m$.

We often say, "$\mathbf{u}(t)$ satisfies the constraint $|u_i(t)| \leq M_i$, $i = 1, 2, \ldots, m$," in place of "$\mathbf{u}(t)$ satisfies the constraint $\Omega$." A particularly important case occurs when $M_1 = M_2 = \cdots = M_m = 1$, as we shall see later on. In this case, the constraint set is the unit hypercube in $R_m$.

**Example 4-7** An example of a time-varying constraint (i.e., not all the $U_t$ are the same) can be obtained by taking

$$U_t = \begin{cases} \{\mathbf{u}\colon |u_i| \leq 1,\ i = 1, 2, \ldots, m \text{ for } t \text{ in } (T_1, T]\} \\ \{\mathbf{u}\colon |u_i| \leq 2,\ i = 1, 2, \ldots, m \text{ for } t \text{ in } (T, T_2)\} \end{cases} \tag{4-217}$$

where $T$ is some given element of $(T_1, T_2)$. Another example can be obtained by letting $M(t)$ be some positive real-valued nonconstant function on $(T_1, T_2)$ and by taking

$$U_t = \{\mathbf{u}\colon |u_i| \leq M(t),\ i = 1, 2, \ldots, m\} \tag{4-218}$$

**Example 4-8** Another important type of constraint is obtained by taking, for all $t$ in $(T_1, T_2)$,

$$U_t = \{\mathbf{u}\colon \|\mathbf{u}\| \leq M\} \tag{4-219}$$

where $M$ is given and $\|\mathbf{u}\|$ denotes the Euclidean norm (see Sec. 3-2) of $\mathbf{u}$. We often say, in this case, that $\mathbf{u}(t)$ satisfies the constraint

$$\|\mathbf{u}(t)\| \leq M \tag{4-220}$$

We shall make use of this type of constraint in Chap. 10. In this case, the constraint set is the hypersphere of radius $M$ in $R_m$.

Now we let $t_0$ be a given element of $(T_1, T_2)$, and we let $\mathbf{x}_0$ be a given element of $R_n$. Then, for $\mathbf{u}$ in $U$,

$$\mathbf{x}(\tau) = \boldsymbol{\phi}(\tau; \mathbf{u}_{(t_0,\tau]}, \mathbf{x}_0) \tag{4-221}$$

is the unique solution of the equation of the system satisfying the initial condition

$$\mathbf{x}(t_0) = \mathbf{x}_0 \tag{4-222}$$

and the function $\hat{J}$ from $R_n \times (T_1, T_2) \times U \times R_n \times (T_1, T_2)$, defined by

$$\hat{J}(\mathbf{x}_0, t_0, \mathbf{u}, \mathbf{x}, t) = K(\mathbf{x}, t) + \int_{t_0}^{t} L[\mathbf{x}(\tau), \mathbf{u}(\tau), \tau]\, d\tau \tag{4-223}$$

where $K$ and $L$ are the functions introduced earlier [Eqs. (4-210) and (4-211)], is a well-defined real number. We have:

***Definition 4-8*** *We say that the control* $\mathbf{u}$ *takes* $\mathbf{x}_0$ *to* $S$ [*or, more precisely,* $(\mathbf{x}_0, t_0)$ *to* $S$], *where* $S$ *is our target set, if the set*

$$\{(\boldsymbol{\phi}(t; \mathbf{u}_{(t_0,t]}, \mathbf{x}_0), t)\colon t \geq t_0\} \tag{4-224}$$

*meets (i.e., intersects)* $S$. *We shall, in this case, say that the motion of the system meets or intersects* $S$. *If* $\mathbf{u}$ *takes* $\mathbf{x}_0$ *to* $S$ *and if* $t_f$ *is the first instant of time after* $t_0$ *when the motion* $\mathbf{x}(t) = \boldsymbol{\phi}(t; \mathbf{u}_{(t_0,t]}, \mathbf{x}_0)$ *enters* $S$ *and if we set*

$$\mathbf{x}_f = \mathbf{x}(t_f) = \boldsymbol{\phi}(t_f; \mathbf{u}_{(t_0,t_f]}, \mathbf{x}_0) \tag{4-225}$$

*then we write* $J(\mathbf{x}_0, t_0, \mathbf{u})$ *in place of* $\hat{J}(\mathbf{x}_0, t_0, \mathbf{u}, \mathbf{x}_f, t_f)$ [*see Eq.* (4-223)], *that is,*

$$J(\mathbf{x}_0, t_0, \mathbf{u}) = \hat{J}(\mathbf{x}_0, t_0, \mathbf{u}, \mathbf{x}_f, t_f) \tag{4-226}$$

$$= K(\mathbf{x}_f, t_f) + \int_{t_0}^{t_f} L[\mathbf{x}(\tau), \mathbf{u}(\tau), \tau]\, d\tau \tag{4-227}$$

$$= K[\boldsymbol{\phi}(t_f; \mathbf{u}_{(t_0,t_f]}, \mathbf{x}_0), t_f] + \int_{t_0}^{t_f} L[\boldsymbol{\phi}(\tau; \mathbf{u}_{(t_0,\tau]}, \mathbf{x}_0), \mathbf{u}(\tau), \tau]\, d\tau \tag{4-228}$$

*We call* $J(\mathbf{x}_0, t_0, \mathbf{u})$ *the value of the performance functional of the system for the control* $\mathbf{u}$ *with respect to the target set* $S$. *We call* $t_f$ *the terminal, or final, time;* $\mathbf{x}_f$ *the terminal, or final, state; and* $K(\mathbf{x}_f, t_f)$ *the terminal cost. If* $\mathbf{u}$ *is an element of* $U$ *which does not take* $\mathbf{x}_0$ *to* $S$, *then we conventionally set*

$$J(\mathbf{x}_0, t_0, \mathbf{u}) = \infty. \tag{4-229}$$

*We call the functional*† $J(\mathbf{x}_0, t_0, \mathbf{u})$ *which maps* $R_n \times (T_1, T_2) \times U$ *into* $R \cup \{\infty\}$‡ *the performance functional of the system.*

Since our initial value $\mathbf{x}_0$ and our initial time $t_0$ will often be fixed, we shall usually write $J(\mathbf{u})$ in place of $J(\mathbf{x}_0, t_0, \mathbf{u})$. Specific examples of performance functionals of practical interest will be given in Chap. 6. We are now ready to give the formal statement of the optimal-control problem.

***Definition 4-9*** *The optimal-control problem for the system of Eqs.* (4-208) *with respect to the target set* $S$, *the performance functional* $J(\mathbf{x}_0, t_0, \mathbf{u})$, *the set* $U$ *of admissible controls, and the initial value* $\mathbf{x}_0$ *at the initial time* $t_0$ *is: Determine the control* $\mathbf{u}$ *in* $U$ *which minimizes the performance functional* $J(\mathbf{u})$.

*In other words, the optimal-control problem is:*

*Given the dynamical system of Eqs.* (4-208), *an initial state* $\mathbf{x}_0$, *an initial time* $t_0$, *a target set* $S$, *and a set* $U$ *of admissible controls. Then find the control* $\mathbf{u}$ *in* $U$ *which takes* $\mathbf{x}_0$ *to* $S$§*—so that* $(\boldsymbol{\phi}(t_f; \mathbf{u}_{(t_0,t_f]}, \mathbf{x}_0), t_f) \in S$*—and which minimizes the performance functional* $J(\mathbf{u}) = J(\mathbf{x}_0, t_0, \mathbf{u})$.

Any control $\mathbf{u}^*$ which provides a solution to the optimal-control problem will be called an *optimal control* (or, simply, *optimal*). We note that optimal controls need not exist; in fact, there may not even be controls $\mathbf{u}$ in $U$ which take $\mathbf{x}_0$ to $S$. We shall have more to say about this problem in Sec. 4-14. On the other hand, there may be more than one optimal control (examples of this are given in Chaps. 6, 8, and 10).

Since finding the maximum of a real-valued function is the same as finding the minimum of the negative of the function, it is clearly sufficient to consider only the minimization of performance functionals. For this reason, we shall *always deal with minima in this book.*

† We use the term *functional* since $J$ depends upon the set of functions $U$; that is, $J$ is a function of functions. (Cf. Sec. 3-16.)

‡ $R \cup \{\infty\}$ is the set of all real numbers together with the symbol $\infty$, which is greater than any real number.

§ That is, which takes $(\mathbf{x}_0, t_0)$ to $S$.

## 4-13 The Control Problem: Important Special Cases

We shall consider several important special cases of the control problem in this section. In particular, we shall discuss:

1. The free, or unconstrained, problem
2. The question of constraints upon the state
3. Various special problems which arise from particular forms of the target set, e.g., the regulator and rendezvous problems

In dealing with these variations of the control problem, we shall use the terminology and notations of Sec. 4-12.

Although the actual inputs to a physical system are invariably constrained, it is often of interest to consider the optimal-control problem in which the constraint set at time $t$, $U_t$ (see Definition 4-7), is *all* of $R_m$ for every $t$. In other words, we often suppose that *any* bounded piecewise continuous function from $(T_1, T_2)$ into $R_m$ is an admissible control. In that case, we say that we are studying the *free*, or *unconstrained*, optimal-control problem. We note that the free control problem may not have a solution, while a similar problem with constraints may, and conversely.

We examined the role of constraints on the value of our control in defining the optimal-control problem; however, we (tacitly) assumed that our state variable could take on *any* value in $R_n$. This leads us to a consideration of control problems in which the state of the system is constrained. In other words, we have:

***Definition 4-10*** *Let $X$ be a closed subset of the state space $R_n$ and suppose that the target set $S$ is contained in $X \times (T_1, T_2)$. Then the optimal-control problem with constrained (or bounded) state is:*

*Given the dynamical system of Eqs. (4-208), an initial state $\mathbf{x}_0$ in $X$, an initial time $t_0$, the target set $S$, and a set $U$ of admissible controls. Then find the control $\mathbf{u}$ in $U$ which takes $\mathbf{x}_0$ to $S$ along a trajectory which lies entirely in $X$ and which gives the minimum of the performance functional $J(\mathbf{u}) = J(\mathbf{x}_0, t_0, \mathbf{u})$ over all such controls.*

We can see that what has been added to the optimal-control problem are the requirements (1) that the initial state be in $X$ and (2) that only controls which generate trajectories lying in $X$ be considered as candidates for an optimal control. We shall *not* consider such problems in this book, and the reader who is interested in these problems should consult Refs. C-4, D-8, and P-5.

We are now going to examine several particular specifications of the target set $S$ and the corresponding control-problem variants which they generate.

We recall that $S$ is a subset of $R_n \times (T_1, T_2)$; that is, the elements of $S$ are pairs $(\mathbf{x}, t)$ consisting of a state $\mathbf{x}$ and a time $t$ in $(T_1, T_2)$. If the set $S$ is of the form

$$S = \bigcup_{t \in (T_1, T_2)} S_t \times \{t\} \tag{4-230}$$

where the $S_t$ are given *nonempty* subsets of $R_n$, then we call the resulting control problem a *free-time problem*. In other words, we are trying to reach a moving set in the state space, and we may meet this moving set at *any* time within the interval of definition $(T_1, T_2)$. A particular case of this is the situation in which $S$ is of the form

$$S = \{(\mathbf{z}(t), t): t \in (T_1, T_2)\} \tag{4-231}$$

where $\mathbf{z}(t)$ is a given $R_n$-valued function on $(T_1, T_2)$; we call this problem the *rendezvous* (or *pursuit*) problem.

On the other hand, if the target set $S$ is of the form

$$S = S_1 \times \{T\} \tag{4-232}$$

where $T$ is a fixed element of $(T_1, T_2)$, then we call the resulting control problem a *fixed-time problem*. In this case, we are trying to reach a given set $S_1$ at a given time $T$. When $T_2 = \infty$, we shall often consider the target set $S_1 \times \{t_0 + T\}$ instead of $S_1 \times \{T\}$ and call the resulting problem a fixed-time problem. To illustrate the formal statement of such problems, we give the following definition:

***Definition 4-11*** *Let $T$ be a given element of $(T_1, T_2)$ and $S_1$ a given subset of $R_n$. Then the fixed-time problem is:*

*Given the dynamical system of Eqs.* (4-208), *an initial state $\mathbf{x}_0$, an initial time $t_0$ with $t_0 < T$, the target set $S_1 \times \{T\}$, and a set $U$ of admissible controls. Then find the control $\mathbf{u}$ in $U$ which takes $\mathbf{x}_0$ to $S$ and which minimizes the performance functional*

$$J(\mathbf{u}) = J(\mathbf{x}_0, t_0, \mathbf{u}) = K[\mathbf{x}(T), T] + \int_{t_0}^{T} L[\mathbf{x}(\tau), \mathbf{u}(\tau), \tau]\, d\tau$$

Now let us suppose that the target set $S$ is of the form

$$S = \{\mathbf{x}^*\} \times (T_1, T_2) \tag{4-233}$$

where $\mathbf{x}^*$ is a *single point*, usually the origin, of $R_n$; then we call the resulting control problem a *fixed-end-point, free-time* problem. In general, if the set of $\mathbf{x}$'s in $R_n$ such that there is a $t$ with $(\mathbf{x}, t)$ in the target $S$ consists of a single point, then we speak of the control problem as a fixed-end-point problem. When the single point $\mathbf{x}^*$ is an equilibrium point of the system,† we say that we are dealing with a *regulator problem*.

Other variations of the control problem will appear in the sequel but should, after this discussion, cause the reader no difficulty in interpretation.

† This means that $\mathbf{f}(\mathbf{x}^*, \mathbf{0}, t) = \mathbf{0}$ for all $t$ in $(T_1, T_2)$, where $\mathbf{f}$ is the function appearing in Eqs. (4-208).

## 4-14 The Set of Reachable States†

Certain qualitative properties of systems play an important role in control theory. In particular, the notions of reachability, controllability, and observability are very important. In this section, we shall discuss the notion of reachability and the concomitant concept of the set of reachable states. In the next section, we shall define the notions of controllability and observability.

Loosely speaking, the term *reachability* means that we can find a control which transfers us from a given state $\mathbf{x}_0$ to another state $\mathbf{x}_1$. Since we are interested in determining, as part of the control problem, the controls which transfer us from our initial state to our target set, we can see that the notion of reachability will play a role in the study of a control problem. Now let us be more precise.

Let us suppose that we are considering a system with transition function $\boldsymbol{\phi}(t;\, \mathbf{u}_{(t_0,t]},\, \mathbf{x}_0)$ and with $U$ as the set of admissible controls satisfying a given constraint $\Omega$ (see Definition 4-7), that is, $\mathbf{u} \in U$ if and only if $\mathbf{u}$ is a piecewise continuous function such that

$$\mathbf{u}(t) \in U_t \tag{4-234}$$

for all $t$ in the interval of definition of the system, where the $U_t$ are given subsets of $R_m$. Then we have:

***Definition 4-12*** *A state $\mathbf{x}_1$ is said to be reachable, or accessible, from the state $\mathbf{x}_0$ at $t_0$ with respect to $U$ if there is an element $\mathbf{u}_1$ of $U$ such that*

$$\boldsymbol{\phi}(t_1;\, \mathbf{u}_{1(t_0,t_1]},\, \mathbf{x}_0) = \mathbf{x}_1 \tag{4-235}$$

*for some finite $t_1$ with $t_1 \geq t_0$. If $A(t;\, \mathbf{x}_0,\, t_0,\, U)$ denotes the subset of $R_n$ which consists of all the states $\mathbf{x}_1$ that are reachable from $\mathbf{x}_0$ at $t_0$ with respect to $U$ at time $t$, that is, if*

$$A(t;\, \mathbf{x}_0,\, t_0,\, U) = \{\mathbf{x}_1\text{: there is a } \mathbf{u}_1 \text{ in } U \text{ such that } \boldsymbol{\phi}(t;\, \mathbf{u}_{1(t_0,t]},\, \mathbf{x}_0) = \mathbf{x}_1\} \tag{4-236}$$

*then we call $A(t;\, \mathbf{x}_0,\, t_0,\, U)$ the set of reachable states at $t$ (from $\mathbf{x}_0$ at $t_0$ with respect to $U$) and we call $\bigcup_{t \geq t_0} A(t;\, \mathbf{x}_0,\, t_0,\, U)$ the set of reachable states (from $\mathbf{x}_0$ at $t_0$ with respect to $U$).*

Now let us suppose that our system is linear, with equations

$$\dot{\mathbf{x}}(t) = \mathbf{A}(t)\mathbf{x}(t) + \mathbf{B}(t)\mathbf{u}(t) \tag{4-237}$$

$$\mathbf{y}(t) = \mathbf{C}(t)\mathbf{x}(t) + \mathbf{D}(t)\mathbf{u}(t) \tag{4-238}$$

† See Refs. H-1 to H-3, L-4, L-7, L-8, L-12, L-14, N-5, and R-7.

and that $\mathbf{\Phi}(t, t_0)$ is the fundamental matrix of Eq. (4-237) (see Sec. 3-20). Then the transition function $\boldsymbol{\phi}$ of our system is given by

$$\boldsymbol{\phi}(t;\, \mathbf{u}_{(t_0,t]},\, \mathbf{x}_0) = \mathbf{\Phi}(t, t_0)\left\{\mathbf{x}_0 + \int_{t_0}^{t} \mathbf{\Phi}(t_0, \tau)\mathbf{B}(\tau)\mathbf{u}(\tau)\, d\tau\right\} \tag{4-239}$$

An examination of Eq. (4-239) leads us to the following theorem:

***Theorem 4-1*** *If the set $U$ of admissible controls for the system of Eqs.* (4-237) *and* (4-238) *is convex (see Sec.* 3-8), *then, for any $\mathbf{x}_0$ and $t_0$, the set $A(t;\, \mathbf{x}_0, t_0, U)$ is also convex. In particular, if $U$ consists of all piecewise continuous functions satisfying the constraint*

$$\mathbf{u}(\tau) \in U_\tau \tag{4-240}$$

*where the $U_\tau$ are all convex, then $A(t;\, \mathbf{x}_0, t_0, U)$ is convex.*

PROOF Suppose that $U$ is convex, and let $\mathbf{x}_0$ be any element of $R_n$ and $t_0$ be any time. We wish to show that if $\mathbf{x}_1$ and $\mathbf{x}_2$ are elements of $A(t;\, \mathbf{x}_0, t_0, U)$ and $r$ and $s$ are real numbers with $r + s = 1$, $r, s \geq 0$, then $r\mathbf{x}_1 + s\mathbf{x}_2$ is in $A(t;\, \mathbf{x}_0, t_0, U)$. But there are controls $\mathbf{u}_1$ and $\mathbf{u}_2$ in $U$ such that

$$\mathbf{x}_1 = \mathbf{\Phi}(t, t_0)\left\{\mathbf{x}_0 + \int_{t_0}^{t} \mathbf{\Phi}(t_0, \tau)\mathbf{B}(\tau)\mathbf{u}_1(\tau)\, d\tau\right\} \tag{4-241}$$

$$\mathbf{x}_2 = \mathbf{\Phi}(t, t_0)\left\{\mathbf{x}_0 + \int_{t_0}^{t} \mathbf{\Phi}(t_0, \tau)\mathbf{B}(\tau)\mathbf{u}_2(\tau)\, d\tau\right\} \tag{4-242}$$

It follows that

$$r\mathbf{x}_1 + s\mathbf{x}_2 = \mathbf{\Phi}(t, t_0)\left\{\mathbf{x}_0 + \int_{t_0}^{t} \mathbf{\Phi}(t_0, \tau)\mathbf{B}(\tau)[r\mathbf{u}_1(\tau) + s\mathbf{u}_2(\tau)]\, d\tau\right\} \tag{4-243}$$

Since $U$ is convex, $r\mathbf{u}_1 + s\mathbf{u}_2$ is in $U$, and so we have shown that $r\mathbf{x}_1 + s\mathbf{x}_2$ is in $A(t;\, \mathbf{x}_0, t_0, U)$. We leave it to the reader to verify the "in particular" part of the theorem.

***Corollary 4-1*** *If $U$ consists of the controls satisfying the constraint*

$$|u_i(\tau)| \leq M_i \qquad i = 1, 2, \ldots, m \tag{4-244}$$

*then $A(t;\, \mathbf{x}_0, t_0, U)$ is convex.*

**Example 4-9** Although all the $A(t;\, \mathbf{x}_0, t_0, U)$ may be convex, it does not follow that $\bigcup_{t \geq t_0} A(t;\, \mathbf{x}_0, t_0, U)$ is convex, as the following simple example shows. Consider the system

$$\begin{bmatrix} \dot{x}_1(t) \\ \dot{x}_2(t) \end{bmatrix} = \begin{bmatrix} 0 & 1 \\ -1 & 0 \end{bmatrix} \begin{bmatrix} x_1(t) \\ x_2(t) \end{bmatrix} + \begin{bmatrix} u_1(t) \\ u_2(t) \end{bmatrix} \tag{4-245}$$

and suppose that $U$ consists only of the function which is identically zero [that is, $u_1(t) = u_2(t) = 0$ for all $t$]. Let $t_0 = 0$ and let $\mathbf{x}_0 = \mathbf{x}(0) = \begin{bmatrix} 0 \\ 1 \end{bmatrix}$. Then $A(t;\, \mathbf{x}_0, 0, U)$ consists of the single point $\begin{bmatrix} \sin t \\ \cos t \end{bmatrix}$ and is, therefore, convex. However, $\bigcup_{t \geq 0} A(t;\, \mathbf{x}_0, 0, U)$ is the set $\left\{\begin{bmatrix} x_1 \\ x_2 \end{bmatrix} : x_1^2 + x_2^2 = 1\right\}$ and is clearly not convex.

***Corollary 4-2*** *If $\mathbf{x}_0 = \mathbf{0}$ and $U$ is convex and contains the function $\mathbf{u}(t) = \mathbf{0}$ for all $t$, then the set of reachable states is convex. In particular, if $U$ is convex, containing $\mathbf{u}(t) = \mathbf{0}$, and the system is linear and time-invariant, then the set of reachable states from $\mathbf{x}_0 = \mathbf{0}$ with respect to $U$† is convex.*

Still more of interest can be deduced from Eq. (4-239). In particular, we have:

***Theorem 4-2*** *Suppose that $U$ consists of all piecewise continuous functions satisfying the constraint*

$$\mathbf{u}(\tau) \in U_\tau \tag{4-246}$$

*where the sets $U_\tau$ are all contained in some sphere $S(\mathbf{0}, M)$ in $R_m$. Then, for each $t > t_0$, the set $A(t; \mathbf{x}_0, t_0, U)$ is bounded (see Sec. 3-6).*

PROOF Suppose that $\mathbf{x}_1$ is in $A(t; \mathbf{x}_0, t_0, U)$. Then there is a $\mathbf{u}$ in $U$ such that

$$\mathbf{x}_1 = \mathbf{\Phi}(t, t_0) \left\{ \mathbf{x}_0 + \int_{t_0}^{t} \mathbf{\Phi}(t_0, \tau)\mathbf{B}(\tau)\mathbf{u}(\tau)\, d\tau \right\} \tag{4-247}$$

Since all the entries in $\mathbf{\Phi}$ and $\mathbf{B}$ are continuous functions on $[t_0, t]$, we may conclude that there are numbers $L$ and $N$‡ such that

$$\|\mathbf{\Phi}(t, t_0)\mathbf{v}\| \le L\|\mathbf{v}\| \tag{4-248}$$

$$\|\mathbf{\Phi}(t_0, \tau)\mathbf{B}(\tau)\mathbf{v}\| \le N\|\mathbf{v}\| \qquad \text{all } \tau \text{ in } [t_0, t] \tag{4-249}$$

for all $\mathbf{v}$ in $R_n$. It follows from Theorem 3-8 that

$$\|\mathbf{x}_1\| \le L\{\|\mathbf{x}_0\| + (t - t_0)N \max_{\tau \in [t_0, t]} \|\mathbf{u}(\tau)\|\} \tag{4-250}$$

$$\le L\{\|\mathbf{x}_0\| + (t - t_0)MN\} \tag{4-251}$$

and hence that

$$A(t; \mathbf{x}_0, t_0, U) \subset \overline{S(\mathbf{0}, L\{\|\mathbf{x}_0\| + (t - t_0)MN\})}\text{§} \tag{4-252}$$

Thus, we have shown that $A(t; \mathbf{x}_0, t_0, U)$ is contained in some sphere in $R_n$ and is, therefore, bounded.

***Corollary 4-3*** *If $U$ consists of the controls satisfying the constraint*

$$|u_i(\tau)| \le M_i \qquad i = 1, 2, \ldots, m \tag{4-253}$$

*then $A(t; \mathbf{x}_0, t_0, U)$ is bounded.*

**Exercise 4-10** Consider the system (4-245) with $u_1(t) = 0$ and $|u_2(t)| \le 1$ for all $t$. Let $t_0 = 0$ and let the initial state be $x_1(0) = x_2(0) = 0$. Show that the set of reachable states $A$ at $t = \pi$ is

$$A = \{(x_1, x_2): x_1^2 + x_2^2 \le 4\}$$

**Exercise 4-11** Consider the system $\dot{x}_1(t) = \dot{x}_2(t)$; $\dot{x}_2(t) = u(t)$; $0 \le u(t) \le 1$, for all $t$. Let $t_0 = 0$ and $x_1(0) = x_2(0) = 0$. Find the set of reachable states at $t = 1$.

† This means that only controls in $U$ are considered.
‡ See Sec. 3-6 and recall that $\|\quad\|$ denotes the Euclidean norm.
§ See Eq. (3-5).

Finally, in the interest of general education, we conclude this section with a theorem which is used quite often in proving the existence of optimal controls. A proof of this theorem can be found in Ref. L-4.

***Theorem 4-3*** *Suppose that, for the system of Eq.* (4-237) *with* $\mathbf{A}(t)$ *and* $\mathbf{B}(t)$ *constant, the set $U$ consists of all measurable*† *functions satisfying the constraint*

$$\mathbf{u}(\tau) \in U_\tau \tag{4-254}$$

*where the sets $U_\tau$ are* (1) *all convex,* (2) *all contained in some sphere* $S(\mathbf{0}, M)$ *in $R_m$, and* (3) *all closed in $R_m$ (see Sec.* 3-4). *Then, for each $t > t_0$, the set $A(t; \mathbf{x}_0, t_0, U)$ is closed.*

## 4-15 Controllability and Observability: Definition

We are going to define the notions of controllability and observability in this section. Roughly speaking, the term *controllability* means that it is possible to drive any state of the system to the *origin* in some finite time, and the term *observability* means that the initial state of the system can be found from a suitable measurement of the output. We shall, in a moment, make these ideas more precise; then we shall, in the next two sections, specialize these concepts to the case of *linear time-invariant systems.* In that case, we shall be able to obtain explicit criteria for controllability and observability. Finally, we shall examine some practical implications of these concepts.‡

Now let us suppose that we are given a dynamical system with equations

$$\dot{\mathbf{x}}(t) = \mathbf{f}[\mathbf{x}(t), \mathbf{u}(t), t] \tag{4-255}$$
$$\mathbf{y}(t) = \mathbf{g}[\mathbf{x}(t), \mathbf{u}(t), t] \tag{4-256}$$

and with $\boldsymbol{\phi}(t; \mathbf{u}_{(t_0,t]}, \mathbf{x}_0)$ as transition function and with $U$ *as the set of all bounded piecewise continuous functions from the interval of definition into* $R_m$. In other words, our admissible controls are *unconstrained.* We then have:

***Definition 4-13 Controllability*** *If the state* $\mathbf{x}_1 = \mathbf{0}$ *is reachable from* $\mathbf{x}_0$ *at $t_0$ (see Definition* 4-12), *then we say that* $\mathbf{x}_0$ *is controllable at (time) $t_0$. In other words,* $\mathbf{x}_0$ *is controllable at $t_0$ if there is a piecewise continuous function* $\mathbf{u}^0$ *such that*

$$\boldsymbol{\phi}(T; \mathbf{u}^0_{(t_0,T]}, \mathbf{x}_0) = \mathbf{0} \tag{4-257}$$

*for some finite $T \geq t_0$. If every state* $\mathbf{x}_0$ *is controllable at $t_0$, then we say that the system is controllable at $t_0$. Finally, if every state* $\mathbf{x}_0$ *is controllable at every time $t_0$ in the interval of definition of the system, then we say that the system is completely controllable (or, simply, controllable).*

† Piecewise continuity is not enough here, and so we require measurability. (See Ref. R-10 for a definition of this concept.)

‡ The reader should consult Refs. B-20, F-2, G-4, K-2, K-5, K-7, K-9, K-26, W-5, and Z-1 for more detailed analyses of these concepts.

***Definition 4-14 Observability*** *We say that a state* $\mathbf{x}_0$ *is observable at* $t_0$ *if, given any control* $\mathbf{u}$, *there is a time* $t_1 \geq t_0$† *such that knowledge of* $\mathbf{u}_{(t_0,t_1]}$ *and the output* $\mathbf{y}_{(t_0,t_1]} = \hat{\mathbf{g}}(\mathbf{x}_0; \mathbf{u}_{(t_0,t_1]})$‡ *is sufficient to determine* $\mathbf{x}_0$. *If every state* $\mathbf{x}_0$ *is observable at* $t_0$, *then we say that the system is observable at* $t_0$. *Finally, if every state* $\mathbf{x}_0$ *is observable at every time* $t_0$ *in the interval of definition of the system, then we say that the system is completely observable (or, simply, observable).*

We shall now show that the concepts of controllability and observability are intrinsic, in the sense that they are preserved under equivalence (see Definition 4-5). More precisely, we have:

***Theorem 4-4*** *Let* $\mathbf{P}$ *be a nonsingular* $n \times n$ *matrix and let* $\mathbf{z}(t) = \mathbf{P}^{-1}\mathbf{x}(t)$. *Then the system*

$$\dot{\mathbf{x}}(t) = \mathbf{f}[\mathbf{x}(t), \mathbf{u}(t), t] \tag{4-258}$$

$$\mathbf{y}(t) = \mathbf{g}[\mathbf{x}(t), \mathbf{u}(t), t] \tag{4-259}$$

*is controllable (or observable) if and only if the equivalent system*

$$\dot{\mathbf{z}}(t) = \mathbf{P}^{-1}\mathbf{f}[\mathbf{P}\mathbf{z}(t), \mathbf{u}(t), t] \tag{4-260}$$

$$\mathbf{y}(t) = \mathbf{g}[\mathbf{P}\mathbf{z}(t), \mathbf{u}(t), t] = \mathbf{h}[\mathbf{z}(t), \mathbf{u}(t), t] \tag{4-261}$$

*is controllable (or observable).*

PROOF Suppose, first, that the system of Eqs. (4-258) and (4-259) is controllable and let $t_0$ be a given time and $\mathbf{z}_0$ a given element of $R_n$. Then there is an $\mathbf{x}_0$ in $R_n$ such that $\mathbf{z}_0 = \mathbf{P}^{-1}\mathbf{x}_0$, and there is an element $\mathbf{u}^0$ in $U$ such that, for some $T \geq t_0$,

$$\boldsymbol{\phi}(T; \mathbf{u}^0_{(t_0,T]}, \mathbf{x}_0) = \mathbf{0} \tag{4-262}$$

where $\boldsymbol{\phi}$ is the transition function of the system of Eqs. (4-258) and (4-259). If we set

$$\mathbf{z}(t) = \mathbf{P}^{-1}\boldsymbol{\phi}(t; \mathbf{u}^0_{(t_0,t]}, \mathbf{x}_0) \tag{4-263}$$

then we can see immediately that $\mathbf{z}(t_0) = \mathbf{z}_0$ and that $\mathbf{z}(T) = \mathbf{P}^{-1}\mathbf{0} = \mathbf{0}$. It follows that the system of Eqs. (4-260) and (4-261) is controllable. Since the argument is reversible, we have established the controllability part of the theorem.

As for the observability part of the theorem, let us suppose that the system of Eqs. (4-258) and (4-259) is observable. Let $t_0$ be a given time and let $\mathbf{z}_0$ be a given element of $R_n$. Let $\mathbf{u}$ be any element of $U$ and let $\mathbf{x}_0 = \mathbf{P}\mathbf{z}_0$. Then there is a time $T \geq t_0$ such that $\mathbf{u}_{(t_0,T]}$ and $\mathbf{y}_{(t_0,T]} = \hat{\mathbf{g}}(\mathbf{x}_0; \mathbf{u}_{(t_0,T]})$ are sufficient to determine $\mathbf{x}_0$. Since $\mathbf{g}[\mathbf{P}\mathbf{z}(t), \mathbf{u}(t), t] = \mathbf{g}[\mathbf{x}(t), \mathbf{u}(t), t]$, it follows that $\mathbf{y}_{(t_0,T]} = \hat{\mathbf{g}}(\mathbf{P}\mathbf{z}_0; \mathbf{u}_{(t_0,T]}) = \hat{\mathbf{h}}(\mathbf{z}_0; \mathbf{u}_{(t_0,T]})$ and, hence, that $\hat{\mathbf{h}}(\mathbf{z}_0; \mathbf{u}_{(t_0,T]})$ and $\mathbf{u}_{(t_0,T]}$ are sufficient to determine $\mathbf{z}_0$. Since this argument is reversible, the theorem is established.

† $t_1$ may depend on $\mathbf{u}$.
‡ See Sec. 4-5.

## 4-16 Controllability for Linear Time-invariant Systems

Now let us consider a linear time-invariant system of the form

$$\dot{\mathbf{x}}(t) = \mathbf{A}\mathbf{x}(t) + \mathbf{B}\mathbf{u}(t) \tag{4-264}$$
$$\mathbf{y}(t) = \mathbf{C}\mathbf{x}(t) + \mathbf{D}\mathbf{u}(t) \tag{4-265}$$

where **A**, **B**, **C**, and **D** are, respectively, $n \times n$, $n \times m$, $p \times n$, and $p \times m$ matrices, and let us attempt to see what insight into the notion of controllability we can come up with in this case.

To begin with, let us suppose that $\mathbf{x}_0$ is controllable at $t_0$. Then there is a $T \geq 0$ such that [see Eq. (4-257), where $T = t_1 - t_0$], for some control $\mathbf{u}$,

$$\mathbf{x}_0 = -\int_{t_0}^{t_0+T} e^{-\mathbf{A}(\tau - t_0)}\mathbf{B}\mathbf{u}(\tau)\,d\tau \tag{4-266}$$

where $e^{\mathbf{A}t}$ is the exponential of $\mathbf{A}t$ (see Sec. 3-21). It follows from Eq. (4-266) and the relation

$$\int_{t_0}^{t_0+T} e^{-\mathbf{A}(\tau - t_0)}\mathbf{B}\mathbf{u}(\tau)\,d\tau = \int_0^T e^{-\mathbf{A}\sigma}\mathbf{B}\mathbf{u}(\sigma)\,d\sigma \tag{4-267}$$

that $\mathbf{x}_0$ is controllable at 0. In other words, when investigating whether or not a state $\mathbf{x}_0$ of a linear time-invariant system is controllable, we may as well suppose that the initial time is 0. Furthermore, we may speak of a controllable state since we have shown that the initial time may be chosen at will. To put it still another way, we have shown that $\mathbf{x}_0$ is controllable at $t_0$ if and only if $\mathbf{x}_0$ is controllable at $t = 0$.

Another important consequence of Eq. (4-266) is the fact that the set of states $\mathbf{x}_0$ which are controllable is a *subspace* of $R_n$.† Theorem 4-5 provides a description of the subspace of controllable states in terms of the system matrix **A** and the matrix **B**. We also observe that if $\mathbf{x}_0$ and $\mathbf{x}_1$ are controllable, then there is a trajectory of the system joining $\mathbf{x}_0$ to $\mathbf{x}_1$. In particular, it is possible to go *from* **0** to any controllable state.

**Exercise 4-12** Prove that the set of states which are controllable at 0 is a subspace of $R_n$.

We let $\mathbf{b}_\beta$, $\beta = 1, 2, \ldots, m$, denote the $n$ vector which is the $\beta$th column of **B**; that is,

$$\mathbf{b}_\beta = \begin{bmatrix} b_{1\beta} \\ b_{2\beta} \\ \cdot \\ \cdot \\ \cdot \\ b_{n\beta} \end{bmatrix} \tag{4-268}$$

† Recall that a subset $W$ of a vector space $V$ is a subspace if (1) $\mathbf{w}_1, \mathbf{w}_2 \in W$ implies that $\mathbf{w}_1 + \mathbf{w}_2 \in W$ and (2) $r \in R$, $\mathbf{w} \in W$ implies that $r\mathbf{w} \in W$. See Sec. 2-5.

where $\mathbf{B} = (b_{i\beta})$, and we let $\mathbf{e}_{\alpha\beta}$ be the $n$ vector, defined by

$$\mathbf{e}_{\alpha\beta} = \mathbf{A}^\alpha \mathbf{b}_\beta \qquad \alpha = 0, 1, \ldots, n-1; \beta = 1, 2, \ldots, m \tag{4-269}$$

where $\mathbf{A}^0 = \mathbf{I}$ and $\mathbf{A}^\alpha = \mathbf{A} \cdot \mathbf{A} \cdots \mathbf{A}$ ($\alpha$ factors) for $\alpha > 0$. Then we have:

***Theorem 4-5*** *The time-invariant linear system of Eqs.* (4-264) *and* (4-265) *is completely controllable if and only if the vectors* $\mathbf{e}_{\alpha\beta}$ *span the entire space* $R_n$, *that is, if and only if every element of* $R_n$ *is a linear combination of the vectors* $\mathbf{e}_{\alpha\beta}$ *(see Sec.* 2-5).

PROOF We shall prove this theorem under the additional assumption that the system matrix $\mathbf{A}$ has *no complex eigenvalues* (a complete proof can be found in Ref. F-2).

First of all, let us suppose that $\xi$ is a controllable (at 0) state of the system. Then there are a $T > 0$ and a control $\mathbf{u}^*$ such that

$$-\xi = \int_0^T e^{-\mathbf{A}s}\mathbf{B}\mathbf{u}^*(s)\, ds \tag{4-270}$$

Since (see Sec. 3-21)

$$e^{-\mathbf{A}s} = \sum_{p=0}^{\infty} (-1)^p \mathbf{A}^p \frac{s^p}{p!}$$

and since every $n \times n$ matrix satisfies its characteristic equation (see Sec. 2-10, Theorem 2-3),

$$e^{-\mathbf{A}s} = \sum_{\alpha=0}^{n-1} \psi_\alpha(s)\mathbf{A}^\alpha \tag{4-271}$$

where the $\psi_\alpha(s)$ are scalar-valued functions of $s$. It follows that

$$-\xi = \sum_{\alpha=0}^{n-1} \int_0^T \psi_\alpha(s)\mathbf{A}^\alpha\mathbf{B}\mathbf{u}^*(s)\, ds \tag{4-272}$$

and hence that

$$-\xi = \sum_{\alpha=0}^{n-1} \sum_{\beta=1}^{m} \left[\int_0^T \psi_\alpha(s) u_\beta^*(s)\, ds\right] \mathbf{A}^\alpha \mathbf{b}_\beta \tag{4-273}$$

since

$$\mathbf{u}^*(s) = \sum_{\beta=1}^{m} u_\beta(s)\mathbf{e}_\beta \tag{4-274}$$

where the $\mathbf{e}_\beta$ are the elements of the natural basis of $R_m$. This proves the "only if" part of the theorem.

On the other hand, let us suppose that $\xi$ is an element of the space $R_n$. Then we shall show that there are a time $T > 0$ and a control $\mathbf{u}^*$ such that

$$-\xi = \int_0^T e^{-\mathbf{A}s}\mathbf{B}\mathbf{u}^*(s)\, ds \tag{4-275}$$

Let us assume for the moment that $T$ is some time $T > 0$ and that there is a real-valued function $\varphi(s)$ such that:

1. $\varphi^{(\alpha)}(0) = \varphi^{(\alpha)}(T) = 0$ for $\alpha = 0, 1, \ldots, n-2$, where $\varphi^{(\alpha)}(s)$ is the $\alpha$th derivative of $\varphi(s)$ and $\varphi^{(0)}(s) = \varphi(s)$.
2. $\varphi(s) > 0$ on $(0, T)$.
3. The matrix $\mathbf{Q}$ given by

$$\mathbf{Q} = \int_0^T \varphi(s)e^{-\mathbf{A}s}\,ds \tag{4-276}$$

is nonsingular.

Since every element of $R_n$ is a linear combination of the $\mathbf{e}_{\alpha\beta}$, we may write

$$-\mathbf{Q}^{-1}\boldsymbol{\xi} = \sum_{\alpha,\beta} a_{\alpha\beta}\mathbf{e}_{\alpha\beta} \tag{4-277}$$

where the $a_{\alpha\beta}$ are real numbers. We claim that the control $\mathbf{u}^*(s)$, with components $u_\beta^*(s)$ given by

$$u_\beta^*(s) = \sum_{\alpha=0}^{n-1} a_{\alpha\beta}\varphi^{(\alpha)}(s) \tag{4-278}$$

drives $\boldsymbol{\xi}$ to the origin in time $T$. Now

$$\int_0^T e^{-\mathbf{A}s}\mathbf{B}\mathbf{u}^*(s)\,ds = \sum_{\alpha=0}^{n-1}\sum_{\beta=1}^{m} a_{\alpha\beta}\left[\int_0^T \varphi^{(\alpha)}(s)e^{-\mathbf{A}s}\,ds\right]\mathbf{b}_\beta \tag{4-279}$$

and so it will be enough to show that

$$\int_0^T \varphi^{(\alpha)}(s)e^{-\mathbf{A}s}\,ds = \mathbf{Q}\mathbf{A}^\alpha \tag{4-280}$$

since Eq. (4-275) would then imply that

$$-\mathbf{Q}\mathbf{Q}^{-1}\boldsymbol{\xi} = -\boldsymbol{\xi} = \int_0^T e^{-\mathbf{A}s}\mathbf{B}\mathbf{u}^*(s)\,ds \tag{4-281}$$

However, Eq. (4-280) is readily established by using induction on $\alpha$, Eq. (4-276), and integration by parts [making use of Eq. (3-278)].

It remains only to show that a suitable function $\varphi(s)$ exists. We claim that the function

$$\varphi(s) = s^{2n}(T-s)^{2n} \tag{4-282}$$

does the job. Clearly, properties 1 and 2 are satisfied. Now let us show that

$$\mathbf{Q} = \int_0^T \varphi(s)e^{-\mathbf{A}s}\,ds \tag{4-283}$$

is nonsingular. Let $\mathbf{J}$ be the Jordan canonical form of $\mathbf{A}$ (see Sec. 2-10). Then there is a nonsingular matrix $\mathbf{P}$ such that $\mathbf{P}^{-1}\mathbf{A}\mathbf{P} = \mathbf{J}$, and so

$$e^{\mathbf{A}(T-s)} = \mathbf{P}^{-1}e^{\mathbf{J}(T-s)}\mathbf{P} \tag{4-284}$$

Since the matrix $e^{\mathbf{A}T}$ is nonsingular, it will be enough to prove that the matrix

$$\mathbf{P}e^{\mathbf{A}T}\mathbf{Q}\mathbf{P}^{-1} = \int_0^T \varphi(s)e^{\mathbf{J}(T-s)}\,ds \tag{4-285}$$

is nonsingular. But (see Sec. 3-22) $e^{\mathbf{J}(T-s)}$ is a triangular matrix with entries $e^{\lambda_i(T-s)}$ on the diagonal, the $\lambda_i$ being the (not necessarily distinct) eigenvalues of **A**. It follows that

$$\det \int_0^T \varphi(s) e^{\mathbf{J}(T-s)}\, ds = \prod_{i=1}^{n} \left[ \int_0^T \varphi(s) e^{\lambda_i s}\, ds \right] \tag{4-286}$$

and hence that the right-hand side of Eq. (4-285) is nonsingular, as

$$\int_0^T \varphi(s) e^{\lambda_i s}\, ds > 0 \tag{4-287}$$

for all $i$. (See Theorem 3-8.) This completes the proof.

Some particularly important corollaries of this theorem are:

***Corollary 4-4*** *The subspace of controllable states of the system of Eqs.* (4-264) *and* (4-265) *is the same as the span of the vectors* $\mathbf{e}_{\alpha\beta}$ *of Eq.* (4-269). *In other words,* $\mathbf{x}_0$ *is controllable if and only if* $\mathbf{x}_0$ *is a linear combination of the vectors* $\mathbf{e}_{\alpha\beta}$.

***Corollary 4-5*** *Let* **G** *be the* $n \times nm$ *matrix, defined by the relation*

$$\mathbf{G} \triangleq [\mathbf{B} \mid \mathbf{AB} \mid \mathbf{A}^2\mathbf{B} \mid \cdots \mid \mathbf{A}^{n-1}\mathbf{B}] \tag{4-288}$$

*or, equivalently, by the relation*

$$\mathbf{G} \triangleq [\mathbf{e}_{01}, \mathbf{e}_{02}, \ldots, \mathbf{e}_{0m} \mid \mathbf{e}_{11}, \mathbf{e}_{12}, \ldots, \mathbf{e}_{1m} \mid \cdots \mid \mathbf{e}_{n-11}, \mathbf{e}_{n-12}, \ldots, \mathbf{e}_{n-1m}] \tag{4-289}$$

*where the* $\mathbf{e}_{\alpha\beta}$ *are given by Eq.* (4-269). *Then the system is completely controllable if and only if the rank of* **G** *is* $n$, *that is,*

$$\text{rank } \mathbf{G} = n \tag{4-290}$$

*or, equivalently, if and only if there is a set of* $n$ *linearly independent column vectors of* **G**.

***Corollary 4-6*** *If* **B** *is an* $n \times 1$ *matrix* **b**, *that is, if the controls are scalar-valued, then the system is completely controllable if and only if the* $n \times n$ *matrix*

$$\mathbf{G} = [\mathbf{b} \mid \mathbf{Ab} \mid \cdots \mid \mathbf{A}^{n-1}\mathbf{b}] \tag{4-291}$$

*is nonsingular, or equivalently, setting*

$$\mathbf{b} = \begin{bmatrix} b_1 \\ b_2 \\ \cdot \\ \cdot \\ \cdot \\ b_n \end{bmatrix}$$

*if and only if*

$$\det \mathbf{G} = \det \begin{bmatrix} b_1 & \sum_{j=1}^{n} \alpha_{1j}^{(1)} b_j & \cdots & \sum_{j=1}^{n} \alpha_{1j}^{(n-1)} b_j \\ b_2 & \sum_{j=1}^{n} \alpha_{2j}^{(1)} b_j & \cdots & \sum_{j=1}^{n} \alpha_{2j}^{(n-1)} b_j \\ \cdots & \cdots & \cdots & \cdots \\ b_n & \sum_{j=1}^{n} \alpha_{nj}^{(1)} b_j & \cdots & \sum_{j=1}^{n} \alpha_{nj}^{(n-1)} b_j \end{bmatrix} \neq 0 \tag{4-292}$$

*where the* $\alpha_{ij}^{(k)}$ *are the entries in the matrix* $\mathbf{A}^k$.

**Corollary 4-7** *If the system matrix* $\mathbf{A}$ *is a diagonal matrix with distinct entries, then the system is completely controllable if and only if* $\mathbf{B}$ *has no zero rows.*

These last two corollaries often provide us with a simple test for the complete controllability of a system.

**Example 4-10** Consider the system

$$\begin{bmatrix} \dot{x}_1 \\ \dot{x}_2 \end{bmatrix} = \begin{bmatrix} 0 & 1 \\ -1 & 0 \end{bmatrix} \begin{bmatrix} x_1 \\ x_2 \end{bmatrix} + \begin{bmatrix} 0 \\ 1 \end{bmatrix} u \tag{4-293}$$

Is it controllable? Well, $\mathbf{B}$ is the one-column vector $\mathbf{b} = \begin{bmatrix} 0 \\ 1 \end{bmatrix}$, and $\mathbf{Ab}$ is the vector $\begin{bmatrix} 1 \\ 0 \end{bmatrix}$. It follows from the theorem that this system is controllable.

**Example 4-11** Consider the system

$$\begin{bmatrix} \dot{x}_1 \\ \dot{x}_2 \end{bmatrix} = \begin{bmatrix} 1 & 1 \\ 0 & 0 \end{bmatrix} \begin{bmatrix} x_1 \\ x_2 \end{bmatrix} + \begin{bmatrix} 1 \\ 0 \end{bmatrix} u \tag{4-294}$$

Is it controllable? Well, $\mathbf{B}$ is the one-column vector $\mathbf{b} = \begin{bmatrix} 1 \\ 0 \end{bmatrix}$, and $\mathbf{Ab}$ is the vector $\begin{bmatrix} 1 \\ 0 \end{bmatrix}$. It follows that $\mathbf{b}$, $\mathbf{Ab}$ do not span $R_2$ and, hence, that the system is *not* controllable.

**Example 4-12** Consider the system

$$\begin{bmatrix} \dot{x}_1 \\ \dot{x}_2 \end{bmatrix} = \begin{bmatrix} 0 & 1 \\ 1 & 2 \end{bmatrix} \begin{bmatrix} x_1 \\ x_2 \end{bmatrix} + \begin{bmatrix} -1 & 1 \\ 0 & 2 \end{bmatrix} \begin{bmatrix} u_1 \\ u_2 \end{bmatrix} \tag{4-295}$$

Is it controllable? Well, $\mathbf{B}$ is the $2 \times 2$ matrix $\begin{bmatrix} -1 & 1 \\ 0 & 2 \end{bmatrix}$, and $\mathbf{A}$ is the $2 \times 2$ matrix $\begin{bmatrix} 0 & 1 \\ 1 & 2 \end{bmatrix}$, so that $\mathbf{AB}$ is the $2 \times 2$ matrix $\begin{bmatrix} 0 & 2 \\ -1 & 5 \end{bmatrix}$. Thus, $\mathbf{G} = [\mathbf{B} \mid \mathbf{AB}]$ is the $2 \times 4$ matrix given by

$$\mathbf{G} = \begin{bmatrix} -1 & 1 & 0 & 2 \\ 0 & 2 & -1 & 5 \end{bmatrix}$$

As the rank of $\mathbf{G}$ is 2, the system is controllable. Observe that the vectors $\begin{bmatrix} -1 \\ 0 \end{bmatrix}$ and $\begin{bmatrix} 1 \\ 2 \end{bmatrix}$ are linearly independent. Also observe that any two-column vectors are linearly independent in this case.

**Exercise 4-13** Show that it is not possible to find a solution of Eq. (4-294) starting at the point $\begin{bmatrix} 0 \\ 1 \end{bmatrix}$ and terminating at the point $\begin{bmatrix} 0 \\ 0 \end{bmatrix}$.

**Exercise 4-14** Which of the following systems are completely controllable?

$$(a)\quad \begin{bmatrix} \dot{x}_1 \\ \dot{x}_2 \end{bmatrix} = \begin{bmatrix} 0 & 1 \\ 0 & 1 \end{bmatrix} \begin{bmatrix} x_1 \\ x_2 \end{bmatrix} + \begin{bmatrix} 1 & -1 \\ 0 & 0 \end{bmatrix} \begin{bmatrix} u_1 \\ u_2 \end{bmatrix}$$

$$(b)\quad \begin{bmatrix} \dot{x}_1 \\ \dot{x}_2 \end{bmatrix} = \begin{bmatrix} 0 & 0 \\ 1 & 1 \end{bmatrix} \begin{bmatrix} x_1 \\ x_2 \end{bmatrix} + \begin{bmatrix} 1 & -1 \\ 0 & 0 \end{bmatrix} \begin{bmatrix} u_1 \\ u_2 \end{bmatrix}$$

$$(c)\quad \begin{bmatrix} \dot{x}_1 \\ \dot{x}_2 \end{bmatrix} = \begin{bmatrix} 0 & 1 \\ 1 & 0 \end{bmatrix} \begin{bmatrix} x_1 \\ x_2 \end{bmatrix} + \begin{bmatrix} 1 & 0 \\ 0 & 0 \end{bmatrix} \begin{bmatrix} u_1 \\ u_2 \end{bmatrix}$$

**Exercise 4-15** What conditions on $b_1$, $b_2$, and $b_3$ ensure the complete controllability of the system

$$\begin{bmatrix} \dot{x}_1 \\ \dot{x}_2 \\ \dot{x}_3 \end{bmatrix} = \begin{bmatrix} \lambda & 1 & 0 \\ 0 & \lambda & 1 \\ 0 & 0 & \lambda \end{bmatrix} \begin{bmatrix} x_1 \\ x_2 \\ x_3 \end{bmatrix} + \begin{bmatrix} b_1 \\ b_2 \\ b_3 \end{bmatrix} u$$

**Exercise 4-16** Consider the *nonlinear* system

$$\begin{aligned} \dot{x}_1 &= \alpha_1 x_2 x_3 + u_1 \\ \dot{x}_2 &= \alpha_2 x_1 x_3 + u_2 \\ \dot{x}_3 &= \alpha_3 x_1 x_2 + u_3 \end{aligned}$$

where $\alpha_1 + \alpha_2 + \alpha_3 = 0$. Can you show that this system is controllable?

## 4-17 Observability for Linear Time-invariant Systems

Again let us consider a linear time-invariant system of the form

$$\dot{\mathbf{x}}(t) = \mathbf{A}\mathbf{x}(t) + \mathbf{B}\mathbf{u}(t) \tag{4-296}$$

$$\mathbf{y}(t) = \mathbf{C}\mathbf{x}(t) + \mathbf{D}\mathbf{u}(t) \tag{4-297}$$

where $\mathbf{A}$, $\mathbf{B}$, $\mathbf{C}$, and $\mathbf{D}$ are, respectively, $n \times n$, $n \times m$, $p \times n$, and $p \times m$ matrices, and let us attempt to see what insight into the notion of observability we can come up with.

Now let us suppose that $\mathbf{x}_0$ is observable at $t_0$; then we claim that $\mathbf{x}_0$ is observable at $t = 0$. To see this, we let $\mathbf{u}$ be a given element of $U$, and we set

$$\mathbf{v}(t) = \mathbf{u}(t - t_0) \tag{4-298}$$

Then $\mathbf{v}$ is an element of $U$, and in view of the observability of $\mathbf{x}_0$ at $t_0$, we may conclude that there is a $T \geq 0$ such that $\mathbf{v}_{(t_0, t_0+T]}$ and $\mathbf{y}_{(t_0,t_0+T]} = \hat{\mathbf{g}}(\mathbf{x}_0; \mathbf{v}_{(t_0,t_0+T]})$ determine $\mathbf{x}_0$. However,

$$\begin{aligned} \mathbf{x}(t) &= \boldsymbol{\phi}(t; \mathbf{v}_{(t_0,t]}, \mathbf{x}_0) \\ &= e^{\mathbf{A}(t-t_0)} \left\{ \mathbf{x}_0 + \int_{t_0}^{t} e^{-\mathbf{A}(\tau - t_0)} \mathbf{B}\mathbf{v}(\tau)\, d\tau \right\} \end{aligned} \tag{4-299}$$

for $t$ in $[t_0, t_0 + T]$, implies that

$$\mathbf{y}(t) = \mathbf{C}e^{\mathbf{A}(t-t_0)} \left\{ \mathbf{x}_0 + \int_{t_0}^{t} e^{-\mathbf{A}(\tau-t_0)} \mathbf{B}\mathbf{v}(\tau)\, d\tau \right\} + \mathbf{D}\mathbf{v}(t) \tag{4-300}$$

for $t$ in $[t_0, t_0 + T]$. If we make the change of variable $s = t - t_0$, $\sigma = \tau - t_0$, then Eq. (4-300) becomes, by virtue of Eq. (4-298),

$$\mathbf{y}(s) = \mathbf{C}e^{\mathbf{A}s}\left\{\mathbf{x}_0 + \int_0^T e^{-\mathbf{A}\sigma}\mathbf{B}\mathbf{u}(\sigma)\,d\sigma\right\} + \mathbf{D}\mathbf{u}(s) \tag{4-301}$$

for $s$ in $[0, T]$. It follows that $\mathbf{u}_{(0,T]}$ and $\mathbf{y}_{(0,T]} = \hat{\mathbf{g}}(\mathbf{x}_0; \mathbf{u}_{(0,T]})$ determine $\mathbf{x}_0$. Thus, it is enough to consider observability at $t = 0$, for linear time-invariant systems. In other words, *a linear time-invariant system is observable if and only if it is observable at* $t = 0$.

Now, if $\mathbf{x}_0$ is observable at 0, then knowledge of $\mathbf{y}^0_{(0,T]} = \hat{\mathbf{g}}(\mathbf{x}_0; \mathbf{0})$ for some $T > 0$ is sufficient to determine $\mathbf{x}_0$. We claim that the converse is true; in other words, we claim that if the free motion of the system, starting from $\mathbf{x}_0$ (see Sec. 3-19), is sufficient to determine $\mathbf{x}_0$, then $\mathbf{x}_0$ is observable. To verify this claim, we merely note that if $\mathbf{u}$ is any element of $U$, then

$$\mathbf{y}(t) = \mathbf{C}\boldsymbol{\phi}(t; \mathbf{u}_{(0,t]}, \mathbf{x}_0) + \mathbf{D}\mathbf{u}(t) \tag{4-302}$$

$$= \mathbf{C}e^{\mathbf{A}t}\left\{\mathbf{x}_0 + \int_0^t e^{-\mathbf{A}\tau}\mathbf{B}\mathbf{u}(\tau)\,d\tau\right\} + \mathbf{D}\mathbf{u}(t) \tag{4-303}$$

$$= \mathbf{y}^0(t) + \mathbf{C}\int_0^t e^{-\mathbf{A}\tau}\mathbf{B}\mathbf{u}(\tau)\,d\tau + \mathbf{D}\mathbf{u}(t) \tag{4-304}$$

It is clear from Eq. (4-304) that if $\mathbf{y}^0_{[0,T]}$ determines $\mathbf{x}_0$, then knowledge of $\mathbf{u}_{(0,T]}$ and $\mathbf{y}_{(0,T]}$ will determine $\mathbf{x}_0$.

Now let us suppose that $\mathbf{x}_0$ and $\mathbf{x}_1$ are two observable states (at $t = 0$) and that $\mathbf{x}_0 \neq \mathbf{x}_1$. If we let $\mathbf{y}_0{}^0(t)$ and $\mathbf{y}_1{}^0(t)$ denote the outputs of our system, starting from $\mathbf{x}_0$ and $\mathbf{x}_1$, respectively, and generated by the control $\mathbf{u}(\tau) = \mathbf{0}$ for all $\tau$, then we know that

$$\mathbf{y}_0{}^0(t) = \mathbf{C}e^{\mathbf{A}t}\mathbf{x}_0 \tag{4-305}$$

$$\mathbf{y}_1{}^0(t) = \mathbf{C}e^{\mathbf{A}t}\mathbf{x}_1 \tag{4-306}$$

We claim that there is a value of $t \geq 0$ (say $T$) for which

$$\mathbf{y}_0{}^0(T) \neq \mathbf{y}_1{}^0(T) \tag{4-307}$$

For, if $\mathbf{y}_0{}^0(t) = \mathbf{y}_1{}^0(t)$ for all $t \geq 0$, then knowledge of $\mathbf{y}_0{}^0$ over *any* interval $(0, t]$ could not determine $\mathbf{x}_0$ and, hence, $\mathbf{x}_0$ would not be observable. It follows that if every state is observable (at $t = 0$), then the mapping which associates with each state $\mathbf{x}$, the corresponding (free) output function $\mathbf{y}(t) = \mathbf{C}e^{\mathbf{A}t}\mathbf{x}$, is one-one (see Sec. 2-4).

We have previously noted that the set of controllable states is a subspace of $R_n$. Is this true for the set of observable states? Let us see. Clearly, if $\mathbf{x}_0$ is observable, then $r\mathbf{x}_0$, where $r \in R$, is observable. Now, suppose that $\mathbf{x}_0$ and $\mathbf{x}_1$ are observable; then we claim that $\mathbf{x}_0 + \mathbf{x}_1$ is observable. In order to verify this claim, we shall first have to take a closer look at what observability means for linear time-invariant systems. We have, in particular, the following theorem:

***Theorem 4-6*** *Let* $\mathbf{c}'_i$, $i = 1, 2, \ldots, p$, *denote the n-column vector whose components are the elements of the ith row of* $\mathbf{C}$, *for example,*

$$\mathbf{c}'_1 = \begin{bmatrix} c_{11} \\ c_{12} \\ \cdot \\ \cdot \\ \cdot \\ c_{1n} \end{bmatrix} \qquad \textit{where } \mathbf{C} = (c_{ij}) \tag{4-308}$$

*and let* $\mathbf{w}_{ki}$, $k = 0, 1, \ldots, n-1$ *and* $i = 1, 2, \ldots, p$, *be the n vector, given by*

$$\mathbf{w}_{ki} = \mathbf{A}'^k \mathbf{c}'_i \tag{4-309}$$

*where* $\mathbf{A}'^k$ *is the transpose [see Eq.* (2-45)] *of* $\mathbf{A}^k$. *Let* $\mathbf{x}$ *be an observable state and let* $r_{ki}$ *be the real number, given by*

$$r_{ki} = \langle \mathbf{w}_{ki}, \mathbf{x} \rangle \tag{4-310}$$

*that is, if* $w_{kij}$ *is the jth component of* $\mathbf{w}_{ki}$ *and* $x_j$ *is the jth component of* $\mathbf{x}$, *then*

$$r_{ki} = \sum_{j=1}^{n} w_{kij} x_j \tag{4-311}$$

*Let* $X_1, X_2, \ldots, X_n$ *be n indeterminates and consider the system of equations*

$$r_{ki} = \sum_{j=1}^{n} w_{kij} X_j \qquad k = 0, 1, \ldots, n-1; i = 1, \ldots, p \tag{4-312}$$

*Then* $\mathbf{x}$ *is the unique solution of the system* (4-312). *Conversely, if an equation of the form* (4-312) *(the* $r_{ki}$ *may be different) has a unique solution* $X_j = y_j$, $j = 1, 2, \ldots, n$, *then the vector* $\mathbf{y}$ *with components* $y_j$ *is an observable state of the system.*

PROOF Since $\mathbf{x}$ is observable, there is a $T > 0$ such that $\mathbf{y}(t) = \mathbf{C}e^{\mathbf{A}t}\mathbf{x}$ over $[0, T]$ determines $\mathbf{x}$ *uniquely*. Now the $i$th component $y_i(t)$ of $\mathbf{y}(t)$ is given by

$$y_i(t) = \sum_{k=0}^{n-1} \psi_k(t) \langle \mathbf{A}'^k \mathbf{c}'_i, \mathbf{x} \rangle \tag{4-313}$$

$$= \sum_{k=0}^{n-1} \psi_k(t) \langle \mathbf{w}_{ki}, \mathbf{x} \rangle \tag{4-314}$$

$$= \sum_{k=0}^{n-1} \psi_k(t) r_{ki} \tag{4-315}$$

where
$$e^{\mathbf{A}t} = \sum_{k=0}^{n-1} \psi_k(t) \mathbf{A}^k$$

in view of the fact that

$$e^{\mathbf{A}t} = \sum_{M=0}^{\infty} \frac{t^M}{M!} \mathbf{A}^M$$

(see Sec. 3-21) and the fact that every $n \times n$ matrix satisfies its characteristic equation.† Now $\mathbf{x}$ is (by definition) a solution of Eq. (4-312). If there were another solution $\mathbf{x}'$ of Eq. (4-312), then it would follow from Eq. (4-315) that $\mathbf{y}(t)$ would also be given by

$$\mathbf{y}(t) = \mathbf{C}e^{\mathbf{A}t}\mathbf{x}' \tag{4-316}$$

on $[0, T]$ and, hence, that $\mathbf{y}_{(0,T]}$ could not determine $\mathbf{x}$ uniquely. Thus, $\mathbf{x}$ is the *unique* solution of Eq. (4-312).

We leave it to the reader to verify the truth of the converse part of the theorem.

***Corollary 4-8*** *If* $\mathbf{x}_0$ *and* $\mathbf{x}_1$ *are observable, then* $\mathbf{x}_0 + \mathbf{x}_1$ *is observable.*

**Exercise 4-17** Prove Corollary 4-8. HINT: Let $\langle \mathbf{w}_{ki}, \mathbf{x}_0 \rangle = r_{ki0}$, $\langle \mathbf{w}_{ki}, \mathbf{x}_1 \rangle = r_{ki1}$; then $\langle \mathbf{w}_{ki}, \mathbf{x}_0 + \mathbf{x}_1 \rangle = r_{ki0} + r_{ki1}$. Of what equation are the components of $\mathbf{x}_0 + \mathbf{x}_1$ a solution? Why are they the unique solution of that equation? An indirect argument should help in answering the last question.

***Corollary 4-9*** *If* $\mathbf{x}$ *is observable, then* $\mathbf{C}e^{\mathbf{A}t}\mathbf{x}$ *over any interval* $[0, T]$, $T > 0$, *determines* $\mathbf{x}$ *uniquely.*

***Corollary 4-10*** *The system is completely observable if and only if every vector* $\mathbf{x}$ *is a linear combination of the vectors* $\mathbf{w}_{ki}$ *(see Sec. 2-5).*

***Corollary 4-11*** *The subspace of observable states is the span of the vectors* $\mathbf{w}_{ki}$. *In other words,* $\mathbf{x}_0$ *is observable if and only if* $\mathbf{x}_0$ *is a linear combination of the vectors* $\mathbf{w}_{ki}$.

***Corollary 4-12*** *Let* $\mathbf{H}$ *be the* $n \times nm$ *matrix, defined by the relation*

$$\mathbf{H} \triangleq [\mathbf{C}' \mid \mathbf{A}'\mathbf{C}' \mid \cdots \mid (\mathbf{A}')^{n-1}\mathbf{C}'] \tag{4-317}$$

*Then the system is completely observable if and only if the rank of* $\mathbf{H}$ *is* $n$ *or, equivalently, if and only if there is a set of* $n$ *linearly independent column vectors of* $\mathbf{H}$. *In particular, if* $\mathbf{C}$ *is a* $1 \times n$ *matrix (row vector)* $\mathbf{c}$, *that is, if the output is scalar-valued, then the system is completely observable if and only if the* $n \times n$ *matrix*

$$\mathbf{H} = [\mathbf{c}' \mid \mathbf{A}'\mathbf{c}' \mid \cdots \mid (\mathbf{A}')^{n-1}\mathbf{c}'] \tag{4-318}$$

*is nonsingular (cf. Corollaries 4-5 and 4-6).*

***Corollary 4-13*** *If the system matrix* $\mathbf{A}$ *is a diagonal matrix with distinct entries, then the system is completely observable if and only if* $\mathbf{C}$ *has no zero columns (cf. Corollary 4-7).*

These last two corollaries often provide us with a simple test for the complete observability of a system.

† By the Cayley-Hamilton theorem (Theorem 2-3).

**Example 4-13** Consider the system

$$\begin{bmatrix}\dot{x}_1\\ \dot{x}_2\end{bmatrix} = \begin{bmatrix}0 & 1\\ 0 & 0\end{bmatrix}\begin{bmatrix}x_1\\ x_2\end{bmatrix} + \begin{bmatrix}0\\ 1\end{bmatrix}u$$
$$y = [1 \quad 0]\begin{bmatrix}x_1\\ x_2\end{bmatrix} = x_1 \tag{4-319}$$

Is it observable? Well, $\mathbf{c}' = \begin{bmatrix}1\\ 0\end{bmatrix}$ and $\mathbf{A}'\mathbf{c}' = \begin{bmatrix}0\\ 1\end{bmatrix}$, so that the system is observable.

**Example 4-14** Consider the system

$$\begin{bmatrix}\dot{x}_1\\ \dot{x}_2\end{bmatrix} = \begin{bmatrix}0 & 1\\ 0 & 0\end{bmatrix}\begin{bmatrix}x_1\\ x_2\end{bmatrix} + \begin{bmatrix}0\\ 1\end{bmatrix}u \tag{4-320}$$
$$y = [0 \quad 1]\begin{bmatrix}x_1\\ x_2\end{bmatrix} = x_2 \tag{4-321}$$

Is it observable? Well, $\mathbf{c}' = \begin{bmatrix}0\\ 1\end{bmatrix}$ and $\mathbf{A}'\mathbf{c}' = \begin{bmatrix}0\\ 0\end{bmatrix}$, so that the system is *not* observable. If we consider a unit mass moving in a frictionless environment and if we let $x_1(t)$ denote the displacement, $x_2(t)$ the velocity, and $u(t)$ the applied force, then $x_1$ and $x_2$ are related by Eq. (4-320). The output $y$ being the velocity could arise from the situation in which we observe the system with a Doppler-shift radar that measures only range rate. From the velocity, we can determine the position, then, only to within an unknown constant. This is an implication of the nonobservability of the system.

**Exercise 4-18** Consider the system

$$\begin{aligned}\dot{x}_1(t) &= a_{11}x_1(t) + a_{12}x_2(t) + b_1u(t)\\ \dot{x}_2(t) &= a_{21}x_1(t) + a_{22}x_2(t) + b_2u(t)\\ y(t) &= c_1x_1(t) + c_2x_2(t)\end{aligned}$$

Derive the conditions for noncontrollability and for nonobservability. Is it possible for the system to be both noncontrollable and nonobservable? Illustrate by an appropriate (nontrivial) choice of numerical values. Construct an analog-computer representation of the system. Explain in terms of the analog-computer block diagram noncontrollability, nonobservability, or both.

**Exercise 4-19** Consider the system, with $\omega \neq 0$,

$$\begin{aligned}\dot{x}_1 &= \alpha x_1 + \omega x_2 + b_1u\\ \dot{x}_2 &= -\omega x_1 + \alpha x_2 + b_2u\\ y &= x_1\end{aligned}$$

Show that the system is controllable unless $b_1 = b_2 = 0$ and that the system is observable.

## 4-18 Practical Implications: Regulating the Output

So far, we have paid little attention to the practical implications of the concepts of controllability and observability. We shall remedy this situation by examining some practical aspects of these concepts in this and the following two sections. In this section, we shall consider the problem of regulating the output.

Let us suppose that our dynamical system is linear and time-invariant and that it is given by

$$\dot{\mathbf{x}}(t) = \mathbf{A}\mathbf{x}(t) + \mathbf{B}\mathbf{u}(t) \tag{4-322}$$

$$\mathbf{y}(t) = \mathbf{C}\mathbf{x}(t) + \mathbf{D}\mathbf{u}(t) \tag{4-323}$$

In engineering practice, we are often interested in taking the output $\mathbf{y}(t)$ to $\mathbf{0}$ and holding it there in the absence of subsequent disturbances; thus, when we shall speak of *regulating the output*, we shall imply that we are interested:

1. In driving the output to zero
2. In determining the control which will maintain the output at zero thereafter

Let the initial time be $t_0 = 0$. Suppose that we have found a control $\mathbf{u}_{(0,T]}$ such that

$$\mathbf{y}(T) = \mathbf{0} \tag{4-324}$$

Suppose that the system is *observable;* this implies that we can compute the initial state $\mathbf{x}_0$ at $t = 0$. Since Eq. (4-322) is linear and time-invariant, Eqs. (4-324) and (4-323) yield the relation

$$\mathbf{C}\left\{e^{\mathbf{A}T}\mathbf{x}_0 + e^{\mathbf{A}T}\int_0^T e^{-\mathbf{A}\tau}\,\mathbf{B}\mathbf{u}(\tau)\,d\tau\right\} + \mathbf{D}\mathbf{u}(T) = \mathbf{0} \tag{4-325}$$

Now suppose that the system is also *controllable:* this means that we can find a $\mathbf{u}_{(0,T]}$ such that $\mathbf{x}(T) = \mathbf{0}$. Moreover, if

$$\mathbf{x}(t) = \mathbf{0} \quad \text{and} \quad \mathbf{u}(t) = \mathbf{0} \quad \text{for all } t > T \tag{4-326}$$

then

$$\mathbf{y}(t) = \mathbf{0} \quad \text{for all } t \geq T \tag{4-327}$$

Therefore, *if the system is completely controllable and completely observable, then we can regulate the output.*

We note that controllability and observability provide us with *sufficient* conditions for the regulation of the output; however, they certainly do not represent *necessary* conditions. To see this, we observe that, in order to regulate the output, it suffices to find a control $\hat{\mathbf{u}}(t)$ for $t > T$ such that the solution of $\dot{\mathbf{x}}(t) = \mathbf{A}\mathbf{x}(t) + \mathbf{B}\hat{\mathbf{u}}(t)$, denoted by $\hat{\mathbf{x}}(t)$ for $t > T$, satisfies the equation

$$\mathbf{C}\hat{\mathbf{x}}(t) + \mathbf{D}\hat{\mathbf{u}}(t) = \mathbf{0} \quad \text{for all } t > T \tag{4-328}$$

Let us denote by $\hat{H}$ the set of all $\hat{\mathbf{x}}(T)$ for which such a control $\hat{\mathbf{u}}(t)$ can be found. If the system is controllable, then we are guaranteed† that we can force the system from $\mathbf{x}_0$ to $\hat{H}$. However, the system may not be controllable, and yet it is conceivable that we can find a control that will force $\mathbf{x}_0$ to $\hat{H}$.

† Because controllability implies that any state is accessible from any other state.

**Example 4-15** To illustrate this point, we consider the simple second-order system

$$\begin{bmatrix} \dot{x}_1(t) \\ \dot{x}_2(t) \end{bmatrix} = \begin{bmatrix} 0 & 1 \\ 0 & -1 \end{bmatrix} \begin{bmatrix} x_1(t) \\ x_2(t) \end{bmatrix} + \begin{bmatrix} 1 \\ 0 \end{bmatrix} u(t) \tag{4-329}$$

$$y(t) = x_1(t) \tag{4-330}$$

It is easy to see that the system is observable but not controllable. Let $\xi_1$ and $\xi_2$ denote the initial values of the state variables $x_1(t)$ and $x_2(t)$. Equation (4-329) is easily integrated to yield

$$\begin{aligned} x_2(t) &= \xi_2 e^{-t} \\ x_1(t) &= \xi_1 - \xi_2 e^{-t} + \xi_2 + \int_0^t u(\tau)\, d\tau \end{aligned} \tag{4-331}$$

Since the system is observable, we can compute $\xi_1$ and $\xi_2$ from the output $y(t)$. Let $T$ be a given time. We wish to find a control such that $y(T) = 0$. Since $y(t) = x_1(t)$, we must make $x_1(T) = 0$; clearly, if $x_1(t) = 0$ for all $t \geq T$, then $y(t) = 0$ for all $t \geq T$. These considerations define the set $\hat{H}$ in the $x_1x_2$ state plane. Clearly, the set $\hat{H}$ is the entire $x_2$ axis.

Let $u(t) = k$ be a constant; then at $t = T$ we must have, from Eq. (4-331),

$$0 = x_1(T) = \xi_1 - \xi_2 e^{-T} + \xi_2 + kT \tag{4-332}$$

which implies that the constant control

$$u(t) = k = \frac{\xi_2 e^{-T} - \xi_1 - \xi_2}{T} \qquad \text{for } 0 < t \leq T \tag{4-333}$$

drives the output to zero at $t = T$. Now the problem is to hold the output at zero or (equivalently) to hold the state on the $x_2$ axis. We observe that, for all $t > T$, the control

$$u(t) = -x_2(t) = -\xi_2 e^{-t} \qquad t > T \tag{4-334}$$

yields [from Eq. (4-329)] $\dot{x}_1(t) = 0$ for all $t > T$ and, hence, $x_1(t) = y(t) = 0$ for all $t > T$. Thus, although the system is not controllable, we can regulate the output. We shall examine an analogous situation in Sec. 4-20, except that we shall require that the control be bounded.

Now suppose that we consider the system with Eq. (4-329) but with output $y(t)$ given by

$$y(t) = x_2(t) \tag{4-335}$$

Then the system (4-329) and (4-335) is neither controllable nor observable. It is easy to see that, in this case, we cannot regulate the output.

Let us now examine some other implications of controllability and observability. We shall see that quite often the fact that the system is not observable or not controllable means that there are more state variables than necessary. To illustrate this point, we consider the physical system shown in Fig. 4-12. The constants $k_1$, $k_2$, $k_3$, and $k_4$ are gains, and $x_1(t)$, $x_2(t)$, $x_3(t)$, and $x_4(t)$ are the outputs of the four integrators. The output $y(t)$ is the sum of the $x_i$ variables, $i = 1, 2, 3, 4$.

It is easy to establish that the vector

$$\mathbf{x}(t) = \begin{bmatrix} x_1(t) \\ x_2(t) \\ x_3(t) \\ x_4(t) \end{bmatrix} \tag{4-336}$$

qualifies as a state vector, and so the physical system of Fig. 4-12 can be represented by the equations

$$\dot{\mathbf{x}}(t) = \mathbf{0x}(t) + \begin{bmatrix} k_1 \\ k_2 \\ k_3 \\ k_4 \end{bmatrix} u(t) \tag{4-337}$$

$$y(t) = [1 \quad 1 \quad 1 \quad 1]\mathbf{x}(t) \tag{4-338}$$

It is left to the reader to verify that the system described by Eqs. (4-337)

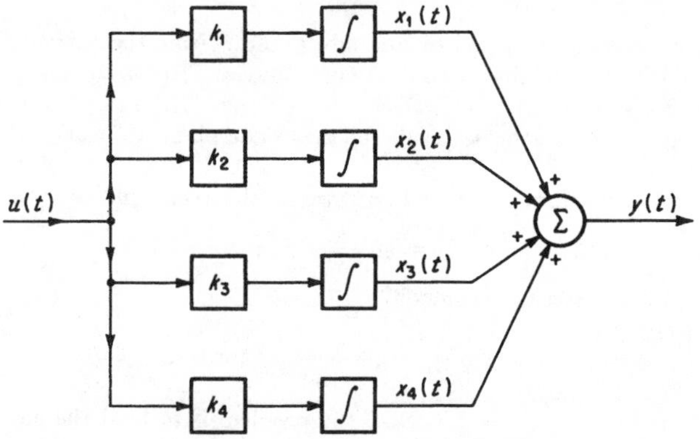


**Fig. 4-12** A physical system simulated on an analog computer.

and (4-338) is neither controllable nor observable. Yet it is easy to establish that the output satisfies the differential equation

$$\dot{y}(t) = ku(t) \tag{4-339}$$

where

$$k = k_1 + k_2 + k_3 + k_4 \tag{4-340}$$

If $k \neq 0$, then it is easy to see that the system

$$\dot{z}(t) = ku(t) \tag{4-341}$$

$$y(t) = z(t) \tag{4-342}$$

is both controllable and observable and, hence, that the output can be regulated and that $z(t) = y(t)$ qualifies as a state.

Clearly, from the input-output point of view, the system (4-337) and (4-338) and the system (4-341) and (4-342) are equivalent. This simple example points out that the choice of the state of the system and especially the dimensionality of the state space are unimportant from the input-output point of view. If, however, the dimensionality of the state space is larger than necessary, then we usually end up with an uncontrollable and unobservable system.

Reference Z-1 contains a rigorous exposition of these ideas; Ref. B-20 provides useful relations for systems treated from the output point of view, using the concept of *output controllability*.

## 4-19 Practical Implications: The Effect of Canceling a Pole with a Zero

We shall show, in this section, that the conventional pole-zero cancellation techniques employed in servomechanism design result in systems which are uncontrollable (i.e., not completely controllable) in the sense of our definition. We consider a system with distinct eigenvalues whose state variables $x_i(t)$ satisfy the differential equations [see Eq. (4-199)]

$$\dot{x}_i(t) = \lambda_i x_i(t) + v_i u(t) \qquad i = 1, 2, \ldots, n \tag{4-343}$$

where $v_i$ is the residue of the transfer function $H(s)$ given by Eq. (4-170) at the $i$th pole.

If we suppose that, in the transfer function $H(s)$ given by Eq. (4-170),

$$\sigma_1 = s_1 \tag{4-344}$$

that is, that there are both a pole and zero at $s = s_1 = \sigma_1$, then the equation for the state variable $x_1(t)$ is

$$\dot{x}_1(t) = s_1 x_1(t) + v_1 u(t) \tag{4-345}$$

where $v_1 = \rho_1$, the residue of $H(s)$ at $s_1 = \lambda_1$. But, by virtue of Eq. (4-202),

$$v_1 = \rho_1 = \frac{b_m(\lambda_1 - \sigma_1)(\lambda_1 - \sigma_2) \cdots (\lambda_1 - \sigma_m)}{(\lambda_1 - \lambda_2)(\lambda_1 - \lambda_3) \cdots (\lambda_1 - \lambda_n)} = 0 \tag{4-346}$$

since $\lambda_1 = s_1 = \sigma_1$. Thus,

$$\dot{x}_i(t) = s_1 x_1(t) \tag{4-347}$$

and so the system is not completely controllable in view of Corollary 4-7.

## 4-20 Practical Implications: An Example

From a practical point of view, controlling a system means that the output can be made to respond in a desired way. With this observation in mind, let us examine the effects of noncontrollability in the very simple example that follows.

Consider the system with (scalar) output $y(t)$ and (scalar) input $u(t)$ which are related by the second-order linear differential equation

$$\ddot{y}(t) + 2\dot{y}(t) + y(t) = \dot{u}(t) + u(t) \tag{4-348}$$

The transfer function $H(s)$ of this system is given by

$$\frac{y(s)}{u(s)} = H(s) = \frac{s+1}{s^2 + 2s + 1} = \frac{s+1}{(s+1)^2} \tag{4-349}$$

Thus, there are two poles and one zero, all at $s = -1$. Following the procedure of Sec. 4-10, we define the state variables $z_1(t)$ and $z_2(t)$ by setting

$$\begin{aligned} z_1(t) &= y(t) \\ z_2(t) &= \dot{y}(t) - u(t) \end{aligned} \tag{4-350}$$

Then

$$\begin{bmatrix} \dot{z}_1(t) \\ \dot{z}_2(t) \end{bmatrix} = \begin{bmatrix} 0 & 1 \\ -1 & -2 \end{bmatrix} \begin{bmatrix} z_1(t) \\ z_2(t) \end{bmatrix} + \begin{bmatrix} 1 \\ -1 \end{bmatrix} u(t) \tag{4-351}$$

and we claim that this system is not controllable. To verify this claim, we form the matrix

$$\mathbf{G} = [\mathbf{b} \mid \mathbf{Ab}] = \begin{bmatrix} 1 & -1 \\ -1 & 1 \end{bmatrix} \tag{4-352}$$

note that its determinant is zero, and apply Corollary 4-6.

Let us now apply the similarity transformation to Eq. (4-351), which puts the system matrix $\mathbf{A}$ into its Jordan canonical form (see Sec. 2-10). Let

$$\mathbf{J} = \begin{bmatrix} -1 & 1 \\ 0 & -1 \end{bmatrix} \tag{4-353}$$

and let

$$\mathbf{P} = \begin{bmatrix} 1 & 0 \\ -1 & 1 \end{bmatrix} \tag{4-354}$$

Then $\mathbf{J}$ is the desired canonical form, and we let

$$\mathbf{x} = \mathbf{P}^{-1}\mathbf{z} \tag{4-355}$$

that is,

$$\begin{aligned} x_1(t) &= z_1(t) = y(t) \\ x_2(t) &= z_1(t) + z_2(t) = y(t) + \dot{y}(t) - u(t) \end{aligned} \tag{4-356}$$

Then the new state variables $x_1(t)$ and $x_2(t)$ satisfy the equation

$$\begin{bmatrix} \dot{x}_1(t) \\ \dot{x}_2(t) \end{bmatrix} = \begin{bmatrix} -1 & 1 \\ 0 & -1 \end{bmatrix} \begin{bmatrix} x_1(t) \\ x_2(t) \end{bmatrix} + \begin{bmatrix} 1 \\ 0 \end{bmatrix} u(t) \tag{4-357}$$

If we set $x_1(0) = \xi_1$ and $x_2(0) = \xi_2$, then the solution of Eq. (4-357) is given by

$$x_1(t) = \xi_1 e^{-t} + \xi_2 t e^{-t} + e^{-t} \int_0^t e^{\tau} u(\tau)\, d\tau \tag{4-358}$$

$$x_2(t) = \xi_2 e^{-t} \tag{4-359}$$

Let us now pose the following problem: *Given initial values $\xi_1$ and $\xi_2$, determine the control which will drive the output $y(t)$ to zero and cause it to remain at zero thereafter.*

So we want to find a control such that, for some $T$,

$$\begin{aligned} y(t) &= 0 \qquad \text{for all } t \geq T \\ \dot{y}(t) &= 0 \qquad \text{for all } t > T \end{aligned} \tag{4-360}$$

This requirement on the output is equivalent to the requirement on the state variables $x_1(t)$ and $x_2(t)$ of Eq. (4-356) that

$$\begin{aligned} x_1(t) &= 0 && \text{for all } t \geq T \\ x_2(t) &= -u(t) && \text{for all } t > T \end{aligned} \tag{4-361}$$

We deduce from Eq. (4-358) that any control $u(t)$ which satisfies the equation

$$-\int_0^T e^{\tau} u(\tau)\, d\tau = \xi_1 + \xi_2 T \tag{4-362}$$

will indeed drive $x_1$ to zero at time $T$, and we deduce from Eq. (4-359) that application of the control

$$u(t) = -\xi_2 e^{-t} \qquad \text{for all } t > T \tag{4-363}$$

will ensure that $\dot{y}(t) = 0$ for all $t > T$. Thus, although the system is not controllable, the objective of driving the output and its time derivative to zero can be attained.

Now let us impose a constraint on the magnitude of the control. In other words, let us suppose that

$$|u(t)| \leq M \qquad \text{for all } t \tag{4-364}$$

(cf. Example 4-6). Again we pose the question of whether or not we can find a control, now satisfying the constraint (4-364), which will drive the output $y(t)$ and its derivative $\dot{y}(t)$ to zero. If we assume that $\xi_1$, $\xi_2$, and $T$ are such that there is a control $u(t)$ with $|u(t)| \leq M$ such that

$$-\int_0^T e^{t} u(t)\, dt = \xi_1 + \xi_2 T \tag{4-365}$$

then $x_1(T)$ will be zero, but it still may be true that $\xi_2$ is such that

$$|\xi_2 e^{-t}| > M \qquad \text{for } T < t \leq T' \tag{4-366}$$

and, hence, that there does not exist a control satisfying the constraint (4-364) and Eq. (4-363). In other words, for such a $\xi_1$ and $\xi_2$, the output and its derivative cannot be forced to zero in a specified time.† Similar problems will be discussed in Sec. 7-13.

## 4-21 Normal Linear Time-invariant Systems‡

We shall discuss the notion of normality in this section. Loosely speaking, a normal system is a system which is controllable with respect to each component $u_1(t), u_2(t), \ldots, u_m(t)$ of the control $\mathbf{u}(t)$. We confine our discussion to the case of a linear time-invariant system (although the concept

† Since $u(t)$ is bounded, there is a time $t^*$ (depending on $\xi_1$ and $\xi_2$) which $T$ must exceed in order for there to be any possibility of solving the problem; the time $T$ may be much larger than $t^*$.

‡ The concept of normality was introduced in Ref. L-4 with regard to time-optimal systems.

of normality can be defined in a more general context). We also remark that the concept of normality will be used extensively in Chap. 6.

Now let us consider a linear time-invariant system with state equation

$$\dot{\mathbf{x}}(t) = \mathbf{A}\mathbf{x}(t) + \mathbf{B}\mathbf{u}(t) \tag{4-367}$$

where $\mathbf{A}$ and $\mathbf{B}$ are, respectively, $n \times n$ and $n \times m$ matrices of real numbers. We let $\mathbf{b}_\beta$, for $\beta = 1, 2, \ldots, m$, denote the $n$ vector which forms the $\beta$th column of $\mathbf{B}$, that is,

$$\mathbf{b}_\beta = \begin{bmatrix} b_{1\beta} \\ b_{2\beta} \\ \cdot \\ \cdot \\ \cdot \\ b_{n\beta} \end{bmatrix} \tag{4-368}$$

where $\mathbf{B} = (b_{i\beta})$. In other words,

$$\mathbf{B} = \left[\begin{array}{c:c:c:c} \uparrow & \uparrow & & \uparrow \\ \mathbf{b}_1 & \mathbf{b}_2 & \cdots & \mathbf{b}_m \\ \downarrow & \downarrow & & \downarrow \end{array}\right] \tag{4-369}$$

We now have:

***Definition 4-15 Normality*** *We say that the system* (4-367) *is normal if each of the systems*

$$\begin{aligned} \dot{\mathbf{x}}(t) &= \mathbf{A}\mathbf{x}(t) + \mathbf{b}_1 u(t) \\ \dot{\mathbf{x}}(t) &= \mathbf{A}\mathbf{x}(t) + \mathbf{b}_2 u(t) \\ &\cdots\cdots\cdots \\ \dot{\mathbf{x}}(t) &= \mathbf{A}\mathbf{x}(t) + \mathbf{b}_m u(t) \end{aligned} \tag{4-370}$$

*is completely controllable* (*see Definition* 4-13), *where the* $\mathbf{b}_\beta$ *are the columns of the matrix* $\mathbf{B}$ *of Eq.* (4-367) [*that is, the* $\mathbf{b}_\beta$ *are given by Eq.* (4-368)].

What is the physical significance of this definition, and how does this definition relate to the controllability of our system $\dot{\mathbf{x}}(t) = \mathbf{A}\mathbf{x}(t) + \mathbf{B}\mathbf{u}(t)$? In order to answer these questions, let us represent our system in the manner illustrated in Fig. 4-13. In Fig. 4-13, each component $u_1(t), u_2(t), \ldots, u_m(t)$ of $\mathbf{u}(t)$ is being operated upon by the corresponding *gain vector* $\mathbf{b}_1, \mathbf{b}_2, \ldots, \mathbf{b}_m$ (respectively) to produce the *vector signals* $\mathbf{b}_1 u_1(t), \mathbf{b}_2 u_2(t), \ldots, \mathbf{b}_m u_m(t)$. These vector signals are added to produce the vector signal $\mathbf{B}\mathbf{u}(t)$. This is simply a graphical representation of the equation

$$\begin{aligned} \mathbf{B}\mathbf{u}(t) &= \left[\begin{array}{c:c:c:c} \uparrow & \uparrow & & \uparrow \\ \mathbf{b}_1 & \mathbf{b}_2 & \cdots & \mathbf{b}_m \\ \downarrow & \downarrow & & \downarrow \end{array}\right] \begin{bmatrix} u_1(t) \\ u_2(t) \\ \cdot \\ \cdot \\ \cdot \\ u_m(t) \end{bmatrix} \\ &= \mathbf{b}_1 u_1(t) + \mathbf{b}_2 u_2(t) + \cdots + \mathbf{b}_m u_m(t) \end{aligned} \tag{4-371}$$

Now we also have indicated $m$ (normally closed) switches, ①, ②, . . . ⓜ, in Fig. 4-13. Each switch, when opened, grounds the corresponding component of $\mathbf{u}(t)$ independently of the other components of $\mathbf{u}(t)$. In other words, if (say) the switch Ⓑ is opened, then $u_\beta(t) = 0$ for all $t$.

Now, if the system $\dot{\mathbf{x}}(t) = \mathbf{A}\mathbf{x}(t) + \mathbf{B}\mathbf{u}(t)$ is completely controllable, then, assuming that all the switches in Fig. 4-13 are *closed*, we can find scalar signals $u_{1(0,T]}, u_{2(0,T]}, \ldots, u_{m(0,T]}$ which reduce any given initial state $\mathbf{x}_0$ to $\mathbf{0}$ at time $T$ (which may depend on $\mathbf{x}_0$).

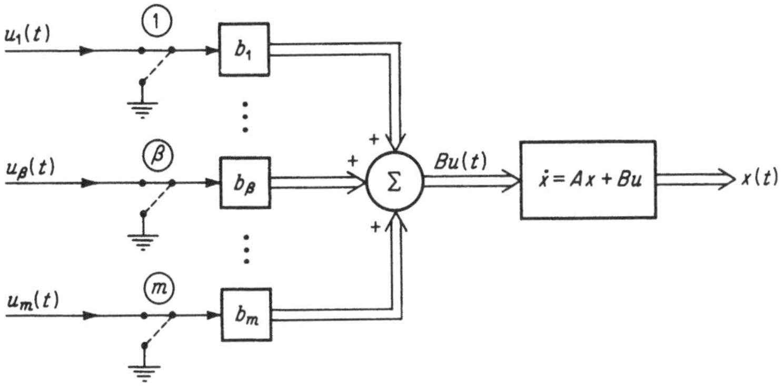


**Fig. 4-13** A setup for the determination of normality of the system $\dot{\mathbf{x}}(t) = \mathbf{A}\mathbf{x}(t) + \mathbf{B}\mathbf{u}(t)$.

Suppose, however, that we open all the switches except the first. Then we have

$$\mathbf{b}_2 u_2(t) = \mathbf{b}_3 u_3(t) = \cdots = \mathbf{b}_m u_m(t) = \mathbf{0} \tag{4-372}$$

and so we can only influence the motion of the system with the scalar control $u_1(t)$. In other words, our system is now described by the equation

$$\dot{x}(t) = \mathbf{A}\mathbf{x}(t) + \mathbf{b}_1 u_1(t) \tag{4-373}$$

since Eq. (4-372) holds. Let us ask the following question:

*Given the system* (4-373) *and the initial state* $\mathbf{x}_0$ *at* $t = 0$, *is there a piecewise continuous scalar function* $u_{1(0,T]}$ *which reduces the initial state* $\mathbf{x}_0$ *to* $\mathbf{0}$*?*

In view of Corollary 4-6, we can see that the answer to this question will be *yes if and only if the system* (4-373) *is controllable or, equivalently, if and only if the* $n \times n$ *matrix* $\mathbf{G}_1$, *given by*

$$\mathbf{G}_1 = [\mathbf{b}_1 \mid \mathbf{A}\mathbf{b}_1 \mid \cdots \mid \mathbf{A}^{n-1}\mathbf{b}_1] \tag{4-374}$$

*is nonsingular.* In a similar way, if we suppose that we open every switch except the $\beta$th, then our system (4-367) will reduce to the system

$$\dot{\mathbf{x}}(t) = \mathbf{A}\mathbf{x}(t) + \mathbf{b}_\beta u_\beta(t) \tag{4-375}$$

and we shall be able to find a control $u_{\beta(0,T]}$ which reduces $\mathbf{x}_0$ to $\mathbf{0}$ if and only if the $n \times n$ matrix

$$\mathbf{G}_\beta = [\mathbf{b}_\beta \mid \mathbf{A}\mathbf{b}_\beta \mid \cdots \mid \mathbf{A}^{n-1}\mathbf{b}_\beta] \tag{4-376}$$

is nonsingular.

*To sum up:* We can see that the term *normality* means that our system is controllable with respect to each component of the control and, also, that normality implies that our system is completely controllable; moreover, we have:

***Theorem 4-7*** *The system* $\dot{\mathbf{x}}(t) = \mathbf{A}\mathbf{x}(t) + \mathbf{B}\mathbf{u}(t)$ *is normal if and only if*

$$\det \mathbf{G}_\beta \neq 0 \qquad \textit{for all } \beta = 1, 2, \ldots, m \tag{4-377}$$

*where* $\mathbf{G}_\beta$ *is the* $n \times n$ *matrix*

$$\mathbf{G}_\beta = [\mathbf{b}_\beta \mid \mathbf{A}\mathbf{b}_\beta \mid \cdots \mid \mathbf{A}^{n-1}\mathbf{b}_\beta] \tag{4-378}$$

*and* $\mathbf{b}_\beta$ *is the* $\beta$th *column of* $\mathbf{B}$.

Finally, we note, as is illustrated in Example 4-16, that not every controllable system is normal.

**Example 4-16** Consider the system

$$\begin{bmatrix} \dot{x}_1(t) \\ \dot{x}_2(t) \end{bmatrix} = \begin{bmatrix} -3 & 0 \\ 0 & -5 \end{bmatrix} \begin{bmatrix} x_1(t) \\ x_2(t) \end{bmatrix} + \begin{bmatrix} 1 & a \\ 1 & 2 \end{bmatrix} \begin{bmatrix} u_1(t) \\ u_2(t) \end{bmatrix} \tag{4-379}$$

where $a$ is some constant. First of all, the system (4-379) is controllable for all values of $a$ because the matrix G, given by

$$\mathbf{G} = \begin{bmatrix} 1 & a & -3 & -3a \\ 1 & 2 & -5 & -10 \end{bmatrix} \tag{4-380}$$

contains the two vectors $\begin{bmatrix} 1 \\ 1 \end{bmatrix}$ and $\begin{bmatrix} -3 \\ -5 \end{bmatrix}$, which are linearly independent. To test for normality, we form the two matrices $\mathbf{G}_1$ and $\mathbf{G}_2$ [see Eq. (4-378)] with

$$\mathbf{G}_1 = \begin{bmatrix} 1 & -3 \\ 1 & -5 \end{bmatrix} \qquad \mathbf{G}_2 = \begin{bmatrix} a & -3a \\ 2 & -10 \end{bmatrix} \tag{4-381}$$

But $\mathbf{G}_1$ is always nonsingular, while $\mathbf{G}_2$ is nonsingular if and only if $a \neq 0$. Thus we conclude that the system (4-379) is normal provided that $a \neq 0$. However, if $a = 0$, then the system (4-379) is *not normal but is still controllable.* It is easy to see that if $a = 0$ and if $u_1(t) = 0$, then the system (4-379) reduces to the system

$$\begin{aligned} \dot{x}_1(t) &= -3x_1(t) \\ \dot{x}_2(t) &= -5x_2(t) + 2u_2(t) \end{aligned} \tag{4-382}$$

and there is no way of changing $x_1(t)$ by means of the control $u_2(t)$ alone.

# Chapter 5

# CONDITIONS FOR OPTIMALITY: THE MINIMUM PRINCIPLE AND THE HAMILTON-JACOBI EQUATION

## 5-1 Introduction

Loosely speaking, we may say that the optimal-control problem is to determine the control $\mathbf{u}^*$ which minimizes a given performance functional $J(\mathbf{u})$.† Our goal in this chapter will be the development of some very general necessary conditions for the optimality of a control $\mathbf{u}^*$. In other words, we shall suppose that we have a solution $\mathbf{u}^*$ to the optimal-control problem, and then we shall attempt to draw some inferences from this. These inferences will provide us with a way of testing whether or not a given control can be a candidate for the solution of our problem. We shall see, in subsequent chapters of the book, that the general conditions developed here are indeed of great value in the solution of practical control problems.

Since the optimal-control problem is a minimization problem, we shall be concerned with determining and characterizing minima. Consequently, the first two parts of this chapter will be devoted to an examination of minimization problems with a view toward obtaining necessary conditions for a minimum. We shall start by studying ordinary minima, i.e., the minima of real-valued functions defined on a Euclidean space $R_n$; then we shall turn our attention to the minima of functionals and, finally, to the minimum principle of Pontryagin.

The results on minimization will be used to obtain a classical (calculus of variations) minimum principle for the free-end-point control problem; this classical minimum principle will be used to motivate our development of the celebrated minimum principle of Pontryagin, which provides us with our desired necessary conditions for optimality. Finally, we shall conclude the chapter with a discussion of sufficiency conditions based upon the Hamilton-Jacobi equation and the principle of optimality.

Before we begin our detailed study of minimization problems, let us briefly comment on the role that necessary conditions play in actually determining

† See Definition 4-9.

solutions. Application of the necessary conditions usually leaves us with a limited number of candidates for a solution to our problem; it is then usually possible to use a process of elimination to determine which of these candidates is the sought-for solution. In fact, it is sometimes the case that there is a *unique* element satisfying the necessary conditions; this element must therefore be the desired solution (provided that a solution exists). These comments will take on a greater significance as the reader progresses through the book.

Suitable references for this chapter are B-5, B-8, B-11, B-15, B-16, B-18, B-22, C-1, C-4 to C-6, D-1, D-3, D-7, D-8, E-2, F-19, G-1, G-2, H-1 to H-3, H-5, K-8, K-11, K-12, K-27, K-28, L-10, M-2, N-5, N-7, P-3, P-5, R-6, R-8, R-9, T-3, W-2, and W-3.

## 5-2 Ordinary Minima†

We shall begin our study of the necessary conditions for optimality with an examination of the theory of ordinary minima. This theory is concerned with the problem of determining the point or points at which some real-valued function $f$, defined on a region in a Euclidean space $R_n$, has a minimum. Here we shall precisely define the notions of minimum and absolute minimum for such a function $f$ and shall describe certain necessary and sufficient conditions for the determination of these minima. In the next section, we shall see what happens when we impose some additional constraints on our problem.

First of all, let us suppose that $f$ is a real-valued function which is defined on all of $R$; then we have:

***Definition 5-1*** *A point $x^*$ in $R$ is said to be a (local) minimum of $f$ if there is an $\epsilon > 0$ such that $|x - x^*| < \epsilon$ implies that*

$$f(x^*) \leq f(x) \tag{5-1}$$

*In other words, $x^*$ is a (local) minimum of $f$ if $f(x^*)$ is no greater than $f(x)$ for all $x$ in a neighborhood of $x^*$ (that is, close to $x^*$).*

We also have:

***Definition 5-2*** *A point $x^*$ in $R$ is said to be an absolute minimum of $f$ if*

$$f(x^*) \leq f(x) \tag{5-2}$$

*for all $x$ in $R$.*

It is clear that an absolute minimum of $f$ is also a (local) minimum of $f$ and that a (local) minimum need *not* be an absolute minimum, as is illustrated in Fig. 5-1.

In practice, we are often interested in real-valued functions $f$ which are defined not necessarily on all of $R$ but rather on an interval [see Eqs. (3-7)]

† See Refs. E-2 and H-5.

contained in $R$. The notions of minimum and absolute minimum are entirely analogous to those given in Definitions 5-1 and 5-2 and are left to the reader to supply.

Bearing these definitions in mind, let us now suppose that $f$ is a real-valued function which is defined and *continuous* on the *closed* interval $[a, b]$. In view of Theorem 3-7, we observe that $f$ *has an absolute minimum on* $[a, b]$. How can we find it?

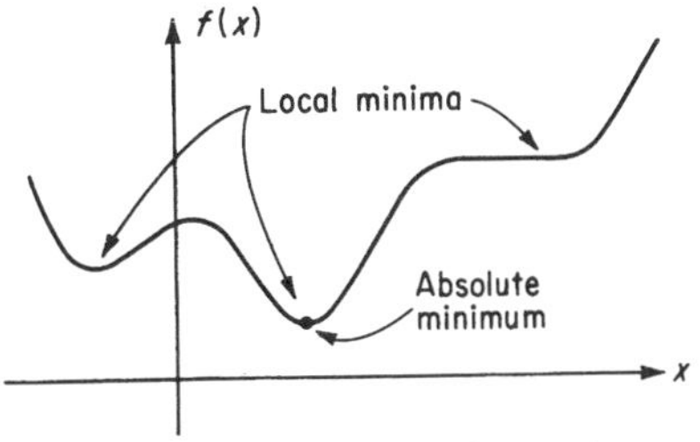


**Fig. 5-1** Types of minima.

We may first note that if $x$ is an interior point of $[a, b]$ [that is, $x$ is in the open interval $(a, b)$] and if the derivative $f'(x)$ exists and is *not* zero, then $x$ *cannot* be a minimum of $f$. This leads us to the necessary condition:

NC1 *If $x^*$ is an interior point of $[a, b]$, if $f'(x^*)$ exists, and if $x^*$ is a minimum of $f$, then*

$$f'(x^*) = \frac{df}{dx}(x^*) = 0 \tag{5-3}$$

A point $x$ at which $f'(x)$ is zero is called an *extremum* of $f$. We may thus conclude that the potential candidates for the absolute minimum of $f$ are the extrema of $f$, the end points $a$ and $b$ of the interval, and the points of $(a, b)$ at which the derivative of $f$ does not exist.

We shall now suppose that $f'(x)$ is a piecewise continuous function† which has only a *finite* number of discontinuities on $[a, b]$. We then have the necessary condition:

NC2 *If $x^*$ is an interior point of $[a, b]$ which is a minimum of $f$, then either*

$$f'(x^*) = 0 \tag{5-4}$$

*or*

$$\lim_{x \to x^*+} f'(x) \geq 0 \quad \textit{and} \quad \lim_{x \to x^*-} f'(x) \leq 0‡ \tag{5-4a}$$

*If $a$ is a minimum of $f$, then*

$$\lim_{x \to a+} f'(x) \geq 0 \tag{5-5}$$

*and if $b$ is a minimum of $f$, then*

$$\lim_{x \to b-} f'(x) \leq 0 \tag{5-6}$$

Various examples of this condition are illustrated in Fig. 5-2.

† This means that $f'(x+)$ and $f'(x-)$ exist for all $x$ but not necessarily $f'(x)$.
‡ See Definition 3-24.

It will be instructive to briefly examine the type of reasoning which leads to necessary condition NC2. Supposing that $x^*$ is an interior point of $[a, b]$ and that $x^*$ is a minimum of $f$, we note that, for $t$ sufficiently small, say for

$$0 \leq |t| < \epsilon \tag{5-7}$$

we have

$$f(x^* + t) \geq f(x^*) \tag{5-8}$$

Let us call $x^* + t$ a *perturbation* of $x^*$. If $f'(x^*)$ exists, then we may, in view of the definition of $f'$, write

$$f(x^* + t) - f(x^*) = tf'(x^*) + o(t) \tag{5-9}$$

where $o(t)$ indicates a correction term which has the property that

$$\lim_{t \to 0} \frac{o(t)}{t} = 0 \tag{5-10}$$

We may thus conclude that $f(x^* + t)$ is *approximately* equal to $f(x^*) + tf'(x^*)$. Moreover, if $f'(x^*)$ were (say) bigger than zero, then, for $t < 0$,

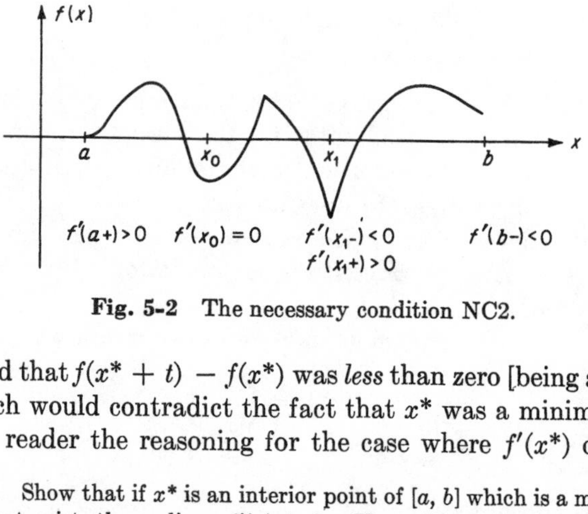


**Fig. 5-2** The necessary condition NC2.

we would find that $f(x^* + t) - f(x^*)$ was *less* than zero [being approximately $tf'(x^*)$], which would contradict the fact that $x^*$ was a minimum of $f$. We leave to the reader the reasoning for the case where $f'(x^*)$ does not exist.

**Exercise 5-1** Show that if $x^*$ is an interior point of $[a, b]$ which is a minimum of $f$ and if $f'(x^*)$ does not exist, then $\lim_{x \to x^*+} f'(x) \geq 0$. Hint: If $t > 0$, then $f(x^* + t) \geq f(x^*)$ implies that $\{f(x^* + t) - f(x^*)\}/t \geq 0$.

**Exercise 5-2** Show that the $o(t)$ term in Eq. (5-9) does not affect our argument that $f'(x^*) > 0$ leads to a contradiction. Hint: If $f'(x^*) > 0$, then there is an $\epsilon > 0$ with $f'(x^*) > \epsilon > 0$. Pick $\delta > 0$, so that $|t| < \delta$ implies that $|o(t)/t| < \epsilon$ and then, for $-\delta < t < 0$, consider the sign of $[f(x^* + t) - f(x^*)]/t$.

We note that the bases of the reasoning used to obtain NC2 are:

1. The fact that small changes (or perturbations) in $x$ cause correspondingly small changes in $f(x)$

2. The fact that, for small changes in $x$, the change in $f(x)$ can be approximated by a term depending upon the derivative of $f(x)$

We can deduce from our derivation of necessary conditions NC1 and NC2 that if $x^*$ is a point at which $f'(x^*)$ is zero and if the derivative of $f$ changes from negative to positive when passing through $x^*$, then $x^*$ is a minimum of $f$. This sufficiency condition for a minimum may be phrased as follows:

SC1 *If $x^*$ is an interior point of $[a, b]$ at which $f'$ is zero* [*that is,* $f'(x^*) = 0$] *and if $f''(x^*) > 0$, then $x^*$ is a (local) minimum of $f$.*

Although there are a variety of sufficiency conditions for a point to be a minimum of $f$, this is the only one which will be used in this book.

Now let us turn our attention to the problem of finding the minimum of a function $g$ of several variables. In other words, $g$ is a real-valued function of a vector $\mathbf{x}$ with components $x_1, x_2, \ldots, x_n$. The question of minimizing such functions is a difficult one, and so we shall only indicate those results which we shall use later on. To begin with, we have:

***Definition 5-3*** *A point $\mathbf{x}^*$ in $R_n$ is said to be a (local) minimum of $g$ if there is an $\epsilon > 0$ such that $\|\mathbf{x} - \mathbf{x}^*\| < \epsilon$† implies that*

$$g(\mathbf{x}^*) \leq g(\mathbf{x}) \tag{5-11}$$

*Again, $\mathbf{x}^*$ is a minimum of $g$ if $g(\mathbf{x}^*)$ is no greater than $g(\mathbf{x})$ for all $\mathbf{x}$ in a neighborhood of $\mathbf{x}^*$.*

We also have:

***Definition 5-4*** *A point $\mathbf{x}^*$ in $R_n$ is said to be an absolute minimum of $g$ if*

$$g(\mathbf{x}^*) \leq g(\mathbf{x}) \tag{5-12}$$

*for all $\mathbf{x}$ in $R_n$.*

We again observe that an absolute minimum of $g$ is also a local minimum of $g$ but that a local minimum of $g$ need *not* be an absolute minimum of $g$.

Now, we are often interested in determining the minima of $g$ in certain subsets of $R_n$ and in studying functions which are not necessarily defined on all of $R_n$ but rather on a subset of $R_n$. In particular, we have:

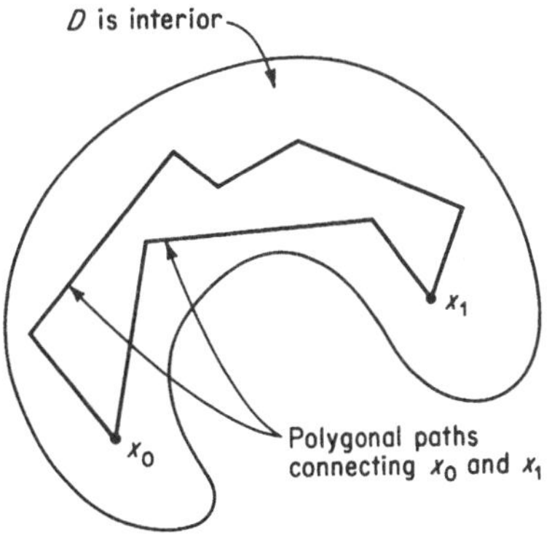


**Fig. 5-3** A typical domain.

***Definition 5-5*** *A bounded open subset $D$ of $R_n$ is called a domain if, given any two points $\mathbf{x}_0$ and $\mathbf{x}_1$ in $D$, there is a piecewise linear function* [*see Eq.* (3-104)] *$\boldsymbol{\lambda}$ from* [0, 1] *into $D$ such that $\boldsymbol{\lambda}(0) = \mathbf{x}_0$ and $\boldsymbol{\lambda}(1) = \mathbf{x}_1$ (that is, there is a polygonal path in $D$ which connects $\mathbf{x}_0$ and $\mathbf{x}_1$, as is illustrated in Fig.* 5-3). *Letting $\partial D$ denote the boundary of $D$ (see Definition* 3-10) *and letting $\hat{D}$ be a subset of $R_n$ such that*

$$D \subset \hat{D} \subset D \cup \partial D \tag{5-13}$$

† Recall that $\| \quad \|$ denotes the Euclidean norm.

*we say that a point* $\mathbf{x}^*$ *of* $\hat{D}$ *is a (local) minimum of* $g$ *in* $\hat{D}$ *if there is an* $\epsilon > 0$ *such that*

$$\hat{\mathbf{x}} \in S(\mathbf{x}^*, \epsilon) \cap \hat{D} \tag{5-14†}$$

*implies that*

$$g(\mathbf{x}^*) \leq g(\hat{\mathbf{x}}) \tag{5-15}$$

*In other words,* $\mathbf{x}^*$ *is a minimum of* $g$ *in* $\hat{D}$ *if* $g(\mathbf{x}^*)$ *is no greater than* $g(\hat{\mathbf{x}})$ *for all* $\hat{\mathbf{x}}$ *in* $\hat{D}$ *which are close to* $\mathbf{x}^*$.

So we shall now suppose that $D$ is a given domain, that $g(\mathbf{x}) = g(x_1, x_2, \ldots, x_n)$ is continuous on the closure $\bar{D}$ $(= D \cup \partial D)$ of $D$, and that $g$ has continuous partial derivatives with respect to each of the components $x_1, x_2, \ldots, x_n$ of $\mathbf{x}$ on $D$. We then have the necessary condition:

NC3 *If a point* $\mathbf{x}^*$ *in* $D$ *is a minimum of* $g$*, then all the partial derivatives of* $g$ *vanish at* $\mathbf{x}^*$*, that is,*

$$\frac{\partial g}{\partial x_1}(\mathbf{x}^*) = \frac{\partial g}{\partial x_2}(\mathbf{x}^*) = \cdots = \frac{\partial g}{\partial x_n}(\mathbf{x}^*) = 0 \tag{5-16}$$

*or, equivalently, the gradient of* $g$ [*see Eqs.* (3-67) *and* (3-68)] *is the zero vector at* $\mathbf{x}^*$.

We may view this necessary condition as follows: Let $\boldsymbol{\alpha}$ be a unit vector with components $\alpha_1, \alpha_2, \ldots, \alpha_n$ and let $f(t)$ be the function defined by

$$f(t) = g(\mathbf{x}^* + t\boldsymbol{\alpha}) \tag{5-17}$$

If $\mathbf{x}^*$ is a minimum of $g$, then 0 is a minimum of $f$, and it follows that

$$f'(0) = \sum_{i=1}^{n} \frac{\partial g}{\partial x_i}(\mathbf{x}^*)\alpha_i = \langle \nabla g(\mathbf{x}^*), \boldsymbol{\alpha} \rangle = 0 \tag{5-18}$$

Noting that $f'(0)$ is the directional derivative of $g$ in the direction $\boldsymbol{\alpha}$ at $\mathbf{x}^*$ [see Eq. (3-80)], we may conclude that necessary condition NC3 is equivalent to the condition that the directional derivative of $g$ vanish at $\mathbf{x}^*$ for every direction $\boldsymbol{\alpha}$.

Let us carry the reasoning of the previous paragraph a step further. Supposing that $g$ has continuous second partial derivatives, we may write, in a neighborhood of 0, the relation

$$f''(t) = \sum_{i=1}^{n} \sum_{j=1}^{n} \frac{\partial^2 g}{\partial x_i \, \partial x_j}(\mathbf{x}^* + t\boldsymbol{\alpha})\alpha_i\alpha_j \tag{5-19}$$

and we may conclude from sufficiency condition SC1 that 0 will be a min-

† See Eqs. (3-4) and (2-3).

imum of $f$ if $f'(0) = 0$ and if

$$f''(0) = \sum_{i=1}^{n} \sum_{j=1}^{n} \frac{\partial^2 g}{\partial x_i \, \partial x_j} (\mathbf{x}^*) \alpha_i \alpha_j > 0 \tag{5-20}$$

The result of this line of thought is the following sufficiency condition:

SC2 *If the directional derivative of $g$ vanishes at $\mathbf{x}^*$ for every direction $\boldsymbol{\alpha}$ and if the inequality*

$$\sum_{i=1}^{n} \sum_{j=1}^{n} \frac{\partial^2 g}{\partial x_i \, \partial x_j} (\mathbf{x}^*) \alpha_i \alpha_j > 0 \tag{5-21}$$

*holds for every direction $\boldsymbol{\alpha}$, then $\mathbf{x}^*$ is a (local) minimum of $g$.*

Now, the term in Eq. (5-21) looks very much like an inner product (see Sec. 2-11). If we let $\mathbf{Q}$ be the symmetric $n \times n$ matrix whose entries are the $(\partial^2 g/\partial x_i \, \partial x_j)(\mathbf{x}^*)$, that is, if $\mathbf{Q}$ is given by the relation

$$\mathbf{Q} = \begin{bmatrix} \frac{\partial^2 g}{\partial x_1^2} (\mathbf{x}^*) & \frac{\partial^2 g}{\partial x_1 \, \partial x_2} (\mathbf{x}^*) & \cdots & \frac{\partial^2 g}{\partial x_1 \, \partial x_n} (\mathbf{x}^*) \\ \frac{\partial^2 g}{\partial x_2 \, \partial x_1} (\mathbf{x}^*) & \frac{\partial^2 g}{\partial x_2^2} (\mathbf{x}^*) & \cdots & \frac{\partial^2 g}{\partial x_2 \, \partial x_n} (\mathbf{x}^*) \\ \cdots & \cdots & \cdots & \cdots \\ \frac{\partial^2 g}{\partial x_n \, \partial x_1} (\mathbf{x}^*) & \frac{\partial^2 g}{\partial x_n \, \partial x_2} (\mathbf{x}^*) & \cdots & \frac{\partial^2 g}{\partial x_n^2} (\mathbf{x}^*) \end{bmatrix} \tag{5-22}$$

then we may define an inner product $Q$ on $R_n$ whose matrix is $\mathbf{Q}$ [see Eq. (2-83)], and we can assert that sufficiency condition SC2 is equivalent to the following condition:

SC3 *If the gradient of $g$ is $\mathbf{0}$ at $\mathbf{x}^*$ and if the matrix $\mathbf{Q}$ of Eq. (5-22) is positive definite* [*see Eq.* (2-84)],† *then $\mathbf{x}^*$ is a minimum of $g$.*

In particular, if $g$ is a function on $R_2$, then the conditions

$$\frac{\partial g}{\partial x_1} (\mathbf{x}^*) = \frac{\partial g}{\partial x_2} (\mathbf{x}^*) = 0 \tag{5-23}$$

$$\frac{\partial^2 g}{\partial x_1^2} (\mathbf{x}^*) > 0 \tag{5-24}$$

$$\frac{\partial^2 g}{\partial x_1^2} (\mathbf{x}^*) \frac{\partial^2 g}{\partial x_2^2} (\mathbf{x}^*) - \left[ \frac{\partial^2 g}{\partial x_1 \, \partial x_2} (\mathbf{x}^*) \right]^2 > 0 \tag{5-25}$$

guarantee that $\mathbf{x}^*$ is a minimum of $g$.

We observe again that the way in which these conditions were obtained was based upon the evaluation of the effect of small changes in $\mathbf{x}$ on $g$ by means of the "derivative" (partial derivatives here) of $g$. We leave it to the reader to examine the results from this point of view.

† See Sec. 2-15.

**Example 5-1** Consider the function $g$ defined on $R_2$ by $g(x, y) = (x^2 + y^2)/2$. Then the origin (0, 0) is a minimum of $g$, since $\partial g/\partial x = x$, $\partial g/\partial y = y$, $\partial^2 g/\partial x\,\partial y = 0$, $\partial^2 g/\partial y^2 = 1$, $\partial^2 g/\partial x^2 = 1$ imply that the sufficiency condition is satisfied at (0, 0). We note that the matrix **Q** of Eq. (5-22) in this case is the identity matrix. The function $g$, viewed as a surface in $R_3$, is a paraboloid of revolution and is illustrated in Fig. 5-4.

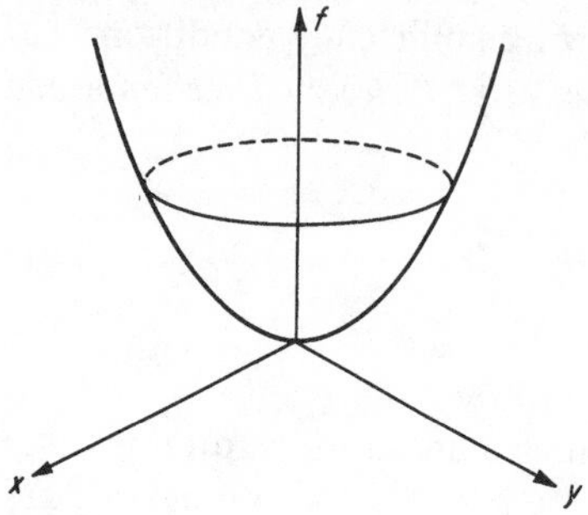


**Fig. 5-4** The paraboloid of Example 5-1.

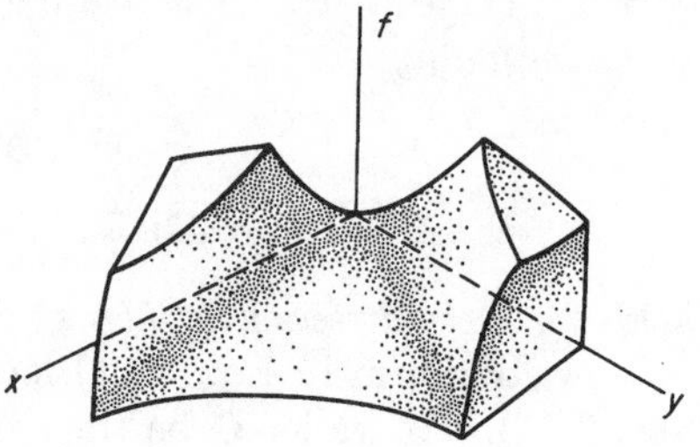


**Fig. 5-5** The hyperbolic paraboloid of Example 5-2. The origin is the "saddle" point.

**Example 5-2** Consider the function $g$ defined on $R_2$ by $g(x, y) = (x^2 - y^2)/2$. Then $\partial g/\partial x$ and $\partial g/\partial y$ vanish at the origin, but the origin is neither a minimum nor a maximum of $g$. Here, the matrix **Q** of Eq. (5-22) is given by

$$\mathbf{Q} = \begin{bmatrix} 1 & 0 \\ 0 & -1 \end{bmatrix}$$

and is neither positive nor negative definite.† The function $g$, viewed as a surface in $R_3$, is a hyperbolic paraboloid and is illustrated in Fig. 5-5. The origin is called a *saddle point* of $g$, and the motivation for this name should be apparent from the figure.

## 5-3 Ordinary Minima with Constraints: A Simple Problem

In the previous section, we considered the problem of determining the minima of a function over a subset of the Euclidean space $R_n$; now we are going to examine the problem of determining the minima of a function subject to certain constraints. In order to motivate the general results which we shall develop in the next section, we shall here consider some special illustrative situations involving functions on $R_2$.

More specifically, let us suppose that $f = f(x, y)$ is a real-valued function defined on $R_2$ and that $h(x)$ is a real-valued function defined on $R$. We may view the equation

$$y = h(x) \qquad x \in R \tag{5-26}$$

as defining a curve in $R_2$,‡ and we note that if we define $g(x, y)$ by setting

$$g(x, y) = y - h(x) \tag{5-27}$$

† See Sec. 2-15.

‡ This curve is the set $\{(x, y): y = h(x)\}$, and, of course, $R_2$ is the $xy$ plane.

then the equation of our curve becomes

$$g(x, y) = y - h(x) = 0 \tag{5-28†}$$

Now we can ask the following question: *Where are the minima of $f(x, y)$ along the curve $y = h(x)$?* In other words, we want to find the points $(x^*, y^*)$ on the curve [that is, $y^* = h(x^*)$] such that if $(x, y)$ is a point *on the curve* which is "near" $(x^*, y^*)$, then

$$f(x^*, y^*) \leq f(x, y) \tag{5-29}$$

Since $y^* = h(x^*)$ and $y = h(x)$, we may write Eq. (5-29) in the more descriptive form

$$f[x^*, h(x^*)] \leq f[x, h(x)] \tag{5-30}$$

We call such a point $(x^*, y^*)$ a *minimum of $f$ subject to the constraint* $y - h(x) = 0$.

Now let us use the perturbation idea to obtain a necessary condition for a constrained [by $y - h(x) = 0$] minimum of $f$. We shall suppose that $(x^*, y^*)$ is such a constrained minimum and that $f$ and $h$ are sufficiently smooth (i.e., differentiable) to validate our procedures. If $(x, y)$ is "near" $(x^*, y^*)$, then we may write

$$x = x^* + \epsilon\xi_1 \qquad y = y^* + \epsilon\xi_2 \tag{5-31}$$

where $\epsilon$ is a small, positive number and $\boldsymbol{\xi} = (\xi_1, \xi_2)$ is a vector. It follows that

$$f(x, y) = f(x^*, y^*) + \epsilon\langle\nabla f(x^*, y^*), \boldsymbol{\xi}\rangle + o(\epsilon) \tag{5-32}$$

where $\nabla f(x^*, y^*)$ is the gradient of $f$ at $(x^*, y^*)$ [see Eq. (3-67)] and $o(\epsilon)$ is a term depending on $\epsilon$, with the property that

$$\lim_{\epsilon \to 0} \frac{o(\epsilon)}{\epsilon} = 0 \tag{5-33}$$

If the point $(x, y)$ is not only "near" $(x^*, y^*)$ but also lies on our curve, then we must have $y = h(x)$, which means, in view of Eqs. (5-31) and of the fact that $y^* = h(x^*)$, that

$$y = h(x) = h(x^* + \epsilon\xi_1) = h(x^*) + \epsilon\xi_2 = y^* + \epsilon\xi_2 \tag{5-34}$$

However, since $x$ is "near" $x^*$, we may write

$$h(x^* + \epsilon\xi_1) = h(x^*) + \epsilon\frac{dh}{dx}(x^*)\xi_1 + o(\epsilon) \tag{5-35}$$

and hence we may conclude that

$$\xi_2 = \frac{dh}{dx}(x^*)\xi_1 \tag{5-36}$$

† See Sec. 3-13. Note that our curve is the set $S(g)$.

[as $\lim_{\epsilon \to 0} o(\epsilon)/\epsilon = 0$]. Thus, for perturbations of $(x^*, y^*)$ *along the curve* $y = h(x)$, Eq. (5-32) becomes

$$f(x, y) = f(x^*, y^*) + \epsilon \left.\frac{\partial f}{\partial x}\right|_{(x^*,y^*)} \xi_1 + \epsilon \left.\frac{\partial f}{\partial y}\right|_{(x^*,y^*)} \left.\frac{dh}{dx}\right|_{x^*} \xi_1 + o(\epsilon) \qquad (5\text{-}37)$$

Since $(x^*, y^*)$ is a minimum of $f$ along the curve, since $\epsilon$ is positive, since $\lim_{\epsilon \to 0} o(\epsilon)/\epsilon = 0$, and since $\xi_1$ may be either positive or negative, we must have

$$\left.\frac{\partial f}{\partial x}\right|_{(x^*,y^*)} + \left.\frac{\partial f}{\partial y}\right|_{(x^*,y^*)} \left.\frac{dh}{dx}\right|_{x^*} = 0 \qquad (5\text{-}38)$$

This equation is thus a necessary condition for $(x^*, y^*)$ to be a minimum of $f$ subject to the constraint $y - h(x) = 0$.

We observe that in deriving Eq. (5-38) we used perturbations along the curve. In other words, once $\xi_1$ was chosen, then $\xi_2$ was determined by Eq. (5-36). Thus, as is consistent with Eq. (5-30), our derivation depended upon only one variable. We see, therefore, that the effect of the constraint was to limit our freedom in choosing a perturbation and, hence, to decrease the dimension of our problem.

**Exercise 5-3** Derive Eq. (5-38) by considering the function $F(x) = f[x, h(x)]$ of one variable. Use the perturbation method.

Now let us look at Eq. (5-38) in a somewhat different way by considering the function

$$g(x, y) = y - h(x) \qquad (5\text{-}39)$$

Our constraint may then be written as

$$g(x, y) = 0 \qquad (5\text{-}40)$$

and we observe that

$$\frac{\partial g}{\partial y} = 1 \qquad (5\text{-}41)$$

is not zero at any point along our constraint curve. Thus, we have

$$\frac{\partial(f, g)}{\partial(x, y)} = \det \begin{bmatrix} \dfrac{\partial f}{\partial x} & \dfrac{\partial f}{\partial y} \\ \dfrac{\partial g}{\partial x} & \dfrac{\partial g}{\partial y} \end{bmatrix} \qquad (5\text{-}42)\dagger$$

$$= \det \begin{bmatrix} \dfrac{\partial f}{\partial x} & \dfrac{\partial f}{\partial y} \\ -\dfrac{dh}{dx} & 1 \end{bmatrix} \qquad (5\text{-}43)$$

$$= \frac{\partial f}{\partial x} + \frac{\partial f}{\partial y}\frac{dh}{dx} \qquad (5\text{-}44)$$

† See Sec. 3-12.

where the term on the left is the Jacobian of $f$ and $g$ with respect to $x$ and $y$ [see Eq. (3-89)]. It follows that Eq. (5-38) may be interpreted as

$$\left.\frac{\partial(f, g)}{\partial(x, y)}\right|_{(x^*,y^*)} = 0 \tag{5-45}$$

What is the significance of this interpretation? For one thing, it means that the system of equations in the variable $p$,

$$\begin{aligned} \left.\frac{\partial f}{\partial x}\right|_{(x^*,y^*)} + p\left.\frac{\partial g}{\partial x}\right|_{(x^*,y^*)} &= 0 \\ \left.\frac{\partial f}{\partial y}\right|_{(x^*,y^*)} + p\left.\frac{\partial g}{\partial y}\right|_{(x,{}^*y^*)} &= 0 \end{aligned} \tag{5-46}$$

has a solution $p = p^*$. In fact, since $g(x, y) = y - h(x)$,

$$p = p^* = -\left.\frac{\partial f}{\partial y}\right|_{(x^*,y^*)} = \text{const} \tag{5-47}$$

is one such solution. We observe that if

$$\begin{aligned} G(x, y) &= f(x, y) + p^*g(x, y) \\ &= f(x, y) + p^*[y - h(x)] \end{aligned} \tag{5-48}$$

where $p^*$ is a solution of Eqs. (5-46), then we may view Eqs. (5-46) with $p^*$ replacing $p$ as a necessary condition for $(x^*, y^*)$ to be a minimum of $G(x, y)$. In other words, if $(x^*, y^*)$ is a minimum of $G(x, y)$, then we know that the partial derivatives of $G$ with respect to $x$ and $y$ must be zero at $(x^*, y^*)$ [see Eq. (5-16)], so that we have

$$\begin{aligned} \left.\frac{\partial G}{\partial x}\right|_{(x^*,y^*)} &= \left.\frac{\partial}{\partial x}\{f(x, y) + p^*g(x, y)\}\right|_{(x^*,y^*)} \\ &= \left.\frac{\partial f}{\partial x}\right|_{(x^*,y^*)} + p^*\left.\frac{\partial g}{\partial x}\right|_{(x^*,y^*)} = 0 \\ \left.\frac{\partial G}{\partial y}\right|_{(x^*,y^*)} &= \frac{\partial}{\partial y}\{f(x, y) + p^*g(x, y)\} \\ &= \left.\frac{\partial f}{\partial y}\right|_{(x^*,y^*)} + p^*\left.\frac{\partial g}{\partial y}\right|_{(x^*,y^*)} = 0 \end{aligned}$$

Thus, we are led to the following conclusion: *In order that $(x^*, y^*)$ be a minimum of $f$ subject to the constraint $g(x, y) = y - h(x) = 0$, it is necessary that there be a number $p^*$ such that the equations*

$$\left.\frac{\partial f}{\partial x}\right|_{(x^*,y^*)} + p^*\left.\frac{\partial g}{\partial x}\right|_{(x^*,y^*)} = 0 \tag{5-49}$$

$$\left.\frac{\partial f}{\partial y}\right|_{(x^*,y^*)} + p^*\left.\frac{\partial g}{\partial y}\right|_{(x^*,y^*)} = 0 \tag{5-50}$$

$$g(x^*, y^*) = y^* - h(x^*) = 0 \tag{5-51}$$

*are satisfied. The number $p^*$ is called a Lagrange multiplier.*

Let us now generalize our problem a bit. Supposing that $f(x, y)$ is a real-valued function defined on $R_2$ and that $g(x, y)$ is a real-valued function on $R_2$ [not necessarily of the form $y - h(x)$], we may ask for necessary conditions for a minimum of $f$ along the curve $g(x, y) = 0$. Well, assuming that $f$ and $g$ are sufficiently smooth (i.e., sufficiently differentiable) and assuming that the norm of the gradient of $g$ is not zero, i.e., that

$$\|\nabla g\| = \left[\left(\frac{\partial g}{\partial x}\right)^2 + \left(\frac{\partial g}{\partial y}\right)^2\right]^{\frac{1}{2}} > 0 \tag{5-52}†$$

we can assert that if $(x^*, y^*)$ is a minimum of $f$ subject to the constraint $g(x, y) = 0$, then there is a number $p^*$, called a *Lagrange multiplier*, such that $x^*$, $y^*$, and $p^*$ are a solution of the system of equations

$$\frac{\partial f}{\partial x} + p\frac{\partial g}{\partial x} = 0 \tag{5-53}$$

$$\frac{\partial f}{\partial y} + p\frac{\partial g}{\partial y} = 0 \tag{5-54}$$

$$g(x, y) = 0 \tag{5-55}$$

that is, that

$$\left.\frac{\partial f}{\partial x}\right|_{(x^*,y^*)} + p^*\left.\frac{\partial g}{\partial x}\right|_{(x^*,y^*)} = 0 \tag{5-56}$$

$$\left.\frac{\partial f}{\partial y}\right|_{(x^*,y^*)} + p^*\left.\frac{\partial g}{\partial y}\right|_{(x^*,y^*)} = 0 \tag{5-57}$$

$$g(x^*, y^*) = 0 \tag{5-58}$$

The verification of this assertion is quite similar to what we have done before, and only the role of the assumption of Eq. (5-52) requires clarification. This assumption [Eq. (5-52)] ensures that the equation $g(x, y) = 0$ can be locally "solved" for either $x$ or $y$. For example, if

$$\left.\frac{\partial g}{\partial y}\right|_{(x^*,y^*)} \neq 0 \tag{5-59}$$

then, in a neighborhood of $x^*$, there is a function $h(x)$ such that

$$y^* = h(x^*) \tag{5-60a}$$

$$0 = g[x, h(x)] \tag{5-60b}$$

$$\frac{dh}{dx} = -\frac{\partial g/\partial x}{\partial g/\partial y} \tag{5-60c}$$

that is, the equations $g(x, y) = 0$ and $y = h(x)$ define the same curve "near" $(x^*, y^*)$. We now leave it to the reader to verify the necessity of Eqs. (5-56) to (5-58).

**Example 5-3** Let $f(x, y) = -xy$ and $g(x, y) = y + xe^x$. What are the potential candidates for minima of $f$ subject to the constraint $g = 0$? Here, we can take $h(x) =$

† Actually, we only require that $\|\nabla g\| > 0$ on the curve $g(x, y) = 0$, as will become apparent.

$-xe^x$ and use Eq. (5-38). Thus, we must solve the equations

$$-y + (-x)(-e^x - xe^x) = 0$$
$$y + xe^x = 0$$

Clearly, (0, 0) and $(-2, +2/e^2)$ are the only solutions of these equations and thus are the only candidates for the minima of $f$ subject to the constraint $g = 0$. Consideration of the function $x^2e^x$ leads us to the conclusion that (0, 0) is *the* minimum of $f$ subject to the constraint $g = 0$ [while $(-2, +2/e^2)$ is a maximum of $f$ subject to our constraint].

**Example 5-4** Let $f(x, y) = -xy$ and let $g(x, y) = x^2 + y^2 - 1$. What are the potential candidates for minima of $f$ subject to the constraint $g = 0$? First of all, we note that the assumption of Eq. (5-52) is not satisfied at (0, 0); however, (0, 0) is not a point on the curve $g(x, y) = 0$, and so we can still apply the necessary conditions of Eqs. (5-56) to (5-58). Thus, setting

$$G(x, y, p) = -xy + p(x^2 + y^2 - 1)$$

we find that we must solve the equations

$$\frac{\partial G}{\partial x} = -y + 2px = 0$$
$$\frac{\partial G}{\partial y} = -x + 2py = 0$$
$$g(x, y) = x^2 + y^2 - 1 = 0$$

It is easy to verify that the solutions of these equations are: $(p^* = \frac{1}{2}, x^* = 1/\sqrt{2}, y^* = 1/\sqrt{2})$, $(p^* = \frac{1}{2}, x^* = -1/\sqrt{2}, y^* = -1/\sqrt{2})$, $(p^* = -\frac{1}{2}, x^* = 1/\sqrt{2}, y^* = -1/\sqrt{2})$, $(p^* = -\frac{1}{2}, x^* = -1/\sqrt{2}, y^* = 1/\sqrt{2})$. Thus, our candidates are the points $(1/\sqrt{2}, 1/\sqrt{2})$, $(-1/\sqrt{2}, -1/\sqrt{2})$, $(1/\sqrt{2}, -1/\sqrt{2})$, and $(-1/\sqrt{2}, 1/\sqrt{2})$.

**Exercise 5-4** Show that the points $(1/\sqrt{2}, 1/\sqrt{2})$ and $(-1/\sqrt{2}, -1/\sqrt{2})$ actually provide the minima in Example 5-4. HINT: Consider the function $-|x|\sqrt{1 - |x|^2}$.

## 5-4 Ordinary Minima with Constraints: Necessary Conditions and the Lagrange Multipliers

Having bolstered our intuition in the previous section, we are now prepared to tackle the problem of determining the minima of a function subject to certain constraints. The development of necessary conditions for this type of problem will lead us to the notion of Lagrange multipliers, which will find frequent application in the sequel. We shall begin with a precise statement of the problem, then we shall present the desired necessary conditions in Theorem 5-1, and finally, we shall briefly discuss these necessary conditions.

So let us now suppose that $D$ is a given domain in $R_n$ (see Definition 5-5), that $f(\mathbf{x}) = f(x_1, x_2, \ldots, x_n)$ is a continuous real-valued function on the closure $\bar{D}$ of $D$,† and that $g_1(\mathbf{x}) = g_1(x_1, x_2, \ldots, x_n)$, $g_2(\mathbf{x}) = g_2(x_1, x_2, \ldots, x_n)$, . . . , $g_m(\mathbf{x}) = g_m(x_1, x_2, \ldots, x_n)$, $m < n$, are continuous real-valued functions on $\bar{D}$. We shall also assume that the $g$'s are independent in the sense that the set of common zeros of the $g$'s, that is,

† See Definition 3-9.

the set of all $\mathbf{x}$ such that

$$g_1(\mathbf{x}) = g_2(\mathbf{x}) = \cdots = g_m(\mathbf{x}) = 0 \tag{5-61}$$

cannot be defined by fewer of the $g$'s.† Then we have:

***Definition 5-6*** *A point $\mathbf{x}^*$ in $\bar{D}$ is said to be a (local) minimum of $f$ subject to the constraints*

$$g_1(\mathbf{x}) = 0,\ g_2(\mathbf{x}) = 0,\ \ldots,\ g_m(\mathbf{x}) = 0 \tag{5-62}$$

*if, firstly,*

$$g_1(\mathbf{x}^*) = 0,\ g_2(\mathbf{x}^*) = 0,\ \ldots,\ g_m(\mathbf{x}^*) = 0 \tag{5-63}$$

*and if, secondly, there is an $\epsilon > 0$ such that*

$$\|\mathbf{x} - \mathbf{x}^*\| < \epsilon \tag{5-64}$$

$$\mathbf{x} \in \bar{D} \tag{5-65}$$

$$g_1(\mathbf{x}) = 0,\ g_2(\mathbf{x}) = 0,\ \ldots,\ g_m(\mathbf{x}) = 0 \tag{5-66}$$

*together imply that*

$$f(\mathbf{x}^*) \le f(\mathbf{x}) \tag{5-67}$$

Consequently, the problem we wish to study is that of determining the minima of $f$ (on $\bar{D}$) subject to the constraints $g_i(\mathbf{x}) = 0$, $i = 1, 2, \ldots, m$.

Let us further assume that $f(x_1, x_2, \ldots, x_n)$ and $g_i(x_1, x_2, \ldots, x_n)$, $i = 1, 2, \ldots, m$, all have continuous partial derivatives with respect to all the $x_j$ on the domain $D$. We shall also require that the set of constraint equations $g_i(\mathbf{x}) = 0$, $i = 1, 2, \ldots, m$, can be locally "solved" for $m$ of the $x_j$ in terms of the $n - m$ remaining ones. This can be guaranteed by supposing (as we shall) that, on $D$, the inequality

$$\sum_{i_1<i_2<\cdots<i_m} \left[\frac{\partial(g_1, g_2, \ldots, g_m)}{\partial(x_{i_1}, x_{i_2}, \ldots, x_{i_m})}\right]^2 > 0 \tag{5-68}$$

holds, where the terms in the summation are the various Jacobians [see Eq. (3-89)] of the $g$'s with respect to subsets of $\{x_1, x_2, \ldots, x_n\}$ containing $m$ distinct elements. We note that the inequality (5-68) holds if and only if the $m$ vectors $\nabla g_1, \nabla g_2, \ldots, \nabla g_m$ are linearly independent at all points of $D$.

**Example 5-5** If $n = 4$ and $m = 2$, so that we are considering two functions $g_1$ and $g_2$ of four variables $x_1$, $x_2$, $x_3$, and $x_4$, then Eq. (5-68) becomes

$$\left[\frac{\partial(g_1, g_2)}{\partial(x_1, x_2)}\right]^2 + \left[\frac{\partial(g_1, g_2)}{\partial(x_1, x_3)}\right]^2 + \left[\frac{\partial(g_1, g_2)}{\partial(x_1, x_4)}\right]^2 + \left[\frac{\partial(g_1, g_2)}{\partial(x_2, x_3)}\right]^2 + \left[\frac{\partial(g_1, g_2)}{\partial(x_2, x_4)}\right]^2 + \left[\frac{\partial(g_1, g_2)}{\partial(x_3, x_4)}\right]^2 > 0$$

Thus, if $\mathbf{x}^* = (x_1^*, x_2^*, x_3^*, x_4^*)$ is a point satisfying the constraint equations

$$g_1(\mathbf{x}^*) = 0 \qquad g_2(\mathbf{x}^*) = 0$$

† See Sec. 3-13.

then one of the Jacobians $\partial(g_1, g_2)/\partial(x_i, x_j)$ must be nonzero at $\mathbf{x}^*$, say

$$\left.\frac{\partial(g_1, g_2)}{\partial(x_1, x_2)}\right|_{\mathbf{x}^*} \neq 0$$

But this means that, in a neighborhood of the 2-vector $(x_3^*, x_4^*)$, we can "solve" for $x_1$, $x_2$ in terms of $x_3$, $x_4$, that is, there are functions $h_1(x_3, x_4)$ and $h_2(x_3, x_4)$ such that

(*a*) $x_1^* = h_1(x_3^*, x_4^*)$, $x_2^* = h_2(x_3^*, x_4^*)$

(*b*) $g_1[h_1(x_3, x_4), h_2(x_3, x_4), x_3, x_4] = 0$
$g_2[h_1(x_3, x_4), h_2(x_3, x_4), x_3, x_4] = 0$

(*c*) $$\frac{\partial h_1}{\partial x_3} = -\frac{\partial(g_1, g_2)/\partial(x_3, x_2)}{\partial(g_1, g_2)/\partial(x_1, x_2)} \qquad \frac{\partial h_1}{\partial x_4} = -\frac{\partial(g_1, g_2)/\partial(x_4, x_2)}{\partial(g_1, g_2)/\partial(x_1, x_2)}$$
$$\frac{\partial h_2}{\partial x_3} = -\frac{\partial(g_1, g_2)/\partial(x_1, x_3)}{\partial(g_1, g_2)/\partial(x_1, x_2)} \qquad \frac{\partial h_2}{\partial x_4} = -\frac{\partial(g_1, g_2)/\partial(x_1, x_4)}{\partial(g_1, g_2)/\partial(x_1, x_2)}$$

The functions $h_1$ and $h_2$ represent the local "solutions" of the equations $g_1 = 0$, $g_2 = 0$ for the two variables $x_1$, $x_2$ in terms of $x_3$ and $x_4$.

**Exercise 5-5** Suppose that $g_1(x_1, x_2, x_3, x_4) = x_1 + x_2 + x_3 + x_4$ and $g_2(x_1, x_2, x_3, x_4) = x_1 - x_2 + x_3^2 + x_4^2$. Can you solve the equations $g_1 = 0$, $g_2 = 0$ for $x_1$ and $x_2$ in terms of $x_3$ and $x_4$ "near" the origin? What about for $x_3$ and $x_4$ in terms of $x_1$ and $x_2$? Exhibit the solutions when they exist.

Under these conditions [Eq. (5-68)], the following theorem provides a necessary condition that $f$ have a minimum subject to the constraints $g_i = 0$, $i = 1, 2, \ldots, m$, at a point $\mathbf{x}^*$ of $D$.

***Theorem 5-1*** *If $\mathbf{x}^*$ is a point of the domain $D$ at which $f$ has a (local) minimum subject to the constraints $g_i = 0$, $i = 1, 2, \ldots, m$, where $f$ and the $g_i$ all have continuous partial derivatives, then there are numbers $p_1^*$, $p_2^*$, $\ldots$, $p_m^*$, called Lagrange multipliers, such that $\mathbf{x}^*$, $p_1^*$, $p_2^*$, $\ldots$, $p_m^*$ are a solution of the following system of $n + m$ equations in the $n + m$ unknowns $x_1, x_2, \ldots, x_n, p_1, p_2, \ldots p_m$:*

$$\frac{\partial f}{\partial x_i} + p_1 \frac{\partial g_1}{\partial x_i} + p_2 \frac{\partial g_2}{\partial x_i} + \cdots + p_m \frac{\partial g_m}{\partial x_i} = 0 \tag{5-69}$$

$$g_j(x_1, x_2, \ldots, x_n) = 0 \tag{5-70}$$

*where $i = 1, 2, \ldots, n$ and $j = 1, 2, \ldots, m$. In other words,*

$$\left.\frac{\partial f}{\partial x_i}\right|_{\mathbf{x}^*} + p_1^* \left.\frac{\partial g_1}{\partial x_i}\right|_{\mathbf{x}^*} + p_2^* \left.\frac{\partial g_2}{\partial x_i}\right|_{\mathbf{x}^*} + \cdots + p_m^* \left.\frac{\partial g_m}{\partial x_i}\right|_{\mathbf{x}^*} = 0 \tag{5-71}$$

$$g_j(x_1^*, x_2^*, \ldots, x_n^*) = 0 \tag{5-72}$$

*for $i = 1, 2, \ldots, n$ and $j = 1, 2, \ldots, m$.*

The proof of this theorem, which will be used frequently in the sequel, is very similar to the argument that was used to derive Eqs. (5-49) to (5-51). Consequently, we shall leave the details to the reader.†

What is the significance of Theorem 5-1? The essential point is that it allows us to replace the problem of determining the minima of $f$ subject to

† See Ref. H-5.

the constraints $g_j = 0, j = 1, 2, \ldots, m$, with the problem of determining the minima of the function

$$G(\mathbf{x}, p_1, p_2, \ldots, p_m) = f(\mathbf{x}) + p_1 g_1(\mathbf{x}) + p_2 g_2(\mathbf{x}) + \cdots + p_m g_m(\mathbf{x}) \tag{5-73}$$

To see this, we let $\mathbf{p}$ and $\mathbf{g}(\mathbf{x})$ be the $m$ vectors with components $p_1, p_2, \ldots, p_m$ and $g_1(\mathbf{x}), g_2(\mathbf{x}), \ldots, g_m(\mathbf{x})$, respectively, so that Eq. (5-73) becomes

$$G(\mathbf{x}, \mathbf{p}) = f(\mathbf{x}) + \langle \mathbf{p}, \mathbf{g}(\mathbf{x}) \rangle \tag{5-74}$$

We then observe that the necessary conditions for a minimum of $G$ at $(\mathbf{x}^*, \mathbf{p}^*)$ are

$$\left.\frac{\partial G}{\partial x_i}\right|_{(\mathbf{x}^*, \mathbf{p}^*)} = \left.\frac{\partial f}{\partial x_i}\right|_{\mathbf{x}^*} + \left\langle \mathbf{p}^*, \left.\frac{\partial \mathbf{g}}{\partial x_i}\right|_{\mathbf{x}^*} \right\rangle = 0 \tag{5-75}$$

$$\left.\frac{\partial G}{\partial p_j}\right|_{(\mathbf{x}^*, \mathbf{p}^*)} = g_j(\mathbf{x}^*) = 0 \tag{5-76}$$

where $i = 1, 2, \ldots, n$ and $j = 1, 2, \ldots, m$ and $\partial \mathbf{g}/\partial x_i$ is the $m$ vector with components $\partial g_j/\partial x_i$.† But Eqs. (5-75) and (5-76) are the same as Eqs. (5-71) and (5-72), which proves our point.

Now we also observe that if we wish to determine the *absolute* minimum of $f$ subject to the constraints $g_j = 0, j = 1, 2, \ldots, m$, *on* $\bar{D}$, then Theorem 5-1 states that the potential candidates for this absolute minimum are the points in $D$ at which Eqs. (5-71) and (5-72) hold *and* the points on the boundary of $D$ which satisfy the constraint equations. This, as we have noted in Sec. 5-2, is typical of minimization problems.

**Exercise 5-6** Determine the minimum of $f(x_1, x_2, \ldots, x_n) = \sum_{i=1}^{n} x_i^2$ subject to the constraint $g(x_1, x_2, \ldots, x_n) = \sum_{i=1}^{n} a_i x_i - 1 = 0$, $a_i > 0$, on the set $x_1 > 0$, $x_2 > 0$, $\ldots, x_n > 0$.

## 5-5 Some Comments

Before we begin our study of functional minima and the variational approach, we should like to review, in a heuristic fashion, the results and techniques that were developed for ordinary minima. We shall have two major aims in this review: first, we should like to indicate the essential ideas behind our treatment of ordinary minima in Secs. 5-2 to 5-4, and second, we should like to provide some intuitive motivation for the methods we shall use in our discussion of functional minima.

In Secs. 5-2 to 5-4, we considered the problem of determining the minima of a real-valued function $f$ defined on a domain $A$ of the Euclidean space $R_n$.

† See Eq. (5-16).

We usually assumed that $f$ was continuous and that the subset $A$ of $R_n$ was compact (see Sec. 3-6). These assumptions guaranteed that there was a solution to our problem, i.e., that a minimum existed. We then sought the potential candidates for the absolute minimum of $f$ on $A$. To aid this search, we developed necessary conditions, based upon perturbation methods, which made use of the fact that the "derivative" represented a best linear approximation to the function (see Sec. 3-12). We found, in fact, that *the local minima of $f$ could occur only at:*

1. *Points on the boundary of $A$ or*
2. *Points at which the "derivative" of $f$ did not exist or*
3. *Interior points of $A$ at which the "derivative" of $f$ vanished (i.e., was zero)*

We called points at which the "derivative" vanished *extrema* of $f$, and the necessary conditions we derived were actually conditions for an extremum.

Now let us consider a generalization of our problem.† We suppose that $V$ is a vector space (see Sec. 2-5) with a distance function $d$ (see Sec. 3-2) which is derived from a norm, $\| \quad \|$, on $V$. In other words, we assume that there is a real-valued function $\|\mathbf{v}\|$ defined on $V$ such that:

1. $\|\mathbf{v}\| \geq 0$ for all $\mathbf{v}$ in $V$ and $\|\mathbf{v}\| = 0$ if and only if $\mathbf{v} = \mathbf{0}$
2. $\|\mathbf{v} + \mathbf{w}\| \leq \|\mathbf{v}\| + \|\mathbf{w}\|$ for all $\mathbf{v}$ and $\mathbf{w}$ in $V$
3. $\|r\mathbf{v}\| = |r| \, \|\mathbf{v}\|$ for all $\mathbf{v}$ in $V$ and for all $r$ in $R$

and we define the distance $d$ on $V$ by setting

$$d(\mathbf{v}, \mathbf{w}) = \|\mathbf{v} - \mathbf{w}\| \tag{5-77}$$

for $\mathbf{v}$, $\mathbf{w}$ in $V$. We leave it to the reader to verify that $d$ defines a distance on $V$ and that this distance is compatible with the notions of sum and product in $V$, in the sense that:

1. Given elements $\mathbf{v}$ and $\mathbf{w}$ in $V$ and sequences $\{\mathbf{v}_k\}$ and $\{\mathbf{w}_k\}$ in $V$ which converge to $\mathbf{v}$ and $\mathbf{w}$, respectively, then $\{\mathbf{v}_k + \mathbf{w}_k\}$ converges to $\mathbf{v} + \mathbf{w}$, that is,

$$d(\mathbf{v}_k + \mathbf{w}_k, \mathbf{v} + \mathbf{w}) \to 0 \tag{5-78}$$

2. Given an element $\mathbf{v}$ in $V$ and an element $r$ in $R$ and sequences $\{\mathbf{v}_k\}$ in $V$ and $\{r_k\}$ in $R$ which converge to $\mathbf{v}$ and $r$, respectively, then $\{r_k\mathbf{v}_k\}$ converges to $r\mathbf{v}$, that is,

$$d(r_k\mathbf{v}_k, r\mathbf{v}) \to 0 \tag{5-79}$$

**Exercise 5-7** Prove that the function $d$ of Eq. (5-77) defines a distance on $V$ which satisfies conditions 1 and 2 of Eqs. (5-78) and (5-79), respectively.

We note that the function spaces $\mathcal{B}$, $\mathcal{P}$, and $\mathcal{C}$ introduced in Sec. 3-15 are vector spaces of this type. If $A$ is a subset of $V$ with the property that, given any sequence $\mathbf{a}_k$, $k = 1, 2, \ldots$, in $A$, there is a subsequence $\mathbf{a}_{k_1}$, $\mathbf{a}_{k_2}$, . . . which converges to an element of $A$ (cf. Sec. 3-6, property C4), then we say that $A$ is *compact*. Finally, supposing that $J$ is a functional

† Compare this discussion with Sec. 3-15.

(i.e., a continuous real-valued function) which is defined on $A$ (which may or may not be compact), we have:

***Definition 5-7*** *A vector* $\mathbf{a}^*$ *in* $A$ *is said to be a (local) minimum of* $J$ *if, for all* $\mathbf{a}$ *in* $A$ *sufficiently "near"* $\mathbf{a}^*$, *the inequality*

$$J(\mathbf{a}^*) \leq J(\mathbf{a}) \tag{5-80}$$

*is satisfied.*

We note then that $J$ has an absolute minimum on $A$ if $A$ is compact (cf. Theorem 3-7), and so we can pose the problem: *What are the potential candidates for this absolute minimum?*

In an effort to answer this question, let us focus our attention on an interior point $\mathbf{a}^*$ of $A$, and let us try to mimic the perturbation methods we have used before. To begin with, we need a notion of derivative, and so we try to approximate $J(\mathbf{a})$ for $\mathbf{a}$ "near" $\mathbf{a}^*$ by a function of the form

$$J(\mathbf{a}^*) + \mathfrak{A}(\mathbf{a} - \mathbf{a}^*) \tag{5-81}$$

where $\mathfrak{A}$ is a (continuous) linear transformation of $V$ into $R$ (cf. Sec. 3-12).† If

$$\lim_{\mathbf{a} \to \mathbf{a}^*} \frac{|J(\mathbf{a}) - J(\mathbf{a}^*) - \mathfrak{A}(\mathbf{a} - \mathbf{a}^*)|}{\|\mathbf{a} - \mathbf{a}^*\|} = 0 \tag{5-82}$$

then *we call the linear transformation* $\mathfrak{A}$ *the derivative of* $J$ *at* $\mathbf{a}^*$, *and we write*

$$\mathfrak{A} = \delta J(\mathbf{a}^*) \tag{5-83}$$

$\delta J(\mathbf{a}^*)$ is often referred to as the *first variation* of $J$ at $\mathbf{a}^*$. We note also that $\delta J(\mathbf{a}^*)$ is again a functional on $V$, and we reiterate that $\delta J(\mathbf{a}^*)$ is a linear transformation from $V$ into $R$.

If $\mathbf{v}$ is some nonzero vector in $V$ and if $\epsilon$ is some real number, then we call $\mathbf{a}^* + \epsilon\mathbf{v}$ a perturbation, or variation, of $\mathbf{a}^*$ in the direction of $\mathbf{v}$. If $\delta J(\mathbf{a}^*)$ exists, then we may write

$$\begin{aligned} J(\mathbf{a}^* + \epsilon\mathbf{v}) &= J(\mathbf{a}^*) + \delta J(\mathbf{a}^*)[\epsilon\mathbf{v}] + o(\epsilon) \\ &= J(\mathbf{a}^*) + \epsilon\delta J(\mathbf{a}^*)\mathbf{v} + o(\epsilon) \end{aligned} \tag{5-84}$$

where $o(\epsilon)$ has the property that

$$\lim_{\epsilon \to 0} \frac{o(\epsilon)}{\epsilon} = 0 \tag{5-85}$$

If $\mathbf{a}^*$ is a minimum of $J$ and if we set

$$F(\epsilon) = J(\mathbf{a}^* + \epsilon\mathbf{v}) \tag{5-86}$$

† That is, $\mathfrak{A}$ is an element of the dual space $V^*$ of $V$ and, hence, is a linear functional on $V$ (see Exercise 2-11).

then $\epsilon = 0$ must be a minimum of the real-valued function $F(\epsilon)$, and we can conclude that

$$\left.\frac{dF(\epsilon)}{d\epsilon}\right|_{\epsilon=0} = 0$$

that is, that

$$\begin{aligned}\lim_{\epsilon\to 0} \frac{|J(\mathbf{a}^* + \epsilon\mathbf{v}) - J(\mathbf{a}^*)|}{\epsilon} &= \lim_{\epsilon\to 0}\left|\frac{\epsilon}{\epsilon}\,\delta J(\mathbf{a}^*)\mathbf{v} + \frac{o(\epsilon)}{\epsilon}\right| \\ &= |\delta J(\mathbf{a}^*)\mathbf{v}| = 0 \qquad \text{(5-87)†}\end{aligned}$$

Since $\mathbf{v}$ can be any nonzero vector in $V$, we must have

$$\delta J(\mathbf{a}^*) = \mathbf{0} \tag{5-88}$$

and this is the exact analog of the necessary conditions that we have derived for functions of one or several variables and will take on more significance in the next several sections. Thus, we may again conclude that the local minima of $J$ can occur only at:

1. Points on the boundary of $A$
2. Points at which the derivative of $J$ does not exist
3. Interior points of $A$ at which the derivative (i.e., the first variation) of $J$ vanishes

**Example 5-6** Let $\varphi(x)$ be a continuously differentiable real-valued function on $R$ and let $J$ be the real-valued function defined on the function space $\mathcal{P}([0, 1], R)$ of all piecewise continuous functions from $[0, 1]$ into $R$‡ by the relation

$$J(u) = \int_0^1 \varphi[u(t)]\,dt \qquad u \in \mathcal{P}([0, 1], R)$$

If $u^*$ is an element of $\mathcal{P}([0, 1], R)$ and if $u^* + \epsilon v$ is a perturbation or variation of $u^*$, then

$$\begin{aligned}J(u^* + \epsilon v) &= \int_0^1 \varphi[u^*(t) + \epsilon v(t)]\,dt \\ &= \int_0^1 \left\{\varphi[u^*(t)] + \epsilon \left.\frac{d\varphi}{dx}\right|_{u^*(t)} v(t) + o(\epsilon)\right\} dt \\ &= J(u^*) + \epsilon \int_0^1 \varphi'[u^*(t)]v(t)\,dt + o(\epsilon)\end{aligned}$$

We may thus conclude that if $u^*$ is a local minimum of $J$, then we must have

$$\int_0^1 \varphi'[u^*(t)]v(t)\,dt = 0$$

for all $v$ in $\mathcal{P}([0, 1], R)$. We note that the transformation $\mathcal{a}$ of $\mathcal{P}([0, 1], R)$ into $R$, given by

$$\mathcal{A}(v) = \int_0^1 \varphi'[u^*(t)]v(t)\,dt$$

is linear and continuous [it is, in fact, $\delta J(u^*)$].

† Recall that $\delta J(\mathbf{a}^*)$ is a linear transformation of $V$ into $R$, so that $\delta J(\mathbf{a}^*)\mathbf{v}$ is a real number whenever $\mathbf{v}$ is in $V$.

‡ See Sec. 3-15.

**Exercise 5-8** Let $\varphi(x, y)$ be a continuously differentiable real-valued function on $R_2$ and let $J$ be the real-valued function defined on the function space $\mathcal{P}([0, 1], R_2)$ [see Eq. (3-136)] by

$$J(\mathbf{u}) = \int_0^1 \varphi[u_1(t), u_2(t)]\,dt \qquad \mathbf{u} \in \mathcal{P}([0, 1], R_2)$$

where $u_1$ and $u_2$ are the components of $\mathbf{u}$. Show, in detail, that if $\mathbf{u}^*$ is a minimum of $J$, then

$$\int_0^1 \langle \nabla\varphi[\mathbf{u}^*(t)], \mathbf{v}(t)\rangle\,dt = 0$$

for all $\mathbf{v}$ in $\mathcal{P}([0, 1], R_2)$, where $\nabla\varphi$ is the gradient of $\varphi$.

We saw in Sec. 5-2 that *the second derivative played an important role in determining whether or not an extremum was actually a minimum.* We propose now to extend those ideas to the more general situation at hand. *First of all, we shall say that* $\mathbf{a}^*$ *is an extremum of* $J$ *if*

$$\delta J(\mathbf{a}^*) = \mathbf{0} \tag{5-89}$$

*Secondly, we shall say that the functional* $J$ *has a second derivative at* $\mathbf{a}^*$ *if there is an inner product* $P$ *(see Sec. 2-11) on* $V$ *such that*

$$J(\mathbf{a}) - J(\mathbf{a}^*) = \delta J(\mathbf{a}^*)(\mathbf{a} - \mathbf{a}^*) + P(\mathbf{a} - \mathbf{a}^*, \mathbf{a} - \mathbf{a}^*) + o(\|\mathbf{a} - \mathbf{a}^*\|^2) \tag{5-90}$$

*for all* $\mathbf{a}$ *"near"* $\mathbf{a}^*$*; the quadratic form induced by* $P$ *(see Sec. 2-11) is a functional on* $V$ *and is called the second derivative, or second variation, of* $J$ *at* $\mathbf{a}^*$*; and we write* $\delta^2 J(\mathbf{a}^*)$ *for this quadratic form, so that*

$$J(\mathbf{a}) - J(\mathbf{a}^*) = \delta J(\mathbf{a}^*)(\mathbf{a} - \mathbf{a}^*) + \delta^2 J(\mathbf{a}^*)(\mathbf{a} - \mathbf{a}^*) + o(\|\mathbf{a} - \mathbf{a}^*\|^2) \tag{5-91}$$

*and*

$$\delta^2 J(\mathbf{a}^*)(\mathbf{a} - \mathbf{a}^*) = P(\mathbf{a} - \mathbf{a}^*, \mathbf{a} - \mathbf{a}^*) \tag{5-92}$$

Now, if $\mathbf{v}$ is some nonzero vector in $V$ and if $\epsilon$ is a small real number, then, assuming $\delta J(\mathbf{a}^*)$ and $\delta^2 J(\mathbf{a}^*)$ exist, we have

$$J(\mathbf{a}^* + \epsilon\mathbf{v}) = J(\mathbf{a}^*) + \epsilon\delta J(\mathbf{a}^*)\mathbf{v} + \epsilon^2\delta^2 J(\mathbf{a}^*)\mathbf{v} + o(\epsilon^2) \tag{5-93}$$

where $o(\epsilon^2)$ has the property that

$$\lim_{\epsilon\to 0} \frac{o(\epsilon^2)}{\epsilon^2} = 0 \tag{5-94}$$

If $\mathbf{a}^*$ is a minimum of $J$, then we know that

$$\delta J(\mathbf{a}^*) = \mathbf{0} \tag{5-95}$$

so that Eq. (5-93) becomes

$$J(\mathbf{a}^* + \epsilon\mathbf{v}) = J(\mathbf{a}^*) + \epsilon^2\delta^2 J(\mathbf{a}^*)\mathbf{v} + o(\epsilon^2) \tag{5-96}$$

Since $\epsilon^2 > 0$ and the $o(\epsilon^2)$ term may be neglected, it follows that

$$J(\mathbf{a}^* + \epsilon\mathbf{v}) - J(\mathbf{a}^*) \geq 0 \tag{5-97}$$

implies that

$$\delta^2 J(\mathbf{a}^*)\mathbf{v} \geq 0 \tag{5-98}$$

for all $\mathbf{v}$ and, hence, that the inner product $P$ of Eq. (5-90) must be *positive* [see Eq. (2-84)]. Equation (5-98) thus represents a further necessary condition that $\mathbf{a}^*$ be a minimum of $J$. The significance of this condition will become clearer in Sec. 5-7.

Finally, if we say that the quadratic form induced by an inner product $P$ on the space $V$ is *positive definite* when there is an $\alpha > 0$ such that

$$P(\mathbf{v}, \mathbf{v}) \geq \alpha \|\mathbf{v}\|^2 \tag{5-99}†$$

for all $\mathbf{v}$ in $V$, then it can be shown that the *positive definiteness of* $\delta^2 J(\mathbf{a}^*)$ *is a sufficient condition for the extremum* $\mathbf{a}^*$ *of* $J$ *to be a local minimum of* $J$.

**Example 5-7** Let $\varphi(x)$ be a twice continuously differentiable real-valued function on $R$ and let $J$ be the real-valued function defined on the function space $\mathcal{P}([0, 1], R)$ of all piecewise continuous functions from $[0, 1]$ into $R$ by

$$J(u) = \int_0^1 \varphi[u(t)]\, dt \qquad u \in \mathcal{P}([0, 1], R)$$

If $u^*$ is in $\mathcal{P}([0, 1], R)$ and if $u^* + \epsilon v$ is a variation of $u^*$, then

$$\begin{aligned} J(u^* + \epsilon v) &= \int_0^1 \varphi[u^*(t) + \epsilon v(t)]\, dt \\ &= \int_0^1 \left\{ \varphi[u^*(t)] + \epsilon \frac{d\varphi}{dx}\bigg|_{u^*(t)} v(t) + \frac{\epsilon^2}{2} \frac{d^2\varphi}{dx^2}\bigg|_{u^*(t)} v^2(t) + o(\epsilon^2) \right\} dt \\ &= J(u^*) + \epsilon \int_0^1 \varphi'[u^*(t)]v(t)\, dt + \frac{\epsilon^2}{2} \int_0^1 \varphi''[u^*(t)]v^2(t)\, dt + o(\epsilon^2) \end{aligned}$$

We may thus conclude that if $u^*$ is a local minimum of $J$, then

$$\int_0^1 \varphi'[u^*(t)]v(t)\, dt = 0$$

for all $v$ in $\mathcal{P}([0, 1], R)$ and

$$\int_0^1 \varphi''[u^*(t)]v^2(t)\, dt \geq 0$$

for all $v$ in $\mathcal{P}([0, 1], R)$. We note that the transformation $\mathcal{A}$, given by

$$\mathcal{A}(v) = \int_0^1 \varphi'[u^*(t)]v(t)\, dt$$

is linear and continuous [it is, in fact, $\delta J(u^*)$] and that the function $P$ from $\mathcal{P}([0, 1], R) \times \mathcal{P}([0, 1], R)$ into $R$, given by

$$P(v, w) = \tfrac{1}{2} \int_0^1 \varphi''[u^*(t)]v(t)w(t)\, dt$$

is an inner product on $\mathcal{P}([0, 1], R)$ [the quadratic form induced by $P$ is $\delta^2 J(u^*)$].

## 5-6 An Example

We shall examine a particular control problem very carefully in this section. Since this problem will represent our first functional minimization

† Compare with Sec. 2-15.

problem, we shall attempt to be very explicit, and we shall try to illustrate and motivate the techniques that we shall use to obtain more general results later on.

Let us begin by stating our problem. We consider the first-order dynamical system with $(-\infty, \infty)$ as interval of definition and with state equation

$$\dot{x}(t) = -x(t) + u(t) \tag{5-100}$$

We suppose that *any* bounded piecewise continuous function from $(-\infty, \infty)$ into $R$ is an admissible control, so that we are dealing with an unconstrained problem (see Sec. 4-13). We let $L(x, u)$ be the continuous function, defined by

$$L(x, u) = \tfrac{1}{2}x^2 + \tfrac{1}{2}u^2 \tag{5-101}$$

and we let $K(x, t)$ be the function given by

$$K(x, t) \equiv 0 \tag{5-102}$$

We suppose that $t_0 = 0$ is our initial time, that

$$x(0) = 1 \tag{5-103}$$

is our initial state, and that our target set $S$ [which is a subset of $R \times (-\infty, \infty)$] consists of the single point $(0, T)$; that is, we want to hit the point $x = 0$ at the fixed time $t = T$ so that the target set $S$ is given by

$$S = \{(x(T), T): x(T) = 0\} \tag{5-104}$$

We are thus dealing with a *fixed-time, fixed-end-point* problem (see Sec. 4-13). Our performance functional $J(u)$ is given by

$$J(u) = \int_0^T [\tfrac{1}{2}x^2(t) + \tfrac{1}{2}u^2(t)]\, dt \tag{5-105}$$

where $x(t)$ is the trajectory of Eq. (5-100), starting from $x(0) = 1$ and generated by the control $u(t)$. Our problem may be stated as follows: *Determine the admissible control $u(t)$ which transfers $x(0) = 1$ to $x(T) = 0$ and which minimizes the performance functional $J(u)$.*

We observe, first of all, that we are here faced with a problem similar to the general problem discussed in the previous section. Our space $V$ is $\mathcal{P}([0, T], R)$ (the set of all piecewise continuous functions on $[0, T]$; see Sec. 3-15), and our subset $A$ is the set of controls $u$ which transfer $x(0) = 1$ to $x(T) = 0$. However, there is an additional constraint, namely, the equation (5-100) of the system. As we shall see, this constraint will be handled in a way which is quite similar to the Lagrange multiplier method of Sec. 5-4. We also note that here *we do not know a priori whether or not a solution to our problem exists.*† We shall, here, simply develop *necessary*

† It is quite difficult in situations of this type to determine whether or not $A$ is compact; even if $A$ is not, the problem may have a solution.

conditions for an optimum and then point out the need for a proof that an optimum actually exists. (In this problem, it does.)

Now let us determine the *necessary conditions* which an optimal control must satisfy. For purposes of exposition, we shall proceed step by step; each step will be numbered, and a brief heuristic explanation of it will accompany the formal development.

*Step 1 Assumption of an Optimum*

In order to develop necessary conditions which an optimal control must satisfy, we assume that there is an optimal control.† So, we *suppose that* $u^*(t)$ *is an optimal control and that* $x^*(t)$ *is the corresponding optimal trajectory.* In other words, $x^*(t)$ and $u^*(t)$ satisfy the following conditions:

1. $$\dot{x}^*(t) = -x^*(t) + u^*(t) \tag{5-106}$$
2. $$x^*(0) = 1 \qquad x^*(T) = 0 \tag{5-107}$$

3. If $u(t)$ is any admissible control whose corresponding trajectory $x(t)$ is such that conditions 1 and 2 are satisfied, i.e., if

$$\begin{aligned} \dot{x}(t) &= -x(t) + u(t) \\ x(0) &= 1 \qquad x(T) = 0 \end{aligned} \tag{5-108}$$

then

$$J^* = J(u^*) = \tfrac{1}{2} \int_0^T [x^{*2}(t) + u^{*2}(t)]\, dt \leq J(u) = \tfrac{1}{2} \int_0^T [x^2(t) + u^2(t)]\, dt \tag{5-109}$$

*We shall use the asterisk superscript to denote optimal quantities throughout the remainder of this chapter and this book.*

We note here that, in view of our definition (see Definition 3-41) of the solution of a differential equation, any piecewise continuous function which differs from $u^*(t)$ only on a countable subset of $[0, T]$ is again an optimal control. Thus, to be strictly correct in our development, we should often have to add the phrase, "except possibly on a countable subset of $[0, T]$," to the statements we make; or else we could speak of equivalence classes, as in Sec. 3-15. However, we shall, having called this point to the reader's attention, slur over these mathematical niceties in what follows.

*Step 2 Perturbation*

As in our treatment of ordinary minima, we shall use a perturbation method to obtain the necessary conditions. Here, we perturb $u^*(t)$ and deduce certain conditions which the perturbation must satisfy.

So we let

$$u(t) = u^*(t) + \epsilon\eta(t) \tag{5-110}$$

† There may be many or none at all.

where $\eta(t)$ is a piecewise continuous function on $[0, T]$, be a perturbation, or variation, of $u^*(t)$. We often write

$$u(t) = u^*(t) + \delta u^*(t) \tag{5-111}$$

and call

$$\delta u^*(t) = \epsilon\eta(t) \tag{5-112}$$

a *variation* of $u^*(t)$. If $x(t)$ is the trajectory generated by $u(t)$, then we may, in view of Axiom 4-3 of the definition of a dynamical system (see Sec. 4-5), write

$$x(t) = x^*(t) + \epsilon\psi(t) \tag{5-113}$$

and we know that the function $\psi(t)$ is bounded on $(0, T]$. In other words, a small variation in the control generates a small variation in the motion of the system. Since $x(t)$ is a solution of Eq. (5-100), we have

$$\dot{x}(t) = -x(t) + u(t) \tag{5-114}$$

$$= -x^*(t) - \epsilon\psi(t) + u^*(t) + \epsilon\eta(t) \tag{5-115}$$

and, in view of Eq. (5-113),

$$\dot{x}(t) = \dot{x}^*(t) + \epsilon\dot{\psi}(t) \tag{5-116}$$

It follows that

$$\dot{\psi}(t) = -\psi(t) + \eta(t) \tag{5-117}$$

or, equivalently, that

$$\frac{d}{dt}\{\delta x^*(t)\} = -\delta x^*(t) + \delta u^*(t) \tag{5-118}$$

where

$$\delta x^*(t) = \epsilon\psi(t) \tag{5-119}$$

is the variation of $x^*(t)$ corresponding to the variation $\delta u^*(t)$ of $u^*(t)$.

Now, we also require that our perturbed control transfer $x(0) = 1$ to $x(T) = 0$, and so we must have

$$\psi(0) = \psi(T) = 0 \tag{5-120}$$

In view of Eq. (5-117), this means that

$$\psi(T) = e^{-T}\int_0^T e^t\eta(t)\,dt = 0 \tag{5-121}$$

so that our perturbation $\epsilon\eta(t)$ is not completely arbitrary. For example, we can let $\eta(t) = e^{-t}\cos(\pi t/T)$, but we cannot let $\eta(t) = e^{-t}\sin(\pi t/T)$.

To sum up, we make a small change $\epsilon\eta(t)$ in the control $u^*(t)$, demanding that $\eta(t)$ be in $\mathcal{P}([0, T], R)$ [that is, that $\eta(t)$ be a piecewise continuous function from $[0, T]$ into $R$] and that $\eta(t)$ satisfy the relation

$$\int_0^T e^t\eta(t)\,dt = 0 \tag{5-122}$$

and we then deduce that the corresponding change $\epsilon\psi(t)$ in the trajectory $x^*(t)$ is the solution of the differential equation

$$\epsilon\dot{\psi}(t) = -\epsilon\psi(t) + \epsilon\eta(t) \qquad \psi(0) = 0 \tag{5-123}$$

These perturbations are illustrated in Fig. 5-6.

*Step 3 Introduction of the Hamiltonian*

Before we calculate the effect of the perturbation on the value of the performance functional $J$, we shall introduce a scalar function $H$, called the *Hamiltonian*, which will facilitate our development of the desired necessary

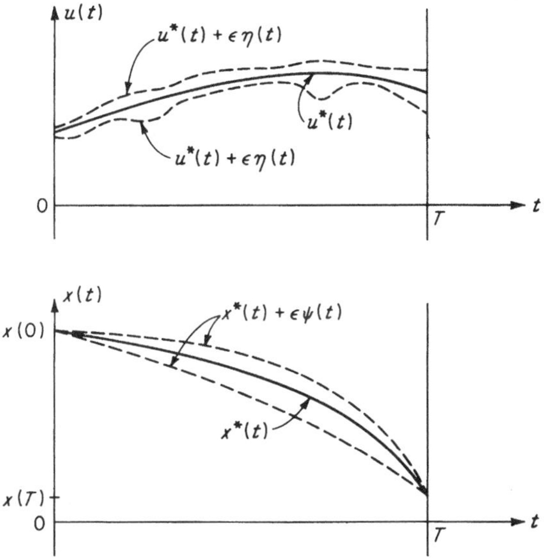


**Fig. 5-6** Perturbations of the control and corresponding perturbations of the trajectory.

conditions. In doing so, we shall also introduce the *costate variable*, which will play a role analogous to that played by the Lagrange multipliers in Sec. 5-4. Although, at this stage, the procedure may seem artificial, the importance of these concepts will become clearer later on.

So, let us suppose that $p(t)$ is some *as yet unspecified* piecewise continuous function; we shall call $p(t)$ the *costate* (or *adjoint* or *Lagrange multiplier*) variable.† Since $x(t)$ and $x^*(t)$ are both solutions of Eq. (5-100) generated by the controls $u(t)$ and $u^*(t)$, respectively, we have

$$-\dot{x}(t) - x(t) + u(t) = 0 \tag{5-124}$$
$$-\dot{x}^*(t) - x^*(t) + u^*(t) = 0 \tag{5-125}$$

† When specified, $p(t)$ will "take care of" the constraint imposed by the equation (5-100) of the system.

It follows that

$$\int_0^T p(t)[-x(t) + u(t) - \dot{x}(t)]\,dt = \int_0^T [-p(t)x(t) + p(t)u(t) - p(t)\dot{x}(t)]\,dt = 0 \quad (5\text{-}126)$$

$$\int_0^T p(t)[-x^*(t) + u^*(t) - \dot{x}^*(t)]\,dt = \int_0^T [-p(t)x^*(t) + p(t)u^*(t) - p(t)\dot{x}^*(t)]\,dt = 0 \quad (5\text{-}127)$$

and hence that

$$J(u) = \int_0^T \left[\frac{x^2(t)}{2} + \frac{u^2(t)}{2} - p(t)x(t) + p(t)u(t) - p(t)\dot{x}(t)\right] dt \quad (5\text{-}128)$$

and that

$$J(u^*) = \int_0^T \left[\frac{x^{*2}(t)}{2} + \frac{u^{*2}(t)}{2} - p(t)x^*(t) + p(t)u^*(t) - p(t)\dot{x}^*(t)\right] dt \quad (5\text{-}129)$$

If we define a scalar-valued function $H(x, p, u)$ on $R \times R \times R$ by setting

$$H(x, p, u) = L(x, u) + p(-x + u) = \frac{x^2}{2} + \frac{u^2}{2} + p(-x + u) \quad (5\text{-}130)$$

then Eqs. (5-128) and (5-129) may be written in the form

$$J(u) = \int_0^T \{H[x(t), p(t), u(t)] - p(t)\dot{x}(t)\}\,dt \quad (5\text{-}131)$$

$$J(u^*) = \int_0^T \{H[x^*(t), p(t), u^*(t)] - p(t)\dot{x}^*(t)\}\,dt \quad (5\text{-}132)$$

The function $H$ is often referred to as the *Hamiltonian function* (or, simply, the *Hamiltonian*) of the problem. We deduce immediately from Eqs. (5-131) and (5-132) that

$$J(u) - J(u^*) = \int_0^T \{H[x(t), p(t), u(t)] - H[x^*(t), p(t), u^*(t)]\}\,dt + \int_0^T p(t)[\dot{x}^*(t) - \dot{x}(t)]\,dt \quad (5\text{-}133)$$

*Step 4 Calculation of $J(u) - J(u^*)$*

We are now ready to calculate the effect of the perturbation upon the value of $J$. We shall do this by making use of the various perturbation equations of step 1, the definition of $H$, a Taylor's series expansion of $H$, and an integration by parts.

First of all we observe that, for $\epsilon$ small,

$$\begin{aligned} H[x(t), p(t), u(t)] &= H[x^*(t), p(t), u^*(t)] \\ &\quad + \frac{\partial H}{\partial x}[x^*(t), p(t), u^*(t)]\,[x(t) - x^*(t)] \\ &\quad + \frac{\partial H}{\partial u}[x^*(t), p(t), u^*(t)]\,[u(t) - u^*(t)] + o(\epsilon) \qquad (5\text{-}134) \\ &= H[x^*(t), p(t), u^*(t)] + \frac{\partial H}{\partial x}[x^*(t), p(t), u^*(t)]\epsilon\psi(t) \\ &\quad + \frac{\partial H}{\partial u}[x^*(t), p(t), u^*(t)]\epsilon\eta(t) + o(\epsilon) \qquad (5\text{-}135) \end{aligned}$$

where $o(\epsilon)$ is a term such that $\lim_{\epsilon \to 0} o(\epsilon)/\epsilon = 0$, since

$$\begin{aligned} x(t) - x^*(t) &= \epsilon\psi(t) \\ u(t) - u^*(t) &= \epsilon\eta(t) \end{aligned} \qquad (5\text{-}136)$$

in view of Eqs. (5-110) and (5-113). Secondly, since

$$\dot{x}^*(t) - \dot{x}(t) = -\epsilon\dot{\psi}(t) \qquad (5\text{-}137)$$

it follows from Eq. (5-133) that

$$\begin{aligned} J(u) - J(u^*) &= \epsilon \int_0^T \left\{ \frac{\partial H}{\partial x}[x^*(t), p(t), u^*(t)]\psi(t) \right. \\ &\quad \left. + \frac{\partial H}{\partial u}[x^*(t), p(t), u^*(t)]\eta(t) \right\} dt - \epsilon \int_0^T p(t)\dot{\psi}(t)\,dt + o(\epsilon) \qquad (5\text{-}138) \end{aligned}$$

**Exercise 5-9** Write Eq. (5-138) in terms of $\delta x^*(t)$, $\delta u^*(t)$, and $(d/dt)\delta x^*(t)$. Why is there no term involving $\partial H/\partial p$ in this equation? Why can we write $o(\epsilon)$ in Eq. (5-138) instead of $\int_0^T o(\epsilon)\,dt$, as would strictly follow from Eq. (5-124)? HINT: Is not $o(\epsilon)$ of the form $\epsilon^2 f(t)$, where $f(t)$ is a *bounded* piecewise continuous function of $t$?

Now the second integral in Eq. (5-138) can be evaluated by parts to give

$$\begin{aligned} -\int_0^T p(t)\dot{\psi}(t)\,dt &= p(0)\psi(0) - p(T)\psi(T) + \int_0^T \dot{p}(t)\psi(t)\,dt \qquad (5\text{-}139) \\ &= \int_0^T \dot{p}(t)\psi(t)\,dt \qquad \text{by Eq. (5-120)} \qquad (5\text{-}140) \end{aligned}$$

Thus, we have

$$\begin{aligned} J(u) - J(u^*) &= \epsilon \left( \int_0^T \left\{ \frac{\partial H}{\partial x}[x^*(t), p(t), u^*(t)] + \dot{p}(t) \right\} \psi(t)\,dt \right. \\ &\quad \left. + \int_0^T \frac{\partial H}{\partial u}[x^*(t), p(t), u^*(t)]\eta(t)\,dt \right) + o(\epsilon) \qquad (5\text{-}141) \end{aligned}$$

where, so far, the function $p(t)$ has not been specified.

Since $u^*$ is an optimal control, we must have

$$J(u) - J(u^*) \geq 0 \qquad (5\text{-}142)$$

Furthermore, since $\epsilon$ may be positive or negative and since the $o(\epsilon)$ term may be neglected (cf., for example, Exercise 5-2), we may conclude that

$$\left( \int_0^T \left\{ \frac{\partial H}{\partial x} [x^*(t), p(t), u^*(t)] + \dot{p}(t) \right\} \psi(t) \, dt + \int_0^T \frac{\partial H}{\partial u} [x^*(t), p(t), u^*(t)] \eta(t) \, dt \right) = 0 \qquad (5\text{-}143)$$

This is the first use we have made of the optimality of $u^*$, and Eq. (5-143) is a *preliminary* form of the desired necessary condition.

*Step 5 A Differential Equation for $p(t)$*

We are now ready to impose some conditions on $p(t)$. In fact, we shall demand that $p(t)$ satisfy a certain differential equation. All that will remain to specify $p(t)$ completely will be the choice of the initial condition $p(0)$.

So we now require that $p(t)$ be a solution of the *linear* differential equation

$$\dot{p}(t) = -\frac{\partial H}{\partial x} [x^*(t), p(t), u^*(t)] \qquad (5\text{-}144)$$

$$= p(t) - x^*(t) \qquad (5\text{-}145)$$

We observe, first of all, that the homogeneous parts of Eqs. (5-145) and (5-106) are adjoint in the sense of Sec. 3-25. Secondly, we note that Eqs. (5-143) and (5-144) imply that

$$\int_0^T \frac{\partial H}{\partial u} [x^*(t), p(t), u^*(t)] \eta(t) \, dt = 0 \qquad (5\text{-}146)$$

whenever $p(t)$ is a solution of Eq. (5-144). Thirdly, we see that $x^*(t)$ is a solution of the differential equation

$$\dot{x}^*(t) = \frac{\partial H}{\partial p} [x^*(t), p(t), u^*(t)] \qquad (5\text{-}147)$$

so that *both* $x^*(t)$ and $p(t)$ are solutions of differential equations involving the Hamiltonian $H$.

*Step 6 A Fundamental Lemma*

We have obtained a relation, Eq. (5-146), which must hold for every piecewise continuous function $\eta(t)$ which satisfies Eq. (5-122). What does this imply about $\partial H[x^*(t), p(t), u^*(t)]/\partial u$? We shall answer this question on the basis of the following lemma:

***Lemma 5-1*** *Let $h(t)$ be a piecewise continuous function on $[0, T]$. Suppose that*

$$\int_0^T h(t)\alpha(t) \, dt = 0 \qquad (5\text{-}148)$$

*for all piecewise continuous functions* $\alpha(t)$ *on* $[0, T]$ *such that*

$$\int_0^T \alpha(t)\, dt = 0 \tag{5-149}$$

*Then* $h(t)$ *is constant on* $[0, T]$; *that is,*

$$h(t) = c \qquad t \in [0, T] \tag{5-150†}$$

PROOF Since $h(t)$ is piecewise continuous, there is a $c$ in $R$ such that

$$\int_0^T h(t)\, dt = cT \tag{5-151}$$

It then follows that

$$\int_0^T [h(t) - c]\, dt = 0 \tag{5-152}$$

so that the function $\alpha(t) = h(t) - c$ is piecewise continuous and satisfies Eq. (5-149). Thus,

$$\int_0^T h(t)[h(t) - c]\, dt = 0 \tag{5-153}$$

and

$$\int_0^T - c[h(t) - c]\, dt = 0 \tag{5-154}$$

which together imply that

$$\int_0^T [h(t) - c]^2\, dt = 0 \tag{5-155}$$

We conclude from Theorem 3-8, part $f$, that

$$h(t) - c = 0 \tag{5-156}$$

(except for $t$ in a countable subset of $[0, T]$), which establishes the lemma.

If we let

$$\eta(t) = e^{-t}\alpha(t) \tag{5-157}$$

where $\alpha(t)$ is *any* piecewise continuous function satisfying Eq. (5-149), then Eq. (5-146) becomes

$$\int_0^T \frac{\partial H}{\partial u}[x^*(t), p(t), u^*(t)]e^{-t}\alpha(t)\, dt = 0 \tag{5-158}$$

and we conclude from the lemma that

$$\frac{\partial H}{\partial u}[x^*(t), p(t), u^*(t)] = e^t c \tag{5-159}$$

where $c$ is a constant.

*Step 7 The Necessary Condition*

We are now ready to give the desired necessary condition. Essentially, we shall show that there is a solution $p^*(t)$ of Eq. (5-145) such that the function $H[x^*(t), p^*(t), u]$ of $u$ has an absolute minimum at $u = u^*(t)$.

† Except possibly for a countable subset $A$ of $[0, T]$.

Let $\hat{p}(t)$ be the solution of Eq. (5-145) with $\hat{p}(0) = 0$ (say); then

$$\hat{p}(t) = -e^t \int_0^t e^{-\tau} x^*(\tau)\, d\tau \tag{5-160}$$

and there is, by virtue of Eq. (5-159), a constant $c^*$ such that

$$\frac{\partial H}{\partial u}\,[x^*(t), \hat{p}(t), u^*(t)] = u^*(t) + \hat{p}(t) = e^t c^* \tag{5-161}$$

In other words,

$$u^*(t) - e^t \int_0^t e^{-\tau} x^*(\tau)\, d\tau = e^t c^* \tag{5-162}$$

or, equivalently,

$$u^*(t) - e^t c^* - e^t \int_0^t e^{-\tau} x^*(\tau)\, d\tau = 0 \tag{5-163}$$

Now we let $p^*(t)$ be the (unique) solution of Eq. (5-145) with $p^*(0) = -c^*$, that is,

$$p^*(t) = -e^t c^* - e^t \int_0^t e^{-\tau} x^*(\tau)\, d\tau \tag{5-164}$$

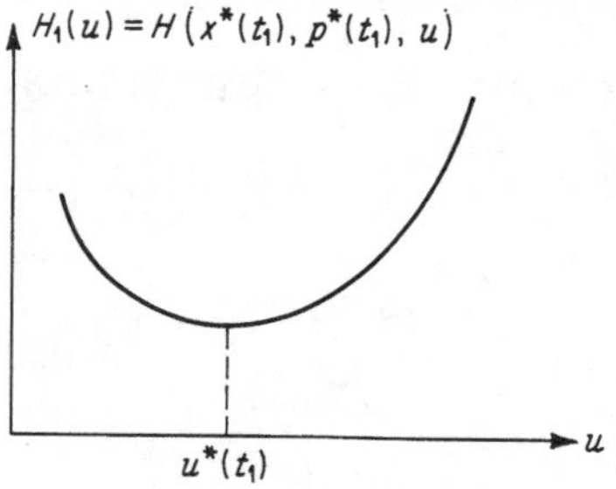

**Fig. 5-7** $H_1(u)$ has an absolute minimum at $u = u^*(t_1)$.

Then *we claim that the function* $H[x^*(t), p^*(t), u]$ *has an absolute minimum as a function of* $u$ *at* $u = u^*(t)$ *for* $t$ *in* $[0, T]$. This claim is to be understood in the following way: Suppose that $t_1$ is some element of $[0, T]$; then $x^*(t_1)$ and $p^*(t_1)$ are well-defined numbers, and we can consider the real-valued function $H_1(u)$ of the real variable $u$, given by

$$H_1(u) = H[x^*(t_1), p^*(t_1), u] \tag{5-165}$$

$$= \frac{x^{*2}(t_1)}{2} + \frac{u^2}{2} + p^*(t_1)[-x^*(t_1) + u] \tag{5-166}$$

and our claim is that $u = u^*(t_1)$ is the absolute minimum of $H_1(u)$. The claim is illustrated in Fig. 5-7.

To verify our claim, we note, first of all, that

$$H[x^*(t), p^*(t), u] = \frac{x^{*2}(t)}{2} + \frac{u^2}{2} + p^*(t)[-x^*(t) + u] \tag{5-167}$$

is a parabola in $u$ which has a unique local minimum which is also its absolute minimum. We next observe that, letting $G(u) = H[x^*(t), p^*(t), u]$,

$$\left.\frac{dG(u)}{du}\right|_{u=u^*(t)} = \frac{\partial H}{\partial u}\,[x^*(t), p^*(t), u^*(t)] \tag{5-168}$$

$$= u^*(t) + p^*(t) \tag{5-169}$$

$$= u^*(t) - e^t c^* - e^t \int_0^t e^{-\tau} x^*(\tau)\, d\tau \tag{5-170}$$

$$= 0 \tag{5-171}$$

in view of Eqs. (5-163) and (5-164). Moreover,

$$\left.\frac{d^2G(u)}{du^2}\right|_{u=u^*(t)} = \frac{\partial^2 H}{\partial u^2}[x^*(t), p^*(t), u^*(t)] = 1 > 0 \qquad (5\text{-}172)$$

so that $u = u^*(t)$ is indeed the minimum of $G(u) = H[x^*(t), p^*(t), u]$ and our claim has been verified.

Thus, we may summarize what we have done in the following theorem, which represents the necessary conditions that we have been seeking for the example.

***Theorem 5-2*** *Let $u^*(t)$ be an admissible control which transfers $x(0) = 1$ to $x(T) = 0$. In order that $u^*(t)$ be optimal, it is necessary that there exist a function $p^*(t)$ such that:*

*a. If $x^*(t)$ is the trajectory corresponding to $u^*(t)$, then $x^*(t)$ and $p^*(t)$ are a solution of the system of equations*

$$\begin{aligned} \dot{x}^*(t) &= \frac{\partial H}{\partial p}[x^*(t), p^*(t), u^*(t)] \\ \dot{p}^*(t) &= -\frac{\partial H}{\partial x}[x^*(t), p^*(t), u^*(t)] \end{aligned} \qquad (5\text{-}173)$$

*b. The function $H[x^*(t), p^*(t), u]$ has an absolute minimum as a function of $u$ at $u = u^*(t)$ for $t$ in $[0, T]$, that is,*

$$\min_u H[x^*(t), p^*(t), u] = H[x^*(t), p^*(t), u^*(t)] \qquad (5\text{-}174)$$

Let us now interpret Theorem 5-2. We are given an optimal control $u^*(t)$ and the corresponding optimal trajectory $x^*(t)$ for all $t$ in $[0, T]$. Then we are guaranteed by part *a* of the theorem that there is a function $p^*(t)$ [which is the solution of Eqs. (5-173)] corresponding to $x^*(t)$ and $u^*(t)$. Thus, at each specific instant of time, say $t_1$, in the interval $[0, T]$, $x^*(t_1)$, $p^*(t_1)$, and $u^*(t_1)$ are three well-defined numbers. Part *b* of the theorem tells us that the *number* $H[x^*(t_1), p^*(t_1), u^*(t_1)]$ is smaller than or equal to the *number* $H[x^*(t_1), p^*(t_1), r]$, where $r$ is *any real number*. This theorem, which is, for our problem, the *minimum principle of Pontryagin*, will be generalized quite a bit in the sequel.

Before reviewing what we have done and before illustrating how the necessary conditions can be used, let us examine the function $H[x^*(t), p^*(t), u^*(t)]$. We have, in particular,

$$\frac{dH[x^*(t), p^*(t), u^*(t)]}{dt} = \left.\frac{\partial H}{\partial x}\right|_* \dot{x}^*(t) + \left.\frac{\partial H}{\partial p}\right|_* \dot{p}^*(t) + \left.\frac{\partial H}{\partial u}\right|_* \dot{u}^*(t) \qquad (5\text{-}175)$$

where $|_*$ indicates that the derivative is to be evaluated at $[x^*(t), p^*(t), u^*(t)]$. It follows, then, from Eqs. (5-168), (5-171), and (5-173), that

$$\frac{dH[x^*(t), p^*(t), u^*(t)]}{dt} = 0 \qquad (5\text{-}176)$$

and, hence, that $H[x^*(t), p^*(t), u^*(t)]$ is *constant*. In other words, the Hamiltonian $H$ for this problem is constant along an optimal trajectory.

Now let us recapitulate what we have done. We began by assuming that there was an optimum; we then proceeded to perturb the control and examine the effect of this perturbation on the trajectory; we next introduced the Hamiltonian $H$ and the costate variable $p$, which later played a role analogous to that of a Lagrange multiplier; after that, we calculated the effect of our perturbation on $J$ in terms of the Hamiltonian; next, we required that our costate be the solution of a linear differential equation; we then proved a lemma about functions with zero integrals; and, finally, we derived the necessary condition, which stated that we could find a suitable costate so that the optimal control minimized the Hamiltonian $H$ as a function of $u$. Thus, we "reduced" our functional-minimization problem to an ordinary minimization problem. Often this "reduction" is of great value in actually solving the control problem.

Now let us see how the necessary conditions of Theorem 5-2 can be used. We observe, first of all, that the theorem states, in effect, that an optimum gives rise to a solution $x^*(t)$, $u^*(t)$, $p^*(t)$ of Eqs. (5-173) and (5-174), that is,

$$\dot{x}^*(t) = \frac{\partial H}{\partial p}[x^*(t), p^*(t), u^*(t)] = -x^*(t) + u^*(t) \tag{5-177}$$

$$\dot{p}^*(t) = -\frac{\partial H}{\partial x}[x^*(t), p^*(t), u^*(t)] = p^*(t) - x^*(t) \tag{5-178}$$

$$\min_u H[x^*(t), p^*(t), u] = H[x^*(t), p^*(t), u^*(t)] \tag{5-179}$$

and

$$x^*(0) = 1 \qquad x^*(T) = 0 \tag{5-180}$$

In view of the definition of $H$, we may write Eq. (5-179) in the form

$$\begin{aligned} &\tfrac{1}{2}x^{*2}(t) + \tfrac{1}{2}u^{*2}(t) + p^*(t)[-x^*(t) + u^*(t)] \\ &\qquad \le \tfrac{1}{2}x^{*2}(t) + \tfrac{1}{2}u^2 + p^*(t)[-x^*(t) + u] \end{aligned} \tag{5-181}$$

for all $u$. Equation (5-181) is equivalent to the relation

$$\tfrac{1}{2}u^{*2}(t) + p^*(t)u^*(t) \le \tfrac{1}{2}u^2 + p^*(t)u \qquad \text{for all } u \tag{5-182}$$

which, in turn, yields the equality

$$u^*(t) = -p^*(t) \tag{5-183}$$

We note that, since

$$\frac{\partial^2 H}{\partial u^2}[x^*(t), p^*(t), u^*(t)] = 1 \tag{5-184}$$

we could also have derived Eq. (5-183) from the relation

$$\frac{\partial H}{\partial u}[x^*(t), p^*(t), u^*(t)] = u^*(t) + p^*(t) = 0 \tag{5-185}$$

Thus, we deduce that an optimum gives rise to a solution of Eqs. (5-177), (5-178), and (5-183), satisfying the boundary conditions of Eqs. (5-180). Therefore, we shall try to find *all* the solutions of these equations. In other words, we want to find all triples of functions $x(t)$, $p(t)$, $u(t)$ such that

$$\dot{x}(t) = -x(t) + u(t) \tag{5-186}$$
$$\dot{p}(t) = p(t) - x(t) \tag{5-187}$$
$$u(t) = -p(t) \tag{5-188}$$

and

$$x(0) = 1 \qquad x(T) = 0 \tag{5-189}$$

So we seek the solutions of the system

$$\begin{aligned} \dot{x}(t) &= -x(t) - p(t) \\ \dot{p}(t) &= p(t) - x(t) \end{aligned} \tag{5-190}$$

with boundary conditions $x(0) = 1$ and $x(T) = 0$. However, the solution of this system is *unique* and is given by

$$\begin{aligned} x(t) &= \frac{1}{2\sqrt{2}} \{x(0)[(\sqrt{2} + 1)e^{-\sqrt{2}t} \\ &\qquad + (\sqrt{2} - 1)e^{\sqrt{2}t}] + p(0)(e^{-\sqrt{2}t} - e^{\sqrt{2}t})\} \\ p(t) &= \frac{1}{2\sqrt{2}} \{x(0)(e^{-\sqrt{2}t} - e^{\sqrt{2}t}) \\ &\qquad + p(0)[(\sqrt{2} - 1)e^{-\sqrt{2}t} + (\sqrt{2} + 1)e^{\sqrt{2}t}]\} \end{aligned} \tag{5-191}$$

where

$$x(0) = 1 \qquad \text{and} \qquad p(0) = \frac{(\sqrt{2} + 1)e^{-\sqrt{2}T} + (\sqrt{2} - 1)e^{\sqrt{2}T}}{e^{\sqrt{2}T} - e^{-\sqrt{2}T}} \tag{5-192}$$

It follows that if there is a solution to our control problem, this solution is *unique and must be of the form*

$$\begin{aligned} u^*(t) = \frac{-1}{2\sqrt{2}} \Bigg\{ e^{-\sqrt{2}t} - e^{\sqrt{2}t} + \frac{(\sqrt{2} + 1)e^{-\sqrt{2}T} + (\sqrt{2} - 1)e^{\sqrt{2}T}}{e^{\sqrt{2}T} - e^{-\sqrt{2}T}} \\ \times [(\sqrt{2} - 1)e^{-\sqrt{2}t} + (\sqrt{2} + 1)e^{\sqrt{2}t}] \Bigg\} \end{aligned} \tag{5-193}$$

Thus, if we had proved that a solution to our problem existed, then it would have to be the control $u^*(t)$ given by Eq. (5-193). We shall see, in later chapters of the book, many other illustrations of the use of necessary conditions in determining optimal controls.

**Exercise 5-10** Show that Eq. (5-191) does indeed define the solution of the system of Eqs. (5-190), satisfying the prescribed boundary conditions. Use the Laplace transform methods of Sec. 3-23.

**Exercise 5-11** In step 3, introduce a function $\hat{H}(x, p, u)$, given by

$$\hat{H}(x, p, u) = -L(x, u) + p(-x + u)$$

in place of the function $H(x, p, u)$. Deduce the following necessary condition for the problem:

Let $u^*(t)$ be an admissible control which transfers $x(0) = 1$ to $x(T) = 0$. In order that $u^*(t)$ be optimal, it is necessary that there exist a function $\hat{p}^*(t)$ such that:

(a) If $x^*(t)$ is the trajectory corresponding to $u^*(t)$, then $x^*(t)$ and $\hat{p}^*(t)$ are a solution of the system of equations

$$\begin{aligned} \dot{x}^*(t) &= \frac{\partial \hat{H}}{\partial p}[x^*(t), \hat{p}^*(t), u^*(t)] \\ \dot{\hat{p}}^*(t) &= -\frac{\partial \hat{H}}{\partial x}[x^*(t), \hat{p}^*(t), u^*(t)] \end{aligned} \tag{5-194}$$

(b) The function $\hat{H}[x^*(t), \hat{p}^*(t), u]$ has an absolute *maximum* as a function of $u$ at $u = u^*(t)$ for $t$ in $[0, T]$, that is,

$$\max_u H[x^*(t), \hat{p}^*(t), u] = H[x^*(t), \hat{p}^*(t), u^*(t)] \tag{5-195}$$

Thus, our necessary conditions could have been developed as a *maximum*, rather than a *minimum*, principle. Show also that calculation of the solutions of Eqs. (5-194) and (5-195) leads to the control $u^*(t)$ of Eq. (5-193).

## 5-7 The Variational Approach to Control Problem 1: Necessary Conditions for a Free-end-point Problem

Having strengthened our intuition with the general comments of Sec. 5-5 and the specific example of Sec. 5-6, we are now prepared to attack the control problem using variational techniques. We shall begin in this section by examining a free-end-point problem with a terminal cost. We shall encounter no difficulties in showing that the optimal control for this problem must *extremize* the Hamiltonian; we shall then introduce the second variation and shall show, *heuristically*, that *the extremum of the Hamiltonian must actually be a minimum.* In the next section, we shall strengthen the necessary conditions to obtain some *local sufficiency conditions* which will be particularly useful in Chap. 9. We shall then briefly examine a fixed-end-point problem and shall encounter certain difficulties which stem from the fact that our variations in the control must satisfy a relation similar to Eq. (5-122). In order to overcome these difficulties, we shall have to reorient our thinking toward the control problem, and we shall thus be led to the minimum principle of Pontryagin. We shall be somewhat more heuristic from now on, referring the reader to the relevant formal proofs in the literature.

Let us now state our problem. We consider the dynamical system

$$\dot{\mathbf{x}}(t) = \mathbf{f}[\mathbf{x}(t), \mathbf{u}(t), t] \tag{5-196}$$

where $\mathbf{x}(t)$ and $\mathbf{f}$ are $n$ vectors and $\mathbf{u}(t)$ is an $m$ vector, with $0 < m \leq n$. We suppose that we are dealing with an unconstrained problem (see Sec. 4-13), so that *any* piecewise continuous function is an admissible control. We suppose that $t_0$ is our initial time, that

$$\mathbf{x}(t_0) = \mathbf{x}_0 \tag{5-197}$$

is our initial state, and that our target set $S$ is of the form $R_n \times \{t_1\}$, where $t_1$ is a *fixed time*, $t_1 > t_0$. In other words, our final state can be any element of $R_n$. We let

$$L(\mathbf{x}, \mathbf{u}, t) \qquad \text{and} \qquad K(\mathbf{x}) \tag{5-198}$$

be sufficiently differentiable real-valued functions, and we consider the performance functional $J(\mathbf{u})$, given by

$$J(\mathbf{u}) = K[\mathbf{x}(t_1)] + \int_{t_0}^{t_1} L[\mathbf{x}(t), \mathbf{u}(t), t]\, dt \tag{5-199}$$

where $\mathbf{x}(t)$ is the trajectory of Eq. (5-196) starting from $\mathbf{x}(t_0) = \mathbf{x}_0$ and generated by the control $\mathbf{u}(t)$. Thus, our problem is: *Determine the admissible control* $\mathbf{u}(t)$ *which minimizes the performance functional* $J(\mathbf{u})$. We note that this is a *fixed-time, free-end-point problem.*

We shall develop necessary conditions entirely similar to those of Theorem 5-2 here, and we shall proceed step by step, just as in Sec. 5-6.

*Step 1 Assumption of an Optimum*

We suppose that $\mathbf{u}^*(t)$ is an optimal control and that $\mathbf{x}^*(t)$ is the corresponding optimal trajectory. In other words, $\mathbf{u}^*(t)$ and $\mathbf{x}^*(t)$ satisfy the following conditions:

1. $$\dot{\mathbf{x}}^*(t) = \mathbf{f}[\mathbf{x}^*(t), \mathbf{u}^*(t), t] \tag{5-200}$$
2. $$\mathbf{x}^*(t_0) = \mathbf{x}_0 \tag{5-201}$$

3. If $\mathbf{u}(t)$ is any admissible control whose corresponding trajectory $\mathbf{x}(t)$ begins at $\mathbf{x}(t_0) = \mathbf{x}_0$, then

$$\begin{aligned} J^* = J(\mathbf{u}^*) &= K[\mathbf{x}^*(t_1)] + \int_{t_0}^{t_1} L[\mathbf{x}^*(t), \mathbf{u}^*(t), t]\, dt \\ \le J(\mathbf{u}) &= K[\mathbf{x}(t_1)] + \int_{t_0}^{t_1} L[\mathbf{x}(t), \mathbf{u}(t), t]\, dt \end{aligned} \tag{5-202}$$

*Step 2 Perturbation*

We now perturb the control $\mathbf{u}^*(t)$ by letting

$$\mathbf{u}(t) = \mathbf{u}^*(t) + \epsilon\boldsymbol{\eta}(t) \tag{5-203}$$

If $\mathbf{x}(t)$ is the trajectory generated by $\mathbf{u}(t)$, then we may write

$$\mathbf{x}(t) = \mathbf{x}^*(t) + \epsilon\boldsymbol{\phi}(t) \tag{5-204}$$

since a small variation in the control generates a small variation in the motion of the system. Since $\mathbf{x}(t)$ is a solution of Eq. (5-196), we have

$$\dot{\mathbf{x}}(t) = \mathbf{f}[\mathbf{x}(t), \mathbf{u}(t), t] \tag{5-205}$$

$$= \mathbf{f}[\mathbf{x}^*(t) + \epsilon\boldsymbol{\phi}(t), \mathbf{u}^*(t) + \epsilon\boldsymbol{\eta}(t), t] \tag{5-206}$$

and, in view of Eq. (5-204),

$$\dot{\mathbf{x}}(t) = \dot{\mathbf{x}}^*(t) + \epsilon\dot{\boldsymbol{\phi}}(t) \tag{5-207}$$

If we expand $\mathbf{f}[\mathbf{x}^*(t) + \epsilon\boldsymbol{\phi}(t), \mathbf{u}^*(t) + \epsilon\mathbf{n}(t), t]$ in a Taylor series, we find that

$$\mathbf{f}[\mathbf{x}^*(t) + \epsilon\boldsymbol{\phi}(t), \mathbf{u}^*(t) + \epsilon\mathbf{n}(t), t] = \mathbf{f}[\mathbf{x}^*(t), \mathbf{u}^*(t), t] + \frac{\partial \mathbf{f}}{\partial \mathbf{x}}\bigg|_* \epsilon\boldsymbol{\phi}(t) + \frac{\partial \mathbf{f}}{\partial \mathbf{u}}\bigg|_* \epsilon\mathbf{n}(t) + \mathbf{o}(\epsilon) \tag{5-208}$$

where $(\partial\mathbf{f}/\partial\mathbf{x})|_*$ is the Jacobian matrix [see Eq. (3-87)] of $\mathbf{f}$ with respect to $\mathbf{x}$ evaluated at $[\mathbf{x}^*(t), \mathbf{u}^*(t), t]$, $(\partial\mathbf{f}/\partial\mathbf{u})|_*$ is the Jacobian matrix of $\mathbf{f}$ with respect to $\mathbf{u}$ evaluated at $[\mathbf{x}^*(t), \mathbf{u}^*(t), t]$, and $\mathbf{o}(\epsilon)$ is a vector such that

$$\lim_{\epsilon\to 0} \frac{\mathbf{o}(\epsilon)}{\epsilon} = \mathbf{0} \tag{5-209}$$

*From now on, we shall use the notation* $|_*$ *to indicate that a function is to be evaluated along the optimal trajectory.* It follows from Eqs. (5-206) to (5-208) that

$$\epsilon\dot{\boldsymbol{\phi}}(t) = \frac{\partial \mathbf{f}}{\partial \mathbf{x}}\bigg|_* \epsilon\boldsymbol{\phi}(t) + \frac{\partial \mathbf{f}}{\partial \mathbf{u}}\bigg|_* \epsilon\mathbf{n}(t) + \mathbf{o}(\epsilon) \tag{5-210}$$

or, equivalently, that

$$\dot{\boldsymbol{\phi}}(t) = \frac{\partial \mathbf{f}}{\partial \mathbf{x}}\bigg|_* \boldsymbol{\phi}(t) + \frac{\partial \mathbf{f}}{\partial \mathbf{u}}\bigg|_* \mathbf{n}(t) + \frac{\mathbf{o}(\epsilon)}{\epsilon} \tag{5-211}$$

Since the trajectory $\mathbf{x}(t)$ begins at $\mathbf{x}(t_0) = \mathbf{x}_0$, we must also have

$$\boldsymbol{\phi}(t_0) = \mathbf{0} \tag{5-212}$$

Now let us suppose that $\boldsymbol{\psi}(t)$ is the solution of the *linear equation*

$$\dot{\boldsymbol{\psi}}(t) = \frac{\partial \mathbf{f}}{\partial \mathbf{x}}\bigg|_* \boldsymbol{\psi}(t) + \frac{\partial \mathbf{f}}{\partial \mathbf{u}}\bigg|_* \mathbf{n}(t) \tag{5-213}$$

with

$$\boldsymbol{\psi}(t_0) = \mathbf{0} \tag{5-214}$$

Then it will follow† that

$$\mathbf{x}(t) = \mathbf{x}^*(t) + \epsilon\boldsymbol{\phi}(t) = \mathbf{x}^*(t) + \epsilon\boldsymbol{\psi}(t) + \mathbf{o}(\epsilon) \tag{5-215}$$

and so, in the calculations that follow, we shall often replace $\mathbf{x}(t)$ by $\mathbf{x}^*(t) + \epsilon\boldsymbol{\psi}(t)$ rather than by the more strictly accurate $\mathbf{x}^*(t) + \epsilon\boldsymbol{\phi}(t)$.

**Exercise 5-12** Suppose that the $\mathbf{o}(\epsilon)$ term in Eq. (5-210) is of the form $\epsilon^2\mathbf{M}(t)$, where $\mathbf{M}(t)$ is a bounded function. Prove Eq. (5-215). HINT: Consider the function $\boldsymbol{\phi}(t) - \boldsymbol{\psi}(t)$ and note that $d[\boldsymbol{\phi}(t) - \boldsymbol{\psi}(t)]/dt = (\partial\mathbf{f}/\partial\mathbf{x})|_* [\boldsymbol{\phi}(t) - \boldsymbol{\psi}(t)] + \epsilon\mathbf{M}(t)$, $\boldsymbol{\phi}(t_0) - \boldsymbol{\psi}(t_0) = \mathbf{0}$; then calculate $\|\mathbf{x}^*(t) + \epsilon\boldsymbol{\phi}(t) - [\mathbf{x}^*(t) + \epsilon\boldsymbol{\psi}(t)]\|$.

† See Ref. P-5, but note that Eq. (5-215) may not be strictly correct without additional assumptions. (Cf. Exercise 5-12.)

*Step 3 Introduction of the Hamiltonian*

We are now ready to introduce the Hamiltonian function and the costate variables for our problem. So we suppose that $\mathbf{p}(t)$ is some *as yet unspecified* piecewise continuous *n-vector-valued* function; we call $\mathbf{p}(t)$ the *costate*, or *adjoint*, vector. Since $\mathbf{x}(t)$ and $\mathbf{x}^*(t)$ are both solutions of Eq. (5-196), we have

$$\mathbf{f}[\mathbf{x}(t), \mathbf{u}(t), t] - \dot{\mathbf{x}}(t) = \mathbf{0} \tag{5-216}$$

$$\mathbf{f}[\mathbf{x}^*(t), \mathbf{u}^*(t), t] - \dot{\mathbf{x}}^*(t) = \mathbf{0} \tag{5-217}$$

It follows that

$$\int_{t_0}^{t_1} \langle \mathbf{p}(t), \mathbf{f}[\mathbf{x}(t), \mathbf{u}(t), t] - \dot{\mathbf{x}}(t)\rangle \, dt = \int_{t_0}^{t_1} \langle \mathbf{p}(t), \mathbf{f}[\mathbf{x}(t), \mathbf{u}(t), t]\rangle \, dt - \int_{t_0}^{t_1} \langle \mathbf{p}(t), \dot{\mathbf{x}}(t)\rangle \, dt = 0 \tag{5-218}$$

$$\int_{t_0}^{t_1} \langle \mathbf{p}(t), \mathbf{f}[\mathbf{x}^*(t), \mathbf{u}^*(t), t] - \dot{\mathbf{x}}^*(t)\rangle \, dt = \int_{t_0}^{t_1} \langle \mathbf{p}(t), \mathbf{f}[\mathbf{x}^*(t), \mathbf{u}^*(t), t]\rangle \, dt - \int_{t_0}^{t_1} \langle \mathbf{p}(t), \dot{\mathbf{x}}^*(t)\rangle \, dt = 0 \tag{5-219}$$

and hence that

$$J(\mathbf{u}) = K[\mathbf{x}(t_1)] + \int_{t_0}^{t_1} \{L[\mathbf{x}(t), \mathbf{u}(t), t] + \langle \mathbf{p}(t), \mathbf{f}[\mathbf{x}(t), \mathbf{u}(t), t]\rangle - \langle \mathbf{p}(t), \dot{\mathbf{x}}(t)\rangle\} \, dt \tag{5-220}$$

and

$$J(\mathbf{u}^*) = K[\mathbf{x}^*(t_1)] + \int_{t_0}^{t_1} \{L[\mathbf{x}^*(t), \mathbf{u}^*(t), t] + \langle \mathbf{p}(t), \mathbf{f}[\mathbf{x}^*(t), \mathbf{u}^*(t), t]\rangle - \langle \mathbf{p}(t), \dot{\mathbf{x}}^*(t)\rangle\} \, dt \tag{5-221}$$

If we define a scalar function $H(\mathbf{x}, \mathbf{p}, \mathbf{u}, t)$ by setting

$$H(\mathbf{x}, \mathbf{p}, \mathbf{u}, t) = L(\mathbf{x}, \mathbf{u}, t) + \langle \mathbf{p}, \mathbf{f}(\mathbf{x}, \mathbf{u}, t)\rangle \tag{5-222}$$

then Eqs. (5-220) and (5-221) may be written in the form

$$J(\mathbf{u}) = K[\mathbf{x}(t_1)] + \int_{t_0}^{t_1} \{H[\mathbf{x}(t), \mathbf{p}(t), \mathbf{u}(t), t] - \langle \mathbf{p}(t), \dot{\mathbf{x}}(t)\rangle\} \, dt \tag{5-223}$$

$$J(\mathbf{u}^*) = K[\mathbf{x}^*(t_1)] + \int_{t_0}^{t_1} \{H[\mathbf{x}^*(t), \mathbf{p}(t), \mathbf{u}^*(t), t] - \langle \mathbf{p}(t), \dot{\mathbf{x}}^*(t)\rangle\} \, dt \tag{5-224}$$

The function $H$ is called the *Hamiltonian* of the problem. We deduce immediately that the change in our performance functional is given by

$$J(\mathbf{u}) - J(\mathbf{u}^*) = \{K[\mathbf{x}(t_1)] - K[\mathbf{x}^*(t_1)]\} + \int_{t_0}^{t_1} \{H[\mathbf{x}(t), \mathbf{p}(t), \mathbf{u}(t), t] - H[\mathbf{x}^*(t), \mathbf{p}(t), \mathbf{u}^*(t), t]\} \, dt + \int_{t_0}^{t_1} \langle \mathbf{p}(t), \dot{\mathbf{x}}^*(t) - \dot{\mathbf{x}}(t)\rangle \, dt \tag{5-225}$$

*Step 4 Calculation of $J(\mathbf{u}) - J(\mathbf{u}^*)$*

We are now ready to calculate the effect of the perturbation upon the value of $J$. Once again we shall do this by making use of the various perturbation equations of step 1, the definition of $H$, a Taylor's series expansion of $H$, and an integration by parts.

First of all, we observe that, for $\epsilon$ small,

$$H[\mathbf{x}(t), \mathbf{p}(t), \mathbf{u}(t), t] = H[\mathbf{x}^*(t), \mathbf{p}(t), \mathbf{u}^*(t), t] + \left\langle \frac{\partial H}{\partial \mathbf{x}}\bigg|_{*\mathbf{p}}, [\mathbf{x}(t) - \mathbf{x}^*(t)] \right\rangle + \left\langle \frac{\partial H}{\partial \mathbf{u}}\bigg|_{*\mathbf{p}}, [\mathbf{u}(t) - \mathbf{u}^*(t)] \right\rangle + o(\epsilon) \quad (5\text{-}226)$$

where $(\partial H/\partial \mathbf{x})|_{*\mathbf{p}}$ is the gradient of $H$ with respect to $\mathbf{x}$ evaluated at $[\mathbf{x}^*(t), \mathbf{p}(t), \mathbf{u}^*(t), t]$ and $(\partial H/\partial \mathbf{u})|_{*\mathbf{p}}$ is the gradient of $H$ with respect to $\mathbf{u}$ evaluated at $[\mathbf{x}^*(t), \mathbf{p}(t), \mathbf{u}^*(t), t]$. In view of Eqs. (5-203) and (5-215), we see that

$$H[\mathbf{x}(t), \mathbf{p}(t), \mathbf{u}(t), t] = H[\mathbf{x}^*(t), \mathbf{p}(t), \mathbf{u}^*(t), t] + \epsilon \left\langle \frac{\partial H}{\partial \mathbf{x}}\bigg|_{*\mathbf{p}}, \boldsymbol{\psi}(t) \right\rangle + \epsilon \left\langle \frac{\partial H}{\partial \mathbf{u}}\bigg|_{*\mathbf{p}}, \boldsymbol{\pi}(t) \right\rangle + o(\epsilon) \quad (5\text{-}227)$$

Secondly, we deduce from Eq. (5-215) that

$$\begin{aligned} K[\mathbf{x}(t_1)] &= K[\mathbf{x}^*(t_1)] + \left\langle \frac{\partial K}{\partial \mathbf{x}}\bigg|_{*}, \mathbf{x}(t_1) - \mathbf{x}^*(t_1) \right\rangle + o(\epsilon) \\ &= K[\mathbf{x}^*(t_1)] + \epsilon \left\langle \frac{\partial K}{\partial \mathbf{x}}\bigg|_{*}, \boldsymbol{\psi}(t_1) \right\rangle + o(\epsilon) \end{aligned} \quad (5\text{-}228)$$

where $(\partial K/\partial \mathbf{x})|_*$ denotes the gradient of $K$ with respect to $\mathbf{x}$ evaluated at $\mathbf{x}^*(t_1)$. Thirdly, we conclude from Eqs. (5-210) and (5-213) that

$$\epsilon \dot{\boldsymbol{\phi}}(t) = \epsilon \dot{\boldsymbol{\psi}}(t) + \epsilon \frac{\partial \mathbf{f}}{\partial \mathbf{x}}\bigg|_{*} [\boldsymbol{\phi}(t) - \boldsymbol{\psi}(t)] + \mathbf{o}(\epsilon) \quad (5\text{-}229)$$

and hence, by virtue of Eq. (5-215), that

$$\epsilon \dot{\boldsymbol{\phi}}(t) = \epsilon \dot{\boldsymbol{\psi}}(t) + \mathbf{o}(\epsilon) \quad (5\text{-}230)$$

It follows that

$$\dot{\mathbf{x}}^*(t) - \dot{\mathbf{x}}(t) = -\epsilon \dot{\boldsymbol{\psi}}(t) + \mathbf{o}(\epsilon) \quad (5\text{-}231)$$

Applying these calculations to Eq. (5-225), we find that

$$J(\mathbf{u}) - J(\mathbf{u}^*) = \epsilon \left\langle \frac{\partial K}{\partial \mathbf{x}}\bigg|_{*}, \boldsymbol{\psi}(t_1) \right\rangle + \epsilon \int_{t_0}^{t_1} \left\langle \frac{\partial H}{\partial \mathbf{x}}\bigg|_{*\mathbf{p}}, \boldsymbol{\psi}(t) \right\rangle dt + \epsilon \int_{t_0}^{t_1} \left\langle \frac{\partial H}{\partial \mathbf{u}}\bigg|_{*\mathbf{p}}, \boldsymbol{\pi}(t) \right\rangle dt - \epsilon \int_{t_0}^{t_1} \langle \mathbf{p}(t), \dot{\boldsymbol{\psi}}(t) \rangle \, dt + o(\epsilon) \quad (5\text{-}232)$$

Now let us evaluate the integral involving $\dot{\boldsymbol{\psi}}(t)$ in Eq. (5-232) by parts. We have

$$\int_{t_0}^{t_1} \langle \mathbf{p}(t), \dot{\boldsymbol{\psi}}(t) \rangle \, dt = \langle \mathbf{p}(t_1), \boldsymbol{\psi}(t_1) \rangle - \langle \mathbf{p}(t_0), \boldsymbol{\psi}(t_0) \rangle - \int_{t_0}^{t_1} \langle \dot{\mathbf{p}}(t), \boldsymbol{\psi}(t) \rangle \, dt \quad (5\text{-}233)$$

and so, noting that $\boldsymbol{\psi}(t_0) = \mathbf{0}$, we can write Eq. (5-232) as

$$J(\mathbf{u}) - J(\mathbf{u}^*) = \epsilon \left\langle \frac{\partial K}{\partial \mathbf{x}}\bigg|_* - \mathbf{p}(t_1), \boldsymbol{\psi}(t_1) \right\rangle + \epsilon \int_{t_0}^{t_1} \left[ \left\langle \frac{\partial H}{\partial \mathbf{x}}\bigg|_{*\mathbf{p}} + \dot{\mathbf{p}}(t), \boldsymbol{\psi}(t) \right\rangle + \left\langle \frac{\partial H}{\partial \mathbf{u}}\bigg|_{*\mathbf{p}}, \boldsymbol{\eta}(t) \right\rangle \right] dt + o(\epsilon) \tag{5-234}$$

where, so far, the $n$ vector $\mathbf{p}(t)$ has not been specified.

Since $\mathbf{u}^*$ is an optimal control, we must have

$$J(\mathbf{u}) - J(\mathbf{u}^*) \geq 0 \tag{5-235}$$

and we deduce, just as in Sec. 5-6, that

$$\left\langle \frac{\partial K}{\partial \mathbf{x}}\bigg|_* - \mathbf{p}(t_1), \boldsymbol{\psi}(t_1) \right\rangle + \int_{t_0}^{t_1} \left[ \left\langle \frac{\partial H}{\partial \mathbf{x}}\bigg|_{*\mathbf{p}} + \dot{\mathbf{p}}(t), \boldsymbol{\psi}(t) \right\rangle + \left\langle \frac{\partial H}{\partial \mathbf{u}}\bigg|_{*\mathbf{p}}, \boldsymbol{\eta}(t) \right\rangle \right] dt = 0 \tag{5-236}$$

*Step 5 A Differential Equation for* $\mathbf{p}(t)$

We are now going to impose some conditions on $\mathbf{p}(t)$. First of all, we shall demand that $\mathbf{p}(t)$ be a solution of the differential equation

$$\dot{\mathbf{p}}(t) = -\frac{\partial H}{\partial \mathbf{x}} [\mathbf{x}^*(t), \mathbf{p}(t), \mathbf{u}^*(t), t] \tag{5-237}$$

$$= -\frac{\partial L}{\partial \mathbf{x}} [\mathbf{x}^*(t), \mathbf{u}^*(t), t] - \left[ \frac{\partial \mathbf{f}}{\partial \mathbf{x}} [\mathbf{x}^*(t), \mathbf{u}^*(t), t] \right]' \mathbf{p}(t) \tag{5-238}$$

where $\partial \mathbf{f}/\partial \mathbf{x}$ is the Jacobian matrix of $\mathbf{f}$ with respect to $\mathbf{x}$ and the prime (′) indicates the transpose of a matrix. We observe that Eq. (5-238) is a *linear* differential equation for $\mathbf{p}(t)$ and that the homogeneous parts of Eqs. (5-238) and (5-213) are adjoint in the sense of Sec. 3-25. Furthermore, we note that Eqs. (5-236) and (5-238) imply that

$$\left\langle \frac{\partial K}{\partial \mathbf{x}}\bigg|_* - \mathbf{p}(t_1), \boldsymbol{\psi}(t_1) \right\rangle + \int_{t_0}^{t_1} \left\langle \frac{\partial H}{\partial \mathbf{u}}\bigg|_{*\mathbf{p}}, \boldsymbol{\eta}(t) \right\rangle dt = 0 \tag{5-239}$$

whenever $\mathbf{p}(t)$ is a solution of Eq. (5-238).

Since $H(\mathbf{x}, \mathbf{p}, \mathbf{u}, t) = L(\mathbf{x}, \mathbf{u}, t) + \langle \mathbf{p}, \mathbf{f}(\mathbf{x}, \mathbf{u}, t) \rangle$ we also have

$$\dot{\mathbf{x}}^*(t) = \mathbf{f}[\mathbf{x}^*(t), \mathbf{u}^*(t), t] = \frac{\partial H}{\partial \mathbf{p}} [\mathbf{x}^*(t), \mathbf{p}(t), \mathbf{u}^*(t), t] \tag{5-240}$$

so that again *both* $\mathbf{x}^*(t)$ *and* $\mathbf{p}(t)$ are solutions of differential equations involving the Hamiltonian $H$.

In contrast to the problem discussed in Sec. 5-6, we can now specify our costate completely because of the presence of the term involving the terminal cost in Eq. (5-239). Thus, we let $\mathbf{p}^*(t)$ be the (unique) solution of Eq. (5-238), with the boundary condition

$$\mathbf{p}^*(t_1) = \frac{\partial K}{\partial \mathbf{x}} [\mathbf{x}^*(t_1)] \tag{5-241}$$

If we let

$$\left.\frac{\partial H}{\partial \mathbf{u}}\right|_* = \frac{\partial H}{\partial \mathbf{u}}[\mathbf{x}^*(t), \mathbf{p}^*(t), \mathbf{u}^*(t), t] \tag{5-242}$$

then we may conclude, from Eqs. (5-239) and (5-241), that

$$\int_{t_0}^{t_1} \left\langle \left.\frac{\partial H}{\partial \mathbf{u}}\right|_*, \mathfrak{n}(t) \right\rangle dt = 0 \tag{5-243}$$

for all piecewise continuous functions $\mathfrak{n}(t)$.

We shall see, in the next step, that Eq. (5-243) implies that $\mathbf{u}^*(t)$ is an *extremum* of $H[\mathbf{x}^*(t), \mathbf{p}^*(t), \mathbf{u}, t]$ viewed as a function of $\mathbf{u}$. We shall then deduce that this extremum must, in fact, be a minimum by introducing the second variation of $J$.

*Step 6 A Fundamental Lemma*

We now prove the following lemma:

***Lemma 5-2*** *Let $\mathbf{h}(t)$ be a piecewise continuous function on $[t_0, t_1]$ and suppose that*

$$\int_{t_0}^{t_1} \langle \mathbf{h}(t), \boldsymbol{\alpha}(t) \rangle\, dt = 0 \tag{5-244}$$

*for all piecewise continuous functions $\boldsymbol{\alpha}(t)$ on $[t_0, t_1]$. Then*

$$\mathbf{h}(t) = \mathbf{0} \qquad t \in [t_0, t_1] \tag{5-245}$$ †

PROOF First of all, we note that Lemma 5-1 implies that $\mathbf{h}(t)$ is constant, i.e., that

$$\mathbf{h}(t) = \mathbf{c} \qquad t \in [t_0, t_1] \tag{5-246}$$

To see this, we observe that if $\alpha_1(t)$ is any piecewise continuous function such that

$$\int_{t_0}^{t_1} \alpha_1(t)\, dt = 0 \tag{5-247}$$

then, setting

$$\boldsymbol{\alpha}^1(t) = \begin{bmatrix} \alpha_1(t) \\ 0 \\ \cdot \\ \cdot \\ \cdot \\ 0 \end{bmatrix} \tag{5-248}$$

we have

$$\langle \mathbf{h}(t), \boldsymbol{\alpha}^1(t) \rangle = h_1(t)\alpha_1(t) \tag{5-249}$$

and

$$\int_{t_0}^{t_1} \langle \mathbf{h}(t), \boldsymbol{\alpha}^1(t) \rangle\, dt = \int_{t_0}^{t_1} h_1(t)\alpha_1(t)\, dt = 0 \tag{5-250}$$

Thus, $h_1(t)$ is a constant $c_1$ on $[t_0, t_1]$ by Lemma 5-1. In a similar way, we can show that each of the components of $\mathbf{h}(t)$ is constant on $[t_0, t_1]$.

However, if $\mathbf{c}$ were not $\mathbf{0}$, then we could consider the function

$$\boldsymbol{\alpha}(t) = \mathbf{c} \neq \mathbf{0} \tag{5-251}$$

† Except possibly for a countable subset $A$ of $[t_0, t_1]$.

and obtain the contradiction

$$\int_{t_0}^{t_1} \langle \mathbf{h}(t), \boldsymbol{\alpha}(t) \rangle \, dt = \int_{t_0}^{t_1} \langle \mathbf{c}, \mathbf{c} \rangle \, dt = \int_{t_0}^{t_1} \|\mathbf{c}\|^2 \, dt > 0 \tag{5-252}$$

This completes the proof of the lemma.

We immediately conclude from the lemma and Eq. (5-243) that

$$\left. \frac{\partial H}{\partial \mathbf{u}} \right|_* = \mathbf{0} \tag{5-253}$$

Therefore, *the function* $H[\mathbf{x}^*(t), \mathbf{p}^*(t), \mathbf{u}, t]$ *has an extremum as a function of* $\mathbf{u}$ *at* $\mathbf{u} = \mathbf{u}^*(t)$ *for* $t$ *in* $[t_0, t_1]$. This conclusion is to be understood in the following way: Suppose that $\hat{t}$ is some element of $[t_0, t_1]$; then $\mathbf{x}^*(\hat{t})$ and $\mathbf{p}^*(\hat{t})$ are well-defined vectors and we can consider the real-valued function $\hat{H}(\mathbf{u})$ of the $m$ vector $\mathbf{u}$, given by

$$\hat{H}(\mathbf{u}) = H[\mathbf{x}^*(\hat{t}), \mathbf{p}^*(\hat{t}), \mathbf{u}, \hat{t}] \tag{5-254}$$

Our conclusion is that $\mathbf{u} = \mathbf{u}^*(\hat{t})$ is an extremum of $\hat{H}(\mathbf{u})$.

*Step 7 The Necessary Condition*

Before we derive the desired necessary condition, let us recall that in Sec. 5-5 we observed, in a general context, that *a necessary condition for an extremum of a functional to be a minimum was the nonnegativeness of the second variation.* [See Eq. (5-98).] Here, we shall make use of that observation. In fact, we shall assume that the second variation $\delta^2 J(\mathbf{u}^*)$ exists and that

$$\delta^2 J(\mathbf{u}^*) > 0 \tag{5-255†}$$

We shall then compute $\delta^2 J(\mathbf{u}^*)$ explicitly and use the formula obtained to derive in a heuristic fashion the analog of Theorem 5-2.

We recall, first of all, that

$$\mathbf{u}(t) = \mathbf{u}^*(t) + \epsilon \boldsymbol{\eta}(t) \tag{5-256}$$

Secondly, we suppose that

$$\mathbf{x}(t) = \mathbf{x}^*(t) + \epsilon \boldsymbol{\psi}(t) + \epsilon^2 \boldsymbol{\xi}(t) + \mathbf{o}(\epsilon^2) \tag{5-257}$$

where $\boldsymbol{\psi}(t)$ is the solution of the linear differential equation

$$\dot{\boldsymbol{\psi}}(t) = \left. \frac{\partial \mathbf{f}}{\partial \mathbf{x}} \right|_* \boldsymbol{\psi}(t) + \left. \frac{\partial \mathbf{f}}{\partial \mathbf{u}} \right|_* \boldsymbol{\eta}(t) \tag{5-258}$$

and

$$\boldsymbol{\psi}(t_0) = \mathbf{0} \qquad \boldsymbol{\xi}(t_0) = \mathbf{0} \tag{5-259}$$

Thirdly, we recall that the change in our performance functional is given by

$$\begin{aligned} J(\mathbf{u}) - J(\mathbf{u}^*) &= K[\mathbf{x}(t_1)] - K[\mathbf{x}^*(t_1)] \\ &\quad + \int_{t_0}^{t_1} \{H[\mathbf{x}(t), \mathbf{p}^*(t), \mathbf{u}(t), t] - H[\mathbf{x}^*(t), \mathbf{p}^*(t), \mathbf{u}^*(t), t]\} \, dt \\ &\quad + \int_{t_0}^{t_1} \langle \mathbf{p}^*(t), \dot{\mathbf{x}}^*(t) - \dot{\mathbf{x}}(t) \rangle \, dt \end{aligned} \tag{5-260}$$

† We shall see later on in our discussion of the minimum principle that this simplifying assumption is not required to obtain our desired result, Theorem 5-3.

where $\mathbf{p}^*(t)$ is the solution of the differential equation

$$\dot{\mathbf{p}}^*(t) = -\left.\frac{\partial H}{\partial \mathbf{x}}\right|_* \tag{5-261}$$

satisfying the boundary condition

$$\mathbf{p}^*(t_1) = \frac{\partial K}{\partial \mathbf{x}}[\mathbf{x}^*(t_1)] \tag{5-262}$$

We propose now to show that the second variation $\delta^2 J(\mathbf{u}^*)(\mathbf{u} - \mathbf{u}^*)$ is given by

$$\delta^2 J(\mathbf{u}^*)(\mathbf{u} - \mathbf{u}^*) = \frac{\epsilon^2}{2}\left\langle \boldsymbol{\psi}(t_1), \left.\frac{\partial^2 K}{\partial \mathbf{x}^2}\right|_* \boldsymbol{\psi}(t_1)\right\rangle + \frac{\epsilon^2}{2}\int_{t_0}^{t_1}\left\langle \begin{bmatrix} \boldsymbol{\psi}(t) \\ \boldsymbol{\eta}(t) \end{bmatrix}, \begin{bmatrix} \left.\frac{\partial^2 H}{\partial \mathbf{x}^2}\right|_* & \left.\frac{\partial}{\partial \mathbf{u}}\left(\frac{\partial H}{\partial \mathbf{x}}\right)\right|_* \\ \left.\frac{\partial}{\partial \mathbf{x}}\left(\frac{\partial H}{\partial \mathbf{u}}\right)\right|_* & \left.\frac{\partial^2 H}{\partial \mathbf{u}^2}\right|_* \end{bmatrix} \begin{bmatrix} \boldsymbol{\psi}(t) \\ \boldsymbol{\eta}(t) \end{bmatrix}\right\rangle dt \tag{5-263}$$

where $(\partial^2 K/\partial \mathbf{x}^2)|_*$ is the Jacobian matrix of $\partial K/\partial \mathbf{x}$ with respect to $\mathbf{x}$ evaluated at $\mathbf{x}^*(t_1)$, $(\partial^2 H/\partial \mathbf{x}^2)|_*$ is the Jacobian matrix of $\partial H/\partial \mathbf{x}$ with respect to $\mathbf{x}$ evaluated at $[\mathbf{x}^*(t), \mathbf{p}^*(t), \mathbf{u}^*(t), t]$, $\left.\frac{\partial}{\partial \mathbf{u}}\left(\frac{\partial H}{\partial \mathbf{x}}\right)\right|_*$ is the Jacobian matrix of $\partial H/\partial \mathbf{x}$ with respect to $\mathbf{u}$ evaluated at $[\mathbf{x}^*(t), \mathbf{p}^*(t), \mathbf{u}^*(t), t]$, $\left.\frac{\partial}{\partial \mathbf{x}}\left(\frac{\partial H}{\partial \mathbf{u}}\right)\right|_*$ is the Jacobian matrix of $\partial H/\partial \mathbf{u}$ with respect to $\mathbf{x}$ evaluated at $[\mathbf{x}^*(t), \mathbf{p}^*(t), \mathbf{u}^*(t), t]$, and $(\partial^2 H/\partial \mathbf{u}^2)|_*$ is the Jacobian matrix of $\partial H/\partial \mathbf{u}$ with respect to $\mathbf{u}$ evaluated at $[\mathbf{x}^*(t), \mathbf{p}^*(t), \mathbf{u}^*(t), t]$.† The following computations, together

† To illustrate the notation, let us suppose that $n = 3$ and $m = 2$; then the right-hand side of Eq. (5-263), when written out explicitly, becomes

$$\frac{\epsilon^2}{2}\left\langle \begin{bmatrix} \psi_1(t_1) \\ \psi_2(t_1) \\ \psi_3(t_1) \end{bmatrix}, \begin{bmatrix} \left.\frac{\partial^2 K}{\partial x_1^2}\right|_* & \left.\frac{\partial^2 K}{\partial x_1\,\partial x_2}\right|_* & \left.\frac{\partial^2 K}{\partial x_1\,\partial x_3}\right|_* \\ \left.\frac{\partial^2 K}{\partial x_2\,\partial x_1}\right|_* & \left.\frac{\partial^2 K}{\partial x_2^2}\right|_* & \left.\frac{\partial^2 K}{\partial x_2\,\partial x_3}\right|_* \\ \left.\frac{\partial^2 K}{\partial x_3\,\partial x_1}\right|_* & \left.\frac{\partial^2 K}{\partial x_3\,\partial x_2}\right|_* & \left.\frac{\partial^2 K}{\partial x_3^2}\right|_* \end{bmatrix} \begin{bmatrix} \psi_1(t_1) \\ \psi_2(t_1) \\ \psi_3(t_1) \end{bmatrix}\right\rangle$$

$$+ \frac{\epsilon^2}{2}\int_{t_0}^{t_1}\left\langle \begin{bmatrix} \psi_1(t) \\ \psi_2(t) \\ \psi_3(t) \\ \eta_1(t) \\ \eta_2(t) \end{bmatrix}, \left[\begin{array}{ccc|cc} \left.\frac{\partial^2 H}{\partial x_1^2}\right|_* & \left.\frac{\partial^2 H}{\partial x_1\,\partial x_2}\right|_* & \left.\frac{\partial^2 H}{\partial x_1\,\partial x_3}\right|_* & \left.\frac{\partial^2 H}{\partial x_1\,\partial u_1}\right|_* & \left.\frac{\partial^2 H}{\partial x_1\,\partial u_2}\right|_* \\ \left.\frac{\partial^2 H}{\partial x_2\,\partial x_1}\right|_* & \left.\frac{\partial^2 H}{\partial x_2^2}\right|_* & \left.\frac{\partial^2 H}{\partial x_2\,\partial x_3}\right|_* & \left.\frac{\partial^2 H}{\partial x_2\,\partial u_1}\right|_* & \left.\frac{\partial^2 H}{\partial x_2\,\partial u_2}\right|_* \\ \left.\frac{\partial^2 H}{\partial x_3\,\partial x_1}\right|_* & \left.\frac{\partial^2 H}{\partial x_3\,\partial x_2}\right|_* & \left.\frac{\partial^2 H}{\partial x_3^2}\right|_* & \left.\frac{\partial^2 H}{\partial x_3\,\partial u_1}\right|_* & \left.\frac{\partial^2 H}{\partial x_3\,\partial u_2}\right|_* \\ \hline \left.\frac{\partial^2 H}{\partial u_1\,\partial x_1}\right|_* & \left.\frac{\partial^2 H}{\partial u_1\,\partial x_2}\right|_* & \left.\frac{\partial^2 H}{\partial u_1\,\partial x_3}\right|_* & \left.\frac{\partial^2 H}{\partial u_1^2}\right|_* & \left.\frac{\partial^2 H}{\partial u_1\,\partial u_2}\right|_* \\ \left.\frac{\partial^2 H}{\partial u_2\,\partial x_1}\right|_* & \left.\frac{\partial^2 H}{\partial u_2\,\partial x_2}\right|_* & \left.\frac{\partial^2 H}{\partial u_2\,\partial x_3}\right|_* & \left.\frac{\partial^2 H}{\partial u_2\,\partial u_1}\right|_* & \left.\frac{\partial^2 H}{\partial u_2^2}\right|_* \end{array}\right] \begin{bmatrix} \psi_1(t) \\ \psi_2(t) \\ \psi_3(t) \\ \eta_1(t) \\ \eta_2(t) \end{bmatrix}\right\rangle dt$$

The indicated blocks in the matrix involving $H$ correspond to the various Jacobian matrices indicated in Eq. (5-263).

with Eqs. (5-261) and (5-262) and the fact that $(\partial H/\partial \mathbf{u})|_* = \mathbf{0}$, prove the validity of Eq. (5-263):

$$K[\mathbf{x}(t_1)] - K[\mathbf{x}^*(t_1)] = \left\langle \frac{\partial K}{\partial \mathbf{x}}\bigg|_*, \epsilon\boldsymbol{\psi}(t_1) + \epsilon^2\boldsymbol{\xi}(t_1)\right\rangle + \frac{1}{2}\left\langle \epsilon\boldsymbol{\psi}(t_1) + \epsilon^2\boldsymbol{\xi}(t_1), \frac{\partial^2 K}{\partial \mathbf{x}^2}\bigg|_* [\epsilon\boldsymbol{\psi}(t_1) + \epsilon^2\boldsymbol{\xi}(t_1)]\right\rangle + o(\epsilon^2) \quad (5\text{-}264)$$

$$= \epsilon\left\langle \frac{\partial K}{\partial \mathbf{x}}\bigg|_*, \boldsymbol{\psi}(t_1)\right\rangle + \epsilon^2\left\langle \frac{\partial K}{\partial \mathbf{x}}\bigg|_*, \boldsymbol{\xi}(t_1)\right\rangle + \frac{\epsilon^2}{2}\left\langle \boldsymbol{\psi}(t_1), \frac{\partial^2 K}{\partial \mathbf{x}^2}\bigg|_* \boldsymbol{\psi}(t_1)\right\rangle + o(\epsilon^2) \quad (5\text{-}265)$$

$$\int_{t_0}^{t_1} \langle \mathbf{p}(t), \dot{\mathbf{x}}^*(t) - \dot{\mathbf{x}}(t)\rangle\, dt = \langle \mathbf{p}(t_1), \mathbf{x}^*(t_1) - \mathbf{x}(t_1)\rangle - \int_{t_0}^{t_1} \langle \dot{\mathbf{p}}(t), \mathbf{x}^*(t) - \mathbf{x}(t)\rangle\, dt \qquad \text{since } \mathbf{x}^*(t_0) = \mathbf{x}(t_0) \quad (5\text{-}266)$$

$$= \epsilon\langle -\mathbf{p}(t_1), \boldsymbol{\psi}(t_1)\rangle + \epsilon^2\langle -\mathbf{p}(t_1), \boldsymbol{\xi}(t_1)\rangle + \epsilon \int_{t_0}^{t_1} \langle \dot{\mathbf{p}}(t), \boldsymbol{\psi}(t)\rangle\, dt + \epsilon^2 \int_{t_0}^{t_1} \langle \dot{\mathbf{p}}(t), \boldsymbol{\xi}(t)\rangle\, dt + o(\epsilon^2) \quad (5\text{-}267)$$

$$H[\mathbf{x}(t), \mathbf{p}^*(t), \mathbf{u}(t), t] - H[\mathbf{x}^*(t), \mathbf{p}^*(t), \mathbf{u}^*(t), t]$$
$$= \left\langle \frac{\partial H}{\partial \mathbf{x}}\bigg|_*, \epsilon\boldsymbol{\psi}(t) + \epsilon^2\boldsymbol{\xi}(t)\right\rangle + \left\langle \frac{\partial H}{\partial \mathbf{u}}\bigg|_*, \epsilon\boldsymbol{\eta}(t)\right\rangle$$
$$+ \frac{1}{2}\left\langle \epsilon\boldsymbol{\psi}(t) + \epsilon^2\boldsymbol{\xi}(t), \frac{\partial^2 H}{\partial \mathbf{x}^2}\bigg|_* [\epsilon\boldsymbol{\psi}(t) + \epsilon^2\boldsymbol{\xi}(t)]\right\rangle + \frac{1}{2}\left\langle \epsilon\boldsymbol{\eta}(t), \frac{\partial^2 H}{\partial \mathbf{u}^2}\bigg|_* \epsilon\boldsymbol{\eta}(t)\right\rangle$$
$$+ \frac{1}{2}\left\langle \epsilon\boldsymbol{\psi}(t) + \epsilon^2\boldsymbol{\xi}(t), \frac{\partial}{\partial \mathbf{u}}\left(\frac{\partial H}{\partial \mathbf{x}}\right)\bigg|_* \epsilon\boldsymbol{\eta}(t)\right\rangle$$
$$+ \frac{1}{2}\left\langle \epsilon\boldsymbol{\eta}(t), \frac{\partial}{\partial \mathbf{x}}\left(\frac{\partial H}{\partial \mathbf{u}}\right)\bigg|_* [\epsilon\boldsymbol{\psi}(t) + \epsilon^2\boldsymbol{\xi}(t)]\right\rangle + o(\epsilon^2) \quad (5\text{-}268)$$

$$= \epsilon\left\langle \frac{\partial H}{\partial \mathbf{x}}\bigg|_*, \boldsymbol{\psi}(t)\right\rangle + \epsilon^2\left\langle \frac{\partial H}{\partial \mathbf{x}}\bigg|_*, \boldsymbol{\xi}(t)\right\rangle + \epsilon\left\langle \frac{\partial H}{\partial \mathbf{u}}\bigg|_*, \boldsymbol{\eta}(t)\right\rangle$$
$$+ \frac{\epsilon^2}{2}\left\langle \begin{bmatrix} \boldsymbol{\psi}(t) \\ \boldsymbol{\eta}(t) \end{bmatrix}, \left[\begin{array}{c:c} \dfrac{\partial^2 H}{\partial \mathbf{x}^2}\bigg|_* & \dfrac{\partial}{\partial \mathbf{u}}\left(\dfrac{\partial H}{\partial \mathbf{x}}\right)\bigg|_* \\ \hdashline \dfrac{\partial}{\partial \mathbf{x}}\left(\dfrac{\partial H}{\partial \mathbf{u}}\right)\bigg|_* & \dfrac{\partial^2 H}{\partial \mathbf{u}^2}\bigg|_* \end{array}\right] \begin{bmatrix} \boldsymbol{\psi}(t) \\ \boldsymbol{\eta}(t) \end{bmatrix}\right\rangle + o(\epsilon^2) \quad (5\text{-}269)$$

Since we have supposed that $\delta^2 J(\mathbf{u}^*) > 0$ and since $\epsilon^2$ is positive, we conclude that

$$\left\langle \boldsymbol{\psi}(t_1), \frac{\partial^2 K}{\partial \mathbf{x}^2}\bigg|_* \boldsymbol{\psi}(t_1)\right\rangle + \int_{t_0}^{t_1} \left\langle \begin{bmatrix} \boldsymbol{\psi}(t) \\ \boldsymbol{\eta}(t) \end{bmatrix}, \left[\begin{array}{c:c} \dfrac{\partial^2 H}{\partial \mathbf{x}^2}\bigg|_* & \dfrac{\partial}{\partial \mathbf{u}}\left(\dfrac{\partial H}{\partial \mathbf{x}}\right)\bigg|_* \\ \hdashline \dfrac{\partial}{\partial \mathbf{x}}\left(\dfrac{\partial H}{\partial \mathbf{u}}\right)\bigg|_* & \dfrac{\partial^2 H}{\partial \mathbf{u}^2}\bigg|_* \end{array}\right] \begin{bmatrix} \boldsymbol{\psi}(t) \\ \boldsymbol{\eta}(t) \end{bmatrix}\right\rangle dt > 0 \quad (5\text{-}270)$$

for all $\boldsymbol{\eta}(t) \neq \mathbf{0}$.

Now let us see where our intuition will lead us. We note, first of all, that, in view of Eq. (5-258) and the condition $\boldsymbol{\psi}(t_0) = \mathbf{0}$, a "small" $\boldsymbol{\eta}(t)$ will generate a "small" $\boldsymbol{\psi}(t)$. However, the converse of this is not true; in other words, *it is possible to construct a "large"* $\boldsymbol{\eta}(t)$ *which generates a "small"* $\boldsymbol{\psi}(t)$. Essentially, the reason for this is that $\boldsymbol{\eta}(t)$ is the input (or control) to a time-varying linear system which is initially at rest and whose output is $\boldsymbol{\psi}(t)$. The idea is illustrated in Fig. 5-8. Thus, we expect that the terms

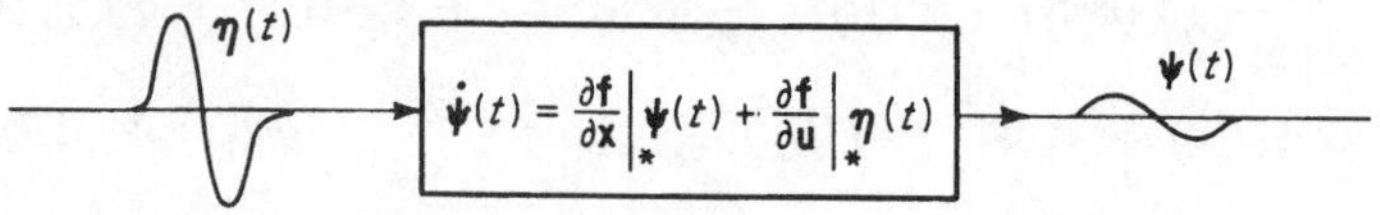


**Fig. 5-8** The control perturbation $\boldsymbol{\eta}(t)$ drives the linear system $\dot{\boldsymbol{\psi}}(t) = (\partial \mathbf{f}/\partial \mathbf{x})|_* \boldsymbol{\psi}(t) + (\partial \mathbf{f}/\partial \mathbf{u})|_* \boldsymbol{\eta}(t)$. "Large" $\boldsymbol{\eta}(t)$ can produce "small" $\boldsymbol{\psi}(t)$.

involving $\boldsymbol{\eta}(t)$ in Eq. (5-270) will have to be positive in order to ensure the validity of that inequality. In particular, since the term

$$\left\langle \boldsymbol{\eta}(t), \left.\frac{\partial^2 H}{\partial \mathbf{u}^2}\right|_* \boldsymbol{\eta}(t) \right\rangle \tag{5-271}$$

is the "most" dependent on $\boldsymbol{\eta}(t)$, we expect that it will be positive for *all* $\boldsymbol{\eta}(t) \neq \mathbf{0}$. Clearly, this can be the case only if the matrix $(\partial^2 H/\partial \mathbf{u}^2)|_*$ is *positive definite*† for $t$ in $[t_0, t_1]$. However, if

$$\left.\frac{\partial^2 H}{\partial \mathbf{u}^2}\right|_* \quad \text{is positive definite} \tag{5-272}$$

*then the extremum of the function* $H[\mathbf{x}^*(t), \mathbf{p}^*(t), \mathbf{u}, t]$ *as a function of* $\mathbf{u}$ *at* $\mathbf{u} = \mathbf{u}^*(t)$ *is, in fact, a minimum.* In effect, then, we have *heuristically* established the following theorem:

***Theorem 5-3*** *Let* $\mathbf{u}^*(t)$ *be an admissible control and let* $\mathbf{x}^*(t)$ *be the trajectory of the system* (5-196) *corresponding to* $\mathbf{u}^*(t)$ *and originating at* $\mathbf{x}_0$ *[that is,* $\mathbf{x}^*(t_0) = \mathbf{x}_0$*]. In order that* $\mathbf{u}^*(t)$ *be optimal, it is necessary that there exist a function* $\mathbf{p}^*(t)$ *such that:*

*a.* $\mathbf{x}^*(t)$ *and* $\mathbf{p}^*(t)$ *are a solution of the system of equations*

$$\begin{aligned} \dot{\mathbf{x}}^*(t) &= \frac{\partial H}{\partial \mathbf{p}} [\mathbf{x}^*(t), \mathbf{p}^*(t), \mathbf{u}^*(t), t] \\ \dot{\mathbf{p}}^*(t) &= -\frac{\partial H}{\partial \mathbf{x}} [\mathbf{x}^*(t), \mathbf{p}^*(t), \mathbf{u}^*(t), t] \end{aligned} \tag{5-273}$$

*satisfying the boundary conditions*

$$\begin{aligned} \mathbf{x}^*(t_0) &= \mathbf{x}_0 \\ \mathbf{p}^*(t_1) &= \frac{\partial K}{\partial \mathbf{x}} [\mathbf{x}^*(t_1)] \end{aligned} \tag{5-274}$$

*b. The function* $H[\mathbf{x}^*(t), \mathbf{p}^*(t), \mathbf{u}, t]$ *has a (possibly local) minimum as a function of* $\mathbf{u}$ *at* $\mathbf{u} = \mathbf{u}^*(t)$ *for* $t$ *in* $[t_0, t_1]$.

† See Sec. 2-15.

Let us now make some comments about this theorem. First of all, we note that we have *not* formally proved it; in point of fact, the formal proof, based upon variational methods, is quite complicated, requires various assumptions (e.g., suitable differentiability), and is, in any case, beyond the scope of this book. Secondly, we observe that, in contrast to Theorem 5-2, we have only shown that $\mathbf{u} = \mathbf{u}^*(t)$ is a local minimum of $H[\mathbf{x}^*(t), \mathbf{p}^*(t), \mathbf{u}, t]$; *it is, in fact, true that* $\mathbf{u} = \mathbf{u}^*(t)$ *is an absolute minimum of* $H[\mathbf{x}^*(t), \mathbf{p}^*(t), \mathbf{u}, t]$, *as we shall see in our discussion of the minimum principle.* We can thus begin to see the need for some additional tools beyond the basic variational ones; actually, in a later section, when we consider a fixed-end-point problem, we shall find it difficult even to show that the optimal control extremizes $H$ and we shall thus be forced to seek a different approach.

Finally, we observe that, along an optimal trajectory, the total derivative of $H$ with respect to time is the same as the partial derivative of $H$ with respect to time, i.e., that

$$\frac{dH[\mathbf{x}^*(t), \mathbf{p}^*(t), \mathbf{u}^*(t), t]}{dt} = \left\langle \frac{\partial H}{\partial \mathbf{x}}\bigg|_*, \dot{\mathbf{x}}^*(t) \right\rangle + \left\langle \frac{\partial H}{\partial \mathbf{p}}\bigg|_*, \dot{\mathbf{p}}^*(t) \right\rangle + \left\langle \frac{\partial H}{\partial \mathbf{u}}\bigg|_*, \dot{\mathbf{u}}^*(t) \right\rangle + \frac{\partial H}{\partial t}\bigg|_* \tag{5-275}$$

$$= \frac{\partial H}{\partial t}\bigg|_* \tag{5-276}$$

in view of Eqs. (5-273) and (5-253); clearly, this implies that *if $H$ does not depend on $t$ explicitly, then $H$ is constant along an optimal trajectory* [compare with Eq. (5-176)].

**Example 5-8** As an illustration of the type of complications which arise in developing a formal proof based on variational methods, we shall show that Eq. (5-270) implies the nonnegative definiteness of $(\partial^2 H/\partial \mathbf{u}^2)|_*$ in the special case where $n = m = 1$, that is, where $x(t)$ and $u(t)$ are scalars. Thus, our system equation is

$$\dot{x}(t) = f[x(t), u(t), t]$$

and the perturbation equation (5-258) becomes

$$\dot{\psi}(t) = a(t)\psi(t) + b(t)\eta(t) \tag{5-277}$$

where we have, for convenience, set $a(t) = (\partial f/\partial x)|_*$ and $b(t) = (\partial f/\partial u)|_*$. With these assumptions, we may write Eq. (5-270) in the form

$$\frac{\partial^2 K}{\partial x^2}\bigg|_* \psi^2(t_1) + \int_{t_0}^{t_1} \left[ \frac{\partial^2 H}{\partial x^2}\bigg|_* \psi^2(t) + 2\frac{\partial^2 H}{\partial x\, \partial u}\bigg|_* \psi(t)\eta(t) + \frac{\partial^2 H}{\partial u^2}\bigg|_* \eta^2(t) \right] dt > 0 \tag{5-278}$$

for all $\eta(t) \neq 0$. We wish to show that Eq. (5-278) implies that

$$\frac{\partial^2 H}{\partial u^2}\bigg|_* \geq 0 \tag{5-279}$$

So let us suppose that $(\partial^2 H/\partial u^2)|_* < 0$ at $t = \hat{t}$; then [by the continuity of $(\partial^2 H/\partial u^2)|_*$] we may assume that, on a small neighborhood $S$ of $\hat{t}$ (that is, on a small interval about $\hat{t}$),

$$\frac{\partial^2 H}{\partial u^2}\bigg|_* < -\alpha \qquad \alpha > 0 \tag{5-280}$$

We now claim that $b(t) \neq 0$ on any subinterval of $S$. For if $b(t) = 0$ on a subinterval $S_1 \subset S$, then, letting $\eta(t) = 0$ on the complement of $S_1$ [that is, $\eta(t) = 0$ for $t \in [t_0, t_1] - S_1$] and letting $\eta(t) = 1$ (say) on $S_1$, we would find that the left-hand side of Eq. (5-278) was negative [since $\psi(t)$ would be identically zero for this choice of $\eta(t)$]. Thus, we may suppose that $b(t) \neq 0$ on $S$. It follows from Eq. (5-277) that

$$\eta(t) = b^{-1}(t)\dot{\psi}(t) - b^{-1}(t)a(t)\psi(t) \tag{5-281}$$

We now restrict ourselves to functions $\eta(t)$ which are zero on the complement of $S$ (that is, on $[t_0, t_1] - S$) and for which the corresponding $\psi$ is zero at the end points of $S$. In other words, if $S = [\hat{t} - \epsilon, \hat{t} + \epsilon]$, then $\eta(t) = 0$ for $t_0 \le t < \hat{t} - \epsilon$ and $\eta(t) = 0$ for $\hat{t} + \epsilon < t \le t_1$ and $\psi(\hat{t} - \epsilon) = \psi(\hat{t} + \epsilon) = 0$. For these functions $\eta(t)$, Eq. (5-278) becomes

$$\int_{\hat{t}-\epsilon}^{\hat{t}+\epsilon} \left\{ \frac{\partial^2 H}{\partial x^2}\bigg|_* \psi^2(t) + 2\frac{\partial^2 H}{\partial x\, \partial u}\bigg|_* \psi(t)[b^{-1}(t)\dot{\psi}(t) - b^{-1}(t)a(t)\psi(t)] + \frac{\partial^2 H}{\partial u^2}\bigg|_* \eta^2(t) \right\} dt > 0 \tag{5-282}$$

We integrate the term involving $\dot{\psi}(t)$ in Eq. (5-282) by parts to obtain the inequality

$$\int_{\hat{t}-\epsilon}^{\hat{t}+\epsilon} \left[ Q(t)\psi^2(t) + \frac{\partial^2 H}{\partial u^2}\bigg|_* \eta^2(t) \right] dt > 0 \tag{5-283}$$

where
$$Q(t) = \frac{\partial^2 H}{\partial x^2}\bigg|_* - \frac{d}{dt}\left[ \frac{\partial^2 H}{\partial x\, \partial u}\bigg|_* b^{-1}(t) \right] - 2\frac{\partial^2 H}{\partial x\, \partial u}\bigg|_* b^{-1}(t)a(t) \tag{5-284}$$

Letting $\varphi(t) = \Phi(t, \hat{t} - \epsilon)$ be the fundamental solution (matrix) of Eq. (5-277) and noting that $\varphi(t) \neq 0$, we have

$$\psi(t) = \varphi(t) \int_{\hat{t}-\epsilon}^{t} \varphi^{-1}(\tau)b(\tau)\eta(\tau)\, d\tau \tag{5-285}$$

Since $\varphi(\hat{t}) \neq 0$ and $b(\hat{t}) \neq 0$, we may suppose that, for suitably small $\epsilon$,

$$\frac{\varphi^2(t)}{b^2(t)} > \beta > 0 \tag{5-286}$$

for $t$ in $[\hat{t} - \epsilon, \hat{t} + \epsilon]$; and, by continuity, we may also suppose that

$$\max_{t \in [\hat{t}-\epsilon, \hat{t}+\epsilon]} |Q(t)\varphi^2(t)| \le M < \infty \tag{5-287}$$

We observe that Eqs. (5-286) and (5-287) remain valid as $\epsilon$ decreases; so let us take $\epsilon$ so small that

$$-\alpha\beta \frac{2\pi^2}{\epsilon} + 2M\epsilon < 0 \tag{5-288}$$

[where $\alpha$ is given in Eq. (5-280)]. We then let

$$\bar{\eta}(t) = b^{-1}(t)\varphi(t)\frac{\pi}{\epsilon}\sin\frac{2\pi(t - \hat{t})}{\epsilon} \tag{5-289}$$

Then the corresponding solution $\bar{\psi}(t)$ of Eq. (5-277) is given by

$$\bar{\psi}(t) = \varphi(t)\int_{\hat{t}-\epsilon}^{t} \frac{\pi}{\epsilon}\sin\frac{2\pi(\tau - \hat{t})}{\epsilon}\, d\tau = \varphi(t)\sin^2\frac{\pi(t - \hat{t})}{\epsilon} \tag{5-290}$$

so that $\bar{\psi}(\hat{t} - \epsilon) = \bar{\psi}(\hat{t} + \epsilon) = 0$. With this choice of $\bar{\eta}(t)$, Eq. (5-283) becomes

$$\int_{\hat{t}-\epsilon}^{\hat{t}+\epsilon} \left[ Q(t)\varphi^2(t)\sin^4\frac{\pi(t - \hat{t})}{\epsilon} \right] dt + \int_{\hat{t}-\epsilon}^{\hat{t}+\epsilon} \left[ \frac{\partial^2 H}{\partial u^2}\bigg|_* \frac{\varphi^2(t)}{b^2(t)}\frac{\pi^2}{\epsilon^2}\sin^2\frac{2\pi(t - \hat{t})}{\epsilon} \right] dt > 0 \tag{5-291}$$

But the left-hand side of Eq. (5-291) is, in view of Eqs. (5-286) and (5-287), *less than*

$$2M\epsilon - \alpha\beta\frac{2\pi^2}{\epsilon}$$

Thus, we have arrived at a contradiction, and so we have established Eq. (5-279). In other words, by assuming Eq. (5-280) to be valid, we were able to find a nonzero perturbation of the control [that is, $\bar{\eta}(t)$] for which Eq. (5-278) did not hold and so were led to the conclusion that Eq. (5-280) could not be valid. (See Ref. G-2.)

## 5-8 The Variational Approach to Control Problem 2: Sufficiency Conditions for the Free-end-point Problem

We are now going to strengthen the necessary conditions of the previous section in order to obtain some local sufficiency conditions which will be used in Chap. 9. Previously, we assumed that there was an optimum control $\mathbf{u}^*$, and we deduced certain conditions that had to be satisfied; now we shall suppose that certain conditions are satisfied for a control $\hat{\mathbf{u}}$, and then we shall deduce that the control $\hat{\mathbf{u}}$ must be (locally) optimal. Our discussion will, by the way, also illustrate the distinction between necessary and sufficient conditions.

We shall again consider the dynamical system

$$\dot{\mathbf{x}} = \mathbf{f}(\mathbf{x}, \mathbf{u}, t) \tag{5-292}$$

where $\mathbf{x}$ and $\mathbf{f}$ are $n$ vectors and $\mathbf{u}$ is an $m$ vector with $0 < m \leq n$. We suppose that $\mathbf{u}$ is unconstrained, that $t_0$ is our initial time, that

$$\mathbf{x}(t_0) = \mathbf{x}_0 \tag{5-293}$$

is our initial state, and that our target set $S$ is of the form $R_n \times \{t_1\}$, where $t_1$ is *fixed* and $t_1 > t_0$. Our performance functional $J(\mathbf{u})$ is given by

$$J(\mathbf{u}) = K[\mathbf{x}(t_1)] + \int_{t_0}^{t_1} L[\mathbf{x}(t), \mathbf{u}(t), t]\, dt \tag{5-294}$$

where $K$ and $L$ are sufficiently differentiable real-valued functions and $\mathbf{x}(t)$ is the trajectory of Eq. (5-292), starting from $\mathbf{x}(t_0) = \mathbf{x}_0$, generated by the control $\mathbf{u}(t)$. Again, our problem is: *Determine the control* $\mathbf{u}(t)$ *which minimizes the performance functional* $J(\mathbf{u})$.

Before presenting the desired sufficiency conditions, let us make some observations about the effects of perturbations on a control $\hat{\mathbf{u}}$ (which is not necessarily optimal). We let $\hat{\mathbf{x}}$ denote the trajectory of Eq. (5-292), starting from $\hat{\mathbf{x}}(t_0) = \mathbf{x}_0$, generated by the control $\hat{\mathbf{u}}$. If $\mathbf{u}$ is a perturbation of $\hat{\mathbf{u}}$ so that

$$\mathbf{u}(t) = \hat{\mathbf{u}}(t) + \epsilon\mathbf{n}(t) \tag{5-295}$$

then the trajectory $\mathbf{x}(t)$ corresponding to $\mathbf{u}(t)$ is given by

$$\mathbf{x}(t) = \hat{\mathbf{x}}(t) + \epsilon\boldsymbol{\psi}(t) + \mathbf{o}(\epsilon) \tag{5-296}$$

where $\boldsymbol{\psi}(t)$ is the solution of the linear differential equation

$$\dot{\boldsymbol{\psi}}(t) = \left.\frac{\partial \mathbf{f}}{\partial \mathbf{x}}\right|_{\wedge} \boldsymbol{\psi}(t) + \left.\frac{\partial \mathbf{f}}{\partial \mathbf{u}}\right|_{\wedge} \mathbf{n}(t) \tag{5-297}$$

with initial condition

$$\boldsymbol{\psi}(t_0) = \mathbf{0} \tag{5-298}$$

and where $|_{\wedge}$ indicates that the quantity in question is to be evaluated along $\hat{\mathbf{x}}(t)$; for example, $(\partial \mathbf{f}/\partial \mathbf{x}) |_{\wedge}$ is the Jacobian matrix of $\mathbf{f}$ with respect to $\mathbf{x}$ evaluated at $[\hat{\mathbf{x}}(t), \hat{\mathbf{u}}(t), t]$. Letting $H(\mathbf{x}, \mathbf{p}, \mathbf{u}, t)$ be the Hamiltonian of the problem, i.e., letting

$$H(\mathbf{x}, \mathbf{p}, \mathbf{u}, t) = L(\mathbf{x}, \mathbf{u}, t) + \langle \mathbf{p}, \mathbf{f}(\mathbf{x}, \mathbf{u}, t) \rangle \tag{5-299}$$

and letting $\hat{\mathbf{p}}(t)$ denote the solution of the linear differential equation

$$\dot{\hat{\mathbf{p}}}(t) = -\frac{\partial H}{\partial \mathbf{x}} [\hat{\mathbf{x}}(t), \hat{\mathbf{p}}(t), \hat{\mathbf{u}}(t), t] \tag{5-300}$$

$$= -\left.\frac{\partial L}{\partial \mathbf{x}}\right|_{\wedge} - \left(\left.\frac{\partial \mathbf{f}}{\partial \mathbf{x}}\right|_{\wedge}\right)' \hat{\mathbf{p}}(t) \tag{5-301}$$

satisfying the boundary condition

$$\hat{\mathbf{p}}(t_1) = \left.\frac{\partial K}{\partial \mathbf{x}}\right|_{\wedge} \tag{5-302}$$

we may deduce, just as in the previous section, that

$$J(\mathbf{u}) - J(\hat{\mathbf{u}}) = \epsilon \int_{t_0}^{t_1} \left\langle \left.\frac{\partial H}{\partial \mathbf{u}}\right|_{\wedge}, \mathbf{n}(t) \right\rangle dt + \frac{\epsilon^2}{2} \left\langle \boldsymbol{\psi}(t_1), \left.\frac{\partial^2 K}{\partial \mathbf{x}^2}\right|_{\wedge} \boldsymbol{\psi}(t_1) \right\rangle$$
$$+ \frac{\epsilon^2}{2} \int_{t_0}^{t_1} \left\langle \begin{bmatrix} \boldsymbol{\psi}(t) \\ \mathbf{n}(t) \end{bmatrix}, \begin{bmatrix} \left.\dfrac{\partial^2 H}{\partial \mathbf{x}^2}\right|_{\wedge} & \left.\dfrac{\partial}{\partial \mathbf{u}}\left(\dfrac{\partial H}{\partial \mathbf{x}}\right)\right|_{\wedge} \\ \left.\dfrac{\partial}{\partial \mathbf{x}}\left(\dfrac{\partial H}{\partial \mathbf{u}}\right)\right|_{\wedge} & \left.\dfrac{\partial^2 H}{\partial \mathbf{u}^2}\right|_{\wedge} \end{bmatrix} \begin{bmatrix} \boldsymbol{\psi}(t) \\ \mathbf{n}(t) \end{bmatrix} \right\rangle dt + o(\epsilon^2) \tag{5-303}$$

where $|_{\wedge}$ indicates that the quantity in question is evaluated along $\hat{\mathbf{x}}$; for example, $(\partial H/\partial \mathbf{u}) |_{\wedge}$ is the gradient of $H$ with respect to $\mathbf{u}$ evaluated at $[\hat{\mathbf{x}}(t), \hat{\mathbf{p}}(t), \hat{\mathbf{u}}(t), t]$. Thus, we have

$$\delta J(\hat{\mathbf{u}})(\mathbf{u} - \hat{\mathbf{u}}) = \epsilon \int_{t_0}^{t_1} \left\langle \left.\frac{\partial H}{\partial \mathbf{u}}\right|_{\wedge}, \mathbf{n}(t) \right\rangle dt \tag{5-304}$$

and

$$\delta^2 J(\hat{\mathbf{u}})(\mathbf{u} - \hat{\mathbf{u}}) = \frac{\epsilon^2}{2} \left\langle \boldsymbol{\psi}(t_1), \left.\frac{\partial^2 K}{\partial \mathbf{x}^2}\right|_{\wedge} \boldsymbol{\psi}(t_1) \right\rangle$$
$$+ \frac{\epsilon^2}{2} \int_{t_0}^{t_1} \left\langle \begin{bmatrix} \boldsymbol{\psi}(t) \\ \mathbf{n}(t) \end{bmatrix}, \begin{bmatrix} \left.\dfrac{\partial^2 H}{\partial \mathbf{x}^2}\right|_{\wedge} & \left.\dfrac{\partial}{\partial \mathbf{u}}\left(\dfrac{\partial H}{\partial \mathbf{x}}\right)\right|_{\wedge} \\ \left.\dfrac{\partial}{\partial \mathbf{x}}\left(\dfrac{\partial H}{\partial \mathbf{u}}\right)\right|_{\wedge} & \left.\dfrac{\partial^2 H}{\partial \mathbf{u}^2}\right|_{\wedge} \end{bmatrix} \begin{bmatrix} \boldsymbol{\psi}(t) \\ \mathbf{n}(t) \end{bmatrix} \right\rangle dt \tag{5-305}$$

We are now ready to prove the following theorem:

***Theorem 5-4*** *Suppose that* $\hat{\mathbf{u}}$ *satisfies the following conditions:*

*a.*
$$\left.\frac{\partial H}{\partial \mathbf{u}}\right|_{\wedge} = \mathbf{0} \qquad \text{for } t \text{ in } [t_0, t_1] \tag{5-306}$$

*b. For* $\mathbf{u}$ *sufficiently near* $\hat{\mathbf{u}}$*, there is a nonnegative piecewise continuous function* $M(t)$ *such that the magnitude of the* $o(\epsilon^2)$ *term in Eq.* (5-303) *satisfies the inequality*

$$|o(\epsilon^2)| \le |\epsilon|^3 \, \|\boldsymbol{\psi}(t_1)\|^3 M(t_1) + |\epsilon|^3 \int_{t_0}^{t_1} \|\mathbf{n}(t)\|^3 M(t)\, dt \tag{5-307}$$

*c. There is a positive constant* $k_1$ *such that*

$$\left\langle \mathbf{v}, \left.\frac{\partial^2 K}{\partial \mathbf{x}^2}\right|_{\wedge} \mathbf{v} \right\rangle \ge k_1 \|\mathbf{v}\|^2 M(t_1) \tag{5-308}$$

*for all vectors* $\mathbf{v}$ *in* $R_n$*, that is, the symmetric matrix* $(\partial^2 K/\partial \mathbf{x}^2)\,|_{\wedge}$ *is positive definite if* $M(t_1) > 0$ *and is positive semidefinite if* $M(t_1) = 0$.†

*d. There is a positive constant* $k_2$ *such that*

$$\left\langle \begin{bmatrix} \mathbf{v} \\ \mathbf{w} \end{bmatrix}, \begin{bmatrix} \left.\dfrac{\partial^2 H}{\partial \mathbf{x}^2}\right|_{\wedge} & \left.\dfrac{\partial}{\partial \mathbf{u}}\left(\dfrac{\partial H}{\partial \mathbf{x}}\right)\right|_{\wedge} \\ \left.\dfrac{\partial}{\partial \mathbf{x}}\left(\dfrac{\partial H}{\partial \mathbf{u}}\right)\right|_{\wedge} & \left.\dfrac{\partial^2 H}{\partial \mathbf{u}^2}\right|_{\wedge} \end{bmatrix} \begin{bmatrix} \mathbf{v} \\ \mathbf{w} \end{bmatrix} \right\rangle \ge k_2 \|\mathbf{w}\|^2 M(t) \tag{5-309}$$

*for all* $\mathbf{v}$ *in* $R_n$*, for all* $\mathbf{w}$ *in* $R_m$*, and for* $t$ *in* $[t_0, t_1]$*, and where the equality may hold only if* $\mathbf{w} = \mathbf{0}$*; this implies that the matrix* $(\partial^2 H/\partial \mathbf{u}^2)\,|_{\wedge}$ *is positive definite for* $t$ *in* $[t_0, t_1]$ *and that the entire* $n + m \times n + m$ *matrix appearing in Eq.* (5-309) *is positive semidefinite.*

*Then* $\hat{\mathbf{u}}$ *is a local minimum of* $J$.

PROOF Let us suppose that condition $b$ is satisfied whenever

$$\|\mathbf{u} - \hat{\mathbf{u}}\| = \max_{t \in [t_0, t_1]} \{\|\mathbf{u}(t) - \hat{\mathbf{u}}(t)\|\} < \delta \tag{5-310}$$

Then in view of condition $a$, we have, for $\mathbf{u}$ satisfying Eq. (5-310),

$$J(\mathbf{u}) - J(\hat{\mathbf{u}}) \ge \delta^2 J(\hat{\mathbf{u}})(\mathbf{u} - \hat{\mathbf{u}}) - |\epsilon|^3 \|\boldsymbol{\psi}(t_1)\|^3 M(t_1) - |\epsilon|^3 \int_{t_0}^{t_1} \|\mathbf{n}(t)\|^3 M(t)\, dt \tag{5-311}$$

Thus, making use of Eq. (5-305) and conditions $c$ and $d$, we conclude that

$$\begin{aligned} J(\mathbf{u}) - J(\hat{\mathbf{u}}) \ge{}& k_1 \frac{\epsilon^2}{2} \|\boldsymbol{\psi}(t_1)\|^2 M(t_1) + k_2 \frac{\epsilon^2}{2} \int_{t_0}^{t_1} \|\mathbf{n}(t)\|^2 M(t)\, dt \\ & - |\epsilon|^3 \|\boldsymbol{\psi}(t_1)\|^3 M(t_1) - |\epsilon|^3 \int_{t_0}^{t_1} \|\mathbf{n}(t)\|^3 M(t)\, dt \end{aligned} \tag{5-312}$$

† See Sec. 2-15.

and hence that

$$J(\mathbf{u}) - J(\hat{\mathbf{u}}) \geq \left[\frac{k_1}{2} - |\epsilon| \, \|\boldsymbol{\psi}(t_1)\|\right] \epsilon^2 \|\boldsymbol{\psi}(t_1)\|^2 M(t_1) + \int_{t_0}^{t_1} \left[\frac{k_2}{2} - |\epsilon| \, \|\mathbf{n}(t)\|\right] \epsilon^2 \|\mathbf{n}(t)\|^2 M(t) \, dt \quad (5\text{-}313)$$

However, in view of the fact that $\boldsymbol{\psi}(t)$ is the solution of Eq. (5-297) with initial condition $\boldsymbol{\psi}(t_0) = \mathbf{0}$, we observe that

$$|\epsilon| \, \|\boldsymbol{\psi}(t_1)\| \leq N|\epsilon| \max_{t \in [t_0, t_1]} \|\mathbf{n}(t)\| = N|\epsilon| \, \|\mathbf{n}\| \quad (5\text{-}314)$$

for some $N > 0$ which does not depend on $\mathbf{n}$. Since $k_1$ and $k_2$ are positive, there is a $\delta_0$ such that

$$0 < \delta_0 \leq \delta \quad (5\text{-}315)$$

and

$$\frac{k_1}{2} > N\delta_0 \qquad \frac{k_2}{2} > \delta_0 \quad (5\text{-}316)$$

It follows that

$$J(\mathbf{u}) - J(\hat{\mathbf{u}}) \geq 0 \quad (5\text{-}317)$$

for all $\mathbf{u}$ such that $\|\mathbf{u} - \hat{\mathbf{u}}\| < \delta_0$. Thus, we have shown that $\hat{\mathbf{u}}$ is a local minimum of $J$.

***Corollary 5-1*** *Suppose that $M(t) \equiv 0$. Then, for $\hat{\mathbf{u}}$ to be a local minimum of $J$, it is sufficient that:*

*a.* $(\partial H/\partial \mathbf{u})|_\wedge = \mathbf{0}$

*b.* $(\partial^2 H/\partial \mathbf{u}^2)|_\wedge$ *be positive definite*

*c. The $n + m \times n + m$ matrix appearing in Eq. (5-309) be positive semidefinite*

***Corollary 5-2*** *Suppose that the dynamical system is linear, i.e., that*

$$\dot{\mathbf{x}}(t) = \mathbf{A}(t)\mathbf{x}(t) + \mathbf{B}(t)\mathbf{u}(t) \qquad \mathbf{x}(t_0) = \mathbf{x}_0 \quad (5\text{-}318)$$

*and suppose that the performance functional $J$ is quadratic, i.e., that*

$$J(\mathbf{u}) = \tfrac{1}{2}\langle \mathbf{x}(t_1), \mathbf{F}\mathbf{x}(t_1)\rangle + \tfrac{1}{2}\int_{t_0}^{t_1} [\langle \mathbf{x}(t), \mathbf{Q}(t)\mathbf{x}(t)\rangle + \langle \mathbf{u}(t), \mathbf{R}(t)\mathbf{u}(t)\rangle] \, dt \quad (5\text{-}319)$$

*where $\mathbf{F}$, $\mathbf{Q}(t)$, and $\mathbf{R}(t)$ are, respectively, $n \times n$, $n \times n$, and $m \times m$ matrices. Then, for $\hat{\mathbf{u}}$ to be a local minimum of $J$, it is sufficient that:*

*a. The matrix $\mathbf{R}(t)$ be positive definite for $t$ in $[t_0, t_1]$*

*b. The matrix $\mathbf{Q}(t)$ be positive semidefinite for $t$ in $[t_0, t_1]$*

*c. The matrix $\mathbf{F}$ be positive semidefinite*

*d. $\hat{\mathbf{u}}$ be a solution of the equation*

$$\mathbf{R}(t)\hat{\mathbf{u}}(t) + \mathbf{B}'(t)\hat{\mathbf{p}}(t) = \mathbf{0} \quad (5\text{-}320)$$

*where $\hat{\mathbf{p}}(t)$ is the solution of the linear differential equation*

$$\dot{\hat{\mathbf{p}}}(t) = -\mathbf{A}'(t)\hat{\mathbf{p}}(t) - \mathbf{Q}(t)\hat{\mathbf{x}}(t) \quad (5\text{-}321)$$

*satisfying the boundary condition*

$$\hat{\mathbf{p}}(t_1) = \mathbf{F}\hat{\mathbf{x}}(t_1) \tag{5-322}$$

$\hat{\mathbf{x}}(t)$ *being the solution of Eq.* (5-318) *corresponding to* $\hat{\mathbf{u}}$.

Corollary 5-2 will be used repeatedly in Chap. 9.

**Exercise 5-13** Prove Corollary 5-2.

We observed in Sec. 5-7 that if $\mathbf{u}^*$ was an optimal control, then $(\partial H/\partial \mathbf{u})\big|_*$ was zero and the matrix $(\partial^2 H/\partial \mathbf{u}^2)\big|_*$ was positive definite. In this section, we supposed that the control $\hat{\mathbf{u}}$ satisfied these two conditions *and* the additional conditions *b*, *c*, and *d* of Theorem 5-4, and then we proved that $\hat{\mathbf{u}}$ had to be a locally optimal control. It is in this sense that *we refer to the sufficiency conditions of Theorem* 5-4 *as being a "strengthening" of the necessary conditions of Sec.* 5-7.

**Exercise 5-14** Consider the first-order system

$$\dot{x}(t) = -x(t) - \frac{x^3(t)}{3} + u(t) \qquad x(0) = x_0$$

and the cost functional

$$J(u) = \tfrac{1}{2}x^2(1) + \tfrac{1}{2}\int_0^1 [x^2(t) + u^2(t)]\,dt$$

Assume that $u^*(t)$ is an optimal control and that $x^*(t)$ is the corresponding optimal trajectory. Let

$$u(t) = u^*(t) + \epsilon\eta(t)$$
$$x(t) = x^*(t) + \epsilon\varphi(t)$$

be perturbations on the control and the trajectory.

(*a*) Show that

$$\dot{x}(t) = -x^*(t) - \tfrac{1}{3}x^{*3}(t) + u^*(t) - \epsilon\varphi(t)[1 + x^{*2}(t)] + \epsilon\eta(t) - \epsilon^2\varphi^2(t)x^*(t) - \frac{\epsilon^3\varphi^3(t)}{3}$$

(*b*) Identify the terms $(\partial \mathbf{f}/\partial \mathbf{x})\big|_*$, $(\partial \mathbf{f}/\partial \mathbf{u})\big|_*$, and $\mathbf{o}(\epsilon)$ of Eq. (5-208) in (*a*).

(*c*) Verify Eq. (5-215) for this particular system.

(*d*) Show that the Hamiltonian for this problem is given by

$$H(x, p, u, t) = \tfrac{1}{2}x^2 + \tfrac{1}{2}u^2 - xp - \tfrac{1}{3}x^3p + up$$

(*e*) Show that the terms in Eq. (5-226) are given for this problem by

$$\frac{\partial H}{\partial \mathbf{x}}\bigg|_{*\mathbf{p}} = x^*(t) - p(t) - x^{*2}(t)p(t)$$

$$\frac{\partial H}{\partial \mathbf{u}}\bigg|_{*\mathbf{p}} = u^*(t) + p(t)$$

$$\begin{aligned}\mathbf{o}(\epsilon) &= \frac{1}{2!}[1 - 2x^*(t)p(t)][x(t) - x^*(t)]^2 + \frac{1}{2!}[u(t) - u^*(t)]^2 - \frac{1}{3!}2p(t)[x(t) - x^*(t)]^3 \\ &= \frac{1}{2!}[1 - 2x^*(t)p(t)]\epsilon^2\phi^2(t) + \frac{1}{2!}\epsilon^2\eta^2(t) - \frac{2}{3!}p(t)\epsilon^3\phi^3(t)\end{aligned}$$

(*f*) Determine the $o(\epsilon)$ terms of Eqs. (5-227), (5-228), and (5-230) for this problem.
(*g*) Write Eq. (5-232) explicitly.
(*h*) Show that $p(t)$ must be a solution of the linear differential equation

$$\dot{p}(t) = -x^*(t) + [1 + x^{*2}(t)]p(t)$$

Deduce that $p^*(t)$ must be the unique solution of this differential equation, satisfying the boundary condition

$$p^*(1) = x^*(1)$$

(*i*) Show that [using Eq. (5-253)] $u^*(t)$ and $p^*(t)$ must be related by

$$u^*(t) = -p^*(t)$$

(*j*) Show that, for this problem, Eq. (5-263) becomes

$$\delta^2 J(u^*)(u - u^*) = \epsilon^2\psi^2(1) + \epsilon^2 \int_0^1 \left\langle \begin{bmatrix} \psi(t) \\ \eta(t) \end{bmatrix}, \begin{bmatrix} 1 - 2x^*(t)p^*(t) & 0 \\ 0 & 1 \end{bmatrix} \begin{bmatrix} \psi(t) \\ \eta(t) \end{bmatrix} \right\rangle dt$$

(*k*) Show that $x^*(t)$ and $p^*(t)$ satisfy the system of equations

$$\dot{x}^*(t) = -x^*(t) - \frac{x^{*3}(t)}{3} - p^*(t)$$
$$\dot{p}^*(t) = -x^*(t) + p^*(t) + x^{*2}(t)p^*(t)$$

with boundary conditions

$$x^*(0) = x_0 \qquad p^*(1) = x^*(1)$$

(*l*) Suppose that $\hat{u}$ is a control for which there is a corresponding $\hat{p}$ such that $\hat{u}$, $\hat{p}$, and $\hat{x}$ satisfy the necessary conditions and, in addition, satisfy the relation

$$\hat{x}(t)\hat{p}(t) \le \tfrac{1}{2} \qquad \text{for } t \text{ in } [0, 1]$$

Show that $\hat{u}$ must be a local minimum of $J$.

**Exercise 5-15** Consider the system

$$\dot{x}_1(t) = -x_2(t) \qquad x_1(0) = \xi_1$$
$$\dot{x}_2(t) = -x_2(t)t^2 + u(t) \qquad x_2(0) = \xi_2$$

and the cost functional

$$J(u) = \tfrac{1}{2} \int_0^1 [x_1^2(t)t^2 + x_2^2(t) + u^2(t)]\, dt$$

Write the following equations explicitly for this problem: (5-208), (5-211), (5-213), (5-215), (5-225), (5-227), (5-228), (5-230), (5-236), (5-237), (5-239), (5-240) to (5-243), (5-263), (5-271), (5-273), (5-303), and (5-309).

**Exercise 5-16** Repeat Exercise 5-15 for the system

$$\dot{x}(t) = -e^{-x^2(t)}x(t) + u(t)x(t) + u(t) \qquad x(0) = x_0$$

and the cost functional

$$J(u) = \tfrac{1}{2} \int_0^1 [x^4(t) + u^2(t)]\, dt$$

## 5-9 The Variational Approach to Control Problem 3: A Fixed-end-point Problem

We are now going to examine a fixed-end-point problem from the variational point of view. We shall arrive at an impasse in trying to show that an optimal control must extremize the Hamiltonian and shall thus have to reorient our thinking toward the control problem.

Let us now state the problem. We consider the dynamical system

$$\dot{\mathbf{x}}(t) = \mathbf{f}[\mathbf{x}(t), \mathbf{u}(t), t] \tag{5-323}$$

where $\mathbf{x}(t)$ and $\mathbf{f}$ are $n$ vectors and $\mathbf{u}(t)$ is an $m$ vector with $0 < m \leq n$. We suppose that we are dealing with an unconstrained problem, that $t_0$ is our initial time, that

$$\mathbf{x}(t_0) = \mathbf{x}_0 \tag{5-324}$$

is our initial state, and that our target set $S$ is of the form

$$S = \{(\mathbf{x}_1, t_1)\} \tag{5-325}$$

where $\mathbf{x}_1$ is a *given* element of $R_n$ and $t_1$ is a *fixed time*, $t_1 > t_0$. In other words, we want our motion to begin at $\mathbf{x}_0$ and to end at a given point $\mathbf{x}_1$. We let $L(\mathbf{x}, \mathbf{u}, t)$ be a sufficiently differentiable real-valued function, and we consider the performance functional $J(\mathbf{u})$ given by

$$J(\mathbf{u}) = \int_{t_0}^{t_1} L[\mathbf{x}(t), \mathbf{u}(t), t]\, dt \tag{5-326}$$

where $\mathbf{x}(t)$ is the trajectory of Eq. (5-323), starting from $\mathbf{x}(t_0) = \mathbf{x}_0$, generated by $\mathbf{u}(t)$. Our problem is: *Determine the admissible control* $\mathbf{u}(t)$ *which transfers* $\mathbf{x}(t_0) = \mathbf{x}_0$ *to* $\mathbf{x}(t_1) = \mathbf{x}_1$ *and which minimizes* $J(\mathbf{u})$ *over all such controls.*

Let us attempt now to develop necessary conditions similar to those of Theorems 5-2 and 5-3 by following the step-by-step procedure of Secs. 5-6 and 5-7.

*Step 1 Assumption of an Optimum*

We suppose that $\mathbf{u}^*(t)$ is an optimal control and that $\mathbf{x}^*(t)$ is the corresponding optimal trajectory. Thus, $\mathbf{u}^*(t)$ and $\mathbf{x}^*(t)$ satisfy the conditions

1. $$\dot{\mathbf{x}}^*(t) = \mathbf{f}[\mathbf{x}^*(t), \mathbf{u}^*(t), t] \tag{5-327}$$
2. $$\mathbf{x}^*(t_0) = \mathbf{x}_0 \qquad \mathbf{x}^*(t_1) = \mathbf{x}_1 \tag{5-328}$$

3. If $\mathbf{u}(t)$ is any admissible control whose corresponding trajectory $\mathbf{x}(t)$ begins at $\mathbf{x}(t_0) = \mathbf{x}_0$ and ends at $\mathbf{x}(t_1) = \mathbf{x}_1$, then

$$J^* = J(\mathbf{u}^*) = \int_{t_0}^{t_1} L[\mathbf{x}^*(t), \mathbf{u}^*(t), t]\, dt \leq J(\mathbf{u}) = \int_{t_0}^{t_1} L[\mathbf{x}(t), \mathbf{u}(t), t]\, dt \tag{5-329}$$

*Step 2 Perturbation*

Here we proceed just as in Sec. 5-7, obtaining the relations

$$\mathbf{u}(t) = \mathbf{u}^*(t) + \epsilon\mathbf{n}(t) \tag{5-330}$$

$$\mathbf{x}(t) = \mathbf{x}^*(t) + \epsilon\boldsymbol{\psi}(t) + o(\epsilon) \tag{5-331}$$

$$\dot{\boldsymbol{\psi}}(t) = \left.\frac{\partial \mathbf{f}}{\partial \mathbf{x}}\right|_* \boldsymbol{\psi}(t) + \left.\frac{\partial \mathbf{f}}{\partial \mathbf{u}}\right|_* \mathbf{n}(t) \qquad \boldsymbol{\psi}(t_0) = \mathbf{0} \tag{5-332}$$

which relate the perturbation of the control to the corresponding change in the trajectory.

However, we also require that our perturbed trajectory terminate at $\mathbf{x}_1$ [that is, that $\mathbf{x}(t_1) = \mathbf{x}_1$]; this implies that

$$\mathbf{0} = \epsilon\boldsymbol{\psi}(t_1) + \mathbf{o}(\epsilon) \tag{5-333}$$

Now, let us observe that $\boldsymbol{\psi}(t)$ does not depend on $\epsilon$, as $\boldsymbol{\psi}(t)$ is a function only of $\boldsymbol{\eta}(t)$ (and the optimal trajectory); on the basis of this observation and Eq. (5-333), we conclude that

$$\boldsymbol{\psi}(t_1) = \mathbf{0} \tag{5-334}$$

and hence that *our variation $\epsilon\boldsymbol{\eta}(t)$ in the control cannot be completely arbitrary.* In fact, if we let $\boldsymbol{\Phi}(t, t_0)$ be the fundamental matrix for the linear system of Eq. (5-332), then $\boldsymbol{\eta}(t)$ must satisfy the relation

$$\mathbf{0} = \int_{t_0}^{t_1} \boldsymbol{\Phi}^{-1}(t, t_0) \left.\frac{\partial \mathbf{f}}{\partial \mathbf{u}}\right|_* \boldsymbol{\eta}(t)\, dt \tag{5-335}$$

[which is analogous to Eq. (5-121)].

Now we can introduce the Hamiltonian $H$ and the costate variable $\mathbf{p}(t)$, just as in Secs. 5-6 and 5-7; that is, the costate variable $\mathbf{p}(t)$ is an arbitrary piecewise continuous function and the Hamiltonian $H$ is given by

$$H(\mathbf{x}, \mathbf{p}, \mathbf{u}, t) = L(\mathbf{x}, \mathbf{u}, t) + \langle \mathbf{p}, \mathbf{f}(\mathbf{x}, \mathbf{u}, t)\rangle \tag{5-336}$$

Arguing in a manner entirely similar to that used before, we may deduce that

$$\int_{t_0}^{t_1} \left\langle \left.\frac{\partial H}{\partial \mathbf{u}}\right|_{*\mathbf{p}}, \boldsymbol{\eta}(t) \right\rangle dt = 0 \tag{5-337}$$

for all $\boldsymbol{\eta}(t)$ satisfying Eq. (5-335) whenever $\mathbf{p}(t)$ is a solution of the linear differential equation

$$\dot{\mathbf{p}}(t) = -\frac{\partial H}{\partial \mathbf{x}}[\mathbf{x}^*(t), \mathbf{p}(t), \mathbf{u}^*(t), t] \tag{5-338}$$

$$= -\frac{\partial L}{\partial \mathbf{x}}[\mathbf{x}^*(t), \mathbf{u}^*(t), t] - \left(\frac{\partial \mathbf{f}}{\partial \mathbf{x}}[\mathbf{x}^*(t), \mathbf{u}^*(t), t]\right)' \mathbf{p}(t) \tag{5-339}$$

However, we are now at an impasse, since it is no longer an easy matter to show that there is a solution $\mathbf{p}^*(t)$ of Eq. (5-339) such that

$$\frac{\partial H}{\partial \mathbf{u}}[\mathbf{x}^*(t), \mathbf{p}^*(t), \mathbf{u}^*(t), t] = \mathbf{0} \tag{5-340}$$

In point of fact, the question of the existence of such a $\mathbf{p}^*(t)$ is quite deep, is intimately related to the notion of controllability and to the concept of a "normal" variational problem, and is, in any case, beyond the scope of this book. (See Ref. M-2.) Although such pronouncements may not seem convincing, we suggest that the reader examine this question a bit and try to see for himself where some of the difficulties lie. He then might profitably consult Refs. B-11 and M-2.

In order to circumvent this problem, we shall reorient our way of thinking about the control problem and shall adopt a somewhat different approach, which will lead us to the minimum principle of Pontryagin. Before developing this different approach in detail, we shall in the next section:

1. Review what we have done so far
2. Make some general comments to aid our intuition
3. Indicate some other substantive reasons (for example, the natural treatment of constrained problems and the removal of severe differentiability requirements on $L$) for adopting a new viewpoint

## 5-10 Comments on the Variational Approach

Our primary aim in this section will be the intuitive explication of the (local) necessary and sufficient conditions that were obtained in Secs. 5-7 and 5-8; consequently, we shall devote the major part of this section to this aim. In the remainder of the section, we shall indicate some of the reasons for adopting an alternative approach to the control problem.

We consider the system

$$\dot{\mathbf{x}}(t) = \mathbf{f}[\mathbf{x}(t), \mathbf{u}(t), t] \tag{5-341}$$

where the state $\mathbf{x}(t)$ is an $n$ vector and the control $\mathbf{u}(t)$ is an $m$ vector with $m \leq n$. We suppose that we are dealing with an unconstrained problem, that $t_0$ is our initial time, that

$$\mathbf{x}(t_0) = \mathbf{x}_0 \tag{5-342}$$

is our initial state, and that our target set $S$ is of the form

$$S = R_n \times \{t_1\} \tag{5-343}$$

where $t_1$ is a *fixed time*, $t_1 > t_0$. In other words, we are dealing with a *free-end-point, fixed-time* problem. We also suppose that our performance functional $J(\mathbf{u})$ is given by

$$J(\mathbf{u}) = \int_{t_0}^{t_1} L[\mathbf{x}(t), \mathbf{u}(t), t]\, dt \tag{5-344}$$

so that $J(\mathbf{u})$ does not involve a terminal cost [that is, $K[\mathbf{x}(t_1)] \equiv 0$]. Thus, our problem is: *Determine the* $\mathbf{u}$ *which minimizes* $J(\mathbf{u})$.

As we are going to illustrate the necessary and sufficient conditions by means of "symbolic plots," we shall have to recall some terminology and some notational conventions. In particular, we remind the reader that:

1. When we write $\mathbf{u}(t)$, we mean the value of the control at $t$, $t \in [t_0, t_1]$, so that $\mathbf{u}(t)$ is an element of $R_m$.

2. When we write $\mathbf{u}_{[t_0, t_1]}$, we mean the entire control function over the interval $[t_0, t_1]$; for the sake of exposition, we often simply write $\mathbf{u}$ in place

of $\mathbf{u}_{[t_0,t_1]}$, that is,

$$\mathbf{u} = \mathbf{u}_{[t_0,t_1]} \tag{5-345}$$

and when dealing with several controls, we often use superscripts to distinguish them; for example,

$$\mathbf{u}^i = \mathbf{u}^i_{[t_0,t_1]} \tag{5-346}$$

for a given value of $i$ would indicate a particular control function and the vector $\mathbf{u}^i(t)$ would be the value of this particular control function at $t$, $t \in [t_0, t_1]$.

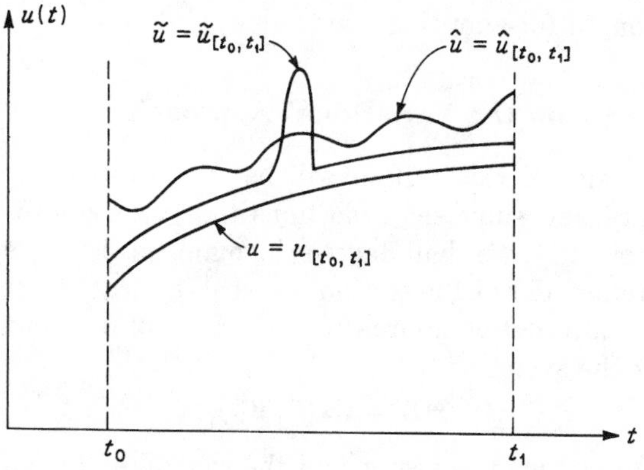


**Fig. 5-9** The control $\hat{\mathbf{u}}$ is nearer to the control $\mathbf{u}$ than the control $\tilde{\mathbf{u}}$ is.

3. When we write $\mathcal{P}([t_0, t_1], R_m)$, we mean the function space† of all piecewise continuous functions from $[t_0, t_1]$ into $R_m$; as this is our set of admissible controls, we shall use $\mathcal{U}$ in place of $\mathcal{P}([t_0, t_1], R_m)$ in the sequel, that is,

$$\mathcal{U} = \{\mathbf{u}: \mathbf{u} \text{ an admissible control}\} = \mathcal{P}([t_0, t_1], R_m) \tag{5-347}$$

and we shall often refer to particular control functions $\mathbf{u}^i$ as points in the function space $\mathcal{U}$.

4. When we say that "$\hat{\mathbf{u}}$ is near $\mathbf{u}$," we mean that the distance, in the function space $\mathcal{U}$, between $\hat{\mathbf{u}}$ and $\mathbf{u}$ is small; we recall that in Sec. 3-15 [Eq. (3-124)], we have defined this distance $d(\hat{\mathbf{u}}, \mathbf{u})$ by setting

$$d(\hat{\mathbf{u}}, \mathbf{u}) = \|\hat{\mathbf{u}} - \mathbf{u}\| = \sup_{t \in [t_0, t_1]} \|\hat{\mathbf{u}}(t) - \mathbf{u}(t)\| \tag{5-348}$$

Thus, the function $\hat{\mathbf{u}}_{[t_0,t_1]}$ in Fig. 5-9 is closer to $\mathbf{u}_{[t_0,t_1]}$ than the function $\tilde{\mathbf{u}}_{[t_0,t_1]}$ is.

Now, if we apply the control $\mathbf{u}^i = \mathbf{u}^i_{[t_0,t_1]}$ to our system, which, at $t = t_0$, is at the known initial state $\mathbf{x}_0$, then we obtain a well-defined trajectory $\mathbf{x}^i = \mathbf{x}^i_{[t_0,t_1]}$ which corresponds to $\mathbf{u}^i$ and, consequently, a well-defined (scalar)

† See Sec. 3-15, particularly Eq. (3-136).

value $J^i$ of our cost. In other words, we have

$$J^i = J(\mathbf{u}^i) = \int_{t_0}^{t_1} L[\mathbf{x}^i(t), \mathbf{u}^i(t), t]\, dt \tag{5-349}$$

Strictly speaking, the cost $J^i$ depends on many things, and we should write

$$J^i = J(\mathbf{u}^i) = J(\mathbf{u}^i; \mathbf{x}_0, t_0, t_1, \mathbf{f}) \tag{5-350}$$

to indicate the dependence of $J^i$ on:

1. The initial state $\mathbf{x}_0$
2. The initial time $t_0$
3. The terminal time $t_1$
4. The system $\dot{\mathbf{x}}(t) = \mathbf{f}[\mathbf{x}(t), \mathbf{u}(t), t]$
5. The control function $\mathbf{u}^i$

However, for *fixed* $\mathbf{x}_0$, $t_0$, $t_1$, *and* $\mathbf{f}$, $J$ is a function only of the element $\mathbf{u}^i$ of $\mathcal{U}$. Thus, we can make a "symbolic plot" of $J(\mathbf{u})$ versus $\mathbf{u}$, as illustrated in Fig.

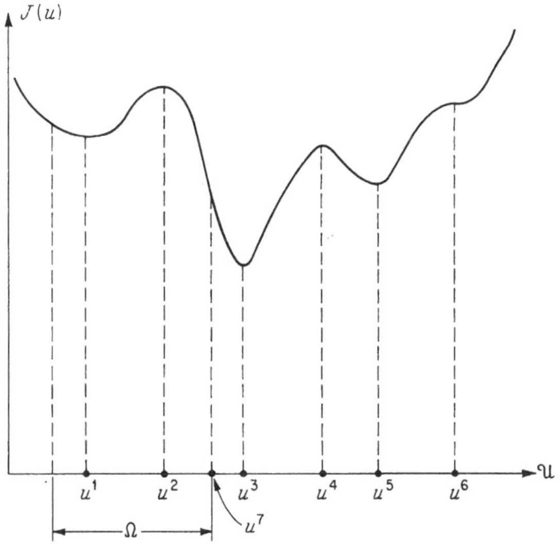


**Fig. 5-10** Symbolic plot of the cost $J(\mathbf{u})$ versus $\mathbf{u}$.

5-10. We point out that the $\mathcal{U}$ "axis" in Fig. 5-10 represents the function space $\mathcal{U}$ and that each "point" $\mathbf{u}$ on the $\mathcal{U}$ "axis" represents a *function* $\mathbf{u}_{[t_0,t_1]}$.

We observe that the functional $J(\mathbf{u})$, which we have plotted in Fig. 5-10, has:

1. An absolute minimum at $\mathbf{u}^3$
2. Relative minima at $\mathbf{u}^1$ and $\mathbf{u}^5$
3. Relative maxima at $\mathbf{u}^2$ and $\mathbf{u}^4$
4. An inflection at $\mathbf{u}^6$

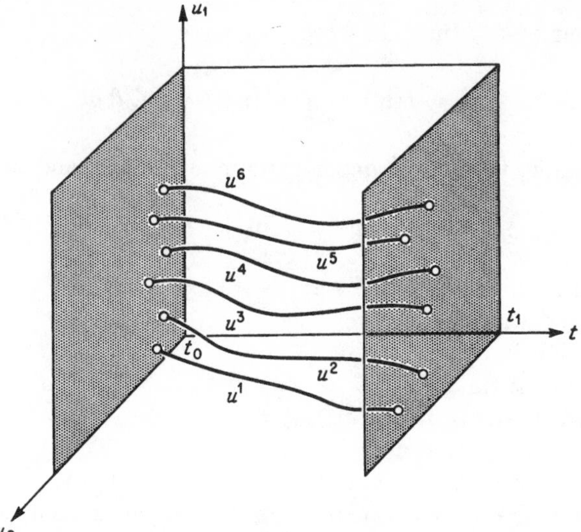


(a)

**Fig. 5-11** (a) The extremal controls $\mathbf{u}^1, \ldots, \mathbf{u}^6$. (b) The states $\mathbf{x}^1, \ldots, \mathbf{x}^6$ generated by the controls of (a). (c) The costates $\mathbf{p}^1, \ldots, \mathbf{p}^6$ that correspond to the controls of (a) and the states of (b).

In other words, $J(\mathbf{u})$ has the following properties:

1. The "derivative" (i.e., the first variation) of $J(\mathbf{u})$ is zero at $\mathbf{u}^1$, $\mathbf{u}^2$, $\mathbf{u}^3$, $\mathbf{u}^4$, $\mathbf{u}^5$, and $\mathbf{u}^6$.
2. The inequalities

$$\begin{aligned} J(\mathbf{u}^1) &< J(\hat{\mathbf{u}}^1) && \text{for all } \hat{\mathbf{u}}^1 \text{ ``near'' } \mathbf{u}^1 \\ J(\mathbf{u}^3) &< J(\hat{\mathbf{u}}^3) && \text{for all } \hat{\mathbf{u}}^3 \neq \mathbf{u}^3 \\ J(\mathbf{u}^5) &< J(\hat{\mathbf{u}}^5) && \text{for all } \hat{\mathbf{u}}^5 \text{ ``near'' } \mathbf{u}^5 \\ J(\mathbf{u}^2) &> J(\hat{\mathbf{u}}^2) && \text{for all } \hat{\mathbf{u}}^2 \text{ ``near'' } \mathbf{u}^2 \\ J(\mathbf{u}^4) &> J(\hat{\mathbf{u}}^4) && \text{for all } \hat{\mathbf{u}}^4 \text{ ``near'' } \mathbf{u}^4 \end{aligned} \tag{5-351}$$

are satisfied.

The control $\mathbf{u}^3$ is *globally optimal*, and the controls $\mathbf{u}^1$ and $\mathbf{u}^5$ are called *locally optimal*.

Let us now examine the behavior of the Hamiltonian $H$ for the problem with the cost functional plotted in Fig. 5-10. We shall first discuss the properties of $H$ from an analytical point of view and then illustrate these properties with "symbolic plots."

The Hamiltonian $H$ for our problem is given by

$$H(\mathbf{x}, \mathbf{p}, \mathbf{u}, t) = L(\mathbf{x}, \mathbf{u}, t) + \langle \mathbf{p}, \mathbf{f}(\mathbf{x}, \mathbf{u}, t) \rangle \tag{5-352}$$

and the canonical system of vector differential equations is given by

$$\dot{\mathbf{x}}(t) = \frac{\partial H}{\partial \mathbf{p}} [\mathbf{x}(t), \mathbf{p}(t), \mathbf{u}(t), t] \tag{5-353}$$

$$\dot{\mathbf{p}}(t) = -\frac{\partial H}{\partial \mathbf{x}} [\mathbf{x}(t), \mathbf{p}(t), \mathbf{u}(t), t] \tag{5-354}$$

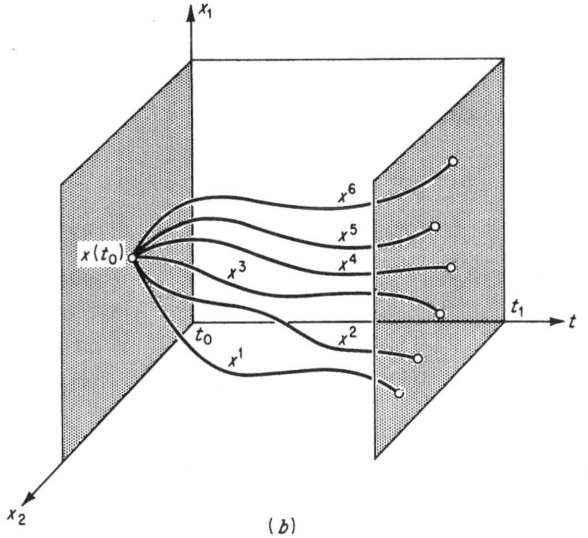

(b)

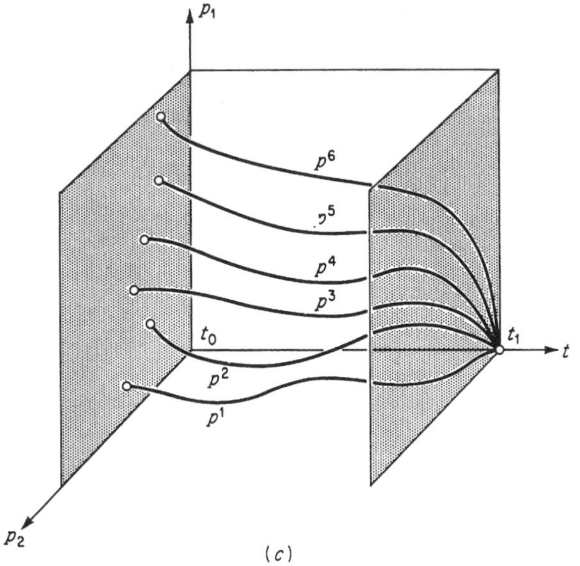

(c)

with boundary conditions

$$\mathbf{x}(t_0) = \mathbf{x}_0 \tag{5-355}$$
$$\mathbf{p}(t_1) = \mathbf{0} \tag{5-356}$$

We shall be particularly interested in the solutions of these equations for the controls $\mathbf{u}^1, \mathbf{u}^2, \ldots, \mathbf{u}^6$ which correspond to the extrema of $J(\mathbf{u})$ in Fig. 5-10.

We suppose that $\mathbf{u}^1, \mathbf{u}^2, \ldots, \mathbf{u}^6$ are plotted in Fig. 5-11$a$ and that the corresponding responses $\mathbf{x}^1, \mathbf{x}^2, \ldots, \mathbf{x}^6$ of our system (5-323), starting from the *same* initial state $\mathbf{x}(t_0) = \mathbf{x}_0$, are plotted in Fig. 5-11$b$. Since

$\partial H/\partial \mathbf{p} = \mathbf{f}(\mathbf{x}, \mathbf{u}, t)$, the responses $\mathbf{x}^1, \mathbf{x}^2, \ldots, \mathbf{x}^6$ do not depend upon the costate. In Fig. 5-11*c*, we plot the solutions $\mathbf{p}^1, \mathbf{p}^2, \ldots, \mathbf{p}^6$ of Eq. (5-354) which correspond to the controls $\mathbf{u}^1, \mathbf{u}^2, \ldots, \mathbf{u}^6$, respectively. In other words, $\mathbf{p}^i = \mathbf{p}^i_{[t_0, t_1]}$ is the solution of the (linear) differential equation

$$\dot{\mathbf{p}}(t) = -\frac{\partial H}{\partial \mathbf{x}}[\mathbf{x}^i(t), \mathbf{p}(t), \mathbf{u}^i(t), t] \tag{5-357}$$

satisfying the boundary condition $\mathbf{p}^i(t_1) = \mathbf{0}$ for $i = 1, 2, \ldots, 6$.

Since the controls $\mathbf{u}^1, \mathbf{u}^2, \ldots, \mathbf{u}^6$ correspond to extrema of $J(\mathbf{u})$, we may deduce that

$$\frac{\partial H}{\partial \mathbf{u}}[\mathbf{x}^i(t), \mathbf{p}^i(t), \mathbf{u}^i(t), t] = \mathbf{0} \tag{5-358}$$

for $i = 1, 2, \ldots, 6$ and $t \in [t_0, t_1]$. Moreover, in view of the fact that $\mathbf{u}^1$, $\mathbf{u}^3$, and $\mathbf{u}^5$ are minima of $J(\mathbf{u})$, that $\mathbf{u}^2$ and $\mathbf{u}^4$ are maxima of $J(\mathbf{u})$, and that $\mathbf{u}^6$ is an inflection of $J(\mathbf{u})$, we may infer that the $m \times m$ matrix $\partial^2 H/\partial \mathbf{u}^2$ has the properties:

1. $\partial^2 H/\partial \mathbf{u}^2$ is positive definite when evaluated at

$$[\mathbf{x}^j(t), \mathbf{p}^j(t), \mathbf{u}^j(t), t] \qquad \text{for } j = 1, 3, 5 \tag{5-359}$$

2. $\partial^2 H/\partial \mathbf{u}^2$ is negative definite when evaluated at

$$[\mathbf{x}^k(t), \mathbf{p}^k(t), \mathbf{u}^k(t), t] \qquad \text{for } k = 2, 4 \tag{5-360}$$

3. $\partial^2 H/\partial \mathbf{u}^2$ is neither positive nor negative definite when evaluated at

$$[\mathbf{x}^6(t), \mathbf{p}^6(t), \mathbf{u}^6(t), t] \tag{5-361}$$

We may also infer that the matrix

$$\left[\begin{array}{c|c} \dfrac{\partial^2 H}{\partial \mathbf{x}^2} & \dfrac{\partial^2 H}{\partial \mathbf{x}\, \partial \mathbf{u}} \\ \hline \dfrac{\partial^2 H}{\partial \mathbf{u}\, \partial \mathbf{x}} & \dfrac{\partial^2 H}{\partial \mathbf{u}^2} \end{array}\right] = \mathbf{Q}(\mathbf{x}, \mathbf{p}, \mathbf{u}, t) \tag{5-362}$$

has the following properties:

1. $\mathbf{Q}$ is positive definite when evaluated at

$$[\mathbf{x}^j(t), \mathbf{p}^j(t), \mathbf{u}^j(t), t] \qquad \text{for } j = 1, 3, 5 \tag{5-363}$$

2. $\mathbf{Q}$ is negative definite when evaluated at

$$[\mathbf{x}^k(t), \mathbf{p}^k(t), \mathbf{u}^k(t), t] \qquad \text{for } k = 2, 4 \tag{5-364}$$

3. $\mathbf{Q}$ is neither positive nor negative definite when evaluated at

$$[\mathbf{x}^6(t), \mathbf{p}^6(t), \mathbf{u}^6(t), t] \tag{5-365}$$

These considerations also allow us to suppose that, for $t \in [t_0, t_1]$, the following inequalities hold:

$$H[\mathbf{x}^1(t), \mathbf{p}^1(t), \mathbf{u}^1(t), t] \le H[\mathbf{x}^1(t), \mathbf{p}^1(t), \mathbf{u}^\alpha(t), t] \qquad \text{for } \alpha = 2, 3, 4, 5, 6 \tag{5-366}$$

$$H[\mathbf{x}^3(t), \mathbf{p}^3(t), \mathbf{u}^3(t), t] \le H[\mathbf{x}^3(t), \mathbf{p}^3(t), \mathbf{u}^\beta(t), t] \qquad \text{for } \beta = 1, 2, 4, 5, 6 \tag{5-367}$$

$$H[\mathbf{x}^5(t), \mathbf{p}^5(t), \mathbf{u}^5(t), t] \le H[\mathbf{x}^5(t), \mathbf{p}^5(t), \mathbf{u}^\gamma(t), t] \qquad \text{for } \gamma = 1, 2, 3, 4, 6 \tag{5-368}$$

$$H[\mathbf{x}^2(t), \mathbf{p}^2(t), \mathbf{u}^2(t), t] \ge H[\mathbf{x}^2(t), \mathbf{p}^2(t), \mathbf{u}^\delta(t), t] \qquad \text{for } \delta = 1, 3, 4, 5, 6 \tag{5-369}$$

$$H[\mathbf{x}^4(t), \mathbf{p}^4(t), \mathbf{u}^4(t), t] \ge H[\mathbf{x}^4(t), \mathbf{p}^4(t), \mathbf{u}^\epsilon(t), t] \qquad \text{for } \epsilon = 1, 2, 3, 5, 6 \tag{5-370}$$

In Fig. 5-12, we plot $H[\mathbf{x}^1(t), \mathbf{p}^1(t), \mathbf{u}(t), t]$ versus $\mathbf{u}(t)$ in order to illustrate the inequality (5-366). The $u_1u_2$ plane in Fig. 5-12 represents the space

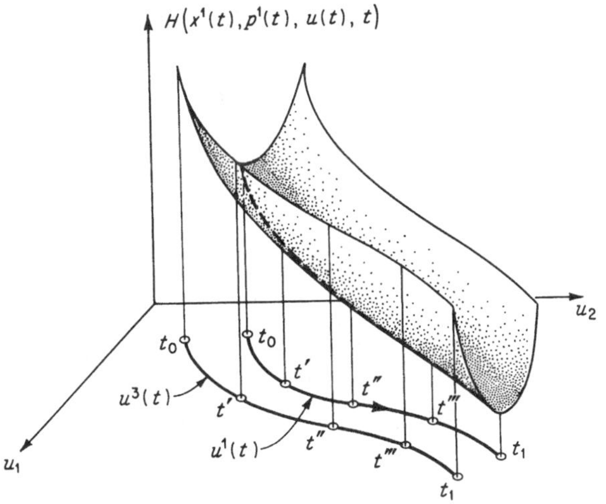


**Fig. 5-12** The Hamiltonian $H[\mathbf{x}^1(t), \mathbf{p}^1(t), \mathbf{u}(t), t]$ has a minimum at $\mathbf{u}^1(t)$ but not at $\mathbf{u}^3(t)$.

$R_m$. We observe that $H[\mathbf{x}^1(t), \mathbf{p}^1(t), \mathbf{u}(t), t]$ has a minimum at $\mathbf{u}(t) = \mathbf{u}^1(t)$ and that, although $\mathbf{u}^3$ is the global minimum of $J(\mathbf{u})$, $H[\mathbf{x}^1(t), \mathbf{p}^1(t), \mathbf{u}(t), t]$ *does not* have a minimum at $\mathbf{u}(t) = \mathbf{u}^3(t)$.

In Fig. 5-13, we show that $H[\mathbf{x}^4(t), \mathbf{p}^4(t), \mathbf{u}(t), t]$ has a maximum at $\mathbf{u}(t) = \mathbf{u}^4(t)$ [see the inequality (5-370)], and in Fig. 5-14, we illustrate the fact that $H[\mathbf{x}^6(t), \mathbf{p}^6(t), \mathbf{u}(t), t]$ has an inflection at $\mathbf{u}(t) = \mathbf{u}^6(t)$.

We hope that these symbolic plots illustrate the point that the necessary conditions of Theorem 5-3 are local in nature (i.e., are satisfied by controls which locally minimize the cost as well as by the globally optimal control)

and also the point that corresponding conditions could be derived for controls which locally maximize the cost. Furthermore, we hope that these symbolic plots emphasize the fact that the necessary conditions do not provide us with enough information to determine the optimal control. For

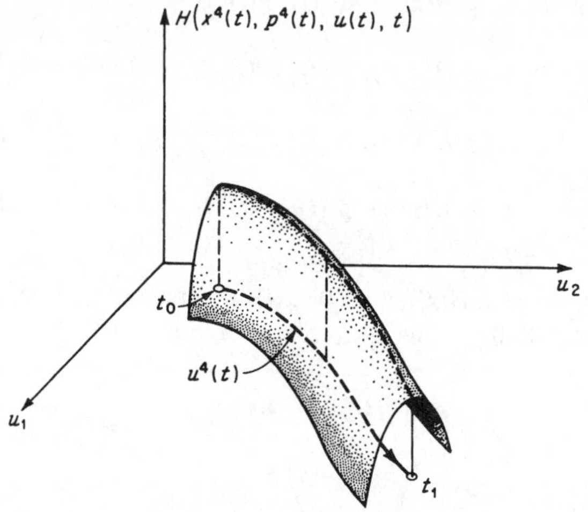


Fig. 5-13 The Hamiltonian $H[\mathbf{x}^4(t), \mathbf{p}^4(t), \mathbf{u}(t), t]$ has a maximum at $\mathbf{u}^4(t)$.

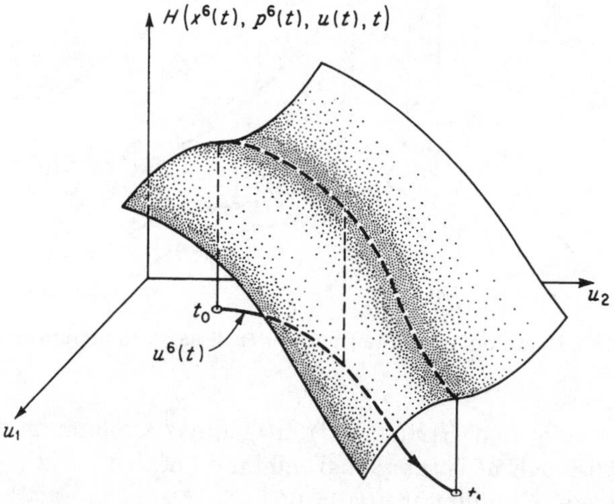


Fig. 5-14 The Hamiltonian $H[\mathbf{x}^6(t), \mathbf{p}^6(t), \mathbf{u}(t), t]$ has an inflection at $\mathbf{u}^6(t)$.

example, the three controls $\mathbf{u}^1$, $\mathbf{u}^3$, and $\mathbf{u}^5$ all satisfy the necessary conditions of Theorem 5-3, and so the fact that $\mathbf{u}^3$ is globally optimal can only be deduced by computing $J(\mathbf{u}^1)$, $J(\mathbf{u}^3)$, and $J(\mathbf{u}^5)$ and then comparing the computed values to find the smallest one.

Let us make one final observation about Fig. 5-10. If we suppose that $\mathbf{u}$ is constrained to be an element of the closed set $\Omega$ shown in Fig. 5-10, then we can observe that

$$J(\mathbf{u}^7) \leq J(\mathbf{u}) \qquad \text{for all } \mathbf{u} \text{ in } \Omega \tag{5-371}$$

However, $\mathbf{u}^7$ is *not* an extremum of $H$; that is, $\frac{\partial H}{\partial \mathbf{u}} [\mathbf{x}^7(t), \mathbf{p}^7(t), \mathbf{u}^7(t), t] \neq \mathbf{0}$, and the matrix $\frac{\partial^2 H}{\partial \mathbf{u}^2} [\mathbf{x}^7(t), \mathbf{p}^7(t), \mathbf{u}^7(t), t]$ need *not* be positive definite. Thus, *if we were seeking the minimum of $J(\mathbf{u})$ over the set $\Omega$, our necessary conditions would not apply and the variational techniques that we have used so far would not be helpful.* The reader should note that $\mathbf{u}^7$ is a boundary point of $\Omega$ and should consider this situation in light of the results obtained in Secs. 5-2 and 5-3 and in light of the comments of Sec. 5-5.

We have already noted that, for a fixed-end-point problem, difficulties arise in developing suitable necessary conditions on the basis of the variational techniques that we used for free-end-point problems. These difficulties stemmed primarily from the fact that we were not sure that there were "enough" variations, since our admissible variations were constrained. [See Eq. (5-335).] We are now going to indicate some further reasons for adopting a different point of view.

First of all, we note that we required $L$ and $\mathbf{f}$ to be "sufficiently" differentiable, and in particular, we required that $\partial L/\partial \mathbf{u}$ be defined. This ruled out such criteria as

$$\int_{t_0}^{t_1} |u(t)|\, dt \tag{5-372}$$

which, as we shall see in Chap. 8, are quite important. So we seek an approach which does not place severe differentiability requirements on $L$ and $\mathbf{f}$.

Secondly, we observed, when examining the behavior of $J(\mathbf{u})$ over $\Omega$ in Fig. 5-10, that our necessary conditions did not apply in the case of a closed constraint set.† Thus, such constraints on the control as

$$|u_i(t)| \leq M \qquad i = 1, 2, \ldots, m \tag{5-373}$$

could not be handled readily. Moreover, if $\Omega$ consists of a *finite* number of points, the variational approach does not work. Since constraints are of great engineering importance, we must try to develop a method of attack which easily takes them into account.

Finally, we note that we did not examine problems in which our target set was neither all of $R_n$ nor a single point. Although such problems are easy to treat using variational techniques *once the desired necessary conditions have been obtained for fixed-end-point problems,* we did not attack them because of

† Such constraints can be handled by variational techniques (see Ref. D-8); however, proofs of the validity of results are quite difficult.

the difficulties involved in the fixed-end-point problem. Thus, we want to develop an approach in which "smooth" target sets can be handled easily.

The minimum principle of Pontryagin will require relatively weak differentiability assumptions, will be well suited to the handling of constraints, and will readily take "smooth" target sets into account. So we turn our attention to it in the next portion of the chapter.

## 5-11 The Minimum Principle of Pontryagin: Introduction

We are now going to study the celebrated minimum principle of Pontryagin (see Ref. P-5). Our three major goals in this study will be:

1. To give a careful, precise statement of the minimum principle
2. To present a *heuristic* proof of the minimum principle based upon the proof given in Ref. P-5 and, in so doing, to provide a suitable introduction to and motivation for that formal proof
3. To interpret the minimum principle in several ways so as to develop a greater intuitive insight into it

In subsequent chapters of the book, we shall show how the minimum principle may be used to obtain the solution of various control problems.

Bearing our major objectives in mind, let us now indicate the path that we shall follow to attain them. First of all, we shall restate the general control problem. Then we shall carefully delineate two special cases of that problem. We shall next give a precise statement of the necessary conditions for optimality which constitute the minimum principle in the two special cases of the control problem. We shall then show, by various changes of variable, how the minimum principle for the general problem can be obtained from the results for the special problems. After doing this, we shall present the heuristic proof based upon the formal proof of Ref. P-5. Finally, we shall interpret and comment on our results and then discuss some sufficient conditions for optimality.

## 5-12 The Minimum Principle of Pontryagin: Restatement of the Control Problem

We are now going to restate the formal control problem (see Definition 4-9) in the form to which the necessary conditions for optimality will apply. The reader may find this section rather formal; however, we must state precisely the assumptions under which the minimum principle is derived.

We suppose that we are given an $n$th-order dynamical system with $(T_1, T_2)$ as interval of definition and with state equation

$$\dot{\mathbf{x}} = \mathbf{f}(\mathbf{x}, \mathbf{u}, t) \tag{5-374}$$

We suppose that $\Omega$ is a given subset of $R_m$ and that our set $U$ of *admissible controls* is the set of all bounded piecewise continuous functions $\mathbf{u}(t)$ on

$(T_1, T_2)$ such that

$$\mathbf{u}(t) \in \Omega \qquad \text{for all } t \text{ in } (T_1, T_2) \tag{5-375}$$

and

$$\mathbf{u}(t-) = \mathbf{u}(t) \qquad \text{for all } t \text{ in } (T_1, T_2) \tag{5-376}$$ †

As usual, we shall denote the transition function of our system by

$$\boldsymbol{\phi}(t; \mathbf{u}_{(t_0,t]}, \mathbf{x}_0) \tag{5-377}$$

We let $L(\mathbf{x}, \mathbf{u}, t)$ be a real-valued function on $R_n \times R_m \times (T_1, T_2)$, and we let $K(\mathbf{x}, t)$ be a real-valued function on $R_n \times (T_1, T_2)$. Finally, we suppose that $S$ is a given subset of $R_n \times (T_1, T_2)$ so that the elements of $S$ are pairs $(\mathbf{x}, t)$ consisting of a state $\mathbf{x}$ and a point $t$ in the interval of definition of the system.

We now make the following assumptions:

CP1 *If $f_1(\mathbf{x}, \mathbf{u}, t), f_2(\mathbf{x}, \mathbf{u}, t), \ldots, f_n(\mathbf{x}, \mathbf{u}, t)$ denote the components of $\mathbf{f}(\mathbf{x}, \mathbf{u}, t)$, then we assume that the functions*

$$f_i(\mathbf{x}, \mathbf{u}, t), \frac{\partial f_i}{\partial \mathbf{x}}(\mathbf{x}, \mathbf{u}, t), \frac{\partial f_i}{\partial t}(\mathbf{x}, \mathbf{u}, t) \qquad i = 1, 2, \ldots, n \tag{5-378}$$

*and the functions*

$$L(\mathbf{x}, \mathbf{u}, t), \frac{\partial L}{\partial \mathbf{x}}(\mathbf{x}, \mathbf{u}, t), \frac{\partial L}{\partial t}(\mathbf{x}, \mathbf{u}, t) \tag{5-379}$$

*are continuous on $R_n \times \bar{\Omega} \times (T_1, T_2)$, where $\bar{\Omega}$ is the closure of $\Omega$ in $R_m$ (see Definition 3-9). (Observe that we do not require that the $f_i$ and $L$ have continuous partial derivatives with respect to the components of $\mathbf{u}$.)*

CP2 *If $S$ is the given subset of $R_n \times (T_1, T_2)$, then we assume that $S$ is of one of the following forms:*

*a. $S = \{\mathbf{x}_1\} \times T$, where $\mathbf{x}_1$ is a fixed element of $R_n$ and $T$ is a subset of $(T_1, T_2)$.*

*b. $S = R_n \times T$, where $T$ is a subset of $(T_1, T_2)$.*

*c. $S = S_1 \times T$, where $S_1$ is a smooth $k$-fold, $1 \le k \le n - 1$, in $R_n$ (see Definition 3-30) and $T$ is a subset of $(T_1, T_2)$.*

*d. $S = \{(\mathbf{g}(t), t): t \in (T_1, T_2)\}$, where $\mathbf{g}(t)$ is a continuously differentiable function from $(T_1, T_2)$ into $R_n$ [so that $\mathbf{g}(t)$ is a curve in $R_n$ with a continuously turning tangent].*

*e. $S$ is a smooth $(k + 1)$-fold, $1 \le k \le n - 1$, in $R_n \times (T_1, T_2)$ which is continuously differentiable in $t$; that is, there are $n - k$ functions $g_1(\mathbf{x}, t)$, $g_2(\mathbf{x}, t), \ldots, g_{n-k}(\mathbf{x}, t)$ on $R_n \times (T_1, T_2)$ such that:*

*1. $S = \{(\mathbf{x}, t): g_1(\mathbf{x}, t) = 0 \text{ and } g_2(\mathbf{x}, t) = 0 \text{ and } \cdots \text{ and } g_{n-k}(\mathbf{x}, t) = 0\}$.*

*2. The functions $g_i(\mathbf{x}, t)$, $\dfrac{\partial g_i}{\partial \mathbf{x}}(\mathbf{x}, t)$, and $\dfrac{\partial g_i}{\partial t}(\mathbf{x}, t)$ are continuous on $R_n \times (T_1, T_2)$ for $i = 1, 2, \ldots, n - k$.*

† See Definition 3-24. We make this assumption in order to avoid having to add the phrase "except possibly on a countable subset" to many of our statements, and we observe that any bounded piecewise continuous function on $(T_1, T_2)$ satisfying Eq. (5-375) is equivalent, in the sense of Sec. 3-15, to a function satisfying this assumption.

3. *The vectors* $\frac{\partial g_i}{\partial \mathbf{x}}(\mathbf{x}, t)$, $i = 1, 2, \ldots, n - k$, *are linearly independent at each point of* $S$.

CP3 *If* $K(\mathbf{x}, t)$ *is the given real-valued function defined on* $R_n \times (T_1, T_2)$, *then we assume that:*

*a. If* $S$ *is of the form a of* CP2, *then* $K(\mathbf{x}, t) \equiv 0$.

*b. If* $S$ *is of the form b of* CP2, *then* $K(\mathbf{x}, t)$ *is independent of* $t$ [*so that* $K(\mathbf{x}, t)$ *equals* $K(\mathbf{x})$] *and the functions* $K(\mathbf{x})$, $\frac{\partial K}{\partial \mathbf{x}}(\mathbf{x})$, $\frac{\partial^2 K}{\partial \mathbf{x}^2}(\mathbf{x})$ *are continuous.*

*c. If* $S$ *is of the form c of* CP2, *then* $K(\mathbf{x}, t)$ *is independent of* $t$ [*so that* $K(\mathbf{x}, t) = K(\mathbf{x})$] *and the functions* $K(\mathbf{x})$, $\frac{\partial K}{\partial \mathbf{x}}(\mathbf{x})$, $\frac{\partial^2 K}{\partial \mathbf{x}^2}(\mathbf{x})$ *are continuous.*

*d. If* $S$ *is of the form d of* CP2, *then* $K(\mathbf{x}, t)$ *is independent of* $\mathbf{x}$ [*so that* $K(\mathbf{x}, t) = K(t)$] *and the functions* $K(t)$, $\frac{dK}{dt}(t)$, *and* $\frac{d^2K}{dt^2}(t)$ *are continuous.*

*e. If* $S$ *is of the form e of* CP2, *then the functions* $K(\mathbf{x}, t)$, $\frac{\partial K}{\partial \mathbf{x}}(\mathbf{x}, t)$, $\frac{\partial K}{\partial t}(\mathbf{x}, t)$, $\frac{\partial^2 K}{\partial \mathbf{x}^2}(\mathbf{x}, t)$, $\frac{\partial^2 K}{\partial \mathbf{x}\, \partial t}(\mathbf{x}, t)$, *and* $\frac{\partial^2 K}{\partial t^2}(\mathbf{x}, t)$ *are continuous.*

Having made these assumptions, we shall call $S$ the *target set* and $K$ the *terminal cost function.* We are now prepared to restate the definitions leading to the formal statement of the control problem. We have (cf. Sec. 4-12):

***Definition 5-8*** *Let* $t_0$ *be an element of* $(T_1, T_2)$ *and let* $\mathbf{x}_0$ *be an element of* $R_n$. *Then we say that the (admissible) control* $\mathbf{u}$ *takes* $\mathbf{x}_0$ *to* $S$ [*or, more precisely,* $(\mathbf{x}_0, t_0)$ *to* $S$], *where* $S$ *is the target set, if the set*

$$\{(\boldsymbol{\phi}(t; \mathbf{u}_{(t_0,t]}, \mathbf{x}_0), t) : t \geq t_0\} \tag{5-380}$$

*meets (i.e., intersects)* $S$. *If* $\mathbf{u}$ *takes* $\mathbf{x}_0$ *to* $S$ *and if* $t_1$ *is the first instant of time after* $t_0$ *when the motion* $\mathbf{x}(t) = \boldsymbol{\phi}(t; \mathbf{u}_{(t_0,t]}, \mathbf{x}_0)$ *enters* $S$ *and if*

$$\mathbf{x}_1 = \mathbf{x}(t_1) = \boldsymbol{\phi}(t_1; \mathbf{u}_{(t_0,t_1]}, \mathbf{x}_0) \tag{5-381}$$

*then the well-defined number* $J(\mathbf{x}_0, t_0, \mathbf{u})$, *given by*

$$J(\mathbf{x}_0, t_0, \mathbf{u}) = K(\mathbf{x}_1, t_1) + \int_{t_0}^{t_1} L[\mathbf{x}(t), \mathbf{u}(t), t]\, dt \tag{5-382}$$

$$= K[\boldsymbol{\phi}(t_1; \mathbf{u}_{(t_0,t_1]}, \mathbf{x}_0), t_1] + \int_{t_0}^{t_1} L[\boldsymbol{\phi}(t; \mathbf{u}_{(t_0,t]}, \mathbf{x}_0), \mathbf{u}(t), t]\, dt \tag{5-383}$$

*is called the value of the performance functional (or, simply, the cost) of the control* $\mathbf{u}$. *We call* $t_1$ *the terminal, or final, time and* $\mathbf{x}_1$ *the terminal, or final, state. If* $\mathbf{u}$ *does not take* $\mathbf{x}_0$ *to* $S$, *then we conventionally set*

$$J(\mathbf{x}_0, t_0, \mathbf{u}) = \infty \tag{5-384}$$

***Definition 5-9*** *The optimal-control problem (or, simply, the control problem) for the system of Eq.* (5-374) *under assumptions* CP1 *to* CP3 *and with* *respect to the target set $S$, the performance functional $J(\mathbf{x}_0, t_0, \mathbf{u})$, the set $U$ of admissible controls, and the initial value $\mathbf{x}_0$ at the initial time $t_0$ is: Determine the control $\mathbf{u}$ in $U$ which minimizes the performance functional $J(\mathbf{x}_0, t_0, \mathbf{u})$. We often drop the explicit dependence of $J$ on $\mathbf{x}_0$ and $t_0$ and write $J(\mathbf{u})$ in place of $J(\mathbf{x}_0, t_0, \mathbf{u})$.*

Having stated the general control problem, we shall now delineate two special cases of it, which we shall call, simply enough, *special problem* 1 and *special problem* 2. We shall simply list the *additional assumptions* which we make in defining the special problems and shall leave the detailed formulation of these problems to the reader.

*Special Problem 1 List of Additional Assumptions*

*a.* The system equation (5-374) does not depend on $t$ explicitly; i.e., the system equation is of the form

$$\dot{\mathbf{x}} = \mathbf{f}(\mathbf{x}, \mathbf{u}) \tag{5-385}$$

*b.* The target set $S$ is of the form

$$S = \{\mathbf{x}_1\} \times (T_1, T_2) \tag{5-386}$$

where $\mathbf{x}_1$ is a *fixed* element of $R_n$. (Thus, special problem 1 is a fixed-end-point, free-time problem.)

*c.* The function $L$ does not depend on $t$ explicitly, so that (in view of form *a* of CP3) the cost functional $J(\mathbf{u}) = J(\mathbf{x}_0, t_0, \mathbf{u})$ is given by

$$J(\mathbf{u}) = \int_{t_0}^{t_1} L[\mathbf{x}(t), \mathbf{u}(t)]\, dt \tag{5-387}$$

(where $t_1$ is not fixed).

*Special Problem 2 List of Additional Assumptions*

*a.* The system equation (5-374) does not depend on $t$ explicitly; i.e., the system equation is of the form

$$\dot{\mathbf{x}} = \mathbf{f}(\mathbf{x}, \mathbf{u}) \tag{5-388}$$

*b.* The target set $S$ is of the form

$$S = S_1 \times (T_1, T_2) \tag{5-389}$$

where $S_1$ is either a smooth $k$-fold in $R_n$ or all of $R_n$. (Thus, special problem 2 is also a free-time problem.)

*c.* The function $L$ does not depend on $t$ explicitly, and the function $K$ is identically zero [that is, $K(\mathbf{x}, t) \equiv 0$], so that the cost functional $J(\mathbf{u}) = J(\mathbf{x}_0, t_0, \mathbf{u})$ is given by

$$J(\mathbf{u}) = \int_{t_0}^{t_1} L[\mathbf{x}(t), \mathbf{u}(t)]\, dt \tag{5-390}$$

We observe that the only difference between the two special problems is the difference between the forms of the target set $S$. In the next section, we shall state the minimum principle for these two problems.

**Exercise 5-17** Formulate special problems 1 and 2 in *full* detail.

## 5-13 The Minimum Principle of Pontryagin

We shall state the minimum principle of Pontryagin for the two special problems in this section. We shall, of course, suppose that all the assumptions of Sec. 5-12 are in force, and we shall also use the notation of Sec. 5-12. Before stating the theorems, we shall introduce some additional terminology and notation. In particular, we have:

***Definition 5-10*** *Let $H(\mathbf{x}, \mathbf{p}, \mathbf{u})$ denote the real-valued function of the $n$ vector $\mathbf{x}$, the $n$ vector $\mathbf{p}$, and the $m$ vector $\mathbf{u}$, given by*

$$H(\mathbf{x}, \mathbf{p}, \mathbf{u}) = L(\mathbf{x}, \mathbf{u}) + \langle \mathbf{p}, \mathbf{f}(\mathbf{x}, \mathbf{u}) \rangle \tag{5-391}$$

*where $\mathbf{f}(\mathbf{x}, \mathbf{u})$ is the function which determines our system [that is, $\mathbf{f}(\mathbf{x}, \mathbf{u})$ is the right-hand side of the state equation] and $L(\mathbf{x}, \mathbf{u})$ is the integrand of our cost functional. We say that $H(\mathbf{x}, \mathbf{p}, \mathbf{u})$ is the Hamiltonian function (or, simply, the Hamiltonian) of our problem and that $\mathbf{p}$ is a costate vector.*

We observe that, in view of our assumption CP1, the functions $H(\mathbf{x}, \mathbf{p}, \mathbf{u})$ and $\dfrac{\partial H}{\partial \mathbf{x}}(\mathbf{x}, \mathbf{p}, \mathbf{u})$ are continuous on $R_n \times R_n \times \bar{\Omega}$, where $\bar{\Omega}$ is the closure of $\Omega$ in $R_m$. Moreover, we note that $\dfrac{\partial H}{\partial \mathbf{p}}(\mathbf{x}, \mathbf{p}, \mathbf{u})$ is well defined and is given by

$$\frac{\partial H}{\partial \mathbf{p}}(\mathbf{x}, \mathbf{p}, \mathbf{u}) = \mathbf{f}(\mathbf{x}, \mathbf{u}) \tag{5-392}$$

Now let us suppose that $\mathbf{x}_0$ is our initial state and that $t_0$ is our initial time. If $\hat{\mathbf{u}}(t)$ is some admissible control and if we let $\hat{\mathbf{x}}(t)$ denote the trajectory of our system starting from $\mathbf{x}_0 = \hat{\mathbf{x}}(t_0)$ generated by $\hat{\mathbf{u}}(t)$, then, for *any* function $\mathbf{p}(t)$,

$$\dot{\hat{\mathbf{x}}}(t) = \frac{\partial H}{\partial \mathbf{p}}[\hat{\mathbf{x}}(t), \mathbf{p}(t), \hat{\mathbf{u}}(t)] = \mathbf{f}[\hat{\mathbf{x}}(t), \hat{\mathbf{u}}(t)] \tag{5-393}$$

In addition, if $\boldsymbol{\pi}$ is *any* $n$ vector, then the *linear* differential equation

$$\begin{aligned}\dot{\mathbf{p}}(t) &= -\frac{\partial H}{\partial \mathbf{x}}[\hat{\mathbf{x}}(t), \mathbf{p}(t), \hat{\mathbf{u}}(t)] \\ &= -\frac{\partial L}{\partial \mathbf{x}}[\hat{\mathbf{x}}(t), \hat{\mathbf{u}}(t)] - \left(\frac{\partial \mathbf{f}}{\partial \mathbf{x}}[\hat{\mathbf{x}}(t), \hat{\mathbf{u}}(t)]\right)' \mathbf{p}(t)\end{aligned} \tag{5-394}$$

has a unique solution $\mathbf{p}(t, \boldsymbol{\pi})$ satisfying the initial condition

$$\mathbf{p}(t_0, \boldsymbol{\pi}) = \boldsymbol{\pi} \tag{5-395}$$

These observations lead us to the following definition:

***Definition 5-11*** *The 2nth-order system of differential equations*

$$\begin{aligned} \dot{\mathbf{x}} &= \frac{\partial H}{\partial \mathbf{p}}(\mathbf{x}, \mathbf{p}, \mathbf{u}) = \mathbf{f}(\mathbf{x}, \mathbf{u}) \\ \dot{\mathbf{p}} &= -\frac{\partial H}{\partial \mathbf{x}}(\mathbf{x}, \mathbf{p}, \mathbf{u}) = -\frac{\partial L}{\partial \mathbf{x}}(\mathbf{x}, \mathbf{u}) - \left(\frac{\partial \mathbf{f}}{\partial \mathbf{x}}(\mathbf{x}, \mathbf{u})\right)' \mathbf{p} \end{aligned} \tag{5-396}$$

*is called the canonical (or Hamiltonian) system associated with our problem. If* $\hat{\mathbf{u}}(t)$ *is an admissible control with* $\hat{\mathbf{x}}(t)$ *as corresponding trajectory, then we say that any solution* $\hat{\mathbf{p}}(t)$ *of the system* (5-394) *corresponds to* $\hat{\mathbf{u}}(t)$ *and* $\hat{\mathbf{x}}(t)$. *In other words,* $\hat{\mathbf{p}}(t)$ *corresponds to* $\hat{\mathbf{u}}(t)$ *if* $\hat{\mathbf{p}}(t)$ *and* $\hat{\mathbf{x}}(t)$ *are a solution of the canonical system.*

We now have:

***Theorem 5-5 The Minimum Principle for Special Problem 1*** *Let* $\mathbf{u}^*(t)$ *be an admissible control which transfers* $(\mathbf{x}_0, t_0)$ *to* $S = \{\mathbf{x}_1\} \times (T_1, T_2)$. *Let* $\mathbf{x}^*(t)$ *be the trajectory [of Eq. (5-385)] corresponding to* $\mathbf{u}^*(t)$, *originating at* $(\mathbf{x}_0, t_0)$, *and meeting S (for the first time) at* $(\mathbf{x}_1, t_1)$ *[that is,* $\mathbf{x}^*(t_1) = \mathbf{x}_1$*]. In order that* $\mathbf{u}^*(t)$ *be optimal (for special problem 1), it is necessary that there exist a function* $\mathbf{p}^*(t)$ *such that:*

*a.* $\mathbf{p}^*(t)$ *corresponds to* $\mathbf{u}^*(t)$ *and* $\mathbf{x}^*(t)$, *so that* $\mathbf{p}^*(t)$ *and* $\mathbf{x}^*(t)$ *are a solution of the canonical system*

$$\begin{aligned} \dot{\mathbf{x}}^*(t) &= \frac{\partial H}{\partial \mathbf{p}}[\mathbf{x}^*(t), \mathbf{p}^*(t), \mathbf{u}^*(t)] \\ \dot{\mathbf{p}}^*(t) &= -\frac{\partial H}{\partial \mathbf{x}}[\mathbf{x}^*(t), \mathbf{p}^*(t), \mathbf{u}^*(t)] \end{aligned} \tag{5-397}$$

*satisfying the boundary conditions*

$$\mathbf{x}^*(t_0) = \mathbf{x}_0 \qquad \mathbf{x}^*(t_1) = \mathbf{x}_1 \tag{5-398}$$

*b. The function* $H[\mathbf{x}^*(t), \mathbf{p}^*(t), \mathbf{u}]$ *has an absolute minimum as a function of* $\mathbf{u}$ *over* $\Omega$ *at* $\mathbf{u} = \mathbf{u}^*(t)$ *for t in* $[t_0, t_1]$; *that is,*

$$\min_{\mathbf{u} \in \Omega} H[\mathbf{x}^*(t), \mathbf{p}^*(t), \mathbf{u}] = H[\mathbf{x}^*(t), \mathbf{p}^*(t), \mathbf{u}^*(t)] \tag{5-399}$$

*or, equivalently,*

$$H[\mathbf{x}^*(t), \mathbf{p}^*(t), \mathbf{u}^*(t)] \leq H[\mathbf{x}^*(t), \mathbf{p}^*(t), \mathbf{u}] \qquad \textit{for all } \mathbf{u} \textit{ in } \Omega \tag{5-400}$$

*c. The function* $H[\mathbf{x}^*(t), \mathbf{p}^*(t), \mathbf{u}^*(t)]$ *is zero for t in* $[t_0, t_1]$; *that is,*

$$H[\mathbf{x}^*(t), \mathbf{p}^*(t), \mathbf{u}^*(t)] = 0 \qquad t \in [t_0, t_1] \tag{5-401}$$

***Theorem 5-6 The Minimum Principle for Special Problem 2*** *Let* $\mathbf{u}^*(t)$ *be an admissible control which transfers* $(\mathbf{x}_0, t_0)$ *to* $S = S_1 \times (T_1, T_2)$. *Let*

*$\mathbf{x}^*(t)$ be the trajectory [of Eq. (5-388)] corresponding to $\mathbf{u}^*(t)$, originating at $(\mathbf{x}_0, t_0)$, and meeting $S$ (for the first time) at $t_1$ [that is, $\mathbf{x}^*(t_1) \in S_1$]. In order that $\mathbf{u}^*(t)$ be optimal (for special problem 2), it is necessary that there exist a function $\mathbf{p}^*(t)$ such that:*

*a. $\mathbf{p}^*(t)$ corresponds to $\mathbf{u}^*(t)$ and $\mathbf{x}^*(t)$, so that $\mathbf{p}^*(t)$ and $\mathbf{x}^*(t)$ are a solution of the canonical system*

$$\dot{\mathbf{x}}^*(t) = \frac{\partial H}{\partial \mathbf{p}}[\mathbf{x}^*(t), \mathbf{p}^*(t), \mathbf{u}^*(t)] \tag{5-402}$$

$$\dot{\mathbf{p}}^*(t) = -\frac{\partial H}{\partial \mathbf{x}}[\mathbf{x}^*(t), \mathbf{p}^*(t), \mathbf{u}^*(t)] \tag{5-403}$$

*satisfying the boundary conditions*

$$\mathbf{x}^*(t_0) = \mathbf{x}_0 \qquad \mathbf{x}^*(t_1) \in S_1 \tag{5-404}$$

*b. The function $H[\mathbf{x}^*(t), \mathbf{p}^*(t), \mathbf{u}]$ has an absolute minimum as a function of $\mathbf{u}$ over $\Omega$ at $\mathbf{u} = \mathbf{u}^*(t)$ for $t$ in $[t_0, t_1]$; that is,*

$$\min_{\mathbf{u} \in \Omega} H[\mathbf{x}^*(t), \mathbf{p}^*(t), \mathbf{u}] = H[\mathbf{x}^*(t), \mathbf{p}^*(t), \mathbf{u}^*(t)] \tag{5-405}$$

*or, equivalently,*

$$H[\mathbf{x}^*(t), \mathbf{p}^*(t), \mathbf{u}^*(t)] \le H[\mathbf{x}^*(t), \mathbf{p}^*(t), \mathbf{u}] \qquad \text{for all } \mathbf{u} \text{ in } \Omega \tag{5-406}$$

*c. The function $H[\mathbf{x}^*(t), \mathbf{p}^*(t), \mathbf{u}^*(t)]$ is zero for $t$ in $[t_0, t_1]$; that is,*

$$H[\mathbf{x}^*(t), \mathbf{p}^*(t), \mathbf{u}^*(t)] = 0 \qquad t \in [t_0, t_1] \tag{5-407}$$

*d. If $S_1$ is a smooth k-fold in $R_n$, then the vector $\mathbf{p}^*(t_1)$ is transversal to $S_1$ at $\mathbf{x}^*(t_1)$; that is,*

$$\langle \mathbf{p}^*(t_1), \mathbf{x} - \mathbf{x}^*(t_1) \rangle = 0 \qquad \text{for all } \mathbf{x} \text{ in } M[\mathbf{x}^*(t_1)] \tag{5-408}$$

*where $M[\mathbf{x}^*(t_1)]$ is the tangent plane of $S_1$ at $\mathbf{x}^*(t_1)$. $M[\mathbf{x}^*(t_1)] = \left\{\mathbf{x}: \left\langle \frac{\partial g_i}{\partial \mathbf{x}}[\mathbf{x}^*(t_1)], \mathbf{x} - \mathbf{x}^*(t_1) \right\rangle = 0 \text{ for } i = 1, 2, \ldots, n-k \right\}$, where $g_1(\mathbf{x}) = 0, g_2(\mathbf{x}) = 0, \ldots, g_{n-k}(\mathbf{x}) = 0$ are the equations of $S_1$. See Sec. 3-13. Thus, $\mathbf{p}^*(t_1)$ is normal to the target set $S_1$ at $\mathbf{x}^*(t_1)$. If $S_1 = R_n$, then the vector $\mathbf{p}^*(t_1)$ is the zero vector; that is, $\mathbf{p}^*(t_1) = \mathbf{0}$.*

We observe, first of all, that the only difference between the two theorems is the additional assertion $d$ of Theorem 5-6, in which the transversality of the costate to the target set is mentioned. We also note that assertion $b$ of each theorem is to be construed in the following way: At each specific instant of time, say $\hat{t}$, in the interval $[t_0, t_1]$, $\mathbf{x}^*(\hat{t})$, $\mathbf{u}^*(\hat{t})$, and $\mathbf{p}^*(\hat{t})$ are three well-defined vectors and part $b$ of the theorems tells us that the *number* $H[\mathbf{x}^*(\hat{t}), \mathbf{p}^*(\hat{t}), \mathbf{u}^*(\hat{t})]$ is smaller than or equal to the *number* $H[\mathbf{x}^*(\hat{t}), \mathbf{p}^*(\hat{t}), \boldsymbol{\omega}]$, *where $\boldsymbol{\omega}$ is any m vector in $\Omega$.*

We have stated these theorems in the form in which they will be used in subsequent chapters. However, in order to be able to handle pathological cases, we should, strictly speaking, include an additional constant $p_0^*$ in our formulation of the theorems. In other words, we should consider the function $\mathcal{H}(\mathbf{x}, \mathbf{p}, \mathbf{u}, p_0) = p_0 L(\mathbf{x}, \mathbf{u}) + \langle \mathbf{p}, \mathbf{f}(\mathbf{x}, \mathbf{u}) \rangle$ instead of the function $H(\mathbf{x}, \mathbf{p}, \mathbf{u})$, and we should assert that there are a nonnegative constant $p_0^*$ (that is, $p_0^* \geq 0$) and a vector $\mathbf{p}^*(t)$ such that

$$\begin{aligned}\dot{\mathbf{p}}^*(t) &= -\frac{\partial \mathcal{H}}{\partial \mathbf{x}}[\mathbf{x}^*(t), \mathbf{p}^*(t), \mathbf{u}^*(t), p_0^*] \\ &= -p_0^* \frac{\partial L}{\partial \mathbf{x}}[\mathbf{x}^*(t), \mathbf{u}^*(t)] - \left(\frac{\partial \mathbf{f}}{\partial \mathbf{x}}[\mathbf{x}^*(t), \mathbf{u}^*(t)]\right)' \mathbf{p}^*(t)\end{aligned}$$

Then the remaining assertions of Theorems 5-5 and 5-6 hold, with $\mathcal{H}$ replacing $H$. The pathological cases occur when $p_0^* = 0$; when $p_0^* \neq 0$, we may choose $p_0^* = 1$.† The control problems we shall examine in subsequent chapters of the book are not pathological. So, having brought this point to the reader's attention, we shall not mention $p_0^*$ again (except in Secs. 5-15 and 5-16, when we present the heuristic proof of the minimum principle, and in Secs. 6-21 and 6-22, when we discuss singular problems).

## 5-14 The Minimum Principle of Pontryagin: Changes of Variable

We are now going to show, by various changes of variable, how the minimum principle for the general control problem can be obtained from Theorems 5-5 and 5-6. We again suppose that all the assumptions of Sec. 5-12 are in force, and we again use the notation of that section. We shall start by removing the restriction of time independence for the system and the cost, then we shall allow the terminal time to be fixed, and finally, we shall treat the problem in which there is a terminal cost. In each case, we shall indicate the additional assumptions that we make and then introduce the change of variable which reduces our problem to one of the special problems. We shall conclude the section with a chart illustrating the various reformulations of the minimum principle for the cases of interest.

### *a. Time-dependent System and Cost*

We suppose that our system equation is of the form

$$\dot{\mathbf{x}}(t) = \mathbf{f}[\mathbf{x}(t), \mathbf{u}(t), t] \tag{5-409}$$

and we assume, to begin with, that our target set $S$ is of the form

$$S = \{\mathbf{x}_1\} \times (T_1, T_2) \tag{5-410}$$

† We can do this because the equation for $\mathbf{p}(t)$ is linear.

where $\mathbf{x}_1$ is a *fixed* element of $R_n$. We also suppose that $L$ is of the form

$$L(\mathbf{x}, \mathbf{u}, t) \tag{5-411}$$

and that our cost functional $J(\mathbf{u})$ is given by

$$J(\mathbf{u}) = \int_{t_0}^{t_1} L[\mathbf{x}(t), \mathbf{u}(t), t]\, dt \qquad t_1 \text{ free} \tag{5-412}$$

As usual, we let $\mathbf{x}_0$ be our initial state and $t_0$ be our initial time.

Now we want to reduce the control problem for the time-dependent system (5-409) with cost (5-412) to one of the special problems—in particular, to special problem 2, as we shall soon see. We shall do this by introducing an additional state variable and state equation and then applying the minimum principle to the new problem.

We let $x_{n+1}$ denote an auxiliary variable, and we consider the $(n + 1)$st-order system

$$\begin{aligned} \dot{\mathbf{x}}(t) &= \mathbf{f}[\mathbf{x}(t), \mathbf{u}(t), x_{n+1}] \\ \dot{x}_{n+1}(t) &= 1 \end{aligned} \tag{5-413}$$

Let us examine the following (auxiliary) control problem for the system (5-413):

Given the initial time $t_0$ and the initial condition $\begin{bmatrix} \mathbf{x}_0 \\ t_0 \end{bmatrix}$ [that is, $x_{n+1}(t_0) = t_0$] in $R_{n+1}$. Let $S = S_1 \times (T_1, T_2)$ be the target set [in $R_{n+1} \times (T_1, T_2)$], where $S_1$ is the *line* (1-fold) given by the equations $x_1 - x_{1,1} = 0$, $x_2 - x_{1,2} = 0, \ldots, x_n - x_{1,n} = 0$, where the $x_{1,j}$ are the components of the vector $\mathbf{x}_1$ of Eq. (5-410). Then determine the admissible control $\mathbf{u}$ which transfers $\left(\begin{bmatrix} \mathbf{x}_0 \\ t_0 \end{bmatrix}, t_0\right)$ to $S_1 \times (T_1, T_2)$ and which, in so doing, minimizes the performance functional $J_1(\mathbf{u})$, given by

$$J_1(\mathbf{u}) = \int_{t_0}^{t_1} L[\mathbf{x}(t), \mathbf{u}(t), x_{n+1}(t)]\, dt \qquad t_1 \text{ free} \tag{5-414}$$

We observe that this problem is equivalent to our original problem (for the system (5-409)] in the sense that *a control is optimal for one if and only if it is optimal for the other*. Furthermore, this new problem is precisely of the form of special problem 2. Thus, we may apply Theorem 5-6 to deduce that if $\mathbf{u}^*(t)$ is optimal, then there are a function $\mathbf{p}^*(t)$ and a function $p^*_{n+1}(t)$ such that the following conditions are satisfied:

1.
$$\begin{aligned} \dot{\mathbf{x}}^*(t) &= \frac{\partial H_1}{\partial \mathbf{p}} [\mathbf{x}^*(t), \mathbf{p}^*(t), \mathbf{u}^*(t), x^*_{n+1}(t), p^*_{n+1}(t)] \\ \dot{\mathbf{p}}^*(t) &= -\frac{\partial H_1}{\partial \mathbf{x}} [\mathbf{x}^*(t), \mathbf{p}^*(t), \mathbf{u}^*(t), x^*_{n+1}(t), p^*_{n+1}(t)] \end{aligned} \tag{5-415}$$

$$\begin{aligned} \dot{x}^*_{n+1}(t) &= \frac{\partial H_1}{\partial p_{n+1}} [\mathbf{x}^*(t), \mathbf{p}^*(t), \mathbf{u}^*(t), x^*_{n+1}(t), p^*_{n+1}(t)] = 1 \\ \dot{p}^*_{n+1}(t) &= -\frac{\partial H_1}{\partial x_{n+1}} [\mathbf{x}^*(t), \mathbf{p}^*(t), \mathbf{u}^*(t), x^*_{n+1}(t), p^*_{n+1}(t)] \end{aligned} \tag{5-416}$$

where $H_1(\mathbf{x}, \mathbf{p}, \mathbf{u}, x_{n+1}, p_{n+1})$ is given by the equation

$$H_1(\mathbf{x}, \mathbf{p}, \mathbf{u}, x_{n+1}, p_{n+1}) = L(\mathbf{x}, \mathbf{u}, x_{n+1}) + \langle \mathbf{p}, \mathbf{f}(\mathbf{x}, \mathbf{u}, x_{n+1})\rangle + p_{n+1} \cdot 1 \quad (5\text{-}417)$$

and where

$$\begin{aligned} \mathbf{x}^*(t_0) &= \mathbf{x}_0 & x^*_{n+1}(t_0) &= t_0 \\ \mathbf{x}^*(t_1) &= \mathbf{x}_1 & x^*_{n+1}(t_1) &= t_1 \end{aligned} \quad (5\text{-}418)$$

2. The function $H_1[\mathbf{x}^*(t), \mathbf{p}^*(t), \mathbf{u}, x^*_{n+1}(t), p^*_{n+1}(t)]$ has an absolute minimum as a function of $\mathbf{u}$ over $\Omega$ at $\mathbf{u} = \mathbf{u}^*(t)$ for $t$ in $[t_0, t_1]$.
3. $H_1[\mathbf{x}^*(t), \mathbf{p}^*(t), \mathbf{u}^*(t), x^*_{n+1}(t), p^*_{n+1}(t)] = 0$ for $t$ in $[t_0, t_1]$.
4. The $n + 1$ vector

$$\begin{bmatrix} \mathbf{p}^*(t_1) \\ p^*_{n+1}(t_1) \end{bmatrix}$$

is transversal (normal) to $S_1$ at $\begin{bmatrix} \mathbf{x}_1 \\ t_1 \end{bmatrix}$, so that

$$p^*_{n+1}(t_1) = 0 \quad (5\text{-}419)$$

Now, we can apply these results in our original problem by simply observing that the auxiliary variable $x_{n+1}$ is, in fact, the time $t$. Thus, for our original problem, we define a function $H(\mathbf{x}, \mathbf{p}, \mathbf{u}, t)$ of the $n$ vector $\mathbf{x}$, the $n$ vector $\mathbf{p}$, the $m$ vector $\mathbf{u}$, and the time $t$ by setting

$$H(\mathbf{x}, \mathbf{p}, \mathbf{u}, t) = L(\mathbf{x}, \mathbf{u}, t) + \langle \mathbf{p}, \mathbf{f}(\mathbf{x}, \mathbf{u}, t)\rangle \quad (5\text{-}420)$$

We call $H$ the *Hamiltonian* of our problem, and we call $\mathbf{p}$ the *costate variable.* We then have:

***Theorem 5-7 The Minimum Principle for the Time-dependent Problem with Fixed End Point*** *Let $\mathbf{u}^*(t)$ be an admissible control which transfers $(\mathbf{x}_0, t_0)$ to $S = \{\mathbf{x}_1\} \times (T_1, T_2)$. Let $\mathbf{x}^*(t)$ be the trajectory [of Eq. (5-409)] corresponding to $\mathbf{u}^*(t)$, originating at $(\mathbf{x}_0, t_0)$, and meeting $S$ for the first time at $t_1$ [that is, $\mathbf{x}^*(t_1) = \mathbf{x}_1$]. In order that $\mathbf{u}^*(t)$ be optimal [for the cost functional (5-412)], it is necessary that there exist a function $\mathbf{p}^*(t)$ such that:*

*a.* *$\mathbf{p}^*(t)$ corresponds to $\mathbf{u}^*(t)$ and $\mathbf{x}^*(t)$ so that $\mathbf{p}^*(t)$ and $\mathbf{x}^*(t)$ are a solution of the canonical system*

$$\dot{\mathbf{x}}^*(t) = \frac{\partial H}{\partial \mathbf{p}}[\mathbf{x}^*(t), \mathbf{p}^*(t), \mathbf{u}^*(t), t] \quad (5\text{-}421)$$

$$\dot{\mathbf{p}}^*(t) = -\frac{\partial H}{\partial \mathbf{x}}[\mathbf{x}^*(t), \mathbf{p}^*(t), \mathbf{u}^*(t), t] \quad (5\text{-}422)$$

*satisfying the boundary conditions*

$$\mathbf{x}^*(t_0) = \mathbf{x}_0 \qquad \mathbf{x}^*(t_1) = \mathbf{x}_1 \quad (5\text{-}423)$$

*b.* *The function $H[\mathbf{x}^*(t), \mathbf{p}^*(t), \mathbf{u}, t]$ has an absolute minimum as a function of $\mathbf{u}$ over $\Omega$ at $\mathbf{u} = \mathbf{u}^*(t)$ for $t$ in $[t_0, t_1]$; that is,*

$$\min_{\mathbf{u}\in\Omega} H[\mathbf{x}^*(t), \mathbf{p}^*(t), \mathbf{u}, t] = H[\mathbf{x}^*(t), \mathbf{p}^*(t), \mathbf{u}^*(t), t] \quad (5\text{-}424)$$

*or, equivalently,*

$$H[\mathbf{x}^*(t), \mathbf{p}^*(t), \mathbf{u}^*(t), t] \leq H[\mathbf{x}^*(t), \mathbf{p}^*(t), \mathbf{u}, t] \qquad \text{for all } \mathbf{u} \text{ in } \Omega \qquad (5\text{-}425)$$

*c. The function* $H[\mathbf{x}^*(t), \mathbf{p}^*(t), \mathbf{u}^*(t), t]$ *satisfies the relations*

$$H[\mathbf{x}^*(t), \mathbf{p}^*(t), \mathbf{u}^*(t), t] = -\int_t^{t_1} \frac{\partial H}{\partial t} [\mathbf{x}^*(\tau), \mathbf{p}^*(\tau), \mathbf{u}^*(\tau), \tau]\, d\tau \qquad (5\text{-}426)$$

$$H[\mathbf{x}^*(t_1), \mathbf{p}^*(t_1), \mathbf{u}^*(t_1), t_1] = 0 \qquad (5\text{-}427)$$

*where $t_1$ is the unspecified terminal time.*

We observe, first of all, that the significant difference between Theorems 5-5 and 5-7 is the difference in the behavior of the Hamiltonians along the optimal trajectory. In the time-independent problem of Theorem 5-5, the Hamiltonian is zero along the optimal path, while in the time-dependent problem at hand, the Hamiltonian varies with time in accordance with Eq. (5-426) along the optimal path.

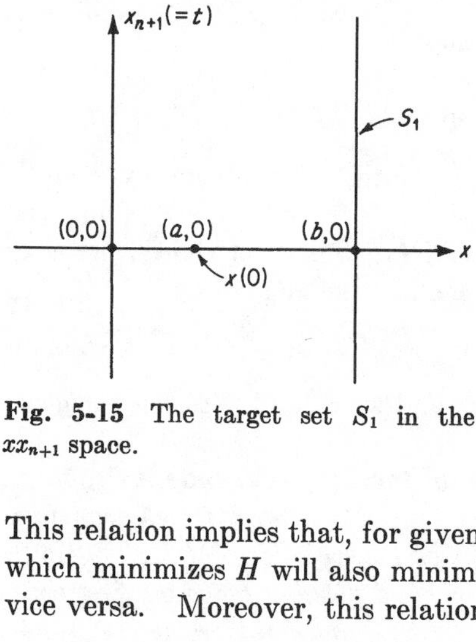

**Fig. 5-15** The target set $S_1$ in the $xx_{n+1}$ space.

We next note that, in view of the fact that $x_{n+1}(t) = t$, the crucial relation which allows us to deduce Theorem 5-7 from conditions 1 to 4, satisfied by the optimum in the new (or auxiliary) problem, is

$$H_1(\mathbf{x}, \mathbf{p}, \mathbf{u}, t, p_{n+1}) = H(\mathbf{x}, \mathbf{p}, \mathbf{u}, t) + p_{n+1} \qquad (5\text{-}428)$$

This relation implies that, for given values of $\mathbf{x}$, $\mathbf{p}$, and $t$, the value of $\mathbf{u}$ in $\Omega$ which minimizes $H$ will also minimize $H_1$ (whatever the value of $p_{n+1}$), and vice versa. Moreover, this relation combined with the fact that

$$\dot{\mathbf{p}}^*_{n+1}(t) = -\frac{\partial H}{\partial t} [\mathbf{x}^*(t), \mathbf{p}^*(t), \mathbf{u}^*(t), t] \qquad p^*_{n+1}(t_1) = 0 \qquad (5\text{-}428a)\dagger$$

enables us to deduce condition $c$ of Theorem 5-7.

In order to illustrate these ideas, let us examine the following simple example:

**Example 5-9** Suppose that our system equation is of the form $\dot{x} = x + ut$, that $t_0 = 0$ is our initial time, and that $x(0) = a$ is our initial state. We let $x_1 = b$ be our desired terminal state [so that $S = \{b\} \times (T_1, T_2)$], and we suppose that our performance functional $J(u)$ is given by

$$J(u) = \int_0^{t_1} (x^2 + t^2)\, dt$$

† As $\partial H_1/\partial x_{n+1} = \partial H(\mathbf{x}, \mathbf{p}, \mathbf{u}, x_{n+1})/\partial x_{n+1}$.

that is, $L(x, u, t) = x^2 + t^2$. Then we introduce the auxiliary variable $x_{n+1}$ and consider the new system $\dot{x} = x + ux_{n+1}$, $\dot{x}_{n+1} = 1$. For this new system, our initial point is

$$\begin{bmatrix} x(0) = a \\ x_{n+1}(0) = 0 \end{bmatrix}$$

and our target set is the *line* $x - b = 0$ in the $xx_{n+1}$ plane (see Fig. 5-15) [that is, our new $S = S_1 \times (T_1, T_2)$, where $S_1$ is the line $x - b = 0$]. Our cost functional $J_1(u)$ is given by

$$J_1(u) = \int_0^{t_1} (x^2 + x_{n+1}^2)\, dt$$

and we have

$$\begin{aligned} H_1(x, p, u, x_{n+1}, p_{n+1}) &= x^2 + x_{n+1}^2 + p(x + ux_{n+1}) + p_{n+1} \\ H(x, p, u, t) &= x^2 + t^2 + p(x + ut) \end{aligned}$$

We now list, side by side, the necessary conditions for the new problem (based on Theorem 5-6) and the original problem (Theorem 5-7):

| *New problem* | *Original problem* |
|---|---|
| (1) $\dot{x}^*(t) = \dfrac{\partial H_1}{\partial p}[x^*(t), p^*(t), u^*(t), x_{n+1}^*(t), p_{n+1}^*(t)]$ <br> $= x^*(t) + u^*(t)x_{n+1}^*(t)$ <br> $\dot{p}^*(t) = -\dfrac{\partial H_1}{\partial x}[x^*(t), p^*(t), u^*(t), x_{n+1}^*(t), p_{n+1}^*(t)]$ <br> $= -2x^*(t) - p^*(t)$ <br> $\dot{x}_{n+1}^*(t) = 1$ <br> $\dot{p}_{n+1}^*(t) = -\dfrac{\partial H_1}{\partial x_{n+1}}[x^*(t), p^*(t), u^*(t), x_{n+1}^*(t), p_{n+1}^*(t)]$ <br> $= -2x_{n+1}^*(t) - p^*(t)u^*(t)$ | (1) $\dot{x}^*(t) = \dfrac{\partial H}{\partial p}[x^*(t), p^*(t), u^*(t), t]$ <br> $= x^*(t) + u^*(t)t$ <br> $\dot{p}^*(t) = -\dfrac{\partial H}{\partial x}[x^*(t), p^*(t), u^*(t), t]$ <br> $= -2x^*(t) - p^*(t)$ |
| (2) $\min_u H_1[x^*(t), p^*(t), u, x_{n+1}^*(t), p_{n+1}^*(t)]$ <br> $= H_1[x^*(t), p^*(t), u^*(t), x_{n+1}^*(t), p_{n+1}^*(t)]$ <br> $= \{x^{*2}(t) + x_{n+1}^{*2}(t) + p^*(t) \times [x^*(t) + u^*(t)x_{n+1}^*(t)]\} + p_{n+1}^*(t)$ | (2) $\min_u H[x^*(t), p^*(t), u, t]$ <br> $= H[x^*(t), p^*(t), u^*(t), t]$ <br> $= x^{*2}(t) + t^2 + p^*(t)[x^*(t) + u^*(t)t]$ |
| (3) $H_1[x^*(t), p^*(t), u^*(t), x_{n+1}^*(t), p_{n+1}^*(t)] = 0$ <br> that is, <br> $x^{*2}(t) + x_{n+1}^{*2}(t) + p^*(t)[x^*(t) + u^*(t)x_{n+1}^*(t)] = -p_{n+1}^*(t)$ | (3) $H[x^*(t), p^*(t), u^*(t), t]$ <br> $= -\int_t^{t_1} \dfrac{\partial H}{\partial t}[x^*(\tau), p^*(\tau), u^*(\tau), \tau]\, d\tau$ <br> $= -\int_t^{t_1} [2\tau + p^*(\tau)u^*(\tau)]\, d\tau$ <br> $H[x^*(t_1), p^*(t_1), u^*(t_1), t_1] = 0$ |
| (4) $p_{n+1}^*(t_1) = 0$ | |

The reader should observe the effect of substituting $x_{n+1}^*(t) = t$ in the conditions for the new problem, as this will lead to the conditions for the original problem.

Now, let us assume that our target set $S$ is of the form

$$S = \{(\mathbf{g}(t), t): t \in (T_1, T_2)\} \tag{5-429}$$

where $\mathbf{g}(t)$ is a continuously differentiable function of $t$ (that is, $S$ is of the form $d$ of assumption CP2, Sec. 5-12). Thus, we are trying to hit a moving point rather than a fixed point and, in so doing, to minimize the cost functional (5-412). We could again introduce the auxiliary variable $x_{n+1}$ and obtain a new problem of the type of special problem 2. We could then deduce the following theorem from Theorem 5-6 in a manner entirely analogous to that used to deduce Theorem 5-7.

***Theorem 5-8 The Minimum Principle for the Time-dependent Problem with Moving End Point*** *Let* $\mathbf{u}^*(t)$ *be an admissible control which transfers* $(\mathbf{x}_0, t_0)$ *to* $S = \{(\mathbf{g}(t), t): t \in (T_1, T_2)\}$. *Let* $\mathbf{x}^*(t)$ *be the trajectory* [*of Eq.* (5-409)] *corresponding to* $\mathbf{u}^*(t)$, *originating at* $(\mathbf{x}_0, t_0)$, *and meeting* $S$ *at* $t_1$ [*that is,* $\mathbf{x}^*(t_1) = \mathbf{g}(t_1)$]. *In order that* $\mathbf{u}^*(t)$ *be optimal* [*for the cost functional* (5-412)], *it is necessary that there exist a function* $\mathbf{p}^*(t)$ *such that:*

*a. Same as statement a* [*Eqs.* (5-421) *and* (5-422)] *of Theorem* 5-7 *except for the boundary conditions* (5-423), *which become*

$$\mathbf{x}^*(t_0) = \mathbf{x}_0 \qquad \mathbf{x}^*(t_1) = \mathbf{g}(t_1) \tag{5-430}$$

*b. Same as statement b* [*Eqs.* (5-424) *and* (5-425)] *of Theorem* 5-7

*c. The function* $H[\mathbf{x}^*(t), \mathbf{p}^*(t), \mathbf{u}^*(t), t]$ *satisfies the relations*

$$H[\mathbf{x}^*(t), \mathbf{p}^*(t), \mathbf{u}^*(t), t] = \left\langle \mathbf{p}^*(t_1), \left.\frac{d\mathbf{g}}{dt}\right|_{t_1} \right\rangle - \int_t^{t_1} \frac{\partial H}{\partial t} [\mathbf{x}^*(\tau), \mathbf{p}^*(\tau), \mathbf{u}^*(\tau), \tau]\, d\tau \tag{5-431}$$

$$H[\mathbf{x}^*(t_1), \mathbf{p}^*(t_1), \mathbf{u}^*(t_1), t_1] = \left\langle \mathbf{p}^*(t_1), \left.\frac{d\mathbf{g}}{dt}\right|_{t_1} \right\rangle \tag{5-432}$$

*where* $t_1$ *is the unspecified terminal time*

We observe that the essential difference between Theorems 5-7 and 5-8 is again the difference in the behavior of the Hamiltonians along the optimal trajectory. We further note that $(d\mathbf{g}/dt)|_{t_1}$ is precisely the "tangent" vector to the curve $\mathbf{g}(t)$ at $t_1$, and so the terminal value of the Hamiltonian (along the optimal trajectory) is the scalar product of the terminal costate and this tangent vector.

Finally, we may assume that our target set $S$ either is a smooth $(k + 1)$-fold, $1 \le k \le n - 1$, in $R_n \times (T_1, T_2)$ which is continuously differentiable in $t$ (that is, $S$ is of the form $e$ of CP2, Sec. 5-12) or is *all* of $R_n \times (T_1, T_2)$. In that case, we may derive, in a manner entirely analogous to that used before, the following theorem:

***Theorem 5-9 The Minimum Principle for the Time-dependent Problem with Moving Target Set*** *Let* $\mathbf{u}^*(t)$ *be an admissible control which transfers* $(\mathbf{x}_0, t_0)$ *to:*

*a. The smooth* $(k+1)$*-fold* $S$ *in* $R_n \times (T_1, T_2)$ *or*

*b. The target set* $S = R_n \times (T_1, T_2)$

*Let* $\mathbf{x}^*(t)$ *be the trajectory [of Eq.* (5-409)] *corresponding to* $\mathbf{u}^*(t)$, *originating at* $(\mathbf{x}_0, t_0)$, *and meeting* $S$ *at* $t_1$. *In order that* $\mathbf{u}^*(t)$ *be optimal [for the cost functional* (5-412)], *it is necessary that there exist a function* $\mathbf{p}^*(t)$ *such that:*

1. *Same as statement a [Eqs.* (5-421) *and* (5-422)] *of Theorem* 5-7 *except for the boundary conditions* (5-423), *which become*

$$\mathbf{x}^*(t_0) = \mathbf{x}_0 \qquad (\mathbf{x}^*(t_1), t_1) \in S \tag{5-433}$$

2. *Same as statement b [Eqs.* (5-424) *and* (5-425)] *of Theorem* 5-7

3 i. *If* $S$ *is a smooth* $(k+1)$*-fold in* $R_n \times (T_1, T_2)$, *then the function* $H[\mathbf{x}^*(t), \mathbf{p}^*(t), \mathbf{u}^*(t), t]$ *satisfies the relations*

$$H[\mathbf{x}^*(t), \mathbf{p}^*(t), \mathbf{u}^*(t), t] = \sum_{i=1}^{n-k} \alpha_i \frac{\partial g_i}{\partial t}[\mathbf{x}^*(t_1), t_1] - \int_t^{t_1} \frac{\partial H}{\partial t}[\mathbf{x}^*(\tau), \mathbf{p}^*(\tau), \mathbf{u}^*(\tau), \tau]\, d\tau \tag{5-434}$$

$$H[\mathbf{x}^*(t_1), \mathbf{p}^*(t_1), \mathbf{u}^*(t_1), t_1] = \sum_{i=1}^{n-k} \alpha_i \frac{\partial g_i}{\partial t}[\mathbf{x}^*(t_1), t_1] \tag{5-435}$$

*where* $t_1$ *is the terminal time; the* $\alpha_i$, $i = 1, 2, \ldots, n-k$, *are constants; and the equations of* $S$ *are*

$$g_1(\mathbf{x}, t) = 0,\ g_2(\mathbf{x}, t) = 0,\ \ldots,\ g_{n-k}(\mathbf{x}, t) = 0 \tag{5-436}$$

ii. *If* $S = R_n \times (T_1, T_2)$, *then the function* $H[\mathbf{x}^*(t), \mathbf{p}^*(t), \mathbf{u}^*(t), t]$ *satisfies the relations*

$$H[\mathbf{x}^*(t), \mathbf{p}^*(t), \mathbf{u}^*(t), t] = -\int_t^{t_1} \frac{\partial H}{\partial t}[\mathbf{x}^*(\tau), \mathbf{p}^*(\tau), \mathbf{u}^*(\tau), \tau]\, d\tau \tag{5-437}$$

$$H[\mathbf{x}^*(t_1), \mathbf{p}^*(t_1), \mathbf{u}^*(t_1), t_1] = 0 \tag{5-438}$$

4 i. *If* $S$ *is a smooth* $(k+1)$*-fold in* $R_n \times (T_1, T_2)$, *then the vector* $\mathbf{p}^*(t_1)$ *is transversal to the smooth* $k$*-fold* $S_{t_1}$ *in* $R_n$, *whose equations are*

$$g_1(\mathbf{x}, t_1) = 0,\ g_2(\mathbf{x}, t_1) = 0,\ \ldots,\ g_{n-k}(\mathbf{x}, t_1) = 0 \tag{5-439}$$

*at* $\mathbf{x}^*(t_1)$; *that is,*

$$\langle \mathbf{p}^*(t_1), \mathbf{x} - \mathbf{x}^*(t_1)\rangle = 0 \qquad \textit{for all } \mathbf{x} \textit{ in } M[\mathbf{x}^*(t_1)] \tag{5-440}$$

*where* $M[\mathbf{x}^*(t_1)]$ *is the tangent plane of* $S_{t_1}$ *at* $\mathbf{x}^*(t_1)$; $M[\mathbf{x}^*(t_1)] = \{\mathbf{x}: \left\langle \frac{\partial g_i}{\partial \mathbf{x}}[\mathbf{x}^*(t_1), t_1], \mathbf{x} - \mathbf{x}^*(t_1)\right\rangle = 0$ *for* $i = 1, 2, \ldots, n-k\}$. *See*

*Sec. 3-13. Thus,* $\mathbf{p}^*(t_1)$ *is normal to* $S_{t_1}$ *at* $\mathbf{x}^*(t_1)$ *and is, therefore, a linear combination of the vectors* $\frac{\partial g_i}{\partial \mathbf{x}}[\mathbf{x}^*(t_1), t_1]$, $i = 1, 2, \ldots, n - k$.

ii. *If* $S = R_n \times (T_1, T_2)$, *then* $\mathbf{p}^*(t_1)$ *is the zero vector; that is,*

$$\mathbf{p}^*(t_1) = \mathbf{0}$$

A particular example illustrating this theorem will appear in Sec. 6-3.

Although the statement of several theorems is rather repetitious, we have done this in order to be able to pinpoint, in subsequent chapters of the book, the precise formulation of the minimum principle that we need.

*b. Fixed Terminal Time*

So far, we have only considered problems in which the terminal time was free; now we shall remove that restriction. We again suppose that our system equation is of the form

$$\dot{\mathbf{x}}(t) = \mathbf{f}[\mathbf{x}(t), \mathbf{u}(t), t] \tag{5-441}$$

and that $\mathbf{x}_0$ is our initial state, with $t_0$ as our initial time. We further assume that our target set $S$ is of one of the forms

$$S = \{\mathbf{x}_1\} \times \{t_1\} = (\mathbf{x}_1, t_1) \tag{5-442}$$
$$S = S_1 \times \{t_1\} \tag{5-443}$$
$$S = R_n \times \{t_1\} \tag{5-444}$$

where $t_1$ is a *fixed* element of $(T_1, T_2)$ with $t_1 > t_0$, $\mathbf{x}_1$ is a given element of $R_n$, and $S_1$ is a smooth $k$-fold, $1 \le k \le n - 1$, in $R_n$. In other words, we want to transfer $\mathbf{x}_0$ to a given subset of $R_n$ at a *specified* time. We also suppose that our cost functional $J(\mathbf{u})$ is given by

$$J(\mathbf{u}) = \int_{t_0}^{t_1} L[\mathbf{x}(t), \mathbf{u}(t), t]\, dt \qquad t_1 \text{ fixed} \tag{5-445}$$

Now we want to reduce the control problem for the system (5-441) with cost functional (5-443) and target set $S$ of Eq. (5-442) [or (5-443) or (5-444)] to one of the special problems, in particular to special problem 1, as we shall soon see. We do this by letting $x_{n+1}$ denote an auxiliary variable and considering the $(n + 1)$st-order system

$$\begin{aligned} \dot{\mathbf{x}}(t) &= \mathbf{f}[\mathbf{x}(t), \mathbf{u}(t), x_{n+1}] \\ \dot{x}_{n+1}(t) &= 1 \end{aligned} \tag{5-446}$$

We can readily verify that the problem of transferring the point $\begin{bmatrix} \mathbf{x}_0 \\ t_0 \end{bmatrix}$ [that is, $x_{n+1}(t_0) = t_0$] to the *point* $\begin{bmatrix} \mathbf{x}_1 \\ t_1 \end{bmatrix}$ [that is, $x_{n+1}(t_1) = t_1$] along a trajectory of Eqs. (5-446) in such a way as to minimize the cost $J_1(\mathbf{u})$ given by

$$J_1(\mathbf{u}) = \int_{t_0}^{T} L[\mathbf{x}(t), \mathbf{u}(t), x_{n+1}(t)]\, dt \qquad T \text{ free} \tag{5-447}$$

is *equivalent* to our original problem [for the system (5-441)] in the sense that a *control is optimal for one if and only if it is optimal for the other.* As the new problem is of the form of special problem 1, we can apply Theorem 5-5 and deduce, in a manner entirely analogous to that used before, the version of the minimum principle applicable to our original problem. We are thus led to the following theorem:

***Theorem 5-10 The Minimum Principle for the Fixed-terminal-time Problem*** *Let* $\mathbf{u}^*(t)$ *be an admissible control which transfers* $(\mathbf{x}_0, t_0)$ *to:*

*a. The target set* $S = \{\mathbf{x}_1\} \times \{t_1\}$ *or*

*b. The target set* $S = S_1 \times \{t_1\}$ *or*

*c. The target set* $S = R_n \times \{t_1\}$

*Let* $\mathbf{x}^*(t)$ *be the trajectory* [*of Eq.* (5-441)] *corresponding to* $\mathbf{u}^*(t)$, *originating at* $(\mathbf{x}_0, t_0)$, *and meeting* $S$ *at* $t_1$.† *In order that* $\mathbf{u}^*(t)$ *be optimal* [*for the cost functional* (5-445)], *it is necessary that there exist a function* $\mathbf{p}^*(t)$ *such that:*

1. *Same as statement a* [*Eqs.* (5-421) *and* (5-422)] *of Theorem* 5-7 *except for the boundary conditions* (5-423), *which become*

$$\begin{array}{lll} \text{i.} & \mathbf{x}^*(t_0) = \mathbf{x}_0 & \mathbf{x}^*(t_1) = \mathbf{x}_1 \end{array} \tag{5-448}$$

$$\begin{array}{lll} \text{ii.} & \mathbf{x}^*(t_0) = \mathbf{x}_0 & \mathbf{x}^*(t_1) \in S_1 \end{array} \tag{5-449}$$

$$\begin{array}{lll} \text{iii.} & \mathbf{x}^*(t_0) = \mathbf{x}_0 & \mathbf{x}^*(t_1)\ \textit{free} \end{array} \tag{5-450}$$

   *according as the target set* $S$ *is of the form a, b, or c.*

2. *Same as statement b* [*Eqs.* (5-424) *and* (5-425)] *of Theorem* 5-7.

4. *The vector* $\mathbf{p}^*(t_1)$ *is:*

   i. *Transversal to* $S_1$ *at* $\mathbf{x}^*(t_1)$ (*see d of Theorem* 5-6)

   ii. *The zero vector; that is,* $\mathbf{p}^*(t_1) = \mathbf{0}$

   *according as the target set* $S$ *is of the form b or c.*

The reader will observe that there is *no* condition 3. In the previous theorems, condition 3 referred to the behavior of the Hamiltonian along the optimal trajectory; however, in the present situation, it is easy to see that the relation

$$H[\mathbf{x}^*(t), \mathbf{p}^*(t), \mathbf{u}^*(t), t] = H[\mathbf{x}^*(t_1), \mathbf{p}^*(t_1), \mathbf{u}^*(t_1), t_1] - \int_{t_1}^{t} \frac{\partial H}{\partial t}[\mathbf{x}^*(\tau), \mathbf{p}^*(\tau), \mathbf{u}^*(\tau), \tau]\, d\tau \tag{5-451}$$

which corresponds to condition 3 is automatically satisfied and provides no additional information.

† We shall treat the three different target sets simultaneously by simply indicating the particular results for each when these results are different.

*c. Terminal Cost Present*

Up to now, we have assumed that our cost functional did not involve a terminal-cost term (i.e., that the function $K$ was zero). We shall now remove that restriction.

We once again suppose that our system equation is of the form

$$\dot{\mathbf{x}}(t) = \mathbf{f}[\mathbf{x}(t), \mathbf{u}(t), t] \tag{5-452}$$

and that $\mathbf{x}_0$ is our initial state, with $t_0$ as our initial time. We assume, to begin with, that our target set $S$ is of the form

$$S = R_n \times (T_1, T_2) \tag{5-453}$$

so that we are dealing with a free-end-point, free-time problem. We suppose that $K(\mathbf{x}, t)$ is a real-valued function on $R_n \times (T_1, T_2)$ such that

$$K(\mathbf{x}, t), \frac{\partial K}{\partial \mathbf{x}}(\mathbf{x}, t), \frac{\partial K}{\partial t}(\mathbf{x}, t), \frac{\partial^2 K}{\partial \mathbf{x}^2}(\mathbf{x}, t), \frac{\partial^2 K}{\partial \mathbf{x}\, \partial t}(\mathbf{x}, t), \frac{\partial^2 K}{\partial t^2}(\mathbf{x}, t) \tag{5-454}$$

are all continuous. We assume then that our cost functional $J(\mathbf{u})$ is given by

$$J(\mathbf{u}) = K[\mathbf{x}(t_1), t_1] + \int_{t_0}^{t_1} L[\mathbf{x}(t), \mathbf{u}(t), t]\, dt \tag{5-455}$$

Now we shall reduce the control problem for the system (5-452) with cost functional (5-455) and target set $S$ of Eq. (5-453) to a problem in which there is *no* terminal cost. We shall then be able to apply Theorem 5-9 to deduce the desired form of the minimum principle.

We first observe that, since $\mathbf{x}(t_0) = \mathbf{x}_0$,

$$K[\mathbf{x}(t_1), t_1] = K(\mathbf{x}_0, t_0) + \int_{t_0}^{t_1} \frac{d}{dt}\{K[\mathbf{x}(t), t]\}\, dt \tag{5-456}$$

However, we know that

$$\frac{d}{dt}\{K[\mathbf{x}(t), t]\} = \left\langle \frac{\partial K}{\partial \mathbf{x}}[\mathbf{x}(t), t], \dot{\mathbf{x}}(t) \right\rangle + \frac{\partial K}{\partial t}[\mathbf{x}(t), t] \tag{5-457}$$

$$= \left\langle \frac{\partial K}{\partial \mathbf{x}}[\mathbf{x}(t), t], \mathbf{f}[\mathbf{x}(t), \mathbf{u}(t), t] \right\rangle + \frac{\partial K}{\partial t}[\mathbf{x}(t), t] \tag{5-458}$$

It follows that the cost $J(\mathbf{u})$ is also given by

$$J(\mathbf{u}) = K(\mathbf{x}_0, t_0) + \int_{t_0}^{t_1} \left\{ L[\mathbf{x}(t), \mathbf{u}(t), t] + \left\langle \frac{\partial K}{\partial \mathbf{x}}[\mathbf{x}(t), t], \mathbf{f}[\mathbf{x}(t), \mathbf{u}(t), t] \right\rangle + \frac{\partial K}{\partial t}[\mathbf{x}(t), t] \right\} dt \tag{5-459}$$

Since $\mathbf{x}_0$ and $t_0$ are given, $K(\mathbf{x}_0, t_0)$ is a known constant; so it is clear that the problem of transferring $(\mathbf{x}_0, t_0)$ to $S$ along a trajectory of Eq. (5-452) in such a way as to minimize the cost $J_1(\mathbf{u})$, given by

$$J_1(\mathbf{u}) = \int_{t_0}^{t_1} \left\{ L[\mathbf{x}(t), \mathbf{u}(t), t] + \left\langle \frac{\partial K}{\partial \mathbf{x}} [\mathbf{x}(t), t], \mathbf{f}[\mathbf{x}(t), \mathbf{u}(t), t] \right\rangle + \frac{\partial K}{\partial t} [\mathbf{x}(t), t] \right\} dt \tag{5-460}$$

is equivalent to our original problem, in the sense that a *control is optimal for one if and only if it is optimal for the other.*

As Theorem 5-9 applies to this new problem, we may deduce that, if $\mathbf{u}^*(t)$ is optimal, then there is a function $\mathbf{p}_1^*(t)$ such that:

1. $$\dot{\mathbf{p}}_1^*(t) = -\frac{\partial H_1}{\partial \mathbf{x}} [\mathbf{x}^*(t), \mathbf{p}_1^*(t), \mathbf{u}^*(t), t] \tag{5-461}$$

$$= -\frac{\partial L}{\partial \mathbf{x}} [\mathbf{x}^*(t), \mathbf{u}^*(t), t] - \frac{\partial}{\partial \mathbf{x}} \left[ \left\langle \frac{\partial K}{\partial \mathbf{x}} (\mathbf{x}, t), \mathbf{f}(\mathbf{x}, \mathbf{u}, t) \right\rangle - \frac{\partial^2 K}{\partial \mathbf{x}\, \partial t} (\mathbf{x}, t) \right] \Bigg|_{[\mathbf{x}^*(t), \mathbf{u}^*(t), t]} - \left( \frac{\partial \mathbf{f}}{\partial \mathbf{x}} [\mathbf{x}^*(t), \mathbf{u}^*(t), t] \right)' \mathbf{p}_1^*(t) \tag{5-462}$$

2. $$\mathbf{p}_1^*(t_1) = \mathbf{0} \tag{5-463}$$

3. $$\min_{\mathbf{u} \in \Omega} H_1[\mathbf{x}^*(t), \mathbf{p}_1^*(t), \mathbf{u}, t] = H_1[\mathbf{x}^*(t), \mathbf{p}_1^*(t), \mathbf{u}^*(t), t] \tag{5-464}$$

where $H_1(\mathbf{x}, \mathbf{p}_1, \mathbf{u}, t)$ is given by

$$\begin{aligned} H_1(\mathbf{x}, \mathbf{p}_1, \mathbf{u}, t) &= \left[ L(\mathbf{x}, \mathbf{u}, t) + \left\langle \frac{\partial K}{\partial \mathbf{x}}, \mathbf{f}(\mathbf{x}, \mathbf{u}, t) \right\rangle + \frac{\partial K}{\partial t} (\mathbf{x}, t) \right] + \langle \mathbf{p}_1, \mathbf{f}(\mathbf{x}, \mathbf{u}, t) \rangle \\ &= L(\mathbf{x}, \mathbf{u}, t) + \left\langle \mathbf{p}_1 + \frac{\partial K}{\partial \mathbf{x}}, \mathbf{f}(\mathbf{x}, \mathbf{u}, t) \right\rangle + \frac{\partial K}{\partial t} (\mathbf{x}, t) \end{aligned} \tag{5-465}$$

Now let us consider the function $H(\mathbf{x}, \mathbf{p}, \mathbf{u}, t)$, given by

$$H(\mathbf{x}, \mathbf{p}, \mathbf{u}, t) = L(\mathbf{x}, \mathbf{u}, t) + \langle \mathbf{p}, \mathbf{f}(\mathbf{x}, \mathbf{u}, t) \rangle \tag{5-466}$$

If we set

$$\mathbf{p}^*(t) = \mathbf{p}_1^*(t) + \frac{\partial K}{\partial \mathbf{x}} [\mathbf{x}^*(t), t] \tag{5-467}$$

then we have, in view of Eqs. (5-462) and (5-458),

$$\begin{aligned} \dot{\mathbf{p}}^*(t) = &-\frac{\partial L}{\partial \mathbf{x}} [\mathbf{x}^*(t), \mathbf{u}^*(t), t] - \left\{ \frac{\partial^2 K}{\partial \mathbf{x}^2} [\mathbf{x}^*(t), t] \mathbf{f}[\mathbf{x}^*(t), \mathbf{u}^*(t), t] \right. \\ &\left. + \left( \frac{\partial \mathbf{f}}{\partial \mathbf{x}} [\mathbf{x}^*(t), \mathbf{u}^*(t), t] \right)' \frac{\partial K}{\partial \mathbf{x}} [\mathbf{x}^*(t), t] + \frac{\partial^2 K}{\partial \mathbf{x}\, \partial t} [\mathbf{x}^*(t), t] \right\} \\ &- \left( \frac{\partial \mathbf{f}}{\partial \mathbf{x}} [\mathbf{x}^*(t), \mathbf{u}^*(t), t] \right)' \left\{ \mathbf{p}^*(t) - \frac{\partial K}{\partial \mathbf{x}} [\mathbf{x}^*(t), t] \right\} \\ &+ \frac{\partial^2 K}{\partial \mathbf{x}^2} [\mathbf{x}^*(t), t] \mathbf{f}[\mathbf{x}^*(t), \mathbf{u}^*(t), t] + \frac{\partial^2 K}{\partial \mathbf{x}\, \partial t} [\mathbf{x}^*(t), t] \end{aligned} \tag{5-468}$$

$$= -\frac{\partial L}{\partial \mathbf{x}}[\mathbf{x}^*(t), \mathbf{u}^*(t), t] - \left(\frac{\partial \mathbf{f}}{\partial \mathbf{x}}[\mathbf{x}^*(t), \mathbf{u}^*(t), t]\right)' \mathbf{p}^*(t) \tag{5-469}$$

$$= -\frac{\partial H}{\partial \mathbf{x}}[\mathbf{x}^*(t), \mathbf{p}^*(t), \mathbf{u}^*(t), t] \tag{5-470}$$

Moreover, since $(\partial K/\partial t)[\mathbf{x}^*(t), t]$ does not depend explicitly on $\mathbf{u}$, it follows from Eq. (5-464) that

$$\min_{\mathbf{u}\in\Omega} H[\mathbf{x}^*(t), \mathbf{p}^*(t), \mathbf{u}, t] = H[\mathbf{x}^*(t), \mathbf{p}^*(t), \mathbf{u}^*(t), t] \tag{5-471}$$

Finally, in view of Eqs. (5-463) and (5-467), we observe that

$$\mathbf{p}^*(t_1) = \frac{\partial K}{\partial \mathbf{x}}[\mathbf{x}^*(t_1), t_1] \tag{5-472}$$

For other forms of the target set $S$, we may argue in an entirely analogous manner and thus deduce the following theorem:

***Theorem 5-11 The Minimum Principle for the Terminal-cost Problem*** *Let* $\mathbf{u}^*(t)$ *be an admissible control which transfers* $(\mathbf{x}_0, t_0)$ *to:*

*a. The target set* $S = R_n \times (T_1, T_2)$

*b. The target set* $S$ *which is a smooth* $(k+1)$*-fold in* $R_n \times (T_1, T_2)$†

*Let* $\mathbf{x}^*(t)$ *be the trajectory [of Eq. (5-452)] corresponding to* $\mathbf{u}^*(t)$*, originating at* $(\mathbf{x}_0, t_0)$*, and meeting* $S$ *at* $t_1$*. In order that* $\mathbf{u}^*(t)$ *be optimal [for the cost functional (5-455)], it is necessary that there exist a function* $\mathbf{p}^*(t)$ *such that:*

1. *Same as statement a [Eqs. (5-421) and (5-422)] of Theorem 5-7 except for the boundary conditions (5-423), which become*

i. $$\mathbf{x}^*(t_0) = \mathbf{x}_0 \qquad \mathbf{x}^*(t_1) \text{ free} \tag{5-473}$$
ii. $$\mathbf{x}^*(t_0) = \mathbf{x}_0 \qquad \mathbf{x}^*(t_1) \in S \tag{5-474}$$

2. *Same as statement b [Eqs. (5-424) and (5-425)] of Theorem 5-7.*
3 i. *If* $S = R_n \times (T_1, T_2)$*, then the function* $H[\mathbf{x}^*(t), \mathbf{p}^*(t), \mathbf{u}^*(t), t]$ *satisfies the relations*

$$H[\mathbf{x}^*(t), \mathbf{p}^*(t), \mathbf{u}^*(t), t] = -\frac{\partial K}{\partial t}[\mathbf{x}^*(t_1), t_1] - \int_t^{t_1} \left\{\frac{\partial H}{\partial t}[\mathbf{x}^*(\tau), \mathbf{p}^*(\tau), \mathbf{u}^*(\tau), \tau] + \frac{\partial^2 K}{\partial t^2}[\mathbf{x}^*(\tau), \tau]\right\} d\tau \tag{5-475}$$

$$H[\mathbf{x}^*(t_1), \mathbf{p}^*(t_1), \mathbf{u}^*(t_1), t_1] = -\frac{\partial K}{\partial t}[\mathbf{x}^*(t_1), t_1] \tag{5-476}$$

† Other cases are left to the reader.

ii. *If $S$ is a smooth $(k+1)$-fold in $R_n \times (T_1, T_2)$, then the function $H[\mathbf{x}^*(t), \mathbf{p}^*(t), \mathbf{u}^*(t), t]$ satisfies the relations*

$$H[\mathbf{x}^*(t), \mathbf{p}^*(t), \mathbf{u}^*(t), t] = \sum_{i=1}^{n-k} \alpha_i \frac{\partial g_i}{\partial t} [\mathbf{x}^*(t_1), t_1] - \frac{\partial K}{\partial t} [x^*(t_1), t_1] - \int_t^{t_1} \left\{ \frac{\partial H}{\partial t} [\mathbf{x}^*(\tau), \mathbf{p}^*(\tau), \mathbf{u}^*(\tau), \tau] + \frac{\partial^2 K}{\partial t^2} [x^*(\tau), \tau] \right\} d\tau \quad (5\text{-}477)$$

$$H[\mathbf{x}^*(t_1), \mathbf{p}^*(t_1), \mathbf{u}^*(t_1), t_1] = \sum_{i=1}^{n-k} \alpha_i \frac{\partial g_i}{\partial t} [\mathbf{x}^*(t_1), t_1] - \frac{\partial K}{\partial t} [\mathbf{x}^*(t_1), t_1] \quad (5\text{-}478)$$

*where $t_1$ is the terminal time; the $\alpha_i$, $i = 1, 2, \ldots, n-k$, are constants; and the equations of $S$ are*

$$g_1(\mathbf{x}, t) = 0, g_2(\mathbf{x}, t) = 0, \ldots, g_{n-k}(\mathbf{x}, t) = 0 \quad (5\text{-}479)$$

4 i. *If $S = R_n \times (T_1, T_2)$, then*

$$\mathbf{p}^*(t_1) = \frac{\partial K}{\partial \mathbf{x}} [\mathbf{x}^*(t_1), t_1] \quad (5\text{-}480)$$

ii. *If $S$ is a smooth $(k+1)$-fold in $R_n \times (T_1, T_2)$, then $\mathbf{p}^*(t_1) - (\partial K/\partial \mathbf{x})[\mathbf{x}^*(t_1), t_1]$ is transversal to the smooth $k$-fold $S_{t_1}$ in $R_n$ whose equations are*

$$g_1(\mathbf{x}, t_1) = 0, g_2(\mathbf{x}, t_1) = 0, \ldots, g_{n-k}(\mathbf{x}, t_1) = 0 \quad (5\text{-}481)$$

*at $\mathbf{x}^*(t_1)$ (cf. Theorem* 5-9).

We observe that the primary effects of the terminal cost occur in the behavior of the Hamiltonian along the optimal trajectory and in the value of the costate at the terminal time.

**Example 5-10** Suppose that $K(\mathbf{x}, t) = \langle \mathbf{c}, \mathbf{x} \rangle$, where $\mathbf{c}$ is a given nonzero vector in $R_n$. Then, for the free-end-point problem [that is, $S = R_n \times (T_1, T_2)$], conditions 3 and 4 of Theorem 5-11 become

3. $$H[\mathbf{x}^*(t), \mathbf{p}^*(t), \mathbf{u}^*(t), t] = -\int_t^{t_1} \frac{\partial H}{\partial t} [\mathbf{x}^*(\tau), \mathbf{p}^*(\tau), \mathbf{u}^*(\tau), \tau]\, d\tau \quad (5\text{-}482)$$

$$H[\mathbf{x}^*(t_1), \mathbf{p}^*(t_1), \mathbf{u}^*(t_1), t_1] = 0 \quad (5\text{-}483)$$

4. $$\mathbf{p}^*(t_1) = \mathbf{c} \quad (5\text{-}484)$$

**Example 5-11** Suppose that $K(\mathbf{x}, t) = \frac{1}{2} \sum_{i=1}^{n} c_i x_i^2$, where the $c_i$ are constants and $x_1, x_2, \ldots, x_n$ are the components of $\mathbf{x}$. Then, for the free-end-point problem [that is, $S = R_n \times (T_1, T_2)$], conditions 3 and 4 of Theorem 5-11 become

3. $$H[\mathbf{x}^*(t), \mathbf{p}^*(t), \mathbf{u}^*(t), t] = -\int_t^{t_1} \frac{\partial H}{\partial t} [\mathbf{x}^*(\tau), \mathbf{p}^*(\tau), \mathbf{u}^*(\tau), \tau]\, d\tau$$

$$H[\mathbf{x}^*(t_1), \mathbf{p}^*(t_1), \mathbf{u}^*(t_1), t_1] = 0 \quad (5\text{-}485)$$

4. $$\mathbf{p}^*(t_1) = \mathbf{C}\mathbf{x}^*(t_1) \quad (5\text{-}486)$$

where $\mathbf{C}$ is the diagonal matrix with entries $c_1, c_2, \ldots, c_n$ on the main diagonal.

Finally, let us note that what we have done is incorporate the terminal cost in the integral to be minimized. On the other hand, we can reduce problems in which there is an integral to be minimized to problems involving only a terminal cost by introducing an auxiliary variable $x_0$ such that

$$\dot{x}_0(t) = L[\mathbf{x}(t), \mathbf{u}(t), t] \qquad x_0(t_0) = 0 \tag{5-487}$$

This is the procedure that is used by Pontryagin and that will be used by us in the next two sections.

*d. Table Summarizing the Results* (pages 306 and 307)

We summarize the various results we have obtained, together with some additional cases, in Table 5-1. We suppose throughout the table that $t_0$ is our initial time and that $\mathbf{x}_0 = \mathbf{x}(t_0)$ is our initial state; therefore, we have omitted explicit reference to these quantities in the table itself. Moreover, we always use the asterisk (*) to indicate optimal quantities. We also make use of the following conventions and notations in the table:

1. Explicit reference to $t$ is often not made; for example, $\mathbf{x}^*$ is written instead of $\mathbf{x}^*(t)$, $\mathbf{p}^*$ instead of $\mathbf{p}^*(t)$.
2. The cost functional $J(\mathbf{u})$ for a particular problem is represented by the particular $K$ and $L$ used to define $J(\mathbf{u})$.
3. $|_*$ is used to indicate that the quantity in question is to be evaluated along the optimal trajectory; for example, $(\partial H/\partial \mathbf{x})\,|_*$ is the gradient of the Hamiltonian $H$ with respect to $\mathbf{x}$ evaluated at $[\mathbf{x}^*(t), \mathbf{p}^*(t), \mathbf{u}^*(t), t]$.
4. $|_{*t_1}$ indicates that the quantity in question is evaluated at the terminal time, using optimal values; for example, $(\partial K/\partial \mathbf{x})\,|_{*t_1}$ is the gradient of $K$ with respect to $\mathbf{x}$ evaluated at $[\mathbf{x}^*(t_1), t_1]$.
5. $H^*$ is the Hamiltonian evaluated along the optimal trajectory, and the notation $H^*[t]$ is used to emphasize particular interest in the behavior of the Hamiltonian as a function of time along the optimal trajectory.

In referring to the table in later portions of the book, we shall often make use of the indicated grid system to locate a particular entry of interest.

## 5-15 Proof of the Minimum Principle: Preliminary Remarks

We are going to present a heuristic proof of the minimum principle based upon the proof given in Ref. P-5. Our primary goals will be to provide a suitable introduction to and motivation for the formal proof and to provide the reader with some justification for accepting the minimum principle. In short, we are going to try to make the minimum principle plausible without giving a formal proof.

In this section, we shall first make some remarks about "proofs" in general; then we shall carefully state the two theorems, relating to special problems 1 and 2 (see Secs. 5-12 and 5-13), which are to be "proved."

The understanding of a theorem and its proof does not consist merely in being able to follow the statements of the theorem and the individual steps in the reasoning of the proof but rather consists in being able to grasp and comprehend the result and the central ideas of the proof as a whole. In short, the entire complex of ideas must be digested and assimilated as part of oneself. This is a highly individual matter and often, in the case of a difficult proof, requires considerable time. Therefore, we suggest (1) that the interested reader consult Ref. P-5 and (2) that he not be dismayed if some of our comments do not seem to "make sense" at first. After all, we are trying to convey our understanding of the proof of the minimum principle, which is no substitute for the reader's own understanding. Moreover, we should, in all fairness, point out that *an understanding of the proof of the minimum principle is not essential to applying the minimum principle intelligently*, and we suggest that those readers who find the "going rough" and who are primarily interested in applications skim over or omit the next section.

Now let us turn our attention to the statement of the theorems to be "proved." We suppose that all the assumptions of Sec. 5-12 are in force, and we use the notation of that section. Furthermore, we deal exclusively with special problems 1 and 2 [see Eqs. (5-385) to (5-390)]. We now have:

***Theorem 5-5P The Minimum Principle for Special Problem 1*** *Let* $\mathbf{u}^*(t)$ *be an admissible control which transfers* $(\mathbf{x}_0, t_0)$ *to* $S = \{\mathbf{x}_1\} \times (T_1, T_2)$. *Let* $\mathbf{x}^*(t)$ *be the trajectory [of Eq. (5-385)] corresponding to* $\mathbf{u}^*(t)$, *originating at* $(\mathbf{x}_0, t_0)$, *and meeting* $S$ *at* $(\mathbf{x}_1, t_1)$. *In order that* $\mathbf{u}^*(t)$ *be optimal (for special problem 1), it is necessary that there exist a nonnegative constant* $p_0^*$ *(that is,* $p_0^* \geq 0$*) and a function* $\mathbf{p}^*(t)$ *such that:*

*a.* $\mathbf{p}^*(t)$ *and* $\mathbf{x}^*(t)$ *are a solution of the canonical system*

$$\dot{\mathbf{x}}^*(t) = \frac{\partial H}{\partial \mathbf{p}} [\mathbf{x}^*(t), \mathbf{p}^*(t), \mathbf{u}^*(t), p_0^*] \tag{5-488}$$

$$\dot{\mathbf{p}}^*(t) = -\frac{\partial H}{\partial \mathbf{x}} [\mathbf{x}^*(t), \mathbf{p}^*(t), \mathbf{u}^*(t), p_0^*] \tag{5-489}$$

*satisfying the boundary conditions*

$$\mathbf{x}^*(t_0) = \mathbf{x}_0 \qquad \mathbf{x}^*(t_1) = \mathbf{x}_1 \tag{5-490}$$

*where the Hamiltonian function* $H(\mathbf{x}, \mathbf{p}, \mathbf{u}, p_0)$ *is given by*

$$H(\mathbf{x}, \mathbf{p}, \mathbf{u}, p_0) = p_0 L(\mathbf{x}, \mathbf{u}) + \langle \mathbf{p}, \mathbf{f}(\mathbf{x}, \mathbf{u}) \rangle \tag{5-491}$$

*b. The function* $H[\mathbf{x}^*(t), \mathbf{p}^*(t), \mathbf{u}, p_0^*]$ *has an absolute minimum as a function of* $\mathbf{u}$ *over* $\Omega$ *at* $\mathbf{u} = \mathbf{u}^*(t)$ *for* $t$ *in* $[t_0, t_1]$; *that is,*

$$\min_{\mathbf{u} \in \Omega} H[\mathbf{x}^*(t), \mathbf{p}^*(t), \mathbf{u}, p_0^*] = H[\mathbf{x}^*(t), \mathbf{p}^*(t), \mathbf{u}^*(t), p_0^*] \tag{5-492}$$

*c. The function* $H[\mathbf{x}^*(t), \mathbf{p}^*(t), \mathbf{u}^*(t), p_0^*]$ *is zero for* $t$ *in* $[t_0, t_1]$; *that is,*

$$H[\mathbf{x}^*(t), \mathbf{p}^*(t), \mathbf{u}^*(t), p_0^*] = 0 \qquad t \in [t_0, t_1] \tag{5-493}$$

**Table 5-1 Summary of Problems and of**

| | Problem | | | | Hamiltonian |
|---|---|---|---|---|---|
| | System | Cost | Time | Target set | |
| 1 | | | | $\mathbf{x}_1$ fixed end point | |
| 2 | | $L = L(\mathbf{x}, \mathbf{u}),\ K = 0$ | $t_1$ free | $S_1$ $k$-fold in $R_n$; $g_i(\mathbf{x}) = 0$ $i = 1, 2, \ldots, n-k$ | |
| 3 | $\dot{\mathbf{x}} = \mathbf{f}(\mathbf{x}, \mathbf{u})$ | $L = L(\mathbf{x}, \mathbf{u}),\ K = K(\mathbf{x})$ | | $R_n$ free end point | $H = H(\mathbf{x}, \mathbf{p}, \mathbf{u}) = L(\mathbf{x}, \mathbf{u}) + \langle \mathbf{p}, \mathbf{f}(\mathbf{x},\mathbf{u})\rangle$ |
| 4 | | $L = L(\mathbf{x}, \mathbf{u}),\ K = 0$ | $t_1$ fixed | $\mathbf{x}_1$ fixed end point | |
| 5 | | $L = L(\mathbf{x}, \mathbf{u}),\ K = K(\mathbf{x})$ | | $S_1$ $k$-fold in $R_n$; $g_i(\mathbf{x}) = 0$ $i = 1, 2, \ldots, n-k$ | |
| 6 | | | | $\mathbf{x}_1$ fixed end point | |
| 7 | | $L = L(\mathbf{x}, \mathbf{u}, t),\ K = 0$ | | $\mathbf{g}(t)$ moving point | |
| 8 | $\dot{\mathbf{x}} = \mathbf{f}(\mathbf{x}, \mathbf{u}, t)$ | $L = L(\mathbf{x}, \mathbf{u}, t),\ K = K(\mathbf{x}, t)$ | $t_1$ free | $S$ $(k+1)$-fold in $R_n \times (T_1, T_2)$; $g_i(\mathbf{x}, t) = 0$ $i = 1, 2, \ldots, n-k$ | $H = H(\mathbf{x}, \mathbf{p}, \mathbf{u}, t) = L(\mathbf{x}, \mathbf{u}, t) + \langle \mathbf{p}, \mathbf{f}(\mathbf{x}, \mathbf{u}, t)\rangle$ |
| 9 | | | | $R_n$ free end point | |
| 10 | | | | $\mathbf{x}_1$ fixed end point | |
| 11 | | $L = L(\mathbf{x}, \mathbf{u}, t),\ K = 0$ | $t_1$ fixed | $S_1$ $k$-fold in $R_n$; $g_i(\mathbf{x}) = 0$ $i = 1, 2, \ldots, n-k$ | |
| 12 | | $L = L(\mathbf{x}, \mathbf{u}, t)$, $K = K(\mathbf{x})$ | | $R_n$ free end point | |
| | A | B | C | D | E |

***Theorem 5-6P The Minimum Principle for Special Problem 2*** *Let* $\mathbf{u}^*(t)$ *be an admissible control which transfers* $(\mathbf{x}_0, t_0)$ *to* $S = S_1 \times (T_1, T_2)$. *Let* $\mathbf{x}^*(t)$ *be the trajectory* [*of Eq.* (5-388)] *corresponding to* $\mathbf{u}^*(t)$, *originating at* $(\mathbf{x}_0, t_0)$, *and meeting* $S$ *at* $t_1$. *In order that* $\mathbf{u}^*(t)$ *be optimal (for special problem 2), it is necessary that there exist a nonnegative constant* $p_0^*$ (*that is,* $p_0^* \geq 0$) *and a function* $\mathbf{p}^*(t)$ *such that:*

*a.* $\mathbf{p}^*(t)$ *and* $\mathbf{x}^*(t)$ *are a solution of the canonical system*

$$\dot{\mathbf{x}}^*(t) = \frac{\partial H}{\partial \mathbf{p}}[\mathbf{x}^*(t), \mathbf{p}^*(t), \mathbf{u}^*(t), p_0^*] \tag{5-494}$$

$$\dot{\mathbf{p}}^*(t) = -\frac{\partial H}{\partial \mathbf{x}}[\mathbf{x}^*(t), \mathbf{p}^*(t), \mathbf{u}^*(t), p_0^*] \tag{5-495}$$

*satisfying the boundary conditions*

$$\mathbf{x}^*(t_0) = \mathbf{x}_0 \qquad \mathbf{x}^*(t_1) \in S_1 \tag{5-496}$$

**the Corresponding Necessary Conditions**

| Necessary conditions (1) | (2) | (3) | (4) | Theorem |
|---|---|---|---|---|
| $\dot{\mathbf{x}}^* = \frac{\partial H}{\partial \mathbf{p}}\Big\|_*$ <br> $\dot{\mathbf{p}}^* = -\frac{\partial H}{\partial \mathbf{x}}\Big\|_*$ | $H^* = H(\mathbf{x}^*, \mathbf{p}^*, \mathbf{u}^*) = \min_{\mathbf{u}\in\Omega} H(\mathbf{x}^*, \mathbf{p}^*, \mathbf{u})$ <br> or <br> $H(\mathbf{x}^*, \mathbf{p}^*, \mathbf{u}^*) \le H(\mathbf{x}^*, \mathbf{p}^*, \mathbf{u})$ for all $\mathbf{u}$ in $\Omega$ | $H^*[t] = H^*[t_1] = 0$ | No condition on $\mathbf{p}^*(t_1)$ | 5-5 |
| | | | $\mathbf{p}^*(t_1)$ normal to $S_1$ at $\mathbf{x}^*(t_1)$ <br> $\mathbf{p}^*(t_1) = \sum_{i=1}^{n-k} \alpha_i \frac{\partial g_i}{\partial \mathbf{x}}\Big\|_{\mathbf{x}^*(t_1)}$ | 5-6 |
| | | | $\mathbf{p}^*(t_1) = \frac{\partial K}{\partial \mathbf{x}}\Big\|_{\mathbf{x}^*(t_1)}$ | 5-6 <br> 5-11 |
| | | $H^*[t] = H^*[t_1] = \text{const}$ | No condition on $\mathbf{p}^*(t_1)$ | 5-10 |
| | | | $\mathbf{p}^*(t_1) = \sum_{i=1}^{n-k} \alpha_i \frac{\partial g_i}{\partial \mathbf{x}}\Big\|_{\mathbf{x}^*(t_1)} + \frac{\partial K}{\partial \mathbf{x}}\Big\|_{\mathbf{x}^*(t_1)}$ | 5-10 <br> 5-11 |
| $\dot{\mathbf{x}}^* = \frac{\partial H}{\partial \mathbf{p}}\Big\|_*$ <br> $\dot{\mathbf{p}}^* = -\frac{\partial H}{\partial \mathbf{x}}\Big\|_*$ | $H^* = H(\mathbf{x}^*, \mathbf{p}^*, \mathbf{u}^*, t) = \min_{\mathbf{u}\in\Omega} H(\mathbf{x}^*, \mathbf{p}^*, \mathbf{u}, t)$ <br> or <br> $H(\mathbf{x}^*, \mathbf{p}^*, \mathbf{u}^*, t) \le H(\mathbf{x}^*, \mathbf{p}^*, \mathbf{u}, t)$ for all $\mathbf{u}$ in $\Omega$ | $H^*[t] = -\int_t^{t_1} \frac{\partial H}{\partial t}\Big\|_* d\tau$ <br> $H^*[t_1] = 0$ | No condition on $\mathbf{p}^*(t_1)$ | 5-7 |
| | | $H^*[t] = \left\langle \mathbf{p}^*(t_1), \frac{d\mathbf{g}}{dt}\Big\|_{*t_1} \right\rangle - \int_t^{t_1} \frac{\partial H}{\partial t}\Big\|_* d\tau$ <br> $H^*[t_1] = \left\langle \mathbf{p}^*(t_1), \frac{d\mathbf{g}}{dt}\Big\|_{*t_1} \right\rangle$ | | 5-8 |
| | | $H^*[t] = H^*[t_1] - \int_t^{t_1} \left[ \frac{\partial H}{\partial t}\Big\|_* + \frac{\partial^2 K}{\partial t^2}\Big\|_* \right] d\tau$ <br> $H^*[t_1] = \sum_{i=1}^{n-k} \alpha_i \frac{\partial g_i}{\partial t}\Big\|_{*t_1} - \frac{\partial K}{\partial t}\Big\|_{*t_1}$ | $\mathbf{p}^*(t_1) = \sum_{i=1}^{n-k} \alpha_i \frac{\partial g_i}{\partial \mathbf{x}}\Big\|_{*t_1} + \frac{\partial K}{\partial \mathbf{x}}\Big\|_{*t_1}$ | 5-11 <br> 5-9 |
| | | $H^*[t] = H^*[t_1] - \int_t^{t_1} \left[ \frac{\partial H}{\partial t}\Big\|_* + \frac{\partial^2 K}{\partial t^2}\Big\|_* \right] d\tau$ <br> $H^*[t_1] = -\frac{\partial K}{\partial t}\Big\|_{*t_1}$ | $\mathbf{p}^*(t_1) = \frac{\partial K}{\partial \mathbf{x}}\Big\|_{*t_1}$ | 5-11 <br> 5-9 |
| | | $H^*[t] = H^*[t_1] - \int_t^{t_1} \frac{\partial H}{\partial t}\Big\|_* d\tau$ | No condition on $\mathbf{p}^*(t_1)$ | 5-10 |
| | | | $\mathbf{p}^*(t_1)$ normal to $S_1$ at $\mathbf{x}^*(t_1)$ <br> $\mathbf{p}^*(t_1) = \sum_{i=1}^{n-k} \alpha_i \frac{\partial g_i}{\partial \mathbf{x}}\Big\|_{\mathbf{x}^*(t_1)}$ | 5-10 |
| | | | $\mathbf{p}^*(t_1) = \frac{\partial K}{\partial \mathbf{x}}\Big\|_{\mathbf{x}^*(t_1)}$ | 5-10 <br> 5-11 |
| F | G | H | I | J |

*where the Hamiltonian function $H(\mathbf{x}, \mathbf{p}, \mathbf{u}, p_0)$ is given by*

$$H(\mathbf{x}, \mathbf{p}, \mathbf{u}, p_0) = p_0 L(\mathbf{x}, \mathbf{u}) + \langle \mathbf{p}, \mathbf{f}(\mathbf{x}, \mathbf{u}) \rangle \tag{5-497}$$

*b. The function $H[\mathbf{x}^*(t), \mathbf{p}^*(t), \mathbf{u}, p_0^*]$ has an absolute minimum as a function of $\mathbf{u}$ over $\Omega$ at $\mathbf{u} = \mathbf{u}^*(t)$ for $t$ in $[t_0, t_1]$; that is,*

$$\min_{\mathbf{u}\in\Omega} H[\mathbf{x}^*(t), \mathbf{p}^*(t), \mathbf{u}, p_0^*] = H[\mathbf{x}^*(t), \mathbf{p}^*(t), \mathbf{u}^*(t), p_0^*] \tag{5-498}$$

*c. The function $H[\mathbf{x}^*(t), \mathbf{p}^*(t), \mathbf{u}^*(t), p_0^*]$ is zero for $t$ in $[t_0, t_1]$; that is,*

$$H[\mathbf{x}^*(t), \mathbf{p}^*(t), \mathbf{u}^*(t), p_0^*] = 0 \qquad t \in [t_0, t_1] \tag{5-499}$$

*d. If $S_1$ is a smooth $k$-fold in $R_n$, then the vector $\mathbf{p}^*(t_1)$ is transversal (normal) to $S_1$ at $\mathbf{x}^*(t_1)$ (see Sec. 3-13); if $S_1 = R_n$, then $\mathbf{p}^*(t_1)$ is the zero vector; that is, $\mathbf{p}^*(t_1) = \mathbf{0}$.*

We observe, first of all, that the difference between these theorems and Theorems 5-5 and 5-6 lies in the additional constant $p_0^*$, which is needed to handle pathological cases (see the discussion at the end of Sec. 5-13). Secondly, we note that the only difference between Theorems 5-5P and 5-6P is the additional assertion *d* of Theorem 5-6P, in which the transversality of the costate to the target set is mentioned. We shall "prove" these two theorems in the next section.

## 5-16 A Heuristic Proof of the Minimum Principle

We are now going to "prove" the minimum principle in a rather intuitive way, making frequent use of geometric ideas. However, before presenting the actual details, let us outline the path we shall follow. As we shall divide this section into subsections, each of which corresponds to a step in the proof, our outline will consist of a list of these subsections with a brief description of their contents.

**1 A Change of Variable** We introduce an auxiliary variable $x_0$ such that $\dot{x}_0(t) = L[\mathbf{x}(t), \mathbf{u}(t)]$, and we replace our problem with an equivalent problem in $R_{n+1}$.

**2 The Principle of Optimality** We show that any portion of an optimal trajectory must be an optimal trajectory, and we interpret this geometrically.

**3 A Small Change in the Initial Condition and Its Consequences** We make a small change in our initial condition, and we see that its future effect is (approximately) described by a certain homogeneous linear system. We view this geometrically as moving a vector along our optimal trajectory, and we construct a "tube" in which our assumed optimal control remains (approximately) optimal.

**4 Moving Hyperplanes** We examine the adjoint system of the homogeneous linear system which describes the effect of a small change in the initial condition. We view this adjoint system geometrically as a hyperplane moving along the optimal trajectory, and we see that our costate will be the normal to a particular hyperplane moving along the optimal trajectory. This particular hyperplane will be a support hyperplane of a convex cone constructed on the basis of the effects of perturbations of the optimal control.

**5 Temporal Variation of the Optimal Control** We make small changes in the time interval over which the optimal control is applied, and we show that their effect may be viewed geometrically as defining a ray through the terminal point.

**6 Spatial Variation of the Optimal Control** We consider controls which differ from the optimal control on a small interval of time and are constant on that interval of time. The effect of these controls leads us to define a cone whose vertex is our terminal point.

**7 The Terminal Cone** We construct a convex cone whose vertex is our terminal point on the basis of the temporal and spatial variations of the optimal control. We show that the elements of this cone correspond to variations of the optimal control.

**8 Proof of Theorem 5-5P** We "prove" Theorem 5-5P. We first show that the ray in the direction of decreasing cost does not meet the interior of our terminal cone. We then determine a hyperplane which separates this ray and our terminal cone. This hyperplane defines our costate, and we proceed to establish Theorem 5-5P by making suitable calculations.

**9 The Transversality Conditions** We establish Theorem 5-6P by suitably modifying our arguments to take the tangent plane to our target set into account.

In essence, we shall follow a step-by-step procedure analogous to that used before in Secs. 5-6 and 5-7 but taking the greater complexity of the problem at hand into account. In other words, we begin by assuming that an optimal control exists. We then replace our problem by an equivalent problem in $R_{n+1}$ to simplify matters (subsection 1). We next develop a property of optimal trajectories known as the principle of optimality (subsection 2). Having completed these preliminaries, we study the effects of perturbations of the optimal control (subsections 3 to 7). Finally, we "prove" the minimum principle (subsections 8 and 9). Basically, our argument consists in:

1. Showing that there are "enough" perturbations of the optimal control to generate a cone (whose vertex is our terminal point) with the property that the ray in the direction of decreasing cost does not meet the interior of the cone

2. Showing that the normal to a hyperplane which separates this ray and the cone can be "moved back" along our optimal trajectory in such a way as to define the costate whose existence is asserted in the theorem

## *1 A Change of Variable*

Let us begin by introducing an auxiliary variable $x_0$ such that

$$\dot{x}_0(t) = L[\mathbf{x}(t), \mathbf{u}(t)] \qquad x_0(t_0) = 0 \tag{5-500}$$

Then we immediately see that

$$J(\mathbf{x}_0, t_0, \mathbf{u}) = \int_{t_0}^{t_1} L[\mathbf{x}(t), \mathbf{u}(t)]\, dt \tag{5-501}$$

$$= \int_{t_0}^{t_1} \dot{x}_0(t)\, dt = x_0(t_1) \tag{5-502}$$

If we consider the $(n + 1)$st-order system

$$\begin{aligned} \dot{x}_0(t) &= L[\mathbf{x}(t), \mathbf{u}(t)] \\ \dot{\mathbf{x}}(t) &= \mathbf{f}[\mathbf{x}(t), \mathbf{u}(t)] \end{aligned} \tag{5-503}$$

then our special problem 1 (see Sec. 5-12) may be rephrased as follows:

*Given the $(n+1)$st-order system (5-503); given the initial time $t_0$ and the initial condition $(0, \mathbf{x}_0)$ [that is, $x_0(t_0) = 0$] in $R_{n+1}$. Let $S = S' \times (T_1, T_2)$ be the target set [in $R_{n+1} \times (T_1, T_2)$] where $S'$ is the line in $R_{n+1}$ through $(0, \mathbf{x}_1)$ and parallel to the $x_0$ axis. $S'$ is given by the equations*

$$x_1 - x_{11} = 0,\ x_2 - x_{12} = 0,\ \ldots,\ x_n - x_{1n} = 0 \tag{5-504}$$

*where the $x_{1j}$ are the components of the vector $\mathbf{x}_1$. Then determine the admissible control $\mathbf{u}$ which transfers $\left(\begin{bmatrix} 0 \\ \mathbf{x}_0 \end{bmatrix}, t_0\right)$ to $S' \times (T_1, T_2)$ and which, in so doing, minimizes $x_0$ at the terminal time.*

In other words, we want to find the trajectory of the system (5-503), starting from the point $(0, \mathbf{x}_0)$, which intersects the line $S'$ in a point with smallest $x_0$ coordinate.

We remind the reader that we wish to hit the specified point $\mathbf{x}_1$ in the state space and that the time of arrival is free.

The geometric situation is illustrated in Fig. 5-16. In Fig. 5-16*a*, we exhibit an optimal (*) trajectory in $R_{n+1}$. We often indicate the state space as the $x_i x_j$ plane, and we shall always view the cost axis ($x_0$) as pointing upward, in order to establish the notions of "above" and "below." Figure 5-16*b* is the "same" as Fig. 5-16*a* except for the fact that we have "compressed" the state space into a single axis. The arrows indicate the motion of a point (in $R_n$ or $R_{n+1}$) as time increases. We shall often exhibit the time axis also; in such a case, one axis will represent the cost, another axis will represent the state space, and the third axis will represent the time. All axes will be labeled, so there should be no difficulty in interpretation.

Since we shall be dealing with vectors in $R_{n+1}$ as well as with vectors in $R_n$, *let us agree to denote a typical element of $R_{n+1}$ by $\mathbf{y}$, and let us suppose that the components of $\mathbf{y}$ are $y_0, y_1, \ldots, y_n$, that is,*

$$\mathbf{y} = \begin{bmatrix} y_0 \\ y_1 \\ \cdot \\ \cdot \\ \cdot \\ y_n \end{bmatrix} \tag{5-505}$$

Then we may define an $(n+1)$-vector-valued function $\mathbf{g}(\mathbf{y}, \mathbf{u})$ with components $g_0, g_1, \ldots, g_n$ by setting

$$\begin{aligned} g_0(y_0, y_1, \ldots, y_n, \mathbf{u}) &= L(y_1, y_2, \ldots, y_n, \mathbf{u}) \\ g_1(y_0, y_1, \ldots, y_n, \mathbf{u}) &= f_1(y_1, y_2, \ldots, y_n, \mathbf{u}) \\ \cdots\cdots\cdots & \cdots\cdots\cdots \\ g_n(y_0, y_1, \ldots, y_n, \mathbf{u}) &= f_n(y_1, y_2, \ldots, y_n, \mathbf{u}) \end{aligned} \tag{5-506}$$

and so we may write the system (5-503) in the form

$$\dot{\mathbf{y}}(t) = \mathbf{g}[\mathbf{y}(t), \mathbf{u}(t)] \tag{5-507}$$

When discussing solutions of the systems (5-503) and (5-507), we shall identify the solution vectors $\mathbf{y}(t)$ and $(x_0(t), \mathbf{x}(t))$ and shall use whichever notation is more convenient to our purposes at hand. We repeat again, for emphasis, *that* $\mathbf{y}$ *will refer to an* $n + 1$ *vector and that* $\mathbf{x}$ *will refer to an* $n$ *vector* throughout this section.

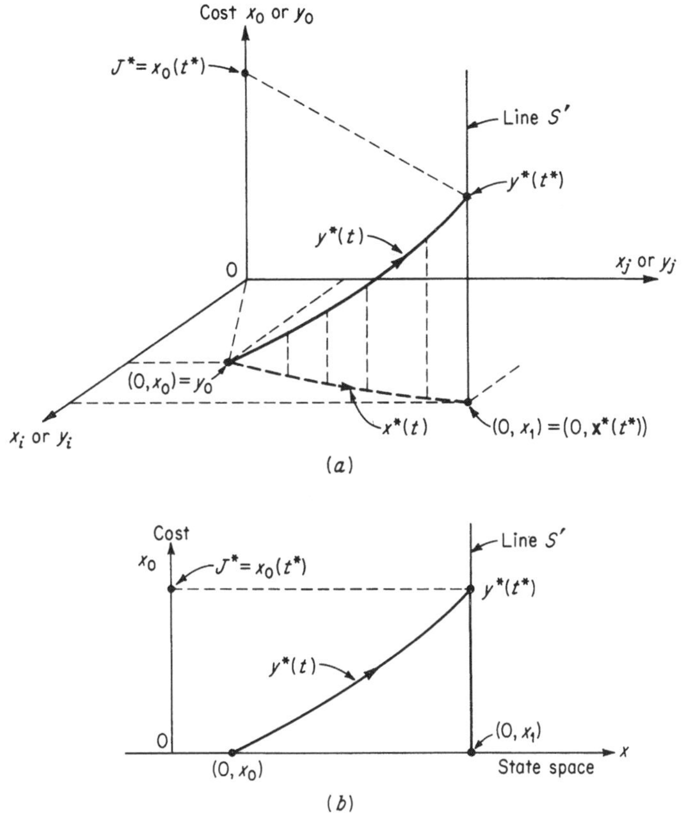


**Fig. 5-16** (*a*) Illustration of the $R_{n+1}$ space. The $x_ix_j$ plane represents the state space, and the $x_0$ axis the cost. The optimal trajectory in $R_{n+1}$ is $\mathbf{y}^*(t)$, and its projection upon the state space $(x_ix_j)$ is $\mathbf{x}^*(t)$. (*b*) In this figure, the state space has been compressed into a single axis (the $x$ axis).

Now, since the minimum principle is a *necessary* condition, we assume that $\mathbf{u}^*(t)$ is an optimal control and that $\mathbf{x}^*(t)$ is the corresponding optimal trajectory (in $R_n$), just as we did in Secs. 5-6 and 5-7. We let $t^*$ denote the terminal time, so that (see Fig. 5-16)

$$\mathbf{x}^*(t^*) = \mathbf{x}_1 \tag{5-508}$$

and we let $x_0^*(t)$ denote the cost along the optimal trajectory, that is,

$$x_0^*(t) = \int_{t_0}^{t} L[\mathbf{x}^*(t), \mathbf{u}^*(t)]\, dt \qquad t_0 \le t \le t^* \tag{5-509}$$

Also, in accordance with our convention, we let $\mathbf{y}^*(t)$ denote the trajectory (in $R_{n+1}$) of (5-507), starting from $(0, \mathbf{x}_0)$ generated by the control $\mathbf{u}^*(t)$ (see Fig. 5-16). Thus, we have

$$\mathbf{y}^*(t_0) = \mathbf{y}_0 = \begin{bmatrix} 0 \\ \mathbf{x}_0 \end{bmatrix} \tag{5-510}$$

$$\mathbf{y}^*(t) = \begin{bmatrix} x_0^*(t) \\ \mathbf{x}^*(t) \end{bmatrix} \tag{5-511}$$

and

$$\mathbf{y}^*(t^*) = \begin{bmatrix} J^* \\ \mathbf{x}_1 \end{bmatrix} = \begin{bmatrix} x_0^*(t^*) \\ \mathbf{x}^*(t^*) \end{bmatrix} \tag{5-512}$$

where we have used $J^*$ to represent the minimum cost, that is,

$$J^* = J(\mathbf{x}_0, t_0, \mathbf{u}^*) \tag{5-513}$$

We know then that the trajectory $\mathbf{y}^*(t)$ in $R_{n+1}$ is optimal for our rephrased problem.

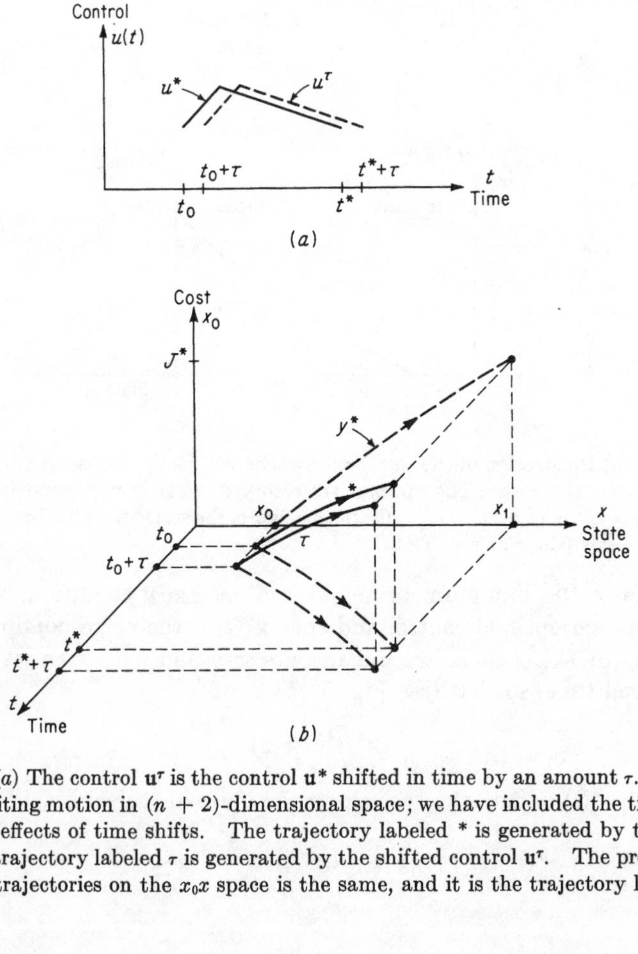


**Fig. 5-17** (a) The control $\mathbf{u}^\tau$ is the control $\mathbf{u}^*$ shifted in time by an amount $\tau$. (b) Here we are exhibiting motion in $(n + 2)$-dimensional space; we have included the time axis to indicate the effects of time shifts. The trajectory labeled * is generated by the control $\mathbf{u}^*$, and the trajectory labeled $\tau$ is generated by the shifted control $\mathbf{u}^\tau$. The projection of the * and $\tau$ trajectories on the $x_0x$ space is the same, and it is the trajectory labeled $\mathbf{y}^*$.

Let us observe that nothing is changed by a translation in time, since both the system and the cost functional do not depend explicitly on the time. In other words, if we let $t_0 + \tau$ be our initial time, then the control $\mathbf{u}^\tau$, defined by setting

$$\mathbf{u}^\tau(s + \tau) = \mathbf{u}^*(s) \qquad s \in [t_0, t^*] \tag{5-514}$$

will be optimal and will generate the "same" trajectory as $\mathbf{u}^*$ (where we view the trajectory as a point set in $R_{n+1}$). This situation is illustrated in Fig. 5-17. In Fig. 5-17*a*, we exhibit the controls $\mathbf{u}^*$ and $\mathbf{u}^\tau$, and in Fig. 5-17*b*, we exhibit the motion in the space which is the product space of the cost, the state space, and the time. The trajectories generated by $\mathbf{u}^*$ and $\mathbf{u}^\tau$ have the same projection on the cost $\times$ state-space "plane" in the figure.

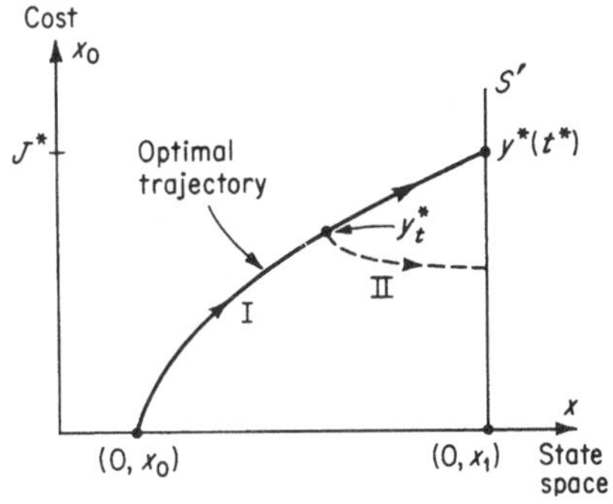


**Fig. 5-18** The trajectory composed of the two segments *I* and *II* cannot occur, because it contradicts the assumption that $\mathbf{u}^*$ is the optimal control.

## 2 *The Principle of Optimality*

We are going to show that *any portion of the optimal trajectory is also an optimal trajectory.*† Let us begin by showing that the control $\mathbf{u}^*$ is optimal for any point on the optimal trajectory. So we suppose that $\mathbf{y}_t^* = \mathbf{y}^*(t) = (x_0^*(t), \mathbf{x}^*(t))$ is a point on our optimal trajectory (with $t_0 < t < t^*$). Then the control $\mathbf{u}^*_{(t,t^*]}$ will transfer this point to $S'$. If $\tilde{\mathbf{u}}$ is an admissible control which transfers $\mathbf{y}_t^*$ to $S'$, then, letting $\tilde{\mathbf{y}}(\tau)$ denote the solution of (5-507) satisfying the initial condition

$$\tilde{\mathbf{y}}(0) = \mathbf{y}_t^* \tag{5-515‡}$$

and letting $\tilde{\tau}$ be the time at which $S'$ is met, we can see that

$$\tilde{y}_0(\tilde{\tau}) \geq J^* = x_0^*(t^*) \tag{5-516}$$

For, if Eq. (5-516) were not satisfied, then the control $\mathbf{u}^*$ would not be optimal, as can readily be seen from Fig. 5-18, in which *II* denotes the $\tilde{\mathbf{y}}(\tau)$ trajectory.

We now assert, as is illustrated in Fig. 5-19, that *all* the trajectories of the system (5-507), starting from $\mathbf{y}_0 = (0, \mathbf{x}_0)$, which *meet* $S'$ must lie "above" the optimal trajectory. We shall verify this assertion by showing that *any portion of the optimal trajectory is also an optimal trajectory* in the following sense:

Suppose that $\mathbf{y}_{t_1}^*$ and $\mathbf{y}_{t_2}^*$ are points on the optimal trajectory, as illustrated in Fig. 5-20, and that $S''$ is the line parallel to the cost axis passing through

† This is often called the principle of optimality. (See Ref. B-5.)

‡ We may start at $\tau = 0$ since there is no explicit time dependence.

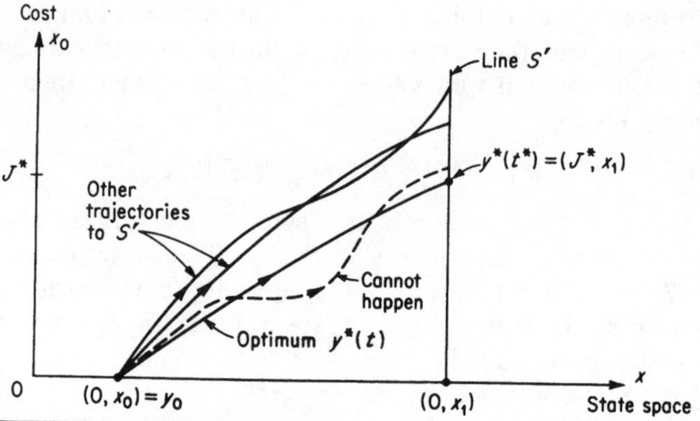


**Fig. 5-19** Illustration of possible and impossible trajectories in $R_{n+1}$.

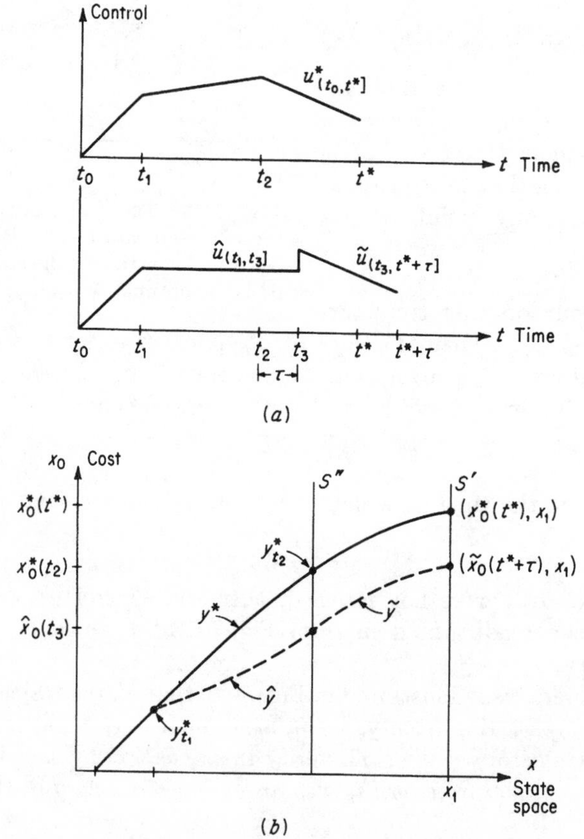


**Fig. 5-20** (*a*) Two controls that transfer the initial state to the line $S'$. The control $\tilde{\mathbf{u}}_{(t_3,t^*+\tau]}$ is the same as the control $\mathbf{u}^*_{(t_2,t^*]}$ except that it is shifted by $\tau$ seconds. (*b*) The trajectory composed of $\hat{\mathbf{y}}$ and $\tilde{\mathbf{y}}$ cannot happen, because it contradicts the optimality of $\mathbf{y}^*$.

$\mathbf{y}^*_{t_2}$. Then the control $\mathbf{u}^*_{(t_1,t_2]}$ transfers $\mathbf{y}^*_{t_1}$ to $S''$, and if $\hat{\mathbf{u}}_{(t_1,t_3]}$ is any admissible control which transfers $\mathbf{y}^*_{t_1}$ *to* $S''$, we must have $\hat{x}_0(t_3) \geq y_0^*(t_2)$, where $t_3$ is the time at which the trajectory $\hat{\mathbf{y}}_{(t_1,t_3]}$ generated by $\hat{\mathbf{u}}$ meets $S''$. In other words, we suppose that $\mathbf{u}^*_{(t_0,t^*]}$ is an optimal control (see Fig. 5-20*a*) and that $\mathbf{y}^*_{(t_0,t^*]}$ is the corresponding optimal trajectory (see Fig. 5-20*b*). We suppose that $t_0 \leq t_1 < t_2 \leq t^*$, and we let

$$\begin{aligned} \mathbf{y}^*_{t_1} &= \mathbf{y}^*(t_1) = \begin{bmatrix} x_0^*(t_1) \\ \mathbf{x}^*(t_1) \end{bmatrix} \\ \mathbf{y}^*_{t_2} &= \mathbf{y}^*(t_2) = \begin{bmatrix} x_0^*(t_2) \\ \mathbf{x}^*(t_2) \end{bmatrix} \end{aligned} \tag{5-517}$$

be two points on the optimal trajectory. We suppose that $S''$ is the line parallel to the cost axis passing through $\mathbf{y}^*_{t_2}$, and we let $\hat{\mathbf{u}}_{(t_1,t_3]}$ be any admissible control (see Fig. 5-20*a*) that transfers $\mathbf{y}^*_{t_1}$ to $S''$; that is, the trajectory $\hat{\mathbf{y}}_{(t_1,t_3]}$ (see Fig. 5-20*b*) has the property that

$$\hat{\mathbf{y}}(t_3) = \begin{bmatrix} \hat{x}_0(t_3) \\ \hat{\mathbf{x}}(t_3) \end{bmatrix} = \begin{bmatrix} \hat{x}_0(t_3) \\ \mathbf{x}^*(t_2) \end{bmatrix} \tag{5-518}$$

or, equivalently, the property that

$$\hat{\mathbf{x}}(t_3) = \mathbf{x}^*(t_2) \tag{5-519}$$

We then claim that

$$\hat{x}_0(t_3) \geq x_0^*(t_2) \tag{5-520}$$

If Eq. (5-520) were not satisfied, then we would have

$$\hat{x}_0(t_3) < x_0^*(t_2) \tag{5-521}$$

If (see Fig. 5-20*a*) $\tau$ is defined by the relation

$$t_3 = t_2 + \tau \tag{5-522}$$

then Eqs. (5-519) and (5-521) become, respectively,

$$\hat{\mathbf{x}}(t_2 + \tau) = \mathbf{x}^*(t_2) \tag{5-523}$$

$$\hat{x}_0(t_2 + \tau) < x_0^*(t_2) \tag{5-524}$$

Now let us suppose that, at $t_3$, we apply the control

$$\tilde{\mathbf{u}}_{(t_3,t^*+\tau]} = \tilde{\mathbf{u}}_{(t_2+\tau,t^*+\tau]} \tag{5-525}$$

as illustrated in Fig. 5-20*a*. In effect, $\tilde{\mathbf{u}}$ is the same as the optimal control $\mathbf{u}^*$ but translated in time; that is,

$$\tilde{\mathbf{u}}(s + \tau) = \mathbf{u}^*(s) \qquad s \in (t_2, t^*] \tag{5-526}$$

The control $\tilde{\mathbf{u}}$ generates a trajectory $\tilde{\mathbf{y}}$, which originates at the point $(\hat{x}_0(t_2 + \tau), \mathbf{x}^*(t_2))$ and which *meets the line* $S'$ at $t^* + \tau$. In other words, if $\tilde{\mathbf{y}}(s+\tau) = (\tilde{x}_0(s+\tau), \tilde{\mathbf{x}}(s+\tau))$ for $s \in (t_2, t^*]$, then

$$\tilde{\mathbf{x}}(t^* + \tau) = \mathbf{x}^*(t^*) = \mathbf{x}_1 \tag{5-527}$$

However, we have

$$\tilde{x}_0(t^* + \tau) = \tilde{x}_0(t_2 + \tau) + \int_{t_2+\tau}^{t^*+\tau} L[\tilde{\mathbf{x}}(s+\tau), \tilde{\mathbf{u}}(s+\tau)]\, d(s+\tau) \tag{5-528}$$

$$= \tilde{x}_0(t_2 + \tau) + \int_{t_2}^{t^*} L[\mathbf{x}^*(s), \mathbf{u}^*(s)]\, ds \tag{5-529}$$

$$= \tilde{x}_0(t_2 + \tau) + x_0^*(t^*) - x_0^*(t_2) \tag{5-530}$$

$$= \hat{x}_0(t_3) + x_0^*(t^*) - x_0^*(t_2) \tag{5-531}$$

In view of Eq. (5-521), we conclude that

$$\tilde{x}_0(t^* + \tau) < x_0^*(t^*) \tag{5-532}$$

which contradicts the optimality of $\mathbf{u}^*$. Thus, our claim (5-520) is established, and so we have shown that optimality of $\mathbf{u}^*$ is equivalent to the geometric property that all the $\mathbf{y}$ trajectories which meet $S'$ must lie "above" the trajectory generated by $\mathbf{u}^*$. (See Fig. 5-19.)

This geometric characterization of optimality will help us to construct a "tube" about the optimal trajectory in which the control $\mathbf{u}^*$ is approximately optimal. Consideration of this "tube" will enable us to overcome the difficulty associated with the requirement that our perturbed trajectories must pass through $\mathbf{x}_1$ (see Sec. 5-9).

### *3 A Small Change in the Initial Condition and Its Consequences*

With a view toward constructing this "tube" in which $\mathbf{u}^*$ is *approximately* optimal, let us now make a very small change in our initial point $(0, \mathbf{x}_0)$ and see how this change "propagates" along the optimal trajectory. In other words, we assume that our new initial point (in $R_{n+1}$) is

$$(0, \mathbf{x}_0) + \epsilon(\nu_0, \boldsymbol{\nu}) + \mathbf{o}(\epsilon) \tag{5-533}$$

where $\epsilon$ is a small positive number and where, as usual, $\mathbf{o}(\epsilon)$ is a vector such that

$$\lim_{\epsilon \to 0} \frac{\|\mathbf{o}(\epsilon)\|}{\epsilon} = 0 \tag{5-534}$$

We observe that the solution of (5-507), starting from (5-533), generated by the control $\mathbf{u}^*$ is of the form

$$(x_0^*(t), \mathbf{x}^*(t)) + \epsilon(\chi_0(t), \boldsymbol{\chi}(t)) + \mathbf{o}(\epsilon) \tag{5-535}$$

where $(\chi_0(t), \boldsymbol{\chi}(t))$ is the solution of the $(n+1)$st-order linear system

$$\dot{\chi}_0(t) = \left\langle \frac{\partial L}{\partial \mathbf{x}}[\mathbf{x}^*(t), \mathbf{u}^*(t)], \boldsymbol{\chi}(t) \right\rangle \tag{5-536}$$

$$\dot{\boldsymbol{\chi}}(t) = \left( \frac{\partial \mathbf{f}}{\partial \mathbf{x}}[\mathbf{x}^*(t), \mathbf{u}^*(t)] \right) \boldsymbol{\chi}(t) \tag{5-537}$$

satisfying the initial conditions

$$\chi_0(t_0) = \nu_0 \tag{5-538}$$
$$\boldsymbol{\chi}(t_0) = \boldsymbol{\nu} \tag{5-539}$$

We also note that the system of Eqs. (5-536) and (5-537) may be written in the matrix form

$$\begin{bmatrix} \dot{\chi}_0(t) \\ \dot{\boldsymbol{\chi}}(t) \end{bmatrix} = \left[\begin{array}{c|cccc} 0 & \left.\dfrac{\partial L}{\partial x_1}\right|_* & \left.\dfrac{\partial L}{\partial x_2}\right|_* & \cdots & \left.\dfrac{\partial L}{\partial x_n}\right|_* \\ \hline \mathbf{0} & & \left.\dfrac{\partial \mathbf{f}}{\partial \mathbf{x}}\right|_* & & \end{array}\right] \begin{bmatrix} \chi_0(t) \\ \boldsymbol{\chi}(t) \end{bmatrix} \tag{5-540}$$

where the $|_*$ indicates that the function in question is to be evaluated along the optimal trajectory. If we let $\mathbf{A}_*(t)$ denote the $n + 1 \times n + 1$ matrix appearing in Eq. (5-540), so that

$$\mathbf{A}_*(t) = \left[\begin{array}{c|cccc} 0 & \left.\dfrac{\partial L}{\partial x_1}\right|_* & \left.\dfrac{\partial L}{\partial x_2}\right|_* & \cdots & \left.\dfrac{\partial L}{\partial x_n}\right|_* \\ \hline \mathbf{0} & & \left.\dfrac{\partial \mathbf{f}}{\partial \mathbf{x}}\right|_* & & \end{array}\right] \tag{5-541}$$

if we let $\boldsymbol{\psi}(t)$ denote the $n + 1$ vector $(\chi_0(t), \boldsymbol{\chi}(t))$, and if we let $\boldsymbol{\xi}$ denote the $n + 1$ vector $(\nu_0, \boldsymbol{\nu})$, then $\boldsymbol{\psi}(t)$ is the solution of the $(n + 1)$st-order *homogeneous* linear system

$$\dot{\boldsymbol{\psi}}(t) = \mathbf{A}_*(t)\boldsymbol{\psi}(t) \tag{5-542}$$

satisfying the initial condition

$$\boldsymbol{\psi}(t_0) = \boldsymbol{\xi} = \begin{bmatrix} \nu_0 \\ \boldsymbol{\nu} \end{bmatrix} \tag{5-543}$$

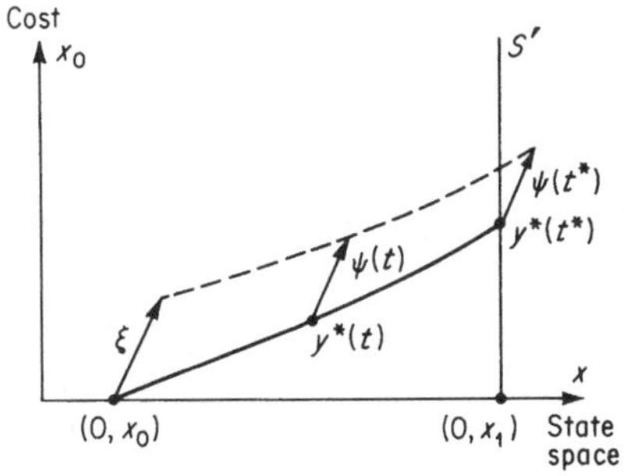


**Fig. 5-21** The result of small perturbations on the initial conditions. The effect of a small perturbation $\epsilon\boldsymbol{\xi}$ can be approximately described by the propagation of a vector $\epsilon\boldsymbol{\psi}(t)$ attached to the optimal trajectory $\mathbf{y}^*(t)$. The small perturbations are exaggerated in this figure for the sake of clarity. The vector $\boldsymbol{\psi}(t)$ is the solution of the linear system (5-542), with the initial condition given by Eq. (5-543).

This means that our new trajectory in $R_{n+1}$ is of the form

$$\mathbf{y}^*(t) + \epsilon\boldsymbol{\psi}(t) + \mathbf{o}(\epsilon) \tag{5-544}$$

We further note that if we view the vector $\boldsymbol{\xi}$ as being "attached" to our initial point $\mathbf{y}_0 = (0, \mathbf{x}_0)$, then, as is illustrated in Fig. 5-21, we may view the vector $\boldsymbol{\psi}(t)$ as being "attached" to the point $\mathbf{y}^*(t)$. Thus, $\boldsymbol{\psi}(t)$ can be considered the result of moving (or transferring) the vector $\boldsymbol{\xi}$ along the optimal trajectory. We also note that the $\mathbf{o}(\epsilon)$ vector in (5-544) depends upon $t$, so that we should write $\mathbf{o}(\epsilon, t)$. However, because of our smoothness assumptions, we can assert that

$$\lim_{\epsilon \to 0} \sup_{t \in [t_0, t_1]} \left\{ \frac{\|\mathbf{o}(\epsilon, t)\|}{\epsilon} \right\} = 0 \tag{5-545}$$

which means that the $t$ dependence of $\mathbf{o}(\epsilon, t)$ can be neglected. The geometric interpretation of this remark appears in Fig. 5-22. As we can see in the figure, the trajectory $\mathbf{y}^*(t) + \epsilon\boldsymbol{\psi}(t) + \mathbf{o}(\epsilon)$ always lies in a narrow band (bounded by the "envelope" curves in Fig. 5-22) about the curve $\mathbf{y}^*(t) + \epsilon\boldsymbol{\psi}(t)$. The interpretation of our remark is, simply, that the

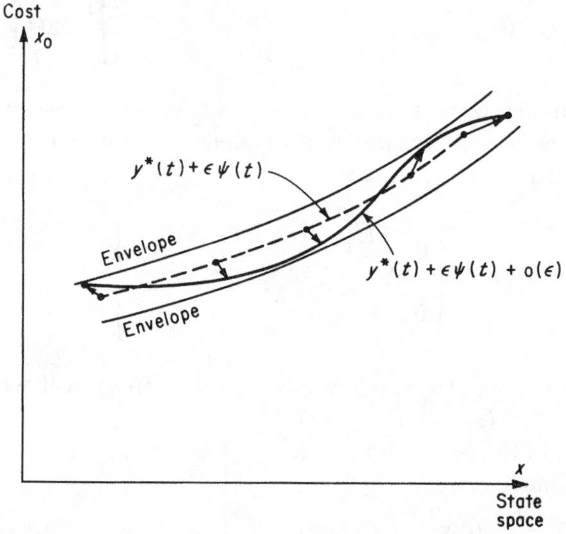


**Fig. 5-22** The "width" of the envelope does not change appreciably with time, and it is approximately $\epsilon^2$.

"width" of this band does not change very much with time; i.e., it is of the order of $\epsilon$ along the entire trajectory.

Letting $\boldsymbol{\Phi}_*(t, t_0)$ denote the $n + 1 \times n + 1$ fundamental matrix of Eq. (5-542) we see that

$$\boldsymbol{\psi}(t) = \boldsymbol{\Phi}_*(t, t_0)\boldsymbol{\xi} \tag{5-546}$$

and that

$$\boldsymbol{\psi}(t^*) = \boldsymbol{\Phi}_*(t^*, t_0)\boldsymbol{\xi} \tag{5-547}$$

Thus, we may view the system (5-542) as defining a linear transformation of the space of $n + 1$ vectors "attached" to $\mathbf{y}_0$ into the space of $n + 1$ vectors "attached" to $\mathbf{y}^*(t)$. Moreover, since the matrix $\boldsymbol{\Phi}_*(t, t_0)$ is *nonsingular*, this linear transformation is nonsingular.

Now, in view of the fact that any portion of our optimal trajectory is an optimal trajectory, we can see that if $t_1$ and $t_2$ are elements of $[t_0, t^*]$, then we can define, in an entirely similar way, a linear transformation of the space of $n + 1$ vectors "attached" to $\mathbf{y}^*(t_1)$ into the space of $n + 1$ vectors "attached" to $\mathbf{y}^*(t_2)$. In fact, if $\boldsymbol{\xi}_1$ is an $n + 1$ vector "attached" to $\mathbf{y}^*(t_1)$, then its image under this linear transformation is precisely the vector

$$\mathbf{\Phi}_*(t_2, t_1)\boldsymbol{\xi}_1 \tag{5-548}$$

Thus, the homogeneous linear system (5-542) may be viewed as describing the future effect of a small change in the optimal trajectory at a given time.

We shall view the small positive number $\epsilon$ as a parameter in the sequel. In other words, we shall, when dealing with specific quantities, assume that $\epsilon$ has been chosen small enough to validate our arguments for those specific quantities. For example, if $\boldsymbol{\xi}$ and $\hat{\boldsymbol{\xi}}$ are $n + 1$ vectors "attached"

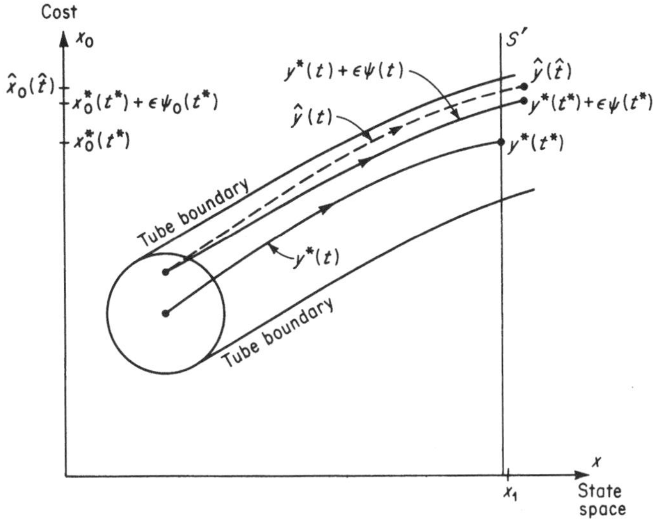


**Fig. 5-23** The trajectory $\hat{\mathbf{y}}(t)$ generated by the admissible control $\hat{\mathbf{u}}$ is "approximately above" the trajectory $\mathbf{y}^*(t) + \epsilon\boldsymbol{\psi}(t)$ because the control $\mathbf{u}^*$ is optimal to within first order in $\epsilon$ in a tube about $\mathbf{y}^*(t)$.

to $\mathbf{y}_0$, then we (tacitly) assume that $\epsilon$ is sufficiently small for both $\mathbf{y}^*(t) + \epsilon\boldsymbol{\psi}(t)$ and $\mathbf{y}^*(t) + \epsilon\hat{\boldsymbol{\psi}}(t)$ to be "close" to $\mathbf{y}^*(t)$, where $\boldsymbol{\psi}(t)$ and $\hat{\boldsymbol{\psi}}(t)$ are solutions of (5-542) satisfying the initial conditions $\boldsymbol{\psi}(t_0) = \boldsymbol{\xi}$ and $\hat{\boldsymbol{\psi}}(t_0) = \hat{\boldsymbol{\xi}}$, respectively.

Since small changes in initial conditions cause correspondingly small changes in the solutions of a differential equation, we can assert that all the trajectories of the system (5-507) which start from the initial point (5-533) and which meet $S'$ to within the first order in $\epsilon$ will lie "approximately above" the trajectory generated by $\mathbf{u}^*$. In other words, if $\hat{\mathbf{u}}$ is an admissible control which generates a trajectory $\hat{\mathbf{y}}$ starting from the point (5-533) with the property that, at some time $\hat{t}$,

$$\hat{\mathbf{x}}(\hat{t}) = \mathbf{x}_1 + \epsilon\hat{\mathbf{x}}_1 \tag{5-549}$$

where $\hat{\mathbf{y}}(t) = [\hat{x}_0(t), \hat{\mathbf{x}}(t)]$

then $\hat{x}_0(\hat{t}) - x_0^*(t^*) - \epsilon\psi_0(t^*) \geq -|o(\epsilon)|$ (5-550)

Thus, to within higher-order terms in $\epsilon$, the control $\mathbf{u}^*$ is optimal from the new initial point (5-533). In fact, as is illustrated in Fig. 5-23, the control $\mathbf{u}^*$ is optimal (to within the first order in $\epsilon$) in a "tube" about the optimal trajectory. The perturbations $\mathbf{u}$ of $\mathbf{u}^*$ which we shall consider will not take us out of this "tube," so that we shall be able to suppose that the cost $J(\mathbf{u})$ is approximately bigger than the cost $J(\mathbf{u}^*)$ for any point in the "tube." We note that this "tube" is, of course, a geometric fiction which we simply use as an aid to our intuition.

## 4 *Moving Hyperplanes*

Now let us examine the adjoint system of the homogeneous linear system $\dot{\boldsymbol{\psi}}(t) = \mathbf{A}_*(t)\boldsymbol{\psi}(t)$ [see Eq. (5-542)] in an effort to see if it has any useful geometric interpretation. (See Sec. 3-25 for the definition of the adjoint system.)

Since the matrix $-\mathbf{A}_*'(t)$ is given by the relation

$$-\mathbf{A}_*'(t) = \left[\begin{array}{c|c} 0 & \mathbf{0} \\ \hline -\left.\dfrac{\partial L}{\partial \mathbf{x}}\right|_* & -\left(\left.\dfrac{\partial \mathbf{f}}{\partial \mathbf{x}}\right|_*\right)' \end{array}\right] \tag{5-551}$$

$$= \left[\begin{array}{c|cccc} 0 & & & \mathbf{0} & \\ \hline -\left.\dfrac{\partial L}{\partial x_1}\right|_* & -\left.\dfrac{\partial f_1}{\partial x_1}\right|_* & -\left.\dfrac{\partial f_2}{\partial x_1}\right|_* & \cdots & -\left.\dfrac{\partial f_n}{\partial x_1}\right|_* \\ -\left.\dfrac{\partial L}{\partial x_2}\right|_* & -\left.\dfrac{\partial f_1}{\partial x_2}\right|_* & -\left.\dfrac{\partial f_2}{\partial x_2}\right|_* & \cdots & -\left.\dfrac{\partial f_n}{\partial x_2}\right|_* \\ \cdots & \cdots & \cdots & \cdots & \cdots \\ -\left.\dfrac{\partial L}{\partial x_n}\right|_* & -\left.\dfrac{\partial f_1}{\partial x_n}\right|_* & -\left.\dfrac{\partial f_2}{\partial x_n}\right|_* & \cdots & -\left.\dfrac{\partial f_n}{\partial x_n}\right|_* \end{array}\right] \tag{5-552}$$

we can immediately see that the adjoint system of (5-542) is of the form

$$\begin{bmatrix} \dot{p}_0(t) \\ \dot{\mathbf{p}}(t) \end{bmatrix} = \left[\begin{array}{c|c} 0 & \mathbf{0} \\ \hline -\left.\dfrac{\partial L}{\partial \mathbf{x}}\right|_* & -\left(\left.\dfrac{\partial \mathbf{f}}{\partial \mathbf{x}}\right|_*\right)' \end{array}\right] \begin{bmatrix} p_0(t) \\ \mathbf{p}(t) \end{bmatrix} \tag{5-553}$$

or, equivalently, of the form

$$\dot{p}_0(t) = 0 \tag{5-554}$$

$$\dot{\mathbf{p}}(t) = -\left.\frac{\partial L}{\partial \mathbf{x}}\right|_* p_0(t) - \left(\left.\frac{\partial \mathbf{f}}{\partial \mathbf{x}}\right|_*\right)' \mathbf{p}(t) \tag{5-555}$$

We may immediately deduce from Eq. (5-554) that $p_0(t)$ is a constant, that is, that $p_0(t) = p_0$ for all $t$. On the other hand, if we recall that the Hamiltonian $H(\mathbf{x}, \mathbf{p}, \mathbf{u}, p_0)$ is defined by

$$H(\mathbf{x}, \mathbf{p}, \mathbf{u}, p_0) = p_0 L(\mathbf{x}, \mathbf{u}) + \langle \mathbf{p}, \mathbf{f}(\mathbf{x}, \mathbf{u}) \rangle \tag{5-556}$$

then we can see that Eq. (5-555) may be written in the form

$$\dot{\mathbf{p}}(t) = -\frac{\partial H}{\partial \mathbf{x}}[\mathbf{x}^*(t), \mathbf{p}(t), \mathbf{u}^*(t), p_0] \tag{5-557}$$

We also observe that if $p_0$ is a constant and $\mathbf{p}(t)$ is a solution of Eq. (5-557), then the scalar product [see Eq. (3-434)]

$$\left\langle \begin{bmatrix} p_0 \\ \mathbf{p}(t) \end{bmatrix}, \boldsymbol{\psi}(t) \right\rangle = \text{const} \tag{5-558}$$

for any solution $\boldsymbol{\psi}(t)$ of Eq. (5-542).

Now, if we let $P_0$ be a hyperplane passing through $\mathbf{y}_0$ with equation

$$\left\langle \begin{bmatrix} \pi_0 \\ \boldsymbol{\pi} \end{bmatrix}, \mathbf{y} - \mathbf{y}_0 \right\rangle = 0 \tag{5-559}†$$

and if we suppose that $\mathbf{y}_0 + \epsilon\boldsymbol{\xi}$ is an $n + 1$ vector, then the solution of (5-553) satisfying the initial condition

$$p_0(t_0) = \pi_0 \qquad \mathbf{p}(t_0) = \boldsymbol{\pi} \tag{5-560}$$

defines a hyperplane $P_t$ with equation

$$\left\langle \begin{bmatrix} \pi_0 \\ \mathbf{p}(t) \end{bmatrix}, \mathbf{y} - \mathbf{y}^*(t) \right\rangle = 0 \tag{5-561}$$

which passes through $\mathbf{y}^*(t)$ and has the same scalar product with the vector $\boldsymbol{\psi}(t)$ "attached" to $\mathbf{y}^*(t)$ as the vector $\boldsymbol{\xi}$ "attached" to $\mathbf{y}_0$ makes with the hyperplane $P_0$. In particular, if $P_0$ is perpendicular to $\boldsymbol{\xi}$, then $P_t$ is perpendicular to $\boldsymbol{\psi}(t)$. The hyperplane $P_t$ can be viewed as the result of moving (or transferring) the hyperplane $P_0$ along the optimal trajectory, as is illustrated in Fig. 5-24.

We observe, in passing, that if $\pi_0 = 0$, then the equation of $P_t$ is of the form

$$\langle \mathbf{p}(t), \mathbf{x} - \mathbf{x}^*(t) \rangle = 0 \tag{5-562}$$

and, consequently, is independent of the "cost" coordinate in $R_{n+1}$, in the sense that $P_t$ is parallel to the $x_0$ axis. This observation will provide a clue to the significance of the pathological situation $p_0^* = 0$.

If we compare Eqs. (5-557) and (5-489), we can see that the costate or adjoint (sic!) variables will correspond to the motion of a particular hyperplane along the optimal trajectory. Thus, the crucial question becomes: *How can we determine the particular hyperplane that will enable us to establish the theorem?* Well, let us begin to answer this question by asking another: *What is a general way of determining hyperplanes passing through a particular*

† Observe that $\begin{bmatrix} \pi_0 \\ \boldsymbol{\pi} \end{bmatrix}$ is normal to the hyperplane $P_0$.

*point* $\mathbf{y}$? If we suppose that $\mathbf{y}$ is a boundary point of a convex set $C$, then we know that there is a support hyperplane of $C$ passing through $\mathbf{y}$ (see Theorem 3-5). So what we propose to do now is to introduce a suitable convex set $C^*_{t_1}$ which has $\mathbf{y}^*(t^*) = (J^*, \mathbf{x}_1)$ as a boundary point and which has a support hyperplane that will be used to prove the theorem. In fact, we shall determine *convex cones* $C_t$ with vertices at $\mathbf{y}^*(t)$ by considering a special class of variations or perturbations of the control $\mathbf{u}^*$. These cones will always lie on the same side of the hyperplanes determined by the costate variables and will be called *cones of attainability*. In a sense, the rays of the cone $C_{t^*}$ will correspond to the perturbations of $\mathbf{u}^*$ that we shall consider.

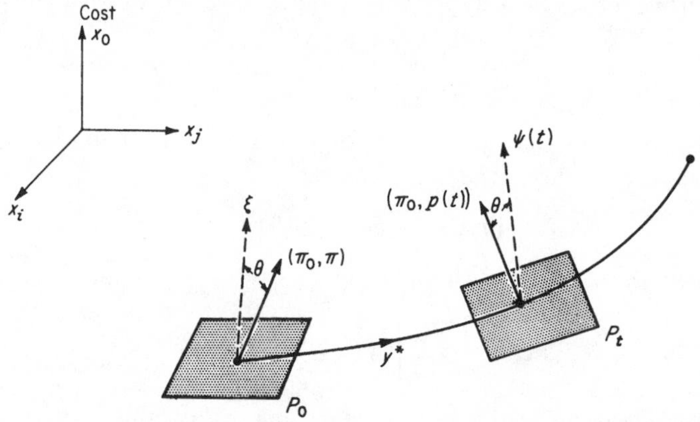


**Fig. 5-24** Illustration of the moving hyperplane $P_t$. The vector $(\pi_0, \pi)$ is perpendicular to the hyperplane $P_0$ at $\mathbf{y}^*(t_0)$. The vector $\xi$ (or $\epsilon\xi$) represents a perturbation on the initial condition $\mathbf{y}^*(t_0)$. At the time $t$, the vector $(\pi_0, \mathbf{p}(t))$ is perpendicular to the hyperplane $P_t$ at $\mathbf{y}^*(t)$. The coordinate system is illustrated at the top of the figure.

So, in the next two subsections, we turn our attention to the description of the class of variations of $\mathbf{u}^*$ with which we shall work.

## 5 *Temporal Variation of the Optimal Control*

The first basic type of variation, which we shall call a *temporal* variation, consists in merely changing the terminal time $t^*$ by a small amount and examining the result of applying the optimal control $\mathbf{u}^*$ over the new time interval. To be more precise, if $\tau$ is an arbitrary real number, then the temporal variation $\mathbf{u}[\tau]$ of $\mathbf{u}^*$ is defined as

$$\mathbf{u}[\tau](t) = \begin{cases} \mathbf{u}^*(t) & t_0 \le t \le t^* + \epsilon\tau \qquad \tau < 0 \\ \mathbf{u}^*(t) & t_0 \le t \le t^* \\ \mathbf{u}^*(t^*) & t^* < t \le t^* + \epsilon\tau \end{cases} \Bigg\} \quad \tau \ge 0 \tag{5-563}$$

Two typical temporal variations of $\mathbf{u}^*$ are illustrated in Fig. 5-25. We note in Fig. 5-25 that $\tau_1 > 0$ and $\tau_2 < 0$.

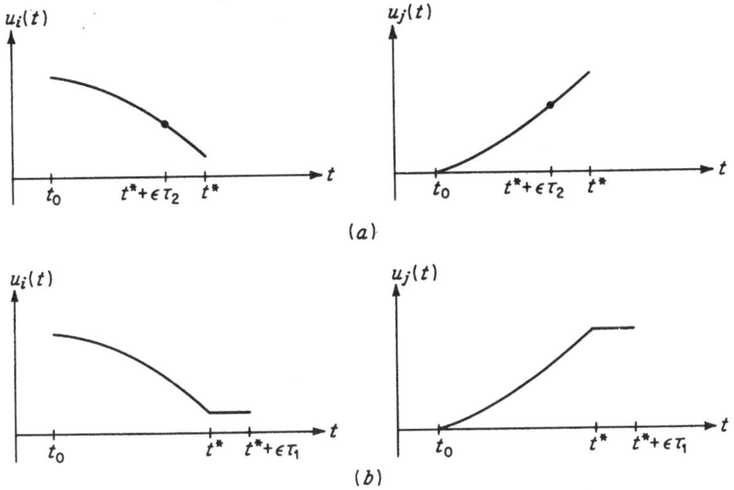


**Fig. 5-25** Typical temporal variations on two components ($u_i$ and $u_j$) of the optimal control $\mathbf{u}^*(t)$. In (a), the time $\tau_2$ is negative, and in (b), the time $\tau_1$ is positive. In (b), the components of $\mathbf{u}[\tau](t)$ are constant over the interval $(t^*, t^* + \epsilon\tau_1]$ and equal to the components of $\mathbf{u}^*(t^*)$.

We observe that if $\epsilon$ is small, then the solution $\mathbf{y}(t) = (x_0(t), \mathbf{x}(t))$ of the system (5-507), starting from $\mathbf{y}_0 = (0, \mathbf{x}_0)$, generated by the control $\mathbf{u}[\tau]$ will have the property that

$$\mathbf{y}(t^* + \epsilon\tau) = \mathbf{y}^*(t^*) + \epsilon\boldsymbol{\delta}[\tau] + \mathbf{o}(\epsilon) \tag{5-564}$$

where $\boldsymbol{\delta}[\tau]$ is an $n + 1$ vector which does not depend on $\epsilon$. In fact, we can show that, in view of our differentiability assumptions and the smallness of $\epsilon$,

$$\boldsymbol{\delta}[\tau] = \mathbf{g}[\mathbf{y}^*(t^*), \mathbf{u}^*(t^*)]\tau \tag{5-565}$$

$$= \begin{bmatrix} L[\mathbf{x}^*(t^*), \mathbf{u}^*(t^*)] \\ \mathbf{f}[\mathbf{x}^*(t^*), \mathbf{u}^*(t^*)] \end{bmatrix} \tau \tag{5-566}$$

$$= \begin{bmatrix} L[\mathbf{x}_1, \mathbf{u}^*(t^*)] \\ \mathbf{f}[\mathbf{x}_1, \mathbf{u}^*(t^*)] \end{bmatrix} \tau \tag{5-567}$$

We note that Eqs. (5-564) and (5-565) can be construed as a precise form of the statement that the "differential" is the best first-order approximation of the change in the function.†

We shall view the vector $\boldsymbol{\delta}[\tau]$ (or $\epsilon\boldsymbol{\delta}[\tau]$) as being "attached" to the point $\mathbf{y}^*(t^*)$, as is illustrated in Fig. 5-26. We can see immediately from Eq. (5-566) that the vectors $\boldsymbol{\delta}[\tau]$, for various values of $\tau$, all lie on the same ray (line) $\vec{\rho}$ passing through the point $\mathbf{y}^*(t^*)$ and that, in fact, these vectors "fill up" that line. Moreover, we can also observe, in view of Eq. (5-566),

† In other words, we are saying that $\mathbf{y}(t^* + \epsilon\tau) - \mathbf{y}^*(t^*) \approx \epsilon\dot{\mathbf{y}}^*(t^*)\tau$.

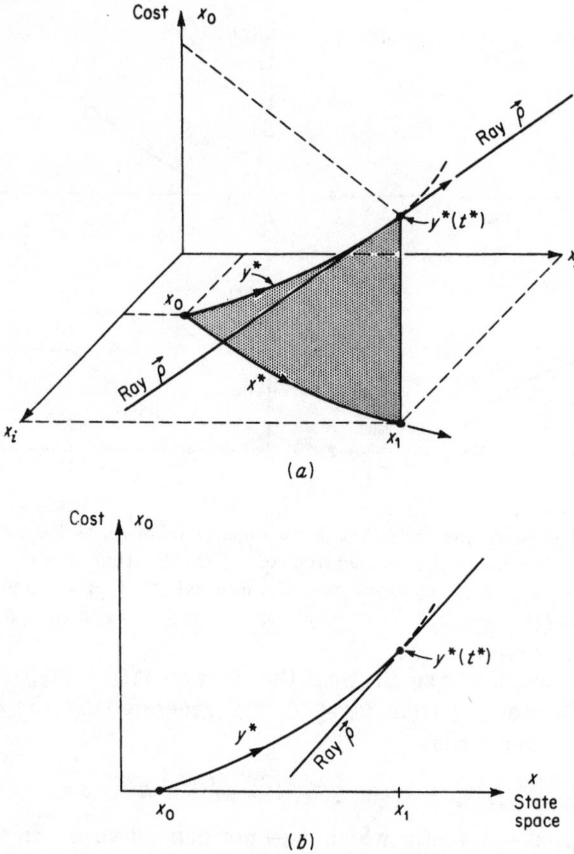


**Fig. 5-26** Illustration of the ray $\vec{\rho}$ generated by the temporal variations on the control. The ray $\vec{\rho}$ passes through the point $\mathbf{y}^*(t^*)$, and its direction is defined by the slope of $\mathbf{y}^*(t)$ at $\mathbf{y}^*(t^*)$.

that $\boldsymbol{\delta}[\tau]$ is a linear function of $\tau$, in that

$$\boldsymbol{\delta}[\alpha_1\tau_1 + \alpha_2\tau_2] = \alpha_1\boldsymbol{\delta}[\tau_1] + \alpha_2\boldsymbol{\delta}[\tau_2] \tag{5-568}$$

for all real numbers $\alpha_1$, $\alpha_2$, $\tau_1$, $\tau_2$. Finally, we can deduce from Eq. (5-566) that, to every vector $\boldsymbol{\delta}$ of the form

$$\boldsymbol{\delta} = \sum_{i=1}^{M} \alpha_i\boldsymbol{\delta}[\tau_i] \qquad M \text{ arbitrary but finite} \tag{5-569}$$

where the $\alpha_i$ are nonnegative and the $\tau_i$ are arbitrary real numbers, there corresponds a perturbation $\mathbf{u}_{\boldsymbol{\delta}}$ of $\mathbf{u}^*$ $\left(\text{in fact, } \mathbf{u}_{\boldsymbol{\delta}} = \mathbf{u}\left[\sum_{i=1}^{M} \alpha_i\tau_i\right]\right)$ such that the vector $\epsilon\boldsymbol{\delta}$ represents the difference (to within the first order in $\epsilon$) between the end point of the perturbed trajectory generated by $\mathbf{u}_\delta$ and the end point of the optimal trajectory.

### 6 *Spatial Variation of the Optimal Control*

The second basic type of variation, which we shall call a *spatial* variation, will be a control that differs from $\mathbf{u}^*$ on a very small interval of time and is *constant* on that interval of time. Physically, such a variation is a small control-energy pulse. More precisely, if $\boldsymbol{\omega}$ is an element of the constraint set $\Omega$ and if $I = (b - \epsilon a, b]$ is a half-open (on the left) subinterval of $[t_0, t^*]$

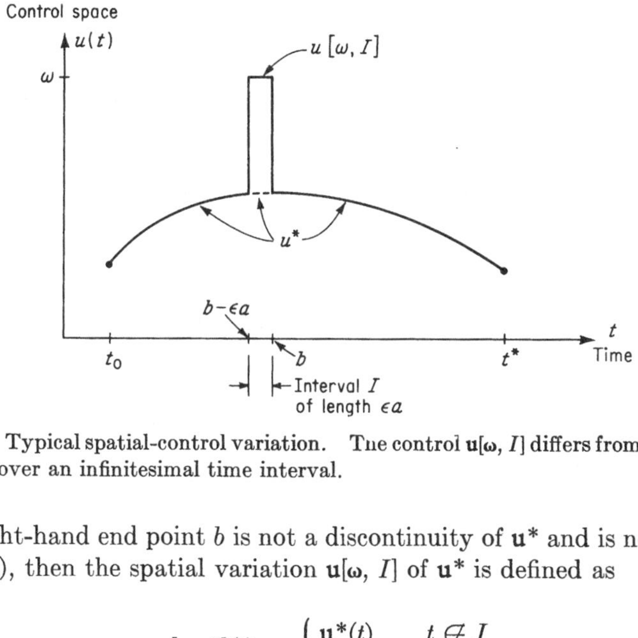


**Fig. 5-27** Typical spatial-control variation. The control $\mathbf{u}[\boldsymbol{\omega}, I]$ differs from the optimal control $\mathbf{u}^*$ over an infinitesimal time interval.

whose right-hand end point $b$ is not a discontinuity of $\mathbf{u}^*$ and is not $t^*$ (that is, $b \neq t^*$), then the spatial variation $\mathbf{u}[\boldsymbol{\omega}, I]$ of $\mathbf{u}^*$ is defined as

$$\mathbf{u}[\boldsymbol{\omega}, I](t) = \begin{cases} \mathbf{u}^*(t) & t \notin I \\ \boldsymbol{\omega} & t \in I \end{cases} \tag{5-570}$$

A typical $\mathbf{u}[\boldsymbol{\omega}, I]$ is illustrated in Fig. 5-27.

Now, what is the effect of applying the control $\mathbf{u}[\boldsymbol{\omega}, I]$ to the system (5-507)? If we denote the solution of the system (5-507), starting from $\mathbf{y}_0 = (0, \mathbf{x}_0)$, generated by the control $\mathbf{u}[\boldsymbol{\omega}, I]$ by $\mathbf{y}(t) = (x_0(t), \mathbf{x}(t))$, then, as is illustrated in Fig. 5-28, $\mathbf{y}(t)$ will coincide with the optimal trajectory for $t_0 \leq t \leq b - \epsilon a$ and will then deviate very slightly from the optimal trajectory. In fact, our differentiability assumptions and the fact that the interval $I$ is small will imply that

$$\mathbf{y}(b) = \mathbf{y}^*(b) + \epsilon\{\dot{\mathbf{y}}(b) - \dot{\mathbf{y}}^*(b)\}a + \mathbf{o}(\epsilon) \tag{5-571}$$

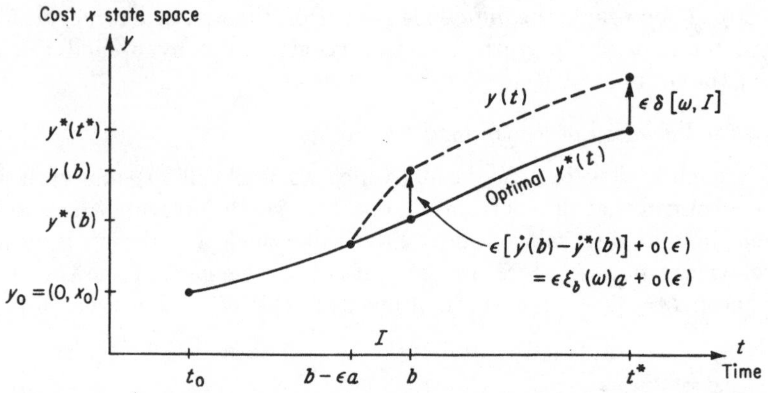


**Fig. 5-28** The effect of a spatial-control variation on the optimal trajectory $\mathbf{y}^*(t)$. At the end of the $I$ interval, the difference between the trajectories is approximately $\epsilon\boldsymbol{\xi}_b(\boldsymbol{\omega})a$. At the terminal time $t^*$, the vector $\epsilon\boldsymbol{\delta}[\boldsymbol{\omega}, I]$ represents the effect of $\epsilon\boldsymbol{\xi}_b(\boldsymbol{\omega})a$ viewed as a perturbation on the "initial" condition at $t = b$.

However, since $\mathbf{y}$ and $\mathbf{y}^*$ are solutions of the system (5-507), we have

$$\dot{\mathbf{y}}(b) = \mathbf{g}[\mathbf{y}(b), \boldsymbol{\omega}] = \begin{bmatrix} L[\mathbf{x}(b), \boldsymbol{\omega}] \\ \mathbf{f}[\mathbf{x}(b), \boldsymbol{\omega}] \end{bmatrix} \tag{5-572}$$

$$\dot{\mathbf{y}}^*(b) = \mathbf{g}[\mathbf{y}^*(b), \mathbf{u}^*(b)] = \begin{bmatrix} L[\mathbf{x}^*(b), \mathbf{u}^*(b)] \\ \mathbf{f}[\mathbf{x}^*(b), \mathbf{u}^*(b)] \end{bmatrix} \tag{5-573}$$

Since $\partial L/\partial \mathbf{x}$ and $\partial \mathbf{f}/\partial \mathbf{x}$ are continuous and since the length $\epsilon a$ of the interval $I$ is small, we have

$$L[\mathbf{x}(b), \boldsymbol{\omega}] = L[\mathbf{x}^*(b), \boldsymbol{\omega}] + \frac{o(\epsilon)}{\epsilon} \tag{5-574}$$

$$\mathbf{f}[\mathbf{x}(b), \boldsymbol{\omega}] = \mathbf{f}[\mathbf{x}^*(b), \boldsymbol{\omega}] + \frac{\mathbf{o}(\epsilon)}{\epsilon} \tag{5-575}$$

and hence, in view of Eq. (5-571), we may deduce that

$$\mathbf{y}(b) = \mathbf{y}^*(b) + \epsilon\{\mathbf{g}[\mathbf{y}^*(b), \boldsymbol{\omega}] - \mathbf{g}[\mathbf{y}^*(b), \mathbf{u}^*(b)]\}a + \mathbf{o}(\epsilon) \tag{5-576}$$

or, equivalently, that

$$x_0(b) = x_0^*(b) + \epsilon\{L[\mathbf{x}^*(b), \boldsymbol{\omega}] - L[\mathbf{x}^*(b), \mathbf{u}^*(b)]\}a + o(\epsilon) \tag{5-577}$$
$$\mathbf{x}(b) = \mathbf{x}^*(b) + \epsilon\{\mathbf{f}[\mathbf{x}^*(b), \boldsymbol{\omega}] - \mathbf{f}[\mathbf{x}^*(b), \mathbf{u}^*(b)]\}a + \mathbf{o}(\epsilon) \tag{5-578}$$

The situation at $b$ is illustrated in Fig. 5-28.

Now the remaining portion of our perturbed trajectory $\mathbf{y}(t)$ is generated by applying the control $\mathbf{u}^*_{(b,\, t^*]}$ to the system, *starting from the point* $\mathbf{y}(b)$. Recalling our discussion of the homogeneous linear system (5-542) and, in particular, the remarks leading to Eq. (5-548), we can see that if we let

$\xi_b(\omega)$ be the vector given by

$$\xi_b(\omega) = \mathbf{g}[\mathbf{y}^*(b), \omega] - \mathbf{g}[\mathbf{y}^*(b), \mathbf{u}^*(b)] \tag{5-579}$$

so that Eq. (5-576) becomes

$$\mathbf{y}(b) = \mathbf{y}^*(b) + \epsilon\xi_b(\omega)a + \mathbf{o}(\epsilon) \tag{5-580}$$

then the trajectory $\mathbf{y}(t)$ for $t > b$ is given by

$$\mathbf{y}(t) = \mathbf{y}^*(t) + \epsilon\mathbf{\Phi}_*(t, b)\xi_b(\omega)a + \mathbf{o}(\epsilon) \tag{5-581}$$

where $\mathbf{\Phi}_*$ is the $n + 1 \times n + 1$ fundamental matrix of the system (5-542). In particular, we have

$$\mathbf{y}(t^*) = \mathbf{y}^*(t^*) + \epsilon\mathbf{\Phi}_*(t^*, b)\xi_b(\omega)a + \mathbf{o}(\epsilon) \tag{5-582}$$

and so, letting $\delta[\omega, I]$ denote the vector $\mathbf{\Phi}_*(t^*, b)\xi_b(\omega)a$, that is, letting

$$\delta[\omega, I] = \mathbf{\Phi}_*(t^*, b)\xi_b(\omega)a = \mathbf{\Phi}_*(t^*, b)\begin{bmatrix} L[\mathbf{x}^*(b), \omega] - L[\mathbf{x}^*(b), \mathbf{u}^*(b)] \\ \mathbf{f}[\mathbf{x}^*(b), \omega] - \mathbf{f}[\mathbf{x}^*(b), \mathbf{u}^*(b)] \end{bmatrix} a \tag{5-583}$$

we can see that $\mathbf{u}[\omega, I]$ generates a trajectory with the property that

$$\mathbf{y}(t^*) = \mathbf{y}^*(t^*) + \epsilon\delta[\omega, I] + \mathbf{o}(\epsilon) \tag{5-584}$$

where the $n + 1$ vector $\delta[\omega, I]$ does not depend on $\epsilon$.

**Exercise 5-18** Verify Eq. (5-580). HINT: $L[\mathbf{x}(b), \omega] = L[\mathbf{x}^*(b), \omega] + \langle \partial L[\mathbf{x}^*(b), \omega]/\partial\mathbf{x}, \mathbf{x}(b) - \mathbf{x}^*(b)\rangle$ + higher-order terms in $\mathbf{x}(b) - \mathbf{x}^*(b)$. Show that $\|\mathbf{x}(b) - \mathbf{x}^*(b)\| = o(\epsilon)$, as the interval $I$ is small.

We shall view the vector $\delta[\omega, I]$ (or $\epsilon\delta[\omega, I]$) as being "attached" to the point $\mathbf{y}^*(t^*)$, as is illustrated in Figs. 5-28 and 5-29. We can see immediately from Eq. (5-583) that the "direction" of the vector $\delta[\omega, I]$ does not depend on the length of the interval $I$. In other words, if (say) $I'$ is the interval $(b - \epsilon a', b]$, then the vectors $\delta[\omega, I]$ and $\delta[\omega, I']$ lie on the same ray through the point $\mathbf{y}^*(t^*)$ and differ only in magnitude. This ray depends only upon the point $b$ and the vector $\omega$, and we shall denote it by $\vec{\rho}\,[\omega, b]$. Thus, $\vec{\rho}\,[\omega, b]$ is [see Eq. (3-22$a$)] the set given by

$$\vec{\rho}\,[\omega, b] = \{\mathbf{y}: \mathbf{y} = \mathbf{y}^*(t^*) + \beta\delta[\omega, I], \beta \geq 0\} \tag{5-585}$$

or, equivalently, by

$$\vec{\rho}\,[\omega, b] = \{\mathbf{y}: \mathbf{y} = \mathbf{y}^*(t^*) + \beta\mathbf{\Phi}_*(t^*, b)\xi_b(\omega), \beta \geq 0\} \tag{5-586}$$

where we recall that

$$\xi_b(\omega) = \begin{bmatrix} L[\mathbf{x}^*(b), \omega] - L[\mathbf{x}^*(b), \mathbf{u}^*(b)] \\ \mathbf{f}[\mathbf{x}^*(b), \omega] - \mathbf{f}[\mathbf{x}^*(b), \mathbf{u}^*(b)] \end{bmatrix} \tag{5-587}$$

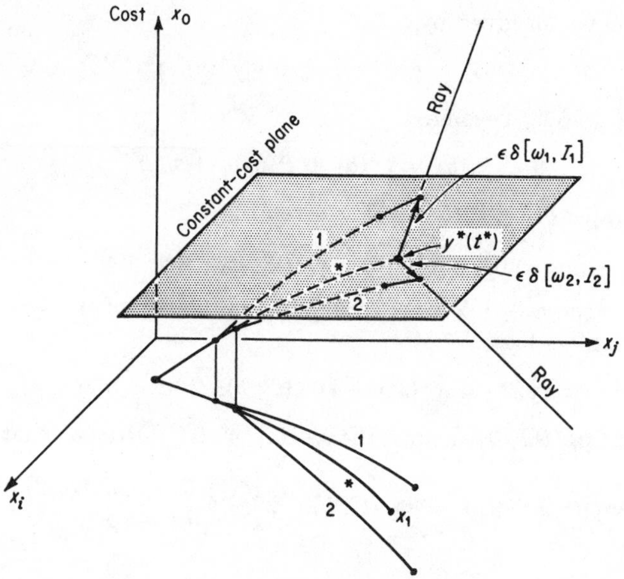


**Fig. 5-29** The effect of two distinct spatial perturbations. The * trajectory is the optimum. The 1-trajectory is due to a perturbation $\mathbf{u}[\omega_1, I_1]$ and generates the vector $\epsilon\boldsymbol{\delta}[\omega_1, I_1]$ at $t = t^*$. The 2-trajectory is due to a perturbation $\mathbf{u}[\omega_2, I_2]$ and generates the vector $\epsilon\boldsymbol{\delta}[\omega_2, I_2]$. The directions of $\boldsymbol{\delta}[\omega_1, I_1]$ and $\boldsymbol{\delta}[\omega_2, I_2]$ define the two indicated rays. Note that both rays originate at $\mathbf{y}^*(t^*)$.

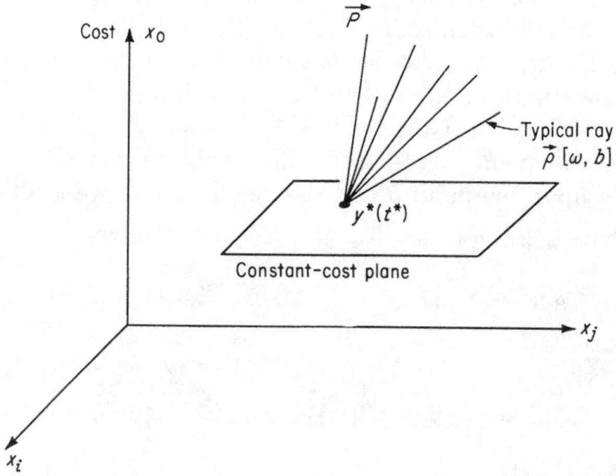


**Fig. 5-30** A typical cone $\vec{P}$ generated by spatial-control variations. The point $\mathbf{y}^*(t^*)$ is the vertex of the cone $\vec{P}$.

Let us denote the cone with vertex $\mathbf{y}^*(t^*)$ formed by all these rays by $\vec{P}$. In other words, $\vec{P}$ is the set of all vectors $\mathbf{y}$ for which there are an $\omega$ in $\Omega$ and a point $b$ in $[t_0, t^*)$ which is not a discontinuity of $\mathbf{u}^*$, such that $\mathbf{y}$ lies on the ray $\vec{\rho}\,[\omega, b]$. We observe that, as is illustrated in Fig. 5-30, the cone $\vec{P}$ need not be convex,† so let us "convexify" it. We do this by considering the convex cover (see Definition 3-21) $\text{co}(\vec{P})$ of $\vec{P}$; that is, we consider the set of all convex combinations of elements of $\vec{P}$ (see Definition 3-20). In other words, $\mathbf{y}$ is an element of $\text{co}(\vec{P})$ if and only if there are real numbers $r_1, r_2, \ldots r_N$ ($N$ may depend on $\mathbf{y}$), with $r_j \geq 0$ for all $j$ and with

$$\sum_{j=1}^{N} r_j = 1$$

and vectors $\mathbf{y}_1, \mathbf{y}_2, \ldots \mathbf{y}_N$ in $\vec{P}$ such that

$$\mathbf{y} = \sum_{j=1}^{N} r_j \mathbf{y}_j \tag{5-588}$$

It is easy to see that the set $\text{co}(\vec{P})$ is a convex cone with vertex $\mathbf{y}^*(t^*)$.

**Exercise 5-19** Show that $\text{co}(\vec{P})$ is a cone with vertex $\mathbf{y}^*(t^*)$. HINT: Suppose that $\mathbf{y} = r\mathbf{y}_1 + s\mathbf{y}_2$, $r, s \geq 0$, $r + s = 1$, and $\mathbf{y}_1, \mathbf{y}_2 \in \vec{P}$. Then $\mathbf{y}_1 = \mathbf{y}^*(t^*) + \beta_1 \boldsymbol{\delta}[\omega_1, I_1]$, $\mathbf{y}_2 = \mathbf{y}^*(t^*) + \beta_2 \boldsymbol{\delta}[\omega_2, I_2]$. We must show that every point on the ray joining $\mathbf{y}$ and $\mathbf{y}^*(t^*)$ is in $\text{co}(\vec{P})$. If $\mathbf{z}$ is on this ray, then $\mathbf{z} = \mathbf{y}^*(t^*) + \beta[\mathbf{y} - \mathbf{y}^*(t^*)]$, $\beta \geq 0$. Show that $\mathbf{z} = \mathbf{y}^*(t^*) + \beta r \beta_1 \boldsymbol{\delta}[\omega_1, I_1] + \beta s \beta_2 \boldsymbol{\delta}[\omega_2, I_2]$. Then use induction.

If we suppose now that $\mathbf{y}$ is an element of $\text{co}(\vec{P})$, then we may write

$$\mathbf{y} = \mathbf{y}^*(t^*) + \epsilon \boldsymbol{\delta} \tag{5-589}$$

where

$$\boldsymbol{\delta} = \sum_{j=1}^{N} \beta_j \boldsymbol{\delta}[\omega_j, I_j] \qquad \beta_j \geq 0 \tag{5-590}$$

and where we view the vector $\boldsymbol{\delta}$ as being "attached" to the point $\mathbf{y}^*(t^*)$. We remark that it can be shown that there is an admissible control $\mathbf{u}_\delta$ such that the trajectory $\mathbf{y}_\delta$ of (5-507), starting from $\mathbf{y}_0 = (0, \mathbf{x}_0)$, generated by $\mathbf{u}_\delta$ has the property that

$$\mathbf{y}_\delta(t^*) = \mathbf{y} + o(\epsilon) \tag{5-591}$$

$$= \mathbf{y}^*(t^*) + \epsilon \boldsymbol{\delta} + o(\epsilon) \tag{5-592}$$

$$= \mathbf{y}^*(t^*) + \epsilon \sum_{j=1}^{N} \beta_j \boldsymbol{\delta}[\omega_j, I_j] + o(\epsilon) \tag{5-593}$$

† This is due to the arbitrary character of $\Omega$ and the nonlinearity of our system. For example, $\Omega$ might consist only of isolated points, in which case the cone $\vec{P}$ might consist only of isolated rays.

The formal proof of this fact appears in Ref. P-5 and requires the introduction of the notation used to describe the control perturbations in Ref. P-5. However, we can see intuitively that, in view of the fact that the system (5-542) is linear and homogeneous, the control $\mathbf{u}_\delta$ will differ from $\mathbf{u}^*$ on a

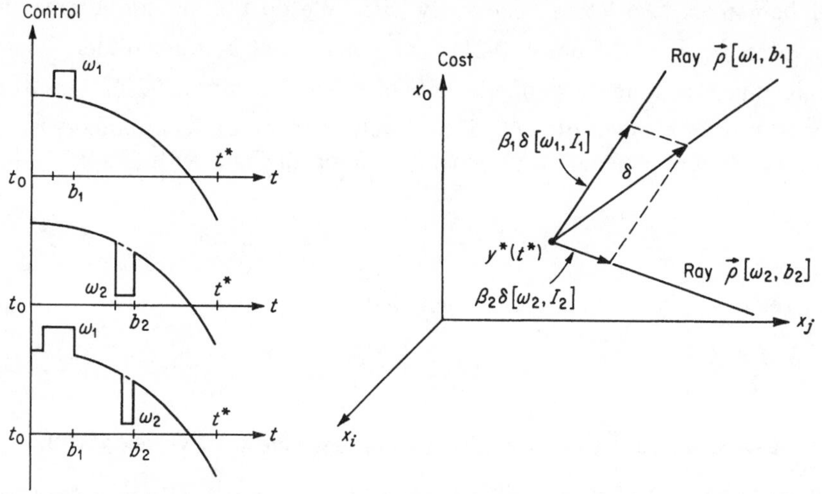


**Fig. 5-31** The construction of the vector $\boldsymbol{\delta}$ [see Eq. (5-594)] from two elements of the rays $\vec{\rho}[\omega_1, b_1]$ and $\vec{\rho}[\omega_2, b_2]$. The bottom control can be used to generate the ray defined by $\boldsymbol{\delta}$ [see Eq. (5-595)].

set of very small subintervals of $[t_0, t^*]$ and will be constant on these subintervals. For example, as is illustrated in Fig. 5-31, if we suppose that

$$\boldsymbol{\delta} = \beta_1\boldsymbol{\delta}[\boldsymbol{\omega}_1, I_1] + \beta_2\boldsymbol{\delta}[\boldsymbol{\omega}_2, I_2] \tag{5-594}$$

where the intervals $I_1$ and $I_2$ are disjoint, then, letting $I_1 = (b_1 - \epsilon a_1, b_1]$ and $I_2 = (b_2 - \epsilon a_2, b_2]$ and supposing that the intervals $I_1' = (b_1 - \epsilon\beta_1 a_1, b_1]$ and $I_2' = (b_2 - \epsilon\beta_2 a_2, b_2]$ are disjoint subintervals of $[t_0, t^*]$, we can readily show that the control $\mathbf{u}_\delta$, defined by setting

$$\mathbf{u}_\delta(t) = \begin{cases} \mathbf{u}^*(t) & t \notin I_1' \cup I_2' \\ \boldsymbol{\omega}_1 & t \in I_1' \\ \boldsymbol{\omega}_2 & t \in I_2' \end{cases} \tag{5-595}$$

will generate a trajectory $\mathbf{y}_\delta(t)$ such that

$$\mathbf{y}_\delta(t^*) = \mathbf{y}^*(t^*) + \epsilon\boldsymbol{\delta} + \mathbf{o}(\epsilon) \tag{5-596}$$

We can also readily verify that the ray of $\text{co}(\vec{P})$ on which $\boldsymbol{\delta}$ lies is given by

$$\{\mathbf{y}: \mathbf{y} = \mathbf{y}^*(t^*) + \epsilon\beta[\boldsymbol{\Phi}_*(t^*, b_1)\boldsymbol{\xi}_{b_1}(\boldsymbol{\omega}_1) + \boldsymbol{\Phi}_*(t^*, b_2)\boldsymbol{\xi}_{b_2}(\boldsymbol{\omega}_2)],\ \beta \geq 0\} \tag{5-597}$$

where the vectors $\boldsymbol{\xi}_{b_i}(\omega_i)$ are defined by Eq. (5-579) and $\boldsymbol{\Phi}_*$ is the fundamental matrix of the system (5-542).

**Exercise 5-20** Show that the control $\mathbf{u}_\delta$ of Eq. (5-595) generates a trajectory $\mathbf{y}_\delta$ satisfying Eq. (5-596). HINT: Use Eq. (5-548) and the transition property of the fundamental matrix.

## 7 *The Terminal Cone*

We shall now combine the effects of the temporal and spatial variations of $\mathbf{u}^*$ in order to construct a convex cone with vertex $\mathbf{y}^*(t^*)$ which will play an important role in our "proof" of the minimum principle.

Let us recall that the vectors $\epsilon\boldsymbol{\delta}[\tau]$ arising from temporal variations of $\mathbf{u}^*$ all lie on a line passing through the point $\mathbf{y}^*(t^*)$. We denote the set of all these vectors $\epsilon\boldsymbol{\delta}[\tau]$ by $\vec{\rho}$, and we claim that the set (see Fig. 5-32)

$$\vec{\rho} + \text{co}(\vec{P}) = \{\mathbf{y}: \mathbf{y} = \mathbf{y}_1 + \mathbf{y}_2,\ \mathbf{y}_1 \in \text{co}(\vec{P}),\ \mathbf{y}_2 \in \vec{\rho}\} \tag{5-598}$$

$$= \left\{\mathbf{y}: \mathbf{y} = \mathbf{y}^*(t^*) + \epsilon \sum_{i=1}^{M} \alpha_i\boldsymbol{\delta}[\tau_i] + \epsilon \sum_{j=1}^{N} \beta_j\boldsymbol{\delta}[\omega_j, I_j] \qquad \alpha_i, \beta_j \geq 0\right\} \tag{5-599}$$

is a convex cone with vertex $\mathbf{y}^*(t^*)$. Since $\vec{\rho}$ is convex by Eq. (5-568) and since $\text{co}(\vec{P})$ is convex by definition, we can see immediately that the set $\vec{\rho} + \text{co}(\vec{P})$ is convex. On the other hand, if

$$\mathbf{y}^*(t^*) + \epsilon \sum_{i=1}^{M} \alpha_i\boldsymbol{\delta}[\tau_i] + \epsilon \sum_{j=1}^{N} \beta_j\boldsymbol{\delta}[\omega_j, I_j]$$

is an element of $\vec{\rho} + \text{co}(\vec{P})$, then a typical element of the ray joining this point to $\mathbf{y}^*(t^*)$ is of the form

$$\mathbf{y}^*(t^*) + \gamma \left\{\mathbf{y}^*(t^*) + \epsilon \sum_{i=1}^{M} \alpha_i\boldsymbol{\delta}[\tau_i] + \epsilon \sum_{j=1}^{N} \beta_j\boldsymbol{\delta}[\omega_j, I_j] - \mathbf{y}^*(t^*)\right\} \qquad \gamma \geq 0 \tag{5-600}$$

or, equivalently, of the form

$$\mathbf{y}^*(t^*) + \epsilon\gamma \sum_{i=1}^{M} \alpha_i\boldsymbol{\delta}[\tau_i] + \epsilon\gamma \sum_{j=1}^{N} \beta_j\boldsymbol{\delta}[\omega_j, I_j] \qquad \gamma \geq 0 \tag{5-601}$$

and hence is an element of $\vec{\rho} + \text{co}(\vec{P})$.

Thus, *the set* $\vec{\rho} + \text{co}(\vec{P})$ *is a convex cone with vertex* $\mathbf{y}^*(t^*)$, *and we denote it by* $C_{t^*}$, *so that*

$$C_{t^*} = \vec{\rho} + \text{co}(\vec{P}) \tag{5-602}$$

The convex cone $C_{t^*}$ is illustrated in Fig. 5-32. The set $C_{t^*}$ in Fig. 5-32 is the set of all vectors emanating from $\mathbf{y}^*(t^*)$ which lie between the two (half) hyperplanes $A$ and $B$, which both contain the ray $\vec{\rho}$.

If we suppose that $\mathbf{y}$ is an element of $C_{t^*}$ then we may write

$$\mathbf{y} = \mathbf{y}^*(t^*) + \epsilon\boldsymbol{\delta} \tag{5-603}$$

where

$$\boldsymbol{\delta} = \sum_{i=1}^{M} \alpha_i \boldsymbol{\delta}[\tau_i] + \sum_{j=1}^{N} \beta_j \boldsymbol{\delta}[\omega_j, I_j] \qquad \alpha_i, \beta_j \geq 0 \tag{5-604}$$

and where we view the vector $\boldsymbol{\delta}$ (or $\epsilon\boldsymbol{\delta}$) as being "attached" to the point $\mathbf{y}^*(t^*)$. We now remark that it can be shown that there is an admissible

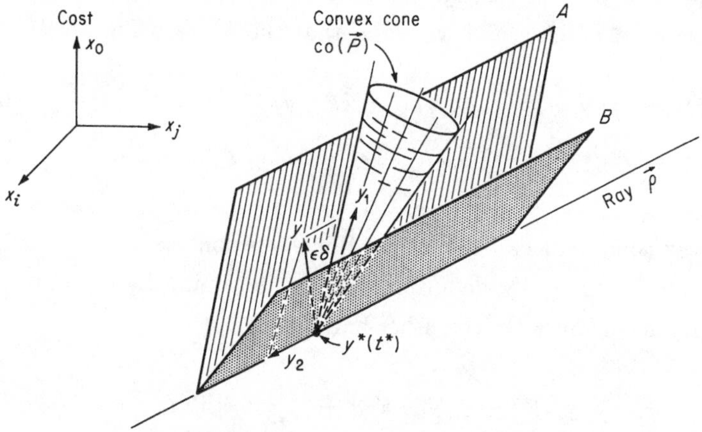


**Fig. 5-32** The generation of the convex cone $\vec{\rho} + \text{co}(\vec{P}) = C_{t^*}$. In this figure, the ray $\vec{\rho}$ is due to the temporal variations (see Fig. 5-26) and the convex cone co$(\vec{P})$ is due to the spatial variations (see Fig. 5-30). The convex cone $C_{t^*}$ with vertex $\mathbf{y}^*(t^*)$ is constructed by linear combinations of elements of $\vec{\rho}$ (such as $\mathbf{y}_2$) and of elements of co$(\vec{P})$ (such as $\mathbf{y}_1$). We shall draw the set $C_{t^*}$ using the two (half) hyperplanes $A$ and $B$ which contain the ray $\vec{\rho}$. Thus, every point in the "trench" defined by $A$ and $B$ is an element of $C_{t^*}$.

control $\mathbf{u}_\delta$ such that the trajectory $\mathbf{y}_\delta$ of (5-507), starting from $\mathbf{y}_0 = (0, \mathbf{x}_0)$, generated by $\mathbf{u}_\delta$ has the property that

$$\mathbf{y}_\delta(t_f) = \mathbf{y} + \mathbf{o}(\epsilon) \tag{5-605}$$
$$= \mathbf{y}^*(t^*) + \epsilon\boldsymbol{\delta} + \mathbf{o}(\epsilon) \tag{5-606}$$

where $t_f$ is the terminal time for the control $\mathbf{u}_\delta$. The formal proof of this fact appears in Ref. P-5. However, we can see intuitively that the control $\mathbf{u}_\delta$ will involve a small change in the terminal time $t^*$ and will also differ from $\mathbf{u}^*$ on a set of very small subintervals of $[t_0, t^*]$ and will be constant on those subintervals. For example, if we suppose that

$$\boldsymbol{\delta} = \boldsymbol{\delta}[\tau] + \boldsymbol{\delta}[\boldsymbol{\omega}, I] \tag{5-607}$$

where $\tau > 0$ and $I = (b - \epsilon a, b]$, then we can readily show that the control $\mathbf{u}_\delta$, defined by setting

$$\mathbf{u}_\delta(t) = \begin{cases} \boldsymbol{\omega} & t \in I \\ \mathbf{u}^*(t) & t \notin I \text{ and } t_0 \leq t < t^* \\ \mathbf{u}^*(t^*) & t^* < t \leq t^* + \epsilon\tau \end{cases} \tag{5-608}$$

will generate a trajectory $\mathbf{y}_\delta(t)$ such that

$$\mathbf{y}_\delta(t^* + \epsilon\tau) = \mathbf{y}^*(t^*) + \epsilon\boldsymbol{\delta} + \mathbf{o}(\epsilon) \tag{5-609}$$

The perturbation $\mathbf{u}_\delta$ is illustrated in Fig. 5-33.

In view of our assertion that any portion of an optimal trajectory is also an optimal trajectory, we can see that if $t$ is any element of $(t_0, t^*]$, then a

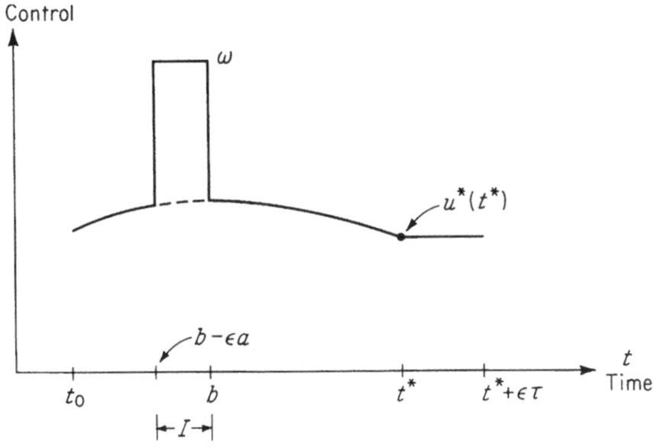


**Fig. 5-33** This control represents a spatial and temporal variation [see Eq. (5-608)]. Such controls will generate elements of the convex cone $C_{t^*}$ illustrated in Fig. 5-32.

convex cone $C_t$ can be defined in a manner entirely similar to that used to define $C_{t^*}$. We shall call the cones $C_t$, for $t$ in $(t_0, t^*]$, *cones of attainability.* If we denote the closure of $C_t$ by $\bar{C}_t$ (see Definition 3-9), then we can readily verify that the set $\bar{C}_t$ is again a convex cone with $\mathbf{y}^*(t)$ as vertex. We also observe that it can be shown that, for $\hat{t} \geq t$,

$$\boldsymbol{\Phi}_*(\hat{t}, t)\bar{C}_t \subset \bar{C}_{\hat{t}} \tag{5-610}†$$

and, in particular, that

$$\boldsymbol{\Phi}_*(t^*, t)\bar{C}_t \subset \bar{C}_{t^*} \tag{5-611}$$

† Usually, these relations will be equalities; however, only the indicated inclusion relationship is needed in the proof of the minimum principle.

It follows from Eq. (5-611) and from the fact that $\boldsymbol{\Phi}_*$ is the fundamental matrix of a linear system [Eq. (5-542)] that

$$\boldsymbol{\Phi}_*(t, t^*)\boldsymbol{\Phi}_*(t^*, t)\bar{C}_t \subset \boldsymbol{\Phi}_*(t, t^*)\bar{C}_{t^*} \tag{5-612}$$

and that

$$\bar{C}_t \subset \boldsymbol{\Phi}_*(t, t^*)\bar{C}_{t^*} \tag{5-613}$$

This relation (5-613) can be used to show that the cones of attainability will always lie on the same side of the moving hyperplane determined by the costate variables.

## *8 Proof of Theorem 5-5P*

We are now ready to "prove" Theorem 5-5P. We begin by asserting that the ray in the direction of decreasing cost emanating from $\mathbf{y}^*(t^*)$ does not meet the interior (see Definition 3-8) of the cone $C_{t^*}$. More precisely, we claim that if $\boldsymbol{\mu}$ is the $n + 1$ vector given by

$$\boldsymbol{\mu} = \begin{bmatrix} -1 \\ 0 \\ 0 \\ \cdot \\ \cdot \\ \cdot \\ 0 \end{bmatrix} \tag{5-614}$$

then the ray $\vec{\mu}$, given by

$$\vec{\mu} = \{\mathbf{y}: \mathbf{y} = \mathbf{y}^*(t^*) + \epsilon\beta\boldsymbol{\mu},\ \beta > 0\} \tag{5-615}$$

does not meet the interior of the cone $C_{t^*}$. The situation is illustrated in Fig. 5-34.

We shall verify this claim about $\vec{\mu}$ by an indirect argument. So let us suppose that the vector

$$\hat{\mathbf{y}} = \mathbf{y}^*(t^*) + \epsilon\hat{\beta}\boldsymbol{\mu} \qquad \hat{\beta} > 0 \tag{5-616}$$

is an interior point (see Definition 3-4) of $C_{t^*}$. This implies that there is a perturbation $\hat{\mathbf{u}}$ of $\mathbf{u}^*$ which generates a trajectory $\hat{\mathbf{y}}(t)$ of (5-507) over the interval $[t_0, \hat{t}]$ with the property that

$$\hat{\mathbf{y}}(\hat{t}) = \mathbf{y}^*(t^*) + \epsilon\hat{\beta}\boldsymbol{\mu} + \mathbf{o}(\epsilon) \tag{5-617}$$

Now, recalling that

$$\hat{\mathbf{y}}(\hat{t}) = \begin{bmatrix} J(\hat{\mathbf{u}}) \\ \hat{\mathbf{x}}(\hat{t}) \end{bmatrix} \qquad \mathbf{y}^*(t^*) = \begin{bmatrix} J^* \\ \mathbf{x}_1 \end{bmatrix} \tag{5-618}$$

and that $\boldsymbol{\mu}$ is given by Eq. (5-614), we can see that Eq. (5-617) may be written in the form

$$J(\hat{\mathbf{u}}) = J^* - \epsilon\hat{\beta} + o(\epsilon) \qquad \hat{\mathbf{x}}(\hat{t}) = \mathbf{x}_1 + \mathbf{o}(\epsilon) \tag{5-619}$$

However, for suitably small $\epsilon$, the control $\hat{\mathbf{u}}$ will not take us out of our "optimal" tube, and so the control $\mathbf{u}^*$ is optimal to within the first order

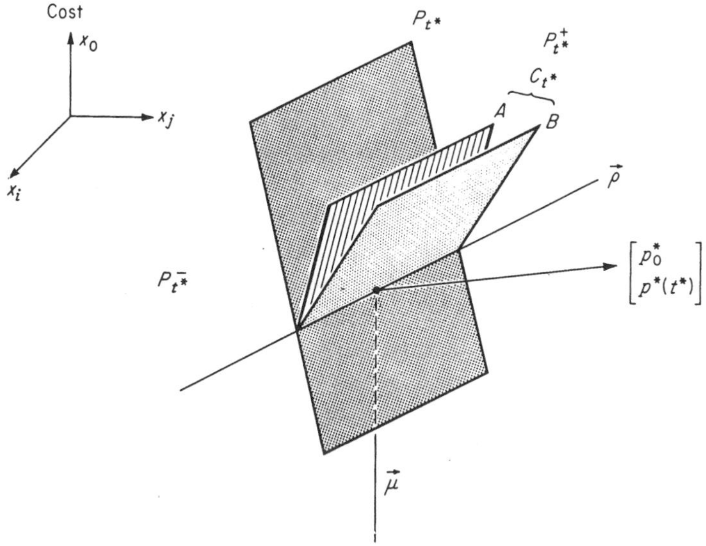


**Fig. 5-34** The hyperplane $P_{t^*}$ shown is a typical support hyperplane to the convex cone $C_{t^*}$, and it separates $C_{t^*}$ and the ray $\vec{\mu}$. Since the ray $\vec{\rho}$ is an element of $C_{t^*}$, the hyperplane $P_{t^*}$ contains the ray $\vec{\rho}$. The vector $\begin{bmatrix} p_0^* \\ \mathbf{p}^*(t^*) \end{bmatrix}$ is normal to $P_t^*$ at the point $\mathbf{y}^*(t^*)$. The convex cone $C_{t^*}$ is on one side $(P_{t^*}{}^+)$ of $P_{t^*}$, and the ray $\vec{\mu}$ is on the other side $(P_{t^*}{}^-)$. The ray $\vec{\mu}$ points in the direction of decreasing cost.

in $\epsilon$. On the other hand, according to Eq. (5-619), the control $\hat{\mathbf{u}}$ leads to *a reduction of order $\epsilon$ in the cost and to a higher-order (in $\epsilon$) change in the terminal point of the system trajectory in $R_n$.* Thus, if $\epsilon$ is very small, we shall obtain a contradiction in that the trajectory $\hat{\mathbf{y}}(t)$ will lie "below" the trajectory $\mathbf{y}^*(t)$ (to within the terms of higher order).† It follows that the vector $\hat{\mathbf{y}}$ cannot be an interior point of $C_{t^*}$. Thus, the ray $\vec{\mu}$ does not meet the interior of $C_{t^*}$.

Since the cone $C_{t^*}$ is convex, its interior $i(C_{t^*})$ is an open convex set in $R_{n+1}$. As the ray $\vec{\mu}$ is a convex set which does not meet the interior $i(C_{t^*})$

† The detailed proof of this key point appears in Ref. P-5 and involves some complicated topological arguments. See, in particular, lemmas 3 and 4 on pp. 94–99 of Ref. P-5. We also note that the relation (5-613) plays a role here.

of $C_{t^*}$, we may deduce from Theorem 3-5 that there is a support hyperplane $P_{t^*}$ of $C_{t^*}$ which separates $C_{t^*}$ and the ray $\vec{\mu}$.† The hyperplane $P_{t^*}$, which is illustrated in Fig. 5-34, passes through the point $\mathbf{y}^*(t^*)$ and has the property that the ray $\vec{\mu}$ lies in the half-space $P_{t^*}^-$ while the cone $C_{t^*}$ is contained in the half-space $P_{t^*}^+$ (see Sec. 3-7). In other words, there is a nonzero $n+1$ vector which we denote by $\begin{bmatrix} p_0^* \\ \mathbf{p}^*(t^*) \end{bmatrix}$, such that the hyperplane $P_{t^*}$ is defined by the equation

$$\left\langle \begin{bmatrix} p_0^* \\ \mathbf{p}^*(t^*) \end{bmatrix}, \mathbf{y} - \mathbf{y}^*(t^*) \right\rangle = 0 \tag{5-620}$$

and, moreover, the following conditions are satisfied:

$$\left\langle \begin{bmatrix} p_0^* \\ \mathbf{p}^*(t^*) \end{bmatrix}, \boldsymbol{\mu} \right\rangle \leq 0 \tag{5-621}$$

$$\left\langle \begin{bmatrix} p_0^* \\ \mathbf{p}^*(t^*) \end{bmatrix}, \boldsymbol{\delta} \right\rangle \geq 0 \tag{5-622}$$

where $\boldsymbol{\mu}$ is (the direction) given by Eq. (5-614) and $\boldsymbol{\delta}$ is any element of the cone $C_{t^*}$. We observe that the vector $\begin{bmatrix} p_0^* \\ \mathbf{p}^*(t^*) \end{bmatrix}$ is normal to the hyperplane $P_{t^*}$ at $\mathbf{y}^*(t^*)$ and that Eq. (5-621) implies that

$$p_0^* \geq 0 \tag{5-623}$$‡

since the vector $\boldsymbol{\mu}$ is given by Eq. (5-614).

We are now going to determine the costate variables on the basis of our discussion of moving hyperplanes in subsection 4. We recall that the system (5-542) is linear and homogeneous and is of the form

$$\dot{\boldsymbol{\psi}}(t) = \mathbf{A}_*(t)\boldsymbol{\psi}(t) \tag{5-624}$$

where $\mathbf{A}_*(t)$ is the $n+1 \times n+1$ matrix given by [see Eq. (5-541)]

$$\mathbf{A}_*(t) = \left[\begin{array}{c:cccc} 0 & \left.\frac{\partial L}{\partial x_1}\right|_* & \left.\frac{\partial L}{\partial x_2}\right|_* & \cdots & \left.\frac{\partial L}{\partial x_n}\right|_* \\ \hdashline \mathbf{0} & & \left(\left.\frac{\partial \mathbf{f}}{\partial \mathbf{x}}\right|_*\right) & & \end{array}\right] \tag{5-625}$$

† According to Theorem 3-5, there is at least one such hyperplane. There may, in fact, be many, and our arguments will apply equally well to any of them. Strictly speaking, this argument is valid only if the interior of $C_{t^*}$ is nonempty. If, however, the interior of $C_{t^*}$ is empty, then the desired result can be obtained by considering the rays $\vec{\mu}(\alpha) = \{\mathbf{y}: \mathbf{y} = \mathbf{y}^*(t^*) + \epsilon\beta\boldsymbol{\mu}, \beta \geq \alpha\}$, $\alpha > 0$, which do not meet $C_{t^*}$ and applying a limiting argument or by using the notions of the footnote on pp. 93 and 94 of Ref. P-5.

‡ We note that if $p_0^* = 0$, then the hyperplane $P_{t^*}$ will contain the ray $\vec{\mu}$ and our problem will be, in a sense, independent of the cost.

This system describes the motion of a vector along the optimal trajectory. We recall also that the system (5-553) which is the adjoint system of (5-624) is also linear and homogeneous and is of the form

$$\begin{bmatrix} \dot{p}_0(t) \\ \dot{\mathbf{p}}(t) \end{bmatrix} = -\mathbf{A}'_*(t) \begin{bmatrix} p_0(t) \\ \mathbf{p}(t) \end{bmatrix} \tag{5-626}$$

We have noted in subsection 4 that this system represents the motion of a hyperplane along the optimal trajectory, and we shall use this fact to determine the desired costate. For convenience, we let $\mathbf{\Phi}_*(t, t_0)$ denote (as before) the fundamental matrix of the system (5-624), and we let $\mathbf{\Psi}_*(t, t_0)$ denote the fundamental matrix of the adjoint system (5-626).

We now construct the costate by finding the hyperplane whose motion along the optimal trajectory leads to $P_{t^*}$. Since the matrix $\mathbf{\Psi}_*(t^*, t_0)$ is nonsingular, we know that there is a nonzero $n + 1$ vector $\begin{bmatrix} \pi_0 \\ \boldsymbol{\pi} \end{bmatrix}$ such that

$$\mathbf{\Psi}_*(t^*, t_0) \begin{bmatrix} \pi_0 \\ \boldsymbol{\pi} \end{bmatrix} = \begin{bmatrix} p_0^* \\ \mathbf{p}^*(t^*) \end{bmatrix} \tag{5-627}$$

If we define an $(n + 1)$-vector-valued function $\begin{bmatrix} p_0^*(t) \\ \mathbf{p}^*(t) \end{bmatrix}$ by setting

$$\begin{bmatrix} p_0^*(t) \\ \mathbf{p}^*(t) \end{bmatrix} = \mathbf{\Psi}_*(t, t_0) \begin{bmatrix} \pi_0 \\ \boldsymbol{\pi} \end{bmatrix} \qquad t \in [t_0, t^*] \tag{5-628}$$

then we can immediately conclude from Eq. (5-554) that

$$p_0^*(t) = p_0^* = \pi_0 \qquad \text{for all } t \tag{5-629}$$

and from Eq. (5-555) that

$$\dot{\mathbf{p}}^*(t) = -\left.\frac{\partial L}{\partial \mathbf{x}}\right|_* p_0^* - \left(\left.\frac{\partial \mathbf{f}}{\partial \mathbf{x}}\right|_*\right)' \mathbf{p}^*(t) \tag{5-630}$$

Since the Hamiltonian $H(\mathbf{x}, \mathbf{p}, \mathbf{u}, p_0)$ is given by

$$H(\mathbf{x}, \mathbf{p}, \mathbf{u}, p_0) = p_0 L(\mathbf{x}, \mathbf{u}) + \langle \mathbf{p}, \mathbf{f}(\mathbf{x}, \mathbf{u}) \rangle \tag{5-631}$$

we can see that Eq. (5-630) may be written in the form

$$\dot{\mathbf{p}}^*(t) = -\frac{\partial H}{\partial \mathbf{x}} [\mathbf{x}^*(t), \mathbf{p}^*(t), \mathbf{u}^*(t), p_0^*] \tag{5-632}$$

We shall now show that $p_0^*$ and $\mathbf{p}^*(t)$ satisfy the requirements of Theorem 5-5P.

We begin by showing that $\mathbf{u}^*$ minimizes the Hamiltonian. Let us suppose that $b$ is a point in $(t_0, t^*]$ which is not a discontinuity of $\mathbf{u}^*$ and that $\boldsymbol{\omega}$ is an element of $\Omega$. We recall [see Eq. (5-579)] that $\boldsymbol{\xi}_b(\boldsymbol{\omega})$ is given by

$$\boldsymbol{\xi}_b(\boldsymbol{\omega}) = \begin{bmatrix} L[\mathbf{x}^*(b), \boldsymbol{\omega}] - L[\mathbf{x}^*(b), \mathbf{u}^*(b)] \\ \mathbf{f}[\mathbf{x}^*(b), \boldsymbol{\omega}] - \mathbf{f}[\mathbf{x}^*(b), \mathbf{u}^*(b)] \end{bmatrix} \tag{5-633}$$

and that $\mathbf{\Phi}_*(t, b)$ is the fundamental matrix of (5-624). Then we know that, for $\epsilon$ small, the vector

$$\epsilon\mathbf{\Phi}_*(t^*, b)\boldsymbol{\xi}_b(\boldsymbol{\omega}) \tag{5-634}$$

belongs to $C_{t^*}$ and hence, in view of (5-622), that

$$\left\langle \begin{bmatrix} p_0^* \\ \mathbf{p}^*(t^*) \end{bmatrix}, \mathbf{\Phi}_*(t^*, b)\boldsymbol{\xi}_b(\boldsymbol{\omega}) \right\rangle \geq 0 \tag{5-635}$$

Since the function $\boldsymbol{\psi}(t) = \mathbf{\Phi}_*(t, b)\boldsymbol{\xi}_b(\boldsymbol{\omega})$ is a solution of the system (5-624), which is adjoint to the system (5-626), we may assert that

$$\left\langle \begin{bmatrix} p_0^* \\ \mathbf{p}^*(b) \end{bmatrix}, \boldsymbol{\xi}_b(\boldsymbol{\omega}) \right\rangle \geq 0 \tag{5-636}$$

or, equivalently, that

$$H[\mathbf{x}^*(b), \mathbf{p}^*(b), \boldsymbol{\omega}, p_0^*] \geq H[\mathbf{x}^*(b), \mathbf{p}^*(b), \mathbf{u}^*(b), p_0^*] \tag{5-637}$$

As $\boldsymbol{\omega}$ can be any element of $\Omega$, we have established (5-492). In other words, we have shown that

$$\min_{\mathbf{u} \in \Omega} H[\mathbf{x}^*(b), \mathbf{p}^*(b), \mathbf{u}, p_0^*] = H[\mathbf{x}^*(b), \mathbf{p}^*(b), \mathbf{u}^*(b), p_0^*] \tag{5-638}$$

which is statement $b$ of the theorem.

Now let us show that the Hamiltonian is zero at the terminal time $t^*$. We recall that, if $\tau$ is an arbitrary real number, then [see Eq. (5-566)]

$$\boldsymbol{\delta}[\tau] = \begin{bmatrix} L[\mathbf{x}^*(t^*), \mathbf{u}^*(t^*)] \\ \mathbf{f}[\mathbf{x}^*(t^*), \mathbf{u}^*(t^*)] \end{bmatrix} \tau \tag{5-639}$$

It follows from Eq. (5-622) that

$$\left\langle \begin{bmatrix} p_0^* \\ \mathbf{p}^*(t^*) \end{bmatrix}, \begin{bmatrix} L[\mathbf{x}^*(t^*), \mathbf{u}^*(t^*)] \\ \mathbf{f}[\mathbf{x}^*(t^*), \mathbf{u}^*(t^*)] \end{bmatrix} \right\rangle \tau \geq 0 \tag{5-640}$$

or, equivalently, that

$$H[\mathbf{x}^*(t^*), \mathbf{p}^*(t^*), \mathbf{u}^*(t^*), p_0^*]\tau \geq 0 \tag{5-641}$$

Since $\tau$ can be either positive or negative, we must have

$$H[\mathbf{x}^*(t^*), \mathbf{p}^*(t^*), \mathbf{u}^*(t^*), p_0^*] = 0 \tag{5-642}$$

Finally, let us show that the function $H[\mathbf{x}^*(t), \mathbf{p}^*(t), \mathbf{u}^*(t), p_0^*]$ is constant on $[t_0, t^*]$. We shall demonstrate this under the additional assumption that the set of points at which $\mathbf{u}^*$ is continuous has the following property: *If* $\hat{t} \in [t_0, t^*]$ *is a point at which* $\mathbf{u}^*$ *is continuous, then* $\mathbf{u}^*$ *is continuous for all* $t$ *in* $[t_0, t^*]$ *sufficiently near* $\hat{t}$.† This assumption will be valid, for example, if $\mathbf{u}^*$ has only a *finite* number of discontinuities.

† The general proof appears in Ref. P-5 and requires a somewhat more complicated argument.

So, let $t_1$ and $t_2$ be elements of $[t_0, t^*]$ such that $t_1 < t_2$ and the function $\mathbf{u}^*(t)$ is continuous on the interval $[t_1, t_2]$. We claim that the function $H[\mathbf{x}^*(t), \mathbf{p}^*(t), \mathbf{u}^*(t), p_0^*]$ is constant on this interval. Since $\mathbf{x}^*(t)$, $\mathbf{p}^*(t)$, and $\mathbf{u}^*(t)$ are continuous on this interval, the sets

$$\{\mathbf{x}^*(t): t \in [t_1, t_2]\} \qquad \{\mathbf{p}^*(t): t \in [t_1, t_2]\} \qquad \{\mathbf{u}^*(t): t \in [t_1, t_2]\} \tag{5-643}$$

are all bounded and, therefore, have compact closures (see Sec. 3-6) $X_1$, $P_1$, and $U_1$, respectively. We observe that the function $H(\mathbf{x}, \mathbf{p}, \mathbf{u}, p_0^*)$ is continuous on the compact set $X_1 \times P_1 \times U_1$ and has, in view of our assumptions, continuous partial derivatives with respect to $\mathbf{x}$ and $\mathbf{p}$ on that compact set. We define a real-valued function $m$ of $\mathbf{x}$ and $\mathbf{p}$ by setting

$$m(\mathbf{x}, \mathbf{p}) = \inf_{\mathbf{u} \in U_1} H(\mathbf{x}, \mathbf{p}, \mathbf{u}, p_0^*) \tag{5-644}$$

Since $\mathbf{u}^*$ is continuous on $[t_1, t_2]$, we may conclude from Eq. (5-638) that

$$m[\mathbf{x}^*(t), \mathbf{p}^*(t)] = H[\mathbf{x}^*(t), \mathbf{p}^*(t), \mathbf{u}^*(t), p_0^*] \tag{5-645}$$

for $t$ in $[t_1, t_2]$, and thus the function $m[\mathbf{x}^*(t), \mathbf{p}^*(t)]$ is continuous. We shall show that the derivative of $m$ is zero and thus complete the proof of Theorem 5-5P.

Now suppose that $t$, $t'$ are distinct points of $[t_1, t_2]$. Then

$$m[\mathbf{x}^*(t'), \mathbf{p}^*(t')] \le H[\mathbf{x}^*(t'), \mathbf{p}^*(t'), \mathbf{u}^*(t), p_0^*] \tag{5-646}$$

and so

$$\begin{aligned} &m[\mathbf{x}^*(t'), \mathbf{p}^*(t')] - m[\mathbf{x}^*(t), \mathbf{p}^*(t)] \\ &\qquad \le H[\mathbf{x}^*(t'), \mathbf{p}^*(t'), \mathbf{u}^*(t), p_0^*] - H[\mathbf{x}^*(t), \mathbf{p}^*(t), \mathbf{u}^*(t), p_0^*] \end{aligned} \tag{5-647}$$

If $t' > t$, then we may deduce from Eq. (5-647) that

$$\frac{m[\mathbf{x}^*(t'), \mathbf{p}^*(t')] - m[\mathbf{x}^*(t), \mathbf{p}^*(t)]}{t' - t} \tag{5-648}$$

is less than or equal to

$$\frac{H[\mathbf{x}^*(t'), \mathbf{p}^*(t'), \mathbf{u}^*(t), p_0^*] - H[\mathbf{x}^*(t), \mathbf{p}^*(t), \mathbf{u}^*(t), p_0^*]}{t' - t} \tag{5-649}$$

However, since $\dot{\mathbf{x}}^*(t)$ and $\dot{\mathbf{p}}^*(t)$ exist and since $\partial H/\partial \mathbf{x}$ and $\partial H/\partial \mathbf{p}$ are continuous, we can see that, as $t' \to t$ from the right, we obtain the relation

$$\frac{d}{ds} m[\mathbf{x}^*(s), \mathbf{p}^*(s)] \Big|_{t+} \le \left\langle \frac{\partial H}{\partial \mathbf{x}} \Big|_*, \dot{\mathbf{x}}^*(t) \right\rangle + \left\langle \frac{\partial H}{\partial \mathbf{p}} \Big|_*, \dot{\mathbf{p}}^*(t) \right\rangle = 0 \tag{5-650}$$

[since $\mathbf{x}^*(t)$ and $\mathbf{p}^*(t)$ satisfy the canonical equations]. In a similar way, we can show that, by letting $t' \to t$ from the left, we have

$$\frac{d}{ds} m[\mathbf{x}^*(s), \mathbf{p}^*(s)] \bigg|_{t-} \geq 0 \tag{5-651}$$

However, the fact that $\partial H/\partial \mathbf{x}$ and $\partial H/\partial \mathbf{p}$ are continuous on $X_1 \times P_1 \times U_1$ and the fact that $\dot{\mathbf{x}}^*(t)$ and $\dot{\mathbf{p}}^*(t)$ exist together imply the existence of the derivative of $m[\mathbf{x}^*(t), \mathbf{p}^*(t)]$, and so we can conclude from Eqs. (5-650) and (5-651) that

$$\frac{d}{ds} m[\mathbf{x}^*(s), \mathbf{p}^*(s)] \bigg|_{t} = 0 \tag{5-652}$$

In view of Eqs. (5-642) and (5-645), we may thus conclude that

$$H[\mathbf{x}^*(t), \mathbf{p}^*(t), \mathbf{u}^*(t), p_0^*] = 0 \tag{5-653}$$

for $t$ in $[t_0, t^*]$.

We have shown that $p_0^* \geq 0$ in Eq. (5-623) and that $\mathbf{p}^*(t)$ is a solution of the canonical system corresponding to $\mathbf{x}^*(t)$ and $\mathbf{u}^*(t)$ in Eq. (5-632). We have shown that $\mathbf{u}^*(t)$ minimizes the Hamiltonian as a function of $\mathbf{u}$ (statement *b* of the theorem) in Eq. (5-637), and finally, we have just demonstrated that the Hamiltonian $H[\mathbf{x}^*(t), \mathbf{p}^*(t), \mathbf{u}^*(t), p_0^*]$ is zero. Thus, we have established Theorem 5-5P.

### 9 *The Transversality Conditions*

The major difference between Theorems 5-5P and 5-6P is the addition of the transversality conditions (statement *d* of Theorem 5-6P). We shall show how the "proof" must be modified in order to establish the transversality conditions.

Let us begin by observing that our arguments depended on the fact that the nonzero vector $\begin{bmatrix} p_0^* \\ \mathbf{p}^*(t^*) \end{bmatrix}$ had the property that

$$\left\langle \begin{bmatrix} p_0^* \\ \mathbf{p}^*(t^*) \end{bmatrix}, \boldsymbol{\mu} \right\rangle \leq 0 \tag{5-654}$$

and the property that

$$\left\langle \begin{bmatrix} p_0^* \\ \mathbf{p}^*(t^*) \end{bmatrix}, \boldsymbol{\delta} \right\rangle \geq 0 \tag{5-655}$$

for all $\boldsymbol{\delta}$ in $C_{t^*}$. In other words, *any* nonzero vector satisfying these two equations would lead to a costate satisfying the requirements of our theorem.

If $S_1$ is our smooth $k$-fold in $R_n$ and if $\mathbf{x}^*(t^*)$ is the terminal point of our optimal trajectory in $R_n$, so that $\mathbf{x}^*(t^*) \in S_1$, then we know that the tangent "plane" $M[\mathbf{x}^*(t^*)]$† of $S_1$ at $\mathbf{x}^*(t^*)$ is well defined and is a $k$-dimensional plane

† If $S_1 = \{\mathbf{x}: g_1(\mathbf{x}) = 0, g_2(\mathbf{x}) = 0, \ldots, g_{n-k}(\mathbf{x}) = 0\}$, then we recall that $M[\mathbf{x}^*(t^*)] = \{\mathbf{x}: \langle (\partial g_i/\partial \mathbf{x})[\mathbf{x}^*(t^*)], \mathbf{x} - \mathbf{x}^*(t^*)\rangle = 0 \text{ for } i = 1, 2, \ldots, n-k\}$. See Sec. 3-13.

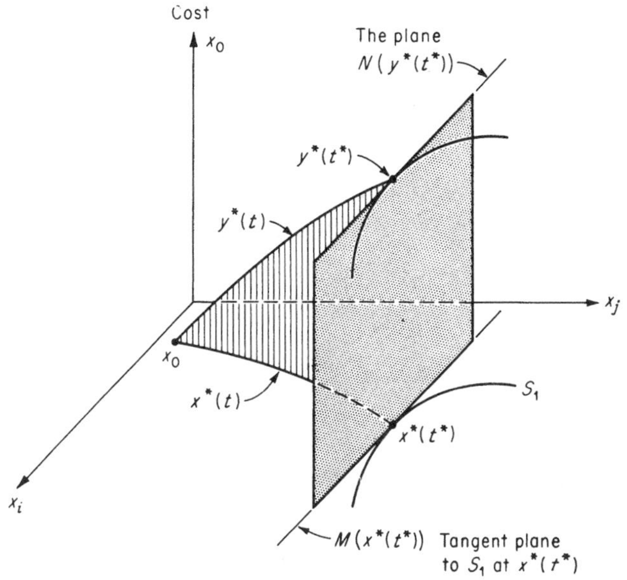


**Fig. 5-35** Illustration of an optimal trajectory to a target set. The target set is $S_1$ and $\mathbf{x}^*(t)$ is the optimal trajectory in $R_n$, while $\mathbf{y}^*(t)$ is the optimal trajectory in $R_{n+1}$. The tangent plane $M[\mathbf{x}^*(*)]$ to $S_1$ at $\mathbf{x}^*(t^*)$ is, in this three-dimensional figure, a straight, line. The plane $N[\mathbf{y}^*(t^*)]$ is the projection of $M[\mathbf{x}^*(t^*)]$ upon the hyperplane of constant cost $x_0 = x_0^*(t^*)$.

in $R_n$ which passes through $\mathbf{x}^*(t^*)$. It will follow that the set of all points $(J^*, \mathbf{x})$, where $J^* = x_0^*(t^*) = J(\mathbf{u}^*)$ and $\mathbf{x} \in M[\mathbf{x}^*(t^*)]$, is a $k$-dimensional plane in $R_{n+1}$ which passes through the terminal point $\mathbf{y}^*(t^*) = (J^*, \mathbf{x}^*(t^*))$ of our optimal trajectory in $R_{n+1}$, as is illustrated in Fig. 5-35. We shall denote this plane by $N[\mathbf{y}^*(t^*)]$, so that

$$N[\mathbf{y}^*(t^*)] = \{\mathbf{y}: \mathbf{y} = (x_0^*(t^*), \mathbf{x}), \mathbf{x} \in M[\mathbf{x}^*(t^*)]\} \tag{5-656}$$

We also observe that every element of $N[\mathbf{y}^*(t^*)]$ may be written in the form

$$\mathbf{y} = \mathbf{y}^*(t^*) + \begin{bmatrix} 0 \\ \hat{\mathbf{x}} \end{bmatrix} \tag{5-657}$$

where $\hat{\mathbf{x}}$ is an $n$ vector with the property that

$$\left\langle \frac{\partial g_i}{\partial \mathbf{x}} [\mathbf{x}^*(t^*)], \hat{\mathbf{x}} \right\rangle = 0 \qquad i = 1, 2, \ldots, n-k \tag{5-658}$$

where $g_1(\mathbf{x}) = 0$, $g_2(\mathbf{x}) = 0$, . . . , $g_{n-k}(\mathbf{x}) = 0$ are the equations of $S_1$. If we let $\hat{M} = \hat{M}[\mathbf{x}^*(t^*)] = \{\hat{\mathbf{x}}$: there is a $\mathbf{y}$ in $N[\mathbf{y}^*(t^*)]$ such that $\mathbf{y} - \mathbf{y}^*(t^*) = (0, \hat{\mathbf{x}})\}$, then we can easily deduce from Eq. (5-658) that $\hat{M}$ is a *subspace* (see Sec. 2-5) of $R_n$ and that the set $\hat{N} = \{\hat{\mathbf{y}} : \hat{\mathbf{y}} = (0, \hat{\mathbf{x}}), \hat{\mathbf{x}} \in \hat{M}\}$ is a subspace of $R_{n+1}$. Moreover, we can see that $\hat{M}$ and $\hat{N}$ are both of dimension $k$. (See Sec. 3-13.) We also note that

$$N[\mathbf{y}^*(t^*)] = \mathbf{y}^*(t^*) + \hat{N} \tag{5-659}$$

The upshot of all this is that the transversality condition $d$ of Theorem 5-6P is equivalent to the requirement that

$$\left\langle \begin{bmatrix} p_0^* \\ \mathbf{p}^*(t^*) \end{bmatrix}, \hat{\mathbf{y}} \right\rangle = 0 \tag{5-660}$$

for all $\hat{\mathbf{y}}$ in $\hat{N}$. (See Fig. 5-36.)

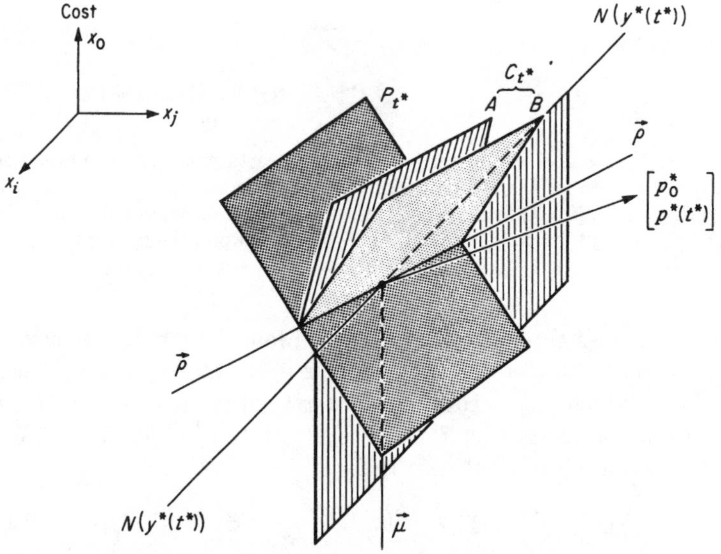


**Fig. 5-36** The hyperplane $P_{t^*}$ is a support hyperplane to the convex cone $C_{t^*}$ and to the convex cone $N[\mathbf{y}^*(t^*)] + \vec{\mu}$, and it separates them. In this three-dimensional figure, the convex cone $N[\mathbf{y}^*(t^*)] + \vec{\mu}$ is the plane defined by the straight line $N[\mathbf{y}^*(t^*)]$ and the ray $\vec{\mu}$ (which points in the direction of decreasing cost). Since the ray $\vec{\rho}$ is an element of $C_{t^*}$ and the line $N[\mathbf{y}^*(t^*)]$ is an element of $N[\mathbf{y}^*(t^*)] + \vec{\mu}$, the hyperplane $P_{t^*}$ must contain the two straight lines $\vec{\rho}$ and $N[\mathbf{y}^*(t^*)]$. Thus, the convex cone $C_{t^*}$ is on one side of $P_{t^*}$ and the convex cone $N[\mathbf{y}^*(t^*)] + \vec{\mu}$ is on the other side. The vector $\begin{bmatrix} p_0^* \\ \mathbf{p}^*(t^*) \end{bmatrix}$ is the normal to the hyperplane $P_{t^*}$ at the point $\mathbf{y}^*(t^*)$. Since $N[\mathbf{y}^*(t^*)]$ is a constant-cost line (see Fig. 5-35), it follows that the $n$ vector $\mathbf{p}^*(t^*)$ is perpendicular to the plane $N[\mathbf{y}^*(t^*)]$ and, hence, perpendicular to the tangent plane $M[\mathbf{x}^*(t^*)]$ of $S_1$ at $\mathbf{x}^*(t^*)$. This is precisely the transversality condition. Thus, knowledge of the target set $S_1$ and of the state $\mathbf{x}^*(t^*)$ completely specifies the direction of the costate $\mathbf{p}^*(t^*)$.

Since $\hat{N}$ is a subspace of $R_{n+1}$, the set $N[\mathbf{y}^*(t^*)]$ is clearly a convex cone with $\mathbf{y}^*(t^*)$ as vertex. If we again let $\boldsymbol{\mu}$ be the $n+1$ vector given by

$$\boldsymbol{\mu} = \begin{bmatrix} -1 \\ 0 \\ 0 \\ \cdot \\ \cdot \\ \cdot \\ 0 \end{bmatrix} \tag{5-661}$$

and if we let $\vec{\mu}$ be the set of all vectors $\beta\boldsymbol{\mu}$, $\beta \geq 0$, then we can see that the set

$$N[\mathbf{y}^*(t^*)] + \vec{\mu} = \{\mathbf{y} : \mathbf{y} = \mathbf{y}^*(t^*) + \hat{\mathbf{y}} + \beta\boldsymbol{\mu},\ \hat{\mathbf{y}} \in \hat{N},\ \beta \geq 0\} \tag{5-662}$$

is a convex cone with vertex at $\mathbf{y}^*(t^*)$. It can be shown by an argument similar to that used previously† that the convex cones $N[\mathbf{y}^*(t^*)] + \vec{\mu}$ and $C_{t^*}$ are separated. Thus, we may deduce from Theorem 3-5 that there is a support hyperplane $P_{t^*}$ of $C_{t^*}$ which separates $C_{t^*}$ and $N[\mathbf{y}^*(t^*)] + \vec{\mu}$. It follows that there is a nonzero $n + 1$ vector $\begin{bmatrix} p_0^* \\ \mathbf{p}^*(t^*) \end{bmatrix}$ such that

$$\left\langle \begin{bmatrix} p_0^* \\ \mathbf{p}^*(t^*) \end{bmatrix}, \hat{\mathbf{y}} + \beta\boldsymbol{\mu} \right\rangle \leq 0 \tag{5-663}$$

for all $\hat{\mathbf{y}}$ in $\hat{N}$ and $\beta \geq 0$, and

$$\left\langle \begin{bmatrix} p_0^* \\ \mathbf{p}^*(t^*) \end{bmatrix}, \boldsymbol{\delta} \right\rangle \geq 0 \tag{5-664}$$

for all $\boldsymbol{\delta}$ in $C_{t^*}$. Since $\hat{N}$ is a subspace of $R_{n+1}$, $\mathbf{0}$ is an element of $\hat{N}$, and so Eq. (5-663) implies that

$$\left\langle \begin{bmatrix} p_0^* \\ \mathbf{p}^*(t^*) \end{bmatrix}, \boldsymbol{\mu} \right\rangle \leq 0 \tag{5-665}$$

On the other hand, letting $\beta = 0$ in Eq. (5-663), we have

$$\left\langle \begin{bmatrix} p_0^* \\ \mathbf{p}^*(t^*) \end{bmatrix}, \hat{\mathbf{y}} \right\rangle \leq 0 \tag{5-666}$$

for all $\hat{\mathbf{y}}$ in $\hat{N}$. As $\hat{N}$ is a subspace of $R_{n+1}$, $\hat{\mathbf{y}} \in \hat{N}$ implies that $-\hat{\mathbf{y}} \in \hat{N}$. Thus, we may conclude that

$$\left\langle \begin{bmatrix} p_0^* \\ \mathbf{p}^*(t^*) \end{bmatrix}, \hat{\mathbf{y}} \right\rangle = 0 \tag{5-667}$$

for all $\hat{\mathbf{y}}$ in $\hat{N}$, which, according to Eq. (5-660), establishes the transversality conditions.

† See Ref. P-5, lemmas 10 and 11.

In the case where the target set $S_1$ is all of $R_n$, we can see that the set $\hat{\mathcal{N}}$ is the entire half-space "below" the hyperplane $y_0 - y_0^*(t^*) = 0$. As the cone $C_{t^*}$ must lie "above" this hyperplane, we can see that the vector $\begin{bmatrix} p_0^* \\ \mathbf{p}^*(t^*) \end{bmatrix}$ must be of the form $\begin{bmatrix} p_0^* \\ \mathbf{0} \end{bmatrix}$ with $p_0^* > 0$; that is, $\mathbf{p}^*(t^*)$ must be the zero vector. Thus, Theorem 5-6P is completely established.

**Exercise 5-21** The purpose of this lengthy exercise is to guide the reader through the various steps of the proof of the minimum principle for a specific optimization problem. In this manner, some of the conceptual difficulties associated with the general proof will be clarified. We suggest that the reader construct good three-dimensional figures in order to enhance his geometric intuition.

Consider the second-order system

$$\dot{x}_1(t) = x_2(t)$$
$$\dot{x}_2(t) = u(t)$$

which, at the initial time $t_0 = 0$, is at the state

$$x_1(0) = 0 \qquad x_2(0) = 0$$

We suppose that the control $u(t)$ is constrained in magnitude by the relation $|u(t)| \leq 1$ for all $t$. Consider the fixed terminal state

$$\mathbf{x}_1 = \begin{bmatrix} 2 \\ 2 \end{bmatrix}$$

and the cost functional (= terminal time)

$$J(u) = \int_0^{t_1} 1\, dt = t_1 \qquad \text{that is, } L(\mathbf{x}, \mathbf{u}) = 1$$

We assert that the control

$$u^*(t) = +1 \qquad \text{for all } t,\ 0 \leq t \leq 2$$

transfers the system from the initial state (0, 0) to the terminal state (2, 2) and minimizes $J(u)$. In other words, the control $u^*(t) = +1$, $t \in [0, 2]$, is a time-optimal control. Clearly,

$$t^* = 2$$

Let

$$\dot{x}_0(t) = 1 \qquad x_0(0) = 0$$

In this case, the cost axis ($x_0$) is the time axis.

(*a*) Refer to subsection 1. Let $\mathbf{y}(t)$ be the 3-vector

$$\mathbf{y}(t) = \begin{bmatrix} x_0(t) \\ x_1(t) \\ x_2(t) \end{bmatrix}$$

Show that

$$\mathbf{y}^*(t) = \begin{bmatrix} t \\ \frac{1}{2}t^2 \\ t \end{bmatrix} \qquad t \in [0, 2]$$

Plot $\mathbf{y}^*(t)$ in the three-dimensional space.

(*b*) Refer to subsection 2. Show that the admissible control

$$\begin{aligned} \hat{u}(t) &= -1 & 0 < t \le 2 \\ \hat{u}(t) &= +1 & 2 < t \le 6 \\ \hat{u}(t) &= 0 & 6 < t \le 8 \end{aligned}$$

transfers (0, 0, 0) to (8, 2, 2). Plot the corresponding $\hat{\mathbf{y}}(t)$ and convince yourself that it is not optimal.

(*c*) Refer to subsection 3. Show that the matrix $\mathbf{A}_*(t)$ of Eq. (5-541) is

$$\mathbf{A}_*(t) = \begin{bmatrix} 0 & 0 & 0 \\ 0 & 0 & 1 \\ 0 & 0 & 0 \end{bmatrix}$$

Show that Eq. (5-542) reduces to the equations

$$\begin{aligned} \dot{\psi}_0(t) &= 0 \\ \dot{\psi}_1(t) &= \psi_2(t) \\ \dot{\psi}_2(t) &= 0 \end{aligned}$$

Show that the fundamental matrix $\mathbf{\Phi}_*(t; t_0)$ is (for $t_0 = 0$)

$$\mathbf{\Phi}_*(t; t_0) = \mathbf{\Phi}_*(t; 0) = \begin{bmatrix} 1 & 0 & 0 \\ 0 & 1 & t \\ 0 & 0 & 1 \end{bmatrix}$$

Show that

$$\begin{aligned} \psi_0(t) &= \xi_0 \\ \psi_1(t) &= \xi_1 + \xi_2 t \\ \psi_2(t) &= \xi_2 \end{aligned}$$

Pick several values of $\xi_0$, $\xi_1$, and $\xi_2$ such that

$$\xi_0^2 + \xi_1^2 + \xi_2^2 \le 1$$

Evaluate $\boldsymbol{\psi}(t)$. Let $t = \frac{1}{2}, 1, \frac{3}{2}, 2$ and plot the vectors $\mathbf{y}^*(t)$ and $\mathbf{y}(t) = \mathbf{y}^*(t) + \epsilon\boldsymbol{\psi}(t)$ for $\epsilon = 0.01$. Convince yourself that $\mathbf{y}(t)$ belongs to a "tube" about $\mathbf{y}^*(t)$ and that $\epsilon$ can regulate the thickness of the tube.

(*d*) Refer to subsection 4. What is the matrix $-\mathbf{A}'_*(t)$ of Eq. (5-551)? Show that Eqs. (5-554) and (5-555) reduce to the equations

$$\begin{aligned} \dot{p}_0(t) &= 0 \\ \dot{p}_1(t) &= 0 \\ \dot{p}_2(t) &= -p_1(t) \end{aligned}$$

What is Eq. (5-556)? Verify Eqs. (5-557) and (5-558). Consider the hyperplane $P_0$ [see Eq. (5-559)] described by the equation

$$y_1 + y_2 + y_3 = 0$$

Suppose that

$$\boldsymbol{\xi} = \begin{bmatrix} 0 \\ 1 \\ 1 \end{bmatrix}$$

Show that the vector

$$\begin{bmatrix} p_0(0) \\ p_1(0) \\ p_2(0) \end{bmatrix} = \begin{bmatrix} 1 \\ 1 \\ 1 \end{bmatrix}$$

is normal to $P_0$ at (0, 0, 0). Show that the scalar product between the vectors

$$\begin{bmatrix} 1 \\ 1 \\ 1-t \end{bmatrix} \quad \text{and} \quad \begin{bmatrix} 0 \\ 1+t \\ 1 \end{bmatrix}$$

is the same as the scalar product between the vectors

$$\begin{bmatrix} 0 \\ 1 \\ 1 \end{bmatrix} \quad \text{and} \quad \begin{bmatrix} 1 \\ 1 \\ 1 \end{bmatrix}$$

What does this mean? Explain in terms of the moving hyperplanes $P_t$.

(*e*) Refer to subsection 5. Show that the temporal control variation is

$$u[\tau](t) = +1 \qquad 0 \le t \le 2 + \epsilon\tau$$

where $\tau$ can be either positive or negative. Show that (since $t^* = 2$)

$$\mathbf{y}(2 + \epsilon\tau) = \begin{bmatrix} 2 + \epsilon\tau \\ 2 + 2\epsilon\tau + \frac{1}{2}\epsilon^2\tau^2 \\ 2 + \epsilon\tau \end{bmatrix}$$

Show that, in Eq. (5-564),

$$\mathbf{y}^*(2) = \begin{bmatrix} 2 \\ 2 \\ 2 \end{bmatrix} \qquad \boldsymbol{\delta}(\tau) = \begin{bmatrix} \tau \\ 2\tau \\ \tau \end{bmatrix} \qquad \mathbf{o}(\epsilon) = \begin{bmatrix} 0 \\ \frac{1}{2}\epsilon^2\tau^2 \\ 0 \end{bmatrix}$$

Verify Eqs. (5-568) and (5-569). Draw the ray $\vec{\rho}$ (in three dimensions) through the point (2, 2, 2).

(*f*) Refer to subsection 6. Consider spatial variations

$$u[\omega, I](t) = \begin{cases} +1 & t \notin I \\ \omega & t \in I \end{cases}$$

Show that we must have

$$-1 \le \omega < +1$$
$$0 < b < 2$$

Let $b = 0.5$, $\epsilon a = 0.1$. Plot the trajectories generated by setting $\omega = 0.9$, $\omega = 0.5$, $\omega = 0$, $\omega = -1$. Repeat for $b = 1.0$, $\epsilon a = 0.1$. Show that

$$\mathbf{y}(b) = \begin{bmatrix} b \\ \frac{1}{2}b^2 - \frac{1}{2}\epsilon^2 a^2 + \frac{1}{2}\epsilon^2\omega a^2 \\ b - \epsilon a + \epsilon a\omega \end{bmatrix}$$

Verify Eq. (5-571) and show that the $\mathbf{o}(\epsilon)$ term in (5-571) is

$$\mathbf{o}(\epsilon) = \begin{bmatrix} 0 \\ \frac{1}{2}\epsilon^2 a^2(\omega - 1) \\ 0 \end{bmatrix}$$

Show that the vector $\boldsymbol{\xi}_b(\omega)$ of Eq. (5-579) is

$$\boldsymbol{\xi}_b(\omega) = \begin{bmatrix} 0 \\ 0 \\ \omega - 1 \end{bmatrix}$$

Evaluate the vectors $\mathbf{o}(\epsilon)$ in Eq. (5-580) and the vector $\mathbf{y}(t^*)$ in Eq. (5-582). Show that the vector $\boldsymbol{\delta}[\omega, I]$ in Eq. (5-583) is

$$\boldsymbol{\delta}[\omega, I] = \begin{bmatrix} 1 & 0 & 0 \\ 0 & 1 & 2-b \\ 0 & 0 & 1 \end{bmatrix} \begin{bmatrix} 0 \\ 0 \\ a\omega - a \end{bmatrix} = a \begin{bmatrix} 0 \\ (b-2)(1-\omega) \\ \omega - 1 \end{bmatrix}$$

Show that a typical ray $\vec{\rho}\,[\omega, b]$ is given by the set of vectors [see Eq. (5-585)]

$$\left\{ \mathbf{y}: \mathbf{y} = \begin{bmatrix} 2 \\ 2 \\ 2 \end{bmatrix} + \beta \begin{bmatrix} 0 \\ (b-2)(1-\omega) \\ \omega - 1 \end{bmatrix}; \beta \geq 0 \right\}$$

Since $-1 \leq \omega < 1$ and $0 < b < 2$, you can construct the set $\text{co}(\vec{P})$; sketch it in a figure. Verify that it is convex.

(*g*) Refer to subsection 7. Show that a typical element of $C_{t^*}$ [see Eq. (5-599)] is of the form

$$\mathbf{y} = \begin{bmatrix} 2 \\ 2 \\ 2 \end{bmatrix} + \alpha\epsilon \begin{bmatrix} \tau \\ 2\tau \\ \tau \end{bmatrix} + \beta\epsilon \begin{bmatrix} 0 \\ (b-2)(1-\omega) \\ \omega - 1 \end{bmatrix} \qquad \alpha \geq 0, \beta \geq 0$$

Sketch $C_{t^*}$. Verify that the vector $\boldsymbol{\delta}$ in Eq. (5-603) can indeed be constructed using Eq. (5-604). Plot a few such vectors $\boldsymbol{\delta}$. Prove Eq. (5-606). Pick a vector $\boldsymbol{\delta}$ and construct a variation of the form (5-608). Evaluate the $\mathbf{o}(\epsilon)$ terms in Eq. (5-609).

(*h*) Refer to subsection 8. Construct the ray $\vec{\mu}$ [see Eq. (5-615)]. Show graphically that $\vec{\mu}$ does not meet the interior of $C_{t^*}$. Let [see Eq. (5-622)]

$$\boldsymbol{\delta} = \alpha \begin{bmatrix} \tau \\ 2\tau \\ \tau \end{bmatrix} + \beta \begin{bmatrix} 0 \\ (b-2)(1-\omega) \\ \omega - 1 \end{bmatrix} \qquad \alpha \geq 0, \beta \geq 0$$

Verify that $p_0^* > 0$ and so set $p_0^* = 1$. Show that Eqs. (5-621) and (5-622) imply that

$$p_1^*(2) \geq -\tfrac{1}{2} \qquad p_2^*(2) \leq 0$$

Show that Eqs. (5-627) and (5-628) imply that

$$\begin{aligned} p_0^*(t) &= 1 = \text{const} \\ p_1^*(t) &= p_1^*(2) = \text{const} \\ p_2^*(t) &= p_2^*(2) + 2p_1^*(2) - p_1^*(2)t \end{aligned}$$

for $t \in [0, 2]$. Show that Eqs. (5-635) to (5-637) and (5-642) are satisfied.

## 5-17 Some Comments on the Minimum Principle

We are now going to comment on the minimum principle in the light of our heuristic proof of it.

To begin with, we observe once again that the minimum principle represents a set of *necessary* conditions for optimality. In fact, if we examine our proof carefully, we can see that these conditions are actually necessary conditions for *local* optimality. In other words, if we suppose that $\hat{\mathbf{u}}$ is a control which transfers our initial point to the target set and if we suppose that for all $\mathbf{u}$ which are near $\hat{\mathbf{u}}$, in the sense that

$$\|\mathbf{u} - \hat{\mathbf{u}}\| = \sup_t \|\mathbf{u}(t) - \hat{\mathbf{u}}(t)\| \leq \delta \qquad (5\text{-}668)\dagger$$

where $\delta$ is a small positive number, the relation

$$J(\mathbf{u}) \geq J(\hat{\mathbf{u}}) \qquad (5\text{-}669)$$

is satisfied, then we can, by letting the parameter $\epsilon$ in our proof be suitably small, show that there are a nonnegative constant $\hat{p}_0$ and a function $\hat{\mathbf{p}}(t)$ such that the various conditions of the minimum principle will be satisfied by $\hat{p}_0$, $\hat{\mathbf{p}}(t)$, $\hat{\mathbf{u}}(t)$, and the trajectory $\hat{\mathbf{x}}(t)$ corresponding to $\hat{\mathbf{u}}(t)$. Thus, the minimum principle is somewhat similar in nature to the condition that the derivative vanish at a local minimum of an ordinary function.

Now let us examine each of the necessary conditions in turn. We begin with (1) the canonical system; then we examine (2) the minimization of the Hamiltonian; next we discuss (3) the behavior of the Hamiltonian along the optimal trajectory; and finally we look at (4) the transversality conditions. The numbers 1, 2, 3, and 4 correspond to the statements in the minimum principle and to columns 1, 2, 3, and 4 under the heading "Necessary conditions" in Table 5-1. We shall number our comments in order to make future references to them quite simple.

*Comment 1*

We observe that the first equation (the $\dot{\mathbf{x}}$ equation) of the canonical system is precisely our original system equation and is actually independent of the costate $\mathbf{p}$. We also note that the second equation (the $\dot{\mathbf{p}}$ equation) of the canonical system describes the motion of a hyperplane (or, to be more exact, the normal to a hyperplane) along the optimal trajectory. There may be many solutions of this equation, each corresponding to the motion of some hyperplane. We show that, as a consequence of optimality, there is a solution of this equation with certain additional and useful properties. We also note (see Definition 5-11) that the canonical system will have solutions along *any* trajectory of the system and not just for an optimal control.

*Comment 2*

The first additional property of the costate $\mathbf{p}^*(t)$, $p_0^*$ is the fact that the optimal control must minimize the Hamiltonian. We repeat once again that this minimization is to be viewed in the following way: At a specific instant of time, say $\hat{t}$, in the interval $[t_0, t^*]$, $\mathbf{x}^*(\hat{t})$, $\mathbf{u}^*(\hat{t})$, and $\mathbf{p}^*(\hat{t})$ are three well-defined vectors and the number $H[\mathbf{x}^*(\hat{t}), \mathbf{p}^*(\hat{t}), \mathbf{u}^*(\hat{t}), p_0^*]$ is smaller than or equal to the number $H[\mathbf{x}^*(\hat{t}), \mathbf{p}^*(\hat{t}), \boldsymbol{\omega}, p_0^*]$, where $\boldsymbol{\omega}$ is any element

† This is just the distance between $\mathbf{u}$ and $\mathbf{u}$ in the function space of piecewise continuous functions (Sec. 3-15).

of the constraint set $\Omega$. We observe also that we have actually shown that, as a function of $\mathbf{u}$, the Hamiltonian has an absolute minimum along the optimal trajectory no matter what the character of the constraint set $\Omega$ may be.

*Comment 3*

We remark that the minimization of the Hamiltonian may be interpreted geometrically as the assertion that all the directions in which we may go from a given point on our optimal trajectory lie on the same side of the hyperplane determined by our costate.

*Comment 4*

Now let us note that we actually verified only that the Hamiltonian was minimized at the points of continuity of $\mathbf{u}^*$; in point of fact, we can really show only that necessary condition 2 must hold for all but the points of some countable subset of the interval $[t_0, t^*]$. As has been our custom, we shall usually slur over this point in the sequel. We also point out that the points at which the Hamiltonian may not be minimized must be discontinuities of $\mathbf{u}^*$.

*Comment 5*

We observe that the form of necessary conditions 1 and 2 does not depend upon the type of target set or upon whether the time is free or fixed. This can readily be seen by glancing at columns F and G of Table 5-1.

*Comment 6*

Necessary condition 3, in which the behavior of the Hamiltonian along the optimal trajectory is described, does indeed depend upon whether or not the time is free. If we examine our proof, we can see that the temporal variations in the control depend upon the fact that the terminal time is free, and hence the conclusion that the Hamiltonian is zero at the terminal time depends upon the fact that the terminal time is free. Geometrically, the temporal variations correspond to the ray $\vec{\rho}$, as illustrated, for example, in Fig. 5-32. If the terminal time is fixed, then there are no temporal variations and the terminal cone $C_{t*}$ will simply be $\text{co}(\vec{P})$ (see Sec. 5-16, subsection 6), as is illustrated in Fig. 5-32. In this case, we have an additional degree of freedom in determining the hyperplane $P_{t*}$, which is compensated for by the fact that we know the terminal time. Thus, in the fixed-time case, because the hyperplane $P_{t*}$ need not contain the ray $\vec{\rho}$ (see Fig. 5-32), we can only show that the Hamiltonian must be constant along the optimal trajectory when our system and our cost do not explicitly depend upon the time.

*Comment 7*

As we shall see in later chapters of the book, the determination of "enough" (that is, $2n$) boundary conditions for the canonical equations is crucial in finding the candidates for the optimal control. Since our initial point $\mathbf{x}_0$ is known, we have $n$ conditions at the initial time. Now, if our end point is fixed, then we also have $n$ conditions on $\mathbf{x}^*(t)$ at the terminal time and we do not require any conditions on our costate. On the other hand, if we are trying to hit a target set, say a smooth $k$-fold, then we have only $n - k$ conditions (the equations of our smooth $k$-fold) on $\mathbf{x}^*(t)$ at the terminal time. However, the transversality conditions provide us with the additional $k$ conditions that we need. In other words, the freer our state is at the terminal time, the more constrained our costate is at the terminal time. This "constraint" on the costate is provided by the transversality conditions and is evidenced by the condition that the hyperplane $P_{t*}$ must contain the plane $N[\mathbf{y}^*(t^*)]$ (see Fig. 5-36). Again, we can see that if the terminal time is fixed, then $P_{t*}$ contains the plane $N[\mathbf{y}^*(t^*)]$ but need not contain the ray $\vec{\rho}$. Thus, there is an additional degree of freedom, which is compensated for by the fact that the terminal time is specified.

*Comment 8*

On occasion, we are faced with problems in which our target set $S$ is only a "piece" of a smooth $k$-fold, as is the case, for example, in Sec. 7-2. We can see that if the point $\mathbf{x}^*(t^*)$ at which our optimal trajectory terminates is a point of $S$ at which a tangent plane may be defined, then the transversality conditions will still hold. On the other hand, if the point $\mathbf{x}^*(t^*)$ is a point of $S$ at which a tangent plane cannot be defined, then there will be no transversality conditions.

*Comment 9*

We have, during the course of our "proof," observed that if $p_0^* = 0$, then the hyperplane $P_{t*}$ contains the "cost" ray $\vec{\mu}$; our problem, therefore, does not explicitly depend upon the cost. Now, if $p_0^* \neq 0$, then the hyperplane $P_{t*}$ is defined by the equation

$$\left\langle \begin{bmatrix} p_0^* \\ \mathbf{p}^*(t^*) \end{bmatrix}, \mathbf{y} - \mathbf{y}^*(t^*) \right\rangle = 0 \tag{5-670}$$

However, when $p_0^* \neq 0$, we can see that Eq. (5-670) is equivalent to the equation

$$\frac{1}{p_0^*} \left\langle \begin{bmatrix} p_0^* \\ \mathbf{p}^*(t^*) \end{bmatrix}, \mathbf{y} - \mathbf{y}^*(t^*) \right\rangle = \left\langle \begin{bmatrix} 1 \\ \dfrac{\mathbf{p}^*(t^*)}{p_0^*} \end{bmatrix}, \mathbf{y} - \mathbf{y}^*(t^*) \right\rangle = 0 \tag{5-671}$$

in that both of these equations define the *same* hyperplane $P_{t^*}$. Thus, in the case where $p_0^* \neq 0$, we may as well suppose that $p_0^* = 1$. Since the control problems that we shall examine in subsequent chapters of the book are not pathological (that is, $p_0^* \neq 0$), we shall always assume that $p_0^* = 1$ and we shall not mention this point (or $p_0^*$) again.

*Comment 10*

Let us suppose that we wished to *maximize* our cost functional rather than minimize it. What changes in our necessary conditions would this lead to? We can see immediately that the form of the canonical system will not change. On the other hand, in condition 2, the minimization of $H$ will be replaced by the maximization of $H$ (that is, "max" will replace "min" in the formula). Moreover, if we examine our "proof," we shall see that, in general, inequalities will be reversed. This will mean, in particular, that the cones of attainability will lie "below" rather than "above" the optimal trajectory. However, the ray $\vec{\rho}$ corresponding to our temporal variations will not change, and so condition 3 (that the Hamiltonian be zero along the optimal trajectory) will not change. Similarly, since the tangent plane to our target set is independent of whether we are minimizing or maximizing, we can see that the transversality conditions will not change. In effect, then, *the only necessary condition which distinguishes between maximization and minimization is condition* 2; *conditions* 1, 3, *and* 4 *may thus be viewed as necessary conditions for an "extremal."*

Most of these comments will take on greater significance in later chapters of the book as we show how the minimum principle may be used to solve various control problems. We suggest also that the reader review Sec. 5-10 in the light of our comments here and of our "proof" of the minimum principle.

## 5-18 Sufficient Conditions for Optimality: Introductory Remarks

We shall discuss some sufficiency conditions for optimality in the remainder of this chapter. These conditions represent a strengthening of the necessary conditions derived earlier and are based upon certain assumptions about the behavior of the cost functional. In effect, we shall show that if the cost functional satisfies a certain partial differential equation and if necessary condition 2 is satisfied over a suitable region, then optimality can be established. Our procedure will combine Bellman's principle of optimality [B-5] and a lemma of Caratheodory in a manner suggested by the work of Kalman [K-8].†

We shall begin in the next section by deriving, under the assumption that the cost functional is sufficiently smooth, an equation describing the behavior

† Additional references that deal with the problem of sufficient conditions from various points of view are C-1, C-4 to C-6, K-27, L-7, L-10, and M-8.

of the cost along an optimal trajectory. We shall then suppose that this equation holds throughout a certain region which is "filled" by trajectories corresponding to a "control" $\mathbf{u}$ which functionally minimizes the Hamiltonian of our problem. After stating a suitable lemma, we shall show that the "control" $\mathbf{u}$ must be optimal.

Before developing these ideas, let us carefully delineate the control problem with which we shall be concerned. We suppose that our system is of the form

$$\dot{\mathbf{x}}(t) = \mathbf{f}[\mathbf{x}(t), \mathbf{u}(t), t] \qquad t \in (T_1, T_2) \tag{5-672}$$

where our set $U$ of admissible controls is the set of all bounded piecewise continuous functions $\mathbf{u}(t)$ on $(T_1, T_2)$ such that

$$\mathbf{u}(t) \in \Omega \qquad \text{for all } t \text{ in } (T_1, T_2) \tag{5-673}$$

with $\Omega$ a given subset of $R_m$ and

$$\mathbf{u}(t-) = \mathbf{u}(t) \qquad \text{for all } t \text{ in } (T_1, T_2) \tag{5-674}$$

We let $L(\mathbf{x}, \mathbf{u}, t)$ be a real-valued function on $R_n \times R_m \times (T_1, T_2)$, and we suppose that $K(\mathbf{x}, t) \equiv 0$, so that our cost functional $J$ is of the form

$$J(\mathbf{u}) = \int_{t_0}^{t_1} L[\mathbf{x}(t), \mathbf{u}(t), t]\, dt \tag{5-675}$$

We suppose that $S$ is our target set, that $S \subset R_n \times (T_1, T_2)$, and that *all the assumptions of Sec. 5-12 are in force.* Previously, in defining a control problem, we fixed the initial time and the initial state; now we shall actually define a family of control problems by considering a region $X$ of $R_n \times (T_1, T_2)$ which contains $S$ and supposing that our initial pair $(\mathbf{x}_0, t_0)$ is in $X$. In other words, we are faced with the following problem:

*Let $X$ be a given region in $R_n \times (T_1, T_2)$ and suppose that $X$ contains the target set $S$. Then, for each element $(\mathbf{x}_0, t_0)$ of $X$, determine the control $\mathbf{u}(\mathbf{x}_0, t_0)$ in $U$ which transfers $(\mathbf{x}_0, t_0)$ to $S$ and which, in so doing, minimizes the performance functional*

$$J(\mathbf{x}_0, t_0, \mathbf{u}) = \int_{t_0}^{t_1} L[\mathbf{x}(t), \mathbf{u}(t), t]\, dt \tag{5-676}$$

As we shall be interested primarily in the behavior of $J(\mathbf{x}_0, t_0, \mathbf{u})$ as a function of $\mathbf{x}_0$ and $t_0$, we shall often omit the subscript 0 and simply write

$$J(\mathbf{x}, t, \mathbf{u}) = \int_{t}^{t_1} L[\mathbf{x}(\tau), \mathbf{u}(\tau), \tau]\, d\tau \tag{5-677}$$

The reasons for our interest will become apparent as we develop the sufficiency conditions in subsequent sections.

## 5-19 Sufficient Conditions for Optimality: An Equation for the Cost

We are now going to derive an equation describing the behavior of the cost along an optimal trajectory, under the assumption that our cost functional is sufficiently smooth.

Let us begin by supposing that $(\mathbf{x}_0, t_0)$ is a given initial pair and that $\hat{\mathbf{u}}$ is an admissible control which transfers $(\mathbf{x}_0, t_0)$ to the target set $S$. As usual, we let $\hat{\mathbf{x}}(t)$ denote the trajectory of Eq. (5-672) originating at $\mathbf{x}_0$ generated by $\hat{\mathbf{u}}$, and we suppose that $t_1$ is the first instant of time after $t_0$ when $\hat{\mathbf{x}}(t)$ meets $S$. Now, if $t \in [t_0, t_1)$, then the control $\hat{\mathbf{u}}$ will also transfer $(\hat{\mathbf{x}}(t), t)$ to $S$ in

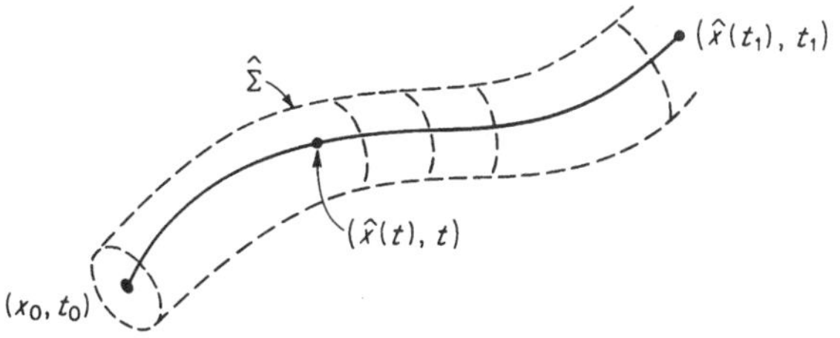


**Fig. 5-37** The region $\hat{\Sigma}$ and the nominal trajectory $(\hat{\mathbf{x}}(t), t)$.

view of the transition property of dynamical systems (see Fig. 5-37), and *so*

$$J[\hat{\mathbf{x}}(t), t, \hat{\mathbf{u}}] = \int_t^{t_1} L[\hat{\mathbf{x}}(\tau), \hat{\mathbf{u}}(\tau), \tau]\, d\tau \tag{5-678}$$

*is a well-defined number for $t$ in $[t_0, t_1)$.*

If we suppose that $\hat{J}(\mathbf{x}, t)$ is a continuously differentiable function defined on a region $\hat{\Sigma} \subset R_n \times (T_1, T_2)$ (see Fig. 5-37) such that

1. $(\hat{\mathbf{x}}(t), t) \in \hat{\Sigma}$ for $t$ in $[t_0, t_1]$ (5-679)
2. $\hat{J}[\hat{\mathbf{x}}(t), t] = J[\hat{\mathbf{x}}(t), t, \hat{\mathbf{u}}]$ for $t$ in $[t_0, t_1]$ (5-680)

then we may deduce that, for $t$ in $[t_0, t_1)$,

$$\frac{d\hat{J}[\hat{\mathbf{x}}(t), t]}{dt} = \left\langle \frac{\partial \hat{J}}{\partial \mathbf{x}} [\hat{\mathbf{x}}(t), t], \dot{\hat{\mathbf{x}}}(t) \right\rangle + \frac{\partial \hat{J}}{\partial t} [\hat{\mathbf{x}}(t), t] \tag{5-681}$$

$$= \left\langle \frac{\partial \hat{J}}{\partial \mathbf{x}} [\hat{\mathbf{x}}(t), t], \mathbf{f}[\hat{\mathbf{x}}(t), \hat{\mathbf{u}}(t), t] \right\rangle + \frac{\partial \hat{J}}{\partial t} [\hat{\mathbf{x}}(t), t] \tag{5-682}$$

and hence, in view of Eqs. (5-678) and (5-682), that

$$-L[\hat{\mathbf{x}}(t), \hat{\mathbf{u}}(t), t] = \left\langle \frac{\partial \hat{J}}{\partial \mathbf{x}} [\hat{\mathbf{x}}(t), t], \mathbf{f}[\hat{\mathbf{x}}(t), \hat{\mathbf{u}}(t), t] \right\rangle + \frac{\partial \hat{J}}{\partial t} [\hat{\mathbf{x}}(t), t] \tag{5-683}$$

or, equivalently, that

$$\frac{\partial \hat{J}}{\partial t} [\hat{\mathbf{x}}(t), t] + L[\hat{\mathbf{x}}(t), \hat{\mathbf{u}}(t), t] + \left\langle \frac{\partial \hat{J}}{\partial \mathbf{x}} [\hat{\mathbf{x}}(t), t], \mathbf{f}[\hat{\mathbf{x}}(t), \hat{\mathbf{u}}(t), t] \right\rangle = 0 \tag{5-684}$$

However, noting that the Hamiltonian $H(\mathbf{x}, \mathbf{p}, \mathbf{u}, t)$ of our problem is given by

$$H(\mathbf{x}, \mathbf{p}, \mathbf{u}, t) = L(\mathbf{x}, \mathbf{u}, t) + \langle \mathbf{p}, \mathbf{f}(\mathbf{x}, \mathbf{u}, t)\rangle \tag{5-685}$$

we may conclude that

$$\frac{\partial \hat{J}}{\partial t}[\hat{\mathbf{x}}(t), t] + H\left[\hat{\mathbf{x}}(t), \frac{\partial \hat{J}}{\partial \mathbf{x}}[\hat{\mathbf{x}}(t), t], \hat{\mathbf{u}}(t), t\right] = 0 \tag{5-686}$$

for $t$ in $[t_0, t_1]$.

Now let us observe, first of all, that such a function $\hat{J}(\mathbf{x}, t)$ may *not* exist. Secondly, even if a $\hat{J}(\mathbf{x}, t)$ can be found, there is *no* guarantee that the function of $t$, $\dfrac{\partial \hat{J}}{\partial \mathbf{x}}[\hat{\mathbf{x}}(t), t]$—the gradient of $\hat{J}(\mathbf{x}, t)$ with respect to $\mathbf{x}$ evaluated at $(\hat{\mathbf{x}}(t), t)$—is a costate which corresponds to $\hat{\mathbf{u}}(t)$ and $\hat{\mathbf{x}}(t)$. In other words, we do not know that

$$\frac{d}{dt}\left\{\frac{\partial \hat{J}}{\partial \mathbf{x}}[\hat{\mathbf{x}}(t), t]\right\} = -\frac{\partial H}{\partial \mathbf{x}}\left[\hat{\mathbf{x}}(t), \frac{\partial \hat{J}}{\partial \mathbf{x}}[\hat{\mathbf{x}}(t), t], \hat{\mathbf{u}}(t), t\right] \tag{5-687}$$

Moreover, we do not know whether or not $\hat{J}(\mathbf{x}, t) = J(\mathbf{x}, t, \hat{\mathbf{u}})$ for points $(\mathbf{x}, t)$ in $\hat{\Sigma}$ which are not on our given system trajectory. Thus, in developing our sufficiency conditions, we shall have to make assumptions which "take care of" these difficulties.

So far, we have considered a control $\hat{\mathbf{u}}$ which transfers $(\mathbf{x}_0, t_0)$ to $S$; now let us turn our attention to an optimal control $\mathbf{u}^*$. In other words, we now *assume that there is an optimal control* $\mathbf{u}^*$ which transfers $(\mathbf{x}_0, t_0)$ to $S$. As usual, we let $\mathbf{x}^*(t)$ denote the corresponding optimal trajectory. We then assert that the control $\mathbf{u}^*$ will also be an optimal control for *any* point $\mathbf{x}^*(t)$ on the optimal trajectory (cf. Sec. 5-16). We know that $\mathbf{u}^*$ will transfer $(\mathbf{x}^*(t), t)$ to $S$, and what we claim is that

$$J[\mathbf{x}^*(t), t, \mathbf{u}^*] \le J[\mathbf{x}^*(t), t, \mathbf{u}] \tag{5-688}$$

for all admissible controls $\mathbf{u}$. However, if Eq. (5-688) were *not* satisfied, then there would be an admissible control $\tilde{\mathbf{u}}$ transferring $(\mathbf{x}^*(t), t)$ to $S$ such that

$$J[\mathbf{x}^*(t), t, \tilde{\mathbf{u}}] < J[\mathbf{x}^*(t), t, \mathbf{u}^*] \tag{5-689}$$

It would then follow that the admissible control $\mathbf{u}^1$, given by

$$\mathbf{u}^1(\tau) = \begin{cases} \mathbf{u}^*(\tau) & \tau \in [t_0, t) \\ \tilde{\mathbf{u}}(\tau) & \tau \ge t \end{cases} \tag{5-690}$$

would transfer $(\mathbf{x}_0, t_0)$ to $S$ and would satisfy the relation

$$J(\mathbf{x}_0, t_0, \mathbf{u}^1) < J(\mathbf{x}_0, t_0, \mathbf{u}^*) \tag{5-691}$$

Since Eq. (5-691) contradicts the optimality of $\mathbf{u}^*$, our claim [Eq. (5-688)] is established. We have, in effect, derived the following principle, which is often referred to as the *principle of optimality:*†

**Principle of Optimality** [B-5] *If* $\mathbf{u}^*$ *is an optimal control and if* $\mathbf{x}^*(t)$, $t \in [t_0, t_1]$, *is the optimal trajectory corresponding to* $\mathbf{u}^*$, *then the restriction of* $\mathbf{u}^*$ *to a subinterval* $[t, t_1]$ *of* $[t_0, t_1]$ *is an optimal control for the initial pair* $(\mathbf{x}^*(t), t)$.

**Exercise 5-22** Derive the following more general form of the principle of optimality: If $\mathbf{u}^*$ is an optimal control and if $\mathbf{x}^*(t)$, $t \in [t_0, t_1]$, is the optimal trajectory corresponding to $\mathbf{u}^*$, then the restriction of $\mathbf{u}^*$ to any subinterval $[\tau_0, \tau_1]$ of $[t_0, t_1]$ (that is, $t_0 \le \tau_0 < \tau_1 \le t_1$) is an optimal control for the initial pair $(\mathbf{x}^*(\tau_0), \tau_0)$ and the target set $(\mathbf{x}^*(\tau_1), \tau_1)$. Heuristically, this version of the principle of optimality is often stated: "Portions of an optimal trajectory are optimal." HINT: See Sec. 5-16, subsection 2.

Now, if $(\mathbf{x}, t)$ is an element of our region $X$ containing $S$ (see Sec. 5-18), then let us denote the minimum (= greatest lower bound) of our cost functional $J(\mathbf{x}, t, \mathbf{u})$ by $J^*(\mathbf{x}, t)$, so that

$$J^*(\mathbf{x}, t) = \min_{\mathbf{u} \in U} J(\mathbf{x}, t, \mathbf{u}) \tag{5-692}$$

where $U$ is our set of admissible controls. Let us assume that

1. $(\mathbf{x}^*(t), t) \in X \qquad \text{for } t \in [t_0, t_1)$ (5-693)
2. $J^*(\mathbf{x}, t)$ is continuously differentiable on $X$ (5-694)

Then, in view of the fact that $\mathbf{u}^*$ is optimal, we may deduce that

$$\frac{\partial J^*}{\partial t}[\mathbf{x}^*(t), t] + H\left[\mathbf{x}^*(t), \frac{\partial J^*}{\partial \mathbf{x}}[\mathbf{x}^*(t), t], \mathbf{u}^*(t), t\right] = 0 \tag{5-695}$$

for $t$ in $[t_0, t_1)$. Thus, under assumptions 1 and 2, Eq. (5-695) is an additional *necessary* condition for optimality.

**Exercise 5-23** Show that if $J^*(\mathbf{x}, t)$ has continuous second-partial derivatives and satisfies the equation

$$\frac{\partial J^*}{\partial t}(\mathbf{x}, t) + H\left[\mathbf{x}, \frac{\partial J^*}{\partial \mathbf{x}}(\mathbf{x}, t), \mathbf{u}^*(t), t\right] = 0 \tag{5-696}$$

throughout $X$, then the function $\mathbf{p}(t) = \dfrac{\partial J^*}{\partial \mathbf{x}}[\mathbf{x}^*(t), t]$ is a costate which corresponds to $\mathbf{u}^*(t)$ and $\mathbf{x}^*(t)$. HINT: Differentiate Eq. (5-695) with respect to $\mathbf{x}$.

## 5-20 A Sufficient Condition for Optimality

We are now going to establish the basic sufficiency condition. We shall first prove the following lemma of Caratheodory (see Refs. C-1 and K-8):

***Lemma 5-3*** *Suppose that for each point* $(\mathbf{x}, t)$ *in* $X$, *the function* $L(\mathbf{x}, \boldsymbol{\omega}, t)$ *has, as a function of* $\boldsymbol{\omega}$, *zero as its unique absolute minimum with respect to all*

† Compare with Sec. 5-16, subsection 2.

$\omega$ *in* $\Omega$ *at* $\omega = \mathbf{u}^\circ(\mathbf{x}, t)$. *In other words, suppose that*

$$0 = L[\mathbf{x}, \mathbf{u}^\circ(\mathbf{x}, t), t] < L(\mathbf{x}, \omega, t) \tag{5-697}$$

*for all* $\omega$ *in* $\Omega$, $\omega \neq \mathbf{u}^\circ(\mathbf{x}, t)$. *Let* $\hat{\mathbf{u}}$ *be an admissible control such that:*

*a.* $\hat{\mathbf{u}}$ *transfers* $(\mathbf{x}_0, t_0)$ *to* $S$.

*b. If* $\hat{\mathbf{x}}(t)$ *is the trajectory corresponding to* $\hat{\mathbf{u}}(t)$, *then*

$$(\hat{\mathbf{x}}(t), t) \in X \tag{5-698}$$

*for* $t$ *in* $[t_0, t_1]$.

*c.* $\hat{\mathbf{u}}$ *satisfies the relation*

$$\hat{\mathbf{u}}(t) = \mathbf{u}^\circ[\hat{\mathbf{x}}(t), t] \tag{5-699}$$

*for* $t$ *in* $[t_0, t_1)$.

*Then* $\hat{\mathbf{u}}$ *is an optimal control relative to the set of controls* $\mathbf{u}$ *that generate trajectories lying entirely in* $X$, *and the cost* $J[\hat{\mathbf{x}}(t), t, \hat{\mathbf{u}}]$ *is zero for* $t$ *in* $[t_0, t_1)$, *that is,*

$$J[\hat{\mathbf{x}}(t), t, ]\hat{\mathbf{u}} = 0 \qquad t \in [t_0, t_1) \tag{5-700}$$

*In other words, if* $\mathbf{u}^1$ *is a control which transfers* $(\mathbf{x}_0, t_0)$ *to* $S$ *such that* $(\mathbf{x}^1(t), t) \in X$, *where* $\mathbf{x}^1(t)$ *is the trajectory corresponding to* $\mathbf{u}^1$, *then*

$$0 = J(\mathbf{x}_0, t_0, \hat{\mathbf{u}}) \leq J(\mathbf{x}_0, t_0, \mathbf{u}^1) \tag{5-701}$$

PROOF In view of properties $a$ to $c$ of $\hat{\mathbf{u}}$, we can see that

$$J(\mathbf{x}_0, t_0, \hat{\mathbf{u}}) = \int_{t_0}^{t_1} L[\hat{\mathbf{x}}(t), \mathbf{u}^\circ[\mathbf{x}(t), t], t]\, dt \tag{5-702}$$

and hence, by virtue of Eq. (5-697), that

$$J(\mathbf{x}_0, t_0, \hat{\mathbf{u}}) = 0 \tag{5-703}$$

Now, if $\mathbf{u}^1$ is any other admissible control transferring $(\mathbf{x}_0, t_0)$ to $S$ such that the trajectory $\mathbf{x}^1(t)$ generated by $\mathbf{u}^1$ lies entirely in $X$, that is, such that $(\mathbf{x}^1(t), t) \in X$, then

$$J(\mathbf{x}_0, t_0, \mathbf{u}^1) = \int_{t_0}^{t_2} L[\mathbf{x}^1(t), \mathbf{u}^1(t), t]\, dt \tag{5-704}\dagger$$

$$\geq \int_{t_0}^{t_2} L[\mathbf{x}^1(t), \mathbf{u}^\circ[\mathbf{x}^1(t), t], t]\, dt \tag{5-705}$$

$$\geq 0 = J(\mathbf{x}_0, t_0, \mathbf{u}) \tag{5-706}$$

and the lemma is established.

We observe that the control $\hat{\mathbf{u}}$ of the lemma may *not* be optimal since there may be controls $\mathbf{u}^2$ which transfer $(\mathbf{x}_0, t_0)$ to $S$ along trajectories which do not lie entirely in $X$, as is illustrated in Fig. 5-38.

Now, the function $L(\mathbf{x}, \mathbf{u}, t)$ of our problem need not have the properties of the lemma; however, we shall, under certain assumptions about the Hamiltonian, be able to replace $L(\mathbf{x}, \mathbf{u}, t)$ with a function $\hat{L}(\mathbf{x}, \mathbf{u}, t)$ which does satisfy the conditions of the lemma and thus obtain the desired suffi-

† Note that the time $t_2$ at which $(\mathbf{x}^1(t), t)$ meets $S$ need not be the same as $t_1$.

ciency conditions. First, we shall need some definitions. In particular, we have:

***Definition 5-12*** *Let $H(\mathbf{x}, \mathbf{p}, \mathbf{u}, t)$ be the Hamiltonian of our problem, so that*

$$H(\mathbf{x}, \mathbf{p}, \mathbf{u}, t) = L(\mathbf{x}, \mathbf{u}, t) + \langle \mathbf{p}, \mathbf{f}(\mathbf{x}, \mathbf{u}, t)\rangle \tag{5-707}$$

*If, for each point $(\mathbf{x}, t)$ in $X$, the function $H(\mathbf{x}, \mathbf{p}, \boldsymbol{\omega}, t)$ has, as a function of $\boldsymbol{\omega}$, a unique absolute minimum with respect to all $\boldsymbol{\omega}$ in $\Omega$ at $\boldsymbol{\omega} = \tilde{\mathbf{u}}(\mathbf{x}, \mathbf{p}, t)$, then we shall say that $H$ is normal relative to $X$ or that our control problem is normal relative to $X$. In that case, we shall call the function $\tilde{\mathbf{u}}(\mathbf{x}, \mathbf{p}, t)$ the H-minimal control relative to $X$.*

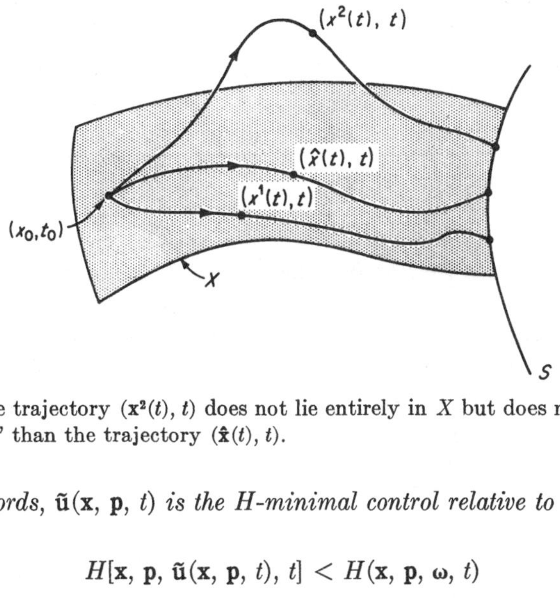


**Fig. 5-38** The trajectory $(\mathbf{x}^2(t), t)$ does not lie entirely in $X$ but does meet $S$ and may be less "costly" than the trajectory $(\hat{\mathbf{x}}(t), t)$.

*In other words, $\tilde{\mathbf{u}}(\mathbf{x}, \mathbf{p}, t)$ is the H-minimal control relative to $X$ if, for each $(\mathbf{x}, t)$ in $X$,*

$$H[\mathbf{x}, \mathbf{p}, \tilde{\mathbf{u}}(\mathbf{x}, \mathbf{p}, t), t] < H(\mathbf{x}, \mathbf{p}, \boldsymbol{\omega}, t) \tag{5-708}$$

*for all $\boldsymbol{\omega}$ in $\Omega$, with $\boldsymbol{\omega} \neq \tilde{\mathbf{u}}(\mathbf{x}, \mathbf{p}, t)$.*

***Definition 5-13*** *If $H$ is normal relative to $X$ and if $\tilde{\mathbf{u}}(\mathbf{x}, \mathbf{p}, t)$ is the H-minimal control relative to $X$, then we shall call the partial differential equation*

$$\frac{\partial \mathcal{J}}{\partial t}(\mathbf{x}, t) + H\left[\mathbf{x}, \frac{\partial \mathcal{J}}{\partial \mathbf{x}}(\mathbf{x}, t), \tilde{\mathbf{u}}\left(\mathbf{x}, \frac{\partial \mathcal{J}}{\partial \mathbf{x}}(\mathbf{x}, t), t\right), t\right] = 0 \tag{5-709}$$

*with boundary condition*

$$\mathcal{J}(\mathbf{x}, t) = 0 \qquad \text{for } (\mathbf{x}, t) \in S \tag{5-710}$$

*the Hamilton-Jacobi equation of our problem relative to $X$.*

We now have the following theorem:

***Theorem 5-12*** *Suppose that $H$ is normal relative to $X$ and that $\tilde{\mathbf{u}}(\mathbf{x}, \mathbf{p}, t)$ is the H-minimal control relative to $X$. Let $\hat{\mathbf{u}}(t)$ be an admissible control such that:*

*a.* $\hat{\mathbf{u}}$ *transfers* $(\mathbf{x}_0, t_0)$ *to* $S$.

*b. If* $\hat{\mathbf{x}}(t)$ *is the trajectory corresponding to* $\hat{\mathbf{u}}(t)$, *then*

$$(\hat{\mathbf{x}}(t), t) \in X \tag{5-711}$$

*for* $t$ *in* $[t_0, t_1]$.

*c. There is a solution* $\hat{J}(\mathbf{x}, t)$ *of the Hamilton-Jacobi equation* (5-709) *satisfying the boundary condition* (5-710) *such that*

$$\hat{\mathbf{u}}(t) = \tilde{\mathbf{u}}\left(\hat{\mathbf{x}}(t), \frac{\partial \hat{J}}{\partial \mathbf{x}}[\hat{\mathbf{x}}(t), t], t\right) \tag{5-712}$$

*for* $t$ *in* $[t_0, t_1]$.

*Then* $\hat{\mathbf{u}}$ *is an optimal control relative to the set of controls* $\mathbf{u}$ *that generate trajectories lying entirely in* $X$, *and*

$$J[\hat{\mathbf{x}}(t), t, \hat{\mathbf{u}}] = \hat{J}[\hat{\mathbf{x}}(t), t] \qquad t \in [t_0, t_1) \tag{5-713}$$

*In other words, if* $\mathbf{u}^1$ *is a control which transfers* $(\mathbf{x}_0, t_0)$ *to* $S$ *in such a way that* $(\mathbf{x}^1(t), t) \in X$ *[where* $\mathbf{x}^1(t)$ *is the trajectory corresponding to* $\mathbf{u}^1$*], then*

$$\hat{J}(\mathbf{x}_0, t_0) = J(\mathbf{x}_0, t_0, \hat{\mathbf{u}}) \leq J(\mathbf{x}_0, t_0, \mathbf{u}^1) \tag{5-714}$$

PROOF Let us consider the function $\hat{L}(\mathbf{x}, \mathbf{u}, t)$, given by

$$\hat{L}(\mathbf{x}, \mathbf{u}, t) = \frac{\partial \hat{J}}{\partial t}(\mathbf{x}, t) + H\left[\mathbf{x}, \frac{\partial \hat{J}}{\partial \mathbf{x}}(\mathbf{x}, t), \mathbf{u}, t\right] \tag{5-715}$$

We claim that $\hat{L}$ satisfies the conditions of Lemma 5-3. To verify this claim, we note first of all that

$$\begin{aligned} \hat{L}\left[\mathbf{x}, \tilde{\mathbf{u}}\left(\mathbf{x}, \frac{\partial \hat{J}}{\partial \mathbf{x}}(\mathbf{x}, t), t\right), t\right] \\ &= \frac{\partial \hat{J}}{\partial t}(\mathbf{x}, t) + H\left[\mathbf{x}, \frac{\partial \hat{J}}{\partial \mathbf{x}}(\mathbf{x}, t), \tilde{\mathbf{u}}\left(\mathbf{x}, \frac{\partial \hat{J}}{\partial \mathbf{x}}(\mathbf{x}, t), t\right), t\right] && \text{(5-716)} \\ &= 0 && \text{(5-717)} \end{aligned}$$

since $\hat{J}$ is a solution of the Hamilton-Jacobi equation. Secondly, we observe that

$$H\left[\mathbf{x}, \frac{\partial \hat{J}}{\partial \mathbf{x}}(\mathbf{x}, t), \tilde{\mathbf{u}}\left(\mathbf{x}, \frac{\partial \hat{J}}{\partial \mathbf{x}}(\mathbf{x}, t), t\right), t\right] < H\left[\mathbf{x}, \frac{\partial \hat{J}}{\partial \mathbf{x}}(\mathbf{x}, t), \boldsymbol{\omega}, t\right] \tag{5-718}$$

if $\boldsymbol{\omega} \neq \tilde{\mathbf{u}}\left(\mathbf{x}, \frac{\partial \hat{J}}{\partial \mathbf{x}}(\mathbf{x}, t), t\right)$, since $H$ is normal and $\tilde{\mathbf{u}}$ is the $H$-minimal control. It follows that

$$0 = \hat{L}\left[\mathbf{x}, \tilde{\mathbf{u}}\left(\mathbf{x}, \frac{\partial \hat{J}}{\partial \mathbf{x}}(\mathbf{x}, t), t\right), t\right] < \hat{L}(\mathbf{x}, \boldsymbol{\omega}, t) \qquad \text{if } \boldsymbol{\omega} \neq \tilde{\mathbf{u}}\left(\mathbf{x}, \frac{\partial \hat{J}}{\partial \mathbf{x}}(\mathbf{x}, t), t\right) \tag{5-719}$$

Now, application of the lemma will show that

$$\int_{t_0}^{t_1} \hat{L}[\hat{\mathbf{x}}(t), \hat{\mathbf{u}}(t), t]\, dt = 0 \tag{5-720}$$

and that

$$\int_{t_0}^{t_2} \hat{L}[\mathbf{x}^1(t), \mathbf{u}^1(t), t]\, dt \geq 0 \tag{5-721}$$

where $\mathbf{u}^1$ is another control which transfers $(\mathbf{x}_0, t_0)$ to $S$ in such a way that $(\mathbf{x}^1(t), t) \in X$. However, we observe that

$$\hat{L}(\mathbf{x}, \mathbf{u}, t) = \frac{\partial \hat{J}}{\partial t}(\mathbf{x}, t) + H\left[\mathbf{x}, \frac{\partial \hat{J}}{\partial \mathbf{x}}(\mathbf{x}, t), \mathbf{u}, t\right] \tag{5-722}$$

$$= \frac{\partial \hat{J}}{\partial t}(\mathbf{x}, t) + L(\mathbf{x}, \mathbf{u}, t) + \left\langle \frac{\partial \hat{J}}{\partial \mathbf{x}}(\mathbf{x}, t), \mathbf{f}(\mathbf{x}, \mathbf{u}, t)\right\rangle \tag{5-723}$$

It follows that

$$\hat{L}[\hat{\mathbf{x}}(t), \hat{\mathbf{u}}(t), t] = L[\hat{\mathbf{x}}(t), \hat{\mathbf{u}}(t), t] + \frac{d}{dt}\hat{J}[\hat{\mathbf{x}}(t), t] \tag{5-724}$$

and that

$$\hat{L}[\mathbf{x}^1(t), \mathbf{u}^1(t), t] = L[\mathbf{x}^1(t), \mathbf{u}^1(t), t] + \frac{d}{dt}\hat{J}[\mathbf{x}^1(t), t] \tag{5-725}$$

[cf. Eq. (5-715)]. We may then conclude that

$$\int_{t_0}^{t_1} \hat{L}[\hat{\mathbf{x}}(t), \hat{\mathbf{u}}(t), t]\, dt = \int_{t_0}^{t_1} L[\hat{\mathbf{x}}(t), \hat{\mathbf{u}}(t), t]\, dt + \int_{t_0}^{t_1} \frac{d\hat{J}}{dt}[\hat{\mathbf{x}}(t), t]\, dt \tag{5-726}$$

$$= J(\mathbf{x}_0, t_0, \hat{\mathbf{u}}) - \hat{J}(\mathbf{x}_0, t_0) + \hat{J}[\hat{\mathbf{x}}(t_1), t_1] \tag{5-727}$$

and that

$$\int_{t_0}^{t_2} \hat{L}[\mathbf{x}^1(t), \mathbf{u}^1(t), t]\, dt = \int_{t_0}^{t_2} L[\mathbf{x}^1(t), \mathbf{u}^1(t), t]\, dt + \int_{t_0}^{t_2} \frac{d\hat{J}}{dt}[\mathbf{x}^1(t), t]\, dt \tag{5-728}$$

$$= \mathbf{J}(\mathbf{x}_0, t_0, \mathbf{u}^1) - \hat{J}(\mathbf{x}_0, t_0) + \hat{J}[\mathbf{x}^1(t_2), t_2] \tag{5-729}$$

However, we know that $(\hat{\mathbf{x}}(t_1), t_1) \in S$ and that $(\mathbf{x}^1(t_2), t_2) \in S$; consequently, in view of our assumption *c* and Eqs. (5-727), (5-729), (5-724), and (5-725), we can see that

$$J(\mathbf{x}_0, t_0, \hat{\mathbf{u}}) - \hat{J}(\mathbf{x}_0, t_0) = 0 \tag{5-730}$$

and that

$$J(\mathbf{x}_0, t_0, \mathbf{u}^1) - \hat{J}(\mathbf{x}_0, t_0) \geq 0 \tag{5-731}$$

Thus, the theorem is established.

We note that this theorem is a local sufficiency condition for optimality, in that all statements are made relative to the region $X$. We also observe, as illustrated in Fig. 5-39, that if $S_1$ is a subset of $S$ containing $(\hat{\mathbf{x}}(t_1), t_1)$ and if $X_1$ is a subregion of $X$ such that $S_1 \subset X_1$ and $(\hat{\mathbf{x}}(t), t) \in X_1$ for $t$ in $[t_0, t_1]$, then $\hat{\mathbf{u}}$ will be an optimal control for the problem with target set $S_1$ relative to the set of controls $\mathbf{u}$ that generate trajectories lying entirely in $X_1$. This is so because the function $\hat{J}(\mathbf{x}, t)$ will satisfy the Hamilton-Jacobi equation on $X_1$ and the boundary condition $\hat{J}(\mathbf{x}, t) = 0$ for $(\mathbf{x}, t) \in S_1$. In other words, restriction of $\hat{J}$ does not change the fact that $\hat{J}$ satisfies the requisite conditions. However, if the region $X$ is enlarged to $X_2$ (say), there is no guarantee that a solution of the Hamilton-Jacobi equation on $X_2$ exists; moreover, even if there is a solution on $X_2$, this solution may not be an extension of $\hat{J}$.

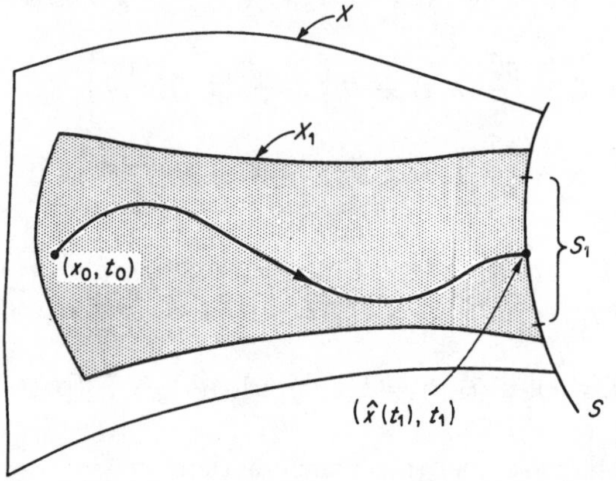


**Fig. 5-39** If $\hat{J}$ satisfies Eqs. (5-709) and (5-710) for $X$ and $S$, then $\hat{J}$ satisfies these equations for $X_1$ and $S_1$ (but *not* conversely).

We further note that if the region $X$ is $R_n \times (T_1, T_2)$ [where $(T_1, T_2)$ is the domain of definition of our system], then the theorem will give a global sufficiency condition for optimality. In other words, we have:

***Theorem 5-13*** *Suppose that $X = R_n \times (T_1, T_2)$, that $H$ is normal relative to $R_n \times (T_1, T_2)$, and that $\tilde{\mathbf{u}}(\mathbf{x}, \mathbf{p}, t)$ is the $H$-minimal control relative to $R_n \times (T_1, T_2)$. Let $\hat{\mathbf{u}}(t)$ be an admissible control such that:*

*a.* *$\hat{\mathbf{u}}$ transfers $(\mathbf{x}_0, t_0)$ to $S$.*

*b.*† *There is a solution $\hat{J}(\mathbf{x}, t)$ of the Hamilton-Jacobi equation* (5-709) *satisfying the boundary condition* (5-710) *such that*

$$\hat{\mathbf{u}}(t) = \tilde{\mathbf{u}}\left(\hat{\mathbf{x}}(t), \frac{\partial \hat{J}}{\partial \mathbf{x}}[\hat{\mathbf{x}}(t), t], t\right) \tag{5-732}$$

*for $t$ in $[t_0, t_1]$.*

*Then $\hat{\mathbf{u}}$ is an optimal control and*

$$J[\hat{\mathbf{x}}(t), t, \hat{\mathbf{u}}] = \hat{J}[\hat{\mathbf{x}}(t), t] \qquad t \in [t_0, t_1) \tag{5-733}$$

We shall comment on these sufficiency conditions in the next section, and examples of their use will appear in subsequent chapters of the book.

† Observe that $(\hat{\mathbf{x}}(t), t)$ is automatically an element of $R_n \times (T_1, T_2)$.

## 5-21 Some Comments on the Sufficiency Conditions

We shall now discuss and interpret geometrically the sufficiency conditions we derived in the previous section, noting that most of our comments will take on greater significance in later chapters of the book.

Let us first observe that there are essentially two major ideas involved in the sufficiency conditions; these are:

1. The notion of normality relative to $X$
2. The concept of the Hamilton-Jacobi partial differential equation for the cost

The assumption of normality is indeed a strengthening of necessary condition 2 of the minimum principle, and the Hamilton-Jacobi equation does indeed represent a requirement on the behavior of the cost. Since normality is relatively easy to understand, we shall concentrate our attention on the Hamilton-Jacobi equation.

We note that the Hamilton-Jacobi equation, *being a partial differential equation*, is often quite difficult (if not impossible) to solve; moreover, a particular solution may only represent the cost along a given trajectory and not throughout the region $X$. For this reason, the Hamilton-Jacobi equation is most often used as a check on the optimality of a control derived from the minimum principle, as we shall see in later chapters of the book.

We have already remarked that Theorem 5-12 is a local sufficiency condition, in that all statements are made relative to the region $X$. For the purposes of proof, we supposed that the region $X$ was specified a priori. Now, in practice, we are usually given an initial pair $(\mathbf{x}_0, t_0)$, and what we do is try to find a suitable region $X$ in which the theorem may be applied. In particular, if $\tilde{\mathbf{u}}(\mathbf{x}, \mathbf{p}, t)$ is the $H$-minimal control and if we know either the optimal cost $J^*(\mathbf{x}, t)$ over a region or a suitable expression $\mathbf{p}^*(\mathbf{x}, t)$ for the costate, then we try to find a region $X$ about $(\mathbf{x}_0, t_0)$ which is "filled" by trajectories generated by controls $\hat{\mathbf{u}}$ such that, along $\hat{\mathbf{x}}(t)$,

$$\hat{\mathbf{u}}(t) = \tilde{\mathbf{u}}\left(\hat{\mathbf{x}}(t), \frac{\partial J^*}{\partial \mathbf{x}}[\hat{\mathbf{x}}(t), t], t\right) \qquad \text{or} \qquad \tilde{\mathbf{u}}(\hat{\mathbf{x}}(t), \mathbf{p}^*[\hat{\mathbf{x}}(t), t], t) \tag{5-734}$$

A typical situation is illustrated in Fig. 5-40.

Now, it often happens that $H$ is normal relative to $R_n \times (T_1, T_2)$ but that a global solution of the Hamilton-Jacobi cannot be found. In such a situation, it is sometimes possible to find disjoint regions $X_1, X_2, \ldots, X_r$ and solutions $J_1, J_2, \ldots, J_r$ of the Hamilton-Jacobi equation such that $X_1 \cup X_2 \cup \cdots \cup X_r = R_n \times (T_1, T_2)$ and the sufficiency conditions apply in each $X_i$. This process of "piecing together" suitable regions $X_i$ often leads to a global proof of optimality. (See, for example, Sec. 7-2.) However, we must usually solve our problem in order to determine the regions $X_i$, and so this procedure is essentially a check on the optimality of a control derived from the minimum principle.

Now let us briefly interpret the results in a somewhat geometric fashion. We first observe that if the cost functional $J(\mathbf{x}, t, \mathbf{u})$ is smooth, then the equation

$$\frac{\partial J}{\partial t}[\mathbf{x}_{\mathbf{u}}(t), t, \mathbf{u}(t)] + H\left[\mathbf{x}_{\mathbf{u}}(t), \frac{\partial J}{\partial \mathbf{x}}[\mathbf{x}_{\mathbf{u}}(t), t, \mathbf{u}(t)], \mathbf{u}(t), t\right] = 0 \quad (5\text{-}735)$$

represents the behavior of $J$ along a trajectory $\mathbf{x}_{\mathbf{u}}(t)$ of our system [see Eq. (5-686)]. Thus, no matter what the control $\mathbf{u}$ which takes us to our target set, the cost must satisfy Eq. (5-735).

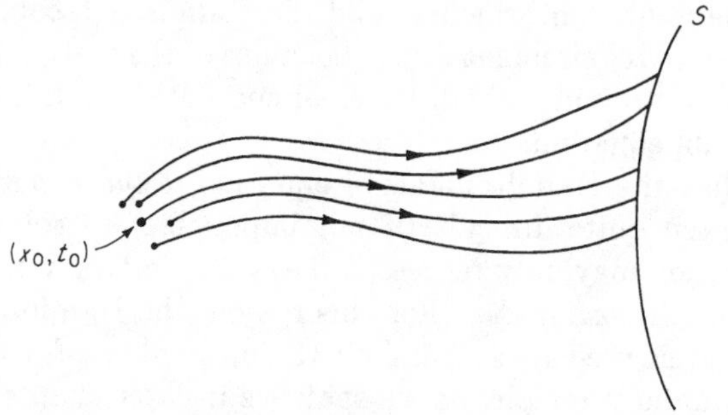


**Fig. 5-40** "Filling" out a region with suitable trajectories.

Now imagine† that $J^*(\mathbf{x}, t)$ is our optimum "cost surface" and that $\mathbf{u}^*$ is the optimal control; then our optimal trajectory (in $R_{n+1}$) defines a curve $j^*(\tau) = J^*[\mathbf{x}^*(\tau), \tau]$ lying on the surface $J^*(\mathbf{x}, t)$. We assert that the "direction" (i.e., the tangent vector) of this curve $j^*(\tau)$ points along the direction of *steepest descent* consistent with the constraints (i.e., the differential equations of motion and the set $\Omega$). In other words, at a point $(\mathbf{x}^*(\tau), \tau)$, we can define a "cone" of tangent vectors

$$\{(\dot{\mathbf{x}}_{\mathbf{u}}(\tau), 1) : \dot{\mathbf{x}}_{\mathbf{u}}(\tau) = \mathbf{f}[\mathbf{x}(\tau), \mathbf{u}, \tau];\ \mathbf{u} \in \Omega\}$$

and the vector $(\dot{\mathbf{x}}^*_{\mathbf{u}^*}(\tau), 1)$ represents the direction of the most rapid decrease in the cost $J^*[\mathbf{x}(t), t]$.‡ This means that

$$\left\langle \begin{bmatrix} \dot{\mathbf{x}}^*_{\mathbf{u}^*}(\tau) \\ 1 \end{bmatrix}, \begin{bmatrix} -\dfrac{\partial J^*}{\partial \mathbf{x}}[\mathbf{x}^*(\tau), \tau] \\ -\dfrac{\partial J^*}{\partial t}[\mathbf{x}^*(\tau), \tau] \end{bmatrix} \right\rangle \le \left\langle \begin{bmatrix} \dot{\mathbf{x}}_{\mathbf{u}}(\tau) \\ 1 \end{bmatrix}, \begin{bmatrix} -\dfrac{\partial J^*}{\partial \mathbf{x}}[\mathbf{x}^*(\tau), \tau] \\ -\dfrac{\partial J^*}{\partial t}[\mathbf{x}^*(\tau), \tau] \end{bmatrix} \right\rangle \quad (5\text{-}736)$$

† Unfortunately, a good geometric figure would be four-dimensional, because one needs at least a plane to represent the state space, one axis for the time, and another axis for the cost.

‡ How does this "cone" relate to the cones of attainability introduced in Sec. 5-16?

for all $\mathbf{u} \in \Omega$, because the $n + 1$ vector

$$\nabla J^*[\mathbf{x}^*(\tau), \tau] = \begin{bmatrix} \dfrac{\partial J^*}{\partial \mathbf{x}} [\mathbf{x}^*(\tau), \tau] \\ \dfrac{\partial J^*}{\partial t} [\mathbf{x}^*(\tau), \tau] \end{bmatrix} \tag{5-737}$$

is the gradient of $J^*$ at $(\mathbf{x}^*(\tau), \tau)$ and it points in the direction of *steepest ascent* (which is opposite to the direction of steepest descent). From Eq. (5-736) and from the definition of the Hamiltonian, we can deduce the necessary condition

$$H\left[\mathbf{x}^*(\tau), \frac{\partial J^*}{\partial \mathbf{x}} [\mathbf{x}^*(\tau), \tau], \mathbf{u}^*(\tau), \tau\right] \leq H\left[\mathbf{x}^*(\tau), \frac{\partial J^*}{\partial \mathbf{x}} [\mathbf{x}^*(\tau), \tau], \mathbf{u}, \tau\right] \tag{5-738}$$

for all $\mathbf{u}$ in $\Omega$.

In substance, then, our sufficiency conditions consist of:

1. The requirement that, over a given region, there is a *unique* direction of steepest descent consistent with the constraints along the optimal cost surface

2. The requirement that the curve along the cost surface generated by the control, which we wish to show is optimal, always has this direction of steepest descent as its direction (or tangent)

If these requirements are met, then our given control will be optimal. A more detailed explication of these ideas, with reference to the time-optimal control problem, will appear in Sec. 6-7.

# Chapter 6

# STRUCTURE AND PROPERTIES OF OPTIMAL SYSTEMS

## 6-1 Introduction

In Chap. 5, we considered the optimal-control problem from a general point of view. The system equations, the constraints, the cost functionals, and the target sets considered were quite general. The results we presented were primarily necessary conditions for optimality and revolved around the very powerful minimum principle.

We can view the developments of the previous four chapters as follows: Chaps. 2 and 3 contained most of the mathematical background necessary for the formulation, understanding, and manipulation of deterministic system problems; Chap. 4 was devoted primarily to concepts, definitions, and properties of dynamical systems; Chap. 5 contained the classical and modern variational results which are basic to the theory of optimal processes. Thus, we have reached, in a sense, a theoretical peak. The purpose of the remainder of the book is to apply the theory to the design of optimal systems and, in so doing, illustrate the logical and computational steps involved in the formulation and solution of optimal-control problems.

The major purpose of this chapter is to provide a buffer between the theoretical development and the specific design problems which we shall consider in Chaps. 7 to 10. Thus, this chapter contains a mixture of geometric concepts, analytical results, and intuitive considerations. We consider optimal-control problems with specific performance criteria in an effort to show how the necessary conditions may be used to derive properties of the optimal control.

The theoretical results of Chap. 5 can be used to derive the conditions that must be satisfied by the optimal control. However, the real problem is to find the optimal control for a given problem. In our search for the optimal control, we can use the necessary conditions provided by the minimum principle in order to isolate the admissible controls that are candidates for optimality. In other words, we start by finding a class of controls, which are called *extremal*, that satisfy all the necessary conditions provided by the minimum principle.

Let us suppose now that we are presented with a problem and that we are able to find the *extremal* controls, i.e., the controls that satisfy the necessary conditions. The question that now arises is: *How can we find the optimal control(s) from the extremal controls?* Well, first of all, we must prove that an optimal control exists. Second, we must find out whether or not the optimal control is unique. Even though the optimal control may be unique, this does not imply that the extremal controls are unique, because the necessary conditions provided by the minimum principle are local in nature. The reader should now appreciate the importance of any information regarding the uniqueness of the extremal controls and of any additional information about special properties of the extremal and optimal controls. It is our purpose in this chapter to present such general results with respect to various performance criteria.

The reader will notice that we shall often sacrifice brevity of exposition for a loose and informal statement of significant results. Also, many proofs are incomplete; in such cases, we either offer references which contain the formal proof or leave the proof to the reader.

This chapter is divided into four major parts: minimum-time problems, minimum-fuel problems, minimum-energy problems, and singular problems.

The minimum-time problem is considered in Secs. 6-2 to 6-10. In a minimum-time problem, we are interested in transferring the state of the system to a target set as quickly as possible.

The minimum-fuel problem is considered in Secs. 6-11 to 6-16. This problem involves the minimization of the consumed fuel required to force an initial state to a target set. Constraints on the response time may also be given.

The minimum-energy problem is considered in Secs. 6-17 to 6-20. Minimum-energy problems arise, for example, when the control is an electrical signal. Again, we may include time constraints in the problem.

In Secs. 6-21 and 6-22, we briefly consider the so-called "singular problem." We shall define singular problems in Secs. 6-3 and 6-13 with respect to the minimum-time and minimum-fuel criteria and, in so doing, motivate the material presented in Secs. 6-21 and 6-22.

We shall conclude this chapter with some general remarks concerning the formulation and solution of optimization problems.

## 6-2 Minimum-time Problems 1: Formulation and Geometric Interpretation

The class of optimization problems for which the sole measure of performance is the minimization of the transition time from an initial state to a target set is called the class of minimum-time (or brachistochrone) problems.

In this section, we shall formulate in a precise manner a time-optimal control problem (Problem 6-1$a$). We shall motivate this mathematical problem by means of a physical example. We shall spend the major part of this section discussing the problem from the geometric point of view. In essence, we shall show that the minimum-time problem reduces to finding:

1. The first instant of time at which the set of reachable states meets the target set
2. The control which accomplishes this

We shall not use the theory of Chap. 5 in this section; this will be done in Sec. 6-3.

***Problem 6-1a Time-optimal Intercept Problem*** *Given the dynamical system [with state* $\mathbf{x}(t)$*, output* $\mathbf{y}(t)$*, and control* $\mathbf{u}(t)$*] defined by the equations*

$$\dot{\mathbf{x}}(t) = \mathbf{f}[\mathbf{x}(t), t] + \mathbf{B}[\mathbf{x}(t), t]\mathbf{u}(t) \tag{6-1}$$
$$\mathbf{y}(t) = \mathbf{h}[\mathbf{x}(t)] \tag{6-2}$$

*Assume that*

$$\begin{aligned} &\mathbf{x}(t) \text{ is an } n\text{-dimensional vector} \\ &\mathbf{y}(t) \text{ is an } m\text{-dimensional vector} \\ &\mathbf{u}(t) \text{ is an } r\text{-dimensional vector} \end{aligned} \tag{6-3}$$

*and that*

$$n \geq r \geq m > 0 \tag{6-4}$$

*Thus,* $\mathbf{f}$ *is an n-vector-valued function,* $\mathbf{B}[\mathbf{x}(t); t]$ *is an* $(n \times r)$*-matrix-valued function, and* $\mathbf{h}$ *is an m-vector-valued function. Moreover, assume that the components* $u_1(t), u_2(t), \ldots, u_r(t)$ *of the control vector* $\mathbf{u}(t)$ *are restricted in magnitude by the inequalities*

$$|u_j(t)| \leq m_j \qquad j = 1, 2, \ldots, r \tag{6-5}$$

*Let* $\mathbf{z}(t)$ *be a vector with m components. Let us agree to call* $\mathbf{z}(t)$ *the desired output. Let*

$$\mathbf{e}(t) = \mathbf{y}(t) - \mathbf{z}(t)$$

*be the error vector.*

*Let* $t_0$ *be the initial time and let* $\mathbf{x}(t_0)$ *be the initial state of the dynamical system.*

*Find the control(s) which:*

1. *Satisfy the constraint* (6-5)
2. *Drive the system in such a way that at the terminal time T*

$$\mathbf{e}(T) \in E \tag{6-6}$$

*where E is some specified subset of* $R_m$

3. *Minimize the response time* $T - t_0$

We shall explain the salient features of this problem in the following example.

**Example 6-1** Suppose that the rocket $A$ is to be controlled so that the missile $B$ will be destroyed, as illustrated in Fig. 6-1. Suppose that at the initial time $t = t_0$ the position of the missile $B$ is $\mathbf{z}(t_0)$ and that its trajectory $\mathbf{z}(t)$ is known for all $t \geq t_0$; the vector $\mathbf{z}(t)$ has three components which define the position of the missile $B$ in space.

Let $\mathbf{y}(t_0)$ be the position of the rocket $A$ at $t = t_0$. The rocket $A$ is carried by an airplane, and at $t = t_0$ the rocket is fired in order to intercept the missile $B$. The equations of motion of the rocket $A$ will be the form (6-1) and (6-2), where the output $\mathbf{y}(t)$ is the position vector of the rocket $A$. The state $\mathbf{x}(t)$ of the rocket $A$ will be its position, velocity, amount of fuel, angle of attack, etc. The control $\mathbf{u}(t)$ will consist of the components of the thrust and any other control scheme that may be used to guide the rocket—such as, for example, the position of the aerodynamic control surfaces. Magnitude constraints on the thrust components and physical limitations on the position of the aerodynamic control surfaces (due to mechanical stops) will be of the type given in Eq. (6-5).

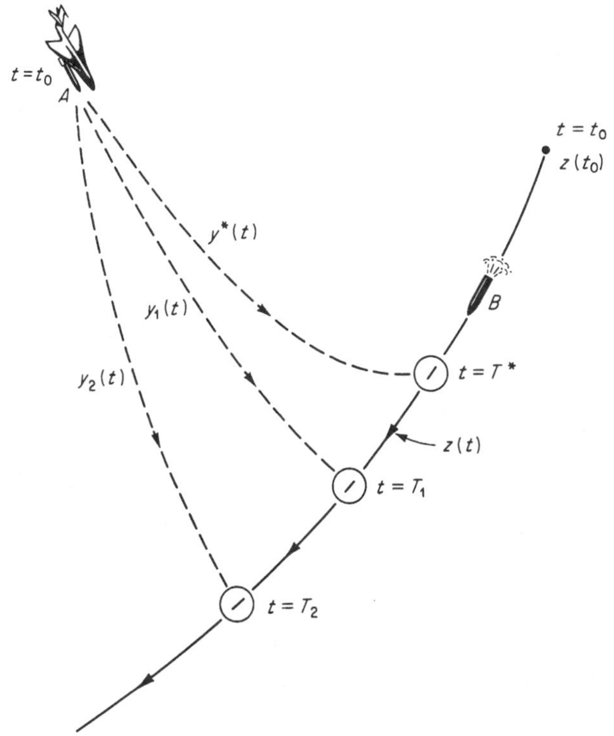


**Fig. 6-1** The problem of intercepting the rocket $B$ by the rocket $A$.

The error vector $\mathbf{e}(t) = \mathbf{y}(t) - \mathbf{z}(t)$ is simply the difference in position between the rocket $A$ and the missile $B$.

The problem is to guide the rocket $A$ so that it will come "near" the missile $B$ in the *shortest possible time.* By "near" we may mean, for example, that the error $\mathbf{e}(T)$ at the terminal time have small norm, i.e., that

$$e_1^2(T) + e_2^2(T) + e_3^2(T) \leq \delta \qquad \delta \text{ small} \tag{6-7}$$

This is a requirement similar to that given by Eq. (6-6), where $E$ is the set defined by $E = \{\mathbf{e}(T)\colon e_1^2(T) + e_2^2(T) + e_3^2(T) \leq \delta\}$.

Let us now discuss Problem 6-1$a$ in the light of Example 6-1. We shall show that the requirements on the error vector can be transformed into requirements in the state space, provided that the system described by Eqs. (6-1) and (6-2) is completely observable.†

First of all, the requirement $\mathbf{e}(T) \in E$ can be changed into a requirement on $\mathbf{y}(T)$. To see this, we note that since $\mathbf{y}(t) = \mathbf{e}(t) + \mathbf{z}(t)$, $\mathbf{e}(T) \in E$ implies that

$$\mathbf{y}(T) \in Y \tag{6-8}$$

where $Y$ is the set defined by the relation

$$Y = \{\mathbf{y}(T) : \mathbf{y}(T) = \mathbf{e}(T) + \mathbf{z}(T); \mathbf{e}(T) \in E\} \tag{6-9}$$

Thus, $Y$ depends upon $E$ and $\mathbf{z}(T)$, and furthermore, $Y$ is uniquely defined by $E$ and $\mathbf{z}(T)$.

We are interested in deriving a relation of the form $\mathbf{x}(T) \in S$, where $S$ is a (target) set in the state space, from the relation (6-8). Since $\mathbf{y}(t) = \mathbf{h}[\mathbf{x}(t)]$ [Eq. (6-2)], we have the relation, at $t = T$,

$$\mathbf{y}(T) = \mathbf{h}[\mathbf{x}(T)] \tag{6-10}$$

Equation (6-10) uniquely defines $\mathbf{y}(T)$ in terms of $\mathbf{x}(T)$. In order to uniquely define a target set $S$ in terms of the set $Y$, there must be a one-to-one correspondence between the state $\mathbf{x}(t)$ and the output $\mathbf{y}(t)$; that is, the statement $\mathbf{y}(t) = \mathbf{h}[\mathbf{x}_1(t)] = \mathbf{h}[\mathbf{x}_2(t)]$ for all $t$ must imply that $\mathbf{x}_1(t) = \mathbf{x}_2(t)$ for all $t$. If the dynamical system described by Eqs. (6-1) and (6-2) is completely observable, then to each $\mathbf{y}(T)$ there corresponds a *unique* state $\mathbf{x}(T)$.‡ It follows that we can define the target set $S$ in the state space by the relation

$$S = \{\mathbf{x}(T) : \mathbf{y}(T) = \mathbf{h}[\mathbf{x}(T)]; \mathbf{y}(T) \in Y\} \tag{6-11}$$

The following lemma summarizes these results:

***Lemma 6-1*** *Consider Problem* 6-1*a. Given a* $\mathbf{z}(t)$, *a time* $T$, *and a set* $E$. *If the system described by Eqs.* (6-1) *and* (6-2) *is completely observable, then the requirement* $\mathbf{e}(T) \in E$ *is equivalent to the requirement*

$$\mathbf{x}(T) \in S \tag{6-12}$$

*where* $S$ *is a well-defined set in the state space, given by*

$$S = \{\mathbf{x}(T) : \mathbf{h}[\mathbf{x}(T)] = \mathbf{e}(T) + \mathbf{z}(T); \mathbf{e}(T) \in E\} \tag{6-13}$$

We illustrate these concepts by the following example.

**Example 6-2** Suppose that $x_1(t)$ and $x_2(t)$ are the components of the state vector and that $y(t) = x_1(t) + x_2(t)$ is the output of a completely observable dynamical system. Let $z(t) = e^{-t}$ be the desired output and let $e(t) = y(t) - z(t)$. Let $E = \{e(T) : |e(T)| \leq 1\}$. The set $S$ is then specified by

$$S = \{(x_1(T), x_2(T)) : |x_1(T) + x_2(T) - e^{-T}| \leq 1\} \tag{6-14}$$

† See Sec. 4-15.

‡ See Theorem 4-6. In essence, if $\mathbf{h}^{-1}(Y)$ is the inverse image of $Y$, then we have: If $\mathbf{x} \in \mathbf{h}^{-1}(Y)$, then $\mathbf{y} \in Y$.

The target set $S$ is a function of the time $T$. The target set $S$ is illustrated in Fig. 6-2$a$ for $T = 0.290$ and in Fig. 6-2$b$ for $T = 0.695$.

The remainder of this section is devoted to interpreting the time-optimal control problem geometrically in terms of sets in the state space which move with time. In order to fix ideas and terminology, we make the following definitions:

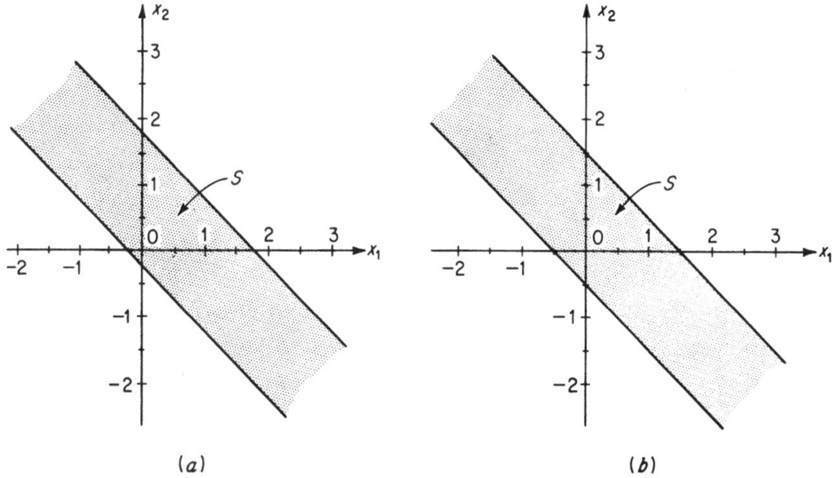


**Fig. 6-2** ($a$) The target set $S$ given by Eq. (6-14) when $T = 0.290$. ($b$) The target set $S$ given by Eq. (6-14) when $T = 0.695$.

***Definition 6-1*** *The constraint set* $\Omega$ *is defined by*

$$\Omega = \{\mathbf{u}(t): |u_j(t)| \leq m_j \qquad j = 1, 2, \ldots, r\} \tag{6-15}$$

*$\Omega$ is a subset (a parallelepiped) of the r-dimensional space of the control variables. When we write $\mathbf{u}(t) \in \Omega$, we mean that at time t, the components of the control vector $\mathbf{u}(t)$ satisfy the inequalities $|u_j(t)| \leq m_j$.*

Let $[t_0, T]$ be a closed interval of time. Let us consider the set of control functions $\mathbf{u}_{[t_0,T]}$ (see Sec. 4-4), with the property that for every $t \in [t_0, T]$, $\mathbf{u}(t) = \mathbf{u}_{[t_0,T]}(t) \in \Omega$.

***Definition 6-2*** *The set of admissible control functions $U_T$ is defined by*

$$U_T = \{\mathbf{u}_{[t_0,T]}: \mathbf{u}_{[t_0,T]}(t) \in \Omega \qquad \textit{for all } t \in [t_0, T]\} \tag{6-16}$$

*The set $U_T$ is a subset of the function space $\mathcal{P}([t_0, T], R_r)$ [where $\mathcal{P}([t_0, T], R_r)$ is the set of all piecewise continuous functions from $[t_0, T]$ to $R_r$; see Sec. 3-15]. Thus, when we write $\mathbf{u}_{[t_0,T]} \in U_T$, we mean that $\mathbf{u}_{[t_0,T]}$ is a control function which at each instant of time satisfies the magnitude constraints imposed.*†

† $U_T$ can be defined in terms of any constraint set $\Omega$ in $R_r$ [not necessarily of the form given by Eq. (6-15)].

Let us now suppose that we conduct the following experiment: We have the system

$$\dot{\mathbf{x}}(t) = \mathbf{f}[\mathbf{x}(t); t] + \mathbf{B}[\mathbf{x}(t), t]\mathbf{u}(t) \tag{6-17}$$

which, at the initial time $t_0$, is at the state $\mathbf{x}(t_0)$. We apply a control $\mathbf{u}_{[t_0,T]} \in U_T$ to the system, and we observe the behavior of the state vector $\mathbf{x}(t)$ for all $t$ in $[t_0, T]$. The entire state trajectory $\mathbf{x}_{[t_0,T]}$ will depend:

1. On the initial time $t_0$
2. On the initial state $\mathbf{x}(t_0)$
3. On the applied control $\mathbf{u}_{[t_0,T]}$

Thus, we write

$$\mathbf{x}_{[t_0,T]} = \mathbf{x}_{[t_0,T]}(\mathbf{x}(t_0), t_0, \mathbf{u}_{[t_0,T]}) \tag{6-18}$$

The state vector $\mathbf{x}(t)$, for some $t \in [t_0, T]$, will be given by

$$\mathbf{x}(t) = \mathbf{x}_{[t_0,T]}(t) = \boldsymbol{\varphi}[\mathbf{x}(t_0), t_0, t, \mathbf{u}_{[t_0,t]}] \tag{6-19}$$

If we apply every element of $U_T$, we shall obtain a set of state trajectories $\mathbf{x}_{[t_0,T]}$. This leads us to the following definition:

***Definition 6-3*†** *By the set of reachable states (at time $t$), we shall mean the subset $A_t$ of the state space, defined by*

$$\begin{aligned} A_t &= A[t; t_0; \mathbf{x}(t_0); U_t] \\ &= \{\mathbf{x}: \textit{there is a } \mathbf{u}_{[t_0,t]} \in U_t \textit{ such that the corresponding solution } \mathbf{x}(t) \\ &\quad \textit{of the system (6-19), starting at } \mathbf{x}(t_0)\textit{, has the property that } \mathbf{x}(t) = \mathbf{x}\}\ddagger \end{aligned} \tag{6-20}$$

*We shall denote the boundary of $A_t$ by $\partial A_t$.*

Let us now *suppose* that the set of reachable states $A_t$ has the following properties:

1. $$A_t \text{ is closed and bounded for every } t \geq t_0 \tag{6-21}$$

(that is, $A_t$ is compact).

2. If $t_1$ and $t_2$ are any values of time such that

$$t_0 \leq t_1 \leq t_2 \tag{6-22}$$

then the inclusion relations

$$\{\mathbf{x}(t_0)\} \subset A_{t_1} \subset A_{t_2} \tag{6-23}$$

hold.

† See also Sec. 4-20.

‡ This definition will also hold for any system of the form $\dot{\mathbf{x}}(t) = \mathbf{f}[\mathbf{x}(t), \mathbf{u}(t), t]$ and any set $U_t$ of admissible controls.

Equations (6-22) and (6-23) imply that the set of reachable states "grows" with increasing time. We illustrate this behavior in Fig. 6-3. The sets $A_{T_1}$, $A_{T_2}$, . . . , where $t_0 \leq T_1 \leq T_2 \leq T_3 \leq T_4 \leq T_5$, are specified by their respective boundaries† $\partial A_{T_1}$, $\partial A_{T_2}$, . . . .

We previously explained that requirements on the error are equivalent to the requirement that the state must reach a target set $S$. This target set $S$ is a subset of the state space and may depend on the time. For this

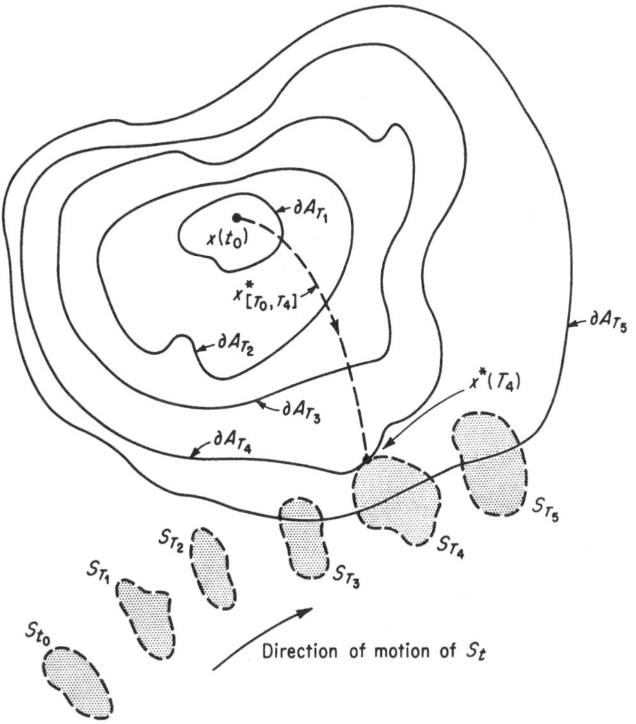


**Fig. 6-3** The sets of reachable states as $t$ increases and the moving target set $S_t$. The point $\mathbf{x}^*(T_4)$ is the first point common to both $A_t$ and $S_t$.

reason, we shall write $S_t$ for the target set to emphasize its time dependence. We let $\partial S_t$ denote the boundary of $S_t$. *We assume that $S_t$ is closed.* In Fig. 6-3, we illustrate the motion of the target set $S_t$ for increasing $t$; we use dashed curves to indicate the boundaries $\partial S_{T_1}$, $\partial S_{T_2}$, $\partial S_{T_3}$, . . . of the target sets $S_{T_1}$, $S_{T_2}$, $S_{T_3}$, . . . , where $t_0 \leq T_1 \leq T_2 \leq T_3 \leq T_4 \leq T_5$.

The basic concepts of minimum-time control can be understood by carefully examining Fig. 6-3. Such an examination will reveal that:

† To compute the boundary $\partial A_T$ of the set $A_T$, we must apply all the elements of $U_T$ to the system (6-17).

1. The sets $A_t$ and $S_t$ do not have any common elements for $t < T_4$. More precisely,

$$A_t \cap S_t = \emptyset \qquad \text{for all } t \in [t_0, T_4) \tag{6-24}$$

where $\emptyset$ is the empty set and $t \in [t_0, T_4)$ means that $t_0 \leq t < T_4$.

2. At $t = T_4$, the sets $A_{T_4}$ and $S_{T_4}$ have precisely a single common point [denoted by $\mathbf{x}^*(T_4)$ in Fig. 6-3].

3. For $t > T_4$, the set $A_t \cap S_t$ contains more than one element of the state space.

Since the set $A_{T_4}$ is the set of reachable states at $T_4$ and is generated by the elements of $U_{T_4}$, it follows that there is an admissible control

$$u^*_{[t_0,T_4]} \in U_{T_4} \tag{6-25}$$

which generates a *unique* state trajectory

$$\mathbf{x}^*_{[t_0,T_4]}$$

such that

$$\mathbf{x}^*_{[t_0,T_4]}(t_0) = \mathbf{x}(t_0) \tag{6-26}$$

$$\mathbf{x}^*_{[t_0,T_4]}(T_4) = \mathbf{x}^*(T_4) \tag{6-27}$$

(The uniqueness part follows from the uniqueness property of the solution of a system of differential equations.) Moreover, it is easy to see from Fig. 6-3 that the state $\mathbf{x}^*(T_4)$ belongs to the boundary of both $A_{T_4}$ and $S_{T_4}$, that is, that

$$\begin{aligned} \mathbf{x}^*(T_4) &\in \partial A_{T_4} \\ \mathbf{x}^*(T_4) &\in \partial S_{T_4} \\ \mathbf{x}^*(T_4) &= \partial A_{T_4} \cap \partial S_{T_4} \end{aligned} \tag{6-28}$$

We can readily see that:

1. The trajectory $\mathbf{x}^*_{[t_0,T_4]}$ is a time-optimal trajectory from the initial state $\mathbf{x}(t_0)$ to the target set $S_t$.

2. The control $\mathbf{u}^*_{[t_0,T_4]}$ (which generates $\mathbf{x}^*_{[t_0,T_4]}$) is a time-optimal control.

3. The response time $T_4 - t_0$ is the minimum possible response time.

These geometric interpretations are powerful tools which can help us understand the basic nature of the time-optimal problem. We summarize these important results in the following theorem:

***Theorem 6-1*** *Let $U_t$ be the set of admissible control functions $\mathbf{u}_{[t_0,t]}$. Let $A_t$ denote the set of reachable states at time $t$ and let $S_t$ be the (moving) target set. Suppose that:*

*a. $A_t$ and $S_t$ are closed and bounded for every $t \geq t_0$.*

*b. If $T_1$ and $T_2$ are any two times such that $t_0 \leq T_1 \leq T_2$, then $\{\mathbf{x}(t_0)\} \subset A_{T_1} \subset A_{T_2}$. Let $Q_t$ denote the subset of the state space defined by*

$$Q_t = A_t \cap S_t \tag{6-29}$$

*Then:*

1. *If*

$$Q_t = \emptyset \qquad \textit{for all } t \geq t_0 \tag{6-30}$$

*a time-optimal control does not exist. This means that the posed problem has no solution.*

2. *If there is a time $T^*$ such that*

$$Q_t = \emptyset \qquad \text{for all } t \in [t_0, T^*) \\ Q_{T^*} \neq \emptyset \tag{6-31}$$

*then the time-optimal control problem has a solution and $T^*$ is the minimum time.*

In the preceding theorem, we have assumed that the sets $A_t$ and $S_t$ are closed for all $t \geq t_0$. We shall now show that *if either $A_t$ or $S_t$ is not closed, then, in general, a solution to the time-optimal problem may not exist.* Let us suppose that $A_t$ is not closed, and let us examine the situation in Fig. 6-3. If $A_t$ is not closed, then it may happen that the state $\mathbf{x}^*(T_4)$ does not belong to $A_{T_4}$. Thus, although the relation

$$\partial A_{T_4} \cap \partial S_{T_4} = \mathbf{x}^*(T_4) \tag{6-32}$$

holds, we have

$$A_{T_4} \cap S_{T_4} = \emptyset \tag{6-33}$$

Now let $\epsilon$ be a very small positive number and consider the set $Q_{T_4+\epsilon}$ given by the relation

$$Q_{T_4+\epsilon} = A_{T_4+\epsilon} \cap S_{T_4+\epsilon} \tag{6-34}$$

As illustrated in Fig. 6-3, the set $Q_{T_4+\epsilon}$ is nonempty. This means that we can find controls that transfer the state $\mathbf{x}(t_0)$ to the target set $S_t$ and that require time $T_4 + \epsilon$, where $\epsilon$ can be arbitrarily small (but not zero). We can find controls that make the time $T_4 + \epsilon$ *arbitrarily close to $T_4$ but never exactly $T_4$* [in view of Eq. (6-33)]. Thus, although an optimal solution does not exist, we can find *$\epsilon$-optimal* solutions which are perfectly satisfactory from the engineering point of view; however, in the case of $\epsilon$-optimal solutions, we cannot use the theory given in Chap. 5 (because all the necessary conditions were derived for the optimal control rather than for $\epsilon$-optimal controls).

The reader can appreciate the difficulties involved if one attempts to solve the time-optimal control problem by using the "brute-force" method of calculating the sets of reachable states and then geometrically trying to determine the first point of contact between $A_t$ and $S_t$. This task is almost impossible if the dimension of the state space is higher than three. In the next section we shall show how use of the minimum principle reduces the time-optimal problem to the solution of a system of differential equations with split boundary conditions; to be sure, such a solution is, in general, very difficult to find, but the difficulties involved are much less than those encountered in computing the sets $A_t$.

**Exercise 6-1** Given the system $\dot{x}(t) = u(t)$ with the initial state $x(0)$. Suppose that $|u(t)| \leq 1$ for all $t$. Let $z(t) = t^2$. It is required to find the control $u(t)$ such that:
(a) $x(T) = z(T)$.
(b) $T$ is minimum.
Determine the set of initial states $\{x(0)\}$ for which no solution exists. Determine the time-optimal control from the initial state $x(0) = 2$.

## 6-3 Minimum-time Problems 2: Application of the Minimum Principle

In the previous section, we demonstrated that requirements on the error (or output) of an observable dynamical system can be expressed in terms of a moving target set in the state space. We indicated the geometric nature of the minimum-time problem, and we noted the difficulties associated with the direct "brute-force" method of solution.

In this section, we shall use the minimum principle of Pontryagin to develop a systematic approach to the solution of the time-optimal problem. This is the first time in the book that we demonstrate in detail the use of the minimum principle and its implications. For this reason, we shall proceed slowly and, sometimes, repeat certain results so that the reader will appreciate the power of the necessary conditions in the solution of optimal problems.

We shall first reformulate Problem 6-1*a* in a slightly different form so that we can use Theorem 5-9 (the minimum principle for the time-dependent problem with moving target set). We next present the results in three sequential steps. In step 1, we state the necessary conditions as they apply to the statement of Problem 6-1*b*. In step 2, we derive a relation between the optimal control, the optimal trajectory, and the corresponding costate; also, we define the notion of a *normal* time-optimal problem and the notion of a *singular* time-optimal problem. In step 3, we demonstrate the procedure that one must follow to determine the time-optimal control(s).

Now let us state the time-optimal problem precisely.

***Problem 6-1b Time-optimal Control to a Moving Target Set*** *Given the system*

$$\dot{x}_i(t) = f_i[\mathbf{x}(t), t] + \sum_{j=1}^{r} b_{ij}[\mathbf{x}(t), t]u_j(t) \qquad i = 1, 2, \ldots n$$

*or, equivalently, in vector form*

$$\dot{\mathbf{x}}(t) = \mathbf{f}[\mathbf{x}(t), t] + \mathbf{B}[\mathbf{x}(t), t]\mathbf{u}(t) \tag{6-35}$$

*Assume that*†

1. $f_i[\mathbf{x}(t), t]$ *and* $b_{ij}[\mathbf{x}(t), t]$ *are continuous in* $\mathbf{x}(t)$ *and* $t$.
2. $\partial f_i[\mathbf{x}(t), t]/\partial x_k(t)$, $\partial f_i[\mathbf{x}(t), t]/\partial t$, $\partial b_{ij}[\mathbf{x}(t), t]/\partial x_k(t)$, $\partial b_{ij}[\mathbf{x}(t), t]/\partial t$ *are continuous in* $\mathbf{x}(t)$ *and* $t$ *for* $i, k = 1, 2, \ldots, n$ *and* $j = 1, 2, \ldots, r$.

† See Sec. 5-12.

*Moreover, assume that the components* $u_1(t), u_2(t), \ldots, u_r(t)$ *are constrained in magnitude by the relation*

$$|u_j(t)| \leq 1 \qquad j = 1, 2, \ldots, r, \text{ for all } t \tag{6-36}$$

*or, more compactly, by*

$$\mathbf{u}(t) \in \Omega$$

*Given a smooth target set S*† *defined by the relations*

$$g_\alpha[\mathbf{x}, t] = 0 \qquad \alpha = 1, 2, \ldots, n - \beta; \beta \geq 1 \tag{6-37}$$

*or, equivalently, by*

$$\mathbf{g}[\mathbf{x}, t] = \mathbf{0} \qquad \mathbf{g} \text{ an } n - \beta \text{ vector with components } g_\alpha$$

*Assume that*

1. $g_\alpha[\mathbf{x}, t]$, $\partial g_\alpha[\mathbf{x}, t]/\partial \mathbf{x}$, $\partial g_\alpha[\mathbf{x}, t]/\partial t$ *are continuous in* $\mathbf{x}$ *and* $t$.
2. *The gradient vectors* $\partial g_\alpha[\mathbf{x}, t]/\partial \mathbf{x}$ *are linearly independent for all* $(\mathbf{x}, t) \in S$.

*Let* $t_0$ *be a given initial time, and let* $\mathbf{x}(t_0)$ *be a given initial state of the system* (6-35).

*Given the cost functional*

$$J(\mathbf{u}) = \int_{t_0}^{T} dt = T - t_0 \qquad T \text{ free} \tag{6-38}$$

*Then determine the control* $\mathbf{u}(t)$ *that:*

i. *Satisfies the constraints* (6-36)

ii. *Forces the state* $\mathbf{x}(t_0)$ *of the system* (6-31) *to the target set S*

iii. *Minimizes the cost functional* $J(\mathbf{u})$ *of Eq.* (6-38)

A few comments on this problem are in order:

1. The continuity assumptions are those stated in Sec. 5-12 for the general control problem.

2. The constraints of Eq. (6-36) differ from the constraints of Eq. (6-5). Since the coefficients $b_{ij}[\mathbf{x}(t), t]$ of the $n \times r$ matrix $\mathbf{B}[\mathbf{x}(t), t]$ have not been exactly specified, we claim that the constants $m_j$ of Eq. (6-5) can be absorbed in the elements of the matrix $\mathbf{B}[\mathbf{x}(t), t]$ with no loss in generality. For example, given the system of equations

$$\dot{x}_i(t) = f_i[\mathbf{x}(t), t] + \sum_{j=1}^{r} b'_{ij}[\mathbf{x}(t), t]u'_j(t) \tag{6-39}$$

with

$$|u'_j(t)| \leq m_j \tag{6-40}$$

we can define new control variables $u_j(t)$ satisfying the constraints

$$|u_j(t)| \leq 1 \tag{6-41}$$

† Of the type $d$ in CP2, Sec. 5-12.

and new coefficients $b_{ij}[\mathbf{x}(t), t]$ by setting

$$u_j(t) = \frac{u'_j(t)}{m_j} \qquad (6\text{-}42)$$
$$b_{ij}[\mathbf{x}(t), t] = b'_{ij}[\mathbf{x}(t), t]m_j$$

to obtain the system of Eq. (6-35). The constraint set defined by Eq. (6-36) is the unit hypercube in $r$-dimensional space.

3. In general, the cost functional to be minimized is of the form

$$J(\mathbf{u}) = \int_{t_0}^{T} L[\mathbf{x}(t), \mathbf{u}(t), t]\, dt \qquad (6\text{-}43)$$

For the time-optimal problem, we set

$$L[\mathbf{x}(t), \mathbf{u}(t), t] = 1 \qquad (6\text{-}44)$$

to obtain the $J(\mathbf{u})$ given by Eq. (6-34). Since $t_0$ is known and since the quantity $T - t_0$ is to be minimized, $T$ must be *free*. We also note that the function $L[\mathbf{x}(t), \mathbf{u}(t), t] = 1$ satisfies all the continuity and differentiability requirements of Sec. 5-12.

Now, the Hamiltonian function $H[\mathbf{x}(t), \mathbf{p}(t), \mathbf{u}(t), t]$ for the system (6-35) and the cost functional (6-38) is, using matrix notation, given by

$$\begin{aligned} H[\mathbf{x}(t), \mathbf{p}(t), \mathbf{u}(t), t] &= 1 + \langle \mathbf{p}(t), \mathbf{f}[\mathbf{x}(t), t] + \mathbf{B}[\mathbf{x}(t), t]\mathbf{u}(t)\rangle \\ &= 1 + \langle \mathbf{p}(t), \mathbf{f}[\mathbf{x}(t), t]\rangle + \langle \mathbf{p}(t), \mathbf{B}[\mathbf{x}(t), t]\mathbf{u}(t)\rangle \\ &= 1 + \langle \mathbf{f}[\mathbf{x}(t), t], \mathbf{p}(t)\rangle + \langle \mathbf{u}(t), \mathbf{B}'[\mathbf{x}(t), t]\mathbf{p}(t)\rangle \qquad (6\text{-}45) \end{aligned}$$

where $\mathbf{p}(t)$ is the costate vector. In terms of the components of the various vectors, we can write the Hamiltonian as

$$\begin{aligned} H &= H[\mathbf{x}(t), \mathbf{p}(t), \mathbf{u}(t), t] \\ &= 1 + \sum_{i=1}^{n} f_i[\mathbf{x}(t), t]p_i(t) + \sum_{j=1}^{r} u_j(t) \left\{ \sum_{i=1}^{n} b_{ij}[\mathbf{x}(t), t]p_i(t) \right\} \qquad (6\text{-}46) \end{aligned}$$

Now suppose that $\mathbf{u}^*(t)$ is a time-optimal control, that $\mathbf{x}^*(t)$ is the resultant time-optimal trajectory, and that $T^*$ is the minimum time. By definition, the optimal quantities must satisfy the relations

$$|u_j^*(t)| \le 1 \qquad j = 1, 2, \ldots, r \qquad (6\text{-}47)$$
$$\mathbf{x}^*(t_0) = \mathbf{x}(t_0) \qquad (6\text{-}48)$$
$$\mathbf{x}^*(T^*) \in S \qquad (6\text{-}49)$$

Equation (6-49) implies [in view of Eq. (6-37)] that

$$\mathbf{g}[\mathbf{x}^*(T^*), T^*] = \mathbf{0} \qquad (6\text{-}50)$$

The material included in step 1 is simply a translation of the statements of Theorem 5-9 as applied to Problem 6-1*b*.

*Step 1*

We shall use the word *optimal* to mean *time-optimal* in all the statements that follow.

Statement 1 of Theorem 5-9 implies that there is an (optimal) costate $\mathbf{p}^*(t)$ corresponding to the optimal control $\mathbf{u}^*(t)$ and optimal trajectory $\mathbf{x}^*(t)$. The existence of $\mathbf{p}^*(t)$ is a necessary condition. It is also necessary that the components $x_k^*(t)$ and $p_k^*(t)$, $k = 1, 2, \ldots, n$, satisfy the *canonical equations*

$$\begin{aligned} \dot{x}_k^*(t) &= \frac{\partial H[\mathbf{x}^*(t), \mathbf{p}^*(t), \mathbf{u}^*(t), t]}{\partial p_k^*(t)} \\ \dot{p}_k^*(t) &= -\frac{\partial H[\mathbf{x}^*(t), \mathbf{p}^*(t), \mathbf{u}^*(t), t]}{\partial x_k^*(t)} \end{aligned} \tag{6-51}$$

for $k = 1, 2, \ldots, n$. But, from Eq. (6-46), we find that

$$\begin{aligned} &H[\mathbf{x}^*(t), \mathbf{p}^*(t), \mathbf{u}^*(t), t] \\ &\quad = 1 + \sum_{i=1}^{n} f_i[\mathbf{x}^*(t), t]p_i^*(t) + \sum_{j=1}^{r} u_j^*(t) \left\{ \sum_{i=1}^{n} b_{ij}[\mathbf{x}^*(t), t]p_i^*(t) \right\} \end{aligned} \tag{6-52}$$

and so the canonical equations (6-51) reduce to

$$\dot{x}_k^*(t) = f_k[\mathbf{x}^*(t), t] + \sum_{j=1}^{r} b_{kj}[\mathbf{x}^*(t), t]u_j^*(t) \tag{6-53}$$

$$\dot{p}_k^*(t) = -\sum_{i=1}^{n} \left\{ \frac{\partial f_i[\mathbf{x}^*(t), t]}{\partial x_k^*(t)} \right\} p_i^*(t) - \sum_{j=1}^{r} u_j^*(t) \sum_{i=1}^{n} \left\{ \frac{\partial b_{ij}[\mathbf{x}^*(t), t]}{\partial x_k^*(t)} \right\} p_i^*(t) \tag{6-54}$$

for $k = 1, 2, \ldots, n$. It is also necessary that [see Eqs. (5-433)]

$$x_k^*(t_0) = x_k(t_0) \tag{6-55}$$

We now make the following observation:

***Remark 6-1*** *The differential equation satisfied by* $\mathbf{p}^*(t)$ *is linear in* $\mathbf{p}^*(t)$. *It follows that if* $\mathbf{p}^*(t)$ *satisfies the differential equation* (6-54), *then so does the vector* $C\mathbf{p}^*(t)$, *where* $C$ *is an arbitrary constant.*

Statement 2 of Theorem 5-9 is

$$H[\mathbf{x}^*(t), \mathbf{p}^*(t), \mathbf{u}^*(t), t] = \min_{\mathbf{u}(t) \in \Omega} H[\mathbf{x}^*(t), \mathbf{p}^*(t), \mathbf{u}(t), t] \tag{6-56}$$

for $t \in [t_0, T^*]$, or, equivalently,

$$H[\mathbf{x}^*(t), \mathbf{p}^*(t), \mathbf{u}^*(t), t] \le H[\mathbf{x}^*(t), \mathbf{p}^*(t), \mathbf{u}(t), t] \tag{6-57}$$

for all $\mathbf{u}(t) \in \Omega$ and $t \in [t_0, T^*]$. In view of Eq. (6-52), we deduce that Eq. (6-57) reduces to the relation

$$\begin{aligned} &1 + \sum_{i=1}^{n} f_i[\mathbf{x}^*(t), t]p_i^*(t) + \sum_{j=1}^{r} u_j^*(t) \left\{ \sum_{i=1}^{n} b_{ij}[\mathbf{x}^*(t), t]p_i^*(t) \right\} \\ &\quad \le 1 + \sum_{i=1}^{n} f_i[\mathbf{x}^*(t), t]p_i^*(t) + \sum_{j=1}^{r} u_j(t) \left\{ \sum_{i=1}^{n} b_{ij}[\mathbf{x}^*(t), t]p_i^*(t) \right\} \end{aligned} \tag{6-58}$$

Since the first two terms are the same on both sides of the inequality, we see that Eq. (6-58) reduces to the inequality

$$\sum_{j=1}^{r} u_j^*(t) \left\{ \sum_{i=1}^{n} b_{ij}[\mathbf{x}^*(t), t]p_i^*(t) \right\} \leq \sum_{j=1}^{r} u_j(t) \left\{ \sum_{i=1}^{n} b_{ij}[\mathbf{x}^*(t), t]p_i^*(t) \right\} \quad (6\text{-}59)$$

for all $\mathbf{u}(t) \in \Omega$ and $t \in [t_0, T^*]$. We shall see in step 2 that Eq. (6-59) is extremely important.

***Remark 6-2*** *If $\mathbf{p}^*(t)$ satisfies Eq.* (6-59), *then so does the vector $C\mathbf{p}^*(t)$, where $C$ is an arbitrary positive constant.*

Statement 3 of Theorem 5-9 [and, in particular, Eq. (5-435)] is

$$H[\mathbf{x}^*(T^*), \mathbf{p}^*(T^*), \mathbf{u}^*(T^*), T^*] = \sum_{\alpha=1}^{n-\beta} e_\alpha \frac{\partial g_\alpha[\mathbf{x}^*(T^*), T^*]}{\partial T^*}$$
$$= \left\langle \mathbf{e}, \frac{\partial \mathbf{g}[\mathbf{x}^*(T^*), T^*]}{\partial T^*} \right\rangle \quad (6\text{-}60)$$

where $\mathbf{e}$ is an $n - \beta$ vector with components $e_1, e_2, \ldots, e_{n-\beta}$. From Eqs. (6-60) and (6-52), we find that

$$1 + \sum_{i=1}^{n} f_i[\mathbf{x}^*(T^*), T^*]p_i^*(T^*) + \sum_{j=1}^{r} u_j^*(T^*) \left\{ \sum_{i=1}^{n} b_{ij}[\mathbf{x}^*(T^*), T^*]p_i^*(T^*) \right\}$$
$$= \sum_{\alpha=1}^{n-\beta} e_\alpha \frac{\partial g_\alpha[\mathbf{x}^*(T^*), T^*]}{\partial T^*} \quad (6\text{-}61)$$

***Remark 6-3*** *If the target set $S$ is not a function of time (i.e., if the target set is a stationary subset of the state space), then*

$$\frac{\partial g_\alpha[\mathbf{x}^*(T^*), T^*]}{\partial T^*} = 0 \qquad \textit{for } \alpha = 1, 2, \ldots, n - \beta \quad (6\text{-}62)$$

*and so Eq.* (6-61) *reduces to*

$$1 + \sum_{i=1}^{n} f_i[\mathbf{x}^*(T^*), T^*]p_i^*(T^*)$$
$$+ \sum_{j=1}^{r} u_j^*(T^*) \left\{ \sum_{i=1}^{n} b_{ij}[\mathbf{x}^*(T^*), T^*]p_i^*(T^*) \right\} = 0 \quad (6\text{-}63)$$

***Remark 6-4*** *If $\mathbf{p}^*(T^*)$ satisfies Eq.* (6-61) *or* (6-63), *then the vector $C\mathbf{p}^*(T^*)$, where $C$ is an arbitrary constant, does not satisfy either Eq.* (6-61) *or Eq.* (6-63). *This is in contradistinction to Remarks* 6-1 *and* 6-3.

Statement 4 of Theorem 5-9 is: The vector $\mathbf{p}^*(T^*)$ must be normal to the target set $S$ at $t = T^*$. The target set $S$ at $t = T^*$ is specified by the

$n - \beta$ equations

$$\begin{aligned} g_1(\mathbf{x}, T^*) &= 0 \\ g_2(\mathbf{x}, T^*) &= 0 \\ \cdots\cdots&\cdots \\ g_{n-\beta}(\mathbf{x}, T^*) &= 0 \end{aligned} \tag{6-64}$$

Let $\mathbf{h}_1(\mathbf{x}, T^*)$, $\mathbf{h}_2(\mathbf{x}, T^*)$, . . . , $\mathbf{h}_{n-\beta}(\mathbf{x}, T^*)$ denote the gradient vectors (each with $n$ components)

$$\mathbf{h}_\alpha[\mathbf{x}, T^*] = \frac{\partial g_\alpha[\mathbf{x}, T^*]}{\partial \mathbf{x}} \qquad \alpha = 1, 2, \ldots, n - \beta \tag{6-65}$$

Then it is necessary that the vector $\mathbf{p}^*(T^*)$ be some linear combination of these gradient vectors at $\mathbf{x}^*(T^*)$; in other words, we must have

$$\mathbf{p}^*(T^*) = \sum_{\alpha=1}^{n-\beta} k_\alpha \mathbf{h}_\alpha[\mathbf{x}^*(T^*), T^*] \tag{6-66}$$

where $k_1, k_2, \ldots, k_{n-\beta}$ are some arbitrary constants.

***Remark 6-5*** *If* $\mathbf{p}^*(T^*)$ *satisfies Eq.* (6-66), *so does the vector* $C\mathbf{p}^*(T^*)$, *where* $C$ *is a constant, because Eq.* (6-66) *specifies a direction rather than a magnitude.*

This concludes the derivation of the necessary conditions for Problem 6-1*b*. We remind the reader that all the statements we have made are of the form "*if* $\mathbf{u}^*(t)$ *is an optimal control and* $\mathbf{x}^*(t)$ *is the optimal state, then it is necessary that there exist a* $\mathbf{p}^*(t)$ *such that Eqs.* (6-51) *to* (6-66) *hold.*"

*Step 2*

In this step we shall derive, from Eq. (6-59), an equation which often relates $\mathbf{u}^*(t)$ to $\mathbf{x}^*(t)$ and $\mathbf{p}^*(t)$ for $t \in [t_0, T^*]$. We shall then "eliminate" the optimal control from all the equations.

Let us define functions $q_1^*(t), q_2^*(t), \ldots, q_r^*(t)$ by the equations

$$q_j^*(t) = \sum_{i=1}^{n} b_{ij}[\mathbf{x}^*(t), t] p_i^*(t) \qquad j = 1, 2, \ldots, r \tag{6-67}$$

or, equivalently, the $r$ vector $\mathbf{q}^*(t)$ by

$$\mathbf{q}^*(t) = \mathbf{B}'[\mathbf{x}^*(t), t]\mathbf{p}^*(t) \tag{6-68}$$

The vector $\mathbf{q}^*(t)$ is thus obtained by a linear transformation (whose matrix is $\mathbf{B}'[\mathbf{x}^*(t), t]$) on the vector $\mathbf{p}^*(t)$.

Using the functions $q_j^*(t)$, Eq. (6-59) becomes

$$\sum_{j=1}^{r} u_j^*(t) q_j^*(t) \le \sum_{j=1}^{r} u_j(t) q_j^*(t) \tag{6-69}$$

for all $|u_j(t)| \leq 1$, $j = 1, 2, \ldots, r$, and every $t \in [t_0, T^*]$. Equation (6-69) means that the function

$$\varphi[\mathbf{u}(t)] = \sum_{j=1}^{r} u_j(t) q_j^*(t) \tag{6-70}$$

attains its absolute minimum at

$$u_j(t) = u_j^*(t) \tag{6-71}$$

We can see that the relation

$$\min_{\mathbf{u}(t) \in \Omega} \varphi[\mathbf{u}(t)] = \min_{\mathbf{u}(t) \in \Omega} \sum_{j=1}^{r} u_j(t) q_j^*(t) = \sum_{j=1}^{r} \{ \min_{|u_j(t)| \leq 1} u_j(t) q_j^*(t) \} \tag{6-72}$$

is true.

***Remark 6-6*** *We can interchange the* "min" *operator and the* "Σ" *operator because the functions* $u_1(t), u_2(t), \ldots, u_r(t)$ *are constrained independently. In other words, if, say, we have* $u_1(t) = +1$, *then* $u_2(t), \ldots, u_r(t)$ *can attain any values consistent with the magnitude constraints.*

Now, it is easy to see that

$$\min_{|u_j(t)| \leq 1} \{u_j(t) q_j^*(t)\} = -|q_j^*(t)| \tag{6-73}$$

The control $u_j^*(t)$ will minimize the function $u_j(t)q_j^*(t)$. It follows, in view of Eq. (6-73), that $u_j^*(t)$ must be the following function of $q_j^*(t)$:

$$\begin{array}{ll} u_j^*(t) = +1 & \text{if } q_j^*(t) < 0 \\ u_j^*(t) = -1 & \text{if } q_j^*(t) > 0 \\ u_j^*(t) \text{ indeterminate} & \text{if } q_j^*(t) = 0 \end{array} \tag{6-74}$$

We can use the signum function† to write Eqs. (6-74) in the more compact form

$$u_j^*(t) = -\operatorname{sgn}\{q_j^*(t)\} = -\operatorname{sgn}\left\{\sum_{i=1}^{n} b_{ij}[\mathbf{x}^*(t), t] p_i^*(t)\right\} \tag{6-75}$$

for $j = 1, 2, \ldots, r$ and $t \in [t_0, T^*]$.

Equation (6-75) relates the components of the time-optimal control $\mathbf{u}^*(t)$ to the state $\mathbf{x}^*(t)$ and to the costate $\mathbf{p}^*(t)$. We note that if $\mathbf{p}^*(t)$ is such

† The signum function, written as sgn { }, is defined as follows: $a = \operatorname{sgn}\{b\}$ means that

$$\begin{array}{ll} a = +1 & \text{if } b > 0 \\ a = -1 & \text{if } b < 0 \\ a \text{ indeterminate} & \text{if } b = 0 \end{array}$$

Note that $\operatorname{sgn}\{ab\} = \operatorname{sgn}\{a\} \operatorname{sgn}\{b\}$.

that Eq. (6-75) holds, then Eq. (6-75) also holds for any vector $c\mathbf{p}^*(t)$, provided that $c > 0$, because sgn $\{c\} = +1$.

We observe that Eq. (6-75) implies that $u_j^*(t)$ is a well-defined function of $\mathbf{x}^*(t)$, $\mathbf{p}^*(t)$, and $t$ as long as the argument of the signum function is nonzero. However, whenever $q_j^*(t) = 0$, $u_j^*(t)$ is not defined by Eq. (6-75).

We shall now distinguish two cases; we shall call the first *normal* and the second *singular*.

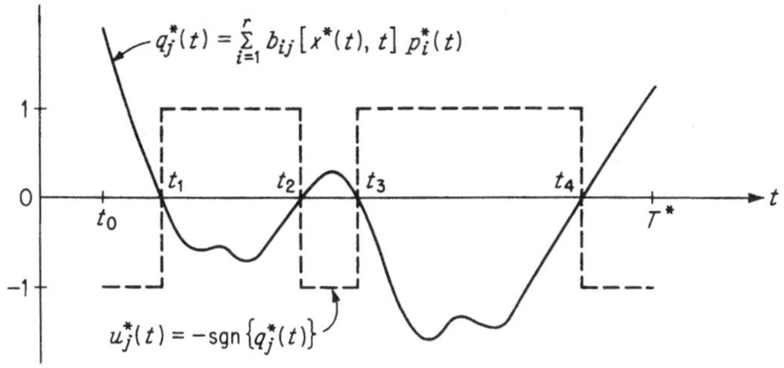


**Fig. 6-4** The function $q_j^*(t)$, which yields the well-defined control $u_j^*(t)$.

***Definition 6-4 Normal Time-optimal Problem*** *Suppose that in the interval* $[t_0, T^*]$, *there is a countable set of times* $t_{1j}, t_{2j}, t_{3j}, \ldots,$

$$t_{\gamma j} \in [t_0, T^*] \qquad \gamma = 1, 2, 3, \ldots ; j = 1, 2, \ldots, r \tag{6-76}$$

*such that*

$$q_j^*(t) = \sum_{i=1}^{n} b_{ij}[\mathbf{x}^*(t), t]p_i^*(t) = \begin{cases} 0 & \textit{if and only if } t = t_{\gamma j} \\ \textit{nonzero otherwise} \end{cases} \tag{6-77}$$

*for all* $j = 1, 2, \ldots, r$.

*Then we shall say that we have a normal time-optimal problem.*

Figure 6-4 illustrates a function $q_j^*(t)$ and the corresponding $u_j^*(t)$ defined by Eq. (6-75). The function $q_j^*(t)$ is zero only at isolated instants of time, and so the time-optimal control is a piecewise constant function with simple jumps. If all the functions $q_j^*(t)$ had the same characteristic, then the time-optimal problem would be normal. It is common terminology to say that the control $u_j^*(t)$ switches at $t = t_{\gamma j}$ and that the *number* of *switchings* of $u_j^*(t)$ is equal to the largest number $\gamma j$ (or $\infty$). Thus, the control $u_j^*(t)$ shown in Fig. 6-4 switches four times and the number of switchings is four.

***Definition 6-5 Singular Time-optimal Problem*** *Suppose that in the interval* $[t_0, T^*]$, *there is one (or more) subinterval* $[T_1, T_2]_j$, $[T_1, T_2]_j \subset [t_0, T^*]$, *such that*

$$q_j^*(t) = \sum_{i=1}^{n} b_{ij}[\mathbf{x}^*(t), t]p_i^*(t) = 0 \qquad \textit{for all } t \in [T_1, T_2]_j \tag{6-78}$$

*Then we say that we have a singular time-optimal problem. We shall call the* $[T_1, T_2]_j$ *interval(s) the singularity interval(s).*

The function $q_j^*(t)$ illustrated in Fig. 6-5 is zero for all $t \in [T_1, T_2]$ and so corresponds to a singular problem. Thus, in a singular time-optimal problem, there is at least one subinterval of time during which the relation $u_j^*(t) = -\operatorname{sgn}\left\{\sum_{i=1}^{n} b_{ij}[\mathbf{x}^*(t), t]p_i^*(t)\right\}$ does not define the time-optimal control as a function of $\mathbf{x}^*(t)$ and $\mathbf{p}^*(t)$. We would like to caution the reader against misunderstanding this statement; it does *not* mean that either the time-optimal control does not exist or the time-optimal control cannot be defined; it simply states that the necessary condition $H[\mathbf{x}^*(t), \mathbf{p}^*(t), \mathbf{u}^*(t), t] \le H[\mathbf{x}^*(t), \mathbf{p}^*(t), \mathbf{u}(t), t]$ does not lead to a well-defined relation between $\mathbf{u}^*(t)$, $\mathbf{x}^*(t)$, $\mathbf{p}^*(t)$, and $t$.

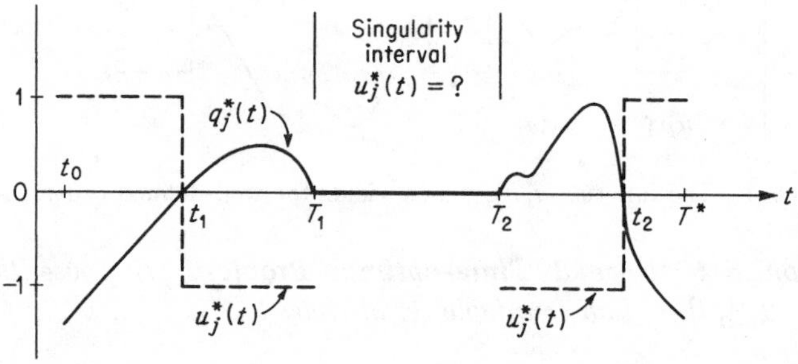


**Fig. 6-5** The function $q_j^*(t)$ shown corresponds to a singular time-optimal problem.

We shall encounter singular problems throughout this chapter.

In the remainder of this section, we shall assume that we are dealing with a normal time-optimal problem. We shall thus be able to substitute Eq. (6-75) into Eqs. (6-52) to (6-54), (6-56) to (6-58), (6-60), and (6-61) and, in so doing, eliminate $\mathbf{u}^*(t)$ from all the necessary conditions. Thus, all the necessary conditions stated in terms of $\mathbf{u}^*(t)$ in step 1 will reduce to necessary conditions which are independent of $\mathbf{u}^*(t)$. As we shall see in step 3, this fact will enable us to determine the time-optimal control.

We shall now state two theorems which summarize these ideas.

***Theorem 6-2 The Bang-Bang Principle*** *Let* $\mathbf{u}^*(t)$ *be a time-optimal control for Problem* 6-1b, *and let* $\mathbf{x}^*(t)$ *and* $\mathbf{p}^*(t)$ *be the corresponding state trajectory and costate. If the problem is normal (see Definition* 6-4), *then the components* $u_1^*(t), u_2^*(t), \ldots, u_r^*(t)$ *of* $\mathbf{u}^*(t)$ *must be defined by the relation*

$$u_j^*(t) = -\operatorname{sgn}\left\{\sum_{i=1}^{n} b_{ij}[\mathbf{x}^*(t), t]p_i^*(t)\right\} \qquad j = 1, 2, \ldots, r \qquad (6\text{-}79)$$

*for* $t \in [t_0, T^*]$.† *Equation* (6-79) *can also be written in the more compact*

† Except at the countable set of switch times $t_{\gamma j}$.

*form*

$$\mathbf{u}^*(t) = -\operatorname{SGN}\{\mathbf{q}^*(t)\} = -\operatorname{SGN}\{\mathbf{B}'[\mathbf{x}^*(t), t]\mathbf{p}^*(t)\} \tag{6-80}$$†

*Thus, if the problem is normal, the components of the time-optimal control are piecewise constant functions of time (i.e., are bang-bang).*

We shall leave it to the reader to verify the following theorem by direct substitution.

***Theorem 6-3 Reduced Necessary Conditions*** *Let* $\mathbf{u}^*(t)$ *be a time-optimal control for Problem* 6-1b. *Let* $\mathbf{x}^*(t)$ *be the state on the time-optimal trajectory, and let* $\mathbf{p}^*(t)$ *be the corresponding costate. Let* $T^*$ *be the minimum time. If the problem is normal (see Definition* 6-4), *then it is necessary that:*

a. *Theorem* 6-2 *be satisfied*

b. *The state* $\mathbf{x}^*(t)$ *and costate* $\mathbf{p}^*(t)$ *satisfy the reduced canonical equations*

$$\dot{x}_k^*(t) = f_k[\mathbf{x}^*(t), t] - \sum_{j=1}^{r} b_{kj}[\mathbf{x}^*(t), t] \operatorname{sgn}\left\{\sum_{i=1}^{n} b_{ij}[\mathbf{x}^*(t), t]p_i^*(t)\right\} \tag{6-81}$$

$$\begin{aligned}\dot{p}_k^*(t) = &-\sum_{i=1}^{n} \frac{\partial f_i[\mathbf{x}^*(t), t]}{\partial x_k^*(t)} p_i^*(t) \\ &+ \left(\sum_{j=1}^{r} \operatorname{sgn}\left\{\sum_{i=1}^{n} b_{ij}[\mathbf{x}^*(t), t]p_i^*(t)\right\}\right) \sum_{i=1}^{n} \frac{\partial b_{ij}[\mathbf{x}^*(t), t]}{\partial x_k^*(t)} p_i^*(t)\end{aligned} \tag{6-82}$$

*for* $k = 1, 2, \ldots, n$ *and for* $t \in [t_0, T^*]$.

c. *The Hamiltonian* (6-52) *along the time-optimal trajectory is given by the equation*

$$\begin{aligned}H[\mathbf{x}^*(t), \mathbf{p}^*(t), \mathbf{u}^*(t), t] = 1 &+ \sum_{i=1}^{n} f_i[\mathbf{x}^*(t), t]p_i^*(t) \\ &- \sum_{j=1}^{r} \left| \sum_{i=1}^{n} b_{ij}[\mathbf{x}^*(t), t]p_i^*(t) \right| \qquad t \in [t_0, T^*]\end{aligned} \tag{6-83}$$

d. *At the terminal time* $T^*$, *the following relation* [*see Eq.* (6-61)] *holds:*

$$\begin{aligned}1 + \sum_{i=1}^{n} f_i[\mathbf{x}^*(T^*), T^*]p_i^*(T^*) - \sum_{j=1}^{r} \left| \sum_{i=1}^{n} b_{ij}[\mathbf{x}^*(T^*), T^*]p_i^*(T^*) \right| \\ = \sum_{\alpha=1}^{n-\beta} e_\alpha \frac{\partial g_\alpha[\mathbf{x}^*(T^*), T^*]}{\partial T^*}\end{aligned} \tag{6-84}$$

† The vector-valued function SGN { } is defined as follows: Let $a_1, a_2, \ldots, a_r$ be the components of the vector $\mathbf{a}$, and let $b_1, b_2, \ldots, b_r$ be the components of the vector $\mathbf{b}$; then

$$\mathbf{a} = \operatorname{SGN}\{\mathbf{b}\}$$

means that

$$a_j = \operatorname{sgn}\{b_j\} \qquad j = 1, 2, \ldots, r$$

*e. At the initial time* $t_0$,

$$\mathbf{x}^*(t_0) = \mathbf{x}(t_0) \tag{6-85}$$

*At the terminal time* $T^*$, *the following relations hold* [*see Eqs.* (6-64) *to* (6-66)]:

$$g_\alpha[\mathbf{x}^*(T^*), T^*] = 0 \qquad \alpha = 1, 2, \ldots, n - \beta; \beta \geq 1 \tag{6-86}$$

$$\mathbf{p}^*(T^*) = \sum_{\alpha=1}^{n-\beta} k_\alpha \frac{\partial g_\alpha[\mathbf{x}^*(T^*), T^*]}{\partial \mathbf{x}^*(T^*)} \tag{6-87}$$

We shall now illustrate the implications of Theorem 6-3. Let us suppose that $n = 3$ and that $r = 2$. As illustrated in Fig. 6-6, the $2 \times 3$ matrix $\mathbf{B}'[\mathbf{x}^*(t), t]$ is associated with a transformation that maps the 3-vector $\mathbf{p}^*(t)$ into the 2-vector $\mathbf{q}^*(t) = \mathbf{B}'[\mathbf{x}^*(t), t]\mathbf{p}^*(t)$. It is easy to see that, in order

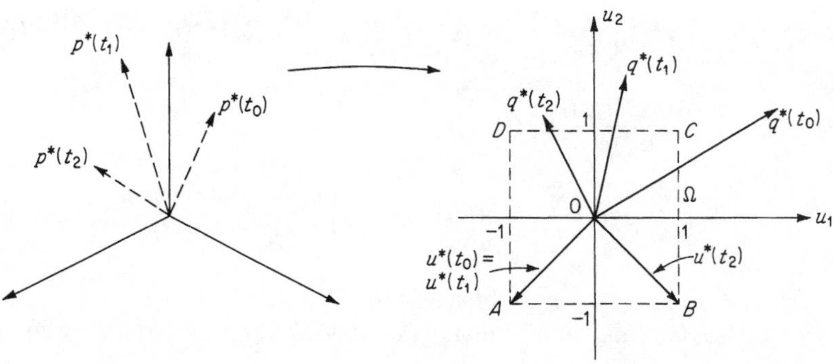


**Fig. 6-6** Geometric interpretation of the fact that $\mathbf{u}^*(t)$ must minimize the scalar product $\langle \mathbf{u}(t), \mathbf{q}^*(t) \rangle$.

to minimize the scalar product $\langle \mathbf{u}^*(t), \mathbf{q}^*(t) \rangle$, the control vector $\mathbf{u}^*(t)$ must attain its maximum magnitude and point in a direction which is roughly opposite to that of the $\mathbf{q}^*(t)$ vector. Thus, whenever $\mathbf{q}^*(t)$ is in the first quadrant, then $\mathbf{u}^*(t)$ must point to the corner $A$ of the constraint square; whenever $\mathbf{q}^*(t)$ is in the second quadrant, $\mathbf{u}^*(t)$ must point to the corner $B$ of the square, and so on.

The reduced necessary conditions provided by Theorems 6-2 and 6-3 lead to a systematic way of deriving the time-optimal control. This exposition will be carried out in step 3. For the time being, we wish to remind the reader that Theorems 6-2 and 6-3 refer to the optimal quantities and represent necessary conditions. Thus, given a control $\mathbf{u}(t)$ and the corresponding trajectory $\mathbf{x}(t)$, if $\mathbf{u}(t)$ violates any one of the necessary conditions, then we are immediately assured that $\mathbf{u}(t)$ is *not* a time-optimal control.

The following example illustrates the results of Theorems 6-2 and 6-3 for a specific system.

**Example 6-3** Consider the dynamical system with state variables $x_1(t)$, $x_2(t)$, $x_3(t)$ and control variables $u_1(t)$ and $u_2(t)$, described by the equations

$$\begin{aligned}\dot{x}_1(t) &= -x_2(t) + x_1(t)u_1(t) \\ \dot{x}_2(t) &= x_1(t) - tx_2^2(t) + u_2(t) \\ \dot{x}_3(t) &= -t^2x_1^3(t) - x_3(t)\end{aligned}$$

Suppose that the control variables $u_1(t)$ and $u_2(t)$ are constrained in magnitude by the inequalities

$$|u_1(t)| \leq 1 \qquad |u_2(t)| \leq 1$$

Suppose that at the initial time $t_0 = 0$, the system is at the initial state

$$x_1(0) = -1 \qquad x_2(0) = -1 \qquad x_3(0) = 10$$

Let us suppose that the target set $S$ is described by the equations

$$\begin{aligned}g_1(x_1, x_2, x_3, t) &= x_3 = 0 \\ g_2(x_1, x_2, x_3, t) &= x_1^2 + x_2^2 - t^2 - 1 = 0\end{aligned}$$

In other words, the target set is the boundary of an expanding circle in the $x_1x_2$ plane, whose radius is $\sqrt{t^2+1}$.

The problem is to find the time-optimal control from the initial state $(-1, -1, 10)$ to the target set. We confess that we do not know whether the time-optimal control exists and, if it exists, whether it is unique. Let us *assume* that it exists, that it is unique, and that the problem is normal, so that we can illustrate the necessary conditions.

The Hamiltonian for this problem is

$$\begin{aligned}H &= H[x_1(t), x_2(t), x_3(t), p_1(t), p_2(t), p_3(t), u_1(t), u_2(t), t] \\ &= 1 - x_2(t)p_1(t) + x_1(t)u_1(t)p_1(t) + x_1(t)p_2(t) - tx_2^2(t)p_2(t) + u_2(t)p_2(t) \\ &\qquad - t^2x_1^3(t)p_3(t) - x_3(t)p_3(t)\end{aligned}$$

The reduced necessary conditions are determined as follows:

From Eq. (6-79), we have

$$\begin{aligned}u_1^*(t) &= -\operatorname{sgn}\{x_1^*(t)p_1^*(t)\} \\ u_2^*(t) &= -\operatorname{sgn}\{p_2^*(t)\}\end{aligned}$$

From Eq. (6-81), we have

$$\begin{aligned}\dot{x}_1^*(t) &= -x_2^*(t) - x_1^*(t)\operatorname{sgn}\{x_1^*(t)p_1^*(t)\} \\ \dot{x}_2^*(t) &= x_1^*(t) - tx_2^{*2}(t) - \operatorname{sgn}\{p_2^*(t)\} \\ \dot{x}_3^*(t) &= -t^2x_1^{*3}(t) - x_3^*(t)\end{aligned}$$

From Eq. (6-82), we have

$$\begin{aligned}\dot{p}_1^*(t) &= +p_1^*(t)\operatorname{sgn}\{x_1^*(t)p_1^*(t)\} - p_2^*(t) + 3t^2x_1^{*2}(t)p_3^*(t) \\ \dot{p}_2^*(t) &= +p_1^*(t) + 2tx_2^*(t)p_2^*(t) \\ \dot{p}_3^*(t) &= +p_3^*(t)\end{aligned}$$

From Eq. (6-84), we find that, at the minimum time $T^*$,

$$\begin{aligned}1 - x_2^*(T^*)p_1^*(T^*) - |x_1^*(T^*)p_1^*(T^*)| + x_1^*(T^*)p_2^*(T^*) - T^*x_2^{*2}(T^*)p_2^*(T^*) \\ - |p_2^*(T^*)| - T^{*2}x_1^{*3}(T^*)p_3^*(T^*) - x_3^*(T^*)p_3^*(T^*) = e_10 + e_22T^*\end{aligned}$$

where $e_2$ is some constant.

At the initial time $t_0 = 0$, we must have

$$x_1^*(0) = -1 \qquad x_2^*(0) = -1 \qquad x_3^*(0) = 10$$

At the minimum time $T^*$, we must have

$$x_3^*(T^*) = 0$$
$$x_1^{*2}(T^*) + x_2^{*2}(T^*) - T^{*2} - 1 = 0$$

and [see Eq. (6-87)]

$$p_1^*(T^*) = k_1 0 + k_2 2x_1^*(T^*)$$
$$p_2^*(T^*) = k_1 0 + k_2 2x_2^*(T^*)$$
$$p_3^*(T^*) = k_1 1 + k_2 0$$

where $k_1$ and $k_2$ are some constants.

*Step 3*

In steps 1 and 2, we stated relations which must be satisfied by the time-optimal control $\mathbf{u}^*(t)$, the resulting state $\mathbf{x}^*(t)$, the corresponding costate $\mathbf{p}^*(t)$, and the minimum time $T^*$. Our problem is, however, to find the time-optimal control, and so the question that arises is the following: *How can we use all these theorems to find the time-optimal control for Problem* 6-1*b?* We shall answer this question in the remainder of this section. We shall develop a systematic procedure and label each step so that the reader can follow the logical development.

*Step 3a Formation of the Hamiltonian*

We start by forming the Hamiltonian function $H[\mathbf{x}(t), \mathbf{p}(t), \mathbf{u}(t), t]$ for the system $\dot{\mathbf{x}}(t) = \mathbf{f}[\mathbf{x}(t), t] + \mathbf{B}[\mathbf{x}(t), t]\mathbf{u}(t)$ and the cost functional $J(\mathbf{u}) = \int_{t_0}^{T} 1\, dt$. The Hamiltonian is [see Eq. (6-45) or (6-46)] given by

$$H[\mathbf{x}(t), \mathbf{p}(t), \mathbf{u}(t), t] = 1 + \langle \mathbf{f}[\mathbf{x}(t), t], \mathbf{p}(t)\rangle + \langle \mathbf{u}(t), \mathbf{B}'[\mathbf{x}(t), t]\mathbf{p}(t)\rangle \quad (6\text{-}88)$$

We write $\mathbf{x}(t)$, $\mathbf{p}(t)$, $\mathbf{u}(t)$ to indicate that eventually these vectors will be functions of time. However, at this step, we do not imply any restrictions either on the values of the vectors $\mathbf{x}(t)$, $\mathbf{p}(t)$, and $\mathbf{u}(t)$ or on the value of time $t$.

*Step 3b Minimization of the Hamiltonian*

The Hamiltonian function $H[\mathbf{x}(t), \mathbf{p}(t), \mathbf{u}(t), t]$ depends on $2n + r + 1$ variables. Let us suppose that we hold $\mathbf{x}(t)$, $\mathbf{p}(t)$, and $t$ constant and that we examine the behavior of the Hamiltonian [which is now a function of $\mathbf{u}(t)$ only, since $\mathbf{x}(t)$, $\mathbf{p}(t)$, and $t$ are held constant] as $\mathbf{u}(t)$ varies over the constraint set $\Omega$. In particular, we wish to find the control which will absolutely minimize the Hamiltonian. For this reason, we shall define an $H$-minimal control as follows:

***Definition 6-6 H-minimal Control*** *The admissible control* $\mathbf{u}^\circ(t)$ *is called H-minimal if it satisfies the relation*

$$H[\mathbf{x}(t), \mathbf{p}(t), \mathbf{u}^\circ(t), t] \le H[\mathbf{x}(t), \mathbf{p}(t), \mathbf{u}(t), t] \quad (6\text{-}89)$$

*for all* $\mathbf{u}(t) \in \Omega$, *all* $\mathbf{x}(t)$, *all* $\mathbf{p}(t)$, *and all* $t$.

If we mimic the development presented in step 2, then we find that the $H$-minimal control $\mathbf{u}^\circ(t)$ for the Hamiltonian function of Eq. (6-88) is given by

$$u_j^\circ(t) = -\operatorname{sgn}\left\{\sum_{i=1}^{n} b_{ij}[\mathbf{x}(t), t]p_i(t)\right\} \qquad j = 1, 2, \ldots, r \tag{6-90}$$

or, more compactly, in vector form by

$$\mathbf{u}^\circ(t) = -\operatorname{SGN}\{\mathbf{B}'[\mathbf{x}(t), t]\mathbf{p}(t)\} \tag{6-91}$$

If we now substitute the $H$-minimal control $\mathbf{u}^\circ(t)$ into Eq. (6-88), we find that

$$\begin{aligned} H[\mathbf{x}(t), \mathbf{p}(t), \mathbf{u}^\circ(t), t] = 1 &+ \langle \mathbf{f}[\mathbf{x}(t), t], \mathbf{p}(t)\rangle \\ &- \langle \operatorname{SGN}\{\mathbf{B}'[\mathbf{x}(t), t]\mathbf{p}(t)\}, \mathbf{B}'[\mathbf{x}(t), t]\mathbf{p}(t)\rangle \end{aligned} \tag{6-92}$$

and hence that

$$H[\mathbf{x}(t), \mathbf{p}(t), \mathbf{u}^\circ(t), t] = 1 + \sum_{i=1}^{n} f_i[\mathbf{x}(t), t]p_i(t) - \sum_{j=1}^{r}\left|\sum_{i=1}^{n} b_{ij}[\mathbf{x}(t), t]p_i(t)\right| \tag{6-93}$$

Note that the right-hand side of Eq. (6-93) is only a function of $\mathbf{x}(t)$, $\mathbf{p}(t)$, and $t$. For this reason, we define the function $H^\circ[\mathbf{x}(t), \mathbf{p}(t), t]$ by the relation

$$H^\circ[\mathbf{x}(t), \mathbf{p}(t), t] = \min_{\mathbf{u}(t)\in\Omega} H[\mathbf{x}(t), \mathbf{p}(t), \mathbf{u}(t), t] \tag{6-94}$$

Once more we remind the reader that these definitions and equations do not explicitly involve trajectories or optimal quantities.

*Step 3c Restricting* $\mathbf{x}(t)$ *and* $\mathbf{p}(t)$

Let us now demand that the (as yet unspecified) vectors $\mathbf{x}(t)$ and $\mathbf{p}(t)$ satisfy the differential equations

$$\dot{\mathbf{x}}(t) = \frac{\partial H^\circ[\mathbf{x}(t), \mathbf{p}(t), t]}{\partial \mathbf{p}(t)} \tag{6-95}$$

$$\dot{\mathbf{p}}(t) = -\frac{\partial H^\circ[\mathbf{x}(t), \mathbf{p}(t), t]}{\partial \mathbf{x}(t)} \tag{6-96}$$

or, equivalently, the differential equations

$$\dot{x}_k(t) = f_k[\mathbf{x}(t), t] - \sum_{j=1}^{r} b_{kj}[\mathbf{x}(t), t] \operatorname{sgn}\left\{\sum_{i=1}^{n} b_{ij}[\mathbf{x}(t), t]p_i(t)\right\} \tag{6-97}$$

$$\begin{aligned} \dot{p}_k(t) = &-\sum_{i=1}^{n}\left\{\frac{\partial f_i[\mathbf{x}(t), t]}{\partial x_k(t)}\right\} p_i(t) \\ &+ \left[\sum_{j=1}^{r} \operatorname{sgn}\left\{\sum_{i=1}^{n} b_{ij}[\mathbf{x}(t), t]p_i(t)\right\}\right] \sum_{i=1}^{n}\left\{\frac{\partial b_{ij}[\mathbf{x}(t), t]}{\partial x_k(t)}\right\} p_i(t) \end{aligned} \tag{6-98}$$

for $k = 1, 2, \ldots, n$. We note that

$$\frac{\partial H^\circ[\mathbf{x}(t), \mathbf{p}(t), t]}{\partial \mathbf{p}(t)} = \frac{\partial H[\mathbf{x}(t), \mathbf{p}(t), \mathbf{u}(t), t]}{\partial \mathbf{p}(t)}\bigg|_{\mathbf{u}(t)=\mathbf{u}^\circ(t)} \tag{6-99}$$

and that

$$\frac{\partial H^\circ[\mathbf{x}(t), \mathbf{p}(t), t]}{\partial \mathbf{x}(t)} = \frac{\partial H[\mathbf{x}(t), \mathbf{p}(t), \mathbf{u}(t), t]}{\partial \mathbf{x}(t)}\bigg|_{\mathbf{u}(t)=\mathbf{u}^\circ(t)} \tag{6-100}$$

*Step 3d An "Experiment"*

Our objective is to find the time-optimal control $\mathbf{u}^*(t)$ that transfers the system $\dot{\mathbf{x}}(t) = \mathbf{f}[\mathbf{x}(t), t] + \mathbf{B}[\mathbf{x}(t), t]\mathbf{u}(t)$ from the given initial state $\mathbf{x}(t_0)$ to the target set $S$. We assume that the problem is normal (see Definition 6-4).

Let us now suppose that we simulate Eqs. (6-97) and (6-98) on an analog computer. At the known initial time $t_0$, we use the known initial values of the state variables $x_1(t_0), x_2(t_0), \ldots, x_n(t_0)$ as initial conditions for the system (6-97). We use some *guessed* values of the initial costate variables $p_1(t_0), p_2(t_0), \ldots, p_n(t_0)$.

Let $q_j(t)$, $j = 1, 2, \ldots, r$, be the functions defined by the relations

$$q_j(t) = \sum_{i=1}^{n} b_{ij}[\mathbf{x}(t), t]p_i(t) \tag{6-101}$$

Let us assume that

$$q_j(t_0) \neq 0 \qquad \text{for all } j = 1, 2, \ldots, r \tag{6-102}$$

Now Eqs. (6-102), (6-101), and (6-90) imply that the numbers

$$u_j^\circ(t_0) = -\operatorname{sgn}\{q_j(t_0)\}$$

are either $+1$ or $-1$. Thus, the solutions of Eqs. (6-97) and (6-98) are well defined, at least for $t$ near $t_0$. We shall denote the solutions of Eqs. (6-97) and (6-98) by

$$\begin{aligned} \mathbf{x}(t) &= \mathbf{x}[t, t_0, \mathbf{x}(t_0), \mathbf{p}(t_0)] \\ \mathbf{p}(t) &= \mathbf{p}[t, t_0, \mathbf{x}(t_0), \mathbf{p}(t_0)] \end{aligned} \tag{6-103}$$

to emphasize their dependence on the *known* initial state $\mathbf{x}(t_0)$ and the *guessed* initial costate $\mathbf{p}(t_0)$.

The "experiment" proceeds as follows: We measure the signals $\mathbf{x}(t)$ and $\mathbf{p}(t)$, and at each instant of time, we also form and measure the signals

$$q_j(t) = \sum_{i=1}^{n} b_{ij}[\mathbf{x}(t), t]p_i(t) \qquad j = 1, 2, \ldots, r \tag{6-104}$$

$$\dot{q}_j(t), \ddot{q}_j(t), \dddot{q}_j(t), \ldots \qquad j = 1, 2, \ldots, r \tag{6-105}$$

$$H^\circ[\mathbf{x}(t), \mathbf{p}(t), t] = 1 + \sum_{i=1}^{n} f_i[\mathbf{x}(t), t]p_i(t) - \sum_{j=1}^{r} |q_j(t)| \tag{6-106}$$

$$g_\alpha[\mathbf{x}(t), t] \qquad \alpha = 1, 2, \ldots, n - \beta \tag{6-107}$$

$$\frac{\partial g_\alpha[\mathbf{x}(t), t]}{\partial t} \qquad \alpha = 1, 2, \ldots, n - \beta \tag{6-108}$$

$$\mathbf{h}_\alpha[\mathbf{x}(t), t] = \frac{\partial g_\alpha[\mathbf{x}(t), t]}{\partial \mathbf{x}(t)} \qquad \alpha = 1, 2, \ldots, n - \beta \tag{6-109}$$

Using the particular (guessed value) $\mathbf{p}(t_0)$, we ask the following questions sequentially for each $t$ in some interval $[t_0, T]$.

**Question 1** If $q_j(t) = 0$, is $\dot{q}_j(t) \neq 0$? If *no*, is $\ddot{q}_j(t) \neq 0$? If *no*, is $\dddot{q}_j(t) \neq 0$? (And so on.) If the answer to question 1 is *yes*, then we ask question 2. If the answer is *no*, then we change the value of $\mathbf{p}(t_0)$ and repeat question 1.

**Question 2** If the answer to question 1 is *yes*, is there a time $T$ such that the relations

$$g_\alpha[\mathbf{x}(T), T] = 0 \tag{6-110}$$

are satisfied for all $\alpha = 1, 2, \ldots, n - \beta$? If the answer to question 2 is *no*, we change $\mathbf{p}(t_0)$ and start all over again. If the answer to question 2 is *yes*, we ask question 3.

**Question 3** If the answer to question 2 is *yes*, are there constants $e_1, e_2, \ldots, e_{n-\beta}$ such that the relation

$$H^\circ[\mathbf{x}(T), \mathbf{p}(T), t] = \sum_{\alpha=1}^{n-\beta} e_\alpha \frac{\partial g_\alpha[\mathbf{x}(T), T]}{\partial T} \tag{6-111}$$

is satisfied? If the answer to question 3 is *no*, we change $\mathbf{p}(t_0)$ and start all over again. If the answer to question 3 is *yes*, we ask question 4.

**Question 4** If the answer to question 3 is *yes*, are there constants $k_1, k_2, \ldots, k_{n-\beta}$ such that the relation

$$\mathbf{p}(T) = \sum_{\alpha=1}^{n-\beta} k_\alpha \frac{\partial g_\alpha[\mathbf{x}(T), T]}{\partial \mathbf{x}(T)} \tag{6-112}$$

is satisfied? If the answer to question 4 is *no*, we change $\mathbf{p}(t_0)$ and start all over again. If the answer to question 4 is *yes*, this implies that we have found a $\mathbf{p}(t_0)$ such that the answer to questions 1 to 4 is *yes;* we *store* this $\mathbf{p}(t_0)$ and start the experiment all over again—until we have found all the vectors $\mathbf{p}(t_0)$ with the property that the answers to questions 1 to 4 are all *yes*. This logical sequence of questions is illustrated in Fig. 6-7.

*Step 3e The Candidates for Time-optimal Control*

We shall now formalize the results of the "experiment" conducted in step 3*d*. Loosely speaking, we define the set $\hat{\mathcal{P}}_0$ to be the set of initial costate vectors $\hat{\mathbf{p}}(t_0)$ corresponding to the given initial state $\mathbf{x}(t_0)$, with the property that the answer to questions 1 to 4 is *yes*. Clearly, $\hat{\mathcal{P}}_0$ is a subset of the $n$-dimensional space $R_n$. We can think of $\hat{\mathcal{P}}_0$ as the "output" of the logical process illustrated in Fig. 6-7. More precisely, we define $\hat{\mathcal{P}}_0$ as follows:

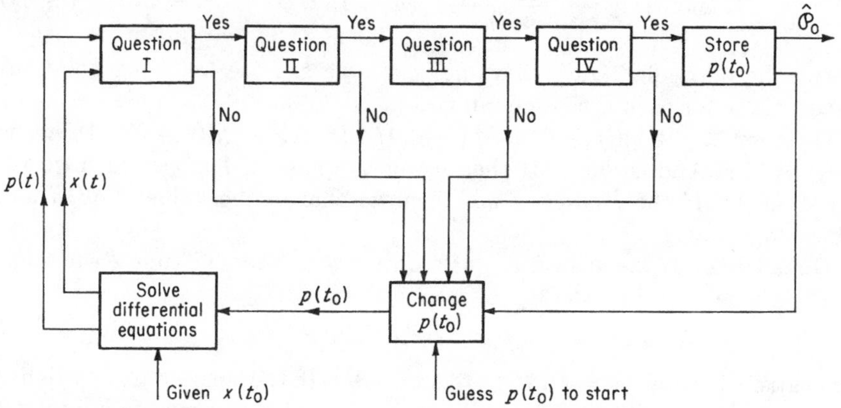


**Fig. 6-7** The logical diagram of the "experiment" which may be used to determine the time-optimal control.

***Definition 6-7*** *Let $\hat{\mathcal{P}}_0$ be the set of initial costates $\hat{\mathbf{p}}(t_0)$ with the following properties:* (1) *For each $\hat{\mathbf{p}}(t_0) \in \hat{\mathcal{P}}_0$, the corresponding solutions of Eqs.* (6-97) *and* (6-98), *denoted by*

$$\begin{aligned} \hat{\mathbf{x}}(t) &= \hat{\mathbf{x}}[t, t_0, \mathbf{x}(t_0), \hat{\mathbf{p}}(t_0)] \\ \hat{\mathbf{p}}(t) &= \hat{\mathbf{p}}[t, t_0, \mathbf{x}(t_0), \hat{\mathbf{p}}(t_0)] \end{aligned} \tag{6-113}$$

*satisfy the relation*

$$\hat{q}_j(t) = \sum_{i=1}^{n} b_{ij}[\hat{\mathbf{x}}(t), t]\hat{p}_i(t) = 0 \qquad j = 1, 2, \ldots, r \tag{6-114}$$

*only at a countable set of times.* (2) *There is a time $\hat{T}$ [which depends on $\mathbf{x}(t_0)$ and $\hat{\mathbf{p}}(t_0)$] such that we can find constants $e_1, e_2, \ldots, e_{n-\beta}$ and $k_1, k_2, \ldots, k_{n-\beta}$ so that the relations*

$$H^o[\hat{\mathbf{x}}(\hat{T}), \hat{\mathbf{p}}(\hat{T}), \hat{T}] = \sum_{i=1}^{n} f_i[\hat{\mathbf{x}}(\hat{T}), \hat{T}]\hat{p}_i(\hat{T}) - \sum_{j=1}^{r} \left| \sum_{i=1}^{n} b_{ij}[\hat{\mathbf{x}}(\hat{T}), \hat{T}]\hat{p}_i(\hat{T}) \right| = \sum_{\alpha=1}^{n-\beta} e_\alpha \frac{\partial g_\alpha[\hat{\mathbf{x}}(\hat{T}), \hat{T}]}{\partial \hat{T}} \tag{6-115}$$

$$g_\alpha[\hat{\mathbf{x}}(\hat{T}), \hat{T}] = 0 \qquad \alpha = 1, 2, \ldots, n-\beta \tag{6-116}$$

$$\hat{\mathbf{p}}(\hat{T}) = \sum_{\alpha=1}^{n-\beta} k_\alpha \frac{\partial g_\alpha[\hat{\mathbf{x}}(\hat{T}), \hat{T}]}{\partial \hat{\mathbf{x}}(\hat{T})} \tag{6-117}$$

*are satisfied.*

We now suggest that the reader go over the statement of Theorem 6-3 and compare Eq. (6-115) with Eq. (6-84), Eq. (6-116) with Eq. (6-86), and Eq. (6-117) with Eq. (6-87). In view of the fact that the functions $\hat{q}_j(t)$ are zero only at a countable set of times and in view of the fact that the differential equations (6-97) and (6-98) are the same as the differential equations (6-81) and (6-82), we deduce the following lemma:

***Lemma 6-2*** *Each solution $\hat{\mathbf{x}}(t)$ and $\hat{\mathbf{p}}(t)$, $t \in [t_0, \hat{T}]$, generated by an element of the set $\hat{\mathcal{P}}_0$ satisfies all the reduced necessary conditions given by Theorem* 6-3.

We have found that the $H$-minimal control (see Definition 6-6) $\mathbf{u}^\circ(t)$ is given by [see Eq. (6-91)]

$$\mathbf{u}^\circ(t) = -\text{ SGN }\{\mathbf{B}'[\mathbf{x}(t), t]\mathbf{p}(t)\}$$

for all $\mathbf{x}(t)$, $\mathbf{p}(t)$, and $t$. If we now evaluate $\mathbf{u}^\circ(t)$, when $\mathbf{x}(t) = \hat{\mathbf{x}}(t)$, $\mathbf{p}(t) = \hat{\mathbf{p}}(t)$, and $t \in [t_0, \hat{T}]$, we find that

$$\mathbf{u}^\circ(t)\Big|_{\substack{\mathbf{x}(t)=\hat{\mathbf{x}}(t)\\ \mathbf{p}(t)=\hat{\mathbf{p}}(t)}} = \hat{\mathbf{u}}^\circ(t) = -\text{ SGN }\{\mathbf{B}'[\hat{\mathbf{x}}(t), t]\hat{\mathbf{p}}(t)\} \qquad t \in [t_0, \hat{T}\} \tag{6-118}$$

If we compare Eqs. (6-118) and (6-91), we immediately deduce, in view of Lemma 6-2, the following lemma:

***Lemma 6-3*** *Each control $\hat{\mathbf{u}}^\circ(t)$ generated by an element of the set $\hat{\mathcal{P}}_0$ satisfies the necessary conditions stated in Theorem* 6-2. *Observe that*

$$H[\hat{\mathbf{x}}(t), \hat{\mathbf{p}}(t), \hat{\mathbf{u}}^\circ(t), t] \leq H[\hat{\mathbf{x}}(t), \hat{\mathbf{p}}(t), \mathbf{u}(t), t] \tag{6-119}$$

*for all $\mathbf{u}(t) \in \Omega$ and $t \in [t_0, \hat{T}]$.*

We shall now discuss the implications of Lemmas 6-2 and 6-3 and, in so doing, explain the usefulness of the necessary conditions in our search for the time-optimal control.

For the sake of concreteness, let us suppose that there are *three* different time-optimal controls from the given initial state $\mathbf{x}(t_0)$ to the given target set $S$; all three time-optimal controls will, by definition, require the same minimum time $T^*$. Let us indicate these (time-optimal) controls by

$$\mathbf{u}_1^*(t),\ \mathbf{u}_2^*(t),\ \mathbf{u}_3^*(t) \qquad t \in [t_0, T^*] \tag{6-120}$$

If we conduct the "experiment" of step 3*d*, then we shall determine a set $\hat{\mathcal{P}}_0$; let us suppose that we can find [from Eq. (6-118)] *five* different controls corresponding to the elements of $\hat{\mathcal{P}}_0$. We shall denote these controls by

$$\hat{\mathbf{u}}_1{}^\circ(t),\ \hat{\mathbf{u}}_2{}^\circ(t),\ \hat{\mathbf{u}}_3{}^\circ(t),\ \hat{\mathbf{u}}_4{}^\circ(t),\ \hat{\mathbf{u}}_5{}^\circ(t) \tag{6-121}$$

and the intervals over which they are defined by

$$[t_0, \hat{T}_1],\ [t_0, \hat{T}_2],\ [t_0, \hat{T}_3],\ [t_0, \hat{T}_4],\ [t_0, \hat{T}_5] \tag{6-122}$$

respectively. We claim that three of the five controls of Eq. (6-121) will be identical to the three time-optimal controls of Eq. (6-120). Again to be specific, we suppose that

$$\begin{aligned} \hat{\mathbf{u}}_1{}^{\circ}(t) &= \mathbf{u}_1^*(t) & \hat{T}_1 &= T^* & t &\in [t_0, T^*] \\ \hat{\mathbf{u}}_2{}^{\circ}(t) &= \mathbf{u}_2^*(t) & \hat{T}_2 &= T^* & t &\in [t_0, T^*] \\ \hat{\mathbf{u}}_3{}^{\circ}(t) &= \mathbf{u}_3^*(t) & \hat{T}_3 &= T^* & t &\in [t_0, T^*] \end{aligned} \tag{6-123}$$

The question that arises is the following: *What is the significance of the controls* $\hat{\mathbf{u}}_4{}^{\circ}(t)$ *and* $\hat{\mathbf{u}}_5{}^{\circ}(t)$*?* Well, these two controls must be locally time-optimal. Since the minimum principle is a local condition, it cannot distinguish between locally optimal and globally optimal controls. For this reason, the set of controls which satisfy all the necessary conditions will consist of both the locally time-optimal and the globally time-optimal controls. The only way to distinguish which of the controls $\hat{\mathbf{u}}_1{}^{\circ}(t), \ldots,$ $\hat{\mathbf{u}}_5{}^{\circ}(t)$ are globally time-optimal is to measure and compare the times $\hat{T}_1, \hat{T}_2, \ldots, \hat{T}_5$ and, in so doing, to find that

$$\begin{aligned} &\hat{T}_1 = \hat{T}_2 = \hat{T}_3 = T^* \\ &\hat{T}_4 > T^* \\ &\hat{T}_5 > T^* \end{aligned} \tag{6-124}$$

It is for this reason that we say that the necessary conditions provide us with all the controls which are candidates for optimality.

In the following section, we shall discuss the results obtained in this section and motivate the subsequent sections of the chapter that deal with the time-optimal control problem.

**Exercise 6-2** Consider the dynamical system of Example 6-3. Using the appropriate theorems of Secs. 5-13 and 5-14, derive the necessary conditions on the optimal control for the following target sets $S$:

(*a*) $S$ is the origin of the three-dimensional state space.

(*b*) $S$ is the state

$$\mathbf{x}_1 = \begin{bmatrix} 1 \\ 1 \\ 1 \end{bmatrix}$$

(*c*) $S$ is the plane $x_1 = 0$.

(*d*) $S$ is the moving point described by the equations

$$x_1 = t \qquad x_2 = t^2 \qquad x_3 = \sin t$$

(*e*) $S$ is the moving plane $x_1 = t + t^2 + t^3$.

**Exercise 6-3** Consider the problem described in Example 6-3. For this particular problem, outline the "experiment" you would perform to determine the candidates for time-optimal control. (Hint: Mimic the development of step 3.) In view of the special form of the equations, can you suggest ways of simplifying the search for the set $\hat{\mathcal{P}}_0$?

**Exercise 6-4** Consider the problem stated in Example 6-3. Suppose now that the control variables $u_1(t)$ and $u_2(t)$ are constrained by the inequalities

$$-1 \le u_1(t) \le +3 \qquad -5 \le u_2(t) \le +2$$

Derive the reduced necessary conditions.

## 6-4 Minimum-time Problems 3: Comments

In the preceding section, we derived necessary conditions for the time-optimal control and suggested an idealized systematic way of determining the candidates for a time-optimal control.

We found (see Theorem 6-2) that if the problem was normal, then the components of the time-optimal control were piecewise constant functions of time. We did not derive any results for the singular problem which will be examined in Sec. 6-21.

Since the components of the time-optimal control must be piecewise constant functions of time for normal problems, we can see that one of the necessary conditions—i.e., the condition $H[\mathbf{x}^*(t), \mathbf{p}^*(t), \mathbf{u}^*(t), t] \leq H[\mathbf{x}^*(t), \mathbf{p}^*(t), \mathbf{u}(t), t]$, $\mathbf{u}(t) \in \Omega$—serves to limit the search for the time-optimal control to the class $|u_j(t)| = 1, j = 1, 2, \ldots, r$. This is perhaps the most useful result derived from the minimum principle, as the other necessary conditions merely provide appropriate boundary and transversality conditions.

The reader has probably noticed† that the Hamiltonian

$$H[\mathbf{x}(t), \mathbf{p}(t), \mathbf{u}(t), t] = 1 + \langle \mathbf{f}[\mathbf{x}(t), t], \mathbf{p}(t) \rangle + \langle \mathbf{u}(t), \mathbf{B}'[\mathbf{x}(t), t]\mathbf{p}(t) \rangle \tag{6-125}$$

and the differential equations

$$\begin{aligned} \dot{\mathbf{x}}(t) &= \frac{\partial H[\mathbf{x}(t), \mathbf{p}(t), \mathbf{u}(t), t]}{\partial \mathbf{p}(t)} \\ \dot{\mathbf{p}}(t) &= -\frac{\partial H[\mathbf{x}(t), \mathbf{p}(t), \mathbf{u}(t), t]}{\partial \mathbf{x}(t)} \end{aligned} \tag{6-126}$$

are completely defined by the system and the cost functional and, thus, are independent of the boundary conditions at $t_0$ and of the target set $S$. In addition, the $H$-minimal control $\mathbf{u}^\circ(t)$ (see Definition 6-6), given by the equation

$$\mathbf{u}^\circ(t) = -\operatorname{SGN}\{\mathbf{B}'[\mathbf{x}(t), t]\mathbf{p}(t)\} \tag{6-127}$$

is (functionally) independent of the imposed boundary conditions. Thus, steps 3*a* to 3*c*, outlined in Sec. 6-3, will be the same for any time-optimal problem. The necessary conditions on the Hamiltonian and the costate variable at the terminal time $T^*$, together with the given initial state and the equations of the target set, merely provide enough boundary conditions to solve the system of $2n$ differential equations satisfied by the state and costate variables.

We indicated in Sec. 6-3 the step-by-step procedure used to determine the controls $\hat{\mathbf{u}}^\circ(t)$, the resultant trajectories $\hat{\mathbf{x}}(t)$, and the corresponding costates $\hat{\mathbf{p}}(t)$ satisfying all the necessary conditions. To distinguish these quantities, we make the following definition:

† This is also reflected by Table 5-1.

***Definition 6-8 Extremal Variables*** *A control* $\hat{\mathbf{u}}^o(t)$ *is called extremal if* $\hat{\mathbf{u}}^o(t)$, *the resultant trajectory* $\hat{\mathbf{x}}(t)$, *and the corresponding costate* $\hat{\mathbf{p}}(t)$ *satisfy all of the necessary conditions* [*i.e., they satisfy Eqs.* (6-113) *and* (6-115) to (6-119)]. *We shall also call* $\hat{\mathbf{x}}(t)$ *and* $\hat{\mathbf{p}}(t)$ *the extremal state and costate trajectories, respectively.*

As we mentioned in Sec. 6-3, in general, we can have many extremal controls. *Each extremal control will then generate either a locally time-optimal or a globally time-optimal trajectory.*† Since an extremal control satisfies all the necessary conditions, we can make the following remarks:

***Remark 6-7*** *If the time-optimal control* $\mathbf{u}^*(t)$ *exists and is unique and if there are no other locally time-optimal controls, then there is one and only one extremal control* $\hat{\mathbf{u}}^o(t)$ *which is, in fact, time-optimal, that is,* $\hat{\mathbf{u}}^o(t) = \mathbf{u}^*(t)$.

It should be clear that in Remark 6-7, the assumption that there are no other locally optimal controls makes the minimum principle *both a necessary and a sufficient condition.*

***Remark 6-8*** *If there are exactly* $m_1$ *distinct time-optimal controls and if there are exactly* $m_2$ *distinct controls which are locally time-optimal (but not globally time-optimal), then there will be exactly* $m_1 + m_2$ *extremal controls.*

***Remark 6-9*** *If a (global) time-optimal control does not exist and if there are* $m_2$ *distinct locally time-optimal controls, then there will be exactly* $m_2$ *extremal controls.*

This remark illustrates that the existence of extremal controls does not necessarily imply the existence of a global time-optimal control.

***Remark 6-10*** *If a time-optimal control exists, we can find it by evaluating the time* $T$ *required by each extremal control and by picking the one which minimizes* $T$.

The above remarks lead us to the conclusion that, when dealing with a time-optimal control problem (such as Problem 6-1*b*), we should know the answers to the following questions:

1. Does a time-optimal control exist?
2. Is the time-optimal control unique?
3. Is the time-optimal problem normal?
4. Is there more information hidden in the necessary conditions for a given system and target set?

Unfortunately, at this time, the answers to such questions are not available for arbitrary nonlinear systems and target sets. Results, however, are known for the class of linear systems. As this class of systems is extremely important, we shall devote several sections to it, deriving additional results which are important from both the theoretical and the practical points of view.

† See also the discussion in Sec. 5-17. Locally time-optimal controls may correspond to saddle points.

## 6-5 Minimum-time Problems 4: Linear Time-invariant Systems†

In this section, we shall specialize the results derived for Problem 6-1*b* to the case of linear time-invariant systems. We shall also assume that the target set is the origin of the state space. For this reason, we shall call the problem the *linear time-optimal regulator problem.* We note also that most of the problems that we shall examine in Chap. 7 are of this type.

We shall first state Problem 6-1*c* so that the reader will know the precise problem that we shall consider. Second, we shall state in Theorem 6-4 the necessary conditions satisfied by a time-optimal control. Third, we shall establish in Theorem 6-6 an explicit test for normality. Fourth, we shall establish in Theorem 6-7 the uniqueness of the time-optimal control for normal problems. Fifth, we shall prove Theorem 6-8 regarding the number of switchings. Finally, we shall establish in Theorem 6-9 the uniqueness of the extremal controls.

The particular special problem we shall consider is the following:

***Problem 6-1c The Time-optimal Regulator Problem for a Linear Time-invariant System*** *Given the dynamical system*

$$\dot{\mathbf{x}}(t) = \mathbf{A}\mathbf{x}(t) + \mathbf{B}\mathbf{u}(t) \tag{6-128}$$

*where*

*a. The n-vector* $\mathbf{x}(t)$ *is the state.*

*b. The system matrix* $\mathbf{A}$ *is an* $n \times n$ *constant matrix.*

*c. The gain matrix* $\mathbf{B}$ *is an* $n \times r$ *constant matrix.*

*d. The r vector* $\mathbf{u}(t)$ *is the control.*

*Assume that the system* (6-128) *is completely controllable and that the components* $u_1(t), u_2(t), \ldots, u_r(t)$ *of* $\mathbf{u}(t)$ *are constrained in magnitude by the relations*

$$|u_j(t)| \le 1 \qquad j = 1, 2, \ldots, r \tag{6-129}$$

*Then, given that at the initial time* $t_0 = 0$, *the initial state of the system* (6-128) *is*

$$\mathbf{x}(0) = \boldsymbol{\xi} \tag{6-130}$$

*find the control* $\mathbf{u}^*(t)$ *that transfers the system* (6-128) *from* $\boldsymbol{\xi}$ *to the origin* $\mathbf{0}$ *in minimum time.*

We let $\lambda_1, \lambda_2, \ldots, \lambda_n$ denote the eigenvalues of the system matrix $\mathbf{A}$, and we let $\mathbf{b}_1, \mathbf{b}_2, \ldots, \mathbf{b}_r$ denote the column vectors of the gain matrix $\mathbf{B}$. In other words,

$$\mathbf{B} = \left[\begin{array}{c:c:c:c} \uparrow & \uparrow & & \uparrow \\ \mathbf{b}_1 & \mathbf{b}_2 & \cdots & \mathbf{b}_r \\ \downarrow & \downarrow & & \downarrow \end{array}\right] \tag{6-131}$$

† Appropriate references for this section are P-5, L-4, B-2, D-2, L-6, and N-2.

We assumed that the system (6-128) is completely controllable.† This, of course, means that there are controls that will transfer the system (6-128) from any initial state $\boldsymbol{\xi}$ to $\mathbf{0}$ in finite time. We also recall that the system (6-128) is completely controllable if and only if the $n \times (rn)$ matrix

$$\mathbf{G} = [\mathbf{B} \mid \mathbf{AB} \mid \mathbf{A}^2\mathbf{B} \mid \cdots \mid \mathbf{A}^{n-1}\mathbf{B}] \tag{6-132}$$

contains $n$ linearly independent column vectors.

If the output $\mathbf{y}(t)$ of our linear system (6-128) is related to the state $\mathbf{x}(t)$ and the control $\mathbf{u}(t)$ by the relation

$$\mathbf{y}(t) = \mathbf{Cx}(t) + \mathbf{Du}(t) \tag{6-133}$$

then we assert that *a control that drives the state to* $\mathbf{0}$ *can be "extended" in such a way as to drive the output to zero and hold it at zero thereafter.* Thus, if at $t = T^*$ we have

$$\mathbf{x}(T^*) = \mathbf{0} \tag{6-134}$$

then

$$\mathbf{y}(T^*) = \mathbf{Du}(T^*) \tag{6-135}$$

and if we "extend" the control by setting

$$\mathbf{u}(t) = \mathbf{0} \qquad \text{for all } t > T^* \tag{6-136}$$

then we have

$$\left.\begin{aligned}\mathbf{x}(t) &= \mathbf{0}\\ \mathbf{y}(t) &= \mathbf{0}\end{aligned}\right\} \qquad \text{for all } t > T^* \tag{6-137}$$

We shall now state the necessary conditions for Problem 6-1c. First of all, we form the Hamiltonian function for the problem,

$$\begin{aligned} H[\mathbf{x}(t), \mathbf{p}(t), \mathbf{u}(t)] &= 1 + \langle \mathbf{Ax}(t), \mathbf{p}(t)\rangle + \langle \mathbf{Bu}(t), \mathbf{p}(t)\rangle \\ &= 1 + \langle \mathbf{Ax}(t), \mathbf{p}(t)\rangle + \langle \mathbf{u}(t), \mathbf{B}'\mathbf{p}(t)\rangle \end{aligned} \tag{6-138}$$

Assuming that a time-optimal control $\mathbf{u}^*(t)$ exists, we may use Theorem 5-5 to obtain the necessary conditions, which are summarized for the sake of completeness in the following theorem:

***Theorem 6-4 Necessary Conditions for Problem 6-1c*** *Let* $\mathbf{u}^*(t)$ *be a time-optimal control that transfers the initial state* $\boldsymbol{\xi}$ *to the origin* $\mathbf{0}$. *Let* $\mathbf{x}^*(t)$ *denote the trajectory of the system* (6-128) *corresponding to* $\mathbf{u}^*(t)$, *originating at* $\boldsymbol{\xi}$ *at* $t_0 = 0$, *and hitting the origin* $\mathbf{0}$ *at the minimum time* $T^*$ [*that is,* $\mathbf{x}^*(0) = \boldsymbol{\xi}$, $\mathbf{x}^*(T^*) = \mathbf{0}$]. *Then there exists a corresponding costate vector* $\mathbf{p}^*(t)$ *such that:*

† See Secs. 4-15 and 4-16.

*a.* $\mathbf{x}^*(t)$ *and* $\mathbf{p}^*(t)$ *are solutions of the canonical equations*

$$\begin{aligned}\dot{\mathbf{x}}^*(t) &= \frac{\partial H[\mathbf{x}^*(t), \mathbf{p}^*(t), \mathbf{u}^*(t)]}{\partial \mathbf{p}^*(t)} \\ &= \mathbf{A}\mathbf{x}^*(t) + \mathbf{B}\mathbf{u}^*(t) \qquad (6\text{-}139)\\ \dot{\mathbf{p}}^*(t) &= -\frac{\partial H[\mathbf{x}^*(t), \mathbf{p}^*(t), \mathbf{u}^*(t)]}{\partial \mathbf{x}^*(t)} \\ &= -\mathbf{A}'\mathbf{p}^*(t) \qquad (6\text{-}140)\end{aligned}$$

*with the boundary conditions*

$$\mathbf{x}^*(0) = \boldsymbol{\xi} \qquad \mathbf{x}^*(T^*) = \mathbf{0} \qquad (6\text{-}141)$$

*b. The relation* [*see Eqs.* (5-400) *and* (6-138)]

$$1 + \langle \mathbf{A}\mathbf{x}^*(t), \mathbf{p}^*(t)\rangle + \langle \mathbf{u}^*(t), \mathbf{B}'\mathbf{p}^*(t)\rangle \le 1 + \langle \mathbf{A}\mathbf{x}^*(t), \mathbf{p}^*(t)\rangle + \langle \mathbf{u}(t), \mathbf{B}'\mathbf{p}^*(t)\rangle \qquad (6\text{-}142)$$

*holds for all admissible* $\mathbf{u}(t)$ *and for* $t \in [0, T^*]$. *Equation* (6-142) *yields, in turn, the relation*

$$\mathbf{u}^*(t) = -\operatorname{SGN}\{\mathbf{q}^*(t)\} = -\operatorname{SGN}\{\mathbf{B}'\mathbf{p}^*(t)\} \qquad (6\text{-}143)$$

*where* $\mathbf{q}^*(t) = \mathbf{B}'\mathbf{p}^*(t)$, *which can be written componentwise* [*in view of Eq.* (6-131)] *in the form*

$$u_j^*(t) = -\operatorname{sgn}\{q_j^*(t)\} = -\operatorname{sgn}\{\langle \mathbf{b}_j, \mathbf{p}^*(t)\rangle\} \qquad j = 1, 2, \ldots, r \qquad (6\text{-}144)$$

*c. The relation* [*see Eqs.* (5-401) *and* (6-138)]

$$H[\mathbf{x}^*(t), \mathbf{p}^*(t), \mathbf{u}^*(t)] = 1 + \langle \mathbf{A}\mathbf{x}^*(t), \mathbf{p}^*(t)\rangle + \langle \mathbf{u}^*(t), \mathbf{B}'\mathbf{p}^*(t)\rangle = 0 \qquad (6\text{-}145)$$

*holds for all* $t \in [0, T^*]$.

In the remainder of this section, we shall derive properties of the time-optimal control (such as uniqueness) from the necessary conditions of Theorem 6-4.

Let us now examine the differential equation for the costate $\mathbf{p}^*(t)$, that is, the equation

$$\dot{\mathbf{p}}^*(t) = -\mathbf{A}'\mathbf{p}^*(t) \qquad (6\text{-}146)$$

We note that Eq. (6-146) is:

1. Homogeneous
2. Time-invariant
3. Adjoint† to Eq. (6-139)
4. Independent of $\mathbf{x}^*(t)$ and $\mathbf{u}^*(t)$

The necessary conditions of Theorem 6-4 do not contain any explicit information regarding either the initial costate $\mathbf{p}^*(0)$ or the terminal costate $\mathbf{p}^*(T)$. Nonetheless, we can determine some properties of $\mathbf{p}^*(t)$ from Eq. (6-145), and an important one is contained in the following lemma:

† See Sec. 3-25.

***Lemma 6-4*** *The costate* $\mathbf{p}^*(t)$ *must be a nonzero vector, that is,*

$$\mathbf{p}^*(t) \neq \mathbf{0} \qquad \text{for } t \in [0, T^*] \tag{6-147}$$

PROOF Suppose that $\mathbf{p}^*(t) = \mathbf{0}$. Then substitution of $\mathbf{p}^*(t) = \mathbf{0}$ into Eq. (6-145) would yield the contradiction $1 = 0$; therefore, $\mathbf{p}^*(t) \neq \mathbf{0}$.

Since Eq. (6-145) must hold for $t \in [0, T^*]$, we can also deduce the following relations:

*a.* For $t = 0$,

$$1 + \langle \mathbf{A}\boldsymbol{\xi}, \mathbf{p}^*(0)\rangle + \langle \mathbf{u}^*(0), \mathbf{B}'\mathbf{p}^*(0)\rangle = 0 \tag{6-148}$$

*b.* For $t = T^*$,

$$1 + \langle \mathbf{u}^*(T^*), \mathbf{B}'\mathbf{p}^*(T^*)\rangle = 0 \tag{6-149}$$

However, we cannot solve Eqs. (6-148) and (6-149) for either $\mathbf{p}^*(0)$ or $\mathbf{p}^*(T^*)$.

Let us agree to denote the *unknown* nonzero (by Lemma 6-4) value of the initial costate by $\boldsymbol{\pi}$; that is, $\boldsymbol{\pi}$ is given by

$$\boldsymbol{\pi} = \mathbf{p}^*(0) \qquad \boldsymbol{\pi} \neq \mathbf{0} \tag{6-150}$$†

The solution of the differential equation (6-146) is then

$$\mathbf{p}^*(t) = e^{-\mathbf{A}'t}\boldsymbol{\pi} \tag{6-151}$$

If we now substitute Eq. (6-151) into Eqs. (6-143) and (6-144), we find that

$$\mathbf{u}^*(t) = -\operatorname{SGN}\{\mathbf{B}'e^{-\mathbf{A}'t}\boldsymbol{\pi}\} \tag{6-152}$$

and hence that

$$u_j^*(t) = -\operatorname{sgn}\{q_j^*(t)\} = -\operatorname{sgn}\{\langle \mathbf{b}_j, e^{-\mathbf{A}'t}\boldsymbol{\pi}\rangle\} \qquad j = 1, 2, \ldots, r \tag{6-153}$$

where

$$q_j^*(t) = \langle \mathbf{b}_j, e^{-\mathbf{A}'t}\boldsymbol{\pi}\rangle \qquad j = 1, 2, \ldots, r \tag{6-154}$$

Now we know the vectors $\mathbf{b}_j$, and we know how to compute the matrix $e^{-\mathbf{A}'t}$ (see Sec. 3-23). Thus, if the problem were normal (see Definition 6-4), then Eq. (6-153) would uniquely‡ define $u_j^*(t)$ in terms of the initial costate $\boldsymbol{\pi}$. For this reason, we now proceed to derive necessary and sufficient conditions for Problem 6-1*c* to be normal. We shall do this by first deriving conditions for Problem 6-1*c* to be singular.

Let us suppose that there is an interval of time $[T_1, T_2]$ such that, for some $j$, the relation

$$q_j^*(t) = \langle \mathbf{b}_j, e^{-\mathbf{A}'t}\boldsymbol{\pi}\rangle = 0 \qquad \text{for all } t \in [T_1, T_2] \tag{6-155}$$

is satisfied. It follows that

$$\dot{q}_j^*(t) = 0; \ \ddot{q}_j^*(t) = 0; \ \dddot{q}_j^*(t) = 0; \ \ldots \qquad \text{for all } t \in [T_1, T_2] \tag{6-156}$$

† The reader should not confuse $\boldsymbol{\pi}$ with the number 3.1415 · · · .

‡ Except at the countable set of switch times.

Since $\mathbf{A}$ and $e^{\mathbf{A}t}$ commute, we deduce that the following relations must be satisfied for all $t \in [T_1, T_2]$:

$$
\begin{aligned}
q_j^*(t) &= \langle e^{-\mathbf{A}t}\mathbf{b}_j, \boldsymbol{\pi}\rangle = 0 \\
\dot{q}_j^*(t) &= \langle \mathbf{A}e^{-\mathbf{A}t}\mathbf{b}_j, \boldsymbol{\pi}\rangle = \langle e^{-\mathbf{A}t}\mathbf{A}\mathbf{b}_j, \boldsymbol{\pi}\rangle = 0 \\
\ddot{q}_j^*(t) &= \langle \mathbf{A}^2e^{-\mathbf{A}t}\mathbf{b}_j, \boldsymbol{\pi}\rangle = \langle e^{-\mathbf{A}t}\mathbf{A}^2\mathbf{b}_j, \boldsymbol{\pi}\rangle = 0 \\
&\cdots\cdots\cdots\cdots\cdots\cdots\cdots\cdots \\
q^{(n-1)*}(t) &= \langle \mathbf{A}^{n-1}e^{-\mathbf{A}t}\mathbf{b}_j, \boldsymbol{\pi}\rangle = \langle e^{-\mathbf{A}t}\mathbf{A}^{n-1}\mathbf{b}_j, \boldsymbol{\pi}\rangle = 0
\end{aligned} \tag{6-157}
$$

Let $\mathbf{G}_j$ be the $n \times n$ matrix defined by the equation

$$
\mathbf{G}_j = \left[\begin{array}{c:c:c:c:c} \uparrow & \uparrow & \uparrow & & \uparrow \\ \mathbf{b}_j & \mathbf{A}\mathbf{b}_j & \mathbf{A}^2\mathbf{b}_j & \cdots & \mathbf{A}^{n-1}\mathbf{b}_j \\ \downarrow & \downarrow & \downarrow & & \downarrow \end{array}\right] \tag{6-158}
$$

Then we can write Eq. (6-157) ass

$$\mathbf{G}_j'e^{-\mathbf{A}'t}\boldsymbol{\pi} = \mathbf{0} \qquad \text{for all } t \in [T_1, T_2] \tag{6-159}$$

or, equivalently,

$$\mathbf{G}_j'(e^{-\mathbf{A}'t}\boldsymbol{\pi}) = \mathbf{0} \qquad \text{for all } t \in [T_1, T_2] \tag{6-160}$$

Since $e^{-\mathbf{A}'t}$ is nonsingular† and $\boldsymbol{\pi} \neq \mathbf{0}$ by Lemma 6-4, there is a nonzero vector $\hat{\boldsymbol{\pi}}$ (that is, $\hat{\boldsymbol{\pi}} \neq \mathbf{0}$) such that

$$\mathbf{G}_j'\hat{\boldsymbol{\pi}} = \mathbf{0} \tag{6-161}$$

and so we conclude that the matrix $\mathbf{G}_j$ must be singular, i.e., that

$$\det \mathbf{G}_j = 0 \tag{6-162}$$

Let us now recapitulate: We have shown that Eq. (6-155) implies Eq. (6-162). But Eq. (6-155) means that the problem is singular. It follows that if the problem is singular, then Eq. (6-162) must hold, or, in other words, Eq. (6-162) is a *necessary condition* for Problem 6-1c to be singular. We shall leave it to the reader to verify that Eq. (6-162) is a *sufficient* condition for the problem to be singular, and so we state the following theorem:

***Theorem 6-5 Necessary and Sufficient Conditions for Problem 6-1c to Be Singular*** *The time-optimal control problem* 6-1c *is singular if and only if, for some $j$, $j = 1, 2, \ldots, r$, the matrix $\mathbf{G}_j$, given by*

$$\mathbf{G}_j = [\mathbf{b}_j \mid \mathbf{A}\mathbf{b}_j \mid \mathbf{A}^2\mathbf{b}_j \mid \cdots \mid \mathbf{A}^{n-1}\mathbf{b}_j] \tag{6-163}$$

*is singular.*

**Exercise 6-5** Prove the "only if" part of Theorem 6-5. HINT: Use the Cayley-Hamilton theorem (Theorem 2-3) and the fact that $\mathbf{G}_j$ is independent of $\mathbf{x}^*(t)$, $\mathbf{u}^*(t)$, $\mathbf{p}^*(t)$, and $t$.

† See Sec. 3-20.

Using Theorem 6-5, we can derive necessary and sufficient conditions for Problem 6-1c to be normal. First of all, if the problem is not singular and if the set of switch times is countable, then the problem is clearly normal. Now, if $\det \mathbf{G}_j \neq 0$ for all $j = 1, 2, \ldots, r$, then the problem is not singular. Moreover, in view of the continuity and differentiability of the elements of $e^{-\mathbf{A}'t}$, we can establish the countability of the set of switch times if the problem is not singular and so deduce the following theorem:

***Theorem 6-6 Necessary and Sufficient Conditions for Problem 6-1c to Be Normal*** *The time-optimal problem* 6-1c *is normal if and only if the* $n \times n$ *matrices* $\mathbf{G}_1, \mathbf{G}_2, \ldots, \mathbf{G}_r$ *given by the equations*

$$\begin{aligned} \mathbf{G}_1 &= [\mathbf{b}_1 \mid \mathbf{A}\mathbf{b}_1 \mid \mathbf{A}^2\mathbf{b}_1 \mid \cdots \mid \mathbf{A}^{n-1}\mathbf{b}_1] \\ \mathbf{G}_2 &= [\mathbf{b}_2 \mid \mathbf{A}\mathbf{b}_2 \mid \mathbf{A}^2\mathbf{b}_2 \mid \cdots \mid \mathbf{A}^{n-1}\mathbf{b}_2] \\ &\cdots\cdots\cdots\cdots\cdots\cdots\cdots \\ \mathbf{G}_r &= [\mathbf{b}_r \mid \mathbf{A}\mathbf{b}_r \mid \mathbf{A}^2\mathbf{b}_r \mid \cdots \mid \mathbf{A}^{n-1}\mathbf{b}_r] \end{aligned} \tag{6-164}$$

*are all nonsingular.*

**Exercise 6-6** Prove Theorem 6-6.

The reader should note that the necessary and sufficient conditions of Theorem 6-6 for the *time-optimal problem* 6-1c *to be normal are identical to the necessary and sufficient conditions for the system* $\dot{\mathbf{x}}(t) = \mathbf{A}\mathbf{x}(t) + \mathbf{B}\mathbf{u}(t)$ *to be normal.*† We can thus see how the normality concept developed in Chap. 4 arises in optimization problems.

We have found the relation between a normal system and a normal time-optimal control problem. Moreover, we can check whether or not a given system is normal. So let us now determine the implications of normality. We shall show that *if the controlled linear system is normal and if a time-optimal control* $\mathbf{u}^*(t)$ *exists, then the time-optimal control is unique.*

The following theorem applies to Problem 6-1c. We recall that in Problem 6-1c the terminal state was the origin of the state space. As we state in Exercise 6-7, the theorem can be generalized to the case where the terminal state $\mathbf{x}(T^*) = \boldsymbol{\theta}$ is any state in the state space.

***Theorem 6-7 Uniqueness of the Time-optimal Control*** *If, in Problem* 6-1c, *the linear system* $\dot{\mathbf{x}}(t) = \mathbf{A}\mathbf{x}(t) + \mathbf{B}\mathbf{u}(t)$ *is normal (or, equivalently, if the time-optimal problem is normal), then the time-optimal control is unique (if it exists).*

PROOF Suppose that $\mathbf{u}_1^*(t)$ and $\mathbf{u}_2^*(t)$ are two distinct time-optimal controls transferring the initial state $\boldsymbol{\xi}$ to $\mathbf{0}$ in the (same) minimum time $T^*$. Let $\mathbf{x}_1^*(t)$ and $\mathbf{x}_2^*(t)$ be the distinct trajectories originating at $\boldsymbol{\xi}$. Then

$$\mathbf{x}_1^*(t) = e^{\mathbf{A}t}\left[\boldsymbol{\xi} + \int_0^t e^{-\mathbf{A}\tau}\mathbf{B}\mathbf{u}_1^*(\tau)\, d\tau\right] \tag{6-165}$$

$$\mathbf{x}_2^*(t) = e^{\mathbf{A}t}\left[\boldsymbol{\xi} + \int_0^t e^{-\mathbf{A}\tau}\mathbf{B}\mathbf{u}_2^*(\tau)\, d\tau\right] \tag{6-166}$$

† See Sec. 4-21.

At $t = T^*$, we must have

$$\mathbf{x}_1^*(T^*) = \mathbf{x}_2^*(T^*) = \mathbf{0} \tag{6-167}$$

Since the matrix $e^{\mathbf{A}t}$ is nonsingular, we find that

$$\int_0^{T^*} e^{-\mathbf{A}t}\mathbf{B}\mathbf{u}_1^*(t)\,dt = \int_0^{T^*} e^{-\mathbf{A}t}\mathbf{B}\mathbf{u}_2^*(t)\,dt \tag{6-168}$$

We shall now use Eq. (6-168) and the necessary conditions to deduce that

$$\mathbf{u}_1^*(t) = \mathbf{u}_2^*(t) \qquad \text{for } t \in [0, T^*] \tag{6-169}$$

Since both $\mathbf{u}_1^*(t)$ and $\mathbf{u}_2^*(t)$ were assumed to be time-optimal, we know that there exist costates $\mathbf{p}_1^*(t) = e^{-\mathbf{A}'t}\boldsymbol{\pi}_1$ and $\mathbf{p}_2^*(t) = e^{-\mathbf{A}'t}\boldsymbol{\pi}_2$ corresponding to $\mathbf{u}_1^*(t)$ and $\mathbf{u}_2^*(t)$, respectively, such that the relations

$$\mathbf{u}_1^*(t) = -\operatorname{SGN}\{\mathbf{B}'e^{-\mathbf{A}'t}\boldsymbol{\pi}_1\} \tag{6-170}$$
$$\mathbf{u}_2^*(t) = -\operatorname{SGN}\{\mathbf{B}'e^{-\mathbf{A}'t}\boldsymbol{\pi}_2\} \tag{6-171}$$

uniquely define (because of the normality) the controls $\mathbf{u}_1^*(t)$ and $\mathbf{u}_2^*(t)$, except possibly at their countable set of switch times. This, in turn, implies that [see Eq. (6-142)] the relation

$$\langle \mathbf{B}\mathbf{u}_1^*(t), \mathbf{p}_1^*(t)\rangle \le \langle \mathbf{B}\mathbf{u}_2^*(t), \mathbf{p}_1^*(t)\rangle \tag{6-172}$$

or, equivalently, the relation

$$\langle \mathbf{B}\mathbf{u}_1^*(t), e^{-\mathbf{A}'t}\boldsymbol{\pi}_1\rangle \le \langle \mathbf{B}\mathbf{u}_2^*(t), e^{-\mathbf{A}'t}\boldsymbol{\pi}_1\rangle \tag{6-173}$$

is an equality whenever $\mathbf{u}_1^*(t) = \mathbf{u}_2^*(t)$ and a strict inequality whenever $\mathbf{u}_1^*(t) \ne \mathbf{u}_2^*(t)$.

If Eq. (6-168) holds, then the relation

$$\left\langle \boldsymbol{\pi}_1, \int_0^{T^*} e^{-\mathbf{A}t}\mathbf{B}\mathbf{u}_1^*(t)\,dt\right\rangle = \left\langle \boldsymbol{\pi}_1, \int_0^{T^*} e^{-\mathbf{A}t}\mathbf{B}\mathbf{u}_2^*(t)\,dt\right\rangle \tag{6-174}$$

or, equivalently, the relation

$$\int_0^{T^*} \langle e^{-\mathbf{A}'t}\boldsymbol{\pi}_1, \mathbf{B}\mathbf{u}_1^*(t)\rangle\,dt = \int_0^{T^*} \langle e^{-\mathbf{A}'t}\boldsymbol{\pi}_1, \mathbf{B}\mathbf{u}_2^*(t)\rangle\,dt \tag{6-175}$$

must be satisfied. From Eqs. (6-175) and (6-173), we immediately deduce that $\mathbf{u}_1^*(t) = \mathbf{u}_2^*(t)$ for $t \in [0, T^*]$.

The reader should note that it was the assumption of normality that allowed us to prove uniqueness. If the problem were not normal, then it would be possible to have $\langle \mathbf{B}\mathbf{u}_1^*(t), e^{-\mathbf{A}'t}\boldsymbol{\pi}_1\rangle = \langle \mathbf{B}\mathbf{u}_2^*(t), \mathbf{e}^{-\mathbf{A}'t}\boldsymbol{\pi}_1\rangle$, although $\mathbf{u}_1^*(t) \ne \mathbf{u}_2^*(t)$.

**Exercise 6-7** Consider Problem 6-1*c*, but suppose the target set is any given state $\boldsymbol{\theta}$ rather than $\mathbf{0}$. Show that the time-optimal control is unique (if it exists), provided that the problem is normal.

We shall next prove a very useful theorem† regarding the "number of switchings" of the time-optimal control. Let us recall that, if the time-optimal problem is normal, then the components of the time-optimal control are piecewise constant functions of time. As illustrated in Fig. 6-4, the control variables jump from $+1$ to $-1$ and from $-1$ to $+1$ at the switch times. The following theorem provides us with an upper bound on the number of switch times, provided that the eigenvalues of the system matrix **A** are real. As we shall see in Chap. 7, this theorem is extremely useful in the design of time-optimal systems.

***Theorem 6-8 Number of Switchings*** *Consider Problem* 6-1c. *Suppose that the system*

$$\dot{\mathbf{x}}(t) = \mathbf{A}\mathbf{x}(t) + \mathbf{B}\mathbf{u}(t) \tag{6-176}$$

*is normal. Suppose that all the eigenvalues* $\lambda_1, \lambda_2, \ldots, \lambda_n$ *of the* $n \times n$ *system matrix* **A** *are real. Let* $u_j^*(t)$, $j = 1, 2, \ldots, r$, *be the components of the (unique) time-optimal control (if it exists). Let* $t_{\gamma j}$ *be the switch times (see Definition* 6-4) *of the piecewise constant functions* $u_j^*(t)$. *Then the maximum value of the number* $\gamma j$ *is at most* $n - 1$, *for all* $j = 1, 2, \ldots, r$. *In other words, each piecewise constant control* $u_j^*(t)$ *can switch (from* $+1$ *to* $-1$ *or from* $-1$ *to* $+1$) *at most* $n - 1$ *times.*

PROOF We shall prove this theorem under the additional assumption that the eigenvalues $\lambda_1, \lambda_2, \ldots, \lambda_n$ are distinct. If two or more eigenvalues are the same, the theorem is still valid and a proof can be found in Ref. P-5 (chap. 3, theorem 10).

We know [see Eq. (6-153)] that the components $u_j^*(t)$ of the time-optimal control are given by the equations

$$u_j^*(t) = -\operatorname{sgn}\{\langle e^{-\mathbf{A}t}\mathbf{b}_j, \boldsymbol{\pi}\rangle\} \qquad \boldsymbol{\pi} \neq \mathbf{0};\ j = 1, 2, \ldots, r \tag{6-177}$$

By the definition of the switch times $t_{\gamma j}$, we have

$$\langle e^{-\mathbf{A}t_{\gamma j}}\mathbf{b}_j, \boldsymbol{\pi}\rangle = 0 \qquad \gamma j = 1, 2, \ldots \tag{6-178}$$

Let $\boldsymbol{\Lambda}$ be the (diagonal) matrix of the eigenvalues; that is,

$$\boldsymbol{\Lambda} = \begin{bmatrix} \lambda_1 & 0 & \cdots & 0 \\ 0 & \lambda_2 & \cdots & 0 \\ \cdots & \cdots & \cdots & \cdots \\ 0 & 0 & \cdots & \lambda_n \end{bmatrix} \tag{6-179}$$

Then $\boldsymbol{\Lambda}$ is related to the system matrix **A** (by the similarity transformation)‡ by the equation

$$\boldsymbol{\Lambda} = \mathbf{P}^{-1}\mathbf{A}\mathbf{P} \tag{6-180}$$

† This theorem was first proved by Bellman [B-4].
‡ See Sec. 2-10.

Now let $e^{\mathbf{\Lambda}t}$ be the diagonal matrix

$$e^{\mathbf{\Lambda}t} = \begin{bmatrix} e^{\lambda_1 t} & 0 & \cdots & 0 \\ 0 & e^{\lambda_2 t} & \cdots & 0 \\ \cdots & \cdots & \cdots & \cdots \\ 0 & 0 & \cdots & e^{\lambda_n t} \end{bmatrix} \tag{6-181}$$

Then we know that† the matrices $e^{\mathbf{\Lambda}t}$ and $e^{\mathbf{A}t}$ are related by

$$e^{\mathbf{\Lambda}t} = \mathbf{P}^{-1}e^{\mathbf{A}t}\mathbf{P} \tag{6-182}$$

From Eq. (6-182), we deduce that

$$e^{-\mathbf{A}t} = \mathbf{P}e^{-\mathbf{\Lambda}t}\mathbf{P}^{-1} \tag{6-183}$$

so that Eq. (6-177) reduces to

$$\begin{aligned} u_j^*(t) &= -\operatorname{sgn}\{\langle \mathbf{P}e^{-\mathbf{\Lambda}t}\mathbf{P}^{-1}\mathbf{b}_j, \boldsymbol{\pi}\rangle\} \\ &= -\operatorname{sgn}\left\{\sum_{k=1}^{n} \rho_{kj}e^{-\lambda_k t}\right\} \end{aligned} \tag{6-184}$$

The constants $\rho_{kj}$ depend on the elements of the matrices $\mathbf{P}$ and $\mathbf{P}^{-1}$, on the components of the vector $\mathbf{b}_j$, and on the components of the vector $\boldsymbol{\pi}$. Clearly, the number of switchings of $u_j^*(t)$ will be the same as the number of zeros of the function

$$g_j^*(t) = \sum_{k=1}^{n} \rho_{kj}e^{-\lambda_k t} \tag{6-185}$$

But this function is a sum of $n$ exponentials. It is not too difficult to see‡ that the number of zeros of a sum of $n$ real exponential functions of time is *at most* $n - 1$, and so the theorem is established.

Although Theorem 6-8 was stated for the origin as target set, it should be clear from the proof that the theorem also applies to cases where the target set is not necessarily the origin of the state space.

The natural question that now arises is the following: *Are there any results available when some of the eigenvalues of* **A** *are complex?* If any of the eigenvalues of **A** are complex,§ then the number of switchings is *finite* but depends on the distance between the initial state $\boldsymbol{\xi}$ and the target set. It is not difficult to see why this happens. For suppose that two of the eigenvalues of **A** are (say)

$$\begin{aligned} \lambda_1 &= \alpha + j\omega \\ \lambda_2 &= \alpha - j\omega \end{aligned} \tag{6-186}$$

Then the argument of the "sgn" function in Eq. (6-177) will contain terms of the form

$$e^{\alpha t}\sin(\omega t + \varphi) \tag{6-187}$$

† See Secs. 3-21 and 3-22 and, in particular, Eqs. (3-298), (3-319), (3-320), and (3-322).
‡ For a formal proof, see the lemma on p. 120 of Ref. P-5.
§ Of course, they must come in complex conjugate pairs since **A** is a real matrix.

But we cannot place an upper bound on the number of zeros of a sinusoidal function unless we specify its domain (i.e., the time interval).

Let us recapitulate what we have done so far. We first stated the conditions for Problem 6-1*c* to be singular (Theorem 6-5) and normal (Theorem 6-6); we saw that the requirement that the time-optimal problem 6-1*c* be normal was equivalent to the requirement that the system $\dot{\mathbf{x}}(t) = \mathbf{A}\mathbf{x}(t) + \mathbf{B}\mathbf{u}(t)$ be normal;† we then established in Theorem 6-7 the uniqueness of the time-optimal control; finally, we found (Theorem 6-8) an upper bound on the number of switchings, under the assumption that **A** had real eigenvalues.

All our results up to now have been concerned with the properties of the time-optimal control. We shall conclude this section with a discussion of the properties of the extremal controls for Problem 6-1*c*. We recall that (see Definition 6-8) an extremal control is a control that satisfies all the necessary conditions. From the statement of Problem 6-1*c* and from Theorem 6-4, we can translate the definition of extremal quantities to give:

***Definition 6-9 Extremal Control for Problem 6-1c*** *The admissible control* $\hat{\mathbf{u}}^o(t)$ *is called extremal if the resulting trajectory* $\hat{\mathbf{x}}(t)$ *and the corresponding costate* $\hat{\mathbf{p}}(t)$ *satisfy the relations*

$$\dot{\hat{\mathbf{x}}}(t) = \mathbf{A}\hat{\mathbf{x}}(t) + \mathbf{B}\hat{\mathbf{u}}^o(t) \tag{6-188}$$

$$\dot{\hat{\mathbf{p}}}(t) = -\mathbf{A}'\hat{\mathbf{p}}(t) \tag{6-189}$$

$$\hat{\mathbf{x}}(0) = \boldsymbol{\xi} \qquad \hat{\mathbf{x}}(\hat{T}) = \mathbf{0} \tag{6-190}$$

$$1 + \langle \mathbf{A}\hat{\mathbf{x}}(t), \hat{\mathbf{p}}(t)\rangle + \langle \hat{\mathbf{u}}^o(t), \mathbf{B}'\hat{\mathbf{p}}(t)\rangle \le 1 + \langle \mathbf{A}\hat{\mathbf{x}}(t), \hat{\mathbf{p}}(t)\rangle + \langle \mathbf{u}(t), \mathbf{B}'\hat{\mathbf{p}}(t)\rangle \tag{6-191}$$

*for all* $\mathbf{u}(t) \in \Omega$ *and* $t \in [0, \hat{T}]$, *and*

$$1 + \langle \mathbf{A}\hat{\mathbf{x}}(t), \hat{\mathbf{p}}(t)\rangle + \langle \hat{\mathbf{u}}^o(t), \mathbf{B}'\hat{\mathbf{p}}(t)\rangle = 0 \tag{6-192}$$

We noted in Sec. 6-4 (see Remark 6-8) that each extremal control corresponded to either a globally or a locally optimal control. We shall now prove an extremely useful theorem which states that for normal systems there is one and only one extremal control (which must then be the time-optimal control).

***Theorem 6-9 Uniqueness of the Extremal Controls for Problem 6-1c*** *Suppose that* $\hat{\mathbf{u}}_1^o(t)$, $0 \le t \le \hat{T}_1$, *and* $\hat{\mathbf{u}}_2^o(t)$, $0 \le t \le \hat{T}_2$, *are two extremal controls for Problem* 6-1*c* (*see Definition* 6-9). *If the system* $\dot{\mathbf{x}}(t) = \mathbf{A}\mathbf{x}(t) + \mathbf{B}\mathbf{u}(t)$ *is normal and if a time-optimal control* $\mathbf{u}^*(t)$ *exists, then*

$$\hat{T}_1 = \hat{T}_2 = T^* \tag{6-193}$$

*and*

$$\hat{\mathbf{u}}_1^o(t) = \hat{\mathbf{u}}_2^o(t) = \mathbf{u}^*(t) \tag{6-194}$$

PROOF The proof of this theorem is similar to the proof of Theorem 6-7, as the reader will soon discover. We start by assuming that the extremal controls $\hat{\mathbf{u}}_1^o(t)$ and $\hat{\mathbf{u}}_2^o(t)$ are distinct and that (say)

$$\hat{T}_1 > \hat{T}_2 \tag{6-195}$$

† See Sec. 4-21.

The extremal trajectories $\hat{\mathbf{x}}_1(t)$ and $\hat{\mathbf{x}}_2(t)$ are given by the equations

$$\hat{\mathbf{x}}_1(t) = e^{\mathbf{A}t}\left[\boldsymbol{\xi} + \int_0^t e^{-\mathbf{A}\tau}\mathbf{B}\hat{\mathbf{u}}_1^{\circ}(\tau)\,d\tau\right] \tag{6-196}$$

$$\hat{\mathbf{x}}_2(t) = e^{\mathbf{A}t}\left[\boldsymbol{\xi} + \int_0^t e^{-\mathbf{A}\tau}\mathbf{B}\hat{\mathbf{u}}_2^{\circ}(\tau)\,d\tau\right] \tag{6-197}$$

But [see Eqs. (6-190)] we have

$$\hat{\mathbf{x}}_1(\hat{T}_1) = \hat{\mathbf{x}}_2(\hat{T}_2) = \mathbf{0} \tag{6-198}$$

and so we deduce that

$$\boldsymbol{\xi} = -\int_0^{\hat{T}_1} e^{-\mathbf{A}t}\mathbf{B}\hat{\mathbf{u}}_1^{\circ}(t)\,dt = -\int_0^{\hat{T}_2} e^{-\mathbf{A}t}\mathbf{B}\hat{\mathbf{u}}_2^{\circ}(t)\,dt \tag{6-199}$$

Corresponding to $\hat{\mathbf{u}}_1^{\circ}(t)$, there is an (extremal) costate $\hat{\mathbf{p}}_1(t)$ which is given by

$$\hat{\mathbf{p}}_1(t) = e^{-\mathbf{A}'t}\hat{\boldsymbol{\pi}}_1 \tag{6-200}$$

Similarly, corresponding to $\hat{\mathbf{u}}_2^{\circ}(t)$, there is a costate $\hat{\mathbf{p}}_2(t)$ given by the equation

$$\hat{\mathbf{p}}_2(t) = e^{-\mathbf{A}'t}\hat{\boldsymbol{\pi}}_2 \tag{6-201}$$

We now form the scalar product $\langle\boldsymbol{\xi}, \hat{\boldsymbol{\pi}}_1\rangle$ to obtain the relation

$$\langle\boldsymbol{\xi}, \hat{\boldsymbol{\pi}}_1\rangle = -\left\langle\hat{\boldsymbol{\pi}}_1, \int_0^{\hat{T}_1} e^{-\mathbf{A}t}\mathbf{B}\hat{\mathbf{u}}_1^{\circ}(t)\,dt\right\rangle = -\left\langle\hat{\boldsymbol{\pi}}_1, \int_0^{\hat{T}_2} e^{-\mathbf{A}t}\mathbf{B}\hat{\mathbf{u}}_2^{\circ}(t)\,dt\right\rangle \tag{6-202}$$

Since $\hat{\boldsymbol{\pi}}_1$ is a constant vector, we deduce that

$$\int_0^{\hat{T}_1} -\langle e^{-\mathbf{A}'t}\hat{\boldsymbol{\pi}}_1, \mathbf{B}\hat{\mathbf{u}}_1^{\circ}(t)\rangle\,dt = \int_0^{\hat{T}_2} -\langle e^{-\mathbf{A}'t}\hat{\boldsymbol{\pi}}_1, \mathbf{B}\hat{\mathbf{u}}_2^{\circ}(t)\rangle\,dt \tag{6-203}$$

and, in view of Eq. (6-200), that

$$\int_0^{\hat{T}_1} -\langle\hat{\mathbf{p}}_1(t), \mathbf{B}\hat{\mathbf{u}}_1^{\circ}(t)\rangle\,dt = \int_0^{\hat{T}_2} -\langle\hat{\mathbf{p}}_1(t), \mathbf{B}\hat{\mathbf{u}}_2^{\circ}(t)\rangle\,dt \tag{6-204}$$

Since $\hat{\mathbf{u}}_1^{\circ}(t)$ is an extremal control, we find from Eq. (6-191) the relation

$$\langle\hat{\mathbf{p}}_1(t), \mathbf{B}\hat{\mathbf{u}}_1^{\circ}(t)\rangle \le \langle\hat{\mathbf{p}}_1(t), \mathbf{B}\hat{\mathbf{u}}_2^{\circ}(t)\rangle \qquad t \in [0, \hat{T}_2] \tag{6-205}$$

Since the system is normal, the above relation will be an equality whenever $\hat{\mathbf{u}}_1^{\circ}(t) = \hat{\mathbf{u}}_2^{\circ}(t)$ and a strict inequality whenever $\hat{\mathbf{u}}_1^{\circ}(t) \neq \hat{\mathbf{u}}_2^{\circ}(t)$. Moreover, since the system is normal, the relation

$$\hat{\mathbf{u}}_1^{\circ}(t) = -\,\mathrm{SGN}\,\{\mathbf{B}'\hat{\mathbf{p}}_1(t)\} \qquad t \in [0, \hat{T}_1] \tag{6-206}$$

will uniquely define $\hat{\mathbf{u}}_1^{\circ}(t)$ except at the countable set of switch times. From Eq. (6-206), we obtain the relation

$$\langle\hat{\mathbf{p}}_1(t), \mathbf{B}\hat{\mathbf{u}}_1^{\circ}(t)\rangle \le 0 \qquad t \in [0, \hat{T}_1] \tag{6-207}$$

The inequality (6-207) is strict whenever at least one of the components of $\hat{\mathbf{u}}_1^{\circ}(t)$ is $+1$ or $-1$. Since $\hat{T}_2 < \hat{T}_1$, we can see that the assumption of normality implies that

$$\int_0^{\hat{T}_1} - \langle \hat{\mathbf{p}}_1(t), \mathbf{B}\hat{\mathbf{u}}_1{}^\circ(t)\rangle \, dt > \int_0^{\hat{T}_2} - \langle \hat{\mathbf{p}}_1(t), \mathbf{B}\hat{\mathbf{u}}_1{}^\circ(t)\rangle \, dt > 0 \tag{6-208}$$

and that

$$\int_0^{\hat{T}_2} - \langle \hat{\mathbf{p}}_1(t), \mathbf{B}\hat{\mathbf{u}}_1{}^\circ(t)\rangle \, dt > \int_0^{\hat{T}_2} - \langle \hat{\mathbf{p}}_1(t), \mathbf{B}\hat{\mathbf{u}}_2{}^\circ(t)\rangle \, dt \tag{6-209}$$

From Eqs. (6-208) and (6-209), we deduce, for distinct $\hat{\mathbf{u}}_1{}^\circ(t)$ and $\hat{\mathbf{u}}_2{}^\circ(t)$, the following relation:

$$\int_0^{\hat{T}_1} - \langle \hat{\mathbf{p}}_1(t), \mathbf{B}\hat{\mathbf{u}}_1{}^\circ(t)\rangle \, dt > \int_0^{\hat{T}_2} - \langle \hat{\mathbf{p}}_1(t), \mathbf{B}\hat{\mathbf{u}}_2{}^\circ(t)\rangle \, dt \tag{6-210}$$

But Eq. (6-204) and the inequality (6-210) are contradictory. It follows that $\hat{\mathbf{u}}_1{}^\circ(t)$ and $\hat{\mathbf{u}}_2{}^\circ(t)$ cannot be distinct, and so $\hat{T}_1 = \hat{T}_2$ and $\hat{\mathbf{u}}_1{}^\circ(t) = \hat{\mathbf{u}}_2{}^\circ(t)$. Since the extremal control is unique and since we have assumed that an optimal control $\mathbf{u}^*(t)$ exists (which is unique in view of the normality), Eqs. (6-193) and (6-194) must hold, and the theorem is established.

This theorem states that there is one and only one extremal control. We derived the theorem under the assumption that the target set was the origin of the state space. Unfortunately, *the theorem is not true if the terminal state is not the origin of the state space.* In other words, if the problem is to transfer the system from a given initial state $\boldsymbol{\xi}$ to a terminal state $\boldsymbol{\theta} \neq \mathbf{0}$ in minimum time, then there may be more than one extremal control. We invite the reader to discover the reason for this by doing the following exercise.

**Exercise 6-8** Try to mimic the proof of Theorem 6-9 when the terminal state is $\boldsymbol{\theta} \neq \mathbf{0}$. Pinpoint the exact place in the logical development where you get stuck. Do you think that, given a terminal state $\boldsymbol{\theta}$, there is a set of initial states such that the extremal controls are unique? If so, can you find this set of initial states?

Let us elaborate a little bit more on one of the points in the proof of Theorem 6-9. Observe that *we have not used the fact that along an extremal trajectory, the Hamiltonian must be zero.* Thus, we could have proved Theorem 6-9 by just using the fact that an extremal control minimizes the Hamiltonian. It is for this reason that in Chap. 7 we shall not use the necessary condition $H[\mathbf{x}^*(t), \mathbf{p}^*(t), \mathbf{u}^*(t)] = 0$ very often.

In the following four sections, we shall continue the development of the available results for Problem 6-1*c*. Specifically, in Sec. 6-6 we shall examine the structure of the open-loop and the closed-loop time-optimal control systems. In Sec. 6-7, we shall introduce the notion of the minimum isochrones, and we shall return once more to the geometric properties of the time-optimal control. In Sec. 6-8, we shall discuss, in a heuristic manner, the problem of existence of the time-optimal control and shall present the most useful existence theorem from the engineering point of view. In Sec. 6-9, we shall present the Hamilton-Jacobi equation for the time-optimal control problem and discuss the difficulties associated with its solution.

## 6-6 Minimum-time Problems 5: Structure of the Time-optimal Regulator and the Feedback Problem

In the previous section, we stated most of the general results that are available for the time-optimal regulator problem. In this section, we shall present an iterative procedure that may be used to determine the time-optimal control. This procedure is similar to the one presented in step 3 of Sec. 6-3. We shall discuss the difficulties associated with this iterative procedure and list references in which various iterative techniques are described. We shall not elaborate on these iterative methods of solution, because we feel that they fall outside the introductory scope of this book. Finally, we shall define the feedback problem and, in so doing, motivate the problems that we shall solve in Chap. 7.

Throughout this section, we shall deal with Problem 6-1*c*. We shall assume that the system $\dot{\mathbf{x}}(t) = \mathbf{A}\mathbf{x}(t) + \mathbf{B}\mathbf{u}(t)$ is normal and that a time-optimal control exists (which is unique by Theorem 6-7). The step-by-step idealized iterative procedure that may be followed to solve Problem 6-1*c* is outlined below.

Given the normal linear time-invariant system

$$\dot{\mathbf{x}}(t) = \mathbf{A}\mathbf{x}(t) + \mathbf{B}\mathbf{u}(t) \qquad \mathbf{x}(0) = \boldsymbol{\xi} \tag{6-211}$$

and given the constraints on the control

$$|u_j(t)| \leq 1 \qquad j = 1, 2, \ldots, r; \text{ for all } t \tag{6-212}$$

To find the time-optimal control that forces $\boldsymbol{\xi}$ to $\mathbf{0}$, proceed as follows:

*Step 1*

Guess an initial costate $\boldsymbol{\pi}$ such that

$$\langle \boldsymbol{\xi}, \boldsymbol{\pi} \rangle \geq 0 \tag{6-213}$$

*Step 2*

Compute the costate $\mathbf{p}(t)$ from the relation

$$\mathbf{p}(t) = e^{-\mathbf{A}'t}\boldsymbol{\pi} \tag{6-214}$$

*Step 3*

Compute the control variables $u_1(t), u_2(t), \ldots, u_r(t)$ from the equations

$$u_j(t) = -\operatorname{sgn} \{\langle \mathbf{b}_j, \mathbf{p}(t) \rangle\} \tag{6-215}$$

where the $\mathbf{b}_j$ are the column vectors of $\mathbf{B}$.

*Step 4*

Solve Eq. (6-211), starting at the initial state $\boldsymbol{\xi}$ and using the controls computed in step 3. Since $\boldsymbol{\pi}$ uniquely defines the control vector and since

$\boldsymbol{\xi}$ uniquely determines the state trajectory (given the control), it is clear that the state, at each instant of time, is a function of both $\boldsymbol{\xi}$ and $\boldsymbol{\pi}$. To emphasize this dependence, we shall denote the solution of Eq. (6-211) by

$$\mathbf{x}_{\xi,\pi}(t) \tag{6-216}$$

*Step 5*

Monitor the solution $\mathbf{x}_{\xi,\pi}(t)$. If there is a time $T$ such that

$$\mathbf{x}_{\xi,\pi}(T) = \mathbf{0} \tag{6-217}$$

then the control computed in step 3 is an extremal control (see Definition 6-9) and, by Theorem 6-9, is time-optimal. If Eq. (6-217) cannot be

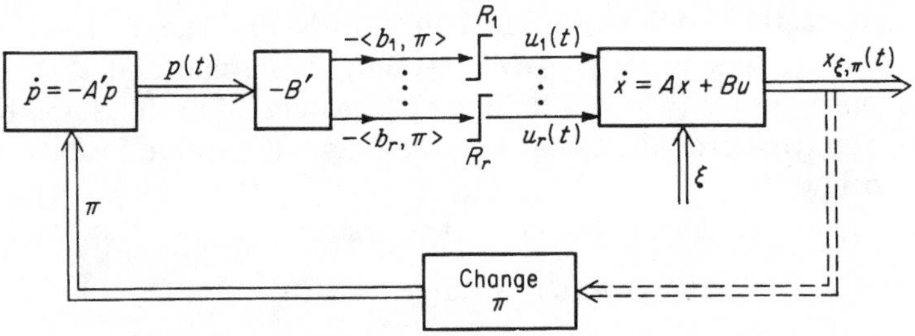


**Fig. 6-8** The "open-loop" structure of the time-optimal problem.

satisfied, then change $\boldsymbol{\pi}$ in step 1 and repeat the procedure until Eq. (6-217) is satisfied.

This procedure is iterative because one must change the initial costate $\boldsymbol{\pi}$ many times to find the time-optimal control. The initial guess and the subsequent values of $\boldsymbol{\pi}$ must be such that Eq. (6-213) is satisfied; Eq. (6-213) is a direct consequence of Eqs. (6-202), (6-203), and (6-208).

The procedure from step 1 to step 5 is illustrated in block-diagram form in Fig. 6-8. We use double arrows to indicate the "flow" of vector signals and single arrows to indicate the "flow" of scalar signals. The "relay" symbol

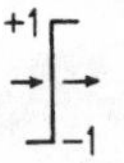


is used to indicate the "sgn" operation.† The outputs of the "relays" $R_1, \ldots, R_r$ are the control variables $u_1(t), \ldots, u_r(t)$. The block labeled "Change $\boldsymbol{\pi}$" monitors $\mathbf{x}_{\xi,\pi}(t)$ and, after a "sufficiently long" time, decides whether or not a new $\boldsymbol{\pi}$ should be introduced to the system.

† Thus, an ideal polarized relay is the engineering realization of the signum function. This convention will be used throughout the book.

There are two questions that arise:

1. How long must one wait before $\boldsymbol{\pi}$ is changed?
2. How do we decide on a new value of $\boldsymbol{\pi}$ based on the previous values used and the "miss distances" of $\mathbf{x}_{\xi,\pi}(t)$ from $\mathbf{0}$?

The answers to these questions are the subject of many investigations;† however, we shall not discuss any iterative techniques, as they are beyond the scope of this book. We would like to comment that these procedures are complex and that they require a large-memory, high-speed computer.

The procedure suggested is an *open-loop* technique. In other words, given an initial state $\boldsymbol{\xi}$, we find a time-optimal control $\mathbf{u}^*(t)$ which will depend on the initial state $\boldsymbol{\xi}$ and on the time $t$; this control will not depend on the instantaneous value of the state $\mathbf{x}(t)$ for $t > 0$. The disadvantages of an open-loop method of control are well known to the engineer. The additional difficulty that is also inherent in this type of computation is the following: Suppose that the eigenvalues of the system matrix $\mathbf{A}$ have negative real parts, i.e., that the system is stable. Then the adjoint dynamical system $\dot{\mathbf{p}}(t) = -\mathbf{A}'\mathbf{p}(t)$ is unstable because the eigenvalues of the matrix $-\mathbf{A}'$ have positive real parts. Thus, a small "error" in the initial costate $\boldsymbol{\pi}$ will be "amplified" as time goes on and may cause the computed control to deviate significantly from the optimal one.‡ For this reason, it is worthwhile to investigate whether or not we can find the time-optimal control $\mathbf{u}^*(t)$ (at time $t$) as a function of the state $\mathbf{x}^*(t)$ (at time $t$). If we can do that, then we shall have derived the optimal *feedback* system which has the property that, at each instant of time, the state completely determines the value of the control.

We claim that there is a "*switching function*" $\mathbf{h}[\mathbf{x}^*(t)]$ such that the time-optimal control $\mathbf{u}^*(t)$ is given by

$$\mathbf{u}^*(t) = -\ \mathrm{SGN}\ \{\mathbf{h}[\mathbf{x}^*(t)]\} \tag{6-218}$$

Equation (6-218) is often called the *time-optimal feedback control law.* We shall now present a *heuristic* justification for Eq. (6-218).

Let $\mathbf{x}^*(t)$ be a state on the optimal trajectory from the initial state $\boldsymbol{\xi}$ to $\mathbf{0}$, and let $\mathbf{p}^*(t)$ denote the value of the costate at that time. Clearly, the relation

$$\mathbf{x}^*(T^*) = \mathbf{0} = e^{\mathbf{A}(T^*-t)}\left[\mathbf{x}^*(t) - \int_t^{T^*} e^{-\mathbf{A}(\tau-t)}\mathbf{B}\ \mathrm{SGN}\ \{\mathbf{B}'\mathbf{p}^*(\tau)\}\ d\tau\right] \tag{6-219}$$

is satisfied. Since the matrix $e^{\mathbf{A}(T^*-t)}$ is nonsingular, we deduce that

$$\mathbf{x}^*(t) = \int_t^{T^*} e^{-\mathbf{A}(\tau-t)}\mathbf{B}\ \mathrm{SGN}\ \{\mathbf{B}'\mathbf{p}^*(\tau)\}\ d\tau \tag{6-220}$$

But

$$\mathbf{p}^*(\tau) = e^{-\mathbf{A}'(\tau-t)}\mathbf{p}^*(t) \qquad \tau \geq t \tag{6-221}$$

† See Refs. N-6, B-1, H-9, E-1, P-1, K-19, G-5, and F-1.

‡ If the adjoint system is simulated on an analog computer, additional errors due to the saturation of amplifiers and due to the amplification of the inherent noise will occur.

and, furthermore, the time $(T^* - t)$ required to force $\mathbf{x}^*(t)$ to $\mathbf{0}$ must be a function of $\mathbf{x}^*(t)$; that is,

$$T^* - t = \alpha[\mathbf{x}^*(t)] \tag{6-222}$$

From Eqs. (6-222) and (6-221), we conclude that Eq. (6-220) represents a relation between $\mathbf{x}^*(t)$ and $\mathbf{p}^*(t)$, which we shall write in the form

$$\mathbf{p}^*(t) = \mathbf{p}^*[\mathbf{x}^*(t)] \qquad \text{(6-223)}\dagger$$

Since $\mathbf{u}^*(t) = -\text{ SGN }\{\mathbf{B}'\mathbf{p}^*(t)\}$, we deduce from Eqs. (6-223) and (6-218) that

$$\mathbf{h}[\mathbf{x}^*(t)] = \mathbf{B}'\mathbf{p}^*[\mathbf{x}^*(t)] \tag{6-224}$$

and, thus, the existence of the switching function has been *heuristically* justified. All the problems we shall solve in Chap. 7 will involve the

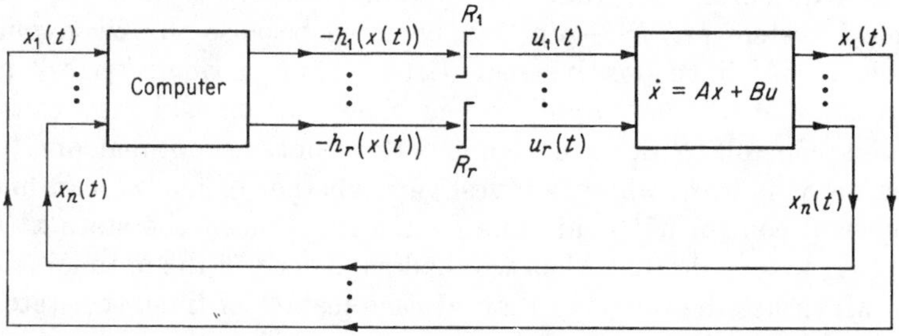


**Fig. 6-9** The feedback structure of a time-optimal control system.

determination of this switching function. As we shall see, this switching function will be defined by the so-called "switch hypersurfaces" in the state space.

We illustrate the block diagram of the optimal feedback system in Fig. 6-9. We note that the state variables $x_1(t), x_2(t), \ldots, x_n(t)$ are measured at each instant of time and are introduced to the subsystem labeled "Computer"; the outputs of the computer are the switching functions $h_1[\mathbf{x}(t)], h_2[\mathbf{x}(t)], \ldots, h_r[\mathbf{x}(t)]$, which are then introduced to the ideal relays $R_1, R_2, \ldots, R_r$ to generate the time-optimal control variables. The determination of the functions $h_1[\mathbf{x}(t)], \ldots, h_r[\mathbf{x}(t)]$ and their engineering realizations is the "heart" of a time-optimal control problem. The primary purposes of Chap. 7 are (1) to present some of the techniques available for determining switching functions and (2) to acquaint the reader with the types of nonlinear-function generators required in the engineering implementation of a time-optimal system.

† This relation is not necessarily unique. In other words, for a given $\mathbf{x}^*(t)$, there may correspond more than one $\mathbf{p}^*(t)$ which will generate the optimal control (see Ref. K-19).

## 6-7 Minimum-time Problems 6: Geometric Properties of the Time-optimal Control

In Sec. 6-2, we discussed the geometric nature of the time-optimal problem in terms of the sets of reachable states. We then jumped from geometric considerations to the analytical results which were obtained from the necessary conditions provided by the minimum principle. In this section, we shall attempt to provide a geometric interpretation of the necessary conditions. We shall limit our discussion to Problem 6-1*c*, and we shall assume that the time-optimal problem is normal according to Definition 6-4. We remark that the material of this section represents a specialization of the comments of Sec. 5-21.

We shall examine the minimum-time surface, and we shall interpret the time-optimal control as the control that causes the state to move in the direction of steepest descent on the minimum-time surface. We shall then be able to establish a correspondence between the costate and the gradient of the minimum-time surface. Our treatment will be heuristic in nature, as we are interested primarily in geometric interpretations of the necessary conditions.

Let $\mathbf{x}$ be a state in our state space, and let us assume that there is a time-optimal control (which is unique) that drives $\mathbf{x}$ to $\mathbf{0}$. We shall denote the unique minimum time required to force $\mathbf{x}$ to $\mathbf{0}$ by

$$T^*(\mathbf{x}) \tag{6-225}$$

We now claim that the minimum time $T^*(\mathbf{x})$ depends only on the state $\mathbf{x}$ and not on the time $t$, that is, that

$$\frac{\partial T^*(\mathbf{x})}{\partial t} = 0 \qquad \text{(6-226)†}$$

This is so because the time invariance of the system $\dot{\mathbf{x}}(t) = \mathbf{A}\mathbf{x}(t) + \mathbf{B}\mathbf{u}(t)$ implies that the minimum time can only be a function of the state. In other words, if $\mathbf{x}$ is the state at $t = 0$ and if the minimum time required to force $\mathbf{x}$ to $\mathbf{0}$ is $T^*(\mathbf{x})$, then, if $\mathbf{x}$ is the state at $t = t_0$, it follows that the time-optimal control will force $\mathbf{x}$ to $\mathbf{0}$ at the time $t_0 + T^*(\mathbf{x})$. [See Eqs. (4-266) and (4-267).]

Since the time required to force $\mathbf{0}$ to $\mathbf{0}$ is zero and since we are concerned with positive time solutions, it is evident that $T^*(\mathbf{x})$ has the properties

$$T^*(\mathbf{x}) = 0 \quad \text{if } \mathbf{x} = \mathbf{0} \tag{6-227}$$
$$T^*(\mathbf{x}) > 0 \quad \text{if } \mathbf{x} \neq \mathbf{0} \tag{6-228}$$

† It has been the experience of the authors that this equation is confusing to students who read it as "the partial derivative of the time $T^*$ with respect to time is zero" and who do not understand the real meaning of a partial derivative. We trust that those who read Sec. 3-12 will not fall into the same trap.

We shall use the notation

$$\frac{\partial T^*(\mathbf{x})}{\partial \mathbf{x}} = \begin{bmatrix} \dfrac{\partial T^*(\mathbf{x})}{\partial x_1} \\ \cdot \\ \cdot \\ \cdot \\ \dfrac{\partial T^*(\mathbf{x})}{\partial x_n} \end{bmatrix} \tag{6-229}$$

in the sequel to denote the gradient of the function $T^*(\mathbf{x})$ with respect to $\mathbf{x}$.

In the sequel, we shall discuss several properties of the function $T^*(\mathbf{x})$. It is useful to think of $T^*(\mathbf{x})$ as the minimum-cost surface for Problem 6-1c and to visualize $T^*(\mathbf{x})$ as illustrated in Fig. 6-10.

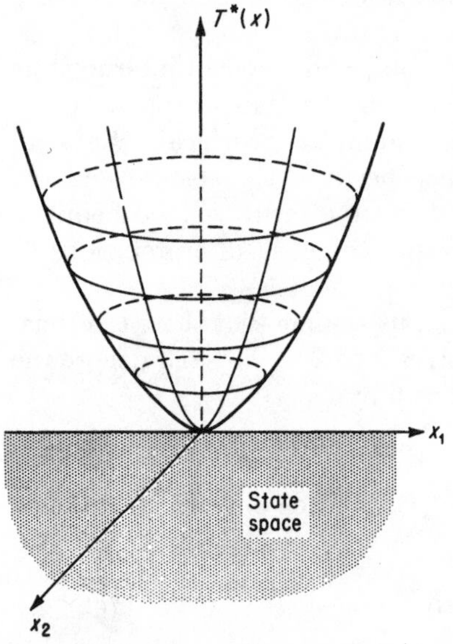


**Fig. 6-10** The minimum-time cost surface $T^*(\mathbf{x})$ as a function of $\mathbf{x}$.

We now define the notion of a minimum isochrone.†

***Definition 6-10 Minimum Isochrones*** *Let $S(\tau)$ be the set of states which can be forced to the origin* $\mathbf{0}$ *in the same minimum time $\tau$, $\tau \geq 0$. We shall call $S(\tau)$ the $\tau$-minimum isochrone. $S(\tau)$ is given by the relation*

$$S(\tau) = \{\mathbf{x}\colon T^*(\mathbf{x}) = \tau; \tau \geq 0\} \tag{6-230}$$

*Let $\hat{S}(\tau)$ denote the set of states which can be forced to the origin by time-optimal controls which require a minimum time less than or equal to $\tau$. More precisely, we have*

$$\hat{S}(\tau) = \{\mathbf{x}\colon T^*(\mathbf{x}) \leq \tau; \tau \geq 0\} \tag{6-231}$$

From Eqs. (6-230) and (6-231), we conclude that $S(\tau)$ is a subset of $\hat{S}(\tau)$. We also assert‡ that $S(\tau)$ is the *boundary* of $\hat{S}(\tau)$ and that $\hat{S}(t)$ *is closed.*

We shall now prove that the set $\hat{S}(\tau)$ is a strictly convex set. Let $\mathbf{x}_1$ and $\mathbf{x}_2$ be two *distinct* states on the $\tau$-minimum isochrone; i.e., let

$$\mathbf{x}_1 \in S(\tau) \qquad \mathbf{x}_2 \in S(\tau) \tag{6-232}$$

In view of the normality, we know that there are a unique time-optimal control $\mathbf{u}_1^*(t) = -\operatorname{SGN}\{\mathbf{B}'\mathbf{p}_1^*(t)\}$ that transfers $\mathbf{x}_1$ to $\mathbf{0}$ and a unique time-optimal control $\mathbf{u}_2^*(t) = -\operatorname{SGN}\{\mathbf{B}'\mathbf{p}_2^*(t)\}$ [distinct from $\mathbf{u}_1^*(t)$] that trans-

† The notion of a minimum isochrone will be used in Chaps. 7, 8, and 10.
‡ For a formal proof, see Ref. S-9.

fers $\mathbf{x}_2$ to $\mathbf{0}$. Thus, the equations

$$\mathbf{x}_1 = \int_0^\tau e^{-\mathbf{A}t}\mathbf{B} \text{ SGN } \{\mathbf{B}'\mathbf{p}_1^*(t)\}\, dt \tag{6-233}$$

$$\mathbf{x}_2 = \int_0^\tau e^{-\mathbf{A}t}\mathbf{B} \text{ SGN } \{\mathbf{B}'\mathbf{p}_2^*(t)\}\, dt \tag{6-234}$$

must be satisfied.

Let $\mathbf{x}$ be a state on the (open) line segment joining $\mathbf{x}_1$ and $\mathbf{x}_2$, as illustrated in Fig. 6-11; more precisely, let $\alpha$ be such that

$$0 < \alpha < 1 \tag{6-235}$$

and consider the state $\mathbf{x}$, defined by

$$\mathbf{x} = \alpha\mathbf{x}_1 + (1 - \alpha)\mathbf{x}_2 \qquad \alpha \in (0, 1) \tag{6-236}$$

From Eqs. (6-236), (6-234), and (6-233), we deduce that

$$\mathbf{x} = \int_0^\tau e^{-\mathbf{A}t}\mathbf{B}[\alpha \text{ SGN } \{\mathbf{B}'\mathbf{p}_1^*(t)\} + (1 - \alpha) \text{ SGN } \{\mathbf{B}'\mathbf{p}_2^*(t)\}]\, dt \tag{6-237}$$

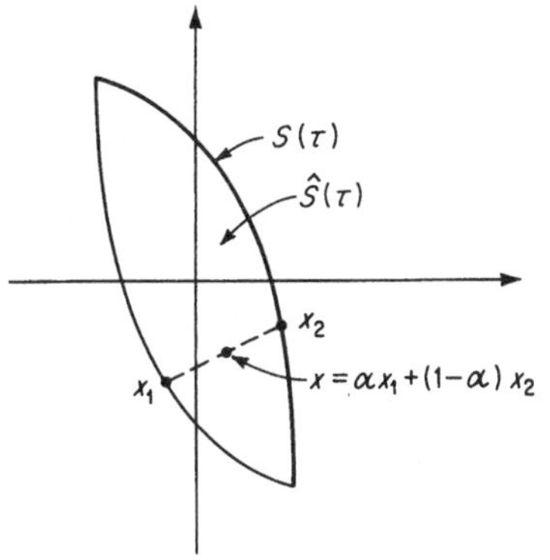


**Fig. 6-11** Illustration of the convexity of $\hat{S}(\tau)$.

Now let $\mathbf{u}^*(t) = -\text{ SGN } \{\mathbf{B}'\mathbf{p}^*(t)\}$ be the time-optimal control that transfers $\mathbf{x}$ to $\mathbf{0}$; let $\tau'$ be the corresponding minimum time [that is, $T^*(\mathbf{x}) = \tau'$]. We shall show that

$$\tau' < \tau \tag{6-238}$$

To do this, we note that

$$\mathbf{x} = \int_0^{\tau'} e^{-\mathbf{A}t}\mathbf{B} \text{ SGN } \{\mathbf{B}'\mathbf{p}^*(t)\}\, dt \tag{6-239}$$

But Eq. (6-237) implies that the control

$$\alpha\mathbf{u}_1^*(t) + (1 - \alpha)\mathbf{u}_2^*(t) = \alpha \text{ SGN } \{\mathbf{B}'\mathbf{p}_1^*(t)\} + (1 - \alpha) \text{ SGN } \{\mathbf{B}'\mathbf{p}_2^*(t)\} \tag{6-240}$$

drives $\mathbf{x}$ to $\mathbf{0}$; however, the control (6-240) is *not* time-optimal, because it is *not* a signum vector. To see this, suppose that at some instant of time $\hat{t}$, we have (say)†

$$\mathbf{u}_1^*(\hat{t}) = \begin{bmatrix} +1 \\ -1 \\ \cdot \\ \cdot \\ \cdot \\ -1 \end{bmatrix} \qquad \mathbf{u}_2^*(\hat{t}) = \begin{bmatrix} -1 \\ -1 \\ \cdot \\ \cdot \\ \cdot \\ -1 \end{bmatrix} \tag{6-241}$$

† Such a time $\hat{t}$ will exist because the controls $\mathbf{u}_1^*(t)$ and $\mathbf{u}_2^*(t)$ are, by assumption, distinct.

Then we can deduce that

$$\alpha \mathbf{u}_1^*(\hat{t}) + (1-\alpha)\mathbf{u}_2^*(\hat{t}) = \begin{bmatrix} 2\alpha - 1 \\ -1 \\ -1 \\ \cdot \\ \cdot \\ \cdot \\ -1 \end{bmatrix} \qquad \alpha \in (0, 1) \tag{6-242}$$

But, since $0 < \alpha < 1$, we have the inequality

$$-1 < 2\alpha - 1 < +1 \tag{6-243}$$

and so the control (6-242) cannot be time-optimal (why?). Since this "non-time-optimal" control forces $\mathbf{x}$ to $\mathbf{0}$ in time $\tau$, it follows that the time-optimal control $\mathbf{u}^*(t)$ requires time $\tau' < \tau$, and Eq. (6-238) is established. Furthermore, since we asserted that $S(\tau)$ is the boundary of $\hat{S}(\tau)$, we conclude that the state $\mathbf{x} = \alpha\mathbf{x}_1 + (1-\alpha)\mathbf{x}_2$, $\alpha \in (0, 1)$, is an element of the interior of $\hat{S}(\tau)$ and so $\hat{S}(\tau)$ is strictly convex.

We summarize these results in the following lemma:

***Lemma 6-5*** *The $\tau$-minimum isochrone $S(\tau)$ is the boundary of the closed and strictly convex set $\hat{S}(\tau)$.*

**Exercise 6-9** Prove that $T^*(\mathbf{x})$ is continuous for Problem 6-1c.
**Exercise 6-10** Prove that $\hat{S}(\tau)$ is closed and that $S(\tau)$ is the boundary of $\hat{S}(\tau)$.

We now wish to point out that the $\tau$-minimum isochrones "grow" with increasing $\tau$. Let us suppose that $\tau_1$ and $\tau_2$ are any two times such that

$$0 < \tau_1 < \tau_2 \tag{6-244}$$

We now claim that

$$\mathbf{0} \subset \hat{S}(\tau_1) \subset \hat{S}(\tau_2) \tag{6-245}$$

**Exercise 6-11** Prove Eq. (6-245).

The inclusion relationship (6-245) implies that the minimum isochrones increase their "distance" from the origin with increasing time. Moreover, this "growth" is smooth, and a heuristic geometric proof is presented to clarify this idea.

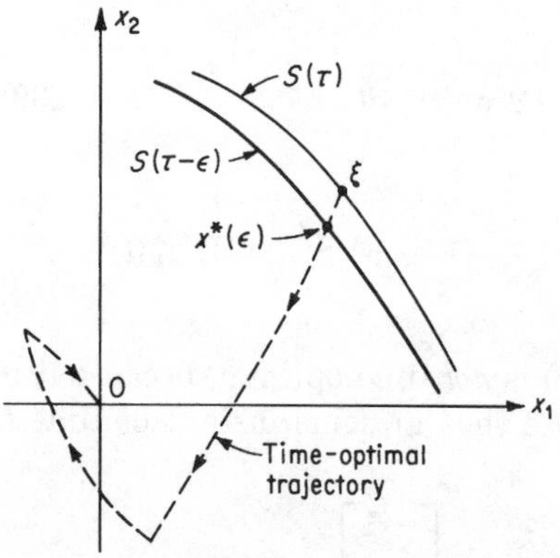


**Fig. 6-12** The minimum isochrones $S(\tau)$ and $S(\tau - \epsilon)$, $\epsilon > 0$. The time-optimal trajectory from $\mathbf{x}^*(\epsilon)$ to $\mathbf{0}$ is a portion of the time-optimal trajectory from $\xi$ to $\mathbf{0}$.

Let us suppose that $\xi$ is an initial state at $t = 0$, and let us suppose that $\tau$ is the time required to force $\xi$ to $\mathbf{0}$ by the time-optimal control $\mathbf{u}^*(t)$, $0 \le t \le \tau$. Thus, $\xi \in S(\tau)$. In Fig. 6-12, we show the time-optimal trajectory $\mathbf{x}^*$ that joins $\boldsymbol{\xi}$ with $\mathbf{0}$. Let $\epsilon > 0$ be a real positive time, and let us consider the state $\mathbf{x}^*(\epsilon)$ at $t = \epsilon$. By the principle of optimality,† we know that the control $\mathbf{u}^*(t)$, for $\epsilon \leq t \leq \tau$, is the time-optimal control that forces $\mathbf{x}^*(\epsilon)$ to $\mathbf{0}$ in the minimum time $\tau - \epsilon$; thus $\mathbf{x}^*(\epsilon) \in S(\tau - \epsilon)$. Since the state trajectory is continuous, we can see that as $\epsilon \to 0$, the state $\mathbf{x}^*(\epsilon)$ tends to $\boldsymbol{\xi}$. If we repeat this experiment for all $\boldsymbol{\xi} \in S(\tau)$, we can see that the isochrones $S(\tau)$ and $S(\tau - \epsilon)$ get "closer and closer together" as $\epsilon \to 0$.

We shall now discuss the geometric properties of the time-optimal control $\mathbf{u}^*(t)$. Let us suppose that $\boldsymbol{\xi}$ is the initial state and that $\boldsymbol{\xi} \in S(\tau)$. As

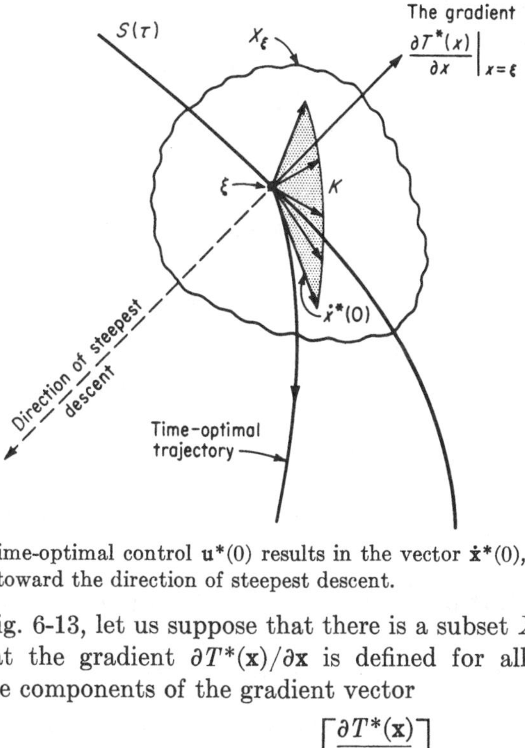


**Fig. 6-13** The time-optimal control $\mathbf{u}^*(0)$ results in the vector $\dot{\mathbf{x}}^*(0)$, which points as much as possible toward the direction of steepest descent.

illustrated in Fig. 6-13, let us suppose that there is a subset $X_\xi$ of the state space such that the gradient $\partial T^*(\mathbf{x})/\partial \mathbf{x}$ is defined for all $\mathbf{x} \in X_\xi$. In other words, the components of the gradient vector

$$\nabla T^*(\mathbf{x}) = \frac{\partial T^*(\mathbf{x})}{\partial \mathbf{x}} = \begin{bmatrix} \dfrac{\partial T^*(\mathbf{x})}{\partial x_1} \\ \cdot \\ \cdot \\ \cdot \\ \dfrac{\partial T^*(\mathbf{x})}{\partial x_n} \end{bmatrix} \tag{6-246}$$

are well-defined functions for all $\mathbf{x} \in X_\xi$. Now the gradient vector at $\mathbf{x} = \boldsymbol{\xi}$, that is, the vector $(\partial T^*(\mathbf{x})/\partial \mathbf{x})|_{\mathbf{x}=\xi}$, defines the direction of steepest change of the function $T^*(\mathbf{x})$ at the point $\boldsymbol{\xi}$. As illustrated in Fig. 6-13.

† See Secs. 5-16 and 5-19.

the gradient vector is normal to the curve $S(\tau)$ at $\mathbf{x} = \boldsymbol{\xi}$, and it points "away from the origin." The direction of the vector $-(\partial T^*(\mathbf{x})/\partial \mathbf{x})|_{\mathbf{x}=\boldsymbol{\xi}}$ (which is the dashed vector in Fig. 6-13) establishes the direction of *steepest descent* on the surface $T^*(\mathbf{x})$ at $\boldsymbol{\xi}$. Thus, if we constructed the cost surface $T^*(\mathbf{x})$ and if we placed a marble at $\boldsymbol{\xi}$, then the marble would start to roll down the surface $T^*(\mathbf{x})$ along the direction of the vector $-(\partial T^*(\mathbf{x})/\partial \mathbf{x})|_{\mathbf{x}=\boldsymbol{\xi}}$.

At $t = 0$, we have the relation

$$\dot{\mathbf{x}}(0) = \mathbf{A}\boldsymbol{\xi} + \mathbf{B}\mathbf{u}(0) \qquad \mathbf{u}(0) \in \Omega \tag{6-247}$$

The direction and magnitude of the vector $\dot{\mathbf{x}}(0)$ clearly depend (1) on the vector $\mathbf{A}\boldsymbol{\xi}$ (which is fixed by the state $\boldsymbol{\xi}$) and (2) on the vector $\mathbf{B}\mathbf{u}(0)$, whose magnitude and direction we can choose as long as we do not violate the constraint $\mathbf{u}(0) \in \Omega$. If we tried all controls $\mathbf{u}(0)$ in $\Omega$, then we would find a set of vectors $\{\dot{\mathbf{x}}(0)\}$ which form a cone $K$. Let us suppose that $K$ is as illustrated in Fig. 6-13. Thus the restriction $\mathbf{u}(0) \in \Omega$ defines a set of directions in Fig. 6-13, and so we cannot make $\dot{\mathbf{x}}(0)$ point along the vector $(-\partial T^*(\mathbf{x})/\partial \mathbf{x})|_{\mathbf{x}=\boldsymbol{\xi}}$. However, there is a vector $\dot{\mathbf{x}}^*(0)$ which points in the direction of steepest descent consistent with the constraints. Let us denote by $\mathbf{u}^*(0)$ the control vector such that

$$\dot{\mathbf{x}}^*(0) = \mathbf{A}\boldsymbol{\xi} + \mathbf{B}\mathbf{u}^*(0) \tag{6-248}$$

Now what property distinguishes $\dot{\mathbf{x}}^*(0)$ from all the other possible $\dot{\mathbf{x}}(0)$ vectors? It is easy to see that $\dot{\mathbf{x}}^*(0)$ satisfies the relation (see Fig. 6-13)

$$\left\langle \dot{\mathbf{x}}^*(0), \frac{\partial T^*(\mathbf{x})}{\partial \mathbf{x}}\bigg|_{\mathbf{x}=\boldsymbol{\xi}} \right\rangle \le \left\langle \dot{\mathbf{x}}(0), \frac{\partial T^*(\mathbf{x})}{\partial \mathbf{x}}\bigg|_{\mathbf{x}=\boldsymbol{\xi}} \right\rangle \tag{6-249}$$

for all $\dot{\mathbf{x}}(0) \in K$. Equivalently, from Eqs. (6-248) and (6-249) we find that, for all $\mathbf{u}(0) \in \Omega$,

$$\left\langle \mathbf{A}\boldsymbol{\xi} + \mathbf{B}\mathbf{u}^*(0), \frac{\partial T^*(\mathbf{x})}{\partial \mathbf{x}}\bigg|_{\mathbf{x}=\boldsymbol{\xi}} \right\rangle \le \left\langle \mathbf{A}\boldsymbol{\xi} + \mathbf{B}\mathbf{u}(0), \frac{\partial T^*(\mathbf{x})}{\partial \mathbf{x}}\bigg|_{\mathbf{x}=\boldsymbol{\xi}} \right\rangle \tag{6-250}$$

or that

$$\left\langle \mathbf{u}^*(0), \mathbf{B}' \frac{\partial T^*(\mathbf{x})}{\partial \mathbf{x}}\bigg|_{\mathbf{x}=\boldsymbol{\xi}} \right\rangle \le \left\langle \mathbf{u}(0), \mathbf{B}' \frac{\partial T^*(\mathbf{x})}{\partial \mathbf{x}}\bigg|_{\mathbf{x}=\boldsymbol{\xi}} \right\rangle \tag{6-251}$$

Let us now recapitulate: We have found that the control vector $\mathbf{u}^*(0)$ that satisfies Eq. (6-251) establishes the direction of steepest descent (consistent with the constraints) along the minimum-time cost surface $T^*(\mathbf{x})$ at the point $\mathbf{x} = \boldsymbol{\xi}$. It should be clear that *the control* $\mathbf{u}^*(0)$ *that satisfies Eq.* (6-251) *also satisfies the relation*

$$\langle \mathbf{u}^*(0), \mathbf{B}'\boldsymbol{\pi} \rangle \le \langle \mathbf{u}(0), \mathbf{B}'\boldsymbol{\pi} \rangle \tag{6-252}$$

*where* $\boldsymbol{\pi}$ *is any outwardly directed normal vector to the minimum isochrone* $S(\tau)$ *at* $\boldsymbol{\xi}$, *that is,*

$$\boldsymbol{\pi} = c \frac{\partial T^*(\mathbf{x})}{\partial \mathbf{x}}\bigg|_{\mathbf{x}=\boldsymbol{\xi}} \qquad c > 0 \tag{6-253}$$

*for any positive constant* $c$.

Now, physically speaking, $\mathbf{u}^*(0)$ must be the time-optimal control at $\mathbf{x} = \boldsymbol{\xi}$ because it causes the state to move in such a way as to maximize the rate of change of the minimum time. If $\mathbf{u}^*(0)$ is the time-optimal control, then we know from the necessary conditions that there is a costate $\mathbf{p}^*(0)$ such that the relation

$$1 + \langle \mathbf{A}\boldsymbol{\xi}, \mathbf{p}^*(0)\rangle + \langle \mathbf{u}^*(0), \mathbf{B}'\mathbf{p}^*(0)\rangle \leq 1 + \langle \mathbf{A}\boldsymbol{\xi}, \mathbf{p}^*(0)\rangle + \langle \mathbf{u}(0), \mathbf{B}'\mathbf{p}^*(0)\rangle \quad \text{for all } \mathbf{u}(0) \in \Omega \qquad (6\text{-}254)$$

or, equivalently, the relation

$$\langle \mathbf{u}^*(0), \mathbf{B}'\mathbf{p}^*(0)\rangle \leq \langle \mathbf{u}(0), \mathbf{B}'\mathbf{p}^*(0)\rangle \qquad \mathbf{u}(0) \in \Omega \qquad (6\text{-}255)$$

is satisfied. Comparing Eqs. (6-255), (6-253), and (6-252), we deduce the following important remark:

***Remark 6-11***† *If* $\mathbf{u}^*(0)$ *is the time-optimal control at* $\mathbf{x}^*(0) = \boldsymbol{\xi}$*, then the initial costate* $\mathbf{p}^*(0)$ *must have the same direction as the gradient vector* $\left.\dfrac{\partial T^*(\mathbf{x})}{\partial \mathbf{x}}\right|_{\mathbf{x}=\boldsymbol{\xi}}$*, if it exists. Another way of saying the same thing is: The initial costate* $\mathbf{p}^*(0)$ *must be the outward normal to the isochrone* $S(\tau)$ *at* $\mathbf{x} = \boldsymbol{\xi}$.

We continue now with a discussion of the important case in which the initial state $\boldsymbol{\xi}$ belongs to a "corner" of a minimum isochrone. Suppose that $\boldsymbol{\xi} \in S(\tau)$, as indicated in Fig. 6-14; let $\boldsymbol{\xi}_1$ be a state near $\boldsymbol{\xi}$ and on the isochrone $S(\tau)$, and let $\boldsymbol{\xi}_2$ be a state also near $\boldsymbol{\xi}$ and also on the isochrone $S(\tau)$. We shall (loosely) say that $\boldsymbol{\xi}_1$ is to the "right" of $\boldsymbol{\xi}$ and that $\boldsymbol{\xi}_2$ is to the "left" of $\boldsymbol{\xi}$. The statement "$\boldsymbol{\xi}$ is at a corner of $S(\tau)$" means that the gradient vector $\partial T^*(\mathbf{x})/\partial \mathbf{x}$ is *not* defined at $\mathbf{x} = \boldsymbol{\xi}$. As illustrated in Fig. 6-14, the gradient vectors

$$\left.\frac{\partial T^*(\mathbf{x})}{\partial \mathbf{x}}\right|_{\mathbf{x}=\boldsymbol{\xi}_1} \qquad \left.\frac{\partial T^*(\mathbf{x})}{\partial \mathbf{x}}\right|_{\mathbf{x}=\boldsymbol{\xi}_2}$$

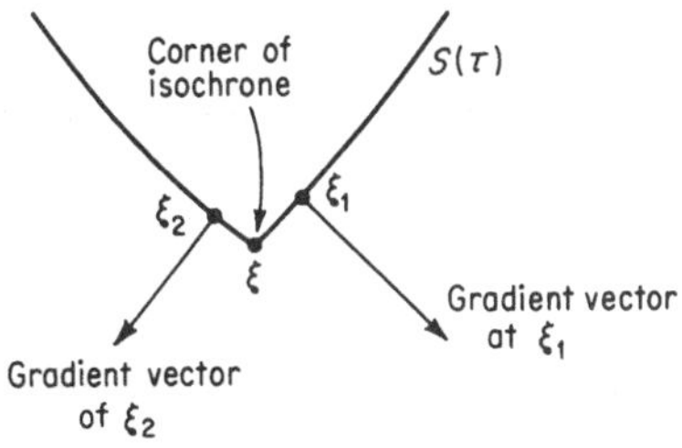


**Fig. 6-14** If $\boldsymbol{\xi}$ is at a corner of a minimum isochrone, then the direction of steepest descent is not defined.

are defined for all $\boldsymbol{\xi}_1$ to the right of $\boldsymbol{\xi}$ and for all $\boldsymbol{\xi}_2$ to the left of $\boldsymbol{\xi}$; however, we note that

$$\lim_{\boldsymbol{\xi}_1 \to \boldsymbol{\xi}} \left\{ \left.\frac{\partial T^*(\mathbf{x})}{\partial \mathbf{x}}\right|_{\mathbf{x}=\boldsymbol{\xi}_1} \right\} \neq \lim_{\boldsymbol{\xi}_2 \to \boldsymbol{\xi}} \left\{ \left.\frac{\partial T^*(\mathbf{x})}{\partial \mathbf{x}}\right|_{\mathbf{x}=\boldsymbol{\xi}_2} \right\} \qquad (6\text{-}256)$$

Thus, at $\mathbf{x} = \boldsymbol{\xi}$ we cannot establish the direction of steepest descent by simply looking at the gradient vector at $\mathbf{x} = \boldsymbol{\xi}$, as the gradient vector is not defined. This means that the time-optimal control cannot be defined at that point by the geometrical argument we presented in our previous

† The geometric interpretation contained in this remark is the starting point for the iterative techniques described in Refs. N-2 and E-1.

discussion. However, if $\mathbf{p}^*(0)$ is the costate corresponding to $\boldsymbol{\xi}$ and $\mathbf{u}^*(0)$, the relation (6-255) still holds. What has happened is that we have lost the correspondence between the initial costate $\mathbf{p}^*(0)$ and the normal to the minimum isochrone.†

In the preceding discussion, we restricted ourselves to initial states on a given minimum isochrone. It is easy to see that the same remarks apply to every state $\mathbf{x}^*(t)$ on a time-optimal trajectory to the origin, simply by invoking the principle of optimality (see Sec. 5-16). We summarize these conclusions in the following remark:

***Remark 6-12*** *Let $\mathbf{x}^*(t)$ be a state on a time-optimal trajectory, and let $\mathbf{p}^*(t)$ be the corresponding costate. Suppose that $\mathbf{x}^*(t) \in S(T)$. Then the costate $\mathbf{p}^*(t)$ is the outward normal to $S(T)$ at the point $\mathbf{x}^*(t)$, provided that the gradient vector $(\partial T^*(\mathbf{x})/\partial \mathbf{x})\,|_{\mathbf{x}=\mathbf{x}^*(t)}$ is defined.*

The following exercise will make the ideas presented in this section somewhat more concrete.

**Exercise 6-12** Consider the system

$$\begin{aligned} \dot{x}_1(t) &= x_2(t) \\ \dot{x}_2(t) &= u(t) \qquad |u(t)| \le 1 \end{aligned} \tag{6-257}$$

We shall prove in Sec. 7-2 that the minimum time $T^*(x_1, x_2)$ required to force any state $(x_1, x_2)$ to $(0, 0)$ is given by

$$T^*(x_1, x_2) = \begin{cases} x_2 + \sqrt{4x_1 + 2x_2^2} & \text{if } x_1 > -\frac{1}{2}x_2|x_2| \\ -x_2 + \sqrt{-4x_1 + 2x_2^2} & \text{if } x_1 < -\frac{1}{2}x_2|x_2| \\ |x_2| & \text{if } x_1 = -\frac{1}{2}x_2|x_2| \end{cases} \tag{6-258}$$

Plot the minimum isochrone $S(4)$. Consider the point

$$\hat{\mathbf{x}} = \begin{bmatrix} 0 \\ \dfrac{4}{1+\sqrt{2}} \end{bmatrix}$$

which belongs to the $S(4)$ minimum isochrone. Compute the gradient vector

$$\left.\frac{\partial T^*(x_1, x_2)}{\mathbf{x}}\right|_{\hat{\mathbf{x}}} = \begin{bmatrix} \dfrac{\partial T^*(x_1, x_2)}{\partial x_1} \\ \dfrac{\partial T^*(x_1, x_2)}{\partial x_2} \end{bmatrix}_{\mathbf{x}=\hat{\mathbf{x}}} \tag{6-259}$$

Plot this gradient vector in your figure as emanating from the point $\hat{\mathbf{x}}$. For all $u$, $|u| \le 1$, compute the set of vectors $\left\{\begin{bmatrix} \dot{x}_1 \\ \dot{x}_2 \end{bmatrix}\right\}$ from Eqs. (6-257) at $\hat{\mathbf{x}}$ and plot them as emanating from $\hat{\mathbf{x}}$ (that is, construct the set $K$ shown in Fig. 6-13). Convince yourself that the time-optimal control must be equal to $-1$ at $\hat{\mathbf{x}}$. Repeat for the point $\tilde{\mathbf{x}} = \begin{bmatrix} 4 \\ 0 \end{bmatrix}$. The

† It is such considerations that make the Hamilton-Jacobi theory weaker, in a sense, than the minimum principle. As long as the minimum-cost surfaces have no "corners," both theoretical approaches are equivalent; it is precisely in problems in which the gradient of the minimum cost is not defined everywhere that the minimum principle of Pontryagin is superior, at least from the applications point of view.

minimum isochrone $S(4)$ you have plotted has two "corners." Compute the gradient vectors on both sides of each corner and convince yourself that the gradient has a jump at the two corners.

## 6-8 Minimum-time Problems 7: The Existence of the Time-optimal Control

In this section, we shall present a discussion of a "useful" theorem (Theorem 6-10) which guarantees the existence of the time-optimal control to the origin for every initial state in the state space.

Questions about the existence of optimal controls from any initial state to any target set are extremely difficult to answer. It is perhaps instructive to examine the specific question of existence of the time-optimal control to the origin from a *heuristic* point of view.

Let us suppose that we are given a completely controllable dynamical system, and let us suppose that the control is constrained in magnitude by the relation $\mathbf{u}(t) \in \Omega$. In view of the assumption of controllability, we know that we can find at least one control which will transfer any initial state $\boldsymbol{\xi}$ to the origin $\mathbf{0}$ in some finite amount of time. However, it may happen that an initial state $\boldsymbol{\xi}$ is sufficiently far from the origin so that the only controls that can force $\boldsymbol{\xi}$ to $\mathbf{0}$ violate the magnitude constraint $\mathbf{u}(t) \in \Omega$. In this case, there are initial states which cannot be forced to $\mathbf{0}$ by controls that satisfy the magnitude constraint.

We can thus see that: *Given a completely controllable dynamical system* $\dot{\mathbf{x}}(t) = \mathbf{f}[\mathbf{x}(t), \mathbf{u}(t)]$ *and a constraint set* $\Omega$, *the n-dimensional state space* $R_n$ *can be divided into two subsets* $\Psi_\Omega$ *and* $R_n - \Psi_\Omega$ *with the following properties:*

1. *If* $\boldsymbol{\xi} \in \Psi_\Omega$, *then there is at least one admissible control that transfers* $\boldsymbol{\xi}$ *to* $\mathbf{0}$ *in some finite time.*
2. *If* $\boldsymbol{\xi} \in R_n - \Psi_\Omega$, *then there are no admissible controls that transfer* $\boldsymbol{\xi}$ *to any element of* $\Psi_\Omega$ *in some finite time (and so* $\boldsymbol{\xi}$ *cannot be forced to* $\mathbf{0}$ *by an admissible control).*

In essence, the constrained controls cannot provide enough "kick" to force states in $R_n - \Psi_\Omega$ into $\Psi_\Omega$ and, subsequently, to the origin.

From a physical point of view, the control $\mathbf{u}(t)$ can either add energy to or subtract energy from the dynamical system. If we think of the state $\mathbf{x} = \mathbf{0}$ as the *zero-energy* state, then we can see that systems with the property that the subset $R_n - \Psi_\Omega$ is nonempty must be, in a sense, unstable. For this reason, we may be reasonably confident that a stable, completely controllable dynamical system will have the property that $\Psi_\Omega = R_n$ and an unstable, completely controllable dynamical system will have the property that $\Psi_\Omega \subset R_n$, $\Psi_\Omega \neq R_n$.

A theorem† which is useful from the verification point of view and which

† Appropriate references which deal with existence theorems will be given in Secs. 6-10 and 6-23.

guarantees the existence of a time-optimal control from *every* initial state to the origin is the following:

***Theorem 6-10*** *Consider the time-optimal control of the controllable system* $\dot{\mathbf{x}} = \mathbf{A}\mathbf{x}(t) + \mathbf{B}\mathbf{u}(t)$, *as described in Problem* 6-1c. *If the eigenvalues of* $\mathbf{A}$ *have nonpositive real parts, then a time-optimal control to the origin exists for any initial state in* $R_n$.

The formal proof of this theorem can be found in Ref. P-5, pages 127–135. We shall indicate the nature of the proof by considering a system with distinct real eigenvalues and a single control variable described by the equations

$$\begin{aligned} \dot{x}_i(t) &= \lambda_i x_i(t) + b_i u(t) \qquad i = 1, 2, \ldots, n \\ |u(t)| &\le 1 \\ \xi_i &= x_i(0) \qquad i = 1, 2, \ldots, n \end{aligned} \tag{6-260}$$

The solution of the system (6-260) for any $u(t)$ is given by

$$x_i(t) = e^{\lambda_i t}\left[\xi_i + \int_0^t e^{-\lambda_i \tau} b_i u(\tau)\, d\tau\right] \tag{6-261}$$

Suppose that we have found an admissible control $\hat{u}(t)$ such that $x_1(\hat{T}) = x_2(\hat{T}) = \cdots = x_n(\hat{T}) = 0$. This implies that the relation

$$\xi_i = -\int_0^{\hat{T}} e^{-\lambda_i t} b_i \hat{u}(t)\, dt \tag{6-262}$$

is satisfied for all $i = 1, 2, \ldots, n$. Since $|\hat{u}(t)| \le 1$, we can then deduce that

$$\begin{aligned} |\xi_i| = \left|\int_0^{\hat{T}} e^{-\lambda_i t} b_i \hat{u}(t)\, dt\right| &\le \int_0^{\hat{T}} e^{-\lambda_i t} |b_i|\, |\hat{u}(t)|\, dt \le \int_0^{\hat{T}} e^{-\lambda_i t} |b_i|\, dt \\ &= -\frac{|b_i|}{\lambda_i}\left(e^{-\lambda_i \hat{T}} - 1\right) \end{aligned} \tag{6-263}$$

for $i = 1, 2, \ldots, n$. Now let us suppose that one of the eigenvalues, say $\lambda_1$, is positive [which means that the system (6-260) is unstable]. From Eq. (6-263), we deduce that

$$|\xi_1| \le -\frac{|b_1|}{\lambda_1}\left(e^{-\lambda_1 \hat{T}} - 1\right) \tag{6-264}$$

which, in turn, yields the relation

$$e^{-\lambda_1 \hat{T}} \le 1 - \frac{\lambda_1 |\xi_1|}{|b_1|} \tag{6-265}$$

Clearly, the relation (6-265) cannot be satisfied for any real positive and finite $\hat{T}$ if the initial state variable $\xi_1$ is such that

$$|\xi_1| \ge \frac{|b_1|}{\lambda_1} \tag{6-266}$$

Thus, if $\lambda_1 > 0$ and if $|\xi_1| \geq |b_1|/\lambda_1$, then we cannot find a $\hat{T}$ such that $x_1(\hat{T}) = 0$, and therefore a time-optimal control does not exist.

Now, if all the eigenvalues $\lambda_i$ are nonpositive, it is easy to show that Eq. (6-263) can be satisfied for *all* $|\xi_i|$ and *all* $i = 1, 2, \ldots, n$, because we can find values of $\hat{T}$ sufficiently large. This, in turn, implies that a time-optimal control exists for all initial states† (why?).

**Example 6-4** Consider the time-optimal control of the system

$$\dot{x}(t) = ax(t) + u(t) \qquad |u(t)| \leq 1 \qquad x(0) = \xi \tag{6-267}$$

If $a \leq 0$, then (by Theorem 6-10) a time-optimal control to the state $x = 0$ exists for all $\xi$. If $a > 0$, then the system is unstable. Let us find the set of initial states $\Psi_\Omega$ for which a time-optimal control exists. If a time-optimal control $u^*(t)$ exists, then $|u^*(t)| = 1$, and so we have

$$\xi = -\int_0^\tau e^{-at}u^*(t)\,dt$$

Thus, we deduce that

$$|\xi| = \left|\int_0^\tau e^{-at}u^*(t)\,dt\right| \leq \int_0^\tau |e^{-at}u^*(t)|\,dt \leq \int_0^\tau |e^{-at}|\,|u^*(t)|\,dt$$
$$= \int_0^\tau e^{-at}\,dt = -\frac{1}{a}(e^{-a\tau} - 1) \tag{6-268}$$

In order for the relation (6-268) to hold for some finite positive $\tau$, it is necessary that

$$|\xi| < \frac{1}{a}$$

Thus, the set of initial states $\Psi_\Omega$ for which a time-optimal control to the origin exists is given by

$$\Psi_\Omega = \left\{\xi: |\xi| < \frac{1}{a} \qquad a > 0\right\} \tag{6-269}$$

If $|\xi| \geq 1/a$, then a time-optimal control to the origin does not exist. In this case, the set $\Psi_\Omega$ is an open set containing the origin.

**Exercise 6-13** Consider the time-optimal control to the origin (0, 0) of the following second-order systems. In each case, the control $u(t)$ is constrained in magnitude by $|u(t)| \leq 1$.

(*a*) $\dot{x}_1(t) = x_1(t) + x_2(t)$
$\quad\;\; \dot{x}_2(t) = -x_1(t) + x_2(t) + u(t)$

(*b*) $\dot{x}_1(t) = x_2(t)$
$\quad\;\; \dot{x}_2(t) = x_2(t) + u(t)$

(*c*) $\dot{x}_1(t) = x_2(t)$
$\quad\;\; \dot{x}_2(t) = x_1(t) + u(t)$

For each system, find the set of initial states $\Psi_\Omega$ for which a time-optimal control to the origin exists. (Assume that if there is an admissible extremal control, then there is an optimal control.)

## 6-9 Minimum-time Problems 8: The Hamilton-Jacobi Equation

In Sec. 5-19, we discussed the behavior of the minimum cost along the optimal trajectory. In Sec. 6-7, we discussed the geometric properties of the time-optimal control for Problem 6-1*c*. In this section, we shall

† See Ref. P-5. We do not deal with existence proofs in this book, because such proofs fall outside its introductory scope.

combine these concepts and examine the Hamilton-Jacobi equation for the time-optimal problem. Our objective will be to illustrate the use of the general results given in Chap. 5 in the time-optimal problem 6-1*c*.

Throughout this section, we shall deal with the time-optimal control of the *normal* system $\dot{\mathbf{x}}(t) = \mathbf{A}\mathbf{x}(t) + \mathbf{B}\mathbf{u}(t)$, where the target set is the origin of the state space. We shall use the following notation: If we are given a state $\mathbf{x}$, then we shall let $T^*(\mathbf{x})$ denote the minimum time required to force $\mathbf{x}$ to $\mathbf{0}$ and we shall let $\mathbf{u}^*$ denote the value of the time-optimal control at the state $\mathbf{x}$.

Our specific objectives in this section are the following:

1. To indicate how the Hamilton-Jacobi equation can be used to test whether or not a given function $T(\mathbf{x})$ can possibly be equal to $T^*(\mathbf{x})$ after we have solved the time-optimal problem
2. To show that difficulties arise if the assumption that a given control is the time-optimal one turns out to be incorrect
3. To indicate the difficulties associated with the determination of the time-optimal control directly from the Hamilton-Jacobi equation

We shall first discuss the use of the Hamilton-Jacobi equation viewed as a necessary condition. From the general theory of Sec. 5-19, we can deduce the following lemma:

***Lemma 6-6*** *Let* $\mathbf{x}^*$ *be a state on a time-optimal trajectory, and let* $\mathbf{u}^*$ *be the value of the time-optimal control at* $\mathbf{x}^*$. *Since* $\partial T^*(\mathbf{x})/\partial t = 0$ *for all* $\mathbf{x}$, *it is necessary that* $T^*(\mathbf{x})$ *satisfy the relation*

$$1 + \left\langle \mathbf{A}\mathbf{x}^*, \left.\frac{\partial T^*(\mathbf{x})}{\partial \mathbf{x}}\right|_{\mathbf{x}=\mathbf{x}^*} \right\rangle + \left\langle \mathbf{u}^*, \mathbf{B}' \left.\frac{\partial T^*(\mathbf{x})}{\partial \mathbf{x}}\right|_{\mathbf{x}=\mathbf{x}^*} \right\rangle = 0 \qquad (6\text{-}270)$$

*provided that* $(\partial T^*(\mathbf{x})/\partial \mathbf{x})|_{\mathbf{x}=\mathbf{x}^*}$ exists.

**Exercise 6-14** Prove Lemma 6-6. HINT: See Eq. (5-695).

This lemma is useful when the time-optimal problem has already been solved and we seek to determine whether or not a given function $T(\mathbf{x})$ can possibly be the expression for the minimum time as a function of the state. If a given function $T(\mathbf{x})$ violates Eq. (6-270) even at a single point, then $T(\mathbf{x})$ can be immediately eliminated as a candidate for the minimum time. We shall illustrate this point by the following example:

**Example 6-5** Suppose that the linear system is described by the equations

$$\begin{bmatrix} \dot{x}_1(t) \\ \dot{x}_2(t) \end{bmatrix} = \begin{bmatrix} 0 & 1 \\ 0 & 0 \end{bmatrix} \begin{bmatrix} x_1(t) \\ x_2(t) \end{bmatrix} + \begin{bmatrix} 0 \\ 1 \end{bmatrix} u(t) \qquad |u(t)| \le 1 \qquad (6\text{-}271)$$

We assert† that the time-optimal control is

$$u^* = -1 \qquad \text{if } \mathbf{x} = \mathbf{x}_\alpha = \begin{bmatrix} 1 \\ 2 \end{bmatrix} \qquad (6\text{-}272)$$

$$u^* = -1 \qquad \text{if } \mathbf{x} = \mathbf{x}_\beta = \begin{bmatrix} 1 \\ 1 \end{bmatrix} \qquad (6\text{-}273)$$

† See Sec. 7-2.

Let us suppose that we have a "hunch" that the expression

$$T(\mathbf{x}) = T(x_1, x_2) = \tfrac{1}{2}x_1^2 + x_2^2 \tag{6-274}$$

is the expression for the minimum time as a function of the state. Observe that this hunch is not unreasonable, because $T(\mathbf{x}) > 0$ for all $\mathbf{x}$, $T(\mathbf{0}) = 0$, and $\lim_{\|\mathbf{x}\| \to \infty} T(\mathbf{x}) = \infty$. We shall now demonstrate how to prove that our hunch is wrong.

First, we calculate the gradient of $T(\mathbf{x})$; we find, from Eq. (6-274), that this gradient is given by

$$\frac{\partial T(\mathbf{x})}{\partial \mathbf{x}} = \begin{bmatrix} \dfrac{\partial T(\mathbf{x})}{\partial x_1} \\ \dfrac{\partial T(\mathbf{x})}{\partial x_2} \end{bmatrix} = \begin{bmatrix} x_1 \\ 2x_2 \end{bmatrix} \tag{6-275}$$

We evaluate it at $\mathbf{x} = \mathbf{x}_\alpha$ and $\mathbf{x} = \mathbf{x}_\beta$ to find that

$$\left.\frac{\partial T(\mathbf{x})}{\partial \mathbf{x}}\right|_{\mathbf{x}=\mathbf{x}_\alpha} = \begin{bmatrix} 1 \\ 4 \end{bmatrix} \tag{6-276}$$

$$\left.\frac{\partial T(\mathbf{x})}{\partial \mathbf{x}}\right|_{\mathbf{x}=\mathbf{x}_\beta} = \begin{bmatrix} 1 \\ 2 \end{bmatrix} \tag{6-277}$$

At $\mathbf{x} = \mathbf{x}_\alpha$, the left-hand side of Eq. (6-270) becomes

$$1 + \left\langle \begin{bmatrix} 0 & 1 \\ 0 & 0 \end{bmatrix} \begin{bmatrix} 1 \\ 2 \end{bmatrix}, \begin{bmatrix} 1 \\ 4 \end{bmatrix} \right\rangle - 1 \left\{ [0 \quad 1] \begin{bmatrix} 1 \\ 4 \end{bmatrix} \right\} = 1 + 2 - 4 = -1 \neq 0 \tag{6-278}$$

We can immediately conclude that the $T(\mathbf{x})$ given by Eq. (6-274) *cannot* be the minimum-time expression, because at $\mathbf{x} = \mathbf{x}_\alpha$ Eq. (6-270) does not hold.

Let us see now what happens if we test $T(\mathbf{x})$ at the point $\mathbf{x} = \mathbf{x}_\beta$. At $\mathbf{x} = \mathbf{x}_\beta$, the left-hand side of Eq. (6-270) becomes

$$1 + \left\langle \begin{bmatrix} 0 & 1 \\ 0 & 0 \end{bmatrix} \begin{bmatrix} 1 \\ 1 \end{bmatrix}, \begin{bmatrix} 1 \\ 2 \end{bmatrix} \right\rangle - 1 \left\{ [0 \quad 1] \begin{bmatrix} 1 \\ 2 \end{bmatrix} \right\} = 1 + 1 - 2 = 0 \tag{6-279}$$

This means that $T(\mathbf{x})$ satisfies the necessary condition of Lemma 6-6 at $\mathbf{x} = \mathbf{x}_\beta$, and on the basis of this test, we could conclude that $T(\mathbf{x})$ *may* be the expression for the minimum time. Nonetheless, the test at $\mathbf{x}_\alpha$ discards this possibility.

Now suppose that we made an error in the determination of the time-optimal control; for example, suppose that we thought that

$$u^* = +1 \qquad \text{if } \mathbf{x} = \mathbf{x}_\beta = \begin{bmatrix} 1 \\ 1 \end{bmatrix} \tag{6-280}$$

Then, instead of Eq. (6-279), we would find that

$$1 + \left\langle \begin{bmatrix} 0 & 1 \\ 0 & 0 \end{bmatrix} \begin{bmatrix} 1 \\ 1 \end{bmatrix}, \begin{bmatrix} 1 \\ 2 \end{bmatrix} \right\rangle + 1 \left\{ [0 \quad 1] \begin{bmatrix} 1 \\ 2 \end{bmatrix} \right\} = 1 + 1 + 2 = 4 \neq 0 \tag{6-281}$$

and we could eliminate $T(\mathbf{x})$ from consideration. Now it is true that $T(\mathbf{x}) \neq T^*(\mathbf{x})$, but the conclusion from Eq. (6-281) cannot be trusted, because the original premise that $u^* = +1$ at $\mathbf{x}_\beta$ is wrong. In other words, if we make a mistake in determining the optimal control, then, in all likelihood, we would eliminate the correct minimum-time expression $T^*(\mathbf{x})$ by this procedure.

From a practical point of view, however, this lemma is not very useful. The reason is that most of the time the engineer is interested in the time-optimal feedback control system rather than in testing whether or not a given $T(\mathbf{x})$ is the same as $T^*(\mathbf{x})$. Nonetheless, the use of the Hamilton-Jacobi equation as a necessary condition is of value in theoretical investigations and in checking whether the results obtained using the minimum principle are correct.

In Chap. 5, we concentrated much more on the Hamilton-Jacobi equation viewed as a sufficient condition. We shall now discuss the problem of solving the Hamilton-Jacobi equation and constructing the time-optimal control. At this point, we feel that the reader should go over the statement of Theorem 5-12 very carefully.

We know that the Hamiltonian for the time-optimal problem 6-1$c$ is

$$H(\mathbf{x}, \mathbf{p}, \mathbf{u}) = 1 + \langle \mathbf{Ax}, \mathbf{p} \rangle + \langle \mathbf{u}, \mathbf{B}'\mathbf{p} \rangle \tag{6-282}$$

Since the system $\dot{\mathbf{x}} = \mathbf{Ax} + \mathbf{Bu}$ is normal, the Hamiltonian $H$ is normal according to Definition 5-12, and so the $H$-minimal control is given by

$$\tilde{\mathbf{u}} = -\operatorname{SGN}\{\mathbf{B}'\mathbf{p}\} \tag{6-283}$$

From Eqs. (6-283) and (6-282), we have

$$\begin{aligned} H(\mathbf{x}, \mathbf{p}, \tilde{\mathbf{u}}) &= 1 + \langle \mathbf{Ax}, \mathbf{p} \rangle - \langle \operatorname{SGN}\{\mathbf{B}'\mathbf{p}\}, \mathbf{B}'\mathbf{p} \rangle \\ &= 1 + \sum_{i=1}^{n} \sum_{k=1}^{n} a_{ik} x_i p_k - \sum_{j=1}^{r} \left| \sum_{i=1}^{n} b_{ij} p_i \right| \end{aligned} \tag{6-284}$$

Now let us consider the partial differential equation (the Hamilton-Jacobi equation)

$$1 + \sum_{i=1}^{n} \sum_{k=1}^{n} a_{ik} x_i \frac{\partial T(\mathbf{x})}{\partial x_i} - \sum_{j=1}^{r} \left| \sum_{i=1}^{n} b_{ij} \frac{\partial T(\mathbf{x})}{\partial x_i} \right| = 0 \tag{6-285}$$

and let us suppose that we were able to find a solution

$$\hat{T}(\mathbf{x}) \tag{6-286}$$

of the partial differential equation (6-285) such that:

1. $$\hat{T}(\mathbf{0}) = 0 \tag{6-287}$$
2. The control vector

$$\hat{\mathbf{u}} = -\operatorname{SGN}\left\{\mathbf{B}' \frac{\partial \hat{T}(\mathbf{x})}{\partial \mathbf{x}}\right\} \tag{6-288}$$

substituted in the linear system yields the system

$$\dot{\mathbf{x}}(t) = \mathbf{Ax}(t) - \mathbf{B} \operatorname{SGN}\left\{\mathbf{B}' \frac{\partial \hat{T}[\mathbf{x}(t)]}{\partial \mathbf{x}(t)}\right\} \tag{6-289}$$

whose solution

$$\hat{\mathbf{x}}(t) \qquad \text{with } \hat{\mathbf{x}}(0) = \boldsymbol{\xi} \tag{6-290}$$

has the property

$$\hat{\mathbf{x}}[\hat{T}(\boldsymbol{\xi})] = \mathbf{0} \tag{6-291}$$

In other words, the solution $\hat{T}(\mathbf{x})$ of the Hamilton-Jacobi equation (6-285) defines a control $\hat{\mathbf{u}}$ [given by Eq. (6-288)] which, in turn, generates a trajectory $\hat{\mathbf{x}}(t)$ which arrives at the origin at the time $\hat{T}(\boldsymbol{\xi})$, where $\boldsymbol{\xi}$ is the given initial state.

If such is the case, then the control $\hat{\mathbf{u}}(t)$ will be time-optimal, at least with respect to neighboring controls; that is, $\hat{\mathbf{u}}(t)$ will be locally time-optimal.

We shall illustrate these ideas by a very simple example. We shall then discuss the shortcomings of this method.

**Example 6-6** Suppose that we are given the first-order system

$$\dot{x}(t) = 2u(t) \qquad |u(t)| \leq 1 \qquad x(0) = \xi \tag{6-292}$$

It is desired to force any initial state $\xi$ to 0 in minimum time. The Hamiltonian for this problem is given by

$$H(x, p, u) = 1 + 2up \tag{6-293}$$

The $H$-minimal control $\tilde{u}$ is given by the relation

$$\tilde{u} = -\operatorname{sgn}\{p\} \tag{6-294}$$

Then, the Hamilton-Jacobi equation for this problem is

$$1 - 2\left|\frac{\partial T(x)}{\partial x}\right| = 0 \tag{6-295}$$

Let us define two regions $X_1$ and $X_2$ of the (one-dimensional) state space as

$$X_1 = \{x: x > 0\} \qquad X_2 = \{x: x < 0\} \tag{6-296}$$

It is trivial to verify that the function

$$\hat{T}(x) = \tfrac{1}{2}|x| \tag{6-297}$$

is the solution to the partial differential equation (6-295) for all $x \in X_1 \cup X_2$, because

$$\frac{\partial \hat{T}(x)}{\partial x} = \tfrac{1}{2} \qquad \text{for } x \in X_1 \tag{6-298}$$

$$\frac{\partial \hat{T}(x)}{\partial x} = -\tfrac{1}{2} \qquad \text{for } x \in X_2 \tag{6-299}$$

Observe that

$$\left.\frac{\partial \hat{T}(x)}{\partial x}\right|_{x=0} \text{ is not defined} \tag{6-300}$$

and that

$$\hat{T}(0) = 0 \tag{6-301}$$

Now consider the control $\hat{u}$, defined by

$$\hat{u} = -\operatorname{sgn}\left\{\frac{\partial \hat{T}(x)}{\partial x}\right\} \tag{6-302}$$

which implies that

$$\hat{u} = -1 \qquad \text{for } x \in X_1 \tag{6-303}$$

$$\hat{u} = +1 \qquad \text{for } x \in X_2 \tag{6-304}$$

$$\hat{u} \text{ not defined} \qquad \text{at } x = 0 \tag{6-305}$$

Let us suppose that $\xi \in X_1$, that is, that $\xi > 0$; by substituting Eq. (6-303) into Eq. (6-292), we find that

$$\dot{\hat{x}}(t) = -2 \tag{6-306}$$

and

$$\hat{x}(t) = \xi - 2t \qquad \xi \in X_1 \tag{6-307}$$

Observe that from Eq. (6-297), we have

$$\hat{T}(\xi) = \tfrac{1}{2}\xi \qquad \xi \in X_1 \tag{6-308}$$

Therefore, we deduce that

$$\hat{x}[\hat{T}(\xi)] = \xi - \xi = 0 \tag{6-309}$$

Moreover, for all $t \in [0, \hat{T}(\xi))$, we have

$$\hat{x}(t) = \xi - 2t > 0 \tag{6-310}$$

which means that

$$\hat{x}(t) \in X_1 \tag{6-311}$$

We thus immediately conclude that the constant control

$$\hat{u} = -1 \tag{6-312}$$

is time-optimal for all $x \in X_1$. Similar reasoning yields the conclusion that the control

$$\hat{u} = +1 \tag{6-313}$$

is time-optimal for all $x \in X_2$.

Note that we have defined the regions $X_1$ and $X_2$ in such a way that $\partial T(x)/\partial x$ was well defined.

We had no particular difficulty in determining the time-optimal control for the above simple system using the Hamilton-Jacobi equation. There are several reasons for this:

1. It was extremely easy to guess a solution to the Hamilton-Jacobi equation that satisfied the boundary conditions.
2. It was extremely easy to define the two regions $X_1$ and $X_2$.
3. It was extremely easy to verify that $\mathbf{x}[\hat{T}(\xi)] = \mathbf{0}$.

If we attempt, however, to determine the time-optimal control for a higher-order system, we shall encounter the following difficulties:†

1. It is almost impossible to determine the solution to the Hamilton-Jacobi equation for systems whose order is higher than 2.
2. For an $n$th-order system, one must subdivide the state space into at least $2n$ regions $X_1, X_2, \ldots, X_{2n}$, which are extremely difficult to specify for systems whose order is higher than 2.

It is for these reasons that, at present, the design of time-optimal feedback systems is mostly carried out using the necessary conditions provided by the minimum principle rather than the sufficient conditions provided by the Hamilton-Jacobi equation.

## 6-10 Minimum-time Problems 9: Comments and Remarks

In the preceding seven sections, we analyzed the time-optimal control problem from several points of view. Our objectives were to interlace the mathematical and geometrical aspects of the problem so that the reader would appreciate their implications.

† The following comments are the outgrowth of the experience of the authors in trying to solve several minimum-time problems.

As the reader has probably noted, the strongest set of theorems were stated with regard to the linear time-invariant time-optimal regulator problem (Problem 6-1*c*). Some of these theorems carry over to the time-optimal control to the origin for the linear time-varying system

$$\dot{\mathbf{x}}(t) = \mathbf{A}(t)\mathbf{x}(t) + \mathbf{B}(t)\mathbf{u}(t) \tag{6-314}$$

provided that the time-optimal problem is normal. In particular, Theorem 6-7 and Exercise 6-7, which state that the time-optimal control is *unique*, still apply† for the time-varying system (6-314). In addition, Theorem 6-9 (which states that if the origin is the terminal state, then the *extremal* controls are *unique*) is also true‡ for the system (6-314). The theorem which deals with the number of switchings (Theorem 6-8) and the most useful existence theorem (Theorem 6-10) have no counterpart for the time-varying system because they are stated in terms of the (constant) eigenvalues of the system matrix **A**.

If we turn our attention to the nonlinear time-optimal problem (such as Problem 6-1*b*), we find, in the current literature, few§ useful results in terms of general characteristics or theorems. The difficulty is not a conceptual one; rather, it is due to the fact that we cannot write analytical expressions for the solutions of nonlinear differential equations.

The theoretical development of the time-optimal control problem will be applied in Chap. 7 to the design of time-optimal systems. Our objective in Chap. 7 will be the determination of the time-optimal control as a function of the state, i.e., the solution of the time-optimal feedback problem. We feel that before the reader can appreciate the general concepts of the time-optimal problem, he should see the explicit solution of some time-optimal problems. Several exercises in Chap. 7 are designed to illustrate the geometric concepts discussed in the preceding sections. In Chap. 7 we shall, whenever possible, present the engineering realization of the time-optimal control laws so that the relative complexity of the required "hardware" can be evaluated. We suggest, at this point, that the reader who is tired of general concepts and theorems skip the remainder of this chapter and go over the material of Chap. 7.

The following exercises are designed to acquaint the reader with the time-optimal control of linear time-varying systems.

**Exercise 6-15** For each of the following systems, determine the necessary and sufficient conditions for the time-optimal problem to be normal. Assume that all the time-varying coefficients are infinitely differentiable functions of time. First refresh your memory by reading Definitions 6-4 and 6-5 and by going over the material preceding Theorem 6-5.

† See Ref. P-5, p. 186.

‡ See Ref. P-5, pp. 186 and 187.

§ Some references which deal with the nonlinear time-optimal problem from the theoretical point of view (and mostly from the point of existence) are H-3, N-5, F-3, L-7, R-8, and W-3.

(a) $\dot{x}_1(t) = x_2(t)$

$\dot{x}_2(t) = +a_1(t)x_1(t) + a_2(t)x_2(t) + b_1(t)u(t) \qquad |u(t)| \leq 1$

(b) $\begin{bmatrix} \dot{x}_1(t) \\ \dot{x}_2(t) \end{bmatrix} = \begin{bmatrix} a_{11}(t) & a_{12}(t) \\ a_{21}(t) & a_{22}(t) \end{bmatrix} \begin{bmatrix} x_1(t) \\ x_2(t) \end{bmatrix} + \begin{bmatrix} b_{11}(t) & b_{12}(t) \\ b_{21}(t) & b_{22}(t) \end{bmatrix} \begin{bmatrix} u_1(t) \\ u_2(t) \end{bmatrix} \qquad \begin{matrix} |u_1(t)| \leq 1 \\ |u_2(t)| \leq 1 \end{matrix}$

(c) $\dot{x}_1(t) = x_2(t) + b_1(t)u(t)$

$\dot{x}_2(t) = x_3(t) + b_2(t)u(t) \qquad |u(t)| \leq 1$

$\dot{x}_3(t) = x_1(t) + b_3(t)u(t)$

(d) $\dot{x}_1(t) = a(t)x_2(t)$

$\dot{x}_2(t) = -a(t)x_1(t) + u(t) \qquad |u(t)| \leq 1$

**Exercise 6-16** Prove Theorem 6-7 for the system $\dot{\mathbf{x}}(t) = \mathbf{A}(t)\mathbf{x}(t) + \mathbf{B}(t)\mathbf{u}(t)$. HINT: See Ref. P-5, page 186.

**Exercise 6-17** Prove Theorem 6-9 for the system $\dot{\mathbf{x}}(t) = \mathbf{A}(t)\mathbf{x}(t) + \mathbf{B}(t)\mathbf{u}(t)$. HINT: See Ref. P-5, pages 186 and 187.

## 6-11 Minimum-fuel Problems 1: Introduction†

We shall study the minimum-fuel problem in this section and in the next five sections. We shall attempt in this study to provide a bridge between the theoretical results of Chap. 5 and the specific designs of simple minimum-fuel systems discussed in Chap. 8.

We note that aerospace problems ranging from simple attitude control to complex rendezvous control are a particularly important source of minimum-fuel problems since we often seek to control aerospace systems by thrusts or torques which are the result of the burning of fuel or the expulsion of mass.

Minimum-fuel problems or, as they are more commonly called, fuel-optimal problems are generally more complicated than time-optimal problems. The additional complexity is present both in the analytical portion of the problem and in the actual design of the fuel-optimal system. We shall see some of the reasons for this additional complexity in Sec. 6-15 and in Chap. 8. At present, let us simply point out that, although conservation of fuel is at the heart of a minimum-fuel design, it is often necessary to consider other quantities (such as response time) in designing a minimum-fuel system.

## 6-12 Minimum-fuel Problems 2: Discussion of the Problem and of the Constraints

We are often interested in the translational and/or rotational motion of a mass. The control is provided by a mechanism that consumes fuel and generates thrusts and/or torques. The consumed fuel may be expelled from the system (for example, by rockets, reaction jets, or internal-combustion engines) or may remain in the system (for example, in the case of a nuclear reactor). At any rate, in every fuel-optimal problem, we shall

† Although there are few references in which the general minimum-fuel problem is examined, there is a good deal of material in Ref. N-3.

make the assumption that the cost of control is related to a decrease in useful mass.

Now let us suppose that the controlled system is described by the equation

$$\dot{\mathbf{x}}(t) = \mathbf{f}[\mathbf{x}(t), \mathbf{u}(t), t] \tag{6-315}$$

where $\mathbf{x}(t)$ is the state vector, $\mathbf{u}(t)$ is the control vector, and $t$ is the time. Let us assume that the control vector $\mathbf{u}(t)$ is a result of the consumption of fuel. Let us agree to denote the rate of change of fuel at time $t$ by $\varphi(t)$; then the total fuel $F$ consumed in the time interval $[t_0, t_1]$ is given by

$$F = \int_{t_0}^{t_1} \varphi(t)\, dt \tag{6-316}$$

We shall measure the rate of flow of fuel as a nonnegative quantity; i.e., we suppose that

$$\varphi(t) \geq 0 \qquad \text{for all } t$$

For most physical systems, there is a relation between the rate of flow of fuel $\varphi$ and the control vector $\mathbf{u}$ of the form

$$\varphi = h(\mathbf{u}) \tag{6-317}$$

This relation can be determined experimentally. To be specific, *we shall assume in this chapter (and in Chap. 8) that Eq. (6-317) takes the form*

$$\varphi = \sum_{j=1}^{r} c_j |u_j| \qquad c_j > 0 \tag{6-318}$$

*where $u_1, u_2, \ldots, u_r$ are the components of the control vector $\mathbf{u}$ and the $c_j$ are positive constants of proportionality.* The relation (6-318) implies that an increase in the rate of flow of fuel leads to an increase in the magnitude of the control vector $\mathbf{u}$. If Eq. (6-318) represents the relation of $\mathbf{u}$ to $\varphi$, we then can see that the fuel consumed in the time interval $[t_0, t_1]$ is given by the relation†

$$F = \int_{t_0}^{t_1} \sum_{j=1}^{r} c_j |u_j(t)|\, dt \tag{6-319}$$

A typical minimum-fuel problem may be stated as follows: We are given the system $\dot{\mathbf{x}}(t) = \mathbf{f}[\mathbf{x}(t), \mathbf{u}(t), t]$, the constraint $\mathbf{u}(t) \in \Omega$, an initial state $\mathbf{x}(t_0)$, and a target set $S$; then we wish to find the admissible control that transfers $\mathbf{x}(t_0)$ to $S$ in such a way that the fuel $F$ is minimized.

If we deal with a problem (such as the control of booster rockets) for which the mass of the fuel $F$ is not negligible compared with the mass of the overall system, then we must include the mass as a state variable. For

† In Chap. 10, we shall consider a different relation between $F$ and $\mathbf{u}$, namely,

$$F = \int_{t_0}^{t_1} \sqrt{u_1^2(t) + \cdots + u_r^2(t)}\, dt$$

example, suppose that the mass of the system at $t = t_0$ is $M(t_0)$; then the rate of change of the mass is proportional to the rate of flow of fuel, and so we can write the differential equation

$$\frac{dM(t)}{dt} = -\sum_{j=1}^{r} c_j|u_j(t)| \tag{6-320}$$

Thus, the mass $M(t)$ at each time $t$ can be calculated from the relation

$$M(t) = M(t_0) - \int_{t_0}^{t} \left\{ \sum_{j=1}^{r} c_j|u_j(\tau)| \right\} d\tau \tag{6-321}$$

In such problems, the control which accomplishes a given control task with minimum fuel has automatically the property of minimizing the mass difference $M(t_0) - M(t_1)$.

If the mass $F$ of the consumed fuel is negligible compared with the mass $M(t_0)$ of the controlled system, then we can reasonably suppose that the system has a constant mass. This supposition decreases the number of state variables by one and simplifies the equations. In all the problems that we shall solve in Chap. 8, we shall make this supposition.

## 6-13 Minimum-fuel Problems 3: Formulation of a Problem and Derivation of the Necessary Conditions

In this section, we shall formulate the fuel-optimal problem in a precise way. We shall derive necessary conditions for fuel optimality using the minimum principle, we shall define the normal and singular fuel-optimal problems, and we shall state the *on-off principle* (Theorem 6-11) for the normal fuel-optimal problem.

The minimum-fuel problem stated below is the analog of the particular time-optimal problem stated as Problem 6-1*b*. In both of these problems (Problem 6-1*b* and Problem 6-2*a*), the controlled system, the control constraints, and the target set are the same. The reader will thus be able to compare the necessary conditions for the two problems. Since many of the necessary conditions are similar and since the logical development is essentially the same as in Sec. 6-3, we shall not present a very detailed development.

***Problem 6-2a Fuel-optimal Control to a Moving Target Set*** *Given the system*

$$\dot{x}_i(t) = f_i[\mathbf{x}(t), t] + \sum_{j=1}^{r} b_{ij}[\mathbf{x}(t), t]u_j(t) \qquad i = 1, 2, \ldots, n$$

*or, equivalently, in vector form,*

$$\dot{\mathbf{x}}(t) = \mathbf{f}[\mathbf{x}(t), t] + \mathbf{B}[\mathbf{x}(t), t]\mathbf{u}(t) \tag{6-322}$$

*Assume that the components* $u_1(t), u_2(t), \ldots, u_r(t)$ *of* $\mathbf{u}(t)$ *are constrained in magnitude by the relation*

$$\left.\begin{aligned} &|u_j(t)| \leq 1 \qquad j = 1, 2, \ldots, r \\ &\text{or, more compactly, by} \\ &\mathbf{u}(t) \in \Omega \end{aligned}\right\} \tag{6-323}$$

*Given a smooth target set* $S$ *defined by Eq.* (6-37) (*see Problem* 6-1*b*). *Suppose that the system* (6-322) *and the target set* $S$ *satisfy the assumptions stated in Problem* 6-1*b*.

*Let* $t_0$ *be a given initial time, and let* $\mathbf{x}(t_0)$ *be a given initial state of the system* (6-322).

*Given the* (*fuel*) *cost functional*

$$J(\mathbf{u}) = \int_{t_0}^{T} \left\{ \sum_{j=1}^{r} c_j |u_j(t)| \right\} dt \qquad c_j > 0 \tag{6-324}$$

*determine the control* $\mathbf{u}(t)$ *that:*

1. *Satisfies the constraints* (6-323)
2. *Forces the system from the state* $\mathbf{x}(t_0)$ *to the target* $S$
3. *Minimizes the fuel functional* $J(\mathbf{u})$ *of Eq.* (6-324)
   i. *If* $T$ *is free*
   ii. *If* $T = T_f$ *is fixed*

We shall now proceed to derive the necessary conditions. We shall first consider the free-time case, i.e., the case in which $T$ is not specified. The appropriate theorem is Theorem 5-9, which provides us with the necessary conditions for a free-time problem to a target set.

First of all, we form the Hamiltonian function for the system (6-322) and the cost functional (6-324); the Hamiltonian is given by the equation

$$H[\mathbf{x}(t), \mathbf{p}(t), \mathbf{u}(t), t] = \sum_{j=1}^{r} c_j |u_j(t)| + \langle \mathbf{f}[\mathbf{x}(t), t], \mathbf{p}(t) \rangle + \langle \mathbf{B}[\mathbf{x}(t), t]\mathbf{u}(t), \mathbf{p}(t) \rangle \tag{6-325}$$

or, equivalently, by the equation

$$\begin{aligned} &H[\mathbf{x}(t), \mathbf{p}(t), \mathbf{u}(t), t] \\ &\quad = \sum_{j=1}^{r} c_j |u_j(t)| + \sum_{i=1}^{n} f_i[\mathbf{x}(t), t] p_i(t) + \sum_{j=1}^{r} \sum_{i=1}^{n} b_{ij}[\mathbf{x}(t), t] u_j(t) p_i(t) \end{aligned} \tag{6-326}$$

We observe that the Hamiltonian for the minimum-fuel problem is *linear* in $u_j(t)$ and $|u_j(t)|$ [for the minimum-time problem, $H$ was linear in $u_j(t)$ only].

Let us assume that $\mathbf{u}^*(t)$ is a fuel-optimal control, that $\mathbf{x}^*(t)$ is the resultant fuel-optimal trajectory, and that $\hat{T}$ is the first time such that

$$\mathbf{x}^*(\hat{T}) \in S \tag{6-327}$$

We use $\hat{T}$, rather than $T^*$, to indicate the time of arrival at $S$, along the fuel-optimal trajectory; we shall retain the notation $T^*$ to indicate the *minimum* possible time of arrival at $S$. It should be clear from the statements of Problems 6-2*a* and 6-1*b* that $\hat{T}$ must satisfy the inequality

$$\hat{T} \geq T^* \tag{6-328}$$

We shall now use the statement of Theorem 5-9 to derive the necessary conditions for our fuel-optimal problem. Statement 1 of Theorem 5-9 guarantees the existence of a costate $\mathbf{p}^*(t)$ corresponding to $\mathbf{u}^*(t)$ and $\mathbf{x}^*(t)$. Now the Hamiltonian (6-326) evaluated at $\mathbf{x}^*(t)$, $\mathbf{p}^*(t)$, and $\mathbf{u}^*(t)$ is

$$H[\mathbf{x}^*(t), \mathbf{p}^*(t), \mathbf{u}^*(t), t] = \sum_{j=1}^{r} c_j|u_j^*(t)| + \sum_{i=1}^{n} f_i[\mathbf{x}^*(t), t]p_i^*(t) + \sum_{j=1}^{r}\sum_{i=1}^{n} b_{ij}[\mathbf{x}^*(t), t]u_j^*(t)p_i^*(t) \tag{6-329}$$

Since $\dot{\mathbf{x}}^*(t) = \partial H/\partial \mathbf{p}^*(t)$ and $\dot{\mathbf{p}}^*(t) = -\partial H/\partial \mathbf{x}^*(t)$, we conclude that $x_k^*(t)$ and $p_k^*(t)$, $k = 1, 2, \ldots, n$, must satisfy the canonical equations

$$\dot{x}_k^*(t) = f_k[\mathbf{x}^*(t), t] + \sum_{j=1}^{r} b_{kj}[\mathbf{x}^*(t), t]u_j^*(t) \tag{6-330}$$

$$\dot{p}_k^*(t) = -\sum_{i=1}^{n} \left\{\frac{\partial f_i[\mathbf{x}^*(t), t]}{\partial x_k^*(t)}\right\} p_i^*(t) - \sum_{j=1}^{r} u_j^*(t) \sum_{i=1}^{n} \left\{\frac{\partial b_{ij}[\mathbf{x}^*(t), t]}{\partial x_k^*(t)}\right\} p_i^*(t) \tag{6-331}$$

for $k = 1, 2, \ldots, n$. The reader should note that the canonical equations (6-330) and (6-331) for this fuel-optimal problem are identical to the canonical equations (6-53) and (6-54) for the time-optimal problem. The reason is that the difference between the Hamiltonians (6-329) and (6-52) is a function of $\mathbf{u}^*(t)$ alone.

Statement 2 of Theorem 5-9 is

$$H[\mathbf{x}^*(t), \mathbf{p}^*(t), \mathbf{u}^*(t), t] \leq H[\mathbf{x}^*(t), \mathbf{p}^*(t), \mathbf{u}(t), t] \tag{6-332}$$

for all $\mathbf{u}(t) \in \Omega$ and $t \in [t_0, \hat{T}]$. From Eq. (6-329), we then deduce that Eq. (6-332) reduces to the relation

$$\begin{aligned}\sum_{j=1}^{r} c_j|u_j^*(t)| &+ \sum_{i=1}^{n} f_i[\mathbf{x}^*(t), t]p_i^*(t) + \sum_{j=1}^{r} u_j^*(t)\left\{\sum_{i=1}^{n} b_{ij}[\mathbf{x}^*(t), t]p_i^*(t)\right\} \\ &\leq \sum_{j=1}^{r} c_j|u_j(t)| + \sum_{i=1}^{n} f_i[\mathbf{x}^*(t), t]p_i^*(t) + \sum_{j=1}^{r} u_j(t)\left\{\sum_{i=1}^{n} b_{ij}[\mathbf{x}^*(t), t]p_i^*(t)\right\}\end{aligned} \tag{6-333}$$

which, in turn, yields the relation

$$\sum_{j=1}^{r} c_j|u_j^*(t)| + \sum_{j=1}^{r} u_j^*(t) \left\{ \sum_{i=1}^{n} b_{ij}[\mathbf{x}^*(t), t]p_i^*(t) \right\}$$
$$\leq \sum_{j=1}^{r} c_j|u_j(t)| + \sum_{j=1}^{r} u_j(t) \left\{ \sum_{i=1}^{n} b_{ij}[\mathbf{x}^*(t), t]p_i^*(t) \right\} \quad (6\text{-}334)$$

for all $\mathbf{u}(t) \in \Omega$ and $t \in [t_0, \hat{T}]$.

For the fuel-optimal problem, statement 3 of Theorem 5-9 becomes

$$\sum_{j=1}^{r} c_j|u_j^*(\hat{T})| + \sum_{i=1}^{n} f_i[\mathbf{x}^*(\hat{T}), \hat{T}]p_i^*(\hat{T}) + \sum_{j=1}^{r} u_j^*(\hat{T}) \left\{ \sum_{i=1}^{n} b_{ij}[\mathbf{x}^*(\hat{T}), \hat{T}]p_i^*(\hat{T}) \right\}$$
$$= \sum_{\alpha=1}^{n-\beta} e_\alpha \frac{\partial g_\alpha}{\partial t} [\mathbf{x}^*(\hat{T}), \hat{T}] \quad (6\text{-}335)$$

where the $e_\alpha$ are constants and $g_\alpha[\mathbf{x}, t] = 0$, $\alpha = 1, 2, \ldots, n - \beta$, are the equations of the target set $S$.

Finally, statement 4 of Theorem 5-9 implies that

$$\mathbf{p}^*(\hat{T}) = \sum_{\alpha=1}^{n-\beta} k_\alpha \mathbf{h}_\alpha[\mathbf{x}^*(\hat{T}), \hat{T}] \quad (6\text{-}336)$$

where $k_1, k_2, \ldots k_{n-\beta}$ are some arbitrary constants and

$$\mathbf{h}_\alpha[\mathbf{x}, \hat{T}] = \frac{\partial g_\alpha[\mathbf{x}, \hat{T}]}{\partial \mathbf{x}} \qquad \alpha = 1, 2, \ldots, n - \beta \quad (6\text{-}337)$$

The reader should note that Eq. (6-336) is identical to Eq. (6-66), which was derived for the minimum-time problem, except that $\hat{T}$ replaces $T^*$.

Up to now, we have simply repeated the logical development of step 1 of Sec. 6-3 for the minimum-fuel problem with unspecified terminal time $T$.

Let us now suppose that the terminal time has been specified a priori to be equal to $T_f$. In this case, we can use Table 5-1, row 11, to determine the necessary conditions for the fixed-time fuel-optimal problem. We leave it to the reader to verify that Eqs. (6-330), (6-331), and (6-334) remain unchanged and that Eq. (6-336) becomes

$$\mathbf{p}^*(T_f) = \sum_{\alpha=1}^{n-\beta} k_\alpha \mathbf{h}_\alpha[\mathbf{x}^*(T_f), T_f] \quad (6\text{-}338)$$

In order for the problem to have a solution, it is necessary that the specified time $T_f$ be greater than or equal to the minimum time $T^*$, that is, that $T_f$ satisfy the inequality

$$T_f \geq T^* \quad (6\text{-}339)$$

We summarize the necessary conditions in the following lemma:

***Lemma 6-7 Necessary Conditions for Problem 6-2a*** *If a fuel-optimal control* $\mathbf{u}^*(t)$ *exists, then the resultant fuel-optimal trajectory* $\mathbf{x}^*(t)$ *and the corresponding costate* $\mathbf{p}^*(t)$ *must satisfy the following conditions:*

*a. If $T$ is free, then the necessary conditions are given by Eqs.* (6-330), (6-331), *and* (6-334) *to* (6-336).

*b. If $T = T_f$ is fixed, then the necessary conditions are given by Eqs.* (6-330), (6-331), (6-334), *and* (6-338).

Let us now examine Eq. (6-334) in more detail; using this equation, we shall (as we have done in step 2 of Sec. 6-3) derive an equation which relates the fuel-optimal control $\mathbf{u}^*(t)$ to the fuel-optimal trajectory $\mathbf{x}^*(t)$ and the corresponding costate $\mathbf{p}^*(t)$. We define functions $q_1^*(t), q_2^*(t), \ldots, q_r^*(t)$ by the relations

$$q_j^*(t) = \sum_{i=1}^{n} b_{ij}[\mathbf{x}^*(t), t]p_i^*(t) \qquad j = 1, 2, \ldots, r \tag{6-340}$$

Thus, the functions $q_j^*(t)$ are the components of the $r$ vector $\mathbf{q}^*(t)$ given by

$$\mathbf{q}^*(t) = \mathbf{B}'[\mathbf{x}^*(t), t]\mathbf{p}^*(t) \tag{6-341}$$†

Using Eq. (6-340), we can see that Eq. (6-334) reduces to the relation

$$\sum_{j=1}^{r} c_j \left\{|u_j^*(t)| + u_j^*(t)\,\frac{q_j^*(t)}{c_j}\right\} \le \sum_{j=1}^{r} c_j \left\{|u_j(t)| + u_j(t)\,\frac{q_j^*(t)}{c_j}\right\} \tag{6-342}$$

for all $|u_j(t)| \le 1$, $j = 1, 2, \ldots, r$, and every $t \in [t_0, \hat{T}]$ or $[t_0, T_f]$. Equation (6-342) means that the function

$$\psi[\mathbf{u}(t)] = \sum_{j=1}^{r} c_j \left\{|u_j(t)| + u_j(t)\,\frac{q_j^*(t)}{c_j}\right\} \tag{6-343}$$

attains its absolute minimum at

$$u_j(t) = u_j^*(t) \tag{6-344}$$

Using Remark 6-6, we find that

$$\begin{aligned}\min_{\mathbf{u}(t)\in\Omega} \psi[\mathbf{u}(t)] &= \min_{\mathbf{u}(t)\in\Omega} \sum_{j=1}^{r} c_j \left\{|u_j(t)| + u_j(t)\,\frac{q_j^*(t)}{c_j}\right\} \\ &= \sum_{j=1}^{r} c_j \left[\min_{|u_j(t)|\le 1} \left\{|u_j(t)| + u_j(t)\,\frac{q_j^*(t)}{c_j}\right\}\right]\end{aligned}$$

† The reader should note that the relations (6-340) and (6-341) are identical to Eqs. (6-67) and (6-68); of course, in Eqs. (6-340) and (6-341) the * quantities refer to fuel-optimal paths, while in Eqs. (6-67) and (6-68) they refer to time-optimal paths.

We claim that

$$\min_{|u_j(t)|\le 1}\left\{|u_j(t)| + u_j(t)\,\frac{q_j^*(t)}{c_j}\right\} = \begin{cases} 0 & \text{if } \left|\dfrac{q_j^*(t)}{c_j}\right| < 1 \\ 1 - \left|\dfrac{q_j^*(t)}{c_j}\right| & \text{if } \left|\dfrac{q_j^*(t)}{c_j}\right| \ge 1 \end{cases} \tag{6-345}$$

Since the minimum is attained when $u_j(t) = u_j^*(t)$, we also conclude that $u_j^*(t)$ is related to $q_j^*(t)$ [and, hence, to $\mathbf{x}^*(t)$ and $\mathbf{p}^*(t)$ by Eq. (6-340)] as follows:

$$\begin{aligned} u_j^*(t) &= 0 && \text{if } -1 < \frac{q_j^*(t)}{c_j} < 1 \\ u_j^*(t) &= +1 && \text{if } \quad \frac{q_j^*(t)}{c_j} < -1 \\ u_j^*(t) &= -1 && \text{if } \quad \frac{q_j^*(t)}{c_j} > +1 \\ 0 \le u_j^*(t) &\le +1 && \text{if } \quad \frac{q_j^*(t)}{c_j} = -1 \\ -1 \le u_j^*(t) &\le 0 && \text{if } \quad \frac{q_j^*(t)}{c_j} = +1 \end{aligned} \tag{6-346}$$

In order to write Eqs. (6-346) in a compact form, we define the *dead-zone function*, denoted by dez { }, as follows:

$$a = \operatorname{dez}\{b\}$$

means that

$$\begin{aligned} a &= 0 && \text{if } |b| < 1 \\ a &= \operatorname{sgn}\{b\} && \text{if } |b| > 1 \\ 0 \le a &\le 1 && \text{if } b = +1 \\ -1 \le a &\le 0 && \text{if } b = -1 \end{aligned} \tag{6-347}$$

Then Eqs. (6-346) can be written in the compact form

$$u_j^*(t) = -\operatorname{dez}\left\{\frac{q_j^*(t)}{c_j}\right\} = -\operatorname{dez}\left\{\frac{1}{c_j}\sum_{i=1}^{n} b_{ij}[\mathbf{x}^*(t), t]p_i^*(t)\right\} \tag{6-348}$$

The formal proof that the control (6-346) indeed minimizes the function $\psi[\mathbf{u}(t)]$ is quite lengthy and does not contribute much to the understanding of the problem. The best way of illustrating Eqs. (6-345) and (6-346) is to plot the possible values of the function $\{|u_j(t)| + u_j(t)[q_j^*(t)/c_j]\}$ for all $u_j(t)$ that satisfy the magnitude constraint $|u_j(t)| \le 1$; this is done in Fig. 6-15. Depending on the value of $u_j(t)$, the value of the function $\{|u_j(t)| + u_j(t)[q_j^*(t)/c_j]\}$ versus $q_j^*(t)/c_j$ will belong to the shaded area. As indicated in Fig. 6-15, the control defined by Eqs. (6-346) yields the minimum value given by Eq. (6-345). On the other hand, the control $u_j(t) = \operatorname{sgn}\{q_j^*(t)/c_j\}$ *maximizes* $\psi[\mathbf{u}(t)]$.

We observe that (6-346) uniquely specifies the magnitude and the polarity of the fuel-optimal control $u_j^*(t)$ in terms of $\mathbf{x}^*(t)$ and $\mathbf{p}^*(t)$ provided that $|q_j^*(t)/c_j| \neq 1$. If $|q_j^*(t)/c_j| = 1$, then we can specify the polarity but *not* the magnitude of the fuel-optimal control. These considerations lead to

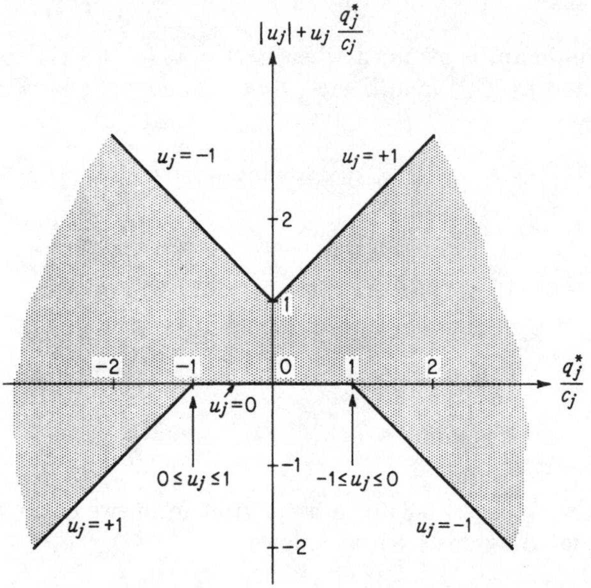


**Fig. 6-15** Plot of the function $|u_j| + u_j(q_j^*/c_j)$ versus $q_j^*/c_j$ for $|u_j| \leq 1$.

the definition of the *normal* fuel-optimal problem and of the *singular* fuel-optimal problem. In particular, we have:

***Definition 6-11 Normal Fuel-optimal Problem*** *Suppose that in the interval* $[t_0, \hat{T}]$ *for the free-time case or in the interval* $[t_0, T_f]$ *for the fixed-time case, there is a countable set of times* $\tau_{1j}, \tau_{2j}, \tau_{3j}, \ldots$, *that is,*

$$\tau_{\gamma j} \in [t_0, \hat{T}] \quad or \quad \tau_{\gamma j} \in [t_0, T_f] \qquad \gamma = 1, 2, \ldots ; j = 1, 2, \ldots, r \tag{6-349}$$

*such that*

$$\left| \frac{q_j^*(t)}{c_j} \right| = \left| \frac{1}{c_j} \sum_{i=1}^{n} b_{ij}[\mathbf{x}^*(t), t] p_i^*(t) \right| = 1$$
$$\textit{if and only if } t = \tau_{\gamma j} \textit{ for all } j = 1, 2, \ldots, r \tag{6-350}$$

*Then we shall say that the fuel-optimal problem is normal. We shall call the times* $\tau_{\gamma j}$ *the switch times.*

Figure 6-16 illustrates a function $q_j^*(t)/c_j$ and the corresponding control $u_j^*(t)$, defined by the equation $u_j^*(t) = -\text{dez}\,\{q_j^*(t)/c_j\}$. We observe that the function $q_j^*(t)/c_j$ is equal to $+1$ or $-1$ only at six isolated instants of time. In this case, the fuel-optimal control is a *piecewise constant* function of time, and its values are $+1$, 0, or $-1$. If all the functions $q_j^*(t)/c_j$, $j = 1, 2, \ldots, r$, had the same characteristic as the one shown in Fig. 6-16, then the fuel-optimal problem would be normal. It is common terminology to call the piecewise constant controls $u_j^*(t)$ *three-level* controls, or *on-off* controls, to indicate that they switch between the three values $+1$, 0, and $-1$. The number of switchings of the control shown in Fig. 6-16 is precisely six.

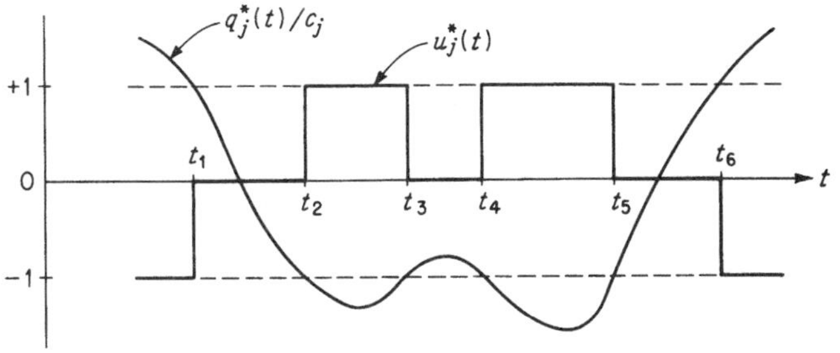


**Fig. 6-16** The function $q_j^*(t)/c_j$ generates the control $u_j^*(t) = -\text{dez}\,\{q_j^*(t)/c_j\}$, which is a three-level piecewise constant function of time.

The reader has probably guessed that the singular fuel-optimal problem arises when $|q_j^*(t)/c_j| = 1$ over a finite interval of time; this is indeed the case, and the precise definition of the singular problem is:

***Definition 6-12 Singular Fuel-optimal Problem*** *Suppose that in the interval* $[t_0, \hat{T}]$ *or in the interval* $[t_0, T_f]$ *there are one or more subintervals* $[T_1, T_2]_j$ *such that*

$$\left|\frac{q_j^*(t)}{c_j}\right| = \left|\frac{1}{c_j}\sum_{i=1}^{n} b_{ij}[\mathbf{x}^*(t), t]p_i^*(t)\right| = 1 \qquad \textit{for all } t \in [T_1, T_2]_j \tag{6-351}$$

*Then we say that we have a singular fuel-optimal problem. We shall call the intervals* $[T_1, T_2]_j$ *singularity intervals.*

The function $q_j^*(t)/c_j$ shown in Fig. 6-17 corresponds to a singular fuel-optimal problem. There are two singularity intervals: $[\tau_2, \tau_3]$ and $[\tau_5, \tau_6]$. With the exception of the switch times $\tau_1$, $\tau_4$, and $\tau_7$, the control $u_j^*(t)$ is uniquely defined by $q_j^*(t)/c_j$ as long as $t \notin [\tau_2, \tau_3]$ and $t \notin [\tau_5, \tau_6]$. For $t \in [\tau_2, \tau_3]$, we know that $u_j^*(t)$ is nonpositive, and for $t \in [\tau_5, \tau_6]$, we know that $u_j^*(t)$ is nonnegative. However, the necessary conditions provided by the minimum principle do not specify $u_j^*(t)$ in terms of $\mathbf{x}^*(t)$ and $\mathbf{p}^*(t)$ during the singularity intervals.

As in Sec. 6-3, we wish to emphasize that these statements do *not* imply that the fuel-optimal control $u_j^*(t)$ does not exist. They simply mean that the necessary condition $H[\mathbf{x}^*(t), \mathbf{p}^*(t), \mathbf{u}^*(t), t] \le H[\mathbf{x}^*(t), \mathbf{p}^*(t), \mathbf{u}(t), t]$ does not result in a unique relation between the fuel-optimal control $\mathbf{u}^*(t)$ and the fuel-optimal trajectory $\mathbf{x}^*(t)$ and the corresponding costate $\mathbf{p}^*(t)$.

We shall discuss fuel-optimal singular problems in Sec. 6-22.

We can now state the following theorem (which is the counterpart of Theorem 6-2) for the normal fuel-optimal problem:

***Theorem 6-11 The On-Off Principle*** *Let $\mathbf{u}^*(t)$ be a fuel-optimal control for Problem 6-2a, and let $\mathbf{x}^*(t)$ and $\mathbf{p}^*(t)$ be the corresponding state trajectory and costate trajectory. If the problem is normal (see Definition 6-11), then*

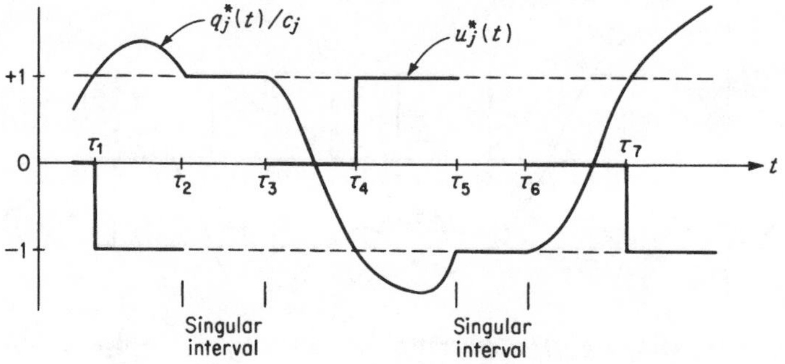


**Fig. 6-17** Illustration ot the singular condition.

*the components $u_1^*(t)$, $u_2^*(t)$, . . . , $u_r^*(t)$ of $\mathbf{u}^*(t)$ must be defined by the relation*

$$u_j^*(t) = -\text{dez}\left\{\frac{1}{c_j}\sum_{i=1}^{n} b_{ij}[\mathbf{x}^*(t), t]p_i^*(t)\right\} \qquad j = 1, 2, \ldots, r \qquad (6\text{-}352)$$

*for $t \in [t_0, T]$ or $t \in [t_0, T_f]$. Equation (6-352) can also be written in the more compact form*

$$\mathbf{u}^*(t) = -\text{DEZ}\,\{\mathbf{C}^{-1}\mathbf{B}'[\mathbf{x}^*(t), t]\mathbf{p}^*(t)\} \qquad (6\text{-}353)\dagger$$

*where*

$$\mathbf{C} = \begin{bmatrix} c_1 & 0 & \cdots & 0 \\ 0 & c_2 & \cdots & 0 \\ \cdots & \cdots & \cdots & \cdots \\ 0 & 0 & \cdots & c_r \end{bmatrix} \qquad (6\text{-}354)$$

*Thus, if the fuel-optimal problem is normal, the components of the fuel-optimal control are piecewise constant functions of time and can switch between the values +1, 0, and −1.*

† The vector-valued function DEZ { } is defined as follows: Let $a_1, a_2, \ldots, a_r$ be the components of the vector $\mathbf{a}$, and let $b_1, b_2, \ldots, b_r$ be the components of the vector $\mathbf{b}$; then

$$\mathbf{a} = \text{DEZ}\,\{\mathbf{b}\}$$

means that $\qquad a_j = \text{dez}\,\{b_j\} \qquad j = 1, 2, \ldots, r$

where the scalar function dez { } is defined in Eqs. (6-347).

If the fuel-optimal problem is normal, we can eliminate the control $\mathbf{u}^*(t)$ from all the necessary conditions stated in Lemma 6-7. In so doing, we would derive reduced necessary conditions for the normal fuel-optimal problem and a theorem which is the counterpart of Theorem 6-3. This we leave as an exercise for the reader.

**Exercise 6-18** Using Theorem 6-11 and Lemma 6-7, derive a theorem stating the reduced (analogous to Theorem 6-3) necessary conditions for the normal fuel-optimal problem.

**Exercise 6-19** Consider the system and the target set described in Example 6-3. Assume that the cost functional to be minimized is $\int_0^T \{|u_1(t)| + |u_2(t)|\}\, dt$, $T$ free. Determine the reduced necessary conditions by assuming that the problem is normal.

Up to now, we have followed the sequence of ideas presented in Sec. 6-3 up to and including Example 6-3. The reader will recall that the remainder of Sec. 6-3 was concerned with the problem of using the necessary conditions to determine the optimal control. We can follow precisely the same reasoning for the fuel-optimal problem. The essential idea is, first, to find the control that absolutely minimizes the Hamiltonian (the $H$-minimal control of Definition 6-6); second, to determine the canonical equations that specify $\dot{\mathbf{x}}(t)$ and $\dot{\mathbf{p}}(t)$; and finally, to determine the initial costate(s) $\mathbf{p}(t_0)$ such that the resulting state, costate, and control satisfy all the necessary conditions. Since these ideas were discussed in great detail in Sec. 6-3, we shall present below only an outline of the essential development.

1. Form the Hamiltonian

$$\begin{aligned} H &= H[\mathbf{x}(t), \mathbf{p}(t), \mathbf{u}(t), t] \\ &= \sum_{j=1}^{r} c_j |u_j(t)| + \langle \mathbf{f}[\mathbf{x}(t), t], \mathbf{p}(t)\rangle + \langle \mathbf{u}(t), \mathbf{B}'[\mathbf{x}(t), t]\mathbf{p}(t)\rangle \end{aligned} \tag{6-355}$$

2. Determine the $H$-minimal control $\mathbf{u}^\circ(t)$ which is given by

$$\mathbf{u}^\circ(t) = -\text{DEZ}\,\{\mathbf{C}^{-1}\mathbf{B}'[\mathbf{x}(t), t]\mathbf{p}(t)\} \tag{6-356}$$

3. Let

$$H^\circ = H^\circ[\mathbf{x}(t), \mathbf{p}(t), t] = H[\mathbf{x}(t), \mathbf{p}(t), \mathbf{u}^\circ(t), t] \tag{6-357}$$

denote the Hamiltonian when $\mathbf{u}(t) = \mathbf{u}^\circ(t)$.

4. Consider the differential equations

$$\dot{\mathbf{x}}(t) = \frac{\partial H^\circ}{\partial \mathbf{p}(t)} \tag{6-358}$$

$$\dot{\mathbf{p}}(t) = -\frac{\partial H^\circ}{\partial \mathbf{x}(t)} \tag{6-359}$$

5. We are given the initial state $\mathbf{x}(t_0)$. Let us guess an initial costate $\mathbf{p}(t_0)$. For the given value of $\mathbf{x}(t_0)$ and the guessed value of $\mathbf{p}(t_0)$, solve

Eqs. (6-358) and (6-359) to obtain the solutions

$$\begin{aligned} \mathbf{x}(t) &= \mathbf{x}[t, t_0, \mathbf{x}(t_0), \mathbf{p}(t_0)] \\ \mathbf{p}(t) &= \mathbf{p}[t, t_0, \mathbf{x}(t_0), \mathbf{p}(t_0)] \end{aligned} \tag{6-360}$$

We can then monitor these solutions and change $\mathbf{p}(t_0)$ until all the necessary conditions are satisfied. As a matter of fact, we can make up a list of questions just as in Sec. 6-3 and construct a logical flow diagram similar to the one given in Fig. 6-7. We leave this task to the reader.

**Exercise 6-20** Formalize the steps involved in the above discussion by describing an "experiment" similar to the one described in step 3*d* of Sec. 6-3. Set up and prove lemmas similar to Lemmas 6-2 and 6-3. Discuss the significance of there being many controls that satisfy the necessary conditions.

*To recapitulate:* We have found that in a normal fuel-optimal problem, the components of the fuel-optimal control are piecewise constant functions of time and can attain (at each instant of time) the value $+1$, or 0, or $-1$. Thus, normal fuel-optimal controls have "coasting" periods ($\mathbf{u} = \mathbf{0}$) and, in this manner, conserve fuel. In the next section, we shall specialize our results to the fuel-optimal control of linear time-invariant systems.

## 6-14 Minimum-fuel Problems 4: Linear Time-invariant Systems

In this section, we shall specialize the results of the previous section to the case of linear time-invariant systems. The precise problem we shall consider will be a *fixed-time* problem; in other words, we shall consider the problem of determining the fuel-optimal control that transfers a *given* initial state $\boldsymbol{\xi}$ to a *given* terminal state $\mathbf{0}$ in a *given* amount of time $T_f$. To motivate the choice of the fixed-time problem, we offer some trivial observations on the free-time problem.

Let us suppose that the dynamical system is described by the differential equation

$$\dot{\mathbf{x}}(t) = \mathbf{A}\mathbf{x}(t) + \mathbf{B}\mathbf{u}(t) \tag{6-361}$$

where $\mathbf{x}(t)$ is the state and $\mathbf{u}(t)$ is the control. Now let us suppose that the initial state is $\mathbf{x}(0) = \boldsymbol{\xi} \neq \mathbf{0}$ and that we apply the control

$$\mathbf{u}(t) = \mathbf{0} \qquad \text{for all } t > 0 \tag{6-362}$$

Then the (unforced) solution $\mathbf{x}^o(t)$ of (6-361) is given by

$$\mathbf{x}^o(t) = e^{\mathbf{A}t}\boldsymbol{\xi} \tag{6-363}$$

Now let us suppose that all the eigenvalues of the matrix $\mathbf{A}$ have negative real parts; this means that the system is stable and that it does not have any integrating capability. In this case, the solution $\mathbf{x}^o(t)$ will approach zero, or more precisely,

$$\lim_{t \to \infty} \mathbf{x}^o(t) = \mathbf{0} \tag{6-364†}$$

It should be clear that if we wish to determine the fuel-optimal control that transfers any initial state $\boldsymbol{\xi}$ to $\mathbf{0}$ for this stable system and if we do not specify the response time, then the fuel-optimal control is $\mathbf{u}^*(t) = \mathbf{0}$ and the response time is infinite. Strictly speaking, the control $\mathbf{u}^*(t) = \mathbf{0}$ for all $t$ is only $\epsilon$-optimal, since it does not transfer $\boldsymbol{\xi}$ to $\mathbf{0}$ in finite time. Nonetheless, it is easy to see that the formulation of such a free-time problem will not lead to a realistic and useful solution. On the other hand, if one or more of the eigenvalues of $\mathbf{A}$ have nonnegative real parts, then the fuel-optimal problem to the origin with unspecified response time is not trivial.

If the fuel-optimal problem is formulated with a specified finite terminal time $T_f$, then the problem can have nontrivial solutions no matter what the eigenvalues of $\mathbf{A}$ are. It is for this reason that the precise statement of Problem 6-2b given below involves a fixed response time.

***Problem 6-2b The Fuel-optimal Regulator Problem for a Linear Time-invariant System with Fixed Response Time*** *Given the dynamical system*

$$\dot{\mathbf{x}}(t) = \mathbf{A}\mathbf{x}(t) + \mathbf{B}\mathbf{u}(t) \tag{6-365}$$

*where*

1. *The n vector* $\mathbf{x}(t)$ *is the state.*
2. *The system matrix* $\mathbf{A}$ *is an* $n \times n$ *constant matrix.*
3. *The gain matrix* $\mathbf{B}$ *is an* $n \times r$ *constant matrix.*
4. *The r vector* $\mathbf{u}(t)$ *is the control.*

*Assume that the system* (6-365) *is completely controllable and that the components* $u_1(t), u_2(t), \ldots, u_r(t)$ *of* $\mathbf{u}(t)$ *are constrained in magnitude by the relations*

$$|u_j(t)| \leq 1 \qquad j = 1, 2, \ldots, r \qquad \text{for all } t \tag{6-366}$$

*Given that at the initial time* $t_0 = 0$, *the initial state of the system* (6-365) *is*

$$\mathbf{x}(0) = \boldsymbol{\xi} \tag{6-367}$$

*Given a terminal state* $\boldsymbol{\theta}$ *(not necessarily* $\mathbf{0}$*); given a real, positive, and finite time* $T_f$ *(greater than the minimum time* $T^*$ *required to force* $\boldsymbol{\xi}$ *to* $\boldsymbol{\theta}$*). Then determine the fuel-optimal control* $\mathbf{u}^*(t)$ *that transfers the system* (6-365) *from* $\boldsymbol{\xi}$ *to* $\boldsymbol{\theta}$ *in time* $T_f$ *and that minimizes the fuel functional*

$$F(\mathbf{u}) = F = \int_0^{T_f} \sum_{j=1}^{r} |u_j(t)|\, dt \tag{6-368‡}$$

In this section, we shall present several theoretical results pertaining to the above problem. We shall proceed in much the same manner as in

† See Sec. 3-26.

‡ We have set the constants $c_j$ equal to 1 [see Eq. (6-324)] to simplify the algebra.

Sec. 6-5 so that the reader will be able to observe the strong similarities and the apparent differences between the two optimal problems. In particular, we shall state the necessary conditions in Theorem 6-12, the sufficient conditions for the fuel-optimal problem to be normal in Theorem 6-13, the uniqueness of the fuel-optimal controls in Theorem 6-14, and the uniqueness of the extremal controls in Theorem 6-15. Thus, our purpose will be to duplicate the development of Sec. 6-5 in deriving the theoretical results for the fuel-optimal problem.

It is very easy to determine the necessary conditions provided by the minimum principle for Problem 6-2*b*. We shall leave it to the reader to verify the following theorem simply by looking over the statement of Lemma 6-7.

***Theorem 6-12 Necessary Conditions for Problem 6-2b*** *Let* $\mathbf{u}^*(t)$ *be a fuel-optimal control that transfers* $\boldsymbol{\xi}$ *to* $\boldsymbol{\theta}$. *Let* $\mathbf{x}^*(t)$ *denote the trajectory of the system* (6-365) *corresponding to* $\mathbf{u}^*(t)$, *originating at* $\boldsymbol{\xi}$ *at* $t_0 = 0$, *and hitting the state* $\boldsymbol{\theta}$ *at the specified time* $T_f$ [*that is,* $\mathbf{x}^*(\mathbf{0}) = \boldsymbol{\xi}$ *and* $\mathbf{x}^*(T_f) = \boldsymbol{\theta}$]. *Then there exists a corresponding costate vector* $\mathbf{p}^*(t)$ *such that:*

*a.* $\mathbf{x}^*(t)$ *and* $\mathbf{p}^*(t)$ *are solutions of the canonical equations*

$$\dot{\mathbf{x}}^*(t) = \mathbf{A}\mathbf{x}^*(t) + \mathbf{B}\mathbf{u}^*(t) \tag{6-369}$$
$$\dot{\mathbf{p}}^*(t) = -\mathbf{A}'\mathbf{p}^*(t) \tag{6-370}$$

*with the boundary conditions*

$$\mathbf{x}^*(0) = \boldsymbol{\xi} \qquad \mathbf{x}^*(T_f) = \boldsymbol{\theta} \tag{6-371}$$

*b. The relation*

$$\sum_{j=1}^{r} |u_j^*(t)| + \langle \mathbf{A}\mathbf{x}^*(t), \mathbf{p}^*(t)\rangle + \langle \mathbf{u}^*(t), \mathbf{B}'\mathbf{p}^*(t)\rangle \leq \sum_{j=1}^{r} |u_j(t)| + \langle \mathbf{A}\mathbf{x}^*(t), \mathbf{p}^*(t)\rangle + \langle \mathbf{u}(t), \mathbf{B}'\mathbf{p}^*(t)\rangle \tag{6-372}$$

*holds for all admissible* $\mathbf{u}(t)$ *and for* $t \in [0, T_f]$. *Equation* (6-372) *yields, in turn, the relation*

$$\mathbf{u}^*(t) = -\operatorname{DEZ}\{\mathbf{q}^*(t)\} = -\operatorname{DEZ}\{\mathbf{B}'\mathbf{p}^*(t)\} \tag{6-373}$$

*or, componentwise,*

$$u_j^*(t) = -\operatorname{dez}\{q_j^*(t)\} = -\operatorname{dez}\{\langle \mathbf{b}_j, \mathbf{p}^*(t)\rangle\} \qquad j = 1, 2, \ldots, r \tag{6-374†}$$

The reader should note that the necessary conditions do not provide us with any explicit information about the boundary values of $\mathbf{p}^*(t)$. In other words, we do not know either $\mathbf{p}^*(0)$ or $\mathbf{p}^*(T_f)$. Furthermore, the reader should note that the necessary conditions do not give us any information

† See Eqs. (6-347) for the definition of the dez { } function.

regarding the value of the Hamiltonian function along the optimal trajectory. Since the system is time-invariant and since the cost functional is not an explicit function of $t$, we know that† the Hamiltonian must be a constant, i.e., that

$$\sum_{j=1}^{r} |u_j^*(t)| + \langle \mathbf{Ax}^*(t), \mathbf{p}^*(t)\rangle + \langle \mathbf{u}^*(t), \mathbf{B}'\mathbf{p}^*(t)\rangle = \gamma = \text{const} \qquad \text{for } t \in [0, T_f] \tag{6-375}$$

but we do *not* know the value of the constant $\gamma$.

We shall now examine the costate vector $\mathbf{p}^*(t)$. We note that $\mathbf{p}^*(t)$ is the solution of the linear homogeneous equation

$$\dot{\mathbf{p}}^*(t) = -\mathbf{A}'\mathbf{p}^*(t) \tag{6-376}$$

which is the same equation as in the time-optimal problem [see Eq. (6-140) in Theorem 6-4]. If we let

$$\boldsymbol{\pi} = \mathbf{p}^*(0) \tag{6-377}$$

denote the initial value of the costate vector, then we know that

$$\mathbf{p}^*(t) = e^{-\mathbf{A}'t}\boldsymbol{\pi} \tag{6-378}$$

Since the fundamental matrix $e^{-\mathbf{A}'t}$ is nonsingular, we can see that the initial condition

$$\boldsymbol{\pi} = \mathbf{0} \tag{6-379}$$

implies that

$$\mathbf{p}^*(t) = \mathbf{0} \qquad \text{for all } t \in [0, T_f] \tag{6-380}$$

which, in turn, implies [in view of Eq. (6-373)] that

$$\mathbf{u}^*(t) = \mathbf{0} \qquad \text{for all } t \in [0, T_f] \tag{6-381}$$

In Sec. 6-5, we deduced (see Theorem 6-6) that a time-optimal problem is normal if and only if the system $\dot{\mathbf{x}}(t) = \mathbf{Ax}(t) + \mathbf{Bu}(t)$ is normal.‡ We shall now derive conditions for the fuel-optimal problem to be normal. As we did in Sec. 6-5, we shall first derive necessary conditions for the fuel-optimal problem to be singular (see Definition 6-12) and translate these into sufficient conditions for the fuel-optimal problem to be normal. The concept of normality will be essential in the derivation of the subsequent uniqueness theorems.

First of all, we have stated in Theorem 6-12 that a fuel-optimal control $\mathbf{u}^*(t)$ must be related to the corresponding costate $\mathbf{p}^*(t)$ by the relation

$$u_j^*(t) = -\text{dez}\,\{\langle \mathbf{b}_j, \mathbf{p}^*(t)\rangle\} \qquad j = 1, 2, \ldots, r \tag{6-382}$$

† See Table 5-1, row 4, column 3.

‡ See Sec. 4-21. Again we remind the reader that the physical significance of a normal system is that every component of the control vector can influence all the state variables.

which means that

$$\begin{aligned} u_j^*(t) &= 0 && \text{if } |\langle \mathbf{b}_j, \mathbf{p}^*(t)\rangle| < 1 \\ u_j^*(t) &= +1 && \text{if } \langle \mathbf{b}_j, \mathbf{p}^*(t)\rangle < -1 \\ u_j^*(t) &= -1 && \text{if } \langle \mathbf{b}_j, \mathbf{p}^*(t)\rangle > +1 \\ 0 \le u_j^*(t) &\le +1 && \text{if } \langle \mathbf{b}_j, \mathbf{p}^*(t)\rangle = -1 \\ -1 \le u_j^*(t) &\le 0 && \text{if } \langle \mathbf{b}_j, \mathbf{p}^*(t)\rangle = +1 \end{aligned} \tag{6-383}$$

For the fuel-optimal problem to be singular, it is necessary that in the control interval $[0, T_f]$, there be at least one subinterval $[t_1, t_2]$ such that for some integer $j$, $|\langle \mathbf{b}_j, \mathbf{p}^*(t)\rangle| = 1$ for all $t \in [t_1, t_2]$. To be specific, let us assume that the relation

$$\langle \mathbf{b}_j, \mathbf{p}^*(t)\rangle = +1 \qquad \text{for all } t \in [t_1, t_2] \tag{6-384}$$

holds, and then let us investigate the implications of Eq. (6-384). Since the function (of time) $\langle \mathbf{b}_j, \mathbf{p}^*(t)\rangle$ is constant, all its time derivatives must be zero. By repeated time differentiation of Eq. (6-384) and by using the fact that $\dot{\mathbf{p}}^*(t) = -\mathbf{A}'\mathbf{p}^*(t)$, we obtain the set of equations

$$\left.\begin{aligned} \langle \mathbf{A}\mathbf{b}_j, \mathbf{p}^*(t)\rangle &= 0 \\ \langle \mathbf{A}^2\mathbf{b}_j, \mathbf{p}^*(t)\rangle &= 0 \\ \cdots&\cdots \\ \langle \mathbf{A}^{n-1}\mathbf{b}_j, \mathbf{p}^*(t)\rangle &= 0 \\ \langle \mathbf{A}^{n}\mathbf{b}_j, \mathbf{p}^*(t)\rangle &= 0 \\ \cdots&\cdots \end{aligned}\right\} \quad \text{for all } t \in [t_1, t_2] \tag{6-385}$$

We pick the first $n$ equations of the set (6-385), and we rewrite them in the following way:

$$\begin{bmatrix} \leftarrow & \mathbf{A}\mathbf{b}_j & \rightarrow \\ \leftarrow & \mathbf{A}^2\mathbf{b}_j & \rightarrow \\ & \cdot & \\ & \cdot & \\ & \cdot & \\ \leftarrow & \mathbf{A}^{n-1}\mathbf{b}_j & \rightarrow \\ \leftarrow & \mathbf{A}^{n}\mathbf{b}_j & \rightarrow \end{bmatrix} \mathbf{p}^*(t) = \mathbf{0} \tag{6-386}$$

Now let us define the $n \times n$ matrix† $\mathbf{G}_j$ as the matrix whose column vectors are $\mathbf{b}_j, \mathbf{A}\mathbf{b}_j, \ldots$, that is,

$$\mathbf{G}_j = [\mathbf{b}_j \mid \mathbf{A}\mathbf{b}_j \mid \cdots \mid \mathbf{A}^{n-1}\mathbf{b}_j] \tag{6-387}$$

Then Eq. (6-386) reduces to the equation

$$\mathbf{G}_j'\mathbf{A}'\mathbf{p}^*(t) = \mathbf{0} \qquad \text{for all } t \in [t_1, t_2] \tag{6-388}$$

† Note that this is the same matrix as the one used in the time-optimal problem [see Eq. (6-158)].

But Eq. (6-384) implies that $\mathbf{p}^*(t) \not\equiv \mathbf{0}$; thus, *for Eq.* (6-388) *to hold, it is necessary that the matrix* $\mathbf{G}_j'\mathbf{A}'$ *be a singular matrix.* Thus, we must have the relation

$$\det (\mathbf{G}_j'\mathbf{A}') = (\det \mathbf{A}) (\det \mathbf{G}_j) = 0 \tag{6-389}$$

Now, Eq. (6-389) can hold either if

$$\det \mathbf{A} = 0 \tag{6-390}$$

and/or if

$$\det \mathbf{G}_j = 0 \tag{6-391}$$

If $\det \mathbf{G}_j = 0$, then $\mathbf{G}_j$ is singular; this means that the system $\dot{\mathbf{x}}(t) = \mathbf{A}\mathbf{x}(t) + \mathbf{B}\mathbf{u}(t)$ is not a normal system (see Definition 4-15). Thus, *if the linear system* $\dot{\mathbf{x}}(t) = \mathbf{A}\mathbf{x}(t) + \mathbf{B}\mathbf{u}(t)$ *is not normal, then the fuel-optimal problem is necessarily singular.*

Now let us suppose that the system is normal, so that $\det \mathbf{G}_j \not\equiv 0$ for all $j$. It is still possible to have a singular fuel-optimal problem if $\det \mathbf{A} = 0$, that is, if the system matrix $\mathbf{A}$ is a singular matrix. Now, $\mathbf{A}$ can be a singular matrix if and only if at least one of the eigenvalues of $\mathbf{A}$ is equal to zero (why?). This means that the system has an integrating capability. We have thus deduced the following theorem (because a nonsingular fuel-optimal problem must be normal):

***Theorem 6-13 Sufficient Conditions for Problem 6-2b to Be Normal or Necessary Conditions for Problem 6-2b to Be Singular*** *A sufficient condition for Problem* 6-2*b to be normal is*

$$\det (\mathbf{G}_j'\mathbf{A}') \not\equiv 0 \qquad \textit{for all } j = 1, 2, \ldots, r \tag{6-392}$$

*where* $\mathbf{G}_j$ *is given by Eq.* (6-387). *On the other hand, for Problem* 6-2*b to be singular, it is necessary that* $\det (\mathbf{G}_j'\mathbf{A}') = 0$ *for some* $j$. *Thus, if the system* $\dot{\mathbf{x}}(t) = \mathbf{A}\mathbf{x}(t) + \mathbf{B}\mathbf{u}(t)$ *is normal and if* $\mathbf{A}$ *is a nonsingular matrix, then Problem* 6-2*b is a normal fuel-optimal problem according to Definition* 6-11.

In general, we cannot prove that Eq. (6-392) is also a necessary condition for the fuel-optimal problem to be normal. As we shall see in Chap. 8, there are systems with the property that the fuel-optimal problem is singular for some initial states but normal for other initial states.

In the time-optimal problem, we demonstrated that if the time-optimal problem is normal, then the time-optimal control is unique (see Theorem 6-7 and Exercise 6-7). This is precisely the case with the minimum-fuel problem. In other words, if the minimum-fuel problem is normal, then the fuel-optimal control (if it exists) is unique. The following theorem states this result in a precise manner. The reader should note that the proof of the theorem proceeds along the same lines as the proof of Theorem 6-7.

***Theorem 6-14 Uniqueness of the Fuel-optimal Control for Problem 6-2b*** *If the fuel-optimal problem* 6-2*b is normal, then the fuel-optimal control* (*if it exists*) *is unique.*

PROOF Let us suppose that $\mathbf{u}^*(t)$ and $\hat{\mathbf{u}}^*(t)$ are two *distinct* fuel-optimal controls transferring the same initial state $\boldsymbol{\xi}$ to $\boldsymbol{\theta}$ in the same time $T_f$. Let $\mathbf{x}^*(t)$ and $\hat{\mathbf{x}}^*(t)$ be the *distinct* trajectories originating at $\boldsymbol{\xi}$. Since both $\mathbf{u}^*(t)$ and $\hat{\mathbf{u}}^*(t)$ are assumed to be fuel-optimal, we have the relation

$$\int_0^{T_f} \sum_{j=1}^{r} |u_j^*(t)|\, dt = \int_0^{T_f} \sum_{j=1}^{r} |\hat{u}_j^*(t)|\, dt \tag{6-393}$$

Now the solutions $\mathbf{x}^*(t)$ and $\hat{\mathbf{x}}^*(t)$ are given by the equations

$$\mathbf{x}^*(t) = e^{\mathbf{A}t}\left[\boldsymbol{\xi} + \int_0^t e^{-\mathbf{A}\tau}\mathbf{B}\mathbf{u}^*(\tau)\, d\tau\right] \tag{6-394}$$

$$\hat{\mathbf{x}}^*(t) = e^{\mathbf{A}t}\left[\boldsymbol{\xi} + \int_0^t e^{-\mathbf{A}\tau}\mathbf{B}\hat{\mathbf{u}}^*(\tau)\, d\tau\right] \tag{6-395}$$

At $t = T_f$, we must have the terminal condition

$$\mathbf{x}^*(T_f) = \hat{\mathbf{x}}^*(T_f) = \boldsymbol{\theta} \tag{6-396}$$

and so we deduce that the (vector) equality

$$\int_0^{T_f} e^{-\mathbf{A}t}\mathbf{B}\mathbf{u}^*(t)\, dt = \int_0^{T_f} e^{-\mathbf{A}t}\mathbf{B}\hat{\mathbf{u}}^*(t)\, dt \tag{6-397}$$

holds. Let $\mathbf{p}^*(t)$ be the costate corresponding to the fuel-optimal control $\mathbf{u}^*(t)$ and trajectory $\mathbf{x}^*(t)$. Let $\boldsymbol{\pi} = \mathbf{p}^*(0)$, so that

$$\mathbf{p}^*(t) = e^{-\mathbf{A}'t}\boldsymbol{\pi} \tag{6-398}$$

From the necessary conditions we know that [see Eq. (6-372)]

$$\begin{aligned}\sum_{j=1}^{r} |u_j^*(t)| &+ \langle \mathbf{A}\mathbf{x}^*(t), \mathbf{p}^*(t)\rangle + \langle \mathbf{B}\mathbf{u}^*(t), \mathbf{p}^*(t)\rangle \\ &\le \sum_{j=1}^{r} |\hat{u}_j^*(t)| + \langle \mathbf{A}\mathbf{x}^*(t), \mathbf{p}^*(t)\rangle + \langle \mathbf{B}\hat{\mathbf{u}}^*(t), \mathbf{p}^*(t)\rangle\end{aligned} \tag{6-399}$$

because Eq. (6-372) holds for all $\mathbf{u}(t)$ and, in particular, for $\hat{\mathbf{u}}^*(t)$. We have assumed that $\mathbf{u}^*(t)$ and $\hat{\mathbf{u}}^*(t)$ are distinct and that the fuel-optimal problem is normal; the normality assumption implies that the inequality in the relation (6-399) is strict whenever

$$\mathbf{u}^*(t) \neq \hat{\mathbf{u}}^*(t) \tag{6-400}$$

For this reason, the relation (6-399) reduces to the strict inequality

$$\sum_{j=1}^{r} |u_j^*(t)| + \langle \mathbf{B}\mathbf{u}^*(t), \mathbf{p}^*(t)\rangle < \sum_{j=1}^{r} |\hat{u}_j^*(t)| + \langle \mathbf{B}\hat{\mathbf{u}}^*(t), \mathbf{p}^*(t)\rangle \tag{6-401}$$

whenever Eq. (6-400) holds.

If we integrate both sides of Eq. (6-401), we find that [for distinct $\mathbf{u}^*(t)$ and $\hat{\mathbf{u}}^*(t)$]

$$\int_0^{T_f} \sum_{j=1}^{r} |u_j^*(t)|\, dt + \int_0^{T_f} \langle \mathbf{B}\mathbf{u}^*(t), \mathbf{p}^*(t)\rangle\, dt$$
$$< \int_0^{T_f} \sum_{j=1}^{r} |\hat{u}_j^*(t)|\, dt + \int_0^{T_f} \langle \mathbf{B}\hat{\mathbf{u}}^*(t), \mathbf{p}^*(t)\rangle\, dt \qquad (6\text{-}402)$$

and that [in view of Eq. (6-393)]

$$\int_0^{T_f} \langle \mathbf{B}\mathbf{u}^*(t), \mathbf{p}^*(t)\rangle\, dt < \int_0^{T_f} \langle \mathbf{B}\hat{\mathbf{u}}^*(t), \mathbf{p}^*(t)\rangle\, dt \qquad (6\text{-}403)$$

Now let us examine Eq. (6-397) and form the scalar product

$$\left\langle \boldsymbol{\pi}, \int_0^{T_f} e^{-\mathbf{A}t}\mathbf{B}\mathbf{u}^*(t)\, dt \right\rangle = \left\langle \boldsymbol{\pi}, \int_0^{T_f} e^{-\mathbf{A}t}\mathbf{B}\hat{\mathbf{u}}^*(t)\, dt \right\rangle \qquad (6\text{-}404)$$

where $\boldsymbol{\pi} = \mathbf{p}^*(0)$. Since $\boldsymbol{\pi}$ is a constant vector and since [from Eq. (6-398)] $\boldsymbol{\pi} = e^{+\mathbf{A}'t}\mathbf{p}^*(t)$, we can see that Eq. (6-404) reduces to the equation

$$\int_0^{T_f} \langle \mathbf{B}\mathbf{u}^*(t), \mathbf{p}^*(t)\rangle\, dt = \int_0^{T_f} \langle \mathbf{B}\hat{\mathbf{u}}^*(t), \mathbf{p}^*(t)\rangle\, dt \qquad (6\text{-}405)$$

Thus, the assumption that $\mathbf{u}^*$ and $\hat{\mathbf{u}}^*$ are distinct and fuel-optimal has led to the two relations (6-403) and (6-405), which are contradictory. It follows that $\mathbf{u}^*(t) = \hat{\mathbf{u}}^*(t)$ for all $t \in [0, T_f]$ (except possibly at the switch times), and so the fuel-optimal control is unique.

The reader should note that it was the assumption that the problem is normal which allowed us to deduce the strict inequality (6-402); if the problem were not normal, then the relation (6-402) would not be a strict inequality, even though $\mathbf{u}^*$ and $\hat{\mathbf{u}}^*$ are distinct.

*Let us now recapitulate:* In this section, we stated a fixed-time fuel-optimal problem (Problem 6-2*b*), we stated the necessary conditions in Theorem 6-12, we derived the sufficient conditions for the fuel-optimal problem to be normal (Theorem 6-13), and we used the normality to derive Theorem 6-14.

It would be nice to know whether or not we are dealing with a problem whose solution is unique (if it exists). However, from the point of view of determining the fuel-optimal control using the necessary conditions, it would be still more useful to know whether or not there is one and only one control that satisfies all the necessary conditions, i.e., a unique extremal control. We discussed the importance of unique extremal controls in the case of time-optimal problems. We shall reiterate these ideas in order to motivate the uniqueness theorem which we shall prove later in this section.

Let us assume that the fuel-optimal control is unique. This, of course, means that the fuel surface has a well-defined *absolute* minimum. However, we do not know whether or not there are any relative minima. If there are many controls which satisfy all of the necessary conditions given in Theorem 6-12, then there may be relative (local) minima in addition to the absolute minimum. In this case, to distinguish the control that corresponds to the absolute minimum, we must compute the cost (fuel) required by every control that satisfies the necessary conditions and compare the costs; naturally, the control with the least cost (fuel) is indeed the required optimum.

The following theorem states that the assumptions which guarantee the uniqueness of the fuel-optimal control also guarantee that there is one and only one control that satisfies the necessary conditions. Clearly, knowledge of such a theorem (coupled with Theorem 6-12) will be of immense help in the long process required to determine the optimal control.

We shall now state the uniqueness theorem in a precise manner. In the statement of the theorem, we shall include for the sake of completeness the necessary conditions.

***Theorem 6-15 Uniqueness of the Extremal Controls for Problem 6-2b*** *Let $T_f$ be a given positive time, let $\boldsymbol{\xi}$ be a given initial state, and let $\boldsymbol{\theta}$ be a given terminal state, not necessarily equal to* $\mathbf{0}$. *Suppose that $\hat{\mathbf{u}}(t)$ and $\tilde{\mathbf{u}}(t)$, $t \in [0, T_f]$, are two admissible extremal controls. More precisely, suppose that:*

*a. The solution $\hat{\mathbf{x}}(t)$ of the system*

$$\dot{\mathbf{x}}(t) = \mathbf{A}\mathbf{x}(t) + \mathbf{B}\hat{\mathbf{u}}(t) \qquad \hat{\mathbf{x}}(0) = \boldsymbol{\xi} \tag{6-406}$$

*and the solution $\tilde{\mathbf{x}}(t)$ of the system*

$$\dot{\mathbf{x}}(t) = \mathbf{A}\mathbf{x}(t) + \mathbf{B}\tilde{\mathbf{u}}(t) \qquad \tilde{\mathbf{x}}(0) = \boldsymbol{\xi} \tag{6-407}$$

*are such that*

$$\hat{\mathbf{x}}(T_f) = \tilde{\mathbf{x}}(T_f) = \boldsymbol{\theta} \tag{6-408}$$

*b. There are a costate $\hat{\mathbf{p}}(t)$ corresponding to $\hat{\mathbf{u}}(t)$ and $\hat{\mathbf{x}}(t)$ and a costate $\tilde{\mathbf{p}}(t)$ corresponding to $\tilde{\mathbf{u}}(t)$ and $\tilde{\mathbf{x}}(t)$ such that*

$$\dot{\hat{\mathbf{p}}}(t) = -\mathbf{A}'\hat{\mathbf{p}}(t) \qquad \hat{\mathbf{p}}(0) = \hat{\boldsymbol{\pi}} \ (\textit{unknown}) \tag{6-409}$$

$$\dot{\tilde{\mathbf{p}}}(t) = -\mathbf{A}'\tilde{\mathbf{p}}(t) \qquad \tilde{\mathbf{p}}(0) = \tilde{\boldsymbol{\pi}} \ (\textit{unknown}) \tag{6-410}$$

*so that*

$$\hat{\mathbf{p}}(t) = e^{-\mathbf{A}'t}\hat{\boldsymbol{\pi}} \tag{6-411}$$

*and*

$$\tilde{\mathbf{p}}(t) = e^{-\mathbf{A}'t}\tilde{\boldsymbol{\pi}} \tag{6-412}$$

*c. The following relations hold (why?) for all $\mathbf{u}(t) \in \Omega$ and $t \in [0, T_f]$:*

$$\sum_{j=1}^{r} |\hat{u}_j(t)| + \langle \mathbf{B}\hat{\mathbf{u}}(t), \hat{\mathbf{p}}(t)\rangle \le \sum_{j=1}^{r} |u_j(t)| + \langle \mathbf{B}\mathbf{u}(t), \hat{\mathbf{p}}(t)\rangle \tag{6-413}$$

$$\sum_{j=1}^{r} |\tilde{u}_j(t)| + \langle \mathbf{B}\tilde{\mathbf{u}}(t), \tilde{\mathbf{p}}(t)\rangle \le \sum_{j=1}^{r} |u_j(t)| + \langle \mathbf{B}\mathbf{u}(t), \tilde{\mathbf{p}}(t)\rangle \tag{6-414}$$

*If the fuel-optimal problem is normal (see Definition* 6-11), *then*

$$\hat{\mathbf{u}}(t) = \tilde{\mathbf{u}}(t) \qquad \text{for all } t \in [0, T_f] \tag{6-415†}$$

*which means that the extremal control is unique.*

PROOF From Eqs. (6-406) to (6-408), we find that

$$\boldsymbol{\theta} = \hat{\mathbf{x}}(T_f) = e^{\mathbf{A}T_f}\left[\boldsymbol{\xi} + \int_0^{T_f} e^{-\mathbf{A}t}\mathbf{B}\hat{\mathbf{u}}(t)\,dt\right] \tag{6-416}$$

and that

$$\boldsymbol{\theta} = \tilde{\mathbf{x}}(T_f) = e^{\mathbf{A}T_f}\left[\boldsymbol{\xi} + \int_0^{T_f} e^{-\mathbf{A}t}\mathbf{B}\tilde{\mathbf{u}}(t)\,dt\right] \tag{6-417}$$

Since the matrix $e^{\mathbf{A}T_f}$ is nonsingular, we deduce the (vector) equality

$$\int_0^{T_f} e^{-\mathbf{A}t}\mathbf{B}\hat{\mathbf{u}}(t)\,dt = \int_0^{T_f} e^{-\mathbf{A}t}\mathbf{B}\tilde{\mathbf{u}}(t)\,dt \tag{6-418}$$

Now we form the scalar product of both sides of Eq. (6-418), first with the constant vector $\hat{\boldsymbol{\pi}}$ and then with the constant vector $\tilde{\boldsymbol{\pi}}$, to derive (how?) the two relations

$$\int_0^{T_f} \langle\hat{\boldsymbol{\pi}}, e^{-\mathbf{A}t}\mathbf{B}\hat{\mathbf{u}}(t)\rangle\,dt = \int_0^{T_f} \langle\hat{\boldsymbol{\pi}}, e^{-\mathbf{A}t}\mathbf{B}\tilde{\mathbf{u}}(t)\rangle\,dt \tag{6-419}$$

$$\int_0^{T_f} \langle\tilde{\boldsymbol{\pi}}, e^{-\mathbf{A}t}\mathbf{B}\hat{\mathbf{u}}(t)\rangle\,dt = \int_0^{T_f} \langle\tilde{\boldsymbol{\pi}}, e^{-\mathbf{A}t}\mathbf{B}\tilde{\mathbf{u}}(t)\rangle\,dt \tag{6-420}$$

But, in view of Eqs. (6-411) and (6-412), we see that Eqs. (6-419) and (6-420) reduce to the relations

$$\int_0^{T_f} \langle\hat{\mathbf{p}}(t), \mathbf{B}\hat{\mathbf{u}}(t)\rangle\,dt = \int_0^{T_f} \langle\hat{\mathbf{p}}(t), \mathbf{B}\tilde{\mathbf{u}}(t)\rangle\,dt \tag{6-421}$$

$$\int_0^{T_f} \langle\tilde{\mathbf{p}}(t), \mathbf{B}\hat{\mathbf{u}}(t)\rangle\,dt = \int_0^{T_f} \langle\tilde{\mathbf{p}}(t), \mathbf{B}\tilde{\mathbf{u}}(t)\rangle\,dt \tag{6-422}$$

Now let us suppose that $\hat{\mathbf{u}}$ and $\tilde{\mathbf{u}}$ are distinct. If the two extremal controls are distinct and if the fuel-optimal problem is normal, then from Eq. (6-413) we deduce that whenever $\hat{\mathbf{u}}(t) \neq \tilde{\mathbf{u}}(t)$, the relation

$$\sum_{j=1}^{r} |\hat{u}_j(t)| + \langle\mathbf{B}\hat{\mathbf{u}}(t), \hat{\mathbf{p}}(t)\rangle < \sum_{j=1}^{r} |\tilde{u}_j(t)| + \langle\mathbf{B}\tilde{\mathbf{u}}(t), \hat{\mathbf{p}}(t)\rangle \tag{6-423}$$

must hold, because Eq. (6-413) holds for all admissible $\mathbf{u}(t)$ and, in particular, for $\mathbf{u}(t) = \tilde{\mathbf{u}}(t)$. Using identical reasoning, we deduce from Eq. (6-415) that the inequality

$$\sum_{j=1}^{r} |\tilde{u}_j(t)| + \langle\mathbf{B}\tilde{\mathbf{u}}(t), \tilde{\mathbf{p}}(t)\rangle < \sum_{j=1}^{r} |\hat{u}_j(t)| + \langle\mathbf{B}\hat{\mathbf{u}}(t), \tilde{\mathbf{p}}(t)\rangle \tag{6-424}$$

must hold. Adding the two inequalities (6-423) and (6-424), we obtain the inequality

$$\langle\mathbf{B}\hat{\mathbf{u}}(t), \hat{\mathbf{p}}(t)\rangle + \langle\mathbf{B}\tilde{\mathbf{u}}(t), \tilde{\mathbf{p}}(t)\rangle < \langle\mathbf{B}\tilde{\mathbf{u}}(t), \hat{\mathbf{p}}(t)\rangle + \langle\mathbf{B}\hat{\mathbf{u}}(t), \tilde{\mathbf{p}}(t)\rangle \tag{6-425}$$

† Except possibly at the switch times.

which is valid whenever $\hat{\mathbf{u}}(t) \neq \tilde{\mathbf{u}}(t)$. Equation (6-425), in turn, yields the inequality

$$\int_0^{T_f} [\langle \mathbf{B}\hat{\mathbf{u}}(t), \hat{\mathbf{p}}(t)\rangle + \langle \mathbf{B}\tilde{\mathbf{u}}(t), \tilde{\mathbf{p}}(t)\rangle]\, dt < \int_0^{T_f} [\langle \mathbf{B}\tilde{\mathbf{u}}(t), \hat{\mathbf{p}}(t)\rangle + \langle \mathbf{B}\hat{\mathbf{u}}(t), \tilde{\mathbf{p}}(t)\rangle]\, dt \tag{6-426}$$

Thus, the assumption of normality and the assumption that $\hat{\mathbf{u}}$ and $\tilde{\mathbf{u}}$ are distinct imply the strict inequality (6-426). On the other hand, the necessary conditions require that the relations (6-421) and (6-422) be satisfied. By adding the right-hand side of Eq. (6-422) to the left-hand side of Eq. (6-421) and by adding the right-hand side of Eq. (6-421) to the left-hand side of Eq. (6-422), we obtain the equality

$$\int_0^{T_f} [\langle \mathbf{B}\hat{\mathbf{u}}(t), \hat{\mathbf{p}}(t)\rangle + \langle \mathbf{B}\tilde{\mathbf{u}}(t), \tilde{\mathbf{p}}(t)\rangle]\, dt = \int_0^{T_f} [\langle \mathbf{B}\tilde{\mathbf{u}}(t), \hat{\mathbf{p}}(t)\rangle + \langle \mathbf{B}\hat{\mathbf{u}}(t), \tilde{\mathbf{p}}(t)\rangle]\, dt \tag{6-427}$$

But the two relations (6-426) and (6-427) are *contradictory*. For this reason, we conclude that $\hat{\mathbf{u}}$ and $\tilde{\mathbf{u}}$ cannot be distinct, and so the uniqueness of the extremal controls has been established.

Both of the uniqueness theorems which we have presented were based upon the assumption that the fuel-optimal problem was normal. We have presented in Theorem 6-13 a *sufficient* condition for the problem to be normal. We wish to reiterate that the condition is not necessary. Thus, it is possible for a problem to be normal even though the sufficiency condition is violated. It may turn out, as we shall see in Chap. 8, that the problem is normal for some initial states but not normal for others.

In the next section, we shall formulate some different types of fuel-optimal problems, which, in a given situation, may represent a superior mathematical formulation of the physical situation. Meanwhile, we invite the reader to examine the following exercises, whose purpose is to illustrate the concepts discussed.

**Exercise 6-21** In Sec. 6-5, we prove (see Theorem 6-8) that if the eigenvalues of the matrix $\mathbf{A}$ are real, then each function $\langle \mathbf{b}_j, \mathbf{p}^*(t)\rangle$ has at most $n - 1$ zeros. Use this theorem to derive an upper bound on the number of switchings for the normal fuel-optimal problem.

**Exercise 6-22** Attempt to show that Eq. (6-392) in Theorem 6-13 is also necessary for Problem 6-2*b* to be normal. Why will this procedure fail? Can you determine any additional assumptions which will make Eq. (6-392) a necessary condition?

**Exercise 6-23** Again consider the linear system $\dot{\mathbf{x}}(t) = \mathbf{A}\mathbf{x}(t) + \mathbf{B}\mathbf{u}(t)$. Suppose that the given initial state $\boldsymbol{\xi}$, the terminal state $\boldsymbol{\theta}$, and the terminal time $T_f$ are related by the equation $\boldsymbol{\xi} = e^{-\mathbf{A}T_f}\boldsymbol{\theta}$. Determine the fuel-optimal control. Describe the physical significance of your answer. Is this fuel-optimal control unique even if $\mathbf{A}$ is a singular matrix? Explain!

**Exercise 6-24** The sufficient conditions stated in Theorem 6-13 were derived for Problem 6-2*b*. Suppose that in the statement of Problem 6-2*b*, we replace the terminal state $\boldsymbol{\theta}$ by an arbitrary target set $S$, and suppose that we consider the fuel-optimal problem of transferring $\xi$ to $S$ in either fixed or free time. Do you think that Eq. (6-392) will still be a sufficient condition for normality? Explain!

**Exercise 6-25** Consider the linear time-varying second-order system

$$\begin{aligned} \dot{x}_1(t) &= x_2(t) \\ \dot{x}_2(t) &= a(t)x_1(t) + b(t)x_2(t) + c(t)u(t) \end{aligned} \tag{6-428}$$

Suppose that $|u(t)| \leq 1$ for all $t$, that $\mathbf{x}(t_0) = \xi$ is given, that the fuel functional is $\int_0^{T_f} |u(t)|\, dt$ ($T_f$ fixed), and that we are given a terminal state $\mathbf{0}$. Can you derive sufficient conditions on $a(t)$, $b(t)$, and $c(t)$ which will imply that the fuel-optimal problem is normal?

## 6-15 Minimum-fuel Problems 5: Additional Formulations and Cost Functionals

In this section, we shall briefly describe several alternative formulations of the minimum-fuel problem. The need for alternative formulations arises because of the intricate relationship between the response time and the consumed fuel. It may happen that in a given physical situation, we may want to fix the response time a priori; in another situation, we may simply wish to specify only an upper bound on the response time; in yet another case, we may wish to put a constraint on the fuel and demand a minimum-time solution. It should be clear that in any given situation, the designer must choose the problem formulation which best fits the physical requirements and which results in the simplest possible control law.

To illustrate these ideas, we shall formulate a sequence of optimal-control problems in which the system, the (control) constraint set, the initial state, and the target set are the same for each problem; we shall then state the properties of each formulation.

*Given the time-invariant system*

$$\dot{\mathbf{x}}(t) = \mathbf{f}[\mathbf{x}(t)] + \mathbf{B}[\mathbf{x}(t)]\mathbf{u}(t) \tag{6-429}$$

*Given the constraint set $\Omega$ in $R_r$*

$$\Omega = \{\mathbf{u}(t): |u_j(t)| \leq 1; j = 1, 2, \ldots, r\} \tag{6-430}$$

*Given the fuel functional*

$$F(\mathbf{u}) = \int_0^T \sum_{j=1}^{r} |u_j(t)|\, dt \tag{6-431}$$

*Given the initial state $\xi$ and a time-invariant target set $S$ (in the state space $R_n$). Then determine the admissible control which transfers $\xi$ to $S$ and which:*

*(Problem* 6-2c*) minimizes the fuel functional $F(\mathbf{u})$ with $T$ free*

*(Problem* 6-2d*) minimizes the fuel functional $F(\mathbf{u})$ with $T = T_f$, $T_f$ fixed*

*(Problem* 6-2e*) minimizes the fuel functional $F(\mathbf{u})$ with $T \leq T_f$, $T_f$ fixed*

(*Problem* 6-2*f*) *minimizes the time T with the constraint* $F(\mathbf{u}) \leq \Phi$, *where* $\Phi$ *is a given positive constant*

(*Problem* 6-2*g*) *minimizes the cost functional*

$$J(\mathbf{u}) = kT + F(\mathbf{u}) = \int_0^T \left\{ k + \sum_{j=1}^{r} |u_j(t)| \right\} dt$$

*where T is free and k is a specified positive constant*

We discussed the free-time problem 6-2*c* and the fixed-time problem 6-2*d* previously. Problems 6-2*e* and 6-2*f* are, in a sense, opposite to each other; in Problem 6-2*e*, we want to minimize the fuel with an upper bound on the response time, while in Problem 6-2*f*, we want to minimize the time with an upper bound on the consumed fuel. Finally, in Problem 6-2*g*, we are interested in minimizing a linear combination of the elapsed time and of the consumed fuel.

The best way of isolating the similarities and differences among these five different problems is to examine the necessary conditions. If we do that, we shall find that, for each problem, the form or shape of the optimal control is the same and that the differences are taken care of by the transversality conditions.

Let us agree to use the subscripts *C*, *D*, *E*, *F*, and *G* for the quantities corresponding to Problems 6-2*c*, 6-2*d*, 6-2*e*, 6-2*f*, and 6-2*g*, respectively. Also, we shall transform each problem into a free-time problem simply by augmenting the *n*-dimensional state space by one variable. (See Sec. 5-14.)

In Problem 6-2*c*, we have a free-time problem, so that we do not need to reformulate the problem.

In Problems 6-2*d* and 6-2*e*, we define the auxiliary variable $x_{n+1}$ by setting

$$x_{n+1}(t) = t \tag{6-432}$$

so that $x_{n+1}$ satisfies the differential equation

$$\dot{x}_{n+1}(t) = 1 \tag{6-433}$$

and the initial condition

$$x_{n+1}(0) = 0 \tag{6-434}$$

Now, in Problem 6-2*d*, we define the target set $S_D$ in $n + 1$ space as

$$S_D = \{(\mathbf{x}, x_{n+1}) : \mathbf{x} \in S \text{ and } x_{n+1} = T_f\} \tag{6-435}$$

In Problem 6-2*e*, we define the target set $S_E$ in $n + 1$ space as

$$S_E = \{(\mathbf{x}, x_{n+1}) : \mathbf{x} \in S \text{ and } x_{n+1} \in [0, T_f]\} \tag{6-436}$$

In Problem 6-2*f*, the fuel is constrained; so let us define the variable $x_{n+1}(t)$ for this problem by setting

$$x_{n+1}(t) = \int_0^t \sum_{j=1}^{r} |u_j(\tau)| \, d\tau \tag{6-437}$$

Here, $x_{n+1}$ satisfies the differential equation

$$\dot{x}_{n+1}(t) = \sum_{j=1}^{r} |u_j(t)| \tag{6-438}$$

and the initial condition

$$x_{n+1}(0) = 0 \tag{6-439}$$

Also, for Problem 6-2*f*, we define the target set $S_F$ in $n + 1$ space as

$$S_F = \{(\mathbf{x}, x_{n+1})\colon \mathbf{x} \in S \text{ and } x_{n+1} \in [0, \Phi]\} \tag{6-440}$$

Problem 6-2*g* is a free-time problem with no additional constraints, so that no reformulation is required.

We shall now write the Hamiltonian functions for each problem. It is easy to see that they are given by the equations

$$H_C = \sum_{j=1}^{r} |u_j(t)| + \langle \mathbf{f}[\mathbf{x}(t)], \mathbf{p}(t)\rangle + \langle \mathbf{B}[\mathbf{x}(t)]\mathbf{u}(t), \mathbf{p}(t)\rangle \tag{6-441}$$

$$H_D = \sum_{j=1}^{r} |u_j(t)| + \langle \mathbf{f}[\mathbf{x}(t)], \mathbf{p}(t)\rangle + \langle \mathbf{B}[\mathbf{x}(t)]\mathbf{u}(t), \mathbf{p}(t)\rangle + 1 \cdot p_{n+1}^D(t) \tag{6-442}$$

$$H_E = \sum_{j=1}^{r} |u_j(t)| + \langle \mathbf{f}[\mathbf{x}(t)], \mathbf{p}(t)\rangle + \langle \mathbf{B}[\mathbf{x}(t)]\mathbf{u}(t), \mathbf{p}(t)\rangle + 1 \cdot p_{n+1}^E(t) \tag{6-443}$$

$$H_F = 1 + \langle \mathbf{f}[\mathbf{x}(t)], \mathbf{p}(t)\rangle + \langle \mathbf{B}[\mathbf{x}(t)]\mathbf{u}(t), \mathbf{p}(t)\rangle + \sum_{j=1}^{r} |u_j(t)| \cdot p_{n+1}^F(t) \tag{6-444}$$ †

$$H_G = k + \sum_{j=1}^{r} |u_j(t)| + \langle \mathbf{f}[\mathbf{x}(t)], \mathbf{p}(t)\rangle + \langle \mathbf{B}[\mathbf{x}(t)]\mathbf{u}(t), \mathbf{p}(t)\rangle \tag{6-445}$$

Next, we shall show that the costate variables $p_{n+1}(t)$ that appear in Eqs. (6-442) to (6-444) are constants. This is so because

$$\dot{p}_{n+1}^D(t) = -\frac{\partial H_D}{\partial x_{n+1}(t)} = 0 \tag{6-446}$$

$$\dot{p}_{n+1}^E(t) = -\frac{\partial H_E}{\partial x_{n+1}(t)} = 0 \tag{6-447}$$

$$\dot{p}_{n+1}^F(t) = -\frac{\partial H_F}{\partial x_{n+1}(t)} = 0 \tag{6-448}$$

so that

$$\begin{aligned} p_{n+1}^D(t) &= k_D \\ p_{n+1}^E(t) &= k_E \\ p_{n+1}^F(t) &= k_F \end{aligned} \tag{6-449}$$

† The term 1 is due to the fact that in Problem 6-2*f*, the cost functional is $\int_0^T 1 \cdot dt$, since we seek a minimum-time solution.

where $k_D$, $k_E$, and $k_F$ are some constants. Thus, the various Hamiltonian functions are given by the following equation (which was written in such a way as to indicate the common elements of each Hamiltonian function):

$$\left.\begin{matrix} H_C \\ H_D \\ H_E \\ H_F \\ H_G \end{matrix}\right\} = \langle \mathbf{f}[\mathbf{x}(t)], \mathbf{p}(t)\rangle + \langle \mathbf{B}[\mathbf{x}(t)]\mathbf{u}(t), \mathbf{p}(t)\rangle + \begin{cases} 0 + \sum_{j=1}^{r} |u_j(t)| \\ k_D + \sum_{j=1}^{r} |u_j(t)| \\ k_E + \sum_{j=1}^{r} |u_j(t)| \\ 1 + k_F \sum_{j=1}^{r} |u_j(t)| \\ k + \sum_{j=1}^{r} |u_j(t)| \end{cases} \tag{6-450}$$

The reader should note that the functional dependence of each Hamiltonian on the state $\mathbf{x}(t)$ and the costate $\mathbf{p}(t)$ is the same for each problem. Moreover, the difference between the five distinct problems is reflected only by the appearance of five different constants in the Hamiltonian functions.

It is easy to see that the control which absolutely minimizes the Hamiltonian (i.e., the $H$-minimal control) is given by the equation

$$\mathbf{u}^\circ(t) = -\text{ DEZ }\{\mathbf{B}'[\mathbf{x}(t)]\mathbf{p}(t)\} \tag{6-451}$$

for Problems 6-2*c*, 6-2*d*, 6-2*e*, and 6-2*g*, and by the equation

$$\mathbf{u}^\circ(t) = -\text{DEZ}\left\{\frac{1}{k_F}\mathbf{B}'[\mathbf{x}(t)]\mathbf{p}(t)\right\} \tag{6-452}$$

for Problem 6-2*f*. Thus, if each problem is normal, then the optimal control (if it exists) must be a three-level (+1, 0, and −1) piecewise constant function of time. This, of course, does not mean that the optimal controls are the same; in general, they will be different. However, for each normal problem, the "shape" of the optimal control is the same.

We shall not discuss here the other necessary conditions. We feel that the reader will benefit much more if he carries out this task by himself; in this case, he will discover that the canonical equations are (functionally) the same and some of the transversality conditions are also the same (e.g., the costate at the terminal time must be normal to $S$). Basically, the optimal controls for these problems will be different because the constants $k_D$, $k_E$, and $k_F$ will be different.

**Exercise 6-26** Consider Problems 6-2*c* to 6-2*g*.

(*a*) Tabulate all the necessary conditions.

(*b*) Comment on the similarities and differences between the necessary conditions.

(*c*) Suppose one of the problems turns out to be singular; do you think that this implies that all of the other problems will be singular? Explain.

(*d*) Show that the optimal control(s) for Problem 6-2*e* will be the same as either the optimal control(s) for Problem 6-2*c* or the optimal control(s) for Problem 6-2*d*.

(*e*) Consider Problem 6-2*g*; do you think that one can find values of the constant $k$ such that the optimal control(s) will be the same as those for Problems 6-2*c* to 6-2*f*?

**Exercise 6-27** Suppose that the system (6-429) is the linear and time-invariant system $\dot{\mathbf{x}}(t) = \mathbf{A}\mathbf{x}(t) + \mathbf{B}\mathbf{u}(t)$, and suppose that the target set is the single state $\mathbf{0}$.

(*a*) Attempt to find sufficient and/or necessary conditions for each problem to be normal.

(*b*) Attempt to find conditions that will guarantee the uniqueness of the optimal control.

(*c*) Try to find conditions that will guarantee the uniqueness of the extremal controls.

We shall now discuss briefly the structure of the optimal feedback control system for each of the above problems; we assume that each problem is normal. In Sec. 6-6, we discussed in a heuristic manner the existence of a "switching function." The same type of arguments can be carried out for minimum-fuel problems. In essence, we assert that for each problem there is a function of the state $\mathbf{x}(t)$ (it may depend also on other variables) which will define the optimal control as long as the problem is normal. To be specific, we assert that:

1. For Problem 6-2*c*, there is a vector-valued switching function $\mathbf{h}_C[\mathbf{x}(t), S]$ such that the optimal control $\mathbf{u}_C^*(t)$ is given by

$$\mathbf{u}_C^*(t) = \text{DEZ}\ \{\mathbf{h}_C[\mathbf{x}(t), S]\} \tag{6-453}$$

and so the feedback control system is time-invariant.

2. For Problem 6-2*d*, the switching function depends on the state $\mathbf{x}(t)$, the target set $S$, and the time $T_f - t$ required to force $\mathbf{x}(t)$ to $S$. Thus, the optimal control $\mathbf{u}_D^*(t)$ is given by

$$\mathbf{u}_D^*(t) = \text{DEZ}\ \{\mathbf{h}_D[\mathbf{x}(t), S, t, T_f]\} \tag{6-454}$$

and the feedback control system is time-varying.

3. For Problem 6-2*e*, the switching function will, in general, depend on the time $T_f - t$ required to force the state $\mathbf{x}(t)$ to $S$. Thus, the optimal control $\mathbf{u}_E^*(t)$ will be given by

$$\mathbf{u}_E^*(t) = \text{DEZ}\ \{\mathbf{h}_E[\mathbf{x}(t), S, t, T_f]\} \tag{6-455}$$

4. For Problem 6-2*f*, the switching function will, in general, depend on the fuel $F(t)$ consumed in the interval $[0, t]$, that is, on

$$F(t) = \int_0^t \sum_{j=1}^{r} |u_j(\tau)|\, d\tau \tag{6-456}$$

and on the upper bound on the fuel $\Phi$. Thus, the optimal control $\mathbf{u}_F^*(t)$ will be given by

$$\mathbf{u}_F^*(t) = \text{DEZ}\ \{\mathbf{h}_F[\mathbf{x}(t), S, F(t), \Phi]\} \tag{6-457}$$

In this case, the system will be time-invariant, although we must keep track of the consumed fuel $F(t)$.

5. For Problem 6-2*g*, the switching function will depend only on the state $\mathbf{x}(t)$ and the target set $S$. Thus, the feedback system will be time-invariant, and for any given value of the constant $k$, the optimal control $\mathbf{u}_G^*(t)$ will be given by

$$\mathbf{u}_G^*(t) = \text{DEZ } \{\mathbf{h}_G[\mathbf{x}(t), S]\} \tag{6-458}$$

We illustrate the structure of the feedback system for Problems 6-2*c* to 6-2*g* in Fig. 6-18; we assume that the problems are normal. In each case, the state variables $x_1(t), x_2(t), \ldots, x_n(t)$ are introduced to the system labeled "Computer"; additional inputs to the computer are the target set $S$,

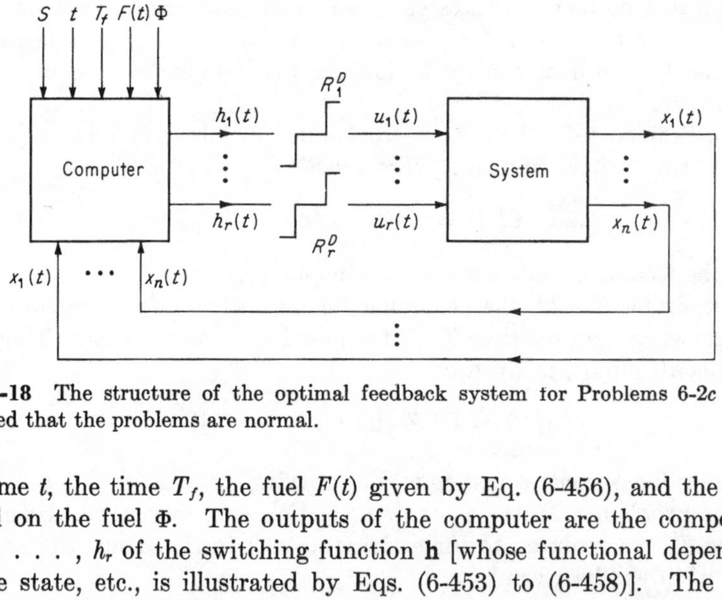


**Fig. 6-18** The structure of the optimal feedback system for Problems 6-2*c* to 6-2*g* provided that the problems are normal.

the time $t$, the time $T_f$, the fuel $F(t)$ given by Eq. (6-456), and the upper bound on the fuel $\Phi$. The outputs of the computer are the components $h_1, h_2, \ldots, h_r$ of the switching function $\mathbf{h}$ [whose functional dependence on the state, etc., is illustrated by Eqs. (6-453) to (6-458)]. The "dez" operation is accomplished by the nonlinear elements $R_1{}^D, R_2{}^D, \ldots, R_r{}^D$, whose input-output characteristics are those of relays with dead zone. This schematic block diagram brings out the structure of the feedback control system. The differences between the different problems are taken care of by the computer, which generates the appropriate switching functions.

In Chap. 8, we shall examine in detail many fuel-optimal problems. We shall illustrate the block diagrams of the optimal control systems, and we shall also comment still further on their properties from the engineering point of view.

In the next section, we shall recapitulate the main points of Secs. 6-11 to 6-15, and we shall present a sequence of exercises. The purpose of these exercises is to let the reader "fill in" the geometric properties of the minimum-fuel problem. We feel that the general geometric concepts of Sec. 5-21 and the discussion of Secs. 6-2 and 6-7 regarding the geometry of the time-optimal problem provide sufficient background information for the reader to tackle such concepts in the fuel-optimal problem himself.

## 6-16 Minimum-fuel Problems 6: Comments

In the preceding five sections, we discussed the minimum-fuel problem. In Sec. 6-15, we presented five distinct formulations of fuel-optimal problems. These formulations reflected the intimate relationship between the time of response and the consumed fuel.

The common characteristic of all the formulations was that the Hamiltonian function was linear in the control variables $u_j(t)$ and $|u_j(t)|$. For this reason, the necessary conditions implied that the optimal control (if it existed) had to be a piecewise constant, three-level ($+1$, $0$, or $-1$) function of time, provided the problem was normal (see Definition 6-11).

From a physical point of view, fuel-optimal controls conserve fuel during the time intervals in which no control is applied. The system is controlled in such a manner as to "fire" whenever a given amount of consumed fuel will result in the most efficient motion. This, of course, is the case whenever the problem is normal. We shall see in Chap. 8 (see Sec. 8-10) that in a specific singular problem, the optimal control is not a three-level one ($+1$, $0$, or $-1$) but a five-level one ($+1$, $k$, $0$, $-k$, $-1$), where $0 < k < 1$.

General existence theorems are not available at present for the various formulated problems.† This is indeed a shortcoming, because one cannot be sure whether or not a solution exists. We shall indicate (by specific examples) in Chap. 8 some of the techniques that one may use. At this point, the reader may wish to bypass the remainder of Chap. 6 and examine the specific fuel-optimal problems of Chap. 8.

**Exercise 6-28** Consider Problem 6-2*c*. Discuss the geometric nature of the problem in a manner analogous to that given in Sec. 6-2. It may be helpful to define the variable $x_0(t) = \int_0^t \sum_{j=1}^{r} |u_j(\tau)|\, d\tau$ and use the concept of reachable states in the new $(n+1)$-dimensional space. Discuss existence and uniqueness of the fuel-optimal controls in terms of the properties of the target set and of the sets of reachable states (in $n + 1$ space).

**Exercise 6-29** Repeat Exercise 6-28 for Problems 6-2*d* to 6-2*g*.

**Exercise 6-30** Once more consider Problem 6-2*c*. Let $F^*(\mathbf{x})$ denote the minimum fuel required to force the state $\mathbf{x}$ to the target set $S$ (the time is free). Define the "minimum isofuel" as the set of states which require the same minimum fuel to be transferred to the target set $S$. Mimic the ideas of Sec. 6-7 to derive geometric properties of the fuel-optimal control.

**Exercise 6-31** Write the Hamilton-Jacobi partial differential equation for each of Problems 6-2*c* to 6-2*g*.

† However, see Ref. C-4.

## 6-17 Minimum-energy Problems 1: Introduction

We shall examine several aspects of the minimum-energy problem in this and the next three sections. The term *minimum energy* reflects the fact that often a quantity related to the energy of an electrical signal is to be minimized in such a problem. We shall give some precise formulations of various minimum-energy problems in Secs. 6-18 and 6-20, and we shall discuss an interesting example in Sec. 6-19. We shall study a particular class of minimum-energy problems in Chap. 9, motivating our choice of a cost functional from the control point of view. In the remainder of this section, we motivate our interest in minimum-energy problems by carefully examining two simple examples.

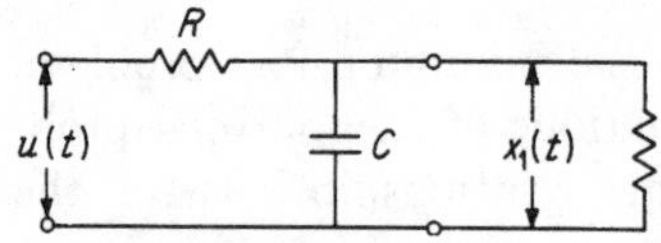


**Fig. 6-19** The $RC$ network considered in Example 6-7.

**Example 6-7** In this example, we are interested in determining a signal which will transfer a given amount of energy to a load. To be more specific, let us examine the simple network shown in Fig. 6-19. We connect a "load" resistor $R_L$ to the output terminals of an $RC$ network. We let $x_1(t)$ denote the voltage across the load resistor $R_L$, and we let $u(t)$ denote the input (or control) voltage. First of all, let us write the differential equation of the network. From elementary circuit-theory concepts, we derive the first-order differential equation

$$RR_LC\dot{x}_1(t) + (R + R_L)x_1(t) = R_Lu(t) \tag{6-459}$$

Let us define the positive constants $a$ and $b$ by setting

$$a = \frac{R + R_L}{RR_LC} \qquad b = \frac{1}{RC} \tag{6-460}$$

Then Eq. (6-459) reduces to the equation

$$\dot{x}_1(t) = -ax_1(t) + bu(t) \tag{6-461}$$

Let us now suppose that at $t = 0$ the capacitor $C$ is not charged, i.e., that

$$x_1(0) = 0 \tag{6-462}$$

The energy $E_L$ delivered to $R_L$ in the time interval $[0, T]$ is given by

$$E_L = \frac{1}{R_L}\int_0^T x_1^2(t)\, dt \tag{6-463}$$

In order to find the energy $E_u$ delivered by the input, we make an energy balance. The energy $E_R$ dissipated in the resistor $R$ is given by

$$E_R = \frac{1}{R}\int_0^T [u(t) - x_1(t)]^2\, dt \tag{6-464}$$

The energy $E_C$ delivered to the capacitor $C$ is given by

$$E_C = C\int_0^T x_1(t)\dot{x}_1(t)\, dt \tag{6-465}$$

[because the instantaneous power delivered to $C$ is the product of the voltage $x_1(t)$ across $C$ and of the current $C\,dx_1(t)/dt$ through $C$]. Thus, the energy $E_u$ delivered by the input is

$$E_u = E_L + E_R + E_C \tag{6-466}$$

If we substitute Eqs. (6-460) and (6-461) into Eq. (6-465), then we find, after some algebraic manipulations, that the input energy $E_u$ is given by

$$E_u = \frac{1}{R}\int_0^T [u^2(t) - x_1(t)u(t)]\,dt \tag{6-467}$$

We can now formulate the following minimum-energy problem: *Given the system (6-461) (i.e., the network of Fig. 6-19), and suppose that $x_1(0) = 0$; then determine the input $u(t)$ which delivers a given amount of energy $E_L = \frac{1}{R_L}\int_0^T x_1^2(t)\,dt$ to the load resistor $R_L$ and which minimizes the input energy $E_u = \frac{1}{R}\int_0^T [u^2(t) - u(t)x_1(t)]\,dt$, where $T$ is fixed.*

In this example, we observe that the quantity

$$\hat{E}_u = \int_0^T u^2(t)\,dt \tag{6-468}$$

is related to (but is *not* proportional to) the energy provided by the input (or control) signal $u(t)$. If the capacitor $C$ were not present, then the quantity $\hat{E}_u$ of (6-468) would be indeed proportional to the input energy.

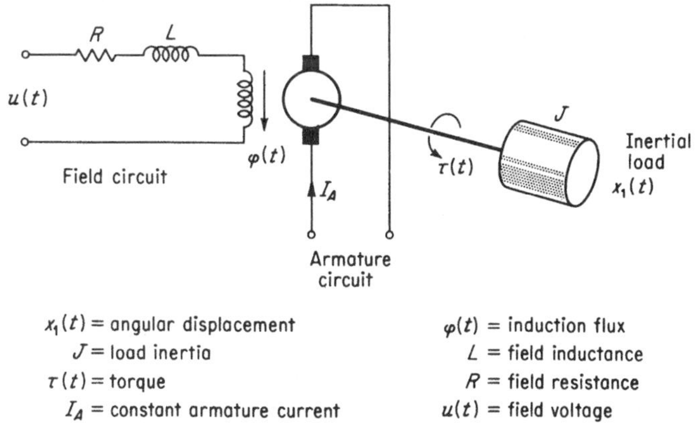


**Fig. 6-20** Schematic of a field-controlled d-c motor.

**Example 6-8** Consider the position control of the field-controlled d-c motor illustrated in Fig. 6-20. The control voltage $u(t)$ is applied to the field circuit; as a result,† the differential equation relating the flux $\varphi(t)$ to the control voltage $u(t)$ is (approximately)

$$L\dot{\varphi}(t) + R\varphi(t) = k_1 u(t) \tag{6-469}$$

where $k_1$ is a constant of proportionality. Let us assume that $I_A$ is the constant armature current. Then the torque $\tau(t)$ is related to the flux $\varphi(t)$ by the equation

$$\tau(t) = k_2 I_A \varphi(t) \tag{6-470}$$

† The detailed derivations of the differential equations can be found, for example, in Ref. G-6, pp. 581–585.

Let us also suppose that there is negligible friction and stiffness on the output shaft. In this case, the output angular displacement $x_1(t)$ satisfies the differential equation

$$J\ddot{x}_1(t) = \tau(t) \tag{6-471}$$

where $J$ is the moment of inertia of the load and rotor. Let us furthermore assume that the field inductance $L$ is negligible, i.e., that

$$L = 0 \tag{6-472}$$

In this case, the system is described by the differential equations

$$\dot{x}_1(t) = x_2(t) \qquad \dot{x}_2(t) = Ku(t) \tag{6-473}$$

where $K$ is a constant of proportionality. The assumption that the field inductance $L$ is negligible allows us to conclude that the quantity

$$E_u = \int_0^T u^2(t)\, dt \tag{6-474}$$

is proportional to the control energy provided by the input signal. For such a system we can easily formulate the minimum-energy problem as follows: *Given the system* (6-473) *with the initial conditions* $x_1(0) = \xi_1$, $x_2(0) = 0$, *determine the control* $u(t)$ *such that the state is forced to the state* $(\theta_1, 0)$ *in a given time* $T$ *and such that the functional* (6-474) *is minimized.*

We shall solve this problem in Sec. 6-19.

The purpose of these two simple examples was to demonstrate that the cost functional

$$J(u) = \int_0^T u^2(t)\, dt \tag{6-475}$$

is often related to the energy of a signal $u(t)$. From the mathematical point of view, the cost functional $J(u)$ assigns a specific number to the cost of control. The fact that the integrand $u^2(t)$ is a quadratic function of $u(t)$ has the obvious significance that *we do not penalize the system very much if* $u(t)$ *is small; on the other hand, we severely penalize the system for large control signals* $u(t)$.

In the next section, we shall formulate and solve a very general minimum-energy problem. We shall first derive the necessary conditions, and we shall then use them to find the optimal control.

## 6-18 Minimum-energy Problems 2: The Linear Fixed-end-point, Fixed-time Problem†

In this section, we shall consider the problem of transferring the initial state $\boldsymbol{\xi}$ of a linear time-varying system to the origin $\mathbf{0}$ in a fixed time $T_f$. We shall assume that the control is not constrained in magnitude. We shall first state the problem, then derive the necessary conditions, and finally determine the optimal control in terms of fundamental matrices.

† Appropriate references for this section are J-6, M-3, L-11, B-17, L-15, L-18, A-6, D-4, F-16, F-17, K-3, K-34, P-8, R-2, R-4, S-1, and T-1.

The precise statement of the problem is:

***Problem 6-3a*** *Given the linear, completely controllable time-varying system*

$$\dot{\mathbf{x}}(t) = \mathbf{A}(t)\mathbf{x}(t) + \mathbf{B}(t)\mathbf{u}(t) \tag{6-476}$$

*where* $\mathbf{x}(t)$ *is the state (an n vector)*

$\mathbf{A}(t)$ *is an* $n \times n$ *system matrix*

$\mathbf{u}(t)$ *is the control (an r vector)*

$\mathbf{B}(t)$ *is an* $n \times r$ *gain matrix*

*Given that at the initial time* $t_0$, *the state is* $\boldsymbol{\xi}$; *that is,*

$$\mathbf{x}(t_0) = \boldsymbol{\xi} \tag{6-477}$$

*Given a terminal time* $T_f$, *and given that* $\mathbf{0}$ *is the desired terminal state. Assume that the control* $\mathbf{u}(t)$ *is unconstrained. Given an* $n + r \times n + r$ *positive definite*† *symmetric matrix* $\mathbf{N}(t)$, *defined by*

$$\mathbf{N}(t) = \left[\begin{array}{c|c} \mathbf{Q}(t) & \mathbf{M}(t) \\ \hline \mathbf{M}'(t) & \mathbf{R}(t) \end{array}\right] \tag{6-478}$$

*where* $\mathbf{Q}(t)$ *is an* $n \times n$ *(positive definite) matrix*

$\mathbf{R}(t)$ *is an* $r \times r$ *(positive definite) matrix*

$\mathbf{M}(t)$ *is an* $n \times r$ *matrix*

*Given the cost functional*

$$\begin{aligned} J(\mathbf{u}) &= \frac{1}{2}\int_{t_0}^{T_f} \left\langle \begin{bmatrix} \mathbf{x}(t) \\ \mathbf{u}(t) \end{bmatrix}, \mathbf{N}(t) \begin{bmatrix} \mathbf{x}(t) \\ \mathbf{u}(t) \end{bmatrix} \right\rangle dt \\ &= \frac{1}{2}\int_{t_0}^{T_f} [\langle \mathbf{x}(t), \mathbf{Q}(t)\mathbf{x}(t)\rangle + \langle \mathbf{u}(t), \mathbf{R}(t)\mathbf{u}(t)\rangle + 2\langle \mathbf{x}(t), \mathbf{M}(t)\mathbf{u}(t)\rangle]\, dt \end{aligned} \tag{6-479}$$

*Then find the control that:*

1. *Transfers the state* $\boldsymbol{\xi}$ *to* $\mathbf{0}$ *in the fixed time* $T_f$, *that is, the control such that*

$$\mathbf{x}(T_f) = \mathbf{0} \tag{6-480}$$

2. *Minimizes the cost functional* $J(\mathbf{u})$ *given by Eq.* (6-479)

It is worthwhile to discuss the choice of the cost functional $J(\mathbf{u})$ at this point. The assumption that $\mathbf{N}(t)$ is positive definite guarantees that the cost functional $J(\mathbf{u})$ is positive, provided that

$$\mathbf{x}(t) \neq \mathbf{u}(t) \neq \mathbf{0} \qquad \text{for all } t \in [t_0, T_f] \tag{6-481}$$

The assumption that $\mathbf{N}(t)$ is positive definite also implies that

$$\mathbf{Q}(t) \text{ and } \mathbf{R}(t) \text{ are positive definite} \tag{6-482}$$

and hence that the inverses

$$\mathbf{N}^{-1}(t),\ \mathbf{Q}^{-1}(t), \text{ and } \mathbf{R}^{-1}(t) \text{ exist} \tag{6-483}$$†

† See Sec. 2-15.

Now let us derive the necessary conditions provided by the minimum principle for this problem. We first form the Hamiltonian function $H$ for the problem:

$$H = H[\mathbf{x}(t), \mathbf{p}(t), \mathbf{u}(t), t] = \tfrac{1}{2}\langle \mathbf{x}(t), \mathbf{Q}(t)\mathbf{x}(t)\rangle + \tfrac{1}{2}\langle \mathbf{u}(t), \mathbf{R}(t)\mathbf{u}(t)\rangle + \langle \mathbf{x}(t), \mathbf{M}(t)\mathbf{u}(t)\rangle + \langle \mathbf{A}(t)\mathbf{x}(t), \mathbf{p}(t)\rangle + \langle \mathbf{B}(t)\mathbf{u}(t), \mathbf{p}(t)\rangle \tag{6-484}$$

We observe that the Hamiltonian is a quadratic function of the control $\mathbf{u}(t)$.

The minimum principle can be used to establish the following theorem:

***Theorem 6-16 Necessary Conditions for Problem 6-3a*** *If $\mathbf{u}^*(t)$ is an optimal control and if $\mathbf{x}^*(t)$ is the resultant optimal trajectory, then there is a corresponding costate $\mathbf{p}^*(t)$ such that:*

*a. The state $\mathbf{x}^*(t)$ and the costate $\mathbf{p}^*(t)$ satisfy the differential equations*

$$\dot{\mathbf{x}}^*(t) = \mathbf{A}(t)\mathbf{x}^*(t) + \mathbf{B}(t)\mathbf{u}^*(t) \tag{6-485}$$
$$\dot{\mathbf{p}}^*(t) = -\mathbf{Q}(t)\mathbf{x}^*(t) - \mathbf{M}(t)\mathbf{u}^*(t) - \mathbf{A}'(t)\mathbf{p}^*(t) \tag{6-486}$$

*and the boundary conditions*

$$\mathbf{x}^*(t_0) = \boldsymbol{\xi} \qquad \mathbf{x}^*(T_f) = \mathbf{0} \tag{6-487}$$

*b. The following relation holds for all $\mathbf{u}(t) \in R_r$ [because $\mathbf{u}(t)$ is not constrained] and $t \in [t_0, T_f]$:*

$$\begin{aligned}\tfrac{1}{2}\langle \mathbf{x}^*(t), \mathbf{Q}(t)\mathbf{x}^*(t)\rangle &+ \tfrac{1}{2}\langle \mathbf{u}^*(t), \mathbf{R}(t)\mathbf{u}^*(t)\rangle + \langle \mathbf{x}^*(t), \mathbf{M}(t)\mathbf{u}^*(t)\rangle \\ &+ \langle \mathbf{A}(t)\mathbf{x}^*(t), \mathbf{p}^*(t)\rangle + \langle \mathbf{B}(t)\mathbf{u}^*(t), \mathbf{p}^*(t)\rangle \\ \le \tfrac{1}{2}\langle \mathbf{x}^*(t), \mathbf{Q}(t)\mathbf{x}^*(t)\rangle &+ \tfrac{1}{2}\langle \mathbf{u}(t), \mathbf{R}(t)\mathbf{u}(t)\rangle + \langle \mathbf{x}^*(t), \mathbf{M}(t)\mathbf{u}(t)\rangle \\ &+ \langle \mathbf{A}(t)\mathbf{x}^*(t),\mathbf{p}^*(t)\rangle + \langle \mathbf{B}(t)\mathbf{u}(t), \mathbf{p}^*(t)\rangle\end{aligned} \tag{6-488}$$

*or, equivalently,*

$$\begin{aligned}\tfrac{1}{2}\langle \mathbf{u}^*(t), \mathbf{R}(t)\mathbf{u}^*(t)\rangle &+ \langle \mathbf{x}^*(t), \mathbf{M}(t)\mathbf{u}^*(t)\rangle + \langle \mathbf{B}(t)\mathbf{u}^*(t), \mathbf{p}^*(t)\rangle \\ &\le \tfrac{1}{2}\langle \mathbf{u}(t), \mathbf{R}(t)\mathbf{u}(t)\rangle + \langle \mathbf{x}^*(t), \mathbf{M}(t)\mathbf{u}(t)\rangle + \langle \mathbf{B}(t)\mathbf{u}(t), \mathbf{p}^*(t)\rangle\end{aligned} \tag{6-489}$$

**Exercise 6-32** Verify Theorem 6-16. HINT: See Table 5-1, row 6.

We shall now use the second necessary condition, i.e., the relation (6-489), to derive an equation which *uniquely* relates the control $\mathbf{u}^*(t)$ to $\mathbf{x}^*(t)$ and $\mathbf{p}^*(t)$. This step is precisely the same as the one that we carried out for the time-optimal problem and for the fuel-optimal problem. However, as the reader will soon discover, we shall not have any singular problems in this case.

Equation (6-489) implies that the function

$$\varphi[\mathbf{u}(t)] = \tfrac{1}{2}\langle \mathbf{u}(t), \mathbf{R}(t)\mathbf{u}(t)\rangle + \langle \mathbf{x}^*(t), \mathbf{M}(t)\mathbf{u}(t)\rangle + \langle \mathbf{B}(t)\mathbf{u}(t), \mathbf{p}^*(t)\rangle \tag{6-490}$$

has an absolute minimum at

$$\mathbf{u}(t) = \mathbf{u}^*(t) \tag{6-491}$$

Since $\mathbf{u}(t)$ is not constrained and since $\varphi[\mathbf{u}(t)]$ is a smooth function of $\mathbf{u}(t)$, we can find the minimum by setting $\partial\varphi[\mathbf{u}(t)]/\partial\mathbf{u}(t) = \mathbf{0}$. But

$$\frac{\partial\varphi[\mathbf{u}(t)]}{\partial\mathbf{u}(t)} = \mathbf{R}(t)\mathbf{u}(t) + \mathbf{M}(t)\mathbf{x}^*(t) + \mathbf{B}'(t)\mathbf{p}^*(t) \tag{6-492}$$

We thus immediately obtain the relation

$$\mathbf{R}(t)\mathbf{u}^*(t) + \mathbf{M}(t)\mathbf{x}^*(t) + \mathbf{B}'(t)\mathbf{p}^*(t) = \mathbf{0} \tag{6-493}$$

and, in view of the nonsingularity of $\mathbf{R}(t)$ [see Eq. (6-483)], we may deduce that

$$\mathbf{u}^*(t) = -\mathbf{R}^{-1}(t)[\mathbf{M}(t)\mathbf{x}^*(t) + \mathbf{B}'(t)\mathbf{p}^*(t)] \tag{6-494}$$

In order to be sure that Eq. (6-494) defines a minimum, we must check whether or not the $r \times r$ matrix $\partial^2\varphi[\mathbf{u}(t)]/\partial\mathbf{u}(t)^2$ is positive definite. By direct computation, we find from Eq. (6-492) that

$$\frac{\partial^2\varphi[\mathbf{u}(t)]}{\partial\mathbf{u}(t)^2} = \mathbf{R}(t) \text{ is positive definite} \tag{6-495}$$

by the assumptions of Problem 6-3*a*. We are thus assured that Eq. (6-494) uniquely defines $\mathbf{u}^*(t)$ in terms of $\mathbf{x}^*(t)$ and $\mathbf{p}^*(t)$.

Since the relations (6-488) and (6-489) are the same, we could deduce Eq. (6-494) by computing the minimum with respect to $\mathbf{u}(t)$ of the function $H[\mathbf{x}^*(t), \mathbf{p}^*(t), \mathbf{u}(t), t]$, because

$$\frac{\partial H[\mathbf{x}^*(t), \mathbf{p}^*(t), \mathbf{u}(t), t]}{\partial\mathbf{u}(t)} = \mathbf{R}(t)\mathbf{u}(t) + \mathbf{M}(t)\mathbf{x}^*(t) + \mathbf{B}(t)\mathbf{p}^*(t) \tag{6-496}$$

and because the matrix

$$\frac{\partial^2 H[\mathbf{x}^*(t), \mathbf{p}^*(t), \mathbf{u}(t), t]}{\partial\mathbf{u}(t)^2} = \mathbf{R}(t) \tag{6-497}$$

is positive definite.

There are three points that we wish to emphasize at this time:

***Remark 6-13*** *It was the assumption that* $\mathbf{R}(t)$ *[rather than* $\mathbf{N}(t)$*] is positive definite which led us to the unique relation* (6-494). *Moreover, the positive definiteness of* $\mathbf{R}(t)$ *guarantees that Eq.* (6-494) *defines the unique minimum (rather than a maximum or saddle point) of* $H[\mathbf{x}^*(t), \mathbf{p}^*(t), \mathbf{u}(t), t]$ *at* $\mathbf{u}^*(t)$.

***Remark 6-14*** *The optimal control* $\mathbf{u}^*(t)$ *is a linear function of* $\mathbf{x}^*(t)$ *and* $\mathbf{p}^*(t)$. *This phenomenon is due to the linearity of the system, the quadratic nature of the cost functional, and the absence of any magnitude constraints on the control* $\mathbf{u}(t)$.

***Remark 6-15*** *The problem is normal because* $\mathbf{u}^*(t)$ *is uniquely defined by* $\mathbf{x}^*(t)$ *and* $\mathbf{p}^*(t)$. *As a matter of fact, whenever*

$$\mathbf{u}^*(t) \neq \mathbf{u}(t) \tag{6-498}$$

*the second necessary condition reduces to the strict inequality*

$$H[\mathbf{x}^*(t), \mathbf{p}^*(t), \mathbf{u}^*(t), t)] < H[\mathbf{x}^*(t), \mathbf{p}^*(t), \mathbf{u}(t), t] \tag{6-499}$$

The next step is to substitute Eq. (6-494) into the canonical equations (6-485) and (6-486) to obtain the system of $2n$ differential equations (the reduced canonical equations)

$$\dot{\mathbf{x}}^*(t) = [\mathbf{A}(t) - \mathbf{B}(t)\mathbf{R}^{-1}(t)\mathbf{M}(t)]\mathbf{x}^*(t) - \mathbf{B}(t)\mathbf{R}^{-1}(t)\mathbf{B}'(t)\mathbf{p}^*(t) \tag{6-500}$$

$$\dot{\mathbf{p}}^*(t) = [-\mathbf{Q}(t) + \mathbf{M}(t)\mathbf{R}^{-1}(t)\mathbf{M}(t)]\mathbf{x}^*(t) + [-\mathbf{A}'(t) + \mathbf{M}(t)\mathbf{R}^{-1}(t)\mathbf{B}'(t)]\mathbf{p}^*(t) \tag{6-501}$$

For the sake of convenience, let us define the $n \times n$ matrices $\mathbf{W}_{11}$, $\mathbf{W}_{12}$, $\mathbf{W}_{21}$, and $\mathbf{W}_{22}$ by setting

$$\begin{aligned} \mathbf{W}_{11}(t) &= \mathbf{A}(t) - \mathbf{B}(t)\mathbf{R}^{-1}(t)\mathbf{M}(t) \\ \mathbf{W}_{12}(t) &= -\mathbf{B}(t)\mathbf{R}^{-1}(t)\mathbf{B}'(t) \\ \mathbf{W}_{21}(t) &= -\mathbf{Q}(t) + \mathbf{M}(t)\mathbf{R}^{-1}(t)\mathbf{M}(t) \\ \mathbf{W}_{22}(t) &= -\mathbf{A}'(t) + \mathbf{M}(t)\mathbf{R}^{-1}(t)\mathbf{B}'(t) \end{aligned} \tag{6-502}$$

We can thus write Eqs. (6-500) and (6-501) in the form

$$\begin{bmatrix} \dot{\mathbf{x}}^*(t) \\ \dot{\mathbf{p}}^*(t) \end{bmatrix} = \begin{bmatrix} \mathbf{W}_{11}(t) & \mathbf{W}_{12}(t) \\ \mathbf{W}_{21}(t) & \mathbf{W}_{22}(t) \end{bmatrix} \begin{bmatrix} \mathbf{x}^*(t) \\ \mathbf{p}^*(t) \end{bmatrix} \tag{6-503}$$

Let $\boldsymbol{\Psi}(t; t_0)$ be the $2n \times 2n$ fundamental matrix for the system (6-503). We partition the $\boldsymbol{\Psi}(t; t_0)$ matrix into four $n \times n$ submatrices as follows:

$$\boldsymbol{\Psi}(t; t_0) = \begin{bmatrix} \boldsymbol{\Psi}_{11}(t; t_0) & \boldsymbol{\Psi}_{12}(t; t_0) \\ \boldsymbol{\Psi}_{21}(t; t_0) & \boldsymbol{\Psi}_{22}(t; t_0) \end{bmatrix} \tag{6-504}$$

We can now write the solution of (6-503) as

$$\mathbf{x}^*(t) = \boldsymbol{\Psi}_{11}(t; t_0)\mathbf{x}^*(t_0) + \boldsymbol{\Psi}_{12}(t; t_0)\mathbf{p}^*(t_0) \tag{6-505}$$

$$\mathbf{p}^*(t) = \boldsymbol{\Psi}_{21}(t; t_0)\mathbf{x}^*(t_0) + \boldsymbol{\Psi}_{22}(t; t_0)\mathbf{p}^*(t_0) \tag{6-506}$$

But it is necessary that $\mathbf{x}^*(t_0) = \boldsymbol{\xi}$ and that $\mathbf{x}^*(T_f) = \mathbf{0}$; therefore, from Eq. (6-505), we find that

$$\boldsymbol{\Psi}_{12}(T_f; t_0)\mathbf{p}^*(t_0) = -\boldsymbol{\Psi}_{11}(T_f; t_0)\boldsymbol{\xi} \tag{6-507}$$

If the matrix $\boldsymbol{\Psi}_{12}(T_f; t_0)$ is nonsingular, i.e., if

$$\det \boldsymbol{\Psi}_{12}(T_f; t_0) \neq 0 \tag{6-508}$$

then we can solve for $\mathbf{p}^*(t_0)$ in Eq. (6-507) and substitute the value

$$\mathbf{p}^*(t_0) = -\boldsymbol{\Psi}_{12}^{-1}(T_f; t_0)\boldsymbol{\Psi}_{11}(T_f; t_0)\boldsymbol{\xi} \tag{6-509}$$

into Eqs. (6-505), (6-506), and (6-494) to find that

$$\mathbf{x}^*(t) = [\boldsymbol{\Psi}_{11}(t; t_0) - \boldsymbol{\Psi}_{12}(t; t_0)\boldsymbol{\Psi}_{12}^{-1}(T_f; t_0)\boldsymbol{\Psi}_{11}(T_f; t_0)]\boldsymbol{\xi} \tag{6-510}$$

$$\mathbf{p}^*(t) = [\boldsymbol{\Psi}_{21}(t; t_0) - \boldsymbol{\Psi}_{22}(t; t_0)\boldsymbol{\Psi}_{12}^{-1}(T_f; t_0)\boldsymbol{\Psi}_{11}(T_f; t_0)]\boldsymbol{\xi} \tag{6-511}$$

$$\begin{aligned} \mathbf{u}^*(t) = -\mathbf{R}^{-1}(t)\{&\mathbf{M}(t)[\boldsymbol{\Psi}_{11}(t; t_0) - \boldsymbol{\Psi}_{12}(t; t_0)\boldsymbol{\Psi}_{12}^{-1}(T_f; t_0)\boldsymbol{\Psi}_{11}(T_f; t_0)] \\ &+ \mathbf{B}'(t)[\boldsymbol{\Psi}_{21}(t; t_0) - \boldsymbol{\Psi}_{22}(t; t_0)\boldsymbol{\Psi}_{12}^{-1}(T_f; t_0)\boldsymbol{\Psi}_{11}(T_f; t_0)]\}\boldsymbol{\xi} \end{aligned} \tag{6-512}$$

The matrix $\Psi_{12}(T_f; t_0)$ is independent of the initial state; it depends on the matrices $\mathbf{A}(t)$, $\mathbf{B}(t)$, $\mathbf{Q}(t)$, $\mathbf{M}(t)$, $\mathbf{R}(t)$, on the initial time $t_0$, and on the terminal time $T_f$. For this reason, we can state the following theorem:

***Theorem 6-17*** *If an optimal control exists and if the matrix $\Psi_{12}(t_f; t_0)$ is nonsingular, then the optimal control is unique and is given by Eq.* (6-512). *Moreover, there is one and only one extremal control.*

PROOF If $\Psi_{12}(T_f; t_0)$ is nonsingular, then Eq. (6-509) implies that there is a unique initial costate $\mathbf{p}^*(t_0)$ corresponding to $\xi$, $t_0$, and $T_f$. Thus, $\mathbf{x}^*(t)$ and $\mathbf{p}^*(t)$ are unique and so is the control $\mathbf{u}^*(t)$. This establishes the uniqueness of the extremal control. If, in addition, an optimal control exists, then it must be $\mathbf{u}^*(t)$ and, hence, unique.

If, on the other hand, the matrix $\Psi_{12}(T_f; t_0)$ is singular, then there may exist many initial costates $\mathbf{p}^*(t_0)$ such that the relation

$$\Psi_{12}(T_f; t_0)\mathbf{p}_i^*(t_0) = -\Psi_{11}(T_f; t_0)\xi \qquad i = 1, 2, \ldots \tag{6-513}$$

is satisfied. In this case, the extremal controls are not unique, and so there may be nonunique optimal solutions together with relatively optimal solutions. To the best of the authors' knowledge, necessary and/or sufficient conditions that imply the nonsingularity of $\Psi_{12}(T_f; t_0)$ are not available at present in the control literature.

There are many directions in which one may go from this point; one could compute the cost $J(\mathbf{u}^*)$ and show that it satisfies the Hamilton-Jacobi equation; one could formulate the problem of going from a given initial state to another state or to a target set; one could specialize the results to time-invariant systems, and so on. In each case, the linearity of the controlled system, the quadratic nature of the cost functional, and the absence of magnitude constraints on the control result in linear equations which can be easily manipulated to derive explicit equations for the necessary conditions. The following exercises illustrate all these points. In the next section, we shall consider the control of a specific second-order system to illustrate the method of determining the optimal control.

**Exercise 6-33** Show that the assumption that the matrix $\mathbf{N}(t)$ given by Eq. (6-478) is positive definite yields a sufficient condition for the local optimality of the extremal control. HINT: See Sec. 5-8.

**Exercise 6-34** Assume that the optimal control $\mathbf{u}^*(t)$ exists and that it is unique. Show that the minimum value $J(\mathbf{u}^*)$ of the cost (6-479) is of the form

$$J(\mathbf{u}^*) = \tfrac{1}{2}\langle \xi, \mathbf{G}(t_0; T_f)\xi \rangle \tag{6-514}$$

where $\mathbf{G}(t_0; T_f)$ is an $n \times n$ matrix independent of $\xi$. Write the equation of $\mathbf{G}(t_0; T_f)$ [HINT: Substitute Eqs. (6-512) and (6-510) into Eq. (6-479) so that $\mathbf{G}(t_0; T_f)$ is specified by the integral of a time-varying matrix]. Show that the assumption that $\mathbf{N}(t)$ is positive definite implies that $\mathbf{G}(t_0; T_f)$ is also positive definite. Show that the minimum cost $J(\mathbf{u}^*)$ is the solution of the Hamilton-Jacobi equation.

**Exercise 6-35** Consider the linear time-invariant system

$$\dot{\mathbf{x}}(t) = \mathbf{A}\mathbf{x}(t) + \mathbf{B}\mathbf{u}(t) \tag{6-515}$$

with the initial state $\mathbf{x}(0) = \boldsymbol{\xi}$. Now let us suppose that we are given a function of time $\mathbf{z}(t)$, a state $\boldsymbol{\theta}$, and a time $T_f$, such that

$$\mathbf{z}(0) = \boldsymbol{\xi} \qquad \mathbf{z}(T_f) = \boldsymbol{\theta} \tag{6-516}$$

We may think of $\mathbf{z}(t)$ as defining a desired path from $\boldsymbol{\xi}$ to $\boldsymbol{\theta}$. The objective is to force the system (6-515) from $\boldsymbol{\xi}$ to $\boldsymbol{\theta}$ and to make $\mathbf{x}(t)$ be near $\mathbf{z}(t)$. For this reason, we define the error $\mathbf{e}(t)$ by setting

$$\mathbf{e}(t) = \mathbf{z}(t) - \mathbf{x}(t) \tag{6-517}$$

and consider the cost functional

$$J_1(\mathbf{u}) = \frac{1}{2}\int_0^{T_f} [\langle \mathbf{e}(t), \mathbf{Q}\mathbf{e}(t)\rangle + \langle \mathbf{u}(t), \mathbf{R}\mathbf{u}(t)\rangle]\, dt \tag{6-518}$$

where both $\mathbf{Q}$ and $\mathbf{R}$ are positive definite constant symmetric matrices and $T_f$ is given. Derive the necessary conditions on the optimal control. Assuming that an optimal control exists, discuss the uniqueness of the optimum control and of the extremal control. Show that one must know $\mathbf{z}(t)$ exactly for all $t \in [0, T_f]$.

**Exercise 6-36** Consider the problem formulated in Exercise 6-35 and suppose that $\mathbf{z}(t)$ is the solution of the known linear homogeneous system

$$\dot{\mathbf{z}}(t) = \mathbf{F}\mathbf{z}(t) \qquad \mathbf{z}(0) = \boldsymbol{\xi} \tag{6-519}$$

such that $\mathbf{z}(T_f) = \boldsymbol{\theta}$. Show that the optimal control $\mathbf{u}^*(t)$ requires knowledge of the matrix $\mathbf{F}$ but does not require knowledge of the entire solution $\mathbf{z}(t)$ for $t \in [0, T_f]$.

## 6-19 Minimum-energy Problems 3: An Example†

In this section, we shall examine several aspects of the minimum-energy control of a second-order linear system. The equations of motion of this system may represent the field-controlled d-c motor described in Example 6-8. We shall derive the optimal-control law for two related performance criteria and explicitly exhibit some of the peculiarities of the solutions.

Let us consider the system

$$\begin{aligned} \dot{x}_1(t) &= x_2(t) \qquad & x_1(0) &= \xi_1 \\ \dot{x}_2(t) &= u(t) \qquad & x_2(0) &= \xi_2 \end{aligned} \tag{6-520}$$

We assume that there are no magnitude constraints on the control $u(t)$. We wish to find the control $u(t)$ which (1) transfers the initial state $(\xi_1, \xi_2)$ to $(0, 0)$ in time $T_f$, that is, we want

$$x_1(T_f) = x_2(T_f) = 0 \tag{6-521}$$

† The material in this section is based, in part, on Ref. G-8.

and (2) minimizes the cost (energy) functional

$$E = E(u) = \frac{1}{2} \int_0^{T_f} u^2(t)\, dt \tag{6-522}$$

where $T_f$ is specified.

In order to find the optimal control, we begin by finding the control that satisfies the necessary conditions, i.e., the extremal control. The first step is to write the Hamiltonian for the system (6-520) and the cost functional (6-522); the Hamiltonian is given by the equation

$$H = \tfrac{1}{2}u^2(t) + x_2(t)p_1(t) + u(t)p_2(t) \tag{6-523}$$

The extremal control must minimize the Hamiltonian; since the Hamiltonian is a quadratic function of $u(t)$, we can find the extremal control by setting $\partial H/\partial u(t) = 0$ and by checking whether or not $\partial^2 H/\partial u(t)^2$ is positive. Since

$$\frac{\partial H}{\partial u(t)} = u(t) + p_2(t) \tag{6-524}$$

and since

$$\frac{\partial^2 H}{\partial u(t)^2} = 1 \tag{6-525}$$

we conclude that the extremal control is given by

$$u(t) = -p_2(t) \qquad t \in [0, T_f] \tag{6-526}$$

The costate variables $p_1(t)$ and $p_2(t)$ must satisfy the differential equations

$$\dot{p}_1(t) = -\frac{\partial H}{\partial x_1(t)} = 0 \tag{6-527}$$

$$\dot{p}_2(t) = -\frac{\partial H}{\partial x_2(t)} = -p_1(t) \tag{6-528}$$

Let

$$\pi_1 = p_1(0) \qquad \pi_2 = p_2(0) \tag{6-529}$$

denote the unknown initial values of the costate variables. We can now solve Eqs. (6-527) and (6-528) to find that

$$\begin{aligned} p_1(t) &= \pi_1 = \text{const} \\ p_2(t) &= \pi_2 - \pi_1 t \end{aligned} \tag{6-530}$$

Thus, the extremal control is given by

$$u(t) = -\pi_2 + \pi_1 t \qquad t \in [0, T_f] \tag{6-531}$$

Equation (6-531) implies that the optimal control (if it exists) must be a linear function of time with $u(0) = -\pi_2$ and slope $\pi_1$. The next step is, therefore, the determination of $\pi_1$ and $\pi_2$ in terms of $\xi_1$, $\xi_2$, and $T_f$.

Let us substitute the control (6-531) into the system equations (6-520) and integrate to obtain the solution

$$\begin{aligned} x_1(t) &= \xi_1 + \xi_2 t - \tfrac{1}{2}\pi_2 t^2 + \tfrac{1}{6}\pi_1 t^3 \\ x_2(t) &= \xi_2 - \pi_2 t + \tfrac{1}{2}\pi_1 t^2 \end{aligned} \tag{6-532}$$

Since it is necessary that $x_1(T_f) = x_2(T_f) = 0$, we find from Eqs. (6-532) and (6-521) that $\pi_1$ and $\pi_2$ are *uniquely* defined by the relations

$$\pi_1 = \frac{6}{T_f^3}(2\xi_1 + \xi_2 T_f) \tag{6-533}$$

$$\pi_2 = \frac{2}{T_f^2}(3\xi_1 + 2\xi_2 T_f) \tag{6-534}$$

The extremal control is then given by

$$u(t) = -\frac{2}{T_f^2}(3\xi_1 + 2\xi_2 T_f) + \frac{6}{T_f^3}(2\xi_1 + \xi_2 T_f)t \tag{6-535}$$

The control (6-535) satisfies the necessary conditions and, furthermore, is a unique function of $\xi_1$, $\xi_2$, $T_f$, and $t$. We also assert that an optimal control exists toı this problem. We can thus conclude that the *optimal control is unique and is given by Eq.* (6-535).

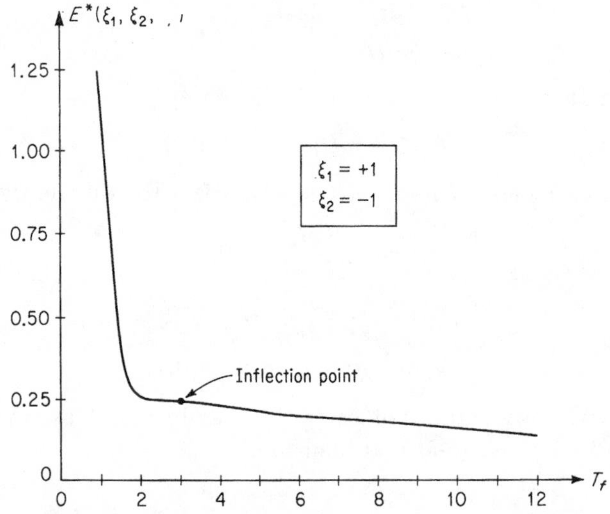


**Fig. 6-21** The plot of the energy $E^*(\xi_1, \xi_2, T_f)$ versus the terminal time $T_f$.

Let us now compute the minimum energy $E^*$ required by the optimal control. To do this, we substitute Eq. (6-535) into Eq. (6-522), we integrate, and we find that

$$E^* = E^*(\xi_1, \xi_2, T_f) = \frac{2}{T_f^3}(3\xi_1^2 + 3\xi_1\xi_2 T_f + \xi_2^2 T_f^2) \tag{6-536}$$

To illustrate the fact that the choice of the terminal time $T_f$ is quite critical, let us consider the initial state $\xi_1 = 1$, $\xi_2 = -1$ and examine the plot of $E^*(1, -1, T_f)$ versus $T_f$, as illustrated in Fig. 6-21. We observe that for values of $T_f < 2$, the energy $E^*$ is quite large; then the energy reaches an inflection point at $T_f = 3$; and for $T_f > 3$ the minimum energy required decreases very slowly. For $T_f$ very large, it is clear that $E^* \approx 2\xi_2^2/T_f$. In Fig. 6-22, we illustrate the optimal trajectories from the initial state $(1, -1)$ to $(0, 0)$. Observe the severe changes that can occur as the terminal time $T_f$ increases. From Fig. 6-21, we know that the energy differential for $T_f = 3$ and $T_f = 10$ is very small; however, the trajectories are radically different (with large overshoots occurring when $T_f$ is large).

The above discussion brings the sensitivity of optimal problems into focus; quite often, small changes in some parameter (say $T_f$) can cause either large or small changes in the minimum cost. A good engineering design must take such effects into account.

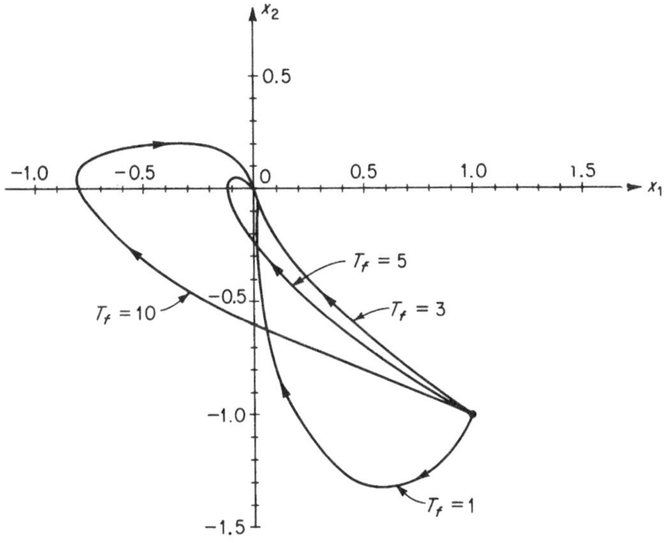


**Fig. 6-22** The behavior of the optimal trajectories originating at $(1, -1)$ as $T_f$ varies.

Next, we shall verify that the minimum energy is positive for all $\xi_1 \neq \xi_2 \neq 0$; to do this, we note that we can write Eq. (6-536) in the scalar-product notation as

$$E^*(\xi_1, \xi_2, T_f) = \left\langle \begin{bmatrix} \xi_1 \\ \xi_2 \end{bmatrix}, \frac{2}{T_f^3} \begin{bmatrix} 3 & \frac{3}{2}T_f \\ \frac{3}{2}T_f & T_f^2 \end{bmatrix} \begin{bmatrix} \xi_1 \\ \xi_2 \end{bmatrix} \right\rangle \tag{6-537}$$

and we observe that the matrix appearing in the scalar product is positive definite for all values of $T_f$ (compare with Exercise 6-34). Thus, $E^* \geq 0$.

If we are willing to accept large values of $T_f$, then we can accomplish the control task with a very small amount of energy. It is easy to see that

$$\lim_{T_f \to \infty} E^*(\xi_1, \xi_2, T_f) = 0 \tag{6-538}$$

and that

$$\lim_{T_f \to \infty} u(t) = 0 \qquad t \in [0, T_f] \tag{6-539}$$

But if the control $u(t)$ is small, then the trajectories exhibit large overshoots (see Fig. 6-22).

We saw in Fig. 6-21 that the $E^*$ versus $T_f$ curve had an inflection point. It is easy to see by setting $\partial E^*(\xi_1, \xi_2, T_f)/\partial T_f = 0$ that an inflection point occurs whenever the relation

$$\xi_2 T_f = -3\xi_1 \tag{6-540}$$

is satisfied. Now we shall demonstrate that if we formulate the minimum-energy problem as a free-terminal-time problem, then the necessary condition that the Hamiltonian function be zero at the terminal time yields this inflection point. Recall that the Hamiltonian is given by [see Eq. (6-523)]

$$H = \tfrac{1}{2}u^2(t) + x_2(t)p_1(t) + u(t)p_2(t) \tag{6-541}$$

Let $T$ be the unknown terminal time; then it is necessary that

$$\tfrac{1}{2}u^2(T) + x_2(T)p_1(T) + u(T)p_2(T) = 0 \tag{6-542}$$

But it is necessary that $x_2(T) = 0$ and that $u(t) = -p_2(t)$ for all $t \in [0, T]$; it follows that $u(T) = -p_2(T)$, and so Eq. (6-542) reduces to the equation

$$\tfrac{1}{2}u^2(T) = 0 \tag{6-543}$$

or, equivalently, to the equation

$$u(T) = 0 \tag{6-544}$$

But Eqs. (6-544) and (6-531) yield the relation

$$-\pi_2 + \pi_1 T = 0 \tag{6-545}$$

From Eqs. (6-545), (6-533), and (6-534), we can conclude that the initial state variables $\xi_1$ and $\xi_2$ and the terminal time $T$ must satisfy the equation

$$\frac{1}{T^2}(6\xi_1 + 2\xi_2 T) = 0 \tag{6-546}$$

Equation (6-546) is satisfied if

$$T = \infty \tag{6-547}$$

or if

$$\xi_2 T = -3\xi_1 \tag{6-548}$$

We thus note that the necessary conditions for the free-time problem have led to Eqs. (6-547) and (6-548). We have shown that the case $T = \infty$ corresponds to the absolute minimum of $E^*$, that is, to $E^* = 0$. We also note that Eq. (6-548) is the same as Eq. (6-540), and so Eq. (6-548) yields an inflection point rather than a minimum; this, of course, can happen, as we have noted that (see Sec. 5-17, comment 10) the necessary condition $H = 0$ corresponds only to extremal points which may be minima, maxima, or inflection and saddle points.

We also remark that the free-time problem does *not* have an optimal solution. The reason is that if $T = \infty$, then $u(t) = 0$ for all $t$, and of course the zero control does not take the initial state $(\xi_1, \xi_2)$ to $(0, 0)$. It is true that we can find optimal solutions for the fixed-time problem for arbitrarily large (but finite) values of $T_f$ and so reduce $E^*$ to an arbitrarily small (but nonzero) value. Nonetheless, the limiting solution cannot satisfy the boundary conditions, and so an optimal solution does not exist. However, an $\epsilon$-optimal solution does exist in this case.

Let us continue our explicit discussion of specific control problems by considering the optimal control of the system (6-520) with respect to a different performance criterion. Our purpose in the remainder of this section is to exhibit the existence of relative minima in some optimization problems.

To be specific, we pose the following optimization problem. We consider the system (6-520), i.e., the system

$$\begin{aligned} \dot{x}_1(t) &= x_2(t) \qquad & x_1(0) &= \xi_1 \\ \dot{x}_2(t) &= u(t) \qquad & x_2(0) &= \xi_2 \end{aligned} \tag{6-549}$$

Let us also consider the cost functional

$$J_1(u) = kT + \tfrac{1}{2}\int_0^T u^2(t)\,dt = \int_0^T [k + \tfrac{1}{2}u^2(t)]\,dt \tag{6-550}$$

where $T$ is free and $k > 0$. We wish to find the control that transfers the system (6-549) from the initial state $(\xi_1, \xi_2)$ to $(0, 0)$ and which minimizes the cost functional (6-550). Clearly, we are dealing with a fixed-end-point, free-terminal-time problem. The cost functional $J_1(u)$ represents† a linear combination of the elapsed time and the consumed energy.

We can solve this problem in two ways. One method of solution is to develop the necessary conditions and proceed in a straightforward manner. The other method of solution is to solve the posed problem assuming that the time $T$ is fixed, plot the cost $J_1$ as a function of the terminal time $T$, and then determine the particular terminal time for which the cost is smallest.

If we attack the problem under the assumption that $T$ is fixed, then the term $kT$ in the cost functional $J_1$ is a known constant, and so the minimization of the cost functional (6-550) with $T$ fixed is equivalent to the minimization of the energy $\frac{1}{2}\int_0^T u^2(t)\,dt$. For this reason, the control (6-535) is still optimal, and we immediately conclude that the minimum value of the cost functional is [see Eq. (6-536)]

$$J_1(\xi_1, \xi_2, T) = kT + \frac{2}{T^3}(3\xi_1^2 + 3\xi_1\xi_2 T + \xi_2^2 T^2) \tag{6-551}$$

† Compare with Problem 6-2*g*.

Clearly, the cost $J_1$ is a function of the terminal time $T$. To find the particular value of the terminal time $T$, say $T^*$, such that

$$J_1(\xi_1, \xi_2, T^*) \leq J_1(\xi_1, \xi_2, T) \tag{6-552}$$

we first determine the extremal points of $J_1(\xi_1, \xi_2, T)$ by setting

$$\frac{\partial J_1(\xi_1, \xi_2, T)}{\partial T} = 0 \tag{6-553}$$

and we find that

$$\frac{2}{\beta^2} - \frac{18\xi_1^2}{T^4} - \frac{12\xi_1\xi_2}{T^3} - \frac{2\xi_2^2}{T^2} = 0 \tag{6-554}$$

where we have set

$$k = \frac{2}{\beta^2} \qquad \beta > 0 \tag{6-555}$$

in order to simplify the equations. From Eq. (6-554), we find, after some algebraic manipulations, that the terminal times at which the function $J_1(\xi_1, \xi_2, T)$ has an extremum are the four solutions of the fourth-order equation

$$T^4 - \beta^2(\xi_2^2T^2 + 6\xi_1\xi_2T + 9\xi_1^2) = 0 \tag{6-556}$$

Let us denote the four roots of Eq. (6-556) by $T_1$, $T_2$, $T_3$, and $T_4$. By taking the square root of both sides of Eq. (6-556), we find that these extremal times are given by

$$\begin{aligned} T_1 &= \tfrac{1}{2}(\beta\xi_2 + \sqrt{\beta^2\xi_2^2 + 12\beta\xi_1}) \\ T_2 &= \tfrac{1}{2}(\beta\xi_2 - \sqrt{\beta^2\xi_2^2 + 12\beta\xi_1}) \\ T_3 &= \tfrac{1}{2}(-\beta\xi_2 + \sqrt{\beta^2\xi_2^2 - 12\beta\xi_1}) \\ T_4 &= \tfrac{1}{2}(-\beta\xi_2 - \sqrt{\beta^2\xi_2^2 - 12\beta\xi_1}) \end{aligned} \tag{6-557}$$

We wish now to isolate the positive terminal times which correspond to the local minima of $J_1(\xi_1, \xi_2, T)$. Thus, the requirement that

$$\frac{\partial^2 J(\xi_1, \xi_2, T)}{\partial T^2} > 0 \tag{6-558}$$

and that

$$T > 0 \tag{6-559}$$

can be used to show that the terminal time $T_i$, $i = 1, 2, 3, 4$, corresponds to a local minimum if and only if

$$0 < T_i < -3\frac{\xi_1}{\xi_2}$$

or

$$T_i > -6\frac{\xi_1}{\xi_2} \tag{6-560}$$

If $T_i$ is such that

$$0 < -3\frac{\xi_1}{\xi_2} < T_i < -6\frac{\xi_1}{\xi_2} \tag{6-561}$$

then that time corresponds to a local maximum.

At this point, let us digress and show that the necessary conditions for the original free-time problem yield Eqs. (6-557). First of all, the Hamiltonian $H_1$ [for the system (6-549) and the cost functional (6-550)] is given by

$$H_1 = \frac{2}{\beta^2} + \frac{1}{2}u^2(t) + x_2(t)p_1(t) + u(t)p_2(t) \tag{6-562}$$

where $k = 2/\beta^2$. The control that minimizes the Hamiltonian is

$$u(t) = -p_2(t) \tag{6-563}$$

and it is easy to see that Eqs. (6-527) to (6-535) remain the same (except that $T_f$ is replaced by $T$). Since we now deal with a free-time problem, the necessary condition that $H_1 = 0$ at $t = T$ yields the relation

$$\frac{2}{\beta^2} + \frac{1}{2}u^2(T) + x_2(T)p_1(T) + u(T)p_2(T) = 0$$

But since $x_2(T) = 0$ and $u(T) = -p_2(T)$, we find that this equation reduces to

$$u(T) = \pm\frac{2}{\beta} \tag{6-564}$$

From Eqs. (6-564), (6-535), (6-533), and (6-534), we find that

$$T^2 = \pm(\beta\xi_2 T + 3\beta\xi_1) \tag{6-565}$$

We can immediately verify that the four roots of Eq. (6-565) are *precisely* the four terminal times $T_1$, $T_2$, $T_3$, and $T_4$ given by Eqs. (6-557). In other words, the necessary conditions imply that the terminal time must be given by Eqs. (6-557).

On the basis of our previous discussion, we can conclude that the necessary condition $H = 0$ provides us only with the terminal times that extremize the cost $J_1(\xi_1, \xi_2, T)$. This is again due to the fact that the necessary condition $H = 0$ for the free-time problem is an extremal condition (see Sec. 5-17, comment 10) rather than a condition for a local minimum.

We shall now proceed to find the optimal solution. The idea is to determine which of the four terminal times $T_1$, $T_2$, $T_3$, and $T_4$ are real, positive, and correspond to local minima of $J_1(\xi_1, \xi_2, T)$. After some algebraic manipulations, we can deduce the following conclusions:

1. If

$$\xi_2 \geq 0 \quad \text{and} \quad \xi_1 \geq 0 \tag{6-566}$$

or if

$$\xi_2 < 0 \quad \text{and} \quad \xi_1 \geq \frac{\beta\xi_2^2}{12} \tag{6-567}$$

then only the time $T_1$ corresponds to a minimum; it follows that the optimal control is unique and that the terminal time is $T_1$.

2. If

$$\xi_2 \leq 0 \qquad \text{and} \qquad \xi_1 \leq 0 \tag{6-568}$$

or if

$$\xi_2 > 0 \qquad \text{and} \qquad \xi_1 \leq -\frac{\beta\xi_2^2}{12} \tag{6-569}$$

then only the time $T_3$ corresponds to a minimum; it follows that the optimal control is unique and that the terminal time is $T_3$.

3. If

$$\xi_2 > 0 \qquad \text{and} \qquad -\frac{\beta\xi_2^2}{12} < \xi_1 < 0 \tag{6-570}$$

then the times $T_1$ and $T_3$ correspond to two local minima, while the time $T_2$ corresponds to a local maximum. In this case, there are two extremal controls, and so there may be two optimal controls.

4. If

$$\xi_2 < 0 \qquad \text{and} \qquad 0 < \xi_1 < \frac{\beta\xi_2^2}{12} \tag{6-571}$$

then the times $T_1$ and $T_3$ correspond to two local minima, while the time $T_4$ corresponds to a local maximum. In this case, there are two extremal controls, and so there may be two optimal controls.

In the last two cases (3 and 4), there is the possibility of nonunique optimal controls. We assert† that there is a set (curve) $\Gamma_\beta$ in the state plane such that if $(\xi_1, \xi_2) \in \Gamma_\beta$, then there are two distinct optimal controls. One of these controls requires a small response time and a large energy expenditure, while the second requires a large response time and a small energy expenditure (why?). The set $\Gamma_\beta$ is defined by the relation

$$\Gamma_\beta = \left\{(\xi_1, \xi_2): \frac{2}{\beta^2} T_1 + \frac{2}{T_1^3}(3\xi_1^2 + 3\xi_1\xi_2 T_1 + \xi_2^2 T_1^2) = \frac{2}{\beta^2} T_3 + \frac{2}{T_3^3}(3\xi_1^2 + 3\xi_1\xi_2 T_3 + \xi_2^2 T_3^2)\right\} \tag{6-572}$$

where $T_1$ and $T_3$ are given by Eqs. (6-557).

We hope that this detailed example has illustrated the steps involved in the determination of the extremal controls and the meaning of the necessary conditions provided by the minimum principle. The following exercises are designed to acquaint the reader with additional properties of simple optimal-control problems.

**Exercise 6-37** Let $\beta = 12$ and $\xi_2 = 2$. Plot $J_1(\xi_1, \xi_2, T)$ versus $T$ for the following values of $\xi_1$: $\xi_1 = -10$, $\xi_1 = -5$, $\xi_1 = -4$, $\xi_1 = -3$, $\xi_1 = -1$, $\xi_1 = 0$, $\xi_1 = 1$, $\xi_1 = 5$. Note the change in the cost function. For the values $\beta = 12$ and $\xi_2 = 2$, determine the value of $\xi_1 = \hat{\xi}_1$ such that $(\hat{\xi}_1, 2) \in \Gamma_\beta$ [see Eq. (6-572)]. Plot the two optimal controls and their corresponding optimal trajectories in the state plane.

† See Ref. G-8.

**Exercise 6-38** Consider the system (6-520) and suppose that the terminal state is $(\theta_1, \theta_2)$. Determine the control that transfers $(\xi_1, \xi_2)$ to $(\theta_1, \theta_2)$ and minimizes the functional (6-522) with $T_f$ fixed. Comment on uniqueness. Now suppose that $\xi_1 = 1$, $\xi_2 = -1$. Find conditions on the terminal state $(\theta_1, \theta_2)$ such that the free-time problem [that is, $T_f$ not specified in Eq. (6-522)] has a solution.

**Exercise 6-39** Consider the system (6-520) and suppose that the terminal state is $(\theta_1, \theta_2)$. Determine the control that transfers $(\xi_1, \xi_2)$ to $(\theta_1, \theta_2)$ and minimizes the cost functional (6-550) with $T$ free. Comment on uniqueness.

**Exercise 6-40** Consider the system

$$\begin{aligned} \dot{x}_1(t) &= x_2(t) & x_1(0) &= \xi_1 \\ \dot{x}_2(t) &= -x_2(t) + u(t) & x_2(0) &= \xi_2 \end{aligned} \tag{6-573}$$

Find the control that transfers $(\xi_1, \xi_2)$ to $(0, 0)$ and minimizes the cost functionals

(a) $$E(u) = \tfrac{1}{2} \int_0^{T_f} u^2(t)\, dt \qquad T_f \text{ fixed} \tag{6-574}$$

(b) $$J_1(u) = \int_0^{T} (k + \tfrac{1}{2}u^2(t))\, dt \qquad T \text{ free; } k > 0 \tag{6-575}$$

**Exercise 6-41** Repeat Exercise 6-40 for the systems

(a) $$\begin{aligned} \dot{x}_1(t) &= x_2(t) \\ \dot{x}_2(t) &= x_2(t) + u(t) \end{aligned} \tag{6-576}$$

(b) $$\begin{aligned} \dot{x}_1(t) &= x_2(t) \\ \dot{x}_2(t) &= -x_1(t) + u(t) \end{aligned} \tag{6-577}$$

(c) $$\begin{aligned} \dot{x}_1(t) &= x_2(t) \\ \dot{x}_2(t) &= -2x_1(t) - 3x_2(t) + u(t) \end{aligned} \tag{6-578}$$

(d) $$\begin{aligned} \dot{x}_1(t) &= x_2(t) \\ \dot{x}_2(t) &= x_1(t) + u(t) \end{aligned} \tag{6-579}$$

**Exercise 6-42** Consider the system (6-520) and suppose that the terminal state is $(0, 0)$. Find the control that transfers $(\xi_1, \xi_2)$ to $(0, 0)$ and minimizes the cost functionals

(a) $$I_2(u) = \tfrac{1}{2} \int_0^{T_f} [x_1^2(t) + u^2(t)]\, dt \qquad T_f \text{ fixed} \tag{6-580}$$

(b) $$I_3(u) = \tfrac{1}{2} \int_0^{T_f} [x_1^2(t) + x_2^2(t) + u^2(t)]\, dt \qquad T_f \text{ fixed} \tag{6-581}$$

(c) $$I_4(u) = \tfrac{1}{2} \int_0^{T_f} [k + x_1^2(t) + u^2(t)]\, dt \qquad T \text{ free; } k > 0 \tag{6-582}$$

**Exercise 6-43** Repeat Exercise 6-42 for the system (6-577).

## 6-20 Minimum-energy Problems 4: Magnitude Constraints on the Control Variables†

In Sec. 6-18, we discussed the problem of optimal control of a linear system with respect to a quadratic performance criterion. We assumed that the components of the control vector were not constrained in magnitude. This assumption enabled us to compute the extremal controls analytically.

† Appropriate references for this section are J-5, L-15 to L-17, J-1, K-24, R-4, and A-6.

In this section, we shall consider Problem 6-3*a* with the additional assumption that the components $u_1(t), u_2(t), \ldots, u_r(t)$ of the control vector $\mathbf{u}(t)$ are constrained in magnitude by the relation

$$|u_j(t)| \leq 1 \qquad j = 1, 2, \ldots, r \tag{6-583}$$

We shall show that the components of the optimal control are *continuous* functions of time and that there are intervals of time during which the components of the optimal control are constant. The continuity of the components of the optimal control contrasts rather sharply with the fact that the components of the optimal control for a normal time-optimal or a normal fuel-optimal problem are piecewise constant. We shall also show that the particular minimum-energy problem at hand is normal.

The precise formulation of our problem is:

***Problem 6-3b*** *Given the linear, completely controllable system*

$$\dot{\mathbf{x}}(t) = \mathbf{A}(t)\mathbf{x}(t) + \mathbf{B}(t)\mathbf{u}(t) \qquad \mathbf{x}(t_0) = \boldsymbol{\xi} \tag{6-584}$$

*and the cost functional*

$$J(\mathbf{u}) = \tfrac{1}{2}\int_{t_0}^{T_f} [\langle \mathbf{x}(t), \mathbf{Q}(t)\mathbf{x}(t)\rangle + \langle \mathbf{u}(t), \mathbf{R}(t)\mathbf{u}(t)\rangle + 2\langle \mathbf{x}(t), \mathbf{M}(t)\mathbf{u}(t)\rangle]\, dt \tag{6-585}$$

*where the matrices* $\mathbf{Q}(t)$, $\mathbf{R}(t)$, *and* $\mathbf{M}(t)$ *satisfy the assumptions stated in Problem* 6-3*a and where the terminal time* $T_f$ *is fixed. Suppose that the control is constrained in magnitude by the relation*

$$|u_j(t)| \leq 1 \qquad j = 1, 2, \ldots, r \tag{6-586}$$

*Then find the control that:*

1. *Satisfies the magnitude constraint* (6-586)
2. *Transfers the system* (6-584) *from the initial state* $\boldsymbol{\xi}$ *to* $\mathbf{0}$ *in time* $T_f$
3. *Minimizes the cost functional* (6-585)

Since the system and the cost functional are the same for both Problem 6-3*a* and Problem 6-3*b*, we can see that the Hamiltonian for Problem 6-3*b* is given by Eq. (6-484), that the canonical equations are given by Eqs. (6-485) and (6-486), and that the boundary conditions are given by Eq. (6-487). The fact that the Hamiltonian must be minimum along the optimal trajectory is stated by Eq. (6-489), which must (for Problem 6-3*b*) hold for all admissible controls. In other words, it is necessary that the relation

$$\begin{aligned}\tfrac{1}{2}\langle \mathbf{u}^*(t), \mathbf{R}(t)\mathbf{u}^*(t)\rangle &+ \langle \mathbf{x}^*(t), \mathbf{M}(t)\mathbf{u}^*(t)\rangle + \langle \mathbf{B}(t)\mathbf{u}^*(t), \mathbf{p}^*(t)\rangle \\ &\leq \tfrac{1}{2}\langle \mathbf{u}(t), \mathbf{R}(t)\mathbf{u}(t)\rangle + \langle \mathbf{x}^*(t), \mathbf{M}(t)\mathbf{u}(t)\rangle + \langle \mathbf{B}(t)\mathbf{u}(t), \mathbf{p}^*(t)\rangle\end{aligned} \tag{6-587}$$

hold for all $\mathbf{u}(t)$ such that $|u_j(t)| \leq 1$, $j = 1, \ldots, r$.

Let us denote the components of the vector $\mathbf{w}^*(t)$, defined by the equation

$$\mathbf{w}^*(t) = \mathbf{R}^{-1}(t)[\mathbf{M}(t)\mathbf{x}^*(t) + \mathbf{B}'(t)\mathbf{p}^*(t)] \tag{6-588}$$

by $w_1^*(t), w_2^*(t), \ldots, w_r^*(t)$.

Then Eq. (6-587) can be written as

$$\tfrac{1}{2}\langle \mathbf{u}^*(t), \mathbf{R}(t)\mathbf{u}^*(t)\rangle + \langle \mathbf{u}^*(t), \mathbf{R}(t)\mathbf{w}^*(t)\rangle \le \tfrac{1}{2}\langle \mathbf{u}(t), \mathbf{R}(t)\mathbf{u}(t)\rangle + \langle \mathbf{u}(t), \mathbf{R}(t)\mathbf{w}^*(t)\rangle \tag{6-589}$$

Now we add the quantity $\frac{1}{2}\langle \mathbf{w}^*(t), \mathbf{R}(t)\mathbf{w}^*(t)\rangle$ to both sides of Eq. (6-589), and we find that

$$\langle [\mathbf{u}^*(t) + \mathbf{w}^*(t)], \mathbf{R}(t)[\mathbf{u}^*(t) + \mathbf{w}^*(t)]\rangle \le \langle [\mathbf{u}(t) + \mathbf{w}^*(t)], \mathbf{R}(t)[\mathbf{u}(t) + \mathbf{w}^*(t)]\rangle \tag{6-590}$$

for all $\mathbf{u}(t)$ such that $|u_j(t)| \le 1, j = 1, 2, \ldots, r$.

We shall now prove that the necessary condition (6-590) implies that

$$\begin{aligned} u_j^*(t) &= -w_j^*(t) && \text{if } |w_j^*(t)| \le 1 \\ u_j^*(t) &= -\operatorname{sgn}\{w_j^*(t)\} && \text{if } |w_j^*(t)| > 1 \end{aligned} \tag{6-591}$$

To prove Eqs. (6-591), we proceed as follows: Let

$$\mathbf{a}(t) = \mathbf{u}(t) + \mathbf{w}^*(t) \tag{6-592}$$

Then Eq. (6-590) implies that the function

$$\psi[\mathbf{u}(t)] = \langle \mathbf{a}(t), \mathbf{R}(t)\mathbf{a}(t)\rangle \tag{6-593}$$

attains its minimum at

$$\mathbf{a}^*(t) = \mathbf{u}^*(t) + \mathbf{w}^*(t) \tag{6-594}$$

Since $\mathbf{R}(t)$ is positive definite for all $t$, the eigenvalues $d_1(t), d_2(t), \ldots, d_r(t)$ of $\mathbf{R}(t)$ are positive.† Let $\mathbf{D}(t)$ be the diagonal matrix of the eigenvalues. We know that there is an orthogonal matrix $\mathbf{P}(t)$ [that is, $\mathbf{P}'(t)\mathbf{P}(t) = \mathbf{I}$] such that

$$\mathbf{D}(t) = \mathbf{P}'(t)\mathbf{R}(t)\mathbf{P}(t) \tag{6-595}$$

But

$$\begin{aligned} \psi[\mathbf{u}(t)] &= \langle \mathbf{a}(t), \mathbf{R}(t)\mathbf{a}(t)\rangle = \langle \mathbf{a}(t), \mathbf{P}(t)\mathbf{D}(t)\mathbf{P}'(t)\mathbf{a}(t)\rangle \\ &= \langle \mathbf{P}'(t)\mathbf{a}(t), \mathbf{D}(t)\mathbf{P}'(t)\mathbf{a}(t)\rangle = \langle \mathbf{b}(t), \mathbf{D}(t)\mathbf{b}(t)\rangle \\ &= \sum_{j=1}^{r} d_j(t)b_j^2(t) \end{aligned} \tag{6-596}$$

where

$$\mathbf{b}(t) = \mathbf{P}'(t)\mathbf{a}(t) \tag{6-597}$$

But since $\mathbf{P}(t)$ and $\mathbf{P}'(t)$ are both orthogonal, we know that

$$\langle \mathbf{b}(t), \mathbf{b}(t)\rangle = \langle \mathbf{a}(t), \mathbf{a}(t)\rangle \tag{6-598}$$

† See Sec. 2-15.

or, equivalently, that

$$\sum_{j=1}^{r} b_j^2(t) = \sum_{j=1}^{r} a_j^2(t) \tag{6-599}$$

Now we establish the relations

$$\begin{aligned}\min_{\mathbf{u}(t)} \psi[\mathbf{u}(t)] &= \min_{\mathbf{a}(t)} \langle \mathbf{a}(t), \mathbf{R}(t)\mathbf{a}(t)\rangle \\ &= \min_{\mathbf{b}(t)=\mathbf{P}'(t)\mathbf{a}(t)} \sum_{j=1}^{r} d_j(t) b_j^2(t) \\ &= \sum_{j=1}^{r} d_j(t) \min_{b_j(t)} b_j^2(t)\end{aligned} \tag{6-600}$$

Equation (6-600) implies that if $\mathbf{a}^*(t)$ minimizes $\langle \mathbf{a}(t), \mathbf{R}(t)\mathbf{a}(t)\rangle$, then the components $b_1^*(t), b_2^*(t), \ldots, b_r^*(t)$ of the vector

$$\mathbf{b}^*(t) = \mathbf{P}'(t)\mathbf{a}^*(t) \tag{6-601}$$

also minimize the scalar product $\langle \mathbf{b}(t), \mathbf{b}(t)\rangle$. In view of Eq. (6-598), we may conclude that the vector $\mathbf{P}(t)\mathbf{P}'(t)\mathbf{a}^*(t) = \mathbf{a}^*(t)$ minimizes the scalar product $\langle \mathbf{a}(t), \mathbf{a}(t)\rangle$. Thus we have established that

$$\begin{aligned}&\text{If} \quad \langle \mathbf{a}^*(t), \mathbf{R}(t)\mathbf{a}^*(t)\rangle \le \langle \mathbf{a}(t), \mathbf{R}(t)\mathbf{a}(t)\rangle \\ &\text{then} \quad \langle \mathbf{a}^*(t), \mathbf{a}^*(t)\rangle \le \langle \mathbf{a}(t), \mathbf{a}(t)\rangle\end{aligned} \tag{6-602}$$

We can reverse our reasoning to establish that

$$\begin{aligned}&\text{If} \quad \langle \mathbf{a}^*(t), \mathbf{a}^*(t)\rangle \le \langle \mathbf{a}(t), \mathbf{a}(t)\rangle \\ &\text{then} \quad \langle \mathbf{a}^*(t), \mathbf{R}(t)\mathbf{a}^*(t)\rangle \le \langle \mathbf{a}(t), \mathbf{R}(t)\mathbf{a}(t)\rangle\end{aligned} \tag{6-603}$$

But we know that

$$\langle \mathbf{a}(t), \mathbf{a}(t)\rangle = \langle [\mathbf{u}(t) + \mathbf{w}^*(t)], [\mathbf{u}(t) + \mathbf{w}^*(t)]\rangle = \sum_{j=1}^{r} [u_j(t) + w_j^*(t)]^2 \tag{6-604}$$

We can thus deduce that

$$\min_{\mathbf{u}(t)} \langle \mathbf{a}(t), \mathbf{a}(t)\rangle = \sum_{j=1}^{r} \min_{|u_j(t)|\le 1} [u_j(t) + w_j^*(t)]^2 \tag{6-605}$$

To minimize the positive quantity $[u_j(t) + w_j^*(t)]^2$, one must set $u_j(t) = -w_j^*(t)$ whenever $|w_j^*(t)| \le 1$, $u_j(t) = +1$ whenever $w_j^*(t) < -1$, and $u_j(t) = -1$ whenever $w_j^*(t) > +1$. Therefore, the control $\mathbf{u}^*(t)$ must be given by Eqs. (6-591), in view of the relations (6-602) and (6-603). Let us now define the sat { } function by the relation

$$\begin{aligned}& x_i = \text{sat}\,\{y_i\} \\ \text{means that} \quad & x_i = y_i \qquad \text{if } |y_i| \le 1 \\ & x_i = \text{sgn}\,\{y_i\} \quad \text{if } |y_i| > 1\end{aligned} \tag{6-606}$$

We can also define the vector-valued function SAT { } as follows:

$$\mathbf{x} = \text{SAT}\,\{\mathbf{y}\}$$
$$\text{means that} \qquad x_i = \text{sat}\,\{y_i\} \qquad \text{for all } i = 1, 2, \ldots, r \tag{6-607}$$

where $x_i$ and $y_i$, $i = 1, 2, \ldots, r$, are the components of $\mathbf{x}$ and $\mathbf{y}$, respectively.

We thus conclude [in view of Eqs. (6-591) and (6-588)] that the optimal control $\mathbf{u}^*(t)$ is uniquely related to the state $\mathbf{x}^*(t)$ and the costate $\mathbf{p}^*(t)$ by the relation

$$\mathbf{u}^*(t) = -\text{SAT}\,\{\mathbf{R}^{-1}(t)[\mathbf{M}(t)\mathbf{x}^*(t) + \mathbf{B}'(t)\mathbf{p}^*(t)]\} \tag{6-608}$$

We observe that the SAT function, unlike the SGN and DEZ functions, provides us with a well-defined $\mathbf{u}^*(t)$. For this reason, the singular problem does not arise in our minimum-energy problem 6-3*b*, provided that the matrix $\mathbf{R}(t)$ is positive definite. It also follows that the components $u_j^*(t)$ of the optimal control are continuous functions of time; this is in contradistinction to the piecewise constant components of the time-optimal control and of the fuel-optimal control.

We have seen in Sec. 6-18 that the optimal control for Problem 6-3*a* (no constraints on the control) is given by

$$\mathbf{u}_A^*(t) = -\mathbf{R}^{-1}(t)[\mathbf{M}(t)\mathbf{x}_A^*(t) + \mathbf{B}'(t)\mathbf{p}_A^*(t)] \tag{6-609}$$

and we have shown that the optimal control for Problem 6-3*b* is given by

$$\mathbf{u}_B^*(t) = -\,\text{SAT}\,\{\mathbf{R}^{-1}(t)[\mathbf{M}(t)\mathbf{x}_B^*(t) + \mathbf{B}'(t)\mathbf{p}_B^*(t)]\} \tag{6-610}$$

We have used the subscripts $A$ and $B$ to distinguish the optimal quantities in the two problems. In Sec. 6-18, we substituted the optimal control (6-609) into the canonical equations, and we were able to derive the optimal control as a function of $\boldsymbol{\xi}$, $t_0$, and $T_f$ [see Eq. (6-512)] under the assumption of the uniqueness of the optimal control.

In the case of a constrained control, we can substitute Eq. (6-608) into the canonical equations (6-485) and (6-486) to find that

$$\begin{aligned} \dot{\mathbf{x}}^*(t) &= \mathbf{A}(t)\mathbf{x}^*(t) - \mathbf{B}(t)\,\text{SAT}\,\{\mathbf{R}^{-1}(t)[\mathbf{M}(t)\mathbf{x}^*(t) + \mathbf{B}'(t)\mathbf{p}^*(t)]\} \\ \dot{\mathbf{p}}^*(t) &= -\mathbf{Q}(t)\mathbf{x}^*(t) - \mathbf{A}'(t)\mathbf{p}^*(t) - \mathbf{M}(t)\,\text{SAT}\,\{\mathbf{R}^{-1}(t)[\mathbf{M}(t)\mathbf{x}^*(t) \\ &\qquad + \mathbf{B}'(t)\mathbf{p}^*(t)]\} \end{aligned} \tag{6-611}$$

Unfortunately, this system of $2n$ differential equations is not linear. For this reason, we cannot obtain an analytical solution, and so we cannot explicitly determine the control $\mathbf{u}_B^*(t)$ as a function of $\boldsymbol{\xi}$, $T_f$, and $t_0$. Rather, we must attempt to find the solution (or solutions) of Eqs. (6-611) that satisfy the boundary conditions $\mathbf{x}^*(t_0) = \boldsymbol{\xi}$ and $\mathbf{x}^*(T_f) = \mathbf{0}$ by an iterative procedure using a computer.

Let us now suppose that we have solved Problem 6-3*a* and that we have found the optimal control $\mathbf{u}_A^*(t)$. It is clear that, in general, the components of the control $\mathbf{u}_A^*(t)$ need not satisfy the magnitude constraints imposed in Problem 6-3*b*. Let us suppose that we consider the control

$$\hat{\mathbf{u}}(t) = \text{SAT}\{\mathbf{u}_A^*(t)\} \tag{6-612}$$

In other words, we pass each component of $\mathbf{u}_A^*(t)$ through a limiter (which performs the sat { } operation), so that the control $\hat{\mathbf{u}}(t)$ does indeed meet the magnitude constraints of Problem 6-3*b*. We can now ask the following question: *Is the control* $\hat{\mathbf{u}}(t)$ *optimal for Problem* 6-3*b?* In other words, is it possible to have

$$\mathbf{u}_B^*(t) = \text{SAT}\{\mathbf{u}_A^*(t)\} \qquad t \in [t_0, T_f] \tag{6-613}$$

In general, the answer is *no*. Equation (6-613) will be true if and only if the components of $\mathbf{u}_A^*(t)$ never violate the magnitude constraints. However, this happens only in small subsets of the state space. For further details, the reader should consult Ref. J-5.

This concludes our treatment of the minimum-energy problem. In Chap. 9, we shall again deal with the design of linear systems with quadratic performance criteria. There we shall deal with the free-end-point case [that is, $\mathbf{x}(T_f)$ not specified]. Meanwhile, the following exercises are designed to illustrate some of the difficulties associated with the solution of minimum-energy problems when the control is constrained in magnitude.

**Exercise 6-44** Consider the system

$$\begin{aligned} \dot{x}_1(t) &= x_2(t) \qquad & x_1(0) &= \xi_1 \\ \dot{x}_2(t) &= u(t) \qquad & x_2(0) &= \xi_2 \end{aligned} \tag{6-614}$$

with

$$|u(t)| \le 1 \tag{6-615}$$

Determine the control that transfers the system (6-614) from $(\xi_1, \xi_2)$ to (0, 0) and that minimizes the cost functional

(*a*) $$J_1(u) = \int_0^{T_f} \tfrac{1}{2}u^2(t)\,dt \qquad T_f \text{ fixed} \tag{6-616}$$

(*b*) $$J_2(u) = \int_0^{T} [k + \tfrac{1}{2}u^2(t)]\,dt \qquad T \text{ free; } k > 0 \tag{6-617}$$

(*c*) $$J_3(u) = \int_0^{T_f} \tfrac{1}{2}[x_1^2(t) + u^2(t)]\,dt \qquad T_f \text{ fixed} \tag{6-618}$$

(*d*) $$J_4(u) = \int_0^{T_f} \tfrac{1}{2}[x_1^2(t) + x_2^2(t) + u^2(t)]\,dt \qquad T_f \text{ fixed} \tag{6-619}$$

(*e*) $$J_5(u) = \int_0^{T} [k + \tfrac{1}{2}x_1^2(t) + \tfrac{1}{2}u^2(t)]\,dt \qquad T \text{ free; } k > 0 \tag{6-620}$$

In parts *a*, *c*, and *d*, plot the optimal control and the optimal trajectories from the initial state (1, 1) for the following values of $T_f$: $T_f = \sqrt{6} + 1$, $T_f = 4$, $T_f = 5$, $T_f = 10$. What happens if $T_f < \sqrt{6} + 1$? Is there a value of $T_f$ such that the optimal control is not a constant in any subinterval of $[0, T_f]$?

**Exercise 6-45** Consider the system

$$\begin{aligned} \dot{x}_1(t) &= x_2(t) \qquad & x_1(0) &= \xi_1 \\ \dot{x}_2(t) &= -x_2(t) + u(t) \qquad & x_2(0) &= \xi_2 \end{aligned} \tag{6-621}$$

with $|u(t)| \leq 1$. Find the controls that transfer $(\xi_1, \xi_2)$ to $(0, 0)$ and that minimize

(*a*) The cost functional (6-617)

(*b*) The cost functional (6-620)

Comment on uniqueness. Plot the optimal trajectories from the initial state (1, 1) for $k = 1$, $k = 5$, $k = 10$, and $k = 100$.

**Exercise 6-46** Consider the system

$$\begin{aligned} \dot{x}_1(t) &= x_2(t) & x_1(0) &= \xi_1 \\ \dot{x}_2(t) &= x_2(t) + u(t) & x_2(0) &= \xi_2 \end{aligned} \qquad (6\text{-}622)$$

with $|u(t)| \leq 1$. Find the control that transfers the system (6-622) from $(\xi_1, \xi_2)$ to $(0, 0)$ and that minimizes the cost functional (6-617). Determine the set of initial states for which no solution exists. Comment on uniqueness.

## 6-21 Singular Problems 1: The Hamiltonian a Linear Function of the Control†

In Sec. 6-3, we discussed the singular time-optimal problem (see Definition 6-5). The reader may recall that the singular time-optimal problem arose when the argument of the sgn { } function was identically zero on a finite interval of time. In this case, the necessary condition

$$H(\mathbf{x}^*, \mathbf{p}^*, \mathbf{u}^*, t) \leq H(\mathbf{x}^*, \mathbf{p}^*, \mathbf{u}, t)$$

did not provide us with any information about the relation between $\mathbf{u}^*$, $\mathbf{x}^*$, and $\mathbf{p}^*$.

In this section and the next, we shall consider singular problems from a more general point of view. Here, we shall consider problems for which the Hamiltonian is a linear function of the control variable $u(t)$. In the next section, we shall examine problems for which the Hamiltonian is a linear function of the control $u(t)$ and of its absolute value $|u(t)|$.

In general, the solution of singular optimization problems is more difficult than the solution of normal optimization problems. This difficulty stems from the fact that the necessary condition that the Hamiltonian is minimized (with respect to the control) along the optimal trajectory does not provide us with a well-defined expression for the optimal control; in the absence of such information, we must manipulate the other necessary conditions in an effort to determine such a well-defined relationship. These ideas will become clearer as our discussion progresses.

Unfortunately, at this time, general results regarding the existence of singular solutions to optimization problems are rather limited. Additional research about the nature, properties, and other characteristics of singular solutions is required. For this reason, we warn the reader that we shall

† Appropriate references for this section and the next are J-6, J-4, W-6, P-2, H-8, R-2, R-4, and A-14. In particular, Ref. J-6 contains an extensive exposition of singular problems, with numerous examples.

merely scratch the surface of this subject in this section. It will become evident later on in this section that singular solutions involve complex computation problems. Our treatment will thus be somewhat loose, and our hope is that we shall provide enough information to the reader so that he will be able to attack a problem which involves singular solutions.

In order to avoid notational complications, we shall consider time-invariant problems with a single control variable; the concepts are the same for problems with many control variables. Furthermore, we shall deal with fixed-end-point problems with either free or fixed terminal time.

Next, we formulate a problem which may have a singular solution.

***Problem 6-4a*** *Given the time-invariant system*

$$\dot{x}_i(t) = f_i[\mathbf{x}(t)] + b_i[\mathbf{x}(t)]u(t) \qquad i = 1, 2, \ldots, n \tag{6-623}$$

*or, in vector notation,*

$$\dot{\mathbf{x}}(t) = \mathbf{f}[\mathbf{x}(t)] + \mathbf{b}[\mathbf{x}(t)]u(t) \tag{6-624}$$

*where the n vector* $\mathbf{x}(t)$ *is the state and where* $u(t)$ *is the scalar control variable. Assume that* $u(t)$ *is constrained in magnitude by the relation*

$$|u(t)| \leq 1 \qquad \text{for all } t \tag{6-625}$$

*Given an initial state* $\mathbf{x}(0) = \boldsymbol{\xi}$ *and a terminal state* $\boldsymbol{\theta}$. *Given the cost functional*

$$J(u) = \int_0^T \{f_0[\mathbf{x}(t)] + b_0[\mathbf{x}(t)]u(t)\}\, dt \tag{6-626}$$

*where* $f_0[\mathbf{x}(t)]$ *and* $b_0[\mathbf{x}(t)]$ *are scalar functions of the state* $\mathbf{x}(t)$ *and where the terminal time T is either free or fixed. Then determine the admissible control that transfers the system* (6-624) *from* $\boldsymbol{\xi}$ *to* $\boldsymbol{\theta}$ *[that is, such that* $\mathbf{x}(T) = \boldsymbol{\theta}$*] and that minimizes the cost functional* $J(u)$.

We note that the scalar control $u(t)$ enters linearly in both the system equation (6-623) and the integrand of the cost functional (6-626).

For the sake of simplifying the algebra, we let

$$\dot{x}_0(t) = f_0[\mathbf{x}(t)] + b_0[\mathbf{x}(t)]u(t) \qquad x_0(0) = 0 \tag{6-627}$$

and we shall write the Hamiltonian $H$ using the scalar†

$$p_0(t) = p_0 = \text{const} \geq 0 \tag{6-628}$$

Thus, the Hamiltonian function $H$ for Problem 6-4$a$ is

$$H = \sum_{i=0}^{n} f_i[\mathbf{x}(t)]p_i(t) + u(t) \sum_{i=0}^{n} b_i[\mathbf{x}(t)]p_i(t) \tag{6-629}$$

† See Secs. 5-15 and 5-16.

We note that the Hamiltonian $H$ is a linear function of $u(t)$. If we define functions $\alpha$ and $\beta$ by setting

$$\alpha[\mathbf{x}(t), \mathbf{p}(t), p_0] = \sum_{i=0}^{n} f_i[\mathbf{x}(t)]p_i(t) \tag{6-630}$$

$$\beta[\mathbf{x}(t), \mathbf{p}(t), p_0] = \sum_{i=0}^{n} b_i[\mathbf{x}(t)]p_i(t) \tag{6-631}$$

then $H$ can be written as

$$H = \alpha[\mathbf{x}(t), \mathbf{p}(t), p_0] + u(t)\beta[\mathbf{x}(t), \mathbf{p}(t), p_0] \tag{6-632}$$

The costate variables $p_i(t)$, $i = 0, 1, 2, \ldots, n$, are given by

$$\dot{p}_i(t) = -\frac{\partial H}{\partial x_i(t)}$$

and so we have, for $i = 0, 1, 2, \ldots, n$, the relation

$$\dot{p}_i(t) = -\sum_{j=0}^{n} p_j(t)\frac{\partial f_j[\mathbf{x}(t)]}{\partial x_i(t)} - u(t)\sum_{j=0}^{n} p_j(t)\frac{\partial b_j[\mathbf{x}(t)]}{\partial x_i(t)} \tag{6-633}$$

We shall now state the necessary conditions for Problem 6-4*a*.

***Theorem 6-18 Necessary Conditions for Problem 6-4a*** *If $u^*(t)$ is an optimal control and if $\mathbf{x}^*(t)$ is the corresponding optimal trajectory, then there are a costate $\mathbf{p}^*(t)$ and a constant $p_0^* \geq 0$ such that:*

*a. For $i = 1, 2, \ldots, n$,*

$$\dot{x}_i^*(t) = f_i[\mathbf{x}^*(t)] + b_i[\mathbf{x}^*(t)]u^*(t)$$
$$\dot{p}_i^*(t) = -\sum_{j=1}^{n} p_j^*(t)\frac{\partial f_j[\mathbf{x}^*(t)]}{\partial x_i^*(t)} - u^*(t)\sum_{j=1}^{n} p_j^*(t)\frac{\partial b_j[\mathbf{x}^*(t)]}{\partial x_i^*(t)} \tag{6-634}$$

*and*

$$\mathbf{x}^*(0) = \boldsymbol{\xi} \qquad \mathbf{x}^*(T) = \boldsymbol{\theta} \tag{6-635}$$

*b. For $t \in [0, T]$ and all $u(t)$ satisfying the constraint $|u(t)| \leq 1$, the following relation holds [see Eqs. (6-630) to (6-632)]:*

$$\alpha[\mathbf{x}^*(t), \mathbf{p}^*(t), p_0^*] + u^*(t)\beta[\mathbf{x}^*(t), \mathbf{p}^*(t), p_0^*] \leq \alpha[\mathbf{x}^*(t), \mathbf{p}^*(t), p_0^*] + u(t)\beta[\mathbf{x}^*(t), \mathbf{p}^*(t), p_0^*] \tag{6-636}$$

*c. If $T$ is free, then*

$$\alpha[\mathbf{x}^*(t), \mathbf{p}^*(t), p_0^*] + u^*(t)\beta[\mathbf{x}^*(t), \mathbf{p}^*(t), p_0^*] = 0 \tag{6-637}$$

*for all $t \in [0, T]$; if $T$ is fixed, then*

$$\alpha[\mathbf{x}^*(t), \mathbf{p}^*(t), p_0^*] + u^*(t)\beta[\mathbf{x}^*(t), \mathbf{p}^*(t), p_0^*] = c = const \tag{6-638}$$

*for all $t \in [0, T]$.*

It is easy to see that Eq. (6-636) implies that

$$u^*(t) = -\operatorname{sgn}\{\beta[\mathbf{x}^*(t), \mathbf{p}^*(t), p_0^*]\} = -\operatorname{sgn}\left\{\sum_{i=0}^{n} b_i[\mathbf{x}^*(t)]p_i^*(t)\right\} \qquad (6\text{-}639)$$

As long as the scalar function† $\beta[\mathbf{x}^*(t), \mathbf{p}^*(t), p_0^*]$ is not zero, Eq. (6-639) provides us with a well-defined relation for the control $u^*(t)$. If, however,

$$\beta[\mathbf{x}^*(t), \mathbf{p}^*(t), p_0^*] = 0 \qquad \text{for all } t \in (t_1, t_2] \qquad (6\text{-}640)$$

where $(t_1, t_2]$ is a subinterval of $[0, T]$, then the sgn { } function is not defined. Indeed, if Eq. (6-640) holds, then the necessary condition (6-636) reduces to

$$u^*(t) \cdot 0 \leq u(t) \cdot 0 \qquad \text{for all } t \in (t_1, t_2] \qquad (6\text{-}641)$$

which is identically satisfied for all $u(t)$ [even those whioh violate the constraint $|u(t)| \leq 1$]. This situation is called *singular*, and we shall examine its implications in this section.

We formalize this discussion in the following definition:

***Definition 6-13 Optimal and Extremal Singular Controls*** *We shall say that Problem 6-4a is singular if the optimal control $u^*(t)$, the resulting trajectory $\mathbf{x}^*(t)$, and the corresponding costate $\mathbf{p}^*(t)$ have the following property: There is at least one (half-open) interval $(t_1, t_2]$ in $[0, T]$ such that*

$$\begin{aligned} \sum_{i=0}^{n} p_i^*(t) b_i[\mathbf{x}^*(t)] &= 0 \qquad \textit{for all } t \in (t_1, t_2] \\ p_0^*(t) &= p_0^* \geq 0 \end{aligned} \qquad (6\text{-}642)$$

*In this case, we shall call the interval $(t_1, t_2]$ a singularity interval, the control function $u^*_{(t_1,t_2]}$ an optimal singular control, and the trajectory $\mathbf{x}^*_{(t_1,t_2]}$ an optimal singular trajectory.*

*If we can find an extremal control $\hat{u}(t)$ (that is, a control that satisfies all the necessary conditions of Theorem 6-18) such that the corresponding state $\hat{\mathbf{x}}(t)$ and costate $\hat{\mathbf{p}}(t)$ have the property that*

$$\begin{aligned} \sum_{i=0}^{n} \hat{p}_i(t) b_i[\hat{\mathbf{x}}(t)] &= 0 \qquad \textit{for all } t \in (t_1, t_2] \\ \hat{p}_0(t) &= \hat{p}_0 \geq 0 \end{aligned}$$

*then we shall call the control $\hat{u}_{(t_1,t_2]}$ an extremal singular control and the trajectory $\hat{\mathbf{x}}_{(t_1,t_2]}$ an extremal singular trajectory.*

Naturally, the existence of an extremal singular control does not necessarily imply that the optimal control is singular. In such cases, we need additional information (e.g., uniqueness) to draw conclusions regarding the optimal control.

† This function is often called the *switching function.*

We shall next examine the implications of singular extremal controls. We shall adopt the point of view that we are given Problem 6-4*a* and that we are trying to find the extremal controls. The first step is to form the Hamiltonian (6-629) and the canonical equations. Let us furthermore suppose that we are dealing with the free-terminal-time case and that we wish to test whether or not it is possible to have extremal singular controls. We carry out this test as follows:

We suppose that†

$$\sum_{i=0}^{n} b_i p_i = 0 \qquad \text{for all } t \in (t_1, t_2] \tag{6-643}$$

Since we deal with a free-terminal-time problem, we must have

$$H = \sum_{i=0}^{n} f_i p_i + u \sum_{i=0}^{n} b_i p_i = 0 \qquad \text{for all } t \in (t_1, t_2] \tag{6-644}$$

Thus, Eqs. (6-643) and (6-644) require that

$$\sum_{i=0}^{n} f_i p_i = 0 \qquad \text{for all } t \in (t_1, t_2] \tag{6-645}$$

But Eq. (6-643) implies that

$$\frac{d^\gamma}{dt^\gamma} \sum_{i=0}^{n} b_i p_i = 0 \qquad \text{for all } \gamma = 1, 2, \ldots \text{ and all } t \in (t_1, t_2] \tag{6-646}$$

Similarly, Eq. (6-645) implies that

$$\frac{d^\gamma}{dt^\gamma} \sum_{i=0}^{n} f_i p_i = 0 \qquad \text{for all } \gamma = 1, 2, \ldots \text{ and all } t \in (t_1, t_2] \tag{6-647}$$

But the canonical equations are

$$\left.\begin{aligned} \dot{x}_i &= f_i + u b_i \\ \dot{p}_i &= -\sum_{j=0}^{n} p_j \frac{\partial f_j}{\partial x_i} - u \sum_{j=0}^{n} p_j \frac{\partial b_j}{\partial x_i} \end{aligned}\right\} \qquad i = 0, 1, 2, \ldots, n \tag{6-648}$$

Let $\gamma = 1$ in Eq. (6-646); then‡

$$\frac{d}{dt} \sum_i b_i p_i = \sum_i \left( \dot{p}_i b_i + p_i \sum_j \frac{\partial b_i}{\partial x_j} \dot{x}_j \right) = 0 \tag{6-649}$$

† From now on, we shall omit the $\mathbf{x}(t)$ dependence of the variables in order to simplify Eqs. (6-643) to (6-675).

‡ Once more, in order to simplify the equations, we shall use $\sum_i$ to mean $\sum_{i=0}^{n}$; furthermore, in Eqs. (6-649) to (6-675), we omit the statement "for all $t \in (t_1, t_2]$."

We substitute Eq. (6-648) into Eq. (6-649), and we find, after some algebraic manipulations, that

$$\sum_i \sum_j \left( f_j p_i \frac{\partial b_i}{\partial x_j} - b_i p_j \frac{\partial f_j}{\partial x_i} \right) + u \sum_i \sum_j \left( p_i b_j \frac{\partial b_i}{\partial x_j} - p_j b_i \frac{\partial b_j}{\partial x_i} \right) = 0 \quad (6\text{-}650)$$

But the coefficient of $u$ in Eq. (6-650) is zero (why?). We thus conclude that Eq. (6-649) implies that

$$\sum_i \sum_j p_i \left( b_j \frac{\partial f_i}{\partial x_j} - f_j \frac{\partial b_i}{\partial x_j} \right) = 0 \quad (6\text{-}651)$$

Next, we let $\gamma = 1$ in Eq. (6-647), and we find that

$$\begin{aligned} 0 &= \frac{d}{dt} \sum_i f_i p_i = \sum_i \left( f_i \dot{p}_i + p_i \sum_j \frac{\partial f_i}{\partial x_j} \dot{x}_j \right) \\ &= \sum_i \sum_j \left( p_i f_j \frac{\partial f_i}{\partial x_j} - p_j f_i \frac{\partial f_j}{\partial x_i} \right) + u \sum_i \sum_j \left( p_i b_j \frac{\partial f_i}{\partial x_j} - p_j f_i \frac{\partial b_j}{\partial x_i} \right) \\ &= u \sum_i \sum_j p_i \left( b_j \frac{\partial f_i}{\partial x_j} - f_j \frac{\partial b_i}{\partial x_j} \right) = 0 \end{aligned} \quad (6\text{-}652)$$

which implies that either

$$u = 0 \quad (6\text{-}653)$$

or

$$\sum_i \sum_j p_i \left( b_j \frac{\partial f_i}{\partial x_j} - f_j \frac{\partial b_i}{\partial x_j} \right) = 0 \quad (6\text{-}654)$$

But Eqs. (6-654) and (6-651) are identical. Therefore, Eq. (6-652) is satisfied for $u \neq 0$. In other words, we cannot conclude that $u = 0$ from Eq. (6-652).

We thus observe that both Eq. (6-643) and Eq. (6-645) lead to the same equation, (6-654). We can thus conclude that Eqs. (6-646) and (6-647) yield the same set of relations for $\gamma = 1$ and, therefore, for all $\gamma = 2, 3, \ldots$.

*Let us now recapitulate:* We have found that, in order to have an extremal singular control, it is necessary (but not sufficient) that the following relations be satisfied for all $t \in (t_1, t_2]$:

$$\sum_i b_i p_i = 0 \quad (6\text{-}655)$$

$$\sum_i f_i p_i = 0 \quad (6\text{-}656)$$

$$\sum_i \sum_j p_i \left( b_j \frac{\partial f_i}{\partial x_j} - f_j \frac{\partial b_i}{\partial x_j} \right) = 0 \quad (6\text{-}657)$$

Now we let $\gamma = 2$ in Eq. (6-646), to find that

$$\frac{d}{dt}\sum_i \sum_j p_i \left( b_j \frac{\partial f_i}{\partial x_j} - f_j \frac{\partial b_i}{\partial x_j} \right) = 0 \tag{6-658}$$

After extensive algebraic manipulations, we find that Eq. (6-658) leads to the relation

$$\begin{aligned}
\sum_i \sum_j \sum_k p_i \Bigg( & -b_j \frac{\partial f_k}{\partial x_j}\frac{\partial f_i}{\partial x_k} + f_j \frac{\partial f_i}{\partial x_k}\frac{\partial b_k}{\partial x_j} + f_k \frac{\partial b_j}{\partial x_k}\frac{\partial f_i}{\partial x_j} \\
& + b_j f_k \frac{\partial^2 f_i}{\partial x_k \, \partial x_j} - f_k \frac{\partial f_j}{\partial x_k}\frac{\partial b_i}{\partial x_j} - f_j f_k \frac{\partial^2 b_i}{\partial x_k \, \partial x_j} \Bigg) \\
& + u \sum_i \sum_j \sum_k p_i \Bigg( f_j \frac{\partial b_k}{\partial x_j}\frac{\partial b_i}{\partial x_k} - b_j \frac{\partial f_k}{\partial x_j}\frac{\partial b_i}{\partial x_k} + b_k \frac{\partial b_j}{\partial x_k}\frac{\partial f_i}{\partial x_j} \\
& \qquad + b_j b_k \frac{\partial^2 f_i}{\partial x_k \, \partial x_j} - b_k \frac{\partial f_j}{\partial x_k}\frac{\partial b_i}{\partial x_j} - f_j f_k \frac{\partial^2 b_i}{\partial x_k \, \partial x_j} \Bigg) = 0
\end{aligned} \tag{6-659}$$

which is of the form

$$\sum_i p_i \psi_{2i}(\mathbf{x}) + u \sum_i p_i \phi_{2i}(\mathbf{x}) = 0 \tag{6-660}$$

In general, the coefficient $\sum_i p_i \phi_{2i}(\mathbf{x})$ of $u$ will not be zero† for all $t \in (t_1, t_2]$. Thus, if

$$\sum_i p_i \phi_{2i}(\mathbf{x}) \not\equiv 0 \tag{6-661}$$

then we can solve Eq. (6-660) for $u$, to find that

$$u = -\frac{\sum_i p_i \psi_{2i}(\mathbf{x})}{\sum_i p_i \phi_{2i}(\mathbf{x})} \tag{6-662}$$

If, on the other hand,

$$\sum_i p_i \phi_{2i}(\mathbf{x}) = 0 \tag{6-663}$$

then Eq. (6-660) reduces to

$$\sum_i p_i \psi_{2i}(\mathbf{x}) = 0 \tag{6-664}$$

Setting $\gamma = 3$ in Eq. (6-646), we find that

$$\frac{d}{dt}\sum_i p_i \psi_{2i}(\mathbf{x}) = 0 \tag{6-665}$$

† To see this, suppose that $b_0(\mathbf{x}), b_1(\mathbf{x}), \ldots, b_n(\mathbf{x})$ are constants (i.e., independent of $\mathbf{x}$); then $\sum_i p_i \phi_{2i}(\mathbf{x}) = \sum_i \sum_j \sum_k p_i b_j b_k \, (\partial^2 f_i / \partial x_k \, \partial x_j)$, which can indeed be nonzero.

The reader can appreciate the notational difficulties involved in the computation of the higher derivatives. For this reason, we shall conceptually outline the results.

Let us suppose that the sequential time differentiation indicated by Eq. (6-646) leads to the set of equations

$$\begin{aligned} \gamma = 1: &\qquad \sum_i p_i\psi_{1i}(\mathbf{x}) = 0 \\ \gamma = 2: &\qquad \sum_i p_i\psi_{2i}(\mathbf{x}) = 0 \\ &\qquad \cdots\cdots\cdots\cdots \\ \gamma = n-1: &\qquad \sum_i p_i\psi_{(n-1)i}(\mathbf{x}) = 0 \end{aligned} \tag{6-666}$$

In other words, in each time derivative the coefficient of $u$ turns out to be zero. It is easy to see that the $n - 1$ relations of Eqs. (6-666) will be linear in the $p_i$ because the differential equations (6-648) are linear in the $p_i$. Of course, $\psi_{1i}(\mathbf{x}) = \sum_j [b_j(\partial f_i/\partial x_j) - f_j(\partial b_i/\partial x_j)]$ in view of Eq. (6-651); $\psi_{2i}(\mathbf{x})$ is the expression in Eqs. (6-666) which, together with Eqs. (6-655) and (6-656), provides us with a total of $n + 1$ equations which are linear in $p_0, p_1, \ldots, p_n$; these equations can be written in the form

$$\begin{bmatrix} f_0(\mathbf{x}) & f_1(\mathbf{x}) & f_2(\mathbf{x}) & \cdots & f_n(\mathbf{x}) \\ b_0(\mathbf{x}) & b_1(\mathbf{x}) & b_2(\mathbf{x}) & \cdots & b_n(\mathbf{x}) \\ \psi_{10}(\mathbf{x}) & \psi_{11}(\mathbf{x}) & \psi_{12}(\mathbf{x}) & \cdots & \psi_{1n}(\mathbf{x}) \\ \psi_{20}(\mathbf{x}) & \psi_{21}(\mathbf{x}) & \psi_{22}(\mathbf{x}) & \cdots & \psi_{2n}(\mathbf{x}) \\ \cdots & \cdots & \cdots & \cdots & \cdots \\ \psi_{(n-1)0}(\mathbf{x}) & \psi_{(n-1)1}(\mathbf{x}) & \psi_{(n-1)2}(\mathbf{x}) & \cdots & \psi_{(n-1)n}(\mathbf{x}) \end{bmatrix} \begin{bmatrix} p_0 \\ p_1 \\ p_2 \\ p_3 \\ \cdot \\ p_n \end{bmatrix} = \begin{bmatrix} 0 \\ 0 \\ 0 \\ 0 \\ \cdot \\ 0 \end{bmatrix} \tag{6-667}$$

which is of the form

$$\mathbf{G}(\mathbf{x}) \begin{bmatrix} p_0 \\ \mathbf{p} \end{bmatrix} = \mathbf{0} \tag{6-668}$$

*Since the* $n + 1$ *vector* $\begin{bmatrix} p_0 \\ \mathbf{p} \end{bmatrix} \neq \mathbf{0}$, *it is necessary that the* $n + 1 \times n + 1$ *matrix* $\mathbf{G}(\mathbf{x})$ *be a singular matrix for all* $t \in (t_1, t_2]$. *Thus, the condition*

$$\det \mathbf{G}(\mathbf{x}) = 0 \qquad \text{for all } t \in (t_1, t_2] \tag{6-669}$$

*is necessary for the existence of a singular extremal control, provided that the repeated time differentiation leads to a set of* $n + 1$ *relations of the form* (6-667).

Now let us suppose† that the repeated time differentiations lead to an equation in which the coefficient of $u$ is nonzero. To be more precise, we assume that the sequential time differentiation indicated by Eq. (6-646)

† It seems that if this is not the case, then there may be an infinite number of optimal controls, although this fact has not been verified.

leads to the set of equations

$$\left.\begin{aligned} \gamma = 1: \qquad & \sum_i p_i\psi_{1i}(\mathbf{x}) = 0 \\ & \cdots\cdots\cdots \\ \gamma = m: \qquad & \sum_i p_i\psi_{mi}(\mathbf{x}) = 0 \end{aligned}\right\} \quad 1 \le m \tag{6-670}$$

$$\gamma = m + 1: \quad \sum_i p_i\psi_{(m+1)i}(\mathbf{x}) + u \sum_i p_i\phi_{(m+1)i}(\mathbf{x}) = 0 \tag{6-671}$$

where

$$\sum_i p_i\phi_{(m+1)i}(\mathbf{x}) \neq 0 \tag{6-672}$$

In such a case, we can solve for $u$ from Eq. (6-671) to find that

$$u = -\frac{\sum_i p_i\psi_{(m+1)i}(\mathbf{x})}{\sum_i p_i\phi_{(m+1)i}(\mathbf{x})} \tag{6-673}$$

*Equation* (6-673) *provides us with a necessary condition on the extremal singular control; in essence, we have found a relation between* $u$, $p_0, p_1, \ldots, p_n$ *and* $x_0, x_1, x_2, \ldots, x_n$. *The magnitude constraint* $|u| \le 1$ *imposes an inequality constraint on the variables of the form*

$$\left| \frac{\sum_i p_i\psi_{(m+1)i}(\mathbf{x})}{\sum_i p_i\phi_{(m+1)i}(\mathbf{x})} \right| \le 1 \tag{6-674}$$

In addition, in view of Eqs. (6-655), (6-656), and (6-670), we must also satisfy the $m + 2$ equations ($m \ge 1$)

$$\begin{aligned} \sum_i p_i f_i(\mathbf{x}) &= 0 \\ \sum_i p_i b_i(\mathbf{x}) &= 0 \\ \sum_i p_i\psi_{1i}(\mathbf{x}) &= 0 \\ \cdots\cdots &\cdots \\ \sum_i p_i\psi_{mi}(\mathbf{x}) &= 0 \end{aligned} \tag{6-675}$$

In any given problem, we must examine whether or not all of the relations are satisfied; if any of the relations are violated, then this represents a violation of the necessary conditions, and so singular extremal controls cannot occur. If the necessary conditions are satisfied, then there may exist singular extremal controls. The following examples illustrate these ideas.

**Example 6-9** Consider the harmonic oscillator

$$\begin{aligned} \dot{x}_1(t) &= x_2(t) \\ \dot{x}_2(t) &= -x_1(t) + u(t) \qquad |u(t)| \le 1 \end{aligned} \tag{6-676}$$

and the cost functional

$$J(u) = \tfrac{1}{2} \int_0^T [x_1^2(t) + x_2^2(t)]\,dt \qquad T \text{ free} \tag{6-677}$$

We wish to test whether or not it is possible to have singular extremal controls for this problem. To do this, we form the Hamiltonian

$$H = \tfrac{1}{2}p_0 x_1^2(t) + \tfrac{1}{2}p_0 x_2^2(t) + x_2(t)p_1(t) - x_1(t)p_2(t) + u(t)p_2(t) \tag{6-678}$$

where

$$p_0 = \text{const} \geq 0 \tag{6-679}$$

$$\dot{p}_1(t) = -\frac{\partial H}{\partial x_1(t)} = -p_0 x_1(t) + p_2(t) \tag{6-680}$$

$$\dot{p}_2(t) = -\frac{\partial H}{\partial x_2(t)} = -p_0 x_2(t) - p_1(t) \tag{6-681}$$

Suppose that

$$p_2(t) = 0 \qquad \text{for all } t \in (t_1, t_2] \tag{6-682}$$

Then, since we deal with a free-terminal-time problem, we must have

$$\tfrac{1}{2}p_0 x_1^2(t) + \tfrac{1}{2}p_0 x_2^2(t) + x_2(t)p_1(t) - x_1(t)p_2(t) = 0 \qquad \text{for all } t \in (t_1, t_2] \tag{6-683}$$

But Eq. (6-682) implies that

$$\dot{p}_2(t) = 0 \qquad \text{for all } t \in (t_1, t_2] \tag{6-684}$$

and so, from Eq. (6-681), we find that

$$p_0 x_2(t) = -p_1(t) \qquad \text{for all } t \in (t_1, t_2] \tag{6-685}$$

If

$$p_0 = 0 \tag{6-686}$$

then Eq. (6-685) implies that

$$p_1(t) = 0 \qquad \text{for all } t \in (t_1, t_2] \tag{6-687}$$

But Eqs. (6-682), (6-686), and (6-687) cannot hold, because the vector

$$\begin{bmatrix} p_0 \\ p_1 \\ p_2 \end{bmatrix}$$

cannot be zero; therefore, $p_0 \neq 0$, and so we set

$$p_0 = 1 \tag{6-688}$$

Then Eq. (6-685) reduces to

$$x_2(t) = -p_1(t) \qquad \text{for all } t \in (t_1, t_2] \tag{6-689}$$

and Eq. (6-683) reduces to

$$\tfrac{1}{2}x_1^2(t) = \tfrac{1}{2}x_2^2(t) \qquad \text{for all } t \in (t_1, t_2] \tag{6-690}$$

which, in turn, implies that

$$x_1(t) = \pm x_2(t) \qquad \text{for all } t \in (t_1, t_2] \tag{6-691}$$

Differentiating Eq. (6-689), we find that

$$\dot{x}_2(t) = -\dot{p}_1(t) = x_1(t) \qquad \text{for all } t \in (t_1, t_2] \tag{6-692}$$

From Eqs. (6-692) and (6-676), we find that

$$u(t) = 2x_1(t) \qquad \text{for all } t \in (t_1, t_2] \tag{6-693}$$

In view of the constraint $|u(t)| \leq 1$ and Eq. (6-691), we conclude that the singular control (6-693) can be extremal if

$$|x_1(t)| \leq \tfrac{1}{2} \qquad |x_2(t)| \leq \tfrac{1}{2} \qquad x_1(t) = \pm x_2(t) \tag{6-694}$$

as illustrated in Fig. 6-23. The two straight lines $x_a$ and $x_b$ shown in Fig. 6-23 are often called singular trajectories, or singular arcs. It is easy to see that if we substitute Eq. (6-693) into the system equations (6-676), we find that

$$\left.\begin{aligned} \dot{x}_1(t) &= x_2(t) \\ \dot{x}_2(t) &= x_1(t) \end{aligned}\right\} \qquad \text{for all } t \in (t_1, t_2] \tag{6-695}$$

It is easy to see that the solution of the system (6-695), starting from an initial state $x_1(0) = \xi_1$, $x_2(0) = \xi_2$, is given by

$$\begin{aligned} x_1(t) &= \tfrac{1}{2}e^t(\xi_1 + \xi_2) + \tfrac{1}{2}e^{-t}(\xi_1 - \xi_2) \\ x_2(t) &= \tfrac{1}{2}e^t(\xi_1 + \xi_2) - \tfrac{1}{2}e^{-t}(\xi_1 - \xi_2) \end{aligned} \tag{6-696}$$

Note that if $\xi_1 = \xi_2$, then the solution (6-696) is unstable; if $\xi_1 = -\xi_2$, then the solution (6-696) is stable. The nature of the solution is indicated in Fig. 6-23, using the arrows on the singular subarcs $x_a$ and $x_b$. Thus, in this problem, depending on the initial conditions $(\xi_1, \xi_2)$ and the terminal state $(\theta_1, \theta_2)$, extremal controls may turn out to be singular.

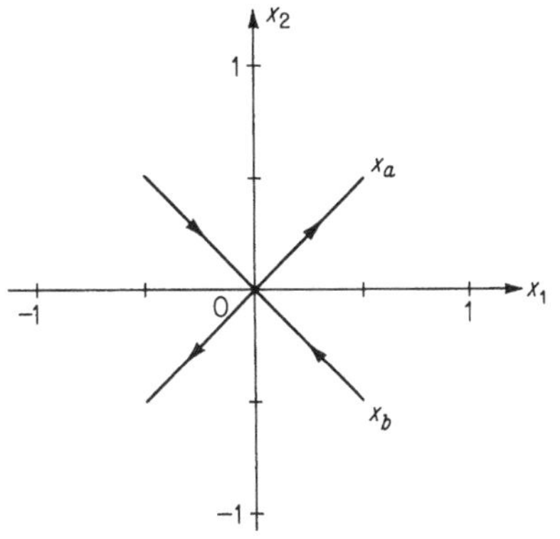


**Fig. 6-23** The singular subarcs $x_a$ and $x_b$.

**Example 6-10** Again we consider the harmonic oscillator

$$\begin{aligned} \dot{x}_1(t) &= x_2(t) \\ \dot{x}_2(t) &= -x_1(t) + u(t) \qquad |u(t)| \leq 1 \end{aligned} \tag{6-697}$$

and we now use the cost functional

$$J(u) = \tfrac{1}{2}\int_0^T x_1^2(t)\,dt \qquad T \text{ free} \tag{6-698}$$

To test whether or not the extremal controls can be singular, we form the Hamiltonian

$$H = \tfrac{1}{2}p_0 x_1^2(t) + x_2(t)p_1(t) - x_1(t)p_2(t) + u(t)p_2(t) \tag{6-699}$$

where

$$p_0 = \text{const} \geq 0 \tag{6-700}$$

$$\dot{p}_1(t) = -\frac{\partial H}{\partial x_1(t)} = -p_0 x_1(t) + p_2(t) \tag{6-701}$$

$$\dot{p}_2(t) = -\frac{\partial H}{\partial x_2(t)} = -p_1(t) \tag{6-702}$$

Suppose that

$$p_2(t) = 0 \qquad \text{for all } t \in (t_1, t_2] \tag{6-703}$$

Then

$$\tfrac{1}{2}p_0 x_1^2(t) + x_2(t)p_1(t) = 0 \qquad \text{for all } t \in (t_1, t_2] \tag{6-704}$$

But

$$\dot{p}_2(t) = 0 \qquad \text{for all } t \in (t_1, t_2] \tag{6-705}$$

and so Eq. (6-702) yields

$$p_1(t) = 0 \qquad \text{for all } t \in (t_1, t_2] \tag{6-706}$$

Therefore, Eq. (6-704) yields

$$p_0 = 1 \tag{6-707}$$

$$x_1(t) = 0 \qquad \text{for all } t \in (t_1, t_2] \tag{6-708}$$

and so

$$\dot{x}_1(t) = 0 \qquad \text{for all } t \in (t_1, t_2] \tag{6-709}$$

Therefore, from Eqs. (6-709) and (6-697), we find that

$$x_2(t) = 0 \qquad \text{for all } t \in (t_1, t_2] \tag{6-710}$$

and thus

$$u(t) = 0 \qquad \text{for all } t \in (t_1, t_2] \tag{6-711}$$

We can thus conclude that the singular control $u(t) = 0$ can be extremal if

$$x_1(t) = x_2(t) = 0 \qquad \text{for all } t \in (t_1, t_2] \tag{6-712}$$

This means that extremal controls will not be singular for any point in the state plane except, possibly, the origin.

These two examples illustrate in a concrete manner the general concepts we have introduced in this section. Example 6-9 is typical because, in many singular problems, we can find the extremal singular control as a function of the state. If we substitute this control into the system differential equations, then we obtain a system of homogeneous equations [such as the system (6-695)] whose solution defines the singular subarcs.

Our discussion up to now has been concerned with the free-terminal-time problem. If we consider the fixed-terminal-time problem, then we know that the Hamiltonian along the optimal trajectory is some (unknown) constant $c$ [see Eq. (6-638)]. We can easily see that, in order to test whether or not it is possible to have a singular extremal control for the fixed-time problem, we can mimic the ideas already presented. We still suppose that Eq. (6-643), that is,

$$\sum_{i=0}^{n} b_i p_i = 0 \qquad \text{for all } t \in (t_1, t_2] \tag{6-713}$$

holds. Now Eq. (6-644) is replaced by the equation

$$H = \sum_{i=0}^{n} f_i p_i + u \sum_{i=0}^{n} b_i p_i = c \qquad \text{for all } t \in (t_1, t_2] \tag{6-714}$$

and so Eq. (6-645) is replaced by the equation

$$\sum_{i=0}^{n} f_i p_i = c \qquad \text{for all } t \in (t_1, t_2] \tag{6-715}$$

However, Eqs. (6-646) and (6-647) must hold; therefore, we can follow the indicated development with the obvious replacement of 0 by $c$ in Eqs. (6-656) and so on. We thus conclude that knowledge of the terminal time $T$ introduces an (unknown) constant $c$ in our equations; on the other hand, if $T$ is not specified, then $c = 0$. In either case, we have a balance of unknown quantities and known relations.

**Exercise 6-47** Consider the completely controllable system

$$\begin{aligned} \dot{x}_1(t) &= a_{11}x_1(t) + a_{12}x_2(t) \\ \dot{x}_2(t) &= a_{21}x_1(t) + a_{22}x_2(t) + u(t) \qquad |u(t)| \le m \end{aligned} \tag{6-716}$$

and the cost functional

$$J(u) = \frac{1}{2} \int_0^T [x_1^2(t) + x_2^2(t)]\, dt \qquad T \text{ free} \tag{6-717}$$

Show that the control

$$u(t) = -(a_{12} + a_{21})x_1(t) - (a_{11} + a_{22})x_2(t) \tag{6-718}$$

can be an extremal singular control. Construct the possible singular extremal trajectories in the plane and show that they are independent of the values of $a_{21}$ and $a_{22}$. Can you explain this behavior? HINT: See Ref. J-6, example 2.

**Exercise 6-48** Consider the system

$$\begin{aligned} \dot{x}_1(t) &= x_2(t) \\ \dot{x}_2(t) &= -x_2(t) - x_2^3(t) + u(t) \qquad |u(t)| \le 1 \end{aligned} \tag{6-719}$$

and the cost functional

$$J(u) = \int_0^T 1 \cdot dt \qquad T \text{ free} \tag{6-720}$$

Let $(\xi_1, \xi_2)$ be the initial state, and let $(\theta_1, \theta_2)$ be the terminal state. Show that there are no extremal singular controls.

**Exercise 6-49** Consider the system

$$\begin{aligned} \dot{x}_1(t) &= x_2(t) \\ \dot{x}_2(t) &= x_3(t) \\ \dot{x}_3(t) &= u(t) \qquad |u(t)| \le 1 \end{aligned} \tag{6-721}$$

and the cost functional

$$J(u) = \frac{1}{2} \int_0^T [x_1^2(t) + x_2^2(t) + x_3^2(t)]\, dt \tag{6-722}$$

Let the initial state be $(\xi_1, \xi_2, \xi_3)$ and the terminal state be (0, 0, 0). Comment on the existence of singular extremal controls for $T$ free and $T$ fixed. Can you find the subset of $R_3$ on which the extremal control must be singular?

## 6-22 Singular Problems 2: The Hamiltonian a Linear Function of the Control and of Its Absolute Value

In Sec. 6-13, we discussed the singular fuel-optimal problem (see Definition 6-12). The reader may recall that the singular fuel-optimal problem arose when the argument of the dez { } function was identically equal to either $+1$ or $-1$ on a finite interval of time. In this section, we shall briefly consider similar types of problems; such problems arise when the Hamiltonian function is linear in the control $u(t)$ and in its absolute value $|u(t)|$.

First, we shall formulate Problem 6-4*b*, whose solution may turn out to be singular. Next, we shall informally discuss the singular case for Problem 6-4*b* and indicate the computations which may be used to determine whether or not the extremal controls are singular.

***Problem 6-4b*** *Given the time-invariant system*

$$\dot{x}_i(t) = f_i[\mathbf{x}(t)] + b_i[\mathbf{x}(t)]u(t) \qquad i = 1, 2, \ldots, n \tag{6-723}$$

*where the n vector* $\mathbf{x}(t)$ *is the state and where* $u(t)$ *is the scalar control variable. Assume that* $u(t)$ *is constrained in magnitude by the relation*

$$|u(t)| \leq 1 \qquad \textit{for all } t \tag{6-724}$$

*Given an initial state* $\mathbf{x}(0) = \boldsymbol{\xi}$ *and a terminal state* $\boldsymbol{\theta}$. *Given the cost functional*

$$J(u) = \int_0^T \{f_0[\mathbf{x}(t)] + b_0[\mathbf{x}(t)]|u(t)|\}\, dt \tag{6-725}$$

*where* $f_0[\mathbf{x}(t)]$ *and* $b_0[\mathbf{x}(t)]$ *are scalar functions of the state and where the terminal time T is either free or fixed. Then determine the admissible control that transfers the system (6-723) from* $\boldsymbol{\xi}$ *to* $\boldsymbol{\theta}$ *[that is, such that* $\mathbf{x}(T) = \boldsymbol{\theta}$*] and that minimizes the cost functional* $J(u)$.

The Hamiltonian function for the system (6-723) and for the cost functional (6-725) is

$$H = p_0 f_0[\mathbf{x}(t)] + p_0 b_0[\mathbf{x}(t)]|u(t)| + \sum_{i=1}^{n} p_i(t) f_i[\mathbf{x}(t)] + u(t) \sum_{i=1}^{n} p_i(t) b_i[\mathbf{x}(t)] \tag{6-726}$$

where

$$p_0 = \text{const} \geq 0 \tag{6-727}$$

The Hamiltonian function will be linear in $|u(t)|$ if

$$p_0 \neq 0 \tag{6-728}$$

and if

$$b_0[\mathbf{x}(t)] \neq 0 \tag{6-729}$$

For this reason, we shall assume that Eqs. (6-728) and (6-729) hold, and we shall set

$$p_0 = 1 \tag{6-730}$$

We now observe that $H$ can be written as

$$H = \left\{f_0[\mathbf{x}(t)] + \sum_{i=1}^{n} p_i(t) f_i[\mathbf{x}(t)]\right\} + b_0[\mathbf{x}(t)] \left\{|u(t)| + u(t) \frac{\sum_{i=1}^{n} p_i(t) b_i[\mathbf{x}(t)]}{b_0[\mathbf{x}(t)]}\right\} \tag{6-731}$$

An extremal control must minimize the Hamiltonian $H$. For this reason, we find the $H$-minimal control which is given by the relation

$$u(t) = -\operatorname{dez}\left\{\frac{\sum_{i=1}^{n} p_i(t) b_i[\mathbf{x}(t)]}{b_0[\mathbf{x}(t)]}\right\} \tag{6-732}$$

where the dez { } function has been defined by Eq. (6-347). The extremal controls for Problem 6-4*b* will be singular if the argument of the dez { } function is +1 or −1 for a finite interval of time, say $(t_1, t_2]$, which is a subset of $[0, T]$. Therefore, if

$$\frac{\sum_{i=1}^{n} p_i(t) b_i[\mathbf{x}(t)]}{b_0[\mathbf{x}(t)]} = \pm 1 \qquad \text{for all } t \in (t_1, t_2] \tag{6-733}$$

then an extremal control may be singular. Basically, Eq. (6-733), together with all the rest of the necessary conditions, can be used to test for the existence of extremal singular controls for Problem 6-4*b*.

The interested reader can derive necessary conditions for the existence of a singular extremal control for Problem 6-4*b* by using the following procedure:

1. Determine the necessary conditions for Problem 6-4*b* and determine the canonical equations.
2. Set, for all $t \in (t_1, t_2]$,

$$\sum_{i=1}^{n} p_i(t) b_i[\mathbf{x}(t)] - b_0[\mathbf{x}(t)] = 0 \tag{6-734}$$

and

$$f_0[\mathbf{x}(t)] + \sum_{i=1}^{n} p_i(t) f_i[\mathbf{x}(t)] = \begin{cases} 0 & \text{for } T \text{ free} \\ \text{const} & \text{for } T \text{ fixed} \end{cases} \tag{6-735}$$

3. Set, for all $t \in (t_1, t_2]$ and $k = 1, 2, \ldots,$

$$\frac{d^k}{dt^k}\left\{\sum_{i=1}^{n} p_i(t) b_i[\mathbf{x}(t)] - b_0[\mathbf{x}(t)]\right\} = 0 \tag{6-736}$$

and

$$\frac{d^k}{dt^k}\left\{f_0[\mathbf{x}(t)] + \sum_{i=1}^{n} p_i(t) f_i[\mathbf{x}(t)]\right\} = 0 \tag{6-737}$$

4. By repeated differentiation, determine the set of equations which are necessary for the existence of extremal singular controls.
5. Replace Eq. (6-734) by

$$\sum_{i=1}^{n} p_i(t) b_i[\mathbf{x}(t)] + b_0[\mathbf{x}(t)] = 0 \tag{6-738}$$

and repeat.

An example which illustrates the existence of optimal singular controls for a fuel-optimal problem is given in Sec. 8-10.

**Exercise 6-50** Consider the system

$$\begin{aligned} \dot{x}_1(t) &= x_2(t) \\ \dot{x}_2(t) &= u(t) \qquad |u(t)| \le 1 \end{aligned} \tag{6-739}$$

and the cost functional ($T$ free, $k \geq 0$)

$$\int_0^T [k + a_1\tfrac{1}{2}x_1^2(t) + a_2\tfrac{1}{2}x_2^2(t) + |u(t)|]\,dt \tag{6-740}$$

Let $(\xi_1, \xi_2)$ be the initial state and let $(0, 0)$ be the terminal state. Test for the possibility of singular extremal controls for the following cases:

| | | |
|---|---|---|
| (*a*) $k > 0$: | $a_1 = a_2 = 0$ | (6-741) |
| (*b*) $k > 0$: | $a_1 = 1 \qquad a_2 = 0$ | (6-742) |
| (*c*) $k > 0$: | $a_1 = 0 \qquad a_2 = 1$ | (6-743) |
| (*d*) $k > 0$: | $a_1 = a_2 = 1$ | (6-744) |
| (*e*) $k = 0$: | $a_1 = 1 \qquad a_2 = 0$ | (6-745) |
| (*f*) $k = 0$: | $a_1 = 0 \qquad a_2 = 1$ | (6-746) |
| (*g*) $k = 0$: | $a_1 = a_2 = 1$ | (6-747) |

**Exercise 6-51** Repeat Exercise 6-50 for each of the following systems ($|u(t)| \leq 1$):

(*a*)
$$\dot{x}_1(t) = x_2(t) \qquad \dot{x}_2(t) = -x_2(t) + u(t) \tag{6-748}$$

(*b*)
$$\dot{x}_1(t) = x_2(t) \qquad \dot{x}_2(t) = x_2(t) + u(t) \tag{6-749}$$

(*c*)
$$\dot{x}_1(t) = x_2(t) \qquad \dot{x}_2(t) = -x_1(t) + u(t) \tag{6-750}$$

(*d*)
$$\dot{x}_1(t) = x_2(t) \qquad \dot{x}_2(t) = x_1(t) + u(t) \tag{6-751}$$

(*e*)
$$\dot{x}_1(t) = x_2(t) \qquad \dot{x}_2(t) = -2x_1(t) - 3x_2(t) + u(t) \tag{6-752}$$

**Exercise 6-52** Consider the system (6-739) and the cost functional

$$J(u) = \int_0^T [1 + e^{-|x_1(t)|}|u(t)|]\,dt \qquad T \text{ free} \tag{6-753}$$

Test for the possibility of extremal singular controls for the terminal state (0, 0).

## 6-23 Some Remarks Concerning the Existence and Uniqueness of the Optimal and Extremal Controls

In the preceding sections of this chapter, we discussed several optimization problems. We indicated that the question of the existence of an optimal control† for a given problem is a very difficult one and must be answered prior to the application of the minimum principle. We wish to caution the reader that statements such as "I know that I am dealing with a physical problem, and therefore, an optimal solution must exist" cannot be used to verify the existence of an optimal control. The reason is that *the first and foremost necessary condition for a mathematical optimization problem to be a faithful representation of a physical optimization problem is that the mathematical problem must have a solution.* In other words, given a mathematical optimization problem, the reader must verify whether or

† The following references deal with the existence problem for various optimization problems: K-14, C-4, F-3, B-18, G-1, P-4, K-16, K-28, H-7, N-5, L-7, and R-8.

not a solution exists, so that he may feel confident that he is dealing with a realistic mathematical model of the physical situation. Quite often, the translation of a physical process and of a set of specifications into their mathematical counterpart is not accurate enough, so that physical reasoning can be on "shaky grounds" regarding the mathematical question of the existence of a solution.

When the engineer formulates an optimization problem, he usually tries to determine the extremal controls, i.e., the controls that satisfy all the necessary conditions. Clearly:†

*The existence of an optimal control implies the existence of at least one extremal control. However, the existence of extremal controls does not necessarily imply the existence of an optimal control.*

Quite often, the nonexistence of an optimal control can be suspected from the fact that one cannot find controls that satisfy all the necessary conditions; in such cases, it may be possible to find controls that satisfy all but one of the necessary conditions. This fact is often a clue to the nonexistence of an optimal control.

Next, we turn to the question of uniqueness of the optimal control. It is easy to see that:

*The existence of an optimal control and the uniqueness of extremal controls imply the uniqueness of the optimal control. However, the uniqueness of the optimal control does not necessarily imply the uniqueness of the extremal controls.*

Recall, for example, that the time-optimal control was unique for a normal linear system for every terminal state $\boldsymbol{\theta}$ (see Theorem 6-7 and Exercise 6-7); however, the extremal controls were proved to be unique only for the terminal state $\boldsymbol{\theta} = \mathbf{0}$ (see Theorem 6-9 and Exercise 6-8). Also:

*The nonuniqueness of an optimal control implies the nonuniqueness of the extremal controls. The converse, however, is not necessarily true.*

Next, let us suppose that for a given optimization problem, we found that an optimum exists for all finite fixed terminal times $T$, and furthermore, let us assume that an optimal control exists for the limiting case $T \to \infty$. Then it is easy to see that:

*The existence of an optimal control for all $T$ (including $T \to \infty$) implies the existence of an optimal control when the terminal time $T$ is left unspecified. The converse statement, however, is not necessarily true.*

Next, let us suppose that an optimal control exists for all fixed terminal states $\boldsymbol{\theta}$ in $R_n$. Let $S$ be a closed target set in $R_n$. Then:

*The existence of an optimal control for all $\boldsymbol{\theta} \in R_n$ implies the existence of an optimal control to a compact target set $S$ which is a subset of $R_n$. The converse statement, again, is not necessarily true.*

The last remark is useful when the target is not smooth so that there is some difficulty in using the transversality conditions. In such problems,

† See also Remarks 6-7 to 6-10.

the safest way is to solve the problem as a fixed-terminal-point problem and then demand that the fixed terminal point belong to the target set. We shall discuss this idea in the following section in greater detail.

## 6-24 Fixed-boundary-condition vs. Free-boundary-condition Problems

In this section, we shall discuss the following points:

1. The solution of free-terminal-time problems by first solving the fixed, but arbitrary, terminal-time problem.
2. The solution of optimal problems to a target set $S$ (in $R_n$) by first solving the fixed, but arbitrary, terminal-state problem.

First of all, let us consider a fixed-terminal-time optimization problem. For simplicity, we consider the time-invariant system

$$\dot{\mathbf{x}}(t) = \mathbf{f}[\mathbf{x}(t), \mathbf{u}(t)] \qquad \mathbf{x}(0) = \boldsymbol{\xi} \tag{6-754}$$

Let $\boldsymbol{\theta}$ be a given terminal state in $R_n$, and let us consider the cost-functional

$$J(\mathbf{u}) = \int_0^T L[\mathbf{x}(t), \mathbf{u}(t)]\, dt \tag{6-755}$$

Let us suppose that we fix the terminal time $T$ and that we find the optimal control $\mathbf{u}^*_{(0,T]}$ which transfers the system (6-754) from the initial state $\boldsymbol{\xi}$ to $\boldsymbol{\theta}$ [that is, $\mathbf{x}^*(T) = \boldsymbol{\theta}$] and which minimizes the cost functional (6-755). Then the minimum value of $J(\mathbf{u})$, denoted by $J^*$, depends on the following quantities:†

1. The initial state $\boldsymbol{\xi}$
2. The terminal state $\boldsymbol{\theta}$
3. The terminal time $T$

For this reason, let us denote the minimum value of the cost functional by

$$J^* = J^*(\boldsymbol{\xi}, \boldsymbol{\theta}, T)$$

Now let us suppose that we always start, at $t = 0$, at the *same* initial state $\boldsymbol{\xi}$, that we wish to hit the *same* terminal state $\boldsymbol{\theta}$, and that we vary the terminal time $T$ so that we can examine the variation of the minimum cost as $T$ changes.

Now let us assume that:

1. If

$$T \in [T_1, T_2] \tag{6-756}$$

(i.e., if $0 \leq T_1 \leq T \leq T_2 \leq \infty$), then we can find an optimal control $\mathbf{u}^*_{(0,T]}$.

2. If

$$T \notin [T_1, T_2] \tag{6-757}$$

then an optimal control does not exist.

† In general, it will also depend on the initial time, which we fix to be 0.

In this case, we can plot the cost $J^*(\boldsymbol{\xi}, \boldsymbol{\theta}, T)$ versus $T$ for $T \in [T_1, T_2]$, holding $\boldsymbol{\xi}$ and $\boldsymbol{\theta}$ fixed. Let us suppose that the plot of $J^*(\boldsymbol{\xi}, \boldsymbol{\theta}, T)$ versus $T$ looks like the one shown in Fig. 6-24$a$; we observe that the function $J^*(\boldsymbol{\xi}, \boldsymbol{\theta}, T)$ has:

1. An absolute minimum at $T_5$ and a relative minimum at $T_7$
2. An absolute maximum at $T_3$ and a relative maximum at $T_6$
3. An inflection point at $T_4$

Now the Hamiltonian function for our problem is

$$H[\mathbf{x}(t), \mathbf{p}(t), \mathbf{u}(t)] = L[\mathbf{x}(t), \mathbf{u}(t)] + \langle \mathbf{f}[\mathbf{x}(t), \mathbf{u}(t)], \mathbf{p}(t) \rangle \tag{6-758}$$

By assumption, for every $T \in [T_1, T_2]$, there are an optimal control $\mathbf{u}^*_{[0,T]}$ and, hence, an optimal trajectory $\mathbf{x}^*_{[0,T]}$ and a corresponding costate $\mathbf{p}^*_{[0,T]}$. For each $T \in [T_1, T_2]$, we have

$$\mathbf{x}^*(T) = \boldsymbol{\theta} \tag{6-759}$$

and so let us consider the function

$$H[\boldsymbol{\theta}, \mathbf{p}^*(T), \mathbf{u}^*(T)] \tag{6-760}$$

In general, given two distinct terminal times

$$\hat{T} \neq \tilde{T} \tag{6-761}$$

in $[T_1, T_2]$, we have the relations

$$\begin{aligned} \mathbf{x}^*(\hat{T}) &= \mathbf{x}^*(\tilde{T}) = \boldsymbol{\theta} \\ \mathbf{p}^*(\hat{T}) &\neq \mathbf{p}^*(\tilde{T}) \\ \mathbf{u}^*(\hat{T}) &\neq \mathbf{u}^*(\tilde{T}) \\ H[\boldsymbol{\theta}, \mathbf{p}^*(\hat{T}), \mathbf{u}^*(\hat{T})] &\neq H[\boldsymbol{\theta}, \mathbf{p}^*(\tilde{T}), \mathbf{u}^*(\tilde{T})] \end{aligned} \tag{6-762}$$

We can see that the value of the Hamiltonian will depend implicitly on the terminal time $T$. If we plot $H[\boldsymbol{\theta}, \mathbf{p}^*(T), \mathbf{u}^*(T)]$ versus $T$, for $T \in [T_1, T_2]$, we shall find a plot similar to that shown in Fig. 6-24$b$. Thus, the Hamiltonian will be zero whenever the cost function $J^*(\boldsymbol{\xi}, \boldsymbol{\theta}, T)$ has an extremum; i.e., in Fig. 6-24$b$, we have

$$H[\boldsymbol{\theta}, \mathbf{p}^*(T_k), \mathbf{u}^*(T_k)] = 0 \qquad \text{for } k = 3, 4, 5, 6, 7 \tag{6-763}$$

Equation (6-763) is one of the necessary conditions for the free-terminal-time problem. We can see that this necessary condition will isolate the extremal points of $J^*(\boldsymbol{\xi}, \boldsymbol{\theta}, T)$, and so additional computations are required to isolate the absolute minimum of $J^*(\boldsymbol{\xi}, \boldsymbol{\theta}, T)$ (in Fig. 6-24$a$, the time $T_5$). We suggest that the reader, at this point, review the proof of the minimum principle (Sec. 5-16) and comments 6 and 10 of Sec. 5-17. He should also look over the material in Sec. 6-19.

The purpose of the above discussion was to point out that we can frequently solve free-terminal-time problems by first solving the fixed-time problem and then examining the minimum cost as a function of the terminal time. In this manner, we are faced with a standard minimization problem and we can use the material of Sec. 5-2.

In the remainder of this section, we shall show how we can frequently solve an optimization problem to a closed target set $S$ by first solving the problem to a terminal state $\boldsymbol{\theta}$ and then demanding that $\boldsymbol{\theta}$ be an element of $S$.

Let us consider the system (6-754) and the cost functional (6-755) with $T$ fixed. Let $S$ be our compact target set in $R_n$. Let us suppose that an

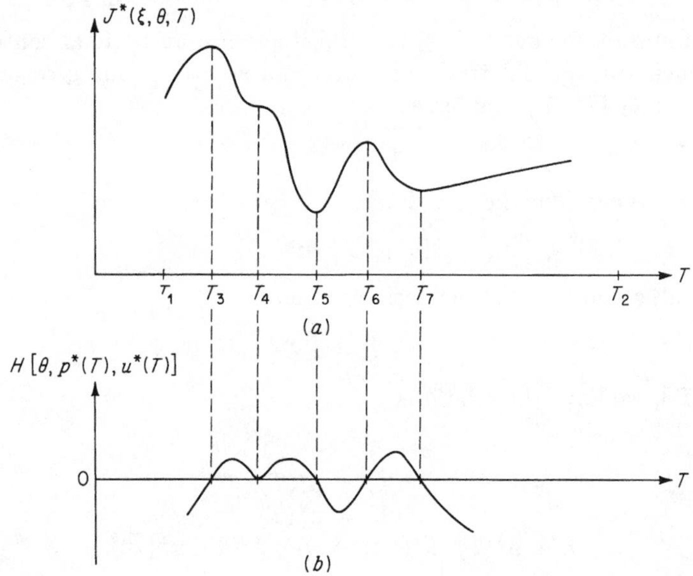


**Fig. 6-24** The typical behavior of $J^*(\boldsymbol{\xi}, \boldsymbol{\theta}, T)$ and $H[\boldsymbol{\theta}, \mathbf{p}^*(T), \mathbf{u}^*(T)]$ as the terminal time $T$ varies.

optimal control exists for all $\boldsymbol{\theta} \in Q$, where $Q$ is a subset of $R_n$ which contains $S$; in other words,

$$S \subset Q \subset R_n \tag{6-764}$$

The cost $J^*(\boldsymbol{\xi}, \boldsymbol{\theta}, T)$ again depends on the initial state $\boldsymbol{\xi}$, the terminal state $\boldsymbol{\theta}$, and the terminal time $T$. If we fix the initial state $\boldsymbol{\xi}$ and the terminal time $T$, then we can plot the cost $J^*(\boldsymbol{\xi}, \boldsymbol{\theta}, T)$ for all $\boldsymbol{\theta}$ in $Q$. If we now demand that the terminal state $\boldsymbol{\theta}$ belongs to the given target set $S$, then we can find, as illustrated in Fig. 6-25, the value of the minimum cost for every $\boldsymbol{\theta}$ in $S$. If we find the cost

$$J^*(\boldsymbol{\xi}, S, T) = \min_{\boldsymbol{\theta} \in S} J^*(\boldsymbol{\xi}, \boldsymbol{\theta}, T) \tag{6-765}$$

then the cost $J^*(\boldsymbol{\xi}, S, T)$ will, of course, be the solution of the control problem to the target set $S$. We can see that the determination of $J^*(\boldsymbol{\xi}, S, T)$

reduces to the minimization of the scalar quantity $J^*(\boldsymbol{\xi}, \boldsymbol{\theta}, T)$, subject to the algebraic constraint $\boldsymbol{\theta} \in S$; therefore, we can use the material given in Secs. 5-3 to 5-5.

We shall now demonstrate that the necessary condition that the costate $\mathbf{p}^*(T)$ be perpendicular to the target set is equivalent to finding the extremal

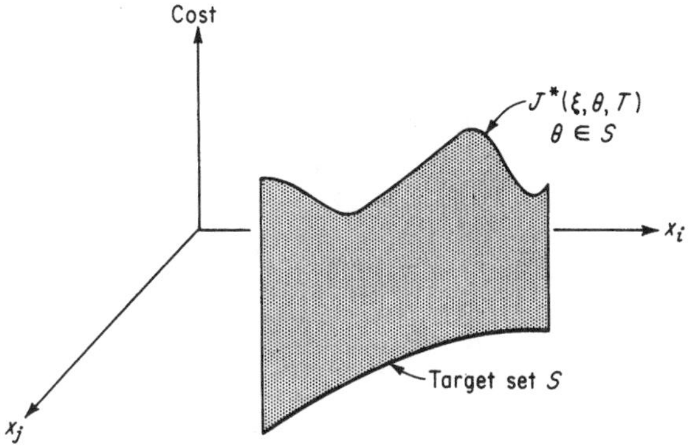


**Fig. 6-25** The cost $J^*(\boldsymbol{\xi}, \boldsymbol{\theta}, T)$ for $\boldsymbol{\theta} \in S$ and for fixed $\boldsymbol{\xi}$ and $T$.

points of $J^*(\boldsymbol{\xi}, \boldsymbol{\theta}, T)$ for $\boldsymbol{\theta} \in S$. To be specific, let us suppose that $S$ is a smooth hypersurface in $R_n$, described by the equation

$$S = \{\mathbf{x}: g(\mathbf{x}) = 0\} \tag{6-766}$$

If $\boldsymbol{\theta} \in S$, then we must have

$$g(\boldsymbol{\theta}) = 0 \tag{6-767}$$

Let us denote, for the sake of simplicity, the cost $J^*(\boldsymbol{\xi}, \boldsymbol{\theta}, T)$ by $\hat{J}^*(\boldsymbol{\theta})$. To find the extremal points of $\hat{J}^*(\boldsymbol{\theta})$ subject to the constraint $g(\boldsymbol{\theta}) = 0$, we can use the techniques of Sec. 5-4. Thus, we let $\lambda$ be a Lagrange multiplier, and we consider the function

$$\hat{J}^*(\boldsymbol{\theta}) + \lambda g(\boldsymbol{\theta}) \tag{6-768}$$

Let us suppose that the points

$$\boldsymbol{\theta}_1^*, \boldsymbol{\theta}_2^*, \ldots, \boldsymbol{\theta}_m^* \tag{6-769}$$

are the extremal (i.e., maxima, minima, saddle points, etc.) points. Then (see Theorem 5-1) there is a number $\lambda_i^*$ corresponding to each $\boldsymbol{\theta}_i$, $i = 1, 2, \ldots, m$, such that $\boldsymbol{\theta}_i^*$ and $\lambda_i^*$ are a solution of the system of $n + 1$ equations (assuming that we deal with a smooth problem)

$$\frac{\partial \hat{J}^*(\boldsymbol{\theta})}{\partial \boldsymbol{\theta}} + \lambda \frac{\partial g(\boldsymbol{\theta})}{\partial \boldsymbol{\theta}} = \mathbf{0} \qquad g(\boldsymbol{\theta}) = 0 \tag{6-770}$$

In other words, for each $i = 1, 2, \ldots, m$, we have

$$\left.\frac{\partial J^*(\boldsymbol{\theta})}{\partial \boldsymbol{\theta}}\right|_{\boldsymbol{\theta}=\boldsymbol{\theta}_i^*} + \lambda_i^* \left.\frac{\partial g(\boldsymbol{\theta})}{\partial \boldsymbol{\theta}}\right|_{\boldsymbol{\theta}=\boldsymbol{\theta}_i^*} = \mathbf{0} \qquad g(\boldsymbol{\theta})\Big|_{\boldsymbol{\theta}=\boldsymbol{\theta}_i^*} = 0 \tag{6-771}$$

Now let us turn our attention to the transversality condition which states that $\mathbf{p}^*(T)$ must be normal to $S$ at $\boldsymbol{\theta}^*$. Since the target set $S$ has the equation

$$g(\mathbf{x}) = 0 \tag{6-772}$$

the gradient vector at $\mathbf{x} = \boldsymbol{\theta}$ is given by

$$\left.\frac{\partial g(\mathbf{x})}{\partial \mathbf{x}}\right|_{\mathbf{x}=\boldsymbol{\theta}} \tag{6-773}$$

Let $c$ be a nonzero constant; then at $\boldsymbol{\theta} = \boldsymbol{\theta}^*$, we must have

$$\mathbf{p}^*(T) = c \left.\frac{\partial g(\mathbf{x})}{\partial \mathbf{x}}\right|_{\mathbf{x}=\boldsymbol{\theta}^*} \tag{6-774}$$

Since we assumed that we are dealing with a smooth problem, we know that (see Sec. 5-19 and Exercise 5-23)

$$\mathbf{p}^*(t) = \left.\frac{\partial J^*(\mathbf{x}, t)}{\partial \mathbf{x}}\right|_{\mathbf{x}^*(t)} \tag{6-775}$$

Thus, for $\mathbf{x}^*(t)$ near $\boldsymbol{\theta}^*$ and for $t$ near $T$, we have the relation

$$\mathbf{p}^*(T) = \left.\frac{\partial \hat{J}^*(\mathbf{x})}{\partial \mathbf{x}}\right|_{\mathbf{x}=\boldsymbol{\theta}^*} \tag{6-776}$$

From Eqs. (6-776) and (6-774), we conclude that

$$\left.\frac{\partial \hat{J}^*(\mathbf{x})}{\partial \mathbf{x}}\right|_{\mathbf{x}=\boldsymbol{\theta}^*} - c \left.\frac{\partial g(\mathbf{x})}{\partial \mathbf{x}}\right|_{\mathbf{x}=\boldsymbol{\theta}^*} = \mathbf{0} \tag{6-777}$$

Since $\boldsymbol{\theta}^*$ is in $S$, we deduce from Eq. (6-772) that

$$g(\mathbf{x})\Big|_{\mathbf{x}=\boldsymbol{\theta}^*} = 0 \tag{6-778}$$

Now let us compare the system of $n + 1$ equations defined by Eqs. (6-777) and (6-778) and the system of $n + 1$ equations defined by Eq. (6-771). We observe that they are the same equations. It follows that, since the points $\boldsymbol{\theta}_1^*, \boldsymbol{\theta}_2^*, \ldots, \boldsymbol{\theta}_m^*$ satisfy Eq. (6-771), they also satisfy Eqs. (6-777) and (6-778) whenever

$$\left.\begin{aligned} \boldsymbol{\theta}^* &= \boldsymbol{\theta}_i^* \\ -c &= \lambda_i^* \end{aligned}\right\} \qquad i = 1, 2, \ldots, m \tag{6-779}$$

We have thus shown that the necessary condition that $\mathbf{p}^*(T)$ be normal to $S$ will isolate all the extremal points of $\hat{J}^*(\boldsymbol{\theta})$, $\boldsymbol{\theta} \in S$. To determine the extremal point which corresponds to the absolute minimum, we must carry out additional computations. At this point, we suggest to the reader that he review subsection 9 of Sec. 5-16 and comments 7, 8, and 10 of Sec. 5-17; in so doing, he will better understand the implications of the transversality conditions.

If the target set $S$ is not smooth (for example, the target set may consist of isolated points)† then we are almost forced to solve the fixed-end-point problem. In such problems, we may not be able to define a tangent plane to the target set, and thus the normal to the target set may not be defined (see comment 8 in Sec. 5-17).

## 6-25 Concluding Remarks

As we stated in the introduction to this chapter, our purpose was to provide a transition from the "theoretical" material of Chap. 5 to the more "applied" material of Chaps. 7 to 10.

In this chapter, we indicated that the first step in the solution of a given problem is to find the extremal controls, i.e., the controls that satisfy the necessary conditions. We pointed out the importance of such theoretical results as the existence and uniqueness of optimal controls, the uniqueness of the extremal controls, the existence of singular controls, and so on. We solved very few specific problems. We shall do this in the remainder of the book. Our purpose in the remaining four chapters of the book is not only to illustrate the theory by means of specific problems but, in addition, to provide a concrete set of examples which illustrate the structure of the optimal feedback system, the type of nonlinearities that must be constructed, the analytical difficulties that arise for higher-order systems, and the relative complexity of the optimal design. Often, we shall suggest suboptimal designs based upon the optimal one. It is our feeling that the advantages and disadvantages of optimal designs can be appreciated only if the engineer is familiar with both the theory and its implications in terms of quantities to be measured, complexity of computing equipment, and trade-off characteristics of optimal systems. Up to now, we have been concerned with general theory; from now on, we shall be concerned with specific systems.

† A simple time-optimal and fuel-optimal problem with such a target set is solved in Ref. A-8.

# Chapter 7
# THE DESIGN OF TIME-OPTIMAL SYSTEMS

## 7-1 Introduction

In this chapter, we shall study in detail the time-optimal control problem for several simple systems. We considered the time-optimal control problem from a general point of view in Secs. 6-1 to 6-10. Here, we shall examine the design of specific time-optimal feedback control systems.

The study of specific time-optimal systems has been a major influence in the development of modern control theory. In 1950, two papers concerned with the efficient use of relays and nonlinear feedback for the improvement of second-order servomechanisms were published in the engineering literature by Hopkin [H-10] and McDonald [M-1]. Heuristic geometric arguments were used to demonstrate that the control was indeed time-optimal. Later, Bushaw [B-23] extended these results to other second-order systems. Many other papers (see, for example, Refs. B-12, B-13, B-24, C-2, D-6, K-1, K-29, K-30, N-1, O-1, P-7, R-5, S-3, and S-8) dealing with the time-optimal control of second- and higher-order systems appeared. The study of second- and third-order systems led to the development of a more mathematically sound theory for the time-optimal problem (see Refs. B-4 and L-2). Meanwhile, time-optimal designs appeared in standard textbooks (see, for example, Refs. T-2, chap. 11; S-6, chaps. 15 to 17; C-3, chap. 9; and G-3, chap. 10). The introduction of the minimum principle enhanced still further the continuing interest in time-optimal control problems.

As we have said, we shall examine in detail many time-optimal systems. Our objectives will be:

1. To illustrate the use of the necessary conditions to obtain a complete design of the time-optimal feedback system
2. To illustrate the nature and importance of such geometric concepts as "switch curves" and "switch surfaces"
3. To illustrate the type of nonlinear computations required to implement the time-optimal feedback system
4. To discuss the relative difficulties involved in the construction of time-optimal controllers

5. To illustrate by specific examples and exercises several theoretical points which were discussed in a general context in preceding chapters

Often, in our exposition, we shall outline proofs of the uniqueness of the extremal controls using geometric arguments rather than invoking Theorem 6-9. We do this because the same type of arguments can be used to show uniqueness for specific problems for which no general theorems can be stated (for example, for nonlinear systems). However, we shall rely heavily on Theorem 6-10, which guarantees the existence of time-optimal controls to the origin.

In many exercises, we shall ask the reader to work out the solutions of problems with "strange" target sets, so that he may become acquainted with some properties of unconventional control systems. It is true that the problems stated in these exercises make very little sense from an engineering point of view; nonetheless, they illustrate several important points which may arise in other, more realistic and complex problems.

Most of the problems we shall consider involve a single control variable $u(t)$. Also, since most of the systems we shall study in this chapter are normal,† the time-optimal control will be piecewise constant. For this reason, we introduce a convenient shorthand notation to indicate the constant values of the piecewise constant control. This shorthand notation is defined below; we remark that it will be used extensively in this and the following chapter.

***Definition 7-1 Control Sequence*** *We shall use the notation*

$$\{u_1, u_2, u_3, \ldots\}$$

*where the* $u_i$ *are constants, to mean*

$$\begin{array}{lll} \textit{Apply the control } u(t) = u_1 & & \textit{for } t_0 < t \leq t_1 \\ \textit{Apply the control } u(t) = u_2 & & \textit{for } t_1 < t \leq t_2 \\ \textit{Apply the control } u(t) = u_3 & & \textit{for } t_2 < t \leq t_3 \\ \cdots\cdots\cdots & & \cdots\cdots\cdots \end{array}$$

*We shall call the set* $\{u_1, u_2, u_3, \ldots\}$ *a control sequence. For example, the statement "use the control sequence* $\{+1, -1\}$*" means apply first the control* $u(t) = +1$ *and then the control* $u(t) = -1$.

Since the systems we shall examine are time-invariant, we shall be able to determine the time-optimal control as a function of the state and thus, in so doing, solve the feedback problem (see Sec. 6-6). Since the time-optimal controls will be piecewise constant functions of time and since the time-optimal control will also be a function of the state, we shall find sets or regions of the state space over which the control is constant. These sets will be separated by curves in two-dimensional space, by surfaces in three-dimensional space, and by hypersurfaces in $n$-dimensional space. The separating sets are called *switch curves, switch surfaces, switch hyper-*

† See Sec. 6-5.

*surfaces,* etc. These ideas will become clearer later on in the chapter. For the moment, we remark that it is the equation and shape of these *switch curves* and *switch surfaces* that determine the nonlinear operations that must be performed.

In every section of this chapter, we shall proceed roughly in the same manner, namely:

1. We shall define the problem precisely.
2. We shall form the Hamiltonian.
3. We shall find the $H$-minimal control (see Sec. 5-3).
4. We shall find the equations of the costate variables.
5. We shall determine the control sequences that are "candidates" for the optimal control.
6. We shall find the control sequences that satisfy the boundary conditions.
7. We shall determine the switch curves, etc., that divide the state space into various regions.
8. We shall prove the validity of the optimal-control law.

The particular problems we shall deal with in the chapter can be classified as follows: Secs. 7-2 to 7-6 deal with the control of systems with real poles (or eigenvalues); Secs. 7-7 to 7-9 deal with the control of systems with complex poles (or eigenvalues); Secs. 7-10 and 7-11 deal with the control of nonlinear systems; finally, Secs. 7-12 to 7-14 deal with the control of systems with zeros in their transfer function. Naturally, there are many other time-optimal problems which have been solved in the control literature;† we hope that the material we have included in this chapter will provide enough background information to enable the interested reader to attack similar time-optimal problems.

Finally, we wish to caution the reader that the criterion of minimum response time is, in many cases, not the most suitable measure of system performance. For example, if one wishes to drive his car from stoplight to stoplight as quickly as possible, one should accelerate as fast as possible and then apply the brakes as hard as possible. This time-optimal driving technique has a number of drawbacks, namely:

1. The dynamics of the car must be known exactly.
2. The acceleration and deceleration must be known exactly, so that the time at which the brakes are applied can be precomputed (otherwise the car might jump the light).
3. Quite a bit of tire rubber will be "peeled" off.
4. Gas consumption will be high.
5. The long arm of the law might reach out with a speeding ticket.

We can see that the minimum-time criterion is not always best and that the system designer should give careful thought to the question of choosing a performance criterion most suitable for his specific application.

† See, for example, Refs. F-4, F-5, O-1, O-2, K-18, K-25, S-5, A-1, D-2, F-6, B-14, F-7, and P-5 (chaps. 1 and 3).

## 7-2 Time-optimal Control of a Double-integral Plant

In this section, we shall solve the time-optimal control problem for the so-called "double-integral" plant. Our exposition will be detailed since the concepts involved will be used again and again in the sequel. We shall solve the problem for the following cases:

1. The terminal state is the origin of the state plane.
2. The target set is an entire axis.
3. The target set is a line segment.

Double-integral plants often represent the motion of inertial loads in a frictionless environment (see also Example 6-9). For example, if we let $m$ be the mass (or moment of inertia) of a body, $y(t)$ be the position (or angular displacement), and $v(t)$ be the applied force (or torque), then, in the absence of friction and gravitational forces, the motion of the body is described by the second-order equation

$$m\ddot{y}(t) = v(t) \tag{7-1}$$

The transfer function $G(s)$ of this system is

$$G(s) = \frac{y(s)}{v(s)} = \frac{1}{ms^2} \tag{7-2}$$

which is why the name "double integral" is used. Let us define the control $u(t)$ by setting $u(t) = v(t)/m$. Then Eq. (7-1) reduces to the equation

$$\ddot{y}(t) = u(t) \tag{7-3}$$

We define the state variables $x_1(t)$ and $x_2(t)$ by

$$\begin{aligned} x_1(t) &= y(t) = \text{output} \\ x_2(t) &= \dot{y}(t) = \text{output rate} \end{aligned} \tag{7-4}$$

and we obtain the system representation

$$\begin{aligned} \dot{x}_1(t) &= x_2(t) \\ \dot{x}_2(t) &= u(t) \end{aligned} \tag{7-5}$$

We assume that the control $u(t)$ is constrained in magnitude by the relation

$$|u(t)| \leq 1 \qquad \text{for all } t \tag{7-6}$$

This magnitude constraint is, as we have commented in Chap. 6, the result of physical limitations on the amount of thrusts or torques which may be realized in practice by equipment.

The first problem we shall solve is the following:

***Problem 7-1*** *Given the system*

$$\begin{aligned} \dot{x}_1(t) &= x_2(t) \\ \dot{x}_2(t) &= u(t) \qquad |u(t)| \leq 1 \end{aligned} \tag{7-7}$$

*find the admissible control that forces the system* (7-7) *from any initial state* $(\xi_1, \xi_2)$ *to the origin* (0, 0) *in the shortest possible time.*

Before we proceed to the solution, we remark that:

1. The system (7-7) is normal, and so singular controls cannot be optimal (see Theorem 6-6).
2. The time-optimal control exists (see Theorem 6-10) and is unique (see Theorem 6-7).
3. Since the terminal state is the origin, the extremal controls are unique (see Theorem 6-9).

Our method of solution will consist of the following steps:

1. We shall find the $H$-minimal control, i.e., the control that minimizes the Hamiltonian.
2. We shall find the equations of the costate variables in terms of their unknown initial values.
3. We shall find the control sequences (see Definition 7-1) which are candidates for time-optimal control.
4. We shall plot the trajectories in the state plane for $u = +1$ and $u = -1$.
5. We shall determine the switch curve.
6. We shall derive Control Law 7-1, which provides the solution to Problem 7-1.
7. We shall draw the schematic diagram of an engineering realization of the control law.

For Problem 7-1, the Hamiltonian function is given by

$$H = 1 + x_2(t)p_1(t) + u(t)p_2(t) \tag{7-8}$$

The $H$-minimal control, i.e., the control which minimizes the Hamiltonian, is given by

$$u(t) = -\operatorname{sgn}\{p_2(t)\} = \Delta = \pm 1 \tag{7-9}$$

The costate variables $p_1(t)$ and $p_2(t)$ satisfy the equations

$$\begin{aligned} \dot{p}_1(t) &= -\frac{\partial H}{\partial x_1(t)} = 0 \\ \dot{p}_2(t) &= -\frac{\partial H}{\partial x_2(t)} = -p_1(t) \end{aligned} \tag{7-10}$$

If we let $\pi_1$ and $\pi_2$ be the initial values of the costate variables, i.e., if we let

$$\pi_1 = p_1(0) \qquad \pi_2 = p_2(0) \tag{7-11}$$

then, from Eqs. (7-10), we find that

$$\begin{aligned} p_1(t) &= \pi_1 = \text{const} \\ p_2(t) &= \pi_2 - \pi_1 t \end{aligned} \tag{7-12}$$

We note that $p_2(t)$ is a straight line in the $p_2(t)$-$t$ plane; the four possible shapes of $p_2(t)$ and the corresponding shapes of the $H$-minimal control $u(t) = -\operatorname{sgn}\{p_2(t)\}$ are shown in Fig. 7-1. From Fig. 7-1, we conclude that the time-optimal control is piecewise constant and can switch, at most,

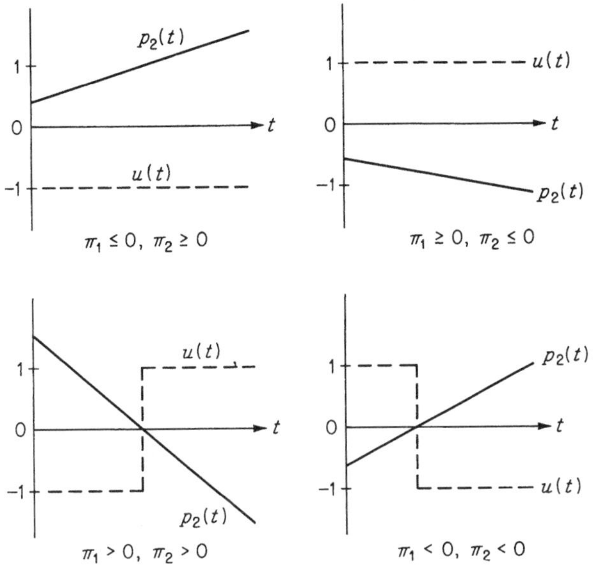


**Fig. 7-1** The four possible "shapes" of $p_2(t) = \pi_2 - \pi_1 t$ and the corresponding controls $u(t) = -\operatorname{sgn}\{\pi_2 - \pi_1 t\}$.

once (compare with Theorem 6-8) because the problem is normal. Thus, we conclude that only the four control sequences (see Definition 7-1)

$$\{+1\},\ \{-1\},\ \{+1,\ -1\},\ \{-1,\ +1\} \tag{7-13}$$

can be candidates for time-optimal control. We note that two of the necessary conditions [namely, Eqs. (7-9) and (7-12)] have helped us to isolate only four possible control sequences.

Since, over a finite interval of time, the time-optimal control is constant, $u(t) = \Delta = \pm 1$, we can solve Eqs. (7-7) using $u(t) = \Delta = \text{const}$ with the initial conditions

$$x_1(0) \triangleq \xi_1 \qquad x_2(0) \triangleq \xi_2 \tag{7-14}$$

to obtain the relations

$$x_2(t) = \xi_2 + \Delta t \tag{7-15}$$
$$x_1(t) = \xi_1 + \xi_2 t + \tfrac{1}{2}\Delta t^2 \tag{7-16}$$

We next eliminate the time $t$, to find that

$$x_1 = \xi_1 + \tfrac{1}{2}\Delta x_2{}^2 - \tfrac{1}{2}\Delta \xi_2{}^2 \qquad (7\text{-}17)$$

where

$$t = \Delta(x_2 - \xi_2) \qquad (7\text{-}18)$$

Equation (7-17) is the equation of the trajectory in the $x_1x_2$ plane originating at $(\xi_1, \xi_2)$ and due to the action of the control $u = \Delta$. These trajectories, which are parabolic, are shown in Fig. 7-2; the solid trajectories are for

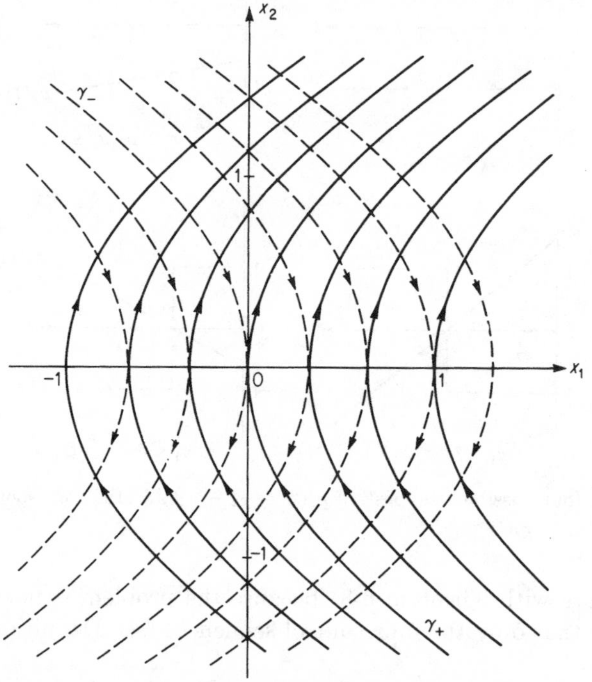


**Fig. 7-2** The forced trajectories in the state plane; the solid trajectories are generated by $u = 1$, and the dashed trajectories are generated by $u = -1$.

$u = \Delta = +1$, whereas the dashed ones are for $u = \Delta = -1$; the arrows indicate the direction of motion for positive time.

Our objective is to drive any initial state to (0, 0), that is, to the origin of the state plane. Since the control must be piecewise constant, we can find the locus of points $(x_1, x_2)$ which can be transferred to (0, 0) using $u = \pm 1$. These two trajectories leading to (0, 0) are labeled $\gamma_+$ and $\gamma_-$ in Fig. 7-2. To be more precise:

***Definition 7-2*** *The $\gamma_+$ curve is the locus of all points $(x_1, x_2)$ which can be forced to (0, 0) by the control $u = +1$. We shall see that*

$$\gamma_+ = \{(x_1, x_2): x_1 = \tfrac{1}{2}x_2{}^2;\ x_2 \le 0\} \qquad (7\text{-}19)$$

***Definition 7-3*** *The $\gamma_-$ curve is the locus of all points $(x_1, x_2)$ which can be forced to* (0, 0) *by the control $u = -1$. We shall see that*

$$\gamma_- = \{(x_1, x_2): x_1 = -\tfrac{1}{2}x_2^2;\ x_2 \geq 0\} \tag{7-20}$$

***Definition 7-4*** *The $\gamma$ curve, called the switch curve, is the union of the $\gamma_+$ and $\gamma_-$ curves. Its equation is*

$$\gamma = \{(x_1, x_2): x_1 = -\tfrac{1}{2}x_2|x_2|\} = \gamma_+ \cup \gamma_- \tag{7-21}$$

The $\gamma$ curve is shown in Fig. 7-3. The $\gamma$ curve divides the state plane into two regions (or sets) $R_-$ and $R_+$.

***Definition 7-5*** *Let $R_-$ be the set of states $(x_1, x_2)$ such that*

$$R_- = \{(x_1, x_2): x_1 > -\tfrac{1}{2}x_2|x_2|\} \tag{7-22}$$

*that is, $R_-$ consists of the points to the right of the switch curve $\gamma$. Let $R_+$ be the set of states $(x_1, x_2)$ such that*

$$R_+ = \{(x_1, x_2): x_1 < -\tfrac{1}{2}x_2|x_2|\} \tag{7-23}$$

*that is, $R_+$ consists of the points to the left of the switch curve $\gamma$.*

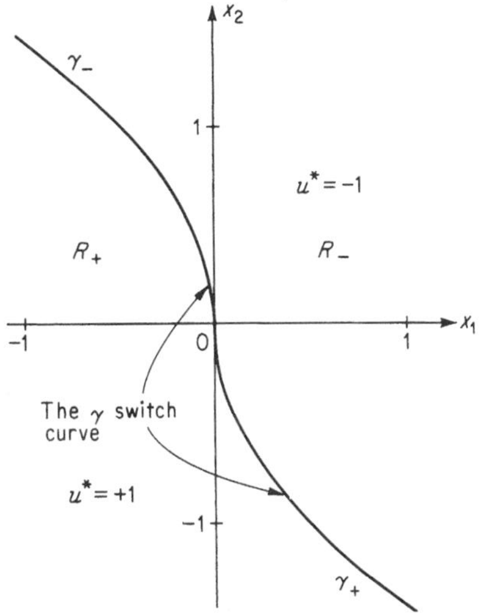


**Fig. 7-3** Illustration of the switch curve for the double-integral plant.

Now we shall prove that *if the initial state $\Xi = (\xi_1, \xi_2)$ belongs to the $\gamma_+$ curve, then $u = +1$ is the time-optimal control.* The arguments are illustrated in Fig. 7-4. Let us consider the four control sequences $\{+1\}$, $\{-1\}$, $\{+1, -1\}$, and $\{-1, +1\}$. By definition, the control sequence $\{+1\}$ results in the trajectory $\Xi 0$, which reaches the origin. The control sequence $\{-1\}$ results in the trajectory $\Xi A$, which never reaches the origin. The control sequence $\{+1, -1\}$ results in trajectories of the type $\Xi BC$, which never reach the origin. The control sequence $\{-1, +1\}$ results in trajectories of the type $\Xi DE$, which never reach the origin. Therefore, if the initial state is on the $\gamma_+$ curve, from all the control sequences which are candidates for minimum-time control, only the sequence $\{+1\}$ can force the state $\Xi$ to 0. Thus, by elimination, it must be time-optimal.

Using analogous arguments, we can show that *if the initial state belongs to the $\gamma_-$ curve, then the time-optimal control is $u = -1$.* Thus, we have derived the time-optimal control law for initial states on the $\gamma$ curve.

Let us now consider an initial state $X$ which belongs to the region $R_-$ (see Definition 7-5). If we use the control sequence $\{+1\}$, the resulting trajectory is $XF$, shown in Fig. 7-4, which never reaches the origin. If we apply the sequence $\{-1\}$, the resulting trajectory $XG$ never reaches the origin. If we apply the sequence $\{+1, -1\}$, the resulting trajectory is of the type $XHI$, which does not reach the origin. However, if we use the sequence $\{-1, +1\}$, then one can reach the origin along the trajectory $XJO$, provided that the transition from the control $u = -1$ to $u = +1$ occurs at the point $J$, that is, at the precise moment that the trajectory crosses the $\gamma$ switch curve. This is true for every state in $R_-$. Thus, by the process of elimination, we have arrived at the conclusion that *the*

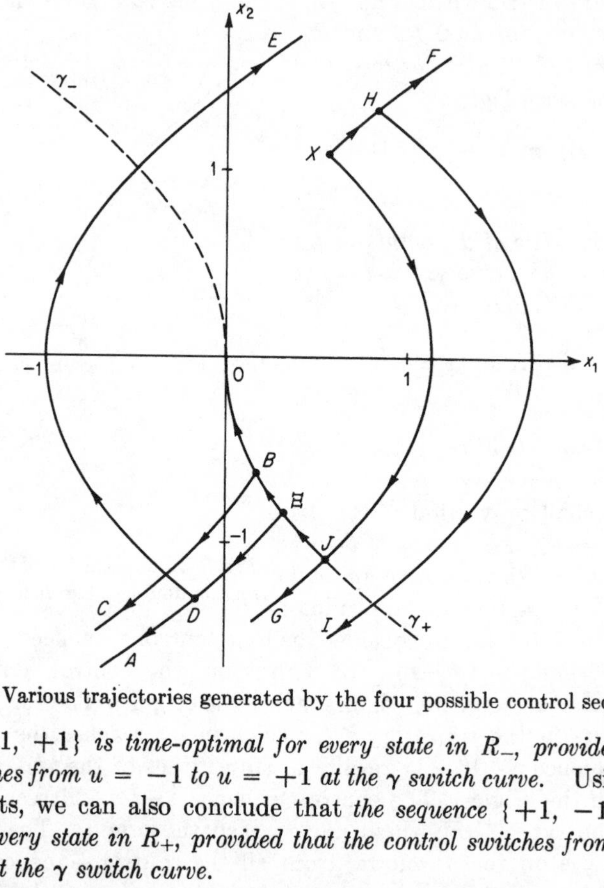


**Fig. 7-4** Various trajectories generated by the four possible control sequences.

*sequence* $\{-1, +1\}$ *is time-optimal for every state in* $R_-$, *provided that the control switches from* $u = -1$ *to* $u = +1$ *at the* $\gamma$ *switch curve.* Using identical arguments, we can also conclude that *the sequence* $\{+1, -1\}$ *is time-optimal for every state in* $R_+$, *provided that the control switches from* $u = +1$ *to* $u = -1$ *at the* $\gamma$ *switch curve.*

We summarize these results in the following control law:

***Control Law 7-1 Solution to Problem 7-1*** *The time-optimal control* $u^*$, *as a function of the state* $(x_1, x_2)$, *is given by*

$$
\begin{aligned}
u^* = u^*(x_1, x_2) &= +1 \qquad \text{for all } (x_1, x_2) \in \gamma_+ \cup R_+ \\
u^* = u^*(x_1, x_2) &= -1 \qquad \text{for all } (x_1, x_2) \in \gamma_- \cup R_-
\end{aligned}
\tag{7-24}
$$

*and, furthermore,* $u^*$ *is unique.*

The above control law assigns a value to the optimal control $u^*$ for every state in the state plane. We remark that the following facts were used to prove this law:

1. The existence of the time-optimal control
2. The unique correspondence of a control sequence to each state, which is equivalent to the uniqueness of the extremal controls
3. The fact that existence of the optimal control, together with the uniqueness of the extremal controls, implies the uniqueness of the optimal control

Next, we shall present an engineering realization of Control Law 7-1. In essence, we can design a nonlinear feedback control system which operates on the state variables $x_1$ and $x_2$ and delivers the correct time-optimal control

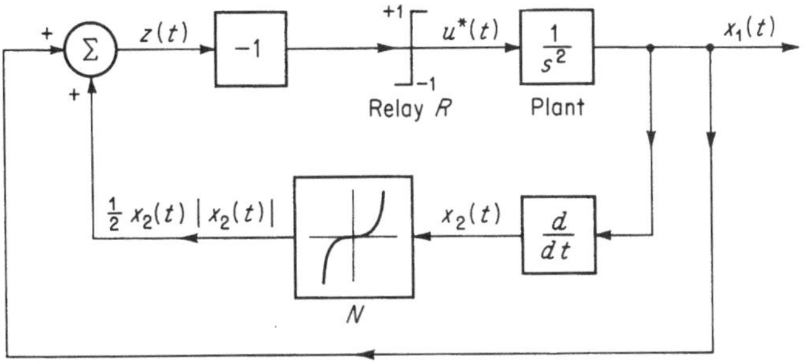


**Fig. 7-5** An engineering realization of the closed-loop time-optimal control system for the double-integral plant.

$u^*$ to the system. Such an engineering realization is illustrated in Fig. 7-5. The state variables $x_1(t)$ (output) and $x_2(t)$ (output rate) are measured at each instant of time. The signal $x_2(t)$ is introduced to the nonlinearity $N$, whose output is $\frac{1}{2}x_2(t)|x_2(t)|$. The signal $z(t)$ is given by

$$z(t) = x_1(t) + \tfrac{1}{2}x_2(t)|x_2(t)| \tag{7-25}$$

and, after a sign reversal, drives the relay $R$. Basically, the relay $R$ is the engineering realization of the signum operation. The relay output $u^*(t)$ is the time-optimal control. The reason is that

$$\text{If } (x_1(t), x_2(t)) \in R_-, \text{ then } -z(t) < 0$$
$$\text{If } (x_1(t), x_2(t)) \in R_+, \text{ then } -z(t) > 0$$

A basic difficulty occurs when the state $(x_1(t), x_2(t))$ is on the $\gamma$ switch curve. From Eqs. (7-25) and (7-21), we can see that

$$\text{If } (x_1(t), x_2(t)) \in \gamma, \text{ then } z(t) = 0$$

In this case, the input signal to the relay is zero; however, since a relay is a physical element with small (but nonzero) inertia, the relay will not switch precisely at the $\gamma$ curve but some distance away, and so, by the time the relay actually switches, the state is no longer on the $\gamma$ curve and the input to the relay is nonzero. This is the reason that we call the system of Fig. 7-2 an engineering realization of the control law. Basically, the inherent physical properties of the equipment remove the ambiguity of the function sgn $\{0\}$, and therefore, the engineering system† will perform in an *almost* optimal manner. Next, we come to the construction of the nonlinearity $N$. First, observe that the input-output characteristics of $N$ are those of the $\gamma$ switch curve. We can construct the nonlinearity $N$ either by using a multiplier and an absolute-value generator [to produce the signal $x_2(t)|x_2(t)|$] or by using biased diodes. We shall let the reader evaluate the effects of approximating $N$ in an exercise.

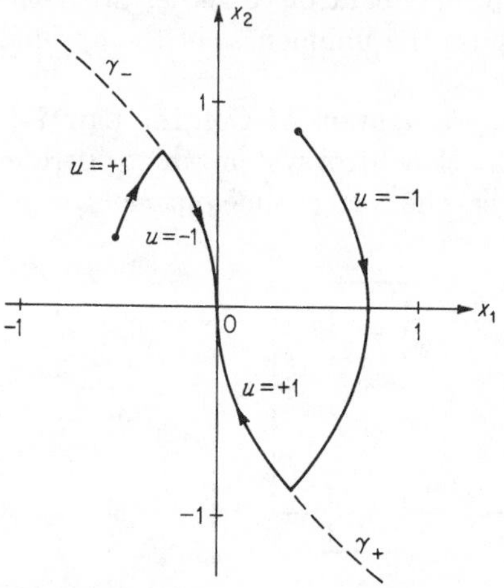


**Fig. 7-6** Two time-optimal trajectories in the state plane.

In Fig. 7-6, we illustrate two time-optimal trajectories to the origin. The arrows on the trajectories indicate the motion of the state for positive time; the time-optimal trajectories are made up of segments of parabolas.

We now turn our attention to the evaluation of the minimum time. For each point $(x_1, x_2)$ in the state plane, there exists a value of time $t^* = t^*(x_1, x_2)$ which represents the minimum time required to force $(x_1, x_2)$ to $(0, 0)$. We shall proceed to evaluate this minimum time $t^*$ as a function of $x_1$, $x_2$ and demonstrate that it is the solution of the Hamilton-Jacobi equation. The method of evaluating $t^*$ is to compute the time required to force $(x_1, x_2)$ to the switch curve (i.e., the $\gamma$ curve) and then compute the time required to go from the point of intersection on the switch curve to the origin $(0, 0)$. Using this technique, we find that the minimum time $t^*$ is given by the relations

$$t^*(x_1, x_2) = t^* = \begin{cases} x_2 + \sqrt{4x_1 + 2x_2^2} & \text{if } x_1 > -\frac{1}{2}x_2|x_2| \\ -x_2 + \sqrt{-4x_1 + 2x_2^2} & \text{if } x_1 < -\frac{1}{2}x_2|x_2| \\ |x_2| & \text{if } x_1 = -\frac{1}{2}x_2|x_2| \end{cases} \tag{7-26}$$

**Exercise 7-1** Derive Eq. (7-26).

† For an extended discussion of the effects of imperfect equipment, see Ref. S-6, chap. 15.

Let us now consider the set of states $(x_1, x_2)$ which can be forced to the origin (0, 0) in the *same minimum time* $t^*$. By manipulating Eq. (7-26) we find that

$$x_1 = \begin{cases} -\frac{1}{2}x_2^2 + \frac{1}{4}(t^* - x_2)^2 & \text{for } x_1 > -\frac{1}{2}x_2|x_2| \\ \frac{1}{2}x_2^2 - \frac{1}{4}(t^* + x_2)^2 & \text{for } x_1 < -\frac{1}{2}x_2|x_2| \\ -\frac{1}{2}x_2 t^* & \text{for } x_1 = -\frac{1}{2}x_2|x_2| \end{cases} \tag{7-27}$$

***Definition 7-6*†** *Let $S(t^*)$ denote the set of states which can be forced to* (0, 0) *in the same minimum time $t^*$. The set $S(t^*)$ is called the $t^*$ minimum*

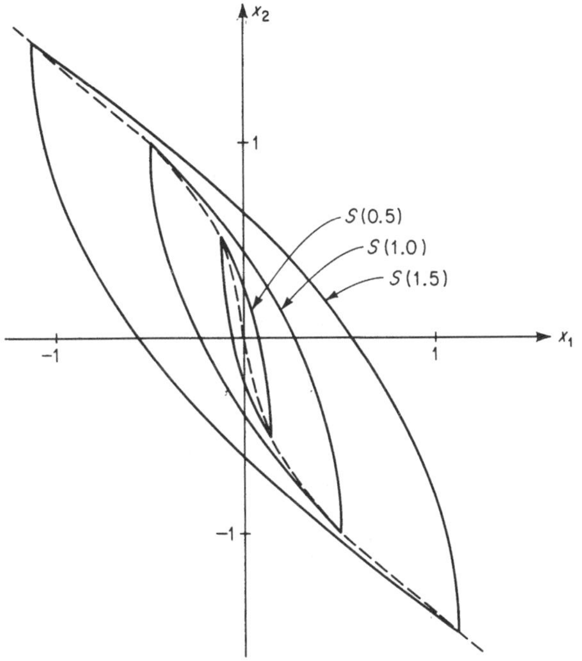


**Fig. 7-7** Three minimum isochrones for the double-integral plant.

*isochrone. A state $(x_1, x_2)$ belongs to $S(t^*)$ if and only if $x_1$, $x_2$, and $t^*$ satisfy Eq.* (7-27).

Each set $S(t^*)$ is a closed curve in the state plane. Figure 7-7 shows three such minimum isochrones. We note that each $S(t^*)$ curve is differentiable everywhere except at the two points at which the $S(t^*)$ curve meets the $\gamma$ switch curve.

From Eq. (7-27), it is easy to establish the following property of the minimum isochrones. Let

$$t_1^* < t_2^* \tag{7-28}$$

† Compare with Definition 6-10.

and consider the $S(t_1^*)$ and $S(t_2^*)$ minimum isochrones. Suppose that $(x_1', x_2) \in S(t_1^*)$, $(x_1'', x_2) \in S(t_2^*)$, and $(x_1''', x_2) \in \gamma$. Then

$$|x_1' - x_1'''| < |x_1'' - x_1'''| \tag{7-29}$$

**Exercise 7-2** Derive Eq. (7-29). The above property, loosely speaking, states that the minimum isochrones "expand" for increasing values of $t^*$.

**Exercise 7-3** Consider the isochrone $S(t^*; x_1, x_2)$ as the boundary of the set $\hat{S}(t^*; x_1, x_2)$ which consists of all states enclosed within $S(t^*; x_1, x_2)$ and on $S(t^*; x_1, x_2)$. Show that the set $\hat{S}(t^*; x_1, x_2)$ is closed, bounded, and convex. Show that the set $\hat{S}(t^*; -x_1, x_2)$ is the set of reachable states from (0, 0) at time $t^*$ with the control constrained by $|u(t)| \leq 1$.

Next, we shall verify that the Hamilton-Jacobi equation† is satisfied along the optimal trajectory. First of all, the minimum cost for our problem is the minimum time $t^*(x_1, x_2)$ required to force the state $(x_1, x_2)$ to (0, 0). Thus

$$J^*(\mathbf{x}, t) = t^*(x_1, x_2) \tag{7-30}$$

and the minimum time is given by Eq. (7-26). In general, the Hamilton-Jacobi equation takes the form

$$\left\{\frac{\partial J^*(\mathbf{x}, t)}{\partial t} + H\left[\mathbf{x}, \frac{\partial J^*(\mathbf{x}, t)}{\partial t}, \mathbf{u}\right]\right\}\Bigg|_{\substack{\mathbf{x}=\mathbf{x}^*\\ \mathbf{u}=\mathbf{u}^*}} = 0 \tag{7-31}$$

For the problem at hand, we have [see Eq. (7-8)]

$$H = 1 + x_2 p_1 + u p_2 \tag{7-32}$$

and so we let

$$p_1 = \frac{\partial t^*}{\partial x_1} \qquad p_2 = \frac{\partial t^*}{\partial x_2} \tag{7-33}$$

Since the time-optimal control $u^*$ is given by

$$\begin{aligned} u^* &= +1 && \text{for all } (x_1, x_2) \in R_+ \\ u^* &= -1 && \text{for all } (x_1, x_2) \in R_- \\ u^* &= -\operatorname{sgn}\{x_2\} && \text{for all } (x_1, x_2) \in \gamma \end{aligned}$$

we may deduce that the Hamilton-Jacobi equation is

$$\begin{aligned} \frac{\partial t^*}{\partial t} + 1 + x_2\frac{\partial t^*}{\partial x_1} - \frac{\partial t^*}{\partial x_2} &= 0 && \text{if } x_1 > -\tfrac{1}{2}x_2|x_2| \\ \frac{\partial t^*}{\partial t} + 1 + x_2\frac{\partial t^*}{\partial x_1} + \frac{\partial t^*}{\partial x_2} &= 0 && \text{if } x_1 < -\tfrac{1}{2}x_2|x_2| \\ \frac{\partial t^*}{\partial t} + 1 + x_2\frac{\partial t^*}{\partial x_1} - \operatorname{sgn}\{x_2\}\frac{\partial t^*}{\partial x_2} &= 0 && \text{if } x_1 = -\tfrac{1}{2}x_2|x_2| \end{aligned} \tag{7-34}$$

† See Secs. 5-18 to 5-21 and 6-9.

But, from Eq. (7-26), we find that

$$\frac{\partial t^*}{\partial t} = 0 \tag{7-35}$$

$$\frac{\partial t^*}{\partial x_1} = \begin{cases} \dfrac{2}{\sqrt{4x_1 + 2x_2^2}} & \text{if } x_1 > -\frac{1}{2}x_2|x_2| \\ -\dfrac{2}{\sqrt{-4x_1 + 2x_2^2}} & \text{if } x_1 < -\frac{1}{2}x_2|x_2| \\ 0 & \text{if } x_1 = -\frac{1}{2}x_2|x_2| \end{cases} \tag{7-36}$$

$$\frac{\partial t^*}{\partial x_2} = \begin{cases} 1 + \dfrac{2x_2}{\sqrt{4x_1 + 2x_2^2}} & \text{if } x_1 > -\frac{1}{2}x_2|x_2| \\ -1 + \dfrac{2x_2}{\sqrt{-4x_1 + 2x_2^2}} & \text{if } x_1 < -\frac{1}{2}x_2|x_2| \\ \operatorname{sgn}\{x_2\} & \text{if } x_1 = -\frac{1}{2}x_2|x_2| \end{cases} \tag{7-37}$$

Consider the set $R_-$ of states $x_1 > -\frac{1}{2}x_2|x_2|$, that is, the points to the right of the switch curve. Then, from Eqs. (7-35) to (7-37), we have

$$1 + \frac{2x_2}{\sqrt{4x_1 + 2x_2^2}} - 1 - \frac{2x_2}{\sqrt{4x_1 + 2x_2^2}} = 0 \tag{7-38}$$

Similarly, for $x_1 < -\frac{1}{2}x_2|x_2|$, that is, for $(x_1, x_2) \in R_+$, we have

$$1 - \frac{2x_2}{\sqrt{-4x_1 + 2x_2^2}} - 1 + \frac{2x_2}{\sqrt{-4x_1 + 2x_2^2}} = 0 \tag{7-39}$$

and, finally, for $x_1 = -\frac{1}{2}x_2|x_2|$, that is, for states on the $\gamma$ curve, we have

$$1 + x_2 0 - \operatorname{sgn}\{x_2\} \operatorname{sgn}\{x_2\} = 1 - 1 = 0 \tag{7-40}$$

Equations (7-38) to (7-40) imply that the Hamilton-Jacobi equation is satisfied for all $(x_1, x_2)$ and, hence, that Control Law 7-1 is indeed time-optimal. However, we note that it was necessary to verify the Hamilton-Jacobi equation separately in each region $R_+$ and $R_-$ and along the switch curve. This is a characteristic property of many time-optimal systems.†

*To summarize:* We have considered the problem of time-optimal control of the double-integral plant from any initial state to the origin. First, we applied the necessary conditions to derive the four control sequences which need to be considered as candidates for the optimal control. Then, we have shown that to each state $(x_1, x_2)$, there corresponds a *unique* control sequence which will drive the state to (0, 0), and so we obtained the time-optimal control law 7-1. We evaluated the minimum time $t^*$ from any state $(x_1, x_2)$ to (0, 0) as a function of $x_1$, $x_2$, and we introduced the concept of a minimum isochrone. Finally, we have shown that the expression for the minimum time $t^*$ is the solution of the Hamilton-Jacobi equation, and

† See also the comments at the end of Sec. 6-9.

so we have provided an alternative proof of the optimality of Control Law 7-1.

The reader should note that we never evaluated the unknown initial values of the costate variables $\pi_1$ and $\pi_2$ in Eqs. (7-12). Thus, we were able to find the time-optimal control as a function of the state solely by using the fact that the optimal control must minimize the Hamiltonian and by using the information provided by the "shape" of $p_2(t)$ (see Fig. 7-1). Also, we did not use the necessary condition that the Hamiltonian must be zero along the time-optimal trajectory; as a matter of fact, we must use this condition to evaluate $\pi_1$ and $\pi_2$, but we shall leave this task to the reader as an exercise at the end of this section.

In the remainder of this section, we shall consider the problem of transferring any initial state to a target set in minimum time. In this manner, we shall illustrate the use of the transversality conditions.

We shall first discuss the problem of reaching the $x_1$ axis from any initial state in minimum time. More precisely:

***Problem 7-2*** *Given the system* (7-5) *and the target set* $S$, *given by*

$$S = \{(x_1, x_2): x_2 = 0, -\infty < x_1 < \infty\}^\dagger \tag{7-41}$$

*Then determine the control* $u(t)$, *subject to the constraint* $|u(t)| \leq 1$, *such that any initial state is forced to* $S$ *in minimum time.*

The fact that we are trying to reach a set rather than a point does not have any effect upon some of the results we have derived by means of the minimum principle. Thus, the Hamiltonian $H$ is the same as the one given by Eq. (7-8); that is,

$$H = 1 + x_2(t)p_1(t) + u(t)p_2(t) \tag{7-42}$$

The control $u(t)$ which minimizes the Hamiltonian is given by Eq. (7-9); that is,

$$u(t) = -\operatorname{sgn}\{p_2(t)\} \tag{7-43}$$

and the equations of the costate variables $p_1(t)$ and $p_2(t)$ are the same as Eqs. (7-10); that is,

$$p_1(t) = \pi_1 \tag{7-44}$$
$$p_2(t) = \pi_2 - \pi_1 t \tag{7-45}$$

The transversality conditions require that at the terminal time $t_1$, the costate vector $\mathbf{p}(t_1)$ be normal to a vector $\mathbf{q}$ which belongs to the tangent hyperplane of the target set $S$. In our problem, the target $S$ is a straight

† Note that the target set $S$ has the property that one can define a unique normal to it at any point.

line, and a vector **q** which may be used is the vector

$$\mathbf{q} = \begin{bmatrix} k \\ 0 \end{bmatrix} \qquad k \neq 0 \tag{7-46}$$

Thus, the condition

$$\langle \mathbf{p}(t_1), \mathbf{q} \rangle = 0 \tag{7-47}$$

implies that

$$kp_1(t_1) + 0p_2(t_1) = 0 \tag{7-48}$$

from which we find that

$$p_1(t_1) = 0 \tag{7-49}$$

But $p_1(t) = \pi_1$ is a constant for all time; hence $\pi_1 = 0$, and it follows that

$$p_2(t) = \pi_2 \tag{7-50}$$

Thus, $p_2(t)$ is a nonzero (why?) constant, either positive or negative, for all time; this and Eq. (7-43) imply that the time-optimal control is $u = +1$ or

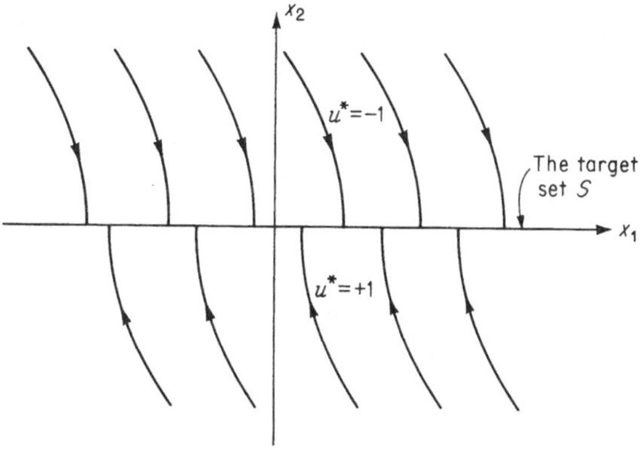


**Fig. 7-8** Time-optimal trajectories to the target set $S$ (the $x_1$ axis).

$-1$ and that no switching can occur. Thus, only the control sequences $\{+1\}$ and $\{-1\}$ can be candidates for time-optimal control. It is easy to verify that the time-optimal control as a function of the state is given by the following control law:

***Control Law 7-2 Solution to Problem 7-2*** *The time-optimal control, as a function of the state $(x_1, x_2)$, is given by*

$$u^* = u^*(x_1, x_2) = -\operatorname{sgn}\{x_2\} \tag{7-51}$$

*and, furthermore, $u^*$ is unique.*

Figure 7-8 illustrates the time-optimal trajectories to the set $S$.

**Exercise 7-4** Verify Control Law 7-2.

**Exercise 7-5** Given any state $(x_1, x_2)$. Determine the minimum time $t_1^*$ required to hit the target set $S$ given by Eq. (7-41). Show that $t_1^*$ is the solution of the Hamilton-Jacobi equation for this problem. Design the time-optimal control system.

Let us now briefly analyze the similarities and differences between the problem of hitting a single point and the problem of hitting this target set. Clearly, the Hamiltonian, the equations of the costate variables, and the equation of the $H$-minimal control (as a function of the costate) are the same for both problems. The only difference is provided by the evaluation of the costate variables at the terminal time. This additional information enabled us to deduce that only the two control sequences $\{+1\}$ and $\{-1\}$ were candidates for the time-optimal control to the set $S$, while the four control sequences $\{+1\}$, $\{-1\}$, $\{+1, -1\}$, and $\{-1, +1\}$ were candidates for the time-optimal control to a single terminal state.

Finally, we shall consider the problem of reaching a closed interval of the $x_1$ axis in minimum time. More precisely:

***Problem 7-3*** *Given the system* (7-5) *and the target set* $S_\alpha$ *defined for* $\alpha > 0$ *by the relation*

$$S_\alpha = \{(x_1, x_2): x_2 = 0, \; -\alpha \le x_1 \le \alpha\} \tag{7-52}$$

*Then determine the control* $u(t)$, *subject to the constraint* $|u(t)| \le 1$, *such that any initial state is forced to* $S_\alpha$ *in minimum time.*

We note that we can define a unique normal to the closed set $S_\alpha$ at every state except the two states $(-\alpha, 0)$ and $(\alpha, 0)$. If the terminal state of the optimal trajectory belongs to the interior of $S_\alpha$, $i(S_\alpha)$, then the transversality conditions will hold. If, on the other hand, the terminal state of the optimal trajectory is either at $(-\alpha, 0)$ or at $(\alpha, 0)$, then we cannot define a unique tangent line to $S_\alpha$ at $(-\alpha, 0)$ or $(\alpha, 0)$, and so we cannot find any transversality conditions. For this reason, we can conclude that *if the terminal state belongs to the interior of* $S_\alpha$, *then only the two control sequences* $\{+1\}$ *and* $\{-1\}$ *can be candidates for time-optimal control; if, on the other hand, the terminal state belongs to the boundary of* $S_\alpha$, *that is,* $\partial S_\alpha$, *then all four control sequences* $\{+1\}$, $\{-1\}$, $\{+1, -1\}$, *and* $\{-1, +1\}$ *can be candidates for optimal control.*

In order to make these considerations more precise, let us define the closed set $G$ shown in Fig. 7-9 and the sets $G_+$ and $G_-$ by the relations

$$\begin{aligned} G &= \{(x_1, x_2): -\tfrac{1}{2}x_2|x_2| - \alpha \le x_1 \le -\tfrac{1}{2}x_2|x_2| + \alpha\} \\ G_+ &= \{(x_1, x_2): (x_1, x_2) \in G \text{ and } x_2 < 0\} \\ G_- &= \{(x_1, x_2): (x_1, x_2) \in G \text{ and } x_2 > 0\} \end{aligned} \tag{7-53}$$

Now, since the set $S_\alpha$ of Problem 7-3 is a subset of the $x_1$ axis (the set $S$ of Problem 7-2) and since Control Law 7-2 provides us with the time-optimal control to the target set $S$, it follows that there is a set of states $\{(x_1, x_2)\}$ such that the control (7-51) is the time-optimal one to the target set $S_\alpha$. It is easy to verify (see Figs. 7-8 and 7-9) that this set $\{(x_1, x_2)\}$ is precisely the set $G$. For this reason,† we conclude that the time-optimal control

† See also the pertinent comments in Secs. 6-23 and 6-24.

for every element of $G$ is given by

$$u^* = u^*(x_1, x_2) = -\operatorname{sgn}\{x_2\} \qquad \text{for all } (x_1, x_2) \in G \tag{7-54}$$

(It is interesting to note the "topological similarity" of the set $G$ to the $\gamma$ switch curve.)

Next, we define the two sets $Q_-$ and $Q_+$ shown in Fig. 7-9 by the relations

$$\begin{aligned} Q_- &= \{(x_1, x_2)\colon x_1 > -\tfrac{1}{2}x_2|x_2| + \alpha\} \\ Q_+ &= \{(x_1, x_2)\colon x_1 < -\tfrac{1}{2}x_2|x_2| - \alpha\} \end{aligned} \tag{7-55}$$

In general, only the four control sequences $\{+1\}$, $\{-1\}$, $\{+1, -1\}$, and $\{-1, +1\}$ can be candidates for time-optimal control. However, if the

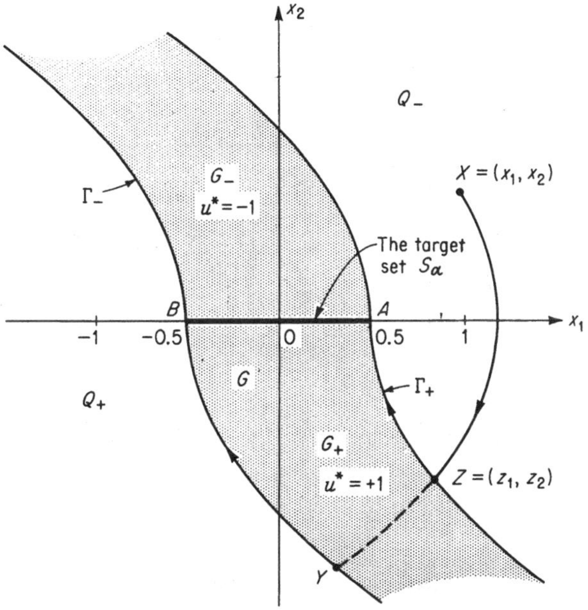


**Fig. 7-9** Illustration of the sets $Q_+$, $Q_-$, $G_+$, $G_-$.

terminal point is in the interior $i(S_\alpha)$ of $S_\alpha$, then only the two control sequences $\{+1\}$, $\{-1\}$ can be time-optimal. Therefore, we need only consider the four control sequences $\{+1\}$, $\{-1\}$, $\{+1, -1\}$, and $\{-1, +1\}$ for elements of $Q_-$ and $Q_+$ and for the two terminal states $(-\alpha, 0)$ and $(+\alpha, 0)$.

Pick any state $X = (x_1, x_2)$ in $Q_-$ and try all control sequences; with the help of Figs. 7-2 and 7-9, it is easy to verify that only the control sequence $\{-1, +1\}$ can force the state $(x_1, x_2) \in Q_-$ to either $(\alpha, 0)$ or $(-\alpha, 0)$. As a matter of fact, we must switch from $u = -1$ to $u = +1$ at the point $Z = (z_1, z_2)$ on the $\Gamma_+$ curve (shown in Fig. 7-9) to reach the state $(\alpha, 0)$ and at the point $Y = (y_1, y_2)$ to reach the state $(-\alpha, 0)$. (See the two trajectories $XZA$ and $XZYB$ in Fig. 7-9.) Now both points $(-\alpha, 0)$ and $(\alpha, 0)$ belong to the target set $S_\alpha$. It is easy to see [see Eq. (7-18)] that from any state in $Q_-$, it takes less time to reach the state $(\alpha, 0)$ than the state $(-\alpha, 0)$.

Similar reasoning can be used to establish that if $(x_1, x_2) \in Q_+$, then only the control sequence $\{+1, -1\}$ can transfer $(x_1, x_2)$ to either $(-\alpha, 0)$ or $(\alpha, 0)$, and that it takes less time to reach the state $(-\alpha, 0)$ than the state $(\alpha, 0)$.

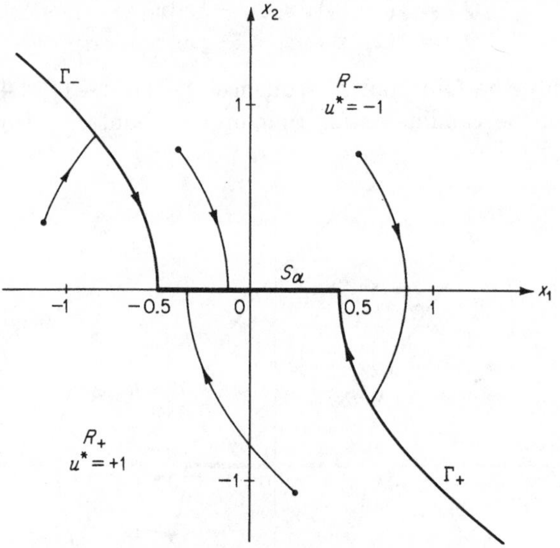


**Fig. 7-10** Time-optimal trajectories to the set $S_\alpha$.

We shall now formalize this discussion as follows: Let $\Gamma_+$ and $\Gamma_-$ be the two curves shown in Fig. 7-9, defined by the relations

$$\Gamma_+ = \{(x_1, x_2): x_1 = \tfrac{1}{2}x_2^2 + \alpha,\ x_2 < 0\} \tag{7-56}$$
$$\Gamma_- = \{(x_1, x_2): x_1 = -\tfrac{1}{2}x_2^2 - \alpha,\ x_2 > 0\} \tag{7-57}$$

Let
$$\Gamma = \Gamma_+ \cup S_\alpha \cup \Gamma_- \tag{7-58}$$

Let $R_-$ be the set of states above the $\Gamma$ curve, that is,

$$R_- = G_- \cup Q_- \tag{7-59}$$

and let $R_+$ be the set of states below the $\Gamma$ curve, that is,

$$R_+ = G_+ \cup Q_+ \tag{7-60}$$

Then the solution to Problem 7-3 is given by the following control law:

***Control Law 7-3 Solution to Problem 7-3*** *The time-optimal control, as a function of the state $(x_1, x_2)$, is given by*

$$\begin{aligned} u^* &= u^*(x_1, x_2) = +1 \qquad \text{for all } (x_1, x_2) \in R_+ \cup \Gamma_+ \\ u^* &= u^*(x_1, x_2) = -1 \qquad \text{for all } (x_1, x_2) \in R_- \cup \Gamma_- \end{aligned} \tag{7-61}$$

*and, furthermore, is unique.*

Some time-optimal trajectories to the set $S_\alpha$ are shown in Fig. 7-10.

**Exercise 7-6** Design a nonlinear system which realizes Control Law 7-3.

**Exercise 7-7** Determine the minimum time $t_\alpha^*$ to the set $S_\alpha$. Show that it is the solution to the Hamilton-Jacobi equation for this problem. Draw the minimum isochrones and show that the curves $\Gamma_+$ and $\Gamma_-$ connect points on the minimum isochrones at which they are not differentiable (i.e., at the points where the minimum isochrones have corners).

**Exercise 7-8** For the double-integral plant, consider the target set $S_\beta$ such that $(x_1, x_2) \in S_\beta$ if $x_1 = 0$, $|x_2| \leq \beta$. Derive the time-optimal control law.

**Exercise 7-9** For the double-integral plant, consider the target set $S_\delta$ such that $(x_1, x_2) \in S_\delta$ if $|x_1| \leq \delta$, $|x_2| \leq \delta$. Derive the time-optimal control law.

A brief comment on the physical interpretation of the state variables might be helpful at this point. Let us consider the double-integral plant with the transfer function $G(s) = 1/s^2$; let $y(t)$ be the plant output, and let $\nu(t)$, $|\nu(t)| \leq M$, be the input to the plant. Suppose that we are given a *reference signal* $r(t)$,

$$r(t) = r_0 + r_1 t + \tfrac{1}{2} r_2 t^2 \tag{7-62}$$

Let $e(t)$ denote the *error signal*

$$e(t) = r(t) - y(t) \tag{7-63}$$

Clearly,

$$\begin{aligned} \dot{e}(t) &= \dot{r}(t) - \dot{y}(t) \\ \ddot{e}(t) &= \ddot{r}(t) - \ddot{y}(t) \end{aligned} \tag{7-64a}$$

But

$$\begin{aligned} \ddot{y}(t) &= \nu(t) \\ \dot{r}(t) &= r_1 + r_2 t \\ \ddot{r}(t) &= r_2 \end{aligned} \tag{7-64b}$$

Therefore, the differential equation of the error signal is

$$\ddot{e}(t) = r_2 - \nu(t) \tag{7-65}$$

Define the control $u(t)$ by the relation

$$u(t) = r_2 - \nu(t) \tag{7-66}$$

to obtain

$$\ddot{e}(t) = u(t) \tag{7-67}$$

If the coefficient $r_2$ is bounded by

$$|r_2| \leq M' < M \tag{7-68}$$

then the control $u(t)$ is bounded by

$$r_2 - M \leq u(t) \leq r_2 + M \tag{7-69}$$

and hence we can define the state variables $x_1(t)$ and $x_2(t)$ by the equations

$$x_1(t) = e(t) \qquad x_2(t) = \dot{e}(t) \tag{7-70}$$

and proceed, using the same techniques as in this section to reduce, say, $e(t)$ and $\dot{e}(t)$ to zero in minimum time. What we wanted to point out is that the techniques presented can be used to design a servomechanism whose output will catch up with a step, ramp, or parabolic reference signal in minimum time. The only difference is that the switch curves will no longer be symmetrical (if $r_2 \neq 0$).

**Exercise 7-10** In Eq. (7-68), let $r_2 = 1$ and $M = 3$, so that Eq. (7-69) reduces to $-2 \leq u(t) \leq +4$. Consider the system (7-67) and find the time-optimal control $u^*(e, \dot{e})$ which reduces any initial error $e(0)$ and error rate $\dot{e}(0)$ to 0 in minimum time.

We shall conclude this section with a series of exercises. Since the equations of the double-integral plant are especially simple, one can use this system to demonstrate many interesting aspects of time-optimal systems without undue computational difficulty (although some of the exercises are definitely nontrivial).

The following six exercises illustrate several aspects of time-optimal behavior when we deal with the single-terminal-state case.

**Exercise 7-11** (This exercise illustrates the relation between the initial state and initial costate.) Consider Problem 7-1. Let $(\pi_1, \pi_2)$ be the initial costate corresponding to the initial state $(\xi_1, \xi_2)$. Use Eqs. (7-8), (7-12), (7-24), and (7-26), together with the necessary condition that the Hamiltonian be zero along the optimal trajectory, to show that

$$\left.\begin{aligned} \pi_1 &= \frac{2}{\sqrt{4\xi_1 + 2\xi_2^2}} \\ \pi_2 &= \frac{2\xi_2}{\sqrt{4\xi_1 + 2\xi_2^2}} + 1 \end{aligned}\right\} \qquad \text{for all } (\xi_1, \xi_2) \in R_-$$

and that
$$\pi_2 = -1 - \xi_2\pi_1 \qquad \text{for all } (\xi_1, \xi_2) \in \gamma_+$$

Determine similar equations for the case $(\xi_1, \xi_2) \in R_+$ and $(\xi_1, \xi_2) \in \gamma_-$. Now think of these equations as defining a nonlinear mapping $\mathfrak{N}$ from the $\xi_1\xi_2$ plane into the $\pi_1\pi_2$ plane [that is, $\pi = \mathfrak{N}(\xi)$]. Construct the image $\mathfrak{N}(R_-)$ of $R_-$ and the image $\mathfrak{N}(R_+)$ of $R_+$ in the $\pi_1\pi_2$ plane. To study a type of "sensitivity" problem, consider the two sets (circles) in the $\xi_1\xi_2$ plane:

$$A_1 = \{(\xi_1, \xi_2): (\xi_1 - 2)^2 + (\xi_2 + 2)^2 \leq 0.04\}$$
$$A_2 = \{(\xi_1, \xi_2): (\xi_1 - 2)^2 + \xi_2^2 \leq 0.04\}$$

and find their images $\mathfrak{N}(A_1)$ and $\mathfrak{N}(A_2)$ in the $\pi_1\pi_2$ plane. Can you draw any conclusions regarding the changes in $(\pi_1, \pi_2)$ arising from small changes in $(\xi_1, \xi_2)$? Next, demonstrate that the vector $\begin{bmatrix} \pi_1 \\ \pi_2 \end{bmatrix}$ is the normal to the minimum isochrone at the point $\begin{bmatrix} \xi_1 \\ \xi_2 \end{bmatrix}$ whenever this normal is defined. Hint: See Eqs. (7-33), (7-36), and (7-37).

**Exercise 7-12** (This exercise illustrates the motion of the trajectories on the minimum-time surface.) First, go over the material in Sec. 6-7. Now consider Problem 7-1. Plot in the state plane the two time-optimal trajectories originating at (2, 0) and (2.2, 0). On the same graph, plot several minimum isochrones using Eq. (7-27). Examine your plot and draw some conclusions about the angles between the tangent vector to the time-optimal trajectory and the tangent line to the minimum isochrone.

**Exercise 7-13** (This exercise is nontrivial, as it requires the construction of an iterative method of solution.) Consider the "experiment" outlined in step 3*d* of Sec. 6-3, and, once more, consider Problem 7-1. Outline an iterative procedure which you may use to find the time-optimal control from any initial state $(\xi_1, \xi_2)$. In other words, suppose that you did not know Control Law 7-1. Start by making an arbitrary guess on $(\pi_1, \pi_2)$, say $(\pi_{10}, \pi_{20})$, and the time $t^*$, say $t_0^*$; then find the control from Eq. (7-9) for $0 \leq t \leq t_0^*$, integrate Eqs. (7-5), and evaluate the state at $t = t_0^*$. On the basis of the "miss distance" from the origin, modify your guesses until, eventually, you determine "almost exactly" the time-optimal control. Verify your "design" for the initial state (2, 0) and state the number of iterations necessary to determine the control which is almost exactly the time-optimal one. Do you think that the results of Exercises 7-11 and 7-12 are helpful in this problem?

**Exercise 7-14** Consider the system (7-7). Prove that there exists a unique time-optimal control which forces *any* initial state $(\xi_1, \xi_2)$ to *any* terminal state $(\theta_1, \theta_2)$. (You may assume the existence of a time-optimal control.)

**Exercise 7-15** (This exercise demonstrates the nonuniqueness of extremal controls when the terminal state is not the origin.) Consider the system (7-7). Let $(\theta_1, \theta_2)$, $\theta_2 \geq 0$, be any given state in the upper half of the state plane. Let $Q_1$ and $Q_2$ be the following two sets in the $x_1x_2$ plane:

$$Q_1 = \{(x_1, x_2): -\tfrac{1}{2}x_2^2 + \theta_1 - \tfrac{1}{2}\theta_2^2 \leq x_1 \leq -\tfrac{1}{2}x_2^2 + \theta_1 + \tfrac{1}{2}\theta_2^2;\ x_2 \geq \theta_2\}$$
$$Q_2 = \{(x_1, x_2): -\tfrac{1}{2}x_2^2 + \theta_1 - \tfrac{1}{2}\theta_2^2 \leq x_1 \leq \tfrac{1}{2}x_2^2 + \theta_1 - \tfrac{1}{2}\theta_2^2;\ 0 \leq x_2 \leq \theta_2\}$$

Show that there are precisely two extremal controls that transfer an initial state $(\xi_1, \xi_2)$ to $(\theta_1, \theta_2)$ provided that $(\xi_1, \xi_2) \in Q_1 \cup Q_2$; otherwise, there is one and only one extremal control. (HINT: Mimic the proof of Theorem 6-9.) Find a similar result for terminal states $(\theta_1, \theta_2)$ such that $\theta_2 \leq 0$. Verify your analytical result by trying the four control sequences of Eq. (7-13), first for the state $(\theta_1, \theta_2) = (1, 1)$ and second for the state $(\theta_1, \theta_2) = (1, -1)$. Determine the unique time-optimal control as a function of the state. Design an engineering realization of your control law.

**Exercise 7-16** (This exercise demonstrates that the minimum-time surface may be only piecewise continuous.) Consider the system (7-7) and the terminal state (1, 1). Let $(\xi_1, \xi_2)$ be any given initial state, and let $t_2^*(\xi_1, \xi_2)$ denote the minimum time required to force $(\xi_1, \xi_2)$ to (1, 1). Find the equation of the minimum time $t_2^*(\xi_1, \xi_2)$. Next, plot the sets of states (minimum isochrones) which require the same minimum time $t_2^*$ to (1, 1) for the following values of $t_2^*$: $t_2^* = 0.5, 1, 2.0, 3.0, 3.9, 4.0, 4.1, 5.0$, and $6.0$. Do you see that the surface $t^*(\xi_1, \xi_2)$ is not continuous? Can you explain why?

The following five exercises deal with the problem of time-optimal control to a target set.

**Exercise 7-17** (This exercise deals with a target set which consists of two isolated points and shows that the time-optimal control need not be unique.) Consider the system (7-7) and the closed, nonconvex set $\hat{S}$ defined by the relation

$$\hat{S} = \{(x_1, x_2): x_1 = 0.2, x_2 = 0 \text{ or } x_1 = -0.2, x_2 = 0\}$$

In other words, the target set $\hat{S}$ is made up of the two isolated states $(-0.2, 0)$ and $(0.2, 0)$. Show that a time-optimal control exists from any given initial state $(\xi_1, \xi_2)$. Also show that:

(*a*) The time-optimal control is unique provided that

$$\xi_1^2 \neq \tfrac{1}{4}\xi_2^4 + 0.2\xi_2^2$$

(*b*) There are precisely two extremal controls for all initial states.

Determine the time-optimal control(s). Plot the minimum isochrones and show that they are *not* boundaries of convex sets. Why are there no transversality conditions?

**Exercise 7-18** Consider the system (7-7) and the target set

$$S_1 = \{(x_1, x_2): x_1 = x_2, |x_1| \leq 1\}$$

Determine the time-optimal control. Repeat for the target set

$$S_2 = \{(x_1, x_2): x_1 = -x_2, |x_1| \leq 1\}$$

Comment on the uniqueness of the optimal and extremal controls.

**Exercise 7-19** (This exercise illustrates that the switch curve is not necessarily a portion of a time-optimal trajectory.) Consider the system (7-7). Suppose that the target set $S$ is the boundary of the unit circle about the origin; that is,

$$S = \{(x_1, x_2): x_1^2 + x_2^2 = 1\}$$

Determine the time-optimal control as a function of the state for initial states outside and inside the circle. (HINT: You should be able to find a parametric rather than an analytical expression for the switch curve.) Plot several minimum isochrones and show that (in contradistinction to Fig. 7-7) the minimum isochrones do not have "corners" at the switch curve. Do you think that this fact implies a one-to-one mapping between the initial costate $(\pi_1, \pi_2)$ corresponding to the initial state $(\xi_1, \xi_2)$?

**Exercise 7-20** (This exercise illustrates the problems that may arise when the target set is not the boundary of a convex set.) Consider the system (7-7). Let $S_1$ and $S_2$ be the two circles

$$S_1 = \{(x_1, x_2): (x_1 - 1)^2 + x_2^2 \leq 1\}$$
$$S_2 = \{(x_1, x_2): (x_1 + 1)^2 + x_2^2 \leq 1\}$$

and let $S = S_1 \cup S_2$ be the target set. Determine the time-optimal control as a function of the state. Comment on the uniqueness of the optimal and extremal controls.

The following exercise shows that one can analyze an optimal design when one of the components is not ideal.

**Exercise 7-21** Consider the control system shown in Fig. 7-5. Suppose that the ideal relay $R$ is replaced by a relay $R_d$ which has a dead zone. In other words, suppose that the input $[-z(t)]$ – output $[u^*(t)]$ characteristics of $R_d$ are

$$u^*(t) = 0 \qquad \text{if } |-z(t)| \leq 0.2$$
$$u^*(t) = \operatorname{sgn}\{-z(t)\} \qquad \text{if } |-z(t)| > 0.2$$

Analyze the effects of this nonideal relay. (HINT: Construct a region about the switch curve.) Is there a steady-state error? Is there a limit cycle? Can you find a way of altering the nonlinearity $N$ so that there will be no steady-state error?

## 7-3 Time-optimal Control of Plants with Two Time Constants

In this section, we shall consider the time-optimal control of a plant whose transfer function has two real poles and no zeros. Basically, the development of ideas is identical to that presented in the preceding section. However, this system has characteristics that are fundamentally different (owing to the lack of any integrating capability) from those of the double-integral plant. Nonetheless, as the reader will soon discover, the structure of the time-optimal control system is basically the same as that of the double-integral plant. This indeed is a common characteristic of all normal second-order systems, as we shall show later in this chapter.

We consider the system described by the second-order differential equation

$$\frac{d^2y(t)}{dt^2} - (\lambda_1 + \lambda_2)\frac{dy(t)}{dt} + \lambda_1\lambda_2 y(t) = Ku(t) \qquad (7\text{-}71)$$

where $y(t)$ is the output, $K$ is a gain constant, $u(t)$ is the control, which we suppose is restricted in magnitude by the relation

$$|u(t)| \leq 1 \qquad (7\text{-}72)$$

and $\lambda_1$, $\lambda_2$ are real, distinct, and nonzero (that is, $\lambda_1 \neq \lambda_2 \neq 0$). In servomechanism practice, the plant described by Eq. (7-71) is said to have the transfer function

$$\frac{y(s)}{u(s)} = G(s) = \frac{1}{(s - \lambda_1)(s - \lambda_2)} \qquad (7\text{-}73)$$

with real poles at $s = \lambda_1$ and $s = \lambda_2$.

Now we shall use two linear transformations to derive a very convenient set of state variables. Let

$$y_1(t) = y(t) \qquad y_2(t) = \dot{y}(t) \qquad (7\text{-}74)$$

The $y_1$ and $y_2$ variables satisfy the system of equations

$$\begin{aligned} \dot{y}_1(t) &= y_2(t) \\ \dot{y}_2(t) &= -\lambda_1\lambda_2 y_1(t) + (\lambda_1 + \lambda_2)y_2(t) + Ku(t) \end{aligned} \qquad (7\text{-}75)$$

Equation (7-75) may be written in matrix form as

$$\begin{bmatrix} \dot{y}_1(t) \\ \dot{y}_2(t) \end{bmatrix} = \begin{bmatrix} 0 & 1 \\ -\lambda_1\lambda_2 & \lambda_1 + \lambda_2 \end{bmatrix} \begin{bmatrix} y_1(t) \\ y_2(t) \end{bmatrix} + \begin{bmatrix} 0 \\ Ku(t) \end{bmatrix} \qquad (7\text{-}76)$$

Note that Eq. (7-76) is of the form

$$\dot{\mathbf{y}}(t) = \mathbf{A}\mathbf{y}(t) + \mathbf{u}(t) \qquad (7\text{-}77)$$

Clearly, the eigenvalues of the matrix $\mathbf{A}$ are $\lambda_1$ and $\lambda_2$. If $\mathbf{\Lambda}$ denotes the diagonal matrix

$$\mathbf{\Lambda} = \begin{bmatrix} \lambda_1 & 0 \\ 0 & \lambda_2 \end{bmatrix} \qquad (7\text{-}78)$$

then† there exists a nonsingular matrix $\mathbf{P}$,

$$\mathbf{P} = \begin{bmatrix} 1 & 1 \\ \lambda_1 & \lambda_2 \end{bmatrix} \qquad (7\text{-}79)$$

with inverse

$$\mathbf{P}^{-1} = \frac{1}{\lambda_2 - \lambda_1}\begin{bmatrix} \lambda_2 & -1 \\ -\lambda_1 & 1 \end{bmatrix} \qquad (7\text{-}80)$$

such that

$$\mathbf{\Lambda} = \mathbf{P}^{-1}\mathbf{A}\mathbf{P} \qquad (7\text{-}81)$$

† See Sec. 2-9.

Now we shall use the similarity transformation;† let us define $\mathbf{z}(t)$ by

$$\mathbf{z}(t) = \mathbf{P}^{-1}\mathbf{y}(t) \tag{7-82}$$

Then the components $z_1(t)$ and $z_2(t)$ of the vector $\mathbf{z}(t)$ satisfy the differential equations

$$\begin{aligned} \dot{z}_1(t) &= \lambda_1 z_1(t) - \frac{K}{\lambda_2 - \lambda_1} u(t) \\ \dot{z}_2(t) &= \lambda_2 z_2(t) + \frac{K}{\lambda_2 - \lambda_1} u(t) \end{aligned} \tag{7-83}$$

It is convenient to define the state variables $x_1(t)$ and $x_2(t)$ by the relations

$$\begin{aligned} x_1(t) &= -\frac{\lambda_1(\lambda_2 - \lambda_1)}{K} z_1(t) \\ x_2(t) &= \frac{\lambda_2(\lambda_2 - \lambda_1)}{K} z_2(t) \end{aligned} \tag{7-84}$$

Then $x_1(t)$ and $x_2(t)$ satisfy the differential equations

$$\begin{aligned} \dot{x}_1(t) &= \lambda_1 x_1(t) + \lambda_1 u(t) \\ \dot{x}_2(t) &= \lambda_2 x_2(t) + \lambda_2 u(t) \end{aligned} \tag{7-85}$$

*We shall use the $x_1(t)$ and $x_2(t)$ state variables exclusively in the rest of this section.* We note that $x_1(t)$ and $x_2(t)$ are related to $y_1(t)$ [or $y(t)$] and $y_2(t)$ [or $\dot{y}(t)$] by known linear transformations. In particular, if $x_1(t) = x_2(t) = 0$, then $y_1(t) = y_2(t) = 0$, and vice versa.

The precise time-optimal problem is as follows:

***Problem 7-4*** *Given the system* (7-85). *Find the admissible control that forces the system* (7-85) *from any initial state* ($\xi_1$, $\xi_2$) *to the origin* (0, 0) *in minimum time.*

We remark that the system (7-85) is normal. We also note (see Theorem 6-10) that a time-optimal control exists and is unique (see Theorem 6-7). The step-by-step procedure we shall follow is identical to the one used in Sec. 7-2. Let us write the Hamiltonian $H$ for this problem. We shall use $p_1(t)$ and $p_2(t)$ to represent the costate variables. We have

$$H = 1 + \lambda_1 x_1(t) p_1(t) + \lambda_2 x_2(t) p_2(t) + u(t)\{\lambda_1 p_1(t) + \lambda_2 p_2(t)\} \tag{7-86}$$

The control $u(t)$ which absolutely minimizes the Hamiltonian $H$ is given by

$$u(t) = -\operatorname{sgn}\{\lambda_1 p_1(t) + \lambda_2 p_2(t)\} \tag{7-87}$$

The costate variables $p_1(t)$ and $p_2(t)$ satisfy the differential equations

$$\begin{aligned} \dot{p}_1(t) &= -\frac{\partial H}{\partial x_1(t)} = -\lambda_1 p_1(t) \\ \dot{p}_2(t) &= -\frac{\partial H}{\partial x_2(t)} = -\lambda_2 p_2(t) \end{aligned} \tag{7-88}$$

† See Sec. 4-9.

Let $\pi_1$ and $\pi_2$ be the initial costate variables; that is,

$$\pi_1 = p_1(0) \qquad \pi_2 = p_2(0) \tag{7-89}$$

Then the solution of Eqs. (7-88) is

$$\begin{aligned} p_1(t) &= \pi_1 e^{-\lambda_1 t} \\ p_2(t) &= \pi_2 e^{-\lambda_2 t} \end{aligned} \tag{7-90}$$

Substituting Eqs. (7-90) into Eq. (7-87), we find that

$$u(t) = -\operatorname{sgn}\{\lambda_1\pi_1 e^{-\lambda_1 t} + \lambda_2\pi_2 e^{-\lambda_2 t}\} \triangleq -\operatorname{sgn}\{q(t)\} \tag{7-91}$$

The function $q(t) = \lambda_1\pi_1 e^{-\lambda_1 t} + \lambda_2\pi_2 e^{-\lambda_2 t}$ has *at most* one zero. Therefore, we conclude that the four control sequences

$$\{+1\}, \{-1\}, \{+1, -1\}, \{-1, +1\} \tag{7-92}$$

are the only candidates for the time-optimal control of this system.† Since the control must be piecewise constant, we solve Eqs. (7-85) using

$$u(t) = \Delta = \pm 1 \tag{7-93}$$

to obtain the solution

$$\begin{aligned} x_1(t) &= (\xi_1 + \Delta)e^{\lambda_1 t} - \Delta \\ x_2(t) &= (\xi_2 + \Delta)e^{\lambda_2 t} - \Delta \end{aligned} \tag{7-94}$$

Eliminating the time $t$ in Eqs. (7-94) and setting

$$\alpha \triangleq \frac{\lambda_2}{\lambda_1} \tag{7-95}$$

we find that

$$x_2(t) = -\Delta + (\xi_2 + \Delta)\left[\frac{x_1(t) + \Delta}{\xi_1 + \Delta}\right]^\alpha \tag{7-96}$$

*In the remainder of this section, we assume that* $0 > \lambda_1 > \lambda_2$. Equation (7-96) describes a trajectory in the $x_1x_2$ plane. The trajectory originates at the state $(\xi_1, \xi_2)$ and evolves as a result of the action of the constant control $u(t) = \Delta$. If the eigenvalues $\lambda_1$ and $\lambda_2$ are negative, then the trajectories generated by $u = -1$, which we shall call the $-1$ *forced trajectories*, tend to the state (1, 1) of the state plane; the trajectories generated by $u = +1$, which we shall call the $+1$ *forced trajectories*, tend to the state $(-1, -1)$ of the state plane. The $-1$ forced trajectories are shown in Fig. 7-11, and the $+1$ forced trajectories are shown in Fig. 7-12.

† See also Theorem 6-8.

We note in passing that Eq. (7-96) is independent of the gain constant $K$; thus, the "shape" of the forced trajectories in the state plane is a function only of the ratio $\lambda_2/\lambda_1$.†

We now proceed to the solution of the time-optimal problem. The reader will find that the approach is very similar to that used in Sec. 7-2.

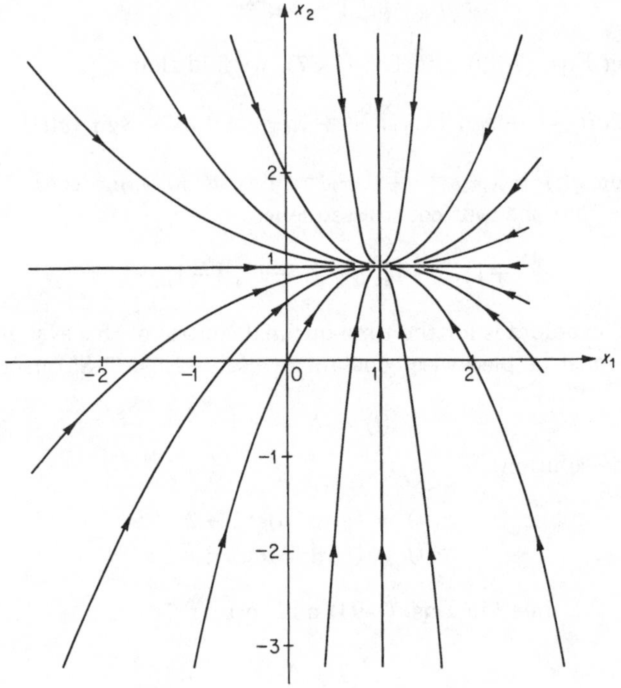


**Fig. 7-11** The −1 forced trajectories; note that as time increases, all trajectories tend to the state (1, 1).

Since the origin of the state plane is the desired terminal state and since we must reach the origin using either the control $u = +1$ or the control $u = -1$, we isolate the two forced trajectories which pass through the origin. We shall denote these trajectories to the origin by $\gamma_+$ and $\gamma_-$. More precisely:

***Definition 7-7*** *The $\gamma_+$ curve is the locus of all states which can be forced to the origin* (0, 0) *by the control sequence* $\{+1\}$ *in positive time. The $\gamma_+$ curve is given by*

$$\gamma_+ = \left\{(x_1, x_2)\colon -1 + (x_2 + 1)\left(\frac{1}{1 + x_1}\right)^{\alpha} = 0;\ x_1 > 0,\ x_2 > 0\right\} \tag{7-97}$$

† This is one of the reasons for working with the $x_1$ and $x_2$ state variables, which, in a sense, are normalized.

*The $\gamma_-$ curve is the locus of all states which can be forced to the origin* (0, 0) *by the control sequence* $\{-1\}$ *in positive time. The $\gamma_-$ curve is given by*

$$\gamma_- = \left\{(x_1, x_2) : 1 + (x_2 - 1)\left(\frac{1}{1 - x_1}\right)^\alpha = 0;\ x_1 < 0,\ x_2 < 0\right\} \tag{7-98}$$

*The $\gamma_+$ and $\gamma_-$ curves are shown in Fig.* 7-13.

Now we shall show that *for any initial state on the $\gamma_+$ curve, the control sequence* $\{+1\}$ *is time-optimal to the origin.* Consider a state on the $\gamma_+$ curve

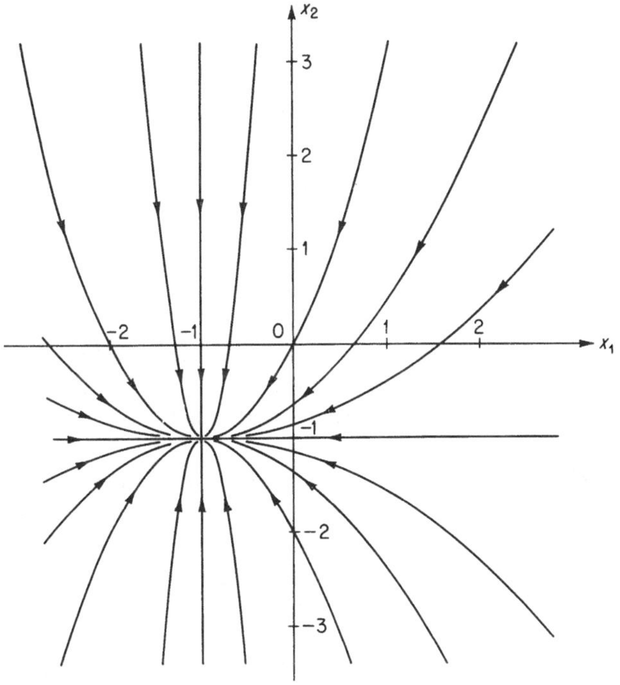


**Fig. 7-12** The +1 forced trajectories; note that as time increases, all trajectories tend to the state $(-1, -1)$.

and consider the four control sequences of Eq. (7-92). Using the shape of the forced trajectories shown in Figs. 7-11 and 7-12, we can immediately conclude that only the control sequence $\{+1\}$ can force this state to the origin. Hence, since the control sequence $\{+1\}$ is the only one that accomplishes the objective, it must be the time-optimal one. Similarly, we can prove that *for any initial state on the $\gamma_-$ curve, the control sequence* $\{-1\}$ *is time-optimal to the origin.* Thus we have derived the control law

$$\begin{aligned} &\text{If } (x_1, x_2) \in \gamma_+, \text{ then } u^* = +1 \\ &\text{If } (x_1, x_2) \in \gamma_-, \text{ then } u^* = -1 \end{aligned} \tag{7-99}$$

***Definition 7-8*** *Let the union of the $\gamma_+$ and the $\gamma_-$ curves be called the $\gamma$ switch curve. The $\gamma$ curve is given by*

$$\gamma = \left\{(x_1, x_2): x_2 = \frac{x_1}{|x_1|}[(1 + |x_1|)^\alpha - 1]\right\} \tag{7-100}$$

[*which is obtained from Eqs.* (7-97) *and* (7-98)].

Let $R_+$ denote the set of states to the right of the $\gamma$ curve, and let $R_-$ denote the set of states to the left of the $\gamma$ curve. Clearly,

$$R_+ = \left\{(x_1, x_2): x_2 < \frac{x_1}{|x_1|}[(1 + |x_1|)^\alpha - 1]\right\} \tag{7-101}$$

$$R_- = \left\{(x_1, x_2): x_2 > \frac{x_1}{|x_1|}[(1 + |x_1|)^\alpha - 1]\right\} \tag{7-102}$$

Now we shall show that *for every state in $R_+$, the time-optimal control sequence is* $\{+1, -1\}$, *provided that the transition (or switch) from $u = +1$ to $u = -1$ occurs at the $\gamma_-$ curve.* To show this, we again consider the four control sequences $\{+1\}$, $\{-1\}$, $\{+1, -1\}$, and $\{-1, +1\}$; the sequences $\{+1\}$ and $\{-1\}$ will generate trajectories which will tend to the states (1, 1) and (−1, −1), respectively, without passing through (0, 0); the control sequence $\{-1, +1\}$ will generate trajectories which do not hit the origin; among all the trajectories generated by the sequence $\{+1, -1\}$, there is one and only one which arrives at the origin, and this trajectory can be generated if and only if we apply the control $u = +1$ until the +1 forced trajectory intersects the $\gamma_-$ curve, at which point we must apply $u = -1$. Similarly, *for every state in $R_-$, the time-optimal control sequence is* $\{-1, +1\}$ *provided that the transition (or switch) from $u = -1$ to $u = +1$ occurs at the $\gamma_+$ curve.* We now summarize the results in the following time-optimal control law:

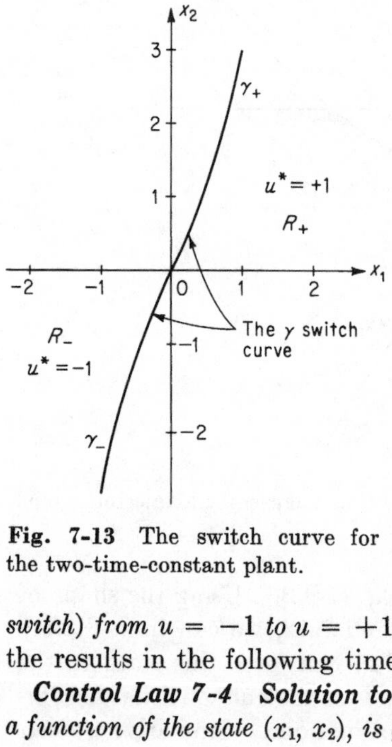


**Fig. 7-13** The switch curve for the two-time-constant plant.

***Control Law 7-4 Solution to Problem 7-4*** *The time-optimal control, as a function of the state* $(x_1, x_2)$, *is given by*

$$\begin{aligned} u^* &= u^*(x_1, x_2) = +1 \qquad \text{for all } (x_1, x_2) \in \gamma_+ \cup R_+ \\ u^* &= u^*(x_1, x_2) = -1 \qquad \text{for all } (x_1, x_2) \in \gamma_- \cup R_- \end{aligned} \tag{7-103}$$

[*where $\gamma_+$, $R_+$, $\gamma_-$, and $R_-$ are given by Eqs.* (7-97), (7-101), (7-98), *and* (7-102), *respectively*]. *Furthermore, the time-optimal control is unique.*

An engineering realization of the above time-optimal control law is shown in Fig. 7-14. The state variables $x_1(t)$ and $x_2(t)$ are measured (or computed). The $x_1(t)$ signal is introduced to the nonlinearity $N$, whose output is

$$a(t) = \frac{x_1}{|x_1|}[(1 + |x_1|)^\alpha - 1] \tag{7-104}$$

The signal $x_2(t)$ is subtracted from $a(t)$ to form the signal $b(t) = a(t) - x_2(t)$. If $b(t)$ is positive, then $(x_1, x_2) \in R_+$; if $b(t)$ is negative, then $(x_1, x_2) \in R_-$; therefore, the signal $b(t)$ has the correct polarity, and hence it may be used to drive the ideal relay $R$ whose output $u^*(t)$ is the time-optimal control.

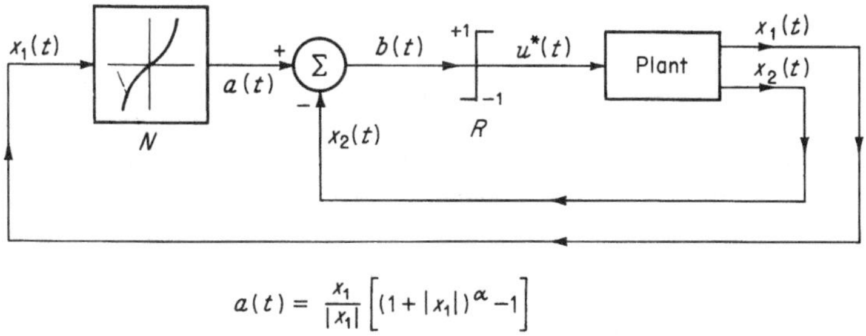


**Fig. 7-14** An engineering realization of the time-optimal system for the two-time-constant plant.

We shall now demonstrate another advantage of using $x_1(t)$ and $x_2(t)$ as state variables. Recall that we originally defined the state variables $y_1(t)$ and $y_2(t)$ by Eqs. (7-74), that is,

$$y_1(t) = y(t) \qquad y_2(t) = \dot{y}(t) \tag{7-105}$$

In other words, we could use the output $y(t)$ and its derivative $\dot{y}(t)$ as the state variables. From Eqs. (7-80), (7-82), and (7-84), we find that $x_1(t)$ and $x_2(t)$ are related to $y_1(t)$ and $y_2(t)$ by

$$\begin{aligned} x_1(t) &= \frac{1}{K}[-\lambda_1\lambda_2 y_1(t) + \lambda_1 y_2(t)] \\ x_2(t) &= \frac{1}{K}[-\lambda_1\lambda_2 y_1(t) + \lambda_2 y_2(t)] \end{aligned} \tag{7-106}$$

Figure 7-15 illustrates, in block-diagram form, the linear transformation necessary to obtain $x_1(t)$ and $x_2(t)$ from $y_1(t)$ and $y_2(t)$. Now let us substitute Eqs. (7-106) into Eq. (7-100) in order to obtain the equation of the $\hat{\gamma}$ switch curve in the $y_1y_2$ plane; the $\hat{\gamma}$ curve is then given by

$$\begin{aligned} &\frac{1}{K}(-\lambda_1\lambda_2 y_1 + \lambda_2 y_2) \\ &\quad = \frac{\frac{1}{K}(-\lambda_1\lambda_2 y_1 + \lambda_1 y_2)}{\left|\frac{1}{K}(-\lambda_1\lambda_2 y_1 + \lambda_1 y_2)\right|}\left\{\left[1 + \left|\frac{1}{K}(-\lambda_1\lambda_2 y_1 + \lambda_1 y_2)\right|\right]^\alpha - 1\right\} \end{aligned} \tag{7-107}$$

It is impossible to manipulate Eq. (7-107) (for arbitrary values of $\alpha$) to obtain $y_2$ *as an explicit function of* $y_1$, or vice versa. This implies that if we measure the state $(y_1, y_2)$ and we wish to test whether or not the state $(y_1, y_2)$ is on the switch curve, we cannot construct a simple nonlinearity into which

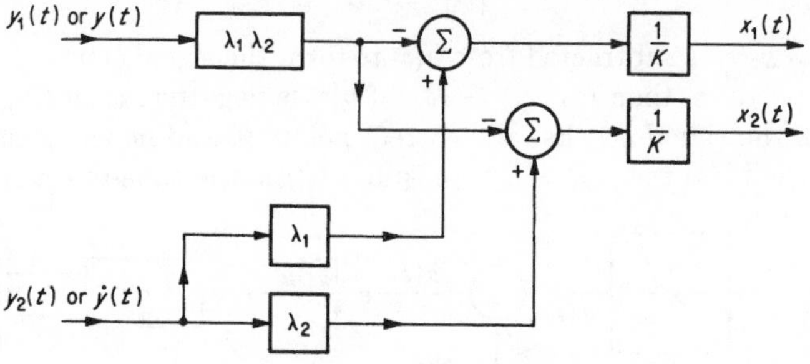


**Fig. 7-15** Block diagram of the linear transformation from the $y_1$ and $y_2$ variables to the $x_1$ and $x_2$ variables.

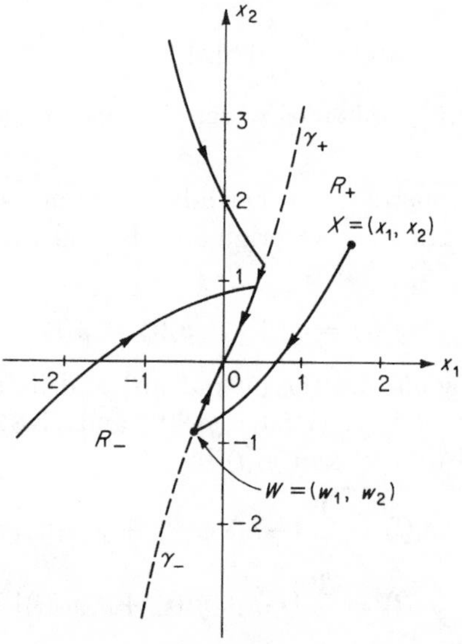


**Fig. 7-16** Several time-optimal trajectories for the two-time-constant plant.

we could introduce (say) the signal $y_1$ and compare the output with $y_2$ to decide what the time-optimal control should be.

Now we shall attempt to evaluate the minimum time $t^*$ required to force any initial state $(x_1, x_2)$ to the origin (0, 0) using the time-optimal control law provided by Eqs. (7-103). Consider an initial state $X = (x_1, x_2)$, as shown in Fig. 7-16, and the time-optimal trajectory $XWO$ to the origin, where $W = (w_1, w_2)$ is on the $\gamma$ curve. Let $u = \Delta^* = \pm 1$ be the control applied during the trajectory $WO$. Let $t_2$ be the time required to go from $W$ to $O$, and let $t_1$ be the time required to go from $X$ to $W$. Then, from Eqs. (7-94) and (7-96), we deduce that

$$t_2 = \frac{1}{\lambda_1} \log \frac{\Delta^*}{w_1 + \Delta^*} = \frac{1}{\lambda_2} \log \frac{\Delta^*}{w_2 + \Delta^*} \tag{7-108}$$

$$0 = -\Delta^* + (w_2 + \Delta^*) \left( \frac{\Delta^*}{w_1 + \Delta^*} \right)^\alpha \tag{7-109}$$

$$\Delta^* = - \operatorname{sgn} \{w_1\} = - \operatorname{sgn} \{w_2\} \tag{7-110}$$

$$t_1 = \frac{1}{\lambda_1} \log \frac{w_1 - \Delta^*}{x_1 - \Delta^*} = \frac{1}{\lambda_2} \log \frac{w_2 - \Delta^*}{x_2 - \Delta^*} \tag{7-111}$$

$$w_2 = \Delta^* + (x_2 - \Delta^*) \left( \frac{w_1 - \Delta^*}{x_1 - \Delta^*} \right)^\alpha \tag{7-112}$$

because, during the trajectory $XW$, we have $u = -\Delta^*$. Let

$$t^* = t_1 + t_2 \tag{7-113}$$

Clearly, $t^*$ is given by

$$t^* = \frac{1}{\lambda_1} \log \left( \frac{\Delta^*}{w_1 + \Delta^*} \, \frac{w_1 - \Delta^*}{x_1 - \Delta^*} \right) \tag{7-114}$$

We want to find $t^*$ as a function of $x_1$ and $x_2$ only, and so we must eliminate $w_1$ from Eq. (7-114). Combining Eqs. (7-109) and (7-112), we find that

$$0 = -\Delta^* + \left[ 2\Delta^* + (x_2 - \Delta^*) \left( \frac{w_1 - \Delta^*}{x_1 - \Delta^*} \right)^\alpha \right] \left( \frac{\Delta^*}{w_1 + \Delta^*} \right)^\alpha \tag{7-115}$$

which provides us with a relationship among $w_1$, $x_1$, and $x_2$. Unfortunately, for arbitrary values of $\alpha$, it is impossible to obtain $w_1$ as an explicit function of $x_1$ and $x_2$ from Eq. (7-115). Thus, we cannot obtain a closed-form expression for $t^*$. However, for specific values of $\lambda_1$ and $\lambda_2$, and hence of $\alpha$, we could obtain the value of $t^*$ from Eqs. (7-114) and (7-115) using a digital computer.

**Exercise 7-22** Let $\lambda_1 = -1$, $\lambda_2 = -2$. Then $\alpha = 2$ and Eq. (7-115) reduces to a quadratic; solve for $w_1$ in terms of $x_1$ and $x_2$ and obtain an analytic expression for $t^*$. Then obtain the equations of the minimum isochrones (see Definition 7-6) and plot the minimum isochrones for the values $t^* = 0.5$, $1.0$, $1.5$, and $2.0$. Show that the expression you have derived for $t^*$ is the solution to the Hamilton-Jacobi equation.

**Exercise 7-23** Consider the system given by Eqs. (7-85) with the control constraint $|u(t)| \leq 1$. Given the target set $S$ described by

$$S = \{(x_1, x_2): x_1 = 0, -\infty < x_2 < \infty\}$$

Show that the control

$$u = \operatorname{sgn} \{x_1\}$$

is time-optimal to $S$.

**Exercise 7-24** Consider Problem 7-4. Determine the relation of the initial costate $(\pi_1, \pi_2)$ corresponding to the initial state $(\xi_1, \xi_2)$. Hint: See Exercise 7-11.

**Exercise 7-25** Consider Problem 7-4. Let $\lambda_1 = -1$ and $\lambda_2 = +1$. Find the set of initial states $\Psi_r$ such that if $(\xi_1, \xi_2) \in \Psi_r$, then a time-optimal control to the origin exists. Determine the time-optimal control. HINT: See Sec. 6-8.

**Exercise 7-26** Repeat Exercise 7-25 for $\lambda_1 = 1$, $\lambda_2 = 2$.

**Exercise 7-27** Consider the system (7-85) with $|u(t)| \leq 1$. Let $\lambda_1 = -1$, $\lambda_2 = -2$. Suppose that $(\theta_1, \theta_2)$ is a given terminal state (not necessarily the origin). Find the set $\Omega$ such that if $(\theta_1, \theta_2) \in \Omega$, then a time-optimal control to $(\theta_1, \theta_2)$ exists from any given initial state $(\xi_1, \xi_2)$. Repeat for:

(*a*) $\lambda_1 = -1$, $\lambda_2 = +1$

(*b*) $\lambda_1 = 1$, $\lambda_2 = 2$

## 7-4 Time-optimal Control of a Third-order Plant with Two Integrators and a Single Time Constant

Previously, we have discussed the time-optimal control problem for second-order systems, and we have shown that the optimal control can be determined as a function of the state by means of a "switch" curve which divides the state plane into two regions. Here, we shall consider the time-optimal control problem for a third-order plant with two integrators and a single time constant. We shall extend the techniques of the previous sections in order to derive the optimal-control law and to obtain an engineering realization of the time-optimal controller (which will be very complicated).

The third-order system we shall examine is described by the differential equation

$$\frac{d^3y(t)}{dt^3} + a\frac{d^2y(t)}{dt^2} = Ku(t) \tag{7-116}$$

where $y(t)$ represents the output, $K$ is a gain constant, and $u(t)$ is the control. We assume that $u(t)$ is constrained in magnitude by the relation

$$|u(t)| \leq 1 \tag{7-117}$$

and that the constant $a$ in Eq. (7-116) is real and positive, that is,

$$a > 0 \tag{7-118}$$

In servomechanism practice, the system described by Eq. (7-116) is said to have the transfer function

$$\frac{y(s)}{u(s)} = G(s) = \frac{K}{s^2(s+a)} \tag{7-119}$$

There are many physical systems which can be described to a good degree of approximation by the transfer function $G(s) = K/s^2(s+a)$. For example, a field-controlled motor system with constant armature current has such a transfer function (see Example 6-9). In this system, the output is angular displacement, the control is the field voltage, the time constant $a$ is a result of the field inductance, and the constraint $|u(t)| \leq 1$ is a consequence of the saturation of the amplifier which provides the field voltage. Another example is provided by problems in torpedo depth control, for which the transfer function $G(s)$ of Eq. (7-119) represents a good approximation to the actual system dynamics. In this system, the output is the depth and the control is the elevator deflection† (limited by mechanical stops).

Let us define the variables $y_1(t)$, $y_2(t)$, and $y_3(t)$ by setting

$$y_1(t) = y(t) \qquad y_2(t) = \dot{y}(t) \qquad y_3(t) = \ddot{y}(t) \tag{7-120}$$

Then the variables $y_1$, $y_2$, $y_3$ satisfy the vector differential equation

$$\begin{bmatrix} \dot{y}_1(t) \\ \dot{y}_2(t) \\ \dot{y}_3(t) \end{bmatrix} = \begin{bmatrix} 0 & 1 & 0 \\ 0 & 0 & 1 \\ 0 & 0 & -a \end{bmatrix} \begin{bmatrix} y_1(t) \\ y_2(t) \\ y_3(t) \end{bmatrix} + \begin{bmatrix} 0 \\ 0 \\ Ku(t) \end{bmatrix} \tag{7-121}$$

which may be written more compactly as

$$\dot{\mathbf{y}}(t) = \mathbf{A}\mathbf{y}(t) + \mathbf{u}(t) \tag{7-122}$$

The eigenvalues of the $3 \times 3$ matrix $\mathbf{A}$ are $0, 0, -a$. Since two of these eigenvalues are the same, we may use a similarity transformation to reduce $\mathbf{A}$ to its Jordan canonical form.‡ Letting $\mathbf{J}(\mathbf{A})$ indicate the Jordan canonical form of $\mathbf{A}$, we have

$$\mathbf{J}(\mathbf{A}) = \begin{bmatrix} 0 & 1 & 0 \\ 0 & 0 & 0 \\ 0 & 0 & -a \end{bmatrix} \tag{7-123}$$

and hence there is a nonsingular matrix $\mathbf{P}$ given by

$$\mathbf{P} = \begin{bmatrix} 1 & 0 & 1/a^2 \\ 0 & 1 & -1/a \\ 0 & 0 & 1 \end{bmatrix} \tag{7-124}$$

with inverse given by

$$\mathbf{P}^{-1} = \begin{bmatrix} 1 & 0 & -1/a^2 \\ 0 & 1 & 1/a \\ 0 & 0 & 1 \end{bmatrix} \tag{7-125}$$

such that the relation

$$\mathbf{J}(\mathbf{A}) = \mathbf{P}^{-1}\mathbf{A}\mathbf{P} \tag{7-126}$$

is satisfied.

We define the column vector $\mathbf{z}(t)$ by setting

$$\mathbf{z}(t) = \mathbf{P}^{-1}\mathbf{y}(t) \tag{7-127}$$

Clearly, the vector $\mathbf{z}(t)$ satisfies the differential equation

$$\dot{\mathbf{z}}(t) = \mathbf{J}(\mathbf{A})\mathbf{z}(t) + \mathbf{P}^{-1}\mathbf{u}(t) \tag{7-128}$$

† See Ref. P-7.
‡ See Sec. 2-10.

or, equivalently,

$$\dot{z}_1(t) = z_2(t) - \frac{K}{a^2}u(t)$$
$$\dot{z}_2(t) = \frac{K}{a}u(t) \tag{7-129}$$
$$\dot{z}_3(t) = -az_3(t) + Ku(t)$$

For simplicity, we define the state variables $x_1(t)$, $x_2(t)$, and $x_3(t)$ by setting

$$x_1(t) = \frac{a^3}{K}z_1(t)$$
$$x_2(t) = \frac{a^2}{K}z_2(t) \tag{7-130}$$
$$x_3(t) = \frac{a}{K}z_3(t)$$

It follows that the $x_i$ satisfy the system of differential equations

$$\dot{x}_1(t) = ax_2(t) - au(t)$$
$$\dot{x}_2(t) = au(t) \tag{7-131}$$
$$\dot{x}_3(t) = -ax_3(t) + au(t)$$

We shall deal exclusively with the state variables $x_1$, $x_2$, $x_3$ in the remainder

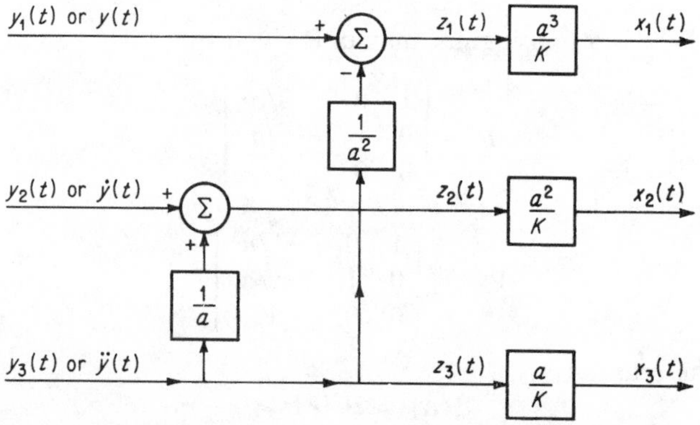


**Fig. 7-17** Block diagram illustrating the linear transformation which generates the $x_1$, $x_2$, and $x_3$ variables from the $y$, $\dot{y}$, and $\ddot{y}$ variables.

of this section. Figure 7-17 illustrates the linear transformation required to generate the $x_i$ from the $y_i$. We note that

$$x_1 = x_2 = x_3 = 0 \quad \text{implies that} \quad y_1 = y_2 = y_3 = 0 \tag{7-132}$$

Now we proceed with the formulation of the time-optimal problem using the system described by Eqs. (7-131).

***Problem 7-5*** *Given the system* (7-131). *Determine the control* [*subject to the constraint* $|u(t)| \leq 1$] *which forces any given initial state* $(\xi_1, \xi_2, \xi_3)$ *to the origin* (0, 0, 0) *in minimum time.*

We remark that the system (7-131) is normal and that the time-optimal control exists and is unique (why?). Once more, the procedure we shall follow to solve this problem is the same as the one presented in Secs. 7-2 and 7-3.

The Hamiltonian $H$ for the minimum-time control of the system described by Eqs. (7-131) is given by the relation

$$H = 1 + ax_2(t)p_1(t) - ax_3(t)p_3(t) + au(t)[-p_1(t) + p_2(t) + p_3(t)] \quad (7\text{-}133)$$

Since $a$ is a positive constant, the control which minimizes $H$ is given by

$$u(t) = -\operatorname{sgn}\{-p_1(t) + p_2(t) + p_3(t)\} \quad (7\text{-}134)$$

The costate variables $p_i(t)$, $i = 1, 2, 3$, satisfy the system of equations

$$\begin{aligned} \dot{p}_1(t) &= -\frac{\partial H}{\partial x_1(t)} = 0 \\ \dot{p}_2(t) &= -\frac{\partial H}{\partial x_2(t)} = -ap_1(t) \\ \dot{p}_3(t) &= -\frac{\partial H}{\partial x_3(t)} = ap_3(t) \end{aligned} \quad (7\text{-}135)$$

Using the familiar notation

$$\pi_i = p_i(0) \qquad i = 1, 2, 3 \quad (7\text{-}136)$$

we solve Eqs. (7-135) to obtain

$$\begin{aligned} p_1(t) &= \pi_1 \\ p_2(t) &= \pi_2 - a\pi_1 t \\ p_3(t) &= \pi_3 e^{at} \end{aligned} \quad (7\text{-}137)$$

The function

$$-p_1(t) + p_2(t) + p_3(t) = -\pi_1 + \pi_2 - a\pi_1 t + \pi_3 e^{at} \quad (7\text{-}138)$$

thus has *at most two zeros* no matter what the values of $\pi_1$, $\pi_2$, and $\pi_3$ are. We can immediately conclude that the only candidates for time-optimal control are the six control sequences

$$\{+1\}, \{-1\}, \{+1, -1\}, \{-1, +1\}, \{+1, -1, +1\}, \{-1, +1, -1\} \quad (7\text{-}139)$$

We next solve Eqs. (7-131) by setting

$$u(t) = \Delta = \pm 1 \quad (7\text{-}140)$$

to obtain the solution

$$\begin{aligned} x_1(t) &= \xi_1 + \xi_2 at + \tfrac{1}{2}\Delta a^2 t^2 - \Delta at \\ x_2(t) &= \xi_2 + \Delta at \\ x_3(t) &= (\xi_3 - \Delta)e^{-at} + \Delta \end{aligned} \tag{7-141}$$

with the usual notation

$$\xi_i = x_i(0) \qquad i = 1, 2, 3 \tag{7-142}$$

Eliminating the time $t$ from Eqs. (7-141), we find that

$$x_1 = \xi_1 - (x_2 - \xi_2) + \tfrac{1}{2}\Delta(x_2^2 - \xi_2^2) \tag{7-143}$$

$$x_3 = (\xi_3 - \Delta)e^{-\Delta(x_2 - \xi_2)} + \Delta \tag{7-144}$$

These two equations determine the trajectory in the three-dimensional state space, which begins at the state $(\xi_1, \xi_2, \xi_3)$ and is generated by the

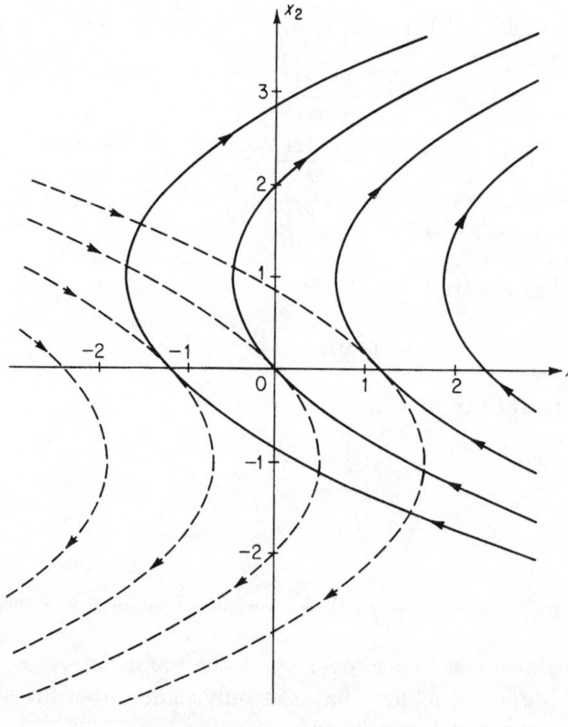


**Fig. 7-18** The projections of the forced trajectories in the $x_1x_2$ plane. The solid trajectories are due to $u = +1$, and the dashed ones are due to $u = -1$; the arrows indicate the motion of the state for positive time.

constant control $u = \Delta$. Equation (7-143) represents the projection of the trajectory onto the $x_1x_2$ plane, and Eq. (7-144) represents its projection onto the $x_2x_3$ plane. We note that the trajectories are independent of the value of $a$.† Figure 7-18 shows the curves generated by Eq. (7-143); the solid curves are for $\Delta = +1$, while the dashed ones are for $\Delta = -1$. Figure 7-19 illustrates the curves generated by Eq. (7-144); again, the solid curves are for $\Delta = +1$, and the dashed ones are for $\Delta = -1$.

Let us consider the two trajectories which pass through the origin (0, 0, 0) of the three-dimensional state space. One of these trajectories is obtained

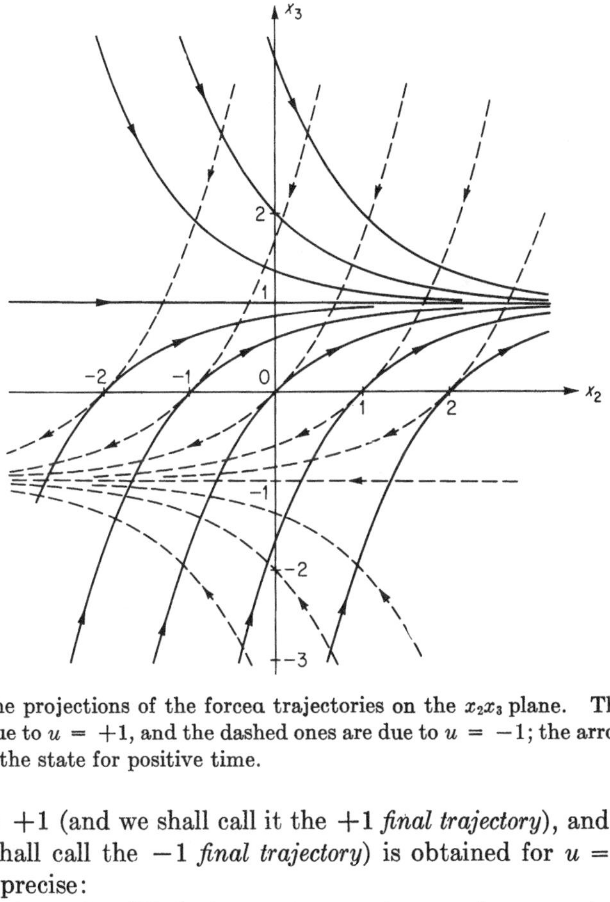


**Fig. 7-19** The projections of the forcea trajectories on the $x_2x_3$ plane. The solid trajectories are due to $u = +1$, and the dashed ones are due to $u = -1$; the arrows indicate the motion of the state for positive time.

for $u = \Delta = +1$ (and we shall call it the $+1$ *final trajectory*), and the other (which we shall call the $-1$ *final trajectory*) is obtained for $u = \Delta = -1$. To be more precise:

***Definition 7-9*** *Let* $\{V_2^+\}$ *denote the set of states* $(x_1, x_2, x_3)$ *which are forced to* (0, 0, 0) *by the control sequence* $\{+1\}$. *The set* $\{V_2^+\}$ *is the* $+1$ *final trajectory and is defined by*

$$\{V_2^+\} = \{(x_1, x_2, x_3) \colon x_1 = \tfrac{1}{2}x_2^2 - x_2,\ x_3 = 1 - e^{-x_2},\ x_2 < 0\} \qquad \text{(7-145)}$$

***Definition 7-10*** *Let* $\{V_2^-\}$ *denote the set of states* $(x_1, x_2, x_3)$ *which are forced to* (0, 0, 0) *by the control sequence* $\{-1\}$. *The set* $\{V_2^-\}$ *is the* $-1$

† This is the reason for the choice of the $x_i$ as state variables in this section.

*final trajectory and is defined by*

$$\{V_2^-\} = \{(x_1, x_2, x_3): x_1 = -\tfrac{1}{2}x_2^2 - x_2,\ x_3 = -1 + e^{x_2},\ x_2 > 0\} \quad (7\text{-}146)$$

***Definition 7-11*** *Let $\{V_2\}$ denote the union of $\{V_2^+\}$ and $\{V_2^-\}$. Using*

$$\Delta^* = -\operatorname{sgn}\{x_2\} \quad (7\text{-}147)$$

*we find that*

$$\{V_2\} = \{(x_1, x_2, x_3): x_1 = \tfrac{1}{2}\Delta^* x_2^2 - x_2,\ x_3 = \Delta^* - \Delta^* e^{-\Delta^* x_2}\} = \{V_2^+\} \cup \{V_2^-\} \quad (7\text{-}148)$$

*It follows from the above definitions that $\{V_2\}$ is a curve in the state space which has the property that any state on $\{V_2\}$ can be forced to the origin by application of the control*

$$u = \Delta^* = -\operatorname{sgn}\{x_2\} \quad (7\text{-}149)$$

The projections of $\{V_2\}$ onto the $x_1x_2$ and $x_2x_3$ planes are shown in Figs. 7-20 and 7-21, respectively.

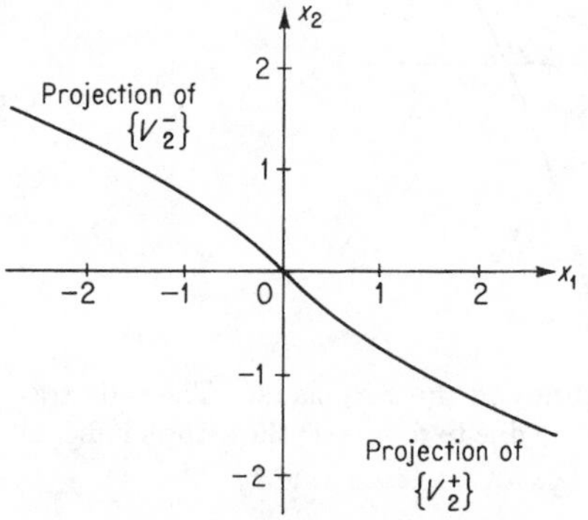


**Fig. 7-20** The projection of the set (or curve) $\{V_2\}$ on the $x_1x_2$ plane.

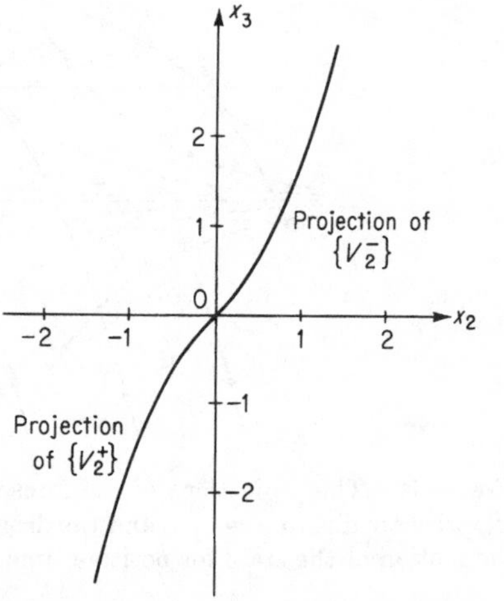


**Fig. 7-21** The projection of the set (or curve) $\{V_2\}$ on the $x_2x_3$ plane.

Consider the set of all states which can be forced to the origin by the control sequence $\{-1, +1\}$. It is clear that the transition from the control $u = -1$ to the control $u = +1$ must occur on the set $\{V_2^+\}$ (since $\{V_2^+\}$ is the set of *all* states which can be driven to the origin by application of the control $u = +1$). If a state $(x_1, x_2, x_3)$ can be forced to the origin by the control sequence $\{-1, +1\}$, then there is a state $(x_{12}, x_{22}, x_{32}) \in \{V_2^+\}$ such that the state $(x_1, x_2, x_3)$ lies on a trajectory, generated by $u = -1$, which terminates at the state $(x_{12}, x_{22}, x_{32}) \in \{V_2^+\}$. We generalize these considerations in the following definition:

***Definition 7-12*** *Let $\{V_1\}$ denote the set of states $(x_1, x_2, x_3)$ which can be forced to the origin by the control sequence $\{-\Delta^*, \Delta^*\}$. Then a state $(x_1, x_2, x_3) \in \{V_1\}$ if and only if*

$$x_{12} = x_1 - (x_{22} - x_2) - \tfrac{1}{2}\Delta^*(x_{22}{}^2 - x_2{}^2) \tag{7-150}$$

$$x_{32} = (x_3 + \Delta^*)e^{\Delta^*(x_{22}-x_2)} - \Delta^* \tag{7-151}$$

*where*

$$(x_{12}, x_{22}, x_{32}) \in \{V_2\} \tag{7-152}$$

*that is [from Eq.* (7-149)],

$$x_{12} = \tfrac{1}{2}\Delta^* x_{22}{}^2 - x_{22} \tag{7-153}$$

$$x_{32} = \Delta^* - \Delta^* e^{-\Delta^* x_{22}} \tag{7-154}$$

*and*

$$\Delta^* = -\operatorname{sgn}\{x_{22}\} \tag{7-155}$$

We note that states belonging to the set $\{V_1\}$ are given in parametric form in terms of states $(x_{12}, x_{22}, x_{32})$ in $\{V_2\}$. We can eliminate $x_{12}, x_{22}, x_{32}$ from Eqs. (7-150) to (7-154) to obtain a relationship between $x_1$, $x_2$, and $x_3$. From Eqs. (7-150) and (7-153), we find that

$$x_{22} = \pm\left[\frac{x_2{}^2}{2} + \Delta^*(x_1 + x_2)\right]^{\frac{1}{2}} \tag{7-156}$$

To eliminate the ambiguity of sign in Eq. (7-156), we use Eq. (7-155), and we then find that

$$x_{22} = -\Delta^*\left[\frac{x_2{}^2}{2} + \Delta^*(x_1 + x_2)\right]^{\frac{1}{2}} \tag{7-157}$$

From Eqs. (7-151) and (7-154), we deduce that

$$x_3 = -\Delta^* + \Delta^* e^{-\Delta^*(x_{22}-x_2)}(2 - e^{-\Delta^* x_{22}}) \tag{7-158}$$

Substituting Eq. (7-157) into Eq. (7-158), we obtain the relation

$$x_3 = -\Delta^* + \Delta^* \exp\left\{\Delta^* x_2 + \left[\frac{x_2{}^2}{2} + \Delta^*(x_1 + x_2)\right]^{\frac{1}{2}}\right\} \times \left(2 - \exp\left\{\left[\frac{x_2{}^2}{2} + \Delta^*(x_1 + x_2)\right]^{\frac{1}{2}}\right\}\right) \tag{7-159}$$

This equation provides a relationship between the variables $x_1, x_2, x_3$, where $(x_1, x_2, x_3) \in \{V_1\}$, and $\Delta^* = \pm 1$. It remains to determine $\Delta^*$ as a function of $x_1$, $x_2$, and $x_3$. Figure 7-22 shows the projection of $\{V_2\}$ on the $x_1x_2$ plane; the equation of this projection is

$$x_1 = -\tfrac{1}{2}\operatorname{sgn}\{x_2\}x_2{}^2 - x_2 \tag{7-160}$$

We observe in Fig. 7-22 that the trajectories which belong to $\{V_1\}$ and which terminate at $\{V_2^+\}$ are generated by $\Delta^* = +1$. On the other hand, the trajectories which belong to $\{V_1\}$ and which terminate at $\{V_2^-\}$ are generated by $\Delta^* = -1$. This means that if a state $(x_1, x_2, x_3)$ belongs to

$\{V_1\}$ and if the two coordinates $x_1$ and $x_2$ satisfy the relation

$$x_1 > -\tfrac{1}{2} \operatorname{sgn} \{x_2\} x_2^2 - x_2 \tag{7-161}$$

then $\Delta^* = +1$. If, however, the two coordinates $x_1$ and $x_2$ satisfy the relation

$$x_1 < -\tfrac{1}{2} \operatorname{sgn} \{x_2\} x_2^2 - x_2 \tag{7-162}$$

then $\Delta^* = -1$. It follows that

$$\Delta^* = \operatorname{sgn} \{x_1 + \tfrac{1}{2} \operatorname{sgn} \{x_2\} x_2^2 + x_2\} \tag{7-163}$$

Thus, we have evaluated $\Delta^*$ in terms of $x_1$ and $x_2$. If we substitute Eq. (7-163) into Eq. (7-159), then we obtain an equation which gives $x_3$ as an explicit function of $x_1$ and $x_2$.

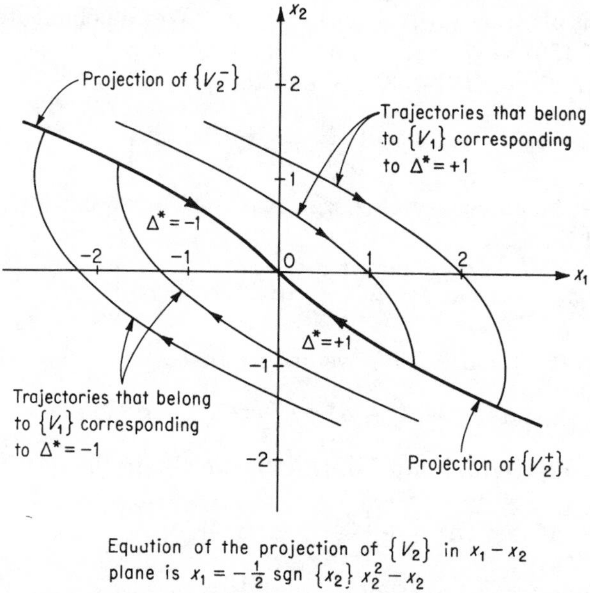


**Fig. 7-22** Projection of the trajectories that belong to $\{V_1\}$.

We note that, in Eq. (7-159), there appear terms of the form

$$\sqrt{\frac{x_2^2}{2} + \Delta^*(x_1 + x_2)} \tag{7-164}$$

Since the variables $x_1$, $x_2$, and $x_3$ must be real, using Eq. (7-163) we obtain the condition

$$\frac{x_2^2}{2} + \operatorname{sgn} \{x_1 + \tfrac{1}{2} \operatorname{sgn} \{x_2\} x_2^2 + x_2\} (x_1 + x_2) \geq 0 \tag{7-165}$$

**Exercise 7-28** Show that Eq. (7-165) holds for all $x_1$ and $x_2$. HINT: Examine the polarities of the functions $x_2^2/2 + (x_1 + x_2)$ and $x_2^2/2 - (x_1 + x_2)$ in the $x_1x_2$ plane.

*To recapitulate:* If the variables $x_1$, $x_2$, and $x_3$ satisfy Eqs. (7-159) and (7-163), then $(x_1, x_2, x_3) \in \{V_1\}$. The set $\{V_1\}$ is a *surface*† in the three-dimensional state space. It is easy to show that the surface $\{V_1\}$ has the following properties:

1. $\mathbf{0} \subset \{V_2\} \subset \{V_1\}$.
2. $\{V_1\}$ is smooth, continuous, and real.
3. $\{V_1\}$ is single-valued; that is, to each pair $(x_1, x_2)$, there corresponds only one $x_3$ for which $(x_1, x_2, x_3) \in \{V_1\}$.
4. $\{V_1\}$ is symmetric about the origin of the state space.
5. The surface $\{V_1\}$ separates the state space into two parts.

We shall make use of the last property in the construction of the time-optimal control policy.

Let us agree to use $\mathbf{x}_1 = (x_{11}, x_{21}, x_{31})$ to indicate that a state belongs to the set $\{V_1\}$. Clearly, if $(x_{11}, x_{21}, x_{31}) \in \{V_1\}$, then

$$x_{31} = -\Delta^* + \Delta^* \exp\left\{\Delta^* x_{21} + \left[\frac{x_{21}^2}{2} + \Delta^*(x_{11} + x_{21})\right]^{\frac{1}{2}}\right\} \times \left(2 - \exp\left\{\left[\frac{x_{21}^2}{2} + \Delta^*(x_{11} + x_{21})\right]^{\frac{1}{2}}\right\}\right) \tag{7-166}$$

where

$$\Delta^* = \operatorname{sgn}\{x_{11} + \tfrac{1}{2}\operatorname{sgn}\{x_{21}\}\, x_{21}^2 + x_{21}\} \tag{7-167}$$

Now consider an arbitrary state $(x_1, x_2, x_3)$ and pass a line parallel to the $x_3$ axis through that state. This line will intersect the surface $\{V_1\}$ at the point $(x_{11}, x_{21}, x_{31})$ for which $x_{11} = x_1$ and $x_{21} = x_2$. The coordinate $x_{31}$ is determined by Eqs. (7-166) and (7-167). This situation is illustrated in Fig. 7-23. *If*

$$x_3 - x_{31} > 0 \tag{7-168}$$

*then we shall say that the state $(x_1, x_2, x_3)$ lies above the surface $\{V_1\}$. If*

$$x_3 - x_{31} < 0 \tag{7-169}$$

*then we shall say that the state $(x_1, x_2, x_3)$ lies below the surface $\{V_1\}$. If*

$$x_3 - x_{31} = 0 \tag{7-170}$$

*then $(x_1, x_2, x_3)$ belongs to $\{V_1\}$.*

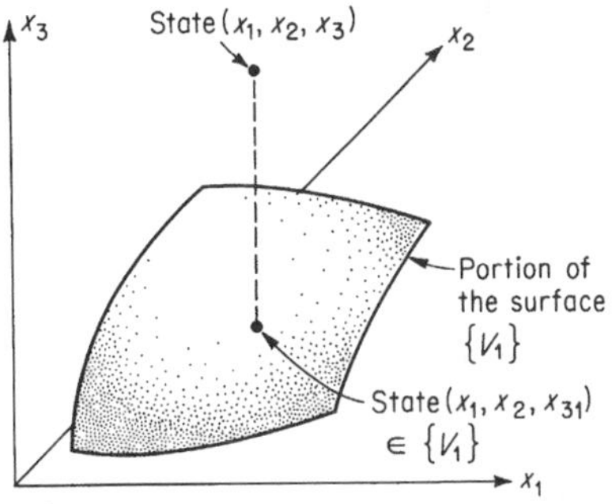


**Fig. 7-23** Illustration of the projection of a state $(x_1, x_2, x_3)$ upon the surface $\{V_1\}$.

Let us now consider states which do not belong to the surface $\{V_1\}$. Suppose that we are given a state $\mathbf{x} = (x_1, x_2, x_3)$ which lies *above* the surface $\{V_1\}$. At the state $\mathbf{x}$, we have two control choices: We can apply either $u = +1$ or $u = -1$. It is easy to verify that the control $u = +1$ results in a trajectory which never intersects the surface $\{V_1\}$, while the control

† Often called the switch surface.

$u = -1$ results in a trajectory which does intersect the surface $\{V_1\}$. Similarly, if a state lies *below* the surface $\{V_1\}$, then the trajectory generated by the control $u = +1$ intersects the surface $\{V_1\}$, while the trajectory generated by the control $u = -1$ does not intersect $\{V_1\}$.

Thus, from the above, we may conclude that: *The control sequence* $\{-1, +1, -1\}$ *drives a state above* $\{V_1\}$ *to the origin, and the control sequence* $\{+1, -1, +1\}$ *drives a state below* $\{V_1\}$ *to the origin. We note in passing that the control sequences* $\{+1, -1, +1\}$ *and* $\{-1, +1, -1\}$ *are both candidates for time-optimal control* [*see Eq.* (7-139)].

Now we shall find the time-optimal control sequence corresponding to each state of the state space. The techniques that will be used are similar to those we have used in the two previous sections; namely, we shall consider the six control sequences of Eq. (7-139), and we shall show that there is only one control sequence which can force a given state to the origin. Thus, we shall prove time optimality by the process of elimination.

Let us consider a state $X$ on the curve $\{V_2^+\}$ described by Definition 7-9. By definition, the control sequence $\{+1\}$ forces $X$ to the origin. Let us examine what happens when we use the control sequence $\{-1\}$ or $\{-1, +1\}$ or $\{-1, +1, -1\}$. Initial application of $u = -1$ will generate a $-1$ forced trajectory. If $Y$ is any point on that trajectory, then it is easy to show that the point $Y$ lies *below* the surface $\{V_1\}$. However, we have shown that, if a state lies *below* $\{V_1\}$, then one must apply the control sequence $\{+1, -1, +1\}$ to drive that state, and hence $Y$, to the origin. It follows that if at $X$ we first apply $u = -1$, then the control sequence which forces $X$ to the origin must be $\{-1, +1, -1, +1\}$. This is *not* an optimal sequence, and so the control sequences $\{-1\}$, $\{-1, +1\}$, and $\{-1, +1, -1\}$ cannot be time-optimal. Let us now consider the remaining two control sequences $\{+1, -1\}$ and $\{+1, -1, +1\}$. Initial application of $u = +1$ forces $X$ to $X'$, with $X' \in \{V_2^+\}$. At $X'$, the control switches to $u = -1$ to generate a $-1$ forced trajectory. If $Y'$ is any point on that trajectory, then $Y'$ lies *below* the surface $\{V_1\}$, and the same type of argument as we used before shows that the control sequences $\{+1, -1\}$ and $\{+1, -1, +1\}$ cannot force $X$ to the origin. Thus, by the process of elimination we conclude that:†

*If a state belongs to* $\{V_2^+\}$, *then* $u = +1$ *is the time-optimal control.*

Similar arguments can be used to show that *if a state belongs to* $\{V_2^-\}$ (*see Definition* 7-10), *then* $u = -1$ *is the time-optimal control.* We can summarize the above conclusions as follows:

***Control Law 7-5a*** *If* $(x_1, x_2, x_3) \in \{V_2\}$, *where* $\{V_2\}$ *is described in Definition* 7-11, *then* $u^* = \Delta^*$ *is the time-optimal control. Note that on* $\{V_2\}$, $\Delta^*$ *is given by Eq.* (7-147).

† We could have bypassed this argument by invoking the uniqueness of extremal controls (see Theorem 6-9). However, we have chosen to outline the geometric proof so that the reader can appreciate the concepts involved.

Next, consider a state $X$ on the surface $\{V_1\}$. For the sake of concreteness, suppose that for that particular state, $\Delta^* = +1$, as determined by Eq. (7-163). Then the time-optimal control sequence is $\{-1, +1\}$. The reasoning is as follows: The control sequence $\{-1, +1\}$ will force $X$ to the origin by definition. The control sequences $\{+1\}$ and $\{-1\}$ generate trajectories that do not hit the origin. The control sequence $\{+1, -1\}$ cannot force $X$ to the origin because it contradicts the assumption that $\Delta^* = +1$. The control sequence $\{+1, -1, +1\}$ is such that $u = +1$ forces $X$ to some $Y$ with $Y$ above $\{V_1\}$. Finally, the control sequence $\{-1, +1, -1\}$ contradicts the assumption $\Delta^* = +1$. Again by elimination, we prove that the time-optimal control sequence must be $\{-1, +1\}$

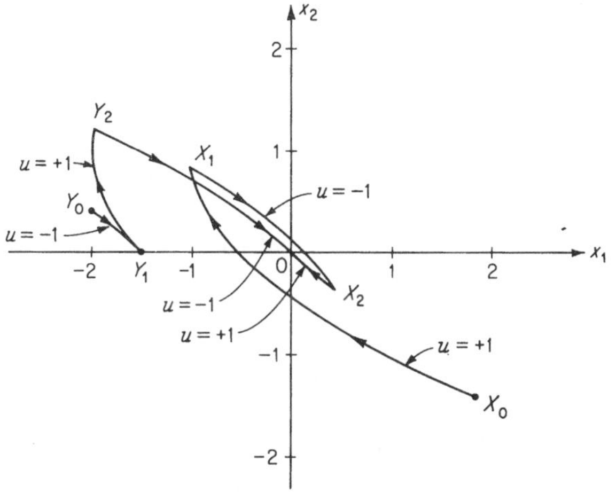


**Fig. 7-24** Projections of time-optimal trajectories on the $x_1x_2$ plane.

provided that the switch from $u = -1$ to $u = +1$ occurs on the curve $\{V_2\}$. We can generalize the above arguments as follows:

***Control Law 7-5b*** *If* $(x_1, x_2, x_3) \in \{V_1\}$, *where* $\{V_1\}$ *is described in Definition* 7-12, *then* $u^* = -\Delta^*$ *is the time-optimal control. Note that on* $\{V_1\}$, $\Delta^*$ *is given by Eq.* (7-163).

We have also shown the following:

***Control Law 7-5c*** *If the state* $(x_1, x_2, x_3)$ *lies above the surface* $\{V_1\}$, *then* $u^* = -1$ *is the time-optimal control. If the state* $(x_1, x_2, x_3)$ *lies below the surface* $\{V_1\}$, *then* $u^* = +1$ *is the time-optimal control.*

Thus, the operation is as follows: Suppose the state lies above the surface $\{V_1\}$; then the control $u = -1$ will generate a trajectory which will intersect $\{V_1\}$, say at $(x_{11}, x_{21}, x_{31})$. At that instant, the control switches from $u = -1$ to $u = +1$, and the resulting trajectory will move along $\{V_1\}$ and will intersect $\{V_2\}$ at $(x_{12}, x_{22}, x_{32})$. If the control switches again from $u = +1$ to $u = -1$, the trajectory will move along $\{V_2\}$ until it hits the origin. As a matter of fact, one can check that at $(x_{11}, x_{21}, x_{31})$, the value of $\Delta^*$ is $-1$, and at $(x_{12}, x_{22}, x_{32})$, $\Delta^*$ is again $-1$. Figures 7-24 and 7-25 illustrate the projections of some time-optimal trajectories.

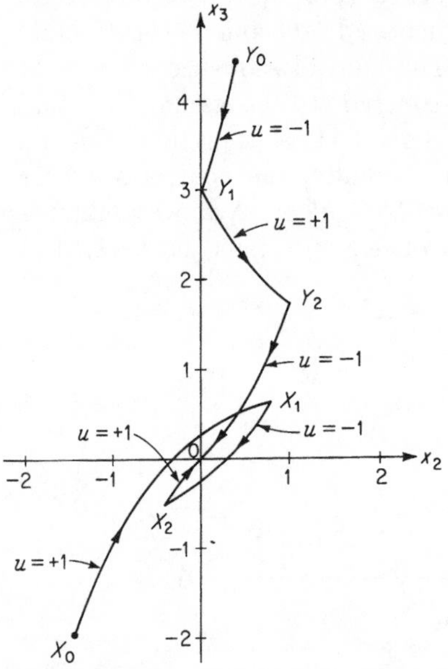


**Fig. 7-25** Projections of time-optimal trajectories on the $x_2x_3$ plane.

Since Control Laws 7-5*a* to 7-5*c* provide us with the time-optimal control for each state $(x_1, x_2, x_3)$, we can design an engineering system which will automatically deliver the time-optimal control to the plant. The compensator suggested here decides whether the state $(x_1, x_2, x_3)$ lies above or below the surface $\{V_1\}$ by computing the projection of $(x_1, x_2, x_3)$ on the surface $\{V_1\}$ (see Fig. 7-23). Thus, the state $(x_1, x_2, x_{31}) \in \{V_1\}$ will be computed, and the difference

$$m(t) = x_3 - x_{31} \tag{7-171}$$

will drive an ideal relay whose output is the time-optimal control $u(t)$. The complete time-optimal system is illustrated in block-diagram form in Fig. 7-26. The state variables $x_1$, $x_2$, and $x_3$ are measured (say using the arrangement of Fig. 7-17). The variables $x_1$ and $x_2$ are used to determine $\Delta^*$, which, according to Eq. (7-163), is given by

$$\Delta^* = \operatorname{sgn}\{a\} = \operatorname{sgn}\{x_1 + x_2 + \tfrac{1}{2}x_2|x_2|\} \qquad \text{(7-172)}\dagger$$

To generate $\Delta^*$, the signal $x_2$ is passed through the nonlinearity $N_1$, whose output is $\frac{1}{2}x_2|x_2|$. The relay $R_1$ performs the signum operation on the signal $a$. The signal $b$, given by the equation

$$b = \tfrac{1}{2}x_2^2 + \Delta^*(x_1 + x_2) \tag{7-173}$$

is obtained by passing $x_2$ through the squaring-type nonlinearity $N_2$, whose output is $\frac{1}{2}x_2^2$, and by adding the result to the signal $\Delta^*(x_1 + x_2)$. The signal $b$ is then introduced into the square-root-type nonlinearity $N_3$ to obtain $\sqrt{b}$. The two nonlinear elements $N_4$ are of exponential type. Thus, by a multitude of linear and nonlinear algebraic operations on the signals $x_1$ and $x_2$, we obtain the signal $x_{31}$. By comparing $x_{31}$ with the measured state variable $x_3$, we obtain the signal $-m(t)$ [see Eq. (7-171)]. The relay $R$

† Note that $\frac{1}{2}$ sgn $\{x_2\}\, x_2^2 = \frac{1}{2}x_2|x_2|$.

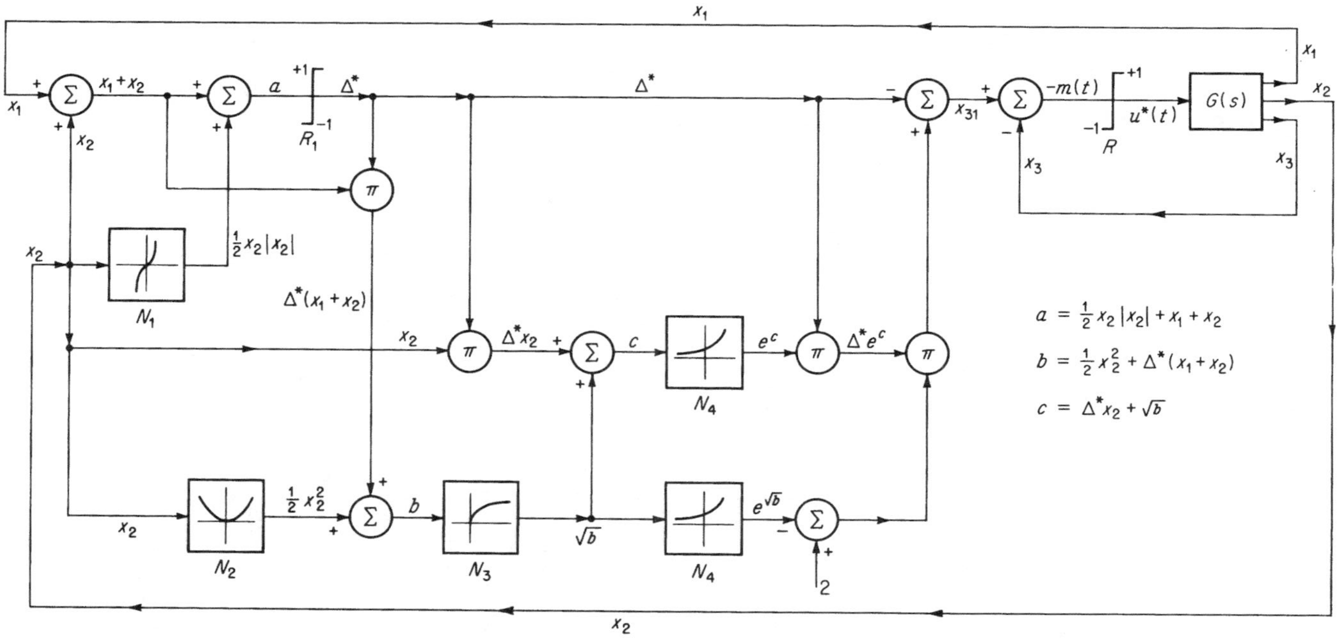


**Fig. 7-26** The time-optimal system for the plant $G(s) = K/s^2(s + a)$.

performs the signum operation on the signal $-m(t)$, and the output of $R$ is the time-optimal control $u^*(t)$ which is used to drive the system $G(s)$.

The system illustrated in Fig. 7-26 is a complex one. It is instructive to compare Fig. 7-26 with Figs. 7-5 and 7-14 to appreciate the complexity of a third-order time-optimal system as compared with the complexity of a second-order system. Yet, for the third-order system considered in this section, it is possible to obtain an explicit equation for the surface $\{V_1\}$ and to design an analog-type nonlinear compensator. As we shall see later on in this chapter, the surface $\{V_1\}$ cannot always be obtained in an explicit form.

In the control literature, the surface $\{V_1\}$ is often called the *switch surface* because the relay switches whenever the state reaches the surface $\{V_1\}$. If we construct or simulate the system of Fig. 7-26, then the unavoidable noise and nonideal equipment will cause the relay to switch when the state is just above or just below the switch surface, so that the actual response of the system will be close to but not exactly the same as the theoretical one. Therefore, an actual system can be expected to exhibit a limit cycle about the origin; the size of the limit cycle will depend upon the amount of deviation of the equipment used in the actual system from the idealizations of the theoretical system.

A final comment regarding the system illustrated in Fig. 7-26 is in order. This engineering realization simply simulates the switch surface $\{V_1\}$; if all the components were ideal, then whenever the state was an element of $\{V_1\}$, the signal $m(t)$ would be zero. Thus, if all our components were ideal, we should include a subsystem which generates the time-optimal control whenever $\mathbf{x} \in \{V_1\}$ and still another subsystem which generates the time-optimal control whenever $\mathbf{x} \in \{V_2\}$. However, we rely on the fact that engineering hardware which is not ideal will never make the state remain either on $\{V_2\}$ or on $\{V_1\}$. Thus, any realization will be slightly suboptimal, and the actual trajectories will be close to, but not exactly on, the switch surface.

In the following section, we shall extend the ideas of this section to systems with an arbitrary number of real poles.

**Exercise 7-29** Consider the so-called "triple-integral" plant with the transfer function

$$G(s) = \frac{1}{s^3} \tag{7-174}$$

and with the state-variable representation

$$\begin{aligned} \dot{x}_1(t) &= x_2(t) \\ \dot{x}_2(t) &= x_3(t) \\ \dot{x}_3(t) &= u(t) \end{aligned} \tag{7-175}$$

Assume that the control is constrained by the relation

$$|u(t)| \le 1 \tag{7-176}$$

Mimic the ideas of this section to find the time-optimal control to the origin. Design an engineering realization of the time-optimal system.

**Exercise 7-30** Consider the unstable system† with the transfer function

$$G(s) = \frac{1}{s^2(s-1)} \tag{7-177}$$

and with the state-variable representation [see Eqs. (7-131)]

$$\begin{aligned} \dot{x}_1(t) &= -x_2(t) + u(t) \\ \dot{x}_2(t) &= -u(t) \\ \dot{x}_3(t) &= x_3(t) - u(t) \end{aligned} \tag{7-178}$$

where the control $u(t)$ is constrained by

$$|u(t)| \le 1 \tag{7-179}$$

Find the set of initial states for which a time-optimal control to the origin (0, 0, 0) exists. Find the time-optimal control law.

## 7-5 Time-optimal Control of Plants with $N$ Real Poles‡

We have developed several techniques for the synthesis of the time-optimal control for second- and third-order systems. Here, we shall examine the time-optimal control problem for an $N$th-order system with $N$ distinct real poles. We shall suppose that the system we shall examine is described by the $N$th-order linear differential equation

$$\frac{d^N y(t)}{dt^N} + a_{N-1}\frac{d^{N-1}y(t)}{dt^{N-1}} + \cdots + a_1\frac{dy(t)}{dt} + a_0 y(t) = Ku(t) \tag{7-180}$$

where $y(t)$ represents the system output, $K$ is a gain constant, and $u(t)$ is the control. We shall also assume that the control $u(t)$ is constrained to satisfy the relation

$$|u(t)| \le 1 \tag{7-181}$$

It is common in servomechanism practice to say that the system described by Eq. (7-180) has the transfer function

$$\frac{y(s)}{u(s)} = G(s) = \frac{K}{s^N + a_{N-1}s^{N-1} + \cdots + a_1 s + a_0} \tag{7-182}$$

We suppose that the function $G(s)$ has $N$ real, negative, and distinct poles, i.e., that $G(s)$ may be written in the form

$$G(s) = \frac{K}{(s+s_1)(s+s_2)\cdots(s+s_N)} \tag{7-183}$$

where the $s_i$ are real and

$$0 < s_1 < s_2 < \cdots < s_N \tag{7-184}$$

† See Sec. 3-26.

‡ The material in this section is based, for the most part, upon Refs. A-1, A-2, S-5, and L-6.

We have discussed the reduction of an $N$th-order system such as (7-180) to a system of $N$ first-order linear differential equations in Sec. 3-24. We shall suppose that this reduction has been made and that a similarity transformation has been used, so that we may view the system described by the transfer function of Eq. (7-183) as having the form

$$\begin{aligned}\dot{z}_1(t) &= -s_1 z_1(t) + r_1 u(t)\\ \dot{z}_2(t) &= -s_2 z_2(t) + r_2 u(t)\\ &\dots\dots\dots\dots\\ \dot{z}_N(t) &= -s_N z_N(t) + r_N u(t)\end{aligned} \tag{7-185}$$

The variables $z_1(t), \ldots, z_N(t)$ are linear combinations of $y(t), \dot{y}(t), \ldots, y^{(N-1)}(t)$, and the constants $r_1, \ldots, r_N$ depend upon the $s_i$ and $K$.

Let us now define a new set of variables, $x_1, x_2, \ldots, x_N$, by setting

$$x_i(t) = z_i(t)\frac{s_i}{r_i} \qquad i = 1, 2, \ldots, N \tag{7-186}$$

Then the $x_i$ satisfy the differential equations

$$\dot{x}_i(t) = -s_i x_i(t) + s_i u(t) \qquad i = 1, 2, \ldots, N \tag{7-187}$$

We shall use the $x_i(t)$ exclusively as state variables in the remainder of this section. The time-optimal control problem we shall consider is:

***Problem 7-6*** *Given the system* (7-187), *find the admissible control which will force any initial state* $(\xi_1, \xi_2, \ldots, \xi_n)$ *to the origin* $(0, 0, \ldots, 0)$ *in minimum time.*

The basic idea in the solution of this problem is to extend the techniques of the preceding section from three dimensions to $N$ dimensions. We shall let $\Delta^*$ denote the value of the time-optimal control during the last trajectory, and then we shall define a series of sets $\{V_{N-1}\}, \{V_{N-2}\}, \ldots, \{V_1\}$ which represent the locus of all points in the state space, with the property that the control must switch from $+1$ to $-1$ (or from $-1$ to $+1$) whenever the trajectory intersects any one of these sets $\{V_{N-1}\}, \{V_{N-2}\}, \ldots, \{V_1\}$. The set $\{V_1\}$ will be the switch hypersurface, and we shall find a parametric set of equations that define it. Our main purpose in this section is to illustrate the computations that must be performed by the time-optimal feedback control system.

The Hamiltonian $H$ for the time-optimal control of the system described by Eq. (7-187) takes the form

$$H = 1 - \sum_{i=1}^{N} s_i x_i(t) p_i(t) + u(t) \sum_{i=1}^{N} s_i p_i(t) \tag{7-188}$$

The control which absolutely minimizes $H$ is given by

$$u(t) = -\operatorname{sgn}\left\{\sum_{i=1}^{N} s_i p_i(t)\right\} \tag{7-189}$$

The costate variables $p_i(t)$ are solutions of the equations

$$\dot{p}_i(t) = -\frac{\partial H}{\partial x_i(t)} = s_i p_i(t) \tag{7-190}$$

for $i = 1, 2, \ldots, N$. Therefore,

$$p_i(t) = \pi_i e^{s_i t} \qquad i = 1, 2, \ldots, N \tag{7-191}$$

where

$$p_i(0) \triangleq \pi_i \qquad i = 1, 2, \ldots, N \tag{7-192}$$

and hence Eq. (7-189) reduces to

$$u(t) = -\operatorname{sgn}\left\{\sum_{i=1}^{N} s_i \pi_i e^{s_i t}\right\} \tag{7-193}$$

It is easy to show that the function $\sum_{i=1}^{N} s_i \pi_i e^{s_i t}$:

1. Has *at most* $N - 1$ zeros for all possible values of $\pi_1, \pi_2, \ldots, \pi_N$
2. Cannot be identically zero over a finite interval of time†

Thus, the time-optimal control is piecewise constant and "switches" *at most* $N - 1$ times. Let

$$u = \Delta = \pm 1 \tag{7-194}$$

and substitute $u = \Delta$ in Eq. (7-187) to obtain the relation

$$x_i(t) = (\xi_i - \Delta)e^{-s_i t} + \Delta \qquad i = 1, 2, \ldots, N \tag{7-195}$$

where

$$\xi_i \triangleq x_i(0) \qquad i = 1, 2, \ldots, N \tag{7-196}$$

It is possible to eliminate the time $t$ from these equations and thus obtain the equation of the trajectory in the $N$-dimensional state space which originates at $(\xi_1, \xi_2, \ldots, \xi_n)$ and evolves owing to the application of the control $u = \Delta$. For example, from Eq. (7-195), using $i = 1$ we find that

$$t = \frac{1}{s_1} \log \frac{\xi_1 - \Delta}{x_1 - \Delta} \tag{7-197}$$

Therefore, substituting Eq. (7-197) into Eq. (7-195) for $i = 2, 3, \ldots, N$, we find that

$$x_k = \Delta + (\xi_k - \Delta)\left(\frac{\xi_1 - \Delta}{x_1 - \Delta}\right)^{s_k/s_1} \qquad k = 2, 3, \ldots, N \tag{7-198}$$

The above $N - 1$ equations describe a trajectory in the $N$-dimensional state space. From Eq. (7-195), we conclude that

$$\lim_{t \to \infty} x_i(t) = \Delta \qquad i = 1, 2, \ldots, N \tag{7-199}$$

† This eliminates the possibility of singular control [as the system (7-187) is clearly normal].

which means that *a trajectory generated by* $u = +1$ *tends to the point* (*or state*) $(1, 1, \ldots, 1)$ *and a trajectory generated by* $u = -1$ *tends to the point* (*or state*) $(-1, -1, \ldots, -1)$ *of the state space.*

We shall now generate a sequence of sets $\{V_{N-1}\}, \{V_{N-2}\}, \ldots, \{V_2\}, \{V_1\}$ analogous to the "switching" curve and surface defined in Sec. 7-4 by generalizing the techniques presented in that section. We shall first give the formal definitions and then discuss their significance.

***Definition 7-13*** *Let* $\{V_{N-1}\}$ *denote the set of states which can be forced to the origin* $\mathbf{0}$ *of the state space by application of the control*

$$u = \Delta^* = \pm 1 \tag{7-200}$$

(*in other words, either by the control sequence* $\{+1\}$ *or by the control sequence* $\{-1\}$). *We shall write* $\mathbf{x}_{N-1}$ *to indicate a state belonging to the set* $\{V_{N-1}\}$; *we shall denote the components of the state vector* $\mathbf{x}_{N-1}$ *by* $x_{i,N-1}$ *for* $i = 1, 2, \ldots, N$. *We denote by* $t_{N-1}$ *the (positive) time required to force* $\mathbf{x}_{N-1}$ *to* $\mathbf{0}$ *using* $u = \Delta^*$. *From Eq.* (7-195), *we find that the* $x_{i,N-1}$ *are given by*

$$x_{i,N-1} = \Delta^* - \Delta^* e^{s_i t_{N-1}} \tag{7-201}$$

*for* $i = 1, 2, \ldots, N$.

***Definition 7-14*** *Let* $\{V_{N-2}\}$ *denote the set of states which can be forced to the set* $\{V_{N-1}\}$ *by application of the control* $u = -\Delta^*$. *We shall write* $\mathbf{x}_{N-2}$ *to indicate a state that belongs to the set* $\{V_{N-2}\}$; *we shall denote the components of the state vector* $\mathbf{x}_{N-2}$ *by* $x_{i,N-2}$ *for* $i = 1, 2, \ldots, N$. *We denote by* $t_{N-2}$ *the (positive) time required to force* $\mathbf{x}_{N-2}$ *to a state* $\mathbf{x}_{N-1} \in \{V_{N-1}\}$. *The* $x_{i,N-2}$ *are given by*

$$x_{i,N-2} = -\Delta^* + (x_{i,N-1} + \Delta^*) e^{s_i t_{N-2}} \tag{7-202}$$

*for* $i = 1, 2, \ldots, N$, *where* $x_{i,N-1}$ *satisfies Eq.* (7-201).

***Definition 7-15*** *Let* $\{V_{N-3}\}$ *denote the set of states which can be forced to the set* $\{V_{N-2}\}$ *by application of the control* $u = \Delta^*$. *We shall write* $\mathbf{x}_{N-3}$ *to indicate a state that belongs to the set* $\{V_{N-3}\}$; *we shall denote the components of the state vector* $\mathbf{x}_{N-3}$ *by* $x_{i,N-3}$ *for* $i = 1, 2, \ldots, N$. *We denote by* $t_{N-3}$ *the (positive) time required to force* $\mathbf{x}_{N-3}$ *to a state* $\mathbf{x}_{N-2} \in \{V_{N-2}\}$. *The* $x_{i,N-3}$ *are given by*

$$x_{i,N-3} = \Delta^* + (x_{i,N-2} - \Delta^*) e^{s_i t_{N-3}} \tag{7-203}$$

*for* $i = 1, 2, \ldots, N$, *where* $x_{i,N-2}$ *satisfies Eq.* (7-202).

We can proceed in a similar way to define as many of these sets as necessary; thus, we have:

***Definition 7-16*** *Let* $\{V_{N-R}\}$ *denote the set of states which can be forced to the set* $\{V_{N-R+1}\}$ *by application of the control* $u = (-1)^{R+1}\Delta^*$. *We shall write* $\mathbf{x}_{N-R}$ *to indicate a state that belongs to the set* $\{V_{N-R}\}$; *we shall denote the components of the state vector* $\mathbf{x}_{N-R}$ *by* $x_{i,N-R}$ *for* $i = 1, 2, \ldots, N$. *We denote by* $t_{N-R}$ *the (positive) time required to force* $\mathbf{x}_{N-R}$ *to a state* $\mathbf{x}_{N-R+1} \in$

$\{V_{N-R+1}\}$. *The* $x_{i,N-R}$ *are given by*

$$x_{i,N-R} = (-1)^{R+1}\Delta^* + [x_{i,N-R+1} - (-1)^{R+1}\Delta^*]e^{s_i t_{N-R}} \tag{7-204}$$

*for* $i = 1, 2, \ldots, N$.

In particular, for $R = N - 1$, we have:

***Definition 7-17*** *Let* $\{V_1\}$ *denote the set of states which can be forced to the set* $\{V_2\}$ *by application of the control* $u = (-1)^N\Delta^*$. *We shall write* $\mathbf{x}_1$ *to indicate a state that belongs to the set* $\{V_1\}$*; we shall denote the components of the state vector* $\mathbf{x}_1$ *by* $x_{i,1}$ *for* $i = 1, 2, \ldots, N$. *We denote by* $t_1$ *the (positive) time required to force* $\mathbf{x}_1$ *to a state* $\mathbf{x}_2 \in \{V_2\}$. *The* $x_{i,1}$ *are given by*

$$x_{i,1} = (-1)^N\Delta^* + [x_{i,2} - (-1)^N\Delta^*]e^{s_i t_1} \tag{7-205}$$

*for* $i = 1, 2, \ldots, N$.

We have thus defined a total of $N - 1$ sets $\{V_1\}, \{V_2\}, \ldots, \{V_{N-2}\}, \{V_{N-1}\}$. It is apparent from the definitions that the equation of $\{V_1\}$ depends upon the equation of $\{V_2\}$, that the equation of $\{V_2\}$ depends upon the equation of $\{V_3\}$, and so on. We wish to emphasize that the definitions imply that *the trajectory originating at any point* $\mathbf{x}_{N-R} \in \{V_{N-R}\}$ *and generated by the control* $u = (-1)^{R+1}\Delta^*$ *will remain on the set* $\{V_{N-R}\}$ *until it hits a point on the set* $\{V_{N-R+1}\}$*;* thus, the sets $\{V_j\}$, $j = 1, 2, \ldots, N - 1$, are made up of families of trajectories.

Now we shall eliminate the variables $x_{i,N-1}, x_{i,N-2}, \ldots, x_{i,2}$ from Eqs. (7-201) to (7-205) and obtain a relationship between the variables $x_{i,1}$ and the times $t_1, t_2, \ldots, t_{N-1}$. To simplify the equations, we define the auxiliary variables $z_1, \ldots, z_{N-1}$ by setting

$$z_j = e^{t_j} \qquad j = 1, 2, \ldots, N-1 \tag{7-206}$$

Using the $z_j$ variables, we see that Eqs. (7-201) to (7-205) become, in a slightly rearranged form,

$$\begin{aligned}
\Delta^* x_{i,N-1} &= 1 - z_{N-1}{}^{s_i} \\
\Delta^* x_{i,N-2} &= -1 + (\Delta^* x_{i,N-1} + 1)z_{N-2}{}^{s_i} \\
\Delta^* x_{i,N-3} &= 1 + (\Delta^* x_{i,N-2} - 1)z_{N-3}{}^{s_i} \\
&\cdots\cdots\cdots\cdots\cdots\cdots\cdots\cdots \\
\Delta^* x_{i,N-R} &= (-1)^{R+1} + [\Delta^* x_{i,N-R+1} - (-1)^{R+1}]z_{N-R}{}^{s_i} \\
&\cdots\cdots\cdots\cdots\cdots\cdots\cdots\cdots \\
\Delta^* x_{i,1} &= (-1)^N + [\Delta^* x_{i,2} - (-1)^N]z_1{}^{s_i}
\end{aligned} \tag{7-207}$$

for $i = 1, 2, \ldots, N$. By successive elimination, we obtain the relations:

*For N odd,*

$$\begin{aligned}
\Delta^* x_{i,1} = -1 + 2z_1{}^{s_i} - 2(z_1z_2)^{s_i} + 2(z_1z_2z_3)^{s_i} - \cdots - 2(z_1z_2 \cdots z_{N-3})^{s_i} \\
+ 2(z_1z_2 \cdots z_{N-2})^{s_i} - (z_1z_2 \cdots z_{N-1})^{s_i}
\end{aligned} \tag{7-208}$$

*For N even,*

$$\Delta^* x_{i,1} = +1 - 2z_1^{s_i} + 2(z_1z_2)^{s_i} - 2(z_1z_2z_3)^{s_i} + \cdots - 2(z_1z_2 \cdots z_{N-3})^{s_i} + 2(z_1z_2 \cdots z_{N-2})^{s_i} - (z_1z_2 \cdots z_{N-1})^{s_i} \quad (7\text{-}209)$$

where $i = 1, 2, \ldots, N$.

Define the variables $w_j$, $j = 1, 2, \ldots, N - 1$, by the equations

$$\begin{aligned} w_1 &= z_1 \\ w_2 &= z_1z_2 \\ w_3 &= z_1z_2z_3 \\ &\cdots\cdots\cdots\cdots \\ w_{N-1} &= z_1z_2z_3 \cdots z_{N-2}z_{N-1} \end{aligned} \quad (7\text{-}210)$$

Thus we obtain for $N$ *odd*, $i = 1, 2, \ldots, N$,

$$\Delta^* x_{i,1} + 1 = 2w_1^{s_i} - 2w_2^{s_i} + \cdots + 2w_{N-2}^{s_i} - w_{N-1}^{s_i} \quad (7\text{-}211)$$

For $N$ *even*, $i = 1, 2, \ldots, N$,

$$\Delta^* x_{i,1} - 1 = -2w_1^{s_i} + 2w_2^{s_i} - \cdots + 2w_{N-2}^{s_i} - w_{N-1}^{s_i} \quad (7\text{-}212)$$

We have specified that the times $t_1, t_2, \ldots, t_{N-1}$ are all positive; this implies that

$$1 < z_j \qquad j = 1, 2, \ldots, N - 1 \quad (7\text{-}213)$$

where the $z_j$ are given by Eq. (7-206). Equation (7-213) and Eqs. (7-210) yield the inequalities

$$1 < w_1 < w_2 < \cdots < w_{N-2} < w_{N-1} \quad (7\text{-}214)$$

*To recapitulate: If a state vector* $\mathbf{x}_1$, *with components* $x_{1,1}, x_{1,2}, \ldots, x_{1,N}$, *belongs to the set* $\{V_1\}$, *then there exist variables* $w_1, w_2, \ldots, w_{N-1}$ *which satisfy Eq.* (7-214) *and a* $\Delta^* = +1$ *or* $\Delta^* = -1$ *such that either Eq.* (7-211) *is satisfied for N odd or Eq.* (7-212) *is satisfied for N even. Conversely, given specific values of* $w_1, w_2, \ldots, w_{N-1}$ *satisfying Eq.* (7-214) *and a value for* $\Delta^*$, *then Eq.* (7-211) *or* (7-212) *defines a state vector* $\mathbf{x}_1$ *which belongs to the set* $\{V_1\}$.

We can use techniques similar to the ones above to determine the equations of the components $x_{i,N-R}$ of a state vector $\mathbf{x}_{N-R}$ which belongs to the set $V_{N-R}$. However, we can obtain these equations by manipulating the equation of the set $\{V_1\}$. If a state $\mathbf{x}_{N-R}$ belongs to the set $\{V_{N-R}\}$, then the motion of this state toward the origin is described by the set of times $t_{N-R}, t_{N-R+1}, \ldots, t_{N-2}, t_{N-1}$. Now, suppose that we set in Eqs. (7-210)

$$w_1 = w_2 = \cdots = w_{N-R-1} = 1 \quad (7\text{-}215)$$

Then we deduce from Eqs. (7-215) and (7-210) that

$$z_1 = z_2 = \cdots = z_{N-R-1} = 1 \quad (7\text{-}216)$$

and from Eqs. (7-216) and (7-206) that

$$t_1 = t_2 = \cdots = t_{N-R-1} = 0 \tag{7-217}$$

It follows that if a state $\mathbf{x}_{N-R}$ belongs to the set $\{V_{N-R}\}$, for $R = 1, 2, \ldots, N-1$, then its components $x_{i,N-R}$, $i = 1, 2, \ldots, N$, must satisfy the equations:

For $N$ *odd* and $i = 1, 2, \ldots, N$,

$$\Delta^* x_{i,N-R} + 1 = 2w_1^{s_i} - 2w_2^{s_i} + \cdots + 2w_{N-2}^{s_i} - w_{N-1}^{s_i} \tag{7-218}$$

For $N$ *even* and $i = 1, 2, \ldots, N$,

$$\Delta^* x_{i,N-R} - 1 = -2w_1^{s_i} + 2w_2^{s_i} - \cdots + 2w_{N-2}^{s_i} - w_{N-1}^{s_i} \tag{7-219}$$

where the $w$ variables satisfy the relationships

$$1 = w_1 = w_2 = \cdots = w_{N-R-1} < w_{N-R} < w_{N-R+1} < \cdots < w_{N-1} \tag{7-220}$$

Actually, we can simplify Eqs. (7-218) to (7-220) to:

For $R$ *even and* $i = 1, 2, \ldots, N$,

$$\Delta^* x_{i,N-R} + 1 = 2w_{N-R}^{s_i} - 2w_{N-R+1}^{s_i} + \cdots + 2w_{N-2}^{s_i} - w_{N-1}^{s_i} \tag{7-221}$$

For $R$ *odd* and $i = 1, 2, \ldots, N$,

$$\Delta^* x_{i,N-R} - 1 = -2w_{N-R}^{s_i} + 2w_{N-R+1}^{s_i} - \cdots + 2w_{N-2}^{s_i} - w_{N-1}^{s_i} \tag{7-222}$$

where the $w_{N-R}, w_{N-R+1}, \ldots, w_{N-1}$ variables satisfy the inequalities

$$1 < w_{N-R} < w_{N-R+1} < \cdots < w_{N-1} \tag{7-223}$$

with $R = 1, 2, \ldots, N-1$.

**Example 7-1** Suppose that the plant is described by the transfer function

$$G(s) = \frac{1}{(s+1)(s+2)(s+3)(s+4)} \tag{7-224}$$

It follows that $N = 4$ (an even number) and $s_1 = 1$, $s_2 = 2$, $s_3 = 3$, $s_4 = 4$ [compare Eq. (7-224) with Eq. (7-183)]. Therefore, if $\mathbf{x}_1$ is a state belonging to the set $\{V_1\}$, the equations satisfied by the components $x_{1,1}$, $x_{2,1}$, $x_{3,1}$, and $x_{4,1}$ of $\mathbf{x}_1$ are [using Eq. (7-212)]

$$\begin{aligned} \Delta^* x_{1,1} - 1 &= -2w_1 + 2w_2 - w_3 \\ \Delta^* x_{2,1} - 1 &= -2w_1^2 + 2w_2^2 - w_3^2 \\ \Delta^* x_{3,1} - 1 &= -2w_1^3 + 2w_2^3 - w_3^3 \\ \Delta^* x_{4,1} - 1 &= -2w_1^4 + 2w_2^4 - w_3^4 \end{aligned} \tag{7-225}$$

where

$$1 < w_1 < w_2 < w_3 \tag{7-226}$$

If $\mathbf{x}_2$ is a state that belongs to the set $\{V_2\}$, then the equations satisfied by the components $x_{1,2}$, $x_{2,2}$, $x_{3,2}$, and $x_{4,2}$ of $\mathbf{x}_2$ are [using $R = 2$, $N = 4$, and hence Eq. (7-221)]

$$\begin{aligned} \Delta^* x_{1,2} + 1 &= 2w_2 - w_3 \\ \Delta^* x_{2,2} + 1 &= 2w_2^2 - w_3^2 \\ \Delta^* x_{3,2} + 1 &= 2w_2^3 - w_3^3 \\ \Delta^* x_{4,2} + 1 &= 2w_2^4 - w_3^4 \end{aligned} \tag{7-227}$$

where

$$1 < w_2 < w_3 \tag{7-228}$$

If $\mathbf{x}_3$ is a state that belongs to the set $\{V_3\}$ then the equations satisfied by the components $x_{1,3}$, $x_{2,3}$, $x_{3,3}$, and $x_{4,3}$ of $\mathbf{x}_3$ are [using $R = 1$, $N = 4$, and Eq. (7-222)]

$$\begin{aligned} \Delta^* x_{1,3} - 1 &= -w_3 \\ \Delta^* x_{2,3} - 1 &= -w_3^2 \\ \Delta^* x_{3,3} - 1 &= -w_3^3 \\ \Delta^* x_{4,3} - 1 &= -w_3^4 \end{aligned} \tag{7-229}$$

where

$$1 < w_3 \tag{7-230}$$

Using the definitions and the equations of the sets $\{V_1\}$, $\{V_2\}$, . . . , $\{V_{N-1}\}$, we may obtain certain important properties of these sets which provide some geometrical insight into the problem.

*Property 1*

The set $\{V_j\}$, $j = 1, 2, \ldots, N-1$, is symmetric about the origin. To see this, we consider the equation of the set $\{V_{N-R}\}$ as given by Eq. (7-221). Setting $\Delta^* = +1$ and considering the variables $(x_{i,N-R})^+$ which satisfy the equations

$$(x_{i,N-R})^+ + 1 = 2w_{N-R}^{s_i} - 2w_{N-R+1}^{s_i} + \cdots + 2w_{N-2}^{s_i} - w_{N-1}^{s_i} \tag{7-231}$$

and setting $\Delta^* = -1$ and considering the variables $(x_{i,N-R})^-$ which satisfy the equations

$$-(x_{i,N-R})^- + 1 = 2w_{N-R}^{s_i} - 2w_{N-R+1}^{s_i} + \cdots + 2w_{N-2}^{s_i} - w_{N-1}^{s_i} \tag{7-232}$$

we see immediately that Eqs. (7-231) and (7-232) imply that

$$(x_{i,N-R})^+ = -(x_{i,N-R})^- \tag{7-233}$$

which means that $\{V_{N-R}\}$ is symmetric about the origin. Physically, this means that on the set $\{V_{N-R}\}$, states which are symmetric about the origin require the same time to reach the origin.†

*Property 2*

In the $N$-dimensional state space, the set $\{V_{N-1}\}$ is a *curve*, or 1-fold;‡ the set $\{V_{N-2}\}$ is a *surface*, or 2-fold; the set $\{V_{N-R}\}$ is an $R$-fold; the set $\{V_1\}$ is a *hypersurface*, or $(N-1)$-fold. Since the set $\{V_1\}$ is a hypersurface, it divides the state space into two parts.

† For example, compare symmetric states on the $\gamma_+$ and $\gamma_-$ curves for the second-order systems considered in Secs. 7-2 and 7-3.

‡ Suppose that in an $n$-dimensional Euclidean space we have $n$ variables $y_1, y_2, \ldots,$ $y_n$ and $m$ parameters $\alpha_1, \alpha_2, \ldots, \alpha_m$, where $m < n$, and suppose that the $y_i$ variables are given by

$$y_i = f_i(\alpha_1, \alpha_2, \ldots, \alpha_m) \qquad i = 1, 2, \ldots, n$$

where the functions $f_i$ are independent. Then we shall say that the points that satisfy the above equations form an $m$-fold in the space. In particular, if $m = 1$, the 1-fold is called a *curve;* if $m = 2$, the 2-fold is called a *surface;* and if $m = n - 1$, the $(n-1)$-fold is called a *hypersurface.* See also Sec. 3-13.

*Property 3*

The sets $\{V_j\}$, $j = 1, 2, \ldots, N-1$, are "smooth" and "continuous." This follows from the fact that these sets were formed by families of smooth and continuous trajectories.

*Property 4*

The sets $\{V_j\}$, $j = 1, 2, \ldots, N-1$, are "infinite in extent," which means that if we set $|x_{1,N-R}| = \infty$ in Eq. (7-221), then, in view of Eq. (7-223), $w_{N-R} = \infty$; it follows that $|x_{k,N-R}| = \infty$ for $k = 2, 3, \ldots, N$.

*Property 5*

The origin **0** is contained in the set $\{V_{N-1}\}$; the set $\{V_{N-1}\}$ is contained in $\{V_{N-2}\}$; and so on. In other words,

$$\mathbf{0} \subset \{V_{N-1}\} \subset \{V_{N-2}\} \subset \cdots \subset \{V_2\} \subset \{V_1\} \tag{7-234}$$

This is a direct consequence of the definition of these sets. We shall make use of property 5 later on in this section.

*Property 6*

We have mentioned that the hypersurface $\{V_1\}$ divides the state space into two parts. It is convenient to establish a notion of "above" and "below"† with respect to the hypersurface $\{V_1\}$. To do this, let us assume that $N$ is odd, so that a state $\mathbf{x}_1 \in \{V_1\}$ satisfies Eq. (7-211), which we write in an expanded form as

$$\Delta^* x_{1,1} + 1 = 2w_1^{s_1} - 2w_2^{s_1} + \cdots + 2w_{N-2}^{s_1} - w_{N-1}^{s_1} \tag{7-235$_1$}$$
$$\Delta^* x_{2,1} + 1 = 2w_1^{s_2} - 2w_2^{s_2} + \cdots + 2w_{N-2}^{s_2} - w_{N-1}^{s_2} \tag{7-235$_2$}$$

. . . . . . . . . . . . . . . . . . . . . . . . . . . . . . . . . . . .

$$\Delta^* x_{N-1,1} + 1 = 2w_1^{s_{N-1}} - 2w_2^{s_{N-1}} + \cdots + 2w_{N-2}^{s_{N-1}} - w_{N-1}^{s_{N-1}} \tag{7-235$_{N-1}$}$$
$$\Delta^* x_{N,1} + 1 = 2w_1^{s_N} - 2w_2^{s_N} + \cdots + 2w_{N-2}^{s_N} - w_{N-1}^{s_N} \tag{7-235$_N$}$$

Suppose that we have a state $\mathbf{x}$,

$$\mathbf{x} = \begin{bmatrix} x_1 \\ x_2 \\ \cdot \\ \cdot \\ \cdot \\ x_{N-1} \\ x_N \end{bmatrix}$$

† As we have done already in Sec. 7-4 for the third-order system.

and we wish to find whether this state $\mathbf{x}$ is above, on, or below the hypersurface $\{V_1\}$. We set

$$x_{1,1} = x_1,\ x_{2,1} = x_2,\ \ldots,\ x_{N-1,1} = x_{N-1} \tag{7-236}$$

in Eqs. (7-235$_1$), (7-235$_2$), . . . , (7-235$_{N-1}$) and solve these $N - 1$ equations to determine the $N - 1$ variables $w_1, w_2, \ldots, w_{N-1}$ and $\Delta^* = +1$ or $\Delta^* = -1$, keeping in mind that the inequalities

$$1 < w_1 < w_2 < \cdots < w_{N-1} \tag{7-237}$$

must be satisfied. We substitute the values of $w_1, \ldots, w_{N-1}$ and $\Delta^*$ in Eq. (7-235$_N$), and we evaluate $x_{N,1}$. We can compare the computed value of $x_{N,1}$ with the last component $x_N$ of the state $\mathbf{x}$. *If*

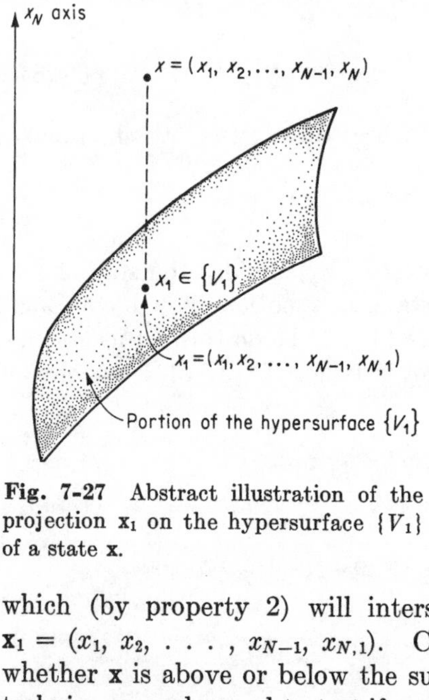


**Fig. 7-27** Abstract illustration of the projection $\mathbf{x}_1$ on the hypersurface $\{V_1\}$ of a state $\mathbf{x}$.

$$x_N - x_{N,1} > 0 \tag{7-238}$$

*then we say that* $\mathbf{x}$ *is above the hypersurface* $\{V_1\}$. *If*

$$x_N - x_{N,1} = 0 \tag{7-239}$$

*then* $\mathbf{x} \in \{V_1\}$, *and if*

$$x_N - x_{N,1} < 0 \tag{7-240}$$

*then we say that* $\mathbf{x}$ *is below* $\{V_1\}$. These ideas are illustrated in Fig. 7-27; as we have done in Sec. 7-4 (Fig. 7-23), we draw a straight line parallel to the $x_N$ axis through the point $\mathbf{x} = (x_1, x_2, \ldots, x_{N-1}, x_N)$, which (by property 2) will intersect the hypersurface $\{V_1\}$ at a point $\mathbf{x}_1 = (x_1, x_2, \ldots, x_{N-1}, x_{N,1})$. Comparison of $x_N$ with $x_{N,1}$ will indicate whether $\mathbf{x}$ is above or below the surface. It should be noted that similar techniques can be used to test if a state $\mathbf{x}_1$ in $\{V_1\}$ also belongs to $\{V_2\}$, etc.

*Property 7*

If $N$ is *odd*, then the state $(1, 1, \ldots, 1, 1)$ is *above* the hypersurface $\{V_1\}$ and the state $(-1, -1, \ldots, -1, -1)$ is *below* the hypersurface $\{V_1\}$. On the other hand, if $N$ is *even*, then the state $(1, 1, \ldots, 1, 1)$ is *below* the hypersurface $\{V_1\}$ and the state $(-1, -1, \ldots, -1, -1)$ is *above* the hypersurface $\{V_1\}$. The complete proof of these statements is omitted. However, we shall illustrate the concepts involved for a third-order system by means of the following example:

**Example 7-2** Consider the third-order system with the transfer function

$$G(s) = \frac{K}{(s + s_1)(s + s_2)(s + s_3)} \qquad 0 < s_1 < s_2 < s_3 \tag{7-241}$$

The equation of the surface $\{V_1\}$† can be found from Eq. (7-211) by using $N = 3$ (an odd number). It is

$$\Delta^* x_{1,1} + 1 = 2w_1^{s_1} - w_2^{s_1} \tag{7-242}$$
$$\Delta^* x_{2,1} + 1 = 2w_1^{s_2} - w_2^{s_2} \tag{7-243}$$
$$\Delta^* x_{3,1} + 1 = 2w_1^{s_3} - w_2^{s_3} \tag{7-244}$$

We shall show that the state point (1, 1, 1) is above the surface $\{V_1\}$. To do this, we shall follow the procedure suggested by property 6. In other words, we shall consider the state point $\mathbf{x}_1 = (1, 1, x_{3,1})$ which belongs to the surface $\{V_1\}$, and we shall demonstrate that $1 - x_{3,1} > 0$, that is, that the point (1, 1, 1) is *above* $\{V_1\}$. The first step is to set $x_{1,1} = 1$ in Eq. (7-242) and $x_{2,1} = 1$ in Eq. (7-243) to obtain the equations

$$\Delta^* + 1 = 2w_1^{s_1} - w_2^{s_1} \tag{7-245}$$
$$\Delta^* + 1 = 2w_1^{s_2} - w_2^{s_2} \tag{7-246}$$

Recall that $\Delta^*$ must be chosen so that the inequalities

$$1 < w_1 < w_2 \tag{7-247}$$

are satisfied.

Suppose that we set $\Delta^* = -1$ in Eqs. (7-245) and (7-246); we then obtain the equations

$$\begin{aligned} 2w_1^{s_1} &= w_2^{s_1} \\ 2w_1^{s_2} &= w_2^{s_2} \end{aligned} \tag{7-248}$$

Clearly,

$$w_2 = 2^{1/s_1} w_1 = 2^{1/s_2} w_1 \tag{7-249}$$

and therefore,

$$2^{1/s_1} = 2^{1/s_2} \tag{7-250}$$

which is impossible, since $0 < s_1 < s_2$; this means that the choice of $\Delta^* = -1$ is not correct. Therefore, in Eqs. (7-245) and (7-246), we must take

$$\Delta^* = +1 \tag{7-251}$$

and, in so doing, obtain the equations

$$2 = 2w_1^{s_1} - w_2^{s_1} \tag{7-252}$$
$$2 = 2w_1^{s_2} - w_2^{s_2} \tag{7-253}$$

or

$$w_2 = [2(w_1^{s_1} - 1)]^{1/s_1} = [2(w_1^{s_2} - 1)]^{1/s_2} \tag{7-254}$$

We note that for arbitrary values of $s_1$ and $s_2$, we cannot solve explicitly for $w_1$ using Eq. (7-254).

The next step is to consider $x_{3,1}$ [defined by Eq. (7-244)], which, using $\Delta^* = +1$, reduces to the relation

$$x_{3,1} = -1 + 2w_1^{s_3} - w_2^{s_3} \tag{7-255}$$

We must show that

$$1 - x_{3,1} > 0 \tag{7-256}$$

or, equivalently, that

$$w_2 > [2(w_1^{s_3} - 1)]^{1/s_3} \tag{7-257}$$

† Since $N = 3$, the hypersurface $\{V_1\}$ is a surface in the three-dimensional state space.

The form of Eqs. (7-254) and (7-257) suggests the definition of a function

$$g(w_1, s) = [2(w_1^s - 1)]^{1/s} \tag{7-258}$$

where this function $g(w_1, s)$ can be viewed as a function of $s$, with $w_1 > 1$ as a parameter. By definition, we must have

$$g(w_1, s_1) = g(w_1, s_2) \tag{7-259}$$

and so we must show that

$$g(w_1, s_3) < g(w_1, s_2) \qquad \text{for } s_3 > s_2 \tag{7-260}$$

It is easy to establish that

$$\lim_{s \to 0} g(w_1, s) = 0 \tag{7-261}$$

$$\lim_{s \to \infty} g(w_1, s) = w_1 > 1 \tag{7-262}$$

and that

$$\frac{d}{ds} g(w_1, s) = \frac{[2(w_1^s - 1)]^{1/s}}{s^2(w_1^s - 1)} [s w_1^s \log w_1 - (w_1^s - 1) \log 2(w_1^s - 1)] \tag{7-263}$$

Since $w_1 > 1$ and $s > 0$, the term $[2(w_1^s - 1)]^{1/s}/s^2(w_1^s - 1)$ in Eq. (7-263) is always positive. We assert that there exists only one $s = \hat{s}$ such that

$$\hat{s} w_1^{\hat{s}} \log w_1 = (w_1^{\hat{s}} - 1) \log 2(w_1^{\hat{s}} - 1) \tag{7-264}$$

which means that

$$\left. \frac{d}{ds} g(w_1, s) \right|_{s = \hat{s}} = 0 \tag{7-265}$$

The above considerations are sufficient to enable us to obtain the shape of the function $g(w_1, s)$. As illustrated in Fig. 7-28, the function $g(w_1, s)$ as a function of $s$ starts at zero, attains a maximum at $s = \hat{s}$, and then drops off, approaching $w_1$ asymptotically as $s \to \infty$. It is clear from this illustration that for any $s_3 > s_2$, $g(w_1, s_3) < g(w_1, s_2)$.

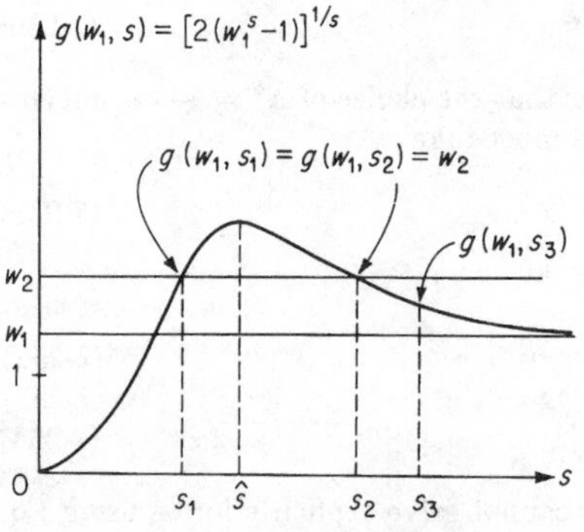


**Fig. 7-28** Plot of the function $g(w_1, s)$ versus $s$.

**Exercise 7-31** Consider the transfer function given by Eq. (7-241), and let $s_1 = 1$, $s_2 = 2$, $s_3 = 3$. Repeat the steps involved in Example 7-2 and show that $w_1 = 3$, $w_2 = 4$, $x_{3,1} = -11$. In this manner, you have verified that the point (1, 1, 1) is *above* the surface $\{V_1\}$, since the point (1, 1, −11) belongs to $\{V_1\}$. Plot the function

$$g(3, s) = [2(3^s - 1)]^{1/s}$$

for $0 < s < \infty$, and verify that its shape is similar to the one illustrated in Fig. 7-28.

We have enumerated a series of properties of the sets $\{V_1\}$, $\{V_2\}$, . . . , $\{V_{N-1}\}$. We shall use these properties to formulate the time-optimal control law for the $N$th-order plant whose transfer function is given by Eq. (7-183) and to sketch the elements of the proof of its optimality.

***Control Law 7-6 Solution to Problem 7-6*** *Given any state* $\mathbf{x}$*; then the time-optimal control* $u^*$ *which forces* $\mathbf{x}$ *to the origin* $\mathbf{0}$ *is defined in the following way:*

*If* $\mathbf{x}$ *is above* $\{V_1\}$, *then* $u^* = (-1)^N$
*If* $\mathbf{x}$ *is below* $\{V_1\}$, *then* $u^* = -(-1)^N$
*If* $\mathbf{x} \in \{V_1\}$, *then* $u^* = (-1)^N\Delta^*$
*If* $\mathbf{x} \in \{V_2\}$, *then* $u^* = -(-1)^N\Delta^*$
. . . . . . . . . . . . . . . . . . . . . . . .
*If* $\mathbf{x} \in \{V_{N-R}\}$, *then* $u^* = (-1)^{R+1}\Delta^*$
. . . . . . . . . . . . . . . . . . . . . . . . .
*If* $\mathbf{x} \in \{V_{N-2}\}$, *then* $u^* = -\Delta^*$
*If* $\mathbf{x} \in \{V_{N-1}\}$, *then* $u^* = \Delta^*$

ELEMENTS OF PROOF First of all, we note that the definitions of the sets $\{V_j\}$ imply that if the state point $\mathbf{x}$ is on the set $\{V_{N-R}\}$, $R = 1, 2, \ldots, N-1$, then the control which forces $\mathbf{x}$ to the origin switches *exactly* $R - 1$ times. In particular, if the state $\mathbf{x}$ belongs to the hypersurface $\{V_1\}$, then the control switches *exactly* $N - 2$ times. Furthermore, we recall that since the system has $N$ real eigenvalues, the time-optimal control can switch, *at most*, $N - 1$ times.

Now consider the state

$$\mathbf{1} = \begin{bmatrix} 1 \\ 1 \\ \cdot \\ \cdot \\ \cdot \\ 1 \end{bmatrix}$$

which (by property 7) is above $\{V_1\}$ for $N$ *odd*. If $N$ is *odd*, the control law states that $u^* = -1$. Let us try to prove this statement. Suppose that at $\mathbf{1}$, we apply the control $u = +1$. Since all trajectories generated by $u = +1$ tend to the point $\mathbf{1}$ [see Eq. (7-199)], we see that if we apply $u = +1$ at $\mathbf{1}$, then the state will remain at $\mathbf{1}$ forever. Therefore, we *must* apply $u = -1$ at $\mathbf{1}$, thus generating a trajectory which hits the hypersurface $\{V_1\}$. If the control switches from $u = -1$ to $u = +1$ at $\{V_1\}$, the total number of switchings is $N - 1$, which clearly does not violate the necessary conditions. Now consider other states $\mathbf{x}$ which are above $\{V_1\}$. If $u = +1$, then $\mathbf{x}$ tends to the state $\mathbf{1}$, and eventually we must switch to $u = -1$ to reach the set $\{V_1\}$. But this policy requires $N$ switchings, which violates the necessary conditions. Therefore, if $N$ is *odd* and the state is above $\{V_1\}$, the control must be $u = -1$. Similar reasoning establishes the fact that if a state $\mathbf{x}$ is above $\{V_1\}$ (and hence by property 5 does not belong to $\{V_2\}, \ldots, \{V_{N-1}\}$), then the control $u = (-1)^N$ forces $\mathbf{x}$ to $\{V_1\}$, and if $\mathbf{x}$ is below $\{V_1\}$, then $u = -(-1)^N$ forces $\mathbf{x}$ to $\{V_1\}$. In either case, the total number of switchings is *exactly* $N - 1$.

The next step is to show that if a state $\mathbf{x}_1$ is on $\{V_1\}$ but not in $\{V_2\}$ (and hence by property 5 not in $\{V_3\}, \ldots, \{V_{N-1}\}$), the control must be $u = (-1)^N\Delta^*$. Suppose for the sake of argument that $N$ is odd and, for the point $\mathbf{x}_1 \in \{V_1\}$, the value of $\Delta^*$ is $\Delta^* = -1$, so that according to the control law we must use $u = +1$ at $\mathbf{x}_1$. Application of $u = +1$ generates a trajectory which follows the hypersurface $\{V_1\}$ and hits $\{V_2\}$ at a point at which the control must switch to $u = -1$, and so on. This control sequence requires *exactly* $N - 2$ switchings. Suppose that at $\mathbf{x}_1$, we apply $u = -1$. The resulting trajectory *will not* follow the hypersurface $\{V_1\}$; as a matter of fact, it will go *below* $\{V_1\}$ (because it will tend to the state $-\mathbf{1}$, which is below $\{V_1\}$), and thus the control must switch from $u = -1$ to $u = +1$ so that the state is brought back to $\{V_1\}$. However, this control sequence requires *exactly* $N$ switchings and hence cannot be a time-optimal one.

We can repeat this procedure and establish that if $\mathbf{x}_{N-R} \in \{V_{N-R}\}$, then the control $u = (-1)^{R+1}\Delta^*$ forces $\mathbf{x}_{N-R}$ to the origin with exactly $R - 1$ switchings, while application of $u = -(-1)^{R+1}\Delta^*$ at $\mathbf{x}_{N-R}$ takes $\mathbf{x}_{N-R}$ above

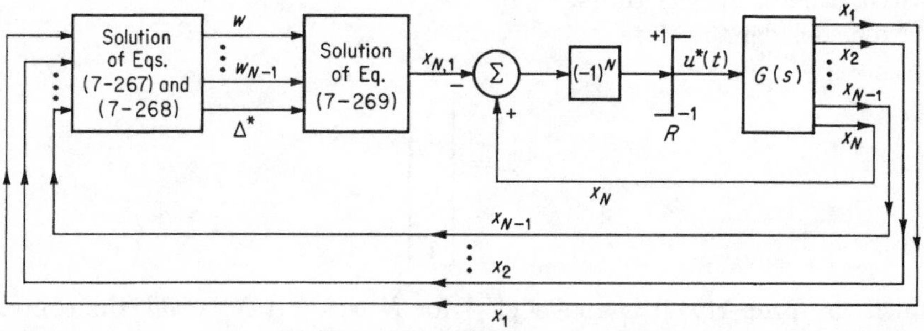


**Fig. 7-29** Block diagram illustrating the computations that must be performed to generate the time-optimal control for the $N$th-order system.

or below $\{V_1\}$; this control procedure cannot be time-optimal since the total number of switchings required is *at least* $N$. This completes the geometric proof.

A direct consequence of the above considerations is that *the control which requires the minimum number of switchings is the time-optimal one.* Of course, this is true for plants with real eigenvalues only.

Figure 7-29 illustrates the computations that must be performed to determine the time-optimal control. The state variables $x_1, x_2, \ldots, x_{N-1}$ and $x_N$ are measured, and the first $N - 1$ of them are used for the computation of $w_1, w_2, \ldots, w_{N-1}$ and $\Delta^*$ from the equations

$$\left.\begin{aligned}
\Delta^* x_1 + 1 &= 2w_1^{s_1} - 2w_2^{s_1} + \cdots + 2w_{N-2}^{s_1} - w_{N-1}^{s_1} \\
\Delta^* x_2 + 1 &= 2w_1^{s_2} - 2w_2^{s_2} + \cdots + 2w_{N-2}^{s_2} - w_{N-1}^{s_2} \\
&\cdots\cdots\cdots\cdots\cdots\cdots\cdots\cdots\cdots \\
\Delta^* x_{N-1} + 1 &= 2w_1^{s_{N-1}} - 2w_2^{s_{N-1}} + \cdots + 2w_{N-2}^{s_{N-1}} - w_{N-1}^{s_{N-1}}
\end{aligned}\right\} \quad \frac{N}{\text{odd}} \tag{7-266a}$$

or

$$\left.\begin{aligned}\Delta^* x_1 - 1 &= -2w_1^{s_1} + 2w_2^{s_1} - \cdots + 2w_{N-2}^{s_1} - w_{N-1}^{s_1} \\ \Delta^* x_2 - 1 &= -2w_1^{s_2} + 2w_2^{s_2} - \cdots + 2w_{N-2}^{s_2} - w_{N-1}^{s_2} \\ &\cdots\cdots\cdots\cdots\cdots\cdots\cdots\cdots \\ \Delta^* x_{N-1} - 1 &= -2w_1^{s_{N-1}} + 2w_2^{s_{N-1}} - \cdots + 2w_{N-2}^{s_{N-1}} - w_{N-1}^{s_{N-1}}\end{aligned}\right\} \frac{N}{\text{even}} \qquad \text{(7-266b)}$$

where $$1 < w_1 < w_2 < \cdots < w_{N-2} < w_{N-1} \qquad \text{(7-267)}$$

The computed values of $w_1, \ldots, w_{N-1}$ and $\Delta^*$ are substituted in the equation

$$\Delta^* x_{N,1} + 1 = 2w_1^{s_N} - 2w_2^{s_N} + \cdots + 2w_{N-2}^{s_N} - w_{N-1}^{s_N} \qquad N \text{ odd} \qquad \text{(7-268a)}$$

or

$$\Delta^* x_{N,1} - 1 = -2w_1^{s_N} + 2w_2^{s_N} - \cdots + 2w_{N-2}^{s_N} - w_{N-1}^{s_N} \qquad N \text{ even} \qquad \text{(7-268b)}$$

and the value of $x_{N,1}$ is obtained. The computed signal $x_{N,1}$ is compared with the measured state variable $x_N$ to form the difference $x_N - x_{N,1}$. If $x_N - x_{N,1}$ is positive, then the measured state is above the hypersurface $\{V_1\}$ (see property 6), and if $x_N - x_{N,1}$ is negative, then the measured state is below $\{V_1\}$. Multiplying the difference $x_N - x_{N,1}$ by $(-1)^N$ delivers a signal with the correct polarity to the relay $R$ whose output $u^*(t)$ is the time-optimal control.

It should be clear that the system illustrated in Fig. 7-29 decides whether the measured state is above or below the hypersurface $\{V_1\}$† and delivers a control which will bring the state to the hypersurface $\{V_1\}$ in minimum time. The usual computational errors and time delays will almost always cause the control to switch either slightly above or below $\{V_1\}$, so that the remaining trajectories will be close to but not exactly on the sets $\{V_1\}$, $\{V_2\}$, $\ldots$, $\{V_{N-1}\}$; thus, the state after $N-1$ switchings will be close to, but not exactly at, the origin. The degree of accuracy will depend upon the speed and accuracy of the computational techniques employed. This is the reason that the engineering realization of Fig. 7-29 does not include logical operations for the sets $\{V_2\}$, $\{V_3\}$, $\ldots$, $\{V_{N-1}\}$.

## 7-6 Some Comments

In the preceding sections, we have examined the time-optimal control problem for two second-order systems, one third-order system, and one $N$th-order system. These systems had the following properties in common:

† The hypersurface $\{V_1\}$ is often called the *switch hypersurface*.

1. The systems were linear, normal, and time-invariant.
2. The transfer function of the system did not contain any zeros.
3. The poles of the transfer function were real and nonpositive (but not necessarily distinct).
4. Control was effected by a single control variable $u(t)$, which was bounded in magnitude [that is, $|u(t)| \leq 1$].
5. The desired terminal state was the origin, which was an equilibrium point of the system, so that upon reaching the origin, it was sufficient merely to shut off the control in order to maintain the system at rest (i.e., at the origin).

The method which we used to obtain the time-optimal control law was almost the same for each of these problems. The following were the essential stages in our synthesis of the optimal control:

1. We first reduced the system differential equation to a set of first-order equations using as our state variables the output signal $y(t)$ and as many of its time derivatives as necessary. For example, the output and its first $N - 1$ time derivatives were used in the case of the $N$th-order system.
2. We then chose a convenient set of state variables by means of a series of nonsingular linear transformations which reduced the system matrix to its Jordan canonical form. In the case of the double-integral system of Sec. 7-2, this reduction was unnecessary since the system matrix was already in Jordan canonical form.
3. We then applied the minimum principle to the system of convenient state variables. We examined the Hamiltonian, determined the equations satisfied by the costate variables, and found the control which absolutely minimized the Hamiltonian. We observed that (in view of the normality) the time-optimal control had to be piecewise constant and could switch at most $N - 1$ times for an $N$th-order system. We summarized the information contained in this observation in terms of control sequences which were candidates for time-optimal control.
4. We next used a process of elimination to determine the time-optimal control. We examined each state in the state space and tried each control sequence which was a candidate for time-optimal control. We found a *unique* control sequence from among the candidates which would force a given state to the origin and thus developed the control law. A by-product of this technique was the notion of *switch curve* for second-order systems, of *switch surface* for the third-order system, and of *switch hypersurface* for $N$th-order systems. In other words, we found a set of states which divided the state space into two distinct regions, in one of which the time-optimal control was $+1$ and in the other of which the time-optimal control was $-1$.
5. Finally, an engineering realization of the time-optimal control law was designed, based upon the simulation of the switch curve (or surface or hypersurface) and the subsequent testing of whether or not the measured

state was above or below the switch curve (or surface or hypersurface). A relay was used to provide the optimal control signal acting as the input to the system.

During the course of determining the time-optimal control for the $N$th-order system, we defined sets $\{V_1\}, \{V_2\}, \ldots, \{V_{N-1}\}$, which included the switch hypersurface and which were generated by trajectories of the system corresponding to application of the constant control $u = \Delta = \pm 1$, as required by the minimum principle.

In point of fact, the sets $\{V_i\}$ are made up of system trajectories if the poles of the system are noncomplex and if the origin is the terminal state. There are many problems (see, for example, Exercise 7-19) such that the switch curve is not a trajectory when the target set is not the origin. Similarly, as we shall see in the next section, the switch curve for a system with complex poles is not entirely made up of trajectories even though the origin is the terminal state.

The following three exercises illustrate the time-optimal control for some systems with noncomplex poles.

**Exercise 7-32** Consider a second-order plant described by the differential equation

$$\frac{d^2y(t)}{dt^2} + a\frac{dy(t)}{dt} = ku(t) \tag{7-269}$$

where

$$a > 0 \qquad k > 0 \qquad |u(t)| \leq 1 \tag{7-270}$$

Such a plant has the transfer function

$$\frac{y(s)}{u(s)} = G(s) = \frac{k}{s(s+a)} \tag{7-271}$$

and is commonly referred to as an "integral-plus-time-constant plant." Using the state variables $x_1(t)$ and $x_2(t)$ which satisfy the vector differential equation

$$\begin{bmatrix} \dot{x}_1(t) \\ \dot{x}_2(t) \end{bmatrix} = \begin{bmatrix} 0 & 0 \\ 0 & -a \end{bmatrix} \begin{bmatrix} x_1(t) \\ x_2(t) \end{bmatrix} + \begin{bmatrix} u \\ -u \end{bmatrix} \tag{7-272}$$

determine the relation between the state variables $x_1(t)$ and $x_2(t)$ and the output state variables $y(t)$ and $\dot{y}(t)$. Show that in order to drive any initial state of the system (7-272) to the origin (0, 0) the time-optimal control (as a function of $x_1$ and $x_2$) is given by the equation

$$u^* = \operatorname{sgn}\left\{\frac{x_1}{|x_1|}(1 - e^{-ax_1}) - x_2\right\} \tag{7-273}$$

**Exercise 7-33** Determine the time-optimal control to the origin for the third-order plant described by the differential equation

$$\frac{d^3y(t)}{dt^3} + (s_1 + s_2)\frac{d^2y(t)}{dt^2} + s_1s_2\frac{dy(t)}{dt} = ku(t) \tag{7-274}$$

where

$$0 < s_1 < s_2 \qquad k > 0 \qquad |u(t)| \leq 1 \tag{7-275}$$

Such a plant has the transfer function

$$\frac{y(s)}{u(s)} = G(s) = \frac{k}{s(s+s_1)(s+s_2)} \tag{7-276}$$

HINT: For this problem, you should be able to get an analytic expression for the switch surface $\{V_1\}$.

**Exercise 7-34** Find the time-optimal control to the origin for the plant

$$\frac{d^4y(t)}{dt^4} = u(t) \qquad |u(t)| \leq 1$$

## 7-7 Time-optimal Control of the Harmonic Oscillator†

Up to this point, we have examined the time-optimal control problem for systems with the property that the eigenvalues of the system matrix were real. In this section, we shall consider a second-order system for which the eigenvalues of the system matrix are imaginary.

Suppose that a system is described by the differential equation

$$\frac{d^2y(t)}{dt^2} + \omega^2 y(t) = Ku(t) \tag{7-277}$$

where $K > 0$ and the control $u(t)$ is assumed to satisfy the relation

$$|u(t)| \leq 1 \tag{7-278}$$

In servomechanism practice, such a system is described by the transfer function

$$\frac{y(s)}{u(s)} = G(s) = \frac{K}{s^2 + \omega^2} \tag{7-279}$$

We define the state variables by setting

$$y_1(t) = y(t) \qquad y_2(t) = \dot{y}(t) \tag{7-280}$$

The vector $\mathbf{y}(t)$ satisfies the matrix differential equation

$$\begin{bmatrix} \dot{y}_1(t) \\ \dot{y}_2(t) \end{bmatrix} = \begin{bmatrix} 0 & 1 \\ -\omega^2 & 0 \end{bmatrix} \begin{bmatrix} y_1(t) \\ y_2(t) \end{bmatrix} + \begin{bmatrix} 0 \\ Ku(t) \end{bmatrix} \tag{7-281}$$

Let us define a more convenient set of state variables $x_1(t)$ and $x_2(t)$ by setting

$$\begin{aligned} x_1(t) &= \frac{\omega}{K} y_1(t) \\ x_2(t) &= \frac{1}{K} y_2(t) \end{aligned} \tag{7-282}$$

Then $x_1$ and $x_2$ (which we shall use in the remainder of this section) satisfy the vector differential equation

$$\begin{bmatrix} \dot{x}_1(t) \\ \dot{x}_2(t) \end{bmatrix} = \begin{bmatrix} 0 & \omega \\ -\omega & 0 \end{bmatrix} \begin{bmatrix} x_1(t) \\ x_2(t) \end{bmatrix} + \begin{bmatrix} 0 \\ u(t) \end{bmatrix} \tag{7-283}$$

† Bushaw [B-23] first obtained the solution to this problem. Other pertinent references are A-5, A-9, B-24, C-2, and K-18.

If we examine the system matrix **A**,

$$\mathbf{A} = \begin{bmatrix} 0 & \omega \\ -\omega & 0 \end{bmatrix} \tag{7-284}$$

then we can easily verify that the eigenvalues of **A** are

$$\lambda_1 = j\omega \qquad \lambda_2 = -j\omega \qquad \text{(7-285)†}$$

This is the first time in this chapter that we have encountered a matrix with complex eigenvalues. As we shall see, this fact significantly changes the form of the switch curve and the method of generating the switch curve.

There are many physical systems for which the state differential equations can be reduced to the form of Eq. (7-283); two of these are considered below.

**Example 7-3** Consider a mass $m$ connected to a linear spring with spring constant $k$, as shown in Fig. 7-30. Let $y(t)$ denote the linear displacement from the equilibrium position (say zero) and let $f(t)$, $|f(t)| \leq F$, be the bounded applied force. Then the differential equation for $y(t)$ is

$$m\frac{d^2y(t)}{dt^2} + ky(t) = f(t) \tag{7-286}$$

Note that Eq. (7-286) can be reduced to Eq. (7-277) [and hence to Eq. (7-283)] by the choice of

$$\omega^2 = \frac{k}{m} \qquad K = \frac{F}{m} \qquad u(t) = \frac{f(t)}{F} \tag{7-287}$$

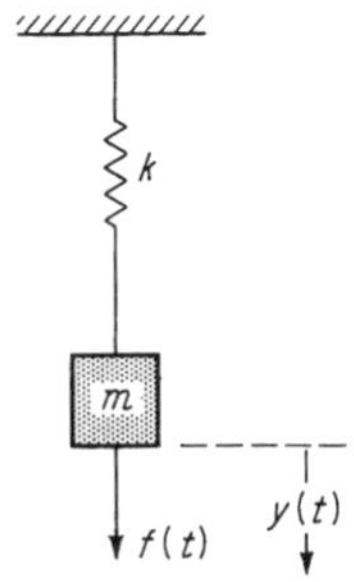

**Fig. 7-30** A mass-spring system.

In the same general category as the mass-spring system are pendulums (for small displacements) and torsion pendulums.

**Example 7-4** In this example, we present the equations of motion for a spinning body in space with a single axis of symmetry. We shall encounter the same system in subsequent sections of the book (cf. Sec. 10-7). Figure 7-31 illustrates such a body with an axis of symmetry. We define the axes 1, 2, 3, fixed on the body, which pass through the center of mass. It is evident from Fig. 7-31 that the 3-axis is the axis of symmetry.

Let $I_3$ denote the moment of inertia about the 3-axis, and let $I$ denote the moment of inertia about the 1 and 2 axes. Let $y_1(t)$, $y_2(t)$, $y_3(t)$ denote the angular velocities about the axes 1, 2, 3, respectively. Suppose that we fix on the body a gas jet delivering a bounded thrust $f(t)$, $|f(t)| \leq F$, so that the distance from the gas jet to the center of mass is $c$ and so that the torque $cf(t)$ controls the angular acceleration $\dot{y}_2(t)$. Then‡ the angular velocities $y_1(t)$, $y_2(t)$, and $y_3(t)$ satisfy the system of differential equations

$$\begin{aligned} I\dot{y}_1(t) &= (I - I_3)y_2(t)y_3(t) \\ I\dot{y}_2(t) &= (I_3 - I)y_1(t)y_3(t) + cf(t) \\ I_3\dot{y}_3(t) &= 0 \end{aligned} \tag{7-288}$$

† Following engineering terminology, we shall use $j$ to represent $\sqrt{-1}$.
‡ See Refs. G-7, A-5, A-9, and F-13.

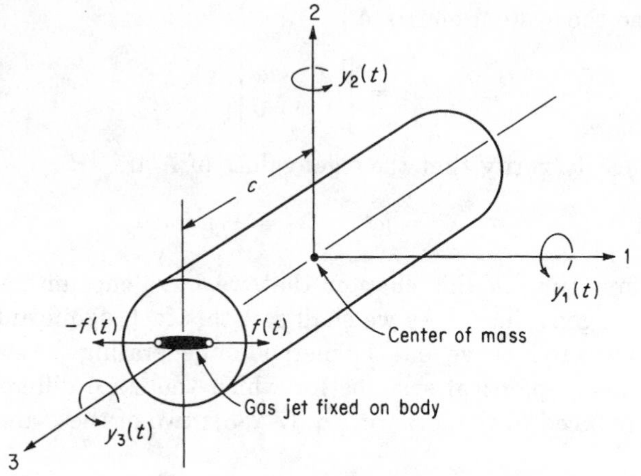


**Fig. 7-31** A symmetric body undergoing simultaneous rotation about the three body axes. The angular velocities are $y_1(t)$, $y_2(t)$, and $y_3(t)$.

From the last equation, we see that

$$y_3(t) = \text{const} \triangleq \eta_3 \tag{7-289}$$

Therefore, if we define

$$\omega = \frac{(I - I_3)\eta_3}{I} \tag{7-290}$$

then we find that the differential equations of the angular velocities $y_1(t)$ and $y_2(t)$ are

$$\begin{aligned} \dot{y}_1(t) &= \omega y_2(t) \\ \dot{y}_2(t) &= -\omega y_1(t) + \frac{c}{I} f(t) \end{aligned} \tag{7-291}$$

If we define

$$K = \frac{c}{I} F \qquad \text{and} \qquad u(t) = \frac{f(t)}{F} \qquad \text{so that } |u(t)| \leq 1 \tag{7-292}$$

then the new variables $x_1(t)$ and $x_2(t)$, defined by setting

$$x_1(t) = \frac{1}{K} y_1(t) \qquad x_2(t) = \frac{1}{K} y_2(t) \tag{7-293}$$

satisfy the differential equations

$$\begin{aligned} \dot{x}_1(t) &= \omega x_2(t) \\ \dot{x}_2(t) &= -\omega x_1(t) + u(t) \end{aligned} \tag{7-294}$$

which are identical to Eq. (7-283).

It is constructive to examine some of the properties of the harmonic oscillator described by Eq. (7-283). First of all, we note that the system matrix **A** given by Eq. (7-284) is a *skew-symmetric matrix.*† If we examine the unforced system [obtained by setting $u(t) = 0$ in Eq. (7-283)]

$$\begin{bmatrix} \dot{x}_1(t) \\ \dot{x}_2(t) \end{bmatrix} = \begin{bmatrix} 0 & \omega \\ -\omega & 0 \end{bmatrix} \begin{bmatrix} x_1(t) \\ x_2(t) \end{bmatrix} \tag{7-295}$$

† See Eqs. (2-99) and (2-100).

we can show that the *fundamental matrix*† $\mathbf{\Phi}(t)$ is given by

$$\mathbf{\Phi}(t) = \begin{bmatrix} \cos \omega t & \sin \omega t \\ -\sin \omega t & \cos \omega t \end{bmatrix} \tag{7-296}$$

It is easy to verify that the fundamental matrix $\mathbf{\Phi}(t)$ is an *orthogonal matrix.*‡ So if we set

$$x_1(0) = \xi_1 \qquad x_2(0) = \xi_2 \tag{7-297}$$

then the solution of Eq. (7-295) is

$$\begin{bmatrix} x_1(t) \\ x_2(t) \end{bmatrix} = \begin{bmatrix} \cos \omega t & \sin \omega t \\ -\sin \omega t & \cos \omega t \end{bmatrix} \begin{bmatrix} \xi_1 \\ \xi_2 \end{bmatrix} \tag{7-298}$$

We thus note that

$$x_1^2(t) + x_2^2(t) = \xi_1^2 + \xi_2^2 = \text{const} \tag{7-299}$$

This means that the trajectories of the unforced system (7-295) are circles in the $x_1x_2$ state plane; these circles have a center at the origin (0, 0) and

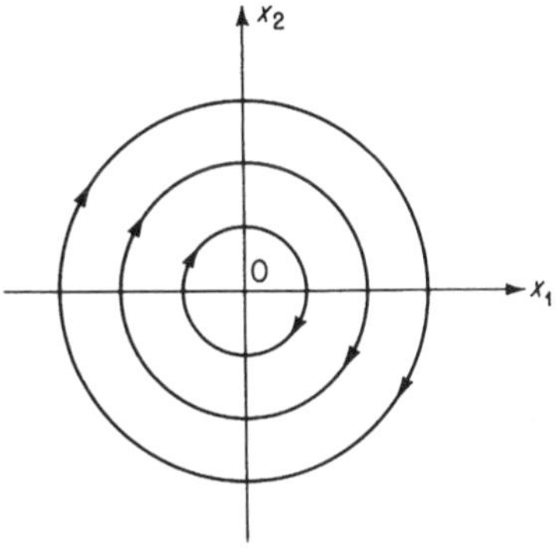


**Fig. 7-32** The trajectories of the harmonic oscillator in the absence of the control (i.e., when $u = 0$).

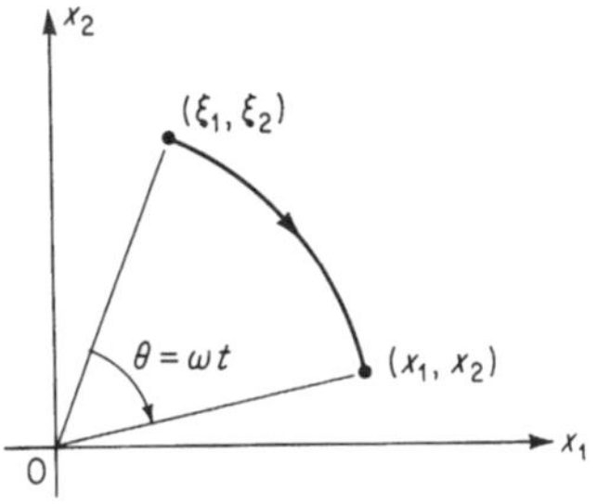


**Fig. 7-33** The angle $\theta = \omega t$ determines the time of travel from $(\xi_1, \xi_2)$ to $(x_1, x_2)$.

radius $\sqrt{\xi_1^2 + \xi_2^2}$, as shown in Fig. 7-32. Note that the state moves in a clockwise sense, as indicated by the arrows. The physical interpretation of this fact is: *The harmonic oscillator is a conservative system.* Of course, this fact is clear from Examples 7-3 and 7-4. The time $t$ required to go from an initial state $(\xi_1, \xi_2)$ to the state $(x_1, x_2)$ along the circular arc can be determined from the equation

$$t = \frac{\theta}{\omega} \tag{7-300}$$

where $\theta$ is the angle shown in Fig. 7-33.

† See Sec. 3-20.
‡ See Eqs. (2-104) to (2-109).

We are now ready to formulate the time-optimal control problem for the harmonic oscillator.

***Problem 7-7*** *Given the system*

$$\begin{bmatrix} \dot{x}_1(t) \\ \dot{x}_2(t) \end{bmatrix} = \begin{bmatrix} 0 & \omega \\ -\omega & 0 \end{bmatrix} \begin{bmatrix} x_1(t) \\ x_2(t) \end{bmatrix} + \begin{bmatrix} 0 \\ u(t) \end{bmatrix} \tag{7-301}$$

*where the control $u(t)$ is constrained to satisfy the relation*

$$|u(t)| \leq 1 \tag{7-302}$$

*determine the admissible control that forces the system* (7-301) *from any initial state* $(\xi_1, \xi_2)$ *to the origin* (0, 0) *in minimum time.*

The basic procedure that we followed in previous sections of this chapter will be used again in this section. However, we shall find that there is no upper bound on the number of switchings of the time-optimal control. For this reason, we cannot isolate the control sequences that are candidates for time-optimal control. However, this apparent lack of information will be compensated for by the additional information provided by the necessary conditions regarding the maximum amount of time for which the optimal control can be constant. Once more we shall be able to find the switch curve for this problem. Owing to the fact that it is more complicated than the ones we found in Secs. 7-2 and 7-3, we shall spend some time discussing suboptimal designs which require "simpler" switch curves. Now we proceed with the solution of the posed problem.

The Hamiltonian for this problem is given by the equation

$$H = 1 + \omega x_2(t)p_1(t) - \omega x_1(t)p_2(t) + u(t)p_2(t) \tag{7-303}$$

The control which absolutely minimizes $H$ is given by

$$u(t) = -\operatorname{sgn}\{p_2(t)\} \tag{7-304}$$

The costate variables $p_1(t)$ and $p_2(t)$ are defined by

$$\begin{aligned} \dot{p}_1(t) &= -\frac{\partial H}{\partial x_1(t)} = \omega p_2(t) \\ \dot{p}_2(t) &= -\frac{\partial H}{\partial x_2(t)} = -\omega p_1(t) \end{aligned} \tag{7-305}$$

which may be written in vector form as

$$\begin{bmatrix} \dot{p}_1(t) \\ \dot{p}_2(t) \end{bmatrix} = \begin{bmatrix} 0 & \omega \\ -\omega & 0 \end{bmatrix} \begin{bmatrix} p_1(t) \\ p_2(t) \end{bmatrix} \tag{7-306}$$

We note that the system (7-306) is of the same form as the system (7-295). This indeed was to be expected from the skew symmetry of the $\mathbf{A}$ matrix. Therefore, using the familiar notation

$$p_1(0) = \pi_1 \qquad p_2(0) = \pi_2 \tag{7-307}$$

we find that

$$\begin{bmatrix} p_1(t) \\ p_2(t) \end{bmatrix} = \begin{bmatrix} \cos \omega t & \sin \omega t \\ -\sin \omega t & \cos \omega t \end{bmatrix} \begin{bmatrix} \pi_1 \\ \pi_2 \end{bmatrix} \tag{7-308}$$

and, in particular, that

$$p_2(t) = -\pi_1 \sin \omega t + \pi_2 \cos \omega t \tag{7-309}$$

Clearly, the function $p_2(t)$ is the sum of two sinusoids, and so $p_2(t)$ is a sinusoid of the form

$$p_2(t) = a \sin (\omega t + \alpha) \tag{7-310}$$

Figure 7-34 shows a typical $p_2(t)$ and the control $u(t)$ obtained from Eq. (7-304). We can immediately draw the following conclusions:

1. The time-optimal control must be piecewise constant and must switch between the two values $u(t) = +1$ and $u(t) = -1$.

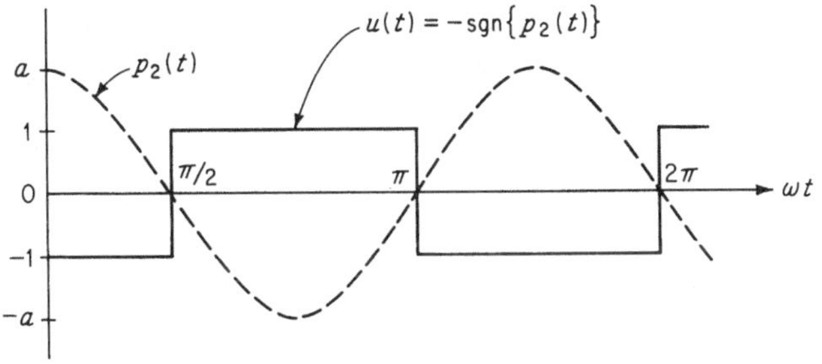


**Fig. 7-34** The sinusoidal costate variable $p_2(t)$ and the control $u(t) = -\operatorname{sgn}\{p_2(t)\}$.

2. The time-optimal control can remain constant for no more than $\pi/\omega$ units of time.

3. There is *no* upper bound on the number of switchings of the time-optimal control.

4. The function $p_2(t)$ cannot be zero over a finite interval of time as this would require $\pi_1 = \pi_2 = 0$, which would in turn imply that $p_1(t) = p_2(t) = 0$ for all $t$ and therefore that the Hamiltonian $H$ would not be zero as required by the minimum principle. Thus, there is no possibility of singular control; i.e., the problem is normal.

The fact that there is no bound on the total number of control switchings for this second-order system stands in marked contrast to the second-order systems (for which the time-optimal control switched *at most* once) considered previously in Secs. 7-2 and 7-3. As we shall see later in this section, the switch curve for systems with complex eigenvalues is different in character from the switch curve for systems with real eigenvalues.

The next step is to find the solution of the system (7-301) using the constant control

$$u = \Delta = \pm 1 \tag{7-311}$$

If the initial state is given by

$$\xi_1 = x_1(0) \qquad \xi_2 = x_2(0) \tag{7-312}$$

then the solution of Eq. (7-301) is

$$\begin{aligned} x_1(t) &= \left(\xi_1 - \frac{\Delta}{\omega}\right)\cos \omega t + \xi_2 \sin \omega t + \frac{\Delta}{\omega} \\ x_2(t) &= -\left(\xi_1 - \frac{\Delta}{\omega}\right)\sin \omega t + \xi_2 \cos \omega t \end{aligned} \tag{7-313}$$

or

$$\begin{aligned} \omega x_1(t) &= (\omega\xi_1 - \Delta)\cos \omega t + \omega\xi_2 \sin \omega t + \Delta \\ \omega x_2(t) &= -(\omega\xi_1 - \Delta)\sin \omega t + \omega\xi_2 \cos \omega t \end{aligned} \tag{7-314}$$

Squaring both sides of the above equations, adding, and rearranging, we obtain the relation

$$[\omega x_1(t) - \Delta]^2 + [\omega x_2(t)]^2 = (\omega\xi_1 - \Delta)^2 + (\omega\xi_2)^2 \tag{7-315}$$

The above equation does not contain the time and is, in fact, the equation of the trajectories in the $\omega x_1$-$\omega x_2$ plane.† Clearly, these trajectories are

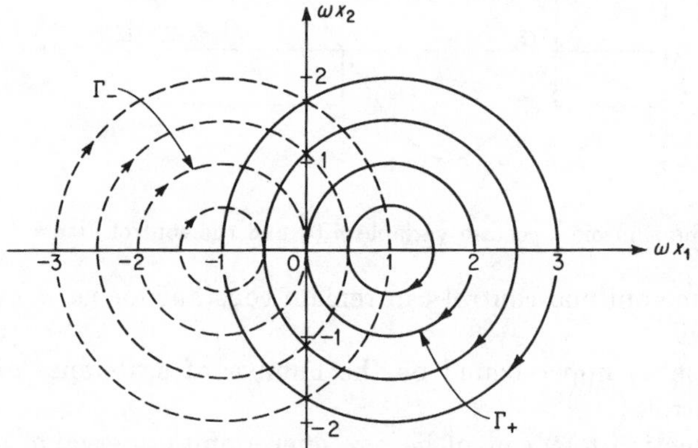


**Fig. 7-35** The solid circles represent the trajectories of the harmonic oscillator when $u = +1$, while the dashed trajectories are due to the control $u = -1$. The motion of the states is in a clockwise direction.

circles in the $\omega x_1$-$\omega x_2$ plane with centers at the point $(\Delta, 0)$. Therefore, if $u = +1$, *the trajectories are circles about the point* $(1, 0)$, *and if* $u = -1$, *the trajectories are circles about the point* $(-1, 0)$. These trajectories are shown in Fig. 7-35.

From Eqs. (7-314), it is easy to obtain an analytical expression for the time required to go from $(\omega\xi_1, \omega\xi_2)$ to $(\omega x_1, \omega x_2)$ when the control is $u = \Delta$;

† We shall use from now on the $\omega x_1$-$\omega x_2$ plane rather than the $x_1x_2$ plane; one could certainly redefine variables $z_1 = \omega x_1$, $z_2 = \omega x_2$, but we feel that this substitution would not add to the clarity of exposition.

the time is given by the expression

$$t = \frac{1}{\omega} \tan^{-1} \frac{(\omega x_1 - \Delta)\omega\xi_2 - \omega x_2(\omega\xi_1 - \Delta)}{(\omega x_2)(\omega\xi_2) + (\omega x_1 - \Delta)(\omega\xi_1 - \Delta)} \tag{7-316}$$

This complex formula has a very simple interpretation in the $\omega x_1$-$\omega x_2$ plane; as is apparent from Fig. 7-36, we may obtain the time by measuring the angle $\theta = \omega t$ of the circular arc from the point $(\omega\xi_1, \omega\xi_2)$ to the point $(\omega x_1, \omega x_2)$, where the circular arc has its center at the point $(\Delta, 0)$.

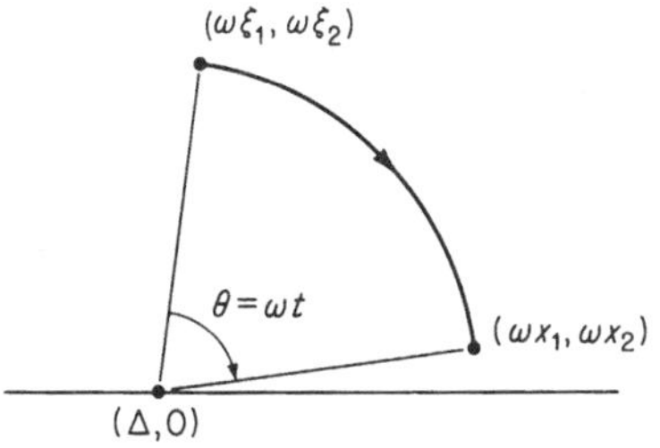


**Fig. 7-36** If $u = \Delta = \pm 1$, then the angle $\theta = \omega t$ determines the time required to force the point $(\omega\xi_1, \omega\xi_2)$ to $(\omega x_1, \omega x_2)$.

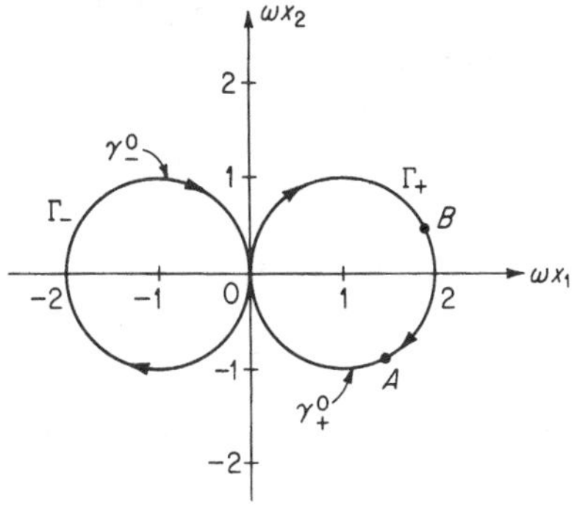


**Fig. 7-37** The two circles $\Gamma_+$ and $\Gamma_-$ represent the two forced trajectories which pass through the origin of the plane. The $\gamma_+{}^0$ semicircle is the lower half of the $\Gamma_+$ circle, while the $\gamma_-{}^0$ semicircle is the upper half of the $\Gamma_-$ circle.

In Fig. 7-35, we indicated the two circles $\Gamma_+$ and $\Gamma_-$ which pass through the origin of the $\omega x_1$-$\omega x_2$ plane; for clarity, Fig. 7-37 shows the circles $\Gamma_+$ and $\Gamma_-$ alone. Clearly, interpreting the circles $\Gamma_+$ and $\Gamma_-$ as sets, we have

$$\Gamma_+ = \{(\omega x_1, \omega x_2) : (\omega x_1 - 1)^2 + (\omega x_2)^2 = 1\} \tag{7-317}$$

and

$$\Gamma_- = \{(\omega x_1, \omega x_2) : (\omega x_1 + 1)^2 + (\omega x_2)^2 = 1\} \tag{7-318}$$

Any state on $\Gamma_+$ can be forced to the origin by the control $u = +1$. In particular, the point (2, 0) can be forced to the origin in exactly $\pi/\omega$ seconds, since the circular arc is exactly $\pi$ radians long. As is clear from Fig. 7-37, use of the control $u = +1$ will force the point $A$ on $\Gamma_+$ to the origin in less than $\pi/\omega$ seconds, while the point $B$ on $\Gamma_+$ requires more than $\pi/\omega$ seconds; in the latter case, the control does not switch for more than $\pi/\omega$ seconds, which is in direct violation of the requirement that the time-optimal control cannot remain constant for more than $\pi/\omega$ seconds.

These considerations lead to the following definitions:

***Definition 7-18*** *Let $\gamma_+{}^0$ be the set of states which can be forced to the origin in no more than $\pi/\omega$ seconds by the control $u = +1$. It is clear that*

$$\gamma_+{}^0 = \{(\omega x_1, \omega x_2) : (\omega x_1 - 1)^2 + (\omega x_2)^2 = 1;\ \omega x_2 < 0\} \tag{7-319}$$

*In other words, as indicated in Fig. 7-37, the $\gamma_+^0$ curve is the portion of the $\Gamma_+$ circle below the $\omega x_1$ axis; so the $\gamma_+^0$ curve is a semicircle.*

***Definition 7-19*** *Let $\gamma_-^0$ be the set of states which can be forced to the origin in no more than $\pi/\omega$ seconds by the control $u = -1$. Clearly,*

$$\gamma_-^0 = \{(\omega x_1, \omega x_2): (\omega x_1 + 1)^2 + (\omega x_2)^2 = 1; \omega x_2 > 0\} \qquad (7\text{-}320)$$

*As indicated in Fig. 7-37, the semicircle $\gamma_-^0$ is the portion of the $\Gamma_-$ circle above the $\omega x_1$ axis.*

Now we shall prove that the $\gamma_+^0$ and $\gamma_-^0$ curves constitute optimal paths to the origin.

***Control Law 7-7a*** *Given a state $(\omega x_1, \omega x_2) \in \gamma_+^0 \cup \gamma_-^0$; then the time-optimal control $u^*$ is unique and is given by*

$$\begin{aligned} u^* &= u^*(\omega x_1, \omega x_2) = +1 \qquad \text{for all } (\omega x_1, \omega x_2) \in \gamma_+^0 \\ u^* &= u^*(\omega x_1, \omega x_2) = -1 \qquad \text{for all } (\omega x_1, \omega x_2) \in \gamma_-^0 \end{aligned}$$

PROOF We could prove this very easily, simply by invoking the existence and uniqueness of the optimal and extremal controls. However, we can also prove it using a geometric argument which we shall now present.

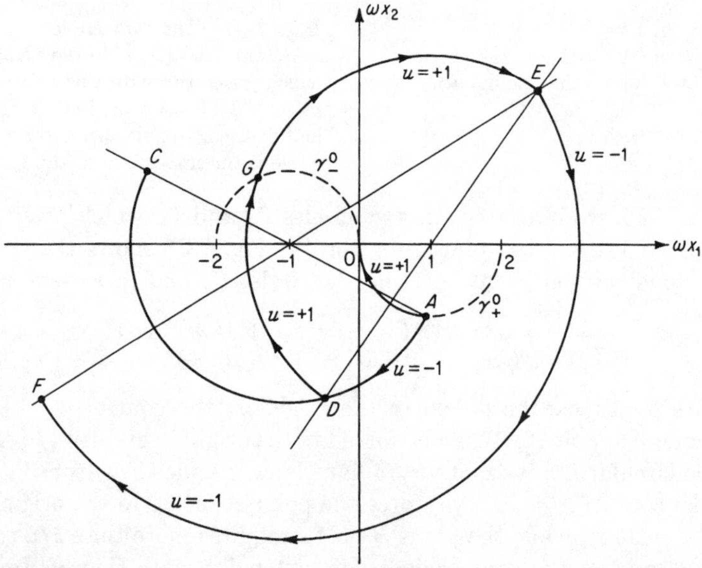


**Fig. 7-38** Illustration of optimal and nonoptimal trajectories.

Consider a point $A$ on $\gamma_+^0$, as indicated in Fig. 7-38. By definition, application of $u = +1$ will force $A$ to $O$ along the circular arc $AO$, which belongs to the $\gamma_+^0$ curve. Now suppose that we apply the control $u = -1$ at $A$; the resulting trajectory will be a circular arc about the center $(-1, 0)$, and the maximum length of that arc is $\pi$ radians (in Fig. 7-38, the arc $AC$ is $\pi$

radians). Clearly, the arc $AC$ does not contain the origin $O$. Now suppose that we switch from $u = -1$ to $u = +1$ at some point $D$ on the arc $AC$; the trajectory due to $u = +1$ will be a circular arc about the center (1, 0) whose maximum length is $\pi$ radians (in Fig. 7-38, the arc $DE$ is exactly $\pi$ radians long). Since the arc $DE$ does not contain the origin, we must switch at $E$ from $u = +1$ to $u = -1$ to obtain the arc $EF$, which again does not contain the origin, and so on. Thus, if at point $A$ we apply $u = -1$ and we follow a control sequence which satisfies the minimum principle, we cannot get to the origin $O$ and, as a matter of fact, we get further and further away from the origin. One more point needs to be clarified: What is wrong with the path $ADGO$, where $G$ is a point on the $\gamma_-^0$ curve? The answer lies in the control sequence required to generate the path $ADGO$, which is of the form $\{-1, +1, -1\}$. It is clear from a reexamination of Fig. 7-34 that the lengths of the arcs $AD$ and $GO$ are arbitrary as long as they do not exceed $\pi$ radians. However, the length of the arc $DG$ must be *exactly* $\pi$ radians, which cannot be true. Therefore, the path $ADGO$ cannot be optimal, since it violates the necessary conditions. The above considerations inevitably lead to the conclusion that for any point $A$ on $\gamma_+^0$, the time-optimal control must be $u = +1$ since there is no other control sequence which satisfies the necessary conditions and forces the point $A$ to the origin. Similar reasoning for points on the $\gamma_-^0$ curve proves the validity of the control law.

***Definition 7-20*** *Let $R_-^1$ denote the set of states which can be forced to the $\gamma_+^0$ curve in no more than $\pi/\omega$ seconds by the control $u = -1$. The set $R_-^1$ is made up of all circular arcs of length $\pi$ radians about the center $(-1, 0)$ which terminate on the $\gamma_+^0$ curve. The set $R_-^1$ is shown in Fig.* 7-39.

***Definition 7-21*** *Let $R_+^1$ denote the set of states which can be forced to the $\gamma_-^0$ curve in no more than $\pi/\omega$ seconds by the control $u = +1$. The set $R_+^1$ is made up of all circular arcs of length $\pi$ radians about the center $(+1, 0)$ which terminate on the $\gamma_-^0$ curve. The set $R_+^1$ is shown in Fig.* 7-39.

***Definition 7-22*** *Let $\gamma_-^1$ denote the set of states which can be forced to the $\gamma_+^0$ curve in exactly $\pi/\omega$ seconds by the control $u = -1$. The $\gamma_-^1$ curve is shown in Fig.* 7-39 *and is a semicircle above the $\omega x_1$ axis with center at $(-3, 0)$ or, more precisely,*

$$\gamma_-^1 = \{(\omega x_1, \omega x_2): (\omega x_1 + 3)^2 + (\omega x_2)^2 = 1;\ \omega x_2 > 0\} \qquad \text{(7-321)}$$

***Definition 7-23*** *Let $\gamma_+^1$ denote the set of states which can be forced to the $\gamma_-^0$ curve in exactly $\pi/\omega$ seconds by the control $u = +1$. The $\gamma_+^1$ curve is shown in Fig.* 7-39 *and is a semicircle below the $\omega x_1$ axis with center at* (3, 0) *or, more precisely,*

$$\gamma_+^1 = \{(\omega x_1, \omega x_2): (\omega x_1 - 3)^2 + (\omega x_2)^2 = 1;\ \omega x_2 < 0\} \qquad \text{(7-322)}$$

On the basis of the above definitions we can derive the following control law:

***Control Law 7-7b*** *Given a state* $(\omega x_1, \omega x_2) \in R_+^1 \cup R_-^1$*; then the time-optimal control* $u^*$ *is unique and is given by*

$$u^* = u^*(\omega x_1, \omega x_2) = +1 \qquad \text{for all } (\omega x_1, \omega x_2) \in R_+^1$$
$$u^* = u^*(\omega x_1, \omega x_2) = -1 \qquad \text{for all } (\omega x_1, \omega x_2) \in R_-^1$$

The proof of the above control law is very similar to the proof of Control Law 7-7*a*; one picks a point in $R_-^1$ and shows that the only control sequence which satisfies the necessary conditions provided by the minimum principle is to apply $u = -1$ until the $\gamma_+^0$ curve is reached and then switch to $u = +1$

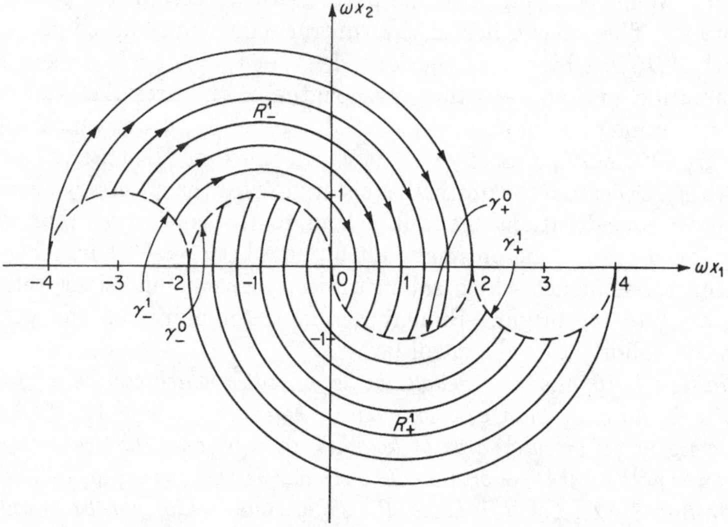


**Fig. 7-39** The semicircles $\gamma_+^0$, $\gamma_-^0$, $\gamma_+^1$, $\gamma_-^1$ and the sets $R_-^1$ and $R_+^1$.

to reach the origin. The interested reader can verify this by constructions similar to the ones of Fig. 7-38.

We generalize these concepts in the following way:

***Definition 7-24*** *Let* $\gamma_+^j$, $j = 1, 2, \ldots$ *, denote the semicircle below the* $\omega x_1$ *axis with center at* $(2j + 1, 0)$ *and unit radius. More precisely,*

$$\gamma_+^j = \{(\omega x_1, \omega x_2) \colon [\omega x_1 - (2j + 1)]^2 + (\omega x_2)^2 = 1; \omega x_2 < 0\} \qquad (7\text{-}323)$$

***Definition 7-25*** *Let* $\gamma_-^j$, $j = 1, 2, \ldots$ *, denote the semicircles above the* $\omega x_1$ *axis with center at* $(-2j - 1, 0)$ *and unit radius. More precisely,*

$$\gamma_-^j = \{(\omega x_1, \omega x_2) \colon [\omega x_1 + (2j + 1)]^2 + (\omega x_2)^2 = 1; \omega x_2 > 0\} \qquad (7\text{-}324)$$

These $\gamma_+^j$ and $\gamma_-^j$ curves are illustrated in Fig. 7-40.

***Definition 7-26*** *Let* $R_-^j$, $j = 1, 2, \ldots$ *, denote the set of states which can be forced to the* $\gamma_+^{j-1}$ *curve in no more than* $\pi/\omega$ *seconds by the control* $u = -1$.

*Let $R_+^j$, $j = 1, 2, \ldots$, denote the set of states which can be forced to the $\gamma_-^{j-1}$ curve in no more than $\pi/\omega$ seconds by the control $u = +1$. The sets $R_-^j$ and $R_+^j$ are illustrated in Fig.* 7-40.

From the definitions above, it follows that *the $\gamma_-^j$ curve is the locus of all states which can be forced in exactly $\pi/\omega$ seconds to the $\gamma_+^{j-1}$ curve by the control $u = -1$, and the $\gamma_+^j$ curve is the locus of all states which can be forced to the $\gamma_-^{j-1}$ curve in exactly $\pi/\omega$ seconds by the control $u = +1$.* In other words, one can generate the $\gamma_-^j$ curve from the $\gamma_+^{j-1}$ curve by rotating every point

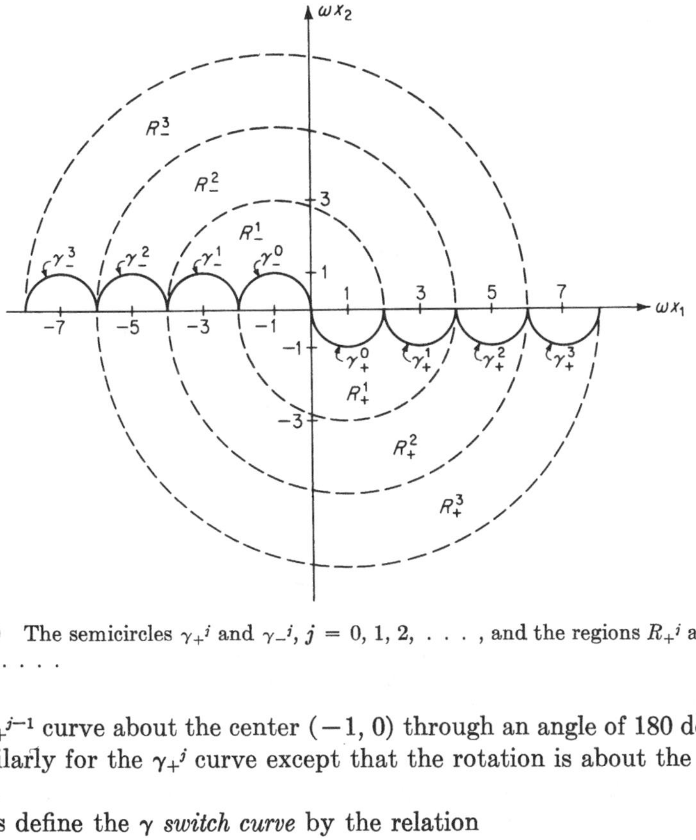


**Fig. 7-40** The semicircles $\gamma_+^j$ and $\gamma_-^j$, $j = 0, 1, 2, \ldots$, and the regions $R_+^j$ and $R_-^j$, $j = 1, 2, \ldots$.

of the $\gamma_+^{j-1}$ curve about the center $(-1, 0)$ through an angle of 180 degrees, and similarly for the $\gamma_+^j$ curve except that the rotation is about the center $(+1, 0)$.

Let us define the $\gamma$ *switch curve* by the relation

$$\gamma = \left[\bigcup_{j=0}^{\infty} \gamma_+^j\right] \cup \left[\bigcup_{j=0}^{\infty} \gamma_-^j\right] \triangleq \gamma_+ \cup \gamma_- \tag{7-325}$$

If we let

$$R_- = \bigcup_{j=1}^{\infty} R_-^j \qquad R_+ = \bigcup_{j=1}^{\infty} R_+^j \tag{7-326}$$

then $R_-$ is the set of points above the $\gamma$ switch curve and $R_+$ is the set of points below the switch curve. We can now state the time-optimal control law for Problem 7-7 as follows:

***Control Law 7-7 Solution to Problem 7-7*** *The time-optimal control, as a function of the state* $(\omega x_1, \omega x_2)$, *is given by*

$$\begin{aligned} u^* &= u^*(\omega x_1, \omega x_2) = +1 \qquad \text{for all } (\omega x_1, \omega x_2) \in R_+ \cup \gamma_+ \\ u^* &= u^*(\omega x_1, \omega x_2) = -1 \qquad \text{for all } (\omega x_1, \omega x_2) \in R_- \cup \gamma_- \end{aligned} \tag{7-327}$$

[*where* $R_+$, $R_-$, $\gamma_+$, $\gamma_-$ *are given by Eqs.* (7-325) *and* (7-326)]. *Furthermore, the time-optimal control is unique.*

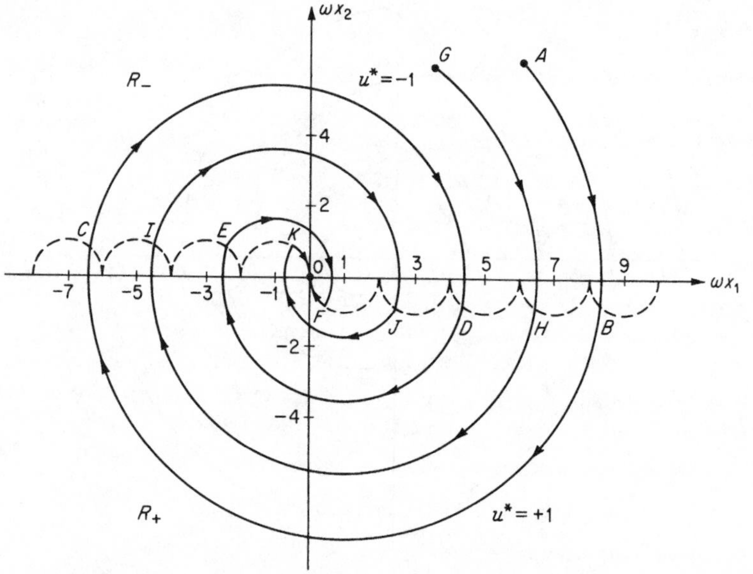


**Fig. 7-41** The $\gamma$ switch curve is represented by the dashed lines. $R_-$ is the region above the switch curve, and $R_+$ is the region below the switch curve. Two time-optimal trajectories are *ABCDEFO* and *GHIJKO*.

We shall omit the proof of this control law because we have already exposed the ideas involved in Control Laws 7-7*a* and 7-7*b*. The proof hinges on the fact that the extremal controls are unique and that an optimal control to the origin exists. Thus, if the initial state is *above* the $\gamma$ switch curve, then the time-optimal control is $u^* = -1$; if the initial state is *below* the $\gamma$ switch curve, then the time-optimal control is $u^* = +1$. Figure 7-41 shows several time-optimal trajectories to the origin.

Let us now examine Fig. 7-41 closely. The $\gamma$ switch curve is indicated with dashed lines, and the two time-optimal trajectories are shown as solid curves. Consider the time-optimal trajectory *ABCDEFO*; the control which generates this trajectory switches exactly five times (at the points

$B$, $C$, $D$, $E$, and $F$ on the switch curve); it is easy to see that† the entire segment $AB$ belongs to the set $R_-^5$, the segment $BC$ belongs to the set $R_+^4$, the segment $CD$ belongs to $R_-^3$, and so on. Moreover, the circular arcs $BC$, $CD$, $DE$, and $EF$ are *exactly* $\pi$ radians long, while the circular arcs $AB$ and $FO$ are less than $\pi$ radians long, which means that the control sequence satisfies the necessary conditions. We also observe that the only portion of the optimal trajectory which belongs to the $\gamma$ switch curve is the arc $FO$, which indeed belongs to the curve $\gamma_+^0$. Considering the other optimal trajectory $GHIJKO$, we also observe that the arc $KO$ is the only one contained in the $\gamma$ switch curve; as a matter of fact, it belongs to the $\gamma_-^0$ curve. Thus, we conclude that *the only portion of the $\gamma$ switch curve which can contain time-optimal trajectories is the union of the $\gamma_-^0$ and $\gamma_+^0$ curves.* In other words, *the $\gamma_+^j$ and $\gamma_-^j$ curves, for $j = 1, 2, \ldots$, are not portions of optimal trajectories.* This is again in marked contrast to the property that the entire switch curve for second-order systems with real eigenvalues‡ is generated by time-optimal trajectories.

There is still another important difference between the harmonic oscillator and the two second-order systems treated in Secs. 7-2 and 7-3: For the latter, the time-optimal control had the property that it required the smallest number of switchings;§ this is not true for the harmonic oscillator. As is evident from Fig. 7-37, the point $B$ on the $\Gamma_+$ circle (which can be forced to the origin by the control $u = +1$) belongs to the set $R_-^1$, and the time-optimal control which forces $B$ to the origin switches once. In other words, *for the harmonic oscillator, the control which requires the minimum number of switchings is not necessarily time-optimal.*

**Exercise 7-35** Show that the constant control $u = +1$ forces the state (1, 1), in the $\omega x_1$-$\omega x_2$ plane, to the origin in $3\pi/2\omega$ seconds, while the time-optimal control requires only $3\pi/4\omega$ seconds.

Since we have derived the time-optimal feedback control, we can provide an engineering realization of the time-optimal controller; such a "design" is illustrated in Fig. 7-42. We observe that the structure of the system of Fig. 7-42 is similar to the structure of the controllers for the other two second-order systems, shown in Figs. 7-5 and 7-14, in the sense that only a single nonlinear element is required. However, the nonlinearity $N$ in Fig. 7-42 (which has the input-output characteristics of the $\gamma$ switch curve) is a very complex one; the engineering realization (say by biased diodes) of the scalloplike input-output characteristic is not easy. For this reason, we shall discuss suboptimal control techniques in Examples 7-5 and 7-6, that is, designs based on the optimal one but with simpler nonlinearities.

† See Definition 7-26.

‡ See Sec. 7-2, Definition 7-4 and Figs. 7-3 and 7-6; and Sec. 7-3, Definition 7-8 and Figs. 7-13 and 7-16.

§ This is also true for the $N$th-order system of Sec. 7-5.

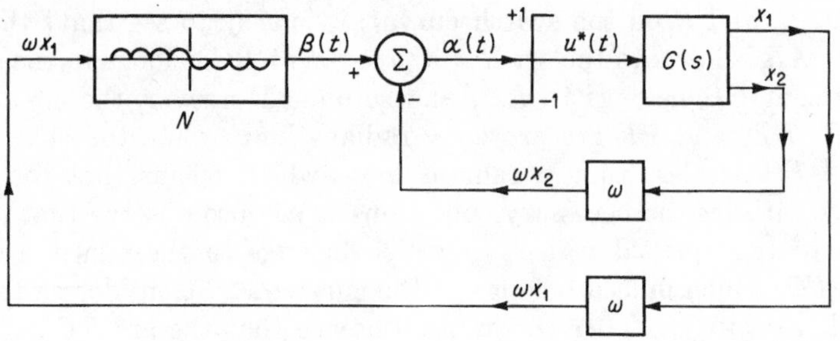


**Fig. 7-42** The block diagram of the time-optimal controller of the harmonic oscillator. The input-output characteristic of the nonlinearity $N$ is the same as the $\gamma$ switch curve.

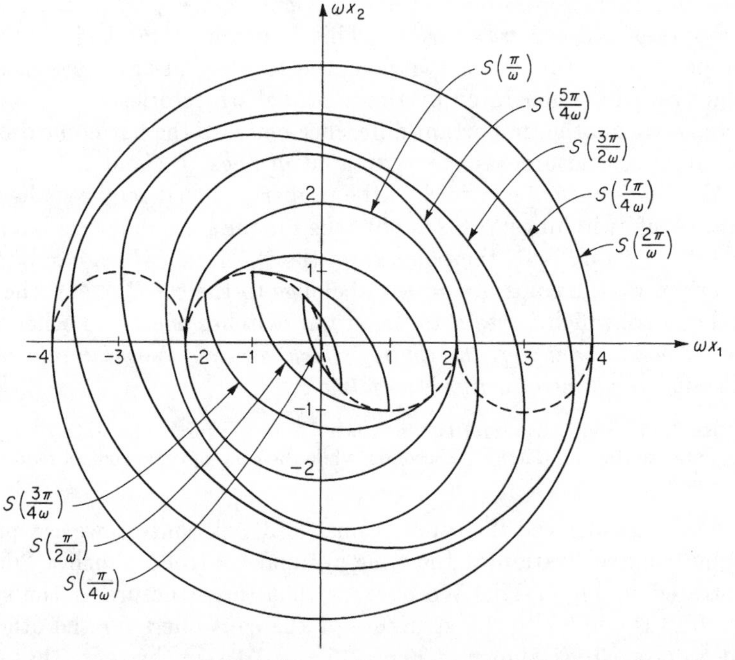


**Fig. 7-43** The minimum isochrones for the harmonic oscillator.

Since the harmonic oscillator is a basic and important system, it is worthwhile to examine its minimum isochrones. As we have done before,† we *define* $S(t^*)$ *to be the set of states which can be forced to the origin in the same minimum time* $t^*$, *and we call* $S(t^*)$ *the* $t^*$ *minimum isochrone.* Figure 7-43 illustrates the minimum isochrones for the harmonic oscillator for various values of $t^*$. Before we discuss the properties of these isochrones, it is instructive to present the method of their construction. In Fig. 7-44, we

† See Definition 7-6.

show a typical time-optimal trajectory $ACO$ in the region $R_+^1$; suppose that the minimum time required to force $A$ to the origin is $t_A^*$. We should like to construct the minimum isochrone $S(t_A^*)$ which passes through the state $A$. In order to do this,

1. We draw a circle of radius 2 with center at $A$, and we find an intersection of this circle with the $\Gamma_+$ circle; this point of intersection is the point $B$. It is easy to show that the minimum time $t_A^*$ can be found from the angle $\omega t_A^*$ (the proof involves the use of congruent triangles in order to show that the angle $\omega t_A^*$ is the sum of the angles $ANC$ and $CMO$).

2. Next, we draw a circular arc of radius 2 with center at $B$ which joins a point $D$ on the $\gamma_+^1$ curve and becomes tangent to the boundary of $R_+^1$

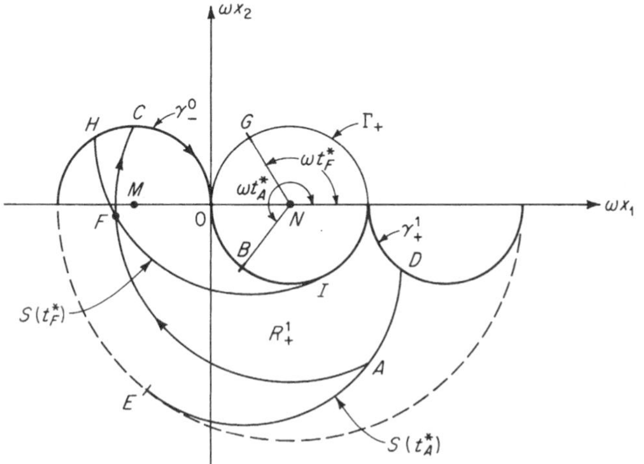


**Fig. 7-44** Construction of the minimum isochrones $S(t_A^*)$ and $S(t_F^*)$.

at the point $E$; then the circular arc $DAE$ is the portion of the isochrone $S(t_A^*)$ in the region $R_+^1$ because, by construction, any point on the arc $DAE$ requires the same minimum time $t_A^*$ to reach the origin.

Now let us pick a different point $F$ in $R_+^1$ with time-optimal trajectory $FCO$, and let us find the minimum isochrone $S(t_F^*)$, where $t_F^*$ is the minimum time to force $F$ to the origin. The same construction as the one outlined above is used to obtain $S(t_F^*)$. The circle of radius 2 about $F$ intersects the $\Gamma_+$ curve at $G$; the time $t_F^*$ is determined from $\omega t_F^*$ (again congruent triangles are used to show that the angle $\omega t_F^*$ is the sum of the angles $FNC$ and $CMO$). The circular arc $HFI$ is drawn with radius 2 and center at $G$, and the arc $HFI$ is the portion of $S(t_F^*)$ in $R_+^1$, where $H$ is on the $\gamma_-^0$ curve and $I$ on the $\gamma_+^0$ curve. The same construction may be employed for any initial state in $R_-^1$.

Figure 7-45 illustrates the technique if the initial state $K$ is in $R_-^2$. The time-optimal trajectory is $KPQO$; the minimum time $t_K^*$ can be found from the angle $\omega t_K^*$ (use of congruent triangles shows that the angle $\omega t_K^*$ equals the sum of the three angles $KMP$, $PNQ$, and $QMO$). To find $S(t_K^*)$, we draw a circle of radius 4 about $K$ and we find its intersection with the $\Gamma_-$ circle, which is the point $L$. The minimum isochrone $S(t_K^*)$ is the arc $SKR$ with center at $L$ and radius 4, where $S$ is the tangent point of $S(t_K^*)$ with the boundary (dashed line) of $R_-^1$ and $R$ is on the $\gamma_+^1$ curve.

These methods can be generalized to construct the minimum isochrone which passes through a given point in the $\omega x_1$-$\omega x_2$ plane. We shall give below a list of directions for constructing these isochrones.

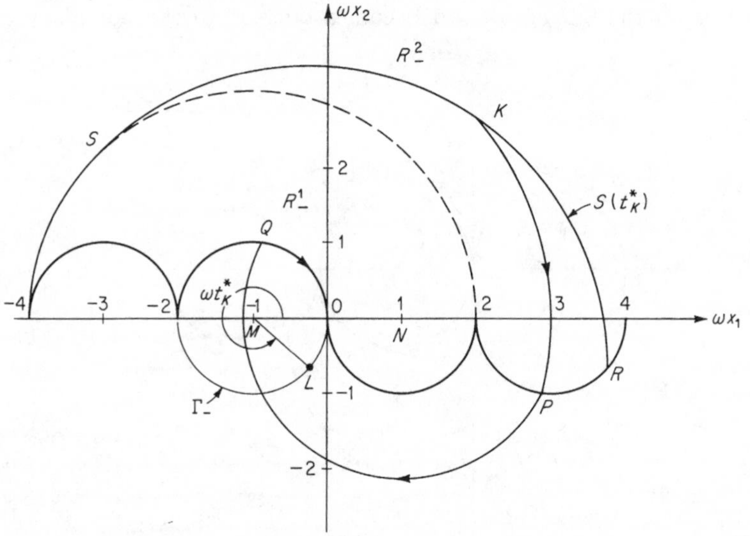


**Fig. 7-45** Construction of the minimum isochrone $S(t_K^*)$, where the point $K$ is in $R_-^2$.

***Definition 7-27*** *Let $C_m$, $m = 2, 4, 6, \ldots$, be the set of states such that*

$$(\omega x_1)^2 + (\omega x_2)^2 \le m^2 \qquad m = 2, 4, 6, \ldots \tag{7-328}$$

*In other words, $C_2$ is the set of points contained within the circle of radius* 2 *with center at the origin, and so on.*

We summarize below the method of constructing the isochrones for states below the switch curve; the isochrones for states above the switch curve can be found by symmetry.

Given a point $A$ below the switch curve:

If $A \in C_2 \cap R_+^1$, then the $S(t_A^*)$ isochrone is a circular arc through $A$ of radius 2 with center on the top half of the $\Gamma_+$ circle.

If $A \in R_+^1 - C_2$,† then the $S(t_A^*)$ isochrone is a circular arc through $A$ of radius 2 with center on the bottom half of the $\Gamma_+$ circle.

† See Eq. (2-4).

If $A \in C_4 \cap R_+^2$, then the $S(t_A^*)$ isochrone is a circular arc through $A$ of radius 4 with center on the top half of the $\Gamma_+$ circle.

If $A \in R_+^2 - C_4$, then the $S(t_A^*)$ isochrone is a circular arc through $A$ of radius 4 with center on the bottom half of the $\Gamma_+$ circle.

So, in general,

If $A \in C_{2j} \cap R_+^j$, $j = 1, 2, \ldots$, then the $S(t_A^*)$ isochrone is a circular arc through $A$ of radius $2j$ with center on the top half of the $\Gamma_+$ circle.

If $A \in R_+^j - C_{2j}$, $j = 1, 2, \ldots$, then the $S(t_A^*)$ isochrone is a circular arc through $A$ of radius $2j$ with center on the bottom half of the $\Gamma_+$ circle.

**Exercise 7-36** Develop your own rules for constructing the $S(t^*)$ minimum isochrone when $t^*$ is given. Plot the $S(7\pi/8\omega)$ and the $S(7\pi/4\omega)$ minimum isochrones.

Now we are able to make some intelligent comments about the minimum isochrones of the harmonic oscillator and to compare them with the minimum isochrones for the double-integral plant (Sec. 7-2, Fig. 7-7).† The reader should refer to Fig. 7-43 in order to visually verify the following conclusions:

1. The minimum isochrones in the set $C_2$, that is, the circle about the origin of radius 2, are composed of *two* circular arcs, symmetric about the $\gamma_-^0$ and $\gamma_+^0$ curves [see, for example, the $S(\pi/4\omega)$, $S(\pi/2\omega)$, $S(3\pi/4\omega)$ minimum isochrones in Fig. 7-43]. We note that these isochrones have "corners" on the $\gamma_-^0$ and $\gamma_+^0$ curves, which means that each minimum isochrone $S(t^*)$, $0 < t^* < \pi/\omega$, is differentiable everywhere except at the union of the $\gamma_-^0$ and $\gamma_+^0$ curves. In this respect, the isochrones $S(t^*)$, $0 < t^* < \pi/\omega$, are "similar" to those of the double-integral system shown in Fig. 7-7.

2. The minimum isochrones $S(i\pi/\omega)$, $i = 1, 2, 3, \ldots$, are circles about the origin of radius $2i$. Clearly, circles are differentiable everywhere, and they do not have "corners."

3. The minimum isochrones $S(t^*)$, $t^* > \pi/\omega$, $t^* \neq i\pi/\omega$, $i = 1, 2, \ldots$, are composed of *four* circular arcs. As indicated by the isochrones $S(5\pi/4\omega)$, $S(3\pi/2\omega)$, $S(7\pi/4\omega)$ in Fig. 7-43, these isochrones are differentiable everywhere; in other words, *all minimum isochrones* $S(t^*)$, $t^* > \pi/\omega$, *do not* have corners at the switch curve or anywhere else,‡ and in that sense, the isochrones for $t^* \geq \pi/\omega$ are "different" from those for $t^* < \pi/\omega$ and hence "different" from the isochrones of systems with real eigenvalues.

4. The centers of all the circular arcs which form the isochrones lie on the circles $\Gamma_+$ and $\Gamma_-$ (see Fig. 7-37). It follows that for large values of $t^*$, the isochrones tend to pure circles. This property enables us to construct good suboptimal systems, as is demonstrated in Examples 7-5 and 7-6.

Now we focus our attention on the problem of suboptimal control, which, in essence, consists in the replacement of the complex nonlinearity $N$ in

† Also for the system of Exercise 7-22 (if you have done it).

‡ This of course implies that we can define a unique support line to every point of $S(t^*)$, $t^* > \pi/\omega$.

Fig. 7-42 by a nonlinearity which is simpler to construct. Since the input-output characteristics of the nonlinearity $N$ of Fig. 7-42 are identical to the switch curve, one may consider the problem of suboptimal control as the problem of approximating the switch curve. The two examples below illustrate two such approximations.

**Example 7-5** Recall that the definition of the $\gamma$ switch curve given by Eq. (7-325) was

$$\gamma = [\bigcup_{j=0}^{\infty} \gamma_+^j] \cup [\bigcup_{j=0}^{\infty} \gamma_-^j] \tag{7-329}$$

Let us consider the curve

$$\hat{\gamma} = \begin{cases} \gamma_+^0 \cup \gamma_-^0 & \text{if } |\omega x_1| \leq 2 \quad (7\text{-}330) \\ \omega x_2 = 0 & \text{if } |\omega x_1| > 2 \quad (7\text{-}331) \end{cases}$$

The $\hat{\gamma}$ curve is illustrated in Fig. 7-46, and it approximates the $\gamma_-^j$ and $\gamma_+^j$ curves, $j = 1, 2, \ldots$, by the $\omega x_1$ axis. *We use the control $u = -1$ for every state above the $\hat{\gamma}$*

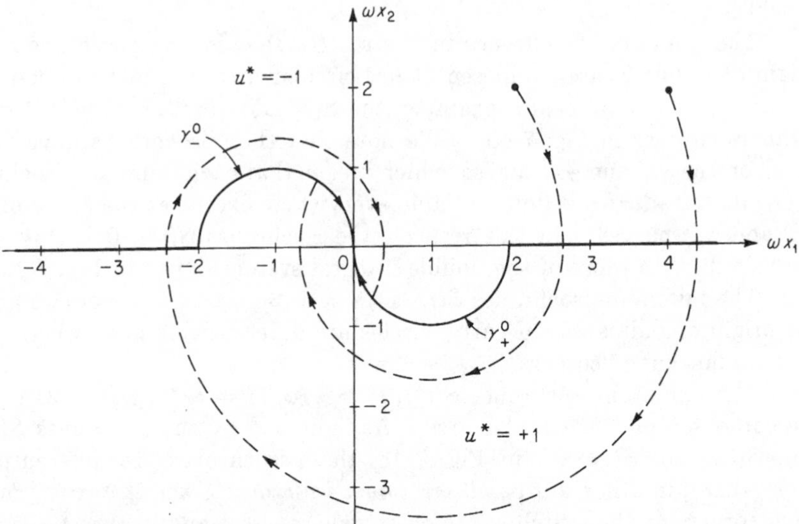


**Fig. 7-46** The switch curve $\hat{\gamma}$ for the suboptimal system of Example 7-5. The control is $u = -1$ above the switch curve and $u = +1$ below it. The dashed curves indicate the trajectories generated by this suboptimal control law.

*curve and the control $u = +1$ for every state below the $\hat{\gamma}$ curve.* Typical trajectories generated by this suboptimal control law are shown as dashed curves in Fig. 7-46. It is clear that this suboptimal control law will force any initial state to the origin; in other words, it is a stable control law.

The engineering realization of this suboptimal control law is identical with that of Fig. 7-42 except that the nonlinearity $N$ of Fig. 7-42 is replaced by the nonlinearity $\hat{N}$ with the input-output characteristic shown in Fig. 7-47. In general, the nonlinearity $\hat{N}$ is easier to construct than the nonlinearity $N$.

Whether or not this suboptimal system is good can only be decided by plotting the isochrones using the $\hat{\gamma}$ curve and comparing them with the minimum isochrones of Fig. 7-43. In order to provide an idea of the difference of the time $\hat{t}$ required by this suboptimal design and the minimum time $t^*$, we present Fig. 7-48.† In Fig. 7-48, the percentage of increase in the response time *when the initial states are on the $\omega x_1$ axis* is plotted. As is evident, there is no difference when $|\omega x_1| \leq 2$, $|\omega x_1| = 4$, $|\omega x_1| = 6$, . . . , because in these regions the $\hat{\gamma}$ curve is identical to the $\gamma$ curve. It is also clear from Fig. 7-48 that the maximum percentage of degradation in time is of the order of 4.3 percent, and for points on the $\omega x_1$ axis sufficiently far from the origin, the time $\hat{t}$ approaches the minimum time $t^*$. This is a consequence of the fact that, for initial states far from the origin, both the minimum isochrones and the suboptimal isochrones tend to the same circles. For many engineering systems, a maximum increase in response time of 4.3 percent is quite acceptable, and so this suboptimal design is often useful.

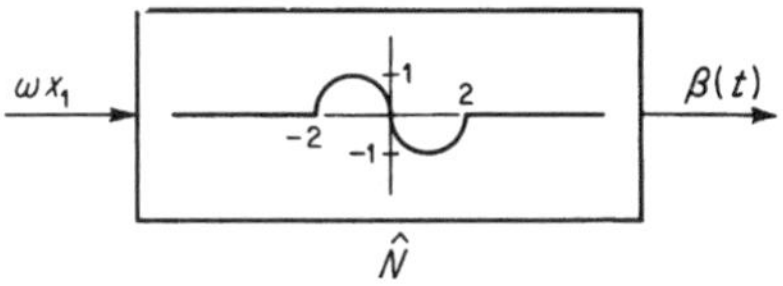


**Fig. 7-47** The input-output characteristics of the nonlinearity $\hat{N}$ required to generate the suboptimal system of Example 7-5.

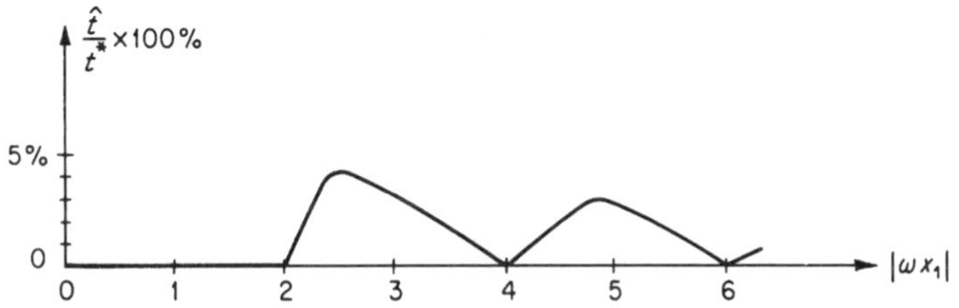


**Fig. 7-48** Plot of $(\hat{t}/t^*) \times 100$ percent versus $|\omega x_1|$.

**Example 7-6** Suppose that we use the following curve $\tilde{\gamma}$ as an approximation to the $\gamma$ switch curve:

$$\tilde{\gamma} = \begin{cases} \gamma_+{}^0 \cup \gamma_-{}^0 & \text{if } |\omega x_1| \leq 1 \\ \omega x_2 = -1 & \text{if } \omega x_1 > 1 \\ \omega x_2 = +1 & \text{if } \omega x_1 < -1 \end{cases} \tag{7-332}$$

The $\tilde{\gamma}$ curve is illustrated in Fig. 7-49. The suboptimal control law is: *Use $u = -1$ for every state above the $\tilde{\gamma}$ curve and $u = +1$ for every state below the $\tilde{\gamma}$ curve.* Typical trajectories generated by this suboptimal control law are shown as dashed curves in Fig. 7-49. This suboptimal control law will also force any initial state to the origin, and so it is a stable control law.

The engineering realization of this suboptimal control law is identical with that of Fig. 7-42 except that the nonlinearity $N$ of Fig. 7-42 is replaced by the nonlinearity $\tilde{N}$ with the input-output characteristic shown in Fig. 7-50. The nonlinearity $\tilde{N}$ is easier to construct than $N$ or $\hat{N}$. However, in general, the system with $\tilde{N}$ will require more time than the system with $\hat{N}$.

The following four exercises deal with the suboptimal and optimal control of the harmonic oscillator.

† Figure 7-48 is fig. C.1 of Ref. K-18, reproduced with the permission of the author.

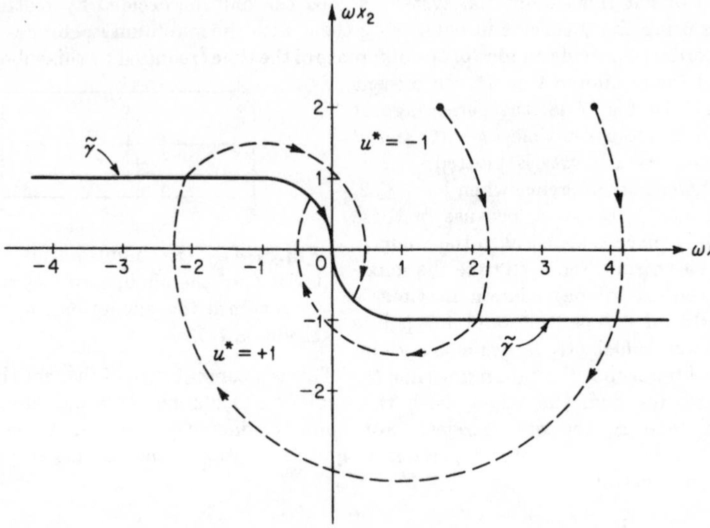


**Fig. 7-49** The switch curve $\tilde{\gamma}$ for the suboptimal system of Example 7-5. The control is $u = -1$ above $\tilde{\gamma}$ and $u = +1$ below $\tilde{\gamma}$. The dashed curves are trajectories generated by this suboptimal control law.

**Exercise 7-37** If $\bar{t}$ is the time required to force any initial state on the $\omega x_1$ axis by the suboptimal system with the nonlinearity $\tilde{N}$, plot $(\bar{t}/t^*) \times 100$ percent versus $|\omega x_1|$ and construct a figure like Fig. 7-48. Show that the maximum of $(\bar{t}/t^*) \times 100$ percent is larger than the maximum of $(\hat{t}/t^*) \times 100$ percent.

**Exercise 7-38** Show that, if the $\gamma$ curve is approximated by the $\omega x_1$ axis (in other words, the suboptimal control law is $u = -\operatorname{sgn}\{\omega x_2\}$), then any initial state *cannot* be forced to the origin but rather can only be forced to the closed segment $|\omega x_1| \leq 1$, $\omega x_2 = 0$.

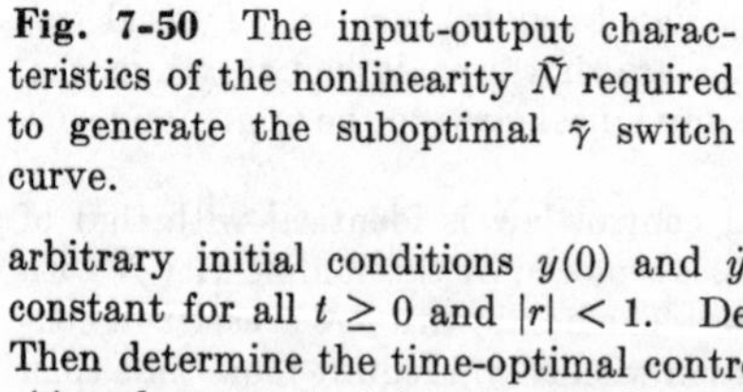

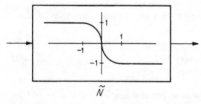

**Fig. 7-50** The input-output characteristics of the nonlinearity $\tilde{N}$ required to generate the suboptimal $\tilde{\gamma}$ switch curve.

**Exercise 7-39** Show that the control law $u = -\operatorname{sgn}\{\omega x_2 + \operatorname{sgn}\{\omega x_1\}\}$ will result in a limit cycle through the points (0, 1) and (0, −1).

**Exercise 7-40** Given the harmonic-oscillator plant with the transfer function

$$G(s) = \frac{1}{s^2 + 1}$$

Let $u(t)$, $|u(t)| \leq 1$, be the control input to the plant, and let $y(t)$ be its output with arbitrary initial conditions $y(0)$ and $\dot{y}(0)$. Given a step reference signal $r(t) = r$, a constant for all $t \geq 0$ and $|r| < 1$. Define the error $e(t)$ by setting $e(t) = r(t) - y(t)$. Then determine the time-optimal control as a function of $e(t)$ and $\dot{e}(t)$ so that the error $e(t)$ and error rate $\dot{e}(t)$ are reduced to zero in minimum time.

The following three exercises deal with the problem of the time-optimal control of the harmonic oscillator when the terminal state $(\theta_1, \theta_2)$ is not necessarily the origin.

**Exercise 7-41** Consider the system (7-301) with the control constrained by Eq. (7-302). Let $(\theta_1, \theta_2)$ be any given terminal state and let $(\xi_1, \xi_2)$ be any given initial state. Show that there is a time-optimal admissible control that transfers the system (7-301) from any initial state $(\xi_1, \xi_2)$ to any terminal state $(\theta_1, \theta_2)$.

**Exercise 7-42** Consider the system (7-301) with $\omega = 1$ and with the control constrained by Eq. (7-302). Suppose that the desired terminal state is the state (1, 1). Find the time-optimal feedback control from any initial state to (1, 1). Construct the switch curves. Comment on the uniqueness of the extremal controls. (HINT: See Exercise 7-15.) Plot some minimum isochrones and comment on the continuity of the minimum-time surface. (HINT: See Exercise 7-16.)

**Exercise 7-43** Repeat Exercise 7-42 for the terminal state (1, −1).

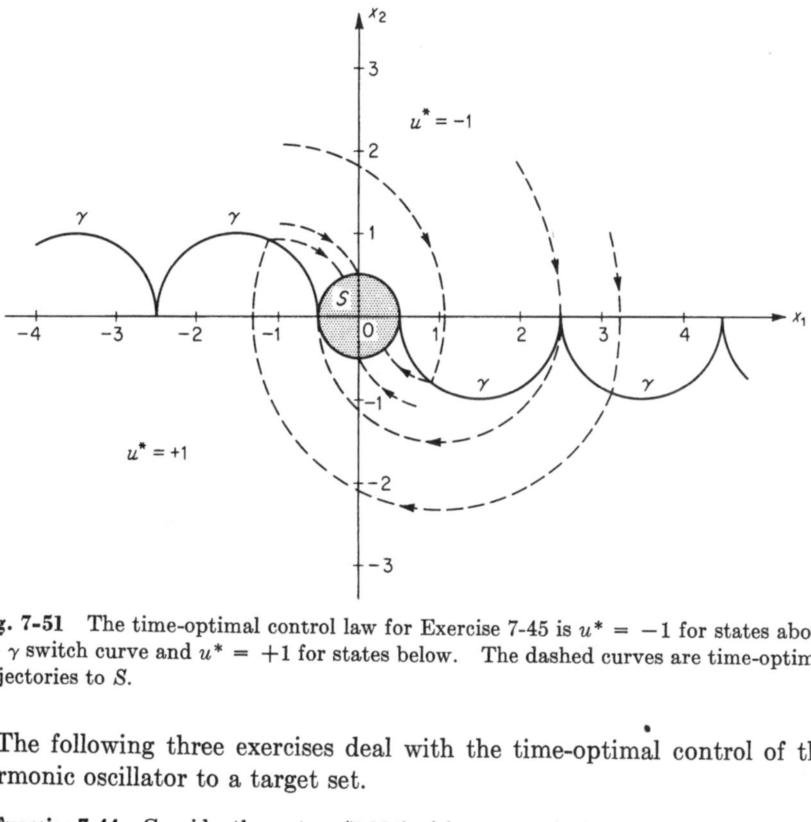


**Fig. 7-51** The time-optimal control law for Exercise 7-45 is $u^* = -1$ for states above the $\gamma$ switch curve and $u^* = +1$ for states below. The dashed curves are time-optimal trajectories to $S$.

The following three exercises deal with the time-optimal control of the harmonic oscillator to a target set.

**Exercise 7-44** Consider the system (7-301) with $\omega = 1$ and with the control constrained by Eq. (7-302). Consider the closed, nonconvex set $\hat{S}$ defined by the relation

$$\hat{S} = \{(x_1, x_2): x_1 = 1, x_2 = 0 \text{ or } x_1 = -1, x_2 = 0\}$$

[In other words, the target set $\hat{S}$ is made up of the two isolated states (1, 0) and (−1, 0).] Show that a time-optimal control to $\hat{S}$ exists from every intial state. Is the time-optimal control unique? Are the extremal controls unique? Determine the time-optimal control law. Plot some minimum isochrones. HINT: See Exercise 7-17.

**Exercise 7-45** Consider the system (7-301) with $\omega = 1$ and with the control constrained by Eq. (7-302). Consider the target set $S$ given by the relation

$$S = \{(x_1, x_2): x_1^2 + x_2^2 \leq (0.5)^2\}$$

Show that the time-optimal control law is the one suggested by Fig. 7-51. (HINT: See Ref. P-5, pages 53 to 58; also, compare with Exercise 7-19.) Show that the switch curve is *not* made up of time-optimal trajectories.

**Exercise 7-46** Repeat Exercise 7-45 for the target sets

$$S_1 = \{(x_1, x_2): x_1^2 + x_2^2 = (1.5)^2\}$$
$$S_2 = \{(x_1, x_2): x_1^2 + x_2^2 = 2^2\}$$

Find the time-optimal control law for initial states inside and outside the circles.

## 7-8 Time-optimal Control of a Stable Damped Harmonic Oscillator†

In the previous section, we discussed physical systems which were described by equations of the harmonic-oscillator type. We assumed that the dynamical systems of Examples 7-3 and 7-4 moved in a frictionless environment. If the same systems are subject to linear friction, then we shall refer to them as damped harmonic oscillators. We shall briefly comment on their time-optimal control in this section.

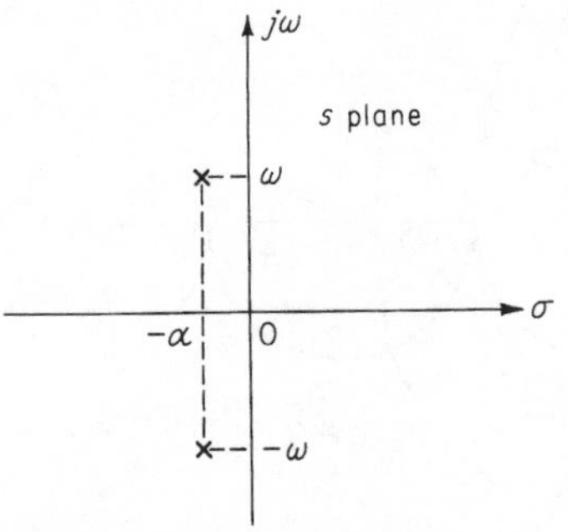


**Fig. 7-52** The s-plane pole configuration of a stable damped harmonic oscillator.

Let us consider a second-order system with the differential equation

$$\frac{d^2y(t)}{dt^2} + 2\alpha\frac{dy(t)}{dt} + (\alpha^2 + \omega^2)y(t) = Ku(t) \tag{7-333}$$

where $y(t)$ is the output, $u(t)$ is the control subject to the usual constraint

$$|u(t)| \leq 1 \qquad \text{for all } t \geq 0 \tag{7-334}$$

$K$ is a positive gain constant, and $\alpha$ and $\omega$ are constants with

$$\alpha > 0 \qquad \omega > 0 \tag{7-335}$$

In servomechanism practice, the system (7-333) is described by the transfer function

$$\frac{y(s)}{u(s)} = G(s) = \frac{K}{(s+\alpha)^2 + \omega^2} \tag{7-336}$$

The experienced reader will recognize that the poles of $G(s)$ are complex conjugates at

$$s = -\alpha \pm j\omega \tag{7-337}$$

Figure 7-52 shows the $s$-plane configuration of $G(s)$. The constant $\omega$ is sometimes called the *natural frequency*, and the constant $\alpha$, the *damping*.

† See Refs. B-24; C-3, pp. 230–234; and K-18.

Let

$$y_1(t) = y(t) \qquad y_2(t) = \dot{y}(t) \tag{7-338}$$

be a set of state variables. The $y_i(t)$ satisfy the vector differential equation

$$\begin{bmatrix} \dot{y}_1(t) \\ \dot{y}_2(t) \end{bmatrix} = \begin{bmatrix} 0 & 1 \\ -(\alpha^2 + \omega^2) & -2\alpha \end{bmatrix} \begin{bmatrix} y_1(t) \\ y_2(t) \end{bmatrix} + \begin{bmatrix} 0 \\ Ku(t) \end{bmatrix} \tag{7-339}$$

It is convenient to define a new set of state variables $x_1(t)$ and $x_2(t)$ by applying a suitable linear transformation, namely,

$$\begin{bmatrix} x_1(t) \\ x_2(t) \end{bmatrix} = \frac{1}{K} \begin{bmatrix} \omega & 0 \\ \alpha & 1 \end{bmatrix} \begin{bmatrix} y_1(t) \\ y_2(t) \end{bmatrix} \tag{7-340}$$

The $x$ variables satisfy the vector differential equation

$$\begin{bmatrix} \dot{x}_1(t) \\ \dot{x}_2(t) \end{bmatrix} = \begin{bmatrix} -\alpha & \omega \\ -\omega & -\alpha \end{bmatrix} \begin{bmatrix} x_1(t) \\ x_2(t) \end{bmatrix} + \begin{bmatrix} 0 \\ u(t) \end{bmatrix} \tag{7-341}$$

**Exercise 7-47** Verify Eq. (7-341).

**Exercise 7-48** Show that if the spinning body of Example 7-4 is subjected to linear friction, then the differential equations of the velocities have the same form as Eq. (7-341).

We note that if in Eq. (7-341) we set $\alpha = 0$, we obtain the differential equation for the undamped harmonic oscillator [see Eq. (7-301)].

Let us examine the system matrix in Eq. (7-341); we note that

$$\begin{bmatrix} -\alpha & \omega \\ -\omega & -\alpha \end{bmatrix} = \begin{bmatrix} -\alpha & 0 \\ 0 & -\alpha \end{bmatrix} + \begin{bmatrix} 0 & \omega \\ -\omega & 0 \end{bmatrix} \tag{7-342}$$

In other words, the system matrix is the sum of a *diagonal matrix* (with elements the real part of the complex poles or eigenvalues) and of a *skew-symmetric matrix* (with elements the imaginary part of the poles or eigenvalues).

Now we formulate the time-optimal control problem for the damped harmonic oscillator.

***Problem 7-8*** *Given the system* (7-341), *determine the control* $u(t)$, *subject to the constraint* $|u(t)| \leq 1$, *which will drive any initial state* $(\xi_1, \xi_2)$ *to the origin* (0, 0) *in minimum time.*

We shall solve this problem by using the techniques of Sec. 7-7. For this reason, we shall not present a detailed exposition of the solution (Control Law 7-8), trusting that the reader can fill in the gaps by referring to Sec. 7-7.

The Hamiltonian $H$ is given by

$$H = 1 - \alpha x_1(t)p_1(t) + \omega x_2(t)p_1(t) - \omega x_1(t)p_2(t) - \alpha x_2(t)p_2(t) + u(t)p_2(t) \tag{7-343}$$

The control which absolutely minimizes the Hamiltonian is

$$u(t) = -\operatorname{sgn}\{p_2(t)\} \tag{7-344}$$

The costate variables $p_1(t)$ and $p_2(t)$ satisfy the vector differential equation

$$\begin{bmatrix} \dot{p}_1(t) \\ \dot{p}_2(t) \end{bmatrix} = \begin{bmatrix} \alpha & \omega \\ -\omega & \alpha \end{bmatrix} \begin{bmatrix} p_1(t) \\ p_2(t) \end{bmatrix} \tag{7-345}$$

The fundamental matrix $\boldsymbol{\Psi}(t)$ for the system (7-345) is

$$\boldsymbol{\Psi}(t) = e^{\alpha t} \begin{bmatrix} \cos \omega t & \sin \omega t \\ -\sin \omega t & \cos \omega t \end{bmatrix} \tag{7-346}$$

Let

$$\mathbf{M}(t) = \begin{bmatrix} \cos \omega t & \sin \omega t \\ -\sin \omega t & \cos \omega t \end{bmatrix} \tag{7-347}$$

Clearly, $\mathbf{M}(t)$ is an orthogonal matrix.† Hence the costate vector $\mathbf{p}(t)$ is given by

$$\mathbf{p}(t) = e^{\alpha t}\mathbf{M}(t)\boldsymbol{\pi} \qquad \boldsymbol{\pi} = \mathbf{p}(0) \tag{7-348}$$

and

$$\|\mathbf{p}(t)\| = e^{\alpha t}\|\boldsymbol{\pi}\| \tag{7-349}$$

This implies that the costate vector $\mathbf{p}(t)$ describes a spiral moving away from the origin in the $p_1p_2$ plane. The function of interest, $p_2(t)$, is given by the relation

$$p_2(t) = e^{\alpha t}(\pi_1 \cos \omega t + \pi_2 \sin \omega t) \tag{7-350}$$

which means that (since $\alpha > 0$) the function $p_2(t)$ is the product of an increasing exponential and a sinusoid. Figure 7-53 illustrates a typical

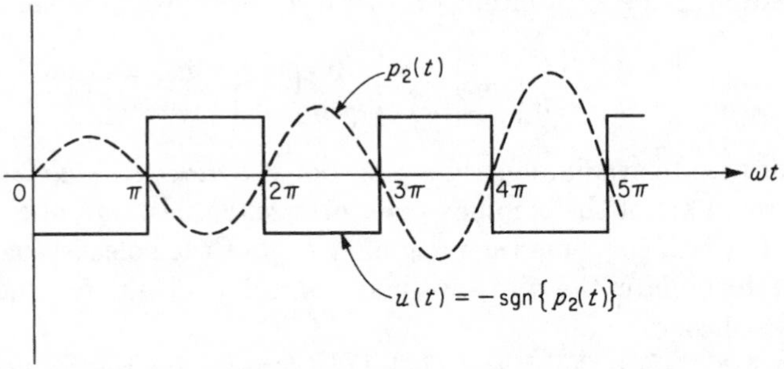


**Fig. 7-53** A typical function $p_2(t)$ and the control generated by Eq. (7-344).

function $p_2(t)$ and the control $u(t)$, defined by Eq. (7-344) as a function of $p_2(t)$. Observe that the control $u(t)$ has the following properties:

1. It must be piecewise constant and must switch between the values $u = +1$ and $u = -1$.
2. It cannot remain constant for more than $\pi/\omega$ units of time.
3. There is no upper bound on the number of switchings.
4. There is no possibility of singular control.

Note that the above necessary conditions on the time-optimal control are identical to those in Sec. 7-7.

† See Eqs. (2-109) to (2-111).

We now proceed to the solution of Eq. (7-341) using the control

$$u = \Delta = \pm 1 \tag{7-351}$$

If we define

$$\xi_1 = x_1(0) \qquad \xi_2 = x_2(0) \tag{7-352}$$

then we find (after extensive algebraic manipulations) that

$$\frac{\omega^2+\alpha^2}{\omega} x_1(t) = \frac{\omega^2+\alpha^2}{\omega} \xi_1 e^{-\alpha t} \cos \omega t + \frac{\omega^2+\alpha^2}{\omega} \xi_2 e^{-\alpha t} \sin \omega t - \Delta \frac{\sqrt{\omega^2+\alpha^2}}{\omega} e^{-\alpha t} \sin(\omega t + \psi) + \Delta \tag{7-353}$$

$$\frac{\alpha^2+\omega^2}{\omega} x_2(t) = -\frac{\alpha^2+\omega^2}{\alpha} \xi_1 e^{-\alpha t} \sin \omega t + \frac{\alpha^2+\omega^2}{\alpha} \xi_2 e^{-\alpha t} \cos \omega t - \Delta \frac{\sqrt{\alpha^2+\omega^2}}{\omega} e^{-\alpha t} \cos(\omega t + \psi) + \Delta \frac{\alpha}{\omega} \tag{7-354}$$

where

$$\psi = \tan^{-1} \frac{\omega}{\alpha} \tag{7-355}$$

We note that

$$\lim_{t\to\infty} x_1(t) = \Delta \qquad \lim_{t\to\infty} x_2(t) = \Delta \frac{\alpha}{\omega} \tag{7-356}$$

The equations (7-353) to (7-355) describe logarithmic spirals which tend to the point $\left(\Delta, \Delta \frac{\alpha}{\omega}\right)$ of the $\left(\frac{\alpha^2+\omega^2}{\omega} x_1, \frac{\alpha^2+\omega^2}{\omega} x_2\right)$ plane. Unfortunately, owing to the transcendental nature of Eqs. (7-353) and (7-354), we cannot eliminate the time in order to obtain the equation of the trajectories. Nevertheless, we can proceed to the following definition:

***Definition 7-28*** *Let $\gamma_+^0$ denote the set of states which can be forced to the origin by the control $u = \Delta = +1$ in no more than $\pi/\omega$ seconds. Let $\gamma_-^0$ denote the set of states which can be forced to the origin by the control $u = \Delta = -1$ in no more than $\pi/\omega$ seconds. Clearly, the $\gamma_+^0$ and $\gamma_-^0$ curves provide paths to the origin which satisfy all the necessary conditions for optimality.*

The shape of the $\gamma_+^0$ and $\gamma_-^0$ curves is illustrated in Fig. 7-54. Using similar arguments to those of Sec. 7-7, we can indeed demonstrate (by the familiar technique of elimination) *that the $\gamma_+^0$ and $\gamma_-^0$ curves are time-optimal paths to the origin.*

We proceed by using reasoning identical to that used in Sec. 7-7 in order to generate a series of curves $\gamma_+^j$ and $\gamma_-^j$ for $j = 1, 2, \ldots$ . We define the $\gamma_-^1$ curve as the set of all points which can be forced by the control $u = -1$ to a point on the $\gamma_+^0$ curve in *exactly* $\pi/\omega$ seconds. Similarly, we define the $\gamma_+^1$ curve as the set of all points which can be forced by the control $u = +1$ to a point on the $\gamma_-^0$ curve in *exactly* $\pi/\omega$ seconds. Thus we have:

*Definition 7-29* *The $\gamma_-^j$ curves, $j = 1, 2, \ldots$, are the sets of points which can be forced by the control $u = -1$ to the $\gamma_+^{j-1}$ curve in exactly $\pi/\omega$ seconds. The $\gamma_+^j$ curves, $j = 1, 2, \ldots$, are the sets of points which can be forced by the control $u = +1$ to the $\gamma_-^{j-1}$ curve in exactly $\pi/\omega$ seconds.*

The $\gamma_+^j$ and $\gamma_-^j$ curves are shown in Fig. 7-54.

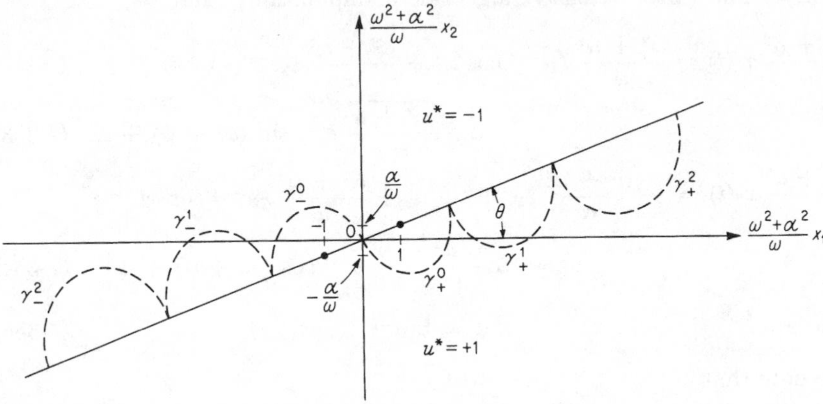


**Fig. 7-54** The $\gamma$ switch curve for the damped harmonic oscillator. The time-optimal control is $u^* = -1$ for states above the switch curve and $u^* = +1$ for states below the switch curve.

*Definition 7-30* *The $\gamma$ switch curve is the union of all the $\gamma_+^j$ and $\gamma_-^j$ curves; in other words,*

$$\gamma = \{\bigcup_{j=0}^{\infty} \gamma_+^j\} \cup \{\bigcup_{j=0}^{\infty} \gamma_-^j\} \tag{7-357}$$

*Control Law 7-8* ***Solution to Problem 7-8*** *The time-optimal control as a function of the state $\left(\frac{\omega^2+\alpha^2}{\omega}x_1, \frac{\omega^2+\alpha^2}{\omega}x_2\right)$ is given by:*

*If $\left(\frac{\omega^2+\alpha^2}{\omega}x_1, \frac{\omega^2+\alpha^2}{\omega}x_2\right)$ is above the $\gamma$ curve, then $u^* = -1$.*

*If $\left(\frac{\omega^2+\alpha^2}{\omega}x_1, \frac{\omega^2+\alpha^2}{\omega}x_2\right)$ is below the $\gamma$ curve, then $u^* = +1$.*

*If $\left(\frac{\omega^2+\alpha^2}{\omega}x_1, \frac{\omega^2+\alpha^2}{\omega}x_2\right)$ belongs to the $\gamma_+^0$ curve, then $u^* = +1$.*

*If $\left(\frac{\omega^2+\alpha^2}{\omega}x_1, \frac{\omega^2+\alpha^2}{\omega}x_2\right)$ belongs to the $\gamma_-^0$ curve, then $u^* = -1$.*

Let us now compare Figs. 7-54 and 7-40 and examine the significance of the constant $\alpha$. It is easy to see that if $\alpha \to 0$, then (the angle) $\theta \to 0$ in Fig. 7-54 and the $\gamma$ curve of Fig. 7-54 will become identical to the $\gamma$ curve of Fig. 7-40. However, for $\alpha > 0$, the size of the $\gamma_+^j$ and $\gamma_-^j$ curves increases with increasing integer values of $j$; this, of course, is the natural consequence of the exponential term $e^{-\alpha t}$ which governs the size of the spiral trajectories.

**Exercise 7-49** Construct approximations to the $\gamma$ curve for generating suboptimal control policies. (HINT: Look over Examples 7-5 and 7-6.) Consider the specific values $\omega = 5$ and $\alpha = 1$, and construct the exact $\gamma$ curve and the approximations you suggested. Evaluate the "goodness" of each of your approximations by plotting the percentage of increase of the response time over the minimum response time.

**Exercise 7-50** Consider the system described by the differential equation

$$\begin{bmatrix} \dot{x}_1(t) \\ \dot{x}_2(t) \end{bmatrix} = \begin{bmatrix} \alpha & \omega \\ -\omega & \alpha \end{bmatrix} \begin{bmatrix} x_1(t) \\ x_2(t) \end{bmatrix} + \begin{bmatrix} 0 \\ u(t) \end{bmatrix} \qquad |u(t)| \leq 1 \tag{7-358}$$

for $\alpha > 0$, $\omega > 0$. This system represents an unstable harmonic oscillator because the eigenvalues (or poles) are at $\alpha + j\omega$ and $\alpha - j\omega$. First, determine the set of initial states for which it is possible to reach the origin (since the system is inherently unstable, there are initial states for which the bounded control cannot provide sufficient force to drive these states to the origin). For this set of controllable states, find the switch curve and the time-optimal control law. HINT: The $\gamma_+^j$ and $\gamma_-^j$ curves for this system diminish in size for increasing values of $j$; see also Ref. C-3, fig. 9-13.

## 7-9 Time-optimal Control of a Harmonic Oscillator with Two Control Variables†

In the preceding sections of this chapter, we examined the time-optimal control problem for dynamical systems with the common property that the control action was accomplished by a single control variable, $u(t)$, subject to the constraint $|u(t)| \leq 1$. In this section, we shall examine the time-optimal control problem for a dynamical system with two control variables, $u_1(t)$ and $u_2(t)$, which are constrained by $|u_1(t)| \leq 1$ and $|u_2(t)| \leq 1$. It is our hope that this simple system will demonstrate to the reader that the techniques which we developed for the single-control-variable case can be readily extended to the case of several control variables. Although the control laws tend to be more complex, the concepts involved present no particular difficulty.

The system we shall examine in this section is the spinning space body already described in Example 7-4. For the sake of completeness, we shall describe the physical body in detail and present its differential equations of motion.

Consider a body in space with a single axis of symmetry,‡ as shown in Fig. 7-55. We define the three principal body axes 1, 2, 3, and we let $x_1(t)$, $x_2(t)$, and $x_3(t)$ denote the *angular velocities* (in radians per second) about

† Appropriate references for this section are A-5, A-9, and P-5, pp. 36–43.

‡ Although the body illustrated in Fig. 7-55 is cylindrical, there are many noncylindrical satellites and reentry vehicles which have a single axis of symmetry, such as the Mercury, Gemini, and Apollo capsules and proposed space stations.

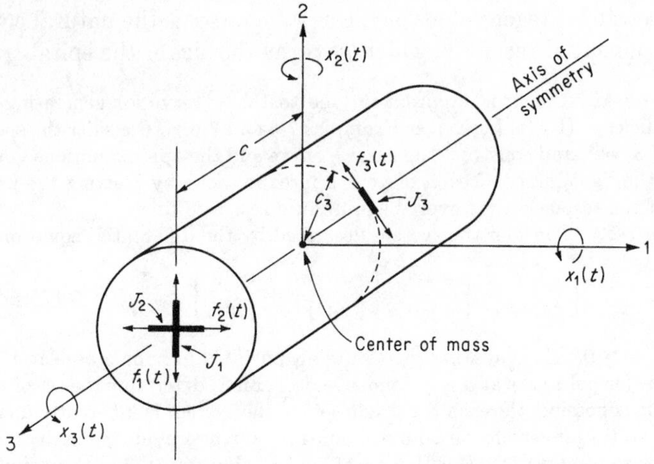


**Fig. 7-55** The space body with a single axis of symmetry.

the three body axes 1, 2, 3, respectively. Let $I_1$, $I_2$, $I_3$ represent the moments of inertia about the 1, 2, 3 axes, respectively. It is well known† that in the absence of external torques, the differential equations satisfied by the three angular velocities $x_1(t)$, $x_2(t)$, and $x_3(t)$ are

$$\begin{aligned} I_1\dot{x}_1(t) &= (I_2 - I_3)x_2(t)x_3(t) \\ I_2\dot{x}_2(t) &= (I_3 - I_1)x_3(t)x_1(t) \\ I_3\dot{x}_3(t) &= (I_1 - I_2)x_1(t)x_2(t) \end{aligned} \tag{7-359}$$

The above equations are the so-called Euler equations of motion. If the body is subject to external torques, then the differential equations of motion are

$$\begin{aligned} I_1\dot{x}_1(t) &= (I_2 - I_3)x_2(t)x_3(t) + \tau_1(t) \\ I_2\dot{x}_2(t) &= (I_3 - I_1)x_3(t)x_1(t) + \tau_2(t) \\ I_3\dot{x}_3(t) &= (I_1 - I_2)x_1(t)x_2(t) + \tau_3(t) \end{aligned} \tag{7-360}$$

where $\tau_1(t)$, $\tau_2(t)$, $\tau_3(t)$ are the torques about the 1, 2, 3 axes, respectively, and represent the components of a torque vector

$$\boldsymbol{\tau}(t) = \begin{bmatrix} \tau_1(t) \\ \tau_2(t) \\ \tau_3(t) \end{bmatrix} \tag{7-361}$$

Let us now examine the body illustrated in Fig. 7-55. Since the 3-axis is the symmetry axis, it follows that the two moments of inertia $I_1$ and $I_2$ are equal; that is,

$$I_1 = I_2 = I \tag{7-362}$$

† Reference G-7, chap. 5.

Now let $J_1$, $J_2$, and $J_3$ be three gas jets mounted on the body, as shown in Fig. 7-55. It is clear that the gas jet $J_3$ produces a torque $\tau_3(t)$ about the 3-axis only, the jet $J_2$ produces a torque $\tau_2(t)$ about the 2-axis only, and the jet $J_1$ produces a torque $\tau_1(t)$ about the 1-axis only. We assume that the three gas jets are fixed on the body and that they can "fire" in both directions.

Let $c$ be the distance between the plane defined by the jets $J_1$ and $J_2$ and the plane of the 1 and 2 axes, and let $c_3$ be the distance from $J_3$ to the 3-axis (for the cylinder of Fig. 7-55, $c$ represents half of its length and $c_3$ its radius). Then, if $f_1(t)$, $f_2(t)$, and $f_3(t)$ represent the thrusts produced by the jets $J_1$, $J_2$, and $J_3$, respectively, it is clear that

$$\tau_1(t) = cf_1(t) \qquad \tau_2(t) = cf_2(t) \qquad \tau_3(t) = c_3f_3(t) \tag{7-363}$$

Using Eqs. (7-363) and (7-362), we find that the equations of motion (7-360) reduce to

$$\dot{x}_1(t) = \frac{I - I_3}{I} x_3(t)x_2(t) + \frac{c}{I} f_1(t) \tag{7-364}$$

$$\dot{x}_2(t) = -\frac{I - I_3}{I} x_3(t)x_1(t) + \frac{c}{I} f_2(t) \tag{7-365}$$

$$\dot{x}_3(t) = \frac{c_3}{I_3} f_3(t) \tag{7-366}$$

Let us examine these equations for a moment. Suppose that at $t = 0$, the three velocities have the values

$$x_1(0) = \xi_1 \qquad x_2(0) = \xi_2 \qquad x_3(0) = \xi_3 \tag{7-367}$$

It is clear that if the thrust $f_3(t)$ is known, then we may integrate Eq. (7-366) to obtain

$$x_3(t) = \xi_3 + \int_0^t \frac{c_3}{I_3} f_3(\tau)\, d\tau \tag{7-368}$$

so that the velocity $x_3(t)$ can be controlled independently of the velocities $x_1(t)$ and $x_2(t)$, and so $x_3(t)$ can be treated as a known function of time. However, since the quantity $x_3(t)$ appears in Eqs. (7-364) and (7-365), the velocity $x_3(t)$ does influence $x_1(t)$ and $x_2(t)$.†

We shall examine a problem of time-optimal velocity control which is motivated by some physical considerations. Suppose that the body represents a reentering capsule and that a constant spin about the symmetry axis is required for aerodynamic stability considerations. We assume that the velocity $x_3(t)$ has been brought to its desired value (say $\xi_3$) using the jet $J_3$ only; but owing to the gyroscopic coupling, the body is now tumbling in space, i.e., the velocities $x_1(t)$ and $x_2(t)$ are not zero, and it is therefore

† This is due, of course, to the familiar gyroscopic effect.

desirable to reduce $x_1(t)$ and $x_2(t)$ to zero in minimum time while maintaining the $x_3(t)$ velocity at its constant value $\xi_3$. Another example is provided by a manned space station with artificial gravity generated by a constant angular velocity about the symmetry axis; in such a case, the reduction of the other two angular velocities $x_1(t)$ and $x_2(t)$ to zero in minimum time might be required.† In both of the above situations, it is desirable to maintain $x_3(t)$ at a constant level and to drive both $x_1(t)$ and $x_2(t)$ to zero.

If we set

$$f_3(t) = 0 \tag{7-369}$$

in Eqs. (7-366) and (7-367), we find that

$$x_3(t) = \xi_3 = \text{const} \tag{7-370}$$

Now suppose that the thrusts $f_1(t)$ and $f_2(t)$ of the jets $J_1$ and $J_2$ are bounded in magnitude by the relations

$$|f_1(t)| \le F \qquad |f_2(t)| \le F \tag{7-371}$$

This means that the jets $J_1$ and $J_2$ are identical. Define

$$\omega = \frac{I - I_3}{I}\,\xi_3 \tag{7-372}$$

$$k = \frac{c}{I}\,F \tag{7-373}$$

$$u_1(t) = \frac{f_1(t)}{F} \tag{7-374}$$

$$u_2(t) = \frac{f_2(t)}{F} \tag{7-375}$$

and substitute in Eqs. (7-364) and (7-365) to obtain

$$\begin{aligned} \dot{x}_1(t) &= \omega x_2(t) + k u_1(t) \\ \dot{x}_2(t) &= -\omega x_1(t) + k u_2(t) \end{aligned} \tag{7-376}$$

In the above system of equations, we can call the angular velocities $x_1(t)$ and $x_2(t)$ the *state variables* and the quantities $u_1(t)$ and $u_2(t)$ the *control variables*, which, owing to Eqs. (7-371), (7-374), and (7-375), are subject to the constraints

$$|u_1(t)| \le 1 \qquad |u_2(t)| \le 1 \qquad \text{for all } t \tag{7-377}$$

We now state the physical problem: *Determine the thrusts $f_1(t)$ and $f_2(t)$ which reduce the angular velocities $x_1(t)$ and $x_2(t)$ to zero in minimum time.* This problem is equivalent to the mathematical problem:

† To eliminate gravity effects which, if severe, might upset the balance of men and equipment.

***Problem 7-9*** *Given the system described by the vector differential equation*

$$\begin{bmatrix} \dot{x}_1(t) \\ \dot{x}_2(t) \end{bmatrix} = \begin{bmatrix} 0 & \omega \\ -\omega & 0 \end{bmatrix} \begin{bmatrix} x_1(t) \\ x_2(t) \end{bmatrix} + k \begin{bmatrix} u_1(t) \\ u_2(t) \end{bmatrix} \tag{7-378}$$

*where*

$$|u_1(t)| \leq 1 \qquad |u_2(t)| \leq 1 \tag{7-379}$$

*Find the controls* $u_1(t)$ *and* $u_2(t)$ *which drive any initial state*

$$x_1(0) = \xi_1 \qquad x_2(0) = \xi_2 \tag{7-380}$$

*to the origin* (0, 0) *in minimum time.*

The reader should compare Eq. (7-378) with Eq. (7-301) and observe that the system matrix is the same. This is the reason for calling the system (7-378) an *undamped harmonic oscillator.*

We shall use the minimum principle to determine the necessary conditions on the optimal controls. The Hamiltonian for the minimum-time control of the system (7-378) is given by the equation

$$H = 1 + \omega x_2(t)p_1(t) - \omega x_1(t)p_2(t) + ku_1(t)p_1(t) + ku_2(t)p_2(t) \tag{7-381}$$

where the costate variables $p_1(t)$ and $p_2(t)$ are solutions of the differential equations

$$\begin{aligned} \dot{p}_1(t) &= -\frac{\partial H}{\partial x_1(t)} = \omega p_2(t) \\ \dot{p}_2(t) &= -\frac{\partial H}{\partial x_2(t)} = -\omega p_1(t) \end{aligned} \tag{7-382}$$

which may be written in the form

$$\begin{bmatrix} \dot{p}_1(t) \\ \dot{p}_2(t) \end{bmatrix} = \begin{bmatrix} 0 & \omega \\ -\omega & 0 \end{bmatrix} \begin{bmatrix} p_1(t) \\ p_2(t) \end{bmatrix} \tag{7-383}$$

Note that Eq. (7-383) is the same as Eq. (7-306), so that the fundamental matrix $\mathbf{\Phi}(t)$ of the system (7-383) is given by Eq. (7-296), which implies that

$$p_1(t) = \pi_1 \cos \omega t + \pi_2 \sin \omega t \tag{7-384}$$

$$p_2(t) = -\pi_1 \sin \omega t + \pi_2 \cos \omega t \tag{7-385}$$

where

$$\pi_1 = p_1(0) \qquad \pi_2 = p_2(0) \tag{7-386}$$

We observe that the controls $u_1(t)$ and $u_2(t)$ which absolutely minimize $H$ are given by (since $k$ is positive)

$$u_1(t) = -\operatorname{sgn}\{p_1(t)\} \tag{7-387}$$

$$u_2(t) = -\operatorname{sgn}\{p_2(t)\} \tag{7-388}$$

Equations (7-384) and (7-385) imply that the functions $p_1(t)$ and $p_2(t)$ are sinusoids of frequency $\omega$; moreover, the function $p_2(t)$ is, to within a constant, the time derivative of the function $p_1(t)$. This fact can be interpreted, in a language very familiar to all electrical engineers, as "the sinusoid $p_2(t)$ is 90 degrees (leading) out of phase with respect to the sinusoid $p_1(t)$."

Figure 7-56 shows two such sinusoids $p_1(t)$ and $p_2(t)$ and the controls $u_1(t)$ and $u_2(t)$ determined by Eqs. (7-387) and (7-388); close examination of Fig. 7-56 reveals the following facts, which are a consequence of the necessary conditions provided by the minimum principle.

1. Each control $u_1(t)$ and $u_2(t)$ is a piecewise constant function. Each control switches between the values $+1$ and $-1$; furthermore, there is *no* upper bound on the number of switchings.

2. Since each sinusoid changes its polarity every $\pi/\omega$ seconds, it follows that each control $u_1(t)$ and $u_2(t)$ can remain constant over an interval of time of at most $\pi/\omega$ seconds of length.

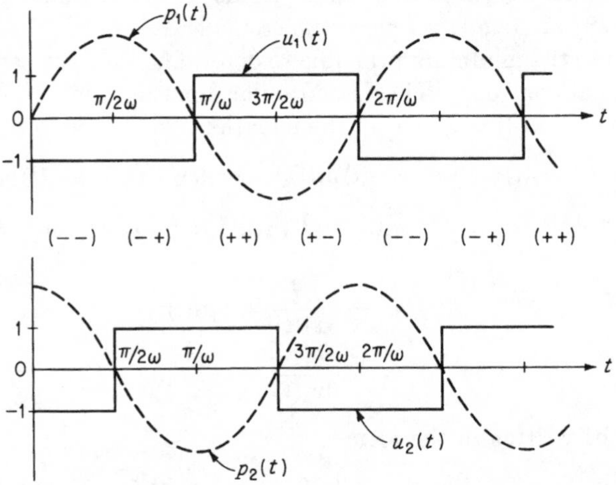


**Fig. 7-56** The two sinusoids $p_1(t)$ and $p_2(t)$ and the two piecewise constant controls generated by $u_1(t) = -\text{sgn}\{p_1(t)\}$ and $u_2(t) = -\text{sgn}\{p_2(t)\}$.

3. If $t_1 \leq t \leq t_2$ is the time interval over which *both* $u_1(t)$ *and* $u_2(t)$ remain constant, then $t_2 - t_1 \leq \pi/2\omega$.

4. If $u_1(t)$ is constant over the time interval $t_1 \leq t \leq t_2$ with $t_2 - t_1 = \pi/\omega$, then the control $u_2(t)$ must switch at exactly $t_3 = t_1 + \frac{1}{2}(t_2 - t_1)$, and vice versa.

5. As is evident from Fig. 7-56, the controls must have the polarity sequence

$$\ldots, (--), (-+), (++), (+-), (--), (-+), (++), (+-), \ldots \tag{7-389}$$

where the symbol $(--)$ means that over a finite interval of time, we have $u_1 = -1$ and $u_2 = -1$, the symbol $(-+)$ means that $u_1 = -1$ and $u_2 = +1$, and so on. The polarity sequence (7-389) also means that the present polarities of the controls fix the subsequent polarities [for example, if at the present time both controls are positive, $(++)$, then after the next switch only the control $u_2$ must switch from $+1$ to $-1$ to obtain $(+-)$ and it is impossible to have immediately after $(++)$ either $(--)$ or $(-+)$]. As we shall see, this information enables us to determine the time-optimal control law very easily by the familiar process of elimination used in all the previous sections of this chapter.

6. There exists no possibility of singular control (why?).

The next step is to solve the system of equations (7-378) using the constant controls

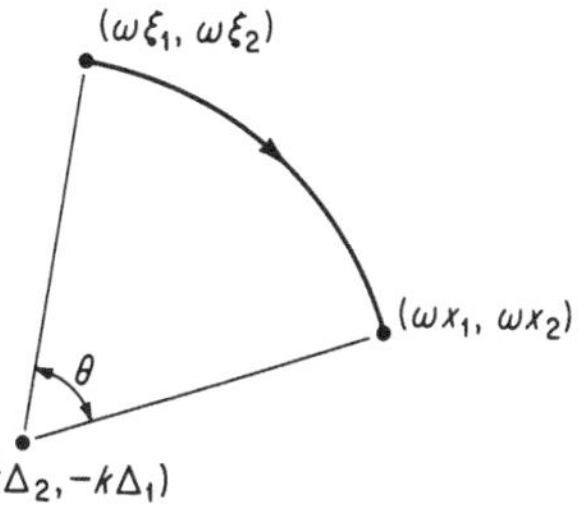


**Fig. 7-57** Angle $\theta$ determines transit time since $\theta = \omega t$.

$$u_1(t) = \Delta_1 = \pm 1 \tag{7-390}$$
$$u_2(t) = \Delta_2 = \pm 1 \tag{7-391}$$

The solution is, in terms of the initial conditions $x_1(0) = \xi_1$, $x_2(0) = \xi_2$,

$$\begin{aligned} \omega x_1(t) &= (\omega\xi_1 - k\Delta_2)\cos\omega t + (\omega\xi_2 + k\Delta_1)\sin\omega t + k\Delta_2 \\ \omega x_2(t) &= -(\omega\xi_1 - k\Delta_2)\sin\omega t + (\omega\xi_2 + k\Delta_1)\cos\omega t - k\Delta_1 \end{aligned} \tag{7-392}$$

Squaring and adding, we obtain the relation

$$[\omega x_1(t) - k\Delta_2]^2 + [\omega x_2(t) + k\Delta_1]^2 = (\omega\xi_1 - k\Delta_2)^2 + (\omega\xi_2 + k\Delta_1)^2 \tag{7-393}$$

The above equation means that *the trajectories are circles with center at the point* $(k\Delta_2, -k\Delta_1)$ *of the* $\omega x_1$-$\omega x_2$ *plane and with radius dependent on the initial conditions.* The motion is always in a clockwise direction. The time $t$ required to go from an initial state $(\omega\xi_1, \omega\xi_2)$ to the state $(\omega x_1, \omega x_2)$ along a circle with center at $(k\Delta_2, -k\Delta_1)$ is given by

$$t = \frac{1}{\omega}\tan^{-1}\frac{(\omega x_1 - k\Delta_2)(\omega\xi_2 + k\Delta_1) - (\omega x_2 + k\Delta_1)(\omega\xi_1 - k\Delta_2)}{(\omega x_2 + k\Delta_1)(\omega\xi_2 + k\Delta_1) + (\omega x_1 - k\Delta_2)(\omega\xi_1 - k\Delta_2)} \tag{7-394}$$

The time of transit $t$ given by Eq. (7-394) can be determined by measuring the angle $\theta$ in Fig. 7-57. Thus, we have

$$t = \frac{\theta}{\omega} \tag{7-395}$$

In the rest of this section, we shall work in the $\omega x_1$-$\omega x_2$ plane, as we have done in Sec. 7-7.

***Definition 7-31*** *Let* $\Gamma_{++}$ *be the set of all states* $(\omega x_1, \omega x_2)$ *which can be forced to the origin* (0, 0) *by the controls* $u_1 = +1$ *and* $u_2 = +1$. *Let* $\Gamma_{+-}$ *be the set of all states* $(\omega x_1, \omega x_2)$ *which can be forced to the origin* (0, 0) *by the controls* $u_1 = +1$ *and* $u_2 = -1$. *Let* $\Gamma_{-+}$ *be the set of all states which can be forced to the origin* (0, 0) *by the controls* $u_1 = -1$ *and* $u_2 = +1$. *Let* $\Gamma_{--}$ *be the set of all states which can be forced to the origin* (0, 0) *by the controls*

*$u_1 = -1$ and $u_2 = -1$. The sets $\Gamma_{++}$, $\Gamma_{+-}$, $\Gamma_{-+}$, and $\Gamma_{--}$ are circles which pass through the origin with centers at $(k, -k)$, $(-k, -k)$, $(k, k)$, and $(-k, k)$, respectively; that is,*

$$\Gamma_{++} = \{(\omega x_1, \omega x_2): (\omega x_1 - k)^2 + (\omega x_2 + k)^2 = 2k^2\} \tag{7-396}$$
$$\Gamma_{+-} = \{(\omega x_1, \omega x_2): (\omega x_1 + k)^2 + (\omega x_2 + k)^2 = 2k^2\} \tag{7-397}$$
$$\Gamma_{-+} = \{(\omega x_1, \omega x_2): (\omega x_1 - k)^2 + (\omega x_2 - k)^2 = 2k^2\} \tag{7-398}$$
$$\Gamma_{--} = \{(\omega x_1, \omega x_2): (\omega x_1 + k)^2 + (\omega x_2 - k)^2 = 2k^2\} \tag{7-399}$$

The four circles $\Gamma_{++}$, $\Gamma_{+-}$, $\Gamma_{-+}$, and $\Gamma_{--}$ are shown in Fig. 7-58, using $k = 1$.

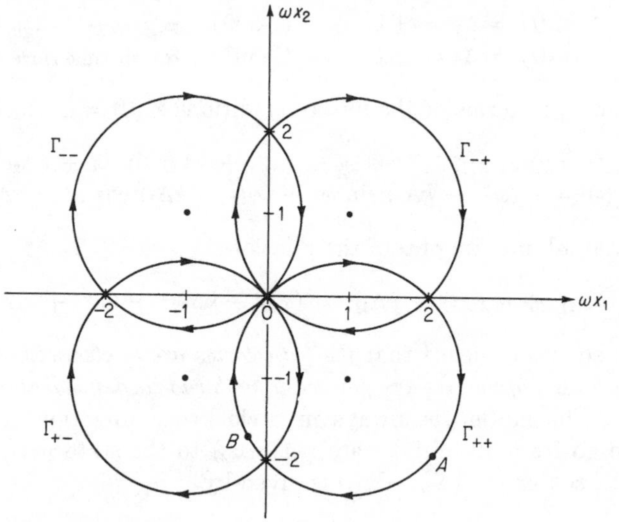


**Fig. 7-58** The four circles $\Gamma_{++}$, $\Gamma_{+-}$, $\Gamma_{-+}$, and $\Gamma_{--}$ which pass through the origin ($k = 1$).

The next step is to determine the portions of these four circles which are generated by controls which satisfy *all* the necessary conditions provided by the minimum principle. Consider the points $A$ and $B$ on the $\Gamma_{++}$ circle of Fig. 7-58. Both $A$ and $B$ can be forced to the origin 0 if $u_1 = +1$ and $u_2 = +1$; the time required to force $A$ to 0 is greater than $\pi/2\omega$ seconds (see Fig. 7-57 for the relation of angles and time), while the time required to force $B$ to 0 is less than $\pi/2\omega$ seconds. Since one of the controls must switch at most every $\pi/2\omega$ seconds, this means that the path $AO$ *cannot* be a time-optimal one. Such considerations lead us to the following definition:

***Definition 7-32*** *Let $\gamma_{++}{}^0$ denote the set of states $(\omega x_1, \omega x_2)$ which can be forced to the origin $(0, 0)$ in less than or equal to $\pi/2\omega$ seconds by the controls $u_1 = +1$, $u_2 = +1$. Let $\gamma_{+-}{}^0$ denote the set of states $(\omega x_1, \omega x_2)$ which can* *be forced to the origin* (0, 0) *in less than or equal to* $\pi/2\omega$ *seconds by the controls* $u_1 = +1$, $u_2 = -1$. *Let* $\gamma_{-+}{}^0$ *denote the set of states* $(\omega x_1, \omega x_2)$ *which can be forced to the origin* (0, 0) *in less than or equal to* $\pi/2\omega$ *seconds by the controls* $u_1 = -1$, $u_2 = +1$. *Let* $\gamma_{--}{}^0$ *denote the set of states* $(\omega x_1, \omega x_2)$ *which can be forced to the origin* (0, 0) *in less than or equal to* $\pi/2\omega$ *seconds by the controls* $u_1 = -1$, $u_2 = -1$.

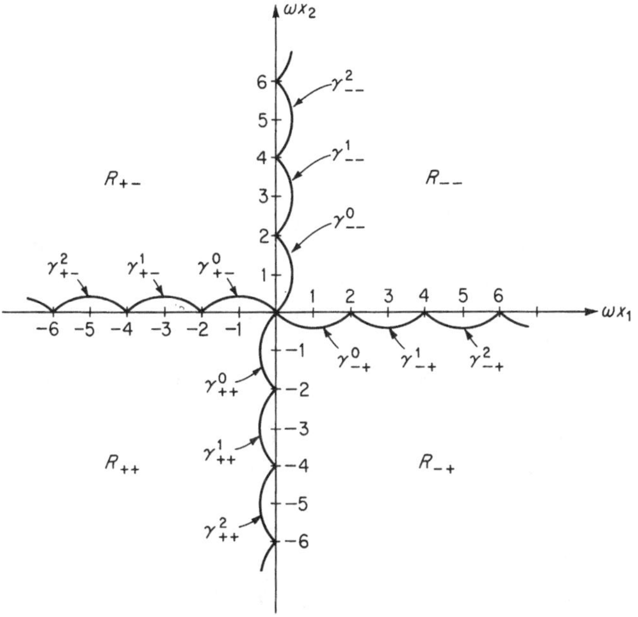


**Fig. 7-59** Illustration of the $\gamma_{++}{}^j$, $\gamma_{+-}{}^j$, $\gamma_{--}{}^j$, and $\gamma_{-+}{}^j$ quarter circles ($u = 1$) and of the four regions $R_{++}$, $R_{+-}$, $R_{--}$, and $R_{-+}$.

*The* $\gamma_{++}{}^0$, $\gamma_{+-}{}^0$, $\gamma_{-+}{}^0$, *and* $\gamma_{--}{}^0$ *sets are quarter circles and portions (or subsets) of the* $\Gamma_{++}$, $\Gamma_{+-}$, $\Gamma_{-+}$, *and* $\Gamma_{--}$ *circles, respectively. These sets are given by*

$$\gamma_{++}{}^0 = \{(\omega x_1, \omega x_2)\colon (\omega x_1 - k)^2 + (\omega x_2 + k)^2 = 2k^2;\quad \omega x_1 \le 0,\ -2k \le \omega x_2 \le 0\} \tag{7-400}$$

$$\gamma_{+-}{}^0 = \{(\omega x_1, \omega x_2)\colon (\omega x_1 + k)^2 + (\omega x_2 + k)^2 = 2k^2;\quad -2k \le \omega x_1 \le 0,\ \omega x_2 \ge 0\} \tag{7-401}$$

$$\gamma_{-+}{}^0 = \{(\omega x_1, \omega x_2)\colon (\omega x_1 - k)^2 + (\omega x_2 - k)^2 = 2k^2;\quad 0 \le \omega x_1 \le 2k,\ \omega x_2 \le 0\} \tag{7-402}$$

$$\gamma_{--}{}^0 = \{(\omega x_1, \omega x_2)\colon (\omega x_1 + k)^2 + (\omega x_2 - k)^2 = 2k^2;\quad 0 \le \omega x_1,\ 0 \le \omega x_2 \le 2k\} \tag{7-403}$$

The quarter circles $\gamma_{++}{}^0$, $\gamma_{+-}{}^0$, $\gamma_{-+}{}^0$, and $\gamma_{--}{}^0$ are shown in Fig. 7-59 for $k = 1$.

We assert that:

***Control Law 7-9a*** *Given a state* $(\omega x_1, \omega x_2) \in \gamma_{++}{}^0 \cup \gamma_{+-}{}^0 \cup \gamma_{-+}{}^0 \cup \gamma_{--}{}^0$; *then the time-optimal control to the origin is given by*

$$\begin{aligned} u_1^* &= +1, u_2^* = +1 && \textit{for all } (\omega x_1, \omega x_2) \in \gamma_{++}{}^0 \\ u_1^* &= +1, u_2^* = -1 && \textit{for all } (\omega x_1, \omega x_2) \in \gamma_{+-}{}^0 \\ u_1^* &= -1, u_2^* = +1 && \textit{for all } (\omega x_1, \omega x_2) \in \gamma_{-+}{}^0 \\ u_1^* &= -1, u_2^* = -1 && \textit{for all } (\omega x_1, \omega x_2) \in \gamma_{--}{}^0 \end{aligned} \tag{7-404}$$

**Exercise 7-51** Prove the validity of Control Law 7-9*a*. HINT: Use the usual technique of picking a point (say on the $\gamma_{++}{}^0$ curve) and show that among *all* controls which satisfy *all* the necessary conditions *only* the controls $u_1 = +1$, $u_2 = +1$ force the state to the origin and hence, by elimination, must be optimal. Similar techniques were used to prove Control Law 7-7*a*.

Since the $\gamma_{++}{}^0$ curve is a time-optimal path to the origin, there must exist states for which a portion of the $\gamma_{++}{}^0$ curve belongs to their time-optimal path to the origin; but since the controls along the $\gamma_{++}{}^0$ curve have the polarities $(++)$, we can conclude from (7-389) that the polarities of the controls which will drive these states to the $\gamma_{++}{}^0$ curve must be $(-+)$, that is, $u_1 = -1$ and $u_2 = +1$. Thus we can find a set of states† which can be forced to the $\gamma_{++}{}^0$ curve by the controls $u_1 = -1$ and $u_2 = +1$ in less than or equal to $\pi/2\omega$ seconds. In particular, we can find the set of states $\gamma_{-+}{}^1$ which can be forced to the $\gamma_{++}{}^0$ curve by the controls $u_1 = -1$, $u_2 = +1$ in *exactly* $\pi/2\omega$ seconds. These considerations lead us to the following definition:

***Definition 7-33*** *Let* $\gamma_{-+}{}^1$ *denote the set of states which can be forced to the* $\gamma_{++}{}^0$ *curve in exactly* $\pi/2\omega$ *seconds by the controls* $u_1 = -1$, $u_2 = +1$.

*Let* $\gamma_{--}{}^1$ *denote the set of states which can be forced to the* $\gamma_{-+}{}^0$ *curve in exactly* $\pi/2\omega$ *seconds by the controls* $u_1 = -1$ *and* $u_2 = -1$.

*Let* $\gamma_{+-}{}^1$ *denote the set of states which can be forced to the* $\gamma_{--}{}^0$ *curve in exactly* $\pi/2\omega$ *seconds by the controls* $u_1 = +1$, $u_2 = -1$.

*Let* $\gamma_{++}{}^1$ *denote the set of states which can be forced to the* $\gamma_{+-}{}^0$ *curve in exactly* $\pi/2\omega$ *seconds by the controls* $u_1 = +1$, $u_2 = +1$.

*The* $\gamma_{++}{}^1$, $\gamma_{-+}{}^1$, $\gamma_{--}{}^1$, *and* $\gamma_{+-}{}^1$ *curves are shown in Fig.* 7-59 *for* $k = 1$.

*It should be clear that:*

*The* $\gamma_{-+}{}^1$ *curve is obtained by a 90-degree counterclockwise rotation of the* $\gamma_{++}{}^0$ *curve about the point* $(k, k)$.

*The* $\gamma_{--}{}^1$ *curve is obtained by a 90-degree counterclockwise rotation of the* $\gamma_{-+}{}^0$ *curve about the point* $(-k, k)$.

*The* $\gamma_{+-}{}^1$ *curve is obtained by a 90-degree counterclockwise rotation of the* $\gamma_{--}{}^0$ *curve about the point* $(-k, -k)$.

*The* $\gamma_{++}{}^1$ *curve is obtained by a 90-degree counterclockwise rotation of the* $\gamma_{+-}{}^0$ *curve about the point* $(k, -k)$.

† Similar to the sets $R_-{}^1$ and $R_+{}^1$ given by Definitions 7-20 and 7-21.

We can generate a whole sequence of these curves, as indicated in the following definition:

***Definition 7-34*** *Let $\gamma_{-+}^{j}$, $\gamma_{--}^{j}$, $\gamma_{+-}^{j}$, and $\gamma_{++}^{j}$, $j = 1, 2, 3, \ldots$, denote curves constructed as follows:*

*The $\gamma_{-+}^{j}$ curve is the curve obtained by a 90-degree counterclockwise rotation of the $\gamma_{++}^{j-1}$ curve about the point $(k, k)$.*

*The $\gamma_{--}^{j}$ curve is the curve obtained by a 90-degree counterclockwise rotation of the $\gamma_{-+}^{j-1}$ curve about the point $(-k, k)$.*

*The $\gamma_{+-}^{j}$ curve is the curve obtained by a 90-degree counterclockwise rotation of the $\gamma_{--}^{j-1}$ curve about the point $(-k, -k)$.*

*The $\gamma_{++}^{j}$ curve is the curve obtained by a 90-degree counterclockwise rotation of the $\gamma_{+-}^{j-1}$ curve about the point $(k, -k)$.*

*The equations of the $\gamma_{-+}^{j}$, $\gamma_{--}^{j}$, $\gamma_{+-}^{j}$, and $\gamma_{++}^{j}$ curves, $j = 1, 2, 3, \ldots$, are*

$$\gamma_{++}^{j} = \{(\omega x_1, \omega x_2): (\omega x_1 - k)^2 + [\omega x_2 + (2j+1)k]^2 = 2k^2; \quad \omega x_1 \le 0; -(2j+2)k \le \omega x_2 \le -2jk\} \tag{7-405}$$

$$\gamma_{+-}^{j} = \{(\omega x_1, \omega x_2): [\omega x_1 + (2j+1)k]^2 + (\omega x_2 + k)^2 = 2k^2; \quad -(2j+2)k \le \omega x_1 \le -2jk; \omega x_2 \ge 0\} \tag{7-406}$$

$$\gamma_{-+}^{j} = \{(\omega x_1, \omega x_2): [\omega x_1 - (2j+1)k]^2 + (\omega x_2 - k)^2 = 2k^2; \quad 2jk \le \omega x_1 \le (2j+2)k; \omega x_2 \le 0\} \tag{7-407}$$

$$\gamma_{--}^{j} = \{(\omega x_1, \omega x_2): (\omega x_1 + k)^2 + [\omega x_2 - (2j+1)k]^2 = 2k^2; \quad \omega x_1 \ge 0; 2jk \le \omega x_2 \le (2j+1)k\} \tag{7-408}$$

The $\gamma_{++}^{j}$, $\gamma_{+-}^{j}$, $\gamma_{-+}^{j}$, and $\gamma_{--}^{j}$ curves are shown in Fig. 7-59.

We can now define four switch curves by setting

$$\begin{aligned} \gamma_{++} &= \bigcup_{j=0}^{\infty} \gamma_{++}^{j} \\ \gamma_{+-} &= \bigcup_{j=0}^{\infty} \gamma_{+-}^{j} \\ \gamma_{-+} &= \bigcup_{j=0}^{\infty} \gamma_{-+}^{j} \\ \gamma_{--} &= \bigcup_{j=0}^{\infty} \gamma_{--}^{j} \end{aligned} \tag{7-409}$$

These four switch curves divide the $\omega x_1$-$\omega x_2$ plane into four regions (or sets) $R_{++}$, $R_{+-}$, $R_{--}$, and $R_{-+}$, as shown in Fig. 7-59. In other words, $R_{++}$ is the set of points to the left of the $\gamma_{++}$ curve and below the $\gamma_{+-}$ curve; $R_{+-}$ is the set of points to the left of the $\gamma_{--}$ curve and above the $\gamma_{+-}$ curve; $R_{--}$ is the set of points to the right of the $\gamma_{--}$ curve and above the $\gamma_{-+}$ curve; $R_{-+}$ is the set of points to the right of the $\gamma_{++}$ curve and below the $\gamma_{-+}$ curve.

We assert that the time-optimal control law is:

***Control Law 7-9 Solution to Problem 7-9*** *The time-optimal control* $\mathbf{u}^*$, *as a function of the state* $(\omega x_1, \omega x_2)$, *is given by*

$$\begin{aligned} u_1^* &= +1, u_2^* = +1 && \text{for all } (\omega x_1, \omega x_2) \in R_{++} \cup \gamma_{++} \\ u_1^* &= +1, u_2^* = -1 && \text{for all } (\omega x_1, \omega x_2) \in R_{+-} \cup \gamma_{+-} \\ u_1^* &= -1, u_2^* = +1 && \text{for all } (\omega x_1, \omega x_2) \in R_{-+} \cup \gamma_{-+} \\ u_1^* &= -1, u_2^* = -1 && \text{for all } (\omega x_1, \omega x_2) \in R_{--} \cup \gamma_{--} \end{aligned} \tag{7-410}$$

*and, furthermore,* $\mathbf{u}^*$ *is unique.*

**Exercise 7-52** Prove the validity of Control Law 7-9.

Figure 7-60 shows several time-optimal trajectories to the origin for $k = 1$; note that the polarity sequence is that of (7-389).

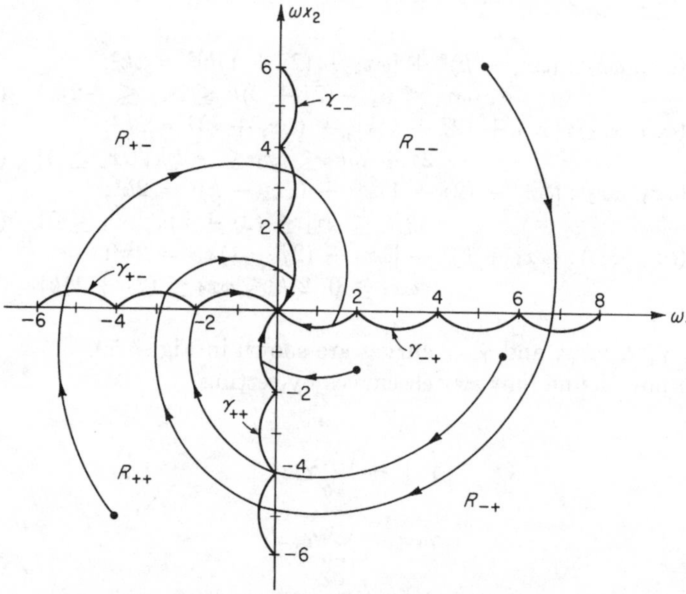


**Fig. 7-60** Several time-optimal trajectories to the origin.

Let us now turn our attention to the engineering realization of Control Law 7-9. A block diagram illustrating the time-optimal feedback control system is shown in Fig. 7-61. The control system measures the signals $\omega x_1$ and $\omega x_2$, decides which region ($R_{++}$, $R_{+-}$, $R_{-+}$, $R_{--}$) the state $(\omega x_1, \omega x_2)$ belongs to, and on the basis of that decision, produces the time-optimal controls $u_1^*(t)$ and $u_2^*(t)$ according to Control Law 7-9. Observe that if the state belongs to either $R_{--}$ or $R_{-+}$, then $u_1^*(t) = -1$ and if the state belongs to either $R_{+-}$ or $R_{++}$, then $u_1^*(t) = +1$, independent of $u_2^*(t)$; on the other hand, if the state belongs to either $R_{+-}$ or $R_{--}$, then $u_2^*(t) = -1$ and if the state belongs to either $R_{++}$ or $R_{-+}$, then $u_2^*(t) = +1$, independent of $u_1^*(t)$. Thus, in Fig. 7-61,

$$\begin{aligned}
&\text{If sgn } \{m(t)\} = +1, \text{ then } (\omega x_1, \omega x_2) \in R_{++} \cup R_{-+} \text{ and } u_2^* = +1 \\
&\text{If sgn } \{m(t)\} = -1, \text{ then } (\omega x_1, \omega x_2) \in R_{+-} \cup R_{--} \text{ and } u_2^* = -1 \\
&\text{If sgn } \{n(t)\} = +1, \text{ then } (\omega x_1, \omega x_2) \in R_{+-} \cup R_{++} \text{ and } u_1^* = +1 \\
&\text{If sgn } \{n(t)\} = -1, \text{ then } (\omega x_1, \omega x_2) \in R_{--} \cup R_{-+} \text{ and } u_1^* = -1
\end{aligned} \tag{7-411}$$

As shown in Fig. 7-61, the signal $\omega x_1$ is introduced to the nonlinearity $N_1$, and the signal $\omega x_2$ is introduced to the nonlinearity $N_2$. The nonlinearities $N_1$ and $N_2$ are identical, and their input-output characteristics are those of the switch curves $\gamma_{-+}$ and $\gamma_{+-}$ (or $\gamma_{--}$ and $\gamma_{++}$).

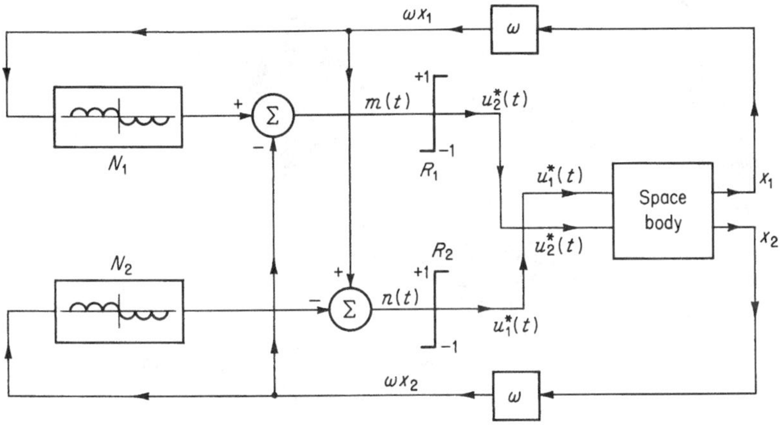


**Fig. 7-61** The block diagram of the time-optimal controller for the spinning space body.

Figure 7-62 illustrates time-optimal manual control. The state variables $\omega x_1$ and $\omega x_2$ are measured and introduced to the $X$ and $Y$ terminals of an oscilloscope; the switch curves have been drawn on the face of the oscilloscope. The man identifies the region in which the state belongs and pushes the appropriate buttons to fire the appropriate gas jets.†

We next turn our attention to the minimum isochrones for this system. Recall that the $S(t^*)$ minimum isochrone is the set of states which require the *same* minimum time $t^*$ to reach the origin. Several of these minimum isochrones are shown‡ in Fig. 7-63, for $k = 1$. We denote by $\hat{S}(t^*)$ the set with $S(t^*)$ as its boundary; that is,

$$\partial \hat{S}(t^*) = S(t^*) \tag{7-412}$$

† We have actually simulated this system on an analog computer, and we have tried our skills at manual time-optimal control. We have found that, after a minute of training, we could control the system optimally, at least for $\omega < 3$ radians/second. The human reaction time is too excessive for higher values of $\omega$.

‡ The method of construction of these isochrones is similar to that presented in Sec. 7-7. Explicit directions for the construction are given in Ref. A-5, pp. 22–23.

It is clear from Fig. 7-63 that the sets $\hat{S}(t^*)$ are *closed* and *convex*. We also note that the sets $\hat{S}(t^*)$ have corners at the $\gamma_{++}^0$, $\gamma_{+-}^0$, $\gamma_{-+}^0$, $\gamma_{--}^0$ curves, which means that in the (open) region (or set) of the state plane, defined by

$$(\omega x_1)^2 + (\omega x_2)^2 < 4k^2 \tag{7-413}$$

each set $\hat{S}(t^*)$, $t^* < \pi/2\omega$, has exactly four *nonregular points*† at the points

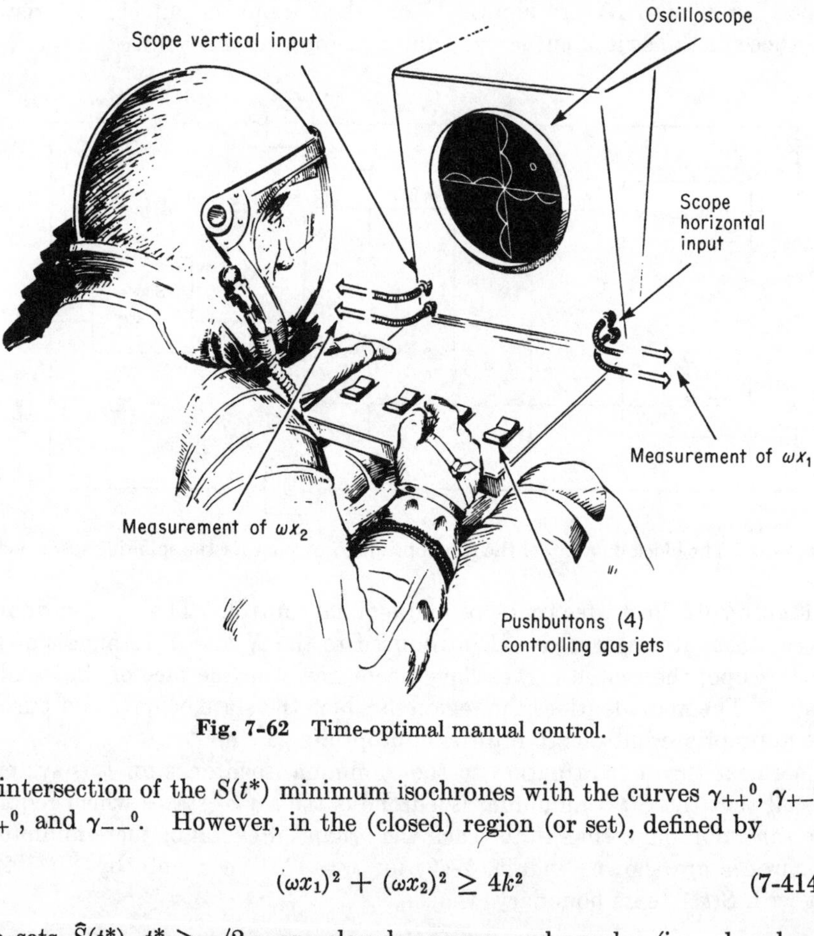


**Fig. 7-62** Time-optimal manual control.

of intersection of the $S(t^*)$ minimum isochrones with the curves $\gamma_{++}^0$, $\gamma_{+-}^0$, $\gamma_{-+}^0$, and $\gamma_{--}^0$. However, in the (closed) region (or set), defined by

$$(\omega x_1)^2 + (\omega x_2)^2 \geq 4k^2 \tag{7-414}$$

the sets $\hat{S}(t^*)$, $t^* \geq \pi/2\omega$, are closed, convex, and regular (i.e., they have no corners). The reader will notice that the isochrones of Fig. 7-63 have properties similar to those of Fig. 7-40.

Finally, it is easy to establish that once the state has been forced to the origin and in the absence of future disturbances, the controls $u_1 = u_2 = 0$ will keep the state at the origin.

† See Definition 3-22.

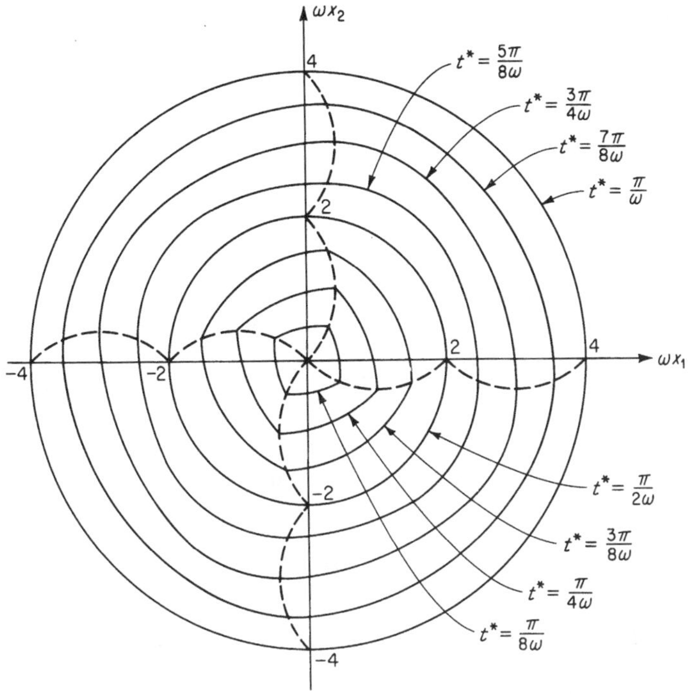


**Fig. 7-63** The minimum isochrones are made of circular arcs. The dashed curves are the switch curves.

**Exercise 7-53** In a manner analogous to that used in the suboptimal system of Example 7-5, consider the following approximation to the $\gamma_{++}$, $\gamma_{+-}$, $\gamma_{-+}$, and $\gamma_{--}$ curves:

$$\begin{aligned}
\hat{\gamma}_{++} &= \begin{cases} \gamma_{++}{}^0 & \text{if } -2k \le \omega x_2 \le 0 \\ \omega x_1 = 0 & \text{if } \omega x_2 < -2k \end{cases} \\
\hat{\gamma}_{+-} &= \begin{cases} \gamma_{+-}{}^0 & \text{if } -2k \le \omega x_1 \le 0 \\ \omega x_2 = 0 & \text{if } \omega x_1 < -2k \end{cases} \\
\hat{\gamma}_{-+} &= \begin{cases} \gamma_{-+}{}^0 & \text{if } 0 \le \omega x_1 \le 2k \\ \omega x_2 = 0 & \text{if } \omega x_1 > 2k \end{cases} \\
\hat{\gamma}_{--} &= \begin{cases} \gamma_{--}{}^0 & \text{if } 0 \le \omega x_2 \le 2k \\ \omega x_1 = 0 & \text{if } \omega x_2 > 2k \end{cases}
\end{aligned} \tag{7-415}$$

Analyze the resulting suboptimal system and demonstrate that it is stable. If $\hat{t}$ represents the response time of this suboptimal system and $t^*$ the minimum response time from the same state, compute and plot $(\hat{t}/t^*) \times 100$ percent for initial states on the $\omega x_1$ axis versus $|\omega x_1|$. Compare your results with those of Fig. 7-48, using $k = 1$.

**Exercise 7-54** Repeat Exercise 7-53 for the following approximation to the switch curves:

$$\begin{aligned}
\tilde{\gamma}_{++} &= \begin{cases} \gamma_{++}{}^0 & \text{if } -k \le \omega x_2 \le 0 \\ \omega x_1 = -k\sqrt{2} + 1 & \text{if } \omega x_2 < -k \end{cases} \\
\tilde{\gamma}_{+-} &= \begin{cases} \gamma_{+-}{}^0 & \text{if } -k \le \omega x_1 \le 0 \\ \omega x_2 = k\sqrt{2} - 1 & \text{if } \omega x_1 < -k \end{cases} \\
\tilde{\gamma}_{-+} &= \begin{cases} \gamma_{-+}{}^0 & \text{if } 0 \le \omega x_1 \le k \\ \omega x_2 = -k\sqrt{2} + 1 & \text{if } \omega x_1 > k \end{cases} \\
\tilde{\gamma}_{--} &= \begin{cases} \gamma_{--}{}^0 & \text{if } 0 \le \omega x_2 \le k \\ \omega x_1 = k\sqrt{2} - 1 & \text{if } \omega x_2 > k \end{cases}
\end{aligned} \tag{7-416}$$

**Exercise 7-55** Consider the system

$$\begin{bmatrix}\dot{x}_1(t)\\ \dot{x}_2(t)\end{bmatrix} = \begin{bmatrix}0 & 1\\ -1 & 0\end{bmatrix}\begin{bmatrix}x_1(t)\\ x_2(t)\end{bmatrix} + \begin{bmatrix}u_1(t)\\ u_2(t)\end{bmatrix} \tag{7-417}$$

where

$$|u_1(t)| \le 1 \qquad |u_2(t)| \le 2 \tag{7-418}$$

Determine the switch curves and the time-optimal control law if the origin is the desired terminal state.

**Exercise 7-56** Consider the system (damped harmonic oscillator with two control variables)

$$\begin{bmatrix}\dot{x}_1(t)\\ \dot{x}_2(t)\end{bmatrix} = \begin{bmatrix}-1 & 1\\ -1 & -1\end{bmatrix}\begin{bmatrix}x_1(t)\\ x_2(t)\end{bmatrix} + \begin{bmatrix}u_1(t)\\ u_2(t)\end{bmatrix} \tag{7-419}$$

where

$$|u_1(t)| \le 1 \qquad |u_2(t)| \le 1 \tag{7-420}$$

Using the results of Sec. 7-8, determine the time-optimal control law if the origin is the terminal state.

The following exercise illustrates the switch curves for systems with real eigenvalues but with two control variables.

**Exercise 7-57** Consider the second-order system

$$\begin{bmatrix}\dot{x}_1(t)\\ \dot{x}_2(t)\end{bmatrix} = \mathbf{A}\begin{bmatrix}x_1(t)\\ x_2(t)\end{bmatrix} + \mathbf{B}\begin{bmatrix}u_1(t)\\ u_2(t)\end{bmatrix} \tag{7-421}$$

where **A** and **B** are $2 \times 2$ matrices. Assume that the control variables are restricted in magnitude by the relations

$$|u_1(t)| \le 1 \qquad |u_2(t)| \le 1 \tag{7-422}$$

Find the time-optimal feedback control law when the origin (0, 0) is the terminal state for the following cases:

(*a*) $\mathbf{A} = \begin{bmatrix}-1 & 0\\ 0 & -2\end{bmatrix}, \mathbf{B} = \begin{bmatrix}1 & 2\\ 1 & 1\end{bmatrix}$ (*b*) $\mathbf{A} = \begin{bmatrix}-1 & 0\\ 0 & -2\end{bmatrix}, \mathbf{B} = \begin{bmatrix}1 & 0\\ 0 & 1\end{bmatrix}$

(*c*) $\mathbf{A} = \begin{bmatrix}0 & 0\\ 0 & 1\end{bmatrix}, \mathbf{B} = \begin{bmatrix}1 & 1\\ 1 & -1\end{bmatrix}$ (*d*) $\mathbf{A} = \begin{bmatrix}0 & 1\\ 0 & -1\end{bmatrix}, \mathbf{B} = \begin{bmatrix}0 & 1\\ 1 & 0\end{bmatrix}$

In each case, check to see whether the system is normal. If the time-optimal control is not unique, explain why in physical terms.

## 7-10 Time-optimal Control of First-order Nonlinear Systems

In the preceding sections of this chapter, we have been concerned with the time-optimal control problem for simple and complex *linear* time-invariant systems. In this section, we shall examine the time-optimal control problem for a first-order nonlinear time-varying system.

Consider the system described by the first-order nonlinear differential equation

$$\dot{x}(t) + g[x(t); t] = Ku(t) \tag{7-423}$$

where $x(t)$ is the (scalar) state of the system, $g[x(t); t]$ is some nonlinear function of the state and of the time, $u(t)$ is the (scalar) control variable, and $K$ is a positive gain constant. We assume that the control $u(t)$ is restricted in magnitude by the relation

$$|u(t)| \leq 1 \qquad \text{for all } t \tag{7-424}$$

**Example 7-7** Consider a mass $m$ moving with velocity $v(t)$ in a medium, and let $f(t)$ be the applied force. Clearly, the differential equation of the velocity $v(t)$ is

$$m\dot{v}(t) + d[v(t)] = f(t) \tag{7-425}$$

where $d[v(t)]$ represents the drag force acting on the body. In the absence of friction, we have

$$d[v(t)] = 0 \tag{7-426}$$

and so

$$\dot{v}(t) = \frac{1}{m} f(t) \tag{7-427}$$

so that the system acts as an integrator. In the case of linear friction, we have

$$d[v(t)] = av(t) \qquad a > 0 \tag{7-428}$$

and the system acts as a linear single-time-constant system. If

$$d[v(t)] = av(t)|v(t)| \qquad a > 0 \tag{7-429}$$

then the drag force is quadratic in nature.

We shall solve the following time-optimal problem:

***Problem 7-10*** *Given the system described by the first-order nonlinear differential equation*

$$\dot{x}(t) + g[x(t); t] = Ku(t) \tag{7-430}$$

*where* $|u(t)| \leq 1$*; determine the control, as a function of the state, which will reduce any initial state* $x(0) = \xi$ *to* 0 *in minimum time, provided that a time-optimal control exists.*

The Hamiltonian for the minimum-time control of the system (7-430) is

$$H = 1 - g[x(t); t]p(t) + Ku(t)p(t) \tag{7-431}$$

where $p(t)$ is the (scalar) costate. But

$$\dot{p}(t) = -\frac{\partial H}{\partial x(t)} = \frac{\partial g[x(t); t]}{\partial x(t)} p(t) \tag{7-432}$$

The control which absolutely minimizes $H$ is given by (since $K$ is a positive constant) the relation

$$u(t) = -\operatorname{sgn}\{p(t)\} \tag{7-433}$$

Let us examine the differential equation satisfied by $p(t)$; from Eq. (7-432), we find that

$$\frac{dp(t)}{p(t)} = h[x(t), t]\, dt \tag{7-434}$$

where we set

$$h[x(t), t] = \frac{\partial g[x(t); t]}{\partial x(t)} \tag{7-435}$$

Integrating both sides of Eq. (7-434), we have

$$\log p(t) = \int_0^t h[x(\tau), \tau] \, d\tau + c \tag{7-436}$$

where $c$ is an integration constant, or

$$p(t) = p(0) \exp \left\{ \int_0^t h[x(\tau), \tau] \, d\tau \right\} \tag{7-437}$$

But, independent of $h[x(t), t]$,

$$\exp \left\{ \int_0^t h[x(\tau), \tau] \, d\tau \right\} > 0 \qquad \text{for } t < \infty \tag{7-438}$$†

which implies that

$$\operatorname{sgn} \{p(t)\} = \operatorname{sgn} \{p(0)\} \tag{7-439}$$

Since $p(0) \neq 0$, the function $p(t)$ can never be zero for any finite $t$. Therefore, the control defined by Eq. (7-433) can be either

$$u(t) = +1 \qquad \text{for all } t > 0 \tag{7-440}$$

or

$$u(t) = -1 \qquad \text{for all } t > 0 \tag{7-441}$$

and it *cannot* switch.

The next step is to determine the time-optimal control as a function of the state.

***Control Law 7-10 Solution to Problem 7-10*** *The time-optimal control that drives any initial state of the system* (7-430) *to zero in minimum time is given by*

$$u^* = - \operatorname{sgn} \{x\} \tag{7-442}$$

*provided that a time-optimal control exists.*

PROOF Suppose that $x(0) > 0$. It is clear that the first time $t^*$ for which $x(t^*) = 0$ is the minimum time required to force $x(t)$ to 0 and that

$$x(t) > 0 \qquad \text{for all } t \text{ with } 0 \leq t < t^* \tag{7-443}$$

Since $x(t)$ is positive, it follows from Eq. (7-442) that $u(t) = -1$. Consider next the two differential equations

$$\dot{x}(t) = -g[x(t); t] - K \tag{7-444}$$

$$\dot{x}(t) = -g[x(t); t] + K \tag{7-445}$$

† It is tacitly assumed that $h[x(\tau), \tau]$ is a bounded function, which is the case for all physical systems representable by Eq. (7-430).

Let $x_-(t)$ be the solution of Eq. (7-444) and let $x_+(t)$ be the solution of Eq. (7-445). We shall show that

$$x_-(t) < x_+(t) \qquad \text{for any } t \text{ with } 0 < t < t^* \tag{7-446}$$

which will prove that the control $u = -1$ is time-optimal. It is clear that at $t = 0$, the slope of the function $x_-(t)$ is smaller than the slope of the function $x_+(t)$ (since $-K < K$); hence there exists an $\epsilon$ such that

$$x_-(t) < x_+(t) \qquad \text{for any } t \text{ with } 0 < t < \epsilon \tag{7-447}$$

as shown in Fig. 7-64. Now suppose that there exists a time $t_1$, $0 < t_1 < t^*$, such that

$$\begin{aligned} x_-(t) &< x_+(t) && \text{for all } t \text{ with } 0 < t < t_1 \\ x_-(t_1) &= x_+(t_1) && \\ x_-(t) &> x_+(t) && \text{for all } t \text{ with } t_1 < t < t_1 + \delta \end{aligned} \tag{7-448}$$

This implies that the slope of $x_+(t)$ at $t_1$ [that is, $\dot{x}_+(t_1)$] is smaller than the slope of $x_-(t)$ at $t_1$ [that is, $\dot{x}_-(t_1)$] or, more precisely, that

$$\dot{x}_-(t_1) > \dot{x}_+(t_1) \tag{7-449}$$

But from Eqs. (7-444) and (7-445), we find that

$$\begin{aligned} \dot{x}_-(t_1) &= -g[x_-(t_1); t_1] - K \\ \dot{x}_+(t_1) &= -g[x_+(t_1); t_1] + K \end{aligned} \tag{7-450}$$

Since $x_+(t_1) = x_-(t_1)$, it follows that

$$\dot{x}_+(t_1) - \dot{x}_-(t_1) = 2K > 0 \tag{7-451}$$

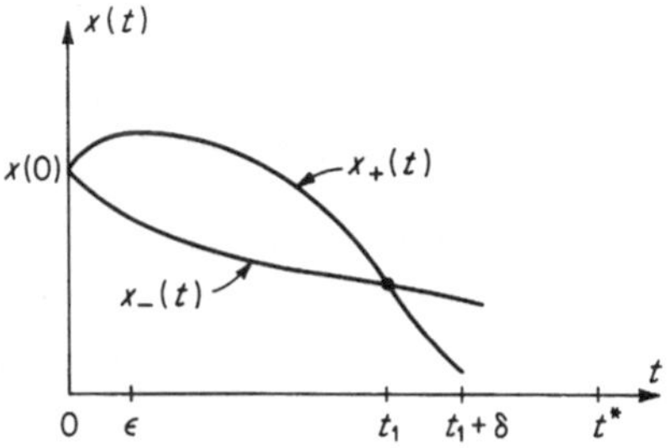


**Fig. 7-64** The solutions $x_+(t)$ and $x_-(t)$ with $x(0) > 0$.

or that

$$\dot{x}_+(t_1) > \dot{x}_-(t_1) \tag{7-452}$$

But Eqs. (7-452) and (7-449) are contradictory; since $t_1$ was arbitrary, we conclude that

$$x_-(t) < x_+(t) \qquad \text{for all } t \tag{7-453}$$

Similar reasoning for $x(0) < 0$ proves the validity of Control Law 7-10.

**Exercise 7-58** Given the system $\dot{x}(t) = u(t)$, $|u(t)| \le 1$, show that the minimum time $t^*$ required to force any initial state $x(0) = \xi$ to zero is given by

$$t^* = |\xi| \tag{7-454}$$

**Exercise 7-59** Given the system

$$\dot{x}(t) = -ax(t) + u(t) \qquad |u(t)| \le 1;\ a > 0 \tag{7-455}$$

show that the minimum time $t^*$ required to force any initial state $x(0) = \xi$ to zero is given by

$$t^* = \frac{1}{a} \log (a|\xi| + 1) \tag{7-456}$$

Show that $t^*$ is the solution of the Hamilton-Jacobi partial differential equation.

**Exercise 7-60** Given the system

$$\dot{x}(t) = bx(t) + Ku(t) \qquad |u(t)| \leq 1; b > 0 \tag{7-457}$$

show that a time-optimal control exists only for initial states $x(0) = \xi$ such that

$$|\xi| < \frac{K}{b} \tag{7-458}$$

In other words, if $|\xi| \geq K/b$, there is *no* control $u(t)$, $|u(t)| \leq 1$, which can force $x(t)$ to zero and hence *no* minimum-time control.

**Exercise 7-61** Given the system

$$\dot{x}(t) = -ax(t)|x(t)| + u(t) \qquad a > 0; |u(t)| \leq 1 \tag{7-459}$$

show that if $x(0) = \xi$, the minimum time to force $\xi$ to 0 is given by

$$t^* = \frac{1}{\sqrt{a}} \tan^{-1} (\sqrt{a}\,|\xi|) \tag{7-460}$$

## 7-11 Time-optimal Control of a Class of Second-order Nonlinear Systems†

In this section, we examine the time-optimal control problem for a class of nonlinear second-order systems described by the differential equation

$$\ddot{y}(t) + f[\dot{y}(t)] = Ku(t) \tag{7-461}$$

where $y(t)$ is the output, $u(t)$ is the control which is constrained by the relation

$$|u(t)| \leq 1 \qquad \text{for all } t \tag{7-462}$$

$K$ is a positive gain constant, and $f[\dot{y}(t)]$ is a nonlinear function of the time derivative $\dot{y}(t)$ of the output $y(t)$.

If

$$\operatorname{sgn} \{f[\dot{y}(t)]\} = \operatorname{sgn} \{\dot{y}(t)\} \qquad \text{for all } \dot{y}(t) \neq 0 \tag{7-463a}$$

then the system is *stable* and if

$$\operatorname{sgn} \{f[\dot{y}(t)]\} = -\operatorname{sgn} \{\dot{y}(t)\} \qquad \text{for all } \dot{y}(t) \neq 0 \tag{7-463b}$$

then the system is *unstable.*

Equation (7-461) is the differential equation of a moving body subject to nonlinear drag forces; if $y(t)$ represents position and $\dot{y}(t)$ velocity, then $f[\dot{y}(t)]$ represents the nonlinear drag force which is a function of the velocity

† Appropriate references for this section are L-7 and L-8.

alone. Clearly, such a function $f[\dot{y}(t)]$ will also satisfy Eq. (7-463*a*) since energy dissipation occurs and the motion is a stable one. To be specific, we shall assume in the remainder of this section that Eq. (7-463*a*) is satisfied.

The state-variable representation of the system satisfying Eq. (7-461) is simple to obtain. We let

$$x_1(t) = y(t) \qquad x_2(t) = \dot{y}(t) \tag{7-464}$$

and so the state differential equations are

$$\begin{aligned} \dot{x}_1(t) &= x_2(t) \\ \dot{x}_2(t) &= -f[x_2(t)] + u(t) \end{aligned} \tag{7-465}$$

The time-optimal control problem which we shall consider is:

***Problem 7-11*** *Given the system* (7-465), *determine the control* $u(t)$, *subject to the constraint* $|u(t)| \leq 1$, *such that any initial state is forced to the origin* (0, 0) *of the* $x_1x_2$ *state plane in minimum time.*

This is the first time we have examined the time-optimal control of a nonlinear second-order system. Our objective in this section is to illustrate how the time-optimal feedback control system can be found by using graphical techniques. As we have done before, we shall use the necessary conditions to show that the extremal controls can switch at most once. As soon as we deduce this conclusion on the number of switchings, we can isolate the four control sequences $\{+1\}$, $\{-1\}$, $\{+1, -1\}$, $\{-1, +1\}$ that can be candidates for time-optimal control. We then mimic the development of the ideas presented in Secs. 7-2 and 7-3 to find the time-optimal control.

We shall outline a list of instructions that one may follow to design the optimal feedback system, working solely with experimental data. We shall illustrate the concepts involved by a specific nonlinear system.

The major point we wish to emphasize is the "structural equivalence" of linear and nonlinear second-order time-optimal systems, provided that the problems are normal. What we mean by "structural equivalence" is that all normal second-order time-optimal systems behave in the same manner. It is only the shape of the switch curve that distinguishes one system from another. Since the nonlinear operations that must be performed by the feedback system are dictated by the shape of the switch curve, we can construct these nonlinearities from experimental data.

Now we proceed with the solution of the posed problem.

The Hamiltonian for this problem is

$$H = 1 + x_2(t)p_1(t) - f[x_2(t)]p_2(t) + u(t)p_2(t) \tag{7-466}$$

The control which absolutely minimizes the Hamiltonian is given by

$$u(t) = -\operatorname{sgn}\{p_2(t)\} \tag{7-467}$$

The costate variables $p_1(t)$ and $p_2(t)$ are the solutions of the differential equations

$$\begin{aligned} \dot{p}_1(t) &= -\frac{\partial H}{\partial x_1(t)} = 0 \\ \dot{p}_2(t) &= -\frac{\partial H}{\partial x_2(t)} = -p_1(t) + \frac{\partial f[x_2(t)]}{\partial x_2(t)} p_2(t) \end{aligned} \tag{7-468}$$

We must assume that

$$\frac{\partial f[x_2(t)]}{\partial x_2(t)} \quad \text{exists for all } x_2(t) \tag{7-469}$$

Let

$$\pi_1 = p_1(0) \qquad \pi_2 = p_2(0)$$

From Eqs. (7-468), we find that

$$p_1(t) = \pi_1 = \text{const} \tag{7-470}$$

Now we come to the point at which we must solve, at least formally, Eqs. (7-468). Let us examine the function $\partial f[x_2(t)]/\partial x_2(t)$, which is a function of $x_2(t)$ and, hence, a function of time; for this reason, let us define $h(t)$ by setting

$$h(t) = \frac{\partial f[x_2(t)]}{\partial x_2(t)} \tag{7-471}$$

Clearly, $h(t)$ is a well-defined [see Eq. (7-469)] but *unknown* function of time. Substituting Eqs. (7-471) and (7-470) into Eqs. (7-468), we next obtain the differential equation

$$\dot{p}_2(t) = -\pi_1 + h(t)p_2(t) \tag{7-472}$$

which can be solved to yield

$$p_2(t) = \exp\left[\int_0^t h(\tau)\, d\tau\right] \left\{\pi_2 - \pi_1 \int_0^t \exp\left[-\int_0^\tau h(\sigma)\, d\sigma\right] d\tau\right\} \tag{7-473}$$

where $\tau$ and $\sigma$ are dummy variables. We shall now prove that the function $p_2(t)$ can be zero *at most* once. For any function $h(t)$, we can immediately conclude that

$$\begin{aligned} a(t) &= \exp\left[\int_0^t h(\tau)\, d\tau\right] > 0 \qquad \text{for } 0 < t < \infty \\ b(\tau) &= \exp\left[-\int_0^\tau h(\sigma)\, d\sigma\right] > 0 \qquad \text{for } 0 < \tau < \infty \\ c(t) &= \int_0^t \exp\left[-\int_0^\tau h(\sigma)\, d\sigma\right] d\tau > 0 \end{aligned} \tag{7-474}$$

and $c(t)$ is monotonically increasing with increasing $t$. If

$$\pi_2 > 0 \qquad \text{and} \qquad \pi_1 \le 0 \tag{7-475a}$$

then $\pi_2 - \pi_1 c(t) > 0$ for all $t$, and hence

$$p_2(t) > 0 \qquad \text{for all } t \geq 0 \tag{7-475b}$$

If
$$\pi_2 = 0 \qquad \text{and} \qquad \pi_1 < 0 \tag{7-476a}$$

then
$$p_2(0) = 0 \qquad \text{and} \qquad p_2(t) > 0 \qquad \text{for all } t > 0 \tag{7-476b}$$

If
$$\pi_2 < 0 \qquad \text{and} \qquad \pi_1 \geq 0 \tag{7-477a}$$

then $\pi_2 - \pi_1 c(t) < 0$ for all $t$; hence

$$p_2(t) < 0 \qquad \text{for all } t \geq 0 \tag{7-477b}$$

If
$$\pi_2 = 0 \qquad \text{and} \qquad \pi_1 > 0 \tag{7-478a}$$

then
$$p_2(0) = 0 \qquad \text{and} \qquad p_2(t) < 0 \qquad \text{for all } t > 0 \tag{7-478b}$$

If
$$\pi_2 > 0 \qquad \text{and} \qquad \pi_1 > 0 \tag{7-479a}$$

then the function $\pi_2 - \pi_1 c(t)$ is monotonically decreasing with increasing $t$, and since $p_2(0) = \pi_2 > 0$, there exists a unique time $t_1 > 0$ such that

$$p_2(t_1) = 0 \tag{7-479b}$$

If
$$\pi_2 < 0 \qquad \text{and} \qquad \pi_1 < 0 \tag{7-480a}$$

then the function $\pi_2 - \pi_1 c(t)$ is monotonically increasing with increasing $t$, and since $p_2(0) = \pi_2 < 0$, there exists a unique time $t_2 > 0$ such that

$$p_2(t_2) = 0 \tag{7-480b}$$

Thus, we have shown that for all possible values of $\pi_1$ and $\pi_2$, the function $p_2(t)$ can be zero at most once;† it follows from Eq. (7-467) that only the four control sequences

$$\{+1\}, \{-1\}, \{+1, -1\}, \{-1, +1\} \tag{7-481}$$

can be candidates for time-optimal control.

The reader will note that for this second-order nonlinear system, the same conclusions have been reached regarding the control sequences that are candidates for time-optimal control as were reached for the linear second-order systems considered in Secs. 7-2 and 7-3. For this reason, the subsequent development of the results and the derivation of the time-optimal control law are identical in reasoning and in concept to the arguments of Secs. 7-2 and 7-3.

The first step is to "solve" the differential equations

$$\begin{aligned} \dot{x}_1(t) &= x_2(t) \\ \dot{x}_2(t) &= -f[x_2(t)] + \Delta \end{aligned} \tag{7-482}$$

where
$$\Delta = \pm 1 \tag{7-483}$$

using
$$x_1(0) = \xi_1 \qquad x_2(0) = \xi_2$$

† The case $\pi_1 = \pi_2 = 0$ is excluded (why?).

From Eqs. (7-482), we observe that the solution $x_2(t)$ is defined by

$$\int \frac{dx_2(t)}{-f[x_2(t)] + \Delta} = t + c \tag{7-484}$$

where $c$ is an integration constant. Of course, we cannot analytically evaluate the integral appearing in Eq. (7-484) for arbitrary functions $f[x_2(t)]$, but the solution can be found either graphically or by using a digital computer. At any rate, for $\Delta = +1$, we can write the solutions of Eqs. (7-482) in the form

$$\begin{aligned} x_2^+(t) &= x_2^+(\xi_2, t) \\ x_1^+(t) &= x_1^+(\xi_1, \xi_2, t) \end{aligned} \tag{7-485}$$

and for $\Delta = -1$,

$$\begin{aligned} x_2^-(t) &= x_2^-(\xi_2, t) \\ x_1^-(t) &= x_1^-(\xi_1, \xi_2, t) \end{aligned} \tag{7-486}$$

In principle, we can eliminate the time $t$ in Eqs. (7-485) and (7-486) and obtain the equations of the trajectories in the state plane,

$$\begin{aligned} x_2^+ &= q^+(x_1^+, \xi_1, \xi_2) \\ x_2^- &= q^-(x_1^-, \xi_1, \xi_2) \end{aligned} \tag{7-487}$$

Of course, we can obtain the trajectories by using graphical techniques (such as the method of isoclines), by using a digital computer, or by simulating the system equations on an analog computer and looking at the trajectories on the screen of an oscilloscope.

We assert that there is a unique trajectory generated by the control $u = +1$ which passes through the origin (0, 0) of the state plane; *we shall denote by $\gamma_+$ the set of states which can be forced to* (0, 0) *by the control $u = +1$. Similarly, we shall denote by $\gamma_-$ the set of states which can be forced to the origin* (0, 0) *by the control $u = -1$.* As we have done before, *we define the $\gamma$ switch curve as the union of the $\gamma_+$ and $\gamma_-$ curves* ($\gamma = \gamma_+ \cup \gamma_-$). The $\gamma$ switch curve divides the state plane into two regions, $R_-$ and $R_+$. *We let $R_-$ denote the set of states to the right of the $\gamma$ switch curve* (more precisely, $R_- = \{(x_1, x_2)\colon \text{if } (x_1', x_2) \in \gamma, \text{ then } x_1 > x_1'\}$). Similarly, *we let $R_+$ denote the set of states to the left of the $\gamma$ switch curve* (more precisely, $R_+ = \{(x_1, x_2)\colon \text{if } (x_1', x_2) \in \gamma, \text{ then } x_1 < x_1'\}$). Using the elimination method of Secs. 7-2 and 7-3, we can deduce the following control law:

***Control Law 7-11 Solution to Problem 7-11*** *The time-optimal control $u^*$, as a function of the state $(x_1, x_2)$, is given by*

$$\begin{aligned} u^* &= u^*(x_1, x_2) = +1 \qquad \textit{for all } (x_1, x_2) \in R_+ \cup \gamma_+ \\ u^* &= u^*(x_1, x_2) = -1 \qquad \textit{for all } (x_1, x_2) \in R_- \cup \gamma_- \end{aligned} \tag{7-488}$$

*and, furthermore, $u^*$ is unique (provided that it exists).*

The $\gamma$ switch curve might look like the one illustrated in Fig. 7-65. As we have done before, we can construct an engineering realization of the time-optimal control law as shown in the block diagram of Fig. 7-66. The input-output characteristics of the nonlinearity $N$ shown in Fig. 7-66 are identical to the $\gamma$ switch curve.

What we have attempted to illustrate in this section is that once one obtains the fact that only the four control sequences of (7-481) can be candidates for time-optimal control or, equivalently, that the time-optimal control can switch *at most* once, one can determine the time-optimal control easily. As a matter of fact, an engineer need not even bother with the actual equations if the construction of the system with the block diagram of Fig. 7-66 is his sole objective. The steps he must follow to design the time-optimal system are summarized below.

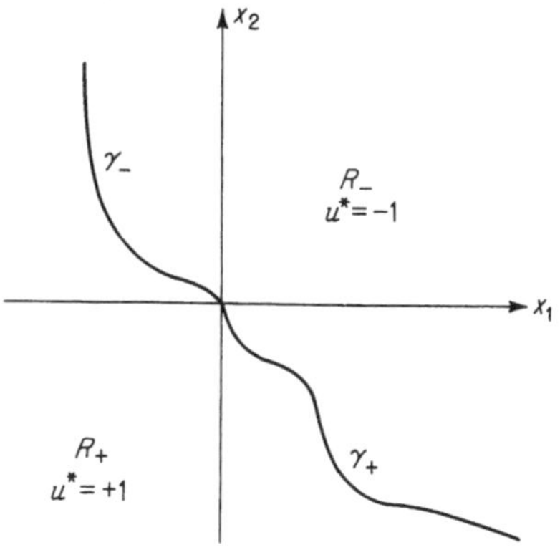


**Fig. 7-65** Possible "shape" of the $\gamma_+$ and $\gamma_-$ curves for a second-order nonlinear system.

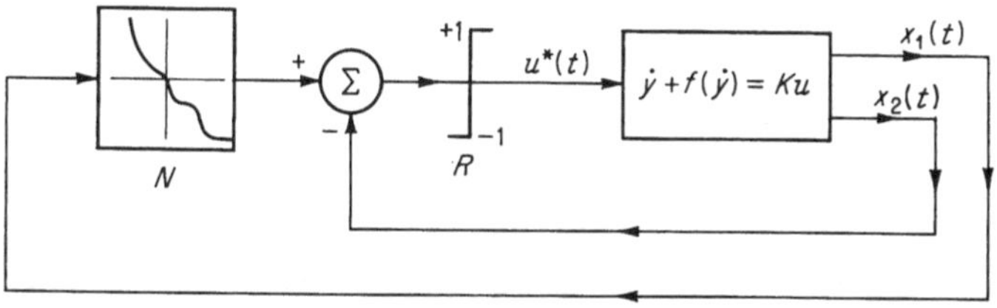


**Fig. 7-66** The block diagram of the time-optimal feedback control system. The input-output characteristics of the nonlinearity $N$ are defined by Fig. 7-65.

*If it can be shown that the time-optimal control can switch at most once, then:*

1. Simulate the system of Eqs. (7-465) on an analog computer.
2. Connect the $x_1(t)$ signal to the $X$ terminal and the $x_2(t)$ signal to the $Y$ terminal of an $X$-$Y$ recorder.
3. Using $u = +1$ and adjusting the initial conditions, determine the trajectory through the origin; this trajectory is the $\gamma_+$ curve.
4. Repeat step 3, using $u = -1$; this will yield the $\gamma_-$ curve.
5. Now you have a graphical plot of the $\gamma$ switch curve.
6. Call the instrument engineer and ask him to construct for you, using a sufficient number of biased diodes, a nonlinear element $N$ which has the input-output characteristics of the $\gamma$ switch curve.
7. Connect the nonlinearity $N$ in the block diagram of Fig. 7-66, and you have the time-optimal system.

In the remainder of this section, we shall present some explicit curves for a specific nonlinear system. Consider a unit mass subject to nonlinear friction; if $y(t)$ is the position and $\dot{y}(t)$ the velocity and if the friction force is a quadratic function of the velocity, then the differential equation is†

$$\ddot{y}(t) + \dot{y}(t)|\dot{y}(t)| = u(t) \qquad |u(t)| \leq 1 \tag{7-489}$$

where $u(t)$ represents the force applied to the mass. We have written the

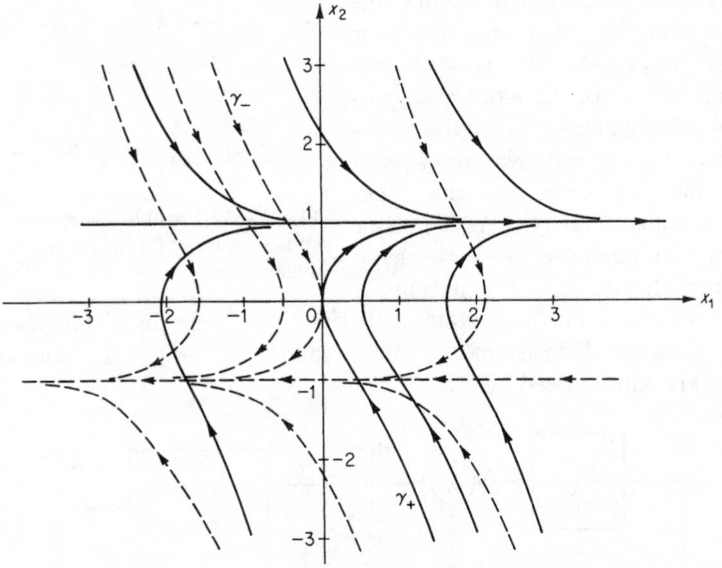


**Fig. 7-67** The trajectories in the state plane of the system $\ddot{y} + \dot{y}|\dot{y}| = u$; the solid trajectories are due to $u = +1$, and the dashed trajectories are due to $u = -1$. The arrows indicate direction of motion of the state.

drag force as $\dot{y}(t)|\dot{y}(t)|$ to indicate that the force is directed oppositely to the direction of motion. Letting

$$x_1(t) = y(t) \qquad x_2(t) = \dot{y}(t)$$

we find that the state variables $x_1(t)$ and $x_2(t)$ satisfy the system of differential equations

$$\begin{aligned} \dot{x}_1(t) &= x_2(t) \\ \dot{x}_2(t) &= -x_2(t)|x_2(t)| + u(t) \end{aligned} \tag{7-490}$$

Figure 7-67 shows the trajectories of this system when $u = +1$ (solid curves) and when $u = -1$ (dashed curves). As is evident from Fig. 7-67, all the trajectories due to $u = +1$ tend to the line $x_2 = +1$ and all the trajectories

† We shall consider the fuel-optimal control of this system in Sec. 8-10.

due to $u = -1$ tend to the line $x_2 = -1$. The trajectories which pass through the origin are labeled as $\gamma_+$ and $\gamma_-$, and they are shown in Fig. 7-68 together with the regions $R_+$ and $R_-$; for time optimality, the control must

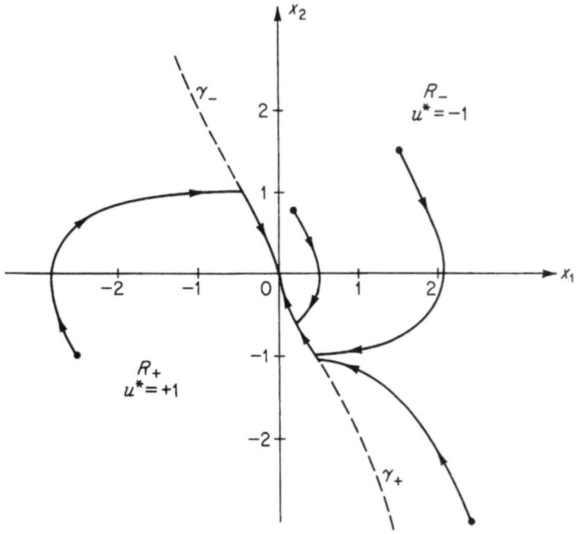


**Fig. 7-68** The $\gamma_+$ and $\gamma_-$ curves for the system $\ddot{y} + \dot{y}|\dot{y}| = u$. The solid curves are time-optimal trajectories to the origin of the state plane.

be $u = +1$ whenever the state is in $R_+$ and $u = -1$ whenever the state is in $R_-$. Several time-optimal trajectories are also illustrated in Fig. 7-68.

**Exercise 7-62** Construct graphically the time-optimal switch curve for the second-order nonlinear system

$$\ddot{y}(t) + [\dot{y}(t)]^3 = u(t) \qquad |u(t)| \leq 1 \tag{7-491}$$

where the state $\dot{y} = y = 0$ is the terminal state.

**Exercise 7-63** Given the system

$$\ddot{y}(t) + f[\dot{y}(t), y(t)] = u(t) \qquad |u(t)| \leq 1 \tag{7-492}$$

Examine the number of switchings of the time-optimal control. Assume that the function $f$ is differentiable with respect to both $\dot{y}(t)$ and $y(t)$.

**Exercise 7-64** Given the system

$$\ddot{y}(t) + Ky(t) + ay^3(t) = u(t) \qquad |u(t)| \leq 1 \tag{7-493}$$

which represents the motion of a nonlinear mass-spring system. Determine the time-optimal control law to the state $\dot{y} = y = 0$ using $K = a = 1$. Hint: The free response of the system (7-493) is oscillatory in nature; hence there is no upper bound on the number of switchings of the time-optimal control. For this reason, use the techniques of Secs. 7-7 to 7-9, as the system (7-493) is a nonlinear oscillator.

**Exercise 7-65** Suggest several techniques for the suboptimal control of the system of Exercise 7-64. Hint: Use techniques similar to those considered in Sec. 7-7, Examples 7-5 and 7-6.

## 7-12 Time-optimal Control of a Double-integral Plant with a Single Zero†

In the preceding sections of this chapter, we developed time-optimal control laws for plants which had the property that their transfer function contained only poles. In other words, we studied plants with the property that the right-hand side of their output differential equation was a function of the control $u(t)$ alone. In this section and in the two subsequent sections, we shall examine the time-optimal control problem for plants with zeros in their transfer function.

We shall first solve two simple examples, and then we shall generalize the results in Sec. 7-14. As we shall see, the time-optimal control of plants with zeros in their transfer function (or plants with *numerator dynamics*, as they are often called) is somewhat different from the time-optimal control of plants which do not contain zeros in their transfer function.

To determine the time-optimal control law for plants with numerator dynamics, we shall proceed, in this section and in the two following ones, as follows:

1. We shall formulate the time-optimal control problem in terms of the output and its time derivatives, and we shall require that the output and its time derivatives be forced to zero in minimum time and be held at zero thereafter.
2. We shall determine the proper state variables for the system, using the techniques of Chap. 4.
3. We shall reformulate the time-optimal control problem in terms of the state variables. We shall show that in the state space there exists a target set $G$ (which may consist of a single state) so that one must force states not in $G$ to $G$ in minimum time and then apply a suitable control to maintain the state in $G$ thereafter.
4. After the set $G$ has been obtained, we shall obtain the time-optimal control law to the set $G$; this will involve the computation of the switch curves, etc.

The reader is warned that the solution of the time-optimal control problem for plants with zeros in their transfer function is much more difficult than the solution of the time-optimal control problem for plants with only poles in their transfer function.

To illustrate these ideas, we consider a plant with output $y(t)$ and control $u(t)$ which are related by the second-order linear differential equation

$$\ddot{y}(t) = \dot{u}(t) + \beta u(t) \tag{7-494}$$

† Appropriate references for Secs. 7-12 to 7-14 are A-4, F-4, F-9, H-6, H-7, H-12, L-9, and S-3.

Assume that the control $u(t)$ is restricted in magnitude, i.e., that

$$|u(t)| \leq 1 \qquad \text{for all } t \tag{7-495}$$

In servomechanism practice, a system described by Eq. (7-494) is said to have the transfer function

$$\frac{y(s)}{u(s)} = H(s) = \frac{s + \beta}{s^2} \tag{7-496}$$

This transfer function contains a single zero at $s = -\beta$ and two poles at $s = 0$ (hence the name double integral with a single zero). We distinguish the following three cases:

*Case 1*

$$\beta > 0 \tag{7-497}$$

Equation (7-497) implies that the zero is in the left half of the $s$ plane, as shown in Fig. 7-69$a$. Such a zero is called a *minimum-phase zero.*

*Case 2*

$$\beta < 0 \tag{7-498}$$

Equation (7-498) implies that the zero is in the right half of the $s$ plane, as shown in Fig. 7-69$b$. Such a zero is called a *non-minimum-phase zero.*

*Case 3*

$$\beta = 0 \tag{7-499}$$

Equation (7-499) implies that the zero is at the origin of the $s$ plane, as shown in Fig. 7-69$c$. The transfer function $H(s)$ of Eq. (7-496) is

$$H(s) = \frac{s}{s^2} \tag{7-500}$$

One may be tempted to replace $H(s)$ by the transfer function

$$H_1(s) = \frac{1}{s} \tag{7-501}$$

and, in this context, to reduce Eq. (7-494) to

$$\dot{y}(t) = u(t) \tag{7-502}$$

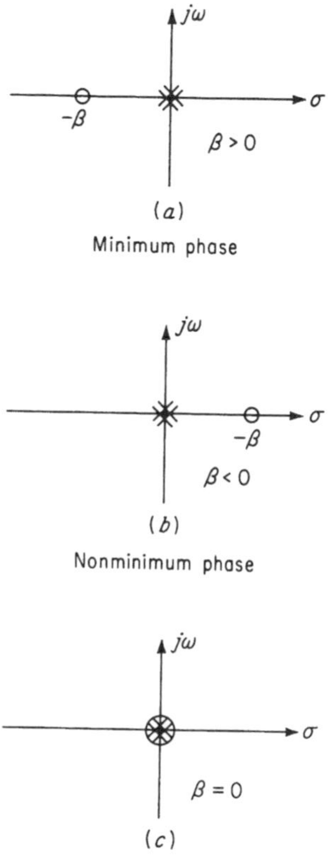


**Fig. 7-69** The $s$-plane pole-and-zero configuration of the transfer function $G(s) = (s + \beta)/s^2$.

and solve the problem using Eq. (7-502). We shall illustrate the pitfalls of this cancellation of the zero and the pole.

The problem we shall consider in this section is the following:

***Problem 7-12a*** *Given the system described by Eq. (7-494), determine the control $u(t)$, satisfying $|u(t)| \leq 1$, which will drive $y$ and $\dot{y}$ to zero in minimum* *time and which will maintain y and ẏ at zero thereafter, in the absence of subsequent disturbances.*

In order to solve this problem correctly, we must define a suitable set of state variables. Following the procedure of Sec. 4-10, we define the state variables $x_1(t)$ and $x_2(t)$ by setting

$$x_1(t) = y(t) - h_0 u(t) \tag{7-503}$$
$$x_2(t) = \dot{y}(t) - h_0 \dot{u}(t) - h_1 u(t) \tag{7-504}$$

The next step is to determine the differential equations satisfied by the state variables. From Eq. (7-503), we obtain

$$\dot{y}(t) = \dot{x}_1(t) + h_0 \dot{u}(t) \tag{7-505}$$

Substituting Eq. (7-505) into Eq. (7-504), we find that

$$\dot{x}_1(t) = x_2(t) + h_1 u(t) \tag{7-506}$$

Differentiating Eq. (7-504), we obtain

$$\ddot{y}(t) = \dot{x}_2(t) + h_0 \ddot{u}(t) + h_1 \dot{u}(t) \tag{7-507}$$

and from Eqs. (7-507) and (7-494),

$$\dot{x}_2(t) = -h_0 \ddot{u}(t) - h_1 \dot{u}(t) + \dot{u}(t) + \beta u(t) \tag{7-508}$$

We note that Eq. (7-506) does not contain any derivatives of $u(t)$, while Eq. (7-508) does; the objective is to choose $h_0$ and $h_1$ so that the equations of $\dot{x}_1(t)$ and $\dot{x}_2(t)$ do not contain any time derivatives of $u(t)$. If, in Eq. (7-508), we set

$$h_0 = 0 \qquad h_1 = 1 \tag{7-509}$$

then Eqs. (7-506) and (7-508) reduce to

$$\begin{aligned} \dot{x}_1(t) &= x_2(t) + u(t) \\ \dot{x}_2(t) &= \beta u(t) \end{aligned} \tag{7-510}$$

and Eqs. (7-503) and (7-504) become

$$\begin{aligned} x_1(t) &= y(t) \\ x_2(t) &= \dot{y}(t) - u(t) \end{aligned} \tag{7-511}$$

We note that the state variable $x_1(t)$ is simply the output $y(t)$, while the state variable $x_2(t)$ is the difference between the output rate $\dot{y}(t)$ and the control $u(t)$.

The problem posed was to drive $y$ and $\dot{y}$ to zero in minimum time and *to maintain them at zero thereafter.*

We shall now assume that the output $y$ and the output rate $\dot{y}$ have been forced to zero, and we shall examine the problem of maintaining them at zero.

Let $t^*$ be the first time such that

$$y(t^*) = \dot{y}(t^*) = 0 \tag{7-512}$$

It follows that if

$$y(t) = \dot{y}(t) = 0 \qquad \text{for all } t > t^* \tag{7-513}$$

then, from Eqs. (7-511), the state variables must satisfy the relation

$$\left.\begin{aligned} x_1(t) &= 0 \\ x_2(t) &= -u(t) \end{aligned}\right\} \qquad \text{for all } t > t^* \tag{7-514}$$

Substituting Eqs. (7-514) into Eqs. (7-510), we find that

$$\left.\begin{aligned} \dot{x}_1(t) &= 0 \\ \dot{x}_2(t) &= -\beta x_2(t) \end{aligned}\right\} \qquad \text{for all } t > t^* \tag{7-515}$$

This implies that

$$\left.\begin{aligned} x_1(t) &= x_1(t^*) = 0 \qquad &(7\text{-}516) \\ x_2(t) &= x_2(t^*)e^{-\beta(t-t^*)} \qquad &(7\text{-}517) \end{aligned}\right\} \qquad \text{for all } t > t^*$$

But, since $x_2(t) = -u(t)$ and since $|u(t)| \leq 1$, we must have

$$|x_2(t^*)| = |u(t^*)| \leq 1 \tag{7-518}$$

and so, from Eq. (7-517), it follows that

$$u(t) = -x_2(t^*)e^{-\beta(t-t^*)} \tag{7-519}$$

for all $t > t^*$.

These considerations lead us to the following conclusions: *In order to drive the output $y$ and the output rate $\dot{y}$ to zero in minimum time, it is sufficient to drive the state variable $x_1$ to zero and the state variable $x_2$ to the closed segment $-1 \leq x_2 \leq +1$ in minimum time. Furthermore, if $t^*$ is this minimum time, then, in order to maintain $y(t) = \dot{y}(t) = 0$ for $t > t^*$, it is necessary to apply the control*

$$u(t) = -x_2(t^*)e^{-\beta(t-t^*)} \qquad t > t^* \tag{7-520}$$

*If $u(t)$ as defined by Eq. (7-520) satisfies the constraint $|u(t)| \leq 1$ for all $t > t^*$, then $y(t) = \dot{y}(t) = 0$ for all $t > t^*$.* However, we have still not completely solved our problem. In order to do so, we shall have to formulate the time-optimal control problem in a more precise fashion, using the state variables rather than the output variables.

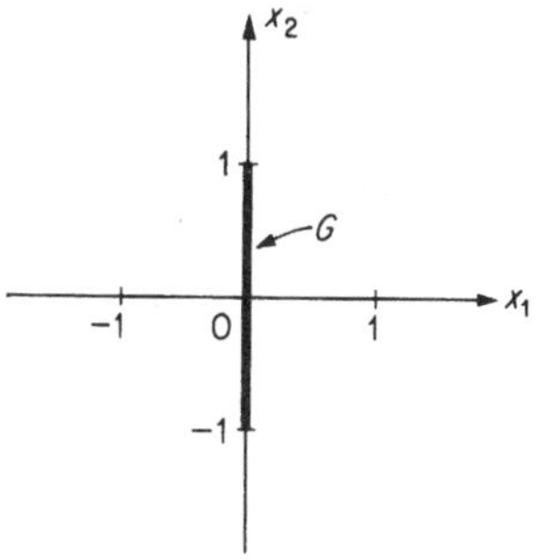


**Fig. 7-70** The target set $G$ in the $x_1x_2$ plane; the set $G$ is the closed interval $-1 \leq x_2 \leq +1$.

***Definition 7-35*** *The target set $G$ is the set defined by the relation*

$$G = \{(x_1, x_2): x_1 = 0, |x_2| \leq 1\} \tag{7-521}$$

*The target set $G$ is shown in Fig. 7-70, and since it is a closed line segment, it is clear that $G$ is closed and convex.*

In terms of the target set $G$, Problem 7-12$a$ is equivalent to:

***Problem 7-12*** *Given the system*

$$\begin{aligned}\dot{x}_1(t) &= x_2(t) + u(t)\\ \dot{x}_2(t) &= \beta u(t)\end{aligned} \tag{7-522}$$

*where*

$$|u(t)| \le 1 \tag{7-523}$$

*Find the control which*

*a. Will force the state to the target set G, given by Eq.* (7-521), *in minimum time*

*b. Will maintain the state in G for all time thereafter*

It should be clear that this problem is equivalent to the time-optimal problem which we originally stated in terms of $y$ and $\dot{y}$. We shall now examine each of the three cases, that is, $\beta > 0$, $\beta < 0$, $\beta = 0$, separately.

*Case 1 Minimum-phase Zero; That Is,* $\beta > 0$

We must control time-optimally the system (7-522) to the target set $G$. To determine the control sequences which are candidates for time-optimal control, we proceed with the familiar steps dictated by the minimum principle. Observe that for $\beta > 0$, the system (7-522) is normal.†

The Hamiltonian $H$ for the time-optimal control of the system (7-522) is

$$H = 1 + x_2(t)p_1(t) + u(t)[p_1(t) + \beta p_2(t)] \tag{7-524}$$

where

$$\dot{p}_1(t) = 0 \qquad \dot{p}_2(t) = -p_1(t) \tag{7-525}$$

or

$$p_1(t) = \pi_1 \qquad p_2(t) = \pi_2 - \pi_1 t \tag{7-526}$$

with

$$\pi_1 = p_1(0) \qquad \pi_2 = p_2(0) \tag{7-527}$$

The control which absolutely minimizes the Hamiltonian is

$$u(t) = -\operatorname{sgn}\{p_1(t) + \beta p_2(t)\} = -\operatorname{sgn}\{\pi_1 + \beta\pi_2 - \beta\pi_1 t\} \tag{7-528}$$

Hence, only the four control sequences

$$\{+1\}, \{-1\}, \{+1, -1\}, \{-1, +1\} \tag{7-529}$$

can be candidates for time-optimal control.

The determination of the switch curves and of the time-optimal control to reach the target set $G$ uses the same reasoning as in Sec. 7-2. First, we solve Eqs. (7-522) using the constant control

$$u = \Delta = \pm 1 \tag{7-530}$$

to obtain

$$\begin{aligned}x_1(t) &= \xi_1 + \xi_2 t + \Delta t + \tfrac{1}{2}\beta t^2\\ x_2(t) &= \xi_2 + \beta\Delta t\end{aligned} \tag{7-531}$$

where

$$\xi_1 = x_1(0) \qquad \xi_2 = x_2(0) \tag{7-532}$$

† See Sec. 6-5, Theorem 6-6, and Sec. 4-21.

Eliminating the time $t$ in Eqs. (7-531), we obtain the trajectory equations (in the $x_1x_2$ state plane)

$$x_1 = \xi_1 + \frac{x_2}{\beta} - \frac{\xi_2}{\beta} + \frac{1}{2}\frac{\Delta}{\beta}x_2{}^2 - \frac{1}{2}\frac{\Delta}{\beta}\xi_2{}^2 \tag{7-533}$$

The trajectories, which are shown in Fig. 7-71 for the value $\beta = 1$, are parabolas with vertex at

$$x_2 = -\Delta \tag{7-534}$$

independent of the value of $\beta$.† Following the reasoning of Sec. 7-2, we may define the switch curves as follows:

***Definition 7-36*** *Let $\gamma_+$ denote the set of states which can be forced to the state* (0, −1) *by the control $u = +1$. Let $\gamma_-$ denote the set of states which can*

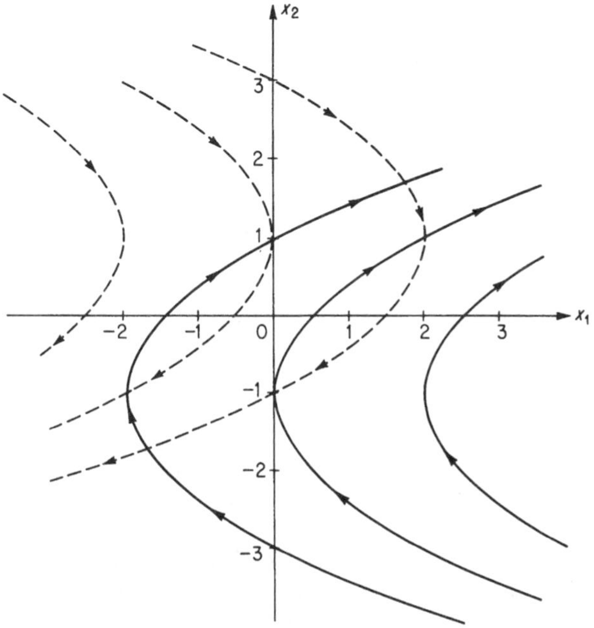


**Fig. 7-71** The trajectories in the $x_1x_2$ plane when $\beta = 1$. The solid trajectories are generated by $u = \Delta = +1$; the dashed trajectories are generated by $u = \Delta = -1$; the arrows indicate the motion of the state for positive time.

*be forced to the state* (0, +1) *by the control $u = -1$. The equations of the $\gamma_+$ and $\gamma_-$ curves can be obtained from Eq.* (7-533), *and they are*

$$\gamma_+ = \left\{(x_1, x_2): x_1 = \frac{1}{2\beta}(x_2 + 1)^2; x_2 \leq -1, \beta > 0\right\} \tag{7-535}$$

$$\gamma_- = \left\{(x_1, x_2): x_1 = -\frac{1}{2\beta}(x_2 - 1)^2; x_2 \geq 1, \beta > 0\right\} \tag{7-536}$$

*The $\gamma_+$ and $\gamma_-$ curves are shown in Fig.* 7-72 *for $\beta = 1$.*

† This is easily seen by computing $\partial x_2/\partial x_1 = \beta/(\Delta x_2 + 1)$, so that $(\partial x_1/\partial x_2)|_{x_2=-\Delta} = 0$.

*We can define the $\gamma$ switch curve as the union of the $\gamma_+$ curve, the target set G given by Eq. (7-521), and the $\gamma_-$ curve; that is,*

$$\gamma = \gamma_+ \cup G \cup \gamma_- \tag{7-537}$$

*The $\gamma$ switch curve separates the state plane into two sets $R_+$ and $R_-$, which are defined below.*

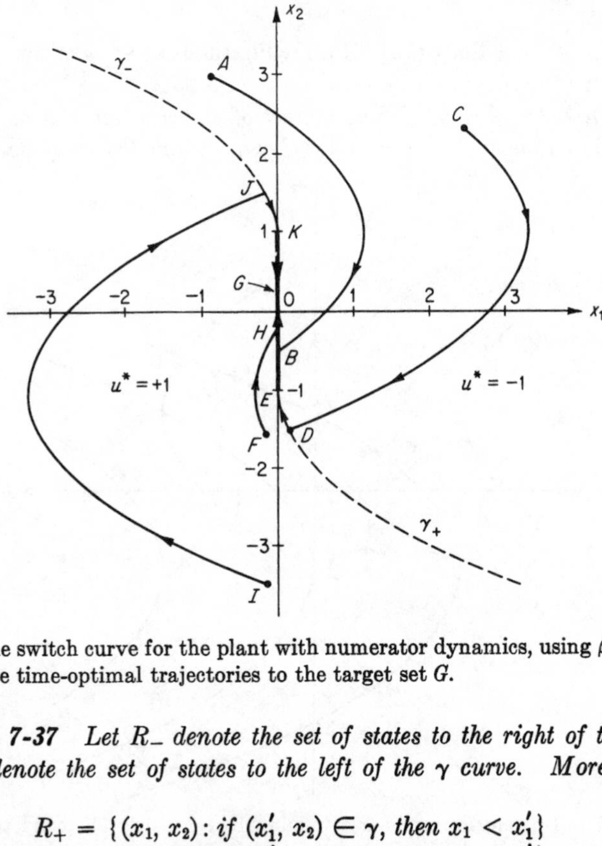


**Fig. 7-72** The switch curve for the plant with numerator dynamics, using $\beta = 1$. The solid curves are time-optimal trajectories to the target set $G$.

***Definition 7-37*** *Let $R_-$ denote the set of states to the right of the $\gamma$ curve and let $R_+$ denote the set of states to the left of the $\gamma$ curve. More precisely,*

$$R_+ = \{(x_1, x_2): \text{if } (x_1', x_2) \in \gamma, \text{ then } x_1 < x_1'\} \tag{7-538}$$
$$R_- = \{(x_1, x_2): \text{if } (x_1', x_2) \in \gamma, \text{ then } x_1 > x_1'\} \tag{7-539}$$

Now we can state the optimal-control law for every state in the state space.

***Control Law 7-12a Solution to Problem 7-12, Part a, $\beta > 0$*** *The time-optimal control $u^*$, as a function of the state $(x_1, x_2)$, that forces any state to the target set G is given by*

$$\begin{aligned} u^* &= u^*(x_1, x_2) = +1 \quad \text{for all } (x_1, x_2) \in R_+ \cup \gamma_+ \\ u^* &= u^*(x_1, x_2) = -1 \quad \text{for all } (x_1, x_2) \in R_- \cup \gamma_- \end{aligned} \tag{7-540}$$

*and, furthermore, $u^*$ is unique.*

**Exercise 7-66** Prove Control Law 7-12*a*. HINT: See Problem 7-3.

Control Law 7-12*a* provides us with the control which will drive the state to the target set $G$ in minimum time or, equivalently, which will drive the output $y$ and the output rate $\dot{y}$ to zero in minimum time. However, it does not provide the solution to the complete problem of maintaining the state in $G$ thereafter or, equivalently, of maintaining $y$ and $\dot{y}$ at zero.

We have found that if $t^*$ is the minimum time, then $(x_1(t^*), x_2(t^*)) \in G$, and that the control $u(t)$ given by Eq. (7-520), that is,

$$u(t) = -x_2(t^*)e^{-\beta(t-t^*)} \qquad t > t^* \tag{7-541}$$

will maintain the state in $G$ as long as $|u(t)| \leq 1$. But since $|x_2(t^*)| \leq 1$ and since $\beta > 0$, it follows that

$$|u(t)| = |x_2(t^*)|\,|e^{-\beta(t-t^*)}| \leq |x_2(t^*)| \leq 1 \tag{7-542}$$

so that

$$|u(t)| \leq 1 \qquad \text{for all } t > t^* \tag{7-543}$$

Hence the control (7-541) will keep the state in $G$ for all $t > t^*$ and, consequently, will keep $y(t) = \dot{y}(t) = 0$ for all $t > t^*$. These considerations provide the *complete* solution to the posed problem. In summary, we have:

***Control Law 7-12 Complete Solution to Problem 7-12 with*** $\beta > 0$ *The optimal control as a function of the state is given by*

$$\begin{aligned} u^* &= u^*(x_1, x_2) = +1 && \text{for all } (x_1, x_2) \in R_+ \cup \gamma_+ \\ u^* &= u^*(x_1, x_2) = -1 && \text{for all } (x_1, x_2) \in R_- \cup \gamma_- \\ u^* &= u^*(x_1, x_2) = -x_2 && \text{for all } (x_1, x_2) \in G \end{aligned} \tag{7-544}$$

Several time-optimal trajectories are shown in Fig. 7-72. Note that there exist initial states which can be forced to the set $G$ with *no* switching of the control (for example, the trajectories $ABO$ and $FHO$); on the other hand, there are initial states which can be forced to $G$ with *one* switching of the control (for example, the trajectories $CDEO$ and $IJKO$).

It is instructive to plot the time-optimal control which forces an initial state to $G$ and holds it in $G$ thereafter. Consider the initial state $A$ in Fig. 7-72; the time-optimal path from $A$ to $G$ is the trajectory $AB$ generated by $u = -1$; the point $B$ is on the set $G$, and for the particular initial state $A$ chosen, we have $B = (0, -0.5)$. Figure 7-73 shows the control $u(t)$, which is equal to 1 in the time interval $0 \leq t \leq t_B$, $t_B$ being the time that the point $B = (0, -0.5)$ is reached. According to Eq. (7-541), the control $u(t)$ must be given by

$$u(t) = 0.5e^{-(t-t_B)} \qquad t > t_B \tag{7-545}$$

since we have been using $\beta = 1$ to plot the figures. Since $u(t) = -x_2(t)$ for $t > t_B$, then we conclude that

$$x_2(t) = -0.5e^{-(t-t_B)} \qquad t > t_B \tag{7-546}$$

which results in the trajectory $BO$, and the origin 0 is reached at $t = \infty$.

Now consider the initial state $C$ in Fig. 7-72. The time-optimal trajectory to the target set $G$ is $CDE$, where $E = (0, -1)$; the control is $u = -1$ along the trajectory $CD$ and $u = +1$ along the trajectory $DE$, as indicated in

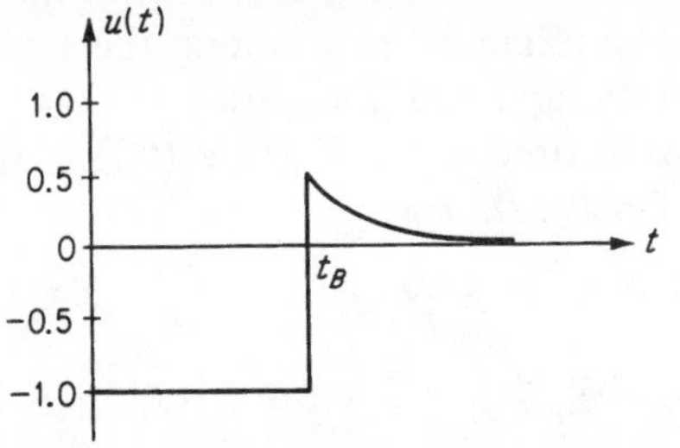


**Fig. 7-73** Plot of the time-optimal control $u(t)$ which generates the time-optimal trajectory $ABO$ of Fig. 7-72. Note that for all $t \geq t_B$ the output $y(t)$ and its derivative $\dot{y}(t)$ are identically zero.

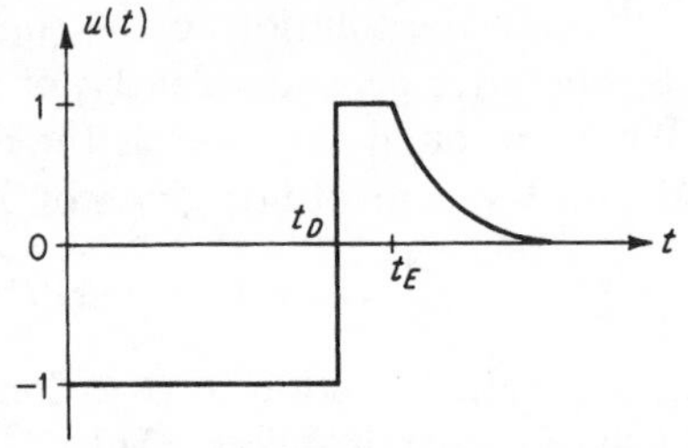


**Fig. 7-74** Plot of the time-optimal control $u(t)$ which generates the time-optimal trajectory $CDEO$ of Fig. 7-72. Note that for all $t \geq t_E$ the output $y(t)$ and its derivative $\dot{y}(t)$ are identically zero.

Fig. 7-74; at $t = t_E$, the state is at $E = (0, -1)$. Thus, for $t > t_E$, the control is

$$u(t) = e^{-(t-t_E)} \qquad t > t_E \tag{7-547}$$

and

$$x_2(t) = -e^{-(t-t_E)} \qquad t > t_E \tag{7-548}$$

so that the control for $t > t_E$ is a decaying exponential, as shown in Fig. 7-74, and the trajectory for $t > t_E$ is $EO$, which belongs to $G$.

**Exercise 7-67** Design an engineering realization of Control Law 7-12.

*Case 2 Non-minimum-phase Zero; That Is, $\beta < 0$*

If $\beta < 0$, the reader can verify that Eqs. (7-524) to (7-533) still hold, since, in the derivation of these equations, no special assumption about $\beta$ was made except that $\beta \neq 0$; therefore, the problem is still normal. Figure 7-75 shows the trajectories of the system when $\beta = -1$. The reader should compare Figs. 7-75 and 7-71 to observe the change in the trajectories when $\beta$ is negative instead of positive. Since the target set $G$ given by Definition 7-35 is independent of $\beta$, we can certainly determine the control which will force any state to the target set $G$ in minimum time. This involves the definition of the switch curves, as shown below.

***Definition 7-38*** *Let $\gamma_+$ denote the set of states which can be forced to the state $(0, -1)$ by the control $u = +1$. Let $\gamma_-$ denote the set of states which can be forced to the state $(0, +1)$ by the control $u = -1$. The $\gamma_+$ and $\gamma_-$ curves are given by*

$$\gamma_+ = \left\{(x_1, x_2): x_1 = \frac{1}{2\beta}(x_2 + 1)^2;\ x_2 \geq -1;\ \beta < 0\right\} \tag{7-549}$$

$$\gamma_- = \left\{(x_1, x_2): x_1 = -\frac{1}{2\beta}(x_2 - 1)^2;\ x_2 \leq 1;\ \beta < 0\right\} \tag{7-550}$$

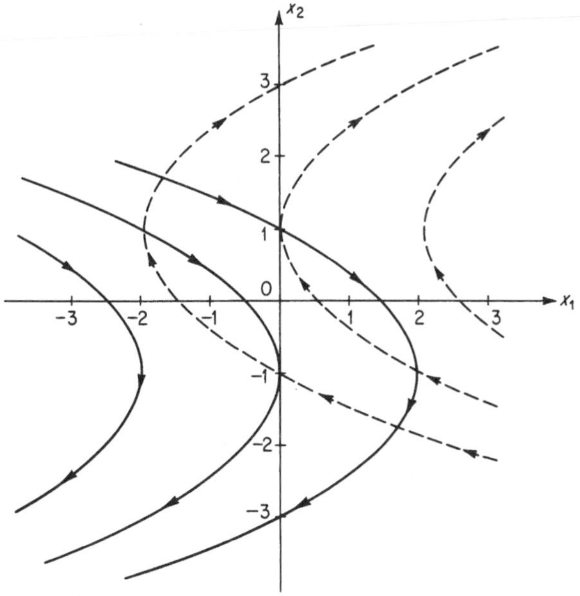


**Fig. 7-75** The trajectories in the $x_1x_2$ plane when $\beta = -1$. The solid trajectories are generated by $u = \Delta = +1$; the dashed trajectories are generated by $u = \Delta = -1$; the arrows indicate the motion of the state for positive time.

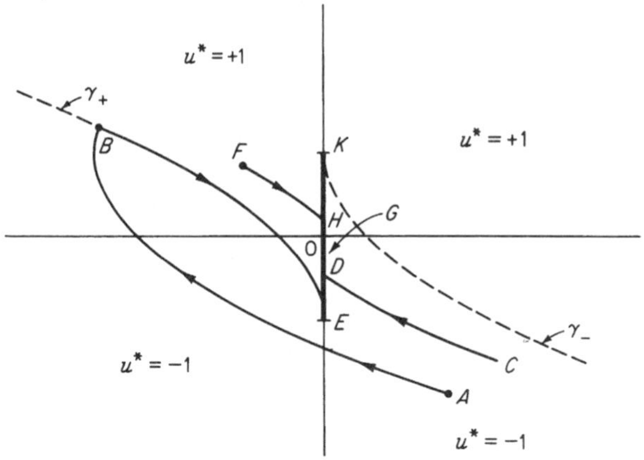


**Fig. 7-76** The switch curves $\gamma_+$ and $\gamma_-$ for the plant $G(s) = (s - 1)/s^2$. The control law indicated will force any initial $y$ and $\dot{y}$ to zero in minimum time. Time-optimal trajectories to the target set $G$ are shown as solid curves.

Note that Definition 7-38 is identical to Definition 7-36; however, since the shape of the trajectories is different for $\beta > 0$ and $\beta < 0$, Eqs. (7-549) and (7-550) are different from Eqs. (7-535) and (7-536). The $\gamma_+$ and $\gamma_-$ curves defined by Eqs. (7-549) and (7-550) are shown in Fig. 7-76, and they were computed using the value $\beta = -1$.

We can define the $\gamma$ curve by

$$\gamma = \gamma_+ \cup G \cup \gamma_- \tag{7-551}$$

Again, the $\gamma$ curve separates the state plane into two sets.

***Definition 7-39*** *Let $R_+$ denote the states above the $\gamma$ curve and let $R_-$ denote the states below the $\gamma$ curve. More precisely,*

$$R_+ = \{(x_1, x_2): \textit{if } (x_1, x_2') \in \gamma, \textit{ then } x_2 > x_2'\} \tag{7-552}$$
$$R_- = \{(x_1, x_2): \textit{if } (x_1, x_2') \in \gamma, \textit{ then } x_2 < x_2'\} \tag{7-553}$$

We now assert that:

***Control Law 7-12b*** *If $\beta < 0$, then the control that forces any given state $(x_1, x_2)$, $(x_1, x_2) \notin G$, to the target set $G$ in minimum time is given by*

$$\begin{aligned} u^* &= u^*(x_1, x_2) = +1 \qquad \textit{for all } (x_1, x_2) \in R_+ \cup \gamma_+ \\ u^* &= u^*(x_1, x_2) = -1 \qquad \textit{for all } (x_1, x_2) \in R_- \cup \gamma_- \end{aligned} \tag{7-554}$$

*and, furthermore, this control is unique.*

**Exercise 7-68** Prove Control Law 7-12*b*.

Figure 7-76 shows several time-optimal trajectories to the target set $G$. Up to now, there has been no significant difference between the plant with the minimum-phase zero ($\beta > 0$) and the plant with the non-minimum-phase zero ($\beta < 0$). The difference that now arises is due to the fact that if $\beta < 0$, there is *no* control which will cause the state to remain in $G$ for *all* time. This is demonstrated below.

Suppose that at $t = t^*$, the state $(x_1(t^*), x_2(t^*)) \in G$, that is, $x_1(t^*) = 0$, $|x_2(t^*)| \le 1$. In order to hold the state in $G$, we must apply the control

$$u(t) = -x_2(t^*)e^{-\beta(t-t^*)} \qquad \text{for } t > t^* \tag{7-555}$$

But since $\beta < 0$, the term $e^{-\beta(t-t^*)}$ is an increasing function of time, and so there exists a time $t_1 \ge t^*$ such that

$$|u(t_1)| = 1 \tag{7-556}$$

The value of $t_1$ is given by the relation

$$t_1 = t^* + \frac{1}{\beta}\log|x_2(t^*)| \tag{7-557†}$$

In the time interval $t^* \le t \le t_1$, the state variables satisfy the equations

$$\left.\begin{aligned} x_1(t) &= 0 \\ x_2(t) &= x_2(t^*)e^{-\beta(t-t^*)} \end{aligned}\right\} \qquad t^* \le t \le t_1 \tag{7-558}$$

† It might appear at first glance that, since $\beta < 0$, $t_1 < t^*$; recall, however, that $|x_2(t^*)| \le 1$, so that $(1/\beta)\log|x_2(t^*)| \ge 0$. If $|x_2(t^*)| = 1$, then $t_1 = t^*$.

so that at $t = t_1$ we have

$$x_1(t_1) = 0 \qquad |x_2(t_1)| = 1 \tag{7-559}$$

It follows that if $|x_2(t^*)| \neq 0$, then *the state can be held in the set $G$ only in the time interval $t^* \leq t \leq t_1$.* This means that *Control Law* 7-12*b forces $y$ and $\dot{y}$ to zero in minimum time $t^*$, but $y$ and $\dot{y}$ can be held at zero only during the finite time interval $t^* \leq t \leq t_1$.*

The question that arises most naturally at this point is: What happens for $t > t_1$? It should be clear that there is *no* control which satisfies the constraint $|u(t)| \leq 1$ and which will cause the state to remain in the target set $G$; therefore, the state will escape from $G$, and for $t > t_1$, Control Law

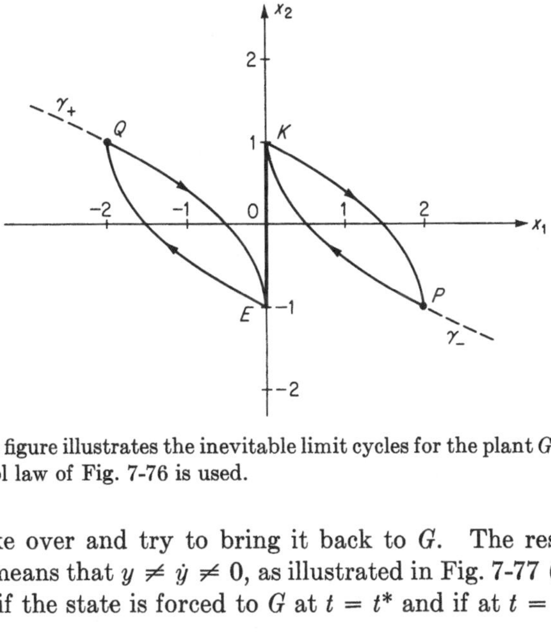


**Fig. 7-77** This figure illustrates the inevitable limit cycles for the plant $G(s) = (s - 1)/s^2$ when the control law of Fig. 7-76 is used.

7-12*b* will take over and try to bring it back to $G$. The result is a limit cycle, which means that $y \neq \dot{y} \neq 0$, as illustrated in Fig. 7-77 (for $\beta = -1$). In Fig. 7-77, if the state is forced to $G$ at $t = t^*$ and if at $t = t^*$ we have

$$x_1(t^*) = 0 \qquad x_2(t^*) > 0 \tag{7-560}$$

then the limit cycle for $t > t_1$ is the closed path $KPKPKP \cdots$. If

$$x_1(t^*) = 0 \qquad x_2(t^*) < 0 \tag{7-561}$$

then the limit cycle for $t > t_1$ is the closed path $EQEQEQ \cdots$. Since the points $E$ and $K$ are specified by

$$E = (0, -1) \qquad K = (0, +1) \tag{7-562}$$

it can be shown that for $\beta = -1$, the points $P$ and $Q$ in Fig. 7-77 are

$$P = (2, -1) \qquad Q = (-2, +1) \tag{7-563}$$

Figure 7-78 illustrates the control which causes this limit cycle. Observe that for $0 < t \le t^*$, the control forces the state to $G$; in the interval $t^* < t \le t_1$, the control is an increasing exponential which holds the state in $G$; for $t > t_1$, the control switches between $+1$ and $-1$ every $T$ seconds, thus generating the limit cycles. If $\beta = -1$, one can calculate the time $T$ in Fig. 7-78 to be exactly 2 seconds.

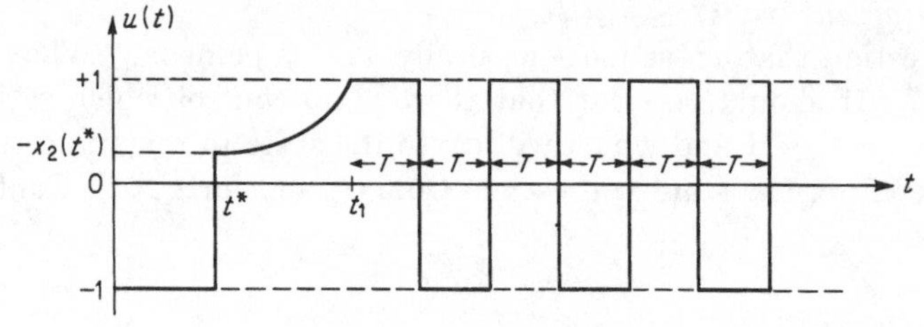


**Fig. 7-78** The shape of the control $u(t)$ which forces $y$ and $\dot{y}$ to zero in minimum time ($t^*$) for the plant $G(s) = (s + \beta)/s^2$, $\beta < 0$. The control $u(t)$ is an increasing exponential during the interval $[t^*, t_1]$; for $t > t_1$ the control is piecewise constant and leads to the limit cycle of Fig. 7-77.

We have shown that if $|x_2(t^*)| \neq 0$, then a limit cycle is unavoidable. Now let us suppose that at $t = t^*$, we have

$$x_1(t^*) = x_2(t^*) = 0 \qquad (7\text{-}564)$$

In this case, if we set

$$u(t) = 0 \qquad \text{for all } t > t^* \qquad (7\text{-}565)$$

then we find that

$$x_1(t) = x_2(t) = 0 \qquad \text{for all } t > t^* \qquad (7\text{-}566)$$

which implies that

$$y(t) = \dot{y}(t) = 0 \qquad \text{for all } t > t^* \qquad (7\text{-}567)$$

as required by the problem.

Now we are faced with the following problem: *We want to drive $y$ and $\dot{y}$ to zero in minimum time and to hold them at zero forever. Control Law* 7-12*b does indeed drive $y$ and $\dot{y}$ to zero in minimum time, but owing to the limit cycle, $y$ and $\dot{y}$ cannot be held at zero forever; but if we are at the origin of the state plane, then the control $u = 0$ will hold $y = \dot{y} = 0$ forever. Therefore, the origin is the only element of $G$ which simultaneously satisfies both objectives of the control problem.* Thus Control Law 7-12*b* is *not* optimal in general, and we must determine another control law which will force any initial state to the origin (0, 0) in minimum time.

Next follow the familiar definitions of the switch curves, the sets in the state plane, and the optimal-control law.

***Definition 7-40*** *Let $\Gamma_+$ denote the set of states which can be forced to the state* (0, 0) *by the control $u = +1$. Let $\Gamma_-$ denote the set of states which can* *be forced to the state* (0, 0) *by the control* $u = -1$. *From Eq.* (7-533), *we obtain*

$$\Gamma_+ = \left\{(x_1, x_2): x_1 = \frac{1}{\beta}(x_2 + \tfrac{1}{2}x_2^2);\ x_2 \geq 0;\ \beta < 0\right\} \tag{7-568}$$

$$\Gamma_- = \left\{(x_1, x_2): x_1 = \frac{1}{\beta}(x_2 - \tfrac{1}{2}x_2^2);\ x_2 \leq 0;\ \beta < 0\right\} \tag{7-569}$$

*The* $\Gamma_+$ *and* $\Gamma_-$ *curves are shown in Fig.* 7-79 (*using* $\beta = -1$). *Let*

$$\Gamma = \Gamma_+ \cup \Gamma_- \tag{7-570}$$

*be the new switch curve.*

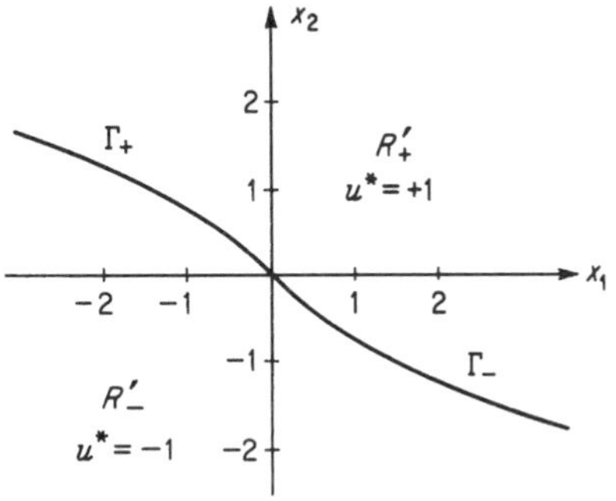


**Fig. 7-79** The switch curves $\Gamma_+$ and $\Gamma_-$.

***Definition 7-41*** *Let* $R'_+$ *denote the set of states above the* $\Gamma$ *curve and let* $R'_-$ *denote the set of states below the* $\Gamma$ *curve; that is,*

$$R'_+ = \left\{(x_1, x_2): x_1 > \frac{1}{\beta}(x_2 + \tfrac{1}{2}x_2|x_2|);\ \beta < 0\right\} \tag{7-571}$$

$$R'_- = \left\{(x_1, x_2): x_1 < \frac{1}{\beta}(x_2 + \tfrac{1}{2}x_2|x_2|);\ \beta < 0\right\} \tag{7-572}$$

We shall now state the solution of Problem 7-12 when $\beta < 0$.

***Control Law 7-12c Solution to Problem 7-12 for*** $\beta < 0$ *The control, as a function of the state* $(x_1, x_2)$, *that forces* $y$ *and* $\dot{y}$ *to zero in minimum time and holds them at zero thereafter is given by*

$$\begin{aligned} u^* &= u^*(x_1, x_2) = +1 \qquad \text{for all } (x_1, x_2) \in R'_+ \cup \Gamma_+ \\ u^* &= u^*(x_1, x_2) = -1 \qquad \text{for all } (x_1, x_2) \in R'_- \cup \Gamma_- \end{aligned} \tag{7-573}$$

*and, furthermore, this control is unique.*

**Exercise 7-69** Prove Control Law 7-12c.
**Exercise 7-70** Design an engineering realization of Control Law 7-12c.

Thus, we have found that in the case of a non-minimum-phase zero, the target set $G$ consists of a single state, the origin of the state plane, and we have developed the time-optimal control law to the origin.

*Case 3* $\beta = 0$

If $\beta = 0$, then the transfer function is

$$H(s) = \frac{s}{s^2} \tag{7-574}$$

and the output $y(t)$ is the solution of the differential equation

$$\ddot{y}(t) = \dot{u}(t) \qquad |u(t)| \leq 1 \tag{7-575}$$

The state variables $x_1(t)$ and $x_2(t)$ are given by Eqs. (7-511), that is,

$$x_1(t) = y(t) \qquad x_2(t) = \dot{y}(t) - u(t) \tag{7-576}$$

and their differential equations are

$$\begin{aligned} \dot{x}_1(t) &= x_2(t) + u(t) \\ \dot{x}_2(t) &= 0 \end{aligned} \tag{7-577}$$

Equations (7-577) are obtained by setting $\beta = 0$ in Eqs. (7-510). The target set $G$ is the same as before; that is,

$$G = \{(x_1, x_2) : x_1 = 0, |x_2| \leq 1\} \tag{7-578}$$

From Eqs. (7-577), we find that if

$$x_1(0) = \xi_1 \qquad x_2(0) = \xi_2 \qquad u(t) = \Delta = \pm 1 \tag{7-579}$$

then

$$\begin{aligned} x_1(t) &= \xi_1 + \xi_2 t + \Delta t \\ x_2(t) &= \xi_2 \end{aligned} \tag{7-580}$$

Since the state variable $x_2(t)$ is a constant and since the objective is to reach the target set $G$ which requires $|x_2| \leq 1$, it is clear that, if we have

$$|x_2(0)| = |\xi_2| > 1 \tag{7-581}$$

then the problem has *no solution*. In other words, *if Eq.* (7-581) *holds, then there is no control which can drive* $y$ *and* $\dot{y}$ *to zero.* If, however,

$$|x_2(0)| = |\xi_2| < 1 \tag{7-582}$$

then it is easy to show that the control

$$u = -\operatorname{sgn}\{x_1\} \tag{7-583}$$

will force $x_1$ and $x_2$ to $G$ in minimum time and that if $(x_1, x_2) \in G$, then the constant control

$$u = -x_2 \tag{7-584}$$

will hold the state in $G$ forever. If

$$|x_2(0)| = |\xi_2| = 1 \tag{7-585}$$

then a solution exists if

$$x_1(0) = \xi_1 < 0 \qquad \text{when } x_2(0) = \xi_2 = +1 \tag{7-586}$$

or if

$$x_1(0) = \xi_1 > 0 \qquad \text{when } x_2(0) = \xi_2 = -1 \tag{7-587}$$

and, in that case, Eq. (7-583) provides us with the time-optimal control to $G$ and Eq. (7-584) with the control which will cause the state to remain in $G$.

To recapitulate: Let $C_+$, $C_-$, and $C$ be the sets of states defined by the relations

$$C_+ = \{(x_1, x_2)\colon -1 < x_2 \leq +1,\, x_1 < 0\} \tag{7-588}$$
$$C_- = \{(x_1, x_2)\colon -1 \leq x_2 < +1,\, x_1 > 0\} \tag{7-589}$$
$$C = C_+ \cup C_- \cup G \tag{7-590}$$

Then *we shall call the set $C$ the set of controllable states.*

***Control Law 7-12d Solution to Problem 7-12 for $\beta = 0$*** *The control, as a function of the state $(x_1, x_2)$, that forces $y$ and $\dot{y}$ to zero and holds them at zero thereafter is unique and is given by*

$$\begin{aligned} u^* &= u^*(x_1, x_2) = +1 && \textit{for all } (x_1, x_2) \in C_+ \\ u^* &= u^*(x_1, x_2) = -1 && \textit{for all } (x_1, x_2) \in C_- \\ u^* &= u^*(x_1, x_2) = -x_2 && \textit{for all } (x_1, x_2) \in G \end{aligned} \tag{7-591}$$

*If $(x_1, x_2) \notin C$ [see Eq. (7-590)], then no control (time-optimal or otherwise) exists which drives $y$ and $\dot{y}$ to zero and maintains them at zero thereafter.*

The set of controllable states and several time-optimal trajectories are shown in Fig. 7-80.

Let us now examine the fallacy of replacing the system

$$\ddot{y}(t) = \dot{u}(t) \qquad |u(t)| \leq 1 \tag{7-592}$$

by the system

$$\dot{y}(t) = u(t) \qquad |u(t)| \leq 1 \tag{7-593}$$

which means that we replace the transfer function $s/s^2$ by the transfer function $1/s$. It is easy to show (see Sec. 7-10) that given the system (7-593), *the control*

$$\begin{aligned} u(t) &= -\operatorname{sgn}\{y(t)\} && y(t) \neq 0 \\ u(t) &= 0 && y(t) = 0 \end{aligned} \tag{7-594}$$

*will reduce $y$ to zero in minimum time and hold it at zero thereafter, no matter what $\dot{y}(t)$ is, since $\dot{y}(t)$ is not a state of the system* (7-593). However, for the system (7-592), we have found that it is not possible to reduce, in general, $y(t)$ to zero if the state variable $x_2(t) = \dot{y}(t) - u(t)$ is such that $|x_2(t)| > 1$ or, equivalently, if $|\dot{y}(t)| > 2$. Thus, if one replaces the system (7-592) by the system (7-593), one might think that one is dealing with a controllable system, which is not true if $|\dot{y}(t)| > 2$, and unexpected stability problems will arise.

Let us briefly summarize the results and conclusions of this section. We have found that if the transfer function contains a *minimum-phase zero* ($\beta > 0$), then there exists a target set $G$ in the state plane and the problem is:

1. To force the state to $G$ in minimum time using a piecewise constant (bang-bang type) control
2. To cause the system to remain in $G$ by a control which is a decaying exponential

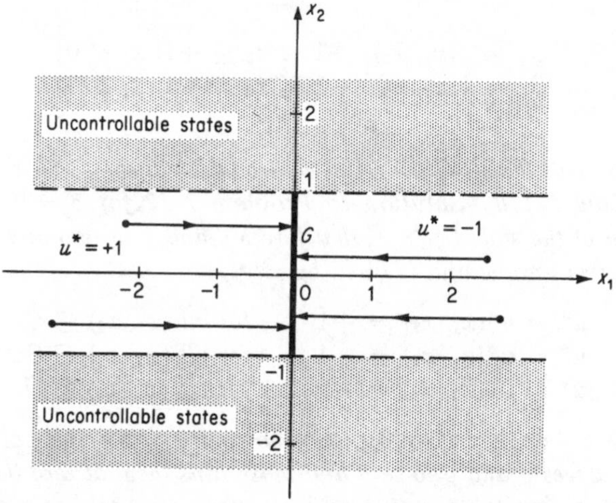


**Fig. 7-80** Time-optimal trajectories to the set $G$ for the plant $G(s) = s/s^2$. Note that only a subset of states is controllable.

If the transfer function contains a *non-minimum-phase zero*, then the problem is to force the state to the origin of the state plane using a piecewise constant (bang-bang type) control. If the zero is on top of the poles ($\beta = 0$), then there exist states which are not controllable.

## 7-13 Time-optimal Control of a Double-integral Plant with Two Zeros

In this section, we consider the problem of the time-optimal control of a plant whose transfer function contains two zeros and two poles. Basically, our reason for studying this system is to provide a specific example of the more general theory of Sec. 7-14. Most of our interest in this section will be focused on the determination of the target set $G$ and on the illustration of the difficulties involved. We shall illustrate the target sets by means of examples. Before we start the specifics, we wish to mention that the more general concepts of Sec. 7-14 are direct extensions of the material of this section.

Consider a plant with output $y(t)$ and control $u(t)$ related by the second-order differential equation

$$\ddot{y}(t) = \ddot{u}(t) + b_1\dot{u}(t) + b_0u(t) \tag{7-595}$$

Assume that the control $u(t)$ is restricted in magnitude by the relation

$$|u(t)| \le 1 \qquad \text{for all } t \tag{7-596}$$

The plant (7-595) is described by the transfer function

$$\frac{y(s)}{u(s)} = H(s) = \frac{s^2 + b_1s + b_0}{s^2} = \frac{(s+\beta_1)(s+\beta_2)}{s^2} \tag{7-597}$$

That is, the zeros are at $s = -\beta_1$ and $s = -\beta_2$, and so

$$b_0 = \beta_1\beta_2 \qquad b_1 = \beta_1 + \beta_2 \tag{7-598}$$

The constants $\beta_1$ and $\beta_2$ are assumed to be real.

The time-optimal control problem we consider is:

*Given the system described by Eq.* (7-595), *find the control* $u(t)$, *satisfying* $|u(t)| \le 1$, *which will drive the output* $y$ *and the output rate* $\dot{y}$ *to zero in minimum time and which will hold them at zero thereafter.*

We define the state variables $x_1(t)$ and $x_2(t)$ by the equations

$$\begin{aligned} x_1(t) &= y(t) - h_0u(t) \\ x_2(t) &= \dot{y}(t) - h_0\dot{u}(t) - h_1u(t) \end{aligned} \tag{7-599}$$

From Eqs. (7-595) and (7-599), we find that the state variables satisfy the differential equations

$$\dot{x}_1(t) = x_2(t) + h_1u(t) \tag{7-600}$$
$$\dot{x}_2(t) = \ddot{u}(t) + b_1\dot{u}(t) + b_0u(t) - h_0\ddot{u}(t) - h_1\dot{u}(t) \tag{7-601}$$

We note that $\dot{x}_1(t)$ is independent of $\dot{u}(t)$ and $\ddot{u}(t)$; in order to make $\dot{x}_2(t)$ independent of $\dot{u}(t)$ and $\ddot{u}(t)$, we set

$$h_0 = 1 \qquad h_1 = b_1 \tag{7-602}$$

so that Eqs. (7-599) reduce to

$$\begin{aligned} x_1(t) &= y(t) - u(t) \\ x_2(t) &= \dot{y}(t) - \dot{u}(t) - b_1u(t) \end{aligned} \tag{7-603}$$

and Eqs. (7-600) and (7-601) reduce to

$$\begin{aligned} \dot{x}_1(t) &= x_2(t) + b_1u(t) \\ \dot{x}_2(t) &= b_0u(t) \end{aligned} \tag{7-604}$$

Suppose that at $t = t^*$, we have

$$y(t^*) = \dot{y}(t^*) = 0 \tag{7-605}$$

and we wish to find the control $u(t)$, $t > t^*$, such that

$$y(t) = \dot{y}(t) = 0 \qquad \text{for all } t > t^* \tag{7-606}$$

Substituting Eqs. (7-605) and (7-606) into Eqs. (7-603), we find that the state variables must satisfy the relationships

$$\left.\begin{aligned} x_1(t) &= -u(t) \\ x_2(t) &= -\dot{u}(t) - b_1 u(t) \end{aligned}\right\} \qquad \text{for } t \geq t^* \tag{7-607}$$

Since we must have

$$x_1(t) = -u(t) \qquad t \geq t^* \tag{7-608}$$

substituting Eq. (7-608) into Eqs. (7-604), we find that

$$\left.\begin{aligned} \dot{x}_1(t) &= -b_1 x_1(t) + x_2(t) \\ \dot{x}_2(t) &= -b_0 x_1(t) \end{aligned}\right\} \qquad t \geq t^* \tag{7-609}$$

Equation (7-609) may be written in the vector form

$$\begin{bmatrix} \dot{x}_1(t) \\ \dot{x}_2(t) \end{bmatrix} = \begin{bmatrix} -b_1 & 1 \\ -b_0 & 0 \end{bmatrix} \begin{bmatrix} x_1(t) \\ x_2(t) \end{bmatrix} \qquad t \geq t^* \tag{7-610}$$

Let $\mathbf{Q}$ be the matrix

$$\mathbf{Q} = \begin{bmatrix} -b_1 & 1 \\ -b_0 & 0 \end{bmatrix} \tag{7-611}$$

It is easy to show that the eigenvalues of the matrix $\mathbf{Q}$ are $-\beta_1$ and $-\beta_2$; that is, *the eigenvalues of* $\mathbf{Q}$ *are the same as the zeros of the transfer function* $H(s)$.

Equation (7-610) is a vector homogeneous differential equation for $t \geq t^*$. Thus, we can solve for $x_1(t)$ in terms of the "initial conditions" $x_1(t^*)$ and $x_2(t^*)$ to obtain the relation

$$\begin{aligned} x_1(t) = \frac{1}{\beta_2 - \beta_1} [x_1(t^*)(-\beta_1 e^{-\beta_1(t-t^*)} + \beta_2 e^{-\beta_2(t-t^*)}) \\ + x_2(t^*)(e^{-\beta_1(t-t^*)} - e^{-\beta_2(t-t^*)})] \qquad t \geq t^* \end{aligned} \tag{7-612}$$

But since $x_1(t) = -u(t)$ for $t \geq t^*$ and since $|u(t)| \leq 1$ for all $t$, we see that for $t \geq t^*$, in view of Eqs. (7-608) and (7-596), we must have

$$\left| \frac{1}{\beta_2 - \beta_1} [x_1(t^*)(-\beta_1 e^{-\beta_1(t-t^*)} + \beta_2 e^{-\beta_2(t-t^*)}) + x_2(t^*)(e^{-\beta_1(t-t^*)} - e^{-\beta_2(t-t^*)})] \right| \leq 1 \tag{7-613}$$

Thus, in order to guarantee that

$$y(t) = \dot{y}(t) = 0 \qquad \text{for all } t \geq t^* \tag{7-614}$$

it is necessary that Eqs. (7-608), (7-610), and (7-613) hold for all $t \geq t^*$.

***Definition 7-42*** *Let $G$ be the set of all states $(x_1(t^*), x_2(t^*))$ which satisfy Eq.* (7-613) *for all $t \geq t^*$. We shall call the set $G$ the target set. We assert that the set $G$ is closed and convex and that the origin is an element of $G$.*†

Since the set $G$ is defined by Eq. (7-613) in a parametric fashion, it is necessary to obtain an expression [which should be an algebraic relationship between $x_1(t^*)$ and $x_2(t^*)$] for the *boundary*‡ of $G$, denoted by $\partial G$. It is necessary to obtain an equation for the boundary of $G$ in order to determine the time-optimal control law to the target set $G$. We shall outline a technique for evaluating this equation.

Define the functions $q_1(t)$, $q_2(t)$ by setting

$$\begin{aligned} q_1(t) &= \frac{1}{\beta_2 - \beta_1} \left(-\beta_1 e^{-\beta_1(t-t^*)} + \beta_2 e^{-\beta_2(t-t^*)}\right) \\ q_2(t) &= \frac{1}{\beta_2 - \beta_1} \left(e^{-\beta_1(t-t^*)} - e^{-\beta_2(t-t^*)}\right) \end{aligned} \tag{7-615}$$

for $t \geq t^*$. Note that

$$q_1(t^*) = 1 \qquad q_2(t^*) = 0 \tag{7-616}$$

Using $q_1(t)$ and $q_2(t)$, we can define the set $G$ by

$$G = \{(x_1, x_2): |x_1 q_1(t) + x_2 q_2(t)| \leq 1 \qquad \text{for all } t \geq t^*\} \tag{7-617}$$

Suppose that the zeros of the transfer function are in the left half of the $s$ plane; i.e., suppose that

$$\beta_2 > \beta_1 > 0 \tag{7-618}$$

Then it is easy to show that

$$\lim_{t \to \infty} q_1(t) = \lim_{t \to \infty} q_2(t) = 0 \tag{7-619}$$

Since each function $q_1(t)$ and $q_2(t)$ is a continuous function of time, it follows that the function

$$q(t) = x_1 q_1(t) + x_2 q_2(t) \tag{7-620}$$

is such that *the function $|q(t)|$ must have a maximum value at some finite time $\hat{t}$*, $\hat{t} \geq t^*$, so that for any $t \neq \hat{t}$, the relation

$$|q(t)| < |q(\hat{t})| \tag{7-621}$$

holds. Since the set $G$ is closed, it follows that a state $(x_1, x_2)$ which belongs to the boundary $\partial G$ of the set $G$ can be determined from the relationship $|q(\hat{t})| = 1$ or, equivalently, from the relation

$$|x_1 q_1(\hat{t}) + x_2 q_2(\hat{t})| = 1 \tag{7-622}$$

† We shall prove this fact in Sec. 7-14 for the general case.

‡ See Definition 3-10.

Let us find the set of $(x_1, x_2)$ which satisfies Eq. (7-622). Since $|q(t)|$ must have a maximum at $t = \hat{t}$, it follows that

$$\left.\frac{\partial |q(t)|}{\partial t}\right|_{t=\hat{t}} = 0 \tag{7-623}$$

and since $q(t)$ will have either a maximum or a minimum at $t = \hat{t}$, we find that

$$\left.\frac{\partial q(t)}{\partial t}\right|_{t=\hat{t}} = x_1 \left.\frac{\partial q_1(t)}{\partial t}\right|_{t=\hat{t}} + x_2 \left.\frac{\partial q_2(t)}{\partial t}\right|_{t=\hat{t}} = 0 \tag{7-624}$$

But from Eqs. (7-615), we can obtain $\partial q_1(t)/\partial t$ and $\partial q_2(t)/\partial t$ and then substitute them in Eq. (7-624) to obtain the relation

$$\hat{t} - t^* = \log \left( \frac{x_1\beta_2^2 - x_2\beta_2}{x_1\beta_1^2 - x_2\beta_1} \right)^{1/(\beta_2 - \beta_1)} \tag{7-625}$$

Substituting Eq. (7-625) into Eq. (7-622), we find, after some algebraic manipulations, that

$$\left| \frac{1}{\beta_2 - \beta_1} \left[ \left( \frac{x_1\beta_2^2 - x_2\beta_2}{x_1\beta_1^2 - x_2\beta_1} \right)^{-\beta_1/(\beta_2-\beta_1)} (-\beta_1 x_1 + x_2) + \left( \frac{x_1\beta_2^2 - x_2\beta_2}{x_1\beta_1^2 - x_2\beta_1} \right)^{-\beta_2/(\beta_2-\beta_1)} (\beta_2 x_1 - x_2) \right] \right| = 1 \tag{7-626}$$

It follows that *if* $\beta_2 > \beta_1 > 0$, *then a state* $(x_1, x_2)$ *belongs to the boundary* $\partial G$ *of the target set* $G$ *if it satisfies Eq.* (7-626).

Now suppose that either one or both of the zeros is in the right half-plane, that is, either one or both of the zeros is *non-minimum-phase.* For the sake of clarity, suppose that

$$\beta_1 < 0 \tag{7-627}$$

Then, from Eqs. (7-615), it follows that

$$\lim_{t\to\infty} |q_1(t)| = \lim_{t\to\infty} |q_2(t)| = \infty \tag{7-628}$$

and hence that the function $q(t) = x_1q_1(t) + x_2q_2(t)$ has the property that

$$\lim_{t\to\infty} |q(t)| = \infty \qquad \text{for any } x_1 \neq 0,\ x_2 \neq 0 \tag{7-629}$$

But we wish to find the set of $(x_1, x_2)$ such that $|q(t)| \leq 1$; therefore, in view of Eq. (7-629), we conclude that if

$$x_1 = 0 \qquad x_2 = 0 \tag{7-630}$$

then

$$|q(t)| = 0 \qquad \text{for all } t \geq t^* \tag{7-631}$$

For this reason, we conclude that *if any of the zeros are non-minimum-phase ones* ($\beta_1 < 0$, $\beta_2 < 0$, *or* $\beta_1 < 0$ *and* $\beta_2 < 0$), *then the target set G consists of a single point,*† *the point* (0, 0), *that is, the origin of the state plane.*

† Note that a set consisting of a single point is both closed and convex.

Finally, let us examine the case

$$\beta_1 = 0 \qquad \text{and} \qquad \beta_2 > 0 \tag{7-632}$$

that is, one zero is at the origin and the other is in the left half-plane. Substituting Eqs. (7-632) into Eqs. (7-615), we find that

$$\begin{aligned} q_1(t) &= e^{-\beta_2(t-t^*)} \\ q_2(t) &= \frac{1}{\beta_2}\left(1 - e^{-\beta_2(t-t^*)}\right) \end{aligned} \tag{7-633}$$

Hence the function

$$q(t) = x_1 q_1(t) + x_2 q_2(t)$$

has the property that

$$\lim_{t \to \infty} |q(t)| = \frac{1}{\beta_2} |x_2| \tag{7-634}$$

But since the set $G$ is the set of all points $(x_1, x_2)$ such that $|q(t)| \leq 1$, Eq. (7-634) implies that

$$|x_2| \leq \beta_2 \tag{7-635}$$

On the other hand, if Eq. (7-635) holds, then the condition $|q(t)| \leq 1$ yields the relation

$$|x_1| \leq 1 \tag{7-636}$$

so that, *if* $\beta_1 = 0$, $\beta_2 > 0$, *then the target set* $G$ *is given by*

$$G = \{(x_1, x_2) \colon |x_1| \leq 1, |x_2| \leq \beta_2\} \tag{7-637}$$

Let us now recapitulate: We have found that if the two zeros are minimum-phase, then the boundary $\partial G$ of the target set $G$ is given by Eq. (7-626). If any of the zeros is a non-minimum-phase zero, then the target set $G$ (and its boundary $\partial G$) is the single point (0, 0). Finally, if one of the zeros cancels a pole and the other zero is a minimum-phase one, then the target set $G$ is given by Eq. (7-637).

Since the first objective is to force the state to the target set $G$ in minimum time, it should be clear that we must use the transversality conditions to find the extremal controls. However, the target sets have, in general, corners; so the reader should be careful in using the transversality conditions.

The following three examples consider the time-optimal problem for specific values of $\beta_1$ and $\beta_2$. We shall present the results fairly rapidly, and we shall leave the details to the reader. Our major objective is to illustrate the shape of the switch curves and of the target set.

**Example 7-8** Consider the system with the transfer function

$$\frac{y(s)}{u(s)} = G(s) = \frac{(s+1)(s+2)}{s^2} \tag{7-638}$$

That is, in Eq. (7-597) we set

$$\beta_1 = 1 \qquad \beta_2 = 2 \qquad b_0 = 2 \qquad b_1 = 3 \tag{7-639}$$

The state variables are [see Eqs. (7-603)]

$$\begin{aligned} x_1(t) &= y(t) - u(t) \\ x_2(t) &= \dot{y}(t) - \dot{u}(t) - 3u(t) \end{aligned} \tag{7-640}$$

and they satisfy the differential equations [see Eqs. (7-604)]

$$\begin{aligned} \dot{x}_1(t) &= x_2(t) + 3u(t) \\ \dot{x}_2(t) &= 2u(t) \end{aligned} \tag{7-641}$$

The boundary of the target set is obtained by substituting $\beta_1 = 1, \beta_2 = 2$ into Eq. (7-626). After some algebraic manipulations, we have

$$\left| -\frac{(x_1 - x_2)^2}{4(2x_1 - x_2)} \right| = 1 \tag{7-642}$$

Equation (7-642) results in the two equations

$$x_2 = x_1 + 2 + 2\sqrt{1 - x_1} \qquad |x_1| \leq 1 \tag{7-643}$$

$$x_2 = x_1 - 2 - 2\sqrt{1 + x_1} \qquad |x_1| \leq 1 \tag{7-644}$$

Thus, a point $(x_1, x_2) \in \partial G$ if it satisfies either Eq. (7-643) or Eq. (7-644). The set $G$ is shown in Fig. 7-81. To determine the time-optimal control to the set $G$, we solve Eqs.

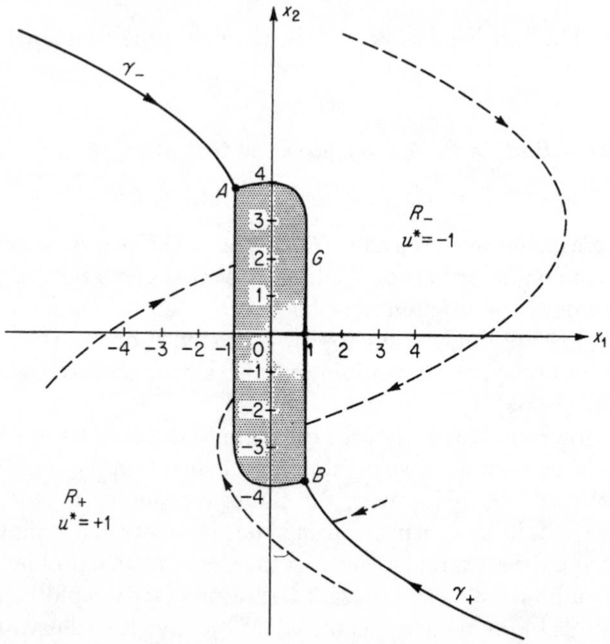


**Fig. 7-81** The target set $G$ for the plant with the transfer function $(s + 1)(s + 2)/s^2$. The switch curves are labeled $\gamma_-$ and $\gamma_+$. The dashed curves represent time-optimal trajectories.

(7-641) using $u = \Delta = \pm 1$ to obtain the equations of the trajectories. Using $x_1(t) = \xi_1$ and $x_2(0) = \xi_2$, we find that

$$\begin{aligned} x_1(t) &= \xi_1 + \xi_2 t + \Delta t^2 + 3\Delta t \\ x_2(t) &= \xi_2 + 2\Delta t \end{aligned} \tag{7-645}$$

Eliminating the time $t$, we obtain the relation

$$x_1 = \xi_1 - \tfrac{1}{4}\Delta\xi_2^2 + \tfrac{1}{4}\Delta x_2^2 + \tfrac{3}{2}x_2 - \tfrac{3}{2}\xi_2 \tag{7-646}$$

Thus the trajectories are parabolas with vertex at

$$x_2 = -\Delta \tag{7-647}$$

Figure 7-81 illustrates the switch curves $\gamma_+$ and $\gamma_-$. The $\gamma_+$ curve is the set of all points which can be forced to the point $B$ of $G$ with $u = +1$, and the $\gamma_-$ curve is the set of all points which can be forced to the point $A$ of $G$ with $u = -1$. From Eqs. (7-643) and (7-644), it is easy to establish that the coordinates of the points $A$ and $B$ are

$$A = (-1, 1 + 2\sqrt{2}) \qquad B = (1, -1 - 2\sqrt{2}) \tag{7-648}$$

Thus, from Eqs. (7-646) and (7-648), we can find the equations of the $\gamma_+$ and $\gamma_-$ curves:

$$\gamma_+ = \{(x_1, x_2): x_1 = \tfrac{1}{4}x_2^2 + \tfrac{3}{2}x_2 - 6.74;\ x_1 \geq 1;\ x_2 \leq -1 - 2\sqrt{2}\} \tag{7-649}$$
$$\gamma_- = \{(x_1, x_2): x_1 = -\tfrac{1}{4}x_2^2 + \tfrac{3}{2}x_2 + 6.74;\ x_1 \leq -1;\ x_2 \geq 1 + 2\sqrt{2}\} \tag{7-650}$$

The $\gamma_+$ curve, the $\gamma_-$ curve, and the set $G$ divide the $x_1x_2$ plane into two sets $R_+$ and $R_-$, as shown in Fig. 7-81. When the state is in $R_+$, the time-optimal control is $u^* = +1$; when the state is in $R_-$, the time-optimal control is $u^* = -1$; when the state is in $G$, then the control is [see Eqs. (7-640)]

$$u^*(t) = -x_1(t) \tag{7-651}$$

The dashed curves in Fig. 7-81 are time-optimal trajectories to the set $G$. It is left to the reader to verify that this control law is indeed time-optimal.

**Example 7-9** Consider the system with the transfer function

$$\frac{y(s)}{u(s)} = G(s) = \frac{(s+1)(s-1)}{s^2} \tag{7-652}$$

That is, in Eq. (7-597), we set

$$\beta_1 = 1 \qquad \beta_2 = -1 \qquad b_0 = -1 \qquad b_1 = 0 \tag{7-653}$$

The system (7-652) has a non-minimum-phase zero (at $s = \beta_1 = 1$).

The state variables are [see Eqs. (7-603)]

$$\begin{aligned} x_1(t) &= y(t) - u(t) \\ x_2(t) &= \dot{y}(t) - \dot{u}(t) \end{aligned} \tag{7-654}$$

and their differential equations are [see Eqs. (7-604)]

$$\begin{aligned} \dot{x}_1(t) &= x_2(t) \\ \dot{x}_2(t) &= -u(t) \end{aligned} \tag{7-655}$$

Since one of the zeros is a non-minimum-phase zero, the target set $G$ consists only of the origin. Since the system (7-655) is almost identical to the system we have examined in Sec. 7-2 [compare Eqs. (7-655) and (7-5)], it is easy to establish that the control law suggested by Fig. 7-82 is the time-optimal one. The equations of the $\gamma_+$ and $\gamma_-$ curves in Fig. 7-82 are

$$\gamma_+ = \{(x_1, x_2): x_1 = -\tfrac{1}{2}x_2^2;\ x_2 \geq 0\} \tag{7-656}$$
$$\gamma_- = \{(x_1, x_2): x_1 = \tfrac{1}{2}x_2^2;\ x_2 \leq 0\} \tag{7-657}$$

The time-optimal control is $u^* = +1$ if the state is in $R_+$ and $u^* = -1$ if the state is in $R_-$. The control is $u^* = 0$ if $x_1 = x_2 = 0$.

**Example 7-10** Consider the system with the transfer function

$$\frac{y(s)}{u(s)} = G(s) = \frac{s(s+1)}{s^2} \tag{7-658}$$

That is, in Eq. (7-597), we set

$$\beta_1 = 0 \qquad \beta_2 = 1 \qquad b_0 = 0 \qquad b_1 = 1 \tag{7-659}$$

The system (7-658) has a zero on top of the two poles at the origin of the $s$ plane.

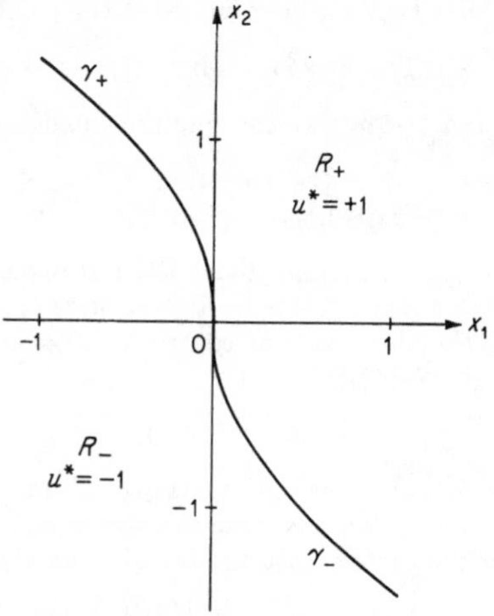


**Fig. 7-82** The target set $G$ for the plant $H(s) = (s+1)(s-1)/s^2$ is the origin. The switch curve is parabolic.

The state variables are [see Eqs. (7-603)]

$$\begin{aligned} x_1(t) &= y(t) - u(t) \\ x_2(t) &= \dot{y}(t) - \dot{u}(t) - u(t) \end{aligned} \tag{7-660}$$

and they satisfy the differential equations [see Eqs. (7-604)]

$$\begin{aligned} \dot{x}_1(t) &= x_2(t) + u(t) \\ \dot{x}_2(t) &= 0 \end{aligned} \tag{7-661}$$

Since $\dot{x}_2(t) = 0$,

$$x_2(t) = x_2(0) = \xi_2 \qquad \text{for all } t \tag{7-662}$$

That is, there is no way that we can change the value of the state variable $x_2(t)$.

The target set $G$ for this problem is given by [see Eq. (7-637)]

$$G = \{(x_1, x_2): |x_1| \le 1, |x_2| \le 1\} \tag{7-663}$$

That is, it is the square shown in Fig. 7-83.

In view of Eq. (7-662), it is immediately apparent that *if* $|\xi_2| > 1$, *there does not exist a control which can drive the state to the target set* $G$. This means that it is impossible to force the output $y(t)$ and output rate $\dot{y}(t)$ to zero and to hold them at zero thereafter.

If $|\xi_2| < 1$, it is possible to force the state to the target set $G$ in minimum time by using the control

$$u^*(t) = -\operatorname{sgn}\{x_1(t)\} \qquad |x_1(t)| > 1 \tag{7-664}$$

If $\xi_2 = 1$ and $x_1(t) > 1$ or if $\xi_2 = -1$ and $x_1(t) < -1$, then the system is again "uncontrollable." If $\xi_2 = 1$ and $x_1(t) < -1$ or if $\xi_2 = -1$ and $x_1(t) > 1$, then the system is "controllable" and the time-optimal control is $u^*(t) = -\operatorname{sgn}\{x_1(t)\}$. If the state

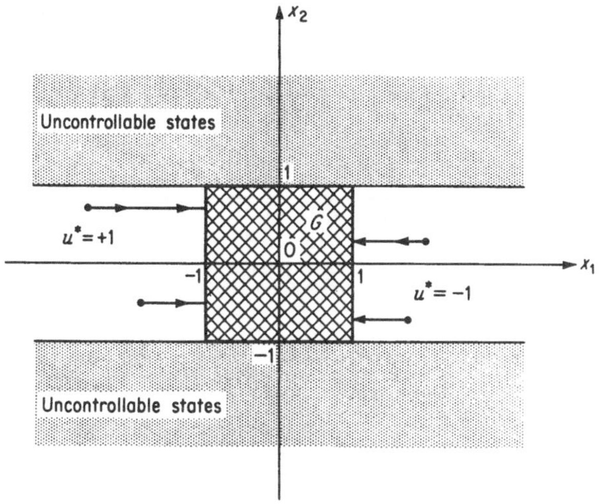


**Fig. 7-83** The target set $G$ for the plant $H(s) = s(s+1)/s^2$ is the unit square.

belongs to $G$, the control must be $u^*(t) = -x_1(t)$. The set of "controllable" states and several time-optimal trajectories to the set $G$ are shown in Fig. 7-83.

## 7-14 General Results on the Time-optimal Control of Plants with Numerator Dynamics

In the preceding two sections, we examined the time-optimal control problem for second-order systems with zeros in their transfer function. We found that the most important step in the analysis of these systems was the determination of the target set $G$; once the target set $G$ was obtained, we could formulate the time-optimal control law to the set $G$ and determine the control for states in the set $G$.

It is our objective to generalize the results obtained in Secs. 7-12 and 7-13 to systems with an arbitrary number of poles and zeros in their transfer function. We shall especially focus our attention on the effect of the zeros in the formulation and solution of the time-optimal problem.

Consider a system with output $y(t)$ and control $u(t)$ related by the $n$th-order time-invariant linear differential equation

$$\begin{aligned}\{D^n + a_{n-1}D^{n-1} + \cdots + a_1D + a_0\}y(t) \\ = \{b_nD^n + b_{n-1}D^{n-1} + \cdots + b_1D + b_0\}u(t)\end{aligned} \tag{7-665}$$

We can write Eq. (7-665) in the form

$$y^{(n)}(t) + \sum_{i=0}^{n-1} a_i y^{(i)}(t) = \sum_{j=0}^{n} b_j u^{(j)}(t) \tag{7-666}$$

where

$$y^{(i)}(t) = \frac{d^i y(t)}{dt^i} \qquad i = 0, 1, \ldots, n$$
$$u^{(j)}(t) = \frac{d^j u(t)}{dt^j} \qquad j = 0, 1, \ldots, n \tag{7-667}$$

It is understood that $y^{(0)}(t) = y(t)$ and that $u^{(0)}(t) = u(t)$.

The system described by Eq. (7-665) has the transfer function

$$H(s) = \frac{b_n s^n + b_{n-1}s^{n-1} + \cdots + b_1 s + b_0}{s^n + a_{n-1}s^{n-1} + \cdots + a_1 s + a_0} \tag{7-668}$$

We assume that the zeros of $H(s)$ are distinct and that they are at

$$s = -\beta_1 \qquad s = -\beta_2 \qquad \cdots \qquad s = -\beta_n \tag{7-669}$$

and that the poles of $H(s)$ are at

$$s = -s_1 \qquad s = -s_2 \qquad \cdots \qquad s = -s_n \tag{7-670}$$

Thus, $H(s)$ can be written in the form

$$H(s) = \frac{b_n(s + \beta_1)(s + \beta_2) \cdots (s + \beta_n)}{(s + s_1)(s + s_2) \cdots (s + s_n)} \tag{7-671}$$

The roots of the polynomial

$$\lambda^n + \frac{b_{n-1}}{b_n}\lambda^{n-1} + \cdots + \frac{b_1}{b_n}\lambda + \frac{b_0}{b_n} \tag{7-672}$$

are at

$$\lambda = -\beta_i \qquad i = 1, 2, \ldots, n \tag{7-673}$$

and the roots of the polynomial

$$\lambda^n + a_{n-1}\lambda^{n-1} + \cdots + a_1\lambda + a_0 \tag{7-674}$$

are at

$$\lambda = -s_i \qquad i = 1, 2, \ldots, n \tag{7-675}$$

Before formulating the time-optimal control problem for the system (7-665), it is instructive to compare this system with the system described by the differential equation

$$\{D^n + a_{n-1}D^{n-1} + \cdots + a_1 D + a_0\}y(t) = b_0 u(t) \tag{7-676}$$

which has the transfer function

$$G(s) = \frac{b_0}{s^n + a_{n-1}s^{n-1} + \cdots + a_1 s + a_0} \tag{7-677}$$

We know that† an appropriate state vector for the system (7-676) is the vector

$$\mathbf{y}(t) = \begin{bmatrix} y(t) \\ \dot{y}(t) \\ \cdot \\ \cdot \\ \cdot \\ y^{(n-2)}(t) \\ y^{(n-1)}(t) \end{bmatrix} \tag{7-678}$$

Suppose that at some time $t = t^*$, we have

$$\mathbf{y}(t^*) = \mathbf{0} \tag{7-679}$$

and suppose that we wish to find a control $u(t)$ such that

$$\mathbf{y}(t) = \mathbf{0} \qquad \text{for all } t > t^* \tag{7-680}$$

Equations (7-679) and (7-680) imply that

$$y^{(n)}(t) = 0 \qquad \text{for all } t > t^*$$

and that

$$\{D^n + a_{n-1}D^{n-1} + \cdots + a_1 D + a_0\}y(t) = 0 \qquad \text{for all } t > t^* \tag{7-681}$$

which implies that the control must be

$$u(t) = 0 \qquad \text{for all } t > t^* \tag{7-682}$$

Now let us turn our attention to the system described by the differential equation (7-665) and suppose that at $t = t^*$, we have

$$\mathbf{y}(t^*) = \mathbf{0} \tag{7-683}$$

Again suppose that we wish to find a control $u(t)$ such that

$$\mathbf{y}(t) = \mathbf{0} \qquad \text{for all } t > t^* \tag{7-684}$$

Employing similar reasoning, we find that the control which will result in Eq. (7-684) must be the solution of the differential equation

$$\{b_n D^n + b_{n-1}D^{n-1} + \cdots + b_1 D + b_0\}u(t) = 0 \qquad \text{for } t > t^* \tag{7-685}$$

† See Sec. 4-9.

It follows that the control $u(t)$ will be of the form

$$u(t) = \sum_{i=1}^{n} \alpha_i(t^*)e^{-\beta_i(t-t^*)} \qquad t \geq t^* \tag{7-686}$$

where the coefficients $\alpha_i(t^*)$ will be functions of $u(t^*), \dot{u}(t^*), \ldots, u^{(n-1)}(t^*)$.

The above considerations lead us to:

***Remark 7-1*** *Given the two systems* (7-665) *and* (7-676). *Suppose that* $\mathbf{y}(t^*) = \mathbf{0}$. *Then, for the system* (7-665), *the control*

$$u(t) = \sum_{i=1}^{n} \alpha_i(t^*)e^{-\beta_i(t-t^*)} \qquad \text{for all } t > t^* \tag{7-687}$$

*and, for the system* (7-676), *the control*

$$u(t) = 0 \qquad \text{for all } t > t^* \tag{7-688}$$

*result in*

$$\mathbf{y}(t) = \mathbf{0} \qquad \text{for all } t > t^* \tag{7-689}$$

We know† that a state vector $\mathbf{x}(t)$ for the system (7-665) can be defined in terms of the output vector $\mathbf{y}(t)$, given by Eq. (7-678), the control $u(t)$, and its $n - 1$ time derivatives as follows:

$$\begin{aligned}
x_1(t) &= y(t) - h_0 u(t) \\
x_2(t) &= \dot{y}(t) - h_0\dot{u}(t) - h_1 u(t) \\
x_3(t) &= \ddot{y}(t) - h_0\ddot{u}(t) - h_1\dot{u}(t) - h_2 u(t) \\
&\cdots\cdots\cdots\cdots\cdots\cdots\cdots\cdots\cdots \\
x_n(t) &= y^{(n-1)}(t) - h_0 u^{(n-1)}(t) - h_1 u^{(n-2)}(t) - \cdots - h_{n-1}u(t)
\end{aligned} \tag{7-690}$$

or, in the abbreviated form,

$$x_i(t) = y^{(i-1)}(t) - \sum_{k=0}^{i-1} u^{(k)}(t)h_{i-k-1} \qquad i = 1, 2, \ldots, n \tag{7-691}$$

The differential equations satisfied by the state variables $x_i(t)$ are‡

$$\begin{bmatrix} \dot{x}_1(t) \\ \dot{x}_2(t) \\ \cdots \\ \dot{x}_{n-1}(t) \\ \dot{x}_n(t) \end{bmatrix} = \begin{bmatrix} 0 & 1 & 0 & \cdots & 0 & 0 \\ 0 & 0 & 1 & \cdots & 0 & 0 \\ \cdots & \cdots & \cdots & \cdots & \cdots & \cdots \\ 0 & 0 & 0 & \cdots & 0 & 1 \\ -a_0 & -a_1 & -a_2 & \cdots & -a_{n-2} & -a_{n-1} \end{bmatrix} \begin{bmatrix} x_1(t) \\ x_2(t) \\ \cdots \\ x_{n-1}(t) \\ x_n(t) \end{bmatrix} + \begin{bmatrix} h_1 \\ h_2 \\ \cdots \\ h_{n-1} \\ h_n \end{bmatrix} u(t) \tag{7-692}$$

† See Sec. 4-10.
‡ See Eq. (4-187).

The constants $h_0, h_1, \ldots, h_{n-1}, h_n$ appearing in Eqs. (7-690) and (7-692) can be found† from the equations

$$\begin{aligned} h_0 &= b_n \\ h_{n-m} &= b_m - \sum_{i=0}^{n-m-1} h_i a_{i+m} \qquad m = 1, 2, \ldots, n-1 \\ h_n &= b_0 - \sum_{i=0}^{n-1} a_i h_i \end{aligned} \tag{7-693}$$

Suppose that at some time $t = t^*$, the vector $\mathbf{y}(t)$ is zero, that is,

$$\mathbf{y}(t^*) = \mathbf{0} \tag{7-694}$$

and suppose that it is desired to find a control such that

$$\mathbf{y}(t) = \mathbf{0} \qquad \text{for all } t > t^* \tag{7-695}$$

Substituting Eq. (7-695) into Eqs. (7-690), we find that

$$\left.\begin{aligned} x_1(t) &= -h_0 u(t) \\ x_2(t) &= -h_0 \dot{u}(t) - h_1 u(t) \\ &\cdots\cdots\cdots\cdots\cdots \\ x_n(t) &= -h_0 u^{(n-1)}(t) - h_1 u^{(n-2)}(t) - \cdots - h_{n-1} u(t) \end{aligned}\right\} \quad \text{for all } t > t^* \tag{7-696}$$

Since

$$x_1(t) = -h_0 u(t) \qquad \text{for all } t > t^* \tag{7-697}$$

we conclude that

$$u(t) = -\frac{1}{h_0} x_1(t) \qquad \text{for all } t > t^* \tag{7-698}$$

Substituting Eq. (7-698) into Eq. (7-692), we find that Eq. (7-692) reduces to the form

$$\dot{\mathbf{x}}(t) = \mathbf{Q}\mathbf{x}(t) \qquad \text{for all } t > t^* \tag{7-699}$$

where $\mathbf{Q}$ is an $n \times n$ real matrix defined by

$$\mathbf{Q} = \begin{bmatrix} -\dfrac{h_1}{h_0} & 1 & 0 & \cdots & 0 \\ -\dfrac{h_2}{h_0} & 0 & 1 & \cdots & 0 \\ \cdots & \cdots & \cdots & \cdots & \cdots \\ -\dfrac{h_{n-1}}{h_0} & 0 & 0 & \cdots & 1 \\ -\dfrac{h_n}{h_0} - a_0 & -a_1 & -a_2 & \cdots & -a_{n-1} \end{bmatrix} \tag{7-700}$$

† See Eqs. (4-189).

Note that $\mathbf{Q}$ is a known matrix since we can compute $h_0, h_1, \ldots, h_n$ from Eqs. (7-693) in terms of the known constants $a_0, a_1, \ldots, a_{n-1}$ and $b_0, b_1, \ldots, b_n$. Equation (7-699) is a linear homogeneous vector differential equation, and we may write the solution $\mathbf{x}(t)$ in terms of the vector $\mathbf{x}(t^*)$ [since Eq. (7-699) holds for $t > t^*$]. The solution is

$$\mathbf{x}(t) = e^{\mathbf{Q}(t-t^*)}\mathbf{x}(t^*) \qquad \text{for all } t > t^* \tag{7-701}$$

where $e^{\mathbf{Q}(t-t^*)}$ is the fundamental matrix† of Eq. (7-699). Let $\mathbf{q}(t)$ denote the first row vector of the fundamental matrix $e^{\mathbf{Q}(t-t^*)}$ and let $q_1(t), q_2(t), \ldots, q_n(t)$ denote the components of the vector $\mathbf{q}(t)$; thus, $q_i(t)$ is the element of the first row and the $i$th column of the matrix $e^{\mathbf{Q}(t-t^*)}$. Thus, for $t > t^*$, we have

$$x_1(t) = \langle \mathbf{q}(t), \mathbf{x}(t^*) \rangle = \sum_{i=1}^{n} q_i(t) x_i(t^*) \tag{7-702}$$

and from Eq. (7-698), we find that

$$u(t) = -\frac{1}{h_0} \sum_{i=1}^{n} q_i(t) x_i(t^*) \qquad t > t^* \tag{7-703}$$

***Remark 7-2*** *Given the system* (7-665) *or its equivalent state representation defined by Eq.* (7-692). *Given that at $t = t^*$, we have*

$$\mathbf{y}(t^*) = \mathbf{0} \tag{7-704}$$

*If, for all $t > t^*$, we apply the control*

$$u(t) = -\frac{1}{h_0} \sum_{i=1}^{n} q_i(t) x_i(t^*) \tag{7-705}$$

*then*

$$\mathbf{y}(t) = \mathbf{0} \qquad \textit{for all } t > t^* \tag{7-706}$$

Let us next investigate briefly a very important property of the matrix $\mathbf{Q}$.

***Lemma 7-1*** *The matrix $\mathbf{Q}$ of Eq.* (7-700) *has the property that*

$$\det(\lambda \mathbf{I} - \mathbf{Q}) = \left(\lambda^n + \frac{b_{n-1}}{b_n}\lambda^{n-1} + \cdots + \frac{b_1}{b_n}\lambda + \frac{b_0}{b_n}\right)(-1)^n \tag{7-707}$$

*which implies that* [*see Eqs.* (7-672) *and* (7-673)] *the eigenvalues of $\mathbf{Q}$ are at*

$$\lambda = -\beta_i \qquad i = 1, 2, \ldots, n \tag{7-708}$$

ELEMENTS OF PROOF The fact that the eigenvalues of $\mathbf{Q}$ are the zeros of the transfer function $H(s)$, given by Eq. (7-671), can be verified by direct computation of $\det(\lambda\mathbf{I} - \mathbf{Q})$ using Eqs. (7-693). There is still another way to prove that the eigenvalues of $\mathbf{Q}$ are $-\beta_1, -\beta_2, \ldots, -\beta_n$. Suppose that the eigenvalues of $\mathbf{Q}$ are $-\gamma_1, -\gamma_2, \ldots, -\gamma_n$. Then, in Eq.

† See Sec. 3-20.

(7-702), we can write $x_1(t)$ as a sum of exponentials of the form

$$x_1(t) = \sum_{i=1}^{n} \delta_i(t^*)e^{-\gamma_i(t-t^*)} \qquad t > t^* \tag{7-709}$$

where the constants $\delta_i(t^*)$ are functions of $x_i(t^*)$ and known constants. From Eq. (7-698), we can write

$$u(t) = \sum_{i=1}^{n} \epsilon_i(t^*)e^{-\gamma_i(t-t^*)} \qquad t > t^* \tag{7-710}$$

But (see Remark 7-1) we know that $\mathbf{y}(t) = \mathbf{0}$ for $t > t^*$ implies that [see Eq. (7-687)]

$$u(t) = \sum_{i=1}^{n} \alpha_i(t^*)e^{-\beta_i(t-t^*)} \qquad t > t^* \tag{7-711}$$

and from Eqs. (7-710) and (7-711), we conclude that

$$\gamma_i = \beta_i \qquad i = 1, 2, \ldots, n \tag{7-712}$$

Let us now formulate the following time-optimal control problem:

*Given the system (7-665) and given that the control $u(t)$ is restricted by the relation*

$$|u(t)| \le 1 \qquad \textit{for all } t \tag{7-713}$$

*find the control $u(t)$ such that*

*a. The output vector $\mathbf{y}(t)$ is driven to zero in minimum time, say $t^*$.*

*b. The output vector is held at zero thereafter.*

Suppose that we have found a control such that $\mathbf{y}(t^*) = \mathbf{0}$, where $t^*$ is the minimum time. From Remark 7-2, we know that $\mathbf{y}(t) = \mathbf{0}$ for all $t > t^*$ requires

$$\begin{aligned} u(t) &= -\frac{1}{h_0} \sum_{i=1}^{n} q_i(t)x_i(t^*) \\ &= -\frac{1}{h_0} \langle \mathbf{q}(t), \mathbf{x}(t^*) \rangle \qquad \text{for all } t > t^* \end{aligned} \tag{7-714}$$

But since $|u(t)| \le 1$ for all $t$, we must also have

$$\left| -\frac{1}{h_0} \langle \mathbf{q}(t), \mathbf{x}(t^*) \rangle \right| \le 1 \qquad \text{for all } t > t^* \tag{7-715}$$

***Definition 7-43*** *Let $G$ denote the set of states $\mathbf{x}$ such that*

$$G = \left\{ \mathbf{x} : \left| -\frac{1}{h_0} \langle \mathbf{q}(t), \mathbf{x} \rangle \right| \le 1 \qquad \textit{for all } t \ge t^* \right\} \tag{7-716}$$

*We shall call the set $G$ the target set.*

***Lemma 7-2*** *The target set $G$, defined by Eq. (7-716), is closed and convex, and the origin $\mathbf{x} = \mathbf{0}$ is an element of $G$.*

ELEMENTS OF PROOF First, we shall prove that $G$ is *closed.* Recall (Definition 3-7) that a set $G$ is closed if every limit point of $G$ is an element of $G$. Consider a sequence $\mathbf{x}_n$ in $G$, $n = 1, 2, \ldots$, that is,

$$\left| -\frac{1}{h_0}\langle \mathbf{q}(t), \mathbf{x}_n\rangle \right| \leq 1 \qquad \text{for all } t \geq t^* \tag{7-717}$$

such that $\mathbf{x}_n$ converges to $\mathbf{x}$, that is,

$$\mathbf{x}_n \to \mathbf{x} \tag{7-718}$$

Suppose that the limit point $\mathbf{x}$ is *not* an element of $G$ (which implies that $G$ is not closed); then there exists a time $\hat{t}$, $\hat{t} \geq t^*$, such that

$$\left| -\frac{1}{h_0}\langle \mathbf{q}(\hat{t}), \mathbf{x}\rangle \right| > 1 + \epsilon \qquad \text{for some } \epsilon > 0 \tag{7-719}$$

But

$$\begin{aligned} \left| -\frac{1}{h_0}\langle \mathbf{q}(\hat{t}), \mathbf{x}\rangle \right| &= \left| -\frac{1}{h_0}\langle \mathbf{q}(\hat{t}), \mathbf{x} - \mathbf{x}_n\rangle + \left(-\frac{1}{h_0}\right)\langle \mathbf{q}(\hat{t}), \mathbf{x}_n\rangle \right| \\ &\leq \left| -\frac{1}{h_0}\langle \mathbf{q}(\hat{t}), \mathbf{x} - \mathbf{x}_n\rangle \right| + \left| -\frac{1}{h_0}\langle \mathbf{q}(\hat{t}), \mathbf{x}_n\rangle \right| \end{aligned} \tag{7-720}$$

where the last step is justified by the triangle inequality (Sec. 2-13, relation N2). Using the Schwarz inequality [Sec. 2-12, Eq. (2-90)] and Eq. (7-717), we obtain the relation

$$\left| -\frac{1}{h_0}\langle \mathbf{q}(\hat{t}), \mathbf{x}\rangle \right| \leq \left| -\frac{1}{h_0} \right| \|\mathbf{q}(\hat{t})\| \, \|\mathbf{x} - \mathbf{x}_n\| + 1 \tag{7-721}$$

where the norms are Euclidean ones (Sec. 2-13). Since $\mathbf{x}_n$ converges to $\mathbf{x}$, we may choose an $n$ such that

$$\|\mathbf{x} - \mathbf{x}_n\| = \frac{\epsilon}{\left| -\frac{1}{h_0} \right| \|\mathbf{q}(\hat{t})\|} \tag{7-722}$$

so that Eq. (7-721) reduces to

$$\left| -\frac{1}{h_0}\langle \mathbf{q}(\hat{t}), \mathbf{x}\rangle \right| \leq \epsilon + 1 \tag{7-723}$$

The two relations (7-719) and (7-723) are contradictory; hence the limit point $\mathbf{x}$ must belong to $G$, and so $G$ is a closed set.

Now we shall prove that $G$ is a convex set. Recall (Definition 3-18) that a set $G$ is convex if for any $\mathbf{x}$ and $\mathbf{y}$ in $G$ and $r + s = 1$, $r \geq 0$, $s \geq 0$, $r\mathbf{x} + s\mathbf{y}$ is in $G$. Consider two points $\mathbf{x}_1$ and $\mathbf{x}_2$ in $G$, that is,

$$\begin{aligned} \left| -\frac{1}{h_0}\langle \mathbf{q}(t), \mathbf{x}_1\rangle \right| &\leq 1 \qquad \text{for all } t \geq t^* \\ \left| -\frac{1}{h_0}\langle \mathbf{q}(t), \mathbf{x}_2\rangle \right| &\leq 1 \qquad \text{for all } t \geq t^* \end{aligned} \tag{7-724}$$

and two real numbers $r$ and $s$ such that

$$r + s = 1 \qquad r \geq 0, s \geq 0 \tag{7-725}$$

We must show that

$$\left| -\frac{1}{h_0} \langle \mathbf{q}(t), r\mathbf{x}_1 + s\mathbf{x}_2 \rangle \right| \leq 1 \qquad \text{for all } t \geq t^* \tag{7-726}$$

But Eq. (7-726) is true because

$$\left| -\frac{1}{h_0} \langle \mathbf{q}(t), r\mathbf{x}_1 + s\mathbf{x}_2 \rangle \right| = \left| -\frac{1}{h_0} r\langle \mathbf{q}(t), \mathbf{x}_1 \rangle + \left(-\frac{1}{h_0}\right) s\langle \mathbf{q}(t), \mathbf{x}_2 \rangle \right|$$
$$\leq r \left| -\frac{1}{h_0} \langle \mathbf{q}(t), \mathbf{x}_1 \rangle \right| + s \left| -\frac{1}{h_0} \langle \mathbf{q}(t), \mathbf{x}_2 \rangle \right| \leq r + s = 1 \qquad \text{for all } t \geq t^* \tag{7-727}$$

To prove that $\mathbf{0}$ is an element of $G$ is very easy since

$$\left| -\frac{1}{h_0} \langle \mathbf{q}(t), \mathbf{0} \rangle \right| = 0 < 1 \qquad \text{for all } t \geq t^* \tag{7-728}$$

***Remark 7-3*** *We can now reformulate the time-optimal control problem for the system* (7-665) *in terms of the state variables as follows:*

*Given the system* (7-692), *where the control* $u(t)$ *is restricted by* $|u(t)| \leq 1$. *Find the control* $u(t)$ *such that*

*a. Any initial state not in the target set* $G$ *is forced to the target set* $G$ *in minimum time [this is equivalent to driving the vector* $\mathbf{y}(t)$ *to* $\mathbf{0}$ *in minimum time].*

*b. Any initial state in the target set* $G$ *is held in* $G$ *for all time [this is equivalent to holding* $\mathbf{y}(t)$ *at* $\mathbf{0}$ *forever].*

In order to completely solve the problem, it is necessary to know the target set $G$. For this reason, we shall examine several properties of the set $G$.

***Lemma 7-3*** *If any of the* $\beta_i$, $i = 1, 2, \ldots, n$, *is a negative number or has a negative real part, then the target set* $G$ *consists of a single point, the origin of the state space. In other words, if*

$$\beta_i < 0 \qquad \textit{or} \qquad \operatorname{Re} \beta_i < 0 \qquad \textit{for any } i \tag{7-729}$$

*then*

$$G = \{\mathbf{x}: \mathbf{x} = \mathbf{0}\} \tag{7-730}$$

*Thus, if the transfer function* $H(s)$ *contains any non-minimum-phase zeros, then the target set is the origin.*

ELEMENTS OF PROOF To be specific, assume that

$$\beta_1 < 0, \beta_2 > 0, \beta_3 > 0, \ldots, \beta_n > 0 \tag{7-731}$$

This means that one of the eigenvalues of the matrix $\mathbf{Q}$ is positive; hence the system (7-699) is unstable. Since the solution of (7-699) will involve the term $e^{-\beta_1(t-t^*)}$, we have

$$\lim_{t\to\infty} \|\mathbf{q}(t)\| = \infty \tag{7-732}$$

which implies that if $\mathbf{x} \neq \mathbf{0}$, then

$$\lim_{t\to\infty} \left| -\frac{1}{h_0} \langle \mathbf{q}(t), \mathbf{x} \rangle \right| = \infty \tag{7-733}$$

If, however, $\mathbf{x} = \mathbf{0}$, then

$$\left| -\frac{1}{h_0} \langle \mathbf{q}(t), \mathbf{0} \rangle \right| = 0 \qquad \text{for all } t \geq t^* \tag{7-734}$$

and Eq. (7-730) follows from Definition 7-43.

Now suppose that all the $\beta_i$ are real and positive, which means that the transfer function $H(s)$ has only minimum-phase zeros; therefore, if

$$0 < \beta_1 < \beta_2 < \cdots < \beta_n \tag{7-735}$$

then the eigenvalues of $\mathbf{Q}$ are negative, and so the system (7-699) is stable. This implies that†

$$\lim_{t\to\infty} \|\mathbf{q}(t)\| = 0 \tag{7-736}$$

It follows that the set $G$ contains more than one element. Unfortunately, the definition of the target set $G$ provided by Eq. (7-716) is not of such a form that we can write an algebraic expression for the boundary of $G$. However, since the set $G$ is closed and convex, we can certainly state the following:

***Definition 7-44*** *Let $\partial G$ denote the boundary of the set $G$. Clearly,*

$$\partial G = \left\{ \mathbf{x} \colon \left| -\frac{1}{h_0} \langle \mathbf{q}(t), \mathbf{x} \rangle \right| = 1 \qquad \textit{for some } t,\ t \geq t^* \right\} \tag{7-737}$$

To find more information about the boundary $\partial G$ of the target set $G$, we proceed in a manner analogous to that used in Sec. 7-13. The function $|-(1/h_0)\langle \mathbf{q}(t), \mathbf{x} \rangle|$ is a continuous and bounded function of time for all $t \geq t^*$. It follows that‡ for a fixed $\mathbf{x}$ in $\partial G$, the function $|-(1/h_0)\langle \mathbf{q}(t), \mathbf{x} \rangle|$ attains an absolute maximum at some time $\hat{t}$, $\hat{t} \geq t^*$. If $\hat{t} > t^*$, then

$$\frac{\partial}{\partial t} \left| -\frac{1}{h_0} \langle \mathbf{q}(t), \mathbf{x} \rangle \right|_{t=\hat{t}} = 0 \tag{7-738}$$

Moreover, the maximum value is 1, since $\mathbf{x}$ is in $\partial G$, so that

$$\left| -\frac{1}{h_0} \langle \mathbf{q}(\hat{t}), \mathbf{x} \rangle \right| = 1 \tag{7-739}$$

† See Sec. 3-26.
‡ See Sec. 3-10.

Equations (7-738) and (7-739) can be used to "eliminate" $\hat{t}$ and obtain an equation for the components of $\mathbf{x}$. Thus:

***Remark 7-4*** *If Eq.* (7-735) *holds, then a point* $\mathbf{x}$ *belongs to* $\partial G$ *if it satisfies Eqs.* (7-738) *and* (7-739).

The reader should be warned that it is, in general, impossible to solve explicitly for $\hat{t}$ in terms of $\mathbf{x}$ from Eq. (7-738) and then substitute in Eq. (7-739). Thus, the determination of the target set $G$ is an extremely difficult affair, and of course one cannot explicitly solve the time-optimal problem without the algebraic equation of the boundary of $G$ (why?).

Up to now, we have assumed that the number of zeros is the same as the number of poles, i.e., that

$$b_n \neq 0 \tag{7-740}$$

In the remainder of this section, we shall define the target set $G$ for the case where

$$b_n = b_{n-1} = \cdots = b_{k+1} = 0 \qquad k = 0, 1, \ldots, n-1 \tag{7-741}$$

This means that the system is described by the differential equation

$$\begin{aligned}\{D^n + a_{n-1}D^{n-1} + \cdots + a_1 D + a_0\}y(t) \\ = \{b_k D^k + b_{k-1}D^{k-1} + \cdots + b_1 D + b_0\}u(t)\end{aligned} \tag{7-742}$$

Substituting Eq. (7-741) into Eqs. (7-693), we find that

$$\begin{aligned} h_0 &= h_1 = h_2 = \cdots = h_{n-k-1} = 0 \\ h_{n-k} &= b_k \\ h_{n-m} &= b_m - \sum_{i=n-k}^{n-m-1} h_i a_{i+m} \qquad m = 1, 2, \ldots, k-1 \\ h_n &= b_0 - \sum_{i=n-k}^{n-1} a_i h_i \end{aligned} \tag{7-743}$$

which implies that Eqs. (7-690) reduce to

$$\begin{aligned} x_1(t) &= y(t) \\ x_2(t) &= \dot{y}(t) \\ &\cdots\cdots\cdots\cdots \\ x_{n-k}(t) &= y^{(n-k-1)}(t) \\ x_{n-k-1}(t) &= y^{(n-k)}(t) - h_{n-k}u(t) \\ x_{n-k-2}(t) &= y^{(n-k+1)}(t) - h_{n-k}\dot{u}(t) - h_{n-k+1}u(t) \\ &\cdots\cdots\cdots\cdots\cdots\cdots\cdots\cdots\cdots \\ x_n(t) &= y^{(n-1)}(t) - h_{n-k}u^{(k-1)}(t) - \cdots - h_{n-1}u(t) \end{aligned} \tag{7-744}$$

The state variables satisfy the differential equations

$$\begin{aligned}
\dot{x}_1(t) &= x_2(t) \\
\dot{x}_2(t) &= x_3(t) \\
&\cdots\cdots\cdots \\
\dot{x}_{n-k-1}(t) &= x_{n-k}(t) \\
\dot{x}_{n-k}(t) &= x_{n-k+1}(t) + h_{n-k}u(t) \\
\dot{x}_{n-k+1}(t) &= x_{n-k+2}(t) + h_{n-k+1}u(t) \\
&\cdots\cdots\cdots \\
\dot{x}_n(t) &= -\sum_{i=0}^{n-1} a_i x_{i+1}(t) + h_n u(t)
\end{aligned} \tag{7-745}$$

Setting

$$y(t) = \dot{y}(t) = \cdots = y^{(n-1)}(t) = 0 \qquad t \geq t^* \tag{7-746}$$

in Eqs. (7-744), we find that

$$\begin{aligned}
x_1(t) = x_2(t) = \cdots = x_{n-k}(t) &= 0 \qquad \text{for all } t > t^* \\
x_{n-k+1}(t) &= h_{n-k}u(t) \qquad \text{for all } t > t^*
\end{aligned} \tag{7-747}$$

Thus, if we substitute

$$u(t) = -\frac{1}{h_{n-k}} x_{n-k+1}(t) \qquad \text{for all } t > t^* \tag{7-748}$$

into the last $k$ equations of Eqs. (7-744), we find that

$$\begin{bmatrix} \dot{x}_{n-k+1}(t) \\ \dot{x}_{n-k+2}(t) \\ \cdots \\ \dot{x}_n(t) \end{bmatrix} = \begin{bmatrix} -\dfrac{h_{n-k+1}}{h_{n-k}} & 1 & 0 & \cdots & 0 \\ -\dfrac{h_{n-k+2}}{h_{n-k}} & 0 & 1 & \cdots & 0 \\ \cdots & \cdots & \cdots & \cdots & \cdots \\ -\dfrac{h_n}{h_{n-k}} - a_{n-k} & -a_{n-k+1} & -a_{n-k+2} & \cdots & -a_{n-1} \end{bmatrix} \begin{bmatrix} x_{n-k+1}(t) \\ x_{n-k+2}(t) \\ \cdots \\ x_n(t) \end{bmatrix}$$

$$\text{for all } t > t^* \tag{7-749}$$

which is a homogeneous vector differential equation of the form

$$\dot{\tilde{\mathbf{x}}}(t) = \tilde{\mathbf{Q}}\tilde{\mathbf{x}}(t) \qquad \text{for } t > t^*$$

where $\tilde{\mathbf{x}}(t)$ is the $k$-dimensional vector whose components are the last $k$ components of the vector $\mathbf{x}(t)$ and $\tilde{\mathbf{Q}}$ is the $k \times k$ matrix appearing in Eq. (7-749). We claim that

$$\det(\lambda\mathbf{I} - \tilde{\mathbf{Q}}) = \left(\lambda^k + \frac{b_{k-1}}{b_k}\lambda^{k-1} + \cdots + \frac{b_1}{b_k}\lambda + \frac{b_0}{b_k}\right)(-1)^k \tag{7-750}$$

so that the $k$ eigenvalues of $\tilde{\mathbf{Q}}$ are the zeros of the system (7-742). Working as before, we find that

$$x_{n-k+1}(t) = \langle \tilde{\mathbf{q}}(t), \tilde{\mathbf{x}}(t^*) \rangle \qquad \text{for all } t > t^* \tag{7-751}$$

where $\tilde{\mathbf{q}}(t)$ is the first row vector of the fundamental matrix $e^{\tilde{\mathbf{Q}}(t-t^*)}$. From Eqs. (7-751) and (7-748) and from the relation $|u(t)| \leq 1$, we find that the relation

$$\left| -\frac{1}{h_{n-k}} \langle \tilde{\mathbf{q}}(t), \tilde{\mathbf{x}}(t^*) \rangle \right| \leq 1 \tag{7-752}$$

must hold for all $t > t^*$.

***Definition 7-45*** *The target set G for the system* (7-745) *is*

$$G = \left\{ \mathbf{x}: \left| -\frac{1}{h_{n-k}} \langle \tilde{\mathbf{q}}(t), \tilde{\mathbf{x}} \rangle \right| \leq 1;\ x_1 = x_2 = \cdots = x_{n-k} = 0; \textit{for all } t \geq t^* \right\} \tag{7-753}$$

*The set G defined above is also closed and convex. The fact that the dimensionality of* $\mathbf{x}$ *is less than that of* $\tilde{\mathbf{x}}$ *helps somewhat, especially if* $k = 1$ *or* $k = 2$.

**Example 7-11** If we have

$$k = 1 \tag{7-754}$$

then Eqs. (7-747) reduce to

$$x_1(t) = x_2(t) = \cdots = x_{n-1}(t) = 0 \qquad \text{for all } t > t^* \tag{7-755}$$

and

$$x_n(t) = -h_{n-1}u(t) \tag{7-756}$$

But if $k = 1$, Eqs. (7-743) reduce to

$$\begin{aligned} h_0 &= h_1 = \cdots = h_{n-2} = 0 \\ h_{n-1} &= b_1 \\ h_0 &= b_0 - a_{n-1}b_1 \end{aligned} \tag{7-757}$$

so that from Eq. (7-756) we obtain the relation

$$u(t) = -\frac{1}{b_1} x_n(t) \tag{7-758}$$

Let us briefly summarize the results of the last three sections. We found that the time-optimal control of plants with minimum-phase zeros in their transfer function is considerably more complex than the time-optimal control of plants with no zeros. We also noted that before we can find the switch curves or the switch hypersurfaces, we must find a target set $G$ in the state space and reduce the time-optimal control problem to driving the state to $G$ in minimum time and then applying a control which forces the state to remain in $G$ thereafter.

If the system is of high order and if the number of zeros is large, the analytical determination of the boundary of the target set $G$ is an extremely laborious task, if at all possible. If, on the other hand, the plant contains a single minimum-phase zero, the target set $G$, as indicated by Example 7-11, is simple.

If the plant contains any non-minimum-phase zeros, then the target set $G$ will always consist of a single point, namely, the origin of the state space.

There will certainly arise circumstances in which the target set $G$ is very complex, and consequently the engineer who is faced with the problem of designing a time-optimal system cannot even determine the boundary of $G$, much less the time-optimal control law. In such cases, a good sub-optimal design is to replace the target set $G$ with a simpler set or even with the origin. The resulting system will not be time-optimal, but it will be much simpler than the time-optimal one.

**Exercise 7-71** For each of the plants listed below, (1) write the equations of the state variables in terms of the output variables and the control and its time derivatives, (2) write the state-variable differential equations, (3) determine the target set $G$ and plot its boundary, (4) discuss whether the system is completely controllable, and (5) determine the time-optimal control law.

(a) $H(s) = \dfrac{s+1}{s(s+2)}$ (b) $H(s) = \dfrac{s}{s(s+2)}$ (c) $H(s) = \dfrac{s+2}{s(s+2)}$

(d) $H(s) = \dfrac{s+3}{s(s+2)}$ (e) $H(s) = \dfrac{s-1}{s(s+2)}$ (f) $H(s) = \dfrac{s-1}{(s+2)(s+4)}$

(g) $H(s) = \dfrac{s}{(s+2)(s+4)}$ (h) $H(s) = \dfrac{s+2}{(s+2)(s+4)}$ (i) $H(s) = \dfrac{s+3}{(s+2)(s+4)}$

(j) $H(s) = \dfrac{s+4}{(s+2)(s+4)}$ (k) $H(s) = \dfrac{s+5}{(s+2)(s+4)}$ (l) $H(s) = \dfrac{s}{s^2+1}$

(m) $H(s) = \dfrac{s+1}{s^2+1}$

**Exercise 7-72** For each of the plants below, find the target set $G$ and its boundary.

(a) $H(s) = \dfrac{(s+1)(s+2)}{s(s+3)}$ (b) $H(s) = \dfrac{(s+1)(s+2)}{(s+4)(s+3)}$

(c) $H(s) = \dfrac{(s+1)(s+2)}{s(s+3)(s+4)}$ (d) $H(s) = \dfrac{(s+1)(s+2)}{(s+3)(s+4)(s+5)}$

## 7-15 Concluding Remarks

In this chapter, we presented techniques which are used in the design of time-optimal feedback control systems for various plants described by simple ordinary differential equations. The approach used was conceptually the same throughout the chapter; we obtained as much information as possible from the minimum principle, and then we used this information together with the shape of the system trajectories in the state space to obtain and prove the time-optimal control law.

Once the time-optimal control law was derived, we knew what control to apply for each state in the state space, and as a result, we could "design" the time-optimal feedback control system. These engineering realizations of the time-optimal control laws used a relay-type element to generate the piecewise constant time-optimal control. We found that for second-order systems with a single control variable, we needed to apply a single nonlinear operation to one of the state variables. The input-output characteristics of the required nonlinearity were identical to the switch curve. The exact equation of the switch curve was a function of the equations of the system trajectories. In Sec. 7-11, we suggested a procedure which could be used to obtain the switch curve experimentally. We found that the switch curve for oscillatory systems was more complex than the switch curve for nonoscillatory systems. We suggested suboptimal designs based on the optimal design in order to simplify the "hardware."

With the exception of the problem solved in Sec. 7-9, we focused our attention on the time-optimal control of normal systems with a single control variable. We showed that if the system is of second order, then the control law reduces to the problem of finding (analytically or experimentally) a switch curve which separates the state plane into two regions. The equation of this switch curve dictates the input-output characteristics of the single nonlinearity required in the construction of the time-optimal feedback control system.

The complexity of the feedback controller may increase rapidly with the increase in the order of the control system. The realization of the control law for the third-order system considered in Sec. 7-4 illustrates this point. As a matter of fact, for systems whose order is higher than three, some iterative procedure must be used to solve the system of transcendental equations that describe the switch hypersurface. This represents a severe drawback in the application of the theory. However, the reader should note that many high-order systems can be approximated in a satisfactory manner by low-order systems, and one may use a controller of the type shown in Fig. 7-66 for the suboptimal control of a high-order system.

Even though the equation of the switch hypersurface is complex, it should be clear that, from a conceptual point of view, the operation of a high-order time-optimal system presents no particular difficulty. Quite often, intimate knowledge of the optimal solution can help the designer to construct an excellent suboptimal system.†

The same comments apply to the design of suboptimal systems for plants which contain more than one zero in their transfer function. The material presented in Secs. 7-13 and 7-14 indicates that for such problems, the determination of the optimal-control law is in itself a difficult task. For this reason, we may often design a good suboptimal system by considering a simpler target set or, still more simply, by specifying that the origin be the target set. In either case, the designer will trade response time for simpler equipment, and the relative merits of each point of view will, of course, depend on the application at hand.

† References that deal with the design of suboptimal systems are, to mention just a few, F-13, K-18, M-6, P-7, and W-1.

# Chapter 8

# THE DESIGN OF FUEL-OPTIMAL SYSTEMS

## 8-1 Introduction

In Secs. 6-11 to 6-16, we considered the general formulation of fuel-optimal problems. In this chapter, we shall derive the optimal-control law for several simple fuel-optimal problems. At this point, the reader should review the pertinent material of Chap. 6 and, especially, the definition of normal fuel-optimal problems (Definition 6-11), the definition of singular fuel-optimal problems (Definition 6-12), the on-off principle (Theorem 6-11), the sufficient conditions for a normal fuel-optimal problem (Theorem 6-13), the uniqueness theorems (Theorems 6-14 and 6-15), and the material of Secs. 6-21 and 6-22 which deals with singular problems.

We remark that the theory of minimum-fuel systems is not so well developed as the theory of minimum-time systems. Also, the design of minimum-fuel systems is more complex than the design of the corresponding minimum-time systems.

Additional complications arise because of the formulation of the problem. If we want to design a system which meets reasonable fuel requirements, then we have a wide choice of performance functionals.† In other words, we may want the response time to be included in our performance functional, as a fuel-optimal system which requires large response times is, from an engineering point of view, impractical. The particular choice of a performance functional combining fuel and response time rests with the designer. An inexperienced designer may have to try many performance criteria, solve the optimal problem for each, and then decide upon a system design. Since it is difficult to predict whether or not a solution will be "good" until the problem is completely solved, we shall, in this chapter, formulate several control problems for the same basic system but using different performance criteria. We shall then be able to compare the different optimal systems and, perhaps more importantly, to determine the type of equipment required for the construction of each. Furthermore, we shall be able to examine the structure of the various optimal systems

† See Sec. 6-15.

with a view to determining whether or not this structure is the same for a wide class of systems. To put it another way, we saw in Chap. 7 that the time-optimal control for a wide variety of second-order systems was the same in the sense that it was determined by a switch curve that divided the state plane into two regions, and we now ask if similar behavior occurs for fuel-optimal systems. We shall, therefore, examine similar second-order systems with various performance criteria in this chapter.

The detailed examples of fuel-optimal systems that we shall consider represent the translational motion of a mass:

1. When there is no friction or drag force
2. When the friction is proportional to the velocity
3. When the friction is a nonlinear function of the velocity

We shall assume that the control force is provided by rockets or gas jets in such a way that the magnitude of the control force is:

1. Bounded
2. Proportional to the rate of flow of fuel or to the rate of flow of gas

Since we shall consider second-order systems (the state variables being the position and velocity of the mass) in order to simplify the exposition, we assume that the mass of fuel, consumed during the control interval, is negligible compared with the mass of the body. In essence, this assumption amounts to viewing our system as one of *constant* mass.†

We shall illustrate the following basic points in this chapter:

1. The fuel-optimal solution does not exist for some systems. Although this may appear somewhat strange, we shall see in Secs. 8-3 and 8-5 that it can happen. Often in such a case, it is possible to find controls whose fuel requirements are arbitrarily close to the ideal minimum fuel, that is, $\epsilon$-optimal controls.

2. The fuel-optimal control is not unique for some systems. In other words, there are many controls which require the same amount of minimum fuel. However, in this case, the response times differ. (See Secs. 8-2 and 8-4.)

3. The formulation and solution of problems in which the response time is specified a priori. The fuel-optimal control is then a function of both the initial state and the specified response time and thus cannot be obtained as a function of the (present) state alone.

4. The formulation and solution of problems in which the response time is specified in terms of the minimum time required to force each state to the origin (see Sec. 8-7).

5. The introduction of a cost functional which is a weighted sum of the consumed fuel and the elapsed time in an effort to overcome the unattractive, from an engineering point of view, aspects of the minimum-fuel solutions when the response time is either free or fixed in advance (see Secs. 8-8

† See Secs. 6-11 and 6-12.

to 8-11). We shall show that the time-optimal control and the fuel-optimal control (with free response time) can be found by taking suitable limits.

6. A technique of suboptimal control which leads to simplification of the optimum controller (see Sec. 8-9).

7. The fact that singular controls can be fuel-optimal when the equations of the system are nonlinear (see Secs. 8-10 and 8-11) and that this situation does not occur when the system is linear.

Just as in Chap. 7, we shall prove certain lemmas in great detail while omitting the proofs of others. We shall try to lead the reader from analytical proofs and constructions to proofs and constructions which are based upon graphical techniques. In so doing, we shall show that quite often the "shapes" of functions, rather than explicit formulas, can be used to determine the optimal control graphically.

Performance criteria involving consumed fuel arise quite frequently in aerospace problems. The fact that the motion of missiles, satellites, and other space vehicles is achieved by the burning of fuel or the exhaust of gas from a jet and the fact that weight is at a premium indicate the need for the inclusion of fuel in the cost functional of a given aerospace mission. In fact, many solutions to minimum-fuel problems have been obtained in this context without the use of the minimum principle.† The techniques employed were, for the most part, a mixture of semiclassical variational techniques and physical reasoning. The reader will find that, often, use of the minimum principle will simplify proofs but, in general, will not alleviate the computational difficulties. References which deal with fuel-optimal problems using the minimum principle are A-3, A-6 to A-8, A-13 to A-15, B-1, B-14, F-12, F-14, F-17, J-6, K-16, K-17, L-1, M-4 to M-6, and N-3.

## 8-2 First-order Linear Systems: The Integrator

In this section and the next, we shall examine the fuel-optimal control problem for linear first-order systems. Although first-order systems are more or less trivial, their study reveals certain important characteristics which are common to almost all fuel-optimal systems.

Before we formulate the problem in precise mathematical language, we shall begin with a description of the physical system. Suppose that we are given a mass $m$ which is undergoing translational motion in a frictionless environment. We let $y(t)$ be the velocity of the mass at time $t$, and we let $u(t)$ be the applied force or thrust. Let us assume that the thrust $u(t)$ is generated by a rocket or gas jet; then it is reasonable to expect that the magnitude of the thrust will be bounded, and so we shall assume that

$$|u(t)| \leq 1 \qquad \text{for all } t \tag{8-1}$$

† Many such problems are treated in Ref. L-13, which also contains an extensive list of references.

Let us also assume that the rate of flow of fuel (or of gas) is proportional to the magnitude of the thrust, that is,

$$\text{Rate of flow of fuel} \sim |u(t)| \tag{8-2}$$

This assumption implies that if we double the rate of flow of fuel, then we are going to get twice the thrust; such an assumption is often either true or a good approximation to physical reality. It follows, then, that the quantity

$$F(u) = F = \int_0^T |u(t)|\, dt \tag{8-3}$$

is proportional to the fuel consumed during the time interval $[0, T]$.

Let us make the further assumption that the mass of the consumed fuel is small compared with the mass $m$ of the moving body.†

If the above assumption holds, then the velocity $y(t)$ of the body satisfies the differential equation

$$m\dot{y}(t) = u(t) \tag{8-4}$$

Suppose that the velocity at $t = 0$ is $y(0)$ and that we are given a desired velocity $y_d$. A reasonable problem is the following: *Find the control thrust $u(t)$ which will change the velocity from $y(0)$ to the desired velocity $y_d$ and which will accomplish this change with a minimum amount of fuel $F$, as measured by Eq. (8-3); furthermore, the (response) time $T$ is not important.*

Let $x(t)$ denote the "error" in the velocity; i.e., let

$$x(t) = y(t) - y_d \tag{8-5}$$

For simplicity, assume that

$$m = 1 \tag{8-6}$$

From Eqs. (8-5) and (8-4), we find that $x(t)$ must be the solution of the equation

$$\dot{x}(t) = u(t) \tag{8-7}$$

since $y_d$ is constant. We shall call the error $x(t)$ the state of (8-7); the name *integrator* in the title refers to the fact that $x(t) = \int u(\tau)\, d\tau$.

Let us now state, in terms of the system (8-7), the precise analog of the physical problem formulated above.

***Problem 8-1*** *Given the system* (8-7) *with the control constrained by Eq.* (8-1)*; given an initial state $x(0) = \xi \neq 0$ and the terminal state $x = 0$. Find the control that transfers $\xi$ to 0 in such a way that the cost functional $F$ of Eq.* (8-3) *is minimum, where $T$ is not specified* (*that is, $T$ is free*).

We observe that Problem 8-1 violates the sufficient condition for a normal fuel-optimal problem (see Theorem 6-13). The reason is that, although

† Such an assumption is often valid for small position or attitude corrections of space vehicles; naturally, it does not apply to the control of large booster rockets, for which most of the mass of the rocket is fuel.

the system $\dot{x}(t) = u(t)$ is a normal system, the system matrix **A** (here the scalar 0) contains a zero eigenvalue. We shall see in the sequel that the nonnormality of Problem 8-1 has as its consequence the nonuniqueness of the optimal controls.

We start our reasoning using the minimum principle. The Hamiltonian $H$ for this problem is given by

$$H = |u(t)| + u(t)p(t) \tag{8-8}$$

The costate $p(t)$ satisfies the equation

$$\dot{p}(t) = -\frac{\partial H}{\partial x(t)} = 0 \tag{8-9}$$

which implies that

$$p(t) = \pi = \text{const} \tag{8-10}$$

The control which absolutely minimizes $H$ is given by

$$\begin{aligned} u(t) &= 0 && \text{if } |p(t)| < 1 && (8\text{-}11)\\ u(t) &= -\operatorname{sgn}\{p(t)\} && \text{if } |p(t)| > 1 && (8\text{-}12)\\ 0 \le u(t) &\le 1 && \text{if } p(t) = -1 && (8\text{-}13)\\ -1 \le u(t) &\le 0 && \text{if } p(t) = +1 && (8\text{-}14) \end{aligned}$$

We note that Eqs. (8-11) and (8-12) uniquely specify the control $u(t)$ as a nonlinear function of $p(t)$; however, Eqs. (8-13) and (8-14) specify the polarity but *not* the magnitude of the control that minimizes $H$. As we shall see shortly, this will result in *nonunique* fuel-optimal controls for the system (8-7).

Let us digress from our usual procedure and begin by performing certain computations which will provide us with a lower bound on the fuel. We first integrate Eq. (8-7) to obtain

$$x(t) = \xi + \int_0^t u(\tau)\,d\tau \tag{8-15}$$

Since we require that at (the *unspecified*) time $T$, $x(T) = 0$, we deduce that

$$\int_0^T u(t)\,dt = -\xi \tag{8-16}$$

which, in turn, implies that†

$$|\xi| = \left|\int_0^T u(t)\,dt\right| \le \int_0^T |u(t)|\,dt = F \tag{8-17}$$

This means that the fuel required to force $\xi$ to 0 cannot be smaller than $|\xi|$, since $F \ge |\xi|$. Thus, if we find a control $u^*(t)$ which requires fuel $F^*$ [that is, $F(u^*) = F^*$],

$$F^* = |\xi| \tag{8-18}$$

† See Eq. (3-108).

then we can conclude that the control $u^*(t)$ is fuel-optimal. We shall see that we can find such a control.

Suppose that $u^*(t)$ is such a fuel-optimal control; then it must satisfy the two equations (assuming $\xi \neq 0$)

$$\int_0^T u^*(t)\, dt = -\xi \tag{8-19}$$

and

$$\int_0^T |u^*(t)|\, dt = |\xi| \tag{8-20}$$

Let us now define the set $V_T^+$ of nonnegative bounded functions $v(t)$ over $[0, T]$ as

$$V_T^+ = \{v(t) : 0 \leq v(t) \leq 1 \qquad \text{for all } t \in [0, T] \text{ and } v(t) \text{ not identically zero}\} \tag{8-21}$$

We now claim that the control (for $\xi \neq 0$)

$$u^*(t) = -\operatorname{sgn}\{\xi\} v(t) \qquad v(t) \in V_T^+ \tag{8-22}$$

satisfies Eqs. (8-19) and (8-20). This is evident, as

$$|u^*(t)| + u^*(t) = (1 - \operatorname{sgn}\{\xi\})v(t) \qquad \text{and} \qquad |\xi| - \xi = (1 - \operatorname{sgn}\{\xi\})|\xi|$$

imply that

$$\int_0^T v(t)\, dt = |\xi|$$

and so there are an infinite number of functions $v(t)$, each requiring a different response time $T$, such that

$$\int_0^T v(t)\, dt = |\xi|$$

We have proved the following:

***Control Law 8-1 Solution to Problem 8-1*** *The fuel-optimal control, as a function of the state $x$, is given by*

$$\begin{aligned} u^*(t) &= -v(t) \qquad \text{for all } x > 0 \\ u^*(t) &= v(t) \qquad \text{for all } x < 0 \end{aligned} \tag{8-23}$$

*where $v(t) \in V_T^+$ [see Eq. (8-21)]; furthermore, the fuel-optimal control is not unique.*

Let us now turn our attention to the minimum principle to see if we can obtain the same results. Since the system (8-7) is time-invariant and since the response time $T$ is not specified, the Hamiltonian $H$ must be identically zero. Since $p(t) = \pi$, this condition implies that

$$H = |u(t)| + u(t)\pi = 0 \qquad \text{for each } t \in [0, T] \tag{8-24}$$

If $u(t) = 0$, then Eq. (8-24) is identically satisfied for any $\pi$; if $u(t) \neq 0$, then Eq. (8-24) requires that [by noting that $|u(t)| = \operatorname{sgn}\{u(t)\}u(t)$]

$$\pi = -\operatorname{sgn}\{u(t)\} = \pm 1 \tag{8-25}$$

But [see Eq. (8-10)] $p(t) = \pi$ for all $t \in [0, T]$, and so it follows that $|p(t)| = 1$ for all $t \in [0, T]$ and, hence, that the minimization of the Hamiltonian yields Eqs. (8-13) and (8-14), which, as we have mentioned, specify the polarity but not the magnitude of the control. If we choose

$$\pi = \operatorname{sgn} \{\xi\} \tag{8-26}$$

then we can conclude [from Eqs. (8-25), (8-26), (8-13), and (8-14)] that the control $u^*(t)$ of Eq. (8-22) forces $\xi$ to 0. Since this control $u^*(t)$ satisfies all the necessary conditions of the minimum principle, it must be fuel-optimal. This result agrees with the conclusions previously derived.

Let us now verify that the minimum fuel as a function of the state is the solution to the Hamilton-Jacobi equation.† In other words, we must show that the function

$$F^*(x) = |x| \tag{8-27}$$

is the solution of the equation

$$\frac{\partial F^*}{\partial t} + |u^*(t)| + u^*(t)\,\frac{\partial F^*}{\partial x} = 0 \tag{8-28}$$

If $x = 0$, then $F^*(x) = 0$ and $u^*(t) = 0$, so that Eq. (8-28) is satisfied; if $x \neq 0$, then

$$\begin{aligned} \frac{\partial F^*}{\partial t} &= 0 & \frac{\partial F^*}{\partial x} &= \operatorname{sgn}\{x\} \\ |u^*(t)| &= v(t) & u^*(t) &= -\operatorname{sgn}\{x\}\, v(t) \\ v(t) &\in V_T{}^+ \end{aligned} \tag{8-29}$$

Substituting Eqs. (8-29) into the left-hand side of Eq. (8-28), we obtain

$$v(t) - \operatorname{sgn}\{x\} \operatorname{sgn}\{x\}\, v(t) = 0 \tag{8-30}$$

Thus, we have verified that $F^* = |x|$ is the minimum fuel and that $u^*(t)$ is a fuel-optimal control.

Whenever the solutions to a given problem are nonunique, the mathematics tend to be somewhat complicated. From the engineering point of view, however, nonunique optimal controls are often a "blessing" rather than a "curse." The reason for this is that, among the nonunique controls, we can usually find one which has additional "desirable" properties. Let us illustrate by means of two examples.

**Example 8-1** Consider the control $u(t)$, defined in terms of the state $x(t)$ by

$$\begin{aligned} u^*(t) &= -\operatorname{sgn}\{x(t)\} && \text{if } |x(t)| \geq 1 \\ u^*(t) &= -x(t) && \text{if } |x(t)| < 1 \end{aligned} \tag{8-31}$$

† See Sec. 5-20.

We leave it to the reader to verify that the control (8-31) is fuel-optimal. Figure 8-1 illustrates the feedback system which generates the control (8-31). The "desirable" property of this fuel-optimal system is that the feedback channel is linear; the limiter (or saturating nonlinearity) in Fig. 8-1 generates the constraint $|u(t)| \leq 1$. The disadvantage of this design is that it takes an infinite time to drive the error $x(t)$ to zero, although the time to drive the error *near* zero is finite.

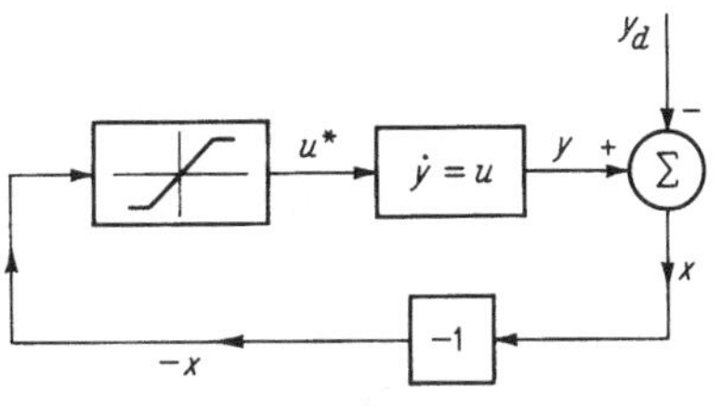


**Fig. 8-1** The fuel-optimal system for the control of Eqs. (8-31).

**Example 8-2** Consider the control defined by

$$\begin{aligned} u^*(t) &= -\operatorname{sgn}\{x(t)\} \qquad &\text{if } |x(t)| \neq 0 \\ u^*(t) &= 0 &\text{if } x(t) = 0 \end{aligned} \tag{8-32}$$

It is easy to establish that the control (8-32) is fuel-optimal [because we can choose $v(t) = +1$ for all $t \in [0, T]$ in Eqs. (8-23)]. The "desirable" property of this fuel-optimal control is that it is also the time-optimal one.† The response time $T$ required by the control (8-32) to force a state $x$ to 0 is $T = |x|$ (= minimum time), and of course the fuel is $|x|$, the smallest possible.

Quite often it is desirable to isolate the fuel-optimal control which results in the smallest possible response time $T$ without going through the process of determining the entire class of fuel-optimal controls. In the remainder of this section, we shall discuss such a technique.

Let us consider the system

$$\dot{x}(t) = u(t) \qquad |u(t)| \leq 1 \tag{8-33}$$

with initial state $x(0) = \xi \neq 0$. Consider the cost functional‡

$$J = \int_0^T [k + |u(t)|]\, dt \tag{8-34}$$

where the response time $T$ is not fixed and

$$k > 0 \tag{8-35}$$

We wish to find the control which drives the system (8-33) from $\xi$ to 0 and minimizes the cost functional $J$ of Eq. (8-34).

The Hamiltonian $H$ for the system (8-33) and the cost (8-34) is

$$H = k + |u(t)| + u(t)p(t) \tag{8-36}$$

Then

$$\dot{p}(t) = -\frac{\partial H}{\partial x(t)} = 0 \tag{8-37}$$

so that

$$p(t) = \pi = \text{const} \tag{8-38}$$

The control $u(t)$ which absolutely minimizes the Hamiltonian of Eq. (8-36) is given by Eqs. (8-11) to (8-14). Since the system (8-33) is time-invariant and since the response time $T$ in the cost functional (8-34) is free, i.e., not

† See Sec. 7-10, Eq. (7-442), and, in particular, Exercise 7-58.
‡ Compare with Problem 6-2*g*.

specified, the Hamiltonian must be zero for all $t \in [0, T]$. If $|\pi| < 1$, then $|p(t)| < 1$ for all $t \in [0, T]$ and Eq. (8-11) requires $u(t) = 0$ for all $t \in [0, T]$; in this case, we deduce that

$$H = k > 0 \tag{8-39}$$

which means that $u(t) = 0$ is not an optimal control.† If $|\pi| = 1$, then $|p(t)| = 1$ for all $t \in [0, T]$ and Eqs. (8-13) and (8-14) require that $u(t) = -\operatorname{sgn}\{\pi\}\, v(t)$, $v(t) \in V_T^+$; in this case,

$$H = k > 0 \tag{8-40}$$

and so the control $u(t) = -\operatorname{sgn}\{\pi\}\, v(t)$ cannot be optimal. If $|\pi| > 1$, then $|p(t)| > 1$ for all $t \in [0, T]$ and the control is $u(t) = -\operatorname{sgn}\{\pi\}$ according to Eq. (8-12); in this case,

$$H = k + 1 - |\pi| = 0 \tag{8-41}$$

which yields (since $k > 0$) the relation

$$|\pi| = k + 1 > 1 \tag{8-42}$$

It follows that the control does not change sign, and so the control

$$u(t) = -\operatorname{sgn}\{\xi\} \tag{8-43}$$

is the extremal control which drives the system (8-33) from $\xi$ to 0. Since the extremal control is unique and since an optimal control exists, Eq. (8-43) must define the optimal control. Note that the control (8-43) is the *time-optimal* control for the system (8-33). Since $|u(t)| = 1$ for all $t \in [0, T]$, it follows that the minimum value of $J$, denoted by $J^*$, is given by

$$J^* = \int_0^T (k + 1)\, dt = (k + 1)T \tag{8-44}$$

But, since the control (8-43) is the time-optimal one, the response time $T$ is the minimum time, and so

$$T = |\xi| \tag{8-45}$$

Therefore, we deduce that

$$J^* = (k + 1)|\xi| \tag{8-46}$$

We also note that

$$\lim_{k \to 0} J^* = |\xi| = F^* \tag{8-47}$$

In other words, by taking the limit as $k \to 0$, the minimum cost $J^*$ converges to the minimum fuel $F^*$. This implies that the control (8-43) is also a fuel-optimal control. Moreover, the control (8-43) requires the least response time—in this case, the minimum time. We shall see in subsequent sections of the book‡ that we can sometimes obtain the fuel-optimal control which

† The word "optimal" is with respect to the cost (8-34).

‡ See Secs. 8-8 to 8-11.

requires the least response time by taking

$$\lim_{k\to 0}\left\{\min_{u(t)}\int_0^T [k + |u(t)|]\, dt\right\} \tag{8-48}$$

*To recapitulate:* In this section, we have shown that there may be many fuel-optimal controls and, in particular, that the time-optimal control may also be fuel-optimal.

**Exercise 8-1** Consider the system $\dot{x}(t) = u(t)$, $|u(t)| \leq 1$. Develop the theory of minimizing the fuel,

$$F = \int_0^T |u(t)|\, dt$$

where the response time $T$ is fixed a priori. Assume that $T$ is greater than the minimum time and that the terminal state is $x(T) = 0$. For each $T$, determine the minimum fuel $F^*(T)$. Plot $F^*(T)$ versus $T$, holding the initial state fixed. Any comments?

## 8-3 First-order Linear Systems: Single-time-constant Case

Let us now consider the translational motion of a unit mass in an environment in which the friction (or drag) force is oppositely directed from the velocity vector and proportional to the velocity. Let $y(t)$ denote the velocity and let $u(t)$ denote the thrust force, which we shall assume bounded by

$$|u(t)| \leq 1 \qquad \text{for all } t \tag{8-49}$$

Under these assumptions, the velocity $y(t)$ satisfies the differential equation

$$\dot{y}(t) = -ay(t) + u(t) \qquad a > 0 \tag{8-50}$$

where $ay(t)$ is the friction-force term.

For the sake of mathematical convenience, we let

$$x(t) = ay(t) \tag{8-51}$$

Then $x(t)$ satisfies the differential equation

$$\dot{x}(t) = -ax(t) + au(t) \qquad a > 0 \tag{8-52}$$

Clearly, $x(t)$ qualifies as a state for the system (8-52). Note that the transfer function of the plant (8-50) is $1/(s + a)$, which is the reason for the terminology *single-time-constant plant.*

The problem we shall solve in this section is:

*Given the system* (8-52) *with the control constrained by Eq.* (8-49). *Given an initial state* $x(0) = \xi$ *and a desired reachable terminal state* $x(T) = \theta \neq \xi$. *Find the control which will force the system* (8-52) *from* $\xi$ *to* $\theta$ *and which will minimize the fuel functional*

$$F(u) = F = \int_0^T |u(t)|\, dt \tag{8-53}$$

*where the response time* $T$ *may or may not be specified.*

The treatment of the basic ideas in this section will be rapid and informal and many of the statements will be offered without proof. The mathematics involved are simple, and so the reader will have no trouble in readily verifying the statements. Our purpose will be to demonstrate that:

1. Quite often, a fuel-optimal solution does not exist; nevertheless, one can find a solution which, for all practical purposes, is almost optimal.

2. Quite often, the fuel-optimal control is identical to the time-optimal control.

The Hamiltonian for the system (8-52) and the cost functional (8-53) is

$$H = |u(t)| - ax(t)p(t) + au(t)p(t) \tag{8-54}$$

The costate $p(t)$ is the solution of the differential equation

$$\dot{p}(t) = -\frac{\partial H}{\partial x(t)} = ap(t) \tag{8-55}$$

so that, if $\pi = p(0)$, then

$$p(t) = \pi e^{at} \tag{8-56}$$

The control which absolutely minimizes $H$ is given by

$$u(t) = 0 \qquad \text{if } |ap(t)| < 1 \tag{8-57}$$

$$u(t) = -\operatorname{sgn}\{ap(t)\} \qquad \text{if } |ap(t)| > 1 \tag{8-58}$$

$$0 \leq u(t) \leq 1 \qquad \text{if } ap(t) = -1 \tag{8-59}$$

$$-1 \leq u(t) \leq 0 \qquad \text{if } ap(t) = +1 \tag{8-60}$$

We note that it is impossible to have†

$$|ap(t)| = 1 \qquad \text{for all } t \in [t_1, t_2] \tag{8-61}$$

For this reason and from Eq. (8-56), we conclude that the only control sequences‡ which can be candidates for fuel-optimal control are

$$\{0\}, \{+1\}, \{-1\}, \{0, +1\}, \{0, -1\} \tag{8-62}$$

It is easy to show that in order to force the system (8-52) from *any* initial state $\xi$ to $\theta$ in some finite time, it is necessary that

$$-1 < \theta < +1 \tag{8-63}$$

This, of course, is due to the lack of integration in the system (8-52) and the constraint (8-49) on the magnitude of the control. From now on, we shall assume that Eq. (8-63) is satisfied.

We shall now examine three cases:

† The posed problem is normal (see Theorem 6-13).

‡ See Definition 7-1.

*Case 1*

Suppose that $\xi$ and $\theta$ are such that

$$\operatorname{sgn}\{\xi\} = \operatorname{sgn}\{\theta\} \qquad |\xi| > |\theta| \qquad \xi \neq \theta \neq 0 \tag{8-64}$$

In this case, the control

$$u(t) = 0 \tag{8-65}$$

will transfer the system (8-52) from $\xi$ to $\theta$ in a time $\hat{T}$ given by

$$\hat{T} = \frac{1}{a} \log \frac{\xi}{\theta} \tag{8-66}$$

[This follows by substituting Eq. (8-65) into Eq. (8-52) and solving the equation.] Clearly, the fuel $F$, given by Eq. (8-53), is *zero* (which is the smallest amount of fuel possible). We note that

$$\lim_{\theta \to 0} \hat{T} = \infty \tag{8-67}$$

Thus, if the response time $T$ is *not* fixed and if $\theta \neq 0$, then the fuel-optimal control is $u(t) = 0$ and the response time is $\hat{T}$. *If, however, $\theta = 0$, then (strictly speaking) a fuel-optimal solution does not exist because, for all finite $t$, we have $x(t) \neq 0$ for $u(t) = 0$.* Nonetheless, it should be clear that one can arrive arbitrarily close to zero, in finite time, using $u(t) = 0$, provided that $T$ is allowed to be large and so *$\epsilon$-optimal* controls exist.

*Case 2*

Suppose that the initial state $\xi$ and the terminal state $\theta$ are such that

$$\operatorname{sgn}\{\xi\} = \operatorname{sgn}\{\theta\} \qquad |\xi| < |\theta| < 1 \tag{8-68}$$

In this case, we claim that the control

$$u(t) = \operatorname{sgn}\{\theta\} \tag{8-69}$$

is the *unique* fuel-optimal control. The time $T^*$ required to force $\xi$ to $\theta$ is given by

$$T^* = \frac{1}{a} \log \frac{\xi - \operatorname{sgn}\{\theta\}}{\theta - \operatorname{sgn}\{\theta\}} \tag{8-70}$$

and the minimum fuel $F^*$ is given by

$$F^* = \int_0^{T^*} 1\, dt = T^* \tag{8-71}$$

Moreover, $T^*$ is the *minimum possible time*† required to force the system from $\xi$ to $\theta$.

† See Sec. 7-10 and, in particular, Exercise 7-59.

In this case, we note that the fuel-optimal control is identical to the time-optimal control. If we solve the problem for $T$ fixed and $T > T^*$, then we shall require more fuel than $F^*$.

*Case 3*

Suppose that the initial state $\xi$ and the terminal state $\theta$ are such that

$$\operatorname{sgn}\{\xi\} = -\operatorname{sgn}\{\theta\} \tag{8-72}$$

If the time $T$ is not specified in advance, it is easy to see that the fuel-optimal control policy is to apply $u(t) = 0$ until the state of the system decays to zero and to then apply $u(t) = \operatorname{sgn}\{\theta\}$ until the state $\theta$ is reached.

As in case 1, one may argue that, from a mathematical point of view, there is no fuel-optimal solution when the response time $T$ is not fixed a priori, but one may find an *$\epsilon$-optimal* solution if $T$ is very large.

**Exercise 8-2** Consider the system

$$\dot{x}(t) = -x(t) + u(t) \qquad |u(t)| \le 1 \tag{8-73}$$

Let $\theta = 0.5$ be the desired terminal state. Let the cost functional be

$$F = \int_0^T |u(t)|\,dt$$

($a$) Suppose that the initial state is $\xi = 1$ and that the response time $T$ is fixed. Plot the minimum of $F$ versus $T$.

($b$) Repeat when $\xi = -1$.

**Exercise 8-3** Consider the system (8-73) and the cost functional

$$J = \int_0^T (k + |u(t)|)\,dt \qquad k > 0 \tag{8-74}$$

where $T$ is not specified. Let the desired terminal state be $\theta = 0$. Show that the optimal control which forces the system (8-73) from the initial state $\xi$ to the terminal state $\theta = 0$ and minimizes $J$ is given [as a function of the state $x(t)$] by

$$\begin{aligned} u^* &= -\operatorname{sgn}\{x(t)\} && \text{if } |x(t)| < k \\ u^* &= 0 && \text{if } |x(t)| \ge k \end{aligned} \tag{8-75}$$

(HINT: Since $T$ is not fixed, use the fact that the Hamiltonian must be zero along the optimal trajectory.) Show that if the initial state $\xi$ is such that $|\xi| < k$, then the control of Eqs. (8-75) is identical to the time-optimal control. Show that the minimum value of the cost $J$ as a function of the state $x$, denoted by $J^*(x)$, is given by

$$\begin{aligned} J^*(x) &= \log\left[|x|^k \frac{(k+1)^{k+1}}{k^k}\right] && \text{if } |x| \ge k \\ J^*(x) &= \log(|x|+1)^{k+1} && \text{if } |x| < k \end{aligned} \tag{8-76}$$

Show that the Hamilton-Jacobi equation for this problem is

$$\begin{aligned} k - x\frac{\partial J^*(x)}{\partial x} &= 0 && \text{if } |x| \ge k \\ k + 1 - x\frac{\partial J^*(x)}{\partial x} - \left|\frac{\partial J^*(x)}{\partial x}\right| &= 0 && \text{if } |x| < k \end{aligned} \tag{8-77}$$

Demonstrate that $J^*(x)$ is indeed a solution to the Hamilton-Jacobi equation. Also show that $\lim_{k \to 0} J^*(x) = 0$.

## 8-4 Fuel-optimal Control of the Double-integral Plant: Formulation†

Let us consider a unit mass undergoing translational motion in the absence of friction. Let $y(t)$ denote the position of the mass and let $u(t)$ denote the thrust force. Assume that the control thrust $u(t)$ is restricted in magnitude by

$$|u(t)| \leq 1 \qquad \text{for all } t \tag{8-78}$$

The position $y(t)$ of this system is the solution of the differential equation

$$\ddot{y}(t) = u(t) \tag{8-79}$$

[Recall that we have examined the time-optimal control of the system (8-79) in Sec. 7-2.]

Suppose that we are given a desired position $y_d$ (a constant) and that our objective is to control the system (8-79) so that at some time $T$, $y(T) = y_d$. Thus, if we define the state variables $x_1(t)$ and $x_2(t)$ by setting

$$x_1(t) = y(t) - y_d \qquad x_2(t) = \dot{y}(t) \tag{8-80}$$

then we can see that the $x_i(t)$ satisfy the differential equations

$$\begin{aligned} \dot{x}_1(t) &= x_2(t) \\ \dot{x}_2(t) &= u(t) \end{aligned} \tag{8-81}$$

We note that the state variable $x_1(t)$ represents the error in position and $x_2(t)$ is the velocity of the mass.

We shall be concerned with the minimization of the cost functional

$$F(u) = F = \int_0^T |u(t)|\, dt \tag{8-82}$$

which we shall call the fuel. The time $T$ will be called the response time. We shall assume that the mass of the consumed fuel is small compared with the mass of the body, so that Eqs. (8-81) are a good approximation to the motion of the body.

In Sec. 8-5, we shall consider the minimization of the cost (8-82) for the system (8-81) with the response time $T$ *not* specified. In Sec. 8-6, we shall consider the minimization of the cost (8-82) for the system (8-81) with the response time either fixed, $T = T_f$, or bounded by a fixed number, $T \leq \hat{T}_f$. In Sec. 8-7, we shall consider the minimization of the cost (8-82) for the system (8-81), and we shall require that the response time $T$ be bounded by a constant multiple of the minimum time corresponding to each state. In Sec. 8-8, we shall consider the minimization of the cost functional

$$J = \int_0^T [k + |u(t)|]\, dt$$

† Appropriate references for Secs. 8-4 to 8-8 are A-3, A-7, A-15, F-12, F-14, K-17, L-1, and M-5.

$T$ not specified, for the system (8-81); in this case, the cost functional is a linear combination of the response time and the consumed fuel.

We shall extensively use several results and formulas obtained in Sec. 7-2 in order to present the material in a condensed form. To facilitate cross referencing, many conclusions will be stated as lemmas.

## 8-5 Fuel-optimal Control of the Double-integral Plant: Free Response Time

In this section, we consider the free-response-time, fuel-optimal control to the origin of the double-integral plant (Problem 8-2). The study of this system is interesting because:

1. There are sets of initial states for which no fuel-optimal solution exists.
2. There are sets of initial states for which many fuel-optimal solutions exist.
3. There are sets of initial states for which the fuel-optimal control is unique.

Our procedure will be similar to the one that we used in Sec. 8-2. First, we shall define the problem; second, we shall find the control sequences that are candidates for fuel-optimal control; third, we shall digress and find an expression for the minimum fuel as a function of the state; and, finally, we shall determine the fuel-optimal control law (whenever a fuel-optimal control exists).

The problem that we shall solve in this section is:

***Problem 8-2*** *Given the system*

$$\begin{aligned} \dot{x}_1(t) &= x_2(t) \\ \dot{x}_2(t) &= u(t) \qquad |u(t)| \le 1 \end{aligned} \tag{8-83}$$

*Find the control which forces the system* (8-83) *from any initial state* $(\xi_1, \xi_2)$ *to the origin* (0, 0) *and which, in so doing, minimizes the fuel*

$$F(u) = F = \int_0^T |u(t)|\, dt \tag{8-84}$$

*where the response time* $T$ *is free (i.e., unspecified).*

The Hamiltonian $H$ for this problem is

$$H = |u(t)| + x_2(t)p_1(t) + u(t)p_2(t) \tag{8-85}$$

The control which absolutely minimizes the Hamiltonian is given by

$$u(t) = 0 \qquad \text{if } |p_2(t)| < 1 \tag{8-86}$$

$$u(t) = -\operatorname{sgn}\{p_2(t)\} \qquad \text{if } |p_2(t)| > 1 \tag{8-87}$$

$$0 \le u(t) \le 1 \qquad \text{if } p_2(t) = -1 \tag{8-88}$$

$$-1 \le u(t) \le 0 \qquad \text{if } p_2(t) = +1 \tag{8-89}$$

The costate variables $p_1(t)$ and $p_2(t)$ are the solutions of the canonical equations

$$\dot{p}_1(t) = -\frac{\partial H}{\partial x_1(t)} = 0 \tag{8-90}$$

$$\dot{p}_2(t) = -\frac{\partial H}{\partial x_2(t)} = -p_1(t) \tag{8-91}$$

It follows that

$$p_1(t) = \pi_1 = \text{const} \tag{8-92}$$

$$p_2(t) = \pi_2 - \pi_1 t \tag{8-93}$$

where

$$\pi_1 = p_1(0) \qquad \pi_2 = p_2(0) \tag{8-94}$$

We note that, if $|p_2(t)| = 1$ for $t \in [t_1, t_2]$, then Eqs. (8-88) and (8-89) specify the polarity but not the magnitude of the control $u(t)$, $t \in [t_1, t_2]$. This singular condition can indeed occur because Problem 8-2 is not normal.† Therefore, we have the lemma

***Lemma 8-1*** *If*

$$\pi_1 = 0 \qquad |\pi_2| = 1 \tag{8-95}$$

*then*

$$|p_2(t)| = 1 \qquad \textit{for all } t \in [0, T] \tag{8-96}$$

*It follows that if $v(t) \in V_T{}^+$, where the set $V_T{}^+$ is* [*see Eq.* (8-21)] *given by*

$$V_T{}^+ = \{v(t)\colon 0 \le v(t) \le 1 \textit{ for all } t \in [0, T] \textit{ and } v(t) \textit{ not identically zero}\} \tag{8-97}$$

*then the control*

$$u(t) = -\operatorname{sgn}\{\pi_2\}\, v(t) \qquad v(t) \in V_T{}^+ \tag{8-98}$$

*is a candidate for the fuel-optimal control.*

PROOF Substituting Eqs. (8-95), (8-92), (8-93), and (8-98) into Eq. (8-85), we find that

$$H = v(t) + x_2(t)0 - v(t) = 0 \qquad \text{for all } t \in [0, T] \tag{8-99}$$

and so the control (8-98) can be extremal.

***Lemma 8-2*** *If*

$$\pi_1 \neq 0 \tag{8-100}$$

*then only the nine control sequences*

$$\{0\}, \{+1\}, \{-1\}, \{+1, 0\}, \{-1, 0\}, \{0, +1\}, \{0, -1\}, \{+1, 0, -1\}, \{-1, 0, +1\} \tag{8-101}$$

*can be candidates for the fuel-optimal control.*

PROOF If $\pi_1 \neq 0$, then $|p_2(t)| = 1$ at most at two isolated times. Since $p_2(t)$ is a linear function of time (see Fig. 7-1), use of Eqs. (8-86) and (8-87) yields the control sequences of (8-101).

† See Definition 6-11 and Theorem 6-13.

We have already calculated the trajectories in the state plane of the system (8-83) when $u = \Delta = \pm 1$. The equations of these trajectories are given by Eqs. (7-14) to (7-18), and the plot of the trajectories appears in Fig. 7-2. If $u(t) = 0$, then the solution of Eqs. (8-83) is given by the equations

$$x_1(t) = \xi_1 + \xi_2 t \qquad x_2(t) = \xi_2 \tag{8-102}$$

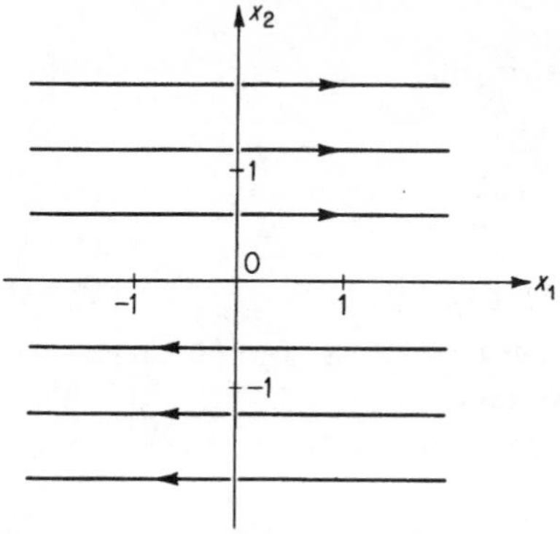


**Fig. 8-2** The trajectories of the system (8-83) when $u(t) = 0$.

The trajectories when $u(t) = 0$ are straight lines and are shown in Fig. 8-2.

Let us digress for a moment and note that we can obtain a lower bound for the fuel $F$ as a function of the initial state; this is possible because of the special form of Eqs. (8-83).†

***Lemma 8-3*** *Let the initial state be* $(\xi_1, \xi_2)$ *and let* $F^*(\xi_1, \xi_2)$ *denote the minimum fuel, if it exists, required to force* $(\xi_1, \xi_2)$ *to* $(0, 0)$ *in some (unspecified) time* $T$. *Then the minimum fuel* $F^*(\xi_1, \xi_2)$ *satisfies the relation*

$$F^*(\xi_1, \xi_2) \geq |\xi_2| \tag{8-103}$$

*It follows that if there is a control* $u^*(t)$ *that forces* $(\xi_1, \xi_2)$ *to* $(0, 0)$ *and requires fuel* $|\xi_2|$, *then* $u^*(t)$ *is optimal and* $F^*(\xi_1, \xi_2) = |\xi_2|$.

PROOF For any $u(t)$, we have $\dot{x}_2(t) = u(t)$, which implies that

$$x_2(t) = \xi_2 + \int_0^t u(\tau)\, d\tau \tag{8-104}$$

Since we must reach the origin at $t = T$, it follows that $x_2(T) = 0$ and, therefore, that

$$|\xi_2| = \left| \int_0^T u(t)\, dt \right|$$

But we can deduce the inequality

$$|\xi_2| = \left| \int_0^T u(t)\, dt \right| \leq \int_0^T |u(t)|\, dt = F \tag{8-105}$$

If $F \geq |\xi_2|$, then it is necessary that the relation

$$F^*(\xi_1, \xi_2) \geq |\xi_2| \tag{8-106}$$

holds.

A few comments are appropriate at this point. First, we want to emphasize that the relationship $F \geq |\xi_2|$ gives us a *lower bound*, $|\xi_2|$, on the fuel. If we can find a control which forces $(\xi_1, \xi_2)$ to $(0, 0)$ with fuel $|\xi_2|$, then we

† We have done the same thing in Sec. 8-2.

are sure that this control is indeed fuel-optimal. Second, although the proof of the lemma does not guarantee that $|\xi_2|$ is the *greatest* lower bound, we shall show in the sequel that this is indeed the case, and so we shall speak of $|\xi_2|$ as the (ideal) minimum fuel. Third, we note that Eq. (8-103) is essentially the same as Eq. (8-18) in Sec. 8-2. Recall that, in Sec. 8-2, it was necessary to force the velocity error to zero with minimum fuel but with no requirements on the position at $t = T$. However, in the problem at hand, we must reduce both the position error and the velocity to zero, if possible, using the fuel $|\xi_2|$. We can immediately sense that some problems will arise. For example, if the initial state of the system (8-83) is at $(\xi_1, 0)$, then we must force the system from $(\xi_1, 0)$ to $(0, 0)$ with *zero* fuel. This implies that $u(t) = 0$ for all $t \in [0, T]$; but if we solve Eqs. (8-83) for $u(t) = 0$ and the initial state $(\xi_1, 0)$, then we shall find that $x_1(t) = \xi_1 \neq 0$ for all $t$. Therefore, a *minimum-fuel solution* does not exist for *the initial state* $(\xi_1, 0)$. As we shall see, there are additional initial states for which, strictly speaking, a fuel-optimal solution does not exist.† Nevertheless, we can find a control whose fuel requirements are arbitrarily close to the (ideal) minimum fuel $F^* = |\xi_2|$. (See Lemma 8-6.)

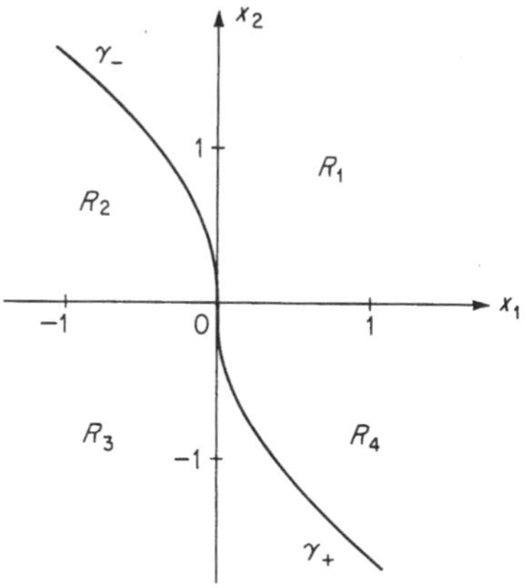


**Fig. 8-3** The $\gamma_+$ curve, the $\gamma_-$ curve, and the $x_1$ axis divide the state plane into the four sets $R_1$, $R_2$, $R_3$, and $R_4$.

To obtain the fuel-optimal solution(s) and to indicate the set of initial states for which no fuel-optimal solutions exist, we divide the $x_1x_2$ state plane into four sets using the $\gamma_+$ and $\gamma_-$ curves (see Definitions 7-2 and 7-3) and the $x_1$ axis. We recall that the $\gamma_+$ and $\gamma_-$ curves were defined by Eqs. (7-19) and (7-20) as

$$\gamma_+ = \{(x_1, x_2): x_1 = \tfrac{1}{2}x_2^2;\ x_2 \leq 0\} \tag{8-107}$$
$$\gamma_- = \{(x_1, x_2): x_1 = -\tfrac{1}{2}x_2^2;\ x_2 \geq 0\} \tag{8-108}$$

***Definition 8-1*** *Let $R_1$, $R_2$, $R_3$, and $R_4$ be the sets (or regions) shown in Fig.* 8-3. *The precise definition of these sets is*

$$R_1 = \{(x_1, x_2): x_2 \geq 0;\ x_1 > x_1',\ \textit{where}\ (x_1', x_2) \in \gamma_-\} \tag{8-109}$$
$$R_2 = \{(x_1, x_2): x_2 > 0;\ x_1 < x_1',\ \textit{where}\ (x_1', x_2) \in \gamma_-\} \tag{8-110}$$
$$R_3 = \{(x_1, x_2): x_2 \leq 0;\ x_1 < x_1',\ \textit{where}\ (x_1', x_2) \in \gamma_+\} \tag{8-111}$$
$$R_4 = \{(x_1, x_2): x_2 < 0;\ x_1 > x_1',\ \textit{where}\ (x_1', x_2) \in \gamma_+\} \tag{8-112}$$

† Recall that this was also the case in Sec. 8-3.

[Note that

$$R_1 \cup R_4 = R_- \qquad R_2 \cup R_3 = R_+ \tag{8-113}$$

where $R_+$ and $R_-$ were defined in Definition 7-5 (see also Fig. 7-3).]

***Lemma 8-4*** *If $(\xi_1, \xi_2) \in \gamma_+$, then the control $u(t) = +1$ is fuel-optimal to the origin and, moreover, is the unique optimal control.*

PROOF Suppose that $\pi_1 \neq 0$, so that Lemma 8-2 holds; then among the control sequences of (8-101), only the control sequence $\{+1\}$ will force the state $(\xi_1, \xi_2) \in \gamma_+$ to the origin. If Lemma 8-1 holds, then the control $u(t)$ must be given by Eq. (8-98). Let $x_1'(t)$ and $x_2'(t)$ be the solutions of Eqs. (8-83) with initial state $(\xi_1, \xi_2) \in \gamma_+$ and with the control given by Eq. (8-98). Clearly, $x_1'(t)$ and $x_2'(t)$ are given by the relations

$$x_2'(t) = \xi_2 + \int_0^t [-\operatorname{sgn}\{\pi_2\}\, v(\tau)]\, d\tau \tag{8-114}$$

$$x_1'(t) = \xi_1 + \xi_2 t + \int_0^t d\tau \int_0^\tau [-\operatorname{sgn}\{\pi_2\}\, v(\sigma)]\, d\sigma \tag{8-115}$$

Let $x_1(t)$ and $x_2(t)$ be the solution of Eqs. (8-83) with $u(t) = +1$, that is,

$$x_2(t) = \xi_2 + \int_0^t 1\, dt \tag{8-116}$$

$$x_1(t) = \xi_1 + \xi_2 t + \int_0^t d\tau \int_0^\tau 1\, d\sigma \tag{8-117}$$

From Eqs. (8-115) and (8-117), we deduce that

$$x_1(t) - x_1'(t) = \int_0^t d\tau \int_0^\tau [1 + \operatorname{sgn}\{\pi_2\}\, v(\sigma)]\, d\sigma \geq 0 \tag{8-118}$$

which means that the trajectory generated by the control of Eq. (8-98) will always be to the left of the $\gamma_+$ curve, and so it will miss the origin. Thus, by the process of elimination, we conclude that if $(\xi_1, \xi_2) \in \gamma_+$, then only the control sequence $\{+1\}$ [that is, the control $u(t) = +1$] forces $(\xi_1, \xi_2)$ to $(0, 0)$. Using Eq. (8-116), we find that the control $u(t) = +1$ requires fuel $|\xi_2|$ to force $(\xi_1, \xi_2)$ to $(0, 0)$ and, therefore, in view of Lemma 8-3, must be optimal.

***Lemma 8-5*** *If $(\xi_1, \xi_2) \in R_4$, then there are many fuel-optimal controls to the origin. In particular, the control sequence $\{0, +1\}$*

*a. Is fuel-optimal*

*b. Has response time $T(\xi_1, \xi_2)$ given by*

$$T(\xi_1, \xi_2) = -\left(\tfrac{1}{2}\xi_2 + \frac{\xi_1}{\xi_2}\right) \tag{8-119}$$

*c. Requires the smallest response time among all the fuel-optimal controls*

PROOF We claim that one can find many functions $v(t) \in V_T^+$ [see Eq. (8-97)] such that the control $u(t) = v(t)$ will force $(\xi_1, \xi_2) \in R_4$ to $(0, 0)$. The reason is that one can find many nonnegative functions $v(t)$ which will satisfy the equations

$$\int_0^T v(\sigma)\, d\sigma = -\xi_2 \qquad \int_0^T d\tau \int_0^\tau v(\sigma)\, d\sigma = -\xi_1 - \xi_2 T \tag{8-120}$$

with $T$ not specified [in other words, $T$ will be, in general, different for each $v(t)$]. Since $(\xi_1, \xi_2) \in R_4$ implies that $\xi_2 < 0$, it is easy to see that

$$\int_0^T v(\sigma)\, d\sigma = |\xi_2|$$

and so every $v(t)$ that satisfies Eq. (8-120) is fuel-optimal (by Lemma 8-3).

Now we consider the control sequences of (8-101); with the aid of Figs. 7-2 and 8-2, we conclude that only the control sequences $\{0, +1\}$ and $\{-1, 0, +1\}$ can force $(\xi_1, \xi_2) \in R_4$ to $(0, 0)$. Figure 8-4 illustrates the trajectory $ABO$ generated by the control sequence $\{0, +1\}$ and the trajectory $ACDO$ generated by the control sequence $\{-1, 0, +1\}$. Let us consider first the trajectory $ABO$; since $u = 0$ during the portion $AB$, no fuel is consumed during that interval; since $B$ is on the $\gamma_+$ curve and since the ordinate of $B$ is $\xi_2$, the fuel consumed during the trajectory $BO$ is $|\xi_2|$, and by Lemma 8-3, we conclude that the control sequence $\{0, +1\}$ is fuel-optimal. The reader should also note that the control sequence $\{0, +1\}$ generates a nonnegative control which satisfies Eqs. (8-120) and that the control corresponding to $\{0, +1\}$ belongs to $V_T^+$. The control sequence $\{-1, 0, +1\}$ and its trajectory $ACDO$ cannot be fuel-optimal; this conclusion follows by inspection of Fig. 8-4 or by actual computation of the fuel. Nevertheless, it is instructive to show that the control sequence $\{-1, 0, +1\}$ is not fuel-optimal by using the necessary condition that the Hamiltonian be zero along the optimal trajectory. To do this, let us suppose that $(\xi_1, \xi_2) \in R_4$ and that the control sequence $\{-1, 0, +1\}$ is fuel-optimal; then, since $u(0) = -1$, we would have [see Eqs. (8-87), (8-92), and (8-93)]

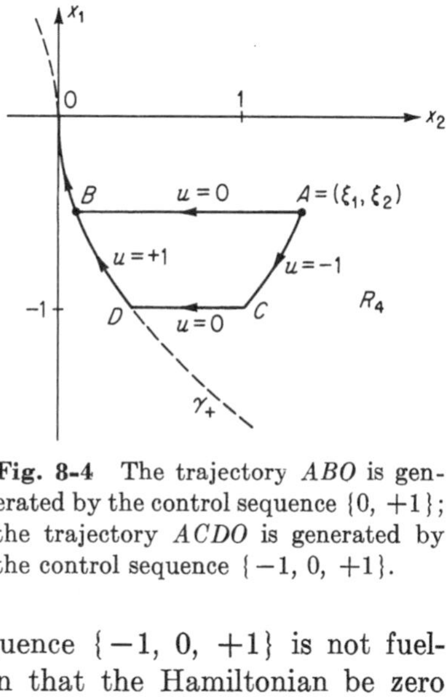


**Fig. 8-4** The trajectory $ABO$ is generated by the control sequence $\{0, +1\}$; the trajectory $ACDO$ is generated by the control sequence $\{-1, 0, +1\}$.

$$p_2(0) = \pi_2 > 1 \qquad p_1(0) = \pi_1 \tag{8-121}$$

Substituting Eqs. (8-121) into Eq. (8-85), we find that the Hamiltonian, at $t = 0$, is given by

$$H\,|_{t=0} = |-1| + \xi_2\pi_1 - \pi_2 \tag{8-122}$$

But $H$ must be identically zero since this is a free-response-time problem and the system (8-83) is time-invariant. But the relation $1 - \pi_2 < 0$ implies that $\xi_2\pi_1 > 0$ [since $(\xi_1, \xi_2) \in R_4$, $\xi_2 < 0$]; therefore, we must have

$$\pi_1 < 0 \qquad \pi_2 > 1 \tag{8-123}$$

But if $\pi_1 < 0$ and $\pi_2 > 1$, then we can see that

$$p_2(t) = \pi_2 - \pi_1 t > \pi_2 > 1 \qquad \text{for all } t > 0 \tag{8-124}$$

If $p_2(t) > 1$ for all $t > 0$, then it can only generate [see Eq. (8-87)] the control sequence $\{-1\}$; hence, it cannot generate the control sequence $\{-1, 0, +1\}$. Thus, we have arrived at a contradiction, and so we conclude that the sequence $\{-1, 0, +1\}$ cannot be fuel-optimal.

It is left as an exercise to the reader to derive Eq. (8-119) and to show that the response time of Eq. (8-119) is the smallest possible. [HINT: $T = \int_0^T dt = \int_0^T (dx_1/dx_1)\, dt = \int_{\xi_1}^0 dx_1/x_2$.] We have thus shown that any control $u(t) = v(t) \in V_T^+$ that satisfies Eqs. (8-120) is fuel-optimal and, therefore, that the control corresponding to the control sequence $\{0, +1\}$ is also fuel-optimal.

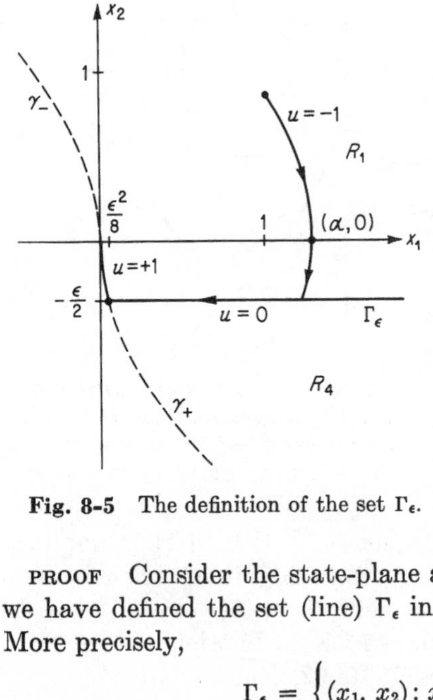


**Fig. 8-5** The definition of the set $\Gamma_\epsilon$.

***Lemma 8-6*** *If $(\xi_1, \xi_2) \in R_1$, then there is no solution to the fuel-optimal problem. However, given any $\epsilon > 0$, there exists a control sequence, namely, $\{-1, 0, +1\}$, which will drive $(\xi_1, \xi_2) \in R_1$ to $(0, 0)$ and for which the fuel $\hat{F}$ is given by*

$$\hat{F} = F^* + \epsilon = |\xi_2| + \epsilon \tag{8-125}$$

*Clearly,* $$\lim_{\epsilon \to 0} \hat{F} = F^* = |\xi_2| \tag{8-126}$$

*Thus, the minimum-fuel solution can be approached as closely as desired but never attained. We call such a control $\epsilon$-fuel-optimal.*

PROOF Consider the state-plane arrangements of Fig. 8-5. In Fig. 8-5, we have defined the set (line) $\Gamma_\epsilon$ in $R_4$ at a distance $\epsilon/2$ from the $x_1$ axis. More precisely,

$$\Gamma_\epsilon = \left\{(x_1, x_2): x_2 = -\frac{\epsilon}{2}, x_1 \geq \frac{\epsilon^2}{8}\right\} \tag{8-127}$$

Now consider a state $(\xi_1, \xi_2)$ in $R_1$. Depending on the position of $(\xi_1, \xi_2)$, it is evident that we can force $(\xi_1, \xi_2)$ to the origin either by the control sequence $\{-1, 0, +1\}$ or by the control sequence $\{-1, +1\}$. If $(\xi_1, \xi_2)$ is transferred to the origin by the control sequence $\{-1, 0, +1\}$, then its fuel requirements are $|\xi_2| + \epsilon/2$ until it reaches the $\Gamma_\epsilon$ curve, zero until it reaches the $\gamma_+$ curve [at the point $(\epsilon^2/8, -\epsilon/2)$], and $\epsilon/2$ to reach the origin along the $\gamma_+$ curve; thus, the total fuel $\hat{F}$ consumed is $\hat{F} = |\xi_2| + \epsilon$. If $(\xi_1, \xi_2)$ is such that the trajectory resulting from the control $u = -1$ intersects the $\gamma_+$ curve between the points $(\epsilon^2/8, -\epsilon/2)$ and $(0, 0)$, then the fuel required will be less than $|\xi_2| + \epsilon$.

In order to show that if $\epsilon = 0$, then there is no fuel-optimal solution, we note that the state $(\xi_1, \xi_2) \in R_1$ is transferred to the state $(\alpha, 0)$ (see Fig. 8-5) by the control $u = -1$ using the minimum amount of fuel possible, namely $|\xi_2|$; any other control requiring the same fuel $|\xi_2|$ will force $(\xi_1, \xi_2)$ to some state $(\beta, 0)$ such that $\beta > \alpha$, and so no fuel-optimal solution exists.

We have derived up to now certain conclusions for states in $\gamma_+ \cup R_4 \cup R_1$. We can certainly use similar arguments for states in $\gamma_- \cup R_2 \cup R_3$. We summarize the results in the following control law:

***Control Law 8-2 Solution to Problem 8-2*** *A fuel-optimal control (there may be many), as a function of the state* $(x_1, x_2)$, *for Problem* 8-2 *is given by*

$$\begin{aligned} u^* &= u^*(x_1, x_2) = +1 && \text{for all } (x_1, x_2) \in \gamma_+ \\ u^* &= u^*(x_1, x_2) = -1 && \text{for all } (x_1, x_2) \in \gamma_- \\ u^* &= u^*(x_1, x_2) = 0 && \text{for all } (x_1, x_2) \in R_2 \cup R_4 \end{aligned}$$

*If* $(x_1, x_2) \in R_1 \cup R_3$, *then there is no fuel-optimal control.*

We remark that the fuel-optimal controls of Control Law 8-2 have the property that they require the least response time and that Control Law 8-2 is a restatement of Lemmas 8-4 and 8-5.

We have seen that, depending on the initial state, there may or may not exist a fuel-optimal solution; furthermore, there are sets of states for which the fuel-optimal control is not unique. In the next section, we shall examine the problem of fuel-optimal control with the response time either fixed or bounded from above.

**Exercise 8-4** Show that if $(\xi_1, \xi_2) \in R_4$, then the response time required by the fuel-optimal control sequence $\{0, +1\}$ is

$$T = -\tfrac{1}{2}\xi_2 - \frac{\xi_1}{\xi_2} \tag{8-128}$$

and if $(\xi_1, \xi_2) \in R_2$, then the response time required by the fuel-optimal control sequence $\{0, -1\}$ is

$$T = \tfrac{1}{2}\xi_2 - \frac{\xi_1}{\xi_2} \tag{8-129}$$

**Exercise 8-5** Consider the state $(1, -1) \in R_4$. Find three different fuel-optimal controls to the origin. Plot the trajectories generated by each fuel-optimal control.

**Exercise 8-6** Show that the minimum fuel $F^*(x_1, x_2) = |x_2|$ required to force $(x_1, x_2) \in \gamma_+ \cup \gamma_- \cup R_2 \cup R_4$ to $(0, 0)$ is the solution to the Hamilton-Jacobi equation.

**Exercise 8-7 Simplified Soft-landing Problem** Consider a body with constant mass $M$; let $x_1(t)$ denote its altitude from the surface of a planet (without atmosphere) and let $x_2(t)$ denote its vertical velocity. Suppose that $g$ is the (constant) gravity of the planet. Let $f(t)$ be the thrust, $-F_M \leq f(t) \leq F_M$, and let

$$F = \int_0^T |f(t)|\, dt$$

be the consumed fuel. Assuming that $F_M > Mg$, determine the fuel-optimal thrust ($T$ is not specified), as a function of the altitude and vertical velocity, which will reduce the altitude and velocity to zero. The equations are

$$\begin{aligned} \dot{x}_1(t) &= x_2(t) \\ M\dot{x}_2(t) &= f(t) - Mg \end{aligned} \tag{8-130}$$

Are there values of initial altitude and velocity which will cause the body to crash? Is there more than one fuel-optimal solution?

**Exercise 8-8** Consider the system (8-83) and the fuel functional (8-84) with $T$ free. Find the optimal solution (if it exists) to the target sets

$$\begin{aligned} S_1 &= \{(x_1, x_2): x_1 = x_2, |x_1| \leq 1\} \\ S_2 &= \{(x_1, x_2): x_1 = -x_2, |x_1| \leq 1\} \\ S_3 &= \{(x_1, x_2): x_1^2 + x_2^2 \leq 1\} \\ S_4 &= \{(x_1, x_2): x_1 = +1, x_2 = 0 \text{ or } x_1 = -1, x_2 = 0\} \end{aligned}$$

**Exercise 8-9** Consider the system (8-83) and the fuel functional (8-84) with $T$ free. Let (1, 1) be the desired terminal state. Determine the fuel-optimal control law and evaluate the minimum fuel as a function of the state. Comment on existence and uniqueness.

## 8-6 Fuel-optimal Control of the Double-integral Plant: Fixed or Bounded Response Time

The results we derived in the preceding section were, from a practical point of view, often unsatisfactory in view of the large response times required by the fuel-optimal control whenever the state was near the $x_1$ axis and in view of the almost infinite response time required by the $\epsilon$-fuel-optimal controls whenever the state was in $R_1$ or $R_3$. For this reason, we shall reconsider the problem of fuel-optimal control of the double-integral plant with:

1. Fixed response time
2. Bounded response time

The procedure we shall adhere to in this section is as follows: After we state the problem, we shall present Lemma 8-7, which states that the fixed-time problem will have a solution if and only if the specified response time is no less than the corresponding minimum time. However, we wish to caution the reader that we are not offering in this section a rigorous proof of the existence of an optimal control. We assert that it does exist; however, the proof involves mathematical concepts (such as measurable functions) which we have not discussed. (See, for example, Ref. C-4 for a flavor of the concepts involved.) For this reason, we shall assume that an optimal solution does exist. We shall then carry out some computations, and we shall find the fuel-optimal control as a function of the initial state and of the specified response time. We now state the problem.

***Problem 8-3*** *Given the system*

$$\begin{aligned} \dot{x}_1(t) &= x_2(t) \\ \dot{x}_2(t) &= u(t) \qquad |u(t)| \leq 1 \end{aligned} \tag{8-131}$$

*Find the control which forces the system* (8-131) *from any initial state* $(\xi_1, \xi_2)$ *to* (0, 0) *and which minimizes the fuel*

$$F = \int_0^{T_f} |u(t)|\, dt \tag{8-132}$$

*where the response time* $T_f$ *is either*

*a. Fixed (i.e., specified) a priori or*

*b. Bounded from above by a fixed time* $\hat{T}_f$*; that is,*

$$T_f \leq \hat{T}_f \tag{8-133}$$

We shall make use of the terminology and results of Secs. 8-5 and 7-2 throughout this section. In particular, we shall use the sets $R_1$, $R_2$, $R_3$, and $R_4$ [see Definition 8-1, Eqs. (8-109) to (8-112), and Fig. 8-3].

Since, in this problem, we specify the three numbers $\xi_1$, $\xi_2$, and $T_f$ (or $\hat{T}_f$), it can happen that the choice of numbers is such that the problem has no solution. To avoid such complication, we state the following lemma:

***Lemma 8-7*** *Problem* 8-3 *has a solution if and only if the specified time* $T_f$ *(or* $\hat{T}_f$*) is not smaller than the minimum time* $t^*$ *corresponding to the given state* $(\xi_1, \xi_2)$*; in other words, for Problem* 8-3 *to have a solution, the relationships [see Eq.* (7-26)]

$$T_f \text{ (or } \hat{T}_f) \geq \xi_2 + \sqrt{4\xi_1 + 2\xi_2^2} \qquad \text{for } (\xi_1, \xi_2) \in R_1 \cup R_4 \tag{8-134}$$

$$T_f \text{ (or } \hat{T}_f) \geq -\xi_2 + \sqrt{-4\xi_1 + 2\xi_2^2} \qquad \text{for } (\xi_1, \xi_2) \in R_2 \cup R_3 \tag{8-135}$$

$$T_f \text{ (or } \hat{T}_f) \geq |\xi_2| \qquad \text{for } (\xi_1, \xi_2) \in \gamma_+ \cup \gamma_- \tag{8-136}$$

*must hold. From now on, we shall assume that Eqs.* (8-134) *to* (8-136) *hold.*

Suppose that we are given an initial state $(\xi_1, \xi_2)$ in $R_1$ and that we require the response time to be an exactly specified number $T_f$. We have seen (see Lemma 8-6) that more than $|\xi_2|$ units of fuel must be used in order to force $(\xi_1, \xi_2) \in R_1$ to (0, 0). We shall now prove the following lemma:

***Lemma 8-8*** *Given* $(\xi_1, \xi_2) \in R_1$ *and* $T_f$*; then the control sequence* $\{-1, 0, +1\}$ *is fuel-optimal.*

PROOF We can use the minimum principle to prove this lemma, and so Eqs. (8-85) to (8-101) hold. Suppose that Eqs. (8-95) hold and that the control (8-98) is optimal; but such a control cannot force $(\xi_1, \xi_2)$ to (0, 0); therefore, Lemma 8-2 must hold. If we examine the nine control sequences of Eq. (8-101), we find (using Figs. 7-2 and 8-2) that only the control sequence $\{-1, 0, +1\}$ can force $(\xi_1, \xi_2)$ to (0, 0); by elimination, it must be fuel-optimal. The reason is that it is the unique extremal control sequence and we asserted the existence of the optimal control.

We have shown that the control sequence $\{-1, 0, +1\}$ is fuel-optimal for $(\xi_1, \xi_2) \in R_1$, given $T_f$. This means that the fuel-optimal control is

$$\begin{aligned} u(t) &= -1 \qquad \text{for } 0 \le t < t_1 \\ u(t) &= 0 \qquad \text{for } t_1 \le t < t_2 \\ u(t) &= +1 \qquad \text{for } t_2 \le t \le T_f \end{aligned} \tag{8-137}$$

Clearly, if we can determine the switch times $t_1$ and $t_2$ in terms of the given quantities $\xi_1$, $\xi_2$, and $T_f$, then we have found the fuel-optimal control. Let $\Xi = (\xi_1, \xi_2)$ be the initial state. Let the points $Z = (z_1, z_2)$ and $W = (w_1, w_2)$ be defined as follows: Let $(x_1(t), x_2(t))$ denote the state variables resulting from the control (8-137) and due to the initial state $(\xi_1, \xi_2)$, and let

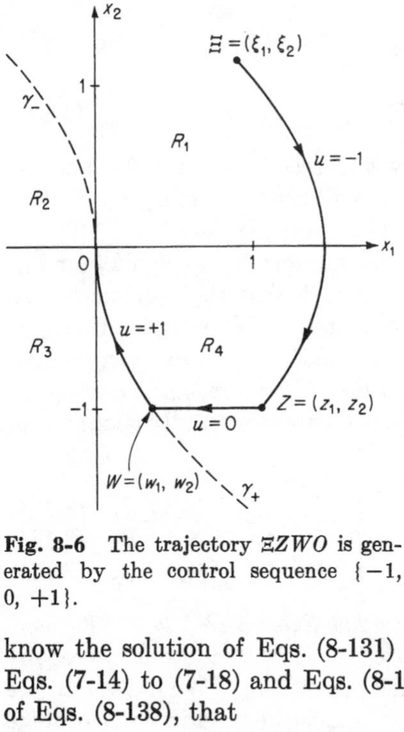


**Fig. 8-6** The trajectory $\Xi ZWO$ is generated by the control sequence $\{-1, 0, +1\}$.

$$\begin{aligned} \Xi &= (\xi_1, \xi_2) = (x_1(0), x_2(0)) \\ Z &= (z_1, z_2) = (x_1(t_1), x_2(t_1)) \\ W &= (w_1, w_2) = (x_1(t_2), x_2(t_2)) \\ O &= (0, 0) = (x_1(T_f), x_2(T_f)) \end{aligned} \tag{8-138}$$

We can immediately conclude that the control (8-137) forces $(\xi_1, \xi_2)$ to $(0, 0)$ if

$$\begin{aligned} Z &= (z_1, z_2) \in R_4 \\ W &= (w_1, w_2) \in \gamma_+ \end{aligned} \tag{8-139}$$

Such a trajectory is illustrated in Fig. 8-6; if it is to be fuel-optimal, then it must satisfy all the necessary conditions of the minimum principle. We know the solution of Eqs. (8-131) when $u(t) = \Delta = \pm 1$ and $u(t) = 0$ [see Eqs. (7-14) to (7-18) and Eqs. (8-102)]; thus, we find, using the definitions of Eqs. (8-138), that

$$\begin{aligned} z_2 &= \xi_2 - t_1 \\ z_1 &= \xi_1 + \xi_2 t_1 - \tfrac{1}{2} t_1^2 \end{aligned} \tag{8-140}$$

$$\begin{aligned} w_2 &= z_2 \\ w_1 &= z_1 + z_2(t_2 - t_1) \end{aligned} \tag{8-141}$$

$$\begin{aligned} 0 &= w_2 + (T_f - t_2) \\ 0 &= w_1 + w_2(T_f - t_2) + \tfrac{1}{2}(T_f - t_2)^2 \end{aligned} \tag{8-142}$$

These six independent equations contain six unknowns, which are $t_1, t_2, z_1, z_2, w_1$, and $w_2$; note that we know $\xi_1$, $\xi_2$, and $T_f$. We are interested in $t_1$ and $t_2$

in order to specify the control of Eqs. (8-137). From Eqs. (8-142) and (8-141), we find that

$$w_1 = \tfrac{1}{2}w_2^2 = \tfrac{1}{2}z_2^2 \tag{8-143}$$
$$t_2 = w_2 + T_f = z_2 + T_f \tag{8-144}$$

From Eqs. (8-140), we obtain the relations

$$t_1 = \xi_2 - z_2 \tag{8-145}$$
$$z_1 = \xi_1 + \tfrac{1}{2}\xi_2^2 - \tfrac{1}{2}z_2^2 \tag{8-146}$$

From Eqs. (8-144) and (8-145), we deduce that

$$t_2 - t_1 = 2z_2 + T_f - \xi_2 \tag{8-147}$$

Substituting Eqs. (8-147), (8-143), and (8-146) into Eq. (8-141), we obtain the relationship

$$z_2^2 + (T_f - \xi_2)z_2 + \xi_1 + \tfrac{1}{2}\xi_2^2 = 0 \tag{8-148}$$

which implies that

$$z_2 = -\frac{T_f - \xi_2}{2} \pm \tfrac{1}{2}[(T_f - \xi_2)^2 - 4\xi_1 - 2\xi_2^2]^{\frac{1}{2}} \tag{8-149}$$

There are two points to clear up: First, we must guarantee that $z_2$ is real and, second, we must choose the correct sign in Eq. (8-149). The fact that $z_2$ is real is an immediate consequence of Lemma 8-7, because Eq. (8-134) implies that $(T_f - \xi_2)^2 - 4\xi_1 - 2\xi_2^2 \geq 0$. In other words, if, by mistake, we demanded a response time $T_f$ smaller than the minimum time, then $z_2$ would not be real. In order to determine the correct sign in Eq. (8-149), we proceed as follows: The times $t_1$, $t_2$, and $T_f$ must be such that

$$0 < t_1 < t_2 < T_f \tag{8-150}$$

which implies that $t_2 - t_1 > 0$ and, therefore [by Eq. (8-147)], that

$$z_2 > -\tfrac{1}{2}(T_f - \xi_2) \tag{8-151}$$

From Eqs. (8-151) and (8-149), we conclude that the + sign must be used in Eq. (8-149); hence

$$z_2 = -\tfrac{1}{2}\{T_f - \xi_2 - [(T_f - \xi_2)^2 - 4\xi_1 - 2\xi_2^2]^{\frac{1}{2}}\} \tag{8-152}$$

Substituting Eq. (8-152) into Eqs. (8-145) and (8-144), we find that

$$t_1 = \tfrac{1}{2}\{T_f + \xi_2 - [(T_f - \xi_2)^2 - 4\xi_1 - 2\xi_2^2]^{\frac{1}{2}}\} \tag{8-153}$$
$$t_2 = \tfrac{1}{2}\{T_f + \xi_2 + [(T_f - \xi_2)^2 - 4\xi_1 - 2\xi_2^2]^{\frac{1}{2}}\} \tag{8-154}$$

which are the required equations of the "switch" times. Thus, we have established the following lemma:

***Lemma 8-9*** *If* $(\xi_1, \xi_2) \in R_1$ *and* $T_f$ *are given, then the control of Eqs.* (8-137) *is unique and fuel-optimal when the "switch" times* $t_1$ *and* $t_2$ *are given by Eqs.* (8-153) *and* (8-154).

It is easy to compute the fuel $\hat{F}(\xi_1, \xi_2, T_f)$ required to force $(\xi_1, \xi_2) \in R_1$ to (0, 0) with response time $T_f$. We leave it to the reader to verify that

$$\hat{F}(\xi_1, \xi_2, T_f) = T_f - [(T_f - \xi_2)^2 - 4\xi_1 - 2\xi_2^2]^{\frac{1}{2}} \tag{8-155}$$

We shall use Eq. (8-155) to establish the following result:

***Lemma 8-10*** *If* $(\xi_1, \xi_2) \in R_1$ *and if it is required to find the fuel-optimal control that forces* $(\xi_1, \xi_2)$ *to* (0, 0) *so that the response time* $T_f$ *is bounded by* $T_f \leq \hat{T}_f$, *then the fuel-optimal control will require* $T_f = \hat{T}_f$.

PROOF It is easy to establish that

$$\frac{\partial \hat{F}(\xi_1, \xi_2, T_f)}{\partial T_f} < 0 \tag{8-156}$$

This means that the fuel increases with decreasing $T_f$; the fuel-optimal control will seek the largest possible $T_f$; therefore, $T_f = \hat{T}_f$.

We have solved Problem 8-3 for initial states in $R_1$. If the initial state is in $R_3$, we can use similar arguments. Now we turn our attention to initial states in $R_4$.

Suppose that $(\xi_1, \xi_2) \in R_4$. Recall (see Lemma 8-5) that if $(\xi_1, \xi_2) \in R_4$, then the control sequence $\{0, +1\}$ will force $(\xi_1, \xi_2)$ to (0, 0) with fuel $F^* = |\xi_2|$ and with response time [see Eq. (8-128)]

$$T = -\tfrac{1}{2}\xi_2 - \frac{\xi_1}{\xi_2} \tag{8-157}$$

Moreover, the response time $T$ is the smallest response time possible using fuel $F^* = |\xi_2|$. If $(\xi_1, \xi_2) \in R_4$ and if we fix the response time $T_f$ so that $T_f < T$, then we must use more fuel than $F^* = |\xi_2|$. On the other hand, if $T_f > T$, then we can always find a nonunique control (see Lemma 8-5) which will force the state to the origin with the minimum fuel $F^* = |\xi_2|$. We formalize these considerations as follows:

***Lemma 8-11*** *Suppose that* $(\xi_1, \xi_2) \in R_4$. *If* $T_f$ *is given and if*

$$T_f < T = -\tfrac{1}{2}\xi_2 - \frac{\xi_1}{\xi_2} \tag{8-158}$$

*then the fuel-optimal control is specified by Eqs.* (8-137), (8-153), *and* (8-154). *If* $T_f$ *is bounded by* $\hat{T}_f$ *and if*

$$T_f \leq \hat{T}_f < T \tag{8-159}$$

*then the fuel-optimal control requires*

$$T_f = \hat{T}_f \tag{8-160}$$

*The minimum fuel in either case is given by Eq.* (8-155).

The proof follows the same lines as the proofs of Lemmas 8-9 and 8-10 and is, therefore, omitted.

***Lemma 8-12*** *If $(\xi_1, \xi_2) \in R_4$ and if $T_f > T$, then there are many fuel-optimal controls to the origin requiring fuel $F^* = |\xi_2|$.*

PROOF This follows from Lemma 8-5.

***Lemma 8-13*** *If $(\xi_1, \xi_2) \in R_4$ and if it is required that $T_f \leq \hat{T}_f$ and if $\hat{T}_f > T$, then there are many fuel-optimal controls to the origin requiring fuel $F^* = |\xi_2|$.*

PROOF This follows from Lemma 8-5.

We shall now summarize the results of Lemmas 8-7 to 8-13 in the following control law. We remark that these lemmas provide us with the solution to the problem whenever the initial state $(\xi_1, \xi_2)$ is in $R_1 \cup R_4$; the results for initial states in $R_2 \cup R_3$ can be obtained by symmetry considerations.

***Control Law 8-3 Solution to Problem 8-3***

*a. $T_f$ fixed*

*Let $(\xi_1, \xi_2)$ be any initial state and let $T_f$ be the fixed response time. If the optimal control $u^*(t)$ exists, then it is given by the following equations:*

*For all $(\xi_1, \xi_2) \in R_1$ and all $T_f$ or for all $(\xi_1, \xi_2) \in R_4$ and all $T_f$ such that $T_f \leq -\frac{1}{2}\xi_2 - \xi_1/\xi_2$*

$$\left.\begin{aligned} u^*(t) &= -1 && \text{for } 0 \leq t < t_1 \\ u^*(t) &= 0 && \text{for } t_1 \leq t < t_2 \\ u^*(t) &= +1 && \text{for } t_2 \leq t \leq T_f \end{aligned}\right\} \quad \textit{unique}$$

*where*

$$\begin{aligned} t_1 &= \tfrac{1}{2}\{T_f + \xi_2 - [(T_f - \xi_2)^2 - 4\xi_1 - 2\xi_2^2]^{\frac{1}{2}}\} \\ t_2 &= \tfrac{1}{2}\{T_f + \xi_2 + [(T_f - \xi_2)^2 - 4\xi_1 - 2\xi_2^2]^{\frac{1}{2}}\} \end{aligned} \tag{8-161a}$$

*For all $(\xi_1, \xi_2) \in R_3$ and all $T_f$ or for all $(\xi_1, \xi_2) \in R_2$ and all $T_f$ such that $T_f \leq \frac{1}{2}\xi_2 - \xi_1/\xi_2$*

$$\left.\begin{aligned} u^*(t) &= +1 && \text{for } 0 \leq t < t_1 \\ u^*(t) &= 0 && \text{for } t_1 \leq t < t_2 \\ u^*(t) &= -1 && \text{for } t_2 \leq t \leq T_f \end{aligned}\right\} \quad \textit{unique}$$

*where*

$$\begin{aligned} t_1 &= \tfrac{1}{2}\{T_f - \xi_2 - [(T_f + \xi_2)^2 + 4\xi_1 - 2\xi_2^2]^{\frac{1}{2}}\} \\ t_2 &= \tfrac{1}{2}\{T_f - \xi_2 + [(T_f + \xi_2)^2 + 4\xi_1 - 2\xi_2^2]^{\frac{1}{2}}\} \end{aligned} \tag{8-161b}$$

*For all $(\xi_1, \xi_2) \in R_4$ and all $T_f$ such that $T_f > -\frac{1}{2}\xi_2 - \xi_1/\xi_2$ or for all $(\xi_1, \xi_2) \in R_2$ and all $T_f$ such that $T_f > \frac{1}{2}\xi_2 - \xi_1/\xi_2$, then the fuel-optimal controls are nonunique* (8-161c)

*For all* $(\xi_1, \xi_2) \in \gamma_+ \cup \gamma_-$, *the fuel-optimal control is unique and is given by*

$$\begin{aligned} u^*(t) &= -\operatorname{sgn}\{\xi_2\} && \text{for } 0 \le t < |\xi_2| \\ u^*(t) &= 0 && \text{for } |\xi_2| \le t \le T_f \end{aligned} \qquad (8\text{-}161d)$$

*b.* $T_f$ *bounded by* $\hat{T}_f$
*Same conclusions as in* (*a*) [*replace* $T_f$ *by* $\hat{T}_f$ *in Eqs.* (8-161*a*) *to* (8-161*d*)].

**Exercise 8-10** Prove Control Law 8-3.
**Exercise 8-11** Consider the initial state $(2, -1) \in R_4$. Plot the fuel-optimal trajectories to the origin for the following values of $T_f$: 2.2, 2.3, 2.4, and 2.5.

We know now the techniques that one must use to find the fuel-optimal trajectory when the response time $T_f$ is given. Let us suppose, once more, that the conditions of Lemma 8-9 or 8-11 are satisfied, so that the fuel-optimal control sequence is $\{-1, 0, +1\}$; the switch times $t_1$ and $t_2$ are given by Eqs. (8-153) and (8-154). We have shown that at the switch time $t_2$, the state $W = (w_1, w_2)$ belongs to the $\gamma_+$ curve. Recall [see Eqs. (8-138)] that at the switch time $t_1$, the state is $Z = (z_1, z_2)$; the value of $z_2$ is given by Eq. (8-152), and the value of $z_1$ is given by Eq. (8-146). Clearly, the state $(z_1, z_2)$ is a function of $\xi_1$, $\xi_2$, and $T_f$. The natural question that arises is: *Does the state* $(z_1, z_2)$ *belong to some (switch) curve?* This question can be answered as follows: Since $(z_1, z_2)$ depends on the three parameters $\xi_1$, $\xi_2$, and $T_f$, if we fix any two of these parameters and vary the third, then we would obtain, from Eqs. (8-152) and (8-146), the locus of the state $(z_1, z_2)$ which will form a curve in the state plane. We shall illustrate this by an example.

**Example 8-3** Suppose that the initial state is $(\xi_1, 0)$, $\xi_1 > 0$, that is, a point on the positive $x_1$ axis. The minimum time $t^*(\xi_1, 0)$ corresponding to this state is [see Eq. (8-134)] given by the relation

$$t^* = 2\sqrt{\xi_1}$$

Suppose we are given a value for $T_f$; then there is a state $(\hat{\xi}_1, 0)$ such that

$$\hat{\xi}_1 = \frac{T_f^2}{4} \qquad (8\text{-}162)$$

which means that the time $T_f$ specified is the minimum time needed to force $(\hat{\xi}_1, 0)$ to $(0, 0)$. Let us now consider the set of states $(\xi_1, 0)$ such that

$$0 \le \xi_1 \le \hat{\xi}_1 \qquad (8\text{-}163)$$

and let us find the fuel-optimal control to the origin which requires the *same* response time $T_f$ to the origin for *every* state $(\xi_1, 0)$. We substitute $\xi_2 = 0$ in Eq. (8-152), and we find that

$$z_2 = -\tfrac{1}{2}T_f + \tfrac{1}{2}(T_f^2 - 4\xi_1)^{\frac{1}{2}} \qquad (8\text{-}164)\dagger$$

Substituting $\xi_2 = 0$ into Eq. (8-146), we find that

$$z_1 = \xi_1 - \tfrac{1}{2}z_2^2 \qquad (8\text{-}165)$$

† Note that Eqs. (8-162) and (8-163) guarantee that $z_2$ is real.

We eliminate $\xi_1$ between Eqs. (8-164) and (8-165) to obtain, after some algebraic manipulations, the relation

$$z_1 = -1.5z_2^2 - z_2T_f \tag{8-166}$$

We know that [see Eqs. (8-139)] $(z_1, z_2) \in R_4$; that is,

$$z_2 < 0 \qquad z_1 > \tfrac{1}{2}z_2^2 \tag{8-167}$$

Thus, Eqs. (8-166) and (8-167) define a curve in the state plane, which we shall call the $\Gamma(T_f)$ "switch" curve, that is,

$$\Gamma(T_f) = \{(z_1, z_2): z_1 = -1.5z_2^2 - z_2T_f;\ z_2 < 0;\ z_1 > \tfrac{1}{2}z_2^2\} \tag{8-168}$$

Figure 8-7 shows such a curve, with $T_f = 3$. Some fuel-optimal trajectories originating on the $x_1$ axis are also shown in Fig. 8-7. Each trajectory requires the same response time $T_f$. Note that the control $u(t) = -1$ is applied until the trajectory hits the $\Gamma(T_f)$ curve; at that time ($t_1$), the control switches to $u(t) = 0$ until the trajectory hits the $\gamma_+$ curve; at that time ($t_2$), the control switches to $u(t) = +1$ to reach the origin. Note, also, that the fuel-optimal trajectory originating at $(\hat{\xi}_1, 0)$ is the time-optimal trajectory (why?).

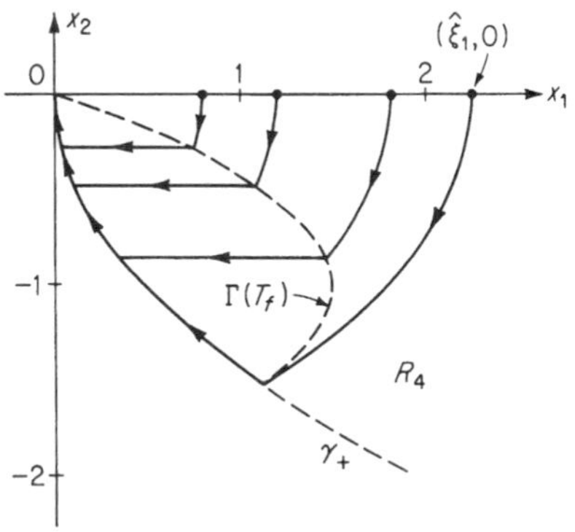


**Fig. 8-7** Illustration of the $\Gamma(T_f)$ switch curve; in this figure, we have used the value $T_f = 3$.

**Exercise 8-12** Compute and plot the consumed fuel as a function of $\xi_1$ for the situation in Example 8-3. Use $T_f = 3$.

**Exercise 8-13** Suppose that the initial state is $(0, \xi_2)$, $\xi_2 > 0$. Compute the locus of $(z_1, z_2)$ and plot the resulting curve for $T_f = 3$.

**Exercise 8-14** Repeat Exercise 8-13 when the initial state $(\xi_1, \xi_2)$ is such that $\xi_1 = \xi_2$, $\xi_2 > 0$.

We conclude this section with the verification of the Hamilton-Jacobi equation† for this problem. The reason for this development is that, up to now, we have been concerned primarily with the optimal control of time-invariant systems with free response time; the system in this section is time-invariant, but the response time is fixed.

First of all, let us suppose that we are given an initial state $(\xi_1, \xi_2)$ in $R_1$ and that the response time $T_f$ is fixed; we know that the fuel-optimal control is

$$\begin{aligned} u^*(t) &= -1 && \text{for } 0 \le t < t_1 \\ u^*(t) &= 0 && \text{for } t_1 \le t < t_2 \\ u^*(t) &= +1 && \text{for } t_2 \le t \le T_f \end{aligned} \tag{8-169}$$

where $t_1$ and $t_2$ are given by Eqs. (8-153) and (8-154), respectively. Let $(x_1, x_2)$ be a state on the fuel-optimal trajectory from $(\xi_1, \xi_2)$ to (0, 0), and let $t$ denote the corresponding time such that

$$0 < t < t_1 \tag{8-170}$$

† See Sec. 5-20.

Let $\hat{F}(x_1, x_2, t)$ denote the fuel required to go from $(x_1, x_2)$ to $(0, 0)$; clearly, $\hat{F}(x_1, x_2, t)$ must be the solution of the Hamilton-Jacobi equation

$$\frac{\partial \hat{F}(x_1, x_2, t)}{\partial t} + |u^*| + x_2 \frac{\partial \hat{F}(x_1, x_2, t)}{\partial x_1} + u^* \frac{\partial \hat{F}(x_1, x_2, t)}{\partial x_2} = 0 \qquad (8\text{-}171)$$

Since $t < t_1$, then, according to Eqs. (8-169),

$$u^* = -1 \qquad (8\text{-}172)$$

Since $(x_1, x_2)$ is a state on the optimal trajectory from $(\xi_1, \xi_2)$ and since $t < t_1$, then

$$\begin{aligned} x_1 &= x_1(t) = \xi_1 + \xi_2 t - \tfrac{1}{2}t^2 \\ x_2 &= x_2(t) = \xi_2 - t \end{aligned} \qquad (8\text{-}173)$$

But the fuel $\hat{F}(x_1, x_2, t)$ is given by the relation

$$\hat{F}(x_1, x_2, t) = \int_t^{t_1} |-1|\, d\tau + \int_{t_1}^{t_2} |0|\, d\tau + \int_{t_2}^{T_f} |1|\, d\tau = t_1 - t + T_f - t_2 \qquad (8\text{-}174)$$

We know $t_1$ and $t_2$ in terms of $\xi_1$, $\xi_2$, and $T_f$. But, from Eqs. (8-173), we also have

$$\begin{aligned} \xi_2 &= x_2 + t \\ \xi_1 &= x_1 - (x_2 + t)t + \tfrac{1}{2}t^2 \end{aligned} \qquad (8\text{-}175)$$

Substituting Eqs. (8-175) into Eqs. (8-153) and (8-154), we obtain $t_1$ and $t_2$ as a function of $x_1$, $x_2$, $t$, and $T_f$, which, substituted in Eq. (8-174), yield

$$\hat{F}(x_1, x_2, t) = T_f - t - (T_f^2 + t^2 + 2tx_2 - 2T_f x_2 - 2T_f t - 4x_1 - x_2^2)^{\frac{1}{2}} \qquad (8\text{-}176)$$

For the sake of simplicity, we define $Y$ as follows:

$$Y = T_f^2 + t^2 + 2tx_2 - 2T_f x_2 - 2T_f t - 4x_1 - x_2^2 \qquad (8\text{-}177)$$

Then we can compute the partial derivatives

$$\frac{\partial \hat{F}(x_1, x_2, t)}{\partial t} = -1 - \frac{t + x_2 - T_f}{\sqrt{Y}} \qquad (8\text{-}178)$$

$$\frac{\partial \hat{F}(x_1, x_2, t)}{\partial x_1} = \frac{2}{\sqrt{Y}} \qquad (8\text{-}179)$$

$$\frac{\partial \hat{F}(x_1, x_2, t)}{\partial x_2} = \frac{x_2 - t + T_f}{\sqrt{Y}} \qquad (8\text{-}180)$$

If we substitute Eqs. (8-172) and (8-178) to (8-180) into the Hamilton-Jacobi equation (8-171), we find that it is indeed satisfied.

**Exercise 8-15** Suppose that $(\xi_1, \xi_2) \in R_1$ and $T_f$ is given; let $(x_1, x_2)$ be a state on the fuel-optimal trajectory at time $t$, $t_1 < t < t_2$. Compute $\hat{F}(x_1, x_2, t)$ and show that it is the solution to the Hamilton-Jacobi equation.

**Exercise 8-16** Repeat Exercise 8-15 if $t_2 < t < T_f$.

We conclude this section with some remarks pertaining to Control Law 8-3. We wish to emphasize that the fuel-optimal control is a function of the initial state $(\xi_1, \xi_2)$ and of the time $T_f$. By methods similar to those used to derive Eq. (8-176), we can find the control as a function of the state $(x_1, x_2)$ and of the remaining time $T_f - t$. In other words, an engineering realization of Control Law 8-3 must contain a "clock" that provides us with the remaining time $T_f - t$. In this sense, the feedback system is a time-varying one although the controlled system and the cost functional are time-invariant. For many applications, this may represent a serious drawback. For this reason, we shall continue our investigation of the double-integral plant, in search of a time-invariant feedback system.

## 8-7 Fuel-optimal Control of the Double-integral Plant: Response Time Bounded by a Multiple of the Minimum Time

In order to motivate the problem posed in this section, we shall comment again on some of the practical disadvantages of the results derived in Secs. 8-5 and 8-6.

The solution to Problem 8-2 in Sec. 8-5 is unsatisfactory from two points of view. From the mathematical point of view, there are no solutions if the initial state is in $R_1$ or in $R_3$. More important, from the engineering point of view, the large response times required when the initial state is near the $x_1$ axis make the solution unattractive.

We must resign ourselves to the fact that in order to decrease the response time, more fuel must be used. For this reason, we posed Problem 8-3 in Sec. 8-6 and required the response time to be either fixed or bounded. We found that the fuel-optimal control in this case is a function of the initial state $(\xi_1, \xi_2)$ and of the response time $T_f$, that is, a function of the three quantities $\xi_1$, $\xi_2$, and $T_f$. If we specify two of these three quantities, then we obtain some sort of switch curve (as in Example 8-3 and Exercises 8-13 and 8-14).

The question that arises is this: *Given an initial state $(\xi_1, \xi_2)$, what is a natural and consistent way of specifying the response time $T_f$?* In particular, suppose that we are given an initial state $(\xi_1, \xi_2)$ and that we have specified a response time $T_f$; now suppose we are given another state $(\xi_1', \xi_2')$ near $(\xi_1, \xi_2)$; how do we pick the new response time $T_f'$ for the state $(\xi_1', \xi_2')$ so that the fuel-optimal trajectories are, in some sense, near each other? All these questions are motivated by physical considerations rather than by mathematical requirements, and they may be summarized as follows: *How can one specify a natural response time $T_f$ for each state?*

There indeed exists a fundamental value of time associated with each state in the state plane, namely, the minimum time $t^*$ required to force that state to the origin. The minimum time $t^*$ takes into account:

1. The fact that the control $u(t)$ is bounded
2. The relative position of the state with respect to the origin

We feel that the choice of the response time $T_f$ must be made by taking into account the minimum time $t^*$. Let us see how we can do that. Suppose that we are given a state $(\xi_1, \xi_2)$ and that we compute the minimum time $t^*(\xi_1, \xi_2)$; then we may require that the response time $T_f$ satisfy the relation

$$T_f = \beta t^*(\xi_1, \xi_2) \tag{8-181}$$

or the relation

$$T_f \leq \hat{T}_f = \beta t^*(\xi_1, \xi_2) \tag{8-182}$$

where $\beta$ is a positive constant such that

$$\beta > 1 \tag{8-183}$$

Of course, for each state $(\xi_1, \xi_2)$, the time $T_f$, given by Eq. (8-181), is a fixed number. However, *if we specify $\beta$ in advance and use the same value of $\beta$ for every initial state $(\xi_1, \xi_2)$, then Eq.* (8-181) *will be a natural way of specifying the response time.*

Let us examine, by means of an example, some of the implications of fixing the response time according to Eq. (8-181).

**Example 8-4** Suppose that we consider the system (8-131) and suppose that the initial state is $(\xi_1, 0) \in R_1$, $\xi_1 > 0$, that is, a point on the positive $x_1$ axis. This is the same assumption as the one we made in Example 8-3. The minimum time $t^*(\xi_1, 0)$ is given by

$$t^* = 2\sqrt{\xi_1} \tag{8-184}$$

Suppose that we require the response time $T_f$ to be

$$T_f = \beta t^* = 2\beta\sqrt{\xi_1} \qquad \beta > 1 \tag{8-185}$$

The reader should note that in Example 8-3 we required the *same* response time for all $(\xi_1, 0)$, while in this example the response time is a function of the initial state. If we substitute $T_f = 2\beta\sqrt{\xi_1}$ and $\xi_2 = 0$ into Eq. (8-152), we obtain, after some algebraic manipulations, the relation

$$z_2 = \sqrt{\xi_1}\,(\sqrt{\beta^2 - 1} - \beta) \tag{8-186}$$

[The reader should compare Eqs. (8-186) and (8-164).] But, since $\xi_2 = 0$, $z_1$ is given by Eq. (8-165), which is

$$z_1 = \xi_1 - \tfrac{1}{2}z_2^2 \tag{8-187}$$

We now eliminate $\xi_1$ between Eqs. (8-186) and (8-187) to obtain

$$z_1 = h_\beta z_2^2 \tag{8-188}$$

where $h_\beta$ is a constant defined by the relation

$$h_\beta = -\frac{1}{2} + \frac{1}{(\sqrt{\beta^2 - 1} - \beta)^2} \tag{8-189}$$

Note that since $\beta > 1$, we have

$$h_\beta > \tfrac{1}{2} \tag{8-190}$$

Since $(z_1, z_2) \in R_4$, we can define the $\Gamma_\beta$ switch curve by

$$\Gamma_\beta = \{(z_1, z_2)\colon z_1 = h_\beta z_2^2;\ z_2 < 0;\ z_1 > \tfrac{1}{2}z_2^2\} \tag{8-191}$$

with $h_\beta$ given by Eq. (8-189). Figure 8-8 illustrates the $\Gamma_\beta$ curve for $h_\beta = 1.5$ (that is, $\beta = 3\sqrt{2}/4 = 1.06$). Figure 8-8 also illustrates some of the fuel-optimal trajectories originating on the $x_1$ axis. Note that $u = -1$ is applied until the trajectory hits $\Gamma_\beta$; at that instant, the control switches from $u = -1$ to $u = 0$. The other control switching, from $u = 0$ to $u = +1$, occurs at the $\gamma_+$ curve. The reader should compare Figs. 8-8 and 8-7 to observe the differences in the operation of the system.

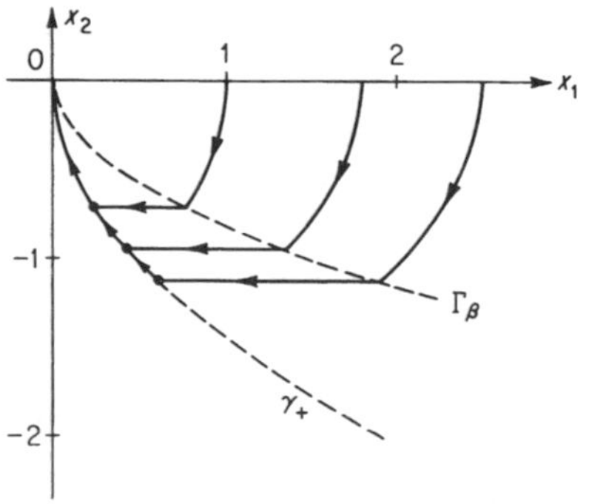


**Fig. 8-8** The $\Gamma_\beta$ switch curve; in this figure, $\beta = 3\sqrt{2}/4$, so that $h_\beta = 1.5$.

**Exercise 8-17** Compute the fuel required to force $(\xi_1, 0)$ using the ideas of Example 8-4. Plot the fuel versus $\xi_1$ for $\beta = 1.5$. Let $\xi_1 = 1$ and plot the fuel versus $\beta$. Repeat for $\xi_1 = 2$.

**Exercise 8-18** Suppose that the initial state is $(0, \xi_2)$, $\xi_2 > 0$. Fix the response time using Eq. (8-181) and find the locus of $(z_1, z_2)$; plot the resulting switch curve with $\beta = 1.2$. Compare with Exercise 8-13.

**Exercise 8-19** Repeat Exercise 8-18 when the initial state $(\xi_1, \xi_2)$ is such that $\xi_1 = \xi_2$, $\xi_2 > 0$. Compare with Exercise 8-14.

Fixing the response time $T_f$ by setting $T_f = \beta t^*(\xi_1, \xi_2)$ has the effect of making the fuel-optimal control a function of $\xi_1$ and $\xi_2$ alone. If we specify a relationship between $\xi_1$ and $\xi_2$ (as we have done in Example 8-4 and Exercises 8-18 and 8-19), then we obtain some sort of switch curve. Note, however, that this switch curve is a function of the "relationship" between $\xi_1$ and $\xi_2$.

From an engineering point of view, we would like to obtain a system which will measure the instantaneous values of the state variables and, on the basis of that information alone, generate the fuel-optimal control. In other words, we would like our system to operate without "remembering" any "relationship" between the initial values of the state variables. All the time-optimal systems we considered had indeed this property. The solution to Problem 8-4, formulated below, will turn out to have this desirable property.

***Problem 8-4*** *Given the system*

$$\begin{aligned} \dot{x}_1(t) &= x_2(t) \\ \dot{x}_2(t) &= u(t) \qquad |u(t)| \le 1 \end{aligned} \tag{8-192}$$

*and the fuel functional*

$$F(u) = F = \int_0^T |u(t)|\, dt \tag{8-193}$$

*Let $(\xi_1, \xi_2)$ be any initial state; if we specify some response time $T_f$, then we can find a fuel-optimal control and a corresponding fuel-optimal trajectory to the origin; let $(x_1(t), x_2(t))$ be a state on this fuel-optimal trajectory at time $t$, $t \in [0, T_f]$, and let $t^*(x_1, x_2)$ be the minimum time associated with the state*

$(x_1(t), x_2(t))$. *We wish to find the fuel-optimal control such that*

$$T_f - t \leq \beta t^*(x_1, x_2) \qquad \beta > 1 \tag{8-194}$$

*for all* $(\xi_1, \xi_2)$ *and all* $t \in [0, T_f]$.

Let us explain the physical significance of Eq. (8-194). Equation (8-194) guarantees that the fuel-optimal control has the property that *every state* $(x_1, x_2)$ *on a fuel-optimal trajectory can be forced to the origin in some time* $T_f - t$ *which is bounded by a multiple of the minimum time* $t^*(x_1, x_2)$ *corresponding to that state.* In a sense, Eq. (8-194) is a time-varying constraint. The significance of this formulation will be clear as we proceed with the solution to Problem 8-4.

Suppose that $(x_1, x_2)$ is a state in $R_4$ [see Eq. (8-112) and Fig. 8-3]. We know that the minimum time $t^*(x_1, x_2)$ corresponding to that state is [see Eq. (8-134)] given by the equation

$$t^*(x_1, x_2) = x_2 + \sqrt{4x_1 + 2x_2^2} \tag{8-195}$$

We also know (see Lemma 8-5) that the state $(x_1, x_2) \in R_4$ can be forced to (0, 0) by the fuel-optimal control sequence $\{0, +1\}$, which requires fuel $|x_2|$ and response time [see Eq. (8-119)]

$$T(x_1, x_2) = -\tfrac{1}{2}x_2 - \frac{x_1}{x_2} \tag{8-196}$$

On the other hand, if $(x_1, x_2) \in R_2$, then the minimum time corresponding to $(x_1, x_2) \in R_2$ is [see Eq. (8-135)] given by

$$t^*(x_1, x_2) = -x_2 + \sqrt{-4x_1 + 2x_2^2} \tag{8-197}$$

We also know (see Lemma 8-12) that the state $(x_1, x_2) \in R_2$ can be forced to (0, 0) by the fuel-optimal control sequence $\{0, -1\}$, which requires fuel $|x_2|$ and response time

$$T(x_1, x_2) = \tfrac{1}{2}x_2 - \frac{x_1}{x_2} \tag{8-198}$$

***Definition 8-2*** *Let* $Q(T)$ *denote the set of states* $(x_1, x_2) \in R_4 \cup R_2$ *which can be forced to the origin by the control sequence* $\{0, +1\}$, *if* $(x_1, x_2) \in R_4$, *or by the control sequence* $\{0, -1\}$, *if* $(x_1, x_2) \in R_2$, *and which require the same response time* $T$.

*From Eqs.* (8-196) *and* (8-198), *we find that*

$$Q(T) = \{(x_1, x_2): x_1 = -x_2T + \tfrac{1}{2}x_2|x_2|;\ (x_1, x_2) \in R_4 \cup R_2\} \tag{8-199}$$

*We also set*

$$\begin{aligned} Q^+(T) &= Q(T) \cap R_4 \\ Q^-(T) &= Q(T) \cap R_2 \end{aligned} \tag{8-200}$$

Figure 8-9 illustrates several of the $Q(T)$ curves. A useful property of the $Q(T)$ curves is provided by the following lemma:

***Lemma 8-14*** *Suppose that* $T_1 < T_2$. *Let* $(x_1, x_2) \in Q(T_1)$, $(x_1', x_2) \in Q(T_2)$. *Then the relation*

$$|x_1'| > |x_1| \tag{8-201}$$

*is satisfied.*

PROOF The proof follows by direct computation using Eq. (8-199). Loosely speaking, this lemma means that the $Q(T)$ curves "expand" with increasing $T$.

Let $t^*$ be some positive number; we know that we can find the set of states for which $t^*$ is the minimum time to the origin and that we have

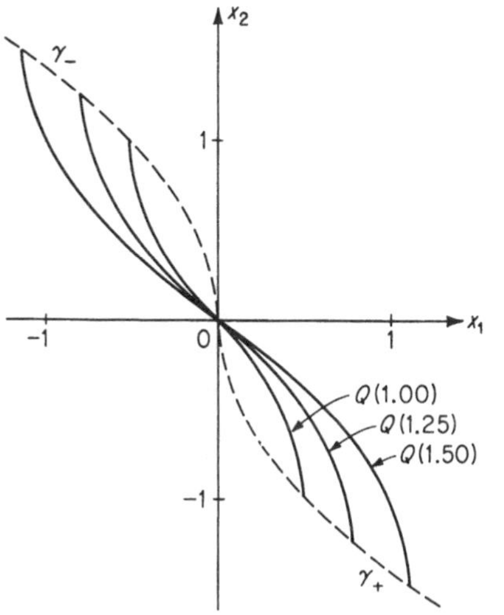


**Fig. 8-9** Some of the $Q(T)$ curves defined by Eq. (8-199); the curves shown in this figure are for $T = 1$, $T = 1.25$, and $T = 1.50$.

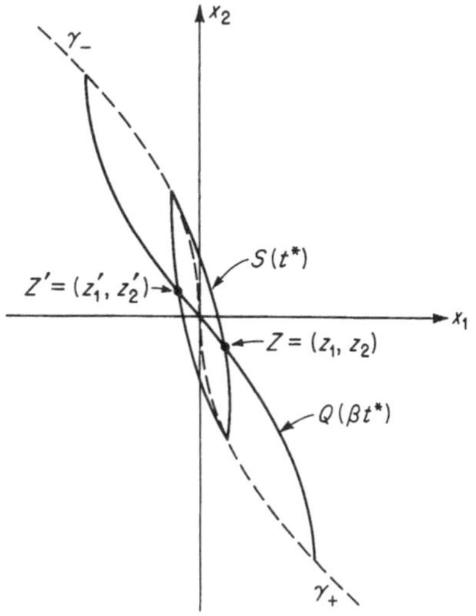


**Fig. 8-10** The points $Z$ and $Z'$ belong to the intersection of $S(t^*)$ and $Q(\beta t^*)$.

denoted this set of states by $S(t^*)$ (see Definition 7-6). The equation of the minimum isochrone $S(t^*)$ is given by the relation [see Eq. (7-27)]

$$\begin{aligned} S(t^*) = \{(x_1, x_2)\colon x_1 &= -\tfrac{1}{2}x_2^2 + \tfrac{1}{4}(t^* - x_2)^2, \text{ if } (x_1, x_2) \in R_4 \cup R_1, \\ \text{or } x_1 &= \tfrac{1}{2}x_2^2 - \tfrac{1}{4}(t^* + x_2)^2, \text{ if } (x_1, x_2) \in R_2 \cup R_3\} \end{aligned} \tag{8-202}$$

Figure 8-10 shows an $S(t^*)$ minimum isochrone. We have also plotted in Fig. 8-10 the $Q(\beta t^*)$ curve, where $\beta > 1$. By setting $T = \beta t^*$ in Eq. (8-197), we deduce that

$$Q(\beta t^*) = \{(x_1, x_2)\colon x_1 = -x_2\beta t^* + \tfrac{1}{2}x_2|x_2|;\ (x_1, x_2) \in R_4 \cup R_2\} \tag{8-203}$$

As illustrated in Fig. 8-10, the sets $Q(\beta t^*)$ and $S(t^*)$ intersect at two points, denoted by $Z = (z_1, z_2)$ and $Z' = (z_1', z_2')$, which are symmetric about the origin. The point $Z = (z_1, z_2)$ is in $R_4$, that is,

$$(z_1, z_2) = S(t^*) \cap Q^+(\beta t^*) = S(t^*) \cap Q(\beta t^*) \cap R_4 \tag{8-204}$$

and the point $Z' = (z_1', z_2')$ is in $R_2$, that is,

$$(z_1', z_2') = S(t^*) \cap Q^-(\beta t^*) = S(t^*) \cap Q(\beta t^*) \cap R_2 \tag{8-205}$$

From Eqs. (8-203) and (8-202), we find that the point $(z_1, z_2)$ is specified by the equation

$$z_1 = -z_2\beta t^* - \tfrac{1}{2}z_2^2 = -\tfrac{1}{2}z_2^2 + \tfrac{1}{4}(t^* - z_2)^2 \tag{8-206}$$

Thus, we conclude that the point $(z_1, z_2)$ can be found from the two equations

$$z_2 = t^*[(1 - 2\beta) + 2\sqrt{\beta(\beta - 1)}] \tag{8-207}$$
$$z_1 = -z_2\beta t^* - \tfrac{1}{2}z_2^2 \tag{8-208}$$

and, owing to the symmetry of $(z_1, z_2)$ and $(z_1', z_2')$, we also have

$$z_2' = -t^*[(1 - 2\beta) + 2\sqrt{\beta(\beta - 1)}] \tag{8-209}$$
$$z_1' = -z_2'\beta t^* + \tfrac{1}{2}z_2'^2 \tag{8-210}$$

Thus, given $\beta$ and $t^*$, the two points $Z = (z_1, z_2)$ and $Z' = (z_1', z_2')$ are completely specified.

We now eliminate the time $t^*$ between Eqs. (8-207) and (8-208) and between Eqs. (8-209) and (8-210) to obtain the relations

$$z_1 = m_\beta z_2^2 \qquad (z_1, z_2) \in R_4 \tag{8-211}$$
$$z_1' = -m_\beta z_2'^2 \qquad (z_1', z_2') \in R_2 \tag{8-212}$$

where $m_\beta$ is given by the equation

$$m_\beta = \frac{\beta}{2\beta - 2\sqrt{\beta(\beta - 1)} - 1} - \frac{1}{2} \tag{8-213}$$

We combine Eqs. (8-211) and (8-212) into a single equation, and we define a curve $\gamma_\beta$ (replacing the $z$'s by $x$'s) as

$$\gamma_\beta = \{(x_1, x_2): x_1 = -m_\beta x_2|x_2|\} \tag{8-214}$$

where $m_\beta$ is given by Eq. (8-213).

Figure 8-11 shows the shape of the $\gamma_\beta$ curve for various values of $\beta > 1$. We shall now establish two lemmas which lead to the control law.

***Lemma 8-15*** *Let $(x_1, x_2) \in \gamma_\beta$ and let $t^*(x_1, x_2)$ be the minimum time associated with $(x_1, x_2)$. If $(x_1, x_2) \in R_4$, then the control sequence $\{0, +1\}$ is fuel-optimal; if $(x_1, x_2) \in R_2$, then the control sequence $\{0, -1\}$ is fuel-optimal; in either case, the response time $T$ is given exactly by the relation*

$$T = \beta t^*(x_1, x_2) \tag{8-215}$$

*for all $(x_1, x_2) \in \gamma_\beta$.*

PROOF This lemma is a direct consequence of the construction of the $\gamma_\beta$ curve. Since the $\gamma_\beta$ curve is the locus of the points $(z_1, z_2)$ and $(z_1', z_2')$, as $t^*$ varies, and since the points $(z_1, z_2)$ and $(z_1', z_2')$ belong both to the isochrone $S(t^*)$ and to the set $Q(\beta t^*)$, the lemma follows from the definition of the $Q(T)$ curves (Definition 8-2).

***Lemma 8-16*** *Given $\beta_1 < \beta_2$; let $(x_1, x_2) \in \gamma_{\beta_1}$ and let $(x_1', x_2) \in \gamma_{\beta_2}$; then $|x_1| < |x_1'|$.*

PROOF The lemma follows by direct computation using Eqs. (8-214) and (8-213).

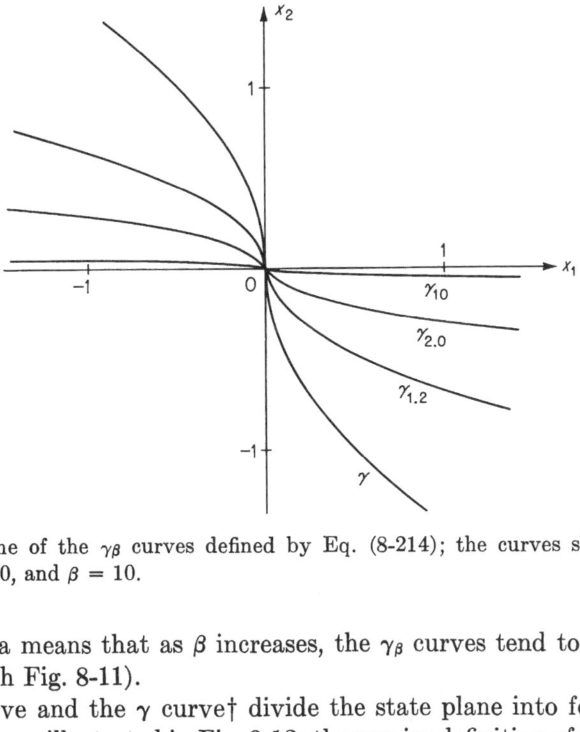


**Fig. 8-11** Some of the $\gamma_\beta$ curves defined by Eq. (8-214); the curves shown are for $\beta = 1.2$, $\beta = 2.0$, and $\beta = 10$.

This lemma means that as $\beta$ increases, the $\gamma_\beta$ curves tend to the $x_1$ axis (compare with Fig. 8-11).

The $\gamma_\beta$ curve and the $\gamma$ curve† divide the state plane into four sets $G_1$, $G_2$, $G_3$, and $G_4$, as illustrated in Fig. 8-12; the precise definition of these sets is

$$G_1 = \{(x_1, x_2): x_1 \geq -\tfrac{1}{2}x_2|x_2| \text{ and } x_1 > -m_\beta x_2|x_2|\} \quad (8\text{-}216)$$
$$G_2 = \{(x_1, x_2): x_1 < -\tfrac{1}{2}x_2|x_2| \text{ and } x_1 \geq -m_\beta x_2|x_2|\} \quad (8\text{-}217)$$
$$G_3 = \{(x_1, x_2): x_1 \leq -\tfrac{1}{2}x_2|x_2| \text{ and } x_1 < -m_\beta x_2|x_2|\} \quad (8\text{-}218)$$
$$G_4 = \{(x_1, x_2): x_1 > -\tfrac{1}{2}x_2|x_2| \text{ and } x_1 \leq -m_\beta x_2|x_2|\} \quad (8\text{-}219)$$

Note that the set $G_1$ includes the $\gamma_-$ curve and that the set $G_3$ includes the $\gamma_+$ curve.

† Recall that $\gamma = \gamma_+ \cup \gamma_-$ [see Definition 7-4 and Eq. (7-21)].

We shall show that Control Law 8-4 provides the solution to Problem 8-4.

***Control Law 8-4 Solution to Problem 8-4*** *The optimal control, as a function of the state* $(x_1, x_2)$, *is given by the equation*

$$\begin{aligned} u^* &= u^*(x_1, x_2) = -1 && \text{for all } (x_1, x_2) \in G_1 \\ u^* &= u^*(x_1, x_2) = +1 && \text{for all } (x_1, x_2) \in G_3 \\ u^* &= u^*(x_1, x_2) = 0 && \text{for all } (x_1, x_2) \in G_2 \cup G_4 \end{aligned} \tag{8-220}$$

*and, furthermore, is unique.*

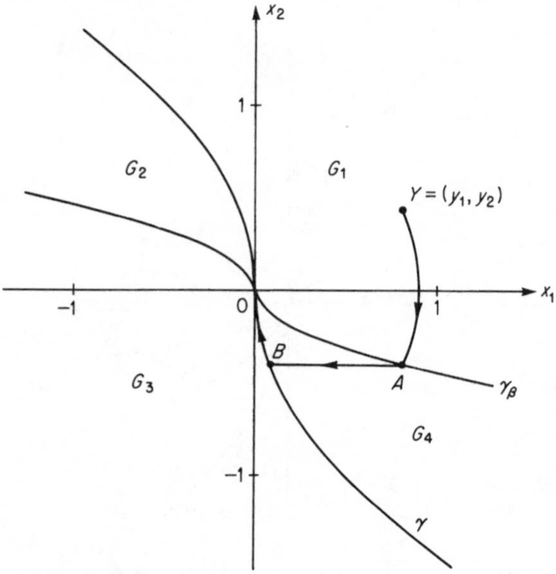


**Fig. 8-12** The division of the state plane into the sets $G_1$, $G_2$, $G_3$, and $G_4$ by the $\gamma$ and $\gamma_\beta$ curves. The trajectory $YABO$ is an optimal one.

PROOF Suppose that $X = (x_1, x_2)$ is a state in $G_4$; let $t_X^*$ be the minimum time corresponding to $X$. We must show that the response time $T_X$ required by the fuel-optimal control sequence $\{0, +1\}$ is such that

$$T_X \leq \beta t_X^* \tag{8-221}$$

Suppose that

$$T_X > \beta t_X^* \tag{8-222}$$

Then there must be a $\beta'$, $\beta' > \beta$, such that

$$T_X = \beta' t_X^* \tag{8-223}$$

and, therefore, a curve $\gamma_{\beta'}$ such that $(x_1, x_2) \in \gamma_{\beta'}$, which, in turn, implies that $\gamma_{\beta'} \in G_4$; but this contradicts Lemma 8-16; therefore, if $(x_1, x_2) \in G_4$, then Eq. (8-221) holds.

Now suppose that $Y = (y_1, y_2)$ is a state in $G_1$, as illustrated in Fig. 8-12. Let $t_Y^*$ be the minimum time associated with $Y$. The control of Eqs. (8-220) results in the trajectory $YABO$ shown in Fig. 8-12. We must show that the response time $T_Y$ is such that

$$T_Y \leq \beta t_Y^* \tag{8-224}$$

Let $t_A$ be the time at which the trajectory hits the $\gamma_\beta$ curve (at the point $A$); since the portion $YA$ of the trajectory $YABO$ is identical to the time-optimal trajectory to the origin, it follows that the minimum time $t_A^*$ corresponding to the point $A$ is

$$t_A^* = t_Y^* - t_A \tag{8-225}$$

But the time required to go from $A$ to the origin along the trajectory $ABO$ is, by definition, $\beta t_A^*$; hence the response time $T_Y$ to go from $Y$ to the origin along $YABO$ is given by

$$T_Y = t_A + \beta(t_Y^* - t_A) = \beta t_Y^* + (1 - \beta)t_A \tag{8-226}$$

Since $\beta > 1$, the term $(1 - \beta)t_A$ is negative, and therefore we deduce that

$$T_Y < \beta t_Y^* \tag{8-227}$$

If the initial state is in $G_2 \cup G_3$, we can use identical reasoning to establish the control law.

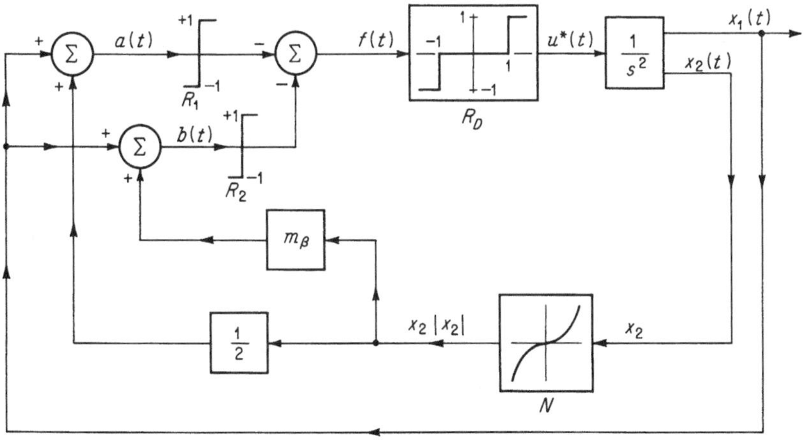


**Fig. 8-13** An engineering realization of Control Law 8-4.

Let us now turn our attention to the engineering realization of Control Law 8-4. The control system must measure the state $(x_1, x_2)$, decide whether the state belongs to $G_1$, to $G_3$, or to $G_2 \cup G_4$, and then deliver the optimal control specified by Eqs. (8-220). The system shown in Fig. 8-13 accomplishes this task. Note that the state variable $x_2$ is introduced to the nonlinearity $N$; the input-output characteristics of $N$ are such that the output is $x_2|x_2|$.† The signal $x_2|x_2|$ is attenuated and added to $x_1$ to produce the signal $a(t)$. The signal $x_2|x_2|$ is also multiplied by the constant gain $m_\beta$ and added to the signal $x_1$ to produce the signal $b(t)$. It is easy to see that

$$\begin{aligned} a(t) > 0 &\quad \text{means} \quad (x_1, x_2) \in G_1 \cup G_4 \\ a(t) < 0 &\quad \text{means} \quad (x_1, x_2) \in G_2 \cup G_3 \\ a(t) = 0 &\quad \text{means} \quad (x_1, x_2) \in \gamma_+ \cup \gamma_- \end{aligned}$$

and that

$$\begin{aligned} b(t) > 0 &\quad \text{means} \quad (x_1, x_2) \in G_1 \cup G_2 \\ b(t) < 0 &\quad \text{means} \quad (x_1, x_2) \in G_3 \cup G_4 \\ b(t) = 0 &\quad \text{means} \quad (x_1, x_2) \in \gamma_\beta \end{aligned}$$

The polarized relays $R_1$ and $R_2$ obtain the sign of $a(t)$ and $b(t)$, since

$$f(t) = -\operatorname{sgn}\{a(t)\} - \operatorname{sgn}\{b(t)\} \tag{8-228}$$

It follows that

$$\begin{aligned} &\text{If } (x_1, x_2) \in G_1, \text{ then } f(t) = -2 \\ &\text{If } (x_1, x_2) \in G_3, \text{ then } f(t) = +2 \\ &\text{If } (x_1, x_2) \in G_2 \cup G_4, \text{ then } f(t) = 0 \end{aligned} \tag{8-229}$$

The input-output characteristics of $R_D$ are those of a perfect relay with a dead zone; that is,

$$\begin{aligned} u^*(t) &= +1 \quad \text{if } f(t) \geq 1 \\ u^*(t) &= 0 \quad \text{if } |f(t)| < 1 \\ u^*(t) &= -1 \quad \text{if } f(t) \leq -1 \end{aligned} \tag{8-230}$$

From Eqs. (8-230) and (8-229), we observe that the system of Fig. 8-13 delivers the control of Eqs. (8-220) and is a time-invariant controller as promised.

We were fortunate in this problem because the equation of the $\gamma_\beta$ curve was similar to the equation of the $\gamma$ curve (both are parabolic). This fact enables the designer to use the same nonlinear element $N$ (in Fig. 8-13) to "generate" both the $\gamma$ and $\gamma_\beta$ curves. We caution the reader that this situation is not always true for other second-order systems; in other words, the equations of the $\gamma$ and $\gamma_\beta$ curves will be, in general, different.

**Exercise 8-20** Show that given any state $(x_1, x_2)$, the response time $T$ required by the control of Eqs. (8-220) is

$$T = \begin{cases} x_2 + (\frac{3}{2} + m_\beta)\sqrt{\dfrac{2x_1 + x_2^2}{1 + 2m_\beta}} & \text{if } (x_1, x_2) \in G_1 \\ \frac{1}{2}|x_2| - \dfrac{x_1}{x_2} & \text{if } (x_1, x_2) \in G_2 \cup G_4 \\ -x_2 + (\frac{3}{2} + m_\beta)\sqrt{\dfrac{-2x_1 + x_2^2}{1 + 2m_\beta}} & \text{if } (x_1, x_2) \in G_3 \end{cases} \tag{8-231}$$

† Note that this is the same nonlinearity as that appearing in Fig. 7-5, which generated the time-optimal control for the same system.

and the fuel $F$ required is

$$F = \begin{cases} x_2 + 2\sqrt{\dfrac{2x_1 + x_2^2}{1 + 2m_\beta}} & \text{if } (x_1, x_2) \in G_1 \\ |x_2| & \text{if } (x_1, x_2) \in G_2 \cup G_4 \\ -x_2 + 2\sqrt{\dfrac{-2x_1 + x_2^2}{1 + 2m_\beta}} & \text{if } (x_1, x_2) \in G_3 \end{cases} \tag{8-232}$$

Let $x_2 = 0$, and plot the fuel $F$ and the time $T$ versus $|x_1|$ for $\beta = 1.0, 1.1, 1.2, 1.5, 2.0, 5.0$, and $10.0$. Also, let $\beta = 1.2$, and plot (for $x_2 = 0$) the ratio $T/t^*$ versus $|x_1|$, where $t^*$ is the minimum time.

## 8-8 Minimization of a Linear Combination of Time and Fuel for the Double-integral Plant

In the previous section, we derived a time-invariant feedback control system which yields the fuel-optimal solution and which guarantees that the response time is always bounded by a constant multiple of the minimum time. If we reexamine the logical development leading to Control Law 8-4, we can see that the analytical expression of the minimum isochrones $S(t^*)$ was used in our calculations. We remarked in Chap. 7 that it is very difficult (if not completely impossible) to find equations for the minimum isochrones; therein lies the major disadvantage of the approach suggested by Problem 8-4.

Our purpose in this section is to formulate and solve an optimization problem using a cost functional which is a weighted linear combination of the response time and the consumed fuel. We shall see that the use of this cost functional leads once more to a time-invariant feedback control system (a desirable characteristic from the practical point of view). As a matter of fact, we shall use the same cost functional in the remaining two sections of this chapter, as we believe it is a good and practical measure of performance.

Our approach in this section will be as follows: First, we shall use the minimum principle to find the control sequences that are candidates for optimal control. We shall then use repeatedly the necessary condition that the Hamiltonian be zero along the optimal trajectory. We shall show that the extremal controls are unique. Finally, we shall derive the equations of the switch curves for the problem and establish the optimal control as a function of the state. We remark at this point that the optimal feedback system turns out to be the one shown in Fig. 8-13; we also hasten to remark that this is a coincidence and is not a general attribute of optimal systems.

The problem we shall solve in this section is the following:

***Problem 8-5*** *Given the system*

$$\begin{aligned} \dot{x}_1(t) &= x_2(t) \\ \dot{x}_2(t) &= u(t) \qquad |u(t)| \le 1 \end{aligned} \tag{8-233}$$

*Find the control which forces the system* (8-233) *from any initial state* $(\xi_1, \xi_2)$ *to* (0, 0) *and which minimizes the cost functional*

$$J(u) = J = \int_0^T [k + |u(t)|]\, dt \tag{8-234}$$

*where*

$$k > 0 \tag{8-235}$$

*and the response time T is not specified.*

We remind the reader that

$$J = kT + \int_0^T |u(t)|\, dt = kT + F \tag{8-236}$$

That is, $J$ is the weighted sum of the response time and the consumed fuel.

The Hamiltonian $H$ for this problem is given by the equation

$$H = k + |u(t)| + x_2(t)p_1(t) + u(t)p_2(t) \tag{8-237}$$

The costate variables $p_1(t)$ and $p_2(t)$ are the solutions of the differential equations

$$\dot{p}_1(t) = -\frac{\partial H}{\partial x_1(t)} = 0 \tag{8-238}$$

$$\dot{p}_2(t) = -\frac{\partial H}{\partial x_2(t)} = -p_1(t) \tag{8-239}$$

Letting

$$\pi_1 = p_1(0) \qquad \pi_2 = p_2(0) \tag{8-240}$$

we find that

$$\begin{aligned} p_1(t) &= \pi_1 = \text{const} \\ p_2(t) &= \pi_2 - \pi_1 t \end{aligned} \tag{8-241}$$

The control which minimizes the Hamiltonian $H$ is given by

$$u(t) = 0 \qquad \text{if } |p_2(t)| < 1 \tag{8-242}$$

$$u(t) = -\operatorname{sgn}\{p_2(t)\} \qquad \text{if } |p_2(t)| > 1 \tag{8-243}$$

$$0 \le u(t) \le +1 \qquad \text{if } p_2(t) = -1 \tag{8-244}$$

$$-1 \le u(t) \le 0 \qquad \text{if } p_2(t) = +1 \tag{8-245}$$

Again we must worry about the possibility of $|p_2(t)| = 1$ for $t$ in some interval, say $[t_1, t_2]$. This resulted in nonunique controls in Sec. 8-5; we shall show now that this situation does not occur in Problem 8-5.

***Lemma 8-17*** *It is not possible for* $|p_2(t)|$ *to be* 1 *for* $t$ *in an interval* $[t_1, t_2]$. *Hence, singular controls cannot be optimal, and so the problem is normal.*

PROOF Suppose that $p_2(t) = -1$ for all $t \in [t_1, t_2]$; then, according to Eq. (8-244), $u(t)$ must be some nonnegative function. At any instant of time $t \in [t_1, t_2]$, the Hamiltonian is then given by

$$H = k + |u(t)| + x_2(t)\pi_1 - u(t) = k + x_2(t)\pi_1 \tag{8-246}$$

But, if $p_2(t) = -1$ for all $t \in [t_1, t_2]$, then

$$\pi_1 = 0 \qquad \text{and} \qquad \pi_2 = -1 \tag{8-247}$$

Substituting $\pi_1 = 0$ into Eq. (8-246), we have (since $k > 0$)

$$H = k > 0 \tag{8-248}$$

But, since the system (8-233) is time-invariant, since the cost functional $J$ of Eq. (8-234) is time-invariant, and since $T$ is not fixed, the Hamiltonian $H$ must be identically zero along an optimal trajectory; clearly, Eq. (8-248) contradicts this requirement. Similar reasoning is employed when $p_2(t) = +1$ for all $t \in [t_1, t_2]$.

Since singular controls cannot be optimal, only the nine control sequences

$$\{0\}, \{+1\}, \{-1\}, \{0, +1\}, \{0, -1\}, \{+1, 0\}, \{-1, 0\}, \{+1, 0, -1\}, \{-1, 0, +1\} \tag{8-249}$$

can be candidates for the optimal control.†

***Lemma 8-18*** *If the origin* (0, 0) *is the desired terminal state, then only the six control sequences*

$$\{+1\}, \{-1\}, \{0, +1\}, \{0, -1\}, \{+1, 0, -1\}, \{-1, 0, +1\} \tag{8-250}$$

*can be candidates for the optimal control.*

PROOF We must show that the control sequences $\{0\}$, $\{+1, 0\}$, and $\{-1, 0\}$ cannot be optimal to the origin. The common characteristic of these sequences is that

$$u(T) = 0 \tag{8-251}$$

Since $x_2(T) = 0$, the Hamiltonian at time $t = T$ is [using $u(T) = 0$] given by

$$H\,|_{t=T} = k + |u(T)| + x_2(T)\pi_1 + u(T)p_2(T) = k > 0 \tag{8-252}$$

This contradicts the requirement that $H = 0$ for all $t \in [0, T]$, and so the lemma is established.

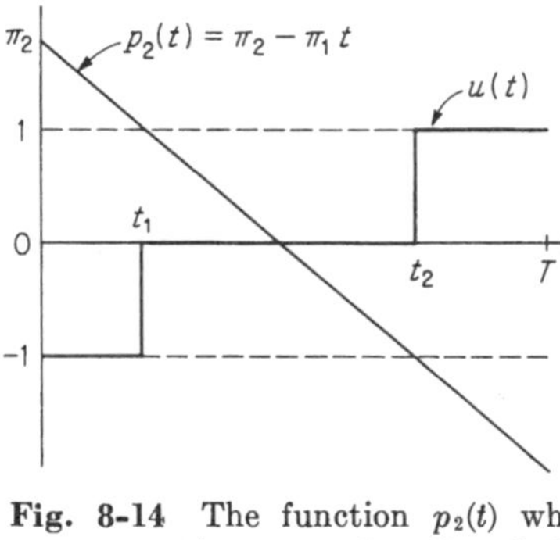


**Fig. 8-14** The function $p_2(t)$ which generates the control $u(t)$ of Eq. (8-253).

We note that the control sequences $\{+1\}$ and $\{0, +1\}$ are subsequences of the control sequence $\{-1, 0, +1\}$ and that the control sequences $\{-1\}$, $\{0, -1\}$ are subsequences of the control sequence $\{+1, 0, -1\}$. By the *principle of optimality,*‡ we can conclude that if a sequence is optimal, then every one of its subsequences must be optimal.

† In this section, the word "optimal" is used with respect to the cost functional $J = \int_0^T [k + |u(t)|]\, dt$.

‡ See Sec. 5-16.

Figure 8-14 shows the function $p_2(t)$ which generates the control sequence $\{-1, 0, +1\}$; that is,

$$\begin{aligned} u(t) &= -1 \qquad 0 \le t < t_1 \\ u(t) &= 0 \qquad t_1 \le t < t_2 \\ u(t) &= +1 \qquad t_2 \le t \le T \end{aligned} \tag{8-253}$$

Clearly, since $p_2(t) = \pi_2 - \pi_1 t$, we must have

$$\pi_2 > 1 \qquad \pi_1 > 0 \tag{8-254}$$

and, in addition, at the switch times $t_1$ and $t_2$

$$p_2(t_1) = 1 = \pi_2 - \pi_1 t_1 \tag{8-255}$$

$$p_2(t_2) = -1 = \pi_2 - \pi_1 t_2 \tag{8-256}$$

Let us now start at some state $(\xi_1, \xi_2)$ and apply the $H$-minimal control (8-253). Let† $(z_1, z_2)$ be the state at $t = t_1$ and let $(w_1, w_2)$ be the state at $t = t_2$. We know that $(w_1, w_2) \in \gamma_+$ and that Eqs. (8-140) to (8-142) hold. We are interested in finding the locus of $(z_1, z_2)$. Since $(w_1, w_2) \in \gamma_+$, we have the relation

$$w_1 = \tfrac{1}{2} w_2^2 \tag{8-257}$$

But Eqs. (8-141) may be written as

$$\begin{aligned} w_2 &= z_2 \\ w_1 &= z_1 + z_2(t_2 - t_1) \end{aligned} \tag{8-258}$$

From Eqs. (8-257) and (8-258), we obtain the relation

$$\tfrac{1}{2} z_2^2 = z_1 + z_2(t_2 - t_1) \tag{8-259}$$

Now we know that the Hamiltonian must be zero for all $t \in [0, T]$. But at $t = t_1$, we have

$$x_1(t_1) = z_1 \qquad x_2(t_1) = z_2 \qquad p_2(t_1) = 1 \qquad u(t_1) \le 0 \tag{8-260}$$

Substituting Eqs. (8-260) into Eq. (8-237), we find that

$$H\,|_{t=t_1} = 0 = k + z_2\pi_1 \tag{8-261}$$

which yields the relation

$$\pi_1 = -\frac{k}{z^2} \tag{8-262}$$

We now subtract Eq. (8-256) from Eq. (8-255) to obtain the equation

$$2 = \pi_1(t_2 - t_1) \tag{8-263}$$

† See Eqs. (8-138).

Substituting Eqs. (8-262) and (8-263) into Eq. (8-259), we find that

$$\tfrac{1}{2}z_2^2 = z_1 - z_2 \frac{2z_2}{k} \tag{8-264}$$

or, equivalently, that

$$z_1 = g_k z_2^2 \tag{8-265}$$

where the constant $g_k$ is defined by the equation

$$g_k = \frac{k+4}{2k} \tag{8-266}$$

Equations (8-265) and (8-266) describe the locus of the states $(z_1, z_2)$ at which the control generated by the control sequence $\{-1, 0, +1\}$ switches from $u = -1$ to $u = 0$. Working in a similar manner, we find that the locus of the states $(z_1', z_2')$ at which the control generated by the control sequence $\{+1, 0, -1\}$ switches from $u = +1$ to $u = 0$ is given by

$$z_1' = -g_k z_2'^2 \tag{8-267}$$

*Let us recapitulate:* We have used *all* the necessary conditions to show that on the extremal trajectory, the state at the switch time $t_1$, that is,

$$x_1(t_1) = z_1$$
$$x_2(t_1) = z_2$$

must satisfy Eq. (8-265), that is, the relation $z_1 = g_k z_2^2$, and that the state at the switch time $t_2$, that is,

$$x_1(t_2) = w_2$$
$$x_2(t_2) = w_1$$

must satisfy Eqs. (8-257) and (8-258), that is, the relations $w_1 = \frac{1}{2}w_2^2$ and $z_2 = w_2$. Since $g_k = (k+4)/2k$ [see Eq. (8-266)], the locus of the states $\{(z_1, z_2)\}$ is completely specified by $k$ and the locus of the states $\{(w_1, w_2)\}$ is completely independent of $k$. For this reason, let us make the following definition:

***Definition 8-3*** *Let $\Gamma_k$ denote the curve defined by the relation*

$$\Gamma_k = \left\{(x_1, x_2): x_1 = -g_k x_2|x_2|;\ g_k = \frac{k+4}{2k}\right\} \tag{8-268}$$

*Figure* 8-15 *shows several $\Gamma_k$ curves for different values of $k$. The $\gamma$ curve and the $\Gamma_k$ curve divide the state plane into the four regions $H_1$, $H_2$, $H_3$, and $H_4$, as shown in Fig.* 8-16. *The precise definitions of $H_1$, $H_2$, $H_3$, and $H_4$ are*

$$H_1 = \{(x_1, x_2): x_1 \geq -\tfrac{1}{2}x_2|x_2| \text{ and } x_1 > -g_k x_2|x_2|\} \tag{8-269}$$
$$H_2 = \{(x_1, x_2): x_1 < -\tfrac{1}{2}x_2|x_2| \text{ and } x_1 \geq -g_k x_2|x_2|\} \tag{8-270}$$
$$H_3 = \{(x_1, x_2): x_1 \leq -\tfrac{1}{2}x_2|x_2| \text{ and } x_1 < -g_k x_2|x_2|\} \tag{8-271}$$
$$H_4 = \{(x_1, x_2): x_1 > -\tfrac{1}{2}x_2|x_2| \text{ and } x_1 \leq -g_k x_2|x_2|\} \tag{8-272}$$

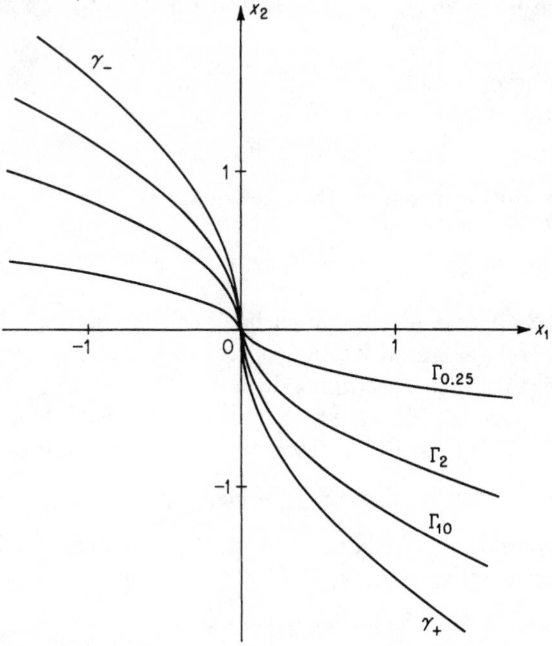


**Fig. 8-15** Some of the $\Gamma_k$ curves; the curves shown are for $k = 0.25$, $k = 2$, and $k = 10$.

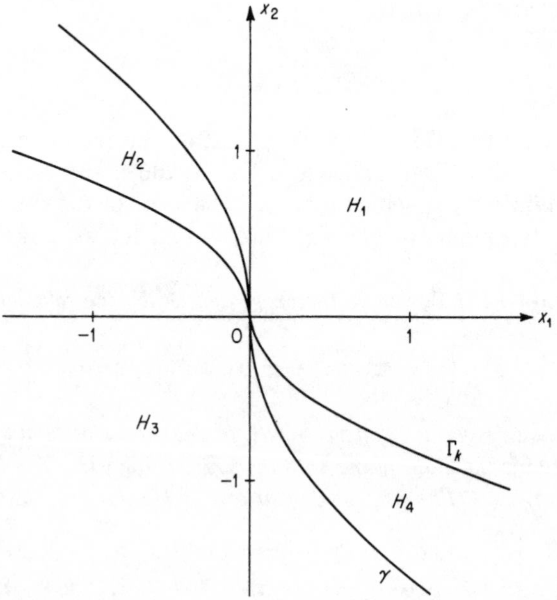


**Fig. 8-16** The division of the state plane into the sets $H_1$, $H_2$, $H_3$, and $H_4$ by the $\gamma$ and $\Gamma_k$ curves.

We claim that the solution to Problem 8-5 is given by the following control law:

***Control Law 8-5 Solution to Problem 8-5*** *The optimal control $u^*$, as a function of the state $(x_1, x_2)$, is given by the equation*

$$
\begin{aligned}
u^* &= u^*(x_1, x_2) = -1 && \text{for all } (x_1, x_2) \in H_1 \\
u^* &= u^*(x_1, x_2) = +1 && \text{for all } (x_1, x_2) \in H_3 \\
u^* &= u^*(x_1, x_2) = 0 && \text{for all } (x_1, x_2) \in H_2 \cup H_4
\end{aligned}
\tag{8-273}
$$

*and, furthermore, $u^*$ is unique.*

**Exercise 8-21** Prove Control Law 8-5. HINT: For existence, assume that an optimal exists for Problem 8-3. For uniqueness, pick states in $H_1$, $H_2$, $H_3$, $H_4$. Try all the six control sequences of Eq. (8-250), and demand that the Hamiltonian be zero along the optimal trajectory.

The engineering realization of Control Law 8-5 is precisely the one illustrated in Fig. 8-13, the only difference being that *the gain $m_\beta$ in Fig.* 8-13 *should be replaced by the gain $g_k = (k + 4)/2k$.* The reason that the two realizations are identical is that the equations of the $\gamma_\beta$ curve [see Eq. (8-214)] and of the $\Gamma_k$ curve [see Eq. (8-268)] have the same form.†

It is easy to establish that

$$\lim_{k \to \infty} g_k = \tfrac{1}{2} \qquad \lim_{k \to 0} g_k \to \infty \tag{8-274}$$

This means (in a somewhat loose terminology) that

$$\lim_{k \to \infty} \Gamma_k \to \gamma \text{ curve} \tag{8-275}$$

$$\lim_{k \to 0} \Gamma_k \to x_1 \text{ axis} \tag{8-276}$$

This confirms our expectations. Since the cost functional is

$$J = \int_0^T [k + |u(t)|]\, dt \tag{8-277}$$

we can see that if $k \to \infty$, then the cost functional weighs only the time, and thus, as $k \to \infty$, Control Law 8-5 reduces to the time-optimal control law (see Control Law 7-1). In other words,

$$
\begin{aligned}
\lim_{k \to \infty} H_1 &= R_- \cup \gamma_- \\
\lim_{k \to \infty} H_3 &= R_+ \cup \gamma_+ \\
\lim_{k \to \infty} H_2 \cup H_4 &= \emptyset \qquad \text{the empty set}
\end{aligned}
\tag{8-278}
$$

On the other hand, if $k \to 0$, then the functional $J$ of Eq. (8-277) reduces to the fuel functional $F$ of Eq. (8-84), and thus Problem 8-5 becomes almost

† This is again a coincidence. The reader should not assume that this situation arises for every second-order system.

equivalent to Problem 8-2; in this case, we have

$$\begin{aligned}\lim_{k\to 0} H_1 &= R_1\\ \lim_{k\to 0} H_3 &= R_3\\ \lim_{k\to 0} H_2 \cup H_4 &= R_2 \cup R_4\end{aligned} \tag{8-279}$$

where $R_1$, $R_2$, $R_3$, and $R_4$ have been defined in Definition 8-1, Eqs. (8-109) to (8-112), and Fig. 8-3. In this case ($k \to 0$), Control Law 8-5 is almost the same as Control Law 8-2. The reader should note that as $k \to 0$, one automatically obtains the fuel-optimal control (whenever it exists) which requires the least response time.† In other words, as $k \to 0$, Control Law 8-5 *does not* bring out the fact that the fuel-optimal controls are not necessarily unique.

**Exercise 8-22** Compute the minimum cost $J^*$ as a function of $(x_1, x_2)$; show that it is the solution to the Hamilton-Jacobi equation. Let $k = 1$ and plot the set of states that require the same cost to the origin (use $J^* = \frac{1}{2}$, 1, and 2). Are these sets the boundaries of convex sets?

**Exercise 8-23** Determine the response time $T$ and the consumed fuel $F$ as a function of the state. Plot the fuel-to-response-time ratio for initial states on the $x_1$ axis, using $k = 1$.

**Exercise 8-24** Determine the relationship of $k$ in Problem 8-5 to $\beta$ in Problem 8-4 so that Control Laws 8-4 and 8-5 are identical.

**Exercise 8-25** Consider the system (8-233) and the cost functional (8-234) with $T$ free. Suppose that the terminal state is $(\theta_1, \theta_2)$, not necessarily the origin. Assuming that an optimum exists, find the optimal control as a function of the state. Comment on the uniqueness of the optimal and extremal controls. Determine the switch curves when $\theta_1 = 1$, $\theta_2 = 1$, using the value $k = 2$. Plot some equal-cost curves. Is the minimum-cost surface a continuous function of the state?

**Exercise 8-26** Consider the system (8-233) and the cost functional (8-234) with $T$ free and $k = 2$. Also consider the target set (circle)

$$S = \{(x_1, x_2): x_1^2 + x_2^2 = 1\}$$

Find the optimal control as a function of the state for states outside and inside the target set (circle) $S$.

**Exercise 8-27** Repeat Exercise 8-26 when the target set is made up of two isolated points, that is,

$$S = \{(x_1, x_2): x_1 = -1, x_2 = 0 \text{ or } x_1 = 1, x_2 = 0\}$$

Comment on the uniqueness of the optimal and extremal controls.

## 8-9 Minimization of a Linear Combination of Time and Fuel for the Integral-plus-time-constant Plant‡

In the previous four sections, we have been concerned with the control of the position and of the velocity of a mass in a frictionless environment. In

† This was also the case in Sec. 8-2.

‡ An appropriate reference for this section is K-17.

this section, we shall consider the control of a mass subject to drag forces. We shall assume that the drag (or friction) force is proportional to the velocity of the body. The performance criterion we shall employ is the same as the one used in Sec. 8-7, that is,

$$J = \int_0^T [k + |u(t)|] \, dt$$

The techniques we shall use are very similar to those developed in Sec. 8-8. Our reasons for solving this problem are twofold. First, we shall see that the control law for this problem is of the same type as Control Law 8-5 in the sense that there are two switch curves that divide the plane into four regions. Second, it is our intent to consider the optimal control of a body subject to *nonlinear* drag forces in the following section; we shall illustrate the fundamental differences between the linear and the nonlinear drag cases, although the performance criterion is the same.

Suppose that we are given a body with unit mass. Let $y(t)$ denote the position of the body and let $u(t)$ be the control force. We suppose that the translational motion of the body occurs in a viscous medium, so that the friction force is proportional to the velocity of the body. The differential equation of the system is

$$\ddot{y}(t) + a\dot{y}(t) = u(t) \tag{8-280}$$

where $a$ is a positive constant (the friction coefficient). We shall assume that the control force $u(t)$ is constrained by the relation

$$|u(t)| \leq 1 \tag{8-281}$$

The transfer function of the plant (8-280) is

$$\frac{y(s)}{u(s)} = G(s) = \frac{1}{s(s + a)} \tag{8-282}$$

and this is the reason for calling the system (8-280) an *integral-plus-time-constant plant.*

Let us define the state variables by the equations

$$x_1(t) = y(t) - y_d \qquad x_2(t) = \dot{y}(t) \tag{8-283}$$

where $y_d$ is a desired position. The problem which we shall solve is:

***Problem 8-6*** *Given the system*

$$\begin{aligned} \dot{x}_1(t) &= x_2(t) \\ \dot{x}_2(t) &= -ax_2(t) + u(t) \qquad |u(t)| \leq 1 \end{aligned} \tag{8-284}$$

*Find the control which forces the system* (8-284) *from any initial state* $(\xi_1, \xi_2)$ *to* (0, 0) *and which minimizes the cost functional*

$$J(u) = J = \int_0^T [k + |u(t)|]\, dt \tag{8-285}$$

*where*

$$k > 0$$

*and the response time* $T$ *is not specified.*

Let us note that the time-optimal control of the system (8-280) was considered in Exercise 7-32.†

The Hamiltonian $H$ for this problem is

$$H = k + |u(t)| + x_2(t)p_1(t) - ax_2(t)p_2(t) + u(t)p_2(t) \tag{8-286}$$

The costate variables $p_1(t)$ and $p_2(t)$ are the solutions of the equations

$$\dot{p}_1(t) = -\frac{\partial H}{\partial x_1(t)} = 0 \tag{8-287}$$

$$\dot{p}_2(t) = -\frac{\partial H}{\partial x_2(t)} = -p_1(t) + ap_2(t) \tag{8-288}$$

We let

$$\pi_1 = p_1(0) \qquad \pi_2 = p_2(0) \tag{8-289}$$

and we solve Eqs. (8-287) and (8-288) to obtain

$$p_1(t) = \pi_1 = \text{const} \tag{8-290}$$

$$p_2(t) = \left(\pi_2 - \frac{\pi_1}{a}\right) e^{at} + \frac{\pi_1}{a} \tag{8-291}$$

The control which absolutely minimizes the Hamiltonian is given by

$$u(t) = 0 \qquad \text{if } |p_2(t)| < 1 \tag{8-292}$$

$$u(t) = -\operatorname{sgn}\{p_2(t)\} \qquad \text{if } |p_2(t)| > 1 \tag{8-293}$$

$$0 \le u(t) \le 1 \qquad \text{if } p_2(t) = -1 \tag{8-294}$$

$$-1 \le u(t) \le 0 \qquad \text{if } p_2(t) = +1 \tag{8-295}$$

***Lemma 8-19*** *The condition* $|p_2(t)| = 1$ *for all* $t \in [t_1, t_2]$ *leads to controls which cannot be fuel-optimal; therefore, there are no singular controls for this problem, and so the problem is normal.*

**Exercise 8-28** Prove Lemma 8-19. HINT: Mimic the steps of the proof of Lemma 8-17 to show that $|p_2(t)| = 1$ for all $t \in [t_1, t_2]$ implies that $H = k > 0$.

Since singular controls cannot occur and since the function $p_2(t)$ is the sum of a constant and of an increasing (in magnitude) exponential, it follows

† The reader should note that the state variables $x_1(t)$ and $x_2(t)$ in Eq. (8-284) are not the same as the state variables used in Eq. (7-272).

from Eq. (8-291) that, for $t \in [0, T]$,

$$\begin{aligned}
&\text{If } \pi_2 = \frac{\pi_1}{a}, \text{ then } p_2(t) \text{ is constant} \\
&\text{If } \pi_2 > \frac{\pi_1}{a}, \text{ then } p_2(t) \text{ is monotonically increasing} \\
&\text{If } \pi_2 < \frac{\pi_1}{a}, \text{ then } p_2(t) \text{ is monotonically decreasing}
\end{aligned} \tag{8-296}$$

We therefore conclude from Eqs. (8-296), (8-292), and (8-293) that only the nine control sequences

$$\{0\},\ \{+1\},\ \{-1\},\ \{0,\ -1\},\ \{0,\ +1\},\ \{+1,\ 0\},\ \{-1,\ 0\},\ \{+1,\ 0,\ -1\},\ \{-1,\ 0,\ +1\} \tag{8-297}$$

can be candidates for optimal control.

***Lemma 8-20*** *If the origin* (0, 0) *is the desired terminal state, then only the six control sequences*

$$\{+1\},\ \{-1\},\ \{0,\ +1\},\ \{0,\ -1\},\ \{+1,\ 0,\ -1\},\ \{-1,\ 0,\ +1\} \tag{8-298}$$

*can be candidates for optimal control.*

**Exercise 8-29** Prove Lemma 8-20. HINT: Mimic the proof of Lemma 8-18.

At this stage of development, it is necessary to solve the system equations (8-284) for the controls

$$u(t) = +1 \qquad u(t) = -1 \qquad u(t) = 0 \tag{8-299}$$

We let

$$\xi_1 = x_1(0) \qquad \xi_2 = x_2(0) \tag{8-300}$$

and we solve Eqs. (8-284) to obtain the equations

$$\left.\begin{aligned} x_2(t) &= \xi_2 e^{-at} \\ x_1(t) &= \xi_1 + \frac{\xi_2}{a}(1 - e^{-at}) \end{aligned}\right\} \quad \text{for } u(t) = 0 \tag{8-301}$$

$$\left.\begin{aligned} x_2(t) &= \left(\xi_2 - \frac{\Delta}{a}\right)e^{-at} + \frac{\Delta}{a} \\ x_1(t) &= \xi_1 + \frac{\Delta}{a}t + \frac{1}{a^2}(\Delta - a\xi_2)e^{-at} + \frac{1}{a}\left(\xi_2 - \frac{\Delta}{a}\right) \end{aligned}\right\} \quad \text{for } u(t) = \Delta = \pm 1 \tag{8-302}$$

If we eliminate the time in Eqs. (8-301) and (8-302), then we shall obtain the equations of the trajectories in the state plane.

***Definition 8-4*** *Let $T^0(\xi_1, \xi_2, x_1, x_2)$ denote the trajectory originating at $(\xi_1, \xi_2)$, generated by the control $u(t) = 0$, and passing through the point $(x_1, x_2)$. The equation of the $T^0(\xi_1, \xi_2, x_1, x_2)$ trajectory is obtained by eliminating the* *time t in Eqs.* (8-301). *Thus, we have the equation*

$$x_1 = \xi_1 + \frac{1}{a}(\xi_2 - x_2) \tag{8-303}$$

*for* $T^0(\xi_1, \xi_2, x_1, x_2)$.

***Definition 8-5*** *Let* $T^+(\xi_1, \xi_2, x_1, x_2)$ *denote the trajectory originating at* $(\xi_1, \xi_2)$, *generated by the control* $u(t) = +1$, *and passing through the point* $(x_1, x_2)$. *From Eqs.* (8-302), *using* $\Delta = +1$, *we deduce that the equation of* $T^+(\xi_1, \xi_2, x_1, x_2)$ *is*

$$x_1 = \xi_1 + \frac{1}{a^2}\log\frac{a\xi_2 - 1}{ax_2 - 1} + \frac{1}{a}(\xi_2 - x_2) \tag{8-304}$$

***Definition 8-6*** *Let* $T^-(\xi_1, \xi_2, x_1, x_2)$ *denote the trajectory originating at* $(\xi_1, \xi_2)$, *generated by the control* $u(t) = -1$, *and passing through the point* $(x_1, x_2)$. *From Eqs.* (8-302), *using* $\Delta = -1$, *we deduce that the equation of* $T^-(\xi_1, \xi_2, x_1, x_2)$ *is*

$$x_1 = \xi_1 - \frac{1}{a^2}\log\frac{a\xi_2 + 1}{ax_2 + 1} + \frac{1}{a}(\xi_2 - x_2) \tag{8-305}$$

Various $T^0$, $T^+$, and $T^-$ trajectories are shown in Figs. 8-17, 8-18, and 8-19, respectively, using the value $a = 1$.

Now we are ready to proceed with the construction of the switch curves and with the formulation of the optimal-control law. We know that (by Lemma 8-20) the control sequence $\{-1, 0, +1\}$ is a candidate for optimal control. The control sequence $\{-1, 0, +1\}$ means that the control is

$$\begin{aligned} u(t) &= -1 \qquad \text{for } 0 \le t < t_1 \\ u(t) &= 0 \qquad \text{for } t_1 \le t < t_2 \\ u(t) &= +1 \qquad \text{for } t_2 \le t \le T \end{aligned} \tag{8-306}$$

Figure 8-20 shows an initial state $\Xi = (\xi_1, \xi_2)$ which can be forced to the origin $O = (0, 0)$ by the control (8-306). We let

$$\begin{aligned} x_1(t_1) &= z_1 \qquad x_2(t_1) = z_2 \qquad Z = (z_1, z_2) \\ x_1(t_2) &= w_1 \qquad x_2(t_2) = w_2 \qquad W = (w_1, w_2) \end{aligned} \tag{8-307}$$

Since $u(t) = -1$ for $t \in [0, t_1)$, we know that the trajectory joining $\Xi = (\xi_1, \xi_2)$ and $Z = (z_1, z_2)$ is $T^-(\xi_1, \xi_2, z_1, z_2)$. Since $u(t) = 0$ for $t \in [t_1, t_2)$, we know that the trajectory joining $Z = (z_1, z_2)$ and $W = (w_1, w_2)$ is $T^0(z_1, z_2, w_1, w_2)$. Finally, since $u(t) = +1$ for $t \in [t_2, T]$, we know that the trajectory joining $W = (w_1, w_2)$ and the origin $O = (0, 0)$ is $T^+(w_1, w_2, 0, 0)$.

If we examine the set of all states for which the control sequence $\{-1, 0, +1\}$ is optimal, then we shall obtain the locus of the points $(w_1, w_2)$ and the locus of the points $(z_1, z_2)$.

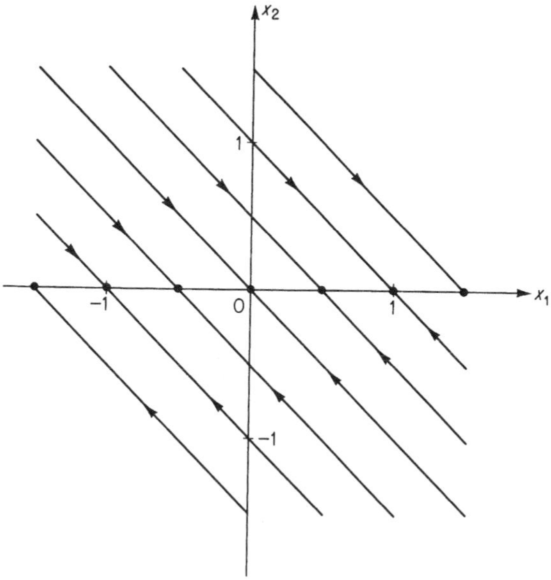


**Fig. 8-17** The $T^0$ trajectories of the system (8-284) when $u(t) = 0$ and $a = 1$.

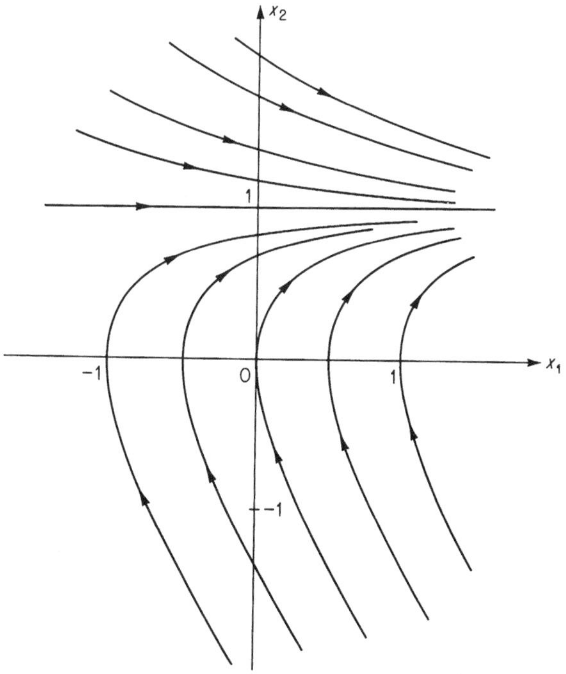


**Fig. 8-18** The $T^+$ trajectories of the system (8-284) when $u(t) = +1$ and $a = 1$.

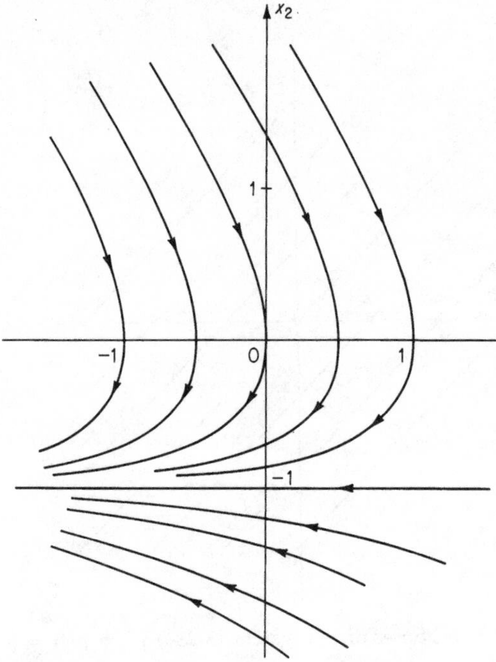


Fig. 8-19 The $T^-$ trajectories of the system (8-284) when $u(t) = -1$ and $a = 1$.

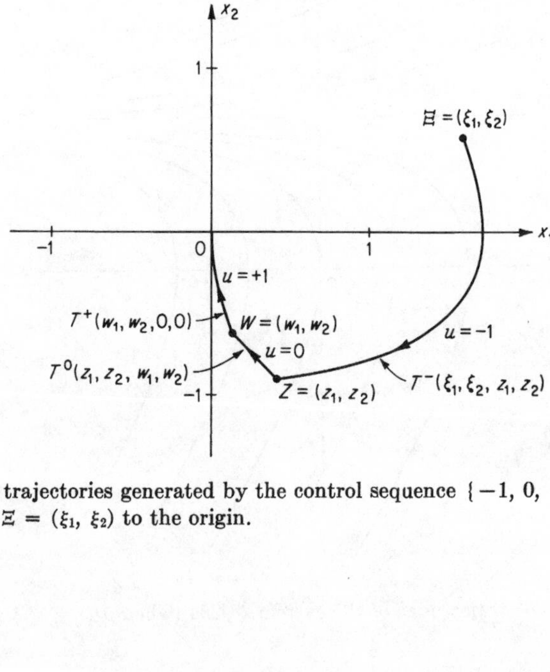


Fig. 8-20 The trajectories generated by the control sequence $\{-1, 0, +1\}$ that forces the initial state $\Xi = (\xi_1, \xi_2)$ to the origin.

Let us determine the locus of $(w_1, w_2)$. Since $(w_1, w_2) \in T^+(w_1, w_2, 0, 0)$, substitution of the values

$$\xi_1 = w_1 \qquad \xi_2 = w_2 \qquad x_1 = 0 \qquad x_2 = 0 \tag{8-308}$$

into Eq. (8-304) will yield the equation

$$w_1 = -\frac{1}{a} w_2 - \frac{1}{a^2} \log (1 - a w_2) \tag{8-309}$$

The physical requirement that $(w_1, w_2)$ be forced to $(0, 0)$ in positive time implies that

$$w_2 \leq 0 \tag{8-310}$$

in Eq. (8-309). If we define the curve $\gamma_+$ by the relation

$$\gamma_+ = \left\{ (x_1, x_2) : x_1 = -\frac{1}{a} x_2 - \frac{1}{a^2} \log (1 - a x_2);\ x_2 \leq 0 \right\} \tag{8-311}$$

then $(w_1, w_2) \in \gamma_+$.

We wish next to determine the locus of the states $(z_1, z_2)$. The states $(w_1, w_2)$ and $(z_1, z_2)$ are connected by the $T^0(z_1, z_2, w_1, w_2)$ trajectory. Thus, substituting the values

$$\xi_1 = z_1 \qquad \xi_2 = z_2 \qquad x_1 = w_1 \qquad x_2 = w_2 \tag{8-312}$$

into Eq. (8-303), we find that

$$w_1 = z_1 + \frac{1}{a} (z_2 - w_2) \tag{8-313}$$

Substituting Eq. (8-309) into Eq. (8-313), we obtain the equation

$$z_1 = -\frac{z_2}{a} - \frac{1}{a^2} \log (1 - a w_2) \tag{8-314}$$

If we can express $w_2$ in terms of $z_2$, then Eq. (8-314) will provide us with the equation of the locus of $(z_1, z_2)$. To obtain such a relationship, we use the information that the Hamiltonian is zero along the optimal trajectory.

At $t = t_1$, we have

$$x_1(t_1) = z_1 \qquad x_2(t_1) = z_2 \qquad p_2(t_1) = +1 \qquad u(t_1) \leq 0 \tag{8-315}$$

Because the control (8-306) switches from $u(t) = -1$ to $u(t) = 0$ at $t = t_1$, $p_2(t_1) = +1$ according to Eqs. (8-292) and (8-293). We now substitute Eqs. (8-315) into Eq. (8-286) to find that, at $t = t_1$, the relation

$$H \big|_{t=t_1} = 0 = k + z_2 \pi_1 - a z_2 \tag{8-316}$$

must hold. At $t = t_2$, we have

$$x_1(t_2) = w_1 \qquad x_2(t_2) = w_2 \qquad p_2(t_2) = -1 \qquad u(t_2) \geq 0 \tag{8-317}$$

Substituting Eqs. (8-317) into Eq. (8-286), we find that

$$H\big|_{t=t_2} = 0 = k + w_2\pi_1 + aw_2 \tag{8-318}$$

We eliminate $\pi_1$ from Eqs. (8-316) and (8-318) to obtain the equation

$$w_2 = \frac{kz_2}{k - 2az_2} \tag{8-319}$$

This is the sought-for relation between $w_2$ and $z_2$. We now substitute Eq. (8-319) into Eq. (8-314), and we find that

$$z_1 = -\frac{z_2}{a} - \frac{1}{a^2}\log\left(1 - \frac{akz_2}{k - 2az_2}\right) \tag{8-320}$$

The requirement that $(z_1, z_2)$ be transferred to $(w_1, w_2)$ in positive time implies that

$$z_2 \leq 0 \tag{8-321}$$

in Eq. (8-320). Thus, if we define the $\Gamma_k^+$ curve by the relation

$$\Gamma_k^+ = \left\{(x_1, x_2): x_1 = -\frac{x_2}{a} - \frac{1}{a^2}\log\left(1 - \frac{akx_2}{k - 2ax_2}\right); x_2 \leq 0\right\} \tag{8-322}$$

then $(z_1, z_2)$ is an element of $\Gamma_k^+$; that is,

$$(z_1, z_2) \in \Gamma_k^+ \tag{8-323}$$

We can repeat the development for the set of initial states $(\xi_1', \xi_2')$ for which the control sequence $\{+1, 0, -1\}$ is optimal. Thus, if the control

$$\begin{aligned} u(t) &= +1 && \text{for } 0 \leq t < t_1' \\ u(t) &= 0 && \text{for } t_1' \leq t < t_2' \\ u(t) &= -1 && \text{for } t_2' \leq t \leq T \end{aligned} \tag{8-324}$$

is optimal and if we let

$$x_1(t_1') = z_1' \qquad x_2(t_1') = z_2' \qquad x_1(t_2') = w_1' \qquad x_2(t_2') = w_2' \tag{8-325}$$

we can find, using similar techniques, that the locus of $(w_1', w_2')$ is given by the equation

$$w_1' = -\frac{1}{a}w_2' + \frac{1}{a^2}\log(1 + aw_2') \tag{8-326}$$

with

$$w_2' \geq 0 \tag{8-327}$$

and that the locus of $(z_1', z_2')$ is given by the equation

$$z_1' = -\frac{z_2'}{a} + \frac{1}{a^2}\log\left(1 + \frac{akz_2'}{k + 2az_2'}\right) \tag{8-328}$$

with

$$z_2' \geq 0 \tag{8-329}$$

We define the $\gamma_-$ curve by the relation

$$\gamma_- = \left\{(x_1, x_2): x_1 = -\frac{1}{a}x_2 + \frac{1}{a^2}\log(1 + ax_2);\ x_2 \geq 0\right\} \tag{8-330}$$

and the $\Gamma_k^-$ curve by the relation

$$\Gamma_k^- = \left\{(x_1, x_2): x_1 = -\frac{x_2}{a} + \frac{1}{a^2}\log\left(1 + \frac{akx_2}{k + 2ax_2}\right);\ x_2 \geq 0\right\} \tag{8-331}$$

and we conclude that

$$(w_1', w_2') \in \gamma_- \tag{8-332}$$
$$(z_1', z_2') \in \Gamma_k^- \tag{8-333}$$

Let $\gamma$ be the switch curve defined by the relation

$$\gamma = \gamma_+ \cup \gamma_- \tag{8-334}$$

Then, from Eqs. (8-311) and (8-330), we deduce that

$$\gamma = \{(x_1, x_2): x_1 = f_1(x_2)\} \tag{8-335}$$

where the function $f_1(x_2)$ is defined by

$$f_1(x_2) = -\frac{x_2}{a} + \operatorname{sgn}\{x_2\}\frac{1}{a^2}\log(1 + a|x_2|) \tag{8-336}$$

Let $\Gamma_k$ be the switch curve defined by the relation

$$\Gamma_k = \Gamma_k^+ \cup \Gamma_k^- \tag{8-337}$$

Then, from Eqs. (8-322) and (8-331), we obtain

$$\Gamma_k = \{(x_1, x_2): x_1 = f_2(x_2)\} \tag{8-338}$$

where

$$f_2(x_2) = -\frac{x_2}{a} + \operatorname{sgn}\{x_2\}\frac{1}{a^2}\log\left(1 + \frac{ak|x_2|}{k + 2a|x_2|}\right) \tag{8-339}$$

The $\gamma$ curve and the $\Gamma_k$ curve separate the state plane into four regions $H_1$, $H_2$, $H_3$, and $H_4$, as shown in Fig. 8-21. The precise definition of these sets is

$$H_1 = \{(x_1, x_2): x_1 \geq f_1(x_2) \text{ and } x_1 > f_2(x_2)\} \tag{8-340}$$
$$H_2 = \{(x_1, x_2): x_1 < f_1(x_2) \text{ and } x_1 \geq f_2(x_2)\} \tag{8-341}$$
$$H_3 = \{(x_1, x_2): x_1 \leq f_1(x_2) \text{ and } x_1 < f_2(x_2)\} \tag{8-342}$$
$$H_4 = \{(x_1, x_2): x_1 > f_1(x_2) \text{ and } x_1 \leq f_2(x_2)\} \tag{8-343}$$

where $f_1(x_2)$ and $f_2(x_2)$ are given by Eqs. (8-336) and (8-339), respectively. Note that the set $H_1$ includes the $\gamma_-$ curve, the set $H_2$ includes the $\Gamma_k^-$ curve, the set $H_3$ includes the $\gamma_+$ curve, and the set $H_4$ includes the $\Gamma_k^+$ curve.

We shall prove the following:

***Control Law 8-6 Solution to Problem 8-6*** *The optimal control $u^*$, as a function of the state $(x_1, x_2)$, is given by*

$$\begin{aligned} u^* &= u^*(x_1, x_2) = -1 \qquad &&\text{for all } (x_1, x_2) \in H_1 \\ u^* &= u^*(x_1, x_2) = +1 &&\text{for all } (x_1, x_2) \in H_3 \\ u^* &= u^*(x_1, x_2) = 0 &&\text{for all } (x_1, x_2) \in H_2 \cup H_4 \end{aligned} \tag{8-344}$$

*and, furthermore, $u^*$ is unique.*

PROOF First of all, we assert that an optimal control exists. Next, we shall use the by now familiar technique of elimination to prove that the control (8-344) is indeed optimal.

Let us suppose that the initial state $(x_1, x_2) \in \gamma_-$ (note that $\gamma_- \subset H_1$). With the aid of Figs. 8-17 to 8-19, we test every control sequence given by (8-298), and we conclude that *only* the control sequence $\{-1\}$ can force $(x_1, x_2) \in \gamma_-$ to (0, 0) along the $\gamma_-$ path. Thus, by elimination, the optimal control is $u(t) = -1$ if $(x_1, x_2) \in \gamma_-$.

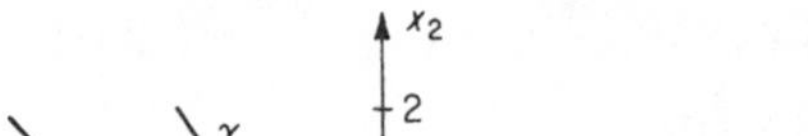

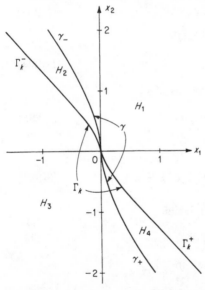

**Fig. 8-21** The switch curves $\gamma_+ \cup \gamma_-$ and $\Gamma_k^+ \cup \Gamma_k^-$ divide the state plane into the regions $H_1$, $H_2$, $H_3$, and $H_4$ $(a = 1)$.

Now let us suppose that $(x_1, x_2) \in H_1 - \gamma_-$;† with the aid of Figs. 8-17 to 8-19, we try each of the control sequences of (8-298), and we conclude that only the control sequence $\{-1, 0, +1\}$ can force the state $(x_1, x_2) \in H_1 - \gamma_-$ to the origin. As a matter of fact, the control must switch from $u(t) = -1$ to $u(t) = 0$ when the trajectory intersects the $\Gamma_k$ curve (it will actually intersect the $\Gamma_k^+$ portion of the $\Gamma_k$ curve), and the control must switch from $u(t) = 0$ to $u(t) = +1$ when the trajectory intersects the $\gamma$ curve (it will actually intersect the $\gamma_+$ curve). The reason that the switchings occur at the $\Gamma_k$ and $\gamma$ curves should be clear from the construction employed.

Let us now suppose that the initial state $(x_1, x_2) \in H_4$. With the aid of Figs. 8-17 to 8-19, we conclude that the state $(x_1, x_2) \in H_4$ can be forced

† $H_1 - \gamma_-$ means the set $H_1$ excluding the $\gamma_-$ curve; see Sec. 2-3.

to (0, 0) either by the control sequence $\{0, +1\}$ or by the control sequence $\{-1, 0, +1\}$. We claim that if $(x_1, x_2) \in H_4$, then the sequence $\{-1, 0, +1\}$ is *not* optimal, and by elimination, the control sequence $\{0, +1\}$ *must* be optimal. Suppose that the control sequence $\{-1, 0, +1\}$ were optimal; then the control would switch from $u(t) = -1$ to $u(t) = 0$ at the $\Gamma_k^+$ curve. But this is impossible, as the following reasoning will show. Consider the initial state $A = (\xi_1, \xi_2) \in H_4$ shown in Fig. 8-22; the trajectory $ABO$ is the trajectory generated by the control sequence $\{0, +1\}$; clearly, the trajectory $AB$ does not intersect the $\Gamma_k^+$ curve (by construction). If we can show that the trajectory $T^-$, generated by the control $u(t) = -1$ and originating at $A = (\xi_1, \xi_2)$, does not intersect the $\Gamma_k^+$ curve, then we can conclude that the control sequence $\{-1, 0, +1\}$ is not optimal. The equation of the trajectory $AB$ is [using Eq. (8-303)]

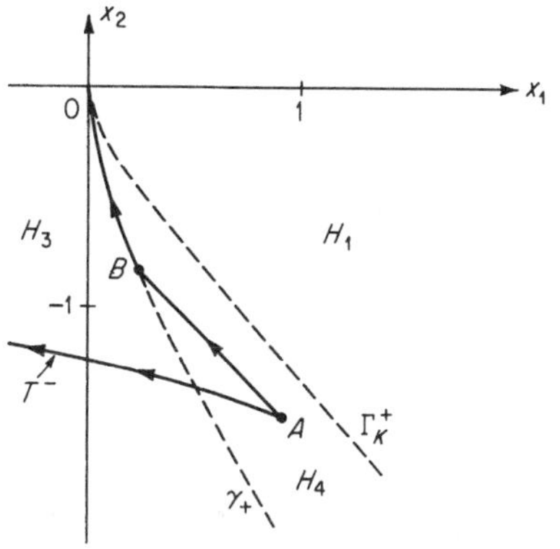

**Fig. 8-22** Illustration of trajectories used in the proof of Control Law 8-6.

$$x_1 = \xi_1 + \frac{\xi_2}{a} - \frac{x_2}{a} \tag{8-345}$$

The trajectory $T^-$, originating at $(\xi_1, \xi_2)$ and due to the control $u(t) = -1$, is [using Eq. (8-305)]

$$x_1 = \xi_1 + \frac{\xi_2}{a} - \frac{1}{a} x_2 - \frac{1}{a^2} \log \frac{a\xi_2 + 1}{ax_2 + 1} \tag{8-346}$$

But, since the trajectories move in positive time, we can show that

$$\frac{a\xi_2 + 1}{ax_2 + 1} > 1$$

which implies that

$$-\frac{1}{a^2} \log \frac{a\xi_2 + 1}{ax_2 + 1} < 0 \tag{8-347}$$

From Eqs. (8-346) and (8-347), we conclude that

$$x_1 < \xi_1 + \frac{\xi_2}{a} - \frac{1}{a} x_2 \tag{8-348}$$

Equations (8-348) and (8-345) imply that the trajectory $T^-$, generated by $u(t) = -1$, is to the left of the trajectory $AB$. Since $AB$ does not intersect the $\Gamma_k^+$ curve, it follows that $T^-$ does not intersect the $\Gamma_k^+$ curve, and so the control sequence $\{-1, 0, +1\}$ is not optimal. Similar arguments can be used to prove Control Law 8-6 for initial states in $H_2 \cup H_3$.

Let us examine the behavior of the optimal system as $k \to 0$ and as $k \to \infty$. We note that the $\gamma$ curve is independent of $k$ [because the function $f_1(x_2)$ given by Eq. (8-336) is independent of $k$]. However, the $\Gamma_k$ curve does depend on $k$ because the function $f_2(x_2)$ given by Eq. (8-339) is dependent on $k$. We note that

$$\lim_{k\to\infty} f_2(x_2) = -\frac{x_2}{a} + \operatorname{sgn}\{x_2\}\frac{1}{a^2}\log(1 + a|x_2|) = f_1(x_2) \quad (8\text{-}349)$$

$$\lim_{k\to 0} f_2(x_2) = -\frac{x_2}{a} \quad (8\text{-}350)$$

Equation (8-349) means that as $k \to 0$, the $\Gamma_k$ curve collapses to the $\gamma$ curve. Therefore, we deduce that

$$\lim_{k\to\infty} \Gamma_k = \gamma \quad (8\text{-}351)$$

$$\lim_{k\to\infty} H_2 \cup H_4 = \emptyset \qquad \text{(the empty set)} \quad (8\text{-}352)$$

and so, as $k \to \infty$, Control Law 8-6 reduces to the time-optimal control law for the system (8-284) (compare with Exercise 7-32).

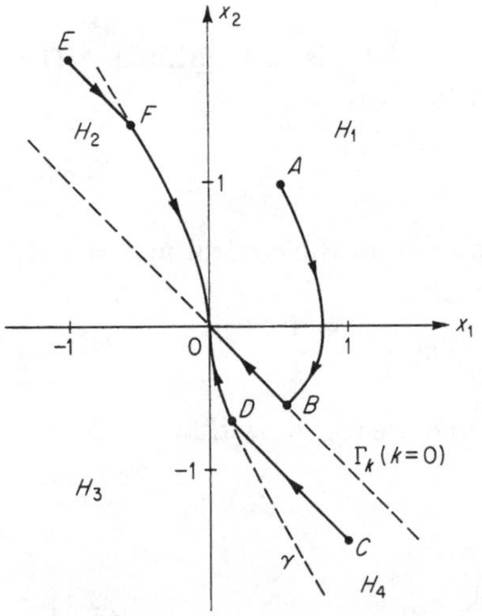


**Fig. 8-23** The limiting behavior of the switch curves as $k \to 0$ and $a = 1$ (the fuel-optimal case with no constraints on the response time). The trajectories $ABO$ and $CDO$ are fuel-optimal to the origin.

As $k \to 0$, the $\Gamma_k$ curve becomes the straight line

$$x_1 = -\frac{x_2}{a} \quad (8\text{-}353)$$

Since, for small values of $k$, the cost functional $J = \int_0^T [k + |u(t)|]\, dt$ weighs mostly the fuel, we claim that as $k \to 0$, we obtain the system which is the fuel-optimal one with no constraints on the response time. Figure 8-23 shows the division of the state plane into the regions $H_1$, $H_2$, $H_3$, and $H_4$ for $k = 0$ and some of the fuel-optimal trajectories. Note that for initial states in $H_1$, the control $u = -1$ is applied until the trajectory hits the $\Gamma_k$ $(k = 0)$ curve, which is itself a trajectory to the origin (see Fig. 8-17). The response time to reach the origin is infinite, but the time required to reach a small neighborhood about the origin is finite.†

† Strictly speaking, this means that a fuel-optimal control does not exist; however, an $\epsilon$-optimal control exists.

Figure 8-24 shows the engineering realization of Control Law 8-6. The signal $x_2$ is introduced to the two nonlinearities $N_1$ and $N_2$. The nonlinearity $N_1$ has the input-output characteristics of the $\gamma$ curve, and its output is the signal $f_1(x_2)$ given by Eq. (8-336). The nonlinearity $N_2$ has the input-output characteristics of the $\Gamma_k$ curve, and its output is the signal

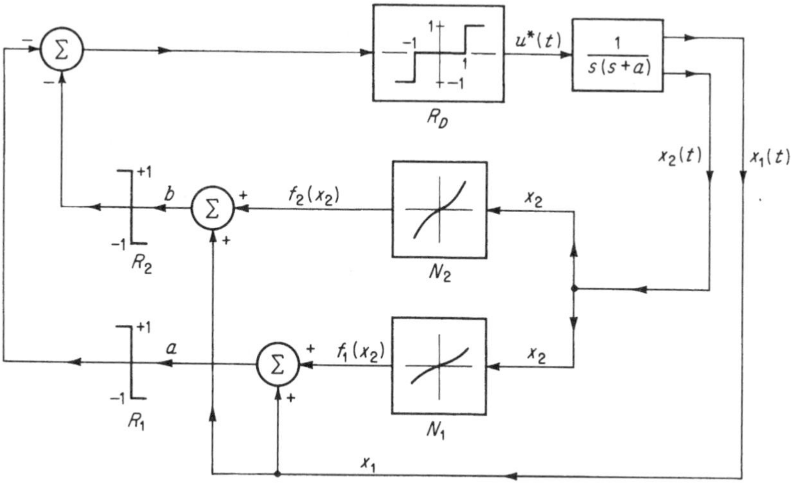


**Fig. 8-24** A realization of Control Law 8-6.

$f_2(x_2)$ given by Eq. (8-339). From Eqs. (8-340) to (8-343), we deduce that the statements

$$\begin{array}{lll} a > 0 & \text{means} & (x_1, x_2) \in H_1 \cup H_4 \\ a < 0 & \text{means} & (x_1, x_2) \in H_2 \cup H_3 \\ b > 0 & \text{means} & (x_1, x_2) \in H_1 \cup H_2 \\ b < 0 & \text{means} & (x_1, x_2) \in H_3 \cup H_4 \end{array} \tag{8-354}$$

are true. The signals $a$ and $b$ are introduced to the relays $R_1$ and $R_2$ with outputs sgn $\{a\}$ and sgn $\{b\}$, respectively; thus,

$$\begin{array}{lll} -\operatorname{sgn}\{a\} - \operatorname{sgn}\{b\} = +2 & \text{means} & (x_1, x_2) \in H_3 \\ -\operatorname{sgn}\{a\} - \operatorname{sgn}\{b\} = -2 & \text{means} & (x_1, x_2) \in H_1 \\ -\operatorname{sgn}\{a\} - \operatorname{sgn}\{b\} = 0 & \text{means} & (x_1, x_2) \in H_2 \cup H_4 \end{array} \tag{8-355}$$

The nonlinear element $R_D$ delivers the optimal control according to Control Law 8-6.

Note that as $k \to \infty$, the gain of the nonlinearity $N_2$ goes to zero (the nonlinearity $N_1$ is not affected) and the system of Fig. 8-24 becomes the time-optimal one. As $k \to 0$, the nonlinearity $N_2$ reduces to a linear gain element (with gain $-1/a$) and the system of Fig. 8-24 reduces to the fuel-optimal system with no constraints on the response time.

If we compare the system of Fig. 8-24 with the system of Fig. 8-13 (which is the realization of Control Laws 8-4 and 8-5 for the double-integral plant), we observe that we require two nonlinearities in the system of Fig. 8-24 and only one in Fig. 8-13.†

It is possible to design a *suboptimal* system using only the nonlinearity $N_1$ in Fig. 8-24. This can be done because we can *approximate* the $\Gamma_k$ curve in Fig. 8-21 by a curve $\hat{\Gamma}_k$ which has the same equation as the $\gamma$ curve. In other words, we know that the $\gamma$ curve is defined by the relation

$$\gamma = \{(x_1, x_2): x_1 = f_1(x_2)\} \tag{8-356}$$

where $f_1(x_2)$ is given by Eq. (8-336). Let us define a new curve $\hat{\Gamma}_k$ by

$$\hat{\Gamma}_k = \{(x_1, x_2): x_1 = \lambda f_1(x_2);\ \lambda > 1\} \tag{8-357}$$

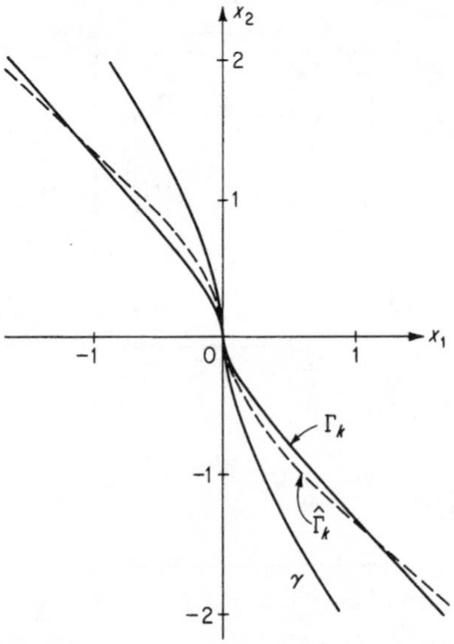

**Fig. 8-25** The approximation of the $\Gamma_k$ switch curve by the $\hat{\Gamma}_k$ curve.

We claim that we can choose the constant $\lambda$ so that the $\hat{\Gamma}_k$ curve is "close"‡ to the $\Gamma_k$ curve. In Fig. 8-25, we show the $\gamma$ curve and the $\Gamma_k$ curve for $a = 1$ and $k = 1$; we have also drawn the $\hat{\Gamma}_k$ curve using the value $\lambda = 2$. If we use the $\gamma$ curve and the $\hat{\Gamma}_k$ curve as the switching curves, the system shown in Fig. 8-26 will be the suboptimal one. The difference between the system of Fig. 8-26 and that of Fig. 8-24 is that, in the former, we use only one nonlinearity ($N_1$) to construct the $\gamma$ and $\hat{\Gamma}_k$ curves.

The choice of the constant $\lambda$ is, of course, facilitated by any additional information at hand. For example, if there is a very high probability that the initial state $(\xi_1, \xi_2)$ will be such that

$$-\Gamma \le \xi_2 \le \Gamma \tag{8-358}$$

then we need only approximate the $\Gamma_k$ curve in the region $-\Gamma \le x_2 \le \Gamma$. The approximation of $\Gamma_k$ by $\hat{\Gamma}_k$ is better for high values of $k$ than for low values of $k$. If the $\hat{\Gamma}_k$ curve is between the $\gamma$ curve and the $\Gamma_k$ curve, then the suboptimal system will require less time and more fuel than the optimal one (why?). If, on the other hand, the $\hat{\Gamma}_k$ curve is entirely in $H_1 \cup H_3$,

† The reason is that the equations of the $\gamma_\beta$ curve in Fig. 8-12 and of the $\Gamma_k$ curve in Fig. 8-16 are of the same type as the equation of the $\gamma$ curve.

‡ In an engineering sense; we do not claim any mathematical rigor here.

then the suboptimal system will require less fuel but more time than the optimal one (why?).

The above comments on the construction of the suggested suboptimal system illustrate, in our opinion, the usefulness of obtaining the optimal design and from it constructing a suboptimal design that may be more "appealing" to the engineer.

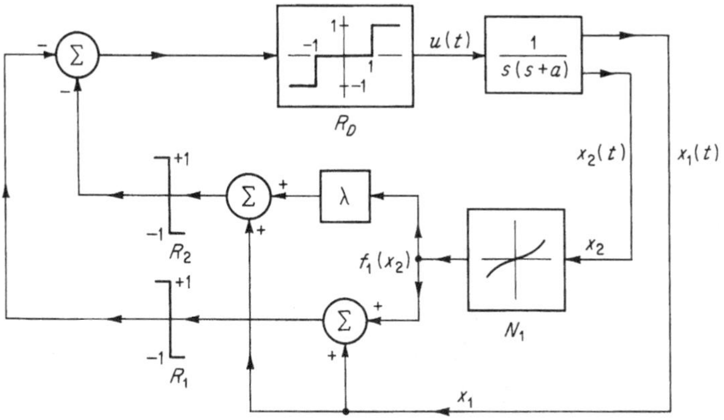


**Fig. 8-26** The realization of the suboptimal control law obtained by replacing the $\Gamma_k$ curve with the $\hat{\Gamma}_k$ curve.

**Exercise 8-30** Consider the approximating $\hat{\Gamma}_k$ curve indicated in Fig. 8-25; that is, use $a = 1$, $k = 1$, and $\lambda = 2$. Consider initial states $(2, \xi_2)$ with $-3 \le \xi_2 \le 3$. Plot, versus $\xi_2$,

(*a*) The ratio of the response time required by the suboptimal system to that of the optimal system

(*b*) The ratio of the consumed fuel required by the suboptimal system to that of the optimal system

(*c*) The ratio of the cost $\int_0^T [1 + |u(t)|]\, dt$ required by the suboptimal system to that of the optimal system

**Exercise 8-31** Consider the system

$$\begin{aligned} \dot{x}_1(t) &= x_2(t) \\ \dot{x}_2(t) &= -x_2(t) + u(t) \qquad |u(t)| \le 1 \end{aligned} \tag{8-359}$$

and the fuel functional $F = \int_0^T |u(t)|\, dt$. Suppose that $T$ is not specified. Derive the fuel-optimal control. Comment on existence and uniqueness. Compare with the results obtained as $k \to 0$ (see Fig. 8-23).

**Exercise 8-32** Repeat Exercise 8-31, but assume that $T$ is fixed or bounded by $\hat{T}_f$. Determine the fuel-optimal control as a function of the initial state $(\xi_1, \xi_2)$ and of the specified response time $\hat{T}_f$.

**Exercise 8-33** Suggest other approximations to the $\Gamma_k$ curve and comment on the suboptimal systems obtained.

## 8-10 Minimization of a Linear Combination of Time and Fuel for a Nonlinear Second-order System†

In Sec. 8-9, we studied the problem of a moving mass subject to linear friction. In this section, we shall consider the problem of controlling a moving mass subject to nonlinear friction forces. We shall use the same performance functional as the one used in Secs. 8-8 and 8-9.

We shall demonstrate that the repeated use and logical interpretation of the necessary conditions provided by the minimum principle allow us to obtain the optimal control as a function of the state, using graphical techniques exclusively. We believe that the techniques used for the solution of this problem illustrate, perhaps better than in any other example in the book, both the step-by-step procedure for using and interpreting the necessary conditions and the fact that we can work with "shapes" of functions rather than with explicit equations.

Let us suppose that a unit mass moves in an environment such that the magnitude of the friction or drag force is proportional to the *square* of the velocity. Let $y(t)$ denote the position of the mass at time $t$ and let $u(t)$ be the control thrust. We assume that

$$|u(t)| \leq 1 \tag{8-360}$$

We also assume that the drag force is proportional to the *square* of the velocity $\dot{y}(t)$ and is oppositely directed to the direction of motion. In this case, the motion of the unit mass is governed by the differential equation

$$\ddot{y}(t) = -a\dot{y}(t)|\dot{y}(t)| + u(t) \tag{8-361}$$

where $a$ is the drag coefficient and

$$a > 0 \tag{8-362}$$

The inclusion of $|\dot{y}(t)|$ in the drag force $a\dot{y}(t)|\dot{y}(t)|$ is necessary in order to guarantee that the drag force is always opposite to the direction of motion. Let $y_d$ be the desired position and define, as before, the state variables

$$x_1(t) = y(t) - y_d \qquad x_2(t) = \dot{y}(t) \tag{8-363}$$

The problem we shall solve is the following:

***Problem 8-7*** *Given the system*

$$\begin{aligned} \dot{x}_1(t) &= x_2(t) \\ \dot{x}_2(t) &= -ax_2(t)|x_2(t)| + u(t) \qquad |u(t)| \leq 1 \end{aligned} \tag{8-364}$$

† An appropriate reference for this section is A-14.

*Find the control which forces the system* (8-364) *from any initial state* $(\xi_1, \xi_2)$ *to* (0, 0) *and which minimizes the cost functional*

$$J(u) = J = \int_0^T [k + |u(t)|]\, dt \tag{8-365}$$

*where*

$$k > 0 \tag{8-366}$$

*and the response time* $T$ *is not specified.*

Recall that we considered the time-optimal control of nonlinear systems of the type $\ddot{y}(t) = -f[\dot{y}(t)] + Ku(t)$ in Sec. 7-11; as a matter of fact, we obtained the time-optimal control law for the system (8-364) as indicated in Fig. 7-68 (using the value $a = 1$). We noted in Sec. 7-11 that the time-optimal control of nonlinear second-order systems was fundamentally the same as the time-optimal control of linear second-order systems. One of the reasons for the consideration of the optimal [with respect to the cost (8-365)] control of the nonlinear system (8-364) is to determine whether or not there is a fundamental difference between linear and nonlinear systems when the fuel enters into the performance functional. As we shall see, there may be a difference. We shall exhibit this difference by showing that singular controls are optimal.

We shall use the minimum principle to derive the necessary conditions for the optimal control. The Hamiltonian $H$ for the system (8-364) with the cost (8-365) is

$$H = k + |u(t)| + x_2(t)p_1(t) - ax_2(t)|x_2(t)|p_2(t) + u(t)p_2(t) \tag{8-367}$$

The costate variables $p_1(t)$ and $p_2(t)$ are specified by the differential equations

$$\dot{p}_1(t) = -\frac{\partial H}{\partial x_1(t)} = 0 \tag{8-368}$$

$$\dot{p}_2(t) = -\frac{\partial H}{\partial x_2(t)} = -p_1(t) + 2a|x_2(t)|p_2(t) \tag{8-369}$$

We let

$$\pi_1 = p_1(0) \qquad \pi_2 = p_2(0) \tag{8-370}$$

and we obtain the equations

$$p_1(t) = \pi_1 = \text{const} \tag{8-371}$$

$$\dot{p}_2(t) = -\pi_1 + 2a|x_2(t)|p_2(t) \tag{8-372}$$

The control which minimizes the Hamiltonian $H$, subject to the constraint $|u(t)| \leq 1$, is given by

$$u(t) = 0 \qquad \text{if } |p_2(t)| < 1 \tag{8-373}$$

$$u(t) = -\operatorname{sgn}\{p_2(t)\} \qquad \text{if } |p_2(t)| > 1 \tag{8-374}$$

$$0 \leq u(t) \leq +1 \qquad \text{if } p_2(t) = -1 \tag{8-375}$$

$$-1 \leq u(t) \leq 0 \qquad \text{if } p_2(t) = +1 \tag{8-376}$$

*Now we shall show that singular extremal controls do exist for this problem, provided that the constant $k$ in the cost functional* (8-365) *is less than one, that is,* $k \leq 1$. To do this,† we suppose that $[t_1, t_2] \subset [0, T]$ and that

$$p_2(t) = +1 \qquad \text{for all } t \in [t_1, t_2] \tag{8-377}$$

This implies that

$$\dot{p}_2(t) = 0 \qquad \text{for all } t \in [t_1, t_2] \tag{8-378}$$

Substituting Eqs. (8-377) and (8-378) into Eq. (8-372), we obtain the relation

$$\pi_1 = 2a|x_2(t)| \qquad \text{for all } t \in [t_1, t_2] \tag{8-379}$$

Since $\pi_1$ and $a$ are constants, this means that

$$x_2(t) = \text{const} \qquad \text{for all } t \in [t_1, t_2] \tag{8-380}$$

One may wonder whether $x_2(t)$ could be a piecewise constant function, for example, the function

$$\begin{aligned} x_2(t) &= \frac{\pi_1}{2a} \qquad &\text{for } t_1 \leq t < t_3 \\ x_2(t) &= -\frac{\pi_1}{2a} \qquad &\text{for } t_3 \leq t \leq t_2 \end{aligned} \tag{8-381}$$

such that Eq. (8-379) still holds; this is impossible because $x_2(t)$ is a state variable and, therefore, must be a continuous function of time.‡ Thus, $x_2(t)$ must be a constant. Equation (8-379) also implies that

$$\pi_1 > 0 \tag{8-382}$$

But since $x_2(t) = \text{const}$ for all $t \in [t_1, t_2]$, we must have

$$\dot{x}_2(t) = 0 \qquad \text{for all } t \in [t_1, t_2] \tag{8-383}$$

and hence from Eqs. (8-364), we deduce that

$$u(t) = ax_2(t)|x_2(t)| \qquad \text{for all } t \in [t_1, t_2] \tag{8-384}$$

Since $x_2(t)$ is constant, Eq. (8-384) implies that

$$u(t) = \text{const} \qquad \text{for all } t \in [t_1, t_2] \tag{8-385}$$

Let us review the logical steps we have taken; we started with [Eq. (8-377)] the supposition that $p_2(t) = +1$ for all $t \in [t_1, t_2]$, and then, following the simple path from Eq. (8-378) to (8-385), we arrived at the conclusion that $u(t)$ must be constant for all $t \in [t_1, t_2]$. If this constant control is to be optimal, it must satisfy all the necessary conditions provided by the minimum principle. One of the requirements is that the optimal control must

† We suggest that the reader review the material in Secs. 6-21 and 6-22.
‡ See Sec. 4-5, Axiom 4-3.

minimize the Hamiltonian. This condition is given by Eq. (8-376). Thus, we conclude that

$$-1 \le u(t) = \text{const} \le 0 \qquad \text{for all } t \in [t_1, t_2] \tag{8-386}$$

in view of Eqs. (8-376) and (8-385). The other necessary condition is that the Hamiltonian (8-367) must be identically zero† for all $t \in [0, T]$; consequently, the Hamiltonian must be zero for all $t \in [t_1, t_2]$. We substitute Eqs. (8-377), (8-379), and (8-384) into Eq. (8-367), and we find that the equation

$$H = 0 = k + |u(t)| + 2ax_2(t)|x_2(t)| - ax_2(t)|x_2(t)| + u(t) \tag{8-387}$$

must hold for all $t \in [t_1, t_2]$. Since $u(t)$ is negative, it is clear that

$$|u(t)| + u(t) = 0 \qquad \text{for all } t \in [t_1, t_2] \tag{8-388}$$

and so, from Eq. (8-387), we deduce the relation

$$-ax_2(t)|x_2(t)| = k \qquad \text{for all } t \in [t_1, t_2] \tag{8-389}$$

From Eq. (8-384), we find that

$$u(t) = -k \qquad \text{for all } t \in [t_1, t_2] \tag{8-390}$$

Thus, *the constant control $u(t) = -k$ satisfies the conditions‡ required by the minimum principle and, thus, is a candidate for optimal control provided that the magnitude restriction $|u(t)| \le 1$ is not violated;* clearly, the control $u(t) = -k$ is a candidate for optimal control *if and only if*

$$k \le 1 \tag{8-391}$$

From Eq. (8-390), we may also conclude that

$$x_2(t) = -\sqrt{\frac{k}{a}} \tag{8-392}$$

We can repeat the entire sequence of arguments from Eq. (8-378) to Eq. (8-392) supposing that $p_2(t) = -1$ for all $t \in [t_1, t_2]$, and we can conclude that the control $u(t) = k$, for all $t \in [t_1, t_2]$, can be a candidate for optimal control. Thus, we established the lemma:

***Lemma 8-21*** *If*

$$k \le 1 \tag{8-393}$$

*and if*

$$|p_2(t)| = 1 \qquad \textit{for all } t \in [t_1, t_2] \tag{8-394}$$

*then the singular control*

$$u(t) = -k \,\text{sgn}\, \{p_2(t)\} \qquad \textit{for all } t \in [t_1, t_2] \tag{8-395}$$

† Because the system (8-364) is time-invariant, the response time $T$ is not fixed and the integrand $k + |u(t)|$ is time-invariant.

‡ We have not verified, as yet, whether or not this control meets the boundary conditions.

*is a candidate for optimal control. This control requires that*

$$\pi_1 = 2\sqrt{ka}\,\operatorname{sgn}\{p_2(t)\} \tag{8-396}$$

*and that*

$$x_2(t) = -\sqrt{\frac{k}{a}}\,\operatorname{sgn}\{p_2(t)\} \tag{8-397}$$

*In the remainder of this section, we shall assume that* $k \le 1$.

Since the controls $u(t) = +1$, $u(t) = -1$, $u(t) = 0$, $u(t) = +k$, and $u(t) = -k$ are candidates for optimal control, it is useful to examine the trajectories of the system (8-364) in the state plane generated by these controls.

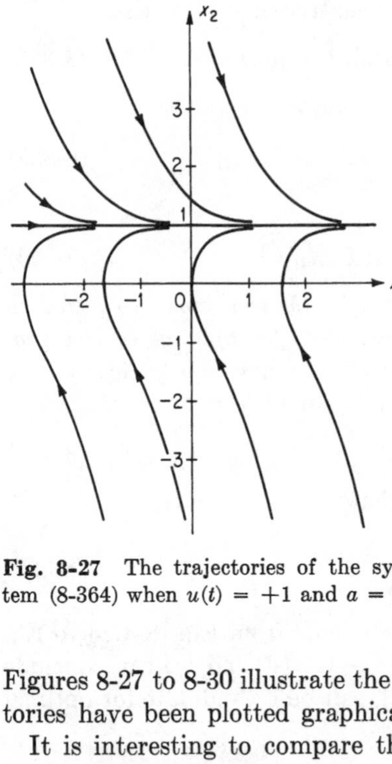


**Fig. 8-27** The trajectories of the system (8-364) when $u(t) = +1$ and $a = 1$.

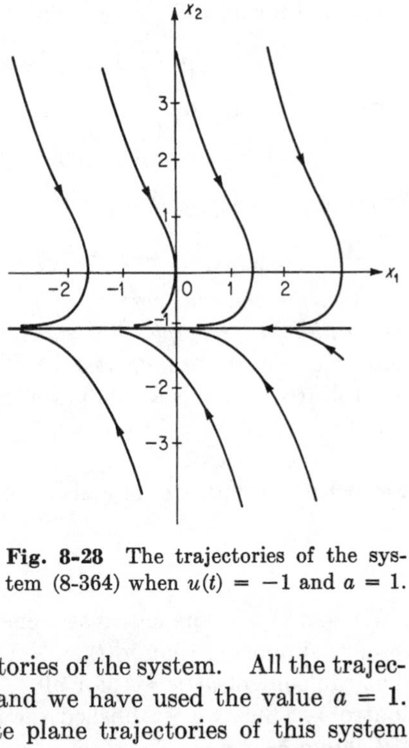


**Fig. 8-28** The trajectories of the system (8-364) when $u(t) = -1$ and $a = 1$.

Figures 8-27 to 8-30 illustrate the trajectories of the system. All the trajectories have been plotted graphically,† and we have used the value $a = 1$.

It is interesting to compare the state plane trajectories of this system with nonlinear friction to the state plane trajectories of the system with linear friction considered in Sec. 8-9. If we compare Fig. 8-27 with Fig. 8-18 and Fig. 8-28 with Fig. 8-19, then we observe that the trajectories are quite similar except for small variations in shape. However, comparison of Figs. 8-29 and 8-17 reveals the essential difference in the characteristics of the two systems.

Since the $H$-minimal control $u(t)$ is a nonlinear function of the costate

† From Eq. (8-364), we find that $dx_1/dx_2 = x_2/(-ax_2|x_2| + u)$, so that the slope of the trajectories is established in the plane.

variable $p_2(t)$, it is desirable to have as much information about $p_2(t)$ as possible. For this reason, we shall establish the following lemma.

***Lemma 8-22*** *The function $p_2(t)$ is continuous for all $t \in [0, T]$, and $p_2(t)$ can be zero at most once.*

PROOF The continuity is a consequence of the fact that $p_2(t)$ is the solution of the differential equation (8-372). The proof that $p_2(t)$ is zero at most once has been presented in Sec. 7-11 and is, therefore, omitted.

***Lemma 8-23*** *The optimal-control sequence to the origin must be of the type $\{. . . , +1\}$ or of the type $\{. . . , -1\}$. In other words, the optimal control at $t = T$ must be either $u(T) = +1$ or $u(T) = -1$ (provided that it exists).*

PROOF The control sequence $\{. . . , 0\}$ cannot be optimal, because $u(T) = x_1(T) = x_2(T) = 0$ substituted into Eq. (8-367) yields $H = k > 0$. The control sequence $\{. . . , +k\}$ cannot be optimal, as substitution of $p_2(T) = -1$, $u(T) = k$, $x_1(T) = x_2(T) = 0$ into Eq. (8-367) will yield $H = k > 0$. The control sequence $\{. . . , -k\}$ cannot be optimal, because substitution of $p_2(T) = +1$, $u(T) = -k$, $x_1(T) = x_2(T) = 0$ into Eq. (8-367) will yield $H = k > 0$. By elimination, the lemma is established.

There is another way that we can prove Lemma 8-23. The argument is based on the shape of trajectories and on the fact that the response time $T$ must be finite for $k > 0$. If we examine Figs. 8-29 and 8-30, then we observe that the trajectories due to the control $u = 0$ and $u = \pm k$ do not hit the origin but that there are two trajectories (one for $u = +1$ and the other for $u = -1$) that hit the origin (see Figs. 8-27 and 8-28).

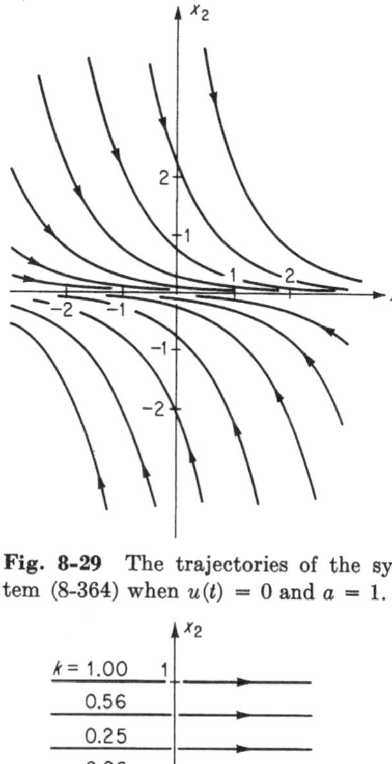


**Fig. 8-29** The trajectories of the system (8-364) when $u(t) = 0$ and $a = 1$.

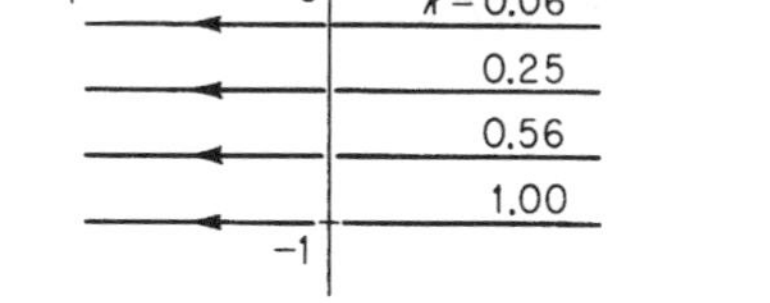


**Fig. 8-30** The trajectories of the system (8-364) when $u(t) = \pm k$ and $a = 1$. The trajectories for $u = k$ are always in the upper half of the plane ($x_2 > 0$), and the trajectories for $u = -k$ are always in the lower half of the plane ($x_2 < 0$).

**Exercise 8-34** Show that, if $u(T) = +1$, then $p_2(T) = -1 - k$, and that, if $u(T) = -1$, then $p_2(T) = 1 + k$.

We shall now outline the subsequent development of this section. In the following two definitions, we shall define (as we did in Sec. 8-9) the various trajectories that arise from the application of the constant controls. We shall then present a sequence of lemmas which isolate the control sequences that can be candidates for optimal control. The proofs of some lemmas are constructive in the sense that we shall define several switch curves from the quantities involved. We shall, however, leave the proofs of many other lemmas as exercises.

Once more, we assert that an optimal control exists. From the sequence of lemmas, we shall conclude that the extremal controls are unique, and so we shall be able to prove Control Law 8-7.

***Definition 8-7*** *Let $T^+(\xi_1, \xi_2, x_1, x_2)$ denote the trajectory originating at $(\xi_1, \xi_2)$ and generated by the control $u(t) = +1$ which passes through the state $(x_1, x_2)$; in view of the positive time requirement, we have (see Fig. 8-27)*

$$\left|\xi_2 - \sqrt{\frac{1}{a}}\right| > \left|x_2 - \sqrt{\frac{1}{a}}\right| \tag{8-398}$$

*All the $T^+$ trajectories are asymptotic (as $t \to \infty$) to the line $x_2 = \sqrt{1/a}$. Let $T^-(\xi_1, \xi_2, x_1, x_2)$ denote the trajectory originating at $(\xi_1, \xi_2)$ and generated by the control $u(t) = -1$ which passes through the state $(x_1, x_2)$; in view of the positive time requirement, we have (see Fig. 8-28)*

$$\left|\xi_2 + \sqrt{\frac{1}{a}}\right| > \left|x_2 + \sqrt{\frac{1}{a}}\right| \tag{8-399}$$

*All the $T^-$ trajectories are asymptotic (as $t \to \infty$) to the line $x_2 = -\sqrt{1/a}$. Let $T^0(\xi_1, \xi_2, x_1, x_2)$ denote the trajectory originating at $(\xi_1, \xi_2)$ and generated by $u(t) = 0$ which passes through the state $(x_1, x_2)$; in view of the positive time requirement (see Fig. 8-29), we have*

$$|\xi_2| > |x_2| \tag{8-400}$$

*Let $T_s^+(\xi_1, \xi_2, x_1, x_2)$ denote the trajectory originating at $(\xi_1, \xi_2)$ and generated by the singular control $u(t) = +k$, $0 < k \leq 1$ (see Fig. 8-30). In view of Lemma 8-21 and the positive time requirement, we have*

$$\xi_2 = x_2 = \sqrt{\frac{k}{a}} \qquad x_1 > \xi_1 \tag{8-401}$$

*Let $T_s^-(\xi_1, \xi_2, x_1, x_2)$ denote the trajectory originating at $(\xi_1, \xi_2)$ and generated by the singular control $u(t) = -k$ (see Fig. 8-30). In view of Lemma 8-21 and the positive time requirement, we have*

$$\xi_2 = x_2 = -\sqrt{\frac{k}{a}} \qquad x_1 < \xi_1 \tag{8-402}$$

***Definition 8-8*** *Let $\gamma_+$ ($\gamma_-$) denote the set of states $(x_1, x_2)$ that can be forced to $(0, 0)$ by the control $u(t) = +1$ [$u(t) = -1$] in positive time. Thus,*

$$\gamma_+ = \{(x_1, x_2): (x_1, x_2) \in T^+(x_1, x_2, 0, 0), x_1 > 0, x_2 < 0\} \quad (8\text{-}403)$$
$$\gamma_- = \{(x_1, x_2): (x_1, x_2) \in T^-(x_1, x_2, 0, 0), x_1 < 0, x_2 > 0\} \quad (8\text{-}404)$$

*Let*
$$\gamma = \gamma_+ \cup \gamma_- \quad (8\text{-}405)$$

*The $\gamma_+$ and $\gamma_-$ curves are shown in Fig.* 8-31. *As shown in Fig.* 8-31, *$S_R$ is the set of states to the right of the $\gamma$ curve and $S_L$ is the set of states to the left of the $\gamma$ curve; that is,*

$$S_R = \{(x_1, x_2): x_1 > x_1' \text{ if } (x_1', x_2) \in \gamma\} \quad (8\text{-}406)$$
$$S_L = \{(x_1, x_2): x_1 < x_1' \text{ if } (x_1', x_2) \in \gamma\} \quad (8\text{-}407)$$

Let us now examine control sequences of the type $\{. . . , +1\}$.

***Lemma 8-24*** *Control sequences of the type $\{. . . , -1, +1\}$, $\{. . . , -k, +1\}$, and $\{. . . , +k, +1\}$ cannot be optimal, and so the control sequence $\{. . . , 0, +1\}$ must be optimal.*

PROOF Suppose that either $\{. . . , -1, +1\}$ or $\{. . . , -k, +1\}$ is optimal, and let $t_1$ be the time at which the control switches from $u = -1$ or $u = -k$ to $u = +1$. Then we must have $p_2(t_1) = -1$ and $p_2(t_1 -) = +1$ [see Eqs. (8-374) and (8-376)], which means that $p_2(t)$ has a discontinuity at $t = t_1$. This violates Lemma 8-22. The control sequence $\{. . . , k, +1\}$ cannot be optimal since the trajectories $T_s^+(\xi_1, \xi_2, x_1, x_2)$ generated by the control $u(t) = +k$ are in the upper half of the state plane and the $\gamma_+$ curve is in the lower half (see Figs. 8-30 and 8-31).

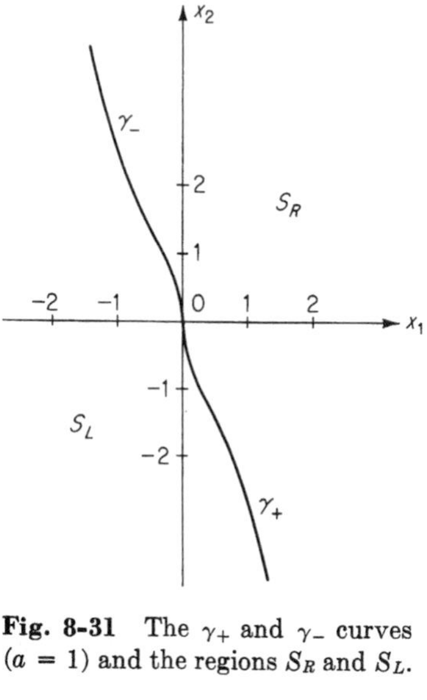


**Fig. 8-31** The $\gamma_+$ and $\gamma_-$ curves ($a = 1$) and the regions $S_R$ and $S_L$.

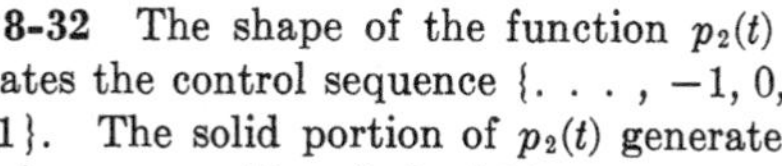

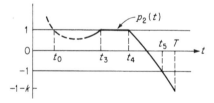

**Fig. 8-32** The shape of the function $p_2(t)$ that generates the control sequence $\{. . . , -1, 0, -k, 0, +1\}$. The solid portion of $p_2(t)$ generates the control sequence $\{0, -k, 0, +1\}$.

***Lemma 8-25*** *If*

$$\pi_1 = 2\sqrt{ka} \quad (8\text{-}408)$$

*then the control sequence $\{0, -k, 0, +1\}$ can be optimal, but the control sequence $\{. . . , -1, 0, -k, 0, +1\}$ cannot be optimal.*

CONSTRUCTIVE PROOF Figure 8-32 shows the shape of the function $p_2(t)$ that generates the control sequence $\{0, -k, 0, +1\}$. The dashed extension of $p_2(t)$ shown is required to generate the sequence $\{\ldots, -1, 0, -k, 0, +1\}$. Let

$$\begin{aligned} x_1(t_5) &= w_1 & x_2(t_5) &= w_2 \\ x_1(t_4) &= s_1 & x_2(t_4) &= s_2 \\ x_1(t_3) &= z_1 & x_2(t_3) &= z_2 \\ x_1(t_0) &= r_1 & x_1(t_0) &= r_2 \end{aligned} \tag{8-409}$$

Clearly,
$$(w_1, w_2) \in \gamma_+ \tag{8-410}$$
and so
$$w_2 < 0 \tag{8-411}$$

But $p_2(t_5) = -1$, and since $\pi_1 = 2\sqrt{ka}$, the condition $H = 0$ yields

$$H = 0 = k + 2\sqrt{ka}\, w_2 - a w_2^2 \tag{8-412}$$

Solving Eq. (8-412) for $w_2$, we find that

$$w_2 = \sqrt{\frac{k}{a}}\,(1 \pm \sqrt{2}) \tag{8-413}$$

and since $w_2 < 0$, we deduce that

$$w_2 = -\sqrt{\frac{k}{a}}\,(\sqrt{2} - 1) \tag{8-414}$$

This completely determines the point $(w_1, w_2)$, because $(w_1, w_2)$ is the intersection of the $\gamma_+$ curve and the line $x_2 = -\sqrt{k/a}\,(\sqrt{2} - 1)$.

Since the state at $t = t_4$ is $(s_1, s_2)$ and since $u(t) = 0$ for all $t \in [t_4, t_5]$, there is a trajectory $T^0(s_1, s_2, w_1, w_2)$ joining $(s_1, s_2)$ to $(w_1, w_2)$. From Eq. (8-400) and the fact that $w_2 < 0$, we deduce the relation

$$s_2 < w_2 < 0 \tag{8-415}$$

At $t = t_4$, $p_2(t_4) = +1$ and the condition $H = 0$ yields the equation

$$H = 0 = k + 2\sqrt{ka}\, s_2 + a s_2^2 \tag{8-416}$$

We solve Eq. (8-416) for $s_2$, and we find that

$$s_2 = -\sqrt{\frac{k}{a}} \tag{8-417}$$

The point $(s_1, s_2)$ is thus completely specified since it is the intersection of the $T^0(s_1, s_2, w_1, w_2)$ trajectory and of the line $x_2 = -\sqrt{k/a}$.

During the time interval $t_3 \le t \le t_4$, the control is singular, that is, $u = -k$, and so the trajectory moves along the straight line $x_2 = -\sqrt{k/a}$ to the point $(s_1, s_2)$. Thus, the points $(z_1, z_2)$ and $(s_1, s_2)$ are joined by the trajectory $T_s^-(z_1, z_2, s_1, s_2)$, and [from Eq. (8-402)] we conclude that

$$z_2 = s_2 = -\sqrt{\frac{k}{a}} < 0 \qquad z_1 > s_1 > 0 \tag{8-418}$$

Suppose that the control sequence $\{. . . , -1, 0, -k, 0, +1\}$ were optimal; then (see Fig. 8-32) there would be a time $t_0 < t_3$ such that $p_2(t_0) = 1$, $\dot{p}_2(t_0) < 0$, and the state would be $(v_1, v_2)$. The state $(v_1, v_2)$ must be transferred to $(z_1, z_2)$ by the trajectory $T^0(v_1, v_2, z_1, z_2)$. But Eqs. (8-418) and (8-400) require that

$$v_2 < z_2 = -\sqrt{\frac{k}{a}} < 0 \tag{8-419}$$

But Eq. (8-372) at $t = t_0$ with $\pi_1 = 2\sqrt{ka}$, $p(t_0) = +1$, $x_2(t_0) = v_2 < 0$ is

$$\dot{p}_2(t_0) = -2\sqrt{ka} - 2av_2 \tag{8-420}$$

Since $\dot{p}_2(t_0) < 0$ (see Fig. 8-32), we deduce from Eq. (8-420) the relation

$$v_2 > -\sqrt{\frac{k}{a}} \tag{8-421}$$

Equations (8-419) and (8-421) are *contradictory;* therefore, the sequence $\{. . . , -1, 0, -k, 0, +1\}$ cannot be optimal, but the sequence $\{0, -k, 0, +1\}$ can be optimal.

***Lemma 8-26*** *If*

$$\pi_1 = 2\sqrt{ka} \tag{8-422}$$

*then the control sequence* $\{0, -1, -k, 0, +1\}$ *can be optimal, but the control sequence* $\{. . . , -1, 0, -1, -k, 0, +1\}$ *cannot be optimal.*

CONSTRUCTIVE PROOF Figure 8-33 shows the shape of the function $p_2(t)$ that generates the control sequence $\{0, -1, -k, 0, +1\}$. The dashed portion of $p_2(t)$ generates the control sequence $\{. . . , -1, 0, -1, -k, 0, +1\}$. We use the notation of Eqs. (8-409), and, in addition, we let

$$x_1(t_2) = q_1 \qquad x_2(t_2) = q_2 \tag{8-423}$$

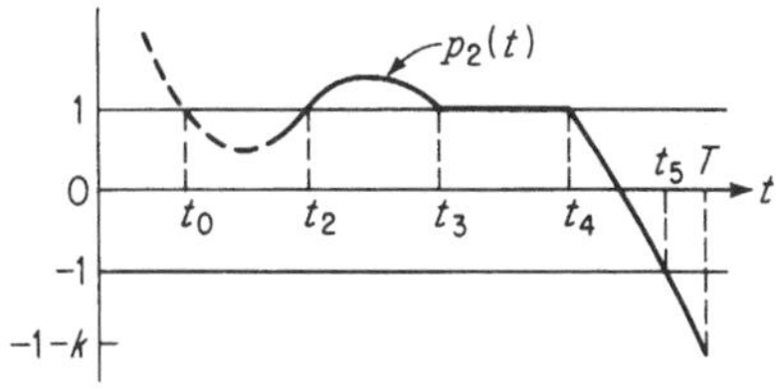


**Fig. 8-33** The shape of the function $p_2(t)$ that generates the control sequence $\{. . . , -1, 0, -1, -k, 0, +1\}$. The solid portion of $p_2(t)$ generates the control sequence $\{0, -1, -k, 0, +1\}$.

Clearly, all the conditions and equations derived in the proof of Lemma 8-25 for the states $(w_1, w_2)$, $(s_1, s_2)$, and $(z_1, z_2)$ continue to hold. Since $u(t) = -1$ for $t \in [t_2, t_3]$, it follows that the state $(q_1, q_2)$ at $t = t_2$ must be transferred to the state $(z_1, z_2)$ by the control $u = -1$, and so $(z_1, z_2)$ and $(q_1, q_2)$ are joined by the $T^-(q_1, q_2, z_1, z_2)$ trajectory. Since [see Eqs. (8-418)] $z_2 = -\sqrt{k/a} > -\sqrt{1/a}$, we must have [see Eq. (8-399)]

$$q_2 > -\sqrt{\frac{k}{a}} \tag{8-424}$$

But at $t = t_2$, $p_2(t_2) = +1$ and $\dot{p}_2(t_2) > 0$. Since $\dot{p}_2(t_2) = -2\sqrt{ka} + a|q_2|p_2(t_2)$, we conclude that

$$|q_2| > \sqrt{\frac{k}{a}} \tag{8-425}$$

From the inequalities (8-424) and (8-425), we conclude that

$$q_2 > \sqrt{\frac{k}{a}} > 0 \tag{8-426}$$

At $t = t_2$, the condition $H = 0$ yields the equation

$$0 = k + 2\sqrt{ka}\, q_2 - aq_2^2$$

which, in turn, implies [in view of Eq. (8-426)] that

$$q_2 = \sqrt{\frac{k}{a}}\,(1 + \sqrt{2}) \tag{8-427}$$

Thus, the state $(q_1, q_2)$ is determined, since it is the intersection of the $T^-(q_1, q_2, z_1, z_2)$ trajectory and the line $x_2 = \sqrt{k/a}\,(1 + \sqrt{2})$.

Now suppose that the control sequence $\{\ldots, -1, 0, -1, -k, 0, +1\}$ were optimal; then (see Fig. 8-33) there would be a time $t_0 < t_2$ such that at $t = t_0$, $x_1(t_0) = r_1$, $x_2(t_0) = r_2$, $p_2(t_0) = 1$, $\dot{p}_2(t_0) < 0$. Since $u(t) = 0$ for $t \in [t_0, t_2]$, it would follow that the state $(r_1, r_2)$ must be joined to the state $(q_1, q_2)$ by the $T^0(r_1, r_2, q_1, q_2)$ trajectory. This and Eqs. (8-400) and (8-427) imply that

$$r_2 > q_2 = \sqrt{\frac{k}{a}}\,(1 + \sqrt{2}) > \sqrt{\frac{k}{a}} > 0 \tag{8-428}$$

But $\dot{p}_2(t_0) = -2\sqrt{ka} + 2a|r_2|p_2(t_0)$; and since $\dot{p}_2(t_0) < 0$ and $p_2(t_0) = 1$, we conclude that

$$|r_2| < \sqrt{\frac{k}{a}} \tag{8-429}$$

But the inequalities (8-428) and (8-429) are *contradictory*. We therefore conclude that the control sequence $\{\ldots, -1, 0, -1, -k, 0, +1\}$ cannot be optimal and that the control sequence $\{0, -1, -k, 0, +1\}$ can be optimal.

***Lemma 8-27*** *If* $\pi_1 = 2\sqrt{ka}$, *then the control sequence* $\{\ldots, -k, 0, -1, -k, 0, +1\}$ *cannot be optimal.*

**Exercise 8-35** Prove Lemma 8-27.

***Lemma 8-28*** *If* $\pi_1 = 2\sqrt{ka}$, *then the control sequences* $\{\ldots, +1, 0, -k, 0, +1\}$ *and* $\{\ldots, +k, 0, -k, 0, +1\}$ *cannot be optimal.*

**Exercise 8-36** Prove Lemma 8-28. HINT: Use Lemma 8-22.

Let us collect our thoughts for a moment and discuss the results of Lemmas 8-25 to 8-28. We have shown that, *if* $\pi_1 = 2\sqrt{ka}$ [*which is the only way that the singular control* $u(t) = -k$ *can be a portion of an optimal control sequence*], *then only the control sequences* $\{0, -k, 0, +1\}$ *and* $\{0, -1, -k, 0, +1\}$ *can be candidates for optimal control. By the principle of optimality, we can conclude that every subsequence of these two optimal sequences is also a candidate for optimal control.*

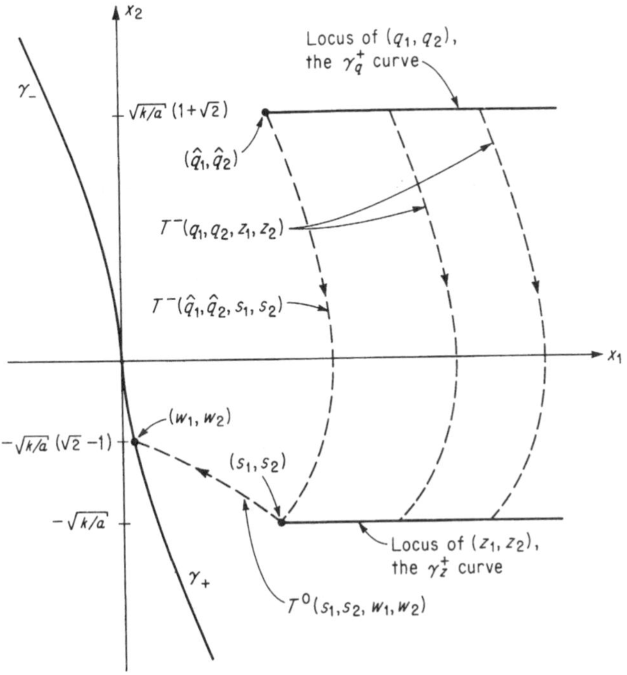


**Fig. 8-34** Illustration of the points $(w_1, w_2)$, $(s_1, s_2)$, and $(q_1, q_2)$ and of the $\gamma_z{}^+$ and $\gamma_q{}^+$ curves.

Let us review some of the definitions and constructions employed in the proof in order to emphasize the graphical techniques involved. We shall use Fig. 8-34 for visualization purposes.

1. First, we pick $\pi_1 = 2\sqrt{ka}$.
2. From Eq. (8-414), we determine that $w_2 = -\sqrt{k/a}\,(\sqrt{2} - 1)$.
3. The point $(w_1, w_2)$ is on the $\gamma_+$ curve.
4. From Eq. (8-417), we determine that $s_2 = -\sqrt{k/a}$.
5. The point $(s_1, s_2)$ is the intersection of the $T^0(s_1, s_2, w_1, w_2)$ trajectory and the line $x_2 = -\sqrt{k/a}$.
6. From Eq. (8-427), we find that $q_2 = \sqrt{k/a}\,(1 + \sqrt{2})$.
7. The point $(q_1, q_2)$ is the intersection of the $T^-(q_1, q_2, z_1, z_2)$ trajectory and the line $x_2 = \sqrt{k/a}\,(1 + \sqrt{2})$.

Let us call the locus of the points $(z_1, z_2)$ the $\gamma_z^+$ curve, that is,

$$\gamma_z^+ = \left\{(x_1, x_2): x_2 = -\sqrt{\frac{k}{a}}, x_1 > s_1\right\} \tag{8-430}$$

Let us examine the locus of the $(q_1, q_2)$ points as $(z_1, z_2)$ varies over $\gamma_z^+$. In particular, let

$$(\hat{q}_1, \hat{q}_2) = \lim_{(z_1, z_2) \to (s_1, s_2)} (q_1, q_2) \tag{8-431}$$

The point $(\hat{q}_1, \hat{q}_2)$ is shown in Fig. 8-34. It is easy to see that $(\hat{q}_1, \hat{q}_2)$ is the intersection of the $T^-(\hat{q}_1, \hat{q}_2, s_1, s_2)$ trajectory and the line

$$x_2 = \sqrt{\frac{k}{a}}\,(1 + \sqrt{2})$$

As $(z_1, z_2)$ varies over the set $\gamma_z^+$, the points $(q_1, q_2)$ describe a curve which we shall call the $\gamma_q^+$ curve. We note that

$$\gamma_q^+ = \left\{(x_1, x_2): x_2 = \sqrt{\frac{k}{a}}\,(1 + \sqrt{2});\ x_1 > \hat{q}_1\right\} \tag{8-432}$$

We shall now establish a sequence of lemmas under the assumption that $\pi_1 \neq 2\sqrt{ka}$. If $\pi_1 \neq 2\sqrt{ka}$, then the singular control $u(t) = -k$ *cannot* be a portion of an optimal control sequence (see Lemma 8-21).

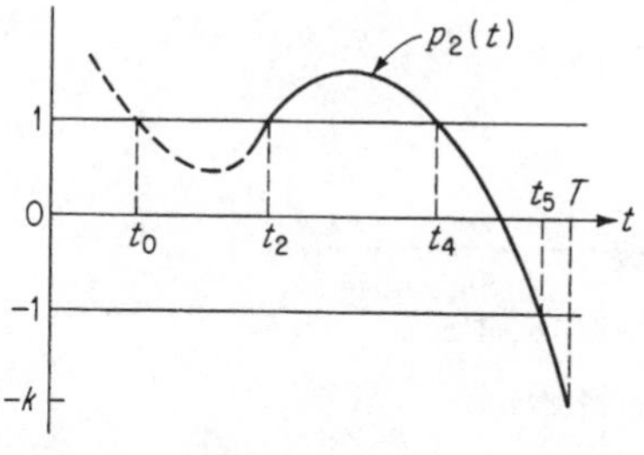


**Fig. 8-35** The shape of the function $p_2(t)$ that generates the control sequence $\{\ldots, -1, 0, -1, 0, +1\}$. The solid portion of $p_2(t)$ generates the control sequence $\{0, -1, 0, +1\}$.

***Lemma 8-29*** *If*

$$\pi_1 > 2\sqrt{ka} \tag{8-433}$$

*then the control sequence* $\{0, -1, 0, +1\}$ *can be optimal but the control sequence* $\{\ldots, -1, 0, -1, 0, +1\}$ *cannot be optimal.*

CONSTRUCTIVE PROOF Figure 8-35 shows the shape of the function $p_2(t)$ that generates the control sequence $\{0, -1, 0, +1\}$. The dashed portion of $p_2(t)$ generates the control sequence $\{\ldots, -1, 0, -1, 0, +1\}$. Let

$$\begin{aligned} m_1 &= x_1(t_5) & m_2 &= x_2(t_5) \\ n_1 &= x_1(t_4) & n_2 &= x_2(t_4) \\ h_1 &= x_1(t_2) & h_2 &= x_2(t_2) \\ g_1 &= x_1(t_0) & g_2 &= x_2(t_0) \end{aligned} \tag{8-434}$$

At $t = t_5$, the control switches from $u = 0$ to $u = +1$ and $u(t) = +1$ for all $t \in [t_5, T]$. This implies that

$$(m_1, m_2) \in \gamma_+ \tag{8-435}$$

and, therefore, that

$$m_2 < 0 \tag{8-436}$$

At $t = t_5$, $p_2(t_5) = -1$; thus, the condition $H = 0$ yields the equation

$$H = 0 = 1 + m_2\pi_1 - am_2^2 \tag{8-437}$$

We find [using Eq. (8-436) to decide on the sign] that

$$m_2 = \frac{1}{2a}(\pi_1 - \sqrt{\pi_1^2 + 4ak}) \qquad \text{(8-438)}\dagger$$

Note that [see Eq. (8-414)] the limiting relations

$$\lim_{\pi_1 \to 2\sqrt{ka}} m_2 = -\sqrt{\frac{k}{a}}(\sqrt{2} - 1) = w_2 \tag{8-439}$$

$$\lim_{\pi_1 \to \infty} m_2 = 0 \tag{8-440}$$

are true. At $t = t_4$, the control switches from $u(t) = -1$ to $u(t) = 0$. Since $u(t) = 0$ for all $t \in [t_4, t_5]$, we deduce the relation

$$n_2 < m_2 < 0 \tag{8-441}$$

At $t = t_4$, $p_2(t_4) = +1$, $\dot{p}_2(t_4) < 0$, and so the equation $\dot{p}_2(t_4) = -\pi_1 + 2a|n_2|p_2(t_4)$ yields the relation

$$n_2 > -\frac{\pi_1}{2a} \tag{8-442}$$

The condition $H = 0$ at $t = t_4$ implies that

$$H = k + n_2\pi_1 + an_2^2 = 0 \tag{8-443}$$

Thus, we find that [using the inequality (8-442) to decide on the sign]

$$n_2 = -\frac{1}{2a}(\pi_1 - \sqrt{\pi_1^2 - 4ak}) \tag{8-444}$$

Note that [see Eq. (8-417)] the limiting relations

$$\lim_{\pi_1 \to 2\sqrt{ka}} n_2 = -\sqrt{\frac{k}{a}} = s_2 \tag{8-445}$$

$$\lim_{\pi_1 \to \infty} n_2 = 0 \tag{8-446}$$

$$\lim_{\pi_1 \to 2\sqrt{ka}} (n_1, n_2) = (s_1, s_2) \tag{8-447}$$

† Note that the inequality (8-433) guarantees that $m_2$ is real.

hold. At $t = t_2$, $p_2(t_2) = 1$ and $\dot{p}_2(t_2) > 0$; thus, the equation $\dot{p}_2(t_2) = -\pi_1 + 2a|h_2|p_2(t_2)$ yields the inequality

$$|h_2| > \frac{\pi_1}{2a} \tag{8-448}$$

Since $u(t) = -1$ for $t \in [t_2, t_4]$ and since

$$x_2(t_4) = n_2 > -\frac{\pi_1}{2a} > -\sqrt{\frac{k}{a}} > -\sqrt{\frac{1}{a}}$$

we know that [see Eq. (8-399)]

$$h_2 > n_2 > -\frac{\pi_1}{2a} \tag{8-449}$$

From the two inequalities (8-448) and (8-449), we conclude that

$$h_2 > \frac{\pi_1}{2a} > \sqrt{\frac{k}{a}} > 0 \tag{8-450}$$

The condition that $H = 0$ at $t = t_2$ yields

$$H = 0 = k + h_2\pi_1 - ah_2^2 \tag{8-451}$$

We solve Eq. (8-451) [using the inequality (8-450) to decide the proper sign], and we find that

$$h_2 = \frac{1}{2a}(\pi_1 + \sqrt{\pi_1^2 + 4ak}) \tag{8-452}$$

Note that

$$\lim_{\pi_1 \to 2\sqrt{ka}} h_2 = \sqrt{\frac{k}{a}}(1 + \sqrt{2}) = \hat{q}_2 \tag{8-453}$$

$$\lim_{\pi_1 \to \infty} h_2 = +\infty \tag{8-454}$$

$$\lim_{\pi_1 \to 2\sqrt{ka}} (h_1, h_2) = (\hat{q}_1, \hat{q}_2) \tag{8-455}$$

To prove that the control sequence $\{\ldots, -1, 0, -1, 0, +1\}$ cannot be optimal, we can mimic the procedure given in the proof of Lemma 8-26 to prove that the control sequence $\{\ldots, -1, 0, -1, -k, 0, +1\}$ was not optimal. For this reason, we leave this part of the proof as an exercise for the reader.

***Lemma 8-30*** *If $\pi_1 > 2\sqrt{ka}$, then control sequences of the type $\{\ldots, +1, 0, -1, 0, +1\}$ cannot be optimal.*

**Exercise 8-37** Prove Lemma 8-30. HINT: Use Lemma 8-28.

***Lemma 8-31*** *If $\pi_1 > 2\sqrt{ka}$, then control sequences of the type $\{\ldots, +1, 0, +1\}$ cannot be optimal.*

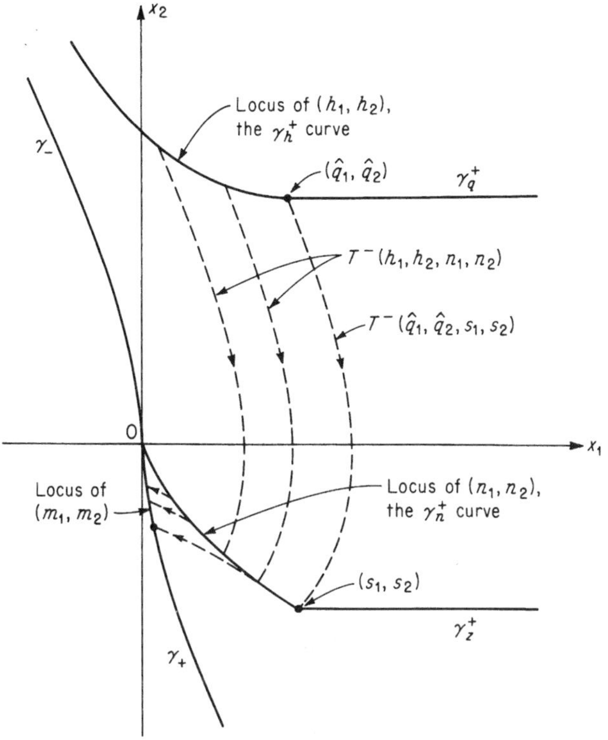


**Fig. 8-36** Illustration of the $\gamma_h^+$ and $\gamma_n^+$ curves.

**Exercise 8-38** Prove Lemma 8-31.

Once more let us see what Lemmas 8-29 to 8-31 and their constructive proofs imply. We shall use Fig. 8-36 for visualization purposes.

1. We pick a value of $\pi_1$ such that $\pi_1 > 2\sqrt{ka}$.
2. From Eq. (8-438), we find that

$$m_2 = \frac{1}{2a}\left(\pi_1 - \sqrt{\pi_1^2 + 4ak}\right)$$

3. The point $(m_1, m_2)$ is on the $\gamma_+$ curve, and so it is the intersection of the $\gamma_+$ curve and the line

$$x_2 = \frac{1}{2a}\left(\pi_1 - \sqrt{\pi_1^2 + 4ak}\right)$$

4. From Eq. (8-444), we find that

$$n_2 = -\frac{1}{2a}\left(\pi_1 - \sqrt{\pi_1^2 - 4ak}\right)$$

5. We draw the $T^0(n_1, n_2, m_1, m_2)$ trajectory passing through $(m_1, m_2)$, and we find its intersection with the line

$$x_2 = -\frac{1}{2a}(\pi_1 - \sqrt{\pi_1^2 - 4ak})$$

The point of intersection is the point $(n_1, n_2)$.

6. From Eq. (8-452), we find that

$$h_2 = \frac{1}{2a}(\pi_1 + \sqrt{\pi_1^2 + 4ak})$$

7. The point $(h_1, h_2)$ is the intersection of the line

$$x_2 = \frac{1}{2a}(\pi_1 + \sqrt{\pi_1^2 + 4ak})$$

and the $T^-(h_1, h_2, n_1, n_2)$ trajectory through the point $(n_1, n_2)$.

Thus, for each value of $\pi_1 > 2\sqrt{ka}$, we can determine graphically the points $(m_1, m_2)$, $(n_1, n_2)$, and $(h_1, h_2)$. As $\pi_1$ varies over the set

$$2\sqrt{ka} < \pi_1 < \infty \tag{8-456}$$

we obtain the loci of the points $(m_1, m_2)$, $(n_1, n_2)$, and $(h_1, h_2)$. We know that

$$(m_1, m_2) \in \left\{(x_1, x_2): (x_1, x_2) \in \gamma_+, -\sqrt{\frac{k}{a}} < x_2 \leq 0\right\} \tag{8-457}$$

***Definition 8-9*** *Let $\gamma_n^+$ denote the locus of the points $(n_1, n_2)$ and let $\gamma_h^+$ denote the locus of the points $(h_1, h_2)$ as $\pi_1$ varies over the set of values $2\sqrt{ka} < \pi_1 < \infty$.*

These sets $\gamma_n^+$ and $\gamma_h^+$ are illustrated in Fig. 8-36.

We conclude our sequence of lemmas with:

***Lemma 8-32*** *If $\pi_1 < 2\sqrt{ka}$, then the control sequence $\{0, +1\}$ can be optimal but control sequences of the type $\{\ldots, -1, 0, +1\}$ and $\{\ldots, +1, 0, +1\}$ cannot be optimal.*

**Exercise 8-39** Prove Lemma 8-32. HINT: Use contradiction.

Up to now, we have defined several curves as indicated in Figs. 8-34 and 8-36. We have called these curves $\gamma_+$, $\gamma_-$, $\gamma_z^+$, $\gamma_q^+$, $\gamma_n^+$, and $\gamma_h^+$. All of these curves are in $S_R$, that is, to the right of the $\gamma_+ \cup \gamma_-$ curve. Since the system is symmetrical, we could prove a sequence of lemmas analogous to Lemmas 8-24 to 8-32 regarding control sequences of the type $\{\ldots, -1\}$. To avoid duplication of effort, we can define the curves $\gamma_z^-$, $\gamma_q^-$, $\gamma_n^-$, and $\gamma_h^-$ to be the curves symmetrical about the origin to the curves $\gamma_z^+$, $\gamma_q^+$, $\gamma_n^+$, and $\gamma_h^+$, respectively.

Recall that each of these curves is a function of the constant $k$, $0 < k \leq 1$. Figure 8-37 shows all these curves, using the values $a = 1$ and $k = 0.25$.

All the curves indicated in Fig. 8-37 were constructed graphically using the techniques described.

As is evident in Fig. 8-37, these curves divide the state plane into six regions (or sets), labeled $H_1$, $H_2$, $H_3$, $H_4$, $H_5$, and $H_6$. We shall not mathematically define these sets but rather use Fig. 8-37, coupled with the remark that

$H_2$ includes the $\gamma_n^-$ curve
$H_3$ includes the $\gamma_q^-$, $\gamma_h^-$, and $\gamma_+$ curves
$H_5$ includes the $\gamma_n^+$ curve
$H_6$ includes the $\gamma_q^+$, $\gamma_h^+$, and $\gamma_-$ curves

The horizontal line segments $\gamma_z^+$ and $\gamma_z^-$ are not included in any of the sets $H_1$ to $H_6$.

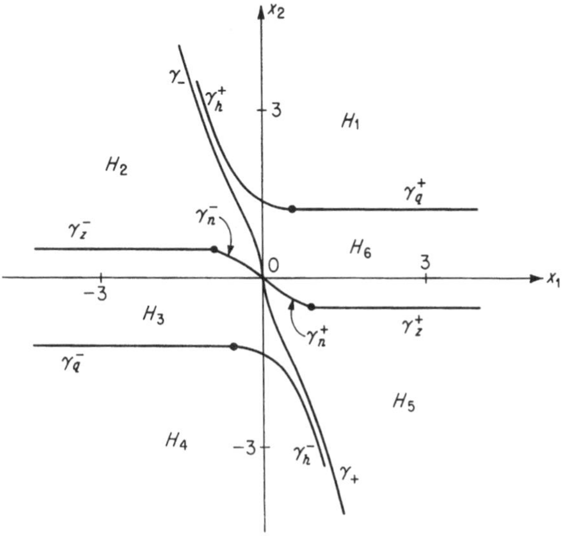


**Fig. 8-37** The division of the state plane into the six regions $H_1$, $H_2$, $H_3$, $H_4$, $H_5$, and $H_6$.

We claim that the solution to the posed problem is:

***Control Law 8-7 Solution to Problem 8-7*** *If* $0 < k \leq 1$, *then the unique optimal control* $u^*$, *as a function of the state* $(x_1, x_2)$, *is given by*

$$\begin{aligned} u^* &= 0 && \text{for all } (x_1, x_2) \in H_1 \cup H_2 \cup H_4 \cup H_5 \\ u^* &= -1 && \text{for all } (x_1, x_2) \in H_6 \\ u^* &= +1 && \text{for all } (x_1, x_2) \in H_3 \\ u^* &= -k && \text{for all } (x_1, x_2) \in \gamma_z^+ \\ u^* &= +k && \text{for all } (x_1, x_2) \in \gamma_z^- \end{aligned} \tag{8-458}$$

PROOF Let $(\xi_1, \xi_2)$ be some initial state. Recall that the construction of the various curves implied that the optimal control sequence to the origin is

$$\begin{array}{ll} \{0, +1\} & \text{if } (\xi_1, \xi_2) \in \gamma_n^+ \\ \{0, -1\} & \text{if } (\xi_1, \xi_2) \in \gamma_n^- \\ \{-k, 0, +1\} & \text{if } (\xi_1, \xi_2) \in \gamma_z^+ \\ \{+k, 0, -1\} & \text{if } (\xi_1, \xi_2) \in \gamma_z^- \\ \{-1, 0, +1\} & \text{if } (\xi_1, \xi_2) \in \gamma_h^+ \\ \{+1, 0, -1\} & \text{if } (\xi_1, \xi_2) \in \gamma_h^- \\ \{-1, -k, 0, +1\} & \text{if } (\xi_1, \xi_2) \in \gamma_q^+ \\ \{+1, +k, 0, -1\} & \text{if } (\xi_1, \xi_2) \in \gamma_q^- \end{array} \tag{8-459}$$

These considerations immediately yield the following:

$$\begin{aligned} &\text{If } (\xi_1, \xi_2) \in H_6, \text{ then } u(t) = -1 \\ &\text{If } (\xi_1, \xi_2) \in H_3, \text{ then } u(t) = +1 \\ &\text{If } (\xi_1, \xi_2) \in \gamma_z^+, \text{ then } u(t) = -k \\ &\text{If } (\xi_1, \xi_2) \in \gamma_z^-, \text{ then } u(t) = +k \end{aligned} \tag{8-460}$$

Now, from Lemmas 8-26 and 8-27, we conclude that the $\gamma_h^+ \cup \gamma_q^+$ curve must be reached by a $T^0$ trajectory. This implies that

$$\text{If } (\xi_1, \xi_2) \in H_1, \text{ then } u(t) = 0 \tag{8-461}$$

From Lemma 8-25, we can conclude that the $\gamma_z^+$ curve must be reached by a $T^0$ trajectory and hence that there is a subset of $H_5$ such that initial states are forced by $u = 0$ to $\gamma_z^+$. On the other hand, Lemma 8-32 implies that there are states that may be transferred to the $\gamma_+$ curve by the control $u(t) = 0$. Thus, we can conclude that

$$\text{If } (\xi_1, \xi_2) \in H_5, \text{ then } u(t) = 0 \tag{8-462}$$

Similar considerations lead to the conclusion that

$$\text{If } (\xi_1, \xi_2) \in H_2, \text{ then } u(t) = 0 \tag{8-463}$$

and the control law is established.

We shall now let $k \to 0$ and examine the limiting behavior of the various curves. This technique will give us the fuel-optimal solution. From Eqs. (8-414), (8-417), and (8-427), we find that

$$\begin{aligned} \lim_{k\to 0} (w_1, w_2) &= (0, 0) \\ \lim_{k\to 0} (s_1, s_2) &= (0, 0) \\ \lim_{k\to 0} (q_1, q_2) &= (0, 0) \end{aligned} \tag{8-464}$$

From Eqs. (8-438), (8-444), and (8-452), we find that

$$\begin{aligned} \lim_{k\to 0} (m_1, m_2) &= (0, 0) \\ \lim_{k\to 0} (n_1, n_2) &= (0, 0) \\ \lim_{k\to 0} (h_1, h_2) &= (\infty, \infty) \end{aligned} \tag{8-465}$$

Thus,
$$\lim_{k\to 0} \gamma_n^+ = \lim_{k\to 0} \gamma_n^- = \emptyset \qquad \text{the empty set} \tag{8-466}$$
$$\lim_{k\to 0} \gamma_z^+ \cup \gamma_z^- = x_1 \text{ axis} \tag{8-467}$$
$$\lim_{k\to 0} \gamma_q^+ \cup \gamma_q^- = x_1 \text{ axis} \tag{8-468}$$
$$\lim_{k\to 0} \gamma_h^+ \cup \gamma_h^- = \gamma_+ \cup \gamma_- \tag{8-469}$$

**Exercise 8-40** Solve independently the problem of fuel-optimal control for the system (8-364) [that is, use the cost $F = \int_0^T |u(t)|\,dt$, $T$ not specified]. Comment on the existence and uniqueness, and compare with the results obtained above as we let $k \to 0$.

**Exercise 8-41** Consider the system (8-364) and the cost functional $F = \int_0^T |u(t)|\,dt$. Suppose that $T$ is fixed. Outline a procedure for determining the optimal control. Are there singular controls in this case?

**Exercise 8-42**† Solve the problem considered in this section for $k > 1$. (HINT: No singular controls can exist; the most general optimal control sequences are $\{0, +1, 0, -1\}$ and $\{0, -1, 0, +1\}$.) Let $k \to \infty$ and verify that your solution tends to the time-optimal solution (given in Sec. 7-11).

**Exercise 8-43** Design an engineering realization of Control Law 8-7 and suggest suboptimal designs. Repeat for the control law that you have found in Exercise 8-42.

## 8-11 Comments and Generalizations

In Sec. 7-11, we discussed the techniques for time-optimal control of a class of nonlinear second-order systems. We concluded that there was no difference between the time-optimal controls of linear and nonlinear second-order systems, except for the "shape" or equation of the switch curve.

If we compare, however, the optimal control‡ for the nonlinear system considered in Sec. 8-10 with the optimal control for the linear systems considered in Secs. 8-8 and 8-9, then we observe that there is indeed a difference between them and that this difference is evidenced by the existence of optimal singular controls for the nonlinear system.

The natural question that arises is this: Is the existence of the optimal singular control a property of the nonlinear system (8-364) or is it a common property of a class of nonlinear systems?

To answer this question, let us consider the nonlinear second-order system

$$\begin{aligned} \dot{x}_1(t) &= x_2(t) \\ \dot{x}_2(t) &= -f[x_2(t)] + u(t) \qquad |u(t)| \le 1 \end{aligned} \tag{8-470§}$$

and the cost functional

$$J = \int_0^T [k + |u(t)|]\,dt \tag{8-471}$$

† The solution is given in Ref. A-14.

‡ With respect to the cost functional $\int_0^T [k + |u(t)|]\,dt$, $0 < k \le 1$.

§ Equations (8-470) are the same as Eqs. (7-465).

with $T$ not specified. Let us assume that the function $f[x_2(t)]$ has the properties

$$f(0) = 0 \tag{8-472}$$

$$\operatorname{sgn}\{f[x_2(t)]\} = \operatorname{sgn}\{x_2(t)\} \tag{8-473}$$

$$g[x_2(t)] = \frac{\partial f[x_2(t)]}{\partial x_2(t)} > 0 \tag{8-474}$$

$$\text{If } |x_2'(t)| < |x_2''(t)|, \text{ then } |f[x_2'(t)]| < |f[x_2''(t)]| \tag{8-475}$$

The system (8-470) can be used to describe the motion of a mass by letting $x_1(t)$ denote the error in position, $x_2(t)$ the velocity, and $f[x_2(t)]$ a drag force which is a nonlinear function of the velocity $x_2(t)$. The assumptions given by Eqs. (8-472) to (8-475) imply, in essence, that the drag force is opposite to the direction of motion and that large velocities create large drag forces.

The Hamiltonian for the system (8-470) and the cost (8-471) is

$$H = k + |u(t)| + x_2(t)p_1(t) - f[x_2(t)]p_2(t) + u(t)p_2(t) \tag{8-476}$$

and so

$$\dot{p}_1(t) = -\frac{\partial H}{\partial x_1(t)} = 0 \tag{8-477}$$

$$\dot{p}_2(t) = -\frac{\partial H}{\partial x_2(t)} = -p_1(t) + g[x_2(t)]p_2(t) \tag{8-478}$$

If $\pi_1 = p_1(0)$, $\pi_2 = p_2(0)$, then

$$p_1(t) = \pi_1 = \text{const} \tag{8-479}$$

$$\dot{p}_2(t) = -\pi_1 + g[x_2(t)]p_2(t) \tag{8-480}$$

The control that minimizes the Hamiltonian $H$ is given by Eqs. (8-373) to (8-376). Singular controls arise if there is a finite time interval $[t_1, t_2]$ such that

$$|p_2(t)| = 1 \qquad \text{for all } t \in [t_1, t_2] \tag{8-481}$$

Suppose that Eq. (8-481) holds; then

$$\dot{p}_2(t) = 0 \qquad \text{for all } t \in [t_1, t_2] \tag{8-482}$$

and from Eq. (8-480), we find that

$$\pi_1 = g[x_2(t)] \operatorname{sgn}\{p_2(t)\} \qquad \text{for all } t \in [t_1, t_2] \tag{8-483}$$

which implies that

$$x_2(t) = \text{const} \qquad \text{for all } t \in [t_1, t_2] \tag{8-484}$$

and that

$$\dot{x}_2(t) = 0 \qquad \text{for all } t \in [t_1, t_2] \tag{8-485}$$

From Eqs. (8-485) and (8-470), we conclude that

$$u(t) = f[x_2(t)] = \text{const} \qquad \text{for all } t \in [t_1, t_2] \tag{8-486}$$

The necessary condition that $H = 0$ for all $t \in [t_1, t_2]$ and Eqs. (8-481) to (8-486) yield

$$H = 0 = k + \operatorname{sgn}\{p_2(t)\}\{x_2(t)g[x_2(t)] - f[x_2(t)]\} \tag{8-487}$$

or

$$x_2(t)g[x_2(t)] - f[x_2(t)] = -\operatorname{sgn}\{p_2(t)\}\,k \tag{8-488}$$

Equation (8-488) can be "solved" to find the constant $x_2(t)$, $t \in [t_1, t_2]$, in terms of $k$. Suppose that the "solution" is

$$x_2(t) = h[-\operatorname{sgn}\{p_2(t)\}\,k] \qquad \text{for all } t \in [t_1, t_2] \tag{8-489}$$

Then from Eq. (8-489), we find that

$$u(t) = f\{h[-\operatorname{sgn}\{p_2(t)\}\,k]\} \qquad \text{for all } t \in [t_1, t_2] \tag{8-490}$$

Clearly, the control (8-490) can be optimal during the time interval $[t_1, t_2]$ provided that the magnitude restriction $|u(t)| \le 1$ is also satisfied; in other words, if

$$|f\{h[-\operatorname{sgn}\{p_2(t)\}\,k]\}| \le 1 \tag{8-491}$$

then the singular control (8-490) is a candidate for optimality.

**Example 8-5** Suppose that in Eqs. (8-470) we set

$$f[x_2(t)] = \tfrac{1}{3}x_2^3(t) \tag{8-492}$$

Then we deduce that

$$g[x_2(t)] = x_2^2(t) \tag{8-493}$$

Equation (8-491) becomes

$$x_2^3(t) - \tfrac{1}{3}x_2^3(t) = -\operatorname{sgn}\{p_2(t)\}\,k \tag{8-494}$$

and the solution is

$$x_2(t) = [-\tfrac{3}{2}\operatorname{sgn}\{p_2(t)\}\,k]^{\frac{1}{3}} \tag{8-495}$$

Equation (8-489) states that

$$u(t) = f[x_2(t)] = \tfrac{1}{3}x_2^3(t) \tag{8-496}$$

We substitute Eq. (8-495) into (8-496) to obtain the relation

$$u(t) = -\tfrac{1}{2}\operatorname{sgn}\{p_2(t)\}\,k \tag{8-497}$$

Since $\operatorname{sgn}\{p_2(t)\} = \pm 1$, we conclude that the singular controls

$$u(t) = -\tfrac{1}{2}k \tag{8-498}$$

and

$$u(t) = \tfrac{1}{2}k \tag{8-499}$$

are candidates for optimal control, provided that

$$0 < k \le 2 \tag{8-500}$$

The above considerations lead us to the conclusion that, in the case of nonlinear systems, optimal singular controls may appear quite often, in contradistinction to their nonoccurrence for linear systems.

**Exercise 8-44** Using the techniques of Sec. 8-10, obtain the optimal-control law to the origin for the system

$$\begin{aligned} \dot{x}_1(t) &= x_2(t) \\ \dot{x}_2(t) &= -\tfrac{1}{3}x_2^3(t) + u(t) \qquad |u(t)| \leq 1 \end{aligned} \tag{8-501}$$

with the cost functional

$$J = \int_0^T [k + |u(t)|]\, dt \qquad T \text{ not specified} \tag{8-502}$$

Treat both the singular control case ($0 < k \leq 2$) and the nonsingular case ($k > 2$).

**Exercise 8-45** Consider the system

$$\begin{aligned} \dot{x}_1(t) &= x_2(t) \\ \dot{x}_2(t) &= -x_2(t) - x_2(t)|x_2(t)| + u(t) \qquad |u(t)| \leq 1 \end{aligned} \tag{8-503}$$

and the cost functional $J = \int_0^T [k + |u(t)|]\, dt$, $T$ not specified. Determine for which values of $k$ singular controls can be optimal. Determine the optimal control to the origin for the singular case.

**Exercise 8-46** Consider the linear system

$$\begin{aligned} \dot{x}_1(t) &= -x_1(t) - u(t) \\ \dot{x}_2(t) &= -2x_2(t) - 2u(t) \qquad |u(t)| \leq 1 \end{aligned} \tag{8-504}$$

and the cost functional $J = \int_0^T [k + |u(t)|]\, dt$, $T$ not specified. Determine the optimal control to the origin. HINT: The time-optimal control of the system (8-504) was treated in Sec. 7-3. There are no singular controls for this problem. The most general optimal control sequences are $\{0, +1, 0, -1\}$ and $\{0, -1, 0, +1\}$. See also Ref. K-17.

**Exercise 8-47** Consider the system (8-504) and the cost functional $F = \int_0^T |u(t)|\, dt$, $T$ fixed a priori. Determine the optimal control to the origin. HINT: Use the techniques of Sec. 8-6.

**Exercise 8-48** Consider the harmonic oscillator

$$\begin{aligned} \dot{x}_1(t) &= x_2(t) \\ \dot{x}_2(t) &= -x_1(t) + u(t) \qquad |u(t)| \leq 1 \end{aligned} \tag{8-505}$$

and the cost functional (fuel) $F = \int_0^T |u(t)|\, dt$, $T$ not specified. Suppose that the initial state is $\xi = (\xi_1, \xi_2)$ and the terminal state is the origin (0, 0). Show that the greatest lower bound $F^*$ on the fuel $F = \int_0^T |u(t)|\, dt$ is $F^* \geq \sqrt{\xi_1^2 + \xi_2^2}$. [HINT: The fundamental matrix $\Phi(t)$ of the system (8-505) is orthogonal; thus,

$$\|\xi\| = \left\| \int_0^T \Phi^{-1}(t) \begin{bmatrix} 0 \\ 1 \end{bmatrix} u(t)\, dt \right\| \leq \int_0^T \left\| \Phi^{-1}(t) \begin{bmatrix} 0 \\ 1 \end{bmatrix} u(t) \right\| dt$$

etc.] Show that, strictly speaking, a fuel-optimal solution does not exist for this problem. Show that given $\epsilon > 0$, there are a $\delta > 0$ and a control of the form

$$\begin{aligned} u(t) &= -\operatorname{sgn}\{x_2\} \qquad &&\text{if } |x_1| \leq \delta \\ u(t) &= 0 &&\text{otherwise} \end{aligned} \tag{8-506}$$

such that the state is forced near the origin with consumed fuel $F^* + \epsilon$. HINT: The time-optimal control of the harmonic oscillator was considered in Sec. 7-7.

**Exercise 8-49** Consider the harmonic oscillator (8-505) with the cost functional $F = \int_0^T |u(t)|\, dt$, $T$ fixed. Use the techniques of Sec. 8-6 to obtain the fuel-optimal control that forces any initial state $(\xi_1, \xi_2)$ to $(0, 0)$, where $\xi_1^2 + \xi_2^2 \leq 4$.

**Exercise 8-50** Consider the harmonic oscillator (8-505) and the cost functional $J = \int_0^T [k + |u(t)|]\, dt$. The origin $(0, 0)$ is the desired terminal state. Determine the optimal control and the switch curves for initial states $(\xi_1, \xi_2)$ such that $\xi_1^2 + \xi_2^2 \leq 1$.

**Exercise 8-51** Consider the system (8-470) with the cost functional (8-471). Suppose that the drag force $f[x_2(t)]$ is such that Eqs. (8-472) to (8-475) are satisfied and that we are given a plot of the function $f[x_2(t)]$ versus $x_2(t)$ but do not know an analytical expression for $f[x_2(t)]$. Outline a method that will enable you to determine the singular control using graphical means exclusively.

# Chapter 9

# THE DESIGN OF OPTIMAL LINEAR SYSTEMS WITH QUADRATIC CRITERIA

## 9-1 Introduction

Our study of the design of time-optimal systems and fuel-optimal systems indicates that there are considerable difficulties in the determination of the optimal-control law and further difficulties in the implementation of the optimal feedback system. Even if the controlled system is linear, it is still almost impossible to obtain general analytical results either in the time-optimal case or in the fuel-optimal case. The reader is probably wondering by now whether there exist systems and criteria for which elegant and general results are available.

In this chapter, we shall obtain analytical results for a particular class of systems with a particular class of performance criteria. In fact, we shall examine linear time-varying systems whose state $\mathbf{x}(t)$, control $\mathbf{u}(t)$, and output $\mathbf{y}(t)$ are related by the system of equations

$$\begin{aligned} \dot{\mathbf{x}}(t) &= \mathbf{A}(t)\mathbf{x}(t) + \mathbf{B}(t)\mathbf{u}(t) \\ \mathbf{y}(t) &= \mathbf{C}(t)\mathbf{x}(t) \end{aligned}$$

The most general form of the cost functional that we shall consider will be

$$\begin{aligned} J(\mathbf{u}) = \tfrac{1}{2}\langle[\mathbf{z}(T) - \mathbf{y}(T)], \mathbf{F}[\mathbf{z}(T) - \mathbf{y}(T)]\rangle \\ + \tfrac{1}{2}\int_{t_0}^{T} \{\langle[\mathbf{z}(t) - \mathbf{y}(t)], \mathbf{Q}(t)[\mathbf{z}(t) - \mathbf{y}(t)]\rangle + \langle\mathbf{u}(t), \mathbf{R}(t)\mathbf{u}(t)\rangle\}\, dt \end{aligned}$$

where $\mathbf{z}(t)$ represents a desired output.

We shall assume that there are no constraints on the magnitude of the components of the control vector $\mathbf{u}(t)$, and we shall also make certain assumptions concerning the matrices $\mathbf{F}$, $\mathbf{Q}(t)$, and $\mathbf{R}(t)$. Under these assumptions, we shall be able to obtain an analytic expression for the optimal control, and we shall find that the optimal feedback system is linear.

We shall now discuss the structure of this chapter, so that the reader can obtain a general idea of its contents.

In Sec. 9-2, we discuss the formulation of the optimum problem. We shall spend some time on motivating the choice of the quadratic cost functional $J(\mathbf{u})$, paying particular attention to its physical significance. Physical considerations will lead us to certain mathematical assumptions on the matrices $\mathbf{F}$, $\mathbf{Q}(t)$, and $\mathbf{R}(t)$. Thus, right from the start, we shall be sure that we are dealing with a meaningful optimization problem from the engineering point of view. In Sec. 9-3, we start the mathematical development for the so-called *state-regulator problem* (i.e., the problem of keeping the state near zero). We proceed in our usual way, using the minimum principle† to derive the necessary conditions on the optimal control. We shall find that the canonical equations relating the state $\mathbf{x}(t)$ to the costate $\mathbf{p}(t)$ are linear. The linearity of the canonical equations allows us to conclude that the state $\mathbf{x}(t)$ and the costate $\mathbf{p}(t)$ are related by an equation of the form $\mathbf{p}(t) = \mathbf{K}(t)\mathbf{x}(t)$, where $\mathbf{K}(t)$ is a real symmetric matrix. Considerations relating to the computation of $\mathbf{K}(t)$ lead us to Control Law 9-1. We then establish that our assumptions allow us to conclude that the local necessary conditions are actually globally sufficient. Thus, we are able to prove that the optimal control exists and is, in addition, unique.

In Sec. 9-4, we discuss, in an informal manner, some of the implications of Control Law 9-1 and comment on the computation of the matrix $\mathbf{K}(t)$ using a digital computer.

In Sec. 9-5, we immediately state Control Law 9-2, which, in essence, states that if the system is time-invariant, if the matrices $\mathbf{Q}(t)$ and $\mathbf{R}(t)$ are constant, and if $T = \infty$, then the resulting optimal feedback system is linear and time-invariant.

In Sec. 9-6, we analyze a very simple first-order optimal regulator. We compute a few time responses so that the reader can gain some physical insight into the operation and properties of optimal regulators.

In Sec. 9-7, we attack the so-called *output-regulator problem;* in this problem, we are concerned with keeping the output $\mathbf{y}(t)$ [rather than the state $\mathbf{x}(t)$] near zero. We obtain Control Law 9-3 by showing that the output-regulator problem can be solved in the same way as the state-regulator problem of Sec. 9-3. We also obtain Control Law 9-4 (which is the counterpart of Control Law 9-2) by specializing the conclusions of Control Law 9-3 to the time-invariant case with $T = \infty$.

In Sec. 9-8, we consider time-invariant single-input–single-output systems and discuss the structure of the optimal feedback system in terms of its closed-loop transfer function and in terms of the poles of the closed-loop system.

In Sec. 9-9, we consider and solve the problem of keeping the output $\mathbf{y}(t)$ near a desired output $\mathbf{z}(t)$; this we call the *optimal tracking problem.* The main results of this section are summarized in Control Law 9-5. We spend some time discussing the results from an engineering point of view. We

† We can actually use the results of Secs. 5-7 and 5-8; in particular, see Corollary 5-2.

find that the feedback structure of the optimal tracking system is identical to that of the optimal output regulator; the feature that distinguishes the tracking system from the output regulator is a feedforward channel which delivers a signal that depends on the desired output $\mathbf{z}(t)$.

Section 9-10 contains *approximate* results that are valid for constant systems, constant desired outputs, and increasingly large values of the terminal time $T$. The approximations indicate that the optimal system behaves approximately as a linear time-invariant feedback system. The lack of precise results for the limiting case $T = \infty$ is due to the current unavailability of a theory which guarantees the existence of an optimal solution as $T \to \infty$.

In Sec. 9-11, we present some trivial (sufficient) conditions on the desired output $\mathbf{z}(t)$ which enable us to transform a tracking problem into a regulator problem.

Section 9-12 contains computed responses of a first-order optimal tracking system. The system is the same as the one analyzed in Sec. 9-6. Figures which illustrate the response of the tracking system to step inputs, delayed step inputs, and sinusoidal inputs are also included.

We conclude the chapter with Sec. 9-13, which contains a very brief discussion of additional results available in the literature.

Throughout this chapter (and especially in Secs. 9-3, 9-5, 9-7, and 9-9), we have relied very heavily on several papers by Kalman. The main references we have used are K-3, K-6, K-8, and K-9. We have omitted in this chapter many details of proofs which can be found in these references.†

Throughout the chapter, we have attempted to illustrate the theory by simple examples. Whenever possible, we have attempted to discuss in a heuristic fashion the engineering implications of the mathematical results. Exercises are spread throughout the chapter. Some of these exercises can be solved only by students who have access to a modern digital computer.

## 9-2 Formulation of the Problem

Suppose that we are given a linear time-varying dynamical system with *state* $\mathbf{x}(t)$, *control input* $\mathbf{u}(t)$, and *output* $\mathbf{y}(t)$. We suppose that the system equations are

$$\begin{aligned} \dot{\mathbf{x}}(t) &= \mathbf{A}(t)\mathbf{x}(t) + \mathbf{B}(t)\mathbf{u}(t) \\ \mathbf{y}(t) &= \mathbf{C}(t)\mathbf{x}(t) \end{aligned} \tag{9-1}$$

We assume that $x_1(t), x_2(t), \ldots, x_n(t)$ are the components of the state vector $\mathbf{x}(t)$, that $u_1(t), u_2(t), \ldots, u_r(t)$ are the components of the control vector $\mathbf{u}(t)$, and that $y_1(t), y_2(t), \ldots, y_m(t)$ are the components of the output vector $\mathbf{y}(t)$. Thus, $\mathbf{A}(t)$ is an $n \times n$ matrix, $\mathbf{B}(t)$ is an $n \times r$ matrix, and

† Other references are K-13, S-1, R-1, and P-8, and the references cited in Secs. 6-17, 6-18, 6-20, and 6-21. See also the book by Merriam [M-7.]

$\mathbf{C}(t)$ is an $m \times n$ matrix. Furthermore, we assume that

$$0 < m \leq r \leq n \tag{9-2}$$

and that

$$\mathbf{u}(t) \text{ is not constrained} \tag{9-3}$$

We now define the objective of the system from a physical point of view. Let $\mathbf{z}(t)$ be a vector with $m$ components $z_1(t), z_2(t), \ldots, z_m(t)$.† *The objective is to control the system* (9-1) *so that the output vector* $\mathbf{y}(t)$ *is "near" the vector* $\mathbf{z}(t)$. Let us agree to call the vector $\mathbf{z}(t)$ the *desired output.* We may then define an *error vector* $\mathbf{e}(t)$ by setting

$$\mathbf{e}(t) = \mathbf{z}(t) - \mathbf{y}(t) \tag{9-4}$$

Thus, loosely speaking, the control objective is: *Find a control* $\mathbf{u}(t)$ *so that the error* $\mathbf{e}(t)$ *is "small."*

We assumed that the control $\mathbf{u}(t)$ was not constrained in magnitude; thus, there may be cases in which the control $\mathbf{u}(t)$ is extremely large. In order to avoid such unrealistic situations (which require extremely large gains in the control loop), we may wish to include in our control objective a statement regarding the fact that "large control signals cost money." *In other words, we would like, on the one hand, to keep the error "small," but, on the other hand, we must not use unnecessarily "large" controls.*

The translation of these physical specifications and requirements into a particular mathematical performance functional is, of course, up to the engineer. In this chapter, we shall choose a particular class of cost functionals, quadratic in nature, which indeed correspond to the physical requirements. It is certainly true that this choice of the cost functionals may not be wise in a given situation; in such a case, the engineer must develop his own performance criterion and work out the problem himself. To aid the engineer in understanding the meaning of the cost functionals we shall consider, we shall present below a discussion of the physical significance of the mathematical terms.

To be precise, let us consider the cost functional

$$J(\mathbf{u}) = \tfrac{1}{2}\langle \mathbf{e}(T), \mathbf{F}\mathbf{e}(T)\rangle + \tfrac{1}{2}\int_{t_0}^{T} [\langle \mathbf{e}(t), \mathbf{Q}(t)\mathbf{e}(t)\rangle + \langle \mathbf{u}(t), \mathbf{R}(t)\mathbf{u}(t)\rangle]\,dt \tag{9-5}$$

where

The terminal time $T$ is specified
$\mathbf{F}$ is a constant $m \times m$ positive semidefinite matrix
$\mathbf{Q}(t)$ is an $m \times m$ positive semidefinite matrix
$\mathbf{R}(t)$ is an $r \times r$ positive definite matrix
(9-6)

The assumptions of (9-6) will be used throughout this chapter.

We shall use the concepts of a positive semidefinite and a positive definite matrix throughout this chapter; we recall (see Sec. 2-15) that a real sym-

† Some or all of the $z_i$ may be equal to zero; note that $m$ is the dimension of the output $\mathbf{y}(t)$.

metric matrix $\mathbf{M}$ is *positive definite* if $\langle \mathbf{a}, \mathbf{Ma} \rangle > 0$ for all $\mathbf{a} \neq \mathbf{0}$ and that a real symmetric matrix $\mathbf{N}$ is *positive semidefinite* if $\langle \mathbf{b}, \mathbf{Nb} \rangle \geq 0$ for all $\mathbf{b} \neq \mathbf{0}$. We note that the zero matrix is clearly positive semidefinite.

Let us now consider each term in the cost functional $J(\mathbf{u})$ given by Eq. (9-5) and see in what way it represents a reasonable mathematical translation of the physical requirements and specifications.

First, consider the term $L_e = \frac{1}{2}\langle \mathbf{e}(t), \mathbf{Q}(t)\mathbf{e}(t) \rangle$, where $\mathbf{Q}(t)$ is positive semidefinite. Clearly, this term is *nonnegative* for all $\mathbf{e}(t)$ and, in particular, is zero when $\mathbf{e}(t) = \mathbf{0}$. If the error is small for all $t \in [t_0, T]$, then the integral of $L_e$ will be small. Since

$$L_e = \tfrac{1}{2}\langle \mathbf{e}(t), \mathbf{Q}(t)\mathbf{e}(t) \rangle = \tfrac{1}{2} \sum_{i=1}^{m} \sum_{j=1}^{m} q_{ij}(t) e_i(t) e_j(t)$$

where the $q_{ij}(t)$ are the entries in $\mathbf{Q}(t)$ and the $e_i(t)$ and $e_j(t)$ are the components of $\mathbf{e}(t)$, it is clear that the cost $L_e$ weighs large errors much more heavily than small errors, and so *the system is penalized much more severely for large errors than for small errors.*†

Next, we consider the term $L_u = \frac{1}{2}\langle \mathbf{u}(t), \mathbf{R}(t)\mathbf{u}(t) \rangle$, with $\mathbf{R}(t)$ positive definite. This term weighs the cost of the control and penalizes the system much more severely for large controls than for small controls. Since $\mathbf{R}(t)$ is positive definite, the cost of control is always positive for all $\mathbf{u}(t) \neq \mathbf{0}$. This term $L_u$ is often called *control power*, and $\int_{t_0}^{T} L_u \, dt$ is often called the *control energy*. The reason for this nomenclature is the following: Suppose that $u(t)$ is a scalar proportional to voltage or current; then $u^2(t)$ is proportional to power and $\int_{t_0}^{T} u^2(t) \, dt$ is proportional to the energy expended in the interval $[t_0, T]$. The requirement that $\mathbf{R}(t)$ be positive definite rather than positive semidefinite is, as we shall see, a condition for the existence of a finite control. We remark that this requirement was also used in Sec. 6-18.

Finally, we turn our attention to the term $\frac{1}{2}\langle \mathbf{e}(T), \mathbf{Fe}(T) \rangle$. This term is often called the *terminal cost*, and its purpose is to guarantee that the error $\mathbf{e}(T)$ at the terminal time $T$ is small. In other words, this term should be included if the value of $\mathbf{e}(t)$ at the terminal time is particularly important; if this is not the case, then we can set $\mathbf{F} = \mathbf{0}$ and rely upon the rest of the cost functional to guarantee that the terminal error is not excessively large.

A question that may arise is the following: Why should the matrices $\mathbf{Q}(t)$ and $\mathbf{R}(t)$ be time-varying rather than constant? If we develop a theory

† If $L_e$ were replaced by $\sum_{i=1}^{m} |q_i(t)e_i(t)|$, then the system would be penalized equally for large or small errors. The use of a functional like $\int_{t_0}^{T} \sum_{i=1}^{m} q_i(t)e_i(t) \, dt$ is not recommended because cancellation of large positive errors by large negative errors may result in a zero cost, although large errors are always present.

for time-varying matrices $\mathbf{Q}(t)$ and $\mathbf{R}(t)$ rather than for constant matrices $\mathbf{Q}$ and $\mathbf{R}$, then we can use the cost functional $J(\mathbf{u})$ for more realistic problems. For example, suppose that at $t = t_0$, the state $\mathbf{x}(t_0)$ is given and hence the output $\mathbf{y}(t_0) = \mathbf{C}(t_0)\mathbf{x}(t_0)$ is known; then it may happen that $\mathbf{z}(t_0)$ is such that the initial error $\mathbf{e}(t_0)$ is very large. The fact that $\mathbf{e}(t_0)$ is large is certainly not the fault of the control system. For this reason, we may want to choose $\mathbf{Q}(t)$ in such a way that large *initial* errors are penalized less severely than subsequent large errors. To do this, we can pick $\mathbf{Q}(t)$ in such a way that, given the times

$$t_0 < t_1 \ll t_2 < T \tag{9-7}$$†

and any constant vector $\mathbf{b}$, the scalar products $\langle \mathbf{b}, \mathbf{Q}(t_1)\mathbf{b}\rangle$ and $\langle \mathbf{b}, \mathbf{Q}(t_2)\mathbf{b}\rangle$ satisfy the relation

$$\langle \mathbf{b}, \mathbf{Q}(t_1)\mathbf{b}\rangle \ll \langle \mathbf{b}, \mathbf{Q}(t_2)\mathbf{b}\rangle \tag{9-8}$$

It is our hope that the preceding discussion of the physical properties of the cost functional $J(\mathbf{u})$ will help the reader understand its properties. We wish to emphasize that whether or not it should be used for a given problem is up to the designer. *Nonetheless, it turns out that this cost functional has two very desirable properties: First, it is mathematically tractable, and second, it leads to optimal feedback systems which are linear.* This is the reason that people say that "quadratic criteria fit linear systems like a glove."

With the physical motivation behind us, we are now ready for the formal statement of the problem.

The problem that we shall consider in this chapter is:

*Given the linear system* (9-1) *and the cost functional* (9-5) *satisfying the assumptions* (9-2) *to* (9-4) *and* (9-6). *Find the optimal control, i.e., the control which will drive the system* (9-1) *so as to minimize the cost functional* (9-5).

Before we attempt the solution of this problem, let us obtain some self-evident but useful results. First of all, if $\mathbf{Q}(t) = \mathbf{F} = \mathbf{0}$, then the optimal control must be $\mathbf{u}(t) = \mathbf{0}$; in this case, the cost is $\frac{1}{2}\int_{t_0}^{T} \langle \mathbf{u}(t), \mathbf{R}(t)\mathbf{u}(t)\rangle\, dt$, and since we do not specify the terminal state $\mathbf{x}(T)$, the cost is minimized if and only if $\mathbf{u}(t) = \mathbf{0}$. To exclude this trivial case, *we shall assume throughout the chapter that* $\mathbf{Q}(t)$ *and* $\mathbf{F}$ *are not both the zero matrix, although we shall allow either* $\mathbf{Q}(t)$ *or* $\mathbf{F}$ *to be the zero matrix individually.*

Now let $\mathbf{\Phi}(t; t_0)$ be the fundamental matrix of the linear system (9-1) and let $\mathbf{x}(t_0)$ be the initial state. Then the state $\mathbf{x}(t)$, for $t \in [t_0, T]$, is given by

$$\mathbf{x}(t) = \mathbf{\Phi}(t; t_0)\left[\mathbf{x}(t_0) + \int_{t_0}^{t} \mathbf{\Phi}^{-1}(\tau; t_0)\mathbf{B}(\tau)\mathbf{u}(\tau)\, d\tau\right] \tag{9-9}$$

† The statement $a \ll b$ means that $b$ is much greater than $a$.

and, hence, the error $\mathbf{e}(t)$ is given by the equation

$$\mathbf{e}(t) = \mathbf{z}(t) - \mathbf{C}(t)\mathbf{\Phi}(t; t_0)\left[\mathbf{x}(t_0) + \int_{t_0}^{t} \mathbf{\Phi}^{-1}(\tau; t_0)\mathbf{B}(\tau)\mathbf{u}(\tau)\, d\tau\right] \quad (9\text{-}10)$$

in view of Eqs. (9-1) and (9-4).

***Lemma 9-1*** *If the initial state $\mathbf{x}(t_0)$ and the desired output $\mathbf{z}(t)$ are related by*

$$\mathbf{z}(t) = \mathbf{C}(t)\mathbf{\Phi}(t; t_0)\mathbf{x}(t_0) \qquad \textit{for all } t \in [t_0, T] \quad (9\text{-}11)$$

*then the optimal control is $\mathbf{u}(t) = \mathbf{0}$ and the minimum cost $J^*$ is zero.*

PROOF Substitute $\mathbf{u}(\tau) = \mathbf{0}$ and Eq. (9-11) into Eq. (9-10) to obtain $\mathbf{e}(t) = \mathbf{0}$.

***Lemma 9-2*** *If $\mathbf{u}(t) \neq \mathbf{0}$, then the cost $J$ must be positive.*

PROOF This follows from the assumptions that $\mathbf{F}$ and $\mathbf{Q}(t)$ are positive semidefinite and that $\mathbf{R}(t)$ is positive definite.

***Lemma 9-3*** *If, for every $\mathbf{u}(t)$, the cost functional $J(\mathbf{u})$ is not defined, that is, if $J(\mathbf{u}) = \infty$ for all $\mathbf{u}(t)$, then an optimal solution does not exist.*

PROOF This is obvious because, if (say) two controls yield an infinite cost, then we cannot distinguish which of the two controls is better.

We conclude this section with the suggestion that the reader review the material of Secs. 6-17 to 6-20. The basic difference between the material of these sections and the material which we shall present in this chapter is: In Chap. 6, we specified either the terminal state or the target set; in this chapter, the terminal state is completely unspecified. We shall show that this small change in the optimal control problem has important consequences from a practical point of view.

## 9-3 The State-regulator Problem†

In this section, we consider the so-called state-regulator problem. Basically, the solution of the state-regulator problem leads to an optimal feedback system with the property that the components of the state vector $\mathbf{x}(t)$ are kept near zero without excessive expenditure of control energy.

We shall now outline the structure of this section. We shall start by applying the necessary conditions provided by the minimum principle to find the extremal controls. We shall show that:

1. The $H$-minimal control turns out to be a linear function of the costate.
2. The canonical equations reduce to a system of $2n$ homogeneous linear differential equations.
3. The transversality condition on the costate $\mathbf{p}(t)$ at the terminal time $T$ specifies that the costate $\mathbf{p}(T)$ be a linear function of the unspecified terminal state $\mathbf{x}(T)$.

Motivated by computational considerations, we shall then show that the extremal control is a linear function of the state. We shall discover that

† We have relied almost exclusively on the results of Ref. K-3 in this section.

the time-varying matrix relating the extremal control to the state can be found by solving a matrix differential equation (the matrix Riccati equation). We shall then prove that the extremal control is at least locally optimal and unique. The existence of the optimal control will be deduced from the fact that the solution of the Hamilton-Jacobi equation is defined everywhere.

Let us consider the linear time-varying system

$$\dot{\mathbf{x}}(t) = \mathbf{A}(t)\mathbf{x}(t) + \mathbf{B}(t)\mathbf{u}(t) \tag{9-12}$$

and the cost functional $J_1$ given by

$$J_1 = \tfrac{1}{2}\langle \mathbf{x}(T), \mathbf{F}\mathbf{x}(T)\rangle + \tfrac{1}{2}\int_{t_0}^{T} [\langle \mathbf{x}(t), \mathbf{Q}(t)\mathbf{x}(t)\rangle + \langle \mathbf{u}(t), \mathbf{R}(t)\mathbf{u}(t)\rangle]\, dt \tag{9-13}$$

Note that if we set $\mathbf{C}(t) = \mathbf{I}$ and $\mathbf{z}(t) = \mathbf{0}$, then $\mathbf{y}(t) = \mathbf{x}(t) = -\mathbf{e}(t)$ and the cost functional $J(\mathbf{u})$ of Eq. (9-5) reduces to the functional $J_1$ of Eq. (9-13). The physical interpretation of $J_1$ is this: *We wish to keep the state near zero without excessive control-energy expenditure.* We still assume that the assumptions given by (9-6) are satisfied.

We shall show in this section that the optimal control is a linear function of the state, i.e., is of the form

$$\mathbf{u}(t) = \mathbf{G}(t)\mathbf{x}(t) \qquad t \in [t_0, T] \tag{9-14}$$

where $\mathbf{G}(t)$ is an $r \times n$ matrix-valued function.

Let us assume that an optimal control exists for any initial state. We can use the minimum principle to obtain the necessary conditions for the optimal control and so derive the extremal controls. The Hamiltonian $H$ for the system (9-12) and the cost $J_1$ of Eq. (9-13) is

$$H = \tfrac{1}{2}\langle \mathbf{x}(t), \mathbf{Q}(t)\mathbf{x}(t)\rangle + \tfrac{1}{2}\langle \mathbf{u}(t), \mathbf{R}(t)\mathbf{u}(t)\rangle + \langle \mathbf{A}(t)\mathbf{x}(t), \mathbf{p}(t)\rangle + \langle \mathbf{B}(t)\mathbf{u}(t), \mathbf{p}(t)\rangle \tag{9-15}$$

The costate vector $\mathbf{p}(t)$ is the solution of the vector differential equation

$$\dot{\mathbf{p}}(t) = -\frac{\partial H}{\partial \mathbf{x}(t)} \tag{9-16}$$

which reduces to

$$\dot{\mathbf{p}}(t) = -\mathbf{Q}(t)\mathbf{x}(t) - \mathbf{A}'(t)\mathbf{p}(t) \tag{9-17}$$

Along the optimal trajectory, we must have†

$$\frac{\partial H}{\partial \mathbf{u}(t)} = \mathbf{0} \tag{9-18}$$

which implies that

$$\frac{\partial H}{\partial \mathbf{u}(t)} = \mathbf{R}(t)\mathbf{u}(t) + \mathbf{B}'(t)\mathbf{p}(t) = \mathbf{0} \tag{9-19}$$

† See Secs. 5-7 and 6-18.

From Eq. (9-19), we deduce that

$$\mathbf{u}(t) = -\mathbf{R}^{-1}(t)\mathbf{B}'(t)\mathbf{p}(t) \tag{9-20}$$

The assumption that $\mathbf{R}(t)$ is positive definite for all $t \in [t_0, T]$ guarantees† the existence of $\mathbf{R}^{-1}(t)$ for all $t \in [t_0, T]$.

We know that the optimal control must minimize the Hamiltonian. The necessary condition $\partial H/\partial\mathbf{u}(t) = \mathbf{0}$ yields only an extremum of $H$ with respect to $\mathbf{u}(t)$. In order for the extremum of $H$ to be a minimum with respect to $\mathbf{u}(t)$, the $r \times r$ matrix $\partial^2 H/\partial\mathbf{u}^2(t)$ must be *positive definite.* But, from Eq. (9-19), we find that

$$\frac{\partial^2 H}{\partial\mathbf{u}^2(t)} = \mathbf{R}(t) \tag{9-21}$$

and hence, since $\mathbf{R}(t)$ was assumed to be positive definite, it follows that the control $\mathbf{u}(t)$ given by Eq. (9-20) does indeed minimize $H$ and, hence, is $H$-minimal.

The next step is to obtain the reduced canonical equations; to do that, we substitute Eq. (9-20) into Eq. (9-12) to obtain the relation

$$\dot{\mathbf{x}}(t) = \mathbf{A}(t)\mathbf{x}(t) - \mathbf{B}(t)\mathbf{R}^{-1}(t)\mathbf{B}'(t)\mathbf{p}(t) \tag{9-22}$$

Equations (9-22) and (9-17) are the reduced canonical equations. Define the matrix $\mathbf{S}(t)$ by setting

$$\mathbf{S}(t) = \mathbf{B}(t)\mathbf{R}^{-1}(t)\mathbf{B}'(t) \tag{9-23}$$

Note that $\mathbf{S}(t)$ is a *symmetric* $n \times n$ *matrix* (why?). Using the matrix $\mathbf{S}(t)$, we can combine the canonical equations (9-22) and (9-17) in the form

$$\begin{bmatrix} \dot{\mathbf{x}}(t) \\ \dot{\mathbf{p}}(t) \end{bmatrix} = \begin{bmatrix} \mathbf{A}(t) & -\mathbf{S}(t) \\ -\mathbf{Q}(t) & -\mathbf{A}'(t) \end{bmatrix} \begin{bmatrix} \mathbf{x}(t) \\ \mathbf{p}(t) \end{bmatrix} \tag{9-24}$$

Equation (9-24) is a system of $2n$ linear time-varying homogeneous differential equations. We know (see Sec. 3-20) that we can obtain a *unique* solution of this system of differential equations provided that we know a total of $2n$ boundary conditions. A total of $n$ boundary conditions is provided by the initial state at $t_0$, $\mathbf{x}(t_0)$. The remaining $n$ boundary conditions are furnished by the transversality conditions,‡ which require [since $\mathbf{x}(T)$ is not specified] that, at the terminal time $T$, the costate $\mathbf{p}(T)$ must satisfy the relation

$$\mathbf{p}(T) = \frac{\partial}{\partial\mathbf{x}(T)} \left[\tfrac{1}{2}\langle\mathbf{x}(T), \mathbf{F}\mathbf{x}(T)\rangle\right] \tag{9-25}$$

Thus, we deduce that

$$\mathbf{p}(T) = \mathbf{F}\mathbf{x}(T) \tag{9-26}$$

† A necessary condition that $\mathbf{M}$ be positive definite is $\det \mathbf{M} > 0$, and so $\mathbf{M}$ is nonsingular. See Theorem 2-4.

‡ See Table 5-1.

Let $\mathbf{\Omega}(t; t_0)$ be the $2n \times 2n$ fundamental matrix for the system (9-24).† If we let $\mathbf{p}(t_0)$ be the (unknown) initial costate, then the solution of Eq. (9-24) is of the form

$$\begin{bmatrix} \mathbf{x}(t) \\ \hdashline \mathbf{p}(t) \end{bmatrix} = \mathbf{\Omega}(t; t_0) \begin{bmatrix} \mathbf{x}(t_0) \\ \hdashline \mathbf{p}(t_0) \end{bmatrix} \tag{9-27}$$

Therefore, at $t = T$, we must have the relation

$$\begin{bmatrix} \mathbf{x}(T) \\ \hdashline \mathbf{p}(T) \end{bmatrix} = \mathbf{\Omega}(T; t) \begin{bmatrix} \mathbf{x}(t) \\ \hdashline \mathbf{p}(t) \end{bmatrix} \tag{9-28}$$

Next, we partition the $2n \times 2n$ matrix $\mathbf{\Omega}(T; t)$ into four $n \times n$ submatrices as follows:

$$\mathbf{\Omega}(T; t) = \begin{bmatrix} \mathbf{\Omega}_{11}(T; t) & \mathbf{\Omega}_{12}(T; t) \\ \mathbf{\Omega}_{21}(T; t) & \mathbf{\Omega}_{22}(T; t) \end{bmatrix} \tag{9-29}$$

Then Eq. (9-28) can be written in the form [using $\mathbf{p}(T) = \mathbf{F}\mathbf{x}(T)$]

$$\mathbf{x}(T) = \mathbf{\Omega}_{11}(T; t)\mathbf{x}(t) + \mathbf{\Omega}_{12}(T; t)\mathbf{p}(t) \tag{9-30}$$

$$\mathbf{p}(T) = \mathbf{\Omega}_{21}(T; t)\mathbf{x}(t) + \mathbf{\Omega}_{22}(T; t)\mathbf{p}(t) = \mathbf{F}\mathbf{x}(T) \tag{9-31}$$

From Eqs. (9-30) and (9-31), we find, after some algebraic manipulations, that

$$\mathbf{p}(t) = [\mathbf{\Omega}_{22}(T; t) - \mathbf{F}\mathbf{\Omega}_{12}(T; t)]^{-1}[\mathbf{F}\mathbf{\Omega}_{11}(T; t) - \mathbf{\Omega}_{21}(T; t)]\mathbf{x}(t) \tag{9-32}$$

provided the indicated inverse exists. Equation (9-32) suggests that the costate $\mathbf{p}(t)$ and the state $\mathbf{x}(t)$ are related by an equation of the form

$$\mathbf{p}(t) = \mathbf{K}(t)\mathbf{x}(t) \tag{9-33}$$

for all $t \in [t_0, T]$. *The matrix* $\mathbf{K}(t)$ *is an* $n \times n$ *time-varying matrix which depends upon the terminal time* $T$ *and the matrix* $\mathbf{F}$ *but does not depend upon the initial state.* In fact,

$$\mathbf{K}(t) = [\mathbf{\Omega}_{22}(T; t) - \mathbf{F}\mathbf{\Omega}_{12}(T; t)]^{-1}[\mathbf{F}\mathbf{\Omega}_{11}(T; t) - \mathbf{\Omega}_{21}(T; t)] \tag{9-34}$$

At this point, a few comments may be helpful. We have not shown that the matrix $[\mathbf{\Omega}_{22}(T; t) - \mathbf{F}\mathbf{\Omega}_{12}(T; t)]$ has an inverse. Let us examine the situation at $t = T$. We know that

$$\mathbf{\Omega}(T; T) = \mathbf{I} \tag{9-35}$$

so that the relations

$$\mathbf{\Omega}_{11}(T; T) = \mathbf{\Omega}_{22}(T; T) = \mathbf{I} \tag{9-36}$$

$$\mathbf{\Omega}_{12}(T; T) = \mathbf{\Omega}_{21}(T; T) = \mathbf{0} \tag{9-37}$$

† Compare also with Eq. (6-503).

hold. Thus, the matrix $[\mathbf{\Omega}_{22}(T; T) - \mathbf{F}\mathbf{\Omega}_{12}(T; T)] = \mathbf{I}$ is nonsingular. Moreover,

$$\mathbf{K}(T) = [\mathbf{\Omega}_{22}(T; T) - \mathbf{F}\mathbf{\Omega}_{12}(T; T)]^{-1}[\mathbf{F}\mathbf{\Omega}_{11}(T; T) - \mathbf{\Omega}_{21}(T; T)] \quad (9\text{-}38)$$

so that, in view of Eqs. (9-36) and (9-37),

$$\mathbf{K}(T) = \mathbf{F} \quad (9\text{-}39)$$

Therefore, Eq. (9-33) holds at $t = T$. It can be shown† that the required inverse exists for all $t$, $t_0 \le t < T$, so that the relation provided by Eq. (9-33) is valid.

Let us now comment on evaluating the matrix $\mathbf{K}(t)$. If the matrices $\mathbf{A}(t)$, $\mathbf{S}(t)$, and $\mathbf{Q}(t)$ are time-varying, then it is impossible, in general, to obtain an analytical expression for the $2n \times 2n$ fundamental matrix $\mathbf{\Omega}(T; t)$. In this case, one must evaluate $\mathbf{K}(t)$ by using (say) a digital computer. If, however, the matrices $\mathbf{A}(t)$, $\mathbf{S}(t)$, and $\mathbf{Q}(t)$ are time-invariant, then the matrix $\mathbf{\Omega}(T; t)$ can be evaluated analytically by using, for example, Laplace transforms; nonetheless, even in that case, the evaluation of the inverse matrix in Eq. (9-32) is an extremely laborious task, especially if the order of the system is high, i.e., if $n$ is a large number.

It seems up to now that the difficulties involved from the computational point of view are almost insurmountable. The natural question that arises is the following: *We have justified the relationship* $\mathbf{p}(t) = \mathbf{K}(t)\mathbf{x}(t)$*; can we find additional properties of the matrix* $\mathbf{K}(t)$ *which, perhaps, will provide us with an alternative means of computing it and which, hopefully, do not involve computation of the inverse of an* $n \times n$ *matrix?*

The answer to this question is *yes,* and we shall devote the remainder of this section to demonstrating this.

What we shall do now is the following: We shall start from the relation $\mathbf{p}(t) = \mathbf{K}(t)\mathbf{x}(t)$, $t \in [t_0, T]$, and differentiate it with respect to time; we shall then substitute the derived equations into the canonical equations, and we shall arrive at the conclusion that the matrix $\mathbf{K}(t)$ must satisfy a certain matrix differential equation.

Suppose that $\mathbf{x}(t)$ and $\mathbf{p}(t)$, the solutions of the canonical equations (9-24), are indeed related by the equation

$$\mathbf{p}(t) = \mathbf{K}(t)\mathbf{x}(t) \qquad t \in [t_0, T] \quad (9\text{-}40)$$

By differentiating with respect to time, we find that

$$\dot{\mathbf{p}}(t) = \dot{\mathbf{K}}(t)\mathbf{x}(t) + \mathbf{K}(t)\dot{\mathbf{x}}(t) \quad (9\text{-}41)$$

But, from Eq. (9-24), we know that

$$\dot{\mathbf{x}}(t) = \mathbf{A}(t)\mathbf{x}(t) - \mathbf{S}(t)\mathbf{p}(t) \quad (9\text{-}42)$$

and that

$$\dot{\mathbf{p}}(t) = -\mathbf{Q}(t)\mathbf{x}(t) - \mathbf{A}'(t)\mathbf{p}(t) \quad (9\text{-}43)$$

† See Ref. K-3.

Substituting Eq. (9-40) into Eq. (9-42), we obtain the relation

$$\dot{\mathbf{x}}(t) = [\mathbf{A}(t) - \mathbf{S}(t)\mathbf{K}(t)]\mathbf{x}(t) \tag{9-44}$$

Substituting Eq. (9-44) into Eq. (9-41), we find that

$$\dot{\mathbf{p}}(t) = [\dot{\mathbf{K}}(t) + \mathbf{K}(t)\mathbf{A}(t) - \mathbf{K}(t)\mathbf{S}(t)\mathbf{K}(t)]\mathbf{x}(t) \tag{9-45}$$

Substituting Eq. (9-40) into Eq. (9-43), we obtain

$$\dot{\mathbf{p}}(t) = [-\mathbf{Q}(t) - \mathbf{A}'(t)\mathbf{K}(t)]\mathbf{x}(t) \tag{9-46}$$

From Eqs. (9-45) and (9-46), we conclude that

$$[\dot{\mathbf{K}}(t) + \mathbf{K}(t)\mathbf{A}(t) - \mathbf{K}(t)\mathbf{S}(t)\mathbf{K}(t) + \mathbf{A}'(t)\mathbf{K}(t) + \mathbf{Q}(t)]\mathbf{x}(t) = \mathbf{0} \tag{9-47}$$

for all $t \in [t_0, T]$. Since Eq. (9-47) must hold for *any* choice of the initial state, since the matrix $\mathbf{K}(t)$ does not depend upon the initial state, and since $\mathbf{x}(t)$ is a solution of the homogeneous equation (9-44), it follows that Eq. (9-47) must hold no matter what the value of $\mathbf{x}(t)$ is. This means that $\mathbf{K}(t)$ must satisfy the matrix differential equation

$$\dot{\mathbf{K}}(t) + \mathbf{K}(t)\mathbf{A}(t) + \mathbf{A}'(t)\mathbf{K}(t) - \mathbf{K}(t)\mathbf{S}(t)\mathbf{K}(t) + \mathbf{Q}(t) = \mathbf{0} \tag{9-48}$$

But [see Eq. (9-23)] $\mathbf{S}(t) = \mathbf{B}(t)\mathbf{R}^{-1}(t)\mathbf{B}'(t)$, and so Eq. (9-48) may be written as

$$\dot{\mathbf{K}}(t) = -\mathbf{K}(t)\mathbf{A}(t) - \mathbf{A}'(t)\mathbf{K}(t) + \mathbf{K}(t)\mathbf{B}(t)\mathbf{R}^{-1}(t)\mathbf{B}'(t)\mathbf{K}(t) - \mathbf{Q}(t) \tag{9-49}$$

Thus, we have established the following lemma.

***Lemma 9-4*** *If* $\mathbf{x}(t)$ *and* $\mathbf{p}(t)$ *are the solutions of the canonical equations* (9-24) *and if* $\mathbf{p}(t) = \mathbf{K}(t)\mathbf{x}(t)$ *for all* $t \in [t_0, T]$ *and all* $\mathbf{x}(t)$, *then* $\mathbf{K}(t)$ *must satisfy Eq.* (9-49).

Now we turn our attention to the boundary conditions at $t = T$; the transversality conditions require that [see Eq. (9-26)]

$$\mathbf{p}(T) = \mathbf{F}\mathbf{x}(T) \tag{9-50}$$

But at $t = T$, we also have the relation

$$\mathbf{p}(T) = \mathbf{K}(T)\mathbf{x}(T) \tag{9-51}$$

and so we conclude that

$$[\mathbf{K}(T) - \mathbf{F}]\mathbf{x}(T) = \mathbf{0} \tag{9-52}$$

for all $\mathbf{x}(T)$† and, therefore, that

$$\mathbf{K}(T) = \mathbf{F} \tag{9-53}$$

[which is the same as Eq. (9-39)].

Equation (9-49) is a matrix differential equation of the *Riccati* type. *By abuse of language, we shall call the matrix differential equation* (9-49) *the*

† Recall that $\mathbf{x}(T)$ was not specified.

*Riccati equation.* Equation (9-53) provides us with the boundary conditions for the Riccati equation, and so, by the existence and uniqueness theorem,† the solution $\mathbf{K}(t)$ of the Riccati equation exists and is unique.

At first glance, it may appear that since $\mathbf{K}(t)$ is an $n \times n$ matrix, Eq. (9-49) represents a system of $n^2$ first-order differential equations. We shall now show that $\mathbf{K}(t)$ is a symmetric matrix.

***Lemma 9-5*** *If* $\mathbf{K}(t)$ *is the solution of the Riccati equation* (9-49) *and if* $\mathbf{K}(T) = \mathbf{F}$, *then* $\mathbf{K}(t)$ *is symmetric for all* $t \in [t_0, T]$, *that is,*

$$\mathbf{K}(t) = \mathbf{K}'(t) \tag{9-54}$$

PROOF We take the transpose of both sides of Eq. (9-49) to find that

$$\left[\frac{d}{dt}\mathbf{K}(t)\right]' = -\mathbf{K}'(t)\mathbf{A}(t) - \mathbf{A}'(t)\mathbf{K}'(t) + \mathbf{K}'(t)\mathbf{B}(t)\mathbf{R}^{-1}(t)\mathbf{B}'(t)\mathbf{K}'(t) - \mathbf{Q}(t) \tag{9-55}$$

since $\mathbf{Q}(t)$ and $\mathbf{B}(t)\mathbf{R}^{-1}(t)\mathbf{B}'(t)$ are symmetric. But, for *any* matrix $\mathbf{K}(t)$, the equation

$$\left[\frac{d}{dt}\mathbf{K}(t)\right]' = \frac{d}{dt}[\mathbf{K}'(t)] \tag{9-56}$$

is true. Using Eq. (9-56), we compare Eqs. (9-55) and (9-49), and we observe that *both* $\mathbf{K}(t)$ *and* $\mathbf{K}'(t)$ are solutions of the *same* differential equation. At $t = T$, we have the boundary condition $\mathbf{K}(T) = \mathbf{F}$. Since $\mathbf{F}$ is symmetric, $\mathbf{F} = \mathbf{F}'$; but $\mathbf{K}'(T) = \mathbf{F}' = \mathbf{F}$, and so we conclude that

$$\mathbf{K}(T) = \mathbf{K}'(T) = \mathbf{F} \tag{9-57}$$

Thus, $\mathbf{K}(t)$ and $\mathbf{K}'(t)$ are solutions of the same differential equation with the same boundary conditions. From the uniqueness of solutions of differential equations, we conclude that $\mathbf{K}(t) = \mathbf{K}'(t)$.

Since $\mathbf{K}(t)$ is symmetric, it follows that the Riccati equation (9-49) represents a system of $n(n+1)/2$ first-order nonlinear time-varying ordinary differential equations.

The control (9-20) is an extremal control. We now claim that it is the unique optimal control. We state this in the following control law. We verify this control law in the remainder of the section.

***Control Law 9-1 Solution to the State-regulator Problem*** *Given the linear system*

$$\dot{\mathbf{x}}(t) = \mathbf{A}(t)\mathbf{x}(t) + \mathbf{B}(t)\mathbf{u}(t) \tag{9-58}$$

*and the cost functional*

$$J_1 = \tfrac{1}{2}\langle \mathbf{x}(T), \mathbf{F}\mathbf{x}(T)\rangle + \tfrac{1}{2}\int_{t_0}^{T} [\langle \mathbf{x}(t), \mathbf{Q}(t)\mathbf{x}(t)\rangle + \langle \mathbf{u}(t), \mathbf{R}(t)\mathbf{u}(t)\rangle]\, dt \tag{9-59}$$

† See Theorem 3-14. See also Ref. K-3 for the proof that $\mathbf{K}(t)$ is defined for all $t < T$.

*where* $\mathbf{u}(t)$ *is not constrained,* $T$ *is specified,* $\mathbf{F}$ *and* $\mathbf{Q}(t)$ *are positive semidefinite, and* $\mathbf{R}(t)$ *is positive definite. Then an optimal control exists, is unique, and is given by the equation*

$$\mathbf{u}(t) = -\mathbf{R}^{-1}(t)\mathbf{B}'(t)\mathbf{K}(t)\mathbf{x}(t) \tag{9-60}$$

*where the* $n \times n$ *symmetric matrix* $\mathbf{K}(t)$ *is the unique solution of the Riccati equation*

$$\dot{\mathbf{K}}(t) = -\mathbf{K}(t)\mathbf{A}(t) - \mathbf{A}'(t)\mathbf{K}(t) + \mathbf{K}(t)\mathbf{B}(t)\mathbf{R}^{-1}(t)\mathbf{B}'(t)\mathbf{K}(t) - \mathbf{Q}(t) \tag{9-61}$$

*satisfying the boundary condition*

$$\mathbf{K}(T) = \mathbf{F} \tag{9-62}$$

*The state of the optimal system is then the solution of the linear differential equation*

$$\dot{\mathbf{x}}(t) = [\mathbf{A}(t) - \mathbf{B}(t)\mathbf{R}^{-1}(t)\mathbf{B}'(t)\mathbf{K}(t)]\mathbf{x}(t) \qquad \mathbf{x}(t_0) = \boldsymbol{\xi} \tag{9-63}$$

We shall first establish that the extremal control $\mathbf{u}(t)$ given by Eq. (9-60) represents at least a local minimum of the cost $J_1$. Recall† that if the matrix

$$\left[\begin{array}{c|c} \dfrac{\partial^2 H}{\partial \mathbf{x}^2(t)} & \dfrac{\partial^2 H}{\partial \mathbf{x}(t)\,\partial \mathbf{u}(t)} \\ \hline \dfrac{\partial^2 H}{\partial \mathbf{u}(t)\,\partial \mathbf{x}(t)} & \dfrac{\partial^2 H}{\partial \mathbf{u}^2(t)} \end{array}\right] \tag{9-64}$$

was positive definite, then the control which made $\partial H/\partial \mathbf{u}(t) = \mathbf{0}$ was, at least locally, optimal. But from Eq. (9-15), we find that

$$\begin{aligned} \frac{\partial H}{\partial \mathbf{x}(t)} &= \mathbf{Q}(t)\mathbf{x}(t) + \mathbf{A}'(t)\mathbf{p}(t) \\ \frac{\partial^2 H}{\partial \mathbf{x}^2(t)} &= \mathbf{Q}(t) \\ \frac{\partial H}{\partial \mathbf{u}(t)} &= \mathbf{R}(t)\mathbf{u}(t) + \mathbf{B}'(t)\mathbf{p}(t) \\ \frac{\partial^2 H}{\partial \mathbf{u}^2(t)} &= \mathbf{R}(t) \\ \frac{\partial^2 H}{\partial \mathbf{x}(t)\,\partial \mathbf{u}(t)} &= \frac{\partial^2 H}{\partial \mathbf{u}(t)\,\partial \mathbf{x}(t)} = \mathbf{0} \end{aligned} \tag{9-65}$$

If we substitute Eqs. (9-65) into Eq. (9-64), we obtain the matrix

$$\left[\begin{array}{c|c} \mathbf{Q}(t) & \mathbf{0} \\ \hline \mathbf{0} & \mathbf{R}(t) \end{array}\right] \tag{9-66}$$

Since $\mathbf{R}(t)$ is positive definite, it follows that if $\mathbf{Q}(t)$ is also positive definite, then the matrix (9-66) is positive definite. If, however, $\mathbf{Q}(t)$ is only posi-

† See Sec. 5-8.

tive semidefinite, then the matrix (9-66) is only positive semidefinite. Since the higher derivatives of $H$ are zero, the assumption,† for this problem, that $\mathbf{R}(t)$ is a positive definite matrix is strong enough to guarantee that the control $\mathbf{u}(t) = -\mathbf{R}^{-1}(t)\mathbf{B}'(t)\mathbf{K}(t)\mathbf{x}(t)$ minimizes, at least locally, the cost. Thus we have established the following lemma:

***Lemma 9-6*** *The control* $\mathbf{u}(t) = -\mathbf{R}^{-1}(t)\mathbf{B}'(t)\mathbf{K}(t)\mathbf{x}(t)$ *yields at least a local minimum for* $J_1$.

The next step is to prove that the extremal control

$$\mathbf{u}(t) = -\mathbf{R}^{-1}(t)\mathbf{B}'(t)\mathbf{K}(t)\mathbf{x}(t)$$

is *unique*. This proof hinges on the fact that the solution $\mathbf{K}(t)$ of the Riccati equation, subject to the boundary condition $\mathbf{K}(T) = \mathbf{F}$, is unique. We shall now prove the following lemma:

***Lemma 9-7*** *If an optimal control exists, then it is unique and is given by Eq.* (9-60).

PROOF We know that the extremal control

$$\mathbf{u}(t) = -\mathbf{R}^{-1}(t)\mathbf{B}'(t)\mathbf{K}(t)\mathbf{x}(t) \tag{9-67}$$

will result in (at least) a local minimum of $J_1$. Since $\mathbf{K}(t)$ is unique, the (locally) optimal control is a unique function of the state. Suppose that there were (say) two optimal controls $\mathbf{u}_1(t)$ and $\mathbf{u}_2(t)$ and two optimal trajectories $\mathbf{x}_1(t)$ and $\mathbf{x}_2(t)$ such that $\mathbf{x}_1(t_0) = \mathbf{x}_2(t_0)$. Then [see Eq. (9-63)] $\mathbf{x}_1(t)$ and $\mathbf{x}_2(t)$ must be two *distinct* solutions of the differential equation

$$\dot{\mathbf{x}}(t) = [\mathbf{A}(t) - \mathbf{B}(t)\mathbf{R}^{-1}(t)\mathbf{B}'(t)\mathbf{K}(t)]\mathbf{x}(t) \tag{9-68}$$

originating at the *same* initial state $\mathbf{x}_1(t_0) = \mathbf{x}_2(t_0)$. But, since $\mathbf{K}(t)$ is unique, the matrix $\mathbf{A}(t) - \mathbf{B}(t)\mathbf{R}^{-1}(t)\mathbf{B}'(t)\mathbf{K}(t)$ is also unique. Since Eq. (9-68) is linear and homogeneous, its solution starting from any initial state is *unique;* hence $\mathbf{x}_1(t) = \mathbf{x}_2(t)$ and $\mathbf{u}_1(t) = \mathbf{u}_2(t)$ for all $t \in [t_0, T]$, and the lemma is established.

We shall next prove a theorem which (1) can be used to show that an optimal control exists and (2) can be used to calculate the minimum cost $J_1^*[\mathbf{x}(t); t]$ as a function of the state $\mathbf{x}(t)$ and of the time $t$.

***Theorem 9-1*** *Given the linear system* (9-58) *and the cost functional* $J_1$ *of Eq.* (9-59). *Let* $J_1^*$ *denote the minimum value of* $J_1$*; then* $J_1^*$ *is given by*

$$J_1^*[\mathbf{x}(t), t] = \tfrac{1}{2}\langle\mathbf{x}(t), \mathbf{K}(t)\mathbf{x}(t)\rangle \tag{9-69}$$

*where* $\mathbf{K}(t)$ *is the symmetric* $n \times n$ *matrix which is the solution of the Riccati equation* (9-61) *satisfying the boundary condition* $\mathbf{K}(T) = \mathbf{F}$. *Moreover, if the optimal control* $\mathbf{u}(t) \neq \mathbf{0}$ *for all states, then* $\mathbf{K}(t)$ *is a positive definite matrix for all* $t$, $t_0 \le t < T$, *and* $\mathbf{K}(T) = \mathbf{F}$ *is positive semidefinite. Since*

† See Corollary 5-2.

*the minimum cost* $J_1^* = \frac{1}{2}\langle \mathbf{x}(t), \mathbf{K}(t)\mathbf{x}(t)\rangle$ *is defined for all* $\mathbf{x}(t)$ *and* $t$, *an optimal control exists and the minimum cost is indeed* $J_1^*$.

PROOF We shall demonstrate that $J_1^*$ is the solution to the Hamilton-Jacobi partial differential equation and that it meets the boundary conditions. The existence of the optimal control will then follow from Theorem 5-12.

First of all, note that at $t = T$, Eq. (9-69) reduces to

$$J_1^*[\mathbf{x}(T), T] = \tfrac{1}{2}\langle \mathbf{x}(T), \mathbf{F}\mathbf{x}(T)\rangle \tag{9-70}$$

which is indeed the terminal cost.

The Hamilton-Jacobi equation† for the system (9-58) and the cost functional $J_1$ is

$$\begin{aligned} \frac{\partial}{\partial t} J_1^*[\mathbf{x}(t), t] + \min_{\mathbf{u}(t)} \Bigg\{ & \tfrac{1}{2}\langle \mathbf{x}(t), \mathbf{Q}(t)\mathbf{x}(t)\rangle + \tfrac{1}{2}\langle \mathbf{u}(t), \mathbf{R}(t)\mathbf{u}(t)\rangle \\ & + \left\langle \mathbf{A}(t)\mathbf{x}(t), \frac{\partial J_1^*[\mathbf{x}(t), t]}{\partial \mathbf{x}(t)} \right\rangle + \left\langle \mathbf{B}(t)\mathbf{u}(t), \frac{\partial J_1^*[\mathbf{x}(t), t]}{\partial \mathbf{x}(t)} \right\rangle \Bigg\} = 0 \end{aligned} \tag{9-71}$$

The control that minimizes the expression in braces is given by

$$\mathbf{u}(t) = -\mathbf{R}^{-1}(t)\mathbf{B}'(t)\,\frac{\partial J_1^*[\mathbf{x}(t), t]}{\partial \mathbf{x}(t)} \tag{9-72}$$

If we substitute Eq. (9-72) into Eq. (9-71), we find that

$$\begin{aligned} \frac{\partial J_1^*}{\partial t} + \tfrac{1}{2}\langle \mathbf{x}(t), \mathbf{Q}(t)\mathbf{x}(t)\rangle + \tfrac{1}{2}\left\langle \mathbf{R}^{-1}(t)\mathbf{B}'(t)\,\frac{\partial J_1^*}{\partial \mathbf{x}(t)}, \mathbf{B}'(t)\,\frac{\partial J_1^*}{\partial \mathbf{x}(t)} \right\rangle \\ + \left\langle \mathbf{A}(t)\mathbf{x}(t), \frac{\partial J_1^*}{\partial \mathbf{x}(t)} \right\rangle - \left\langle \mathbf{B}(t)\mathbf{R}^{-1}(t)\mathbf{B}'(t)\,\frac{\partial J_1^*}{\partial \mathbf{x}(t)}, \frac{\partial J_1^*}{\partial \mathbf{x}(t)} \right\rangle = 0 \end{aligned} \tag{9-73}$$

But, if $J_1^* = \frac{1}{2}\langle \mathbf{x}(t), \mathbf{K}(t)\mathbf{x}(t)\rangle$, then

$$\frac{\partial J_1^*}{\partial t} = \tfrac{1}{2}\langle \mathbf{x}(t), \dot{\mathbf{K}}(t)\mathbf{x}(t)\rangle \tag{9-74}$$

and

$$\frac{\partial J_1^*}{\partial \mathbf{x}(t)} = \mathbf{K}(t)\mathbf{x}(t) \tag{9-75}$$

Substituting Eqs. (9-74) and (9-75) into Eq. (9-73), we obtain the equation

$$\begin{aligned} \tfrac{1}{2}\langle \mathbf{x}(t), \dot{\mathbf{K}}(t)\mathbf{x}(t)\rangle + \tfrac{1}{2}\langle \mathbf{x}(t), \mathbf{Q}(t)\mathbf{x}(t)\rangle + \langle \mathbf{A}(t)\mathbf{x}(t), \mathbf{K}(t)\mathbf{x}(t)\rangle \\ - \tfrac{1}{2}\langle \mathbf{B}(t)\mathbf{R}^{-1}(t)\mathbf{B}'(t)\mathbf{K}(t)\mathbf{x}(t), \mathbf{K}(t)\mathbf{x}(t)\rangle = 0 \end{aligned} \tag{9-76}$$

But, since $\mathbf{K}(t)$ is symmetric, we can write

$$\begin{aligned} \langle \mathbf{A}(t)\mathbf{x}(t), \mathbf{K}(t)\mathbf{x}(t)\rangle &= \tfrac{1}{2}\langle \mathbf{A}(t)\mathbf{x}(t), \mathbf{K}(t)\mathbf{x}(t)\rangle + \tfrac{1}{2}\langle \mathbf{A}(t)\mathbf{x}(t), \mathbf{K}(t)\mathbf{x}(t)\rangle \\ &= \tfrac{1}{2}\langle \mathbf{x}(t), \mathbf{A}'(t)\mathbf{K}(t)\mathbf{x}(t)\rangle + \tfrac{1}{2}\langle \mathbf{x}(t), \mathbf{K}(t)\mathbf{A}(t)\mathbf{x}(t)\rangle \end{aligned} \tag{9-77}$$

† See Sec. 5-20.

Substituting Eq. (9-77) into Eq. (9-76) and combining all the terms into a single scalar product, we deduce that

$$\tfrac{1}{2}\langle \mathbf{x}(t), [\dot{\mathbf{K}}(t) + \mathbf{K}(t)\mathbf{A}(t) + \mathbf{A}'(t)\mathbf{K}(t) - \mathbf{K}(t)\mathbf{B}(t)\mathbf{R}^{-1}(t)\mathbf{B}'(t)\mathbf{K}(t) + \mathbf{Q}(t)]\mathbf{x}(t)\rangle = 0 \quad (9\text{-}78)$$

Clearly, if $\mathbf{K}(t)$ satisfies the Riccati equation (9-61), then the matrix in the brackets is the zero matrix and Eq. (9-78) holds, and vice versa.

To show that $\mathbf{K}(t)$ is *positive definite*, we proceed as follows: Suppose that at $t = t' < T$, the matrix $\mathbf{K}(t')$ is *not* positive definite; then there exists an $\mathbf{x}(t')$ such that $\frac{1}{2}\langle \mathbf{x}(t'), \mathbf{K}(t')\mathbf{x}(t')\rangle \leq 0$; clearly, this violates Lemma 9-2, and so $\mathbf{K}(t)$ must be positive definite for all $t$, $t_0 \leq t < T$.

The minimum cost $J_1^* = \frac{1}{2}\langle \mathbf{x}(t), \mathbf{K}(t)\mathbf{x}(t)\rangle$ is a scalar function whose domain is the $(n + 1)$-dimensional space which is the product space of the $n$-dimensional state space and of the reals (time). The cost surface defined by $J_1^*$ is smooth and has a unique gradient everywhere (why?).

We also remark that the function $J_1^*[\mathbf{x}(t), t] = \frac{1}{2}\langle \mathbf{x}(t), \mathbf{K}(t)\mathbf{x}(t)\rangle$ is the solution to the Hamilton-Jacobi equation for all $\mathbf{x}(t) \in R_n$ and all $t \in [t_0, T]$. We can thus invoke the results of Sec. 5-20 to show that an optimal control exists.

We now claim that we have completely verified Control Law 9-1. We have shown that an optimum exists and that the unique extremal control (9-60) corresponds to a minimum of the functional $J_1(\mathbf{u})$. It should be clear that the optimal control $\mathbf{u}(t)$ must be given by Eq. (9-60).

In the following section, we shall comment on several aspects of Control Law 9-1.

## 9-4 Discussion of the Results and Examples

In Sec. 9-3, we derived the optimal-control law for the state-regulator problem. In this section, we shall discuss the implications of Control Law 9-1 from a physical point of view.

### *Structure of the Optimal Feedback System*

Figure 9-1 shows the structure of the optimal feedback system. Since the optimal control is $\mathbf{u}(t) = -\mathbf{R}^{-1}(t)\mathbf{B}'(t)\mathbf{K}(t)\mathbf{x}(t)$, the state $\mathbf{x}(t)$ is operated on by the linear transformation $\mathbf{K}(t)$ and then by the linear transformation $-\mathbf{R}^{-1}(t)\mathbf{B}'(t)$ to generate the control. The feedback system is, thus, linear and time-varying. Since $\mathbf{R}(t)$ and $\mathbf{B}(t)$ are known matrices, it follows that the matrix $\mathbf{K}(t)$ governs the behavior of the system. We shall often call the matrix $\mathbf{K}(t)$ the *gain* matrix.

The response $\mathbf{x}(t)$ of the optimal system is the solution of the differential equation

$$\dot{\mathbf{x}}(t) = \mathbf{G}(t)\mathbf{x}(t) \quad (9\text{-}79)$$

where [see Eq. (9-63)] the $n \times n$ matrix $\mathbf{G}(t)$ is given by

$$\mathbf{G}(t) = \mathbf{A}(t) - \mathbf{B}(t)\mathbf{R}^{-1}(t)\mathbf{B}'(t)\mathbf{K}(t) \tag{9-80}$$

Figure 9-2 shows the simulation of the optimal system $\dot{\mathbf{x}}(t) = \mathbf{G}(t)\mathbf{x}(t)$.

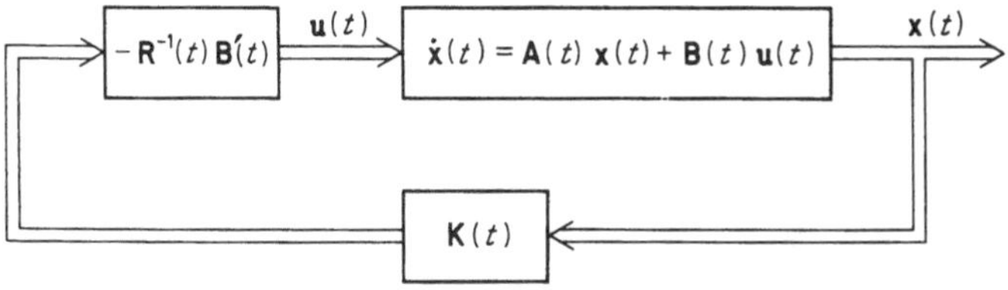


**Fig. 9-1** The structure of the optimal state-regulating system. The "thickened" channels indicate transmission of vector-valued signals.

*Computation of the Gain Matrix* $\mathbf{K}(t)$

The positive definite matrix $\mathbf{K}(t)$, $t \in [t_0, T]$, is the solution of the matrix Riccati equation

$$\dot{\mathbf{K}}(t) = -\mathbf{K}(t)\mathbf{A}(t) - \mathbf{A}'(t)\mathbf{K}(t) + \mathbf{K}(t)\mathbf{B}(t)\mathbf{R}^{-1}(t)\mathbf{B}'(t)\mathbf{K}(t) - \mathbf{Q}(t) \tag{9-81}$$

with the boundary condition

$$\mathbf{K}(T) = \mathbf{F} \tag{9-82}$$

The matrix Riccati equation is *nonlinear*, and for this reason, we usually cannot obtain closed-form solutions; therefore, we must compute $\mathbf{K}(t)$ using a digital computer. Such a computation can be based, for example, on an approximation of $\dot{\mathbf{K}}(t)$. Recall that

$$\frac{d}{dt}\mathbf{K}(t) = \lim_{\Delta \to 0} \frac{\mathbf{K}(t+\Delta) - \mathbf{K}(t)}{\Delta} \tag{9-83}$$

so that we can solve Eq. (9-81) approximately by using the formula

$$\mathbf{K}(t+\Delta) \approx \mathbf{K}(t) + \Delta\{-\mathbf{K}(t)\mathbf{A}(t) - \mathbf{A}'(t)\mathbf{K}(t) + \mathbf{K}(t)\mathbf{B}(t)\mathbf{R}^{-1}(t)\mathbf{B}'(t)\mathbf{K}(t) - \mathbf{Q}(t)\} \tag{9-84}$$

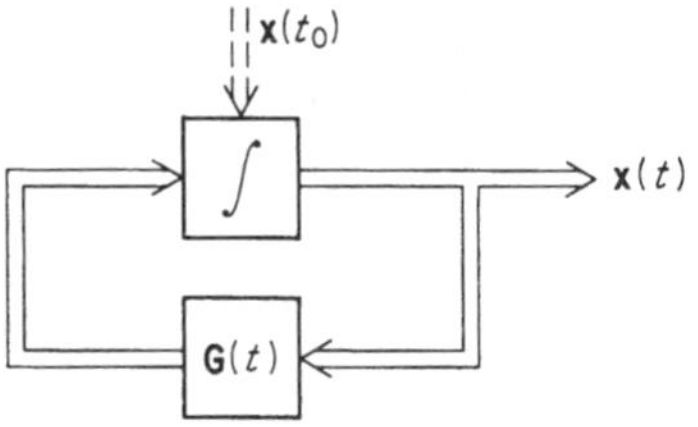


**Fig. 9-2** The simulation of the optimal state regulator given by Eq. (9-79).

We solve the Riccati equation *backward in time*, using a small negative $\Delta$ and setting $\mathbf{K}(T) = \mathbf{F}$. A computation such as the one indicated by Eq. (9-84) can be done extremely rapidly by a modern digital computer. Naturally, the smaller the value of $\Delta$, the better the approximation.

The important thing to realize is that the gain matrix $\mathbf{K}(t)$ is *independent* of the state, so that once the system and the cost $J_1$ have been specified, $\mathbf{K}(t)$ can be computed before the optimal system starts to operate.

Let us now illustrate the theory by means of a simple example.

**Example 9-1** Consider the second-order system (the double-integral plant)

$$\begin{aligned}\dot{x}_1(t) &= x_2(t)\\ \dot{x}_2(t) &= u(t)\end{aligned} \tag{9-85}$$

and the cost functional

$$J_1 = \tfrac{1}{2}[x_1^2(3) + 2x_2^2(3)] + \tfrac{1}{2}\int_0^3 [2x_1^2(t) + 4x_2^2(t) + 2x_1(t)x_2(t) + \tfrac{1}{2}u^2(t)]\,dt \tag{9-86}$$

For the system (9-85), we have

$$\mathbf{A}(t) = \begin{bmatrix}0 & 1\\ 0 & 0\end{bmatrix} \qquad \mathbf{B}(t) = \begin{bmatrix}0\\ 1\end{bmatrix} \tag{9-87}$$

For the cost (9-86), we have the relations

$$\mathbf{F} = \begin{bmatrix}1 & 0\\ 0 & 2\end{bmatrix} \qquad \mathbf{Q}(t) = \begin{bmatrix}2 & 1\\ 1 & 4\end{bmatrix} \qquad \mathbf{R}(t) = \tfrac{1}{2} \qquad t_0 = 0 \qquad T = 3 \tag{9-88}$$

Let $\mathbf{K}(t)$ be the $2 \times 2$ matrix

$$\mathbf{K}(t) = \begin{bmatrix}k_{11}(t) & k_{12}(t)\\ k_{12}(t) & k_{22}(t)\end{bmatrix} \tag{9-89}$$

The (scalar) optimal control is then given by the equation

$$\begin{aligned}u(t) &= -\mathbf{R}^{-1}(t)\mathbf{B}'(t)\mathbf{K}(t)\mathbf{x}(t)\\ &= -2[0 \quad 1]\begin{bmatrix}k_{11}(t) & k_{12}(t)\\ k_{12}(t) & k_{22}(t)\end{bmatrix}\begin{bmatrix}x_1(t)\\ x_2(t)\end{bmatrix}\\ &= -2[k_{12}(t)x_1(t) + k_{22}(t)x_2(t)]\end{aligned} \tag{9-90}$$

The Riccati equation is

$$\begin{aligned}\begin{bmatrix}\dot{k}_{11}(t) & \dot{k}_{12}(t)\\ \dot{k}_{12}(t) & \dot{k}_{22}(t)\end{bmatrix} = &-\begin{bmatrix}k_{11}(t) & k_{12}(t)\\ k_{12}(t) & k_{22}(t)\end{bmatrix}\begin{bmatrix}0 & 1\\ 0 & 0\end{bmatrix} - \begin{bmatrix}0 & 0\\ 1 & 0\end{bmatrix}\begin{bmatrix}k_{11}(t) & k_{12}(t)\\ k_{12}(t) & k_{22}(t)\end{bmatrix}\\ &+\begin{bmatrix}k_{11}(t) & k_{12}(t)\\ k_{12}(t) & k_{22}(t)\end{bmatrix}\begin{bmatrix}0\\ 1\end{bmatrix} 2[0 \quad 1]\begin{bmatrix}k_{11}(t) & k_{12}(t)\\ k_{12}(t) & k_{22}(t)\end{bmatrix} - \begin{bmatrix}2 & 1\\ 1 & 4\end{bmatrix}\end{aligned} \tag{9-91}$$

and the boundary conditions at $t = 3$ are

$$\begin{bmatrix}k_{11}(3) & k_{12}(3)\\ k_{12}(3) & k_{22}(3)\end{bmatrix} = \begin{bmatrix}1 & 0\\ 0 & 2\end{bmatrix} \tag{9-92}$$

We perform the indicated matrix multiplications in Eq. (9-91), and we obtain the three differential equations and corresponding boundary conditions

$$\begin{aligned}\dot{k}_{11}(t) &= 2k_{12}^2(t) - 2 & k_{11}(3) &= 1\\ \dot{k}_{12}(t) &= -k_{11}(t) + 2k_{12}(t)k_{22}(t) - 1 & k_{12}(3) &= 0\\ \dot{k}_{22}(t) &= -2k_{12}(t) + 2k_{22}^2(t) - 4 & k_{22}(3) &= 2\end{aligned} \tag{9-93}$$

If we solve Eqs. (9-93) for $k_{12}(t)$ and $k_{22}(t)$, we shall be able to generate the optimal control $u(t)$ using Eq. (9-90).

*Is Controllability† Necessary?*

In Sec. 9-3, we did not require that the system be controllable. Since the system state equation is

$$\dot{\mathbf{x}}(t) = \mathbf{A}(t)\mathbf{x}(t) + \mathbf{B}(t)\mathbf{u}(t) \tag{9-94}$$

the use of the cost functional $J_1$ implies that we wish to keep the state $\mathbf{x}(t)$ *near* zero; since controllability implies that we can find a control that drives the state to zero [and, hence, a control that drives $\mathbf{x}(t)$ *near* zero], one may wonder whether or not the system (9-94) should be controllable in order that the optimal solution be given by Control Law 9-1. However, it is not necessary that the system (9-94) be controllable, because the contribution of the uncontrollable states to the cost functional is always finite provided that the control interval $[t_0, T]$ is finite. We shall see in the next section that we shall require controllability as $T \to \infty$ to ensure that the cost is finite. Meanwhile, we demonstrate that we can find the optimal control for two uncontrollable systems.

**Example 9-2** Consider the "most uncontrollable" system

$$\dot{\mathbf{x}}(t) = \mathbf{A}(t)\mathbf{x}(t) + \mathbf{B}(t)\mathbf{u}(t) \tag{9-95}$$

with

$$\mathbf{B}(t) = \mathbf{0} \qquad \text{for all } t \tag{9-96}$$

and the cost functional $J_1$. Clearly, the value of $\langle \mathbf{x}(t), \mathbf{Q}(t)\mathbf{x}(t)\rangle$ is independent of $\mathbf{u}(t)$; we should expect then that the optimal control $\mathbf{u}(t)$ should be zero so that the term $\langle \mathbf{u}(t), \mathbf{R}(t)\mathbf{u}(t)\rangle$ does not contribute to the cost. This is indeed the case, because the optimal control is given by $\mathbf{u}(t) = -\mathbf{R}^{-1}(t)\mathbf{B}'(t)\mathbf{K}(t)\mathbf{x}(t)$. Since $\mathbf{B}'(t) = \mathbf{0}$, it follows that $\mathbf{u}(t) = \mathbf{0}$.

**Example 9-3** Consider the second-order *uncontrollable* system

$$\begin{aligned} \dot{x}_1(t) &= ax_2(t) + u(t) \\ \dot{x}_2(t) &= bx_2(t) \end{aligned} \tag{9-97}$$

and the cost functional

$$J_1 = \tfrac{1}{2}\int_0^T [x_1^2(t) + hx_2^2(t) + u^2(t)]\,dt \tag{9-98}$$

We leave it to the reader to verify that the optimal control is given by

$$u(t) = -k_{11}(t)x_1(t) - k_{12}(t)x_2(t) \tag{9-99}$$

and that the Riccati equation yields the three differential equations

$$\begin{aligned} \dot{k}_{11}(t) &= k_{11}^2(t) - 1 & k_{11}(T) &= 0 \\ \dot{k}_{12}(t) &= -ak_{11}(t) - bk_{12}(t) + k_{11}(t)k_{12}(t) & k_{12}(T) &= 0 \\ \dot{k}_{22}(t) &= -2ak_{12}(t) - 2bk_{22}(t) + k_{12}^2(t) - h & k_{22}(T) &= 0 \end{aligned} \tag{9-100}$$

Now let us examine the system (9-97). Since the state variable $x_2(t)$ cannot be influenced by the control, the term $hx_2^2(t)$ in the cost functional $J_1$ cannot be influenced by the control. Thus, we should expect that the optimal control will be independent of the value of $h$. This is indeed the case, because the control [see Eq. (9-99)] depends only on $k_{11}(t)$ and $k_{12}(t)$, the equations for $\dot{k}_{11}(t)$ and $\dot{k}_{12}(t)$ are [see Eqs. (9-100)] independent of $k_{22}(t)$,

† See Secs. 4-15 and 4-16.

and $h$ appears only in the equation for $\dot{k}_{22}(t)$. However, although $x_2(t)$ cannot be controlled, the control *must* depend on $x_2(t)$. The reason for this is that, for $a \neq 0$, the value of $x_2(t)$ influences the value of $x_1(t)$; in other words, if $a \neq 0$, then $k_{12}(t) \neq 0$. If, however, $a = 0$, then the control should not depend on $x_2(t)$ and so $k_{12}(t)$ should turn out to be zero. This is indeed the case, because, if $a = 0$, then [see Eqs. (9-100)]

$$\dot{k}_{11}(t) = k_{11}^2(t) - 1 \qquad \text{independent of } k_{12}(t) \tag{9-101}$$

and

$$\dot{k}_{12}(t) = k_{12}(t)[k_{11}(t) - b] \tag{9-102}$$

which is integrated to yield

$$k_{12}(t) = k_{12}(0) \exp \int_0^t [k_{11}(\tau) - b]\, d\tau \tag{9-103}$$

But, since $k_{12}(T) = 0$ and since $\exp \int_0^T [k_{11}(\tau) - b]\, d\tau \neq 0$, this implies that $k_{12}(0) = 0$, and, hence, that

$$k_{12}(t) = 0 \qquad \text{for all } t \in [t_0, T] \tag{9-104}$$

The two examples demonstrate that an optimal control exists for uncontrollable systems and behaves as would be expected from physical reasoning.

**Exercise 9-1** Consider the system (9-97) and the cost functional

$$J_1 = \tfrac{1}{2}[x_1^2(T) + gx_1(T)x_2(T) + mx_2^2(T)] + \tfrac{1}{2}\int_0^T [x_1^2(t) + nx_1(t)x_2(t) + hx_2^2(t) + \tfrac{1}{2}u^2(t)]\, dt \tag{9-105}$$

Show that the optimal control will be a function of $x_1(t)$ alone if and only if

$$a = g = n = 0 \tag{9-106}$$

Verify that your mathematical answer agrees with your physical intuition.

We wish to emphasize that the optimal feedback system will turn out to be linear but time-varying as long as the control interval $[t_0, T]$ is finite. This will be the case even if the system and the cost functional are time-invariant, i.e., even if the matrices $\mathbf{A}(t)$, $\mathbf{B}(t)$, $\mathbf{Q}(t)$, and $\mathbf{R}(t)$ are constant matrices. The property of linearity is an appealing one; however, the engineering construction of time-varying functions must be done by a digital computer.

The engineer may not quarrel with the necessity of time-varying controllers for time-varying systems. However, he may like time-invariant controllers for time-invariant systems. In Chaps. 7 and 8, we saw that we could obtain time-invariant controllers if the response time $T$ was not specified a priori. Unfortunately, we cannot let $T$ be unspecified in the cost functional $J_1(\mathbf{u})$, because we have not specified the terminal state (why?). We shall show in the next section that if we let $T \to \infty$, then we obtain a time-invariant controller for a time-invariant system and cost functional.

## 9-5 The State-regulator Problem: Time-invariant Systems; $T = \infty$

Let us suppose that the matrices $\mathbf{A}(t)$, $\mathbf{B}(t)$, $\mathbf{R}(t)$, and $\mathbf{Q}(t)$ are all constant matrices, so that both the linear system and the cost functional are time-invariant. Furthermore, we suppose that $\mathbf{F} = \mathbf{0}$. Clearly, the gain matrix $\mathbf{K}(t)$ is still the solution of the Riccati equation and now satisfies the boundary condition $\mathbf{K}(T) = \mathbf{0}$.

It is worthwhile to ask the following question: *Under what circumstances is the matrix $\mathbf{K}(t)$ constant?* Clearly, if the gain matrix $\mathbf{K}(t)$ is constant, then the resulting optimal system will be linear and time-invariant; from the engineering point of view, such an optimal system would indeed be simple to construct.

The following control law states the precise conditions under which the matrix $\mathbf{K}(t)$ is a constant.

***Control Law 9-2*** *Given the controllable linear time-invariant system*

$$\dot{\mathbf{x}}(t) = \mathbf{A}\mathbf{x}(t) + \mathbf{B}\mathbf{u}(t) \tag{9-107}$$

*and the cost functional*

$$\hat{J}_1 = \tfrac{1}{2}\int_0^\infty [\langle \mathbf{x}(t), \mathbf{Q}\mathbf{x}(t)\rangle + \langle \mathbf{u}(t), \mathbf{R}\mathbf{u}(t)\rangle]\, dt \tag{9-108}$$

*where $\mathbf{u}(t)$ is not constrained, $\mathbf{Q}$ is a positive definite matrix, and $\mathbf{R}$ is a positive definite matrix.*

*Then an optimal control exists, is unique, and is given by the equation*

$$\mathbf{u}(t) = -\mathbf{R}^{-1}\mathbf{B}'\hat{\mathbf{K}}\mathbf{x}(t) \tag{9-109}$$

*where $\hat{\mathbf{K}}$ is the (constant) $n \times n$ positive definite matrix which is the solution of the nonlinear matrix algebraic equation*

$$-\hat{\mathbf{K}}\mathbf{A} - \mathbf{A}'\hat{\mathbf{K}} + \hat{\mathbf{K}}\mathbf{B}\mathbf{R}^{-1}\mathbf{B}'\hat{\mathbf{K}} - \mathbf{Q} = \mathbf{0} \tag{9-110}$$

*In this case, the optimal trajectory is the solution of the linear time-invariant homogeneous system*

$$\dot{\mathbf{x}}(t) = \mathbf{G}\mathbf{x}(t) \qquad \mathbf{x}(0) \textit{ given} \tag{9-111}$$

*where $\mathbf{G}$ is defined by*

$$\mathbf{G} = \mathbf{A} - \mathbf{B}\mathbf{R}^{-1}\mathbf{B}'\hat{\mathbf{K}} \tag{9-112}$$

*The minimum cost $\hat{J}_1^*$ is given by*

$$\hat{J}_1^*[\mathbf{x}(t)] = \tfrac{1}{2}\langle \mathbf{x}(t), \hat{\mathbf{K}}\mathbf{x}(t)\rangle \tag{9-113}$$

Let us discuss some of the differences between the assumptions of Control Laws 9-1 and 9-2.

First of all, we require that the system (9-107) be completely controllable. This means that the matrix

$$[\mathbf{B} \mid \mathbf{A}\mathbf{B} \mid \cdots \mid \mathbf{A}^{n-1}\mathbf{B}] \tag{9-114}$$

must contain $n$ linearly independent column vectors.† The requirement of controllability guarantees that the minimum cost is finite; it is quite conceivable that if the system were uncontrollable and unstable, then the cost functional would be infinite for all controls since the control interval is infinite. In this case, one cannot distinguish the optimal control from any other control.

Another difference is that we use $\mathbf{F} = \mathbf{0}$ in Control Law 9-2. Thus, we do not consider any terminal cost of the type $\lim_{T\to\infty} \frac{1}{2}\langle \mathbf{x}(T), \mathbf{F}\mathbf{x}(T)\rangle$. A terminal cost at $T = \infty$ does not make much engineering sense since the engineer is always interested in the response of a system in finite time. We let $T \to \infty$ for the following reasons: *First, we wish to guarantee that the state stays near zero after an initial transient interval and, second, we wish to avoid the (somewhat arbitrary) specification of a large terminal time $T$.* In the latter case, we could not guarantee that the state would be near zero for $t > T$.

Now let us discuss the matrix $\hat{\mathbf{K}}$. Kalman has shown that‡ the assumptions of controllability and $\mathbf{F} = \mathbf{0}$ imply that $\lim_{T\to\infty} \mathbf{K}(t)$ exists, is unique, and is $\hat{\mathbf{K}}$, that is, that

$$\lim_{T\to\infty} \mathbf{K}(t) = \hat{\mathbf{K}} \tag{9-115}$$

where $\hat{\mathbf{K}}$ is the positive definite matrix which is the solution of the algebraic equation (9-110).

A useful interpretation of the matrix $\hat{\mathbf{K}}$ is the following: Consider the Riccati equation

$$\dot{\mathbf{K}}(t) = -\mathbf{K}(t)\mathbf{A} - \mathbf{A}'\mathbf{K}(t) + \mathbf{K}(t)\mathbf{B}\mathbf{R}^{-1}\mathbf{B}'\mathbf{K}(t) - \mathbf{Q} \tag{9-116}$$

with the boundary condition $\mathbf{K}(T) = \mathbf{0}$. Think of the time $T$ as the "starting time" and of $\mathbf{K}(T)$ as the "initial condition"; then the matrix $\hat{\mathbf{K}}$ can be thought of as the "steady-state" solution of the Riccati equation as the time decreases. As $T \to \infty$, the "transient" due to the "initial condition" $\mathbf{K}(T) = \mathbf{0}$ dies down and the steady-state interval, during which $\mathbf{K}(t) = \hat{\mathbf{K}}$, grows without bound, as illustrated in Fig. 9-3, that is, $\mathbf{K}(t) \to \hat{\mathbf{K}}$ for all finite $t$.

Since the optimal control is given by $\mathbf{u}(t) = -\mathbf{R}^{-1}\mathbf{B}'\hat{\mathbf{K}}\mathbf{x}(t)$, the optimal trajectory is the solution of the homogeneous equation

$$\dot{\mathbf{x}}(t) = \mathbf{G}\mathbf{x}(t) \tag{9-117}$$

where

$$\mathbf{G} = \mathbf{A} - \mathbf{B}\mathbf{R}^{-1}\mathbf{B}'\hat{\mathbf{K}} \tag{9-118}$$

† See Sec. 4-16.

‡ See Ref. K-6, theorem (7-7), p. 33; or Ref. K-3, proposition (6.6) and theorem (6-7), pp. 112–113.

***Lemma 9-8*** *The eigenvalues of the matrix* $\mathbf{G} = \mathbf{A} - \mathbf{B}\mathbf{R}^{-1}\mathbf{B}'\hat{\mathbf{K}}$ *must have negative real parts, and so the optimal system* (9-117) *is stable.*†

PROOF Suppose that one or more eigenvalues of **G** has a nonnegative real part; then some of the state variables would not go to zero, and hence the cost $J_1^*$ would be infinite.

We thus observe that the positive definite matrix $\hat{\mathbf{K}}$ must be such that the eigenvalues of **G** have negative real parts, although one or more of the eigenvalues of **A** may have a nonnegative real part. In other words,

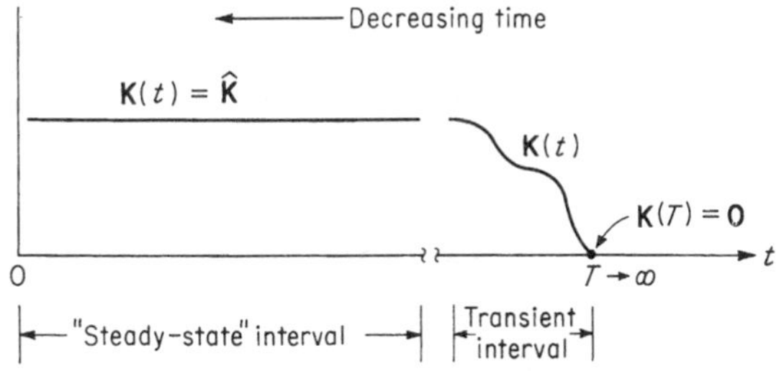


**Fig. 9-3** A loose interpretation of the constant matrix $\hat{\mathbf{K}}$. As $T \to \infty$, the "transient interval" tends to infinity and the "steady-state interval" occupies all finite times.

although the controlled system is unstable, the optimal system must be strictly stable.

We shall illustrate these concepts by a simple example.

**Example 9-4** Consider the controllable system

$$\begin{aligned} \dot{x}_1(t) &= x_2(t) \\ \dot{x}_2(t) &= u(t) \end{aligned} \tag{9-119}$$

We know‡ that Eqs. (9-119) are the state-variable representation of the system with output $x_1(t)$ and input $u(t)$ having the transfer function

$$\frac{x_1(s)}{u(s)} = G(s) = \frac{1}{s^2} \tag{9-120}$$

That is, the system (9-119) is the so-called double-integral plant. Let us now consider the cost functional

$$\hat{J}_1 = \frac{1}{2} \int_0^\infty [x_1^2(t) + 2bx_1(t)x_2(t) + ax_2^2(t) + u^2(t)]\, dt \tag{9-121}$$

where we assume that

$$a - b^2 > 0 \tag{9-122}$$

For this problem, we have

$$\mathbf{A} = \begin{bmatrix} 0 & 1 \\ 0 & 0 \end{bmatrix} \qquad \mathbf{B} = \begin{bmatrix} 0 \\ 1 \end{bmatrix} \qquad \mathbf{Q} = \begin{bmatrix} 1 & b \\ b & a \end{bmatrix} \qquad \mathbf{R} = 1 \tag{9-123}$$

Equation (9-122) guarantees that **Q** is positive definite.

† See Sec. 3-26.
‡ See, for example, Sec. 7-2.

From Eqs. (9-123) and (9-109), we find that the optimal control is given by

$$u(t) = -\hat{k}_{12}x_1(t) - \hat{k}_{22}x_2(t) \tag{9-124}$$

From Eqs. (9-123) and (9-110), we deduce that

$$-\begin{bmatrix}\hat{k}_{11} & \hat{k}_{12}\\ \hat{k}_{12} & \hat{k}_{22}\end{bmatrix}\begin{bmatrix}0 & 1\\ 0 & 0\end{bmatrix} - \begin{bmatrix}0 & 0\\ 1 & 0\end{bmatrix}\begin{bmatrix}\hat{k}_{11} & \hat{k}_{12}\\ \hat{k}_{12} & \hat{k}_{22}\end{bmatrix} + \begin{bmatrix}\hat{k}_{11} & \hat{k}_{12}\\ \hat{k}_{12} & \hat{k}_{22}\end{bmatrix}\begin{bmatrix}0\\ 1\end{bmatrix}[0 \quad 1]\begin{bmatrix}\hat{k}_{11} & \hat{k}_{12}\\ \hat{k}_{12} & \hat{k}_{22}\end{bmatrix} - \begin{bmatrix}1 & b\\ b & a\end{bmatrix} = \begin{bmatrix}0 & 0\\ 0 & 0\end{bmatrix} \tag{9-125}$$

Performing the indicated matrix multiplications, we obtain the three *algebraic* equations

$$\hat{k}_{12}^2 = 1 \tag{9-126}$$
$$-\hat{k}_{11} + \hat{k}_{12}\hat{k}_{22} - b = 0 \tag{9-127}$$
$$-2\hat{k}_{12} - \hat{k}_{22}^2 - a = 0 \tag{9-128}$$

The solutions of the above equations are

$$\hat{k}_{12} = \pm 1 \tag{9-129}$$
$$\hat{k}_{11} = \hat{k}_{12}\hat{k}_{22} - b \tag{9-130}$$
$$\hat{k}_{22} = \pm\sqrt{a + 2\hat{k}_{12}} \tag{9-131}$$

In order to remove the ambiguity of sign, we must use the fact that $\hat{\mathbf{K}}$ is positive definite, which implies that the relations†

$$\hat{k}_{11} > 0 \tag{9-132}$$
$$\hat{k}_{11}\hat{k}_{22} - \hat{k}_{12}^2 > 0 \tag{9-133}$$

must hold.

Equations (9-132) and (9-133) imply that

$$\hat{k}_{22} > 0 \tag{9-134}$$

From Eqs. (9-134) and (9-131), we deduce that

$$\hat{k}_{22} = \sqrt{a + 2\hat{k}_{12}} \tag{9-135}$$

We claim that the value

$$\hat{k}_{12} = +1 \tag{9-136}$$

will make the $\hat{\mathbf{K}}$ matrix positive definite. To see this, suppose that we let

$$\hat{k}_{12} = -1 \tag{9-137}$$

Then, from Eqs. (9-137) and (9-135), we must conclude that

$$\hat{k}_{22} = \sqrt{a - 2} \tag{9-138}$$

Since $\hat{k}_{22}$ must be real, we must have $a > 2$. From Eqs. (9-130), (9-132), (9-137), and (9-138), we obtain

$$\hat{k}_{11} = -\hat{k}_{22} - b = -\sqrt{a - 2} - b > 0 \tag{9-139}$$

and so the relation

$$b < -\sqrt{a - 2} < 0 \tag{9-140}$$

must hold. Equations (9-133), (9-129), (9-139), and (9-138) yield the inequality

$$-(a - 2) - b\sqrt{a - 2} > 1 \tag{9-141}$$

† See Sec. 2-15.

or, equivalently, the inequality

$$b^2 > \frac{(a-1)^2}{a-2} = \frac{(a-2+1)^2}{a-2} = a + \frac{1}{a-2} > a \tag{9-142}$$

and so we deduce that

$$b^2 > a \tag{9-143}$$

But the inequalities (9-143) and (9-122) are contradictory; therefore, the value $\hat{k}_{12} = -1$ cannot make $\hat{\mathbf{K}}$ positive definite, and so the solution must be

$$\begin{aligned} \hat{k}_{12} &= +1 \\ \hat{k}_{22} &= \sqrt{a+2} \\ \hat{k}_{11} &= \sqrt{a+2} - b \end{aligned} \tag{9-144}$$

The optimal control is thus given by [see Eq. (9-124)]

$$u(t) = -x_1(t) - \sqrt{a+2}\, x_2(t) \tag{9-145}$$

and so, for this problem, $u(t)$ is independent of $b$.

The block diagram of the optimal system is shown in Fig. 9-4. In Fig. 9-4, the output of the plant $1/s^2$ is $x_1(t)$; since $x_2(t) = dx_1(t)/dt$, we indicate the differentiation by passing $x_1(t)$ through the block indicated by $s$ (which is the conventional way of indicating a derivative using Laplace transforms). Figure 9-5 is obtained by a block-diagram transformation of Fig. 9-4, and shows that the optimal system contains a feedback compensator with the transfer function $(1 + s\sqrt{a+2})$. The closed-loop transfer function is

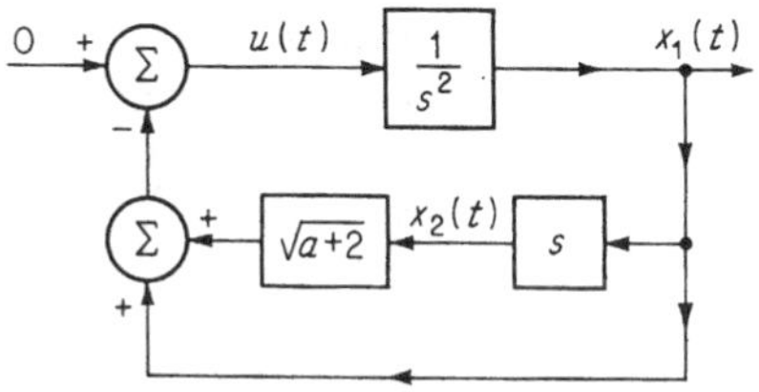


**Fig. 9-4** The optimal state regulator for the system of Example 9-4.

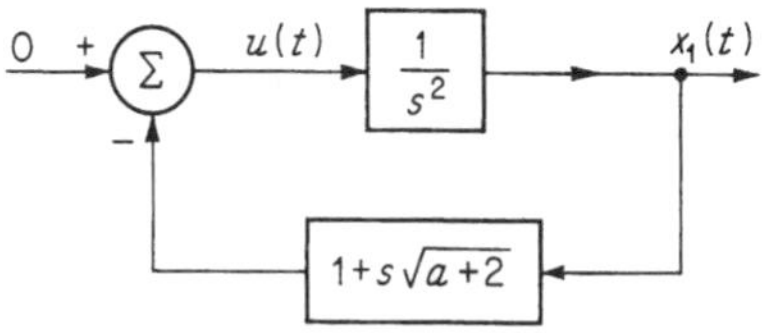


**Fig. 9-5** The feedback system derived by block-diagram transformations from the system of Fig. 9-4.

$$G^*(s) = \frac{1/s^2}{1 + (1 + s\sqrt{a+2})/s^2} = \frac{1}{s^2 + s\sqrt{a+2} + 1} \tag{9-146}$$

Thus, the poles of the optimal closed-loop feedback system are at

$$s = -\frac{\sqrt{a+2}}{2} \pm \frac{1}{2} j \sqrt{a-2} \tag{9-147}$$

Figure 9-6 shows the locus of the poles of the optimal feedback system as $a$ varies from 0 to $\infty$. At $a = 0$, the poles are at $s = -\sqrt{2}/2 \pm j(\sqrt{2}/2)$, which corresponds to the case [see Eq. (9-121)] where the signal $x_2(t)$ (that is, the derivative of the output) is not weighed in the cost functional. As $a$ increases, the poles of the optimal system tend to the real axis of the $s$ plane; the net effect is that the system response becomes less oscillatory and more sluggish. For $a > 2$, both poles are real and negative. Thus, the more we weigh the derivative of the output, the less oscillatory the system.

A few more comments on the solution of the algebraic equations (9-126) to (9-128) are in order. Note that although $\hat{k}_{11}$ does not appear in the expression for the optimal control, we must solve for $\hat{k}_{11}$ in order to obtain the solution of the algebraic equations that results in a positive definite $\hat{\mathbf{K}}$ matrix. The reader should not underestimate the difficulty of isolating the proper solution of these quadratic equations. It is often easier to solve the Riccati differential equation and determine its "steady-state" solution than to isolate the positive definite soution of the algebraic equation (9-110).

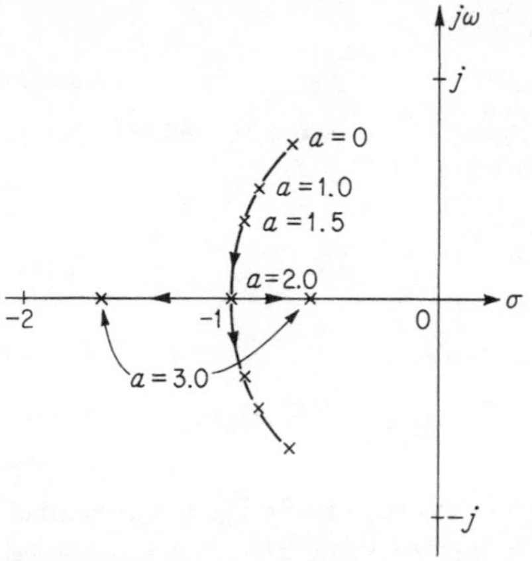


**Fig. 9-6** The locus of the poles of the optimal closed-loop system as $a$ varies.

## 9-6 Analysis of a First-order System

The preceding four sections contain the theory of the state-regulator problem. We illustrated the implications of this theory by means of a few simple examples. In this section, we shall examine a first-order system and shall further illustrate the response of the optimal system by presenting a few time plots. Although the system is very simple, the results illustrate the properties of optimal systems, and we feel that the reader will gain some additional insight into their behavior.

Consider the system

$$\dot{x}(t) = ax(t) + u(t) \tag{9-148}$$

and the cost functional

$$J_1 = \tfrac{1}{2}fx^2(T) + \tfrac{1}{2}\int_0^T [qx^2(t) + ru^2(t)]\,dt \tag{9-149}$$

We assume that

$$f \geq 0 \qquad q > 0 \qquad r > 0 \tag{9-150}$$

Using Control Law 9-1, we conclude that the optimal control is given by

$$u(t) = -\frac{1}{r}k(t)x(t) \tag{9-151}$$

where the scalar $k(t)$ is the solution of the Riccati equation

$$\dot{k}(t) = -2ak(t) + \frac{1}{r}k^2(t) - q \tag{9-152}$$

with the boundary condition

$$k(T) = f \tag{9-153}$$

From Eq. (9-152), we find that

$$\int_{k(t)}^{f} \frac{dk(t)}{(1/r)k^2(t) - 2ak(t) - q} = \int_t^T d\tau \tag{9-154}$$

which can be integrated to obtain

$$k(t) = r\frac{\beta + a + (\beta - a)\dfrac{f/r - a - \beta}{f/r - a + \beta}e^{2\beta(t-T)}}{1 - \dfrac{f/r - a - \beta}{f/r - a + \beta}e^{2\beta(t-T)}} \tag{9-155}$$

where
$$\beta = \sqrt{\frac{q}{r} + a^2} \tag{9-156}$$

Thus, for this system, we can solve the first-order Riccati differential equation analytically.

The optimal trajectory is the solution of the first-order time-varying equation

$$\dot{x}(t) = \left[a - \frac{1}{r}k(t)\right]x(t) \qquad x(0) \text{ given} \tag{9-157}$$

Thus,
$$x(t) = x(0)\exp\int_0^t\left[a - \frac{1}{r}k(\tau)\right]d\tau \tag{9-158}$$

Figure 9-7 shows the structure of the optimal system. We generate $k(t)$ by simulating the Riccati equation; the initial condition $k(0)$ is obtained

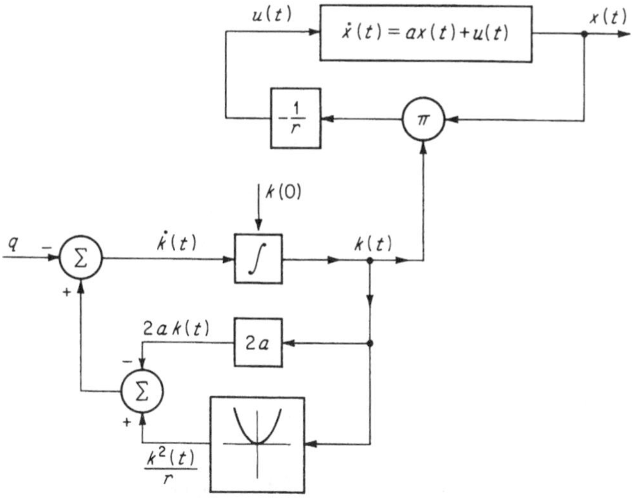


**Fig. 9-7** The optimal state regulator for the system $\dot{x}(t) = ax(t) + u(t)$. The "gain" $k(t)$ is generated by a simulation of Eq. (9-152). The initial value $k(0)$ is precomputed.

from Eq. (9-155) using $t = 0$. The signal $k(t)$ is multiplied by the state $x(t)$ to generate the optimal control.

Figure 9-8$a$ shows the state $x(t)$ of the optimal system for $a = -1$, $f = 0$, $T = 1$, $x(0) = 1$, and $q = 1$, using $r$ as a parameter. Observe that when $r$ is small (which means that the cost of the control is not important), the state $x(t)$ is forced to zero rapidly; when $r$ is large (which means that the cost of the control is very important), the state $x(t)$ decays slowly.

Figure 9-8*b* shows the optimal control $u(t)$, with $r$ as a parameter. Observe that as $r$ decreases, the control becomes quite large in the initial part of the control interval [0, 1]; as $r \to 0$, the optimal control approaches an impulse at $t = 0$.

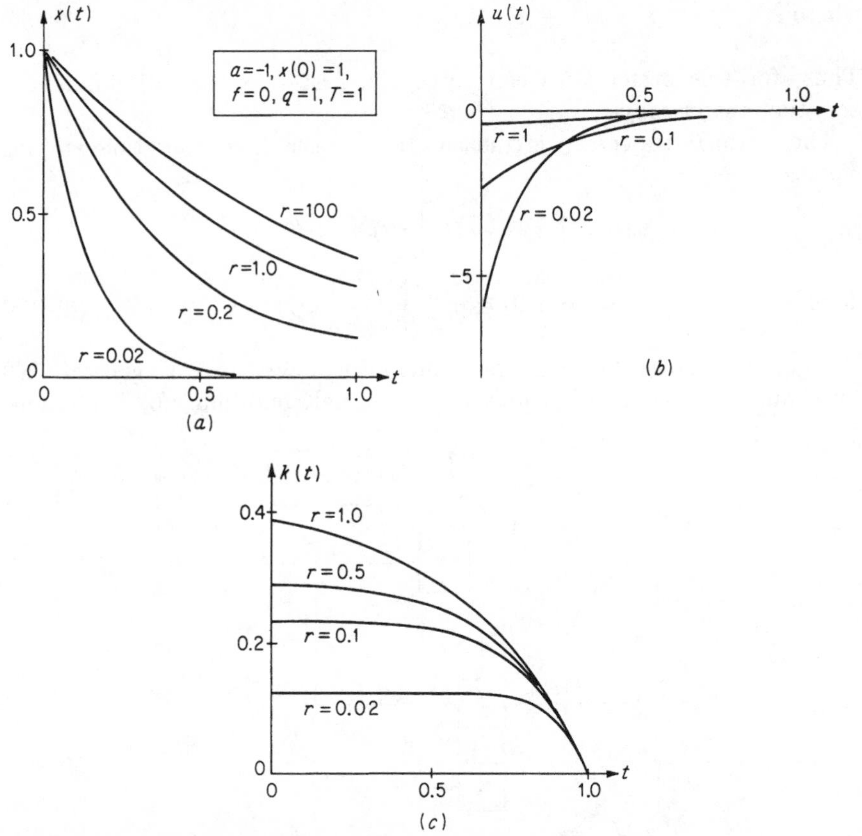


**Fig. 9-8** (*a*) The response of the optimal regulator when $a = -1$ (stable case). (*b*) The optimal control . (*c*) The solution $k(t)$ of the Riccati equation.

Figure 9-8*c* shows the solution $k(t)$ of the Riccati equation, using $r$ as a parameter. Note that as $r$ decreases, $k(t)$ is almost a constant during the initial part of the control interval; for small values of $r$, the "time-varying nature" of $k(t)$ is exhibited only in the latter part of the control interval. As $r$ increases, $k(t)$ becomes a true "time-varying gain."

Figure 9-9 shows the solution $k(t)$ of the Riccati equation for different values of the terminal time $T$ and with the two boundary conditions $k(T) = 0$ and $k(T) = 1$ (that is, $f = 0$ and $f = 1$). The other values of the constants are $a = -1$, $q = r = 1$. Figure 9-9 illustrates that the function $k(t)$

approaches the same "steady-state value backward in time" as $T$ increases, independent of the terminal conditions. This fact can also be analytically

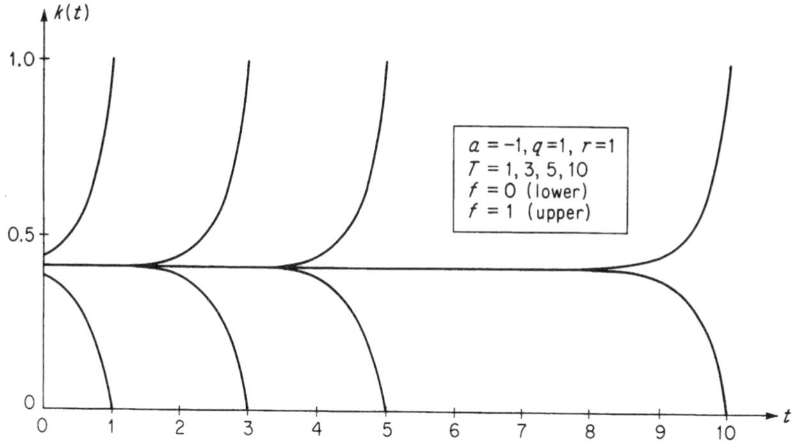


**Fig. 9-9** The solution $k(t)$ of the Riccati equation for various values of the terminal time $T$.

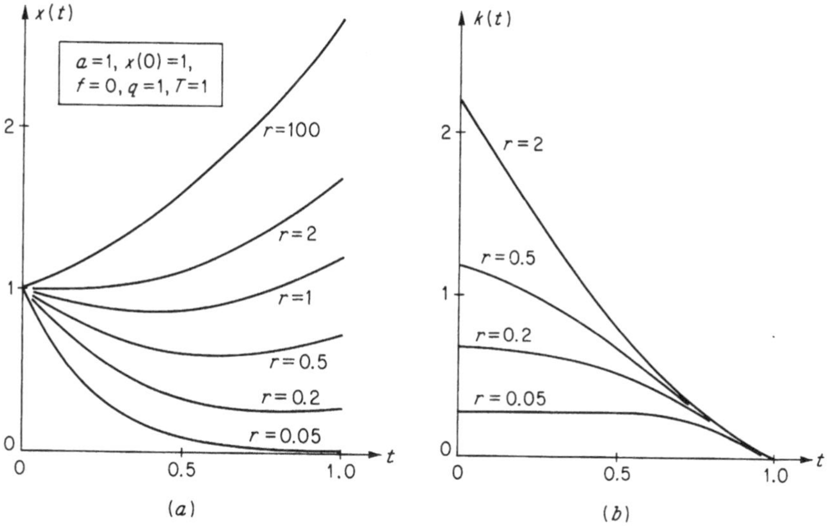


**Fig. 9-10** (*a*) The response of the optimal regulator when $a = +1$ (unstable case). (*b*) The solution $k(t)$ of the Riccati equation when $a = +1$.

verified by using Eq. (9-156), which yields (since $\beta$ is positive) the relation

$$\lim_{T\to\infty} k(t) = r(\beta + a) = ar + r\sqrt{\frac{q}{r} + a^2} \tag{9-159}$$

The results of Figs. 9-8 and 9-9 were obtained for $a = -1$, which means that the system $\dot{x}(t) = ax(t) + u(t)$ is *stable*. Now suppose that $a = +1$, which means that the controlled system is *unstable*. Figure 9-10*a* shows the state $x(t)$ with $r$ as a parameter and using the values $a = 1$, $x(0) = 1$, $f = 0$, and $q = 1$. The response of this system is very interesting because it tends to illustrate some of the idiosyncrasies of optimal systems. First of all, note that for large values of $r$, the state regulator does a very bad job of regulating the state; the reason is that it is cheaper for the system to have a large error in the state over this finite interval of time than to use large control. As the value of $r$ decreases, the state $x(t)$ starts to decrease from its initial value $x(0)$, reaches a minimum, and then increases during the latter part of the control interval. This behavior is due to the fact that $k(T) = 0$, which implies that the optimal control must go to zero as $t \to T$. Figure 9-10*b* shows the behavior of the function $k(t)$ as $r$ varies. Once more, note that $k(t)$ is almost a constant for small values of $r$ at the beginning of the control interval [0, 1].

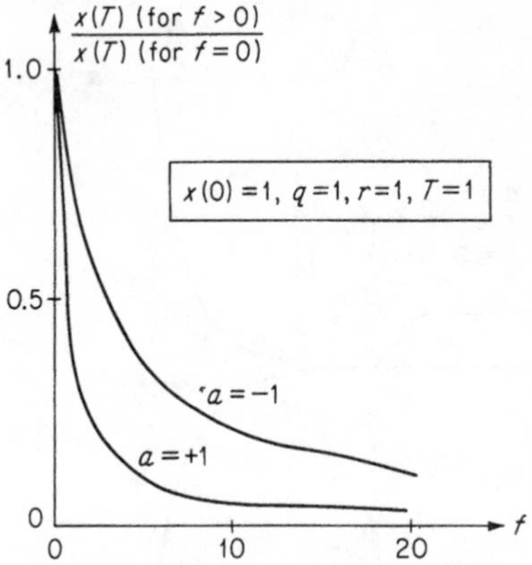

**Fig. 9-11** The effect of $f$ upon the terminal state $x(T)$ of the optimal regulator when $a = -1$ (stable case) and $a = +1$ (unstable case).

Next, let us examine the effect of the boundary condition $k(T) = f$ upon the terminal value $x(T)$ for both the stable ($a = -1$) and the unstable ($a = +1$) systems. In Fig. 9-11, we plot the ratio

$$\frac{x(T)\ (\text{for } f > 0)}{x(T)\ (\text{for } f = 0)} \tag{9-160}$$

versus $f$, using, in both cases, the values $x(0) = q = r = T = 1$. From Fig. 9-8*a*, we know that, when $f = 0$ and $a = -1$, $x(1) = 0.2184$; and from Fig. 9-10*a*, we know that, when $f = 0$ and $a = +1$, $x(1) = 1.2244$. The conclusion that we may draw from Fig. 9-11 is: *Inclusion of the terminal cost* $\frac{1}{2}fx^2(T)$ *leads to a better regulator in the unstable case than in the stable case, especially for small values of* $f$. In either case, $\lim_{f\to\infty} x(T) = 0$, but $x(T) \neq 0$ for any finite $f$.

Let us now turn our attention to the case $T = \infty$. Using Control Law 9-2, we find that the optimal control is given by

$$u(t) = -\frac{1}{r}\hat{k}x(t) \tag{9-161}$$

where $\hat{k}$ is the positive root of the algebraic equation

$$\frac{1}{r}\hat{k}^2 - 2a\hat{k} - q = 0 \tag{9-162}$$

and so $\hat{k}$ is given by the equation

$$\hat{k} = ar + r\sqrt{\frac{q}{r} + a^2} \tag{9-163}$$

Comparing Eqs. (9-163) and (9-159), we observe that

$$\hat{k} = \lim_{T\to\infty} k(t) \qquad \textit{for all } f \geq 0 \tag{9-164}$$

If $a = -1$ and $q = r = 1$, then $\hat{k} = -1 + \sqrt{2} \approx 0.414$, which is indeed the "steady-state value" of $k(t)$ in Fig. 9-9. If $a = +1$, $q = 1$, and $r = 0.05$, then $\hat{k} \approx 0.278$, which is the "steady-state value" of $k(t)$ in Fig. 9-10*b* for $r = 0.05$.

**Exercise 9-2** Consider the system $\dot{\mathbf{x}}(t) = \mathbf{A}\mathbf{x}(t) + \mathbf{B}\mathbf{u}(t)$ and the cost functional

$$J_1 = \tfrac{1}{2}\langle \mathbf{x}(T), \mathbf{F}\mathbf{x}(T)\rangle + \tfrac{1}{2}\int_0^T [\langle \mathbf{x}(t), \mathbf{Q}\mathbf{x}(t)\rangle + \langle \mathbf{u}(t), \mathbf{R}\mathbf{u}(t)\rangle]\, dt \tag{9-165}$$

Suppose that $\mathbf{F}$ is positive definite and that $\mathbf{F}$ satisfies the algebraic equation

$$-\mathbf{FA} - \mathbf{A}'\mathbf{F} + \mathbf{FBR}^{-1}\mathbf{B}'\mathbf{F} - \mathbf{Q} = \mathbf{0} \tag{9-166}$$

Show that the minimum cost is $J_1^* = \frac{1}{2}\langle \mathbf{x}(t), \mathbf{F}\mathbf{x}(t)\rangle$ for all $t$ and $T$, either finite or infinite. Show that the optimal control is a time-invariant linear function of the state. Can you explain what Eq. (9-166) implies?

**Exercise 9-3** Consider the system with input $u(t)$ and output $y(t)$ and with the transfer function

$$\frac{y(s)}{u(s)} = G(s) = \frac{1}{s^2 + a_1 s + a_0} \tag{9-167}$$

Let $x_1(t) = y(t)$ and $x_2(t) = \dot{y}(t)$, and consider the cost functional

$$\hat{J}_1 = \tfrac{1}{2}\int_0^\infty [x_1^2(t) + q x_2^2(t) + r u^2(t)]\, dt \qquad q > 0;\ r > 0 \tag{9-168}$$

Determine the optimal control and construct the block diagram (similar to the one shown in Fig. 9-5) of the optimal system. In each of the following cases

$$a_0 = 0 \qquad a_1 = 1 \tag{9-169a}$$
$$a_0 = 0 \qquad a_1 = -1 \tag{9-169b}$$
$$a_0 = 2 \qquad a_1 = 3 \tag{9-169c}$$
$$a_0 = -1 \qquad a_1 = 0 \tag{9-169d}$$
$$a_0 = 1 \qquad a_1 = 0 \tag{9-169e}$$

(*a*) Plot the locus of the poles of the optimal closed-loop system for $r = 1$ and $r = 0.1$ as $q$ varies.

(*b*) Plot the locus of the poles of the optimal closed-loop system for $q = 1$ and $q = 0.1$ as $r$ varies.

(*c*) State your conclusions about the character of the system (i.e., oscillatory, sluggish, etc.).

The following exercises are for students who have access to a digital computer and know how to program it.

**Exercise 9-4** Consider the first-order system $\dot{x}(t) = (2e^{-t} - 1)x(t) + u(t)$ and the cost functional $\frac{1}{2}\int_0^2 [x^2(t) + u^2(t)]\,dt$. Suppose that $x(0) = 0$. Plot the optimal $x(t)$ and $u(t)$. Repeat for the system $\dot{x}(t) = (\sin 10t)x(t) + u(t)$, $x(0) = 0$.

**Exercise 9-5** Consider the system

$$\begin{aligned}\dot{x}_1(t) &= x_2(t)\\ \dot{x}_2(t) &= u(t)\end{aligned} \tag{9-170}$$

with

$$x_1(0) = 1 \qquad x_2(0) = 0 \tag{9-171}$$

and the cost functional

$$J_1 = \tfrac{1}{2}\int_0^T [x_1^2(t) + 0.1x_2^2(t) + 0.1u^2(t)]\,dt \tag{9-172}$$

Plot $x_1(t)$ and $x_2(t)$ versus $t$ in the interval $0 \le t \le 1$ for $T = 1$, $T = 2$, $T = 5$, $T = 10$, and $T = \infty$. Determine the time-varying eigenvalues $\lambda_1(t)$ and $\lambda_2(t)$ of the optimal system; plot $\lambda_1(t)$ and $\lambda_2(t)$ (both real and imaginary parts if necessary) versus $t$ in the interval $0 \le t \le 1$ for $T = 1$, and compare them with the (constant) eigenvalues of the optimal system when $T = \infty$.

**Exercise 9-6** Consider the third-order system

$$\begin{aligned}\dot{x}_1(t) &= x_2(t)\\ \dot{x}_2(t) &= x_3(t)\\ \dot{x}_3(t) &= -ax_3(t) + u(t)\end{aligned} \tag{9-173}$$

with $x_1(0) = 1$, $x_2(0) = x_3(0) = 0$. Consider the cost functional

$$\hat{J}_1 = \tfrac{1}{2}\int_0^\infty [x_1^2(t) + 0.1x_2^2(t) + 0.1u^2(t)]\,dt \tag{9-174}$$

Determine the optimal control. Plot the response $x_1(t)$ of the optimal system for $a = 0.1$, $a = 1$, $a = 10$, and $a = 100$. Compare the $x_1(t)$ you have just determined with the $x_1(t)$ you found in Exercise 9-5 for $T = \infty$. Any conclusions?

## 9-7 The Output-regulator Problem

Up to now, we have been concerned with the problem of making all the components of the state vector $\mathbf{x}(t)$ small. In this section, we shall be concerned with the problem of making the components of the output vector $\mathbf{y}(t)$ small. In the following section, we shall briefly discuss the output-regulator problem for single-input–single-output systems.

In this section, we shall show that if the controlled dynamical system is observable, then we can reduce the output-regulator problem to the state-regulator problem. In essence, the concept of observability allows us to deduce the uniqueness of the optimal control.

Let us consider the linear time-varying system

$$\begin{aligned}\dot{\mathbf{x}}(t) &= \mathbf{A}(t)\mathbf{x}(t) + \mathbf{B}(t)\mathbf{u}(t)\\ \mathbf{y}(t) &= \mathbf{C}(t)\mathbf{x}(t)\end{aligned} \tag{9-175}$$

where

$$\begin{array}{l}\text{The state } \mathbf{x}(t) \text{ is an } n\text{-column vector} \\ \text{The control } \mathbf{u}(t) \text{ is an } r\text{-column vector} \\ \text{The output } \mathbf{y}(t) \text{ is an } m\text{-column vector} \\ \qquad 0 < m \leq r \leq n \end{array} \tag{9-176}$$

We now consider the cost functional

$$J_2 = \tfrac{1}{2}\langle \mathbf{y}(T), \mathbf{F}\mathbf{y}(T)\rangle + \tfrac{1}{2}\int_{t_0}^{T} [\langle \mathbf{y}(t), \mathbf{Q}(t)\mathbf{y}(t)\rangle + \langle \mathbf{u}(t), \mathbf{R}(t)\mathbf{u}(t)\rangle]\, dt \tag{9-177}$$

We assume that

$$\begin{array}{l}\text{The system (9-175) is completely observable}\dagger \\ \text{The control } \mathbf{u}(t) \text{ is not constrained} \\ \mathbf{R}(t) \text{ is positive definite} \\ \mathbf{Q}(t) \text{ is positive semidefinite} \\ \mathbf{F} \text{ is positive semidefinite} \\ T \text{ is fixed}\end{array} \tag{9-178}$$

We wish to determine the optimal control, i.e., the control that minimizes the cost $J_2$.

The use of the cost functional $J_2$ indicates our desire to bring and keep the *output* $\mathbf{y}(t)$ near zero without using an excessive amount of control energy. The presence of the terminal cost $\frac{1}{2}\langle \mathbf{y}(T), \mathbf{F}\mathbf{y}(T)\rangle$ implies that, for $\mathbf{F} \neq \mathbf{0}$, we wish to guarantee that the output at the terminal time $T$ is also small.

We shall solve this problem, called the output-regulator problem, by reducing it to a problem similar to the state-regulator problem. We shall then use the results of Sec. 9-3 to obtain the optimal-control law. To do this, we substitute $\mathbf{y}(t) = \mathbf{C}(t)\mathbf{x}(t)$ into Eq. (9-177) to obtain the relation

$$\begin{aligned} J_2 = {} & \tfrac{1}{2}\langle \mathbf{x}(T), \mathbf{C}'(T)\mathbf{F}\mathbf{C}(T)\mathbf{x}(T)\rangle \\ & + \tfrac{1}{2}\int_{t_0}^{T} [\langle \mathbf{x}(t), \mathbf{C}'(t)\mathbf{Q}(t)\mathbf{C}(t)\mathbf{x}(t)\rangle + \langle \mathbf{u}(t), \mathbf{R}(t)\mathbf{u}(t)\rangle]\, dt \end{aligned} \tag{9-179}$$

If we compare the cost $J_2$ of Eq. (9-179) with the cost $J_1$ of Eq. (9-13), then we observe that their form is identical and the only difference between them is that the matrices $\mathbf{F}$ and $\mathbf{Q}(t)$ in Eq. (9-13) are replaced by the matrices $\mathbf{C}'(T)\mathbf{F}\mathbf{C}(T)$ and $\mathbf{C}'(t)\mathbf{Q}(t)\mathbf{C}(t)$, respectively, in Eq. (9-179). If we can show that $\mathbf{C}'(T)\mathbf{F}\mathbf{C}(T)$ and $\mathbf{C}'(t)\mathbf{Q}(t)\mathbf{C}(t)$ are positive semidefinite whenever $\mathbf{F}$ and $\mathbf{Q}(t)$ are positive semidefinite, then the correspondence between $J_1$ and $J_2$ will be complete.

***Lemma 9-9*** *If* $\mathbf{F}$ *and* $\mathbf{Q}(t)$ *are positive semidefinite and if the system* (9-175) *is completely observable, then the matrices* $\mathbf{C}'(T)\mathbf{F}\mathbf{C}(T)$ *and* $\mathbf{C}'(t)\mathbf{Q}(t)\mathbf{C}(t)$ *are positive semidefinite.*

PROOF First of all, since $\mathbf{F}$ and $\mathbf{Q}(t)$ are symmetric, the matrices $\mathbf{C}'(T)\mathbf{F}\mathbf{C}(T)$ and $\mathbf{C}'(t)\mathbf{Q}(t)\mathbf{C}(t)$ are symmetric. If the system (9-175) is

† See Sec. 4-15, Definition 4-14, and Sec. 4-17.

observable, then $\mathbf{C}'(t)$ cannot be zero for all $t \in [t_0, T]$. If $\mathbf{Q}(t)$ is positive semidefinite, then $\langle \mathbf{y}(t), \mathbf{Q}(t)\mathbf{y}(t)\rangle \geq 0$ for *all* $\mathbf{y}(t)$; therefore, if $\mathbf{y}(t) = \mathbf{C}(t)\mathbf{x}(t)$, then $\langle \mathbf{C}(t)\mathbf{x}(t), \mathbf{Q}(t)\mathbf{C}(t)\mathbf{x}(t)\rangle \geq 0$ for all $\mathbf{C}(t)\mathbf{x}(t)$. But observability implies that each output $\mathbf{y}(t)$ is generated by a unique state $\mathbf{x}(t)$, and so we conclude that $\langle \mathbf{x}(t), \mathbf{C}'(t)\mathbf{Q}(t)\mathbf{C}(t)\mathbf{x}(t)\rangle \geq 0$ for all $\mathbf{x}(t)$ and that $\mathbf{C}'(t)\mathbf{Q}(t)\mathbf{C}(t)$ is positive semidefinite. An identical argument establishes that $\mathbf{C}'(T)\mathbf{F}\mathbf{C}(T)$ is also positive semidefinite.

We can now use the results of Sec. 9-3 to establish the following control law:

***Control Law 9-3 Solution to the Output-regulator Problem*** *Given the linear observable system*

$$\begin{aligned} \dot{\mathbf{x}}(t) &= \mathbf{A}(t)\mathbf{x}(t) + \mathbf{B}(t)\mathbf{u}(t) \\ \mathbf{y}(t) &= \mathbf{C}(t)\mathbf{x}(t) \end{aligned} \tag{9-180}$$

*and the cost functional*

$$J_2 = \tfrac{1}{2}\langle \mathbf{y}(T), \mathbf{F}\mathbf{y}(T)\rangle + \tfrac{1}{2}\int_{t_0}^{T} [\langle \mathbf{y}(t), \mathbf{Q}(t)\mathbf{y}(t)\rangle + \langle \mathbf{u}(t), \mathbf{R}(t)\mathbf{u}(t)\rangle]\, dt \tag{9-181}$$

*Assume that* $\mathbf{u}(t)$ *is not constrained, that* $\mathbf{F}$ *and* $\mathbf{Q}(t)$ *are positive semidefinite, that* $\mathbf{R}(t)$ *is positive definite, and that* $T$ *is fixed.*

*Then the optimal control exists, is unique, and is given by*

$$\mathbf{u}(t) = -\mathbf{R}^{-1}(t)\mathbf{B}'(t)\mathbf{K}(t)\mathbf{x}(t) \tag{9-182}$$

*where the* $n \times n$ *symmetric and positive definite matrix* $\mathbf{K}(t)$ *is the solution of the Riccati equation*

$$\dot{\mathbf{K}}(t) = -\mathbf{K}(t)\mathbf{A}(t) - \mathbf{A}'(t)\mathbf{K}(t) + \mathbf{K}(t)\mathbf{B}(t)\mathbf{R}^{-1}(t)\mathbf{B}'(t)\mathbf{K}(t) - \mathbf{C}'(t)\mathbf{Q}(t)\mathbf{C}(t) \tag{9-183}$$

*with the boundary condition*

$$\mathbf{K}(T) = \mathbf{C}'(T)\mathbf{F}\mathbf{C}(T) \tag{9-184}$$

*The optimal state is then the solution of the differential equation [with* $\mathbf{x}(t_0)$ *given]*

$$\dot{\mathbf{x}}(t) = [\mathbf{A}(t) - \mathbf{B}(t)\mathbf{R}^{-1}(t)\mathbf{B}'(t)\mathbf{K}(t)]\mathbf{x}(t) \tag{9-185}$$

*The minimum cost* $J_2^*$ *is given by*

$$J_2^*[\mathbf{x}(t), t] = \tfrac{1}{2}\langle \mathbf{x}(t), \mathbf{K}(t)\mathbf{x}(t)\rangle \tag{9-186}$$

*for all* $\mathbf{x}(t)$ *and* $t$.

Let us now comment on the implications of Control Law 9-3.

The structure of the optimal output regulator is shown in Fig. 9-12. Note that *the optimal control is a function of the state* $\mathbf{x}(t)$ *rather than of the output* $\mathbf{y}(t)$. *This fact may appear disappointing to the reader, since he might have expected the optimal control to be a function of the output rather than of the state.* Since the dimension of the state vector is, in general, higher than the dimension of the output, this means that the optimal system is quite complicated. However, the fact that the system is observable guarantees that we can compute the state from knowledge of the output. *If the system were not observable, then we could not compute the state from the output, and so we could not construct the optimal control.*

A little thought should convince the reader that the optimal control *must* be a function of the state, since the state at each instant of time contains all the information necessary to predict both the future states and the outputs and since the optimal control has the property that it must utilize all the information in the most "efficient" manner. We also observe that the minimum cost $J_2^*$ is a function of the state [see Eq. (9-186)] rather than of the output. The output alone cannot determine the minimum cost [although the cost $J_2$ was formulated in terms of $\mathbf{y}(t)$] because the output does not, in general, contain all the information that governs the future evolution of the process.

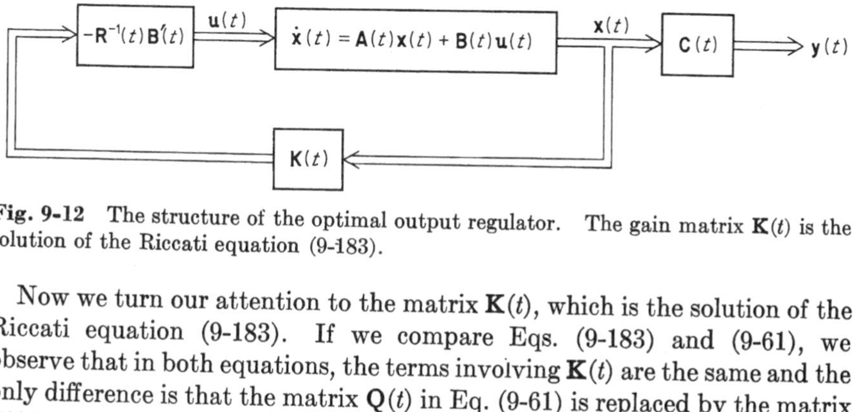


**Fig. 9-12** The structure of the optimal output regulator. The gain matrix $\mathbf{K}(t)$ is the solution of the Riccati equation (9-183).

Now we turn our attention to the matrix $\mathbf{K}(t)$, which is the solution of the Riccati equation (9-183). If we compare Eqs. (9-183) and (9-61), we observe that in both equations, the terms involving $\mathbf{K}(t)$ are the same and the only difference is that the matrix $\mathbf{Q}(t)$ in Eq. (9-61) is replaced by the matrix $\mathbf{C}'(t)\mathbf{Q}(t)\mathbf{C}(t)$ in Eq. (9-183). Naturally, the solutions $\mathbf{K}(t)$ will be different for the state-regulator problem and for the output-regulator problem. Nevertheless, since $\mathbf{K}(t)$ is the solution of the same type of Riccati equation in both problems, all the comments of Sec. 9-4 concerning the computation of $\mathbf{K}(t)$ still apply.

Finally, we note that (just as in Sec. 9-3) controllability of the linear system (9-180) is not required as long as the terminal time $T$ is finite.†

Let us now specialize our results to the case of linear time-invariant systems and cost functionals with $T = \infty$. We mimic the results of Sec. 9-5 to obtain the following control law:

***Control Law 9-4*** *Given the controllable and observable linear time-invariant system*

$$\begin{aligned} \dot{\mathbf{x}}(t) &= \mathbf{A}\mathbf{x}(t) + \mathbf{B}\mathbf{u}(t) \\ \mathbf{y}(t) &= \mathbf{C}\mathbf{x}(t) \end{aligned} \tag{9-187}$$

† See the comments on controllability in Sec. 9-4.

*and the cost functional*

$$\hat{J}_2 = \frac{1}{2} \int_0^\infty [\langle \mathbf{y}(t), \mathbf{Q}\mathbf{y}(t)\rangle + \langle \mathbf{u}(t), \mathbf{R}\mathbf{u}(t)\rangle]\, dt \tag{9-188}$$

*Assume that* $\mathbf{u}(t)$ *is not constrained, that* $\mathbf{Q}$ *is positive definite, and that* $\mathbf{R}$ *is positive definite. Then the optimal control exists, is unique, and is given by*

$$\mathbf{u}(t) = -\mathbf{R}^{-1}\mathbf{B}'\hat{\mathbf{K}}\mathbf{x}(t) \tag{9-189}$$

*where the* $n \times n$ *constant, symmetric, and positive definite matrix* $\hat{\mathbf{K}}$ *is the solution of the nonlinear time-invariant matrix algebraic equation*

$$-\hat{\mathbf{K}}\mathbf{A} - \mathbf{A}'\hat{\mathbf{K}} + \hat{\mathbf{K}}\mathbf{B}\mathbf{R}^{-1}\mathbf{B}'\hat{\mathbf{K}} - \mathbf{C}'\mathbf{Q}\mathbf{C} = \mathbf{0} \tag{9-190}$$

*In this case, the state of the optimal system is the solution of the linear time-invariant homogeneous equation [with* $\mathbf{x}(0)$ *given]*

$$\dot{\mathbf{x}}(t) = [\mathbf{A} - \mathbf{B}\mathbf{R}^{-1}\mathbf{B}'\hat{\mathbf{K}}]\mathbf{x}(t) \triangleq \mathbf{G}\mathbf{x}(t) \tag{9-191}$$

*The minimum cost* $\hat{J}_2^*$ *is given by*

$$\hat{J}_2^* = \frac{1}{2}\langle \mathbf{x}(t), \hat{\mathbf{K}}\mathbf{x}(t)\rangle \tag{9-192}$$

*Moreover, the eigenvalues of the matrix*

$$\mathbf{G} = [\mathbf{A} - \mathbf{B}\mathbf{R}^{-1}\mathbf{B}'\hat{\mathbf{K}}] \tag{9-193}$$

*have negative real parts.*

We remind the reader that the system (9-187) is *controllable* if and only if the matrix

$$[\mathbf{B} \mid \mathbf{A}\mathbf{B} \mid \mathbf{A}^2\mathbf{B} \mid \cdots \mid \mathbf{A}^{n-1}\mathbf{B}] \tag{9-194}$$

contains $n$ linearly independent column vectors and that the system (9-187) is *observable* if and only if the matrix

$$[\mathbf{C}' \mid \mathbf{A}'\mathbf{C}' \mid (\mathbf{A}')^2\mathbf{C}' \mid \cdots \mid (\mathbf{A}')^{n-1}\mathbf{C}'] \tag{9-195}$$

contains $n$ linearly independent column vectors.

**Example 9-5** Consider the system

$$\begin{aligned} \dot{x}_1(t) &= x_2(t) \\ \dot{x}_2(t) &= u(t) \\ y(t) &= x_1(t) \end{aligned} \tag{9-196}$$

This is the same system as the one we considered in Example 9-4; its transfer function is $y(s)/u(s) = 1/s^2$. Let us consider the cost functional

$$\hat{J}_2 = \frac{1}{2} \int_0^\infty [y^2(t) + ru^2(t)]\, dt \tag{9-197}$$

We apply Control Law 9-4, and we find that the optimal control is

$$u(t) = -\frac{1}{r}[\hat{k}_{12}x_1(t) + \hat{k}_{22}x_2(t)] \tag{9-198}$$

The matrix algebraic equation (9-190) yields, in this case, the three equations

$$\frac{1}{r}\hat{k}_{12}^2 = 1 \tag{9-199}$$

$$-\hat{k}_{11} + \frac{1}{r}\hat{k}_{12}\hat{k}_{22} = 0 \tag{9-200}$$

$$-2\hat{k}_{12} + \frac{1}{r}\hat{k}_{22}^2 = 0 \tag{9-201}$$

Since $\hat{\mathbf{K}}$ is positive definite, we must have the inequalities

$$\hat{k}_{11} > 0 \qquad \hat{k}_{22} > 0 \qquad \hat{k}_{11}\hat{k}_{22} - \hat{k}_{12}^2 > 0 \tag{9-202}$$

Thus, the correct solutions of Eqs. (9-199) to (9-201) are

$$\begin{aligned} \hat{k}_{12} &= r^{\frac{1}{2}} \\ \hat{k}_{22} &= r^{\frac{3}{4}}\sqrt{2} \\ \hat{k}_{11} &= r^{\frac{1}{4}}\sqrt{2} \end{aligned} \tag{9-203}$$

Substituting Eqs. (9-203) into Eq. (9-198), we find that the optimal control is given by the relation

$$\begin{aligned} u(t) &= -r^{-\frac{1}{2}}x_1(t) - r^{-\frac{1}{4}}\sqrt{2}\,x_2(t) \\ &= -r^{-\frac{1}{2}}y(t) - r^{-\frac{1}{4}}\sqrt{2}\,\dot{y}(t) \end{aligned} \tag{9-204}$$

The block diagram of the optimal system (using Laplace transform notation) is shown in Fig. 9-13. The optimal control is generated by passing the output signal $y(t)$ through a feedback compensator with the transfer function

$$F(s) = r^{-\frac{1}{2}} + r^{-\frac{1}{4}}\sqrt{2}\,s \tag{9-205}$$

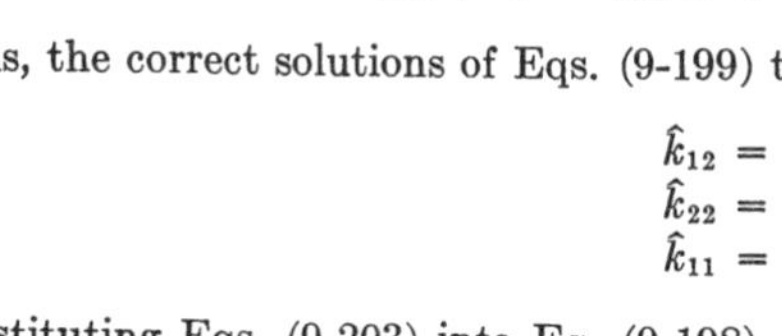

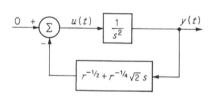

**Fig. 9-13** The optimal closed-loop output regulator for the double-integral plant.

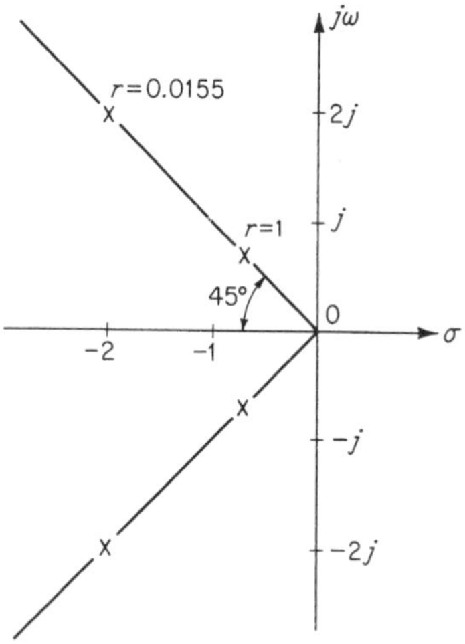


**Fig. 9-14** The locus of the poles of the closed-loop system of Fig. 9-13 as $r$ varies. The positions of the poles for $r = 0.0155$ and $r = 1$ are indicated.

The transfer function $G^*(s)$ of the optimal closed-loop feedback system is

$$G^*(s) = \frac{1/s^2}{1 + (r^{-\frac{1}{2}} + r^{-\frac{1}{4}}\sqrt{2}\,s)/s^2} = \frac{1}{s^2 + \sqrt{2}\,r^{-\frac{1}{4}}s + r^{-\frac{1}{2}}} \tag{9-206}$$

The poles of the optimal closed-loop transfer function $G^*(s)$ are at

$$s = -\frac{\sqrt{2}}{2}r^{-\frac{1}{4}}(1 \pm j) \tag{9-207†}$$

Figure 9-14 shows the locus of the poles of the closed-loop system as $r$ varies. The locus

† $j = \sqrt{-1}$.

is a straight line at 45 degrees from the $j\omega$ axis. The remarkable property of this system is that the *damping ratio*† $\zeta$ of the optimal system is

$$\zeta = 0.707 \tag{9-208}$$

for every value of $r$. Changes in $r$ simply increase or decrease the gain of the system; when $r$ is small, the gain is large, and when $r$ is large, the gain is small.

**Exercise 9-7** Consider the linear system

$$\begin{aligned}\dot{\mathbf{x}}(t) &= \mathbf{A}(t)\mathbf{x}(t) + \mathbf{B}(t)\mathbf{u}(t)\\ \mathbf{y}(t) &= \mathbf{C}(t)\mathbf{x}(t) + \mathbf{D}(t)\mathbf{u}(t)\end{aligned}$$

and the cost functional $J_2$ given by Eq. (9-181). Determine the optimal control and the minimum value of $J_2$. State clearly all your assumptions.

**Exercise 9-8** Consider the system $\dot{\mathbf{x}}(t) = \mathbf{A}\mathbf{x}(t) + \mathbf{B}\mathbf{u}(t)$, $\mathbf{y}(t) = \mathbf{C}\mathbf{x}(t)$, and the cost functional

$$J_2 = \tfrac{1}{2}\langle \mathbf{y}(T), \mathbf{F}\mathbf{y}(T)\rangle + \tfrac{1}{2}\int_0^T [\langle \mathbf{y}(t), \mathbf{Q}\mathbf{y}(t)\rangle + \langle \mathbf{u}(t), \mathbf{R}\mathbf{u}(t)\rangle]\, dt$$

where $T$ is fixed and $T < \infty$. Can you derive conditions such that $\mathbf{K}(t) = \text{const}$ for all $t \in [0, T]$? HINT: See Exercise 9-2.

## 9-8 The Output-regulator Problem for Single-input–Single-output Systems

In this section, we shall specialize the results of the previous section to systems with a single control input $u(t)$ and a single output $y(t)$. Our major objectives are:

1. To indicate the structure of the optimal feedback system
2. To present the equations that must be solved

We shall carry out this development for systems with only poles in their transfer function. We shall indicate the concepts involved for systems with both poles and zeros by means of a simple example.

First, we consider the system described by the $n$th-order linear time-invariant differential equation

$$\{D^n + a_{n-1}D^{n-1} + \cdots + a_1 D + a_0\}y(t) = u(t) \tag{9-209}$$

The transfer function $G(s)$ of this system is

$$\frac{y(s)}{u(s)} = G(s) = \frac{1}{s^n + a_{n-1}s^{n-1} + \cdots + a_1 s + a_0} \tag{9-210}$$

We define‡ the state $\mathbf{x}(t)$ for the system (9-209) by setting

$$x_i(t) = \frac{d^{i-1}}{dt^{i-1}}\, y(t) \qquad i = 1, 2, \ldots, n \tag{9-211}$$

† *Damping ratio* is a term very familiar to control engineers. If a pair of complex poles (or eigenvalues) are at $s = -\alpha_n \pm j\omega_n$, $\alpha_n \geq 0$, $\omega_n \geq 0$, then the damping ratio $\zeta$ is defined as

$$\zeta = \sin \tan^{-1} \frac{\omega_n}{\alpha_n}$$

‡ See Sec. 4-9.

Then the state equations of the system (9-209) are

$$\begin{bmatrix} \dot{x}_1(t) \\ \dot{x}_2(t) \\ \cdots \\ \dot{x}_n(t) \end{bmatrix} = \begin{bmatrix} 0 & 1 & 0 & \cdots & 0 \\ 0 & 0 & 1 & \cdots & 0 \\ \cdots & \cdots & \cdots & \cdots & \cdots \\ -a_0 & -a_1 & -a_2 & \cdots & -a_{n-1} \end{bmatrix} \begin{bmatrix} x_1(t) \\ x_2(t) \\ \cdots \\ x_n(t) \end{bmatrix} + \begin{bmatrix} 0 \\ 0 \\ \cdots \\ 1 \end{bmatrix} u(t) \tag{9-212}$$

and the output equation takes the form

$$y(t) = [1 \quad 0 \quad \cdots \quad 0] \begin{bmatrix} x_1(t) \\ x_2(t) \\ \cdot \\ \cdot \\ \cdot \\ x_n(t) \end{bmatrix} \tag{9-213}$$

Equation (9-212) is of the form

$$\dot{\mathbf{x}}(t) = \mathbf{A}\mathbf{x}(t) + \mathbf{b}u(t) \tag{9-214}$$

and Eq. (9-213) is of the form

$$y(t) = \mathbf{C}\mathbf{x}(t) \tag{9-215}$$

where $\mathbf{C}$ is a $1 \times n$ matrix (i.e., a row vector).

Let

$$\hat{J}_2 = \tfrac{1}{2} \int_0^\infty [y^2(t) + ru^2(t)]\, dt \tag{9-216}$$

be the cost functional under consideration. We now apply Control Law 9-4. If we let $\hat{k}_{ij}$ denote the elements of the matrix $\hat{\mathbf{K}}$, then the optimal control is given by

$$\begin{aligned} u(t) &= -\frac{1}{r} [\hat{k}_{1n}x_1(t) + \hat{k}_{2n}x_2(t) + \cdots + \hat{k}_{nn}x_n(t)] \\ &= -\frac{1}{r} \left[ \sum_{j=1}^{n} \hat{k}_{jn}x_j(t) \right] \\ &= -\frac{1}{r} \left[ \sum_{j=1}^{n} \hat{k}_{jn} \frac{d^{j-1}}{dt^{j-1}} y(t) \right] \end{aligned} \tag{9-217}$$

The structure of the optimal feedback system is shown in Fig. 9-15 using Laplace transform notation. The output $y(t)$ is introduced to the feedback compensator with the transfer function

$$F(s) = \hat{k}_{nn}s^{n-1} + \cdots + \hat{k}_{2n}s + \hat{k}_{1n} \tag{9-218}$$

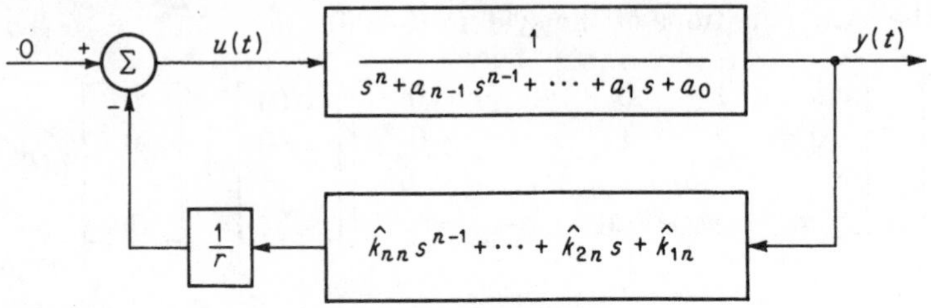


**Fig. 9-15** The optimal closed-loop output regulator when the transfer function of the controlled plant contains only poles.

to produce the optimal control. The transfer function $G^*(s)$ of the optimal closed-loop system is

$$G^*(s) = \frac{G(s)}{1 + (1/r)G(s)F(s)}$$
$$= \frac{1}{s^n + \left(a_{n-1} + \frac{\hat{k}_{nn}}{r}\right)s^{n-1} + \cdots + \left(a_1 + \frac{\hat{k}_{2n}}{r}\right)s + \left(a_0 + \frac{\hat{k}_{1n}}{r}\right)} \tag{9-219}$$

Thus, the poles (or eigenvalues) of the optimal closed-loop system are the $n$ solutions of the algebraic equation

$$s^n + \left(a_{n-1} + \frac{\hat{k}_{nn}}{r}\right)s^{n-1} + \cdots + \left(a_1 + \frac{\hat{k}_{2n}}{r}\right)s + \left(a_0 + \frac{\hat{k}_{1n}}{r}\right) = 0 \tag{9-220}$$

The values of $\hat{k}_{1n}, \ldots, \hat{k}_{nn}$ must be evaluated by solving the equation

$$-\hat{\mathbf{K}}\mathbf{A} - \mathbf{A}'\hat{\mathbf{K}} + \frac{1}{r}\hat{\mathbf{K}}\mathbf{b}\mathbf{b}'\hat{\mathbf{K}} - \mathbf{C}'\mathbf{C} = \mathbf{0} \tag{9-221}$$

where $\mathbf{A}$ is the matrix appearing in Eq. (9-212),

$$\mathbf{b} = \begin{bmatrix} 0 \\ 0 \\ \cdot \\ \cdot \\ \cdot \\ 1 \end{bmatrix}$$

and $\mathbf{C} = [1 \quad 0 \quad 0 \quad \cdots \quad 0]$.

The location of the closed-loop poles will, of course, depend on the weighting factor $r$ which appears in the cost functional (9-216). It has been shown by Kalman [K-10] that the asymptotic behavior of the poles of $G^*(s)$ [see Eq. (9-219)] can be described as follows: *As $r \to 0$, the poles of the optimal closed-loop system approach a stable Butterworth configuration.*† As we have done before, we can plot the loci of the closed-loop poles as $r$ varies. We

† For a discussion of Butterworth filters see, for example, Ref. W-4.

remark that the value of the loop gain is a (nonlinear) function of the parameter $r$. The smaller the value of $r$, the higher the loop gain; the larger the value of $r$, the smaller the loop gain.

From a practical point of view, the construction of the feedback compensator $F(s)$ [see Fig. 9-15 and Eq. (9-218)] must be done in an approximate manner. The reason is that a system with the transfer function $F(s)$ is an all-zero network. A suboptimal system can be obtained by replacing $F(s)$ by a network with the transfer function

$$\tilde{F}(s) = \frac{F(s)}{Q(s)}$$

where the roots of $Q(s)$ are as far as possible, in the left half of the $s$ plane, from the roots of $F(s)$.

We could apply the theory to a system whose transfer function contains $m$ zeros and $n$ poles, with $m < n$. We shall leave this task to the reader and shall illustrate the general theory by means of a simple example.

**Example 9-6** Consider the system† with output $y(t)$ and control $u(t)$ related by

$$\ddot{y}(t) = \dot{u}(t) + \beta u(t) \tag{9-222}$$

This system has the transfer function

$$\frac{y(s)}{u(s)} = H(s) = \frac{s+\beta}{s^2} \tag{9-223}$$

We define the state variables by setting

$$\begin{aligned} x_1(t) &= y(t) \\ x_2(t) &= \dot{y}(t) - u(t) \end{aligned} \tag{9-224}$$

Then the system equations are

$$\begin{aligned} \dot{x}_1(t) &= x_2(t) + u(t) \\ \dot{x}_2(t) &= \beta u(t) \\ y(t) &= x_1(t) \end{aligned} \tag{9-225}$$

We wish to find the control which will minimize the cost $\hat{J}_2$ given by Eq. (9-216). We apply Control Law 9-4, and we find that [see Eq. (9-189)]

$$u(t) = -\frac{1}{r}[(\hat{k}_{11} + \beta\hat{k}_{12})x_1(t) + (\hat{k}_{12} + \beta\hat{k}_{22})x_2(t)] \tag{9-226}$$

From Eq. (9-190), we obtain the algebraic equations

$$\begin{aligned} \hat{k}_{11}^2 + 2\beta\hat{k}_{11}\hat{k}_{12} + \beta^2\hat{k}_{12}^2 - r &= 0 \\ \beta\hat{k}_{11}\hat{k}_{22} + \beta\hat{k}_{12}^2 + \beta^2\hat{k}_{12}\hat{k}_{22} + \hat{k}_{11}\hat{k}_{12} - r\hat{k}_{11} &= 0 \\ \hat{k}_{12}^2 + 2\beta\hat{k}_{12}\hat{k}_{22} + \beta^2\hat{k}_{22}^2 - 2r\hat{k}_{12} &= 0 \end{aligned} \tag{9-227}$$

Figure 9-16 shows the block diagram of the optimal system (using Laplace transform notation). Since $x_2(t) = \dot{y}(t) - u(t)$, we obtain the state variable $x_2(t)$ by subtracting the control $u(t)$ from $\dot{y}(t)$. Naturally, the block diagram of Fig. 9-16 can be reduced further by standard block-diagram transformations.

† We examined the time-optimal control of this system in Sec. 7-12.

From Eqs. (9-227), we find that

$$\begin{aligned}(\hat{k}_{11} + \beta\hat{k}_{12})^2 &= r \\ (\hat{k}_{12} + \beta\hat{k}_{22})^2 &= 2r\hat{k}_{12}\end{aligned} \tag{9-228}$$

Since $\hat{k}_{11} > 0$, $\hat{k}_{22} > 0$, $\hat{k}_{11}\hat{k}_{22} - \hat{k}_{12}^2 > 0$, we deduce that

$$\hat{k}_{11} + \beta\hat{k}_{12} = \sqrt{r} \tag{9-229}$$

$$\hat{k}_{12} + \beta\hat{k}_{22} = \sqrt{2r\hat{k}_{12}} \tag{9-230}$$

From Eqs. (9-229), (9-230), and (9-227), we find that $\hat{k}_{12}$ is the solution of a fourth-order algebraic equation, which we shall not attempt to solve here.

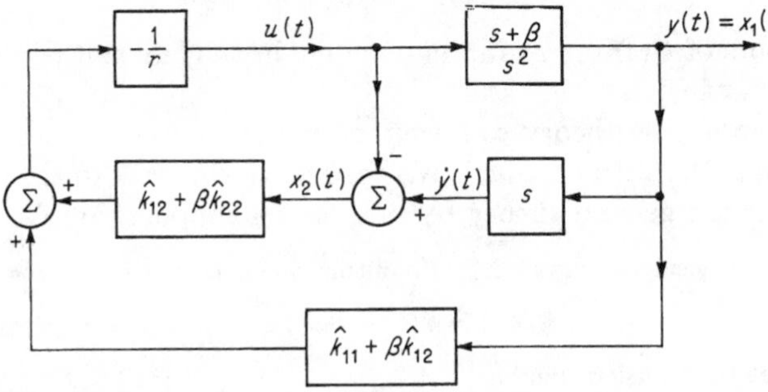


**Fig. 9-16** The optimal closed-loop output regulator for the plant with the transfer function $(s + \beta)/s^2$.

An interesting aspect of the problem arises when $\beta = 0$, that is, when the system (9-225) is not controllable. If the system is not controllable, then Control Law 9-4 should not be used. To illustrate the nonsense that arises when we attempt to use Control Law 9-4, we blindly substitute $\beta = 0$ into Eqs. (9-227) to obtain

$$\hat{k}_{11}^2 = r \tag{9-231}$$

$$\hat{k}_{11}\hat{k}_{12} = r\hat{k}_{11} \tag{9-232}$$

$$\hat{k}_{12}^2 = 2r\hat{k}_{12} \tag{9-233}$$

Since $\hat{k}_{11} > 0$ and $r > 0$, we see from Eq. (9-231) that

$$\hat{k}_{11} = \sqrt{r} > 0 \tag{9-234}$$

Equation (9-232) leads to the conclusion

$$\hat{k}_{12} = r \tag{9-235}$$

and Eq. (9-233) results in the relations

$$\hat{k}_{12} = 0 \qquad \text{or} \qquad \hat{k}_{12} = 2r \tag{9-236}$$

Clearly, Eqs. (9-235) and (9-236) are contradictory. This ridiculous answer arises because we have tried to use Control Law 9-4 for an uncontrollable system.

**Exercise 9-9** For each of the following systems, $y(t)$ represents the output and $u(t)$ the control; the cost functional is

$$\hat{J}_2 = \frac{1}{2}\int_0^\infty [y^2(t) + ru^2(t)]\,dt \tag{9-237}$$

For each system, determine the optimal control, determine the poles of the optimal-control system, and plot the locus of these poles as $r$ varies.

(*a*) $\ddot{y}(t) + \dot{y}(t) = u(t)$ (*b*) $\ddot{y}(t) - \dot{y}(t) = u(t)$

(*c*) $\ddot{y}(t) + 3\dot{y}(t) + 2y(t) = u(t)$ (*d*) $\ddot{y}(t) - y(t) = u(t)$

(*e*) $\ddot{y}(t) + y(t) = u(t)$ (*f*) $\dddot{y}(t) = u(t)$

(*g*) $\dddot{y}(t) + \ddot{y}(t) = u(t)$

Is there a common characteristic of the loci of the poles of the second-order systems (*a*) to (*e*)?

**Exercise 9-10** Consider the system with output $y(t)$ and input $u(t)$ related by the differential equation

$$\ddot{y}(t) + \dot{y}(t) = \alpha\dot{u}(t) + u(t) \tag{9-238}$$

and with the cost functional given by Eq. (9-237). Determine and plot the locus of the poles of the optimal closed-loop system as $r$ varies and as $\alpha$ varies. What happens when $\alpha = 1$?

## 9-9 The Tracking Problem†

In the previous sections, we examined the state- and output-regulator problems. In this section, we shall focus our attention on the tracking problem. Recall that we discussed in detail the physical aspects of the tracking problem in Sec. 9-2 and that we presented its formulation; we shall repeat it here for the sake of completeness.

In this section, we shall proceed in a manner similar to that of Sec. 9-3. We shall use the minimum principle to determine the extremal controls. We shall show that:

1. The $H$-minimal control is a linear function of the costate.
2. The reduced canonical equations are linear but not homogeneous (as the desired output will appear as a forcing term).
3. We shall find that the costate $\mathbf{p}(t)$ is related to the state $\mathbf{x}(t)$ by an expression of the form $\mathbf{p}(t) = \mathbf{K}(t)\mathbf{x}(t) - \mathbf{g}(t)$; we shall show that the matrix $\mathbf{K}(t)$ must be the solution of the Riccati equation and that $\mathbf{g}(t)$ must be the solution of a linear differential equation which involves the desired output.

We shall leave the proofs of existence and uniqueness to the reader, and we shall concentrate on the discussion of the properties of the optimal tracking system.

We suppose that we are given a linear observable system with state $\mathbf{x}(t)$, output $\mathbf{y}(t)$, and control $\mathbf{u}(t)$, described by the equations

$$\begin{aligned}\dot{\mathbf{x}}(t) &= \mathbf{A}(t)\mathbf{x}(t) + \mathbf{B}(t)\mathbf{u}(t)\\ \mathbf{y}(t) &= \mathbf{C}(t)\mathbf{x}(t)\end{aligned} \tag{9-239}$$

Suppose that the vector $\mathbf{z}(t)$ is the *desired output;* we assume that the dimension of $\mathbf{z}(t)$ is equal to the dimension of the output $\mathbf{y}(t)$. Our objective is to control the system in such a way as to make the output $\mathbf{y}(t)$ be "near"

† This problem is treated in Ref. K-8.

$\mathbf{z}(t)$ without excessive control-energy expenditure. For this reason, we define the *error vector* $\mathbf{e}(t)$ by setting

$$\mathbf{e}(t) = \mathbf{z}(t) - \mathbf{y}(t) \tag{9-240}$$

The cost functional $J_3$ which we shall minimize is given by

$$J_3 = \tfrac{1}{2}\langle \mathbf{e}(T), \mathbf{F}\mathbf{e}(T)\rangle + \tfrac{1}{2}\int_{t_0}^{T} [\langle \mathbf{e}(t), \mathbf{Q}(t)\mathbf{e}(t)\rangle + \langle \mathbf{u}(t), \mathbf{R}(t)\mathbf{u}(t)\rangle]\, dt \tag{9-241}$$

with the (usual) assumptions that $T$ is specified, that $\mathbf{R}(t)$ is positive definite, and that $\mathbf{Q}(t)$ and $\mathbf{F}$ are positive semidefinite.

Since the error can be expressed as a function of $\mathbf{z}(t)$ and $\mathbf{x}(t)$, that is, since

$$\mathbf{e}(t) = \mathbf{z}(t) - \mathbf{C}(t)\mathbf{x}(t) \tag{9-242}$$

we can, by substituting Eq. (9-242) into Eq. (9-241), obtain $J_3$ as a function of $\mathbf{x}(t)$, $\mathbf{u}(t)$, and $\mathbf{z}(t)$. We are now ready to use the minimum principle and thus obtain the necessary conditions. Our method of approach will be very similar to that used in Sec. 9-3, and many proofs will be omitted or left as exercises for the reader.

The Hamiltonian for the tracking problem is given by

$$\begin{aligned} H = {} & \tfrac{1}{2}\langle [\mathbf{z}(t) - \mathbf{C}(t)\mathbf{x}(t)], \mathbf{Q}(t)[\mathbf{z}(t) - \mathbf{C}(t)\mathbf{x}(t)]\rangle \\ & + \tfrac{1}{2}\langle \mathbf{u}(t), \mathbf{R}(t)\mathbf{u}(t)\rangle + \langle \mathbf{A}(t)\mathbf{x}(t), \mathbf{p}(t)\rangle + \langle \mathbf{B}(t)\mathbf{u}(t), \mathbf{p}(t)\rangle \end{aligned} \tag{9-243}$$

The condition $\partial H/\partial \mathbf{u}(t) = \mathbf{0}$ yields the equation

$$\frac{\partial H}{\partial \mathbf{u}(t)} = \mathbf{0} = \mathbf{R}(t)\mathbf{u}(t) + \mathbf{B}'(t)\mathbf{p}(t) \tag{9-244}$$

and so we deduce the relation

$$\mathbf{u}(t) = -\mathbf{R}^{-1}(t)\mathbf{B}'(t)\mathbf{p}(t) \tag{9-245}$$

Since $\mathbf{R}(t)$ is positive definite, the control (9-245) minimizes $H$; that is, the control (9-245) is $H$-minimal.

The condition $\dot{\mathbf{p}}(t) = -\partial H/\partial \mathbf{x}(t)$ yields

$$\dot{\mathbf{p}}(t) = -\mathbf{C}'(t)\mathbf{Q}(t)\mathbf{C}(t)\mathbf{x}(t) - \mathbf{A}'(t)\mathbf{p}(t) + \mathbf{C}'(t)\mathbf{Q}(t)\mathbf{z}(t) \tag{9-246}$$

From Eqs. (9-245) and (9-239), we obtain the relation

$$\dot{\mathbf{x}}(t) = \mathbf{A}(t)\mathbf{x}(t) - \mathbf{B}(t)\mathbf{R}^{-1}(t)\mathbf{B}'(t)\mathbf{p}(t) \tag{9-247}$$

We define, for the sake of simplicity, the matrices

$$\mathbf{S}(t) = \mathbf{B}(t)\mathbf{R}^{-1}(t)\mathbf{B}'(t) \tag{9-248}$$

$$\mathbf{V}(t) = \mathbf{C}'(t)\mathbf{Q}(t)\mathbf{C}(t) \tag{9-249}$$

$$\mathbf{W}(t) = \mathbf{C}'(t)\mathbf{Q}(t) \tag{9-250}$$

We now combine Eqs. (9-246) and (9-247) to obtain the reduced canonical equations

$$\begin{bmatrix} \dot{\mathbf{x}}(t) \\ \dot{\mathbf{p}}(t) \end{bmatrix} = \begin{bmatrix} \mathbf{A}(t) & -\mathbf{S}(t) \\ -\mathbf{V}(t) & -\mathbf{A}'(t) \end{bmatrix} \begin{bmatrix} \mathbf{x}(t) \\ \mathbf{p}(t) \end{bmatrix} + \begin{bmatrix} \mathbf{0} \\ \mathbf{W}(t)\mathbf{z}(t) \end{bmatrix} \tag{9-251}$$

This is a system of $2n$ linear time-varying differential equations. The vector $\mathbf{W}(t)\mathbf{z}(t)$ acts as the "forcing function." At $t = t_0$, $n$ boundary conditions are provided by the initial state

$$\mathbf{x}(t_0) = \boldsymbol{\xi} \tag{9-252}$$

At $t = T$, $n$ boundary conditions are provided by the transversality condition

$$\begin{aligned} \mathbf{p}(T) &= \frac{\partial}{\partial \mathbf{x}(T)} [\tfrac{1}{2}\langle \mathbf{e}(T), \mathbf{F}\mathbf{e}(T)\rangle] \\ &= \mathbf{C}'(T)\mathbf{F}\mathbf{C}(T)\mathbf{x}(T) - \mathbf{C}'(T)\mathbf{F}\mathbf{z}(T) \end{aligned} \tag{9-253}$$

Let $\boldsymbol{\Phi}(t; t_0)$ be the $2n \times 2n$ fundamental matrix for the system (9-251). We can now derive the relation

$$\begin{bmatrix} \mathbf{x}(T) \\ \mathbf{p}(T) \end{bmatrix} = \boldsymbol{\Phi}(T; t) \left\{ \begin{bmatrix} \mathbf{x}(t) \\ \mathbf{p}(t) \end{bmatrix} + \int_t^T \boldsymbol{\Phi}^{-1}(\tau; t) \begin{bmatrix} \mathbf{0} \\ \mathbf{W}(\tau)\mathbf{z}(\tau) \end{bmatrix} d\tau \right\} \tag{9-254}$$

If we substitute the expression for $\mathbf{p}(T)$ given by Eq. (9-253) into Eq. (9-254) and if we proceed as in Sec. 9-3, then we find that the state $\mathbf{x}(t)$ and the costate $\mathbf{p}(t)$ are (linearly) related by the equation

$$\mathbf{p}(t) = \mathbf{K}(t)\mathbf{x}(t) - \mathbf{g}(t) \tag{9-255}$$

Just as in Sec. 9-3, we proceed to derive the properties of the $n \times n$ matrix $\mathbf{K}(t)$ and of the $n$-column vector $\mathbf{g}(t)$, because additional properties of $\mathbf{K}(t)$ and of $\mathbf{g}(t)$ may be valuable from the computational point of view. To derive these properties, we start by differentiating Eq. (9-255), and we find that

$$\dot{\mathbf{p}}(t) = \dot{\mathbf{K}}(t)\mathbf{x}(t) + \mathbf{K}(t)\dot{\mathbf{x}}(t) - \dot{\mathbf{g}}(t) \tag{9-256}$$

From Eqs. (9-247), (9-248), and (9-255), we obtain

$$\dot{\mathbf{x}}(t) = [\mathbf{A}(t) - \mathbf{S}(t)\mathbf{K}(t)]\mathbf{x}(t) + \mathbf{S}(t)\mathbf{g}(t) \tag{9-257}$$

and so Eq. (9-256) reduces to

$$\dot{\mathbf{p}}(t) = [\dot{\mathbf{K}}(t) + \mathbf{K}(t)\mathbf{A}(t) - \mathbf{K}(t)\mathbf{S}(t)\mathbf{K}(t)]\mathbf{x}(t) + \mathbf{K}(t)\mathbf{S}(t)\mathbf{g}(t) - \dot{\mathbf{g}}(t) \tag{9-258}$$

om Eqs. (9-246), (9-249), (9-250), and (9-255), we find that

$$\dot{\mathbf{p}}(t) = [-\mathbf{V}(t) - \mathbf{A}'(t)\mathbf{K}(t)]\mathbf{x}(t) + \mathbf{A}'(t)\mathbf{g}(t) + \mathbf{W}(t)\mathbf{z}(t) \tag{9-259}$$

As long as an optimal solution exists, Eqs. (9-258) and (9-259) must hold for all $\mathbf{x}(t)$, $\mathbf{z}(t)$, and $t$. Therefore, we conclude that:

1. The $n \times n$ matrix $\mathbf{K}(t)$ must satisfy the matrix differential equation

$$\dot{\mathbf{K}}(t) = -\mathbf{K}(t)\mathbf{A}(t) - \mathbf{A}'(t)\mathbf{K}(t) + \mathbf{K}(t)\mathbf{S}(t)\mathbf{K}(t) - \mathbf{V}(t) \qquad (9\text{-}260)$$

2. The $n$-column vector $\mathbf{g}(t)$ must satisfy the vector differential equation

$$\dot{\mathbf{g}}(t) = [\mathbf{K}(t)\mathbf{S}(t) - \mathbf{A}'(t)]\mathbf{g}(t) - \mathbf{W}(t)\mathbf{z}(t) \qquad (9\text{-}261)$$

The boundary conditions are derived as follows: From Eq. (9-255), we know that

$$\mathbf{p}(T) = \mathbf{K}(T)\mathbf{x}(T) - \mathbf{g}(T) \qquad (9\text{-}262)$$

and from Eq. (9-253), we know that

$$\mathbf{p}(T) = \mathbf{C}'(T)\mathbf{F}\mathbf{C}(T)\mathbf{x}(T) - \mathbf{C}'(T)\mathbf{F}\mathbf{z}(T) \qquad (9\text{-}263)$$

Since Eqs. (9-262) and (9-263) must hold for all $\mathbf{x}(T)$ and $\mathbf{z}(T)$, we conclude that

$$\mathbf{K}(T) = \mathbf{C}'(T)\mathbf{F}\mathbf{C}(T) \qquad (9\text{-}264)$$

and that

$$\mathbf{g}(T) = \mathbf{C}'(T)\mathbf{F}\mathbf{z}(T) \qquad (9\text{-}265)$$

Thus, the boundary conditions for the differential equations (9-260) and (9-261) are completely specified at the (specified) terminal time $T$, and so the differential equations (9-260) and (9-261) can be solved to obtain $\mathbf{K}(t)$ and $\mathbf{g}(t)$ for all $t \in [t_0, T]$.

If we substitute Eq. (9-255) into Eq. (9-245), we can obtain the extremal control as a function of $\mathbf{K}(t)$ and $\mathbf{g}(t)$. The extremal trajectory will then be the solution of Eq. (9-257), starting from the given initial state $\mathbf{x}(t_0) = \boldsymbol{\xi}$.

We now assert that an optimal control exists and that the extremal control is unique. Thus, we can derive the following control law:

***Control Law 9-5 Solution to the Tracking Problem*** *Given the linear observable system*

$$\begin{aligned}\dot{\mathbf{x}}(t) &= \mathbf{A}(t)\mathbf{x}(t) + \mathbf{B}(t)\mathbf{u}(t)\\ \mathbf{y}(t) &= \mathbf{C}(t)\mathbf{x}(t)\end{aligned} \qquad (9\text{-}266)$$

*Given the desired output* $\mathbf{z}(t)$ *and the error* $\mathbf{e}(t) = \mathbf{z}(t) - \mathbf{y}(t)$. *Given the cost functional*

$$J_3 = \tfrac{1}{2}\langle\mathbf{e}(T), \mathbf{F}\mathbf{e}(T)\rangle + \tfrac{1}{2}\int_{t_0}^{T} [\langle\mathbf{e}(t), \mathbf{Q}(t)\mathbf{e}(t)\rangle + \langle\mathbf{u}(t), \mathbf{R}(t)\mathbf{u}(t)\rangle]\, dt \qquad (9\text{-}267)$$

*where* $\mathbf{u}(t)$ *is unconstrained,* $T$ *is specified,* $\mathbf{R}(t)$ *is positive definite, and* $\mathbf{F}$ *and* $\mathbf{Q}(t)$ *are positive semidefinite. Then the optimal control exists, is unique, and is given by*

$$\mathbf{u}(t) = \mathbf{R}^{-1}(t)\mathbf{B}'(t)[\mathbf{g}(t) - \mathbf{K}(t)\mathbf{x}(t)] \qquad (9\text{-}268)$$

*The* $n \times n$ *real, symmetric, and positive definite matrix* $\mathbf{K}(t)$ *is the solution of the Riccati-type matrix differential equation*

$$\dot{\mathbf{K}}(t) = -\mathbf{K}(t)\mathbf{A}(t) - \mathbf{A}'(t)\mathbf{K}(t) + \mathbf{K}(t)\mathbf{B}(t)\mathbf{R}^{-1}(t)\mathbf{B}'(t)\mathbf{K}(t) - \mathbf{C}'(t)\mathbf{Q}(t)\mathbf{C}(t) \qquad (9\text{-}269)$$

*with the boundary condition*

$$\mathbf{K}(T) = \mathbf{C}'(T)\mathbf{F}\mathbf{C}(T) \tag{9-270}$$

*The vector* $\mathbf{g}(t)$ *(with n components) is the solution of the linear vector differential equation*

$$\dot{\mathbf{g}}(t) = -[\mathbf{A}(t) - \mathbf{B}(t)\mathbf{R}^{-1}(t)\mathbf{B}'(t)\mathbf{K}(t)]'\mathbf{g}(t) - \mathbf{C}'(t)\mathbf{Q}(t)\mathbf{z}(t) \tag{9-271}$$

*with the boundary condition*

$$\mathbf{g}(T) = \mathbf{C}'(T)\mathbf{F}\mathbf{z}(T) \tag{9-272}$$

*The optimal trajectory is the solution of the linear differential equation*

$$\dot{\mathbf{x}}(t) = [\mathbf{A}(t) - \mathbf{B}(t)\mathbf{R}^{-1}(t)\mathbf{B}'(t)\mathbf{K}(t)]\mathbf{x}(t) + \mathbf{B}(t)\mathbf{R}^{-1}(t)\mathbf{B}'(t)\mathbf{g}(t) \tag{9-273}$$

*starting at the (known) initial state* $\mathbf{x}(t_0) = \boldsymbol{\xi}$. *The minimum value* $J_3^*$ *of the cost* $J_3$ *is given by*

$$J_3^* = \tfrac{1}{2}\langle \mathbf{x}(t), \mathbf{K}(t)\mathbf{x}(t)\rangle - \langle \mathbf{g}(t), \mathbf{x}(t)\rangle + \varphi(t) \tag{9-274}$$

*where*
$$\dot{\varphi}(t) = -\tfrac{1}{2}[\langle \mathbf{z}(t), \mathbf{Q}(t)\mathbf{z}(t)\rangle - \langle \mathbf{g}(t), \mathbf{B}(t)\mathbf{R}^{-1}(t)\mathbf{B}'(t)\mathbf{g}(t)\rangle]$$
$$\varphi(T) = \langle \mathbf{z}(T), \mathbf{K}(T)\mathbf{z}(T)\rangle$$

*and is defined for all* $\mathbf{x}(t)$ *and* $t \in [t_0, T]$.

**Exercise 9-11** Show that the assumptions on $\mathbf{R}(t)$, $\mathbf{Q}(t)$, and $\mathbf{F}$ guarantee that the extremal control given by Eq. (9-268) yields (at least) a local minimum of $J_3$. Hint: See Eqs. (9-64) to (9-66).

**Exercise 9-12** Derive Eq. (9-274) using the Hamilton-Jacobi equation. Hint: See Theorem 9-1.

**Exercise 9-13** Show that an optimal control exists and that the control $\mathbf{u}(t)$ given by Eq. (9-268) is optimal and, furthermore, that it is unique.

Let us now discuss the implications of Control Law 9-5. First, let us examine the "gain" matrix $\mathbf{K}(t)$. We note that *the Riccati equation* (9-269) *and the boundary condition* (9-270) *are independent of the desired output* $\mathbf{z}(t)$. This means that the "gain" matrix $\mathbf{K}(t)$ is completely specified once the system, the cost, and the terminal time $T$ are specified. If we compare Eqs. (9-269) and (9-270) with Eqs. (9-183) and (9-184) (see Control Law 9-3, which provided the solution to the output-regulator problem), we note that they are identical. This means that *the feedback structure of the optimal tracking system is the same as the feedback structure of the optimal output-regulator system.* This fact is further evidenced by a comparison of Eqs. (9-273) and (9-185); in both differential equations, the system matrix $[\mathbf{A}(t) - \mathbf{B}(t)\mathbf{R}^{-1}(t)\mathbf{B}'(t)\mathbf{K}(t)]$ is the same. This means that *the eigenvalues of the optimal closed-loop tracking system are identical to the eigenvalues of the optimal closed-loop output-regulator system and are independent of the desired output* $\mathbf{z}(t)$. At the risk of being repetitious, we wish to emphasize that *the "gain" matrix* $\mathbf{K}(t)$ *is a function only of the matrices* $\mathbf{A}(t)$, $\mathbf{B}(t)$, $\mathbf{C}(t)$, $\mathbf{F}$, $\mathbf{Q}(t)$, $\mathbf{R}(t)$ *and of the terminal time* $T$.

The essential difference between the optimal tracking system and the optimal output-regulator system is provided by the vector $\mathbf{g}(t)$ [compare Eqs. (9-268) and (9-182)]. We can think of the vector $\mathbf{g}(t)$ as the forcing function to the system [see Eq. (9-273)]

$$\dot{\mathbf{x}}(t) = [\mathbf{A}(t) - \mathbf{B}(t)\mathbf{R}^{-1}(t)\mathbf{B}'(t)\mathbf{K}(t)]\mathbf{x}(t) + \mathbf{B}(t)\mathbf{R}^{-1}(t)\mathbf{B}'(t)\mathbf{g}(t) \qquad (9\text{-}275)$$

Thus, $\mathbf{g}(t)$ tends to counterbalance the regulator "features" of the system.

Let us now examine in more detail the differential equation of $\mathbf{g}(t)$, that is, the equation

$$\dot{\mathbf{g}}(t) = -[\mathbf{A}(t) - \mathbf{B}(t)\mathbf{R}^{-1}(t)\mathbf{B}'(t)\mathbf{K}(t)]'\mathbf{g}(t) - \mathbf{C}'(t)\mathbf{Q}(t)\mathbf{z}(t) \qquad (9\text{-}276)$$

The boundary condition, given at $t = T$, is $\mathbf{g}(T) = \mathbf{C}'(T)\mathbf{F}\mathbf{z}(T)$. First of all, we note that the system (9-276) is the *adjoint* to the system (9-275). Thus, if $\mathbf{\Phi}(t; t_0)$ is the fundamental matrix for the system (9-275) and if $\mathbf{\Psi}(t; t_0)$ is the fundamental matrix for the system (9-276), then we know that

$$\mathbf{\Psi}'(t; t_0)\mathbf{\Phi}(t; t_0) = \mathbf{I} \qquad (9\text{-}277)$$

Another implication of the adjoint relationship is that the eigenvalues of the matrix $-[\mathbf{A}(t) - \mathbf{B}(t)\mathbf{R}^{-1}(t)\mathbf{B}'(t)\mathbf{K}(t)]'$ are the negatives of the eigenvalues of the matrix $[\mathbf{A}(t) - \mathbf{B}(t)\mathbf{R}^{-1}(t)\mathbf{B}'(t)\mathbf{K}(t)]$. Thus, the dynamical behavior of the system (9-275) is also independent of the desired output $\mathbf{z}(t)$. For this reason, we can think of the desired output $\mathbf{z}(t)$ as the forcing function to the dynamical system that generates the signal $\mathbf{g}(t)$ [see Eq. (9-276)].

Let $\mathbf{\Psi}(t; t_0)$ be the fundamental matrix for the system (9-276). We immediately conclude that

$$\mathbf{C}'(T)\mathbf{F}\mathbf{z}(T) = \mathbf{g}(T) = \mathbf{\Psi}(T; t)\left[\mathbf{g}(t) - \int_t^T \mathbf{\Psi}^{-1}(\tau; t)\mathbf{C}'(\tau)\mathbf{Q}(\tau)\mathbf{z}(\tau)\, d\tau\right] \qquad (9\text{-}278)$$

or, for all $t \in [t_0, T]$, that

$$\mathbf{g}(t) = \mathbf{\Psi}^{-1}(T; t)\mathbf{g}(T) + \int_t^T \mathbf{\Psi}^{-1}(\tau; t)\mathbf{C}'(\tau)\mathbf{Q}(\tau)\mathbf{z}(\tau)\, d\tau \qquad (9\text{-}279)$$

The implication of Eq. (9-279) is this: *In order to evaluate* $\mathbf{g}(t)$, $t \in [t_0, T]$, *we must know* $\mathbf{z}(\tau)$ *for all* $\tau \in [t, T]$. In other words, *in order to evaluate the present value of* $\mathbf{g}(t)$, *we must know all the future values of the desired output* $\mathbf{z}(\tau)$. Since the optimal control is given by the expression

$$\mathbf{u}(t) = -\mathbf{R}^{-1}(t)\mathbf{B}(t)[\mathbf{K}(t)\mathbf{x}(t) - \mathbf{g}(t)] \qquad (9\text{-}280)$$

we further conclude that *the present value of the optimal control depends upon the future values of the desired output.*

At this point, the frustrated engineer might accuse the optimal control of "double-crossing" him because it involves knowledge of the future and, as such, is often† *unrealizable.* Nevertheless, it is easy to see that the optimal control‡ must be a function of the future values of $\mathbf{z}(t)$. The reason is that the optimal control will utilize all the available information. Certainly, *the formulation of our control problem did not include the requirement of realizability.*

At this point, let us ask the following question: *Is there an easy way of incorporating the requirement of realizability in the mathematical formulation of the optimal problem?* The answer is *no.* To see this, we let $t$ be the *present,* $[t_0, t)$ be the *past,* and $(t, T]$ be the *future;* then, the *present* control $\mathbf{u}(t)$ can only affect the *future* response of the system, and $\mathbf{u}(t)$, clearly, cannot change the *past.* The *past* behavior of the system is summarized by the *present* state $\mathbf{x}(t)$. Since the *present* state $\mathbf{x}(t)$ will determine a portion of the *future* response, the *present* optimal $\mathbf{u}(t)$ must be a function of the *present* state $\mathbf{x}(t)$. But the action of the *present* $\mathbf{u}(t)$ must be such as to minimize *future* errors. These *future* errors will depend on the *future* values of $\mathbf{z}(\tau)$, $\tau \in (t, T]$. Therefore, we conclude that the *present* optimal control $\mathbf{u}(t)$ must depend on $\mathbf{z}(\tau)$, for all $\tau \in (t, T]$. In other words, *if we do not know the future exactly, then we cannot be expected to react now in a precise optimal manner.*

Once this somewhat philosophical point has been clarified, we can proceed in two ways. We can either use a *predicted value for our future desired output* or formulate the problem in a *stochastic* manner (i.e., minimizing a function of an expected error). If we use our "best" deterministic estimate, we shall obtain an optimal system which is as good as our estimate. If, on the other hand, we formulate stochastically an essentially deterministic problem, then we would obtain a system which will be optimal "on the average"; this does not guarantee that, in any *one* experiment, the system response will be satisfactory.

Now that we have accepted the fact that we must know the desired output a priori, let us comment on the computational aspects of determining $\mathbf{K}(t)$ and $\mathbf{g}(t)$. Since the matrix $\mathbf{K}(t)$ is independent of $\mathbf{z}(t)$, we can compute it a priori for all $t \in [t_0, T]$. Knowledge of $\mathbf{K}(t)$ and of $\mathbf{z}(t)$ will enable us to compute $\mathbf{g}(t)$, backward in time, for all $t \in [t_0, T]$. We can either store $\mathbf{g}(t)$ on tape or compute the initial value $\mathbf{g}(t_0)$ using the formula

$$\mathbf{g}(t_0) = \mathbf{\Psi}'^{-1}(T; t_0)\mathbf{C}'(T)\mathbf{F}\mathbf{z}(T) + \int_{t_0}^{T} \mathbf{\Psi}'^{-1}(\tau; t_0)\mathbf{C}'(\tau)\mathbf{Q}(\tau)\mathbf{z}(\tau)\, d\tau \quad (9\text{-}281)$$

† There are, of course, many cases for which the future values of $\mathbf{z}$ are known. Consider, for example, the control of an antenna used to obtain telemetry data from a satellite, and suppose that the satellite has been in orbit for a sufficiently long time for its orbit to be known with great accuracy. In this case, one knows that the satellite will rise above the horizon at (say) 4:13 A.M. and that it will disappear at (say) 4:27 A.M. Clearly, $\mathbf{z}(t)$ is known, and the optimal-control system for the tracking radar can be realized. It is easy to see that the antenna should start to move before 4:13 A.M., so that it will "match" the satellite path when the satellite rises over the horizon.

‡ The statements that follow apply equally well to all optimal tracking systems, and they are not restricted to those considered in this section.

We can then use this computed value of $\mathbf{g}(t_0)$ as the initial condition to the system (9-271) and generate the solution $\mathbf{g}(t)$ forward in time. Figure 9-17 illustrates the dynamical system which generates $\mathbf{g}(t)$, once $\mathbf{g}(t_0)$ has been precomputed. Figure 9-18 shows the structure of the overall optimal system; we use a vector feedback channel around the integrators in order to emphasize the adjoint property of the two dynamical systems.

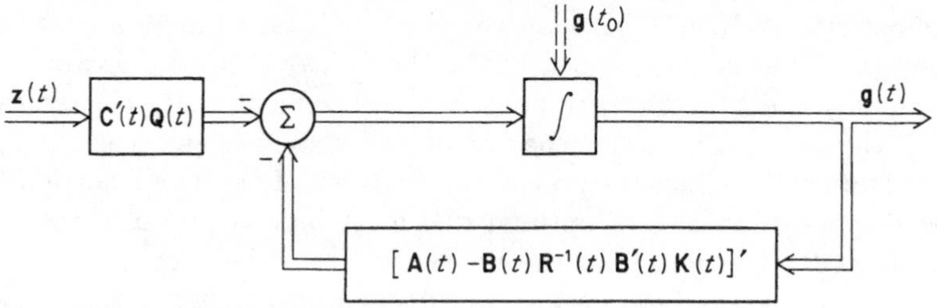


**Fig. 9-17** The simulation of the system that produces $\mathbf{g}(t)$. It is assumed that $\mathbf{g}(t_0)$ and $\mathbf{K}(t)$ have been precomputed.

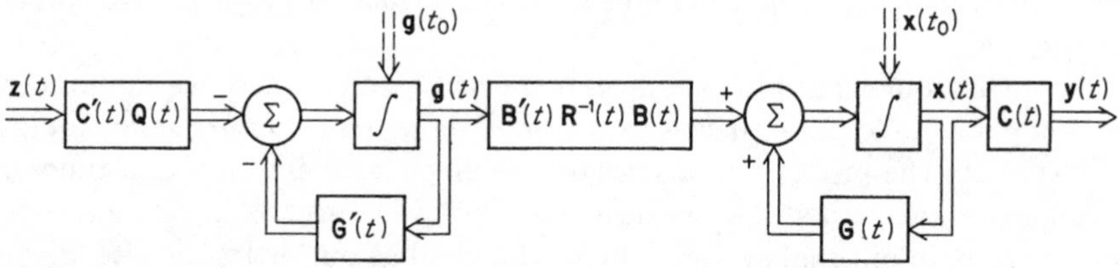


**Fig. 9-18** The simulation of the optimal tracking system based on Eqs. (9-274) and (9-276). The matrix $\mathbf{G}(t)$ is given by $\mathbf{G}(t) = \mathbf{A}(t) - \mathbf{B}(t)\mathbf{R}^{-1}(t)\mathbf{B}'(t)\mathbf{K}(t)$. It is assumed that $\mathbf{g}(t_0)$ and $\mathbf{K}(t)$ have been precomputed.

Since the mapping defined by Eq. (9-281) is linear, it is easy to establish the following theorem:

***Theorem 9-2*** *Let*

$$\mathbf{g}_\beta(t_0) = \mathbf{\Psi}^{-1}(T; t_0)\mathbf{C}'(T)\mathbf{F}\mathbf{z}_\beta(T) + \int_{t_0}^{T} \mathbf{\Psi}^{-1}(\tau; t_0)\mathbf{C}'(\tau)\mathbf{Q}(\tau)\mathbf{z}_\beta(\tau)\, d\tau \qquad \beta = 1, 2, \ldots, N \quad (9\text{-}282)$$

*If*

$$\mathbf{z}(\tau) = \sum_{\beta=1}^{N} \gamma_\beta \mathbf{z}_\beta(\tau) \qquad \tau \in [t_0, T] \quad (9\text{-}283)$$

*where $\gamma_\beta$ are any constants, then the initial-condition vector $\mathbf{g}(t_0)$, corresponding to the signal $\mathbf{z}_{[t_0,T]}$, is given by*

$$\mathbf{g}(t_0) = \sum_{\beta=1}^{N} \gamma_\beta \mathbf{g}_\beta(t_0) \quad (9\text{-}284)$$

PROOF The theorem follows upon substitution of Eq. (9-283) into Eq. (9-281) and application of Eq. (9-282).

The implications of this theorem are as follows: Suppose that we know that the desired output signals $\mathbf{z}(t)$ are always linear combinations of known signals $\mathbf{z}_\beta(t)$; then we can precompute the initial vectors $\mathbf{g}_\beta(t_0)$ and generate $\mathbf{g}(t_0)$ using Eq. (9-283), provided that no stability problems arise.

In the following section, we shall discuss certain approximate relations for time-invariant systems as the terminal time $T$ becomes increasingly large.

The following exercises, which can be done by mimicking the development of this section, will provide some additional theoretical results.

**Exercise 9-14** Consider the linear observable system

$$\begin{aligned}\dot{\mathbf{x}}(t) &= \mathbf{A}(t)\mathbf{x}(t) + \mathbf{B}(t)\mathbf{u}(t) + \boldsymbol{\omega}(t)\\ \mathbf{y}(t) &= \mathbf{C}(t)\mathbf{x}(t)\end{aligned} \tag{9-285}$$

where $\boldsymbol{\omega}(t)$ is a known disturbance vector. Let $\mathbf{z}(t)$ be the desired output, let $\mathbf{e}(t_0) = \mathbf{z}(t) - \mathbf{y}(t)$, and consider the cost functional $J_3$ of Eq. (9-267). Show that the optimal control is given by

$$\mathbf{u}(t) = \mathbf{R}^{-1}(t)\mathbf{B}'(t)[\mathbf{h}(t) - \mathbf{K}(t)\mathbf{x}(t)] \tag{9-286}$$

where $\mathbf{K}(t)$ is the solution of the Riccati equation (9-269) with the boundary condition (9-270) and where the vector $\mathbf{h}(t)$ is the solution of the linear differential equation

$$\dot{\mathbf{h}}(t) = -[\mathbf{A}(t) - \mathbf{B}(t)\mathbf{R}^{-1}(t)\mathbf{B}'(t)\mathbf{K}(t)]'\mathbf{h}(t) - \mathbf{C}'(t)\mathbf{Q}(t)\mathbf{z}(t) + \mathbf{K}(t)\boldsymbol{\omega}(t) \tag{9-287}$$

with the boundary condition

$$\mathbf{h}(T) = \mathbf{C}'(T)\mathbf{Q}(T)\mathbf{z}(T) \tag{9-288}$$

Determine, using the Hamilton-Jacobi equation, the equation for the minimum cost. Now suppose that $\mathbf{z}(t) = \mathbf{0}$ for all $t \in [t_0, T]$. Can you generate $\mathbf{h}(t)$ without knowing the disturbance $\boldsymbol{\omega}(\tau)$, for all $\tau \in [t, T]$?

**Exercise 9-15** Consider the linear observable system

$$\begin{aligned}\dot{\mathbf{x}}(t) &= \mathbf{A}(t)\mathbf{x}(t) + \mathbf{B}(t)\mathbf{u}(t) + \boldsymbol{\omega}(t)\\ \mathbf{y}(t) &= \mathbf{C}(t)\mathbf{x}(t) + \mathbf{D}(t)\mathbf{u}(t) + \mathbf{v}(t)\end{aligned} \tag{9-289}$$

where $\boldsymbol{\omega}(t)$ and $\mathbf{v}(t)$ are known disturbance vectors. Let $\mathbf{z}(t)$ be the desired output, let $\mathbf{e}(t) = \mathbf{z}(t) - \mathbf{y}(t)$, and consider the cost $J_3$ of Eq. (9-267). Determine the optimal control and find the equation of the minimum cost.

## 9-10 Approximate Relations for Time-invariant Systems

In this section, we shall examine the tracking problem for linear time-invariant systems, constant desired output vectors, and increasingly large $T$. All the results that we shall derive are *approximate* in nature and are valid for very large values of the terminal time $T$. Unfortunately, at this time, the theory which deals with the limiting case $T = \infty$ is not available.

Consider the observable and controllable linear time-invariant system

$$\begin{aligned}\dot{\mathbf{x}}(t) &= \mathbf{A}\mathbf{x}(t) + \mathbf{B}\mathbf{u}(t)\\ \mathbf{y}(t) &= \mathbf{C}\mathbf{x}(t)\end{aligned} \tag{9-290}$$

Let $\hat{\mathbf{z}}$ be a *constant* vector which represents the desired output, so that the error $\mathbf{e}(t)$ is given by

$$\mathbf{e}(t) = \hat{\mathbf{z}} - \mathbf{y}(t) = \hat{\mathbf{z}} - \mathbf{C}\mathbf{x}(t) \tag{9-291}$$

Let us consider the cost functional

$$\hat{J}_3 = \frac{1}{2} \int_0^T [\langle \mathbf{e}(t), \mathbf{Q}\mathbf{e}(t)\rangle + \langle \mathbf{u}(t), \mathbf{R}\mathbf{u}(t)\rangle]\, dt \tag{9-292}$$

and let us assume that the constant matrices $\mathbf{Q}$ and $\mathbf{R}$ are positive definite. Clearly, the results of Control Law 9-5 can be used for the solution of this special problem, as long as $T$ is specified and finite.

We know that† as $T \to \infty$, the "gain" matrix $\mathbf{K}(t)$ of the Riccati equation (9-269) tends to the constant positive definite matrix $\hat{\mathbf{K}}$. Recall that the matrix $\hat{\mathbf{K}}$ is the positive definite solution of the algebraic matrix equation

$$-\hat{\mathbf{K}}\mathbf{A} - \mathbf{A}'\hat{\mathbf{K}} + \hat{\mathbf{K}}\mathbf{B}\mathbf{R}^{-1}\mathbf{B}'\hat{\mathbf{K}} - \mathbf{C}'\mathbf{Q}\mathbf{C} = \mathbf{0} \tag{9-293}$$

Let the terminal time $T$ be *exceedingly large* but finite. Let $T_1$ and $T_2$ be two other *very large* times such that

$$0 \ll T_1 \ll T_2 \ll T < \infty \tag{9-294}$$

We have interpreted the matrix $\hat{\mathbf{K}}$ as the "steady-state" solution of Eq. (9-269) attained when the Riccati equation is solved backward in time. Since $T_2 \ll T$, let us *approximate* the matrix $\mathbf{K}(t)$ by the matrix $\hat{\mathbf{K}}$ for all $t \in [0, T_2]$. In other words, we let

$$\mathbf{K}(t) \approx \hat{\mathbf{K}} \qquad t \in [0, T_2] \tag{9-295}$$

Using this approximation and Control Law 9-5, we find that [see Eq. (9-273)]

$$\dot{\mathbf{x}}(t) \approx (\mathbf{A} - \mathbf{B}\mathbf{R}^{-1}\mathbf{B}'\hat{\mathbf{K}})\mathbf{x}(t) + \mathbf{B}\mathbf{R}^{-1}\mathbf{B}'\mathbf{g}(t) \qquad t \in [0, T_2] \tag{9-296}$$

and that [see Eq. (9-271)]

$$\dot{\mathbf{g}}(t) \approx (\mathbf{A} - \mathbf{B}\mathbf{R}^{-1}\mathbf{B}'\hat{\mathbf{K}})'\mathbf{x}(t) - \mathbf{C}'\mathbf{Q}\hat{\mathbf{z}} \qquad t \in [0, T_2] \tag{9-297}$$

Let $\mathbf{G}$, $\mathbf{S}$, and $\mathbf{W}$ be the matrices

$$\mathbf{G} = \mathbf{A} - \mathbf{B}\mathbf{R}^{-1}\mathbf{B}'\hat{\mathbf{K}} \qquad \mathbf{S} = \mathbf{B}\mathbf{R}^{-1}\mathbf{B}' \qquad \mathbf{W} = \mathbf{C}'\mathbf{Q} \tag{9-298}$$

Then Eqs. (9-296) and (9-297) reduce to

$$\dot{\mathbf{x}}(t) \approx \mathbf{G}\mathbf{x}(t) + \mathbf{S}\mathbf{g}(t) \qquad t \in [0, T_2] \tag{9-299}$$

$$\dot{\mathbf{g}}(t) \approx -\mathbf{G}'\mathbf{g}(t) - \mathbf{W}\hat{\mathbf{z}} \qquad t \in [0, T_2] \tag{9-300}$$

We know that (see Control Law 9-4) the eigenvalues of $\mathbf{G}$ have negative

† By Control Law 9-4.

real parts; therefore, the eigenvalues of $-\mathbf{G}'$ have positive real parts. Hence, the system (9-300) is unstable. Nevertheless, since $\mathbf{F} = \mathbf{0}$ (because $\hat{J}_3$ contains no terminal cost), it follows that $\mathbf{g}(T) = \mathbf{0}$ and that the vector

$$\mathbf{g}(T_2) = \int_{T_2}^{T} \mathbf{\Psi}^{-1}(\tau;\, T_2)\mathbf{C}'\mathbf{Q}\hat{\mathbf{z}}\, d\tau \neq \mathbf{0} \tag{9-301}$$

is finite.

For the approximate system (9-300), the fundamental matrix is $e^{-\mathbf{G}'(t-t_0)}$ and so we deduce that

$$\mathbf{g}(T_2) \approx e^{-\mathbf{G}'(T_2-t)} \left[\mathbf{g}(t) - \int_t^{T_2} e^{\mathbf{G}'(\tau-t)}\mathbf{W}\hat{\mathbf{z}}\, d\tau\right] \qquad t \in [0, T_2] \tag{9-302}$$

or, equivalently, that

$$\mathbf{g}(t) \approx e^{\mathbf{G}'T_2}e^{-\mathbf{G}'t}\mathbf{g}(T_2) + \int_t^{T_2} e^{\mathbf{G}'(\tau-t)}\mathbf{W}\hat{\mathbf{z}}\, d\tau \qquad t \in [0, T_2] \tag{9-303}$$

Since the desired output $\hat{\mathbf{z}}$ is a constant vector, we find that

$$\begin{aligned}\int_t^{T_2} e^{\mathbf{G}'(\tau-t)}\mathbf{W}\hat{\mathbf{z}}\, d\tau &= \left(\int_t^{T_2} e^{\mathbf{G}'\tau}\, d\tau\right) e^{-\mathbf{G}'t}\, \mathbf{W}\hat{\mathbf{z}} \\ &= -(\mathbf{G}')^{-1}\mathbf{W}\hat{\mathbf{z}} + (\mathbf{G}')^{-1}e^{\mathbf{G}'T_2}e^{-\mathbf{G}'t}\mathbf{W}\hat{\mathbf{z}}\end{aligned} \tag{9-304}$$

In the derivation of Eq. (9-304), we have used the facts:

1. That $(\mathbf{G}')^{-1}$ exists, since all the eigenvalues of $\mathbf{G}'$ have negative real parts:
2. That $(\mathbf{G}')^{-1}$ and $e^{\mathbf{G}'}$ commute (why?)

We now substitute Eq. (9-304) into Eq. (9-303), and we find that

$$\mathbf{g}(t) \approx -(\mathbf{G}')^{-1}\mathbf{W}\hat{\mathbf{z}} + e^{\mathbf{G}'T_2}[(\mathbf{G}')^{-1}e^{-\mathbf{G}'t}\mathbf{W}\hat{\mathbf{z}} + e^{-\mathbf{G}'t}\mathbf{g}(T_2)] \tag{9-305}$$

for all $t \in [0, T_2]$.

Now suppose that $T_1$ is such that

$$0 \ll T_1 \ll T_2 \tag{9-306}$$

and that

$$t \in [0, T_1] \tag{9-307}$$

Since all the eigenvalues of $\mathbf{G}'$ have negative real parts, it follows that

$$e^{\mathbf{G}'T_2} \approx \mathbf{0} \tag{9-308}$$

and that

$$\mathbf{g}(t) \approx -(\mathbf{G}')^{-1}\mathbf{W}\hat{\mathbf{z}} = \hat{\mathbf{g}} \qquad t \in [0, T_1] \tag{9-309}$$

Equation (9-309) implies that $\mathbf{g}(t) \approx \hat{\mathbf{g}}$, that is, that $\mathbf{g}(t)$ is approximately constant for small values of $t$. All the indicated approximations become better and better as $T$, $T_2$, and $T_1$ become larger and larger.

Unfortunately, at this time, a rigorous exposition of the limiting case ($T = \infty$) is not available. Nevertheless, the indicated approximations can be used with confidence for an engineering system as long as the terminal time is very large. Figure 9-19 illustrates the structure of the "approximate" tracking system.

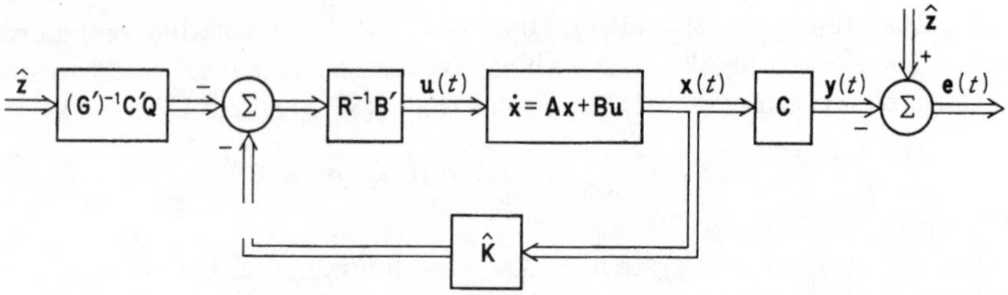


**Fig. 9-19** The structure of the approximate optimal time-invariant system when the desired output is a constant vector $\hat{\mathbf{z}}$. The matrix $\mathbf{G}$ is given by $\mathbf{G} = \mathbf{A} - \mathbf{B}\mathbf{R}^{-1}\mathbf{B}'\hat{\mathbf{K}}$.

## 9-11 Tracking Problems Reducible to Output-regulator Problems

We stated in Sec. 9-10 that a rigorous treatment of the limiting case $T = \infty$ was not available for the optimal tracking problem. We recall, however, that the limiting case $T = \infty$ was well developed for both the optimal state-regulator problem and the optimal output-regulator problem. The question that arises is this: *Are there any tracking problems which are equivalent to output-regulator problems?* The answer is *yes*.

In this section, we shall derive sufficient conditions on the desired output which will enable us to transform a tracking problem into an output-regulator problem. We shall present the main ideas for a single-input–single-output system.

Let $\mathcal{L}(D)$ be the differential operator

$$\mathcal{L}(D) = \{D^n + a_{n-1}D^{n-1} + \cdots + a_1 D + a_0\} \qquad D = \frac{d}{dt} \tag{9-310}$$

Let $y(t)$ be the (scalar) output and let $u(t)$ be the (scalar) input of a system described by the relation

$$\mathcal{L}(D)y(t) = u(t) \tag{9-311}$$

Let $z(t)$ be the desired output and let

$$e(t) = z(t) - y(t) \tag{9-312}$$

denote the error. The optimum tracking problem is to find the control that minimizes the cost $\hat{J}_3$ given by

$$\hat{J}_3 = \tfrac{1}{2}\int_0^\infty [qe^2(t) + ru^2(t)]\,dt \qquad q > 0;\ r > 0 \tag{9-313}$$

The theory of Sec. 9-9 cannot be used to solve this problem because (as we have mentioned in Sec. 9-10) a rigorous theory for the case $T = \infty$ is not available. We shall now proceed to find conditions on $z(t)$ which will be sufficient to guarantee that the problem can be solved; this can be done simply by transforming the tracking problem into an output-regulator problem.

To obtain the (sufficient) conditions on $z(t)$, we apply the differential operator $\mathfrak{L}(D)$ to both sides of Eq. (9-312) and, in so doing, find that

$$\begin{aligned}\mathfrak{L}(D)e(t) &= \mathfrak{L}(D)[z(t) - y(t)] \\ &= \mathfrak{L}(D)z(t) - \mathfrak{L}(D)y(t)\end{aligned}$$

Suppose that $z(t)$ is (at least) $n$ times differentiable and, furthermore, that $z(t)$ is such that

$$\mathfrak{L}(D)z(t) = \{D^n + a_{n-1}D^{n-1} + \cdots + a_1D + a_0\}z(t) = 0 \tag{9-314}$$

for all $t$.

From Eqs. (9-311) and (9-314), we find that

$$\mathfrak{L}(D)e(t) = -u(t) \qquad \text{for all } t \tag{9-315}$$

We can interpret Eq. (9-315) as the input-output relationship for a linear time-invariant system with "input" $-u(t)$ and "output" $e(t)$. Since the system (9-315) is controllable and observable (why?), we can use Control Law 9-4 to find the control that minimizes the cost $\hat{J}_3$ given by Eq. (9-313). Thus, if we define state variables $x_i(t)$ by setting $x_i(t) = D^{i-1}e(t)$, $i = 1, 2, \ldots, n$, then the system (9-315) can be put into the form required by Control Law 9-4. Thus, the optimal control will be a linear combination of the error and its first $n - 1$ time derivatives.

We present two examples to illustrate these simple ideas. In each case, the tracking problem is reducible to a regulator problem because $z(t)$ satisfies Eq. (9-314).

**Example 9-7** Suppose that the system (9-311) is

$$\ddot{y}(t) + 3\dot{y}(t) + 2y(t) = u(t) \tag{9-316}$$

Then the signal $z(t)$ given by the equation

$$z(t) = \alpha_1 e^{-t} + \alpha_2 e^{-2t} \tag{9-317}$$

(where $\alpha_1$ and $\alpha_2$ are arbitrary constants) is such that

$$\mathfrak{L}(D)z(t) = \ddot{z}(t) + 3\dot{z}(t) + 2z(t) = 0 \tag{9-318}$$

for all $t$.

**Example 9-8** Suppose that the system (9-311) is such that

$$\frac{d^n}{dt^n}y(t) = u(t) \tag{9-319}$$

Then the signal $z(t)$ given by the equation

$$z(t) = \alpha_0 + \alpha_1 t + \alpha_2 t^2 + \cdots + \alpha_{n-1}t^{n-1} \tag{9-320}$$

(where $\alpha_0, \alpha_1, \ldots, \alpha_{n-1}$ are arbitrary constants) is such that

$$\mathfrak{L}(D)z(t) = \frac{d^n}{dt^n}z(t) = 0 \tag{9-321}$$

for all $t$.

The following exercises illustrate several aspects of optimal tracking problems.

**Exercise 9-16** Interpret the results of this section using transfer-function ideas. What is the physical significance of Eq. (9-314)?

**Exercise 9-17** Let $\mathcal{L}(D)$ and $\mathfrak{M}(D)$ be the linear time-invariant differential operators $(m < n)$

$$\mathcal{L}(D) = \{D^n + a_{n-1}D^{n-1} + \cdots + a_1 D + a_0\} \tag{9-322}$$

$$\mathfrak{M}(D) = \{b_m D^m + b_{m-1}D^{m-1} + \cdots + b_1 D + b_0\} \tag{9-323}$$

and consider the linear controllable and observable system

$$\mathcal{L}(D)y(t) = \mathfrak{M}(D)u(t) \tag{9-324}$$

Let $z(t)$ be the desired output and let $e(t) = z(t) - y(t)$. Determine nontrivial sufficiency conditions on $z(t)$ such that the minimization of the cost (9-313) can be accomplished using the theory of output-regulator systems.

**Exercise 9-18** (This exercise requires a digital computer.) Consider the system

$$\begin{aligned} \dot{x}_1(t) &= x_2(t) & x_1(0) &= 0 \\ \dot{x}_2(t) &= -x_1(t) + u(t) & x_2(0) &= 0 \\ y(t) &= x_1(t) \end{aligned} \tag{9-325}$$

Let $z(t) = \hat{z}$, a constant, for all $t \in [0, T]$, and let $e(t) = \hat{z} - y(t)$. It is desired to minimize the cost

$$J_3 = \tfrac{1}{2}fe^2(T) + \tfrac{1}{2}\int_0^T [e^2(t) + ru^2(t)]\,dt \tag{9-326}$$

Using Control Law 9-5, determine the optimal control. Let $k_{11}(t)$, $k_{12}(t)$, and $k_{22}(t)$ denote the elements of the $2 \times 2$ gain matrix $\mathbf{K}(t)$, and let $g_1(t)$ and $g_2(t)$ be the components of the vector $\mathbf{g}(t)$.

(*a*) Let $f = 0$, $r = 0.1$, $\hat{z} = 1$. Plot $e(t)$, $k_{11}(t)$, $k_{12}(t)$, $k_{22}(t)$, $g_1(t)$, $g_2(t)$ versus $t$ for $T = 1, 2, 5, 10$, and 20.

(*b*) Let $f = 0$, $\hat{z} = 1$, $T = 2$. Repeat part *a* for $r = 10, 1, 0.1, 0.01$, and 0.001.

(*c*) Let $r = 0.1$, $\hat{z} = 1$, $T = 2$. Repeat part *a* for $f = 0, 1, 5, 10, 50$, and 100.

(*d*) Can you find a $z(t)$ such that the tracking problem reduces to a regulator problem? Is $\hat{z}$ such a $z(t)$?

**Exercise 9-19** Repeat Exercise 9-18, but replace the system (9-325) by the system

$$\begin{aligned} \dot{x}_1(t) &= x_2(t) & x_1(0) &= 0 \\ \dot{x}_2(t) &= x_1(t) + u(t) & x_2(0) &= 0 \\ y(t) &= x_1(t) \end{aligned} \tag{9-327}$$

## 9-12 Analysis of a First-order Tracking System

In Sec. 9-6, we presented some computed responses of a first-order regulator system. In this section, we shall present some computed responses of a first-order tracking system which illustrate some of the properties of optimal tracking systems. We consider the same system as the one in Sec. 9-6, i.e., the system

$$\begin{aligned} \dot{x}(t) &= ax(t) + u(t) \\ y(t) &= x(t) \end{aligned} \tag{9-328}$$

The desired output is $z(t)$, the error is $e(t) = z(t) - y(t) = z(t) - x(t)$, and the cost is given by

$$J_3 = \tfrac{1}{2}fe^2(T) + \tfrac{1}{2}\int_0^T [qe^2(t) + ru^2(t)]\,dt \qquad (9\text{-}329)$$

where

$$f \geq 0 \qquad q > 0 \qquad r > 0 \qquad (9\text{-}330)$$

Using Control Law 9-5, we find that the optimal control is given by

$$u(t) = \frac{1}{r}[g(t) - k(t)x(t)] \qquad (9\text{-}331)$$

The scalar $k(t)$ is the solution of the first-order Riccati equation

$$\dot{k}(t) = -2ak(t) + \frac{1}{r}k^2(t) - q \qquad (9\text{-}332)$$

with the boundary condition

$$k(T) = f \qquad (9\text{-}333)$$

The scalar $g(t)$ is the solution of the first-order linear equation

$$\dot{g}(t) = -\left[a - \frac{1}{r}k(t)\right]g(t) - z(t) \qquad (9\text{-}334)$$

with the boundary condition

$$g(T) = fz(T) \qquad (9\text{-}335)$$

Figure 9-20*a* shows the response of the optimal system to a step input. We have computed the response using $r$ as a parameter, $a = -1$ (stable case), $x(0) = 0$, $f = 0$, $q = 1$, $T = 1$, and $z(t) = +1$ for all $t \in [0, 1]$. We observe that the tracking capabilities of the system improve as $r$ decreases. We also note that the error starts to increase near the end of the control interval. This behavior is due to the fact that $f = 0$. Since $f = 0$, $g(T) = k(T) = 0$, and hence $u(T) = 0$. Since the control decreases with increasing $t$ and since the system is stable, we deduce that the state starts to decay toward zero.

Figure 9-20*b* illustrates the behavior of $g(t)$, $t \in [0, 1]$. Note that as $r$ decreases, $g(t)$ remains almost a constant near the start of the control interval and then $g(t)$ decreases to zero (because $f = 0$). We shall not present any plots of $k(t)$, because $k(t)$ is independent of $z(t)$ and its behavior has already been illustrated in Fig. 9-10*b*.

Figure 9-20*c* shows the behavior of the optimal control. Observe that $u(t)$ decreases to zero; this accounts for the increase of the error in Fig. 9-20*a* for $t$ near $T$.

Figure 9-21*a*, *b*, and *c* illustrates the response of the optimal system when $a = 0$ (i.e., the system is a pure integrator), $x(0) = 0$, $f = 0$, $q = 1$,

$T = 1$, and $z(t) = 1$ for all $t \in [0, 1]$. We observe that there is no increase in the error for $t$ near $T$ although the optimal control decreases to zero. This is due to the integrating capabilities of the system.

Figure 9-22*a*, *b*, and *c* illustrates the response of the optimal system when $a = +1$ (unstable case), $x(0) = 0, f = 0, q = 1, T = 1$, and $z(t) = +1$

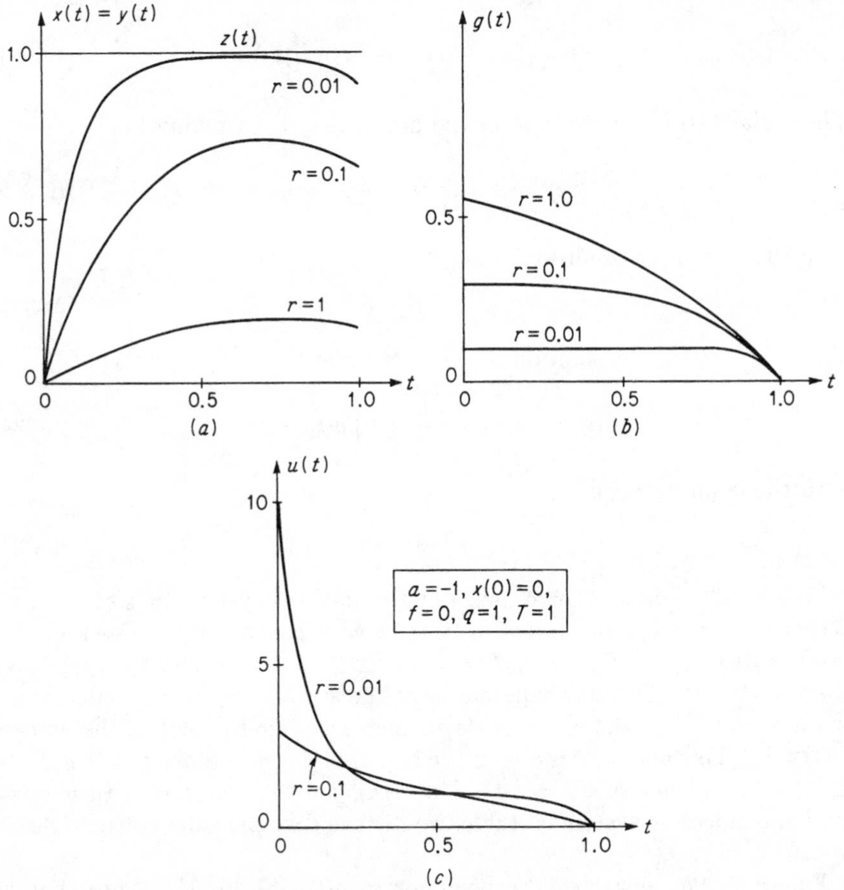


**Fig. 9-20** (*a*) The response of the optimal system when $a = -1$ (stable case) and $z(t) = 1$. (*b*) The function $g(t)$ when $a = -1$ and $z(t) = 1$. (*c*) The optimal control when $a = -1$ and $z(t) = 1$.

for all $t \in [0, 1]$. Since this system is unstable, we observe that the output exceeds the desired output for $t$ near $T$ (why?). We also note that when $r = 0.01$, the optimal control $u(t)$ becomes negative (see Fig. 9-22*c*) in order to counteract the instability of the system; this causes the error to remain small.

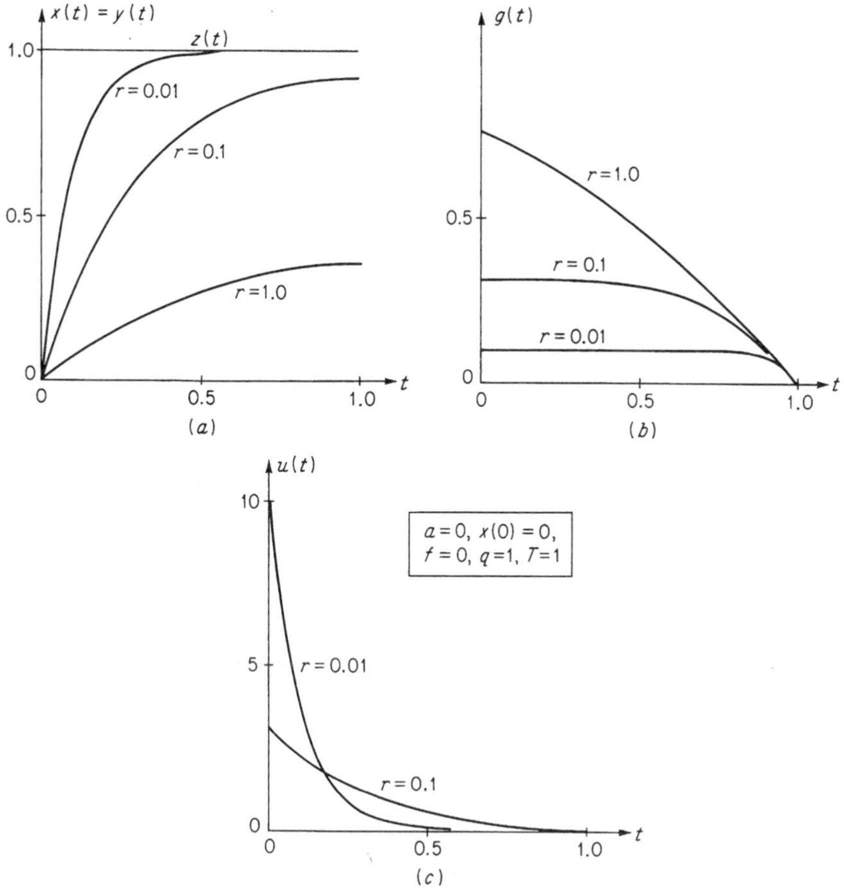


**Fig. 9-21** (*a*) The response of the optimal system when $a = 0$ (i.e., when the system is an integrator) and $z(t) = 1$. (*b*) The function $g(t)$ when $a = 0$ and $z(t) = 1$. (*c*) The optimal control when $a = 0$ and $z(t) = 1$.

Figure 9-23*a* illustrates the anticipatory nature of the optimal system. We have computed the responses by using $a = 0, f = 0, q = 1$, and $T = 5$. The desired output is

$$z(t) = \begin{cases} 0 & \text{for } 0 \leq t < 2.5 \\ 1 & \text{for } 2.5 \leq t \leq 5.0 \end{cases} \tag{9-336}$$

The initial state is $x(0) = 1$. When $r = 0.01$, the system acts as a regulator in the time interval [0, 1.5]; however, the system "knows" that a step is "coming" at $t = 2.5$, so that the output starts to increase at about $t = 2$. Thus, by the time the step has arrived, the error is small for $t > 2.5$. When $r = 1$, the same behavior occurs. Figure 9-23*b* illustrates the behavior of $g(t)$ when $z(t)$ is given by Eq. (9-336).

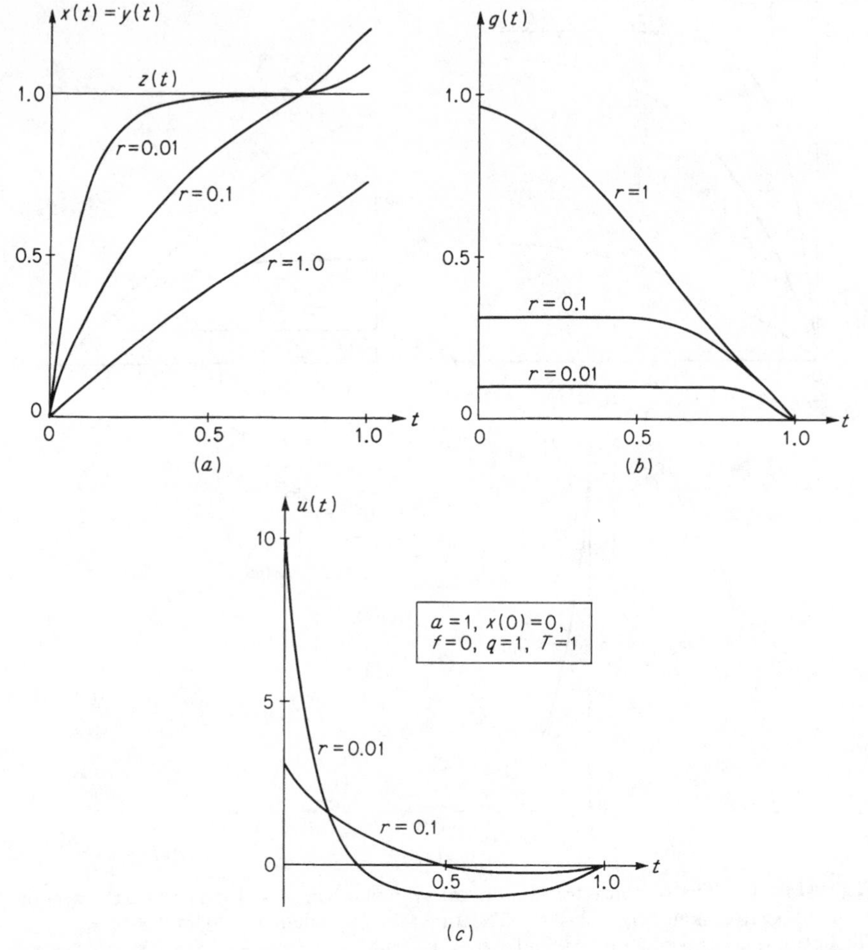


**Fig. 9-22** (*a*) The response of the optimal system when $a = +1$ (unstable case) and $z(t) = 1$. (*b*) The function $g(t)$ when $a = +1$ and $z(t) = 1$. (*c*) The optimal control when $a = +1$ and $z(t) = 1$.

Figure 9-24 shows the response of the optimal system when the desired output $z(t)$ is a sine wave. We have computed these responses for $a = 0$, $f = 0$, $q = 1$, $x(0) = 1$, and $T = 5$. The equation of $z(t)$ was

$$z(t) = 2 \sin (4t) \qquad t \in [0, 5] \tag{9-337}$$

For small values of $r$, we note that, after an initial transient stage, the output is sinusoidal with no detectable phase shift. As $r$ increases, the system behaves like a regulator, and its tracking capabilities are degraded.

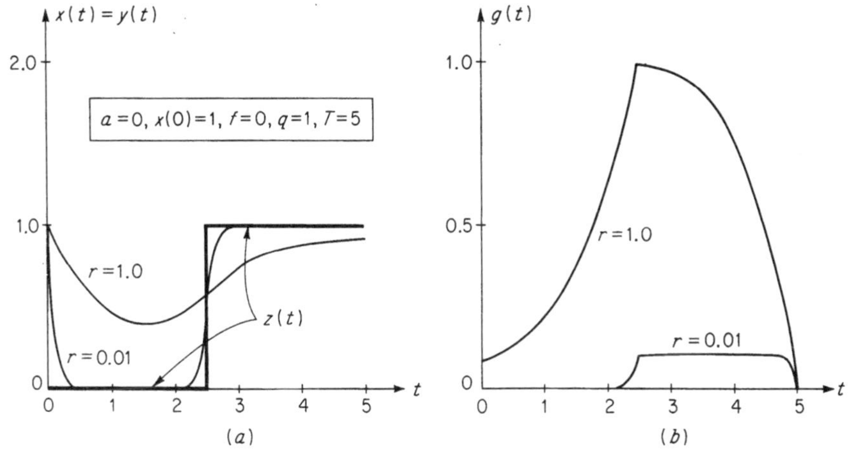


**Fig. 9-23** (*a*) The response of the optimal system when $a = 0$ and $z(t)$ is a delayed step input. (*b*) The function $g(t)$ when $a = 0$ and when $z(t)$ is a delayed step input.

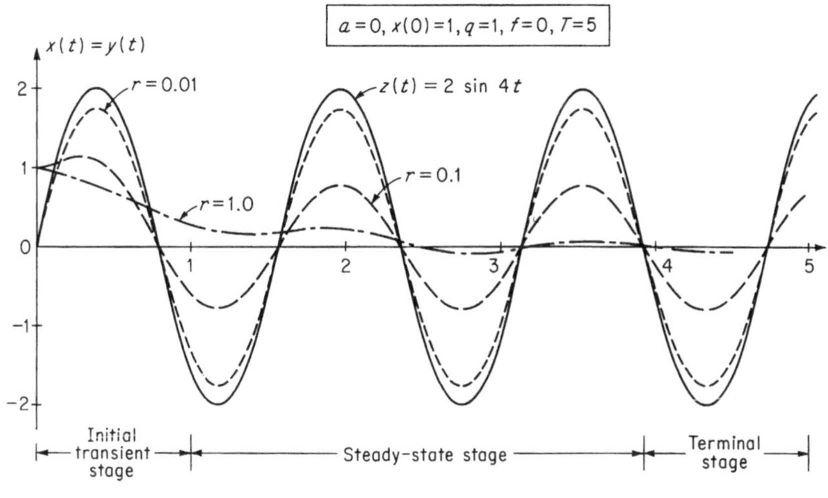


**Fig. 9-24** The response of the optimal system when $a = 0$ and $z(t) = 2 \sin (4t)$.

Figure 9-25 shows the error $e(t)$, the optimal control $u(t)$, and the function $g(t)$ when $a = -1$, $x(0) = 0$, $f = 0$, $q = 1$, $r = 0.1$, $T = 10$, and

$$z(t) = \hat{z} = 1.0 \qquad t \in [0, 10] \tag{9-338}$$

We shall use the results of Fig. 9-25 to verify the approximate relationships we obtained in Sec. 9-10. Note that, during the steady-state stage, we have

$$\begin{aligned} u(t) &\approx 0.91 \\ e(t) &\approx 0.09 \\ g(t) &\approx 0.31 \end{aligned} \tag{9-339}$$

Since the time $T = 10$ is large, we find that $k(t) \approx \hat{k}$, where $\hat{k}$ is the positive solution of the algebraic equation (since $r = 0.1$, $a = -1$, $q = 1$)

$$10\hat{k}^2 + 2\hat{k} - 1 = 0 \tag{9-340}$$

and so

$$\hat{k} = 0.23 \tag{9-341}$$

The matrices **G**, **S**, and **W** given by Eqs. (9-298) are now scalars and are given by

$$\begin{aligned} \mathbf{G} &= -1 - 10\hat{k} = -3.3 \\ \mathbf{S} &= 10 \\ \mathbf{W} &= 1 \end{aligned} \tag{9-342}$$

Equation (9-309) states that $\mathbf{g}(t) \approx -(\mathbf{G}')^{-1}\mathbf{W}\hat{\mathbf{z}}$, and so, for this problem,

$$g(t) \approx -(-3.3)^{-1} \times 1 \times 1 = 0.315 \tag{9-343}$$

We observe that this approximate value of $g(t)$ agrees with the value of $g(t)$ in Fig. 9-25 and in Eqs. (9-339).

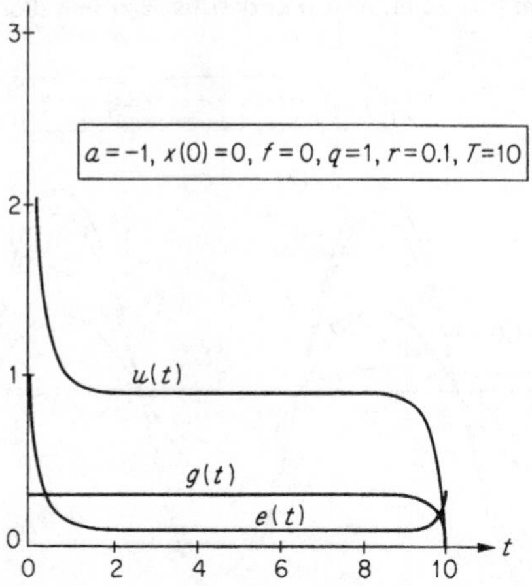


**Fig. 9-25** Illustration of the steady-state quantities when $a = -1$ and $z(t) = +1$ for $t \in [0, 10]$.

If we let $T$ increase without bound, then the "steady-state interval" shown in Fig. 9-25 will also increase without bound. Since both $u(t)$ and $e(t)$ are constant and nonzero over this interval, it is easy to see that $\lim_{T\to\infty} J_3^* = \infty$, and thus, strictly speaking, an optimal solution does *not* exist.

We conclude this section by commenting on the relative weighting of the error vs. the control. In general, to obtain a good tracking system, one should weight the error about 50 to 100 times more than the control. Such rules of thumb play a very important role in engineering and, without doubt, will be developed further as additional experience is obtained regarding the response of optimal systems.

## 9-13 Some Comments

The theory and results of the preceding sections of this chapter are both elegant from the mathematical point of view and useful from the engineering point of view, as the optimal feedback system is linear. In this section, we shall make some comments indicating some additional results which can be found in the literature.

Let us consider the single-input–single-output regulator systems which we examined in Sec. 9-8. The design of (cascade or feedback) compensators has been the main subject of conventional control theory; a conventional servo design is usually based on the transient response and on the frequency response of the closed-loop system. One naturally wonders whether it is possible to translate the properties of an optimal time-invariant feedback system into frequency-domain characteristics.

Reference K-10 contains a rigorous exposition of the frequency-domain properties of a linear system which is optimal with respect to some quadratic cost function. This reference considers the so-called "inverse problem"† and, in particular, shows that the traditional criteria of "goodness" (namely, moderate overshoot, high loop gain, and flat frequency response) correspond to the properties of an optimal output-regulator system.

Before the state-space approach gained in popularity, optimal systems were designed using Laplace and Fourier transforms. The transition from the time domain to the frequency domain can be accomplished using *Parseval's theorem*. This point of view is contained in chap. 2 of Ref. N-8 and in chap. 2 of Ref. C-3. Unfortunately, a reference which contains a rigorous exposition of the relative merits, conceptual and computational, of each approach is not available at this time.

In Ref. M-7, the design of optimal linear systems using the dynamic-programming approach is considered. This reference contains examples of application of the theory to nontrivial engineering problems.

We now focus our attention on the relaxation of some of the assumptions we have made throughout this chapter. Let us consider, for example, the linear time-invariant system

$$\dot{\mathbf{x}}(t) = \mathbf{A}\mathbf{x}(t) + \mathbf{B}\mathbf{u}(t) \tag{9-344}$$

and the cost functional

$$J_1 = \frac{1}{2} \int_0^T [\langle \mathbf{x}(t), \mathbf{Q}\mathbf{x}(t) \rangle + \langle \mathbf{u}(t), \mathbf{R}\mathbf{u}(t) \rangle] \, dt \tag{9-345}$$

One of our assumptions was that $\mathbf{R}$ was a positive definite matrix. Suppose that we set

$$\mathbf{R} = \mathbf{0} \tag{9-346}$$

† The inverse problem is: Given a system and a control law, find the cost functional which is minimized.

in Eq. (9-345) and require that $\mathbf{Q}$ be positive definite. If $\mathbf{R} = \mathbf{0}$, we do not penalize the system for its control-energy expenditure. The optimal control, in this case, will try to bring the state to zero as fast as possible; since we have not placed any magnitude restrictions on $\mathbf{u}(t)$, the optimal control will turn out to be impulsive. The impulsive nature of the control is also borne out by the mathematics. To see this, we form the Hamiltonian $H$ for the system (9-344) and the cost (9-345) with $\mathbf{R} = \mathbf{0}$; the Hamiltonian is given by

$$H = \tfrac{1}{2}\langle \mathbf{x}(t), \mathbf{Q}\mathbf{x}(t)\rangle + \langle \mathbf{A}\mathbf{x}(t), \mathbf{p}(t)\rangle + \langle \mathbf{B}\mathbf{u}(t), \mathbf{p}(t)\rangle \tag{9-347}$$

Since $H$ is linear in $\mathbf{u}(t)$ and since $\mathbf{u}(t)$ is unconstrained, we see that the minimum of $H$ is attained as $\mathbf{u}(t) \to \pm\infty$ [whenever $\mathbf{p}(t) \neq \mathbf{0}$]. Clearly, in order to obtain a meaningful engineering solution, we must place magnitude constraints upon the components of the control vector.

If we formulate the problem with magnitude constraints on the control, then the solutions are no longer nice and linear (even if $\mathbf{R} \neq \mathbf{0}$). In such problems, the optimal control turns out to be either strictly continuous (with intervals of time during which some of the control variables are constant) or piecewise constant.

Quite often, the optimal control turns out to be singular. Reference W-6 considers the system

$$\dot{\mathbf{x}}(t) = \mathbf{A}\mathbf{x}(t) + \mathbf{b}u(t) \tag{9-348}$$

with the (scalar) control restricted in magnitude by $|u(t)| \leq m$. The cost functional to be minimized is of the form $\frac{1}{2}\int_0^T \langle \mathbf{x}(t), \mathbf{Q}\mathbf{x}(t)\rangle\, dt$. Reference W-6 contains conditions on $\mathbf{A}$, $\mathbf{Q}$, and $\mathbf{b}$ which, if satisfied, imply that the optimal control is singular. Otherwise, the optimal control is piecewise constant (bang-bang).

In conclusion, we wish to point out that any violation of the assumptions that we have made in this chapter [and, especially, the assumption that $\mathbf{u}(t)$ is not constrained] will lead in general to a more complicated solution and to a nonlinear feedback system.

Chapter 10

# OPTIMAL - CONTROL PROBLEMS WHEN THE CONTROL IS CONSTRAINED TO A HYPERSPHERE

## 10-1 Introduction

In Chap. 9, we considered the optimal control of linear systems with quadratic cost functionals. We derived analytical expressions for the optimal control under the assumption that the control was not constrained in magnitude. In this chapter, we shall obtain analytical expressions for the optimal control for a class of nonlinear systems and for a variety of performance criteria under the assumption that the control is constrained in magnitude.

The basic assumption throughout this chapter is that the control vector $\mathbf{u}(t)$ is constrained to a hypersphere. In other words, we shall assume that the components $u_1(t), u_2(t), \ldots, u_r(t)$ of the control vector $\mathbf{u}(t)$ are constrained by the relation

$$\|\mathbf{u}(t)\| = \sqrt{u_1^2(t) + \cdots + u_r^2(t)} \leq m \qquad \text{for all } t$$

Our objectives in this chapter are the following:

1. We wish to illustrate the minimization of the Hamiltonian when the control is constrained to satisfy $\|\mathbf{u}(t)\| \leq m$.

2. We wish to show that optimal controls can be found by using standard tools of functional analysis (such as the Schwarz inequality) rather than by using the minimum principle.

3. We wish to indicate, by means of a specific example, the value of optimal-control theory at the preliminary-design stage.

We now discuss the structure of the chapter. In Sec. 10-2, we briefly comment on the physical and mathematical implications of the constraint $\|\mathbf{u}(t)\| \leq m$.

In Sec. 10-3, we formulate a time-optimal control problem. We show that the constraint $\|\mathbf{u}(t)\| \leq m$ leads to a time-optimal control $\mathbf{u}^*(t)$ whose components $u_1^*(t), u_2^*(t), \ldots, u_r^*(t)$ are continuous (rather than piecewise constant) functions of time.

In Sec. 10-4, we derive a very simple expression for the time-optimal control for a special class of nonlinear systems. The time-optimal control is derived by use of the Schwarz inequality rather than by use of the minimum principle. In Sec. 10-5, we discuss the assumptions and techniques used in Sec. 10-4.

In Sec. 10-6, we consider the optimal control of a class of systems, called *norm-invariant,* with respect to several performance criteria. For each criterion, we derive the corresponding optimal control using the Schwarz inequality.

In Sec. 10-7, we use the results of Secs. 7-7, 7-9, and 10-6 to illustrate the value of the theory of optimal control at the preliminary-design stage of a system. We consider the time-optimal velocity control of a spinning body. A basic restriction on the rate of flow of fuel and the freedom of placing the gas jets on the body results in three constraint sets for the control vector. We evaluate the performance of the system for each constraint set, and we draw some interesting conclusions which may be of value to the design engineer.

We conclude the chapter with Sec. 10-8, which contains some comments and some suggestions for further reading.

We wish to caution the reader that the class of systems we shall consider in this chapter is a special one, and so the applicability of the results is rather limited. However, the techniques that we shall expose are interesting, and they demonstrate that the control engineer should always be alert to the possibility that there may be "direct methods" for the solution of a given problem which may be used instead of or in conjunction with the necessary conditions provided by the minimum principle.

## 10-2 Discussion of the Constraint $\|\mathbf{u}(t)\| \le m$

In all the time-optimal and fuel-optimal problems that we have examined, we assumed that the components $u_1(t), u_2(t), \ldots, u_r(t)$ of the control vector $\mathbf{u}(t)$ were constrained in magnitude by relations of the type

$$|u_j(t)| \le m \qquad j = 1, 2, \ldots, r \tag{10-1}$$

The implications of such a magnitude constraint on the control vector are:

1. The control vector must belong to a hypercube in $r$-dimensional space.
2. Each component of the control vector acts independently of the other components. For example, the fact that at some instant of time $u_1(t) = +m$ does not prevent the remaining components $u_2(t), \ldots, u_r(t)$ from attaining any value consistent with Eq. (10-1).

Let us now suppose that the components of the control vector $\mathbf{u}(t)$ are constrained in magnitude by the relation

$$\sqrt{u_1^2(t) + u_2^2(t) + \cdots + u_r^2(t)} = \|\mathbf{u}(t)\| \le m \tag{10-2}$$

The norm $\|\mathbf{u}(t)\|$ is the Euclidean length of the vector $\mathbf{u}(t)$. The implications of such a constraint are:

1. The control vector must belong to a hypersphere (of radius $m$) in $r$-dimensional space.

2. The components of the control vector are now dependent. For example, if at some time $t'$, we have $u_1(t') = m$, then $u_2(t') = u_3(t') = \cdots = u_r(t') = 0$.

The constraint $\|\mathbf{u}(t)\| \leq m$ arises quite often when the control vector $\mathbf{u}(t)$ represents a thrust force which is generated by a gimbaled rocket or gas jet, i.e., a thrust source which is free to rotate in space. Quite often, the constraint $\|\mathbf{u}(t)\| \leq m$ may be an artificial one and may be used in lieu of the constraint $|u_j(t)| \leq m$. In this case, one *approximates* the hypercube by the inscribed hypersphere. It should be clear that if $\|\mathbf{u}(t)\| \leq m$, then all the components $u_j(t)$ of $\mathbf{u}(t)$ automatically satisfy the constraint $|u_j(t)| \leq m$.

We shall see in this chapter that the "smoothing" of the control constraint set from a hypercube to a hypersphere leads to solutions which are "smooth." For example, time-optimal and fuel-optimal control laws are not determined by means of switch curves.

## 10-3 Formulation of a Time-optimal Problem

In this section, we shall formulate a time-optimal problem with the constraint $\|\mathbf{u}(t)\| \leq m$ so that the reader can obtain an idea of the structure of the time-optimal controller for this type of constraint. The problem which we consider is the following:

*Given the system*

$$\dot{\mathbf{x}}(t) = \mathbf{g}[\mathbf{x}(t); t] + \mathbf{B}(t)\mathbf{u}(t) \tag{10-3}$$

*where the control is constrained by*

$$\|\mathbf{u}(t)\| \leq m \qquad \textit{for all } t \tag{10-4}$$

*Given an initial time $t_0$ and an initial state $\mathbf{x}(t_0)$. Determine the control which will force the system* (10-3) *to the origin, i.e., the state $\mathbf{x} = \mathbf{0}$, in the minimum possible time.*

We start by using the minimum principle. The Hamiltonian $H$ for this problem is

$$H = 1 + \langle \mathbf{g}[\mathbf{x}(t); t], \mathbf{p}(t)\rangle + \langle \mathbf{B}(t)\mathbf{u}(t), \mathbf{p}(t)\rangle \tag{10-5}$$

The costate vector $\mathbf{p}(t)$ is the solution of the system of differential equations

$$\dot{p}_i(t) = -\frac{\partial H}{\partial x_i(t)} = -\sum_{j=1}^{n} p_j(t)\,\frac{\partial g_j[\mathbf{x}(t); t]}{\partial x_i(t)} \tag{10-6}$$

where $i = 1, 2, \ldots, n$. Let $\mathbf{p}^*(t)$ be the solution of Eq. (10-6) that corresponds to the optimal trajectory $\mathbf{x}^*(t)$ and the optimal control $\mathbf{u}^*(t)$, $\|\mathbf{u}^*(t)\| \le m$. By the minimum principle, it is necessary that

$$H[\mathbf{x}^*(t), \mathbf{p}^*(t), \mathbf{u}^*(t), t] \le H[\mathbf{x}^*(t), \mathbf{p}^*(t), \mathbf{u}(t), t] \tag{10-7}$$

and so we find that

$$\begin{aligned} 1 + \langle \mathbf{g}[\mathbf{x}^*(t); t], \mathbf{p}^*(t)\rangle + \langle \mathbf{B}(t)\mathbf{u}^*(t), \mathbf{p}^*(t)\rangle \\ \le 1 + \langle \mathbf{g}[\mathbf{x}^*(t); t], \mathbf{p}^*(t)\rangle + \langle \mathbf{B}(t)\mathbf{u}(t), \mathbf{p}^*(t)\rangle \end{aligned} \tag{10-8}$$

From Eq. (10-8), we deduce the relation

$$\langle \mathbf{u}^*(t), \mathbf{B}'(t)\mathbf{p}^*(t)\rangle \le \langle \mathbf{u}(t), \mathbf{B}'(t)\mathbf{p}^*(t)\rangle \tag{10-9}$$

Equation (10-9) means that, whatever $\mathbf{p}^*(t)$ is, the time-optimal control must minimize the scalar product

$$\langle \mathbf{u}^*(t), \mathbf{B}'(t)\mathbf{p}^*(t)\rangle \tag{10-10}$$

We claim that the time-optimal control must be given by

$$\mathbf{u}^*(t) = -m \frac{\mathbf{B}'(t)\mathbf{p}^*(t)}{\|\mathbf{B}'(t)\mathbf{p}^*(t)\|} \qquad \text{if } \|\mathbf{B}'(t)\mathbf{p}^*(t)\| \ne 0 \tag{10-11}$$

If, however, the relation

$$\|\mathbf{B}'(t)\mathbf{p}^*(t)\| = 0 \tag{10-12}$$

holds, then $\mathbf{B}'(t)\mathbf{p}^*(t) = \mathbf{0}$ and Eq. (10-9) is satisfied for all $\mathbf{u}(t)$. This is the familiar singular case, and so, if Eq. (10-12) holds, we cannot obtain any information regarding $\mathbf{u}^*(t)$ from Eq. (10-9).

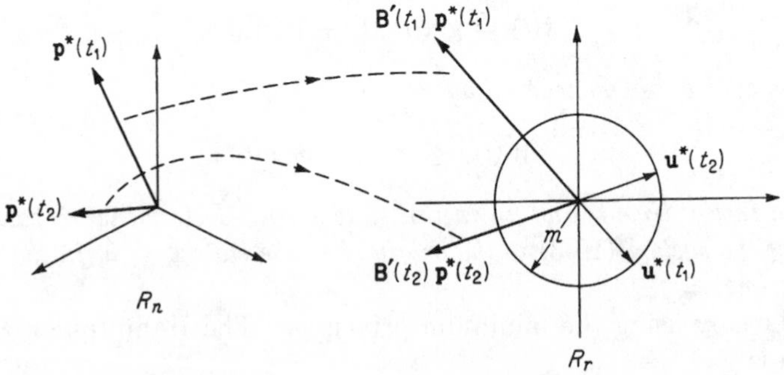


**Fig. 10-1** Graphical interpretation of Eq. (10-11).

Figure 10-1 shows a geometric interpretation of Eq. (10-11). Suppose that $t_1$ is a time such that the $n$-dimensional vector $\mathbf{p}^*(t_1) \ne \mathbf{0}$ and $\mathbf{B}'(t_1) \ne \mathbf{0}$. The matrix $\mathbf{B}'(t_1)$ corresponds to a linear transformation from the $n$-dimensional space of the costate variables to the $r$-dimensional space of the control variables. Thus, the vector $\mathbf{p}^*(t_1)$ is mapped into the vector $\mathbf{B}'(t_1)\mathbf{p}^*(t_1)$. Clearly, to minimize the scalar product $\langle \mathbf{u}(t_1), \mathbf{B}'(t_1)\mathbf{p}^*(t_1)\rangle$, the vector $\mathbf{u}^*(t_1)$ must be oppositely directed to the vector $\mathbf{B}'(t_1)\mathbf{p}^*(t_1)$ and, furthermore, $\mathbf{u}^*(t_1)$ must be as large as possible [which means that $\|\mathbf{u}^*(t_1)\| = m$]. Similarly, this must occur for any time $t$. Thus, the implication of Eq. (10-11) is that *the time-optimal control must be oppositely directed to the* $\mathbf{B}'(t)\mathbf{p}^*(t)$ *vector and the magnitude of* $\mathbf{u}^*(t)$ *must be as large as possible, that is,* $\|\mathbf{u}^*(t)\| = m$.

If we substitute Eq. (10-11) into Eq. (10-3) (and drop the * for convenience), we find that

$$\dot{\mathbf{x}}(t) = \mathbf{g}[\mathbf{x}(t); t] - m\mathbf{B}(t)\frac{\mathbf{B}'(t)\mathbf{p}(t)}{\|\mathbf{B}'(t)\mathbf{p}(t)\|} \tag{10-13}$$

Equations (10-13) and (10-6) are the canonical equations for the problem. For a specific system, we can solve the canonical equations and determine the time-optimal control, provided that it exists and that it is unique.

A very important property of the time-optimal control

$$\mathbf{u}^*(t) = -\frac{m\mathbf{B}'(t)\mathbf{p}^*(t)}{\|\mathbf{B}'(t)\mathbf{p}^*(t)\|}$$

is that *its components* $u_1^*(t), u_2^*(t), \ldots, u_r^*(t)$ *are continuous functions of time.* This immediately follows from the continuity of $\mathbf{p}^*(t)$ (why?) and of $\mathbf{B}'(t)$. This property is in direct contrast with the *piecewise constancy* of the components of the time-optimal control vector when the constraint set was the hypercube.† If $\mathbf{u}^*(t)$ is a continuous vector-valued function, then $\dot{\mathbf{x}}^*(t)$ is also continuous. This means that the time-optimal trajectory $\mathbf{x}^*(t)$ has no "corners," and so we should not expect any "switch curves" in the state space. Rather, the time-optimal control will turn out to be a nonlinear, but smooth, function of the state of the system.

We shall illustrate, by means of the following example, the concepts involved and the computational difficulties that may occur.

**Example 10-1** Consider the time-optimal control of the system

$$\begin{bmatrix} \dot{x}_1(t) \\ \dot{x}_2(t) \end{bmatrix} = \begin{bmatrix} -1 & 0 \\ 0 & -2 \end{bmatrix} \begin{bmatrix} x_1(t) \\ x_2(t) \end{bmatrix} + \begin{bmatrix} 1 & 1 \\ 0 & 1 \end{bmatrix} \begin{bmatrix} u_1(t) \\ u_2(t) \end{bmatrix} \tag{10-14}$$

with the constraint

$$\sqrt{u_1^2(t) + u_2^2(t)} \le 1 \tag{10-15}$$

We assume that $(\xi_1, \xi_2)$ is the initial state and $(0, 0)$ is the terminal state. The Hamiltonian for this problem is given by

$$H = 1 - x_1(t)p_1(t) - 2x_2(t)p_2(t) + [u_1(t) + u_2(t)]p_1(t) + u_2(t)p_2(t) \tag{10-16}$$

† See Theorem 6-2.

The control which minimizes the Hamiltonian is given by the equation

$$\mathbf{u}(t) = -\frac{\begin{bmatrix}1 & 0\\ 1 & 1\end{bmatrix}\begin{bmatrix}p_1(t)\\ p_2(t)\end{bmatrix}}{\left\|\begin{bmatrix}1 & 0\\ 1 & 1\end{bmatrix}\begin{bmatrix}p_1(t)\\ p_2(t)\end{bmatrix}\right\|} \tag{10-17}$$

We perform the indicated operations, and we find that

$$\begin{aligned} u_1(t) &= -\frac{p_1(t)}{\sqrt{p_1^2(t) + [p_1(t) + p_2(t)]^2}} \\ u_2(t) &= -\frac{p_1(t) + p_2(t)}{\sqrt{p_1^2(t) + [p_1(t) + p_2(t)]^2}} \end{aligned} \tag{10-18}$$

The costate variables are the solutions of the differential equations

$$\begin{aligned} \dot{p}_1(t) &= -\frac{\partial H}{\partial x_1(t)} = p_1(t) \\ \dot{p}_2(t) &= -\frac{\partial H}{\partial x_2(t)} = 2p_2(t) \end{aligned} \tag{10-19}$$

Let $\pi_1$ and $\pi_2$ be the (unknown) initial costates corresponding to the initial state $(\xi_1, \xi_2)$; then the solution of Eq. (10-19) is given by

$$\begin{aligned} p_1(t) &= \pi_1 e^t \\ p_2(t) &= \pi_2 e^{2t} \end{aligned} \tag{10-20}$$

We substitute Eqs. (10-20) and (10-18) into Eq. (10-14), and since the equations are linear, we integrate to find that $x_1(t)$ and $x_2(t)$ are given by the equations

$$x_1(t) = e^{-t}\left(\xi_1 - \int_0^t e^\tau \frac{2\pi_1 e^\tau + \pi_2 e^{2\tau}}{\sqrt{2\pi_1^2 e^{2\tau} + \pi_2^2 e^{4\tau} + 2\pi_1\pi_2 e^{3\tau}}}\, d\tau\right) \tag{10-21}$$

$$x_2(t) = e^{-2t}\left(\xi_2 - \int_0^t e^{2\tau} \frac{\pi_1 e^\tau + \pi_2 e^{2\tau}}{\sqrt{2\pi_1^2 e^{2\tau} + \pi_2^2 e^{4\tau} + 2\pi_1\pi_2 e^{3\tau}}}\, d\tau\right) \tag{10-22}$$

where $\xi_1$ and $\xi_2$ are the initial values of the state variables at $t = 0$. Let $t^*$ be the minimum time required to force the initial state $(\xi_1, \xi_2)$ to the origin $(0, 0)$. We substitute

$$x_1(t^*) = x_2(t^*) = 0 \qquad t = t^* \tag{10-23}$$

into Eqs. (10-21) and (10-22) and, thus, determine the relationship between $\xi_1$, $\xi_2$, $\pi_1$, $\pi_2$, and $t^*$. We can obtain an additional relationship between $\pi_1$, $\pi_2$, and $t^*$ by using the fact that $H = 0$ for all $t \in [0, t^*]$ and, in particular, that $H = 0$ at $t = t^*$. Thus, substituting Eqs. (10-23), (10-20), and (10-18) into Eq. (10-16), we find that

$$\sqrt{2\pi_1^2 e^{2t^*} + \pi_2^2 e^{4t^*} + 2\pi_1\pi_2 e^{3t^*}} = 1 \tag{10-24}$$

If we could evaluate analytically the integrals that appear in Eqs. (10-21) and (10-22), then we would be able to obtain an analytical expression for the extremal control. In many cases this is possible, but in general, the integrals must be evaluated using a computer. In the following section, we shall discuss a class of systems for which analytical expressions for the time-optimal controls can be derived easily.

This example illustrates that the determination of the extremal controls reduces to the evaluation of some integrals. This is indeed a common characteristic of such systems. The analytical difficulties arise from the fact that the integrals are somewhat messy. Thus, the analytical solution of the two-point boundary-value problem is often impossible.

## 10-4 Analytical Determination of the Time-optimal Control for a Class of Nonlinear Systems†

In Sec. 10-3, we formulated a time-optimal control problem to illustrate the use of the minimum principle when the control is constrained in magnitude by the relation $\|\mathbf{u}(t)\| \leq m$, for all $t$. In this section, we shall derive an analytical expression for the time-optimal control for a class of nonlinear systems. The class of nonlinear systems is a very restricted one; however, the expression for the time-optimal control turns out to be an extremely simple function of the state, and it is derived in a direct manner without using the necessary conditions.

The results, restricted or not, are important from the theoretical point of view. It turns out that there is (at least) one engineering system which falls in this category; we shall examine it in detail later on in this chapter.

The problem we shall solve in this section is the following:

***Problem 10-1*** *Given the system*

$$\dot{\mathbf{x}}(t) = \mathbf{f}[\mathbf{x}(t); t] + \mathbf{u}(t) \qquad \mathbf{x}(t_0) = \boldsymbol{\xi} \tag{10-25}$$

*Assume that:*

*a. The system is controllable.*

*b. The control vector* $\mathbf{u}(t)$ *and the state vector* $\mathbf{x}(t)$ *have the same dimension, say* $n$.

*c. The control vector* $\mathbf{u}(t)$ *is restricted in magnitude by the relation*

$$\|\mathbf{u}(t)\| = \sqrt{u_1^2(t) + \cdots + u_n^2(t)} \leq m \tag{10-26}$$

*for all* $t \geq t_0$.

*d. The vector-valued function* $\mathbf{f}[\mathbf{x}(t); t]$ *has the property that*

$$\langle \mathbf{f}[\mathbf{x}(t); t], \mathbf{x}(t) \rangle = h[\|\mathbf{x}(t)\|, t] \tag{10-27}$$

*for all* $\mathbf{x}(t)$ *and all* $t \geq t_0$*; the function* $h[\|\mathbf{x}(t)\|; t]$ *is some scalar function of the norm* $\|\mathbf{x}(t)\|$ *[that is, of the Euclidean length of the state vector* $\mathbf{x}(t)$*] and of the time* $t$.

*Determine the control* $\mathbf{u}^*(t)$ *which will force the system* (10-25) *from any given initial state* $\mathbf{x}(t_0) = \boldsymbol{\xi}$ *to the origin* $\mathbf{0}$ *of the state space in the shortest possible time.*

We postpone, for the moment, the physical interpretation of the problem and of the assumptions and proceed with certain trivial but very important observations.

***Lemma 10-1*** *Suppose that* $\mathbf{u}(t)$ *is a control which reduces the norm* $\|\mathbf{x}(t)\|$ *of the solution* $\mathbf{x}(t)$ *of the system* (10-25) *from* $\|\mathbf{x}(t_0)\| = \|\boldsymbol{\xi}\|$ *to* $\|\mathbf{x}(t^*)\| = 0$ *in the least possible time* $t^*$. *Then* $\mathbf{u}(t) = \mathbf{u}_{\xi}^*(t)$*; that is, the control* $\mathbf{u}(t)$ *is time-optimal.*

PROOF This is obvious, because $\mathbf{x}(t^*) = \mathbf{0}$ if and only if $\|\mathbf{x}(t^*)\| = 0$.

† This section is based upon the results of Ref. A-11.

Throughout this chapter, we shall deal with the time derivative of the length of a vector. For this reason, it is helpful to establish the following lemma:

***Lemma 10-2*** *Let* $\mathbf{w}(t)$ *be a vector with components* $w_1(t), w_2(t), \ldots, w_n(t)$. *Then*

$$\frac{d}{dt}\|\mathbf{w}(t)\| = \frac{\langle \dot{\mathbf{w}}(t), \mathbf{w}(t)\rangle}{\|\mathbf{w}(t)\|} = \left\langle \dot{\mathbf{w}}(t), \frac{\mathbf{w}(t)}{\|\mathbf{w}(t)\|}\right\rangle \tag{10-28}$$

PROOF Equation (10-28) is obtained by the direct computation

$$\begin{aligned}\frac{d}{dt}\|\mathbf{w}(t)\| &= \frac{d}{dt}\Big[\sum_{i=1}^{n} w_i^2(t)\Big]^{\frac{1}{2}} \\ &= \tfrac{1}{2}\Big[\sum_{i=1}^{n} w_i^2(t)\Big]^{-\frac{1}{2}} 2\sum_{i=1}^{n} \dot{w}_i(t)w_i(t) \\ &= \frac{\langle \dot{\mathbf{w}}(t), \mathbf{w}(t)\rangle}{\|\mathbf{w}(t)\|}\end{aligned} \tag{10-29}$$

Now let $\mathbf{x}(t)$ be the solution of Eq. (10-25). Using Eqs. (10-28) and (10-25), we find that the rate of change of $\|\mathbf{x}(t)\|$ can be found from the relation

$$\frac{d}{dt}\|\mathbf{x}(t)\| = \frac{\langle \dot{\mathbf{x}}(t), \mathbf{x}(t)\rangle}{\|\mathbf{x}(t)\|} = \frac{\langle \mathbf{f}[\mathbf{x}(t); t], \mathbf{x}(t)\rangle}{\|\mathbf{x}(t)\|} + \frac{\langle \mathbf{u}(t), \mathbf{x}(t)\rangle}{\|\mathbf{x}(t)\|} \tag{10-30}$$

By assumption $d$ of Problem 10-1 and, in particular, by substituting Eq. (10-27) into Eq. (10-30), we find that

$$\frac{d}{dt}\|\mathbf{x}(t)\| = \frac{h[\|\mathbf{x}(t)\|; t]}{\|\mathbf{x}(t)\|} + \left\langle \mathbf{u}(t), \frac{\mathbf{x}(t)}{\|\mathbf{x}(t)\|}\right\rangle \tag{10-31}$$

Equation (10-31) is a first-order differential equation which defines the behavior of the norm of $\mathbf{x}(t)$, $\|\mathbf{x}(t)\|$, as a function of the control $\mathbf{u}(t)$. In order to see more clearly the implications of Eq. (10-31), we let

$$y(t) = \|\mathbf{x}(t)\| \tag{10-32}$$

$$a[y(t); t] = \frac{h[\|\mathbf{x}(t)\|; t]}{\|\mathbf{x}(t)\|} \tag{10-33}$$

$$v(t) = \left\langle \mathbf{u}(t), \frac{\mathbf{x}(t)}{\|\mathbf{x}(t)\|}\right\rangle \tag{10-34}$$

so that Eq. (10-31) reduces to the equation

$$\dot{y}(t) = a[y(t); t] + v(t) \tag{10-35}$$

where $y(t_0) = \|\mathbf{x}(t_0)\| = \|\boldsymbol{\xi}\| > 0$.

Our objective is to solve Problem 10-1. However, we note that *driving any initial state to zero in minimum time is equivalent to reducing the length of the initial state to zero in minimum time, and vice versa.* By Lemma 10-1, we know that the control that reduces the length of the state vector to zero in minimum time is also the control that provides the solution to Problem 10-1. Equation (10-35) is a first-order differential equation. It follows that, if we find the $v(t)$ which drives the system (10-35) from any initial state $y(t_0) > 0$ to zero in minimum time and if we can relate this $v(t)$ to the vector $\mathbf{u}(t)$ using Eq. (10-34), then this procedure will give us the time-optimal $\mathbf{u}^*(t)$. This is the procedure we shall now follow.

In order to formulate a meaningful time-optimal problem for the system

$$\dot{y}(t) = a[y(t); t] + v(t) \qquad y(t_0) > 0 \tag{10-36}$$

we must have magnitude constraints on $v(t)$. These are obtained as follows: Since $v(t) = \langle \mathbf{u}(t), \mathbf{x}(t)/\|\mathbf{x}(t)\| \rangle$, it follows by the Schwarz inequality† that

$$|v(t)| = \left| \left\langle \mathbf{u}(t), \frac{\mathbf{x}(t)}{\|\mathbf{x}(t)\|} \right\rangle \right| \leq \|\mathbf{u}(t)\| \left\| \frac{\mathbf{x}(t)}{\|\mathbf{x}(t)\|} \right\| \leq m \tag{10-37}$$

because $\|\mathbf{u}(t)\| \leq m$ and because $\mathbf{x}(t)/\|\mathbf{x}(t)\|$ is a vector of unit length. Thus, the scalar $v(t)$ is constrained in magnitude by the relation

$$|v(t)| \leq m \tag{10-38}$$

But we have already solved for the time-optimal control of a first-order system in Sec. 7-10. Using the results of Sec. 7-10 and noting that $y(t) = \|\mathbf{x}(t)\| \geq 0$, we conclude that the function

$$v^*(t) = -m \tag{10-39}$$

drives any initial $y(t_0) > 0$ to zero in minimum time (if this is possible). It follows that the time-optimal control vector $\mathbf{u}^*(t)$ must be such that

$$\left\langle \mathbf{u}^*(t), \frac{\mathbf{x}^*(t)}{\|\mathbf{x}^*(t)\|} \right\rangle = v^*(t) = -m \tag{10-40}$$

where $\mathbf{x}^*(t)$ is the state of the time-optimal system at time $t$. We can immediately conclude that *the time-optimal control* $\mathbf{u}^*(t)$ *must be given by the equation*

$$\mathbf{u}^*(t) = -m \frac{\mathbf{x}^*(t)}{\|\mathbf{x}^*(t)\|} \tag{10-41}$$

*and that the state* $\mathbf{x}^*(t)$ *is the solution of the differential equation*

$$\dot{\mathbf{x}}(t) = \mathbf{f}[\mathbf{x}(t); t] - m \frac{\mathbf{x}(t)}{\|\mathbf{x}(t)\|} \tag{10-42}$$

*starting from the known initial state* $\mathbf{x}(t_0) = \boldsymbol{\xi}$ *(provided that a solution exists).*

Clearly, the time-optimal control $\mathbf{u}^*(t)$ given by Eq. (10-41) is *unique;* the uniqueness is a consequence of the uniqueness of the solutions of Eq.

† Recall that the Schwarz inequality states that $|\langle \mathbf{v}, \mathbf{w} \rangle| \leq \|\mathbf{v}\| \, \|\mathbf{w}\|$. (See Sec. 2-12.)

(10-42) and the fact that there is one and only one $\mathbf{u}^*(t)$ which satisfies Eq. (10-40).

We summarize these results in the following control law:

***Control Law 10-1 Solution to Problem 10-1*** *The time-optimal control which is the solution of Problem* 10-1 *is given by*

$$\mathbf{u}^*(t) = -m \frac{\mathbf{x}^*(t)}{\|\mathbf{x}^*(t)\|} \tag{10-43}$$

*for all* $t \geq t_0$. *Moreover,* $\mathbf{u}^*(t)$ *is unique (provided that it exists).*

Equation (10-43) implies that the time-optimal control is as large as possible $[\|\mathbf{u}^*(t)\| = m]$ and is oppositely directed to the state $\mathbf{x}^*(t)$.

We shall discuss the significance of the assumptions in Problem 10-1 and of Control Law 10-1 in the following section.

## 10-5 Discussion of the Results

In Sec. 10-4, we stated a time-optimal problem (Problem 10-1) and we found its solution (Control Law 10-1). In this section, we shall discuss the technique used to derive Control Law 10-1 and the importance of the assumptions of Problem 10-1.

The reader has probably noticed that our method of solution deviated from the standard steps provided by the minimum principle, which were outlined in Sec. 10-3. The basic idea in Sec. 10-4 was to examine the first-order differential equation satisfied by the Euclidean norm $\|\mathbf{x}(t)\|$ [see Eqs. (10-30) and (10-31)] and to determine the control that reduced $\|\mathbf{x}(t)\|$ to zero in minimum time. The assumptions of Problem 10-1 had the following basic implications:

1. The homogeneous part $h[\|\mathbf{x}(t)\|;\, t]/\|\mathbf{x}(t)\|$ of Eq. (10-31) was independent of the components $x_i(t)$ of the state $\mathbf{x}(t)$ and was only a function of $\|\mathbf{x}(t)\|$.

2. The "control" $v(t) = \langle \mathbf{u}(t), \mathbf{x}(t)/\|\mathbf{x}(t)\|\rangle$ was constrained in magnitude by $|v(t)| \leq m$; thus, the constraints on the "control" were independent of the state and of the time.

If we examine Eq. (10-31) carefully, we observe that the time-optimal control $\mathbf{u}(t) = -m\mathbf{x}(t)/\|\mathbf{x}(t)\|$ has the following property: *The time-optimal control is such that the time rate of change* $\frac{d}{dt}\|\mathbf{x}(t)\|$ *of the norm* $\|\mathbf{x}(t)\|$ *is the smallest possible for each time t and each state* $\mathbf{x}(t)$. The verification of this property is very easy. Suppose that $\hat{\mathbf{u}}(t)$ is a control vector which is admissible and distinct from the time-optimal control; that is,

$$\|\hat{\mathbf{u}}(t)\| \leq m \qquad \hat{\mathbf{u}}(t) \neq -\frac{m\mathbf{x}(t)}{\|\mathbf{x}(t)\|} \tag{10-44}$$

Using the Schwarz inequality, we find that

$$\left\langle \hat{\mathbf{u}}(t), \frac{\mathbf{x}(t)}{\|\mathbf{x}(t)\|} \right\rangle > -m \tag{10-45}$$

from Eqs. (10-44), (10-31), and (10-40).

We can interpret these results in the following way: *The control which is "locally optimal," i.e., the control which locally causes the norm $\|\mathbf{x}(t)\|$ to decrease as fast as possible, turns out to be globally optimal, and vice versa.* Of course, this conclusion is valid only for Problem 10-1 and *does not represent a general property of time-optimal systems.*

We shall illustrate this point by means of an example. In this example, we consider the time-optimal control of a system which satisfies some, but not all, of the assumptions stated in Problem 10-1.

**Example 10-2** Consider the controllable system (harmonic oscillator)

$$\begin{aligned} \dot{x}_1(t) &= x_2(t) \\ \dot{x}_2(t) &= -x_1(t) + u(t) \end{aligned} \tag{10-46}$$

Assume that the scalar control $u(t)$ is constrained in magnitude by the relation

$$|u(t)| \leq 1 \tag{10-47}$$

Let us mimic the development of Sec. 10-4 to discover the implications.

We compute $\frac{d}{dt}\|\mathbf{x}(t)\|$ for the system (10-46), and we find that

$$\begin{aligned} \frac{d}{dt}\|\mathbf{x}(t)\| &= \frac{x_2(t)x_1(t) - x_1(t)x_2(t)}{\|\mathbf{x}(t)\|} + \frac{x_2(t)u(t)}{\|\mathbf{x}(t)\|} \\ &= u(t)\frac{x_2(t)}{\sqrt{x_1^2(t) + x_2^2(t)}} \end{aligned} \tag{10-48}$$

Now, the system (10-46) is such that Eq. (10-27) is satisfied because $h[\|\mathbf{x}(t)\|; t] = 0$ for all $\mathbf{x}(t)$ and all $t$. Thus, assumptions $a$ and $d$ of Problem 10-1 are satisfied, but assumptions $b$ and $c$ are violated.

Let us now try to find the "locally optimal" control. In other words, we would like to find the control $u'(t)$ which makes $\frac{d}{dt}\|\mathbf{x}(t)\|$ as negative as possible at each instant of time. From Eq. (10-28) and in view of Eq. (10-47), we find that this "locally optimal" control is given by the equation

$$u'(t) = -\operatorname{sgn}\left\{\frac{x_2(t)}{\sqrt{x_1^2(t) + x_2^2(t)}}\right\} = -\operatorname{sgn}\{x_2(t)\} \tag{10-49}$$

We claim that this "locally optimal" control $u'(t)$ is not identical to the global time-optimal control $u^*(t)$. Recall that we have derived the global time-optimal control $u^*(t)$ for the system (10-46) in Sec. 7-7 [see Eq. (7-283), with $\omega = 1$]; we have found (see Fig. 7-41) that the global time-optimal control is $u^*(t) = -1$ whenever the state is above the $\gamma$ switch curve and that $u^*(t) = +1$ whenever the state is below the $\gamma$ switch curve. Since the $\gamma$ switch curve of Fig. 7-41 is *not* the $x_1$ axis, we conclude that the "locally optimal" control $u'(t)$ is *not* identical to the global time-optimal control

$u^*(t)$ (which was unique). The shaded region in Fig. 10-2 is the set of states for which the "locally optimal" control $u'(t) = -\text{sgn}\{x_2(t)\}$ differs from the global time-optimal control $u^*(t)$. We have plotted in Fig. 10-2 the true time-optimal trajectory $T_1$ and the trajectory $T_2$ generated by the control $u'(t)$. Note that the trajectories are identical until the point $Q$ is reached. After the point $Q$, the nonoptimal trajectory $T_2$ is closer to the origin than $T_1$, until the trajectories hit the $x_1$ axis. However, by the time the initial state reaches the point $P$ along the trajectory $T_2$, the same initial state has already arrived at the origin along $T_1$. With ideal equipment, the trajectory $T_2$ would terminate at $P$. However, in any engineering realization, the chattering of the relay would cause a state on $T_2$ to eventually arrive at the origin along the zigzag path indicated in Fig. 10-2. Loosely speaking, the control $u'(t)$ is "shortsighted" because it does not take into account the future effects of the present control (and this is why it is

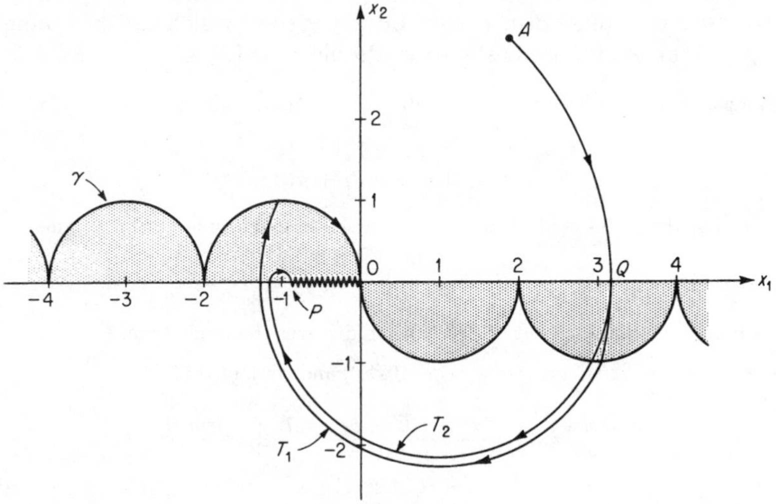


**Fig. 10-2** The trajectory $T_1$ is the time-optimal one. The trajectory $T_2$ is generated by the control $u'(t) = -\text{sgn}\{x_2(t)\}$.

only "locally optimal"). Nonetheless, the trajectories due to $u^*(t)$ and $u'(t)$ are almost identical for states far from the origin, and the superiority of $u^*(t)$ is exhibited only for states near the origin.

The derivation of the time-optimal control by reducing the $n$th-order system to a first-order system presented no particular difficulties. We claim that it is extremely difficult to derive the time-optimal control using the minimum principle. To see this, we mimic the steps of Sec. 10-3, and we find that [see Eq. (10-11)] the time-optimal control must be given by the equation

$$\mathbf{u}^*(t) = -m \frac{\mathbf{p}^*(t)}{\|\mathbf{p}^*(t)\|} \tag{10-50}$$

where $\mathbf{p}^*(t)$ is the costate vector corresponding to the time-optimal control $\mathbf{u}^*(t)$ and the time-optimal trajectory $\mathbf{x}^*(t)$. The canonical equations for Problem 10-1 reduce to

$$\begin{aligned}\dot{x}_i^*(t) &= f_i[\mathbf{x}^*(t); t] - \frac{p_i^*(t)}{\|\mathbf{p}^*(t)\|} \\ \dot{p}_i^*(t) &= -\sum_{j=1}^{n} p_j^*(t) \frac{\partial f_i[\mathbf{x}^*(t); t]}{\partial x_i^*(t)}\end{aligned} \tag{10-51}$$

for $i = 1, 2, \ldots, n$. Now, since the time-optimal control is given by the equation

$$\mathbf{u}^*(t) = -m \frac{\mathbf{x}^*(t)}{\|\mathbf{x}^*(t)\|} \tag{10-52}$$

it follows that the solutions of the canonical equations must satisfy the relation

$$\frac{\mathbf{p}^*(t)}{\|\mathbf{p}^*(t)\|} = \frac{\mathbf{x}^*(t)}{\|\mathbf{x}^*(t)\|} \tag{10-53}$$

in view of Eqs. (10-50) and (10-52). It is even difficult to verify that Eq. (10-53) represents the relationship between the state and the costate, and it is extremely difficult, if not impossible, to arrive at Eq. (10-53) by solving Eqs. (10-51).

**Exercise 10-1** Using the assumptions of Problem 10-1, verify that the solutions $\mathbf{p}^*(t)$ and $\mathbf{x}^*(t)$ related by Eq. (10-53) satisfy Eqs. (10-51).

There is still another simplification that the direct method of Sec. 10-4 has to offer. If we attempted to solve the posed problem 10-1 using the minimum principle, then we would find the extremal control. As we noted in Chap. 6, we must prove the existence of an optimal control and verify that the extremal control is unique to arrive at the conclusion that it is time-optimal. Since we deal with a nonlinear time-varying system, it is extremely difficult to verify uniqueness. We emphasize that all these steps have been bypassed in the solution that we presented for Problem 10-1.

Finally, we wish to comment on the engineering realization of the time-optimal control $\mathbf{u}^*(t)$. Since the $i$th component $u_i^*(t)$ is given by

$$u_i^*(t) = -m \frac{x_i^*(t)}{\|\mathbf{x}^*(t)\|} \tag{10-54}$$

it follows that we must measure the components $x_i^*(t)$ of the state vector and that we must divide each component by $\sqrt{x_1^{*2}(t) + \cdots + x_n^{*2}(t)}$. Reference S-7 describes a circuit which can be used to compute the approximate square root of the sum of the squares of an arbitrary number of variables. Thus, the generation of the time-optimal control does not require an extremely complicated system from the practical point of view.

In the following section, we shall consider the control of so-called *norm-invariant systems*. Norm-invariant systems constitute the class of nonlinear (or linear) systems of the form $\dot{\mathbf{x}}(t) = \mathbf{f}[\mathbf{x}(t); t] + \mathbf{u}(t)$, where $\mathbf{f}[\mathbf{x}(t); t]$ has the property that $\langle \mathbf{f}[\mathbf{x}(t); t], \mathbf{x}(t) \rangle = 0$ for all $\mathbf{x}(t)$ and all $t$. Thus, norm-invariant systems represent a special class of the type of systems which we considered in Problem 10-1.

**Exercise 10-2** Consider the time-optimal control to the origin of the system

$$\dot{\mathbf{x}}(t) = \mathbf{f}[\mathbf{x}(t); t] + \mathbf{B}[\mathbf{x}(t); t]\mathbf{u}(t) \qquad \|\mathbf{u}(t)\| \leq m \tag{10-55}$$

Suppose that the system (10-55) satisfies all the assumptions of Problem 10-1. In addition, suppose that the $n \times n$ matrix $\mathbf{B}[\mathbf{x}(t); t]$ is orthogonal for all $\mathbf{x}(t)$ and all $t$. Show that the time-optimal control to the origin is given by

$$\mathbf{u}^*(t) = -m\mathbf{B}'[\mathbf{x}^*(t); t] \frac{\mathbf{x}^*(t)}{\|\mathbf{x}^*(t)\|} \tag{10-56}$$

HINT: See Sec. 2-14.

**Exercise 10-3** Consider Problem 10-1. Given an initial state $\mathbf{x}(t_0) = \xi$, show that the time-optimal control to the target set $S$,

$$S = \{\mathbf{x}: \|\mathbf{x}\| = \mathbf{r}\} \tag{10-57}$$

is given by the relations

$$u^*(t) = \begin{cases} -m \dfrac{\mathbf{x}^*(t)}{\|\mathbf{x}^*(t)\|} & \text{if } r < \|\xi\| \\ +m \dfrac{\mathbf{x}^*(t)}{\|\mathbf{x}^*(t)\|} & \text{if } r > \|\xi\| \end{cases} \tag{10-58}$$

**Exercise 10-4** Consider the linear system

$$\dot{\mathbf{x}}(t) = [k\mathbf{I} + \mathbf{S}]\mathbf{x}(t) + \mathbf{u}(t) \qquad \|\mathbf{u}(t)\| \leq m \tag{10-59}$$

where $k$ is an arbitrary constant, $\mathbf{I}$ is the identity matrix, and $\mathbf{S}$ is a skew-symmetric matrix, that is,

$$\mathbf{S} = -\mathbf{S}' \tag{10-60}$$

(a) Show that this system satisfies all the assumptions of Problem 10-1.

(b) If $\mathbf{x}(0) = \xi$ is the initial state, show that the state of the system (10-59) can be forced to the origin if and only if

$$\|\xi\| < \frac{m}{k} \qquad \text{for } k > 0 \tag{10-61}$$

and that it can be forced to the origin for all $k \leq 0$.

(c) If $k < m/\|\xi\|$, show that the minimum time $t^*$ required to force $\xi$ to $\mathbf{0}$ is given by the equation

$$t^* = -\frac{1}{k} \log \left(1 + \frac{k\|\xi\|}{m}\right) \tag{10-62}$$

(d) Derive the time-optimal control $\mathbf{u}^*(t) = -m[\mathbf{x}^*(t)/\|\mathbf{x}^*(t)\|]$ by solving the canonical equations. HINT: The fundamental matrix for the system (10-59) is $e^{[k\mathbf{I}+\mathbf{S}]t} = e^{kt}\mathbf{\Phi}(t)$, where $\mathbf{\Phi}(t)$ is an orthogonal matrix.

(e) Show that the time $t^*$ of Eq. (10-62) is the solution of the Hamilton-Jacobi equation for all $\mathbf{x}$.

## 10-6 The Optimal Control of Norm-invariant Systems†

In this section, we shall consider the control of a class of systems which we shall call *norm-invariant.* Loosely speaking, we say that a system is norm-invariant if it has the property that the Euclidean length of the state vector is constant whenever the control is zero. We first make some precise definitions, and we then proceed with the formulation and solution of optimal problems for norm-invariant systems with respect to several performance criteria. We shall obtain the optimal control, not by using the minimum principle, but rather by using the Schwarz inequality.

***Definition 10-1*** *The dynamical system*

$$\dot{\mathbf{x}}(t) = \mathbf{g}[\mathbf{x}(t); t] + \mathbf{u}(t) \tag{10-63}$$

*is called norm-invariant if the solution* $\mathbf{x}(t)$ *of the homogeneous system*

$$\dot{\mathbf{x}}(t) = \mathbf{g}[\mathbf{x}(t); t] \tag{10-64}$$

*has the property that*

$$\frac{d}{dt}\|\mathbf{x}(t)\| = 0 \tag{10-65}$$

*for all* $\mathbf{x}(t)$ *and all* $t$. *An equivalent way of defining the notion of norm invariance for a system of the type* (10-63) *can be obtained as follows: We know that* [*see Eq.* (10-28)]

$$\frac{d}{dt}\|\mathbf{x}(t)\| = \frac{\langle\dot{\mathbf{x}}(t), \mathbf{x}(t)\rangle}{\|\mathbf{x}(t)\|} \tag{10-66}$$

*and so, in view of Eqs.* (10-65) *and* (10-64), *we may conclude that the system* $\dot{\mathbf{x}}(t) = \mathbf{g}[\mathbf{x}(t); t] + \mathbf{u}(t)$ *is norm-invariant if and only if* $\mathbf{g}[\mathbf{x}(t); t]$ *has the property that*

$$\langle\mathbf{g}[\mathbf{x}(t); t], \mathbf{x}(t)\rangle = 0 \tag{10-67}$$

*for all* $\mathbf{x}(t)$ *and all* $t$.

**Example 10-3** The homogeneous system

$$\begin{aligned} \dot{x}_1(t) &= t^4 \sin\{x_1(t)x_2(t)\}\, x_2^2(t)x_1^3(t) \\ \dot{x}_2(t) &= -t^4 \sin\{x_1(t)x_2(t)\}\, x_2(t)x_1^4(t) \end{aligned} \tag{10-68}$$

is a norm-invariant system because the equation

$$\dot{x}_1(t)x_1(t) + \dot{x}_2(t)x_2(t) = t^4 \sin\{x_1(t)x_2(t)\}\, x_2^2(t)x_1^4(t) - t^4 \sin\{x_1(t)x_2(t)\}\, x_2^2(t)x_1^4(t) = 0 \tag{10-69}$$

holds.

Let us now ask the following question: *Given the linear homogeneous system*

$$\dot{\mathbf{x}}(t) = \mathbf{S}(t)\mathbf{x}(t) \tag{10-70}$$

† The material of this section is based, for the most part, upon Refs. A-10 and A-12.

*what are conditions on the system matrix* $\mathbf{S}(t)$ *which ensure that the system* (10-70) *is norm-invariant?* We claim that the matrix $\mathbf{S}(t)$ must be a skew-symmetric matrix, i.e., that

$$\mathbf{S}(t) = -\mathbf{S}'(t) \tag{10-71}$$

To see this, we form the scalar product of Eq. (10-67), which, for the system (10-70), is

$$\langle \mathbf{x}(t), \mathbf{S}(t)\mathbf{x}(t)\rangle = 0 \tag{10-72}$$

Since $\langle \mathbf{x}(t), \mathbf{S}(t)\mathbf{x}(t)\rangle = \langle \mathbf{S}'(t)\mathbf{x}(t), \mathbf{x}(t)\rangle$ for all $\mathbf{S}(t)$, we conclude that $\mathbf{S}(t)$ must be such that

$$\langle [\mathbf{S}(t) + \mathbf{S}'(t)]\mathbf{x}(t), \mathbf{x}(t)\rangle = \mathbf{0} \tag{10-73}$$

for all $\mathbf{x}(t)$ and all $t$. Therefore, Eq. (10-71) must hold.

In order to distinguish a linear norm-invariant system from a nonlinear norm-invariant system, we make the following definition:

***Definition 10-2*** *We shall call the linear norm-invariant system*

$$\dot{\mathbf{x}}(t) = \mathbf{S}(t)\mathbf{x}(t) \qquad \mathbf{S}(t) = -\mathbf{S}'(t) \tag{10-74}$$

*self-adjoint. The term "self-adjoint" arises from the fact that the adjoint system to* (10-74) *is* $\dot{\mathbf{z}}(t) = -\mathbf{S}'(t)\mathbf{z}(t) = \mathbf{S}(t)\mathbf{z}(t)$ *and, hence, the system* (10-74) *is its own adjoint.*

**Example 10-4** The linear system

$$\begin{bmatrix} \dot{x}_1(t) \\ \dot{x}_2(t) \\ \dot{x}_3(t) \end{bmatrix} = \begin{bmatrix} 0 & \sin t & t^3 \\ -\sin t & 0 & -e^t \\ -t^3 & e^t & 0 \end{bmatrix} \begin{bmatrix} x_1(t) \\ x_2(t) \\ x_3(t) \end{bmatrix} \tag{10-75}$$

is norm-invariant and self-adjoint.

After these preliminary definitions, we are now ready to formulate several optimal problems; in order to eliminate repetitious statements about assumptions, etc., we shall formulate all the optimal problems in a single statement.

***Problem 10-2*** *Given the controllable norm-invariant system*

$$\dot{\mathbf{x}}(t) = \mathbf{g}[\mathbf{x}(t); t] + \mathbf{u}(t) \qquad \mathbf{x}(0) = \boldsymbol{\xi} \tag{10-76}$$

*Assume that the dimension of* $\mathbf{u}(t)$ *is the same as the dimension of* $\mathbf{x}(t)$ *and that* $\|\mathbf{u}(t)\| \leq m$ *for all* $t$. *Then determine the control which forces the system* (10-76) *from the initial state* $\boldsymbol{\xi}$ *to* $\mathbf{0}$ *and which minimizes the following cost functionals:*

*Problem 10-2a. Minimum time*

$$J_A = \int_0^T dt = T \qquad T \text{ free} \tag{10-77}$$

*Problem 10-2b. Minimum fuel*

$$J_B = \int_0^T \|\mathbf{u}(t)\| \, dt \qquad T \text{ free} \tag{10-78}$$

*Problem 10-2c. Minimization of a linear combination of elapsed time and consumed fuel*

$$J_C = \int_0^T \{k + \|\mathbf{u}(t)\|\} \, dt \qquad T \text{ free; } k > 0 \tag{10-79}$$

*Problem 10-2d. Minimum energy*

$$J_D = \int_0^T \tfrac{1}{2}\|\mathbf{u}(t)\|^2 \, dt \qquad T \text{ fixed} \tag{10-80}$$

*Problem 10-2e. Minimization of a linear combination of elapsed time and expended energy*

$$J_E = \int_0^T \{k + \tfrac{1}{2}\|\mathbf{u}(t)\|^2\} \, dt \qquad T \text{ free; } k > 0 \tag{10-81}$$

We shall find the optimal control for each of the above problems. As we mentioned before, we shall derive the optimal controls using the Schwarz inequality. The proofs are relatively easy, because we can examine the first-order differential equation satisfied by the norm of $\mathbf{x}(t)$. Since the system (10-76) is norm-invariant, we deduce that $\|\mathbf{x}(t)\|$ is the solution of the first-order differential equation

$$\frac{d}{dt}\|\mathbf{x}(t)\| = \left\langle \mathbf{u}(t), \frac{\mathbf{x}(t)}{\|\mathbf{x}(t)\|} \right\rangle = v(t) \tag{10-82}$$

Since $\|\mathbf{u}(t)\| \leq m$, we also conclude that $v(t)$ is constrained by the relation

$$|v(t)| = \left| \left\langle \mathbf{u}(t), \frac{\mathbf{x}(t)}{\|\mathbf{x}(t)\|} \right\rangle \right| \leq m \tag{10-83}$$

We shall use these two equations repeatedly in the remainder of this section to derive the optimal-control laws for each individual problem. Basically, the procedure will be as follows: We shall first establish the greatest lower bound for each cost functional and then show that we can find one or more controls which attain this greatest lower bound; these controls will then be optimal. Then we shall clear up questions of uniqueness.

The mathematics involved in the various proofs are not complicated, because we shall use the two first-order relations (10-82) and (10-83). In our opinion, the important concepts that the reader should grasp are the techniques of establishing the greatest lower bounds using standard inequalities and the uniqueness considerations involved.

First, we turn our attention to Problem 10-2*a*. Since the system (10-76) is norm-invariant and since norm-invariant systems are only a special class of the systems considered in Problem 10-1, we can state the following control law:

***Control Law 10-2a* Solution to Problem 10-2a** *The unique time-optimal control $\mathbf{u}_A^*(t)$, that is, the control which minimizes the cost $J_A$ of Eq.* (10-77), *is given by*

$$\mathbf{u}_A^*(t) = -m \frac{\mathbf{x}_A^*(t)}{\|\mathbf{x}_A^*(t)\|} \tag{10-84}$$

*where $\mathbf{x}_A^*(t)$ is the solution of Eq.* (10-76), *with $\mathbf{u}(t) = \mathbf{u}_A^*(t)$. The minimum value $J_A^*$ of the cost $J_A$, that is, the minimum time $t^*$, required to force $\boldsymbol{\xi}$ to $\mathbf{0}$ is given by the equation*

$$J_A^* = t^* = \frac{\|\boldsymbol{\xi}\|}{m} \tag{10-85}$$

PROOF Equation (10-84) follows by Control Law 10-1. Equation (10-85) is derived as follows: We substitute Eq. (10-84) into Eq. (10-82), and we find that

$$\frac{d}{dt}\|\mathbf{x}_A^*(t)\| = -m \tag{10-86}$$

We integrate Eq. (10-86), and we deduce that

$$\|\mathbf{x}_A^*(t)\| = \|\boldsymbol{\xi}\| - mt \tag{10-87}$$

But at $t = t^*$, we must have $\|\mathbf{x}_A^*(t^*)\| = 0$, and so we can conclude that the minimum time is $t^* = \|\boldsymbol{\xi}\|/m$.

We now proceed to the solution of Problem 10-2*b*. As we shall see, the fuel-optimal control is not unique.

***Control Law 10-2b* Solution to Problem 10-2b** *Let $\mathcal{B}$ be the set of nonnegative scalar functions $\beta(t)$ defined as*

$$\mathcal{B} = \{\beta(t): 0 \le \beta(t) \le m \text{ for all } t, \text{ and for every } \beta(t) \text{ there is a } T_\beta \text{ such that } \int_0^{T_\beta} \beta(t)\, dt = \|\boldsymbol{\xi}\|\} \tag{10-88}$$

*Then the control*

$$\mathbf{u}_B^*(t) = -\beta(t) \frac{\mathbf{x}_B^*(t)}{\|\mathbf{x}_B^*(t)\|} \qquad \beta(t) \in \mathcal{B} \tag{10-89}$$

*is a fuel-optimal control; in other words, the control $\mathbf{u}_B^*(t)$ minimizes the cost $J_B$ given by Eq.* (10-78). *Since $\beta(t)$ is not unique, the fuel-optimal controls are not unique. The minimum fuel $J_B^*$ required to force $\boldsymbol{\xi}$ to $\mathbf{0}$ is given by*

$$J_B^* = \|\boldsymbol{\xi}\| \tag{10-90}$$

*Furthermore, if we have*

$$\beta(t) = m \qquad \text{for all } t \tag{10-91}$$

*then* [*see Eq.* (10-84)] *we can deduce that*

$$\mathbf{u}_B^*(t) = -m \frac{\mathbf{x}_B^*(t)}{\|\mathbf{x}_B^*(t)\|} = \mathbf{u}_A^*(t) \tag{10-92}$$

*which implies that the time-optimal control is also a fuel-optimal control.*

PROOF Let us examine Eq. (10-82), which states that

$$\frac{d}{dt}\|\mathbf{x}(t)\| = \left\langle \mathbf{u}(t), \frac{\mathbf{x}(t)}{\|\mathbf{x}(t)\|} \right\rangle \tag{10-93}$$

We integrate this equation to find that

$$\|\mathbf{x}(t)\| = \|\boldsymbol{\xi}\| + \int_0^t \left\langle \mathbf{u}(\tau), \frac{\mathbf{x}(\tau)}{\|\mathbf{x}(\tau)\|} \right\rangle d\tau \tag{10-94}$$

Let $T$ be the time such that $\|\mathbf{x}(T)\| = 0$. Then

$$\|\boldsymbol{\xi}\| = -\int_0^T \left\langle \mathbf{u}(t), \frac{\mathbf{x}(t)}{\|\mathbf{x}(t)\|} \right\rangle dt \tag{10-95}$$

But from Eq. (10-95) we can deduce the relations

$$\begin{aligned} |\|\boldsymbol{\xi}\|| = \|\boldsymbol{\xi}\| &= \left| -\int_0^T \left\langle \mathbf{u}(t), \frac{\mathbf{x}(t)}{\|\mathbf{x}(t)\|} \right\rangle dt \right| \\ &\le \int_0^T \left| \left\langle \mathbf{u}(t), \frac{\mathbf{x}(t)}{\|\mathbf{x}(t)\|} \right\rangle \right| dt \\ &\le \int_0^T \|\mathbf{u}(t)\| \left\| \frac{\mathbf{x}(t)}{\|\mathbf{x}(t)\|} \right\| dt = \int_0^T \|\mathbf{u}(t)\|\, dt = J_B \end{aligned} \tag{10-96}$$

Thus, we have shown that

$$J_B \ge \|\boldsymbol{\xi}\| \tag{10-97}$$

which means that the initial norm $\|\boldsymbol{\xi}\|$ is the greatest lower bound of the cost $J_B$. Clearly, if a control yields this lower bound, $\|\boldsymbol{\xi}\|$, then this control is optimal.

Next, we show that any control $\mathbf{u}_B^*(t)$ specified by Eqs. (10-88) and (10-89) is optimal. Since

$$\|\mathbf{u}_B^*(t)\| = \beta(t) \tag{10-98}$$

by substituting $\mathbf{u}_B^*(t)$ into Eq. (10-95), we find that

$$\begin{aligned} \|\boldsymbol{\xi}\| &= \int_0^{T_\beta} \beta(t) \left\langle \frac{\mathbf{x}_B^*(t)}{\|\mathbf{x}_B^*(t)\|}, \frac{\mathbf{x}_B^*(t)}{\|\mathbf{x}_B^*(t)\|} \right\rangle dt \\ &= \int_0^{T_\beta} \beta(t)\, dt = \int_0^{T_\beta} \|\mathbf{u}_B^*(t)\|\, dt \end{aligned} \tag{10-99}$$

where $T_\beta$ is the time required to force $\boldsymbol{\xi}$ to $\mathbf{0}$ given a $\beta(t)$; that is, $T_\beta$ is such that

$$\mathbf{x}_B^*(T_\beta) = \mathbf{0} \tag{10-100}$$

Equation (10-99) indicates that the control $\mathbf{u}_B^*(t)$ is such that $J_B$ attains its greatest lower bound $\|\boldsymbol{\xi}\|$ [see Eq. (10-97)], and so $\mathbf{u}_B^*(t)$ is fuel-optimal; furthermore, the minimum cost (fuel) is given by the equation

$$J_B^* = \|\boldsymbol{\xi}\| \tag{10-101}$$

It is easy to establish by direct substitution that the time-optimal control $\mathbf{u}_A^*(t)$ is also fuel-optimal. Since the time-optimal control is unique and since the minimum time $t^*$ is given by (see Control Law 10-2*a*) $t^* = \|\boldsymbol{\xi}\|/m$, we conclude that the response time $T_\beta$ associated with each $\beta(t)$ in $\mathcal{B}$ satisfies the inequality

$$T_\beta > t^* = \frac{\|\boldsymbol{\xi}\|}{m} \qquad \text{for } \beta(t) \neq m \tag{10-102}$$

Finally, we shall verify that all the fuel-optimal controls are given by Eq. (10-89). Note that Eq. (10-89) implies that a fuel-optimal control must:

1. Be directed oppositely to the state vector
2. Have arbitrary magnitude $\beta(t)$ as long as this magnitude does not exceed $m$

Let $\mathbf{R}(t)$ be an *orthogonal* matrix† and let us consider the class of controls defined by the relations

$$\begin{aligned} \hat{\mathbf{u}}(t) &= -\beta(t)\mathbf{R}(t)\frac{\hat{\mathbf{x}}(t)}{\|\hat{\mathbf{x}}(t)\|} \qquad & \mathbf{R}(t) &\neq \mathbf{I} \\ \beta(t) &\in \mathcal{B} & \mathbf{R}'(t)\mathbf{R}(t) &= \mathbf{I} \end{aligned} \tag{10-103}$$

In other words, we consider the controls $\hat{\mathbf{u}}(t)$ which are obtained from the controls $\mathbf{u}_B^*(t)$ by an *orthogonal* transformation. Since $\beta(t)$ and $\mathbf{R}(t)$ are, in a sense, arbitrary, it is easy to see that the set of controls $\hat{\mathbf{u}}(t)$ is identical with the set of controls which take the state $\boldsymbol{\xi}$ to $\mathbf{0}$ and satisfy the relation $\|\mathbf{u}(t)\| \leq m$.

The controls $\hat{\mathbf{u}}(t)$ and $\mathbf{u}_B^*(t)$ have the same magnitude, i.e.,

$$\|\hat{\mathbf{u}}(t)\| = \|\mathbf{u}_B^*(t)\| = \beta(t) \tag{10-104}$$

since

$$\|\mathbf{R}(t)\mathbf{x}(t)\| = \|\mathbf{x}(t)\|$$

and so the only difference between $\hat{\mathbf{u}}(t)$ and $\mathbf{u}_B^*(t)$ is that $\hat{\mathbf{u}}(t)$ is *not* oppositely directed to the state vector. Let $T_\beta$ be the response time such that $\mathbf{x}_B^*(T_\beta) = \mathbf{0}$ and let $\hat{T}_\beta$ be the response time such that $\hat{\mathbf{x}}(\hat{T}_\beta) = \mathbf{0}$. We shall show that

$$\int_0^{\hat{T}_\beta} \|\hat{\mathbf{u}}(t)\|\, dt > \int_0^{T_\beta} \|\mathbf{u}_B^*(t)\|\, dt \tag{10-105}$$

and, hence, that $\hat{\mathbf{u}}(t)$ is not fuel-optimal. However, in view of Eq. (10-104), we can see that we need only prove the inequality

$$\hat{T}_\beta > T_\beta \tag{10-106}$$

If we substitute the control $\mathbf{u}_B^*(t)$ into Eq. (10-93), we obtain the equation

$$\frac{d}{dt}\|\mathbf{x}_B^*(t)\| = -\beta(t) \qquad \|\mathbf{x}_B^*(0)\| = \|\boldsymbol{\xi}\| \tag{10-107}$$

† See Sec. 2-14.

If we substitute the control $\hat{\mathbf{u}}(t)$ into Eq. (10-93), we obtain the equation

$$\frac{d}{dt}\|\hat{\mathbf{x}}(t)\| = -\beta(t)\frac{\langle \mathbf{R}(t)\hat{\mathbf{x}}(t), \hat{\mathbf{x}}(t)\rangle}{\|\hat{\mathbf{x}}(t)\|^2} \qquad \|\hat{\mathbf{x}}(0)\| = \|\boldsymbol{\xi}\| \tag{10-108}$$

By the Schwarz inequality and by the assumption that $\mathbf{R}(t) \neq \mathbf{I}$, we deduce that

$$\langle \mathbf{R}(t)\hat{\mathbf{x}}(t), \hat{\mathbf{x}}(t)\rangle < \|\hat{\mathbf{x}}(t)\|^2 \tag{10-109}$$

and, therefore, that

$$\frac{d}{dt}\|\hat{\mathbf{x}}(t)\| > \beta(t) \qquad \|\hat{\mathbf{x}}(0)\| = \|\boldsymbol{\xi}\| \tag{10-110}$$

From Eqs. (10-107) and (10-110) and by an identical argument to that given in Sec. 7-10, we conclude that

$$\hat{T}_\beta > T_\beta \tag{10-111}$$

In essence, the rate of change of $\|\hat{\mathbf{x}}(t)\|$ is greater than the rate of change of $\|\mathbf{x}_B^*(t)\|$ for each $t$. Thus, Eq. (10-105) has been verified, and so we conclude that the fuel-optimal controls must belong to the class defined by $\mathbf{u}_B^*(t)$.

The reader should note that the proof we presented for Control Law 10-2*b* is very similar to the technique we used to obtain the fuel-optimal controls for the integrator plant in Sec. 8-2.

We next turn our attention to Problem 10-2*c*, which involves the minimization of the cost $J_C = \int_0^T [k + \|\mathbf{u}(t)\|]\, dt$. The cost $J_C$ can be interpreted as a linear combination of elapsed time and consumed fuel. We shall show† that the control $\mathbf{u}_C^*(t)$ which minimizes the cost $J_C$ is the time-optimal control $\mathbf{u}_A^*(t)$. More precisely, we have:

***Control Law 10-2c Solution to Problem 10-2c*** *Let $\mathbf{u}_C^*(t)$ be the control that minimizes the cost $J_C$ given by Eq.* (10-79). *We let $J_C^*$ denote the minimum value of $J_C$. Then the unique optimal control $\mathbf{u}_C^*(t)$ is given by*

$$\mathbf{u}_C^*(t) = -m\frac{\mathbf{x}_C^*(t)}{\|\mathbf{x}_C^*(t)\|} \tag{10-112}$$

*and the minimum cost $J_C^*$ is given by*

$$J_C^* = \|\boldsymbol{\xi}\|\left(1 + \frac{k}{m}\right) \tag{10-113}$$

*Equation* (10-112) *means that the optimal control $\mathbf{u}_C^*(t)$ is identical to the time-optimal control $\mathbf{u}_A^*(t)$ [compare Eqs.* (10-113) *and* (10-84)].

PROOF From Eq. (10-96), we know that

$$\int_0^T \|\mathbf{u}(t)\|\, dt \geq \|\boldsymbol{\xi}\| \tag{10-114}$$

† We arrived at the same conclusion in Sec. 8-2.

Since $t^* = \|\boldsymbol{\xi}\|/m$ is the minimum time, we also have (since $k > 0$)

$$kT \geq kt^* = \frac{k\|\boldsymbol{\xi}\|}{m} \tag{10-115}$$

From Eqs. (10-114) and (10-115), we find that

$$\begin{aligned} J_C &= \int_0^T [k + \|\mathbf{u}(t)\|]\, dt = kT + \int_0^T \|\mathbf{u}(t)\|\, dt \\ &\geq \frac{k\|\boldsymbol{\xi}\|}{m} + \|\boldsymbol{\xi}\| = \|\boldsymbol{\xi}\|\left(1 + \frac{k}{m}\right) \end{aligned} \tag{10-116}$$

Equation (10-116) means that the quantity $\|\boldsymbol{\xi}\|(1 + k/m)$ is the greatest lower bound of the cost $J_C$. We shall show first, by direct substitution, that the control $\mathbf{u}_C^*(t)$ given by Eq. (10-112) yields this greatest lower bound. Since $\|\mathbf{u}_C^*(t)\| = m$, we deduce that

$$J_C^* = \int_0^T [k + \|\mathbf{u}_C^*(t)\|]\, dt = \int_0^{t^* = \|\boldsymbol{\xi}\|/m} \{k + m\}\, dt = \|\boldsymbol{\xi}\|\left(1 + \frac{k}{m}\right) \tag{10-117}$$

Next, we must show that the optimal control $\mathbf{u}_C^*(t)$ is unique. To do this, we can consider the control $\hat{\mathbf{u}}(t)$ given by Eqs. (10-103), i.e., the control which we have used in the proof of Control Law 10-2*b*. Clearly, the response time $\hat{T}_\beta$ required by $\hat{\mathbf{u}}(t)$ is such that $\hat{T}_\beta > t^* = \|\boldsymbol{\xi}\|/m$. In addition, we have shown that $\int_0^{\hat{T}_\beta} \|\hat{\mathbf{u}}(t)\|\, dt \geq \|\boldsymbol{\xi}\|$. We can then immediately conclude that

$$\int_0^{\hat{T}_\beta} [k + \|\hat{\mathbf{u}}(t)\|]\, dt > \|\boldsymbol{\xi}\|\left(1 + \frac{k}{m}\right) = J_C^* \tag{10-118}$$

and so, since the class of controls $\hat{\mathbf{u}}(t)$ is identical to the class of controls with the property $\|\mathbf{u}(t)\| \leq m$, the uniqueness of $\mathbf{u}_C^*(t)$ is established.

Next, we shall obtain the solution to Problem 10-2*d*. We shall call the optimal control which minimizes the cost $J_D$ the *energy-optimal control.* We now state the following control law:

***Control Law 10-2d Solution to Problem 10-2d*** *Let $\mathbf{u}_D^*(t)$ be the energy-optimal control, i.e., the control that minimizes the cost $J_D = \frac{1}{2}\int_0^T \|\mathbf{u}(t)\|^2\, dt$, with $T$ fixed. We let $J_D^*$ denote the minimum value of $J_D$. Then the unique energy-optimal control $\mathbf{u}_D^*(t)$ is given by the equation*

$$\mathbf{u}_D^*(t) = -\frac{\|\boldsymbol{\xi}\|}{T}\frac{\mathbf{x}_D^*(t)}{\|\mathbf{x}_D^*(t)\|} \tag{10-119}$$

*provided that the specified response time $T$ is such that $T \geq t^* = \|\boldsymbol{\xi}\|/m$ (which is necessary for the problem to have a solution).*

*For this problem, the minimum cost $J_D^*$ is given by the relation*

$$J_D^* = \frac{\|\boldsymbol{\xi}\|^2}{2T} \tag{10-120}$$

PROOF From the proof of Control Law 10-2*b*, we know that

$$\|\boldsymbol{\xi}\| \leq \int_0^T \|\mathbf{u}(t)\| \, dt \tag{10-121}$$

Now let us recall† the following inequality: For any two piecewise continuous functions $a(t)$ and $b(t)$, the relation

$$\left[ \int_0^T a(t)b(t) \, dt \right]^2 \leq \left[ \int_0^T a^2(t) \, dt \right] \left[ \int_0^T b^2(t) \, dt \right] \tag{10-122}$$

is true. If we let

$$a(t) = \|\mathbf{u}(t)\| \qquad b(t) = 1 \tag{10-123}$$

then we find from Eq. (10-122) that

$$\left[ \int_0^T \|\mathbf{u}(t)\| \, dt \right]^2 \leq T \left[ \int_0^T \|\mathbf{u}(t)\|^2 \, dt \right] \tag{10-124}$$

From Eqs. (10-124) and (10-121), we immediately conclude that

$$J_D = \tfrac{1}{2} \int_0^T \|\mathbf{u}(t)\|^2 \, dt \geq \frac{\|\boldsymbol{\xi}\|^2}{2T} \tag{10-125}$$

It is trivial to show by direct substitution that the control $\mathbf{u}_D^*(t)$ given by Eq. (10-119) has the property

$$\tfrac{1}{2} \int_0^T \|\mathbf{u}_D^*(t)\|^2 \, dt = \tfrac{1}{2} \int_0^T \frac{\|\boldsymbol{\xi}\|^2}{T^2} \, dt = \frac{\|\boldsymbol{\xi}\|^2}{2T} \tag{10-126}$$

This means that the control $\mathbf{u}_D^*(t)$ yields the greatest lower bound $\|\boldsymbol{\xi}\|^2/2T$ of $J_D$, and so $\mathbf{u}_D^*(t)$ is energy-optimal.

Now we must show that $\mathbf{u}_D^*(t)$ is unique. To prove uniqueness, we first note that the inequality (10-124) is a strict equality if and only if $\|\mathbf{u}(t)\|$ is a constant for all $t \in [0, T]$. It follows that, if $\|\hat{\mathbf{u}}(t)\| \neq$ const for $t \in [0, T]$, then we have the strict inequality

$$T \int_0^T \|\hat{\mathbf{u}}(t)\|^2 \, dt > \left[ \int_0^T \|\hat{\mathbf{u}}(t)\| \, dt \right]^2$$

which, in turn, implies that

$$\tfrac{1}{2} \int_0^T \|\hat{\mathbf{u}}(t)\|^2 \, dt > \frac{\|\boldsymbol{\xi}\|^2}{2T} \tag{10-127}$$

and so $\hat{\mathbf{u}}(t)$ cannot be optimal. We leave the trivial case where $\|\hat{\mathbf{u}}(t)\|$ is constant to the reader. Thus, the uniqueness is established.

Since the energy-optimal control $\mathbf{u}_D^*(t)$ is a function of the initial state $\boldsymbol{\xi}$ and of the time $T$ and since the minimum energy $J_D^* = \|\boldsymbol{\xi}\|^2/2T$ decreases as $T$ increases, we look for the optimal control that minimizes a linear combination of the elapsed time and of the expended energy (Problem

† See Sec. 3-15, Eq. (3-148).

10-2e). We claim that the control that minimizes the cost $J_E$ given by Eq. (10-81) is provided by the following control law:

***Control Law 10-2e Solution to Problem 10-2e*** *Let* $\mathbf{u}_E^*(t)$ *be the control that minimizes the cost*

$$J_E = \int_0^T [k + \tfrac{1}{2}\|\mathbf{u}(t)\|^2]\, dt$$

*Let* $J_E^*$ *denote the minimum value of* $J_E^*$. *Then, given* $k$, *the control* $\mathbf{u}_E^*(t)$ *is unique and is given by*

$$\mathbf{u}_E^*(t) = -m \frac{\mathbf{x}_E^*(t)}{\|\mathbf{x}_E^*(t)\|} \qquad \text{if } k \geq \frac{m^2}{2} \tag{10-128}$$

$$\mathbf{u}_E^*(t) = -\sqrt{2k} \frac{\mathbf{x}_E^*(t)}{\|\mathbf{x}_E^*(t)\|} \qquad \text{if } k < \frac{m^2}{2} \tag{10-129}$$

*The minimum cost* $J_E^*$ *is given by*

$$J_E^* = \frac{\|\boldsymbol{\xi}\|}{m} (k + \tfrac{1}{2}m^2) \qquad \text{if } k \geq \frac{m^2}{2} \tag{10-130}$$

$$J_E^* = \sqrt{2k}\, \|\boldsymbol{\xi}\| \qquad \text{if } k < \frac{m^2}{2} \tag{10-131}$$

*The response time required to force* $\boldsymbol{\xi}$ *to* $\mathbf{0}$ *is given by*

$$T = t^* = \frac{\|\boldsymbol{\xi}\|}{m} \qquad \text{if } k \geq \frac{m^2}{2} \tag{10-132}$$

$$T = \frac{\|\boldsymbol{\xi}\|}{\sqrt{2k}} \qquad \text{if } k < \frac{m^2}{2} \tag{10-133}$$

**Exercise 10-5** Prove Control Law 10-2e. HINT: First, prove that the optimal control must have constant magnitude, using Eqs. (10-121) and (10-124) and using the fact that Eq. (10-124) is an equality if and only if $\|\mathbf{u}(t)\| = \text{const.}$ Second, prove that the optimal control must be of the form $-\alpha(\mathbf{x}(t)/\|\mathbf{x}(t)\|)$ and determine $J_E(\alpha)$. Set $\partial J_E(\alpha)/\partial\alpha = 0$ to find that $\alpha = \sqrt{2k}$ or $\alpha = m$, depending on the magnitude of $k$. See also Sec. 6-24.

We hope that the reader has found these "inequality-type" proofs interesting. Our purpose was to demonstrate that, in special cases, optimal-control laws can be derived without having to use the minimum principle. The proofs which use the tools of functional analysis (such as the Schwarz inequality) have quite often the advantage of not obscuring the physical aspects of the problems, as is often the case with the introduction of the costate variables.

We conclude this section with an example of a physical system which is norm-invariant.

**Example 10-5**† Consider a nonsymmetric body in space. This body may be a satellite or any other space vehicle. Define three principal axes 1, 2, 3 through the center of

† We have considered such a physical system in Example 7-4 and in Sec. 7-9.

mass of this body. Assume that these axes are fixed on the body. Let $I_1$, $I_2$, and $I_3$ be the three principal moments of inertia of the body about the axes 1, 2, 3, respectively. As indicated in Fig. 10-3, let $y_1(t)$, $y_2(t)$, and $y_3(t)$ be the three angular velocities about the axes 1, 2, 3, respectively. We would like to emphasize that this is a body-fixed

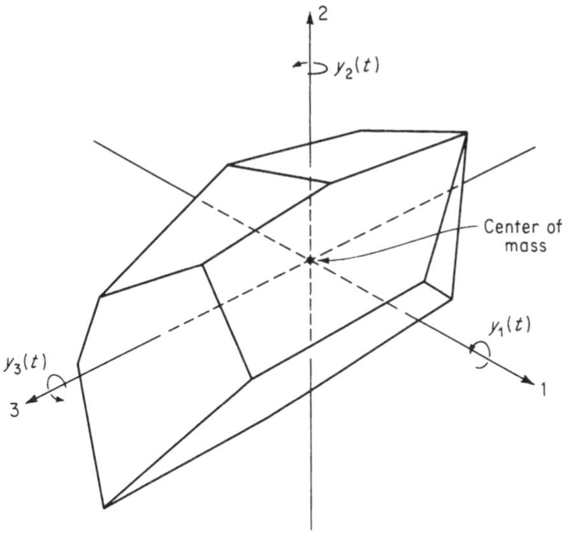


**Fig. 10-3** An asymmetrical body with the three body axes 1, 2, 3 through the center of mass.

coordinate system and that all the measurements are taken with respect to this body-fixed coordinate system. In the absence of any external torques on the body, the equations of motion are†

$$\begin{aligned} I_1\dot{y}_1(t) &= (I_2 - I_3)y_2(t)y_3(t) \\ I_2\dot{y}_2(t) &= (I_3 - I_1)y_3(t)y_1(t) \\ I_3\dot{y}_3(t) &= (I_1 - I_2)y_1(t)y_2(t) \end{aligned} \tag{10-134}$$

The differential equations of motion are nonlinear. The terms $(I_i - I_j)y_i(t)y_j(t)$ are due to the gyroscopic coupling effect. If the body were completely symmetrical, i.e., if $I_1 = I_2 = I_3$, then the nonlinear terms would be zero and the equations of motion would be "uncoupled."

Next, we proceed to show that the system (10-134) is norm-invariant according to Definition 10-1. At first glance, it may appear that this is not so, because if we compute the rate of change of the magnitude of the velocity vector $\mathbf{y}(t)$, we find that

$$\begin{aligned} \frac{d}{dt}\|\mathbf{y}(t)\| &= \frac{d}{dt}\sqrt{y_1^2(t) + y_2^2(t) + y_3^2(t)} \\ &= \left(\frac{I_2 - I_3}{I_1} + \frac{I_3 - I_1}{I_2} + \frac{I_1 - I_2}{I_3}\right)\frac{y_1(t)y_2(t)y_3(t)}{\|\mathbf{y}(t)\|} \neq 0 \end{aligned} \tag{10-135}$$

This means that the magnitude of the velocity vector changes with time. Thus, although the velocity vector $\mathbf{y}(t)$ is indeed a state vector for the system (10-134), $\mathbf{y}(t)$ is not the invariant quantity.

† The system of equations (10-134) is commonly called the Euler equations of motion. See Ref. G-7, chap. 5.

From the physical point of view, we know that the space body shown in Fig. 10-3 conserves its angular momentum. For this reason, we let $\mathbf{x}(t)$ be the *angular-momentum vector* which is defined by the transformation

$$\mathbf{x}(t) = \begin{bmatrix} x_1(t) \\ x_2(t) \\ x_3(t) \end{bmatrix} = \begin{bmatrix} I_1 & 0 & 0 \\ 0 & I_2 & 0 \\ 0 & 0 & I_3 \end{bmatrix} \begin{bmatrix} y_1(t) \\ y_2(t) \\ y_3(t) \end{bmatrix} \tag{10-136}$$

In terms of the components of the angular-momentum vector, Eqs. (10-134) reduce to

$$\begin{aligned} \dot{x}_1(t) &= \frac{I_2 - I_3}{I_2 I_3} x_2(t) x_3(t) \\ \dot{x}_2(t) &= \frac{I_3 - I_1}{I_3 I_1} x_3(t) x_1(t) \\ \dot{x}_3(t) &= \frac{I_1 - I_2}{I_1 I_2} x_1(t) x_2(t) \end{aligned} \tag{10-137}$$

Now we compute $\frac{d}{dt} \|\mathbf{x}(t)\|$, and we find that

$$\frac{d}{dt} \|\mathbf{x}(t)\| = \frac{d}{dt} \sqrt{x_1^2(t) + x_2^2(t) + x_3^2(t)} = 0 \tag{10-138}$$

and so the system (10-137) is norm-invariant according to Definition 10-1.

Equations (10-137) describe the motion of the body in the absence of any applied torques. A torque vector $\mathbf{u}(t)$ can be generated by means of gas jets, reaction wheels, gravity-gradient arrangements, and so on. At any rate, if $\mathbf{u}(t)$ is a control torque, the equations of motion become

$$\begin{aligned} \dot{x}_1(t) &= \frac{I_2 - I_3}{I_2 I_3} x_2(t) x_3(t) + u_1(t) \\ \dot{x}_2(t) &= \frac{I_3 - I_1}{I_3 I_1} x_3(t) x_1(t) + u_2(t) \\ \dot{x}_3(t) &= \frac{I_1 - I_2}{I_1 I_2} x_1(t) x_2(t) + u_3(t) \end{aligned} \tag{10-139}$$

We can immediately conclude that, if the constraints on the control torque $\mathbf{u}(t)$ are of the form

$$\|\mathbf{u}(t)\| = \sqrt{u_1^2(t) + u_2^2(t) + u_3^2(t)} \leq m \tag{10-140}$$

then all the theory we have developed in this section can be used to find the optimal torque.

We can, as a matter of fact, interpret the physical implications of the control laws we have obtained. For example, we know from Control Law 10-2*a* that the time-optimal control torque $\mathbf{u}^*(t)$ must be given by

$$\mathbf{u}^*(t) = -m \frac{\mathbf{x}^*(t)}{\|\mathbf{x}^*(t)\|} \tag{10-141}$$

This means that, *in order to reduce the angular-momentum vector* $\mathbf{x}(t)$ *to zero in the shortest possible time, the torque vector* $\mathbf{u}(t)$ *must point in the opposite direction to the angular-momentum vector* $\mathbf{x}(t)$ *and the torque* $\mathbf{u}(t)$ *must be as large as possible.* This indeed makes sense from the physical point of view.

In the next section, we shall discuss the time-optimal control of a space body with a single axis of symmetry ($I_1 = I_2 \neq I_3$), and we shall compare three different means of time-optimal control.

**Exercise 10-6** Consider the linear self-adjoint system (see Definition 10-2)

$$\begin{aligned} \dot{\mathbf{x}}(t) &= \mathbf{S}(t)\mathbf{x}(t) + \mathbf{u}(t) \qquad & \mathbf{x}(t_0) &= \boldsymbol{\xi} \\ \mathbf{S}(t) &= -\mathbf{S}'(t) & \|\mathbf{u}(t)\| &\le m \end{aligned} \tag{10-142}$$

Prove Control Laws 10-2*a* to 10-2*e* using the minimum principle exclusively. HINT: For all the proofs, use repeatedly the fact that the fundamental matrix $\mathbf{\Phi}(t;\, t_0)$ for the system (10-142) is orthogonal for all $t$. Many of the proofs can be found in Ref. A-12.

**Exercise 10-7** Suppose that the spinning body considered in Example 10-5 is subject to friction torques. Let $\mathbf{x}(t)$ be the angular-momentum vector and let $\mathbf{u}(t)$, $\|\mathbf{u}(t)\| \le m$, be the control-torque vector. If the equations of motion are

$$\begin{aligned} \dot{x}_1(t) &= -\alpha(t)x_1(t) + \frac{I_2 - I_3}{I_2 I_3} x_2(t)x_3(t) + u_1(t) \\ \dot{x}_2(t) &= -\alpha(t)x_2(t) + \frac{I_3 - I_1}{I_3 I_1} x_3(t)x_1(t) + u_2(t) \\ \dot{x}_3(t) &= -\alpha(t)x_3(t) + \frac{I_1 - I_2}{I_1 I_2} x_1(t)x_2(t) + u_3(t) \end{aligned} \tag{10-143}$$

where $\alpha(t) > 0$ is the friction coefficient, then show that the time-optimal control can be found using the methods of Sec. 10-4. What is the physical implication of the fact that the friction coefficient $\alpha(t)$ is the same for all three equations?

**Exercise 10-8** Consider the linear time-invariant self-adjoint system

$$\begin{aligned} \dot{\mathbf{x}}(t) &= \mathbf{S}\mathbf{x}(t) + \mathbf{u}(t) \qquad & \mathbf{x}(0) &= \boldsymbol{\xi} \\ \mathbf{S} &= -\mathbf{S}' & \|\mathbf{u}(t)\| &\le m \end{aligned} \tag{10-144}$$

and consider the cost functional

$$J(\mathbf{u}) = \frac{1}{2} \int_0^T [\|\mathbf{x}(t)\|^2 + r\|\mathbf{u}(t)\|^2]\, dt \qquad r > 0 \tag{10-145}$$

where $T$ is fixed and $\mathbf{x}(T)$ is not specified. Determine the optimal control. What happens as $T \to \infty$?

## 10-7 Time-optimal Velocity Control of a Rotating Body with a Single Axis of Symmetry†

In this section, we shall apply the theory developed in Secs. 7-7, 7-9, and 10-6 to the problem of controlling the velocity of a spinning space body with a single axis of symmetry. Our purpose is twofold: First, we wish to indicate the use of optimal-control theory in the preliminary stages of design, and second, we wish to explicitly show the effects upon the performance of the system of changing the constraint on the control.

In Secs. 7-7 and 7-9, we derived the equations of motion for the space body with a single axis of symmetry; we shall repeat the equations in this section for the sake of completeness. Suppose that the body described in Example 10-5 has a single axis of symmetry. Let us assume that the axis of symmetry is the 3-axis. This means that the moments of inertia $I_1$ and $I_2$ about the 1-axis and the 2-axis, respectively, are equal, i.e., that

$$I = I_1 = I_2 \neq I_3 \tag{10-146}$$

† The material in this section is based on Refs. A-5 and A-9.

If we substitute Eq. (10-146) into Eqs. (10-134), we find that the angular velocities $y_1(t)$, $y_2(t)$, and $y_3(t)$ satisfy the differential equations

$$\begin{aligned} I\dot{y}_1(t) &= (I - I_3)y_3(t)y_2(t) \\ I\dot{y}_2(t) &= (I_3 - I)y_3(t)y_1(t) \\ I_3\dot{y}_3(t) &= 0 \end{aligned} \tag{10-147}$$

Now let us suppose that by some means we generate a torque vector $\mathbf{u}(t)$, so that the equations of motion become

$$\begin{aligned} I\dot{y}_1(t) &= (I - I_3)y_3(t)y_2(t) + u_1(t) \\ I\dot{y}_2(t) &= (I_3 - I)y_3(t)y_1(t) + u_2(t) \\ I_3\dot{y}_3(t) &= u_3(t) \end{aligned} \tag{10-148}$$

We are now faced with the task of generating the torques by some system. We can generate torques by gas jets, reaction wheels, etc. Let us suppose that we have decided to use gas jets. The next problem that we face is: *Where should we place the gas jets on the body?* If we examine Eqs. (10-148), we observe that the "spin" velocity $y_3(t)$ (which is the angular velocity of the body about its axis of symmetry) is only a function of the torque component $u_3(t)$ because

$$y_3(t) = y_3(0) + \frac{1}{I_3}\int_0^t u_3(\tau)\,d\tau \tag{10-149}$$

Clearly, the velocity $y_3(t)$ is not a function of $y_1(t)$ and $y_2(t)$; however, $y_1(t)$ and $y_2(t)$ do depend on $y_3(t)$. It is often desirable† to control the spin velocity $y_3(t)$ independent of the other velocities. For this reason, let us agree that: *A jet $J_3$, which generates a thrust $\mathbf{f}_3(t)$, will be placed on the body in such a manner that the torque vector $\mathbf{u}_3(t)$ produced by the thrust $\mathbf{f}_3(t)$ is of the form*

$$\mathbf{u}_3(t) = \begin{bmatrix} 0 \\ 0 \\ u_3(t) \end{bmatrix} \tag{10-150}$$

*In other words, the jet $J_3$ must generate a torque about the symmetry axis only; this implies that the thrust $\mathbf{f}_3(t)$ must be in a plane perpendicular to the symmetry axis (the 3-axis).*

Next, we consider the problem of generating the torques $u_1(t)$ and $u_2(t)$. We shall propose three different schemes of generating these two torques. Later on in this section, we shall compare the three schemes and draw some interesting conclusions.

† Suppose that the space body is a manned space station; then we can establish a constant artificial gravity by maintaining a constant spin velocity and keeping the other velocities $y_1(t)$ and $y_2(t)$ at zero so that the space station will not tumble. If the space body is a reentry vehicle, then a spin velocity about the symmetry axis is desirable because it increases aerodynamic stability.

In the following definitions, the components of a vector are measured along the 1-, 2-, 3-axes of the body. The axes 1, 2, 3 define a right-hand coordinate system, so that all the rules of establishing directions in a right-hand coordinate system are followed.

*Scheme A Single Fixed Jet*

A single gas jet $J^A$ is fixed on the body (see Fig. 10-4*a*) so that the thrust vector $\mathbf{f}^A(t)$ has components

$$\mathbf{f}^A(t) = \begin{bmatrix} f_1{}^A(t) \\ 0 \\ 0 \end{bmatrix} \tag{10-151}$$

and produces a torque vector $\mathbf{u}^A(t)$ with components

$$\mathbf{u}^A(t) = \begin{bmatrix} 0 \\ u_2{}^A(t) \\ 0 \end{bmatrix} = \begin{bmatrix} 0 \\ \alpha f_1{}^A(t) \\ 0 \end{bmatrix} \tag{10-152}$$

Thus, the thrust $\mathbf{f}^A(t)$ generates a torque $u_2{}^A(t) = \alpha f_1{}^A(t)$ about the 2-axis only. We also assume that the thrust is constrained in magnitude by the relation

$$\|\mathbf{f}^A(t)\| = |f_1{}^A(t)| \le F^A \tag{10-153}$$

*Scheme B Two Fixed Jets*

Two gas jets $J^{B1}$ and $J^{B2}$ are fixed on the body (see Fig. 10-4*b*) so that their respective thrust vectors $\mathbf{f}^{B1}(t)$ and $\mathbf{f}^{B2}(t)$ have components

$$\mathbf{f}^{B1}(t) = \begin{bmatrix} f_1{}^{B1}(t) \\ 0 \\ 0 \end{bmatrix} \qquad \mathbf{f}^{B2}(t) = \begin{bmatrix} 0 \\ f_2{}^{B2}(t) \\ 0 \end{bmatrix} \tag{10-154}$$

and produce torque vectors $\mathbf{u}^{B1}(t)$ and $\mathbf{u}^{B2}(t)$ with components

$$\mathbf{u}^{B1}(t) = \begin{bmatrix} 0 \\ u_2{}^{B1}(t) \\ 0 \end{bmatrix} = \begin{bmatrix} 0 \\ \alpha f_1{}^{B1}(t) \\ 0 \end{bmatrix} \qquad \mathbf{u}^{B2}(t) = \begin{bmatrix} u_1{}^{B2}(t) \\ 0 \\ 0 \end{bmatrix} = \begin{bmatrix} \alpha f_2{}^{B2}(t) \\ 0 \\ 0 \end{bmatrix} \tag{10-155}$$

Thus, the thrust $\mathbf{f}^{B1}(t)$ generates a torque $u_2{}^{B1}(t) = \alpha f_1{}^{B1}(t)$ about the 2-axis only, and the thrust $\mathbf{f}^{B2}(t)$ generates a torque $u_1{}^{B2}(t) = \alpha f_2{}^{B2}(t)$ about the 1-axis only. We also assume that the thrusts are constrained in magnitude by the relations

$$\begin{aligned} \|\mathbf{f}^{B1}(t)\| &= |f_1{}^{B1}(t)| \le F^B \\ \|\mathbf{f}^{B2}(t)\| &= |f_2{}^{B2}(t)| \le F^B \end{aligned} \tag{10-156}$$

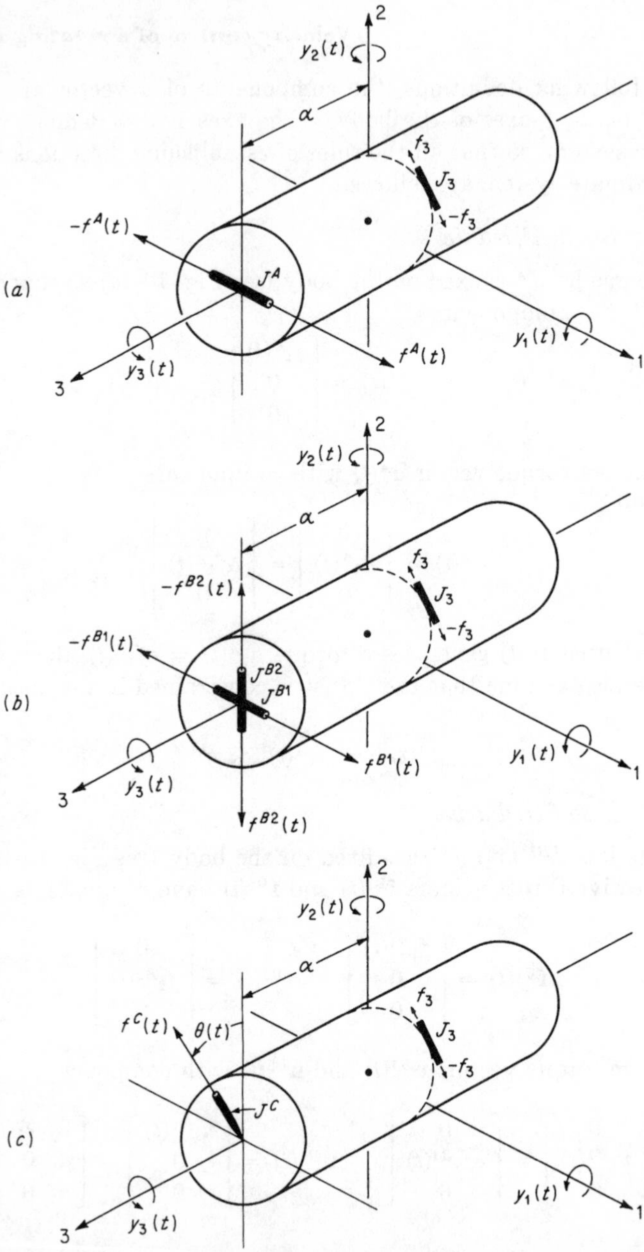


**Fig. 10-4** (*a*) The physical arrangement of the jets on the body for scheme $A$. The jet $J^A$ is fixed on the body, and it can "fire" in both directions so as to generate a torque about the 2-axis only. (*b*) The physical arrangement of the jets on the body for scheme $B$. The jets $J^{B1}$ and $J^{B2}$ are fixed on the body. The jet $J^{B1}$ generates a torque about the 2-axis only, and the jet $J^{B2}$ generates a torque about the 1-axis only. (*c*) The physical arrangement of the jets on the body for scheme $C$. The jet $J^C$ is gimbaled and is allowed to rotate in a plane parallel to the 1-2 plane. The angle $\theta(t)$ specifies the direction of the thrust vector $\mathbf{f}^C(t)$.

*Scheme C Single Gimbaled Jet*

A single jet $J^C$ is placed on the body (see Fig. 10-4c) and is allowed to rotate in a plane perpendicular to the 3-axis. The thrust vector $\mathbf{f}^C(t)$ produced by the jet $J^C$ has components

$$\mathbf{f}^C(t) = \begin{bmatrix} f_1{}^C(t) \\ f_2{}^C(t) \end{bmatrix} \tag{10-157}$$

The torque vector $\mathbf{u}^C(t)$ has components

$$\mathbf{u}_C(t) = \begin{bmatrix} u_1{}^C(t) \\ u_2{}^C(t) \end{bmatrix} = \begin{bmatrix} \alpha f_2{}^C(t) \\ \alpha f_1{}^C(t) \end{bmatrix} \tag{10-158}$$

Thus, the gimbaled jet can produce torques about the 1-axis and the 2-axis simultaneously. We assume that the thrust vector $\mathbf{f}^C(t)$ is constrained in magnitude by the relation

$$\|\mathbf{f}^C(t)\| = \sqrt{f_1{}^C(t)^2 + f_2{}^C(t)^2} \leq F^C \tag{10-159}$$

Let us now discuss the control problem from a physical point of view. Suppose that, prior to $t = 0$, we have "fired" the jet $J_3$ in order to bring the spin velocity to a desired value—say $\hat{y}_3 \neq 0$—so that at $t = 0$ we have

$$y_3(0) = \hat{y}_3 \tag{10-160}$$

Owing to the gyroscopic coupling we shall have, in general, some nonzero initial values $y_1(0)$ and $y_2(0)$ for the other two angular velocities. In the absence of any external torques for $t > 0$, the body will tumble in space. It is then desired to determine the control torque that will reduce the velocities $y_1(t)$ and $y_2(t)$ to zero; in other words, it is desired to reduce the tumbling motion of the body to a pure spinning motion (at the desired spin velocity $\hat{y}_3$). In order to have a meaningful optimal problem, we must have a performance criterion. We shall consider the *minimum-time* criterion. Thus, the physical problem is: *Reduce the initial velocities $y_1(0)$ and $y_2(0)$ to zero in minimum time and maintain the spin velocity $y_3(t)$ at its desired value $\hat{y}_3$*, that is,

$$y_3(t) = y_3(0) = \hat{y}_3 = \text{const} \qquad \text{for all } t > 0 \tag{10-161}$$

If we substitute $y_3(t) = \hat{y}_3 = \text{const}$ into Eqs. (10-148), then we can determine the differential equations of the velocities $y_1(t)$ and $y_2(t)$. It is more convenient, however, to deal with angular momenta; for this reason, let us define the *angular momenta* $x_1(t)$ and $x_2(t)$ by the equations

$$x_1(t) = Iy_1(t) \qquad x_2(t) = Iy_2(t) \tag{10-162}$$

From Eqs. (10-162), (10-161), and (10-148), we find that the angular momenta satisfy the equations

$$\begin{aligned} \dot{x}_1(t) &= \omega x_2(t) + u_1(t) \\ \dot{x}_2(t) &= -\omega x_1(t) + u_2(t) \end{aligned} \tag{10-163}$$

where $\omega$ is a constant given by

$$\omega = \frac{I - I_3}{I} \hat{y}_3 \qquad \text{radians/second} \tag{10-164}$$

The *two*-dimensional angular-momentum vector $\mathbf{x}(t) = \begin{bmatrix} x_1(t) \\ x_2(t) \end{bmatrix}$ is the projection of the three-dimensional angular-momentum vector

$$\begin{bmatrix} x_1(t) \\ x_2(t) \\ I_3\hat{y}_3 \end{bmatrix}$$

on the 1-2 plane. The *two*-dimensional torque vector $\mathbf{u}(t) = \begin{bmatrix} u_1(t) \\ u_2(t) \end{bmatrix}$ is the projection of the three-dimensional torque vector

$$\begin{bmatrix} u_1(t) \\ u_2(t) \\ 0 \end{bmatrix}$$

on the 1-2 plane. As we mentioned before, *we wish to find the control vector* $\mathbf{u}(t)$ *that will drive the angular-momentum vector* $\mathbf{x}(t)$ *to* $\mathbf{0}$ *in minimum time.*

From the definitions of the three schemes of control, we conclude that we must solve for the time-optimal control to the origin for the following systems and constraints (we omit all the superscripts to avoid confusion):

*Scheme A Single Fixed Jet*

$$\begin{aligned} \dot{x}_1(t) &= \omega x_2(t) \\ \dot{x}_2(t) &= -\omega x_1(t) + u_2(t) \qquad |u_2(t)| \le F^A\alpha \end{aligned} \tag{10-165}$$

*Scheme B Two Fixed Jets*

$$\begin{aligned} \dot{x}_1(t) &= \omega x_2(t) + u_1(t) \qquad |u_1(t)| \le F^B\alpha \\ \dot{x}_2(t) &= -\omega x_1(t) + u_2(t) \qquad |u_2(t)| \le F^B\alpha \end{aligned} \tag{10-166}$$

*Scheme C Single Gimbaled Jet*

$$\begin{aligned} \dot{x}_1(t) &= \omega x_2(t) + u_1(t) \\ \dot{x}_2(t) &= -\omega x_1(t) + u_2(t) \qquad \sqrt{u_1^2(t) + u_2^2(t)} \le F^C\alpha \end{aligned} \tag{10-167}$$

We derived the time-optimal control for the system (10-165) in Sec. 7-7. We derived the time-optimal control for the system (10-166) in Sec. 7-9. We derived the time-optimal control for the system (10-167) (which is self-adjoint and norm-invariant) in Sec. 10-6. Our objective is to compare the three time-optimal solutions.

Any comparison of different schemes of control must be done fairly. If we examine Eqs. (10-165) to (10-167), we see that the controlled system is the same but the control constraints on the torque vector $\mathbf{u}(t)$ are different. In order to obtain a fair comparison, we must determine some relation between $F^A$, $F^B$, and $F^C$ (which are the bounds on the magnitude of the thrusts generated by the gas jets). We shall establish such a relation among $F^A$, $F^B$, and $F^C$ by considering the rate of flow of fuel.

We assumed in Chap. 8 that *the rate of flow of fuel $\varphi$ through a gas jet is proportional to the magnitude of the generated thrust produced by that jet.* If we let $\beta$ be a positive constant of proportionality, then we deduce that for the three proposed schemes, we have the relations [see Eqs. (10-153), (10-156), and (10-159)]

$$\varphi^A = \beta\|\mathbf{f}^A(t)\| = \beta|f_1{}^A(t)| \leq \beta F^A \tag{10-168}$$

$$\begin{aligned}\varphi^B &= \beta[\|\mathbf{f}^{B1}(t)\| + \|\mathbf{f}^{B2}(t)\|] \\ &= \beta[|f_1{}^{B1}(t)| + |f_2{}^{B2}(t)|] \leq 2\beta F^B\end{aligned} \tag{10-169}$$

$$\varphi^C = \beta\|\mathbf{f}^C(t)\| = \beta\sqrt{f_1{}^C(t)^2 + f_2{}^C(t)^2} \leq \beta F^C \tag{10-170}$$

Since, for time-optimal control, the jets must produce the maximum thrust for all $t$, we conclude that the rate of flow of fuel will attain its upper bound and, therefore, that

$$\varphi^A = \beta F^A \qquad \varphi^B = 2\beta F^B \qquad \varphi^C = \beta F^C \tag{10-171}$$

A fair way of comparing the three schemes of control is to demand that *the rate of flow of fuel be the same for each scheme,* i.e., that

$$\varphi^A = \varphi^B = \varphi^C \tag{10-172}$$

which, in turn, yields the required relation

$$F \triangleq F^A = 2F^B = F^C \tag{10-173}$$

Using Eq. (10-173), we can specify the constraint set of the torque vector $\mathbf{u}(t)$ for each scheme of control. The constraint sets are

$$\left.\begin{aligned}
&\textit{Scheme A}\\
&\qquad \mathbf{u}(t) \in \Omega_A \qquad \Omega_A = \{\mathbf{u}(t);\ u_1(t) = 0,\ |u_2(t)| \leq \alpha F\}\\
&\textit{Scheme B}\\
&\qquad \mathbf{u}(t) \in \Omega_B \qquad \Omega_B = \left\{\mathbf{u}(t):\ |u_1(t)| \leq \frac{\alpha F}{2},\ |u_2(t)| \leq \frac{\alpha F}{2}\right\}\\
&\textit{Scheme C}\\
&\qquad \mathbf{u}(t) \in \Omega_C \qquad \Omega_C = \{\mathbf{u}(t):\ \sqrt{u_1{}^2(t) + u_2{}^2(t)} \leq \alpha F\}
\end{aligned}\right\} \tag{10-174}$$

The sets $\Omega_A$, $\Omega_B$, and $\Omega_C$ are shown in Fig. 10-5. *Since the sets $\Omega_A$ and $\Omega_B$ are both subsets of $\Omega_C$, we should expect that the time-optimal control of scheme C will be "better" than the time-optimal control of schemes A and B.* By the word "better," we mean that given an initial state $\mathbf{x}(0)$, that is, an initial angular momentum, then the minimum time required by the gimbaled-jet scheme [to reduce $\mathbf{x}(0)$ to $\mathbf{0}$] is smaller than the minimum time required by either the single-fixed-jet or the two-fixed-jet scheme. On the other hand, *since the set $\Omega_A$ is not contained in $\Omega_B$, we cannot say that the two-fixed-jet scheme will always be "better" than the single-fixed-jet scheme; it may be reasonable to expect, however, that scheme B will be "better" than scheme A for most of the initial states, because the set $\Omega_B$ "contains" a large portion of $\Omega_A$.*

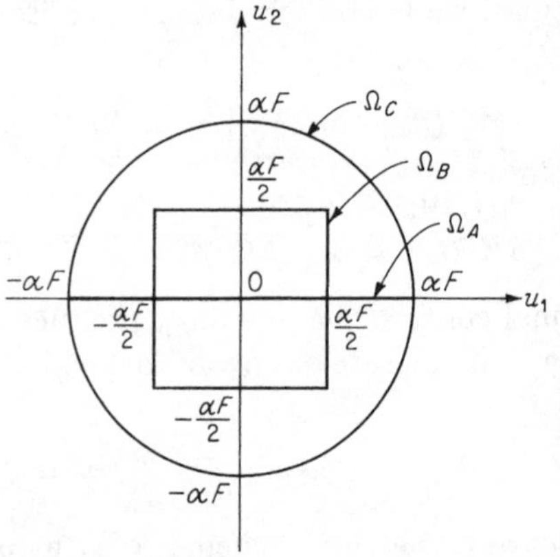


**Fig. 10-5** The three constraint sets $\Omega_A$, $\Omega_B$, and $\Omega_C$.

We shall verify these intuitive arguments by actually plotting some *minimum isochrones* for each scheme. For the various plots, we shall use the value

$$\alpha F = 2 \tag{10-175}$$

First, we state the time-optimal control laws for each scheme (assuming $\alpha F = 2$).

*Scheme A*†

The time-optimal torque vector $\mathbf{u}(t)$ [see Eqs. (10-165)] is given by

$$\begin{aligned} u_1(t) &= 0 \\ u_2(t) &= \begin{cases} -2 & \text{if } (\omega x_1, \omega x_2) \in R_- \\ +2 & \text{if } (\omega x_1, \omega x_2) \in R_+ \end{cases} \end{aligned} \tag{10-176}$$

where $R_-$ is the set of points above the $\gamma$ switch curve and $R_+$ is the set of points below the $\gamma$ switch curve, as shown in Fig. 10-6*a*.

† See Fig. 7-41.

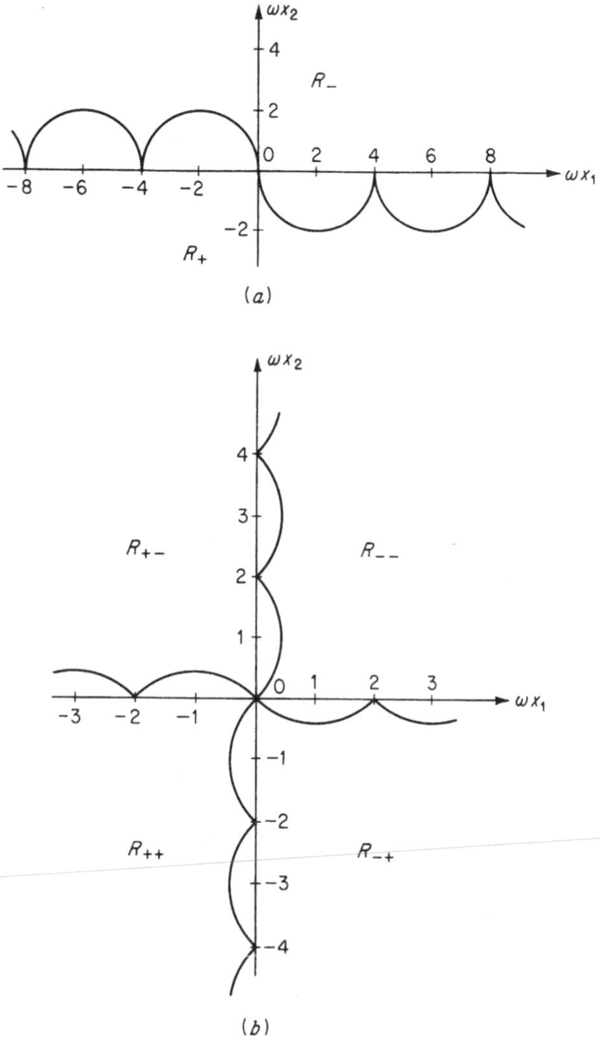


**Fig. 10-6** (*a*) The sets $R_-$ and $R_+$. (*b*) The sets $R_{++}$, $R_{+-}$, $R_{-+}$, and $R_{--}$.

*Scheme B*†

The time-optimal torque vector $\mathbf{u}(t)$ [see Eqs. (10-166)] is given by

$$
\begin{aligned}
u_1(t) &= \begin{cases} -1 & \text{if } (\omega x_1,\ \omega x_2) \in R_{-+} \cup R_{--} \\ +1 & \text{if } (\omega x_1,\ \omega x_2) \in R_{+-} \cup R_{++} \end{cases} \\
u_2(t) &= \begin{cases} -1 & \text{if } (\omega x_1,\ \omega x_2) \in R_{+-} \cup R_{--} \\ +1 & \text{if } (\omega x_1,\ \omega x_2) \in R_{-+} \cup R_{++} \end{cases}
\end{aligned}
\qquad (10\text{-}177)
$$

where the sets $R_{-+}$, $R_{+-}$, $R_{++}$, and $R_{--}$ are shown in Fig. 10-6*b*.

† See Fig. 7-60.

*Scheme C*†

The time-optimal torque vector $\mathbf{u}(t)$ [see Eqs. (10-167)] is given by

$$\mathbf{u}(t) = -2\frac{\mathbf{x}(t)}{\|\mathbf{x}(t)\|} = -2\frac{\omega\mathbf{x}(t)}{\|\omega\mathbf{x}(t)\|} \tag{10-178}$$

or, equivalently, by

$$\begin{aligned} u_1(t) &= -\frac{2x_1(t)}{\sqrt{x_1^2(t) + x_2^2(t)}} \\ u_2(t) &= -\frac{2x_2(t)}{\sqrt{x_1^2(t) + x_2^2(t)}} \end{aligned} \tag{10-179}$$

If we let $\theta(t)$ be the angle between the jet $J^C$ (see Fig. 10-4*c*) and the 2-axis, then from Eqs. (10-157), (10-158), and (10-179) we deduce that

$$\theta(t) = \tan^{-1}\frac{x_2(t)}{x_1(t)} = \tan^{-1}\frac{y_2(t)}{y_1(t)} \tag{10-180}$$

The minimum time $t_C^*$ required by the control (10-178) to force a state $(\omega x_1, \omega x_2)$ to (0, 0) is given by [see Eq. (10-85)] the relation

$$t_C^* = \frac{1}{\alpha F}\sqrt{(\omega x_1)^2 + (\omega x_2)^2} = \tfrac{1}{2}\sqrt{(\omega x_1)^2 + (\omega x_2)^2} \tag{10-181}$$

Equation (10-181) implies that the minimum isochrones are circles of radius $2t_C^*$ in the $\omega x_1$-$\omega x_2$ plane.

We shall now plot some minimum isochrones for each scheme and compare them in order to decide which scheme is "better."

Let $\hat{S}_A(t^*)$, $\hat{S}_B(t^*)$, $\hat{S}_C(t^*)$ denote the set of states that can be forced to the origin in time less than or equal to $t^*$ by the three time-optimal schemes $A$, $B$, and $C$. Let $S_A(t^*)$, $S_B(t^*)$, $S_C(t^*)$ be the boundaries of $\hat{S}_A(t^*)$, $\hat{S}_B(t^*)$, $\hat{S}_C(t^*)$, respectively. The curves $S_A(t^*)$, $S_B(t^*)$, and $S_C(t^*)$ are the minimum isochrones. Figure 10-7 shows the minimum isochrones $S_A(\pi/4\omega)$, $S_B(\pi/4\omega)$, and $S_C(\pi/4\omega)$; Fig. 10-8 shows the minimum isochrones $S_A(\pi/2\omega)$, $S_B(\pi/2\omega)$, $S_C(\pi/2\omega)$. We observe that

$$\begin{aligned} \hat{S}_A\left(\frac{\pi}{4\omega}\right) &\subset \hat{S}_C\left(\frac{\pi}{4\omega}\right) \\ \hat{S}_B\left(\frac{\pi}{4\omega}\right) &\subset \hat{S}_C\left(\frac{\pi}{4\omega}\right) \end{aligned} \tag{10-182}$$

and that

$$\begin{aligned} \hat{S}_A\left(\frac{\pi}{2\omega}\right) &\subset \hat{S}_C\left(\frac{\pi}{2\omega}\right) \\ \hat{S}_B\left(\frac{\pi}{2\omega}\right) &\subset \hat{S}_C\left(\frac{\pi}{2\omega}\right) \end{aligned} \tag{10-183}$$

† See Eq. (10-84).

If we plot all the minimum isochrones, we shall find that

$$\hat{S}_A(t^*) \subset \hat{S}_C(t^*) \qquad \hat{S}_B(t^*) \subset \hat{S}_C(t^*) \tag{10-184}$$

for all $t^* > 0$. Equation (10-184) implies that, given any state $(\omega x_1, \omega x_2)$, the minimum times $t_A^*$, $t_B^*$, and $t_C^*$ required by the schemes $A$, $B$, and $C$ to

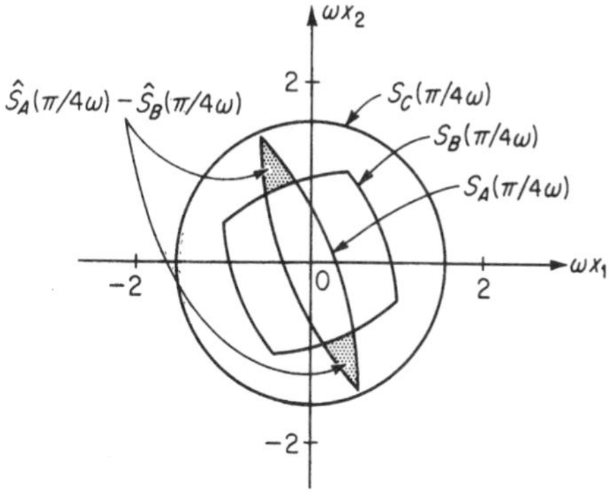


**Fig. 10-7** The minimum isochrones $S_A(t^*)$, $S_B(t^*)$, and $S_C(t^*)$ for $t^* = \pi/4\omega$.

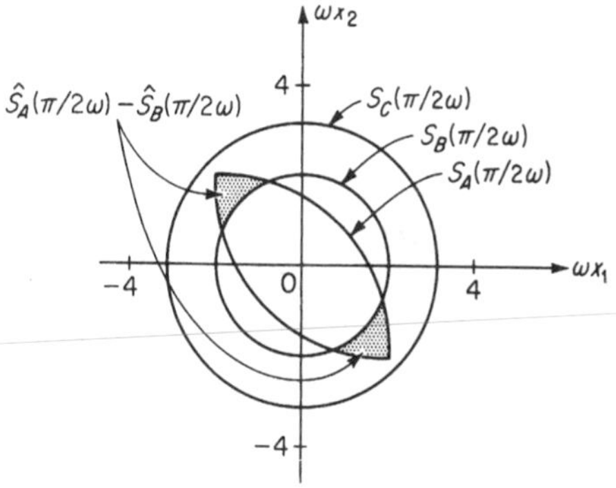


**Fig. 10-8** The minimum isochrones $S_A(t^*)$, $S_B(t^*)$, and $S_C(t^*)$ for $t^* = \pi/2\omega$.

force $(\omega x_1, \omega x_2)$ to (0, 0) satisfy the inequalities

$$t_C^* < t_A^* \qquad t_C^* < t_B^* \tag{10-185}$$

In this sense, scheme $C$ is "*better*" than either scheme $A$ or scheme $B$.

The comparison of schemes $A$ and $B$ is not as straightforward. If we examine Figs. 10-7 and 10-8, we conclude that initial states in the crosshatched regions [or, more precisely, the sets $\hat{S}_A(\pi/4\omega) - \hat{S}_B(\pi/4\omega)$ and $\hat{S}_A(\pi/2\omega) - \hat{S}_B(\pi/2\omega)$] can be forced to the origin faster by scheme $A$ than by scheme $B$. In general, *if* $(\omega x_1, \omega x_2) \in \hat{S}_A(t^*) - \hat{S}_B(t^*)$, *then scheme A is "better" than scheme B; if* $(\omega x_1, \omega x_2) \in \hat{S}_B(t^*) - \hat{S}_A(t^*)$, *then scheme B* *is "better" than scheme A.* However, the area (or measure) of $\hat{S}_A(t^*)$ is (in general) less than the area (or measure) of $\hat{S}_B(t^*)$. For this reason, we conclude that, *on the average*, scheme $B$ is "better" than scheme $A$.

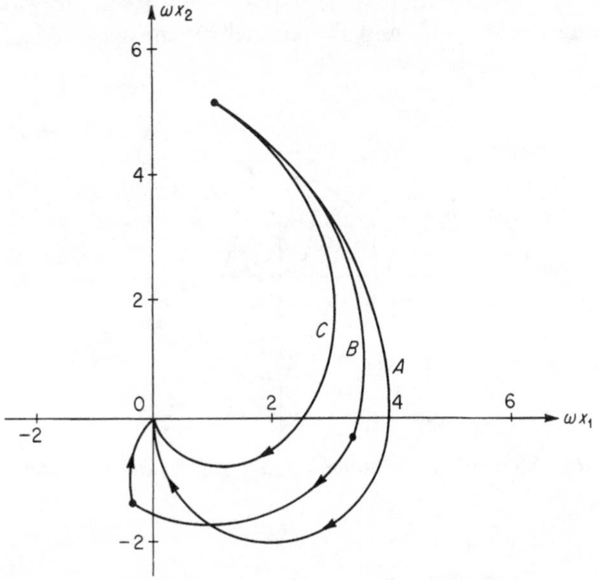


**Fig. 10-9** Three time-optimal trajectories ($A$, $B$, $C$) to the origin generated by schemes $A$, $B$, and $C$, respectively.

In Fig. 10-9, we have plotted the three time-optimal trajectories (from the same initial point to the origin) generated by each scheme. The associated minimum times $t_A^*$, $t_B^*$, and $t_C^*$ turn out to be

$$\begin{aligned} t_A^* &= \frac{4.19}{\omega} \quad \text{seconds} \\ t_B^* &= \frac{4.10}{\omega} \quad \text{seconds} \\ t_C^* &= \frac{2.65}{\omega} \quad \text{seconds} \end{aligned} \tag{10-186}$$

Clearly, scheme $C$ is superior (timewise) for this initial state.

We have established that the gimbaled-jet scheme is better than the fixed-jet schemes with respect to the time of response. There are, however, other considerations that we must take into account if we are faced with the problem of designing the velocity-control system for a given space body. For example, we must know the problems associated with the engineering realization of each time-optimal scheme, the fuel consumed by each scheme, etc. In the remainder of this section, we shall discuss the relative merits and disadvantages of each scheme of control. Such knowledge will enable the designer to make an intelligent decision on the basis of complete information about the properties of each control scheme.

### 1 *Engineering Realization*

Since the objective is to reduce the angular velocities $y_1(t)$ and $y_2(t)$ to zero, we must be able to measure these two velocities. The measurement of the velocities is necessary for all three schemes.

Schemes $A$ and $B$ require the computation of the quantities $\omega x_1(t) = I\omega y_1(t)$ and $\omega x_2(t) = I\omega y_2(t)$ because the switch curves (see Fig. 10-6*a* and *b*) are defined in the $\omega x_1$-$\omega x_2$ plane. Thus, we must compute the constant $\omega = [(I - I_3)/I]\hat{y}_3$, which means that we must measure the spin velocity $\hat{y}_3$.

Using these measurements, the block diagram of the time-optimal control system for scheme $A$ is similar to the one illustrated in Fig. 7-42; the signal $\alpha(t)$ in Fig. 7-42 will control the firing of the fixed jet $J^A$ (see Fig. 10-4*a*).

The block diagram of the time-optimal control system for scheme $B$ is similar to the one illustrated in Fig. 7-61; the signal $m(t)$ in Fig. 7-61 will control the firing of the fixed jet $J^{B1}$, and the signal $n(t)$ in Fig. 7-61 will control the firing of the fixed jet $J^{B2}$ (see Fig. 10-4*b*).

The gimbaled-jet control scheme (scheme $C$) is basically different. First of all, the jet $J^C$ (see Fig. 10-4*c*) must generate its maximum thrust at all times. Thus, the only task to be performed by the time-optimal control system is to point the jet $J^C$ correctly. This means that [see Eq. (10-180)] the jet $J^C$ must be rotated in such a way that the angle $\theta(t)$ is

$$\theta(t) = \tan^{-1} \frac{y_2(t)}{y_1(t)}$$

We can immediately see that we need not compute the signals $\omega x_1(t)$ and $\omega x_2(t)$, which implies that we do not have to measure the spin velocity $\hat{y}_3$. Thus, the time-optimal control system for scheme $C$ must:

1. Measure the velocities $y_1(t)$ and $y_2(t)$
2. Compute the angle $\theta(t) = \tan^{-1}[y_2(t)/y_1(t)]$
3. Use a servomotor to rotate $J^C$

We can now compare the three schemes from the point of view of the required control system. Schemes $A$ and $B$ require the measurement of $\hat{y}_3$ and the use of nonlinear function generators to generate the switch curves. Scheme $C$ requires one simple function generator to compute $\theta(t)$ and a high-performance servo for pointing the jet $J^C$. It is difficult to say which control system is more reliable or cheaper to construct. This evaluation will depend on the specific system at hand.

### 2 *Fuel Consumption*

We are comparing three time-optimal schemes of control. We recall that the basis of comparison was that the rate of flow of fuel $\varphi$ was the same for each scheme. Since the consumed fuel is proportional to $\int_0^T \varphi \, dt$ and since scheme $C$ requires the least time to force any initial velocity to zero, we deduce that *the fuel consumed by the gimbaled-jet scheme is less than the fuel consumed by the fixed-jet schemes.* As a matter of fact, we know from Control Law 10-2*b* that the time-optimal control of scheme $C$ is also fuel-optimal. Therefore, we conclude that: *With respect to fuel consumption, scheme C is "better" than either scheme A or scheme B.*

### 3 *Time-varying Spin Velocity*

In the formulation of the physical problem, we assumed that the spin velocity $y_3(t)$ was at its desired value $\hat{y}_3$ at the initial time $t = 0$. This means that first we establish the desired spin velocity and only then do we drive the other two velocities to zero. This assumption had the mathematical consequence that the system of differential equations (10-163) satisfied by the angular momenta $x_1(t)$ and $x_2(t)$ was linear and time-invariant.

Now suppose that we change the problem in the following manner: Let us assume that at the initial time $t = 0$, we measure the three angular velocities $y_1(0)$, $y_2(0)$, and $y_3(0)$. Suppose that the initial spin velocity $y_3(0)$ is not the desired one, i.e., that

$$y_3(0) \neq \hat{y}_3 \tag{10-187}$$

and that

$$y_1(0) \neq 0 \qquad y_2(0) \neq 0 \tag{10-188}$$

The spin velocity $y_3(t)$ can be controlled independently by the jet $J_3$. Since $y_3(t)$ is given by Eq. (10-149), we can define $\omega(t)$ by setting

$$\omega(t) = \frac{I - I_3}{I} y_3(t) \tag{10-189}$$

The differential equations of the angular momenta $x_1(t)$ and $x_2(t)$ are then

$$\begin{aligned} \dot{x}_1(t) &= \omega(t)x_2(t) + u_1(t) \\ \dot{x}_2(t) &= -\omega(t)x_1(t) + u_2(t) \end{aligned} \tag{10-190}$$

Since these equations are time-varying, we cannot use the results of Secs. 7-7 and 7-9 to determine the switch curves for the time-optimal control. To determine the time-optimal torques for scheme $A$ or $B$, we must determine some sort of time-varying switch curves which depend on $\omega(t)$. On the other hand, if we use scheme $C$, then the time-optimal control is independent of $\omega(t)$, and so Eq. (10-180) [which specifies the angle $\theta(t)$ in terms of the angular velocities] still holds, even though $\omega(t)$ is a function of time. The reason is, of course, that the time-varying system (10-190) is norm-invariant, and so Control Laws 10-2*a* and 10-2*b* apply.

The above considerations mean that the gimbaled-jet control system can be used to drive the angular velocities $y_1(0)$ and $y_2(0)$ to zero in minimum time and with minimum fuel and to simultaneously control the spin velocity.

To accomplish the same task with either scheme $A$ or scheme $B$, we must construct time-varying nonlinear function generators and require, as before, more time and more fuel. The construction of a time-varying nonlinear function generator is far more difficult than the construction of a servo which points the jet $J^C$ at the required angle $\theta(t) = \tan^{-1} [y_2(t)/y_1(t)]$.

*To recapitulate:* We have found that, among the three schemes of time-optimal control, scheme $C$ requires the least time and the least amount of fuel to drive any initial velocities $y_1(0)$ and $y_2(0)$ to zero; moreover, the time-optimal control system for scheme $C$ is independent of the spin velocity (which is not necessarily constant), while the time-optimal control systems for schemes $A$ and $B$ do depend on the spin velocity.

We resist the temptation of recommending any scheme to the design engineer. Our objective is merely to indicate that the theory of optimal control can be used in the preliminary-design stage of a system and serve as a guide to the intelligent choice of a particular control scheme. The solution of a well-formulated optimal problem is a mathematical one; the selection of a constraint set, however, is very often up to the engineer. For this reason, the engineer should always try to obtain the properties of the optimal-control system for various possible control-constraint sets.

**Exercise 10-9** Consider the system

$$\begin{aligned} \dot{x}_1(t) &= x_2(t) + u_1(t) \\ \dot{x}_2(t) &= -x_1(t) + u_2(t) \end{aligned} \tag{10-191}$$

Suppose that the control vector $\mathbf{u}(t)$ belongs to the three different constraint sets $\Omega_1$, $\Omega_2$, and $\Omega_3$ defined by

$$\Omega_1 = \{\mathbf{u}(t): \sqrt{u_1^2(t) + u_2^2(t)} \le 1\} \tag{10-192}$$
$$\Omega_2 = \{\mathbf{u}(t): |u_1(t)| \le 1;\ |u_2(t)| \le 1\} \tag{10-193}$$
$$\Omega_3 = \{\mathbf{u}(t): \sqrt{u_1^2(t) + u_2^2(t)} \le \sqrt{2}\} \tag{10-194}$$

Suppose that at $t = 0$, the initial state of the system is $(\xi_1, 0)$. Let $t_1^*$, $t_2^*$, and $t_3^*$ be the minimum times required to force $(\xi_1, 0)$ to $(0, 0)$ when $\mathbf{u}(t) \in \Omega_1$, $\mathbf{u}(t) \in \Omega_2$, and $\mathbf{u}(t) \in \Omega_3$, respectively. Plot $t_1^*/t_2^*$ and $t_2^*/t_3^*$ versus $|\xi_1|$.

Suppose that the consumed fuel is measured by

$$F_1 = \int_0^{t_1*} dt \qquad F_2 = \int_0^{t_2*} 2\,dt \qquad F_3 = \int_0^{t_3*} \sqrt{2}\,dt \tag{10-195}$$

Plot $F_1/F_2$ and $F_2/F_3$ versus $|\xi_1|$.

## 10-8 Suggestions for Further Reading

In this chapter, we demonstrated that the Schwarz inequality could be used to obtain the solution for a class of optimal problems. The use of functional analysis (rather than the theory of the minimum principle) in the investigation of optimal-control problems has received some attention in the literature. The interested reader will find such expositions in Refs. K-21, K-22, K-31 to K-33, and S-2.

The use of gimbaled rockets for the propulsion of missiles and other spacecraft leads to constraints of the type $\|\mathbf{u}(t)\| \leq m$ on the control variable $\mathbf{u}(t)$. Reference L-13 considers such aerospace problems and includes additional references. Reference M-4 also deals with constraints of the type $\|\mathbf{u}(t)\| \leq m$ and considers a minimum-fuel mid-course guidance scheme; it also suggests an iterative procedure for the computation of the optimal control.

The technique of "local optimization" [i.e., the technique of choosing the control in such a way as to make the rate of change of $\|\mathbf{x}(t)\|$ as negative as possible] is appealing from the design point of view. We demonstrated in Example 10-2 that this technique has the disadvantage that the state is not driven to the origin. Nonetheless, this idea of "local optimization" perhaps can be used as a tool for the design of suboptimal control systems. The interested reader can also consult Ref. K-4 for some additional results.

# REFERENCES


A-1 Athanassiades, M., and O. J. M. Smith: Theory and Design of High Order Bang Bang Control Systems, *IRE Trans. Autom. Control*, vol. AC-6, pp. 125–134, 1961.

A-2 Athanassiades, M.: Bang-Bang Control for Tracking Systems, *IRE Trans. Autom. Control*, vol. AC-7, pp. 77–78, 1962.

A-3 Athanassiades, M.: On Optimal Linear Control Systems Which Minimize the Time Integral of the Absolute Value of the Control Function (Minimum-fuel Control Systems), *MIT Lincoln Lab. Rept.* 22G-4, Lexington, Mass., May, 1962.

A-4 Athanassiades, M., and P. L. Falb: Time Optimal Control for Plants with Numerator Dynamics, *IRE Trans. Autom. Control*, vol. AC-7, pp. 46–47, 1962.

A-5 Athanassiades, M., and P. L. Falb: Time Optimal Velocity Control of a Spinning Space Body, *MIT Lincoln Lab. Rept.* 22G-8, Lexington, Mass., July, 1962.

A-6 Athanassiades, M.: Optimal Control for Linear Time Invariant Plants with Time-, Fuel-, and Energy Constraints, *IEEE Trans. Appl. Ind.*, vol. 81, pp. 321–325, 1963.

A-7 Athans, M.: Minimum Fuel Feedback Control Systems: Second Order Case, *IEEE Trans. Appl. Ind.*, vol. 82, pp. 8–17, 1963.

A-8 Athans, M.: Time- and Fuel-optimal Attitude Control, *MIT Lincoln Lab. Rept.* 22G-9, Lexington, Mass., May, 1963.

A-9 Athans, M., P. L. Falb, and R. T. Lacoss: Time Optimal Velocity Control of a Spinning Space Body, *IEEE Trans. Appl. Ind.*, vol. 83, pp. 206–214, 1963.

A-10 Athans, M., P. L. Falb, and R. T. Lacoss: Time-, Fuel-, and Energy-optimal Control of Nonlinear Norm-invariant Systems, *IEEE Trans. Autom. Control*, vol. AC-8, pp. 196–202, 1963.

A-11 Athans, M., and P. L. Falb: Time-optimal Control for a Class of Nonlinear Systems, *IEEE Trans. Autom. Control*, vol. AC-8, p. 379, 1963.

A-12 Athans, M., P. L. Falb, and R. T. Lacoss: On Optimal Control of Self Adjoint Systems, *IEEE Trans. Appl. Ind.*, vol. 83, pp. 161–166, 1964.

A-13 Athans, M.: Minimum Fuel Control of Second Order Systems with Real Poles, *IEEE Trans. Appl. Ind.*, vol. 83, pp. 148–153, 1964.

A-14 Athans, M., and M. D. Canon: Fuel-optimal Singular Control of a Nonlinear Second Order System, *preprints 1964 Joint Autom. Control Conf.*, pp. 245–255, June, 1964.

A-15 Athans, M.: Fuel-optimal Control of a Double Integral Plant with Response Time Constraints, *IEEE Trans. Appl. Ind.*, vol. 83, pp. 240–246, 1964.

B-1 Balakrishnan, A. V., and L. W. Neustadt (eds.): "Computing Methods in Optimization Problems," Academic Press Inc., New York, 1964.

B-2 Bass, R. W.: Optimal Feedback Control System Design by the Adjoint System, *Aeronca Tech. Rept.* 60-22A, Baltimore, Md., June, 1960.

B-3 Beckenbach, E. F., and R. Bellman: "Inequalities," Springer-Verlag OHG, Berlin, 1961.

B-4 Bellman, R., I. Glicksberg, and O. Gross: On the Bang-Bang Control Problem, *Quart. Appl. Math.*, vol. 14, pp. 11–18, 1956.

B-5 Bellman, R.: "Dynamic Programming," Princeton University Press, Princeton, N.J., 1957.

B-6 Bellman, R.: "Introduction to Matrix Analysis," McGraw-Hill Book Company, New York, 1960.

B-7 Bellman, R. (ed.): "Mathematical Optimization Techniques," University of California Press, Berkeley, Calif., 1963.

B-8 Berkovitz, L. D.: Variational Methods in Problems of Control and Programming, *J. Math. Anal. Appl.*, vol. 3, pp. 145–169, 1961.

B-9 Birkhoff, G., and S. MacLane: "A Survey of Modern Algebra," rev. ed., The Macmillan Company, New York, 1958.

B-10 Birkhoff, G., and G. C. Rota: "Ordinary Differential Equations," Ginn and Company, Boston, 1962.

B-11 Bliss, G.: "Calculus of Variations," Mathematical Association of America, The Open Court Publishing Company, LaSalle, Ill., 1925.

B-12 Bogner, I.: An Investigation of the Switching Criteria for Higher Order Contactor Servomechanisms, *Cook Res. Lab. Rept.* PR-16-9, 1953.

B-13 Bogner, I., and L. F. Kazda: An Investigation of the Switching Criteria for Higher Order Servomechanisms, *Trans. AIEE*, pt. II, vol. 73, pp. 118–127, 1954.

B-14 Boyadjieff, G., et al.: Some Applications of the Maximum Principle to Second Order Systems, Subject to Input Saturation, Minimizing Error, and Effort, *J. Basic Eng.*, vol. 86, pp. 11–22, 1964.

B-15 Breakwell, J. V.: The Optimization of Trajectories, *J. SIAM*, vol. 7, pp. 215–247, 1959.

B-16 Breakwell, J. V., J. L. Speyer, and A. E. Bryson: Optimization and Control of Nonlinear Systems Using the Second Variation, *J. SIAM Control*, ser. A, vol. 1, pp. 193–223, 1963.

B-17 Breakwell, J. V., and Y. C. Ho: On the Conjugate Point Condition for the Control Problem, *Harvard Univ. Cruft Lab. Tech. Rept.* 441, Cambridge, Mass., March, 1964.

B-18 Bridgland, T. F.: On the Existence of Optimal Feedback Controls, *J. SIAM Control*, ser. A, vol. 1, pp. 261–274, 1963.

B-19 Brockett, R. W.: "The Invertibility of Dynamical Systems with Application to Control," Ph.D. dissertation, Case Institute of Technology, Cleveland, May, 1963.

B-20 Brockett, R. W., and M. Mesarovic: The Reproducibility of Multivariable Systems, *preprints 1964 Joint Autom. Control Conf.*, pp. 481–486, June, 1964.

B-21 Brockett, R. W.: Poles, Zeroes, and Feedback: State Space Interpretation, *IEEE Trans. Autom. Control*, vol. AC-10, pp. 129–135, 1965.

B-22 Bryson, A. E., and W. F. Denham: A Steepest-ascent Method for Solving Optimum Programming Problems, *J. Appl. Mech.*, ser. E, vol. 29, pp. 247–257, 1962.

B-23 Bushaw, D. W.: Differential Equations with a Discontinuous Forcing Term, *Stevens Inst. Technol. Experimental Towing Tank Rept.* 469, Hoboken, N.J., January, 1953.

B-24 Bushaw, D. W.: Optimal Discontinuous Forcing Terms, in S. Lefschetz (ed.), "Contributions to the Theory of Nonlinear Oscillations," vol. 4, pp. 29–52, Princeton University Press, Princeton, N.J., 1958.

C-1 Caratheodory, C.: "Variationsrechnung und partielle Differentialgleichungen erster Ordnung," B. G. Teubner Verlagsgesellschaft, mbH, Leipzig, 1935.

C-2 Chandaket, P., C. T. Leondes, and E. C. Deland: Optimum Non-linear Bang-Bang Control Systems with Complex Roots, *Trans. AIEE*, pt. II, vol. 80, pp. 82–102, 1961.

C-3 Chang, S. S. L.: "Synthesis of Optimum Control Systems," McGraw-Hill Book Company, New York, 1961.

C-4 Chang, S. S. L.: An Extension of Ascoli's Theorem and Its Applications to the Theory of Optimal Control, *New York Univ. Dept. Elec. Eng. Tech. Rept.* 400-51, New York, 1962.

C-5 Chang, S. S. L.: Sufficient Condition for Optimal Control of Linear Systems with Nonlinear Cost Functions, *preprints 1964 Joint Autom. Control Conf.*, pp. 295–296, June, 1964.

C-6 Chzhan, Sy-In: On Sufficient Conditions for an Optimum, *Appl. Math. Mech.*, vol. 25, pp. 1420–1423, 1961.

C-7 Coddington, E. A., and N. Levinson: "Theory of Ordinary Differential Equations," McGraw-Hill Book Company, New York, 1955.

D-1 Denham, W. F.: Steepest-ascent Solution of Optimal Programming Problems, Harvard Summer Program, 1963, "Optimization of Dynamic Systems," *Raytheon Co. Space and Inform. Systems Div. Rept.*, Bedford, Mass., 1963.

D-2 Desoer, C. A.: The Bang Bang Servo Problem Treated by Variational Techniques, *Inform. and Control*, vol. 2, pp. 333–348, 1959.

D-3 Desoer, C. A.: Pontryagin's Maximum Principle and the Principle of Optimality, *J. Franklin Inst.*, vol. 271, pp. 361–367, 1961.

D-4 Dem'yanov, V. F., and V. V. Khomenyuk: The Solution of a Linear Problem in Optimal Control, *Automation and Remote Control*, vol. 24, pp. 1068–1070, 1964.

D-5 Dieudonne, J.: "Foundations of Modern Analysis," Academic Press Inc., New York, 1960.

D-6 Doll, H. G., and T. M. Stout: Design and Analogue Computer Analysis of an Optimum Third-order Non-linear Servomechanism, *Trans. ASME*, vol. 79, pp. 513–525, 1957.

D-7 Dreyfus, S. E.: Dynamic Programming and the Calculus of Variations, *J. Math. Anal. Appl.*, vol. 1, pp. 228–239, 1960.

D-8 Dreyfus, S. E.: Variational Problems with Inequality Constraints, *J. Math. Anal. Appl.*, vol. 4, pp. 297–308, 1962.

D-9 Dreyfus, S. E., and J. R. Elliot: An Optimal Linear Feedback Guidance Scheme, *J. Math. Anal. Appl.*, vol. 8, pp. 364–386, 1964.

D-10 Dubovitskii, A. Y., and A. A. Milyutin: Certain Optimality Problems for Linear Systems, *Automation and Remote Control*, vol. 24, pp. 1471–1481, 1964.

E-1 Eaton, J. H.: An Iterative Solution to Time Optimal Control, *J. Math. Anal. Appl.*, vol. 5, pp. 329–344, 1962.

E-2 Edelbaum, T.: Theory of Maxima and Minima, in G. Leitman (ed.), "Optimization Techniques," Academic Press Inc., New York, 1962.

E-3 Eggleston, H. G.: "Convexity," Cambridge University Press, London, 1958.

F-1 Fadden, E. J., and E. G. Gilbert: Computational Aspects of the Time-optimal Control Problem, in A. V. Balakrishnan and L. W. Neustadt (eds.), "Computing Methods in Optimization Problems," Academic Press Inc., New York, 1964.

F-2 Falb, P. L., and M. Athans: A Direct Proof of the Criterion for Complete Controllability of Time-invariant Linear Systems, *IEEE Trans. Autom. Control*, vol. AC-9, pp. 189–190, 1964.

F-3 Filippov, A. F.: On Certain Questions in the Theory of Optimal Control, *J. SIAM Control*, ser. A, vol. 1, pp. 76–84, 1962.

F-4 Flügge-Lotz, I., and A. A. Frederickson: Contactor Control of Higher-order Systems Whose Transfer Functions Contain Zeros, *Stanford Univ. Div. Eng. Mech. Tech. Rept.* 119, Stanford, Calif., 1959.

F-5 Flügge-Lotz, I., and T. Ishikawa: Investigation of Third-order Contactor Control Systems with Two Complex Poles without Zeros, *NASA Tech. Note* D248, 1960.

F-6 Flügge-Lotz, I.: Synthesis of Third-order Contactor Control Systems, *Proc. First IFAC Congr.*, pp. 390–397, 1961.

F-7 Flügge-Lotz, I., and Mih Yin: On the Optimum Response of Third-order Contactor Control Systems, *Stanford Univ. Div. Eng. Mech. Tech. Rept.* 125, Stanford, Calif., 1960.

F-8 Flügge-Lotz, I., and H. Halkin: Pontryagin's Maximum Principle and Optimal Control, *Stanford Univ. Div. Eng. Mech. Tech. Rept.* 130, Stanford, Calif., 1961.

F-9 Flügge-Lotz, I., and T. Ishikawa: Investigation of Third-order Contactor Control Systems with Zeros in Their Transfer Functions, *NASA Tech. Note* D719, 1961.

F-10 Flügge-Lotz, I., and Mih Yin: The Optimum Response of Second-order Velocity-controlled Systems with Contactor Control, *J. Basic Eng.*, vol. 83, pp. 59–64, 1961.

F-11 Flügge-Lotz, I., and H. A. Titus: The Optimum Response of Full Third-order Systems with Contactor Control, *J. Basic Eng.*, 1962, pp. 554–558.

F-12 Flügge-Lotz, I., and H. Marbach: The Optimal Control of Some Attitude Control Systems for Different Performance Criteria, *J. Basic Eng.*, vol. 85, pp. 165–176, 1963.

F-13 Flügge-Lotz, I., and M. D. Maltz: Attitude Stabilization Using a Contactor Control System with a Linear Switching Criterion (to be published in *Automatica*).

F-14 Foy, W. H.: Fuel Minimization in Flight Vehicle Attitude Control, *IEEE Trans. Autom. Control*, vol. AC-8, pp. 84–88, 1963.

F-15 Friedland, B.: A Minimum Response-time Controller for Amplitude and Energy Constraints, *IRE Trans. Autom. Control*, vol. AC-7, pp. 73–74, 1962.

F-16 Friedland, B.: The Structure of Optimum Control Systems, *J. Basic Eng.*, vol. 84, pp. 1–12, 1962.

F-17 Friedland, B.: The Design of Optimum Controllers for Linear Processes with Energy Limitations, *J. Basic Eng.*, vol. 85, pp. 181–196, 1963.

F-18 Friedland, B.: Optimum Control of an Aerodynamically Unstable Booster, *General Precision Aerospace Res. Rept.* 63-RC-8, Little Falls, N.J., November, 1963.

F-19 Friedman, A.: Optimal Control for Hereditary Processes, *Arch. Rational Mech. Anal.*, vol. 15, pp. 396–416, 1964.

G-1 Gambill, R. A.: Generalized Curves and the Existence of Optimal Controls, *J. SIAM Control*, ser. A, vol. 1, pp. 246–260, 1963.

G-2 Gelfand, I. M., and S. V. Fomin: "Calculus of Variations," Prentice-Hall, Inc., Englewood Cliffs, N.J., 1963.

G-3 Gibson, J. E.: "Nonlinear Automatic Control," McGraw-Hill Book Company, New York, 1963.

G-4 Gilbert, E. G.: Controllability and Observability in Multivariable Control Systems, *J. SIAM Control*, ser. A, vol. 1, pp. 128–151, 1963.

G-5 Gilbert, E. G.: The Application of Hybrid Computers to the Iterative Solution of Optimal Control Problems, in A. V. Balakrishnan and L. W. Neustadt (eds.), "Computing Methods in Optimization Problems," Academic Press Inc., New York, 1964.

G-6 Gille, J. C., et al.: "Feedback Control Systems," McGraw-Hill Book Company, New York, 1959.

G-7 Goldstein, H.: "Classical Mechanics," Addison-Wesley Publishing Company, Inc., Reading, Mass., 1959.

G-8 Gottlieb, G.: "Energy Optimum Systems with Constraints on the Control and the Response Time," M.S. thesis, Massachusetts Institute of Technology, Cambridge, Mass., May, 1964.

H-1 Halkin, H.: Liapounov's Theorem on the Range of a Vector Measure and Pontryagin's Maximum Principle, *Arch. Rational Mech. Anal.*, vol. 10, pp. 296–304, 1962.

H-2 Halkin, H.: The Principle of Optimal Evolution, in J. P. LaSalle and S. Lefschetz (eds.), "Nonlinear Differential Equations and Nonlinear Mechanics," Academic Press Inc., New York, 1963.

H-3 Halkin, H.: On the Necessary Condition for Optimal Control of Nonlinear Systems, *J. Anal. Math.*, vol. 12, pp. 1–82, 1963.

H-4 Halmos, P. R.: "Finite Dimensional Vector Spaces," 2d ed., D. Van Nostrand Company, Inc., Princeton, N.J., 1958.

H-5 Hancock, H.: "The Theory of Maxima and Minima," Ginn and Company, Boston, 1907; also Dover Publications Inc., New York, 1960.

H-6 Harvey, C. A.: Determining the Switching Criterion for Time Optimal Control, *J. Math. Anal. Appl.*, vol. 5, pp. 245–257, 1962.

H-7 Harvey, C. A., and E. B. Lee: On the Uniqueness of Time-optimal Control for Linear Processes, *J. Math. Anal. Appl.*, vol. 5, pp. 258–268, 1962.

H-8 Hermes, H., and G. Haynes: On the Nonlinear Control Problem with the Control Appearing Linearly, *J. SIAM Control*, ser. A, vol. 1, pp. 185–205, 1963.

H-9 Ho, Y. C.: A Successive Approximation Technique for Optimal Control Systems Subject to Input Saturation, *J. Basic Eng.*, vol. 84, pp. 33–40, 1962.

H-10 Hopkin, A. M.: A Phase Plane Approach to the Design of Saturating Servomechanisms, *Trans. AIEE*, vol. 70, pp. 631–639, 1950.

H-11 Howard, D. R., and Z. V. Rekasius: Determination of Reachable Zone Boundaries, *Aerospace Corp. Rept.* ATN-64(4540-70)-1, Los Angeles, Calif., June, 1964.

H-12 Hutchinson, C. E.: Minimum Time Control of a Linear Combination of State Variables, *Stanford Univ. Electron. Lab. Systems Theory Lab. Tech. Rept.* 6311-1, Stanford, Calif., August, 1963.

J-1 Jen-Wei, C.: A Problem in the Synthesis of Optimal Systems Using Maximum Principle, *Automation and Remote Control*, vol. 22, pp. 1170–1176, 1962.

J-2 Jen-Wei, C.: Synthesis of Relay Systems from the Minimum Integral Quadratic Deviation, *Automation and Remote Control*, vol. 22, pp. 1463–1469, 1962.

J-3 Johnson, C. D., and J. E. Gibson: Singular Solutions in Problems of Optimal Control, *IEEE Trans. Autom. Control*, vol. AC-8, pp. 4–14, 1963.

J-4 Johnson, C. D., and W. M. Wonham: On a Problem of Letov in Optimal Control, *preprints 1964 Joint Autom. Control Conf.*, pp. 317–325, June, 1964.

J-5 Johnson, C. D., and J. E. Gibson: Optimal Control of a Linear Regulator with Quadratic Index of Performance and Fixed Terminal Time, *IEEE Trans. Autom. Control*, vol. AC-9, pp. 355–360, 1964.

J-6 Johnson, C. D.: Singular Solutions in Problems of Optimal Control, in C. T. Leondes (ed.), "Advances in Control Systems: Theory and Applications," vol. II, Academic Press Inc., New York, 1965.

K-1 Kalman, R. E.: Analysis and Design Principles of Second and Higher Order Saturating Servomechanisms, *Trans. AIEE*, pt. II, vol. 74, pp. 294–310, 1955.

K-2 Kalman, R. E.: On the General Theory of Control Systems, *Proc. First IFAC Congr.*, pp. 481–493.

K-3 Kalman, R. E.: Contributions to the Theory of Optimal Control, *Bol. Soc. Mat. Mex.*, vol. 5, pp. 102–119, 1960.

K-4 Kalman, R. E., and J. E. Bertram: Control Systems Analysis and Design via the Second Method of Lyapunov, *J. Basic Eng.*, vol. 82, pp. 371–393, 1960.

K-5 Kalman, R. E., et al.: Controllability of Linear Dynamical Systems, in "Contributions to Differential Equations," vol. 1, John Wiley & Sons, Inc., New York, 1962.

K-6 Kalman, R. E., et al.: Fundamental Study of Adaptive Control Systems, *Wright-Patterson Air Force Base Tech. Rept.* ASD-TR-61-27, vol. 1, April, 1962.

K-7 Kalman, R. E.: Canonical Structure of Linear Dynamical Systems, *Proc. Natl. Acad. Sci. U.S.*, vol. 48, pp. 596–600, 1962.

K-8 Kalman, R. E.: The Theory of Optimal Control and the Calculus of Variations, in R. Bellman (ed.), "Mathematical Optimization Techniques," University of California Press, Berkeley, Calif., 1963.

K-9 Kalman, R. E.: Mathematical Description of Linear Dynamical Systems, *J. SIAM Control*, ser. A, vol. 1, pp. 152–192, 1963.

K-10 Kalman, R. E.: When Is a Linear Control System Optimal? *J. Basic Eng.*, vol. 86, pp. 51–60, 1964.

K-11 Kelley, H. J.: Guidance Theory and Extremal Fields, *IRE Trans. Autom. Control*, vol. AC-7, pp. 75–81, 1962.

K-12 Kelley, H. J.: Method of Gradients, in G. Leitman (ed.), "Optimization Techniques," Academic Press Inc., New York, 1962.

K-13 Kirillova, L. S.: The Problem of Optimizing the Final State of a Controlled System, *Automation and Remote Control*, vol. 23, pp. 1485–1494, 1963.

K-14 Kirillova, L. S.: An Existence Theorem in Terminal Control Problems, *Automation and Remote Control*, vol. 24, pp. 1071–1074, 1964.

K-15 Kipiniak, W.: "Dynamic Optimization and Control," John Wiley & Sons, Inc., New York, 1961.

K-16 Kishi, F. H.: The Existence of Optimal Controls for a Class of Optimization Problems, *IEEE Trans. Autom. Control*, vol. AC-8, pp. 173–175, 1963.

K-17 Kleinman, D. L.: "Fuel-optimal Control of Second and Third Order Systems with Different Time Constraints," M.S. thesis, Massachusetts Institute of Technology, Cambridge, Mass., June, 1963.

K-18 Knudsen, H.: "Maximum Effort Control for an Oscillatory Element," M.S. thesis, University of California, Berkeley, Calif., February, 1960.

K-19 Knudsen, H. K.: An Iterative Procedure for Computing Time-optimal Controls, *IEEE Trans. Autom. Control*, vol. AC-9, pp. 23–30, 1964.

K-20 Kolmogorov, A., and S. Fomin: "Elements of the Theory of Functions and Functional Analysis," vol. 1, "Metric and Normed Spaces," Graylock Press, Rochester, N.Y., 1957.

K-21 Kranc, G. M., and P. E. Sarachik: An Application of Functional Analysis to the Optimum Control Problem, *J. Basic Eng.*, vol. 85, pp. 143–150, 1963.

K-22 Krasovskii, N. N.: On the Theory of Optimum Regulation, *Automation and Remote Control*, vol. 18, pp. 1005–1016, 1958.

K-23 Krasovskii, N. N.: On the Theory of Optimum Control, *Appl. Math. Mech.*, vol. 23, pp. 899–919, 1959.

K-24 Krasovskii, N. N., and A. M. Letov: The Theory of Analytical Design of Controllers, *Automation and Remote Control*, vol. 23, pp. 649–656, 1962.

K-25 Kreindler, E.: Contributions to the Theory of Time-optimal Control, *J. Franklin Inst.*, vol. 275, pp. 314–344, 1963.

K-26 Kreindler, E., and P. E. Sarachik: On the Concept of Controllability and Observability of Linear Systems, *IEEE Trans. Autom. Control*, vol. AC-9, pp. 129–136, 1964.

K-27 Krotov, V. F.: Methods for Solving Variational Problems on the Basis of the Sufficient Conditions for an Absolute Minimum. I., *Automation and Remote Control*, vol. 23, pp. 1473–1484, 1963.

K-28 Krotov, V. F.: Methods for Solving Variational Problems. II. Sliding Regimes, *Automation and Remote Control*, vol. 24, pp. 539–553, 1963.

K-29 Kuba, R. E., and L. F. Kazda: A Phase Space Method for the Synthesis of Nonlinear Servomechanisms, *Trans. AIEE*, pt. II, pp. 282–290, 1956.

K-30 Kuba, R. E., and L. F. Kazda: The Design and Performance of a Model Second Order Nonlinear Servomechanism, *IRE Trans. Autom. Control*, vol. AC-1, pp. 43–48, 1958.

K-31 Kulikowski, R.: On Optimum Control with Constraints, *Bull. Polish Acad. Sci., Ser. Tech. Sci.*, vol. 7, pp. 385–394, 1959.

K-32 Kulikowski, R.: Synthesis of a Class of Optimum Control Systems, *Bull. Polish Acad. Sci., Ser. Tech. Sci.*, vol. 7, pp. 663–671, 1959.

K-33 Kulikowski, R.: Optimizing Processes and Synthesis of Optimizing Automatic Control Systems with Nonlinear Invariable Elements, *Proc. First IFAC Congr.*, pp. 473–477.

K-34 Kurtsveil, Y.: The Analytical Design of Control Systems, *Automation and Remote Control*, vol. 22, pp. 593–599, 1961.

L-1 Ladd, H. O.: Minimum Fuel Control of a Second Order Linear Process with a Constraint on Time to Run, *Raytheon Missile and Space Div., Rept.* BR-2113, Bedford, Mass., November, 1962.

L-2 LaSalle, J. P.: Time Optimal Control Systems, *Proc. Natl. Acad. Sci. U.S.*, vol. 45, pp. 573–577, 1959.

L-3 LaSalle, J. P.: Time Optimal Control, *Bol. Soc. Mat. Mex.*, vol. 5, pp. 120–124, 1960.

L-4 LaSalle, J. P.: The Time-optimal Control Problem, in "Contributions to Differential Equations," vol. V, pp. 1–24, Princeton University Press, Princeton, N.J., 1960.

L-5 LaSalle, J. P.: The Bang-Bang Principle, *Proc. First IFAC Congr.*, pp. 493–497.

L-6 Lee, E. B.: Mathematical Aspects of the Synthesis of Linear Minimum Response Time Controllers, *IRE Trans. Autom. Control*, vol. AC-5, pp. 283–289, 1960.

L-7 Lee, E. B., and L. Markus: Optimal Control for Nonlinear Processes, *Arch. Rational Mech. Anal.*, vol. 8, pp. 36–58, 1961.

L-8 Lee, E. B., and L. Markus: Synthesis of Optimal Control for Nonlinear Processes with One Degree of Freedom, *Inst. Math. Acad. Sci. Ukrainian SSR*, Kiev, 1961.

L-9 Lee, E. B.: On the Time-optimal Regulation of Plants with Numerator Dynamics, *IRE Trans. Autom. Control*, vol. AC-6, pp. 351–352, 1961.

L-10 Lee, E. B.: A Sufficient Condition in the Theory of Optimal Control, *J. SIAM Control*, ser. A, vol. 1, pp. 241–245, 1963.

L-11 Lee, E. B.: Geometric Properties and Optimal Controllers for Linear Systems, *IEEE Trans. Autom. Control*, vol. AC-8, pp. 379–381, 1963.

L-12 Lee, E. B.: On the Domain of Controllability for Linear Systems, *IEEE Trans. Autom. Control*, vol. AC-8, pp. 172–173, 1963.

L-13 Leitmann, G. (ed.): "Optimization Techniques with Applications to Aerospace Systems," Academic Press Inc., New York, 1962.

L-14 LeMay, J. L.: Recoverable and Reachable Zones to Control Systems with Linear Plants and Bounded Controller Outputs, *preprints 1964 Joint Autom. Control Conf.*, pp. 305–312, June, 1964.

L-15 Letov, A. M.: Analytical Controller Design I, *Automation and Remote Control*, vol. 21, pp. 303–306, 1960.

L-16 Letov, A. M.: Analytical Controller Design II, *Automation and Remote Control*, vol. 21, pp. 389–393, 1960.

L-17 Letov, A. M.: The Analytical Design of Control Systems, *Automation and Remote Control*, vol. 22, pp. 363–372, 1961.

L-18 Litovchenko, I. A.: Isoperimetric Problem in Analytic Design, *Automation and Remote Control*, vol. 22, pp. 1417–1423, 1962.

M-1 McDonald, D. C.: Nonlinear Techniques for Improving Servo Performance, *Proc. Natl. Electron. Conf.*, vol. 6, pp. 400–421, 1950.

M-2 McShane, E. J.: On Multipliers for Lagrange Problems, *Am. J. Math.*, vol. 61, pp. 809–819, 1939.

M-3 Meditch, J. S.: Synthesis of a Class of Linear Feedback Minimum Energy Controls, *IEEE Trans. Autom. Control*, vol. AC-8, pp. 376–378, 1963.

M-4 Meditch, J. S., and L. W. Neustadt: An Application of Optimum Control to Midcourse Guidance, *Proc. Second IFAC Congr.*, Paper 427, Basle, 1963.

M-5 Meditch, J. S.: On Minimal Fuel Satellite Attitude Controls, *IEEE Trans. Appl. Ind.*, vol. 83, pp. 120–128, 1964.

M-6 Meditch, J. S.: On the Problem of Optimal Thrust Programming for a Lunar Soft Landing, *preprints 1964 Joint Autom. Control Conf.*, pp. 233–238, June, 1964.

M-7 Merriam, C. W.: "Optimization Theory and the Design of Feedback Control Systems," McGraw-Hill Book Company, New York, 1964.

M-8 Miele, A.: The Calculus of Variations in Applied Aerodynamics and Flight Mechanics, in G. Leitman (ed.), "Optimization Techniques," Academic Press Inc., New York, 1962.

N-1 Neiswander, R. S., and R. H. MacNeal: Optimization of Non-linear Control Systems by Means of Non-linear Feedbacks, *Trans. AIEE*, pt. II, vol. 72, pp. 262–272, 1953.

N-2 Neustadt, L. W.: Synthesizing Time-optimal Control Systems, *J. Math. Anal. Appl.*, vol. 1, pp. 484–493, 1960.

N-3 Neustadt, L. W.: Time-optimal Control Systems with Position and Integral Limits, *J. Math. Anal. Appl.*, vol. 3, pp. 406–427, 1961.

N-4 Neustadt, L. W.: Minimum Effort Control Systems, *J. SIAM Control*, ser. A, vol. 1, pp. 16–31, 1962.

N-5 Neustadt, L. W.: The Existence of Optimum Controls in the Absence of Convexity Conditions, *J. Math. Anal. Appl.*, vol. 7, pp. 110–117, 1963.

N-6 Neustadt, L. W.: On Synthesizing Optimal Controls, *Proc. Second IFAC Congr.*, Paper 421, Basle, 1963.

N-7 Neustadt, L. W.: Optimization, a Moment Problem, and Nonlinear Programming, *Aerospace Corp. Rept.* TDR-169-(3540-10)-TN-1, El Segundo, Calif., July, 1963.

N-8 Newton, G. C., L. A. Gould, and J. F. Kaiser: "Analytical Design of Linear Feedback Controls," John Wiley & Sons, Inc., New York, 1957.

O-1 Oldenburger, R.: Optimum Non-linear Control, *Trans. ASME*, vol. 79, pp. 527–546, 1957.

O-2 Oldenburger, R., and G. Thompson: Introduction to Time Optimal Control of Stationary Linear Systems, *Automatica*, vol. 1, pp. 177–205, 1963. (Contains 58 references, mostly on time-optimal control.)

P-1 Paiewonsky, B.: Time-optimal Control of Linear Systems with Bounded Control, in "International Symposium on Nonlinear Differential Equations and Nonlinear Mechanics," Academic Press Inc., New York, 1963.

P-2 Paraev, Y. I.: On Singular Control in Optimal Processes That Are Linear with Respect to the Control Inputs, *Automation and Remote Control*, vol. 23, pp. 1127–1134, 1962.

P-3 Pars, L. A.: "An Introduction to the Calculus of Variations," John Wiley & Sons, Inc., New York, 1962.

P-4 Pittel, B. G.: Some Problems of Optimum Control. I, *Automation and Remote Control*, vol. 24, pp. 1078–1091, 1964.

P-5 Pontryagin, L. S., V. Boltyanskii, R. Gamkrelidze, and E. Mishchenko: "The Mathematical Theory of Optimal Processes," Interscience Publishers, Inc., New York, 1962.

P-6 Pontryagin, L. S.: "Ordinary Differential Equations," Addison-Wesley Publishing Company, Inc., Reading, Mass., 1962.

P-7 Preston, J. L.: Non-linear Control of Saturating Third-order Servomechanism, *MIT Servo Lab. Tech. Mem.* 6897-TM-14, Cambridge, Mass., 1954.

P-8 Pryakhin, N. S.: The Problem of Analytical Regulator Design, *Automation and Remote Control*, vol. 24, pp. 1075–1077, 1964.

R-1 Rekasius, Z. V.: A General Performance Index for Analytical Design of Control Systems, *IRE Trans. Autom. Control*, vol. AC-6, pp. 217–222, 1961.

R-2 Rekasius, Z. V., and T. C. Hsia: On an Inverse Problem in Optimal Control, *preprints 1964 Joint Autom. Control Conf.*, pp. 313–316, June, 1964.

R-3 Repin, Yu. M., and V. I. Tret'yakov: The Analytical Design of Controls Based on Electronic Analog Devices, *Automation and Remote Control*, vol. 24, pp. 674–679, 1963.

R-4 Rohrer, R. A., and M. Sobral: Optimal Singular Solution for Linear, Multi-input Systems, *Univ. Illinois Coordinated Science Lab. Rept.* R-199, Urbana, Ill., April, 1964.

R-5 Rose, N. J.: Optimum Switching Criteria for Discontinuous Controls, *IRE Natl. Conv. Record*, pp. 61–66, 1956.

R-6 Rozonoer, L. I.: L. S. Pontryagin's Maximum Principle in the Theory of Optimum Systems I, II, III, *Automation and Remote Control*, vol. 20, I, pp. 1288–1302; II, pp. 1405–1421; III, pp. 1517–1532, 1960.

R-7 Roxin, E.: Reachable Zones in Autonomous Differential Systems, *Bol. Soc. Mat. Mex.*, vol. 5, pp. 125–135, 1960.

R-8 Roxin, E.: The Existence of Optimal Controls, *Mich. Math. J.*, vol. 9, pp. 109–119, 1962.

R-9 Roxin, E.: A Geometric Interpretation of Pontryagin's Maximum Principle, in J. P. LaSalle and S. Lefschetz (eds.), "Nonlinear Differential Equations and Nonlinear Mechanics," Academic Press Inc., New York, 1963.

R-10 Rudin, W.: "Principles of Mathematical Analysis," 2d ed., McGraw-Hill Book Company, New York, 1964.

S-1 Salukvadze, M. E.: On the Analytical Design of an Optimal Controller, *Automation and Remote Control*, vol. 24, pp. 409–417, 1963.

S-2 Sarachik, P. E., and G. M. Kranc: On Optimal Control of Systems with Multinorm Constraints, *Proc. Second IFAC Congr.*, Paper 423, Basle, 1963.

S-3 Schmidt, S. F.: The Analysis and Design of Continuous and Sampled Data Feedback Control Systems with a Saturation Type Non-linearity, *NASA Tech. Note* D-20, 1959.

S-4 Simmons, G. F.: "Introduction to Topology and Modern Analysis," McGraw-Hill Book Company, New York, 1963.

S-5 Smith, F. B.: Time-optimal Control of Higher-order Systems, *IRE Trans. Autom. Control*, vol. AC-6, pp. 16–21, 1961.

S-6 Smith, O. J. M.: "Feedback Control Systems," McGraw-Hill Book Company, New York, 1958.

S-7 Stern, T. E., and R. M. Lerner: A Circuit for the Square Root of the Sum of the Squares, *Proc. IEEE*, vol. 51, pp. 593–596, 1963.

S-8 Stout, T. M.: Effects of Friction in an Optimum Relay Servomechanism, *Trans. AIEE*, pt. II, vol. 72, pp. 329–336, 1953.

S-9 Sun-Jian and Hang King-ching: Analysis and Synthesis of Time-optimal Control Systems, submitted to the Second IFAC Congress, Basle, September, 1963.

T-1 Tou, J. T.: "Optimum Design of Digital Control Systems," Academic Press Inc., New York, 1962.

T-2 Truxal, J. G.: "Automatic Feedback Control System Synthesis," McGraw-Hill Book Company, New York, 1955.

T-3 Troitskii, V. A.: The Mayer Bolza Problem of the Calculus of Variations and the Theory of Optimum Systems, *J. Appl. Math. Mech.*, vol. 25, pp. 994–1010, 1961.

V-1 Vulikh, B. Z.: "Introduction to Functional Analysis," Addison-Wesley Publishing Company, Inc., Reading, Mass., 1963.

W-1 Wang, P. K. C.: Analytical Design of Electrohydraulic Servomechanisms with Near Time-optimal Response, *IEEE Trans. Autom. Control*, vol. AC-8, pp. 15–27, 1963.

W-2 Warga, J.: Relaxed Variational Problems, *J. Math. Anal. Appl.*, vol. 4, pp. 111–128, 1962.

W-3 Warga, J.: Necessary Conditions for Minimum in Relaxed Variational Problems, *J. Math. Anal. Appl.*, vol. 4, pp. 129–145, 1962.

W-4 Weinberg, L.: "Network Analysis and Synthesis," McGraw-Hill Book Company, New York, 1962.

W-5 Weiss, L., and R. E. Kalman: Contributions to Linear System Theory, *RIAS Tech. Rept.* 64-9, Baltimore, Md., April, 1964.

W-6 Wonham, W. M., and C. D. Johnson: Optimal Bang-Bang Control with Quadratic Index of Performance, *J. Basic Eng.*, vol. 86, pp. 107–115, 1964.

Z-1 Zadeh, L. A., and C. A. Desoer: "Linear System Theory: The State Space Approach," McGraw-Hill Book Company, New York, 1963.

# INDEX